AF758419

ENZYME INHIBITION IN DRUG DISCOVERY AND DEVELOPMENT

ENZYME INHIBITION IN DRUG DISCOVERY AND DEVELOPMENT

The Good and the Bad

Edited by

CHUANG LU

ALBERT P. LI

WILEY

A JOHN WILEY & SONS, INC., PUBLICATION

Copyright © 2010 by John Wiley & Sons, Inc. All rights reserved.

Published by John Wiley & Sons, Inc., Hoboken, New Jersey
Published simultaneously in Canada

No part of this publication may be reproduced, stored in a retrieval system, or transmitted in any form or by any means, electronic, mechanical, photocopying, recording, scanning, or otherwise, except as permitted under Section 107 or 108 of the 1976 United States Copyright Act, without either the prior written permission of the Publisher, or authorization through payment of the appropriate per-copy fee to the Copyright Clearance Center, Inc., 222 Rosewood Drive, Danvers, MA 01923, (978) 750-8400, fax (978) 750-4470, or on the web at www.copyright.com. Requests to the Publisher for permission should be addressed to the Permissions Department, John Wiley & Sons, Inc., 111 River Street, Hoboken, NJ 07030, (201) 748-6011, fax (201) 748-6008, or online at http://www.wiley.com/go/permission.

Limit of Liability/Disclaimer of Warranty: While the publisher and author have used their best efforts in preparing this book, they make no representations or warranties with respect to the accuracy or completeness of the contents of this book and specifically disclaim any implied warranties of merchantability of fitness for a particular purpose. No warranty may be created or extended by sales representatives or written sales materials. The advice and strategies contained herein may not be suitable for your situation. You should consult with a professional where appropriate. Neither the publisher nor author shall be liable for any loss of profit or any other commercial damages, including but not limited to special, incidental, consequential, or other damages.

For general information on our other products and services or for technical support, please contact our Customer Care Department within the United States at (800) 762-2974, outside the United States at (317) 572-3993 or fax (317) 572-4002.

Wiley also publishes its books in a variety of electronic formats. Some content that appears in print may not be available in electronic formats. For more information about Wiley products, visit our web site at www.wiley.com.

Library of Congress Cataloging-in-Publication Data:

Enzyme inhibition in drug discovery and development : the good and the bad / [edited by] Chuang Lu, Albert P. Li.
p. ; cm.
Includes bibliographical references and index.
ISBN 978-0-470-28174-1 (cloth)
1. Enzyme inhibitors. 2. Drug development. 3. Drugs–Metabolism. I. Lu, Chuang. II. Li, A. P.
[DNLM: 1. Enzyme Inhibitors–pharmacology. 2. Drug Discovery–methods. 3. Enzyme Inhibitors–chemical synthesis. QU 143 E605 2009]
QP601.5.E58 2009
615′.19–dc22

2009014027

Printed in the United States of America

10 9 8 7 6 5 4 3 2 1

CONTENTS

PREFACE

Development of pharmaceuticals is a complex science that involves collaborative activities of scientists from several different areas, such as: biochemistry to validate drug targets, medicinal chemistry to synthesize compounds; pharmacology to test the compounds' efficacy; drug metabolism and pharmacokinetics (DMPK) to ensure that the compounds have "drugable" properties, such as appropriate absorption, distribution, metabolism, and excretion (ADME) profiles; formulation science to design appropriate drug products; toxicology to characterize the safety profile in animals and help estimate the safety margin for humans; and last but not least, clinical sciences to ensure adequate pharmacokinetic, efficacy, and safety profiles in humans. Marketing and finance departments are involved right from the start and throughout the process in projecting the market on the basis of the drug's properties and the existing competition. During the costly and lengthy drug development process, the management team continually evaluates the progress of individual programs, sets priorities for development, and allocates resources to maximize the possibility of success given the company's position and market conditions.

Diseases are often associated with the elevation or repression of certain proteins that could be used as drug targets. In general, drug discovery starts by manipulating the drug target proteins (a receptor, kinase, or enzyme) with an agonist or, more commonly, with an antagonist or inhibitor. Such inhibition can generally be termed "good inhibition," because it leads to a desired outcome—that is, suppression of a target and, in turn, a treatment for a disease. More often than not, other proteins unrelated to the target are also inhibited, potentially leading to adverse effects. This additional, unwanted inhibition can be termed "bad inhibition." The inhibited proteins may be receptors, metabolizing enzymes, drug transporters, and so on. For the purposes of this book, only bad inhibition related to DMPK is covered. Just as good inhibition can lead to bad effects (e.g., related to mechanism-based adverse effects), bad inhibition can have a positive side as well. For example, drugs known to affect nontarget proteins may be used in select cases to cause mixed-type positive effects or in combination with appropriate drugs to complement their shortcomings. These attributes will be discussed in many of the chapters. In this high-throughput era, individual specialities are sometimes too focused on delivering in their own area. An overall understanding of the drug discovery and development process allows each scientist to look beyond his or her specialty at the bigger picture and to mold and integrate his or her research to fill the grand need of the product from all departments' perspectives. The goal of this book is to provide in-depth information to these specialized scientists on the latest science, strategy, and approaches in enzyme inhibition as applied to drug discovery and drug development, as well as to provide value to those who are interested in learning the science of fields other than their own.

This book represents the first of its kind in terms of its inclusion of both pharmacologic and pharmacokinetic aspects of enzyme inhibition, with chapters written by over 50 leading scientists in their fields, from both academia and industry (major pharmaceutical and biotechnology companies). The first part of this book gives an overall review of the drug discovery processes, including chapters on drug discovery strategy, medicinal chemistry, analytical chemistry, drug metabolism, pharmacokinetics, and safety biomarker assessment. The second part of this book discusses the manipulation of drug-metabolizing enzymes and transporters and their mostly adverse consequences, such as drug–drug interactions. The third part of this book reviews the inhibition of several major drug target pathways, such as the GPCR pathway, the NFκB pathway, and the ion channel pathway. This book will allow DMPK scientists to learn and appreciate target biology in drug discovery and, conversely, will allow discovery biologists and medicinal chemists to have a broader understanding of DMPK. It is also intended for readers with a general interest in the field—for instance, students in pharmacology, clinical development staff, those in marketing, and regulatory practitioners, as well as pharmaceutical executives, because it offers the opportunity to learn in a single treatise the basic science of drug discovery and development.

Chuang Lu
Albert P. Li

Cambridge, Massachusetts
Columbia, Maryland
June 2009

CONTRIBUTORS

Elena Afonina, Sequoia Pharmaceuticals, Inc., Gaithersburg, Maryland

Suresh K. Balani, Department of DMPK, Millennium Pharmaceuticals, Inc., Cambridge, Massachusetts

Tonika Bohnert, Department of DMPK, Biogen Idec, Inc., Cambridge, Massachusetts

Cuiping Chen, Pharmacokinetics, Depomed, Inc., Menlo Park, California

Wei Chen, Department of Oncology Biochemistry, Millennium Pharmaceuticals, Inc., Cambridge, Massachusetts

Anjaneya Chimalakonda, Drug Metabolism and Pharmacokinetics, Bristol Myers Squibb, Princeton, New Jersey

Maria Almira Correia, Department of Cellular and Molecular Pharmacology, University of California, San Francisco, San Francisco, California

Lenny Dang, Department of Discovery Oncology, Millennium Pharmaceuticals, Inc., Cambridge, Massachusetts

Michael Eissenstat, Sequoia Pharmaceuticals, Inc., Gaithersburg, Maryland

Amy Elder, Galenea, Inc., Cambridge, Massachusetts

Eric R. Fedyk, Department of Drug Safety Evaluation, Millennium Pharmaceuticals, Inc., Cambridge, Massachusetts

Stephen S. Ferguson, CellzDirect, Invitrogen, Durham, North Carolina

Michael B. Fisher, Boehringer Ingelheim Pharmaceuticals Inc., Ridgefield, Connecticut

Aleksandra Galetin, School of Pharmacy and Pharmaceutical Sciences, University of Manchester, Manchester, United Kingdom

Liang-Shang Gan, Department of DMPK, Biogen Idec, Inc., Cambridge, Massachusetts

Indranath Ghosh, Galenea, Inc., Cambridge, Massachusetts

David J. Greenblatt, Department of Pharmacology and Experimental Therapeutics, Tufts University School of Medicine and Tufts Medical Center, Boston Massachusetts

Thomas Guenthner, University of Illinois at Chicago, Chicago, Illinois

Sergei V. Gulnik, Sequoia Pharmaceuticals, Inc., Gaithersburg, Maryland

Geraldine Harriman, Galenea, Inc., Cambridge, Massachusetts

Nicola Hewitt, Kaly-Cell, Parc d'Innovation, Illkirch, France

J. Brian Houston, School of Pharmacy and Pharmaceutical Sciences, University of Manchester, Manchester, United Kingdom

Ping Kang, Pharmacokinetics, Dynamics, and Metabolism Department, Pfizer Global Research and Development, San Diego, California

Edward L. LeCluyse, CellzDirect, Invitrogen, Durham, North Carolina

Frank Lee, Department of DMPK, Millennium Pharmaceuticals, Inc., Cambridge, Massachusetts

Albert P. Li, In Vitro ADMET Laboratories, Columbia, Maryland

Mingxiang Liao, Department of DMPK, Millennium Pharmaceuticals, Inc., Cambridge, Massachusetts

Xingrong Liu, Drug Metabolism and Pharmacokinetics, Roche Palo Alto, Palo Alto, California

Yi Liu, East Coast Research and Early Development, Johnson & Johnson Pharmaceutical Research and Development, L.L.C., Pennsylvania

Chuang Lu, Department of DMPK, Millennium Pharmaceuticals, Inc., Cambridge, Massachusetts

Zheng Lu, Department of DMPK, Millennium Pharmaceuticals, Inc., Cambridge, Massachusetts

Gang Luo, Covance Laboratories Inc., Madison, Wisconsin

Wenyan Miao, Wyeth Inflammation, Wyeth Pharmaceuticals, Inc., Cambridge, Massachusetts

Gerald T. Miwa, Nextcea Inc., Lexington, Massachusetts

Lisa L. von Moltke, Department of Clinical Pharmacology, Millennium Pharmaceuticals, Inc., Cambridge, Massachusetts

Bernard P. Murray, Drug Metabolism Department, Gilead Sciences, Foster City, California

Molly Nix, School of Design, Carnegie Mellon University, Pittsburgh, Pennsylvania

R. Scott Obach, Pfizer, Inc., Groton, Connecticut

Ning Qin, East Coast Research and Early Development, Johnson & Johnson Pharmaceutical Research and Development, L.L.C., Pennsylvania

Richard Ridgewell, Covance Laboratories Inc., Madison, Wisconsin

Amin Rostami-Hodjegan, Department of Systems Pharmacology The Medical School, University of Sheffield, United Kingdom

Jasminder Sahi, CellzDirect, Invitrogen, Durham, North Carolina

Michael W. Sinz, Bristol Myers Squibb, Wallingford, Connecticut

Wen Chyi Shyu, Discovery Medicine and Clinical Pharmacology, Bristol Myers Squibb, Princeton, New Jersey

Jing-Tao Wu, Department of DMPK, Millennium Pharmaceuticals, Inc., Cambridge, Massachusetts

Lijun Wu, Resolvyx Pharmaceuticals, Inc., Bedford, Massachusetts

Cindy Q. Xia, Department of DMPK, Millennium Pharmaceuticals, Inc., Cambridge, Massachusetts

Hua Yang, Department of DMPK, Millennium Pharmaceuticals, Inc., Cambridge, Massachusetts

Johnny J. Yang, Department of DMPK, Millennium Pharmaceuticals, Inc., Cambridge, Massachusetts

Ping Zhao, Sonus Pharmaceuticals, Bothell, Washington. *Current Address*: Office of Clinical Pharmacology, Center for Drug Evaluation and Research, The Food and Drug Administration, Silver Spring, Maryland

PART I

DRUG DISCOVERY APPROACHES AND TECHNOLOGIES

1

THE DRUG DISCOVERY PROCESS

GERALD T. MIWA

1.1 INTRODUCTION

The discovery and development of drugs is an inefficient process. Only one of approximately 10,000 compounds synthesized reaches the marketplace, and this requires approximately 10–15 years and $800,000,000 in R&D expenditures (Khosla and Keasling, 2003). The objective of drug discovery should be to identify a compound that will prove to be safe and effective against the intended disease with minimal attrition due to toxicities, inadequate exposure, or unsuccessful translation of target modulation to clinical benefit against the disease. Drug discovery, the process used to select a compound for drug development, is common in most pharmaceutical and biotechnology companies. This process is composed of the following phases: target identification and validation, hit identification, lead optimization, and development candidate nomination (Fig. 1.1).

Target identification and validation is the process used to identify and confirm that the modulation of a biological target will produce a desired therapeutic effect. The methods employed for this phase are mainly biological. The confirmation or validation of the utility of the target in modifying a human disease is critical and remains the greatest potential issue of this phase because of its implication in clinical failures due to inadequate efficacy. Hit identification is the process employed to initially identify molecules that interact with the target. Both biological and chemical methods are used to identify hits. The methods employed, although more comprehensive than in the past, may still not identify all possible hits. Usually, more stringent selection criteria are

Enzyme Inhibition in Drug Discovery and Development, edited by Lu and Li
Copyright © 2010 John Wiley & Sons, Inc.

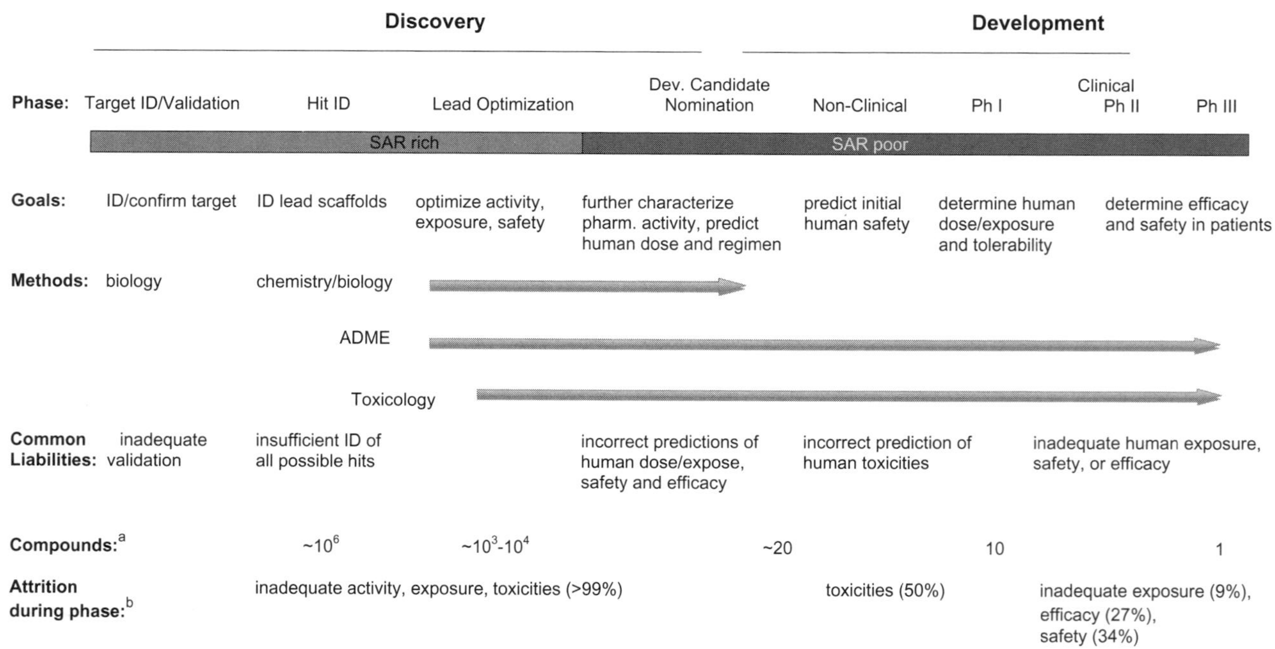

[a] adapted from Khosla 2003
[b] major causes of attrition during a phase and (percentage of total attrition due to these causes), adapted from Kola and Landis 2004

Figure 1.1. Goals and challenges in the processes of drug discovery and development.

implemented to screen hits in order to identify lead compounds that have the potential to improve further with structure modifications. Lead optimization is the chemical structure–activity optimization process that identifies the best possible drug-like molecule. Usually absorption, distribution, metabolism and excretion (ADME) and toxicology assays are added to the biological and chemical methods during lead optimization. Development candidate nomination is the process used to further characterize the potential exposure, efficacy, and safety of the nomination candidate and to judge if the molecule is suitable for drug development. The most common liabilities remaining from this phase are inaccurate predictions of human exposure, safety and efficacy. Although the objectives are common among companies engaged in this process, the methods, issues, and acceptance standards vary. Nevertheless, the best measure for the success of drug discovery is the demonstration of a compound's effectiveness and safety in patients. This chapter describes the elements of the drug discovery process and comments on its successfulness and areas currently under evaluation to improve success.

1.2 TARGET IDENTIFICATION AND VALIDATION

Target identification is the process used to identify potential therapeutic targets amenable to modulation by drug molecules, antibodies, aptamers, or gene modulators such as siRNAs and antisense oligonucleotides. The identification of potential targets for therapeutic intervention by drugs has greatly improved in the past two decades. Prior to the mid-1980s, drug targets were identified by the serendipitous discovery of active agents such as the penicillins and the benzodiazepines, through the symptomatic changes in disease models in animals such as cardiovascular drugs or through activity in suitable *in vitro* systems such as for anti-infective drugs. Biochemical targets were identified through their postulated relevance in pathways thought to be involved in disease processes such as HMG-CoA reductase in cholesterol biosynthesis and coronary heart disease (Tobert, 2003), and the H2-receptor in gastric acid secretion and gastric ulceration. In rare cases, targets could be identified and validated in humans through existing genetic mutations in the human population such as the deficiencies in 5-alpha reductase (Johnson et al., 1986) which led to the 5-alpha reductase inhibitor class of drugs for benign prostatic hypertrophy and propecia.

During the past two decades, greater knowledge and newer biological methods have permitted the mining of patient samples and animal models of diseases to elucidate the probable genes implicated in the disease etiology. Techniques such as gene expression profiling and comparative genomics have been valuable in identifying potential targets. For example, the capability to identify the overexpression of HER2 in the diseased tissues of some breast cancer patients was used to identify HER2 as a new target for metastatic breast cancer and led to the discovery of Herceptin (Chang, 2007).

Single-nucleotide polymorphisms (SNP) genotype–phenotype correlations from patient samples have also yielded many more, albeit clinically less validated, potential targets for drug design (Wunber et al., 2006). In addition, greater knowledge of the role of specific receptors, cell cycle regulators, enzymes, and other proteins in biochemical pathways has led to the identification of potential drug targets at the organism and biochemical levels.

The complete sequencing of the human genome ushered in the potential to vastly increase the number of drug targets. It has been estimated that there are as many as 10,000 potential drug target genes in the human genome, which is comprised of approximately 30,000 to 50,000 genes (Venter et al., 2001; IHGSC, 2001). In addition, this may underestimate the total number of possible protein targets because of splice variants and post-translational modifications (Pillutla et al., 2002). More recently, nonreceptor or noncatalytic protein targets such as protein–protein interactions and nonprotein targets such as DNA and mRNA have also been recognized. This is an enormous increase over the approximately 43 protein targets for the 100 best selling drugs (Zambrowicz and Sands, 2003) and is the basis for optimism for the future of the pharmaceutical industry. Of course, many, perhaps most, of the proteins will not be suitable for pharmaceutical development. Since there are thousands of potential protein targets, a growing effort has been made to develop high-thoughput screens to identify potential drug targets through ligand–protein (Pellecchia et al., 2004) and protein–protein interactions with protein microarrays (McBeath and Schreiber, 2000) or cell microarrays (Bailey et al., 2002). This and other technology for screening large protein libraries is comprehensively reviewed by Pillutla et al. (2002).

Target validation is the assessment and evidence supporting the linkage of a target to the etiology of the disease or pathology and the amelioration of the disease by modulation of the target. In other words, it is the convincing demonstration that modulation of the target, usually through inhibition, results in amelioration of the disease. Target validation is critical and, arguably, the most important step in translating a new potential target into a viable drug target because of its role in achieving efficacy in patients, currently one of the major reasons for drug attrition in the clinic (Kola and Landis, 2004).

The most accepted criteria for target validation during drug discovery are based on three categories: (1) demonstration of the target protein expression or mRNA in relevant cell types or in the target tissues from animal models or patients, (2) demonstration that modulation of the target in cell systems results in the desired functional affect, and (3) demonstration that the target has a causal role in producing the disease phenotype in animal models and/or patients (Windler et al., 2003).

The demonstration of protein expression is usually accomplished in diseased or target tissues by *in situ* hybridization or immunocytochemistry. For example, the localization of orexin peptides and receptors in the hypothalamus provided evidence for its role in the regulation of feeding (Sakurai et al., 1998). *In situ* hybridization permits gene detection only at the transcriptional or mRNA level, whereas immunocytochemistry identifies and locates the protein expression product but requires appropriate antibodies.

Today, the association of the target protein with diseased or target tissue is rarely considered sufficient for target validation. The functional association of the target with disease modification is also required. The demonstration that modulation of the target results in the desired functional affect can often be explored or demonstrated through inhibition of the target by transgenic (Tornell and Snaith, 2002) and gene knockout mice, small molecule inhibitors, antisense oligonucleotides, and small interfering RNA (siRNA). Antisense oligonucleotides inhibit gene expression through complementary binding to single-stranded DNA or RNA. They have been used to characterize the function of genes as well as potential drugs. In contrast, siRNAs inhibit gene expression through complementary binding to mRNA followed by catalytic destruction of the mRNA (Natt, 2007; Behike, 2006; Hammond et al., 2000). Consequently, siRNAs have rapidly eclipsed antisense oligonucleotides as a tool for inhibiting gene expression and target validation (Szymkowski, 2003). Scholarly reviews of techniques and examples of target identification and validation can be found in Natt (2007), Behike (2006), Pillutla et al. (2002), and Ohlstein et al. (2000). Proteomic methods have been less successful in identifying or validating potential drug targets because analytical techniques for exploring the entire proteome have not been adequate (Kopec et al., 2005). However, the recent development of cell microarrays (Bailey et al., 2002) may provide more feasible methods for quickly assessing disease-based changes in the proteome.

Despite the enormous advances in the technologies for examining the genome and proteome in cell-based and animal models of diseases and for validating potential drug targets with these methods, there remains considerable uncertainty about the prospective translation of these findings to human diseases (Williams, 2003). Today, target validation still ultimately depends on retrospective verification from clinical studies while the lack of efficacy in patients continues to be a major source of drug attrition. Consequently, the current performance of target validation has been criticized (Sams-Dodd, 2005). This had led to more holistic approaches for target validation such as the confirmation of activity in cell systems (Kenakin, 2003) and target deconvolution (Terstappen et al., 2007). Target deconvolution is the identification of both target and off-target proteins and their functional roles affected by a compound in mammalian cells or model organisms such as zebrafish or nematodes. Both target and off-target proteins are identified by a growing battery of techniques that include the use of the compound as an immobilized ligand for binding and isolating these proteins through affinity chromatography and through the gene expression changes caused by the compound. A more integrated preclinical and clinical paradigm has also been advocated for drug discovery (Schadt et al., 2003).

1.3 HIT IDENTIFICATION

Up to this point in the discovery process, the focus has been on the target and not the drug. Following target validation sufficient to create confidence that the target is involved in the etiology of a disease and that modulation of the target will result in the reduction of the phenotypic expression of the disease, the search for a drug molecule is initiated. Hit identification is the first stage of this process. For small molecules, the goal is to identify the core molecular structure that demonstrates promise for further structure optimization.

There are a number of strategies for identifying hits. With the advent of large compound libraries through combinatorial synthesis, proprietary compound collections, and natural products and the increased feasibility of screening provided by automated analysis and data informatic methods, high-throughput screening (HTS) of large compound libraries against potential targets emerged as a major source of compound hits. This approach was eagerly pursued, especially by large pharmaceutical companies, during the 1990s. Although the random screening of large combinatorial libraries is less common today, many companies continue to use HTS of focused compound libraries as a source of hits. Focused libraries are subsets of chemical libraries that are chosen based on a company's proprietary knowledge of the target structure or by other chemoinformatic criteria (Goodnow, 2006). In addition to the use of smaller, more focused compound libraries to screen for chemical hits, a growing effort to identify higher-quality hits through further characterization has been adopted by some pharmaceutical companies. The most common approach is the characterization of desirable drug-like properties using assays that can be easily implemented in a high-capacity and rapid-turnaround format such as those for drug absorption, metabolism, drug interactions, and cardiac ion channel modulation. For example, *in vitro* assays for oral absorption such as the Caco-2 cell system, metabolic stability such as the hepatic liver microsomal system, the potential for drug interactions with CYP inhibition assays, and the potential for cardiac QT interval prolongation with hERG binding have been moved to the end of the hit identification stage to provide higher-quality characterization to the leads they generate (Bleicher et al., 2003).

Computational or modeling techniques have also been employed as a source of hits (Balakin et al., 2006). For example, ligand-based design employs known binding ligands to map out the structural binding features of the target. This computational method does not require knowledge of the structure of the binding domain of the target but does depend on a comprehensive structure–activity relationship with known ligands. A more refined computational method, structure-based design, can be employed when the three-dimensional structure of the ligand-binding domain of the target protein is known. Modeling of ligands docking to the target provides a means for identifying potential hits (Kitchen et al., 2004). Actual HTS screening with focused compound libraries can then be used to confirm these potential hits or identify new hits.

An alternative or complementary approach to HTS with compound libraries is fragment-based lead design (FBLD). This method has been gaining popularity in the last decade (Hajduk and Greer, 2007; Everts, 2008) and is based on measuring the binding of smaller chemical fragments or functional groups to target proteins rather than the functional assay of the interaction of the larger molecules containing more chemical fragments employed with HTS. The strategy of FBLD is to identify individual chemical functional groups (such as amino, carboxylate, carbonyl, aryl, etc.) that bind to different sites on the target and then combine them in a single molecule in order to create compounds with higher binding affinities. One strategic advantage of FBLD over HTS is that a smaller library of compounds is required to define a greater fraction of the total number of possible chemical scaffolds (Hajduk and Greer, 2007). For example, it has been estimated that for scaffolds of MW 500, there are 10^{62} possible scaffolds while the usual HTS screen is limited to about 10^6 compounds leaving a vast amount of chemical space unscreened. In contrast, the smaller size of FBLD fragments, MW < 200, results in approximately 10^7 possible combinations requiring a much smaller fragment library of 10^4–10^5 to give much better screening coverage. The low coverage of all possible scaffolds possible with HTS may be a significant contributor to the disappointing success of this approach in identifying clinical candidates (Drews, 2000).

Although the functional assays of HTS are often not sensitive enough to detect the lower affinity binding of fragments, a number of general methods have been applied to successfully screen FBLD libraries such as NMR detection of the chemical shifts produced from fragment binding to the target protein, X-ray crystallographic detection of ligand binding, and surface plasmon resonance, which measures the change in protein surface refractive index upon ligand binding. To date, FBLD has yielded several clinical candidates from a number of companies such as Abbott, Astex Therapeutics, Novartis, Plexxikon, and Sunesis (Everts, 2008).

1.4 LEAD OPTIMIZATION

Hit identification, through the screening of compound libraries, FBLD, or other sources, provides the entry point for the drug discovery phase known as lead optimization (LO). Lead optimization is a systematic effort to maximize the pharmacological and drug exposure properties of a lead candidate while minimizing its toxicity and drug–drug interaction potential through structure modifications (refer to Chapter 2 in this book). Often, some assessment of certain acute toxicity markers and factors that can contribute to inadequate exposure or overexposure is either conducted systematically or conducted occasionally during LO or as part of the compound characterization stage prior to the nomination of a compound to development. The most common toxicity and exposure assays employed during this stage are: hERG binding for potential adverse cardiac ion channel binding, cytotoxicity assays as predictors of organ toxicities, limited AMES assays to identify direct acting genotoxins, the evaluation of chemically reactive metabolites (Baillie, 2006), the identification of off target receptors with the NOVA screen (Kramer et al., 2007; Sasseville et al., 2004), the identification of major metabolic pathways (refer to Chapter 5 in this book), and the evaluation of drug–drug interaction potential through the identification of enzymes and transporters responsible for the clearance of the compound, the evaluation of polymorphic CYPs such as 2C9, 2C19, and 2D6 as primary clearance enzymes, the evaluation of the induction potential for CYPs involved in drug clearance (refer to Chapter 8 in this book) and the inhibition of these enzymes and drug transporters (refer to Chapters 7, 15, and 17 in this book; Balani et al., 2005). In some cases, acute single or repeat dose *in vivo* toxicology studies are conducted, usually based on the observation of organ toxicities from previous compounds in the class or LO program.

The first step in LO is the identification of the pharmacophore or component groups on the lead molecule responsible for the desired interaction with the target. Often, the pharmacophore is only a small portion of the entire molecule, and other portions are either (a) providing the framework to support the pharmacophore for either favorable or less favorable binding to the target or (b) extraneous and have no effect on binding of the pharmacophore. Identification of the pharmacophore, framework, and extraneous portions of the lead molecule requires systematic structural modifications, most commonly through selective deletion, of portions of the lead molecule to define these components. Lead optimization can then be pursued through selective additions, deletions, and functional group modifications of the lead molecule to maximize activity. Of course, it is also essential to optimize the biopharmaceutical and exposure or pharmacokinetic properties while minimizing the toxicity potential during LO. If LO is conducted *in vivo*, the structure can also be directly optimized for pharmacokinetic properties such as bioavailability, systemic exposure, distribution to target organs or tissues, and duration of action while minimizing the potential for drug–drug interactions via induction or inhibition of clearance enzymes (refer to Chapter 6 in this book). The systematic structure modifications lead to an understanding of the relationship between chemical structure and the activity, exposure, and safety properties of the lead. This structure–activity relationship (SAR) is often the most time-consuming phase of LO, requiring hundreds to thousands of compounds to optimize these properties.

1.5 CRITERIA FOR THE DEVELOPMENT CANDIDATE

During the last phase of LO, when one or a few compounds are judged to possess the optimal properties of exposure, activity, and safety, they are usually more fully characterized to verify their potential as drug development candidates. The most common discovery criterion for drug development is that there is compelling enough evidence to merit confidence that the compound will have the desired efficacy, exposure, and safety properties in the clinic. The specific criteria in each of these areas are compound-specific and depend on factors such as the seriousness of the disease, the properties of previous development candidates, the strength of the patent position, and the needed ease of administration of the dose and dose regimen to ensure compliance. In addition, there are a number of nondiscovery criteria that also influence the decision to advance a compound into development such as the competition from other drugs in development and existing marketed drugs, the anticipated manufacturing requirements and cost of producing drug product, the projected development times, the anticipated development and regulatory issues, and the market potential and other commercial issues.

Although the majority of development candidates arise from the drug discovery process or candidates that are licensed in from other companies, there are other, less common, sources for drug candidates. Drug metabolites have been a source for development compounds (Fura, 2006), and some notable examples have gone on to become drugs such as oxazepam, pravastatin, and fexofenadine. Side effects or off-target effects observed in the clinic have led to new or increased indications for some drugs such as Viagra, marketed for erectile dysfunction although originally developed for angina and hypertension, and Zyban, marketed for smoking cessation although originally developed as an antidepressant.

1.6 PERSPECTIVE

Pharmaceutical companies pursue similar processes for discovering new drugs. Although each company has its own specific criteria and metrics for measuring success during drug discovery, the most appropriate metric should be the success in meeting safety and efficacy criteria in patients with minimal compound attrition. By this measure the pharmaceutical industry has not become more successful and actually may have become less successful during the past decade in its efforts to discovery new drugs (Mathieu, 2007). As an industry, pharmaceutical companies have greatly advanced in their knowledge and accessibility to new drug targets and the technology for screening for hits. During the last two decades, the addition of screening assays relevant to drug disposition and drug–drug interactions in humans and the optimization of ADME properties during hit identification and lead optimization has had remarkable success in reducing the attrition from inadequate drug exposure during clinical development (Kola and Landis, 2004). The process and methods employed for lead optimization of pharmacological activity have not seen as dramatic an improvement in speed or scope, but their success in identifying development candidates has been maintained.

The greatest challenges facing successful drug discovery are those closest to meeting the primary objectives for a new drug: safety and efficacy. Only 1 of 10 compounds entering clinical development is approved as a new drug. The main contributors to this 90% attrition during clinical development are inadequate efficacy (27%) and inadequate safety (34%) (Fig. 1.1). The absence of better models and methods to predict drug efficacy for many new disease targets and to predict toxicities in humans portend that these will continue to be the major sources for drug attrition during development. This is recognized by the industry, and there is a growing emphasis on better target validation and predictive toxicity screens that can be implemented earlier in drug discovery. Solutions to these needs will be crucial to improving the success of the discovery process.

REFERENCES

Bailey SN, Wu RZ, Sabatini DM. Applications of transfected cell microarrays in high-throughput drug discovery. *Drug Discovery Today* 2002;7:16.

Baillie T. Future of toxicology-metabolic activation and drug design: challenges and opportunities in chemical toxicology. *Chem Res Toxicol* 2006;19:889–893.

Balakin KV, Kozintsev AV, Kiselyov AS, Savchuk NP. Rational design approaches to chemical libraries for hit identification. *Curr Drug Discovery Technol* 2006;3: 49–65.

Balani SK, Miwa GT, Gan L-S, Wu J-T, Lee FW. Strategy of utilizing *in vitro* and *in vivo* ADME tools for lead optimization and drug candidate selection. *Curr Topics Med Chem* 2005;5:1033–1038.

Behike MA. Progress towards *in vivo* use of siRNAs. *Mol Ther* 2006;13:644–670.

Bleicher KH, Bohm H-J, Muller K, Alanine AI. Hit and lead generation: beyond high-throughput screening. *Nat Rev: Drug Discovery* 2003;2:369–378.

Chang JC. HER2 inhibition: From discovery to clinical practice. *Clin Cancer Res* 2007;13:1–3.

Drews J. Drug discovery: a historical perspective. *Science* 2000;287:1960–1964.

Everts S. Piece by piece. *Chem Eng News* July 21, 2008:15–23.

Fura A. Role of pharmacologically active metabolites in drug discovery and development. *Drug Discovery Today* 2006;11:133–142.

Goodnow RA Jr. Hit and lead identification: integrated technology-based approaches. *Drug Discovery Today: Technologies* 2006;3:367–375.

Hajduk PJ, Greer J. A decade of fragment-based drug design: strategic advances and lessons learned. *Nat Rev: Drug Discovery* 2007;6:211–219.

Hammond S, Bernstein E, Beach D, Hannon G. An RNA-directed nuclease mediates post-transcriptional gene silencing in *Drosophila* cells. *Nature* 2000;404:293–296.

IHGSC International Human Genome Sequencing Consortium. Initial sequencing and analysis of the human genome. *Nature* 2001;409:860–921.

Johnson L, George FW, Neaves WB, Rosenthal IM, Christensen RA, Decristoforo A, Schweikert HU, Sauer MV, Leshin M, Griffin JE. Characterization of the testicular abnormality in 5 alpha-reductase deficiency. *Clin Endocrinol Metab* 1986;63: 1091–1099.

Kenakin T. Predicting therapeutic value in the lead optimization phase of drug discovery. *Nat Rev: Drug Discovery* 2003;2:429–438.

Khosla C, Keasling JD. Metabolic engineering for drug discovery and development. *Nat Rev: Drug Discovery* 2003;2:1019–1024.

Kitchen DB, Decornez H, Furr JR, Bajorath J. Docking and scoring in virtual screening for drug discovery: methods and applications. *Nat Rev: Drug Discovery* 2004;3: 935–949.

Kola I, Landis J. Can the pharmaceutical industry reduce attrition rates? *Nature Rev: Drug Discovery* 2004;3:711–715.

Kopec KK, Bozyczko-Coyne D, Williams M. Target identification and validation in drug discovery: the role of proteomics. *Biochem Pharmacol* 2005;69:1133–1139.

Kramer JA, Sagartz JE, Morris DL. The application of discovery toxicology and pathology towards the design of safer pharmaceutical lead candidates. *Nat Rev: Drug Discovery* 2007;6:636–649.

Mathieu MP, editor. Statistics on drug development: costs/complexity, development time, success rates. In: *Parexel's Bio/Pharmaceutical R&D Statistical Sourcebook 2007/2008*. Waltham, MA: PAREXEL International Corp.; 2007, p. 229.

McBeath G, Schreiber SL. Printing proteins as microarrays for high throughput function determination. *Science* 2000;289:1760–1763.

Natt F. siRNAs in drug discovery: target validation and beyond. *Curr Opin Mol Ther* 2007;9:242–247.

Ohlstein EH, Ruffolo RR, Jr, Elliott JD. Drug discovery in the Next Millennium. *Annu Rev Pharmacol Toxicol* 2000;40:177–191.

Pellecchia M, Becattini B, Crowell KJ, Fattorusso R, Forino M, Fragai M, Jung D, Mustelin T, Tautz L. NMR-based techniques in the hit identification and optimization processes. *Expert Opin Ther Targets* 2004;8:597–611.

Pillutla RC, Fisher PB, Blume AJ, Godstein NI. Target validation and drug discovery using genomic and protein–protein interaction technologies. *Expert Opin Ther Targets* 2002;6:517–531.

Sakurai T, Amemiya A, Ishii M, Matsuzaki I, Chemelli RM, Tanaka H, Williams SC, Richardson JA, Kozlowski GP, Wilson S, Arch JR, Buckingham RE, Haynes AC, Carr SA, Annan RS, McNulty DE, Liu WS, Terrett JA, Elshourbagy NA, Bergsma DJ, Yanagisawa M. Orexins and oresin receptors: a family of hypothalamic neuropeptides and G protein-coupled receptors that regulate feeding behavior. *Cell* 1998;92:573–585.

Sams-Dodd F. Target-based drug discovery: Is something wrong? *Drug Discovery Today* 2005;10:139–147.

Sasseville VG, Lane JH, Kadambi VJ, Bouchard P, Lee FW, Balani, SK, Miwa GT, Smith PF, Alden CL. Testing paradigm for prediction of development-limiting barriers and human drug toxicity. *Chem-Biol Int* 2004;150:9–25.

Schadt EE, Monks SA, Friend SH. A new paradigm for drug discovery: integrating clinical, genetic, genomic and molecular phenotype data to identify drug targets. *Biochem Soc Trans* 2003;31:437–443.

Szymkowski DE. Target validation joins the pharma fold. *Targets* 2003;2:8–9.

Terstappen GC, Schlupen C, Raggiaschi R, Gaviraghi G. Target deconvolution strategies in drug discovery. *Nat Rev: Drug Discovery* 2007;6:891–903.

Tobert JA. Lovastatin and beyond: the history of the HMG-CoA reductase inhibitors. *Nat Rev Drug Discovery* 2003;2:517–526.

Tornell J, Snaith M. Transgenic systems in drug discovery: from target identification to humanized mice. *Drug Discovery Today* 2002;7:461–470.

Venter JC, Adams MD, Myers EW. The sequence of the human genome. *Science* 2001;291:1304–1349.

Williams M. Target validation. *Curr Opin Pharmacol* 2003;3:571–577.

Windler H. Target validation requirements in the pharmaceutical industry. *Targets* 2003;2:69–71.

Wunber T, Hendrix M, Hillisch A, Lobell M, Meir H, Schmeck C, Wild H, Hinzen B. Improving the hit-to-lead process: data-driven assessment of drug-like and lead-like screening hits. *Drug Discovery Today* 2006;11:175–180.

Zambrowicz BP, Sands AT. Knockouts model the 100 best-selling drugs—will they model the next 100? *Nat Rev Drug Discovery* 2003;2:38–51.

2

MEDICINAL CHEMISTRY OF THE OPTIMIZATION OF ENZYME INHIBITORS

GERALDINE HARRIMAN, AMY ELDER, AND INDRANATH GHOSH

2.1 INTRODUCTION

Over the years the increased knowledge of the function of enzymes in signaling pathways has led to the identification of new targets and the discovery of new therapeutic agents directed at these targets. Enzymes have been well-exploited as medicinal targets, leading to discoveries that have greatly improved the quality of life of patients. Recently highlighted are the advancements within the kinase field. Several new molecular entities have entered the clinic, and certain inhibitors have been approved as new drugs for important therapeutic indications. (Fischer, 2004; Klebl and Mueller, 2005).

In the search for new medicinal agents, it is important to appropriately modulate the therapeutically targeted enzyme while being mindful not to also modulate enzymes in non-related biological processes. In this chapter, we will describe certain types of enzyme inhibitors in addition to highlighting some of the "off-target" enzymes or proteins that are considered when creating selective enzyme inhibitors that have good drug-like properties (DLPs). A variety of strategies are used by medicinal chemists to create molecules that are effective enzyme inhibitors but also avoid modulating metabolizing enzymes (such as the cytochrome P450s), being degraded (e.g., amidases) or sequestered by proteins (e.g., serum albumin) or interacting undesirably with other targets (e.g., the hERG channel). These are the topics that will be reviewed in this chapter.

Enzyme Inhibition in Drug Discovery and Development, edited by Lu and Li
Copyright © 2010 John Wiley & Sons, Inc.

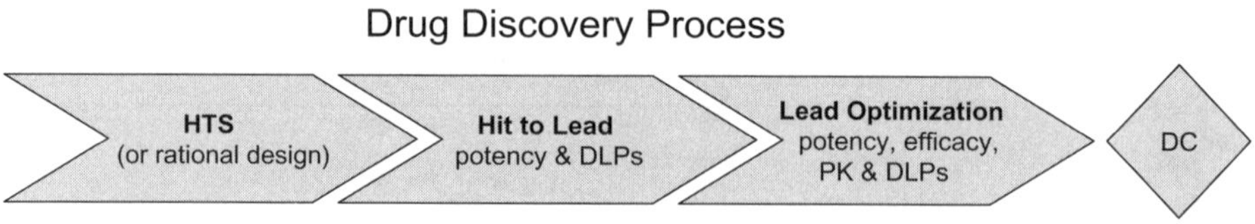

Figure 2.1. The process utilized by many drug discovery companies.

There are many approaches to the identification of novel chemical matter that initiate the involvement of medicinal chemistry. One common approach is to subject a well-characterized protein (enzyme, receptor, etc.) to a chemical library and "screen" the contents of this compound collection for small molecules that can modulate its activity (e.g., enzyme inhibition). This exercise is referred to as a high throughput screening (HTS) campaign (Colas, 2008; Dunlop et al., 2008; Kale et al., 2008; Mayr and Fuerst, 2008; Wunder et al., 2008). From this exercise, small molecules that affect protein function (i.e., screening hits) are selected for advancement into the next phase of the drug discovery process (Fig. 2.1). In the next phase, known as hit-to-lead (HTL), a battery of structural variants or analogs of each hit are synthesized, and their potency against the same biological target is determined. Through this iterative process, structure–activity relationships (SAR) are developed and drug-like properties (DLPs) are determined (solubility, selectivity against related and nonrelated proteins, stability against drug metabolism enzymes, etc.) (Lipinski, 2000; Li, 2001, 2005; Yamashita and Hashida, 2003; Wang et al., 2007; Gleeson, 2008). Key issues identified in the HTL stage are addressed with the synthesis and evaluation of additional analogs of the "lead" or best structures in the lead optimization (LO) stage. It is in this phase that desirable properties are optimized, and undesirable properties are eliminated or minimized, culminating in the identification of a development candidate (DC) which is then moved forward into preclinical development. These optimized properties usually include significant potency against the primary biological target, acceptable DLPs, significant efficacy in a pharmacological model of disease, and a pharmacokinetic profile that will translate into acceptable efficacious exposure in human. Simultaneous optimization of all these properties to a single DC can be quite a feat, because each one of these parameters involves its own unique structure–activity relationship. Some of the strategies used in the optimization of small molecules as potential therapeutics will be discussed in this chapter.

2.2 TYPES OF ENZYME INHIBITORS

Many drugs are enzyme inhibitors, inhibiting enzyme processes in the body by inhibiting either binding of the substrate or catalysis. Enzymes can also be activated by small molecules, but this is beyond the scope of this chapter

(Goode et al., 2008; Guertin and Grimsby, 2006; Evgenov et al., 2006; Thakur et al., 2007; Milne et al., 2007; Putt et al., 2006). Enzyme inhibitors can be broadly classified as reversible or irreversible depending on their binding (Robertson, 2005; Cleland, 2007; Copeland, 2007; Johnson, 2007; Hammes, 2008; Morrison, 2007; Pliska, 2007; Stojan, 2005; Tipton, 2007; Varfolomeev, 2005). Certain drugs are designed as reversible inhibitors because of the potential for lower side effects associated with a reversible inhibitor relative to an irreversible inhibitor. This chapter will focus on reversible inhibitors and their design and optimization. Reversible inhibitors can be further classified into competitive, uncompetitive, or noncompetitive inhibitors, depending on how an inhibitor binds to the target enzyme. Competitive inhibitors typically bind at the active site and compete with the substrate; the binding of each is mutually exclusive. This has been the primary objective for many medicinal chemistry programs; however, uncompetitive and noncompetitive inhibitors have also been successfully pursued. This chapter will describe examples from all of these classes of enzyme inhibitors.

2.2.1 Competitive Inhibition and Transition-State Mimetics

One method to design competitive enzyme inhibitors is based on a rational-drug design process which takes advantage of known structures of substrates and mimics these in the design of inhibitors. The premise for this strategy is that the inhibitor can bind in the active site in a similar way to that of the substrate without forming product, while inhibiting the binding of the substrate. As a result, many inhibitors exhibit a striking resemblance to enzyme substrates, and "compete" for their binding sites. This approach has been used successfully in the kinase inhibitor field where many compounds are competitive with the kinase substrate ATP (Liu and Gray, 2006).

Certain classes of molecules, such as the aminoquinazolines, have been shown to competitively bind to the ATP binding pocket (Fig. 2.2) of kinases and inhibit their function. Examples of competitive inhibitors are the marketed drug Gefitinib, and its analog Erlotinib (Fig. 2.3) (Clark et al., 2005), which are both ATP competitive EGFR (epidermal growth factor receptor) kinase inhibitors. (Laufer et al., 2005; Crespo et al., 2008). These inhibitors are structurally similar to ATP and make similar key hydrogen bond interactions with the kinase. Figure 2.4 illustrates the binding of Erlotinib in the ATP binding pocket of EGFR, forming key interactions with the kinase at the ATP binding site.

Many examples of transition-state mimetics exist in the aspartic protease field where the inhibitors have been designed to mimic the proposed transition-state structure during cleavage of the enzyme substrate (Fig. 2.5). (Bursavich and Rich, 2002; Yokokawa et al., 2008) Inhibitors in the aspartic protease field have been designed to target diseases such as AIDS, Alzheimer's disease, and hypertension. Ritonavir was at least partially designed based on the structure of the proposed transition-state hydrolysis intermediate of the

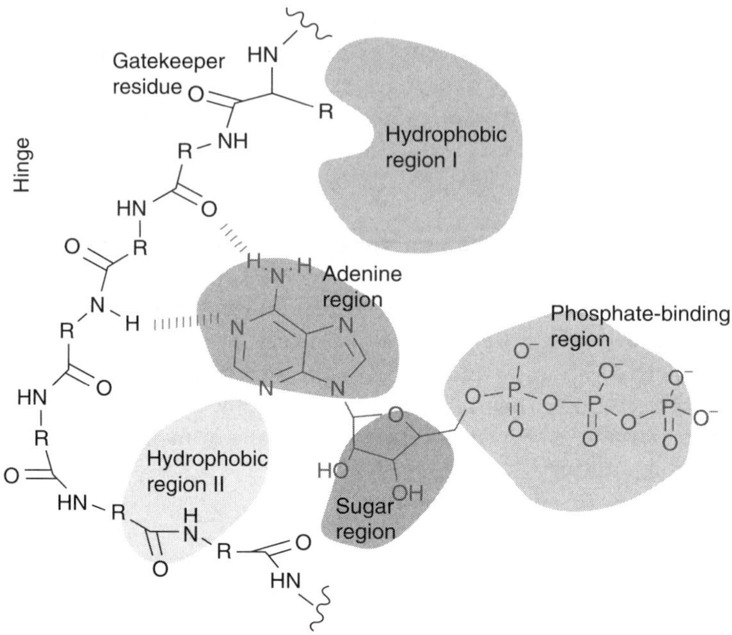

Figure 2.2. ATP binding in the active site, illustrating the important hydrogen bonds made between the backbone of the protein (the hinge region) and its substrate ATP. See the insert for color representation of this figure.

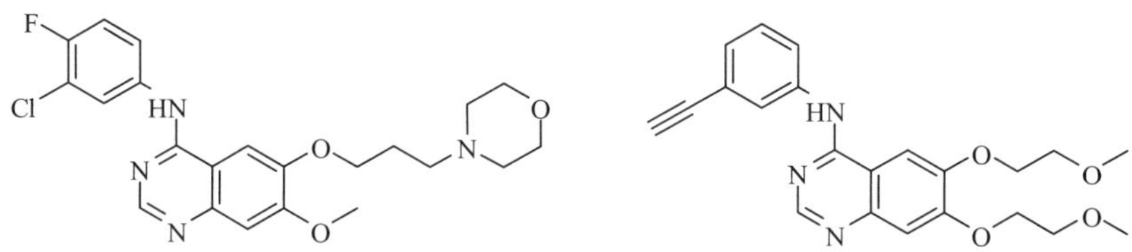

Figure 2.3. Structure of Gefitinib and Erlotinib, both ATP competitive inhibitors of EGFR.

viral polypeptide substrate by HIV protease and the symmetry of the protease (Fig. 2.6). (Kempf et al., 1998) Initial attempts to design inhibitors of HIV proteases included key transition state mimetics (TS mimetics) which lacked the scissile peptide bond but mimicked the hydrolysis of the HIV polypeptide substrate by the protease (Boudes and Geiger, 1996; Kempf et al., 1998). A significant number of transition state mimetic HIV protease inhibitors have been advanced into the clinic or approved as drugs, including Saquinavir, Ritonavir, Indinavir, Amprenavir, Lopinavir, and Nelfinavir (Mastrolorenzo et al., 2007; Nguyen et al., 2008)

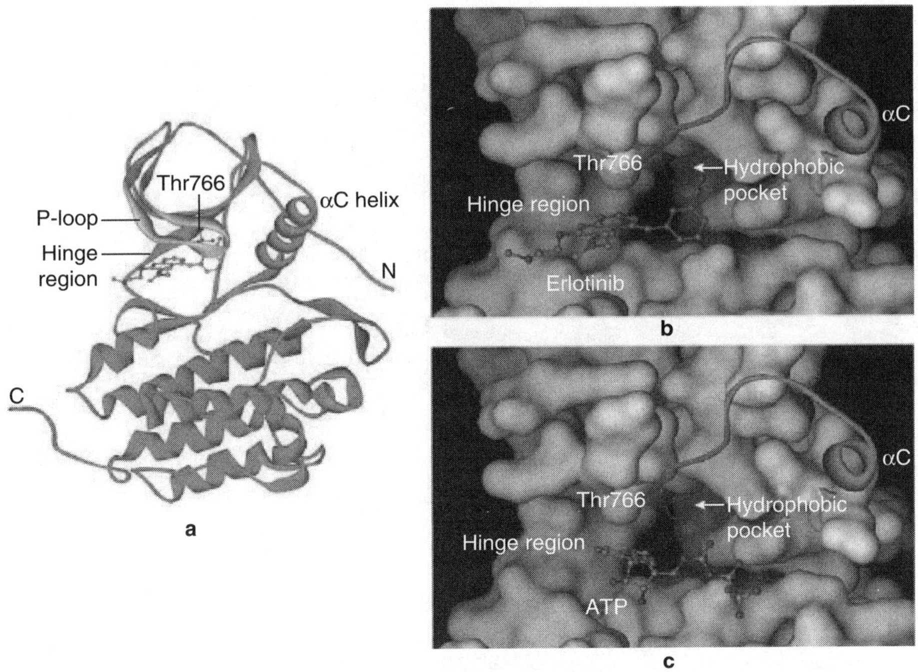

Figure 2.4. Erlotinib bound in the ATP binding pocket. See the insert for color representation of this figure.

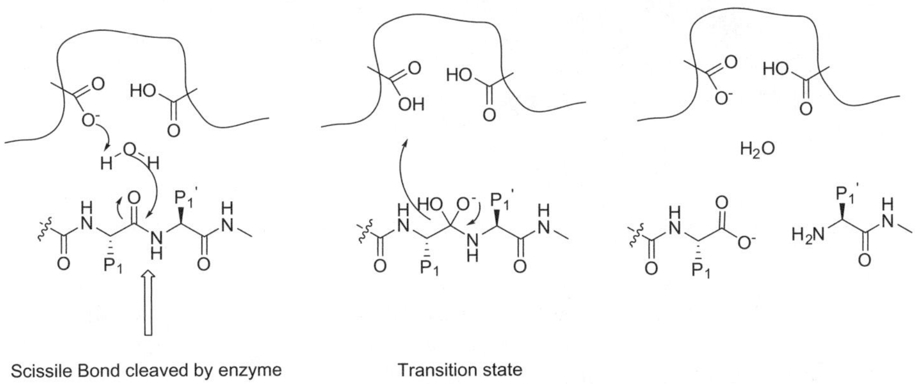

Figure 2.5. Proposed cleavage of peptide substrate by aspartic protease.

Transition-state mimetic inhibitors also take advantage of the better binding resulting from what has been referred to as an induced fit model. This model was proposed by Koshland to explain the unexpectedly high enzyme affinity exhibited by TS mimetics (Thomas and Koshland, 1960) More recently, compounds in the HIV protease field have used the growing body of knowledge around the TS analogs to solve some of the limitation within this field of

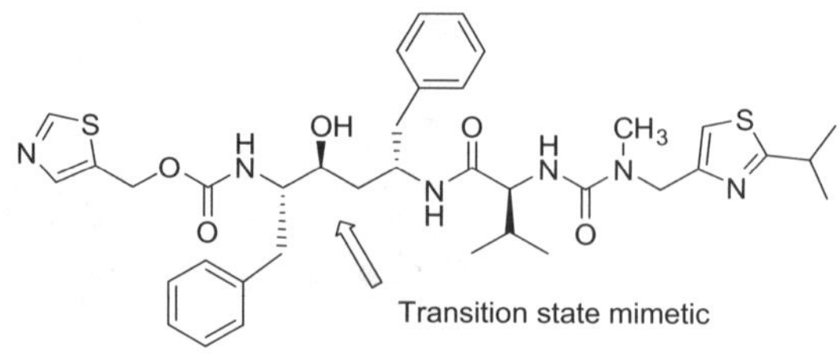

Figure 2.6. Ritonavir.

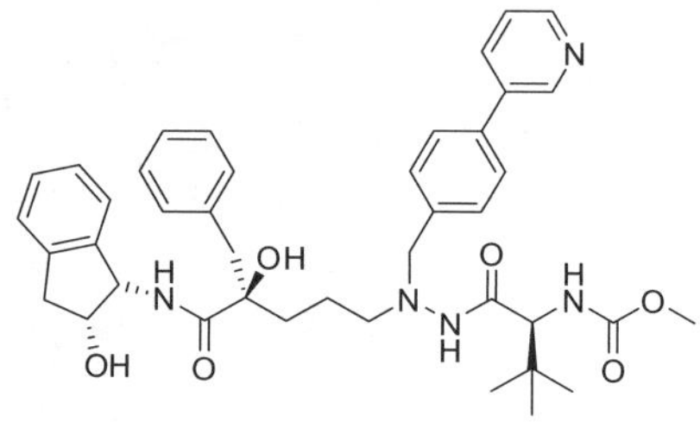

Figure 2.7. "Shielded inhibitor" of HIV protease.

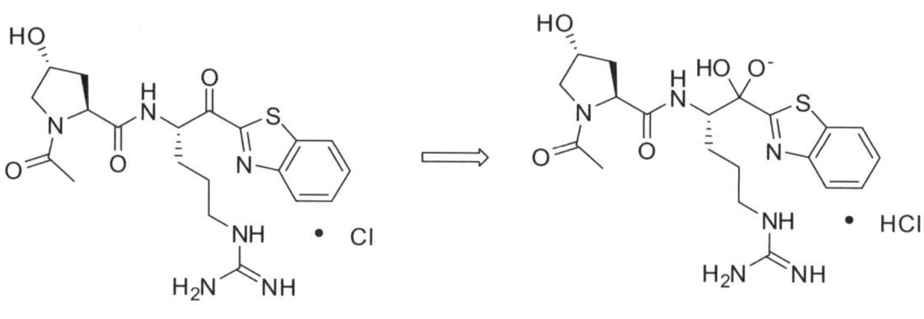

Figure 2.8. Tryptase inhibitor utilizing an activated ketone as a TS mimetic.

inhibitors, including the low oral bioavailability, poor aqueous solubility, high protein binding, and overall poor pharmacokinetic properties. Ekergren and co-workers described inhibitors containing the traditional alcohol-based transition-state mimetic with a twist, converting the secondary alcohol to a tertiary alcohol to create "shielded inhibitors" (Fig. 2.7) (Ekegren et al., 2005, 2006a,b; Wu et al., 2008). Pleasingly, a set of these inhibitors exhibited increased permeability and excellent human microsomal stability which has the potential to improve overall pharmacokinetic properties. TS mimics have also been used as inhibitors of the serine protease, human mast cell tryptase. TS tryptase inhibitors have been designed with the presence of a heterocyclic activated ketone as the transition-state mimetic. This ketone, when hydrated, forms the key transition structure that mimics hydrolysis of the substrate (Fig. 2.8)

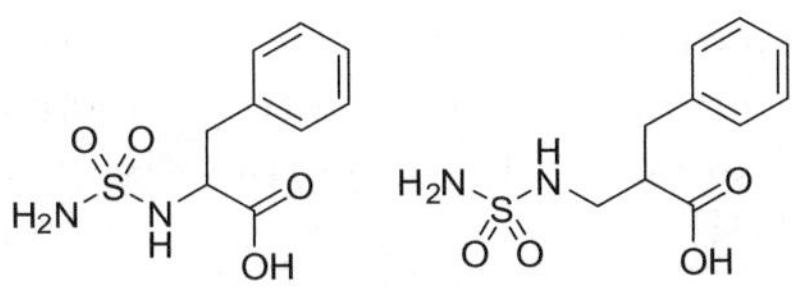

Figure 2.9. Carboxypeptidase A inhibitors.

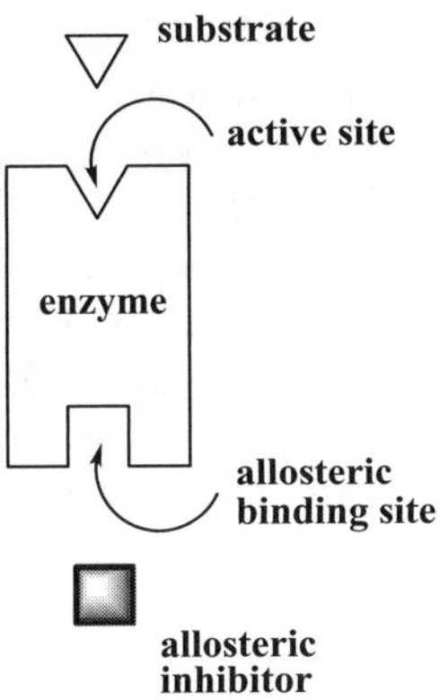

Figure 2.10. Inhibition of an enzyme by an allosteric inhibitor.

(Costanzo et al., 2003). In addition, novel transition state mimetics have been discovered for use in zinc proteases. Dae Park and colleagues discovered that the sulfonyl urea based inhibitors can act as novel transition state analogs to produce inhibitors of carboxypeptidase A (Fig. 2.9) (Park et al., 2002).

2.2.2 Noncompetitive/Allosteric Inhibitors

Noncompetitive inhibitors bind to a site that is remote from the active site (Fig. 2.10). The binding of a noncompetitive inhibitor to the enzyme elicits a conformational change within the enzyme. This conformational change no longer allows the substrate to bind to the enzyme in a fully productive manner. During this process, both the substrate and inhibitor may remain capable of binding to the enzyme at the same time to create a nonproductive complex. In contrast to competitive inhibitors, a noncompetitive inhibitor cannot be competed out of the enzyme by the substrate and can still effectively inhibit the enzyme under high substrate concentrations. At a simple mechanistic level, noncompetitive and allosteric inhibition appears to be the same and the terms are often used interchangeably.

Although drug discovery programs have been very successful in identifying competitive inhibitors and excellent drugs, for some therapeutic targets, these inhibitors may not provide the target selectivity required for closely related receptor subtypes and may produce undesirable off-target side effects. In this respect, allosteric inhibitors are particularly attractive and can potentially address many of the undesirable properties associated with competitive inhibitors (Kenakin, 2004).

2.2.3 Protein Kinase Inhibition: A Good Example of Allosteric Inhibition

The vast majority of protein kinase inhibitors (PKIs) reported to date are ATP-competitive and bind at the ATP binding site, which is common to all kinases and highly conserved. Thus, it was believed that developing selective small-molecule PKIs would be a challenging endeavor. Although selective inhibitors competitive with ATP have been identified (Fischer, 2004; Parang and Sun, 2004; Barnett et al., 2005) for specific protein kinases, it remains to be seen whether side effects or resistance will appear with these inhibitors in long-term clinical studies (Cowan-Jacob et al., 2004).

An alternative way to avoid target promiscuity in the design of kinase inhibitors is to look for noncompetitive inhibitors, with the aim of obtaining interactions remote from the ATP site. In this way, regions outside the ATP binding pocket can be targeted for designing allosteric inhibitors that cause significant conformational changes in the target kinase. These compounds are likely to be more selective within the kinase family with fewer off-target side effects. Thus, there is growing interest in identifying new classes of allosteric PKIs that do not directly compete with ATP (Bogoyevitch and Fairlie, 2007), and some of these novel small molecule inhibitors are discussed below.

BIRB 796 (Fig. 2.11) is a potent selective inhibitor of p38 mitogen-activated protein kinase (MAPK), a Ser/Thr protein kinase (Pargellis et al., 2002; Regan et al., 2002, 2003a,b; Mol et al., 2004) It was developed from a diaryl urea (Fig. 2.11), which was identified from a high-throughput screening campaign. The allosteric mode of inhibition of this class of compounds was evident from the crystal structure of the urea with human p38 MAPK, which showed the compound bound to the enzyme at a site remote from the ATP binding region (Fig. 2.11) (Regan et al., 2002). In the case of the diaryl urea inhibitor and BIRB 796, the mechanism of inhibition involves a large conformational change in the enzyme which blocks productive binding of ATP (Regan et al., 2002).

Interestingly, BIRB 796 appears to be kinetically "ATP-competitive" in its inhibition despite its allosteric mechanism of action. However, there are other allosteric protein kinase inhibitors that show noncompetitive kinetics with respect to ATP (Martinez et al., 2002; Davidson et al., 2004; Bogoyevitch and Fairlie, 2007) Two examples of such allosteric PKIs are shown in Fig. 2.12. Compounds PD098059 and PD318088 are inhibitors of MEK1/2 (Favata et al., 1998; Sebolt-Leopold et al., 1999). A crystal structure of MEK1 with PD318088 in the presence of Mg^{2+} and ATP indicates that PD318088 does not inhibit ATP binding; however, in the presence of PD318088 the enzyme

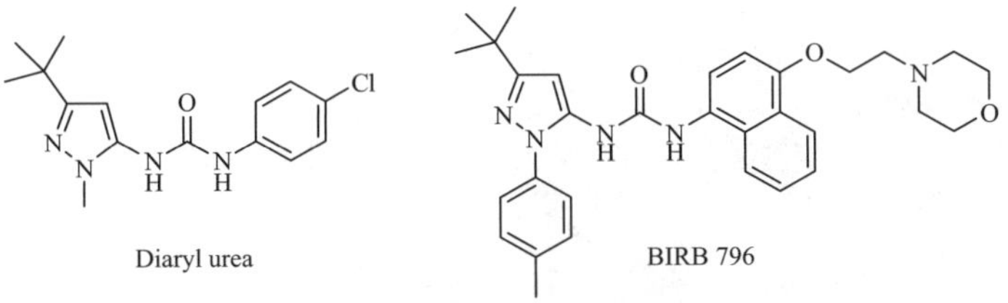

Diaryl urea BIRB 796

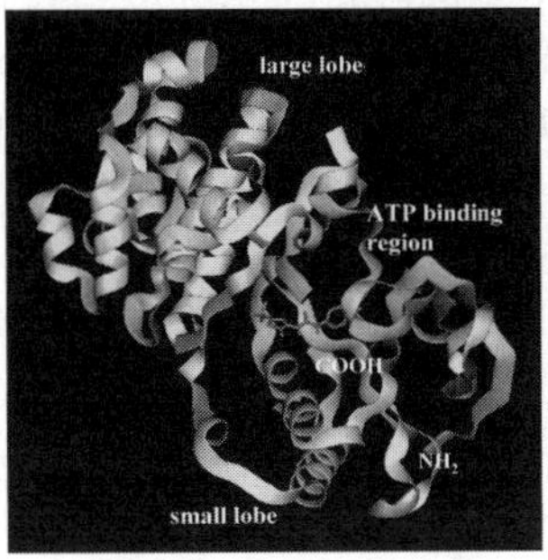

Figure 2.11. Diaryl urea and BIRB 796 are allosteric inhibitors of p38 MAPK. Also shown is the crystal structure of human p38 MAP kinase and diaryl urea (blue). See the insert for color representation of this figure.

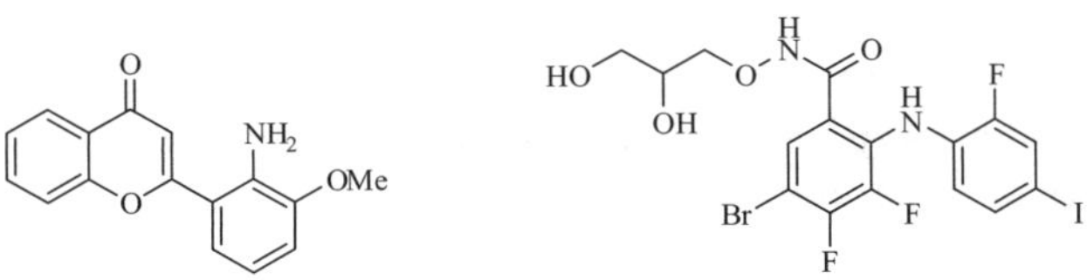

PD098059 PD318088

Figure 2.12. Compounds PD098059 and PD318088 are non-ATP competitive allosteric inhibitors of MEK1/2.

undergoes significant conformational changes in the kinase activation loop and N-terminal lobe resulting in inhibition of the kinase (Ohren et al., 2004).

2.3 IMPROVEMENT OF DRUG-LIKE PROPERTIES (DLPs)

Drugs encounter numerous hurdles in the form of processes in the human body that may prevent them from reaching their final targets. In the drug discovery process, it is the job of the medicinal chemist to identify likely hurdles and design molecules that either overcome or avoid these obstacles. The goal of this process is to obtain development candidates that can effectively and selectively reach their intended biological target *in vivo*, elicit the desired effect, at the same time not affect other targets, which would result in

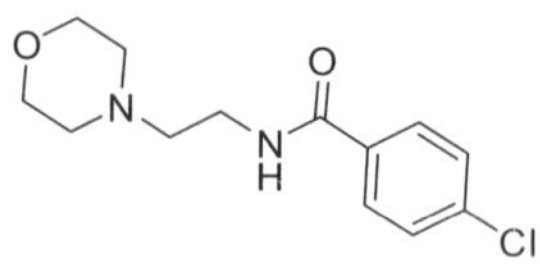

Figure 2.13. Aurorex.

potential undesired effects. Such undesired effects may be driven by the molecule's interaction with transporters, metabolizing enzymes, absorption mechanisms, and so on. This part of the chapter will address a few examples of medicinal chemistry strategies to overcome these hurdles.

2.3.1 Avoiding Enzyme-Related Associated Side Effects

The area of antidepressants offers one good example of a drug discovery process that has resulted in safer and more efficacious drugs. The monoamine oxidase inhibitors were first discovered in the 1950s and 1960s, but exhibited severe side effects including cardiovascular side effects, hypertensive crisis, hyperpyrexia, and psychosis. These side effects were believed to be partly due to the poor selectivity of binding to the Apo-A and B enzyme and the irreversible nature of the monoamine oxidase inhibition. Improvements in selectivity and reversibility resulted in the identification of reversible inhibitors of monoamine oxidase type A (RIMAs), examples of these include Aurorex (Fig. 2.13), which was the first RIMA (Lotufo-Neto et al., 1999). Aurorex no longer requires close dietary supervision by a physician, which is still needed for older monoamine oxidase inhibitors.

2.3.2 Improving Drug-like Properties; Strategies to Improve Plasma Stability

Beyond inhibitor selectivity, many programs have struggled with the issue of stability in plasma, which ultimately will limit the amount of compound that reaches the site of action. Within the antiviral field, the human cytomegalovirus protease has proven to be a difficult target to drug. Borthwick and coworkers have described successful efforts to overcome the problems of both poor solubility and poor plasma stability of their inhibitors (Borthwick et al., 2003). Modification to the lactam nitrogen in the inhibitors resulted in increased stability in human plasma and whole-cell antiviral activity (Table 2.1). One strategy utilized increased steric hindrance around the lactam and the second included the deactivation of the carbonyl of the lactam, making it less susceptible to enzymatic hydrolysis. This was achieved by substituting the lactam nitrogen with different groups, including a cyclopropyl carbonyl (with various substitution patterns) as well as electron-donating and electron-withdrawing aryl and heteroaryl groups. The best plasma stability was observed with an electron-withdrawing heterocyclic group that resulted in human plasma

TABLE 2.1. Human Plasma Stability of Lactams

	Tested in Fresh Human Plasma at 100 μM			
		Human Plasma Stability $t_{1/2}$ (h)		Human Plasma Stability $t_{1/2}$ (h)
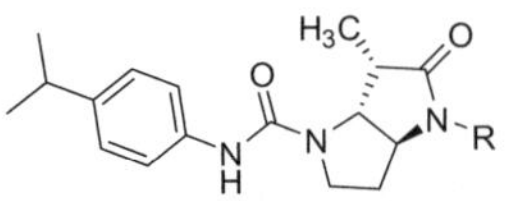	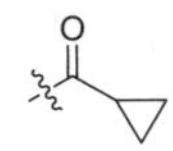	0.5	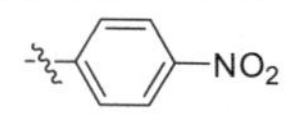	>24
	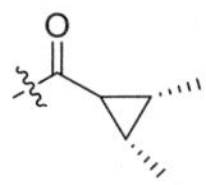	1.5	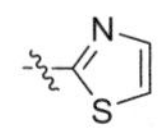	>24
	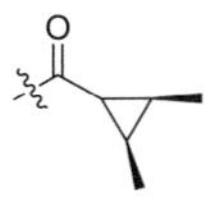	6.0	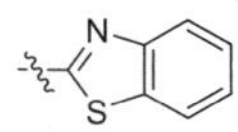	>24
	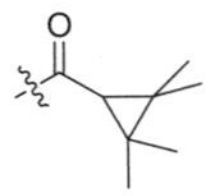	16.0		

stability of >24 h, however, acceptable antiviral activity was not achieved except with a benzothiazole substitution.

2.3.3 Improving Drug-like Properties; Strategies to Avoid Potential Drug–Drug Interactions

Cytochrome P450 enzymes are the major metabolizing enzymes in the human body. Inhibition of P450 enzymes is undesirable because of the potential risk of side effects due to drug–drug interactions. Overcoming this challenge in drug discovery continues to be a major challenge for medicinal chemists due to the promiscuous nature of the cytochrome P450 enzymes. Companies have developed *in vitro* HTS methods to help predict drug-drug interactions early in drug discovery. In addition, computational models have been developed to identify structural features that are likely to inhibit the five major cytochrome P450 isozymes (1A2, 2D6, 2C9, 2C19, and 3A4) (Yan and Caldwell, 2001). Developing SAR around compounds that exhibit P450 inhibition continues to be a challenge for medicinal chemists. However, successful examples using true rational drug design strategies have been reported in the literature, and a few examples will be shown.

Nicotinic acid (niacin) has recently been shown to bind to the GPCR receptor, GPR109A (Raghavan et al., 2008). Niacin is well known for its ability to raise HDL-cholesterol levels, the "good cholesterol," and is the leading marketed drug for this purpose. A group at Merck was interested in identifying a series of compounds that bound to the GPCR receptor GPR109A, in hopes of identifying other potent niacin receptor agonists. The group screened their library and identified an initial anthranilic acid-based hit (Fig. 2.14). This hit

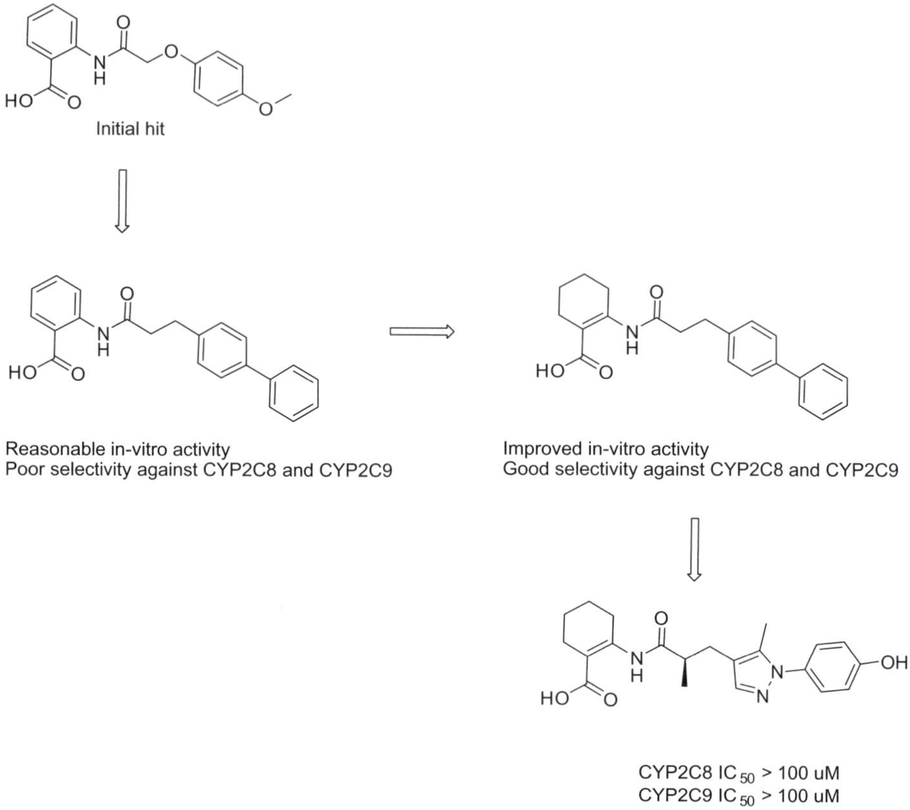

Figure 2.14. Optimization of anthranilic acid hit to a tetrahydro anthranilic acid derivatives.

was elaborated to a biphenyl anthranilic acid derivative that showed undesirable activity against the CYP2C8 and CYP2C9 P450 subtypes. Replacements of the anthranilic acid with a tetrahydroanthranilic acid proved to be acceptable with improved *in vitro* activity against the niacin receptor and an improved overall profile for the cytochrome P450 isoforms, CYP2C8 and CYP2C9.

Medicinal chemists involved in the discovery of human gonadotropin-releasing hormone receptor antagonists (*h*-GnRH-R) identified a series of potent uracil-containing analogs, which also exhibited potent inhibition of cytochrome CYP3A4. SAR work showed that the incorporation of an acid functionality could separate activity at the *h*-GnRH-R from that at CYP3A4 (Fig. 2.15) (Chen et al., 2008). Kim and co-workers used a similar strategy of incorporation of an acidic carboxylic acid to overcome inhibition of CYP3A4 in the dipeptidyl peptidase IV (DPP-IV) serine protease (Fig. 2.16) (Jun et al., 2008).

Data-mining has been used to design computational models to predict structural features that will inhibit specific P450 enzymes. A variety of differ-

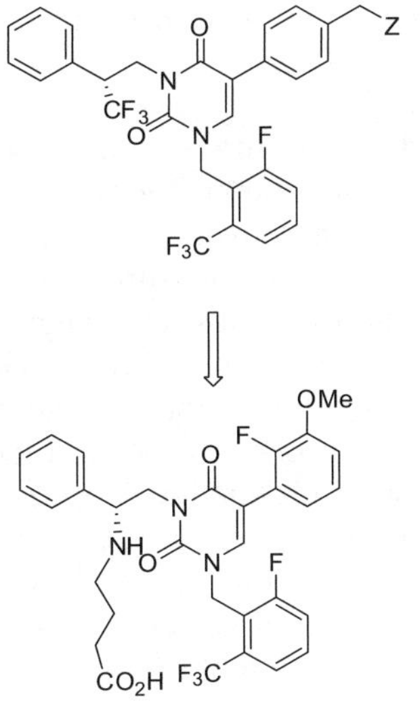

Z	GnRH K_i (nM)	CYP3A4 IC_{50} (uM)
-OH	69	0.80
$-NH_2$	1100	9.1
$-NHCH_2C(O)OEt$	110	0.35
-OCHC(O)OH	>10,000	20
$-NHCH_2CH_2C(O)OH$	>10,000	31

GnRH
K_i = 2.2 nM

CYP3A4
IC_{50} = 36 uM

Figure 2.15. Optimization of the uracil core to differentiate GnRH antagonism and CYP3A4 inhibition.

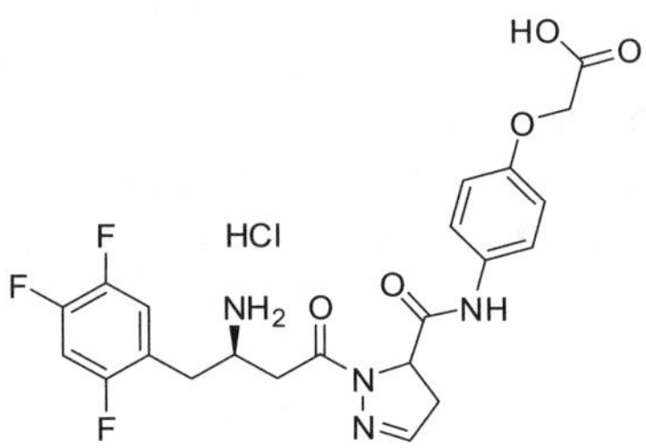

DPP-IV
K_i = 0.51 μM

CYP3A4
K_i = 59.3 μM

Figure 2.16. Dipeptidyl peptidase IV inhibitor.

ent computational models have been developed and used as predictors to filter compounds with structural features predicted to inhibit specific cytochrome P450's. Multiple groups have used Quantitative Structure Activity Relationship (QSAR) models using different statistical approaches, including partial least squares (PLS), multiple linear regression (MLR), classification and regression trees (CART) and Bayesian neuronal network (BNN) to name a few (Chohan et al., 2005; Burton et al., 2006; Olsen et al., 2006; Hudelson et al., 2008; McMasters et al., 2007; Jensen et al., 2007).

2.3.4 Improving Drug-like Properties; Strategies to Improve Intestinal Absorption and Cell Permeability

Absorption of compounds is crudely described as the movement of compounds into the bloodstream and can greatly affect the ultimate bioavailability of a compound. Multiple routes of administration can deliver compounds to the bloodstream, including intravenous, intramuscular, or subcutaneous, however, most common for human therapeutics and the most challenging is oral delivery. The delivery of compounds orally involves the additional challenges that a compound must undergo dissolution, solubilize, and be absorbed across the gutwall. In order for compounds to be absorbed into the bloodstream, they must cross the gastrointestinal wall, which is lined with epithelial cells. Compounds that target the brain have the additional challenge of crossing the blood–brain barrier (BBB). Compound ionic charge, size, and overall polar surface area as well as the presence or absence of certain functional groups can effect how well compounds cross the gut wall and/or the BBB. In addition, the presence of transporter at either site can decrease the ultimate absorption of compounds. Absorption remains a major focus for medicinal chemists since compounds must be absorbed before an *in vivo* pharmacological response can be elicited. For this reason, there exist a number of *in vitro* and *in silico* predictive models of drug absorption across the gut wall and BBB permeability. *In vitro* tools include the PAMPA assay (parallel artificial membrane permeation assay), which is a predictor of passive diffusion only and is a measure of permeability across an artificial membrane without interference from other factors. The PAMPA assay does not predict paracellular transport, active uptake, or efflux by transporters. As a result, the PAMPA assay allows the ability to rank order compounds based solely on passive permeability through an artificial membrane. Additional *in vitro* models for predicting absorption include cell-based assays employing either Caco-2, the MDCK cell line with and without MDR1 transfection, or the bovine brain microvessel endothelial cell (BBMEC) (Terasaki et al., 2003; Garberg et al., 2005; Irvine et al., 1999). The Caco-2 assay evaluates compound permeability due to transcellular passive diffusion as well as active and paracellular transport. This cell line is a cultured human colonic cell line that expresses transporters present in the gut wall, and for this reason it can serve as a very good predictor of physiological drug absorption. Results from the Caco-2 assay provide information about the compounds' effects on transporters, including the P-glycoprotein transporter. The MDCK cell line has gained popularity as a better predictor of brain penetration than the Caco-2 cell line. In addition, the MDCK cell line is a better predictor for efflux for compounds that are recognized and effluxed by the P-glycoprotein (P-gp) versus the Caco-2 cell line (which has the presence of multiple transporters). The topic of transporters is discussed within this text. Some examples of the use of these *in vitro* tools to drive medicinal chemistry to identify compounds with improved drug-like properties will be highlighted.

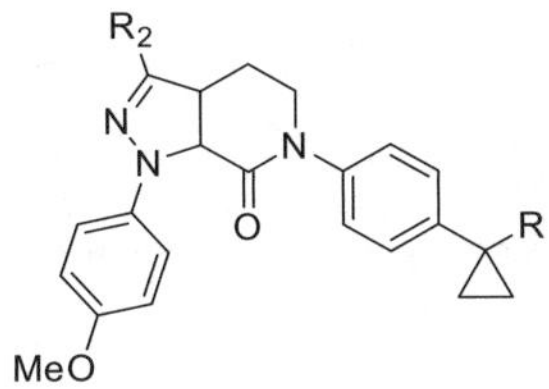

R_2	R	Caco-2 P_{app} (nm/s)
$-CF_3$	$-CH_2NHMe$	26
$-SO_2Me$	$-CH_2NHMe$	1
$-CONH_2$	$-CH_2NHMe$	1
-CN	$-CH_2NHMe$	6

R_2	R	Caco-2 P_{app} (nm/s)
$-CF_3$	$-CH_2NMe_2$	85
$-SO_2Me$	$-CH_2NMe_2$	8
$-CONH_2$	$-CH_2NMe_2$	5
-CN	$-CH_2NMe_2$	46

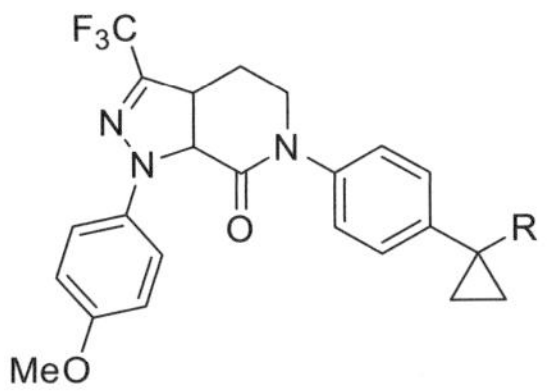

R	Caco-2 P_{app} (nm/s)
$-CO_2Et$	not detected
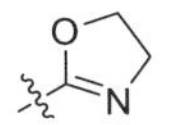	145
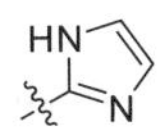	57

Figure 2.17. Factor Xa inhibitors with improved permeability.

Qiao and co-workers at BMS describe work to discover efficacious factor Xa inhibitors (Qiao et al., 2009). Modification of functional groups greatly affected the permeability of the compounds based on evaluation using the Caco-2 assay. Modification of groups such as sulfonyl and amides, which are known to reduce permeability due to the higher polar surface area, showed a reduced permeability as measured using the Caco-2 assay (Fig. 2.17). However, the presence of a trifluoromethyl group greatly increased permeability. Subsequent modifications to the pyrazolodihydropyridone core were investigated. The group also achieved improvements in permeability through heterocyclic isosteric replacements of esters and acids. An additional example has been shown in the design of novel noradrenaline reuptake inhibitors. Fish and co-workers designed molecules with modification to a key amide that showed modulation of permeability and a decrease in efflux by the P-gp transporter using the MDCK cell line (Fig. 2.18). (Fish et al., 2008). This decrease in efflux is attributed to the increased lipophilicity and the altered H-bond capacity of the amide.

In the field of phosphatase inhibitors a continued challenge has been the ability to obtain permeable, orally bioavailable compounds. Key recognition binding motifs utilized in many phosphatase inhibitors are polar functionalities that mimic the phosphate present on the substrate. For this reason, many

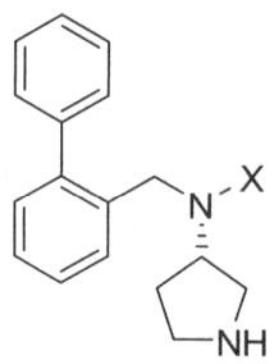

X	MDCK-MDR1 AB/BA	MDCK-MDR1 ER
-COi-Pr	11/68	6.1
$-SO_2Me$	10/51	5.1
-C(O)OEt	22/44	2.0

Figure 2.18. Novel noradrenaline reuptake inhibitors.

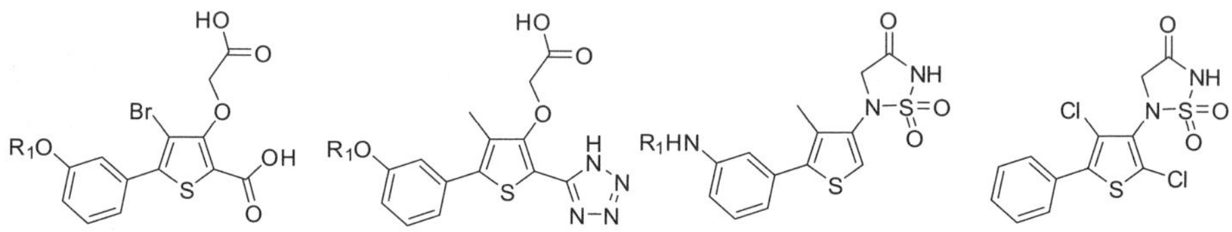

R_1 = cyclic benzylsulfonamide

PAMPA (Pe x 10^{-6} cm/s)			
0	0.22	0.19	3.54

Figure 2.19. PTP1B inhibitors.

phosphatase inhibitors contain an acidic carboxylic acid (Fig. 2.19). Inhibitors designed for PTP1B (protein tyrosine phosphatase 1B) by a group at Wyeth Research show the effect of a diacid on cellular permeability. Compounds with this functional group showed decreased passive permeability based on measurements using the PAMPA assay (Wan et al., 2007). However, replacing the acid with a well-known acid isosteric replacement, the tetrazole, slightly improves the passive permeability. Subsequent optimization and replacement of the acid with a cyclic thiadiazolidine improved the passive permeability to an acceptable level.

In silico models can also be used to predict compound absorption based on lipophilicity, H-bond capacity, molecular size, polar surface area (PSA), and quantum properties. (Borchardt et al., 2004). Many models have been generated to predict compound absorption and ability to penetrate the BBB (Clark, 2005), in addition, the well known Lipinski rules help predict certain properties of compounds that could improve absorption. (Lipinski, 2000) As with the models that have been generated for the prediction of P450 inhibition, these models provide guidance for compound selection and evaluation. Ultimately, advancing a compound into *in vivo* pharmacokinetic studies provides the most accurate readout of compound absorption and bioavailability.

2.3.5 Improving Drug-like Properties; Strategies to Avoid hERG Binding

Among the many safety issues considered during drug discovery research and development, it is likely that no other off-target event has received more recent attention than binding to the hERG channel (*h*uman *E*ther-a-go-go *R*elated *G*ene or *KCNH2)*. The hERG channel is a potassium-gated ion channel, which is expressed in the heart and nervous system and found in tetrameric pores on the cell surface. Compounds that bind or interact with the hERG channel have been shown to affect repolarization of the heart following a depolarization event. This is often described as QT prolongation, an event that affects the time between ventricular depolarization (Q point) and ventricular repolarization (end of T wave) (Roden, 2004). Affecting or prolonging the QT interval leads to problematic endpoints such as ventricular arrhythmias, torsade de pointes (Webster et al., 2002), and ventricular fibrillation (Casis et al., 2006; Kamiya et al., 2006; Sanguinetti and Tristani-Firouzi, 2006).

Over the past decade, many important drugs have been removed from the market due to their cardiotoxicity. Due to the hydrophobic nature of the hERG channel, it has been shown to effectively bind a diverse set of pharmaceutically important compounds. Reports in the literature have illustrated that certain drug design strategies can be utilized to avoid undesired interactions with the hERG channel. Some successful strategies employed include standard SAR to eliminate hERG activity as well as several computational approaches (Pearlstein et al., 2003a,b). Here we review four approaches that have been successfully used to attenuate hERG binding affinity: modifying hydrophobic interactions with key amino acid residues in the channel, decreasing molecular lipophilicity of hERG ligands, decreasing basicity of the ligands, and creation of zwitterionic molecules that disfavor hERG binding (Jamieson et al., 2006).

Several examples appear in the literature that support an approach which considers the interactions made between the ligand and the hERG channel that utilizes rational drug design approaches (Pearlstein et al., 2003). These reports describe the necessary interactions between hydrophobic groups found on the ligands interacting with hERG's tyrosine 562. Disruption of, or decreases in, these hydrophobic–hydrophobic interactions have opened an avenue for some researchers to obtain acceptable *in vitro* selectivity windows. Figure 2.20

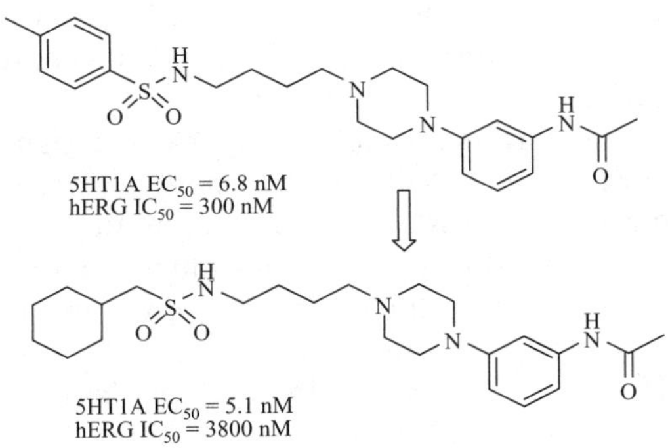

Figure 2.20. Modulating hydrophobic interactions to decrease hERG affinity.

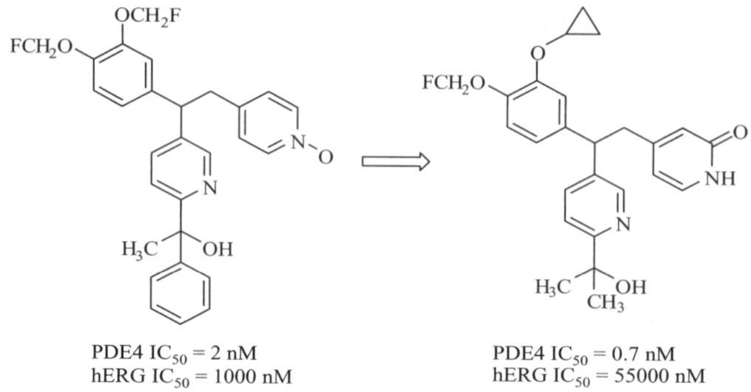

Figure 2.21. Modifications to lipophilicity to modulate hERG binding.

illustrates this approach. Replacement of the *p*-tolyl group of the 5HT1A agonist with a cyclohexylmethylene allowed for retention of agonist activity against 5HT1A while decreasing hERG affinity (Becker et al., 2006).

Other strategies include decreasing the lipophilicity of the hERG ligands. It is believed that it is critical for the hERG ligands to access the pore and bind to the hydrophobic residues next to the pore surface. The residues found near the top of the channel pore favor the binding of hydrophobic compounds. Many groups have found that decreasing the lipophilicity of hERG ligands can decrease their affinity for the channel. An example of this is shown below, where a series of PDE4 inhibitors were optimized and this strategy was employed to effectively decrease the affinity of this series of inhibitors for the hERG channel (Figure 2.21) (Friesen et al., 2003). Replacement of a phenyl with a small methyl substituent in addition to the modification of the pyridine to the pyridone allowed for a sufficient decrease in lipophilicity to obtain a 50-fold decrease in hERG affinity while improving potency against PDE4A.

The strategy of adding polar functional groups to hERG ligands has also been effective at attenuating the affinity of molecules for this channel (Figure 2.22). This approach is especially effective when basic nitrogen-containing hERG ligands are converted to their corresponding zwitterion. The most well known example of this strategy is the transformation of terfenadine to fexofenadine. Terfenadine, an antihistamine, was removed from the market in 1997 as a result of reports of cardiac arrhythmias. The addition of the carboxylate, which results in a zwitterionic structure, removes a significant amount of hERG affinity from the molecule. Fexofenadine is a currently marketed antihistamine with no reported hERG-associated liabilities.

A series of KDR (kinase domain receptor) inhibitors were developed utilizing the pyrazolopyrimidine scaffold. Initial lead compound in this series exhibited high affinity for KDR but with insufficient *in vitro* selectivity against the hERG channel. In Fig. 2.23, modifications of the basicity of an appended piperazine substituent increased the window of affinity to the desired target over the hERG channel (Dinges et al., 2007). This approach also proved

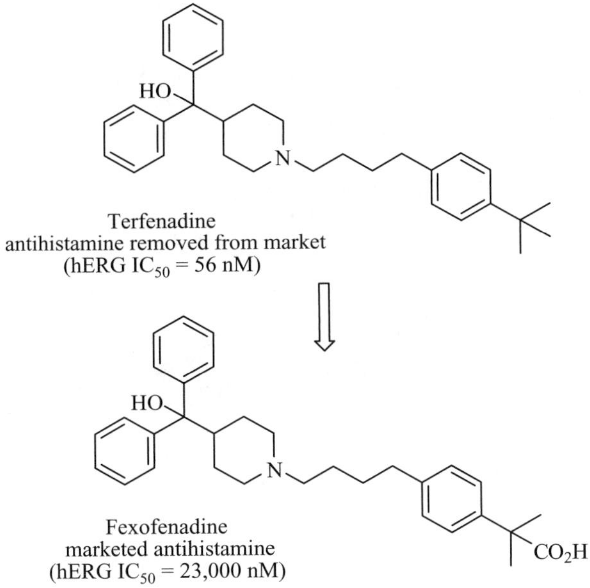

Figure 2.22. Zwitterions modulate hERG binding.

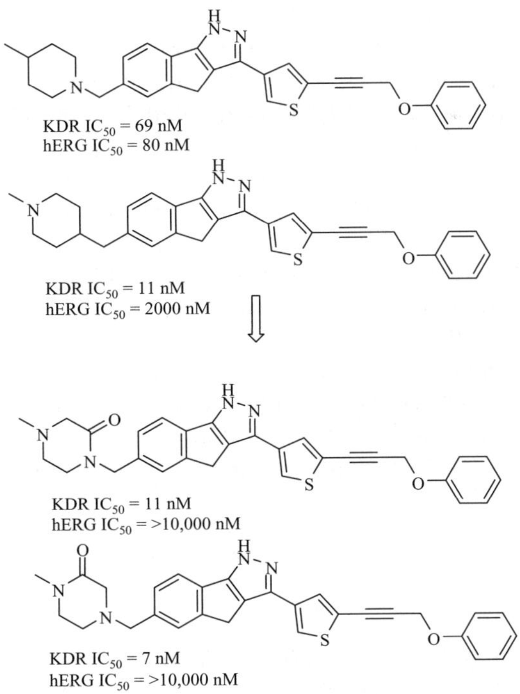

Figure 2.23. Modifications of basicity to modulate hERG binding.

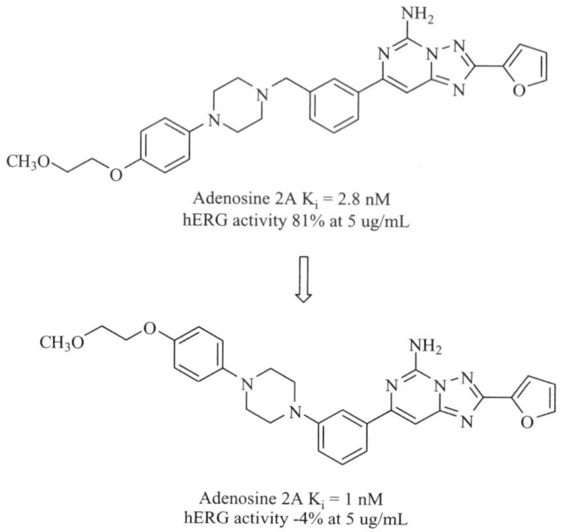

Figure 2.24. Decrease of the basicity of the piperazine nitrogen led to a lower affinity for the hERG channel.

fruitful in the advancement of certain piperazine-containing adenosine A2 antagonists. In this body of work, the reported selectivity of A2 over hERG greatly improved when the benzylic piperazine's basicity was decreased. The modified aniline-type analog proved to be equally potent against adenosine A2, but with no affinity to the hERG channel at 5 μg/mL (Matasi et al., 2005; Slee et al., 2008) (Fig. 2.24).

Within this chapter, we have provided a glimpse into the process of the discovery and optimization of new drug entities, including the process from identification of a hit to the conversion to a drug candidate. Many aspects need to be addressed between identification of the initial hit and advancement of a compound with drug-like properties into preclinical studies. A more comprehensive summary would require a full book, and many good reviews have been published for this sole reason. The goal of this chapter is to provide a brief overview of the challenges and difficulties of developing compounds that can advance into the clinic and to highlight many of the aspects that need to be addressed. Many of these concepts will be more fully examined in other chapters within this book.

REFERENCES

Barnett SF, Bilodeau MT, Lindsley CW. The Akt/PKB family of protein kinases: a review of small molecule inhibitors and progress towards target validation. *Curr Top Med Chem* 2005;5(2):109–125.

Becker OM, Dhanoa DS, Marantz Y, Chen D, Shacham S, Cheruku S, Heifetz A, Mohanty P, Fichman M, Sharadendu A, Nudelman R, Kauffman M, Noiman S. An integrated in silico 3D model-driven discovery of a novel, potent, and selective amidosulfonamide 5-HT1A agonist (PRX-00023) for the treatment of anxiety and depression. *J Med Chem* 2006;49(11):3116–3135.

Bogoyevitch MA, Fairlie DP. A new paradigm for protein kinase inhibition: blocking phosphorylation without directly targeting ATP binding. *Drug Discovery Today* 2007;12(15–16):622–633.

Borchardt RT, Kerns EH, Lipinski CA, Thakker DR, Wang B, Editors. *Pharmaceutical Profiling in Drug Discovery for Lead Selection. (Proceedings of the Workshop held 19–21 May 2003 in Whippany, New Jersey.)* [In: Biotechnol.: Pharm. Aspects 2004, 1)].

Borthwick AD, Davies DE, Ertl PF, Exall AM, Haley TM, Hart GJ, Jackson DL, Parry NR, Patikis A, Trivedi N, Weingarten GG, Woolven JM. Design and synthesis of pyrrolidine-5,5′-trans-lactams (5-oxo-hexahydropyrrolo[3,2-b]pyrroles) as novel mechanism-based inhibitors of human cytomegalovirus protease. 4. Antiviral activity and plasma stability. *J Med Chem* 2003;46(21):4428–4449.

Boudes P, Geiger JM. HIV protease inhibitors: general review. *Therapie* 1996; 51(3):319–325.

Bursavich MG, Rich DH. Designing non-peptide peptidomimetics in the 21st century: inhibitors targeting conformational ensembles. *J Med Chem* 2002;45(3):541–558.

Burton J, Ijjaali I, Barberan O, Petitet F, Vercauteren DP, Michel A. Recursive partitioning for the prediction of cytochromes P450 2D6 and 1A2 inhibition: importance of the quality of the dataset. *J Med Chem* 2006;49(21):6231–6240.

Casis O, Olesen S-P, Sanguinetti MC. Mechanism of action of a novel human ether-a-go-go-related gene channel activator. *Mol Pharmacol* 2006;69(2):658–665.

Chen C, Chen Y, Pontillo J, Guo Z, Huang CQ, Wu D, Madan A, Chen T, Wen J, Xie Q, Tucci FC, Rowbottom M, Zhu Y-F, Wade W, Saunders J, Bozigian H, Struthers RS. Potent and orally bioavailable zwitterion GnRH antagonists with low CYP3A4 inhibitory activity. *Bioorg Med Chem Lett* 2008;18(11):3301–3305.

Chohan KK, Paine SW, Mistry J, Barton P, Davis AM. A rapid computational filter for cytochrome P450 1A2 inhibition potential of compound libraries. *J Med Chem* 2005;48(16):5154–5161.

Clark DE. Computational prediction of blood–brain barrier permeation. *Annu Rep Med Chem* 2005;40:403–415.

Clark J, Cools J, Gilliland DG. EGFR inhibition in non-small cell lung cancer: resistance, once again, rears its ugly head. *PLoS Med* 2005;2(3):e75.

Cleland WW. Enzyme kinetics: steady state. *Handbook of Proteins* 2007;1:365–369.

Colas P. High-throughput screening assays to discover small-molecule inhibitors of protein interactions. *Curr Drug Discovery Technol* 2008;5(3):190–199.

Copeland RA. Enzymology methods. *Handbook of Proteins* 2007;2:1007–1011.

Costanzo MJ, Yabut SC, Almond HR, Jr., Andrade-Gordon P, Corcoran TW, De Garavilla L, Kauffman JA, Abraham WM, Recacha R, Chattopadhyay D, Maryanoff BE. Potent, small-molecule inhibitors of human mast cell tryptase. Antiasthmatic action of a dipeptide-based transition-state analogue containing a benzothiazole ketone. *J Med Chem* 2003;46(18):3865–3876.

Cowan-Jacob SW, Guez V, Fendrich G, Griffin JD, Fabbro D, Furet P, Liebetanz J, Mestan J, Manley PW. Imatinib (STI571) resistance in chronic myelogenous leukemia: molecular basis of the underlying mechanisms and potential strategies for treatment. *Mini Rev Med Chem* 2004;4(3):285–299.

Crespo A, Zhang X, Fernandez A. Redesigning kinase inhibitors to enhance specificity. *J Med Chem* 2008;51(16):4890–4898.

Davidson W, Frego L, Peet GW, Kroe RR, Labadia ME, Lukas SM, Snow RJ, Jakes S, Grygon CA, Pargellis C, Werneburg BG. Discovery and characterization of a substrate selective p38alpha inhibitor. *Biochemistry* 2004;43(37):11658–11671.

Dinges J, Albert DH, Arnold LD, Ashworth KL, Akritopoulou-Zanze I, Bousquet PF, Bouska JJ, Cunha GA, Davidsen SK, Diaz GJ, Djuric SW, Gasiecki AF, Gintant GA, Gracias VJ, Harris CM, Houseman KA, Hutchins CW, Johnson EF, Li H, Marcotte PA, Martin RL, Michaelides MR, Nyein M, Sowin TJ, Su Z, Tapang PH, Xia Z, Zhang HQ. 1,4-Dihydroindeno[1,2-*c*]pyrazoles with acetylenic side chains as novel and potent multitargeted receptor tyrosine kinase inhibitors with low affinity for the hERG ion channel. *J Med Chem* 2007;50(9):2011–2029.

Dunlop J, Bowlby M, Peri R, Tawa G, LaRocque J, Soloveva V, Morin J. Ion channel screening. *Comb Chem High Throughput Screening* 2008;11(7):514–522.

Ekegren JK, Ginman N, Johansson SA, Wallberg H, Larhed M, Samuelsson B, Unge T, Hallberg A. Microwave-accelerated synthesis of P1′-extended HIV-1 protease inhibitors encompassing a tertiary alcohol in the transition-state mimicking scaffold. *J Med Chem* 2006a;49(5):1828–1832.

Ekegren JK, Gising J, Wallberg H, Larhed M, Samuelsson B, Hallberg A. Variations of the P2 group in HIV-1 protease inhibitors containing a tertiary alcohol in the transition-state mimicking scaffold. *Org Biomol Chem* 2006b;4(16):3040–3043.

Ekegren JK, Unge T, Safa MZ, Wallberg H, Samuelsson B, Hallberg A. A new class of HIV-1 protease inhibitors containing a tertiary alcohol in the transition-state mimicking scaffold. *J Med Chem* 2005;48(25):8098–8102.

Evgenov OV, Pacher P, Schmidt PM, Hasko G, Schmidt HHHW, Stasch J-P. NO-independent stimulators and activators of soluble guanylate cyclase: discovery and therapeutic potential. *Nat Rev Drug Discovery* 2006;5(9):755–768.

Favata MF, Horiuchi KY, Manos EJ, Daulerio AJ, Stradley DA, Feeser WS, Van Dyk DE, Pitts WJ, Earl RA, Hobbs F, Copeland RA, Magolda RL, Scherle PA, Trzaskos JM. Identification of a novel inhibitor of mitogen-activated protein kinase kinase. *J Biol Chem* 1998;273(29):18623–18632.

Fischer PM. The design of drug candidate molecules as selective inhibitors of therapeutically relevant protein kinases. *Curr Med Chem* 2004;11(12):1563–1583.

Fish PV, Barta NS, Gray DLF, Ryckmans T, Stobie A, Wakenhut F, Whitlock GA. Derivatives of (3*S*)-*N*-(biphenyl-2-ylmethyl)pyrrolidin-3-amine as selective noradrenaline reuptake inhibitors: Reducing P-gp mediated efflux by modulation of H-bond acceptor capacity. *Bioorg Med Chem Lett* 2008;18(15):4355–4359.

Friesen RW, Ducharme Y, Ball RG, Blouin M, Boulet L, Cote B, Frenette R, Girard M, Guay D, Huang Z, Jones TR, Laliberte F, Lynch JJ, Mancini J, Martins E, Masson P, Muise E, Pon DJ, Siegl PKS, Styhler A, Tsou NN, Turner MJ, Young RN, Girard Y. Optimization of a tertiary alcohol series of phosphodiesterase-4 (PDE4) inhibitors: structure-activity relationship related to PDE4 inhibition and human ether-a-go-go related gene potassium channel binding affinity. *J Med Chem* 2003;46(12):2413–2426.

Garberg P, Ball M, Borg N, Cecchelli R, Fenart L, Hurst RD, Lindmark T, Mabondzo A, Nilsson JE, Raub TJ, Stanimirovic D, Terasaki T, Oeberg JO, Oesterberg T. *In vitro* models for the blood-brain barrier. *Toxicol in Vitro* 2005;19(3):299–334.

Gleeson MP. Generation of a set of simple, interpretable ADMET rules of thumb. *J Med Chem* 2008;51(4):817–834.

Goode DR, Totten RK, Heeres JT, Hergenrother PJ. Identification of promiscuous small molecule activators in high-throughput enzyme activation screens. *J Med Chem* 2008;51(8):2346–2349.

Guertin KR, Grimsby J. Small molecule glucokinase activators as glucose lowering agents: a new paradigm for diabetes therapy. *Curr Med Chem* 2006;13(15): 1839–1843.

Hammes GG. How do enzymes really work? *J Biol Chem* 2008;283(33):22337–22346.

Hudelson MG, Ketkar NS, Holder LB, Carlson TJ, Peng C-C, Waldher BJ, Jones JP. High confidence predictions of drug–drug interactions: predicting affinities for cytochrome P450 2C9 with multiple computational methods. *J Med Chem* 2008; 51(3):648–654.

Irvine JD, Takahashi L, Lockhart K, Cheong J, Tolan JW, Selick HE, Grove JR. MDCK (Madin-Darby canine kidney) cells: a tool for membrane permeability screening. *J Pharm Sci* 1999;88(1):28–33.

Jamieson C, Moir EM, Rankovic Z, Wishart G. Medicinal chemistry of hERG optimizations: highlights and hang-ups. *J Med Chem* 2006;49(17):5029–5046.

Jensen BF, Vind C, Padkjr SB, Brockhoff PB, Refsgaard HHF. *In silico* prediction of cytochrome P450 2D6 and 3A4 inhibition using gaussian kernel weighted *k*-nearest neighbor and extended connectivity fingerprints, including structural fragment analysis of inhibitors versus noninhibitors. *J Med Chem* 2007;50(3):501–511.

Johnson KA. Enzyme kinetics: transient phase. *Hand Proteins* 2007;1:370–376.

Jun MA, Park WS, Kang SK, Kim KY, Kim KR, Rhee SD, Bae MA, Kang NS, Sohn S-K, Kim SG, Lee JO, Lee DH, Cheon HG, Kim SS, Ahn JH. Synthesis and biological evaluation of pyrazoline analogues with [beta]-amino acyl group as dipeptidyl peptidase IV inhibitors. *Eur J Med Chem* 2008;43(9):1889–1902.

Kale DL, Chaturvedi SC, Patil D, Kakde RB. High throughput screening (HTS) technique in drug discovery. *Pharma Rev* 2008;6(32):46–48, 50–51.

Kamiya K, Niwa R, Mitcheson JS, Sanguinetti MC. Molecular determinants of hERG channel block. *Mol Pharmacol* 2006;69(5):1709–1716.

Kempf DJ, Sham HL, Marsh KC, Flentge CA, Betebenner D, Green BE, McDonald E, Vasavanonda S, Saldivar A, Wideburg NE, Kati WM, Ruiz L, Zhao C, Fino L, Patterson J, Molla A, Plattner JJ, Norbeck DW. Discovery of Ritonavir, a potent inhibitor of HIV protease with high oral bioavailability and clinical efficacy. *J Med Chem* 1998;41(4):602–617.

Kenakin T. Allosteric modulators: the new generation of receptor antagonist. *Mol Interv* 2004;4(4):222–229.

Klebl BM, Mueller G. Second-generation kinase inhibitors. *Expert Opin Ther Targets* 2005;9(5):975–993.

Laufer SA, Domeyer DM, Scior TRF, Albrecht W, Hauser DRJ. Synthesis and biological testing of purine derivatives as potential ATP-competitive kinase inhibitors. *J Med Chem* 2005;48(3):710–722.

Li AP. Screening for human ADME/Tox drug properties in drug discovery. *Drug Discovery Today* 2001;6(7):357–366.

Li AP. Preclinical *in vitro* screening assays for drug-like properties. *Drug Discovery Today: Technologies* 2005;2(2):179–185.

Lipinski CA. Changes in the profiles of drug properties: an experimental, computational, and informatics perspective. *Book of Abstracts, 219th ACS National Meeting, San Francisco, CA, March 26–30, 2000*, CINF-020.

Liu Y, Gray NS. Rational design of inhibitors that bind to inactive kinase conformations. *Nat Chem Biol* 2006;2(7):358–364.

Lotufo-Neto F, Trivedi M, Thase ME. Meta-analysis of the reversible inhibitors of monoamine oxidase type A moclobemide and brofaromine for the treatment of depression. *Neuropsychopharmacology* 1999;20(3):226–247.

Martinez A, Alonso M, Castro A, Perez C, Moreno FJ. First non-ATP competitive glycogen synthase kinase 3 beta (GSK-3beta) inhibitors: thiadiazolidinones (TDZD) as potential drugs for the treatment of Alzheimer's disease. *J Med Chem* 2002; 45(6):1292–1299.

Mastrolorenzo A, Rusconi S, Scozzafava A, Barbaro G, Supuran CT. Inhibitors of HIV-1 protease: current state of the art 10 years after their introduction. From antiretroviral drugs to antifungal, antibacterial and antitumor agents based on aspartic protease inhibitors. *Curr Med Chem* 2007;14(26):2734–2748.

Matasi JJ, Caldwell JP, Zhang H, Fawzi A, Higgins GA, Cohen-Williams ME, Varty GB, Tulshian DB. 2-(2-Furanyl)-7-phenyl[1,2,4]triazolo[1,5-c]pyrimidin-5-amine analogs as adenosine A2A antagonists: the successful reduction of hERG activity. *Bioorg Med Chem Lett* 2005;15(16):3675–3678.

Mayr LM, Fuerst P. The future of high-throughput screening. *J Biomol Screening* 2008;13(6):443–448.

McMasters DR, Torres RA, Crathern SJ, Dooney DL, Nachbar RB, Sheridan RP, Korzekwa KR. Inhibition of recombinant cytochrome P450 isoforms 2D6 and 2C9 by diverse drug-like molecules. *J Med Chem* 2007;50(14):3205–3213.

Milne JC, Lambert PD, Schenk S, Carney DP, Smith JJ, Gagne DJ, Jin L, Boss O, Perni RB, Vu CB, Bemis JE, Xie R, Disch JS, Ng PY, Nunes JJ, Lynch AV, Yang H, Galonek H, Israelian K, Choy W, Iffland A, Lavu S, Medvedik O, Sinclair DA, Olefsky JM, Jirousek MR, Elliott PJ, Westphal CH. Small molecule activators of SIRT1 as therapeutics for the treatment of type 2 diabetes. *Nature (London, United Kingdom)* 2007;450(7170):712–716.

Mol CD, Fabbro D, Hosfield DJ. Structural insights into the conformational selectivity of STI-571 and related kinase inhibitors. *Curr Opin Drug Discovery Dev* 2004;7(5): 639–648.

Morrison JF. Enzyme activity: reversible inhibition. *Hand Proteins* 2007;1:482–490.

Nguyen J-T, Hamada Y, Kimura T, Kiso Y. Design of potent aspartic protease inhibitors to treat various diseases. *Arch Pharmazie (Weinheim, Germany)* 2008;341(9):523–535.

Ohren JF, Chen H, Pavlovsky A, Whitehead C, Zhang E, Kuffa P, Yan C, McConnell P, Spessard C, Banotai C, Mueller WT, Delaney A, Omer C, Sebolt-Leopold J, Dudley DT, Leung IK, Flamme C, Warmus J, Kaufman M, Barrett S, Tecle H, Hasemann CA. Structures of human MAP kinase kinase 1 (MEK1) and MEK2 describe novel noncompetitive kinase inhibition. *Nat Struct Mol Biol* 2004; 11(12):1192–1197.

Olsen L, Rydberg P, Rod TH, Ryde U. Prediction of activation energies for hydrogen abstraction by cytochrome P450. *J Med Chem* 2006;49(22):6489–6499.

Parang K, Sun G. Design strategies for protein kinase inhibitors. *Curr Opin Drug Discovery Dev* 2004;7(5):617–629.

Pargellis C, Tong L, Churchill L, Cirillo PF, Gilmore T, Graham AG, Grob PM, Hickey ER, Moss N, Pav S, Regan J. Inhibition of p38 MAP kinase by utilizing a novel allosteric binding site. *Nat Struct Biol* 2002;9(4):268–272.

Park JD, Kim DH, Kim S-J, Woo J-R, Ryu SE. Sulfamide-based inhibitors for carboxypeptidase A. Novel type transition state analogue inhibitors for zinc proteases. *J Med Chem* 2002;45(24):5295–5302.

Pearlstein RA, Vaz RJ, Kang J, Chen X-L, Preobrazhenskaya M, Shchekotikhin AE, Korolev AM, Lysenkova LN, Miroshnikova OV, Hendrix J, Rampe D. Characterization of HERG potassium channel inhibition using CoMSiA 3D QSAR and homology modeling approaches. *Bioorg Med Chem Lett* 2003a;13(10): 1829–1835.

Pearlstein R, Vaz R, Rampe D. Understanding the structure-activity relationship of the human ether-a-go-go-related gene cardiac K+ channel. A model for bad behavior. *J Med Chem* 2003b;46(11):2017–2022.

Pliska VK. Substrate binding to enzymes. *Hand Proteins* 2007;1:417–426.

Putt KS, Chen GW, Pearson JM, Sandhorst JS, Hoagland MS, Kwon J-T, Hwang S-K, Jin H, Churchwell MI, Cho M-H, Doerge DR, Helferich WG, Hergenrother PJ. Small-molecule activation of procaspase-3 to caspase-3 as a personalized anticancer strategy. *Nat Chem Biol* 2006;2(10):543–550, S543/1–S543/29.

Qiao JX, King SR, He K, Wong PC, Rendina AR, Luettgen JM, Xin B, Knabb RM, Wexler RR, Lam PYS. Highly efficacious factor Xa inhibitors containing [alpha]-substituted phenylcycloalkyl P4 moieties. *Bioorg Med Chem Lett* 2009;19(2): 462–468.

Raghavan S, Tria GS, Shen HC, Ding F-X, Taggart AK, Ren N, Wilsie LC, Krsmanovic ML, Holt TG, Wolff MS, Waters MG, Hammond ML, Tata JR, Colletti SL. Tetrahydro anthranilic acid as a surrogate for anthranilic acid: application to the discovery of potent niacin receptor agonists. *Bioorg Med Chem Lett* 2008;18(11): 3163–3167.

Regan J, Breitfelder S, Cirillo P, Gilmore T, Graham AG, Hickey E, Klaus B, Madwed J, Moriak M, Moss N, Pargellis C, Pav S, Proto A, Swinamer A, Tong L, Torcellini C. Pyrazole urea-based inhibitors of p38 MAP kinase: from lead compound to clinical candidate. *J Med Chem* 2002;45(14):2994–3008.

Regan J, Capolino A, Cirillo PF, Gilmore T, Graham AG, Hickey E, Kroe RR, Madwed J, Moriak M, Nelson R, Pargellis CA, Swinamer A, Torcellini C, Tsang M, Moss N. Structure-activity relationships of the p38alpha MAP kinase inhibitor 1-(5-*tert*-butyl-2-*p*-tolyl-2*H*-pyrazol-3-yl)-3-[4-(2-morpholin-4-yl-ethoxy)naph- thalen-1-yl]urea (BIRB 796). *J Med Chem* 2003a;46(22):4676–4686.

Regan J, Pargellis CA, Cirillo PF, Gilmore T, Hickey ER, Peet GW, Proto A, Swinamer A, Moss N. The kinetics of binding to p38MAP kinase by analogues of BIRB 796. *Bioorg Med Chem Lett* 2003b;13(18):3101–3104.

Robertson JG. Mechanistic basis of enzyme-targeted drugs. *Biochemistry* 2005;44(15):5561–5571.

Roden DM. Drug-induced prolongation of the QT interval. Reply. *N Engl J Med* 2004;350(25):2620–2621.

Sanguinetti MC, Tristani-Firouzi M. hERG potassium channels and cardiac arrhythmia. *Nature (London, United Kingdom)* 2006;440(7083):463–469.

Sebolt-Leopold JS, Dudley DT, Herrera R, Van Becelaere K, Wiland A, Gowan RC, Tecle H, Barrett SD, Bridges A, Przybranowski S, Leopold WR, Saltiel AR. Blockade of the MAP kinase pathway suppresses growth of colon tumors *in vivo*. *Nat Med* 1999;5(7):810–816.

Slee DH, Moorjani M, Zhang X, Lin E, Lanier MC, Chen Y, Rueter JK, Lechner SM, Markison S, Malany S, Joswig T, Santos M, Gross RS, Williams JP, Castro-Palomino JC, Crespo MI, Prat M, Gual S, Diaz J-L, Jalali K, Sai Y, Zuo Z, Yang C, Wen J, O'Brien Z, Petroski R, Saunders J. 2-Amino-*N*-pyrimidin-4-ylacetamides as A2A receptor antagonists: 2. Reduction of hERG activity, observed species selectivity, and structure–activity relationships. *J Med Chem* 2008;51(6):1730–1739.

Stojan J. Enzyme inhibitors. *Enzymes and Their Inhibition:* 2005;149–169.

Terasaki T, Ohtsuki S, Hori S, Takanaga H, Nakashima E, Hosoya K-I. New approaches to *in vitro* models of blood–brain barrier drug transport. *Drug Discovery Today* 2003;8(20):944–954.

Thakur CS, Jha BK, Dong B, Das Gupta J, Silverman KM, Mao H, Sawai H, Nakamura AO, Banerjee AK, Gudkov A, Silverman RH. Small-molecule activators of RNase L with broad-spectrum antiviral activity. *Proc Nat Acad Sci USA* 2007;104(23): 9585–9590.

Thomas JA, Koshland DE, Jr. Stereochemistry of enzyme, substrate, and products during b-amylase action. *J Biol Chem* 1960;235:2511–2517.

Tipton KF. Enzymes: irreversible inhibition. *Hand Proteins* 2007;1:490–503.

Varfolomeev SD. Enzymatic catalysis: kinetics, active site structure, bioinformatics. *Chem Biol Kinetics New Horizons* 2005;2:151–186.

Wan Z-K, Follows B, Kirincich S, Wilson D, Binnun E, Xu W, Joseph-McCarthy D, Wu J, Smith M, Zhang Y-L, Tam M, Erbe D, Tam S, Saiah E, Lee J. Probing acid replacements of thiophene PTP1B inhibitors. *Bioorg Med Chem Lett* 2007;17(10):2913–2920.

Wang J, Urban L, Bojanic D. Maximising use of *in vitro* ADMET tools to predict *in vivo* bioavailability and safety. *Expert Opin Drug Metab Toxicol* 2007;3(5): 641–665.

Webster R, Leishman D, Walker D. Towards a drug concentration effect relationship for QT prolongation and torsades de pointes. *Curr Opin Drug Discovery Dev* 2002;5(1):116–126.

Wu X, Oehrngren P, Ekegren JK, Unge J, Unge T, Wallberg H, Samuelsson B, Hallberg A, Larhed M. Two-carbon-elongated HIV-1 protease inhibitors with a tertiary-alcohol-containing transition-state mimic. *J Med Chem* 2008;51(4):1053–1057.

Wunder F, Kalthof B, Mueller T, Hueser J. Functional cell-based assays in microliter volumes for ultra-high throughput screening. *Comb Chem High Throughput Screening* 2008;11(7):495–504.

Yamashita F, Hashida M. *In silico* prediction and control of ADME properties. *Drug Delivery System* 2003;18(1):22–29.

Yan Z, Caldwell GW. Metabolism profiling, and cytochrome P450 inhibition & induction in drug discovery. *Curr Top Med Chem (Hilversum, Netherlands)* 2001;1(5): 403–425.

Yokokawa F, Maibaum J. Recent advances in the discovery of non-peptidic direct renin inhibitors as antihypertensives: new patent applications in years 2000–2008. *Expert Opin Ther Patents* 2008;18(6):581–602.

3

BIOANALYTICAL TECHNOLOGIES IN DRUG DISCOVERY

Jing-Tao Wu

3.1 INTRODUCTION

The majority of the ADME information we generate on a compound, such as pharmacokinetics, metabolic stability, enzyme inhibition/induction, permeability, protein binding, and so on, rely on the quantitative measurement of a compound, its metabolites, or some surrogates or markers in biological matrices. Since radio-labeled compounds are generally not available in the drug discovery stage, most of these measurements are conducted using cold compounds. Traditionally, the analytics of quantitatively measuring analytes in biological matrices, namely bioanalysis, is a low-throughput approach with medium sensitivity and specificity. The introduction and routine implementation of liquid chromatography–tandem mass spectrometry (LC/MS/MS) in drug discovery laboratories revolutionized the bioanalysis and turned it into a technology with much higher throughput, sensitivity, and specificity. This technology advancement enabled ADME screening to be routinely incorporated into the drug discovery process. For the past two decades, bioanalytical technology has evolved rapidly to meet the ever-increasing needs.

There are some unique and difficult challenges associated with bioanalysis. First, biological matrixes are highly complex and contain numerous components that can potentially cause interference with the detection of the analyte. With LC/MS/MS methods, these matrix components can also potentially alter the quantitative results by ionization suppression, namely, matrix effect

Enzyme Inhibition in Drug Discovery and Development, edited by Lu and Li
Copyright © 2010 John Wiley & Sons, Inc.

(Matuszewski et al., 1999). Considering that the analyte concentrations are usually at low or even subnanomolar levels while the concentrations of some of the matrix components are at the millimolar levels, matrix complexity forms a formidable challenge for bioanalysis. Also, metabolites of the parent compound are frequently formed either *in vivo* or *in vitro*. These metabolites are often structural analogs or conjugates of the parent compounds, and therefore they may cause interference with the quantitation of the parent compound (Jemal and Xia, 1999). In addition, analyte stability in matrix is often a concern since there are enzymes in the biological matrixes that may catalyze the degradation of the molecule. To make it even more challenging, bioanalysis requires highly quantitative results with well-defined accuracy and precision and good day-to-day, batch-to-batch reproducibility. Finally, with the broad role of bioanalysis in drug discovery and development, the volume of the work is significant and therefore the bioanalytical methodology must meet the throughput requirement.

In this chapter, we will first describe the bioanalytical practices that are commonly used in the pharmaceutical industry today to tackle the challenges discussed above. We will then focus on some new and, in the author's opinion, highly promising technologies that are emerging and will likely shape the way we do bioanalysis in the coming years.

3.2 THE CURRENT COMMONLY USED BIOANALYTICAL PRACTICES

A large variety of bioanalytical tools and processes are available and are used by the pharmaceutical industry. Because of the specific needs and strategy of each company as well as the experience and preference of the scientists who perform the bioanalytical work, there are no standard bioanalytical practices across the industry. However, some common or similar tools and process do exist. This section will describe these most commonly used practices in bioanalysis.

Unarguably, the core technology used in today's bioanalytical work is based on LC/MS/MS. This chapter will only review LC/MS/MS-based bioanalytical practices. It should be recognized that other analytical methodologies such as liquid chromatography with fluorescent/UV detection is occasionally still used in certain types of bioanalytical applications with the distinct advantage of cost reduction. Before the biological samples are analyzed on a LC/MS/MS system, they are subject to a sample preparation procedure to selectively enrich the analyte of interest and remove the matrix components. We will discuss the sample preparation and LC/MS/MS analysis steps separately.

3.2.1 Sample Preparation

Because of the complexity of the biological matrices, biological samples need to be cleaned up before analysis. There are three types of commonly used sample preparation methods. They are protein precipitation, liquid–liquid extraction, and solid-phase extraction. This section will only provide a brief discussion on these methods along with their pros and cons. A detailed description of these methods can be found elsewhere (Wells, 2003).

3.2.1.1 Protein Precipitation (PPT). Protein precipitation is the most commonly used method in drug discovery due to its simplicity and broad applicability. In this method, an organic solvent such as acetonitrile is added to the biological samples, usually at a volume ratio of 3–6:1. The mixture is vortexed followed by centrifugation. The supernatant can be dried down and reconstituted or it can be injected directly after dilution with aqueous solvent to reduce the organic strength. Because of the relatively nonselective nature, this method involves minimal method development work and usually gives good extraction recovery for compounds with a large variety of chemical structures. The disadvantage of this method, due to the same nonselective nature, is frequently associated with ion suppression and lack of robustness for batches of large size.

PPT is generally the method of choice for sample preparation during the early screening stage of drug discovery due to its simplicity and broad applicability. It is also a good method to consider if multiple analytes or metabolites with diversified structures need to be analyzed from the same sample. After a lead compound is selected as a potential development candidate, a more refined method is usually developed to provide the ruggedness and reproducibility required.

3.2.1.2 Liquid–Liquid Extraction (LLE). LLE is another commonly used sample preparation technique that selectively separates the analyte from matrix components by differential partitioning between the aqueous sample matrix and an immiscible solvent such as methyl-*tert*-butylether (MTBE), ethyl acetate, diethyl ether, or hexane. The addition of a pH buffer to the sample prior to extraction is generally important to keep the analyte in its un-ionized form. After the addition of the immiscible solvent, the mixture is thoroughly vortexed to allow the analyte and sample matrix components to distribute between the aqueous and organic phases. The analyte in its un-ionized form will preferentially distribute into the organic solvent (the analyte must be soluble in the organic solvent used). Hydrophilic components in the matrix will stay in the aqueous phase. The organic phase is then isolated, evaporated, and reconstituted in mobile phase before injection. LLE is generally applicable to a wide range of compounds, and the extracts are generally clean since most of the water-soluble inorganic salts and proteins are removed during extraction. Although LLE is a relatively low-cost extraction method from a laboratory supply perspective, the procedures are often tedious and labor-intensive.

3.2.1.3 Solid-Phase Extraction (SPE). In an SPE method, a biological sample containing the analyte of interest flows through a solid-phase bed and the analyte is adsorbed onto the surface of the solid while the matrix components go to waste. A wash step is then usually employed to further clean up the samples on the solid before the elution step where the analyte is eluted off the solid surface to a collection plate. Compared to PPT and LLE, SPE offers the best selectivity through the selection of (a) the sorbent and the type and (b) the strength of the wash solvent and the elution solvent. SPE is also more readily amenable for parallel processing with automation. Some of the key disadvantages include (a) increased complexity of the selection of the sorbent and the selection of the washing and elution solvent and (b) the manipulation of pH of the samples. It also has a relatively higher cost compared to the other two methods.

3.2.1.4 Selection of the Extraction Method. There is no fixed rule on extraction method selection. The method selected should fit for the purpose of the applications and should also take into account the experience of the bioanalyst and the laboratory and automation equipment available. Generally speaking, PPT is usually preferred for discovery screening studies where the batch size of the samples is relatively small and the quality of the data does not need to meet regulatory standards (although carefully optimized PPT methods have been routinely used to support regulated studies). This is mostly because of the wide applicability of the PPT methods and the simplicity in method development. When a molecule enters into development stage, an LLE or SPE method becomes more desirable. The resources invested in developing these more selective methods are readily justifiable by the large number of samples and the need to meet regulatory standards.

3.2.1.5 Automation. Because of the multistep handling of the samples during sample preparation, sample preparation is a well-recognized bottleneck in bioanalysis. Tremendous efforts have been made in this area to improve the throughput by automation. Microplates, typically in a 96-well format, have become the gold standard for bioanalytical sample processing. The use of these 96-well plates allows for parallel processing by hand and more importantly by liquid handlers. The use of liquid handlers for automated sample preparation is getting more and more popular. Some of the popular liquid handlers most commonly used in bioanalytical laboratories include MultiPROBE (PerkinElmer Life Sciences), Quadra 96 (Tomtec), Microlab (Hamilton), and Tecan Genesis (Tecan USA). A more detailed review and comparison for these liquid handlers can be found elsewhere (Wells, 2003).

3.2.2 LC/MS Analysis

Liquid chromatography with UV or florescence detection traditionally was the standard tool for bioanalysis. Its key drawbacks include lack of specificity, long method development and sample analysis time, and often suboptimal sensitivity. The introduction of LC/MS technology was a quantum leap because it provided the unparalleled specificity, speed, and sensitivity in bioanalysis. It is also a key enabling factor that ADME screening has now been routinely incorporated into the drug discovery process.

The gold standard in mass spectrometry for bioanalysis is the triple-quadrupole mass spectrometer (Rolando and Sablier, 2003). Although the single quadrupole, ion trap, or time-of-flight mass spectrometers were occasionally also used for bioanalysis, they generally only fit some specific bioanalytical applications due to their different limitations and, therefore, will not be discussed in detail in this chapter. The triple quadrupole mass spectrometer is a tandem mass spectrometer with two quadrupole mass analyzers and a collision cell (originally also configured as a quadrupole) in between. During bioanalysis, the triple quadrupole mass spectrometer is commonly operated in the multiple reaction monitoring (MRM) mode. In this mode, the first mass analyzer is set to a fixed mass-to-charge ratio (m/z) so that only compounds with this specific m/z can pass through. Once the analyte passes through the first mass analyzer, it enters the collision cell where it will be fragmented through collision with an inert gas such as nitrogen or argon. The second mass analyzer is then set to a fixed m/z that is corresponding to a specific fragment of the analyte. In the MRM mode, for any compounds that can cause interference with the analyte, they must have the same m/z as the analyte and they must also form a fragment that has the same m/z as that of the analyte. Because of this multistage gating mechanism, the chances for an endogenous component to cause interference with the analyte are significantly reduced. However, as will be discussed a little later, the use of a tandem mass spectrometer does not guarantee the specificity of the assay, particularly when it comes to interferences from metabolites.

Most of the molecules generated in the drug discovery process are nonvolatile, and the samples for ADME studies are usually in liquid phase. There had not been a straightforward ionization method to ionize nonvolatile compound from liquid matrixes until the introduction of atmospheric pressure ionization (API) methods in the late 1980s. Two most common forms of API methods are electrospray ionization (ESI) and atmospheric pressure chemical ionization (APCI) (Kebarle and Tang, 1993). These API methods provide a simple way to ionize a large variety of structurally diversified compounds directly out of a liquid flow and thus making them an almost ideal interface between LC and MS. Although these API methods are predominantly used for LC/MS/MS, they also carry two potential caveats for bioanalysis. First, these methods, particularly ESI, are readily subject to ion suppression (King et al., 2000). When the analyte is eluting with matrix components, these matrix components may potentially suppress the signal of the analyte, resulting in a quantitative bias. This is often called the matrix effect in bioanalysis. It is one of the most common issues in bioanalysis, particularly during the discovery stage when stable isotope-labeled internal standards are not available. The second caveat is associated with metabolite interference via collision-induced dissociation (CID) in the ion source region (Wells and McLuckey, 2003).

When a metabolite with a labile chemical bond (such as a glucuronide conjugate) is ionized, the collision with the gases in the ion source region may break down the labile bond and thus potentially convert the metabolite back into the parent compound, resulting in potential interference. The matrix effects and metabolite conversion in the ion source are among the most important factors for evaluation during the development of a bioanalytical assay. This is why chromatographic separations prior to mass spectrometric detection are still an essential step in bioanalysis regardless of the high specificity offered by the mass spectrometer. By carefully optimizing chromatography, the analyte can be separated from matrix components that cause the ion suppression. Also, metabolites with labile chemical bonds can be separated from the analyte in retention time, thus avoiding interference. As an essential step, unfortunately, chromatography has thus become a bottleneck in bioanalysis in many cases.

3.3 EMERGING NEW BIOANALYTICAL TECHNOLOGIES

Bioanalytical technology has evolved rapidly in the past 20 years because ADME science is fully engaged into the drug discovery process. In this section, we will review some of the emerging and highly promising technologies in the bioanalytical area. Some of these technologies are still in the early phase of development and implementation, but in the author's opinion they are in the right direction to meet the ever-increasing demands for bioanalysis. These new technology developments are discussed in this section under three categories. They are (a) technologies to increase throughput, (b) technologies to enhance assay performance, and (c) technologies to create new capability.

3.3.1 Higher Throughput

As ADME screening is incorporated into the drug discovery process, both the workload and the turnaround requirement have increased significantly, thereby posing strong demands for higher-throughput bioanalytical technologies. As discussed in the previous section, sample preparation and chromatographic separations are the two most common bottlenecks in bioanalysis and a lot of new developments have been made in technologies to address these two bottlenecks.

3.3.1.1 High-Speed Separations. One straightforward approach to increase the bioanalytical throughput is to reduce the chromatographic runtime. It is important to note that this goal should be achieved without sacrificing the quality of the chromatographic separations. This is one of the most active areas in bioanalytical technology development. Two technologies have thus far shown to be most promising in this area and will be discussed in detail below.

3.3.1.1.1 Ultrahigh Pressure Using LC Columns with Small Particle Sizes. The use of ultrahigh pressure and LC columns with small particle sizes was pioneered by the group at the University of North Carolina led by Jorgenson (MacNair et al., 1997). More recently, Waters Corporation was among the first companies to commercialize the technology with less elevated but practically more manageable backpressure (Mazzeo et al., 2005). According to the Van Deemter curves, higher separation efficiency, as measured by the plate height, can be achieved by using particles with smaller sizes. Also, separation efficiency is well-maintained at much higher flow rate with particles of smaller sizes. Therefore, it is possible to use particles of smaller sizes with higher linear flow rate to achieve faster separations with good separation efficiency. The use of high flow rate and small particle sizes is associated with a high backpressure that requires modification of conventional HPLC equipment. One of most popular versions of these modified HPLC equipment is the ACQUITY UPLC (Ultra-Performance Liquid Chromatography) system from Waters Corporation.

There have been a few publications in the literature on the use of the UPLC system for higher-throughput bioanalytical applications (Yu et al., 2006; Plumb et al., 2008; Wang et al., 2006; O'Connor et al., 2006). One good comparison of the performance of UPLC and HPLC in bioanalysis was demonstrated in the separation of a drug mixture in rat plasma (Yu et al., 2006). In this work, a mixture of ibuprofen, alprazolam, naproxen, prednisolone, and diphenhydramine in rat plasma were analyzed on the UPLC/MS/MS and the HPLC/MS/MS systems. As shown in Fig. 3.1, while maintaining comparable chromatographic resolution, the UPLC/MS/MS offered a threefold decrease in analysis time and up to tenfold increase in detected peak height, which translated into a five- to tenfold improvement in the lower limit of quantitation for four out of the five compounds. Also, a repeat injection of 963 plasma samples resulted in highly reproducible retention time and measured concentration, which demonstrated the robustness of the system for routine bioanalytical applications.

The ultrahigh-pressure separations employing columns with small particle sizes are quickly gaining popularity in the bioanalytical area. They offer distinct advantage in speed and sensitivity with simple instrumentation. The high flow rates associated with these separations require a robust ion source on the mass spectrometer or the use of a flow splitter. The narrow chromatographic peak width also requires high sampling speed on the mass spectrometer. Since the nonscanning MRM mode is routinely used in bioanalysis, this is generally not a problem except for multiple-component assays measuring a large number of analytes at the same time.

3.3.1.1.2 Monolithic LC Columns. Monolithic columns consisting of one piece of silica or an organic polymer has generated a lot of interest as a potential high speed separation tool (Tanaka et al., 2001). Compared to conventional particulate columns, monolithic columns have a unique bi-porous

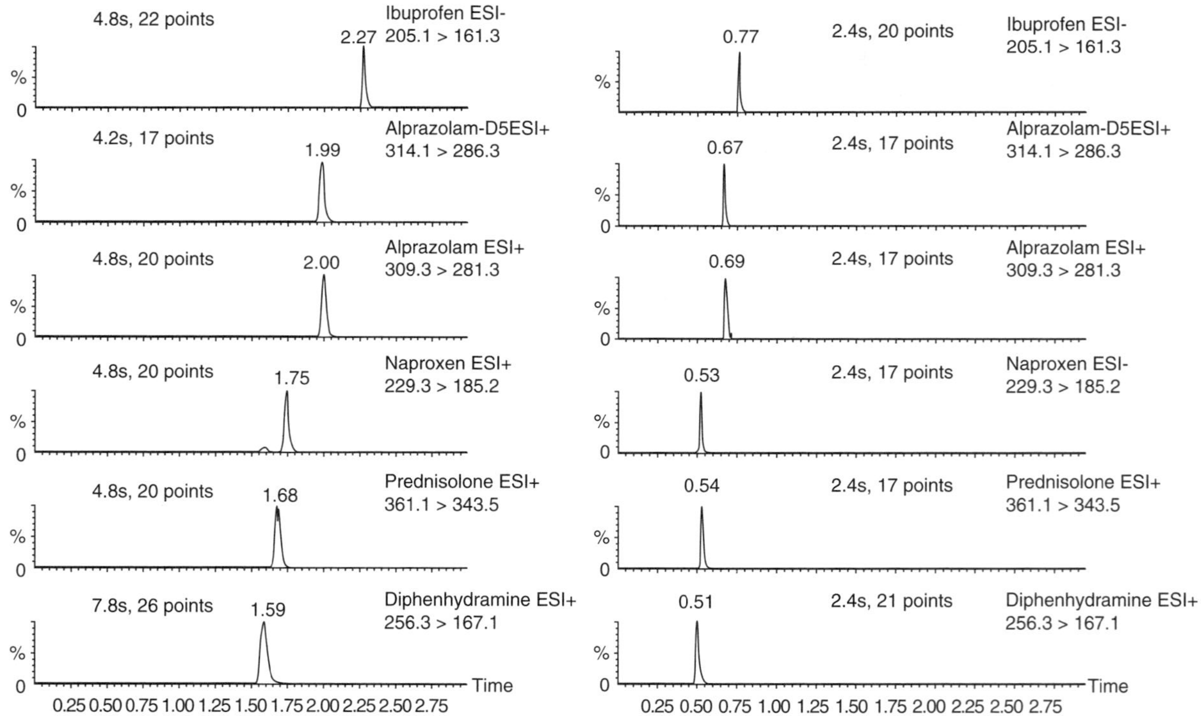

Figure 3.1. Comparison of MRM chromatograms on HPLC/MS/MS and UPLC/MS/MS for a mixture of test compounds in rat plasma extracts. The left panel is from HPLC/MS/MS and the right panel is from UPLC/MS/MS. Adapted from Yu et al. (2006). Reproduced with permission of John Wiley & Sons, Ltd.

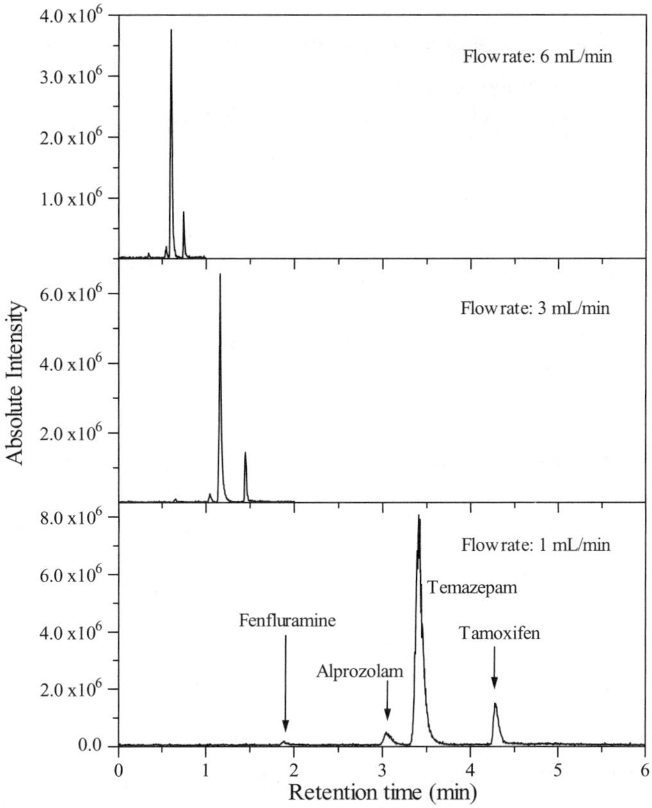

Figure 3.2. Total ion chromatograms of the separation of rat plasma extract containing a four-component mixture. Separations were performed on the monolithic column with a flow rate of 1, 3, and 6 mL/min, respectively. A linear gradient from 20% to 80% solvent B was used. Adapted from Wu et al. (2001). Reproduced with permission of John Wiley & Sons, Ltd.

structure. The smaller pores with low-nanometer diameter on the silica skeleton provide the surface area needed for capacity. The larger pores with low-micrometer diameter reduce the flow resistance to allow a high flow rate with reduced backpressure. Like LC columns with small particle sizes, monolithic columns can maintain high separation efficiency at high linear flow rate, resulting in high-speed separations with high chromatographic resolving power.

The use of monolithic columns for high-throughput bioanalysis has been reported by several laboratories (Plumb et al., 2001; Wu et al., 2001; Huang et al., 2006; Zhou et al., 2005; Peng et al., 2003; Hsieh et al., 2002). The advantages of using these columns for bioanalysis were demonstrated in the analysis of a mixture of fenfluramine, alprozolam, temazepam, and tamoxifen in rat plasma (Wu et al., 2001). In this work, the separation was performed on a monolithic column at flow rates of 1, 3, and 6 mL/min, respectively (Fig. 3.2).

TABLE 3.1. Performance of the Monolithic Column at Different Flow Rates for the Separation of a Four-Component Plasma Extract

Figure of Merit Flow Rate (mL/min)	Number of Plates per Meter ($\times 10^5$)			Resolution[a]			Signal/Noise		
	1	3	6	1	3	6	1	3	6
Fenfluramine	0.87	0.85	0.69	14.2	14.7	13.5	45	65	70
Alprozolam	1.15	1.23	1.28	3.7	3.6	3.4	150	150	100
Temazepam	1.46	1.30	1.22	3.7	3.6	3.4	240	172	258
Tamoxifen	4.54	4.65	3.43	10.5	10.8	8.1	250	340	226

[a]Resolution was calculated using the most adjacent peak pair.

Source: Adapted from Wu et al. (2001). Reproduced with permission of John Wiley & Sons, Ltd.

As shown in the figure, by increasing the flow rate from the normally used 1 mL/min to an elevated 6 mL/min, the chromatographic run time was reduced accordingly from 6 min to 1 min. The performance of these separations is summarized in Table 3.1. The number of separation plates, resolution, and signal/noise ratio were well-maintained at the high separation speed.

Monolithic columns offer the speed and simplicity for high-throughput bioanalysis. The choices of columns, in terms of column dimension and column chemistry, are relatively limited due to technical challenges in column preparation.

3.3.1.2 Online Extraction. The two high-speed separation methods discussed above have the potential to alleviate the bottleneck of chromatographic separations. Similar efforts have also been made to address the bottleneck in sample preparation. While automation using robotic liquid handlers is playing a more and more significant role in bioanalytical sample preparation, online extraction based on column switching has generated considerable interest, largely because this technology only involves conventional HPLC equipment with which most of the bioanalytical scientists are familiar.

A schematic diagram of a column switching setup is shown in Fig. 3.3. In this technology, samples in their biological fluids are directly injected onto an extraction column or cartridge with no or minimal pretreatment. The analyte is retained on the extraction column using different mechanisms based on the nature of the extraction column while sample matrix components are washed to waste. The retained analyte is then eluted off the extraction column onto a separation column via column switching. The analyte is then further separated from matrix components on the separation column and then detected by the mass spectrometer. There have been numerous publications on the use of online extraction for bioanalysis (Wu, et al., 2000; Jemal et al., 1998; Ingelse et al., 2008; Herman, 2005; Wu, 2001). One of the most popular modes is the use of a high-flow—or sometimes called "turbulent-flow"—online extraction

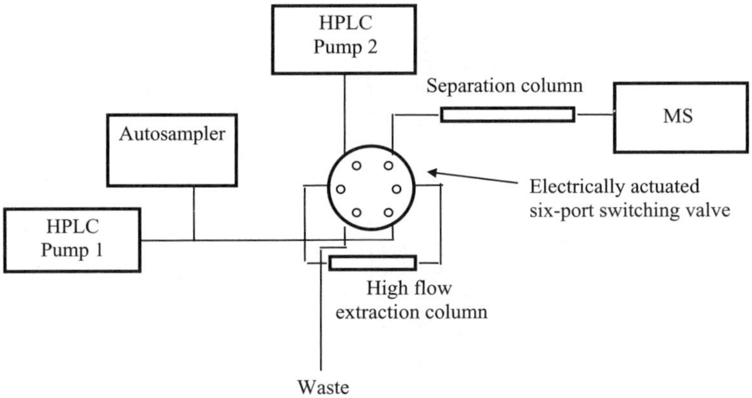

Figure 3.3. A representative schematic diagram of a column switching system.

method. In this mode, an extraction column with large particle sizes (30–60 µm in diameter) is operated at a high linear flow rate during the online extraction. The use of this high linear flow rate improves the effectiveness and robustness of the extraction.

Figure 3.4 shows the MRM chromatograms for a multiple-component assay using this high-flow online extraction method to support a 14-in-1 cassette dosing pharmacokinetic screening (Wu et al., 2000). Not only could all the analytes be well retained using online extraction, good separation efficiency was also achieved on the separation column via column switching. In this work, one extraction column could be routinely used for 200–300 plasma sample injections with good reproducibility. The precision and accuracy, as well as the lower limit of quantitation, were comparable with conventional offline sample preparation methods.

The use of online extraction with column switching can effectively remove the sample extraction step with either a commercial or simple custom-built systems. A potential drawback of this technique is usually associated with the overall robustness of the system with the direct handling of untreated biological fluids. The stability of the analyte in the biological matrix is also a potential concern since the unextracted samples will be placed in the autosampler for hours before analysis.

3.3.1.3 Parallel Separations. In addition to the high-speed and online extraction methods mentioned above, another trend in high-throughput bioanalysis is to go parallel. A most common form of the parallel mode is to multiplex chromatographic separations since it is often one of the key bottlenecks in bioanalysis. One of the simplest modes of parallel separations is to stagger the injections so that the analyte on one column elutes at a retention time when there is no analyte eluting on the other column (Wu, 2001; Jemal et al., 2001; Korfmacher et al., 1999; King et al., 2002). A more sophisticated

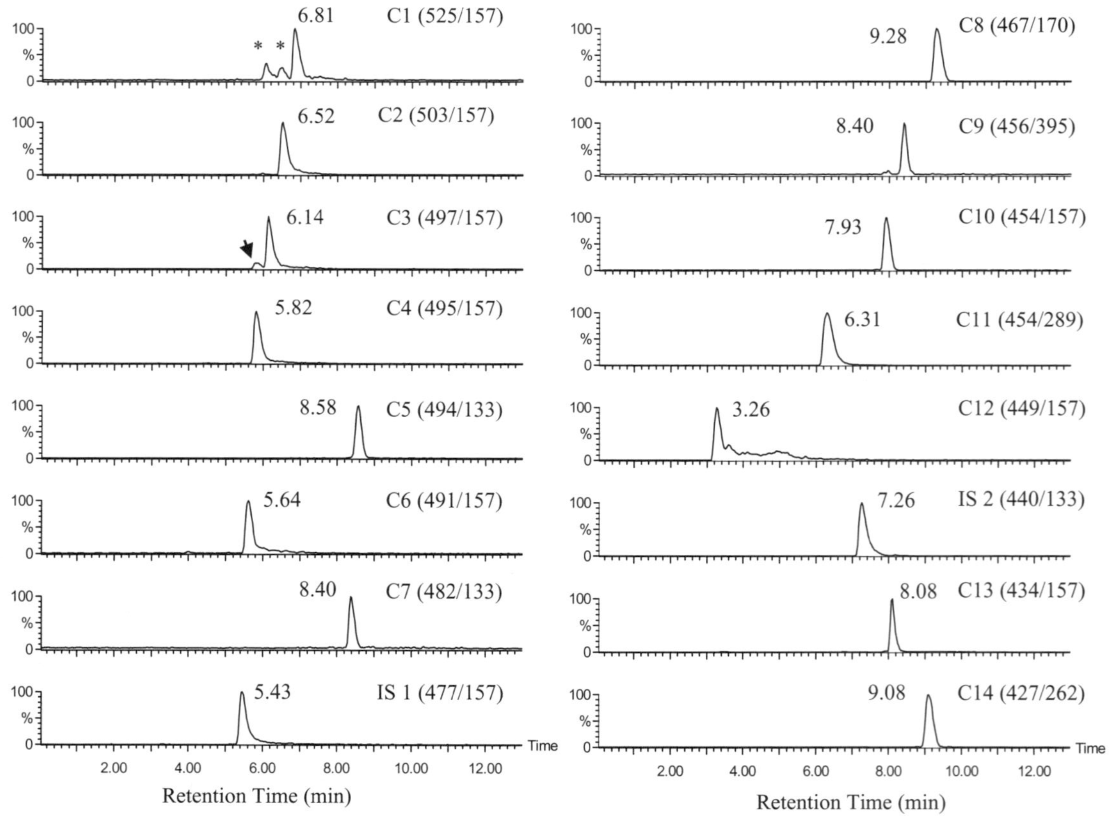

Figure 3.4. MRM chromatograms of a plasma sample collected 1 h after intravenous dosing of 14 compounds to a dog. The plasma sample was directly injected onto an online extraction column switching system. Adapted from Wu et al. (2000). Reproduced with permission of the American Chemical Society.

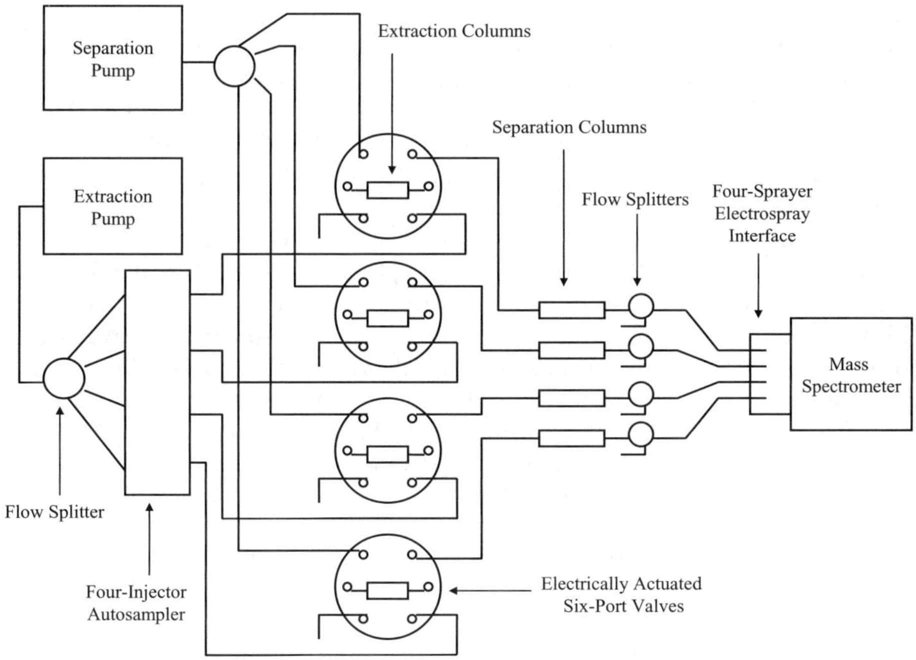

Figure 3.5. Schematic diagram of the four-way parallel extraction, parallel separation, and pseudo-parallel mass spectrometric detection system. Adapted from Deng et al. (2001). Reproduced with permission of John Wiley & Sons, Ltd.

mode, however, involved the parallel online extraction, parallel separation, and pseudo-parallel mass spectrometric detection (Deng et al., 2001). A schematic diagram of the instrument setup is shown in Fig. 3.5. In this work, an autosampler equipped with four injectors made four injections of unextracted plasma samples simultaneously onto four high-flow online extraction columns connected to four electrically actuated valves. After the extraction step, when the analyte was retained on the extraction column and matrix components were washed to waste, the four valves switched simultaneously and eluted the analyte onto four separation columns. The effluent from the four columns was introduced into the mass spectrometer simultaneously. The mass spectrometer was equipped with an indexed four-sprayer electrospray interface that allowed for the pseudo-simultaneous and indexed detection of the effluent from the four parallel separations. Although sophisticated in instrumentation, the system enabled a fourfold throughput improvement for extraction and separation without having to modify the analytical method itself.

3.3.1.4 Ambient Ionization. Ambient ionization refers to a class of ionization methods that were developed recently to ionize samples under ambient conditions in their native environment (Cooks et al., 2006). The unique feature of this ionization method is that samples can be ionized directly without any sample preparation. Although several forms of the ambient ionization methods are available, two of these methods have drawn most of the attention. They are the DESI (Desorption ElectroSpray Ionization) and the DART (Direct Analysis in Real Time) methods (Takats et al., 2004; Cody et al., 2005). The use of these techniques in different application areas has been reported (Chen et al., 2005; Petucci et al., 2007).

Because of the fact that no sample preparation is required for these ambient ionization methods, they appear to be extremely attractive for high-throughput bioanalysis. However, the application of these methods in the bioanalytical area is still in the very early stage due to the quantitative nature of these applications and the complexity of the sample matrixes. Two recent publications reported extensive evaluation of DART as a potential high-throughput tool for bioanalysis (Zhao et al., 2008; Yu et al., 2009). DART was chosen in these works because of its superb ability to handle ion suppression of biological matrixes possibly due to its preferential ionization of low-molecular-weight species over large molecules such as proteins in the matrix.

With refinement and optimization of the sample introduction process, the signal reproducibility of the DART/MS/MS system has shown to be satisfactory for quantitative applications of biological samples with CV% in the single digit. With the current instrument setting, the reported lower limit of quantitation is generally one to two orders of magnitude higher than that from conventional LC/MS/MS, but it is still very useful for many types of bioanalytical applications. The ability of the DART method to handle untreated biological matrixes, as measured by matrix effects, is shown in Fig. 3.6 as an example (Yu et al., 2009). Although matrix ion suppression still exists, this has shown great promise considering that there was no sample preparation and chromatography. A comparison of the DART data with those from the conventional LC/MS/MS method for a mouse PK study is shown in Table 3.2. Clearly, DART was able to provide comparable results as LC/MS/MS methods for certain types of applications.

A major drawback for using the DART system for bioanalysis is that metabolites with labile chemical bonds may likely break down in the ion source and cause potential interference with parent compound quantitation. Therefore, the DART system is most suitable for either early-stage screening studies or late-stage compounds whose major metabolite profiles are known. In addition to its potential as a high-throughput bioanalytical tool, DART may also have the potential as a clinical diagnostic tool due to its simplicity in sample manipulation and the fast readout of the results.

3.3.2 Enhanced Assay Performance

In addition to the tremendous amount of the effort made on throughput improvement, the reliability of the bioanalytical assay, particularly for compounds at the late discovery stage, has also been an area where some impor-

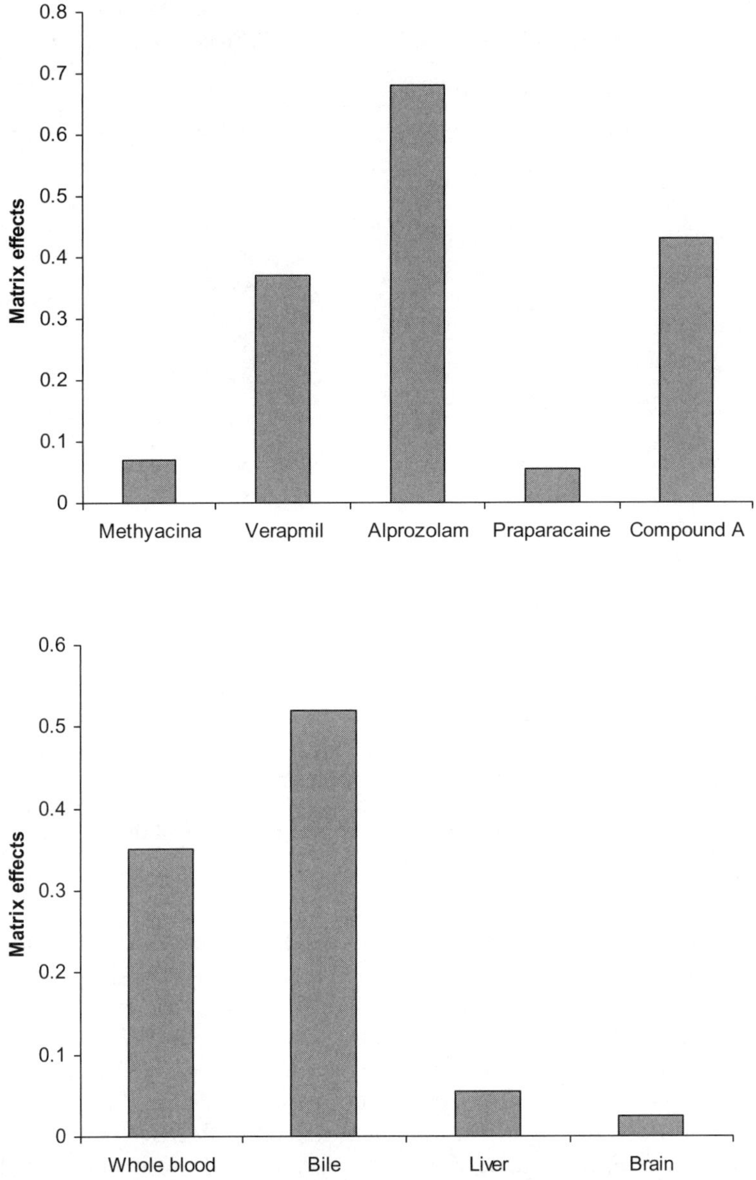

Figure 3.6. Evaluation of matrix effect with DART/MS/MS without extraction and chromatography. **Top panel:** Matrix effect of test compounds in plasma. **Lower panel:** Matrix effect of verapamil in different tissue homogenates. Matrix effect is defined as the ratio of analyte peak height in matrix over that in neat solvent. A matrix effect value of 1 is no matrix effect. Adapted from Yu et al. (2009). Reproduced with permission of the American Chemical Society.

TABLE 3.2. Concentration Comparison Between the DART/MS/MS and LC/MS/MS Methods for a PK Study in Mouse on a Proprietary Compound

	Concentration (nM)		
Sample	LC/MS/MS	DART/MS/MS	%MPD
30 min (A)	3990	4170	2.2
30 min (B)	3930	4412	5.8
30 min (C)	2300	2918	11.8
1 h (A)	3500	4297	10.2
1 h (B)	4200	5846	16.4
1 h (C)	2480	3011	9.7
2 h (A)	3500	4430	11.7
2 h (B)	1890	2340	10.6
2 h (C)	2900	3073	2.9
4 h (A)	2940	2674	−4.7
4 h (B)	3480	3672	2.7
4 h (C)	4510	4766	2.8
8 h (A)	3110	2851	−4.3
8 h (B)	1900	2111	5.3
8 h (C)	3270	3130	−2.2
16 h (A)	BQL	BQL	NA[a]
16 h (B)	2	BQL	NA
16 h (C)	BQL	BQL	NA
24 h (A)	BQL	BQL	NA
24 h (B)	BQL	BQL	NA
24 h (C)	1.39	BQL	NA

[a]NA, not available.

Source: Adapted from Yu et al. (2009). Reproduced with permission of the American Chemical Society.

tant technology improvements are being made. This section will highlight two key technologies in this area.

3.3.2.1 High-Field Asymmetric Waveform Ion Mobility Spectrometry (FAIMS). FAIMS is a technology that was developed in the late 1990s by Purves and Guevremont (1999). Its principle of operation is based on separating ions by differences in ion mobility under high electric field. This unique capability allows FAIMS to separate compounds that are isobaric or that are difficult to separate by HPLC. Because of this unique feature, this technique has drawn more attention recently in the bioanalytical area to remove interferences and improve the reliability of the assay (Kapron et al., 2005; Hatsis et al., 2007; Xia et al., 2008). The FAIMS is now a commercially available add-on device to some of the standard triple-quadrupole mass spectrometers. One good example of this unique capability is demonstrated in removing the interference caused by the ion source fragmentation of a coeluting N-oxide metabolite (Hatsis and Kapron, 2008). As shown in Fig. 3.7, the N-oxide metabolite caused an increase in the peak area of the parent compound in a

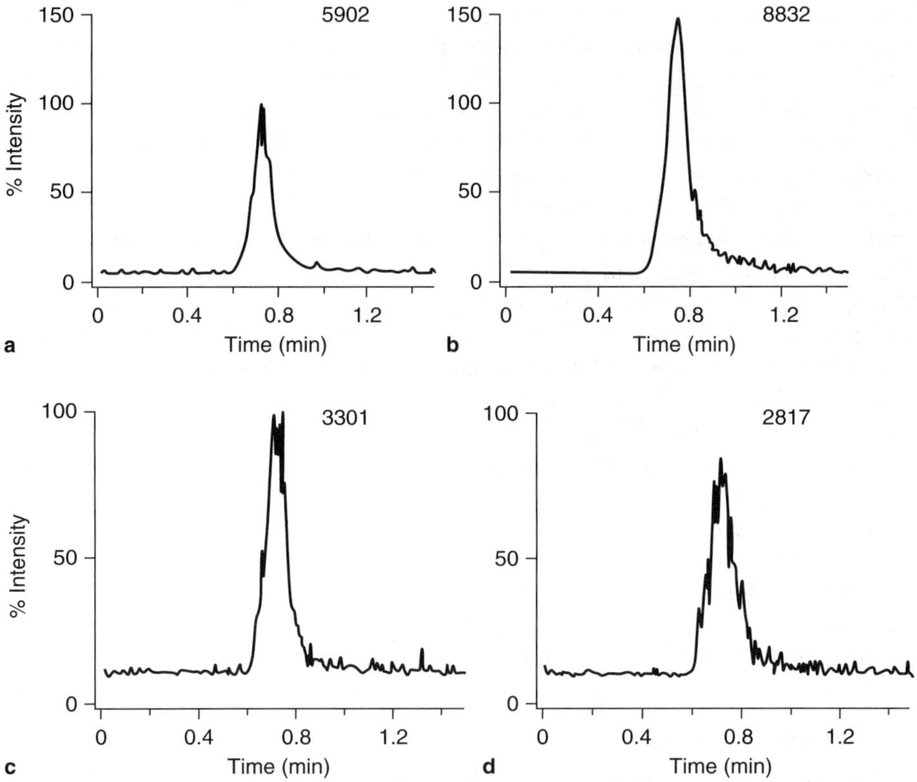

Figure 3.7. Use of FAIMS to remove the contribution of in-source decay of a metabolite to the precursor ion signal. **(a)** Representative chromatogram of an amine drug. **(b)** Chromatogram of the amine drug in the presence of its *N*-oxide metabolite. **(c)** Chromatogram of the amine drug obtained with FAIMS. **(d)** Chromatogram of the amine drug in the presence of its *N*-oxide metabolite obtained with FAIMS. The numbers in each chromatogram refer to the signal obtained for the amine drug. Adapted from Hatsis and Kapron (2008). Reproduced with permission of John Wiley & Sons, Ltd.

conventional LC/MS/MS assay. After employing the FAIMS, the interference of the N-oxide on the parent compound was completely removed. Therefore, the specificity of the assay was improved without the need to optimize the chromatography to separate the metabolite from the parent.

It is not expected that FAIMS will become a mainstream standard add-on equipment for LC/MS/MS systems for routine bioanalysis. However, it is a potentially useful tool for a bioanalytical scientist to solve issues associated with assay specificity.

3.3.2.2 High-Resolution MS. As discussed earlier, triple quadrupole mass spectrometers have become the standard platform for bioanalysis. Traditionally, triple quadrupole mass spectrometers have been operated at unit mass resolution during these applications. More recently, ThermoFinnigan introduced their Quantum product line, which offers enhanced mass resolution on a triple quadrupole instrument. With this enhanced resolution, the mass spectrometer can work at a resolution of 0.2-Da peak width, FWHM (full width at half-maximum) while substantially maintaining the ion transmission efficiency. In bioanalysis, this enhanced resolution will translate into increased specificity and sensitivity by removing or reducing interferences or chemical background (Xu et al., 2003; Yang et al., 2002; Pucci et al., 2006; Jemal and Ouyang, 2003). An example of the effective use of this enhanced resolution is shown for the quantitation of nefazodone in human urine (Fig. 3.8) (Jemal and Ouyang, 2003). When the instrument was operated at unit mass resolution (0.7 FWHM), there was considerable chemical noises at the lower concentration (30 pg/mL). When the enhanced resolution (0.2 FWHM) was used, the chemical noise background was significantly reduced, resulting in significantly improved signal-to-noise ratio.

The enhanced resolution capability is a useful tool to increase the sensitivity of the assay if the quantitation limit is chemistry noise limited. It is important to recognize that the enhanced resolution capability may not be able to remove interferences from drug-related species such as metabolites because these interferences may share the same exact masses.

3.3.3 New Capability

In addition to technologies to improve the throughput and performance of the bioanalytical assays, some recent development in technology is also potentially enabling bioanalysis to answer questions that conventional bioanalytical methods are not able to address. This section will cover two new technologies that offer these new capabilities.

3.3.3.1 Nanospray. Nanospray refers to electrospray at low flow rates, typically ≤1 μL/min. From the first report on nanospray by Wilm and Mann (1994), significant interest has been generated in this area. Nanospray is more than simply a reduction of electrospray flow, because it offers some unique advantages over conventional electrospray. One of the most notable advantages is its reduced ion suppression and increased tolerance for salt contamination compared to conventional electrospray (Gangl et al., 2001). This is a highly attractive feature for high-throughput bioanalysis because it becomes potentially feasible to eliminate chromatographic separation. Based on this concept, a chip-based nanospray device was developed and became commercially available. It has been used for different types of bioanalytical applications from high-throughput assays for *in vitro* samples to regulatory-compliant assays (Van Pelt et al., 2003; Wickremsinhe et al., 2005). Taking advantage of the reduced ion suppression, Hatsis et al. (2007) have combined nanospray with FAIMS to speed up separations from the liquid phase to the gas phase.

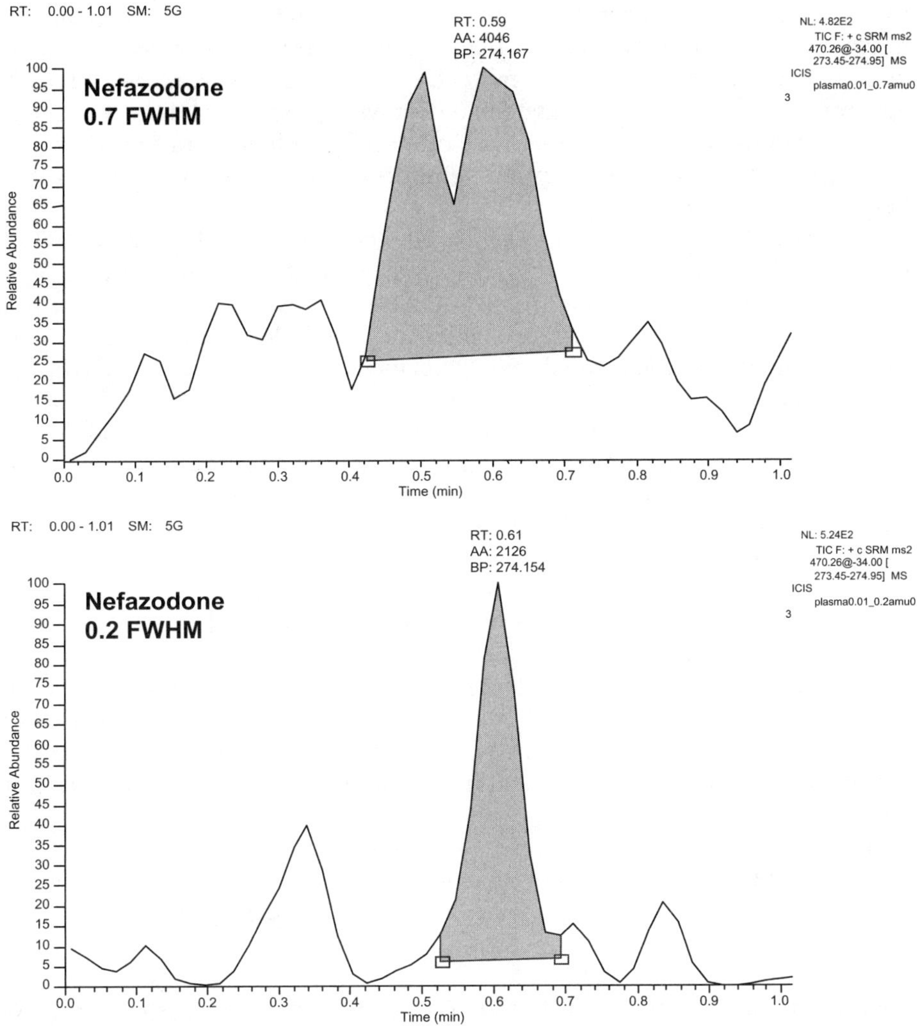

Figure 3.8. Comparison of the cleanliness of the SRM chromatograms obtained at FWHM settings of 0.70 and 0.20 FWHM from a 30 pg/mL nefazodone sample in human urine, with 60 fg injected onto column: upper panel at 0.70 FWHM; lower panel at 0.20 FWHM. Adapted from Jemal and Ouyang (2003). Reproduced with permission of John Wiley & Sons, Ltd.

Although often being viewed as a potential high-throughput tool, the most intriguing application of nanospray in the bioanalytical area is the potential as a semi-equimolar response ionization technique. The qualitative and quantitative information on metabolite formation plays an important role in drug discovery because it provides potential means to reduce metabolic clearance, understand the pharmacokinetics–pharmacodynamics relationship, and alert us of any chemical toxicity. Although the conventional electrospray technique has been very successful for bioanalysis, the ionization efficiency of compounds may differ significantly and therefore a pure reference standard is generally required for each analyte in bioanalysis. At the early discovery stage, the pure standards of metabolites are often not available, and therefore our capability of acquiring the quantitative information on metabolites has often been hindered at this stage.

Compared to conventional electrospray, the initial droplets generated in nanospray are much smaller. The smaller droplets increase the surface area and reduce the diffusion time for a solvated species to the droplet surface where they can be ionized. Also, smaller initial droplets reduce the number of columbic explosions, which is a mechanism for differed electrospray response due to preferential species enrichment during the process. Therefore, it is possible to achieve more uniform or equimolar response for compounds with different structures in nanospray. This equimolar response phenomenon was first reported by Karas and co-workers and was more recently demonstrated by Henion, Hop, and Wu (Schmidt et al., 2003; Henion, 2004; Hop et al., 2005; Valaskovic et al., 2006). The evaluation of this semi-equimolar response phenomenon for metabolite quantitation was demonstrated in Fig. 3.9, where the relative response for these six compounds and their metabolites were compared in conventional electrospray and nanospray at different flow rate (Valaskovic et al., 2006). In all these cases, it was observed that nanospray at low flow rate always generated more uniform response between the parent compound and the metabolite. The largest difference in response between the parent and the metabolite observed in nanospray is about 2 in this set of test compounds.

The equimolar response feature of nanospray ionization still needs to be fully validated with compound sets of larger sizes, but it certainly offers the potential capability to answer some challenging questions in drug discovery.

3.3.3.2 Mass Spectrometry-Based Tissue Imaging. Mass spectrometry-based tissue imaging has generated a lot of interest recently (Chaurand et al., 2004). Compared to the information acquired in conventional bioanalysis, which includes only the identity and quantity of the drug, tissue imaging also provides distribution of the drug in the tissues. In addition to the information provided by whole-body autoradiography, which only provides the distribution and quantity of all drug-related material, mass spectrometry-based tissue imaging can provide molecular-specific information and thus can differentiate the parent molecule from each metabolite. Another advantage of mass spectrometry-based imaging is that there is no need for radio-labeled compound which facilitates acquiring the distribution data in the discovery stage.

The MALDI (Matrix-Assisted Laser Desoprtion Ionization) technique has often been used as the method of choice for mass spectrometry-based tissue

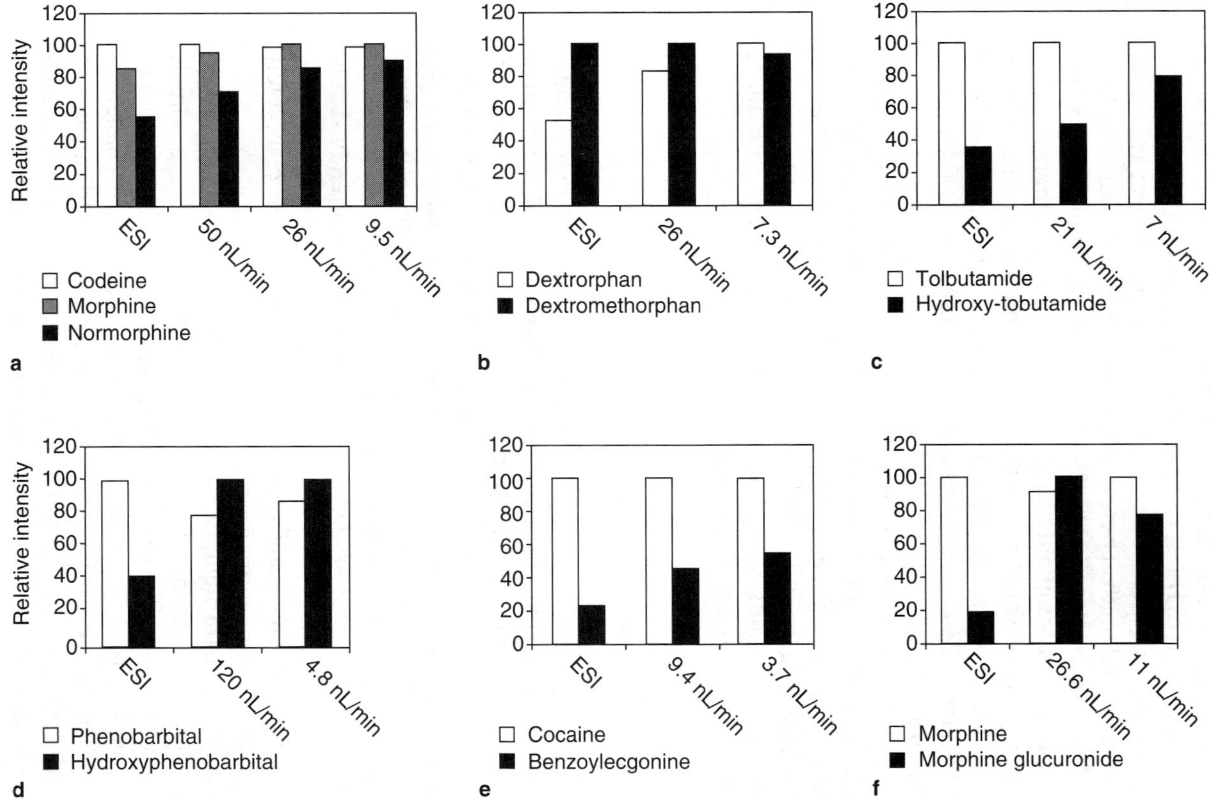

Figure 3.9. Results (**A–F**) of the equimolar response study for a variety of diverse compounds and their metabolites (as shown in Fig. 3.6), respectively, for both conventional ESI (5 µL/min) and nanospray (flow given in nL/min). Each bar graph shows the relative average signal intensities obtained from full-scan MS (average 12 scans). Each compound set was spiked into LLE plasma (2.5 µM final concentration each). Analysis was carried out by conventional ESI (5 µL/min) and ultra-low flow nanospray (a 5-µm-i.d., short-taper, metal-coated fused-silica emitter). Note that the commonly observed trend toward equimolar response as flow rate is reduced. Adapted from Valaskovic et al. (2006). Reproduced with permission of John Wiley & Sons, Ltd.

imaging because of its high sensitivity and excellent spatial resolution (Karas and Hillenkamp, 1988). MALDI is a soft ionization method that uses a matrix solution to form co-crystals with the analyte before being subject to a laser beam with a certain wavelength. These crystals absorb energy from the laser beam and transfer them to the analyte for desorption and ionization. While MALDI imaging has been frequently used for large molecules, there are also a few interesting reports on small molecule applications. Korfmacher and co-workers used MALDI imaging to directly measure clozapine in rat brain (Hsieh et al., 2006). Troendle et al. (1999) reported the use of MALDI imaging for paclitaxol in rat liver.

More recently, DESI has been reported to be used for tissue imaging instead of MALDI. As discussed in Section 3.3.1.4, a key advantage of DESI is that the samples can be analyzed at ambient under intact conditions. Kertesz et al. (2008) have reported the use of DESI for tissue imaging using propranolol as an example. The results were found to be in agreement with those generated by whole-body autoradiography.

3.4 CONCLUSIONS AND FUTURE DIRECTIONS

Bioanalytical technology has been evolving rapidly in the past two decades largely due to the new development in mass spectrometry and separations sciences. The detection sensitivity and ion source robustness of mass spectrometers have improved significantly over the years. It may become possible in the near future that a dilute-and-shoot approach can be routinely employed with biological samples such as plasma, thus significantly reducing the workload for sample preparation and minimize effects associated with sample matrixes. The new development in high-speed separations has already significantly shortened chromatographic runs, and new technology is on the horizon to move the separation from liquid phase to gas phase or even completely eliminate separation for some applications.

There is no doubt that throughput has been the key theme of these new developments, but it is worth pointing out that the new trend is gradually shifting toward enabling technologies that offers new capabilities. Instead of measuring systemic exposure in conventional bioanalysis, future challenges in bioanalysis will be in the areas such as measuring drug distribution and unbound drug concentration at the target tissue, measuring metabolite levels without reference standards, and reliably measuring various types of safety and efficacy biomarkers in the clinic. The latter one is another fast-growing area that is not covered in this chapter, but bioanalysis will play an important role in it.

The fast growth of bioanalytical technology has clearly been driven by the increasing role of ADME science in drug discovery. On the other hand, the advancement in bioanalytical technology has also successfully enabled ADME science to be engaged more and more into the drug discovery process. It is expected that this healthy cycle will continue on as ADME science continues to play a more and more important role in drug discovery.

REFERENCES

Chaurand P, Schwartz SA, Caprioloi RM. Profiling and imaging proteins in tissue sections by MS. *Anal Chem* 2004;76:87A–93A.

Chen H, Talaty NN, Takats Z, Cooks RG. Desorption electrospray ionization mass spectrometry for high-throughput analysis of pharmaceutical samples in the ambient environment. *Anal Chem* 2005;77:6915–6927.

Cody RB, Laramee JA, Durst HD. Versatile new ion source for the analysis of materials in open air under ambient conditions. *Anal Chem* 2005;77:2297–2302.

Cooks RG, Ouyang Z, Takats Z, Wiseman JM. Ambient mass spectrometry. *Science* 2006;311:1566–1570.

Deng Y, Zeng H, Unger SE, Wu J-T. Multiple-sprayer tandem mass spectrometry with parallel high flow extraction and parallel separation for high-throughput quantitation in biological fluids. *Rapid Commun Mass Spectrom* 2001;15:1634–1640.

Gangl ET, Annan A, Spooner N, Vorous P. Reduction of signal suppression effects in ESI-MS using a nanosplitting device. *Anal Chem* 2001;73:5635–5644.

Hatsis P, Brockman AH, Wu J-T. Evaluation of high-field asymmetric waveform ion mobility spectrometry coupled to nanoelectrospray ionization for bioanalysis in drug discovery. *Rapid Commun Mass Spectrom* 2007;21:2295–2300.

Hatsis P, Kapron JT. A review on the application of high-field asymmetric waveform ion mobility spectrometry (FAIMS) in drug discovery. *Rapid Commun Mass Spectrom* 2008;22:735–738.

Henion JD. Paper presented at the 52nd ASMS Conference on Mass Spectrometry and Allied Topics, 2004.

Herman JL. The use of turbulent flow chromatography and the isocratic focusing effect to achieve on-line cleanup and concentration of neat biological samples for low-level metabolite analysis. *Rapid Commun Mass Spectrom* 2005;19:696–700.

Hop CECA, Chen Y, Yu LJ. Uniformity of ionization response of structurally diverse analytes using a chip-based nanoelectrospray ionization source. *Rapid Commun Mass Spectrom* 2005;19:3139–3142.

Hsieh Y, Wang G, Wang Y, Chackalamannil S, Brisson J-M, Ng K, Korfmacher WA. Simultaneous determination of a drug candidate and its metabolite in rat plasma samples using ultrafast monolithic column high-performance liquid chromatography/tandem mass spectrometry. *Rapid Commun Mass Spectrom* 2002;16:944–950.

Hsieh Y, Casale R, Fukuda E, Chen J, Knemeyer I, Wingate J, Morrison R, Korfmacher WA. Matrix-assisted laser desorption/ionization imaging mass spectrometry for direct measurement of clozapine in rat brain tissue. *Rapid Commun Mass Spectrom* 2006;20:965–972.

Huang M-Q, Mao Y, Jemal M, Arnold M. Increased productivity in quantitative bioanalysis using a monolithic column coupled with high-flow direct-injection liquid chromatography/tandem mass spectrometry. *Rapid Commun Mass Spectrom* 2006;20:1709–1714.

Ingelse BA, Vogel G, Botterblom M, Nanninga D, Ooms B. Direct injection of whole blood for liquid chromatography/tandem mass spectrometry analysis to support single-rodent pharmacokinetic studies. *Rapid Commun Mass Spectrom* 2008;22: 834–840.

Jemal M, Ouyang Z. Enhanced resolution triple-quadrupole mass spectrometry for fast quantitative bioanalysis using liquid chromatography/tandem mass spectrometry: investigations of parameters that affect ruggedness. *Rapid Commun Mass Spectrom* 2003;17:24–38.

Jemal M, Xia Y-Q. The need for adequate chromatographic separation in the quantitative determination of drugs in biological samples by high performance liquid chromatography with tandem mass spectrometry. *Rapid Commun Mass Spectrom* 1999;13:97–106.

Jemal M, Xia Y-Q, Whigan DB. The use of high-flow high performance liquid chromatography coupled with positive and negative ion electrospray tandem mass spectrometry for quantitative bioanalysis via direct injection of the plasma/serum samples. *Rapid Commun Mass Spectrom* 1998;12:1389–1399.

Jemal M, Huang M, Mao Y, Whigan D, Powell ML. Increased throughput in quantitative bioanalysis using parallel-column liquid chromatography with mass spectrometric detection. *Rapid Commun Mass Spectrom* 2001;15:994–999.

Kapron JT, Jemal M, Duncan G, Kolakowski B, Purves R. Removal of metabolite interference during liquid chromatography/tandem mass spectrometry using high-field asymmetric waveform ion mobility spectrometry. *Rapid Commun Mass Spectrom* 2005;19:1979–1983.

Karas M, Hillenkamp F. Laser desorption ionization of proteins with molecular masses exceeding 10,000 daltons. *Anal Chem* 1988;60:2299–2301.

Kebarle P, Tang L. From ions in solution to ions in the gas phase. *Anal Chem* 1993;65:972A–986A.

Kertesz V, Van Berkel GJ, Vavrek M, Koeplinger KA, Schneider BB, Covey TR. Comparison of drug distribution images from whole-body thin tissue sections obtained using desorption electrospray ionization tandem mass spectrometry and autoradiography. *Anal Chem* 2008;80:5168–5177.

King R, Bonfiglio R, Fernandez-Metzler C, Miller-Stein C, Olah T. Mechanistic investigation of ionization suppression in electrospray ionization. *J Am Soc Mass Spectrom* 2000;11:942–950.

King RC, Miller-Stein C, Magiera DJ, Brann J. Description and validation of a staggered parallel high performance liquid chromatography system for good laboratory practice level quantitative analysis by liquid chromatography/tandem mass spectrometry. *Rapid Commun Mass Spectrom* 2002;16:43–52.

Korfmacher WA, Veals J, Dunn-Meynell K, Zhang X, Tucker G, Cox KA, Lin C-C. Demonstration of the capabilities of a parallel high performance liquid chromatography tandem mass spectrometry system for use in the analysis of drug discovery plasma samples. *Rapid Commun Mass Spectrom* 1999;13:1991–1998.

MacNair JE, Patel KD, Jorgenson JW. Ultrahigh-pressure reversed-phase capillary liquid chromatography: isocratic and gradient elution using columns packed with 1.0-μm particles. *Anal Chem* 1997;71:700–708.

Matuszewski BK, Costanzer ML, Chavez-Eng CM. Matrix effect in quantitative LC/MS/MS analyses of biological fluids: a method for determination of Finasteride in human plasma at picogram per milliliter concentrations. *Anal Chem* 1999;70: 882–889.

Mazzeo JR, Neue UD, Kele M, Plumb RS. A new separation technique takes advantage of sub-2-μm porous particles. *Anal Chem* 2005;77:460A–467A.

O'Connor D, Mortishire-Smith R, Morrison D, Davies A, Dominguez M. Ultra-performance liquid chromatography coupled to time-of-flight mass spectrometry for robust, high-throughput quantitative analysis of an automated metabolic stability assay, with simultaneous determination of metabolic data. *Rapid Commun Mass Spectrom* 2006;20:851–857.

Peng SX, Barbone AG, Ritchie DM. High-throughput cytochrome P450 inhibition assays by ultrafast gradient liquid chromatography with tandem mass spectrometry using monolithic columns. *Rapid Commun Mass Spectrom* 2003;17:509–518.

Petucci C, Diffendal J, Kaufman D, Mekonnen B, Terefenko G, Musselman B. Direct analysis in real time for reaction monitoring in drug discovery. *Anal Chem* 2007;79:5064–5070.

Plumb R, Dear G, Mallett D, Ayrton J. Direct analysis of pharmaceutical compounds in human plasma with chromatographic resolution using an alkyl-bonded silica rod column. *Rapid Commun Mass Spectrom* 2001;15:986–993.

Plumb RS, Potts WB III, Rainville PD, Alden PG, Shave DH, Baynham G, Mazzeo JR. Addressing the analytical throughput challenges in ADME screening using rapid ultra-performance liquid chromatography/tandem mass spectrometry methodologies. *Rapid Commun Mass Spectrom* 2008;22:2139–2152.

Pucci V, Bonelli F, Monteagudo E, Laufer R. Enhanced mass resolution method development, validation and assay application to support preclinical studies of a new drug candidate. *Rapid Commun Mass Spectrom* 2006;20:1240–1246.

Purves RW, Guevremont R. Electrospray ionization high-field asymmetric waveform ion mobility spectrometry–mass spectrometry. *Anal Chem* 1999;71:2346–2357.

Rolando C, Sablier M. Multiquadrupoles. In: *The Encyclopedia of Mass Spectrometry: Theory and Ion Chemistry*, Armentrout PB, editor. Elsevier, 2003.

Schmidt A, Karas M, Dulcks R. Effect of different solution flow rates on analyte ion signals in nano-ESI MS, or: when does ESI turn into nano-ESI? *J Am Soc Mass Spectrom* 2003;14:492–500.

Takats Z, Wiseman JM, Gologan B, Cooks RG. Mass spectrometry sampling under ambient conditions with desorption electrospray ionization. *Science* 2004;306: 471–473.

Tanaka N, Kobayashi H, Nakanishi K, Minakuchi H, Ishizuka N. Monolithic LC columns: a new type of chromatographic support could lead to higher separation efficiencies. *Anal Chem* 2001;73:421A–429A.

Troendle FJ, Reddick CD, Yost RA. Detection of pharmaceutical compounds in tissue by matrix-assisted laser desorption/ionization and laser desorption/chemical ionization tandem mass spectrometry with a quadrupole ion trap. *J Am Soc Mass Spectrom* 1999;10:1315–1321.

Valaskovic GA, Utley L, Lee MS, Wu J-T. Ultra-low flow nanospray for the normalization of conventional liquid chromatography/mass spectrometry through equimolar response: standard-free quantitative estimation of metabolite levels in drug discovery. *Rapid Commun Mass Spectrom* 2006;20:1087–1096.

Van Pelt CK, Zhang S, Fung E, Chu I, Liu T, Li C, Korfmacher WA, Henion J. A fully automated nanoelectrospray tandem mass spectrometric method for analysis of Caco-2 samples. *Rapid Commun Mass Spectrom* 2003;17:1573–1578.

Wang G, Hsieh Y, Cui X, Cheng K-C, Korfmacher WA. Ultra-performance liquid chromatography/tandem mass spectrometric determination of testosterone and its metabolites in *in vitro* samples. *Rapid Commun Mass Spectrom* 2006;20:2215–2221.

Wells DA. *High Throughput Bioanalytical Sample Preparation*. Amsterdam: Elsevier, 2003.

Wells JM, McLuckey SA. Collisional activation and dissociation: methodology. In: *The Encyclopedia of Mass Spectrometry: Theory and Ion Chemistry*, Armentrout PB, editor. Elsevier, 2003.

Wickremsinhe ER, Ackermann BL, Chaudhary AK. Validating regulatory-compliant wide dynamic range bioanalytical assays using chip-based nanoelectrospray tandem mass spectrometry. *Rapid Commun Mass Spectrom* 2005;19:47–56.

Wilm MS, Mann M. Electrospray and Taylor–Cone theory, Dole's beam of macromolecules at last? *Int J Mass Spectrom Ion Processes* 1994;136:167–180.

Wu J-T. The development of a staggered parallel separation liquid chromatography/tandem mass spectrometry system with on-line extraction for high-throughout screening of drug candidates in biological fluids. *Rapid Commun Mass Spectrom* 2001;15:73–81.

Wu J-T, Zeng H, Qian M, Brogdon BL, Unger SE. Direct plasma sample injection in multiple-component LC–MS–MS assays for high-throughput pharmacokinetic screening. *Anal Chem* 2000;72:61–67.

Wu J-T, Zeng H, Deng Y, Unger SE. High-speed liquid chromatography/tandem mass spectrometry using a monolithic column for high-throughput bioanalysis. *Rapid Commun Mass Spectrom* 2001;15:1113–1119.

Xia Y-Q, Wu ST, Jemal M. LC-FAIMS-MS/MS for quantification of a peptide in plasma and evaluation of FAIMS global selectivity from plasma components. *Anal Chem* 2008;80:7137–7143.

Xu X, Veals J, Korfmacher WA. Comparison of conventional and enhanced mass resolution triple-quadrupole mass spectrometers for discovery bioanalytical applications. *Rapid Commun Mass Spectrom* 2003;17:832–837.

Yang L, Amad M, Winnik WM, Schoen AE, Schweingruber H, Mylchreest I, Rudewicz PJ. Investigation of an enhanced resolution triple quadrupole mass spectrometer for high-throughput liquid chromatography/tandem mass spectrometry assays. *Rapid Commun Mass Spectrom* 2002;16:2060–2066.

Yu K, Little D, Plumb R, Smith B. High-throughput quantification for a drug mixture in rat plasma—a comparison of ultra performance liquid chromatography/tandem mass spectrometry with high-performance liquid chromatography/tandem mass spectrometry. *Rapid Commun Mass Spectrom* 2006;20:544–552.

Yu S, Crawford E, Tice J, Musselman B, Wu J-T. Bioanalysis without sample cleanup or chromatography: the evaluation and initial implementation of direct analysis in real time ionization mass spectrometry for the quantification of drugs in biological matrixes. *Anal Chem* 2009;81:193–202.

Zhao Y, Lam M, Wu D, Mak R. Quantification of small molecules in plasma with direct analysis in real time tandem mass spectrometry, without sample preparation and liquid chromatographic separation. *Rapid Commun Mass Spectrom* 2008;22: 3217–3224.

Zhou S, Zhou H, Larson M, Miller DL, Mao D, Jiang X, Weng N. High-throughput biological sample analysis using on-line turbulent flow extraction combined with monolithic column liquid chromatography/tandem mass spectrometry. *Rapid Commun Mass Spectrom* 2005;19:2144–2150.

4

SAFETY BIOMARKERS IN DRUG DEVELOPMENT: EMERGING TRENDS AND IMPLICATIONS

ERIC R. FEDYK

4.1 INTRODUCTION

Biomarker is a general term that is utilized by many scientific disciplines and with differing definitions. The National Institutes of Health defined medical biomarkers as "A characteristic that is objectively measured and evaluated as an indicator of normal biologic processes, pathogenic processes, or pharmacologic responses to a therapeutic intervention" (Biomarkers Definitions Working Group, 2001). This general definition is indeed an accurate representation of the vast spectrum of types of biomarkers. Common examples include pharmacodynamic biomarkers, efficacy biomarkers, safety biomarkers, disease biomarkers, surrogate endpoints and bridging biomarkers (Table 4.1). While biomarkers such as fever have been utilized by clinicians for centuries, the new millennia has brought renewed interest this field due to the potential to personalize medicine and/or reduce the cost of drug discovery and development. The use of biomarkers in drug discovery and development is such a large field that insightful review generally requires focusing on specific types of biomarkers and disciplines. This chapter will focus on safety biomarkers, characteristics that are objectively measured and evaluated as indicators of pathogenic processes or toxicologic responses to a therapeutic intervention, because this subdiscipline is currently the least defined and offers the greatest opportunity to improve drug development and medicine.

Enzyme Inhibition in Drug Discovery and Development, edited by Lu and Li
Copyright © 2010 John Wiley & Sons, Inc.

TABLE 4.1. Types of Biomarkers

Type	Description
Pharmacodynamic biomarkers	Used to understand the relationship between dose and pharmacologic activity of a test agent on the target *in vivo*.
Efficacy biomarkers	Correlate with the desired effect of a treatment, but unvalidated (unlike surrogate endpoints).
Safety biomarkers	Indicate potentially harmful effects of a drug in cell-based, preclinical or clinical studies.
Disease biomarkers	Indicate the presence or likelihood of a particular disease.
Surrogate endpoints	Accepted as a substitute for desired outcome because it can be predictive before meaningful clinical endpoints.
Bridging biomarkers	Can be used in both preclinical and clinical studies. Can be disease, efficacy, or toxicity biomarkers.

4.2 THE OPPORTUNITY FOR MODERNIZATION OF PRECLINICAL SAFETY ASSESSMENT

In 1962, congressional amendments to the Food, Drug, and Cosmetic Act created for the first time a requirement that drugs be scientifically shown to be effective before they could be marketed (a requirement for safety had been in effect since 1938). During the 1960s–1980s, drug developers, the academic community, and regulators worked to develop and refine ways to design, conduct, and analyze randomized controlled clinical trials that could produce the needed evidence. Many important advances in pharmacotherapy (e.g., cardiovascular therapies, psychiatric drugs, anti-infectives, and cancer treatments) were introduced during this era. However, the evidence generated in drug development programs was still somewhat limited. For example, dose–response information was usually scanty, often few women were studied, data on long-term use (even for chronically administered drugs) were lacking, evaluations of subgroups such as patients with renal or hepatic insufficiency were not conducted, and data on drug–drug interactions were not available. From the mid-1980s through the 1990s, as an increasing number of drug therapies became available, the United States Food and Drug Administration (FDA) as well as the international regulatory community established the expectation that such information would be obtained during drug development programs. Therefore, modern development programs usually are much more extensive and contain many more clinical studies and patient exposures than was usual in 1960–1985.

Despite these significant advances, recalcitrant issues remain that generate uncertainty about the performance of drugs that are new to the market. Data from long-term use, for example, are still usually limited due to the escalating cost of developing therapeutics (Fig. 4.1), in terms of time (6–20

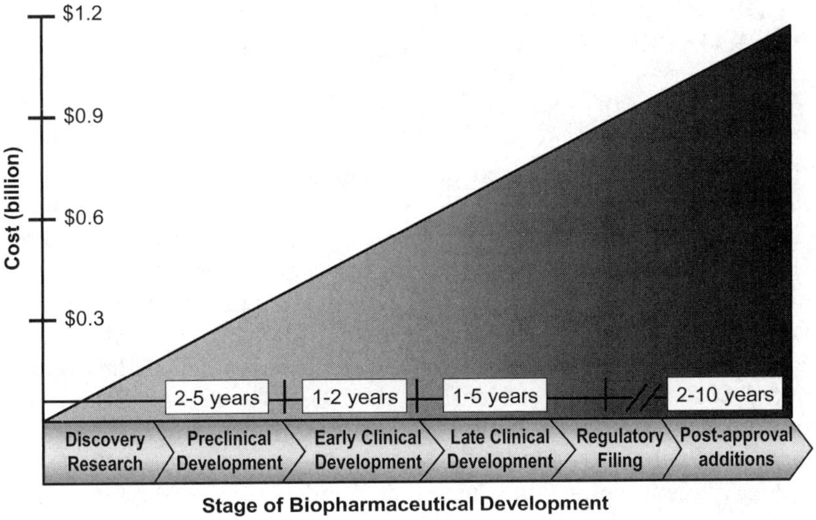

Figure 4.1. The relative cost and timelines to develop a biopharmaceutical.

years) and money (0.8–1.7 billion dollars) (DiMasi et al., 2003; Adams and Brantner, 2006; Development TTCFD, 2006). Secondly, current drug development programs cannot detect drug-related adverse outcomes that represent a small increase in frequency of a problem that is already common in the treated population (e.g., ischemic cardiovascular events). Current methodologies to predict the occurrence of rare (1 in >1000 subjects) and sometimes catastrophic side effects are impractical, cost-prohibitive, and consequently rarely practiced. A number of constraints have thwarted well-intentioned attempts to make the results of clinical trials more widely applicable, and hence the patients enrolled in trials do not reflect the full range of the population or treatment situations that occur in practice. New safety issues are consequently identified only after drugs enter the market, with the withdrawals of rofecoxib and valdecoxib for cardiovascular toxicity being notable examples (Solomon, 2006; Hermann and Ruschitzka, 2007). New issues have emerged as well. Increased pressure upon regulatory agencies to accelerate the drug approval process by shortening review cycles, coupled with aggressive marketing techniques targeting the consumer (Wood, 2006), have nevertheless led to early release, immediate uptake, and widespread use of many new drugs, with a general expectation that their performance is well understood for this patient population. Increasing recognition of this problem has since led to calls for larger trials and longer patient exposures prior to drug marketing which, conversely, can be incompatible with accelerating the drug review process and lowering the cost of developing therapeutics.

4.3 A DECREASE IN THE EFFICIENCY OF DEVELOPMENT OF PHARMACEUTICALS

The productivity of the pharmaceutical industry has been steadily decreasing the last couple of decades because successful development of novel drugs has decreased, despite increased investment in pharmaceutical research and development (Administration USFaD, 2004a, 2006b). Recent data indicates that 2004 represented a 20-year low in introductions of new chemical entities (NMEs) worldwide. The same phenomenon has been observed in the United States, where the submission rate of new drug applications for NMEs has shown a downward trend in the past decade (Administration USFaD, 2004a,b). Not surprisingly, the investment needed per successful NME has risen to an estimated $0.8–1.7 billion (DiMasi et al., 2003; Adams and Brantner, 2006; Development TTCF, 2006). This cost is driven by the high rate of clinical failure, estimated at 70–90% of candidates (Administration USFaD, 2004b). The rising percentage of late-stage clinical failures, now 50% of compounds tested in phase 3 trials, is of particular concern. These late-stage failures affect the cost of successful drug development disproportionately since the cost of each stage of development grows progressively larger (Fig. 4.1). Costly failures such as these may discourage investment in more innovative, risky approaches, as well as in therapeutics for diseases that represent smaller markets. Additionally, the need to recoup this investment during the period of market exclusivity, prior to the introduction of generic copies, is an incentive for aggressive marketing techniques (Wood, 2006). Rapid market uptake also entails a large number of individuals who have already been exposed by the time a drug problem is discovered after marketing.

Thus, rising societal demands for greater certainty about the outcomes of drug therapy are occurring at a time when the pharmaceutical industry is experiencing difficulty controlling the cost and quality of drug development. These concurrent trends generated significant concern on a federal level, given the number of medical conditions that currently have unsatisfactory or no therapeutic options. The FDA, with its dual roles of protecting and promoting health, is charged with implementing policies that ensure that the benefits of new products will outweigh their risks, while simultaneously promoting innovations that can improve health. The challenges inherent to this mission drove the genesis of the Critical Path Initiative.

4.4 THE CRITICAL PATH INITIATIVE

Expectations have been high that significant public investment in biomedical research would produce an explosion of new therapies for previously untreatable or inadequately treated diseases. President Richard Nixon declared a war on cancer in his State of the Union address in 1971 and stated that the same kind of concentrated effort that split the atom and took man to the moon should be turned toward conquering cancer. The failure of this and similar efforts to achieve these lofty goals has prompted extensive speculation on the cause of this problem. In 2004, the FDA published a White Paper entitled "Innovation or Stagnation: Challenges and Opportunities on the Critical Path to Medical Product Development" (Administration USFaD, 2004a). This document acknowledges that a combination of factors has likely led to the current issues with development of therapeutics. The Critical Path also illuminated an important and generally unrecognized problem: the lagging science of drug development.

Drug development can be conceptualized as a process leading from basic research through a series of developmental steps to a commercial product. The FDA White Paper identified the "Critical Path" as a process beginning with identification of a drug candidate during the stage of Discovery Research (Fig. 4.1) and culminating in a marketing approval. Along the path to marketing, the product is subjected to a series of evaluations to predict its safety and effectiveness and to enable its mass production. There has been very little change in the science of the development process, despite extensive investment in basic biomedical science over the past four decades. Many of the newer and more sophisticated scientific tools used in identifying drug targets and lead compounds during discovery research have yet to be employed in the preclinical and clinical development stages (Fig. 4.1). This is particularly true for safety biomarkers. There are relatively few new safety biomarkers in conventional use; most widely used safety biomarkers have been recognized for decades (Table 4.2). The high rate of failure in the clinic due to safety

TABLE 4.2. Examples of Conventional Safety Biomarkers

Safety Biomarker	Type	Target Tissue
Body temperature	Clinical	Multiple
Stool consistency	Clinical	Intestine, colon
Alanine transferase	Serum chemistry	Liver
Gamma glutamyl transpeptidase	Serum chemistry	Biliary system
Conjugated bilirubin	Urine test	Liver, biliary system
pH	Urine test	Digestive system, urinary tract
Red blood cell indices	Hematology	Bone marrow, vasculature
White blood cell count	Hematology	Bone marrow, vasculature
Platelet count	Coagulation	Bone marrow, vasculature
Activated clotting time	Coagulation	Vasculature
Organ weights	Gross pathology	Multiple
Organ appearance	Gross pathology	
Hyperplasia	Histology	Multiple
Hypertrophy	Histology	Multiple
QT prolongation	Electrocardiogram	Cardiac
PR interval prolonation	Electrocardiogram	Cardiac

issues suggests that we need to improve our cache of biomarkers and that the safety biomarkers of yesteryear are insufficient for evaluating the therapeutics of tomorrow. Reducing these clinical failures will also, in turn, increase the overall efficiency of developing therapeutics.

4.5 THE CRITICAL PATH OPPORTUNITIES LIST

FDA's 2004 Critical Path White Paper generated considerable discussion and debate among academics, clinicians, the biopharmaceutical and medical device industries and patient advocacy groups. Over 100 groups submitted comments on the paper. After extensive consultation with numerous stakeholders, FDA issued the "Critical Path Report and List" in 2006 (Administration USFaD, 2006b). This report highlighted the scientific areas that would have the greatest impact on improving the development process: development and utilization of biomarkers; modernizing clinical trial methodologies and processes; the aggressive use of bioinformatics, including disease modeling and trial simulation; and improvement in manufacturing technologies. It also contained the "2006 Critical Path Opportunities List," 76 discrete projects that, if completed, could improve product development and subsequent use. A number of these projects are now being undertaken, many in partnership with the FDA (Administration USFaD, 2007).

Development of new biomarkers was identified as the highest priority for scientific effort. Genomic, proteomic, and metabolomic technologies, as well as advanced imaging techniques, offer tremendous opportunity for generating new biomarkers that can reflect the state of health or disease at the molecular level (Woodcock, 2007; Wagner, 2008). The greatest need and opportunity in biomarker development exists in safety, in part because previous activity focused primarily on pharmacodynamics, efficacy, and surrogate endpoint biomarkers (Table 4.1). Prediction of adequate safety is an essential part of drug development. Currently, preclinical safety testing involves traditional *in vitro* and animal toxicology studies. Animal toxicology tests are very useful for assessing safety for initial human testing; however, they often fail to uncover the types of toxicities seen after widespread human exposure. New technologies, such as gene expression assays (Mardis, 2008) in whole-cell or animal systems, proteomics (Petricoin et al., 2006), or metabolomics (van Ravenzwaay et al., 2007), may provide much greater insight into the whole spectrum of pharmacologic effects of a candidate drug. Such technologies may also be useful in comparing the candidate's effects [both pharmacologic ("On-target") and nonspecific ("Off-target") effects] to those of other drugs in its class or other drugs intended for similar uses (Caskey, 2007). Drug developers are currently using such technologies in preclinical safety assessment; and the clinical utilization and, ultimately, implications of such findings are being assessed.

4.6 IDEAL ATTRIBUTES OF SAFETY BIOMARKERS

A theoretical description of the ideal attributes of a toxicity biomarker would include high specificity for the toxicity of interest. The biomarker would only be expressed in the target tissue (e.g., kidney) or within a specific cell type within a given tissue (e.g., renal tubules) and not expressed in nontarget tissues in various types of pathologies. The biomarker would also exhibit specificity for various mechanisms of toxicity—for example, necrosis and not apoptosis, hypertrophy, and so on. The ideal biomarker would also exhibit high sensitivity, such that minimal injury has occurred prior to detection, perhaps by exhibiting low levels of basal expression and robust and immediate release upon tissue injury. The amount of release would, moreover, be proportionate to the degree of injury. Measuring the biomarker would require noninvasive procedures—for example, imaging technologies or detection from easily accessible samples such as saliva, urine, blood, and so on. The biomarker should be robust in the detection is rapid, simple, inexpensive and for all toxicology species, including humans (i.e., a bridging biomarker). Lastly, the stability of the ideal biomarker would be sufficiently long to permit easy detection and yet short enough such that reversibility of an injury could be detected by monitoring levels of that same biomarker. A real-world example of a safety biomarker that illustrates many of these attributes is serum troponin for drug-induced cardiac toxicity (Wallace et al., 2004).

4.7 SAFETY BIOMARKERS IMPROVE IDENTIFICATION OF THE BEST CANDIDATE THERAPEUTICS

The emerging concept of "personalized medicine" is a paradigm for improving the overall quality of patient care and is defined as the right drug, for the right patient, at the right dose and at the right time (Woodcock, 2007). The utilization of novel safety biomarkers to improve drug development can be framed in a similar paradigm; selecting the right therapeutic candidates, the right subjects, the right dose and time. Safety biomarkers are increasingly utilized to select the best therapeutic candidates. Well-established, validated, and qualified safety biomarkers are utilized early in development to screen and prioritize lead therapeutic candidates with respect to the potential risk of adverse events. This improves the productivity of drug development by first identifying potential failures early and then re-allocating downstream resources exclusively to the most promising candidate therapeutics. Failures late in development are particularly devastating to productivity because the cost of drug development increases with each stage (Fig. 4.1). Biomarkers can aid in early decision making on whether to drop a drug from consideration or to move it through trials. A conventional example of such a screen is the human Ether-a-go-go Related Gene (hERG) assay (Meyer et al., 2007). hERG encodes the $K_v11.1$ potassium ion channel responsible for the repolarizing I_{Kr}

current in the cardiac action potential. The hERG potassium ion channel is sensitive to drug binding, which can result in (a) decreased channel function and (b) the acquired long QT syndrome and fatal cardiac arrhythmia, due to repolarization disturbances of the cardiac action potential. The hERG channel contains a relatively large inner vestibule that provides more space for many different drug classes to bind and block this potassium channel. Some of the most common drugs that can cause QT prolongation include antiarrhythmics (especially Class 1A and Class III), anti-psychotic agents, and certain antibiotics (including quinolones and macrolides). The severity and frequency of consequences with compounds that bind to the hERG channel has convinced many pharmaceutical companies to assay effects on this channel during early preclinical drug development (Fig. 4.1), lead optimization in some instances, to prioritize candidate therapeutics with respect to their potential to cause cardiac arrhythmia (Meyer et al., 2007).

4.8 SAFETY BIOMARKERS IMPROVE IDENTIFICATION OF THE MOST APPROPRIATE SUBJECTS

The current problems with predicting and evaluating drug safety are often due to extreme variability of the human disease response. Safety biomarkers can be utilized to identify subjects that are likely to respond adversely to a drug. Omission of these subjects from clinical trials increases the likelihood of enrolling subjects that will respond favorably and in so doing can reduce the overall cost of the trial by reducing the size of a study population required to obtain a statistically significant effect. Identification of subject CYP2D6 and CYP2C19 genotypes is predictive of drug metabolism enzyme activity and can be used to select subjects for clinical trials of therapeutics that are metabolized by these gene products.

Cytochrome P450 isoenzymes are a group of heme-containing enzymes involved in oxidative metabolism of a number of drug classes and xenobiotics. The P450 genes are polymorphic, which have a functional significance for drug metabolism as certain allelic variants exhibit either altered activity or complete absence of enzymatic activity. The CYP2D6 gene, encoding for debrisoquine hydroxylase, has at least 70 allelic variants resulting in four phenotypic types: poor metabolizers with gene inactivation of both alleles, intermediate metabolizers with one reduced activity allele and one null allele, extensive metabolizers with at least one functional allele, and ultrarapid metabolizers with excess enzymatic activity due to multiple copies (3–13) of functional alleles from gene duplication (Johansson et al., 1993; Sachse et al., 1997; Griese et al., 1998). The enzyme encoded by CYP2D6 metabolizes many antidepressants, antipsychotics, antiarrhythmics, opiates, antiemetics, and beta-adrenergic receptor blocker drugs. The CYP2C19 gene, encoding the enzyme *S*-mephenytoin hydroxylase, has two major variant alleles that result in enzyme deficiency (Goldstein, 2001). Differences in drug metabolism due to CYP450 phenotype can impact plasma levels of both the active moiety (drug or drug metabolite) and toxic metabolites. The enzyme encoded by CYP2C19 metabolizes compounds from the classes of anticonvulsants, proton pump inhibitors, anticoagulants, benzodiazepines and antimalarials. Diagnostic assays that identify a patient's CYP2D6 and CYP2C19 genotype from a whole blood sample are currently utilized for this express purpose.

Poor metabolizers treated with drugs that are extensively metabolized by these isoenzymes are at increased risk for an excessive or prolonged therapeutic effect and for toxicity, while ultrarapid metabolizers may not achieve therapeutic plasma levels with standard dosing. In the case of prodrugs that require enzymatic action to become therapeutic, the opposite phenomenon occurs. The distribution of functional, reduced function, and nonfunctional CYP2D6 alleles shows racial/ethnic differences (Bradford, 2002; Bjornsson et al., 2003; Solus et al., 2004). Functional CYP2D6 alleles are predominant in European Caucasians, while reduced function alleles have a high frequency in Asian and African populations and their descendants. Nonfunctional alleles are present at the highest frequency in European Caucasians, but are present at the lowest frequency in Asians. The frequency of the ultrarapid metabolizer phenotype is highest in Saudi Arabian and Ethiopian populations (Ingelman-Sundberg, 1999). The two most common allelic variants of CYP2C19 (*2 and *3) result in a nonfunctional enzyme. The frequency of poor metabolizers is highest in Asian populations (approximately 15–30%) (Bjornsson et al., 2003).

4.9 SAFETY BIOMARKERS IMPROVE IDENTIFICATION OF THE OPTIMAL DOSE

The relationship between the dose of a drug and desired effect is often bimodal; low doses are insufficient for efficacy, moderate doses may be optimal and higher doses detrimental due to toxicity. The therapeutic window is an index for estimating the dosage which can treat disease effectively while staying within the safety range (Fig. 4.2). More specifically, it is the range between the ED50 and the starting point of LD50 curve. The inherent genetic and environmental heterogeneity within a human population can influence the pharmacologic and toxicologic components of the therapeutic index to the extent that there may not be a single dose that is optimal for each subject within a population. There is a need hence, for safety biomarkers that guide dosing of individual subjects. Current safety biomarkers used for this purpose are often the end product of the toxicity itself, a change from an irreversible cellular event, for example release of Alanine Transferase from necrotic hepatocytes. The unmet need is for antecedent safety biomarkers; markers which precede and predict a toxicological response which provide the opportunity to intervene and minimize toxicity experienced by the subject. Recent improvement in the comprehension of pathogenic mechanisms and advancements in systems biology provide an unprecedented opportunity to identify antecedent

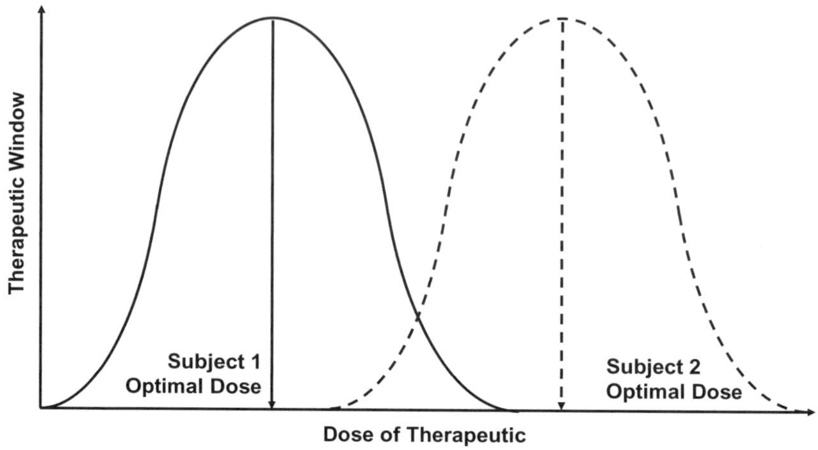

Figure 4.2. Safety biomarkers can define optimal dosing for subjects.

candidate safety biomarkers *in silico*. These candidates can then be tested empirically to assess potential utility as safety biomarkers. This activity will yield antecedent biomarkers that are useful in guiding dosing of individual subjects.

4.10 SAFETY BIOMARKERS CAN ASSIST IN IDENTIFYING THE OPTIMAL DOSE AT A SPECIFIC TIME

The inherent environmental heterogeneity during the course of a subject's therapy also influences pharmacologic and toxicologic components of the therapeutic index. Stress, concomitant medications (i.e., drug–drug interations), diet, and infections are a few of numerous variables that can shift the optimal dose of a therapeutic at a given point in time (Fig. 4.3). Monitoring levels of a predictive safety biomarker and understanding the relationship of the biomarker to the toxicity can assist in adjusting the dosing to within the optimal exposure at that point in the subject's therapy. If the therapeutic modulates the level of a causal safety biomarker to a level that presents a safety hazard, then the investigator could reduce the dose to minimize the likelihood of that toxicity. An intriguing example is the level of serum immunoglobulin with respect to immunosuppression and the predisposition of subjects to bacterial infection and/or malignancy.

Levels of serum immunoglobulin are inversely related to exposure to immunosuppressive therapeutics; a decrease in levels of serum immunoglobulin by a therapeutic is indicative of immunosuppression that predisposes subjects to infection and/or malignancy. A limitation of using of serum immunoglobulins as a predictive safety biomarker is the inherent delay in detecting modulation by a therapeutic due to two confounding factors. A steady-state

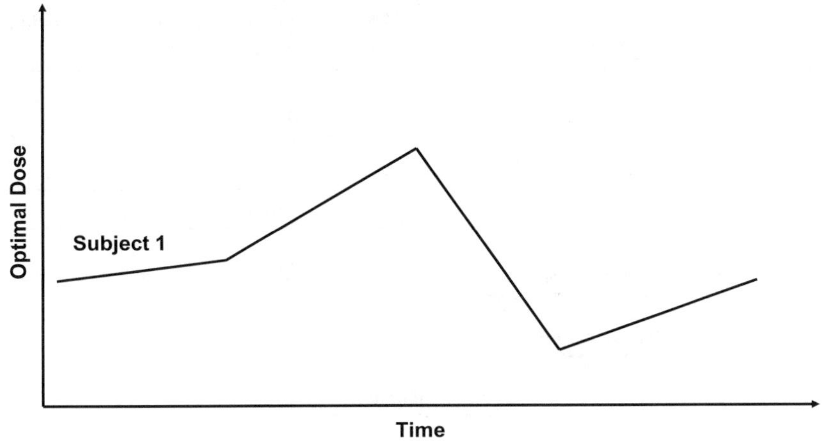

Figure 4.3. Safety biomarkers can define optimal dosing for subject at a given point in time.

level of serum immunoglobulins exists prior to exposure to a therapeutic; hence, an immunosuppressant is characterized by a reduction in signal. A decrease from this homeostatic level is primarily determined by two variables, the rate of production and the rate of catabolism, and these two rates are vastly different. Production can be inhibited with minutes of exposure to the therapeutic. The half-life of a serum immunoglobulin, namely gamma (IgG), is 27 days. This long serum half-life delays detection of therapeutics that inhibit production because detecting a 50% reduction in these steady-state IgGs will require 27 days, assuming that production is completely inhibited by the therapeutic. Calculating an IC_{50} for this potential effect further requires measuring less than full inhibition of production by a therapeutic; hence two months of dosing with a therapeutic could elapse before a robust biomarker signal is detected, depending on the sensitivity of the detection assay. The long half-life of the protein could be circumvented by measuring a less stable intermediate, such as mRNA with a half-life of two days. Innovative new safety biomarkers such as these could provide more timely feedback on the initial toxic activity and would greatly reduce the risk of infection and the need for supportive therapy.

4.11 STRATEGIES FOR IDENTIFYING AND DEVELOPING SAFETY BIOMARKERS

Biomarkers are often identified and developed in parallel with the candidate therapeutic (Fig. 4.4) because a particular candidate therapeutic may have a recognized liability and/or partnering these two activities is an efficient use of time and resources. Pharmacology experiments designed to characterize efficacy of a candidate therapeutic are also excellent opportunities for

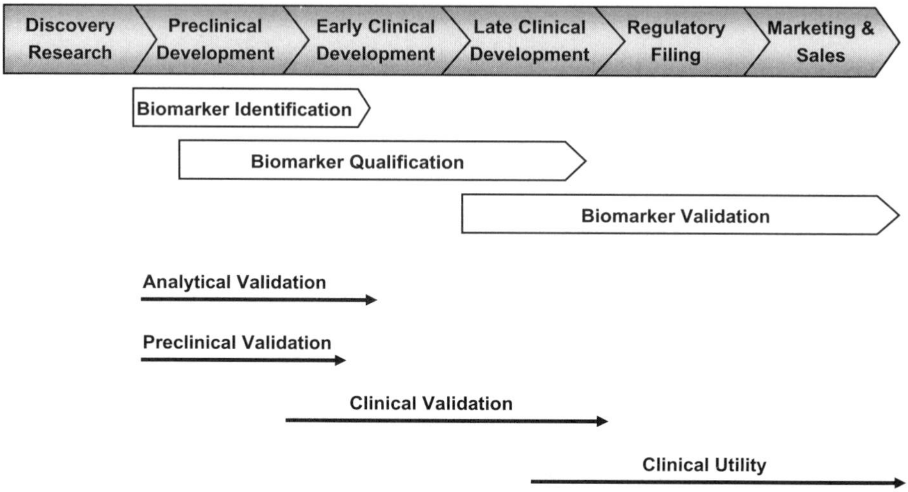

Figure 4.4. Development of safety biomarkers in biopharmaceutical development.

developing pharmacodynamic and/or efficacy biomarkers. They also can be valuable indicators of toxicity, both mechanistic (on-target) and chemical (off-target). The stage of drug development at which a biomarker is developed also heavily influences the selection strategy for discovery, validation, and qualification. Early identification of biomarkers permits abundant opportunities for co-development with the candidate therapeutic—for example, nonclinical efficacy and toxicity assessment, Phase I, II, and III clinical trials, and post-filing (Fig. 4.4). Conversely, identification of biomarkers later in development limits opportunity for co-development, perhaps only Phase III trials and/or post-filing and may require repeating resource-intensive investigations (e.g., nonclinical toxicology, clinical studies, etc.). Safety biomarkers can be particularly challenging in this respect because the need for these biomarkers is often recognized later in the drug development process than for pharmacodynamic, efficacy, and disease biomarkers. One strategy for overcoming this challenge of a "late-start" is to validate a well-developed biomarker for this new toxicity, thereby skipping assay development and analytical validation of the qualification stage of development. More often, however, a suitable pre-existing safety biomarker assay is not been suitable and one must consider "Top-down" versus "Bottom-up" strategies for identifying a suitable safety biomarker(s).

4.12 "TOP-DOWN" VERSUS "BOTTOM-UP" BIOMARKER IDENTIFICATION STRATEGIES

Identification strategies utilizing unbiased, high-content and high-sample throughput technologies are commonly referred to as "Top-down" approaches.

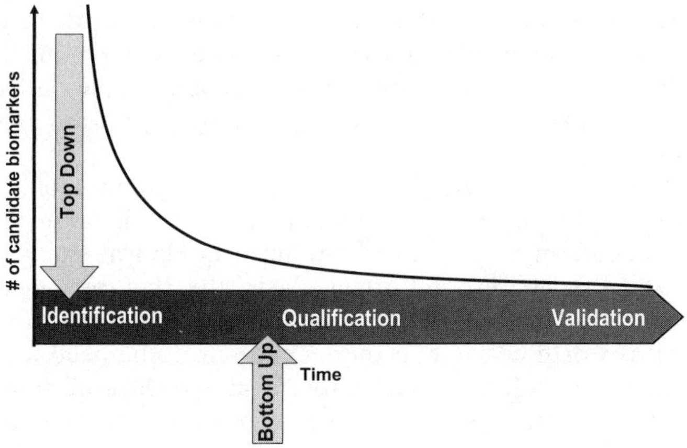

Figure 4.5. "Top-down" versus "bottom-up" biomarker discovery strategies.

This strategy consists of a broad screening of molecules on a sufficiently high content platform in order to identify molecular patterns associated with a phenotype of interest (Fig. 4.5). A defining characteristic is that a high-content molecular platform and often large numbers of samples is utilized in lieu of existing data and hypothesis-driven science. This approach is often utilized when the activity of potential therapeutic is poorly understood. A weakness of this strategy is that it provides correlative data that may not indicate the underlying mechanism of the adverse phenotype. Additional prospective studies are consequently required to determine the likely predictive utility (and any mechanistic relevance) of signatures discovered following these types of exploratory analyses. These "Top-down" strategies are typically employed early in preclinical development and with sufficient opportunity for additional prospective investigations.

An alternative strategy is hypothesis-driven science that utilizes knowledge of the candidate therapeutic and pathology to guide functional investigations that test causality. This "Bottom-up" strategy can be more efficient than a "Top-down" strategy if causality is proven because it minimizes the size and number of identification investigations necessary for qualification (Fig. 4.5). "Bottom-up" strategies are typically utilized later in preclinical development or with second- and third-generation therapeutic programs, in response to an adverse phenotype (i.e., issue management). A weakness in with "Bottom-up" strategy is that it is limited by the content of the existing knowledgebase and is subject to bias in these data, which can prevent identification of the best biomarker for the given issue. Safety biomarkers that directly reflect a toxicologic mechanism often possess the highest predictive clinical utility due to their proximal relationship to the pathology and due to a higher likelihood of evolutionary conservation of this critical activity across species.

A wide variety of profiling technology platforms is utilized to identify safety biomarkers of adverse events. Evaluations of DNA, RNA, protein, or metabolic profiles in preclinical models can enable us to identify signatures that can be refined and eventually used for safety assessment in clinical development (Glas et al., 2006). Examples of more common patient stratification biomarkers with varying degrees of utility include (a) critical DNA variants in genes encoding proteins involved in a drug's mechanism of action and (b) critical protein levels in specimens as indicated by immunohistochemistry. Similarly, transcriptomic, proteomic, and metabolomic signatures identified in preclinical models can be refined and utilized in clinical studies to identify patients most likely to respond adversely to the therapeutic intervention. While genomic, proteomic, and metabolomic standards are currently being established, it is worth noting that there continues to be need for integrating innovative new technologies that outperform the current standards, with next-generation sequencing being a recent example (Mardis, 2008). These "Nextgen" sequencing systems will be similar to existing unbiased, open-content, mass spectrometry, and NMR-based systems used to analyze proteomic and metabolic signatures, with the notable exception that all of the profiled moieties will be identifiable upon generation of the original expression profile signature, whereas additional effort may be required to identify moieties via mass spectrometry. These platform-based solutions for the thorough interrogation of DNA, RNA, proteins, and/or metabolites in a given biological specimen are powerful new tools for the discovery of novel safety biomarkers.

Appropriate study design, careful selection of the most informative and reproducible molecular signatures, and validation of the assays used to assess those biomarker profiles in subsequent clinical studies are all critical components that can maximize the chances of predicting adverse events in human patients. Another primary component is the relevance of the preclinical models from which the signatures are derived. A potential limitation is the identification and utilization of models that mimic the pathways encountered clinically, whether animals models or human *in vitro* and/or *ex vivo* systems. There are species-dependent metabolic and toxic effects, and traditional animal models can provide misleading data. A prototypical example is troglitazone-induced hepatotoxicity in humans, which is not observed in the rat. The variability observed between hepatocytes from different human donors, along with the high cost associated with obtaining these cultures, has severely limited the utility of human hepatocyte cultures. Searching for continual improvements in the predictability of model systems is critical for improving the safety of biopharmaceuticals and, hence, the efficiency of the biopharmaceutical development.

4.13 QUALIFICATION OF SAFETY BIOMARKERS

Biomarker qualification entails validating the analytical performance, the preclinical relevance, and the clinical relevance of the potential safety biomarker.

Analytical and preclinical validation often begin during the identification phase and proceed in parallel well into the biomarker qualification phase (Fig. 4.4). Analytical validation is the design and assessment of a detection assay for the potential safety biomarker. Key analytical performance criteria, such as sensitivity and specificity, are assessed to determine if they fulfill criteria required for clinical validation. Preclinical validation is an assessment of the utility of the potential safety biomarker in preclinical model systems. Utility is assessed with respect to a number of different variables. If the potential safety biomarker was discovered in an *in vitro* rat system, for example, is a similar phenomenon observed *in vivo* in the rat, the dog, and/or nonhuman primates? Is the potential safety biomarker representative of the effect of the specific therapeutic or a group of therapeutics of similar toxic activity, or dissimilar toxic activity (toxicologically active and inactive versions), but of a similar structure? Is the potential safety biomarker representative of a specific pathway of toxicity—for instance, a biomarker of cellular necrosis and not apoptosis? Clinical validation proceeds only if the safety biomarker(s) is representative of toxicity in preclinical species and the performance characteristics of the analytical assay(s) surpass the criteria set for initiation of clinical validation.

The analytical performance of a high-content technology platform, such as microarrays or mass spectrometry, utilized to identify the biomarker is often insufficient for clinical validation due to the rigor of diagnostic standards (Glas et al., 2006). Identification technologies also tend to be newer technology platforms that may be incompatible with existing practices and/or infrastructure in clinics throughout the world. Clinical validation and utility (Fig. 4.4) often require reconfiguration of the detection assay with a technology platform that offers enhanced analytical performance and/or is more amenable to clinics worldwide, such as quantitative polymerase chain reaction or enzyme-linked immunoabsorbent assays (Glas et al., 2006). Reconfiguration also entails additional rigorous analytical validation.

4.14 A SAFETY BIOMARKER QUALIFICATION PROCESS

The absence of practical processes for the clinical validation of a potential safety biomarker has severely limited the use of novel biomarkers in drug development and the clinic. The validity of preclinical and clinical biomarkers was traditionally settled by debate, consensus, and the passage of time. This process was too inefficient, often requiring a decade before a consensus was reached. Recognizing the urgent need for accelerated application of biomarkers in drug development, the FDA outlined a process for accurate, comprehensive, and aggressive qualification of biomarkers (Goodsaid and Frueh, 2007). The process is driven by the Interdisciplinary Pharmacogenomic Review Group (IPRG) to allow contributions of expertise from different FDA Centers, such as the Center for Drug Evaluation

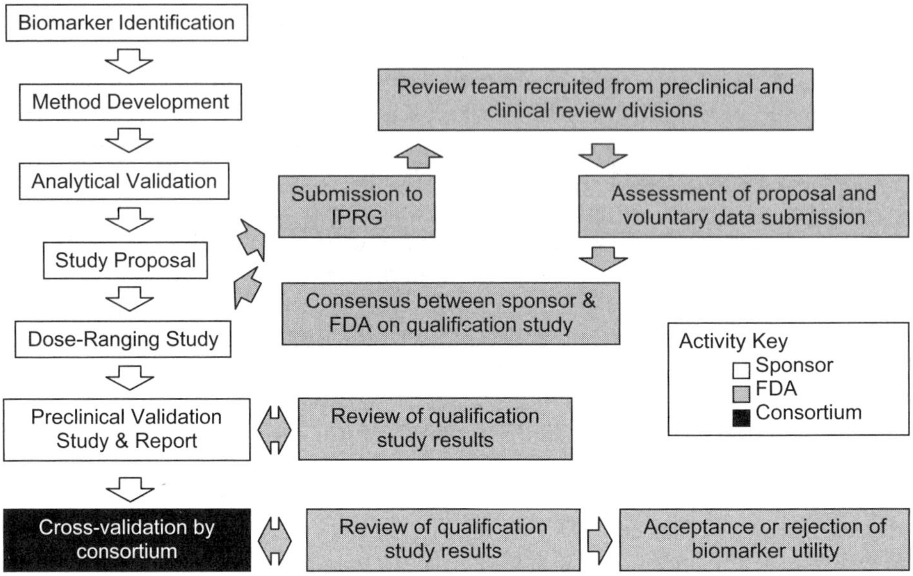

Figure 4.6. Process map for biomarker qualification.

and Research (CDER), the Center for Devices and Radiological Health, and the National Center for Toxicological Research, as well as across clinical divisions and from nonclinical toxicology reviewers in CDER (Goodsaid and Frueh, 2006, 2007). The responsibilities of IPRG include creation of a specific review function for the assessment of biomarker qualification data sets: the IPRG Biomarker Qualification Review Team. The IPRG Biomarker Qualification Review Team will evaluate study protocols and review study results for the qualification of novel biomarkers of drug safety, using appropriate preclinical, clinical, and statistical considerations (Fig. 4.6). The team will then develop recommendations and guidance for the submission of biomarker data, assess the original biomarker context proposal through voluntary data submission, and then evaluate the qualification study protocol together with the sponsor to reach a consensus protocol (Fig. 4.6). This process map reflects the expectation of a true partnership between sponsors and the FDA in the critical steps in this process of initial evaluation, qualification protocol draft, and data review (Goodsaid and Frueh, 2006, 2007). The qualification process must also be coordinated with the review of drug submissions at the FDA, since biomarker development is often a component of an overall drug development program. The decision to accept, reject, or amend the Qualification Data Report will be made by the IPRG Biomarker Qualification Review Team, which will ultimately draft a recommendation for the appropriate clinical division regarding the approval or rejection of the qualification submission.

4.15 THE CRITICAL PATH INSTITUTE AND THE PREDICTIVE SAFETY TESTING CONSORTIUM

The ultimate validity and utility of a safety biomarker is furthermore determined by the extent to which the biotherapeutic development community uses it. The qualification process outlined by the FDA incorporates an interlaboratory review involving consortium of potential users. Reaching a consensus position between these parties also requires a review structure within regulatory agencies that can evaluate qualification data for these biomarkers and determine whether these biomarkers can be qualified. The FDA and other stakeholders established a consortia for this purpose called the Critical Path Institute (C-Path) (Administration USFaD, 2005). C-Path is envisioned as a neutral third party, without financial support from the regulated industry. Because of C-Path's neutral funding and its mission to focus on process, not products, FDA can actively participate in the work without concerns about conflicts of interest. C-Path's strategy is to invite stakeholders to join consortia in which they can work with the FDA to improve the process of medical product development (Administration USFaD, 2005). The FDA agreed to participate under a Memorandum of Understanding (http://www.C-Path.org). The first and essential requirement is that there are champions for the project within the FDA. Also, there must be two or more companies willing to work together on the project, and there must be a source of external funding that is independent of commercial interests. The projects focus on precompetitive aspects of drug development, specifically preclinical toxicology.

The first consortium formed by C-Path, the Predictive Safety Testing Consortium (PSTC), was announced in March 2006 by Secretary of Health and Human Services Michael Leavitt (Administration USFaD, 2006a). The PSTC entails 16 global pharmaceutical companies that are sharing their preclinical methods and data for tests of nephrotoxicity, hepatotoxicity, vasculitis, muscle toxicity, and carcinogenicity. In this consortium, methods developed by one company that appear to best predict drug toxicity are verified by experiments performed by a second company. Hundreds of scientists from pharmaceutical companies, the FDA, and the European Medicines Agency (EMEA) participate in the meetings and discussions as advisors. The methods that are cross-validated by the companies are expected to eventually provide the scientific basis for regulatory guidance to be issued by the FDA and the EMEA (Administration USFaD, 2006a).

As of this writing, the US Congress is considering the FDA Revitalization Act. This legislation would create the Reagan–Udall Institute, a foundation established to advance FDA's mission and engage in critical path research. Legislation being introduced in the House of Representatives would authorize funding for the FDA to create multiple critical path public–private partnerships. The European Commission is in the final stages of approving the Innovative Medicines Initiative, a similar partnership among the pharmaceutical industry, the European Union, and academia that would conduct research relevant to drug development, with funding divided equally between government and industry. A systematic approach to improving the process of developing therapeutics will ultimately become part of the fabric of translational research.

4.16 SUMMARY AND CONCLUSIONS

The use of new safety biomarkers combined with active surveillance tools in the health-care environment has the potential to revolutionize approaches to drug safety, moving from detection and enumeration to prediction, prevention, and active management of drug toxicity in the future. To achieve this potential, drug safety must be integrated further into the overall process of developing drugs. Steps are being taken for more and better data to be generated earlier in selection of the best clinical candidate drugs using improved nonclinical model systems that promise to be more informative and predictive of human outcome. Technologies are delivering on improved approaches to monitor all major toxicities seen in animals in accelerated human trials. The changes in regulatory policy, new technologies, and platform reduction approaches being implemented in drug development are also critical contributors. Consortia and partnerships among industry representatives, together with an open dialogue with regulatory authorities, are critical to the efficient acceptance and implementation of these new models, endpoints, and new approaches in early human clinical trials. Taken together, it would appear that drug development is on the verge of a significant transformation.

REFERENCES

Adams CP, Brantner VV. Estimating the cost of new drug development: Is it really 802 million dollars? *Health Aff (Millwood)* 2006;25:420–428.

Administration USFaD. Innovation or stagnation: challenge and opportunity on the critical path to new medical products. 2004. http://www.fda.gov/ScienceResearch/SpecialTopics/CriticalPathInitiative/CriticalPathOpportunitiesReports/ucm077262.htm. http://www.fda.gov/oc/initiatives/criticalpath/whitepaper.html.

Administration USFaD. FDA announces partnership with Critical Path Institute to conduct essential research to spur Medical Innovation Safety Consortium Study and Community Pharmacy Network among first projects. 2005. http://www.fda.gov/NewsEvents/Newsroom/PressAnnouncements/2005/ucm108532.htm.

Administration USFaD. Predictive safety testing consortium. 2006a. http://www.fda.gov/NewsEvents/Newsroom/PressAnnouncements/2006/ucm108617.htm.

Administration USFaD. Innovation or stagnation: critical path opportunities report and list. 2006b. http://www.fda.gov/downloads/ScienceResearch/SpecialTopics/CriticalPathInitiative/CriticalPathOpportunitiesReports/UCM077258.pdf.

Administration USFaD. Critical path opportunities initiated during 2006. 2007. http://www.fda.gov/ScienceResearch/SpecialTopics/CriticalPathInitiative/CriticalPathOpportunitiesReports/ucm077251.htm.

Biomarkers Definitions Working Group. Biomarkers and surrogate endpoints: preferred definitions and conceptual framework. *Clin Pharmacol Ther* 2001;69:89–95.

Bjornsson TD, Wagner JA, Donahue SR, Harper D, Karim A, Khouri MS, Murphy WR, Roman K, Schneck D, Sonnichsen DS, Stalker DJ, Wise SD, Dombey S, Loew C. A review and assessment of potential sources of ethnic differences in drug responsiveness. *J Clin Pharmacol* 2003;43:943–967.

Bradford LD. CYP2D6 allele frequency in European Caucasians, Asians, Africans and their descendants. *Pharmacogenomics* 2002;3:229–243.

Caskey CT. The drug development crisis: efficiency and safety. *Annu Rev Med* 2007;58:1–16.

Development TTCFD. Average cost to develop a new biotechnology product is $1.2 billion. 2006. http://csdd.tufts.edu/NewsEvents/NewsArticle.asp?newsid=69.

DiMasi JA, Hansen RW, Grabowski HG. The price of innovation: new estimates of drug development costs. *J Health Econ* 2003;22:151–185.

Glas AM, Floore A, Delahaye LJ, Witteveen AT, Pover RC, Bakx N, Lahti-Domenici JS, Bruinsma TJ, Warmoes MO, Bernards R, Wessels LF, Van't Veer LJ. Converting a breast cancer microarray signature into a high-throughput diagnostic test. *BMC Genomics* 2006;7:278.

Goldstein JA. Clinical relevance of genetic polymorphisms in the human CYP2C subfamily. *Br J Clin Pharmacol* 2001;52:349–355.

Goodsaid F, Frueh F. Process map proposal for the validation of genomic biomarkers. *Pharmacogenomics* 2006;7:773–782.

Goodsaid F, Frueh F. Biomarker qualification pilot process at the US Food and Drug Administration. *Aaps J* 2007;9:E105–E108.

Griese EU, Zanger UM, Brudermanns U, Gaedigk A, Mikus G, Morike K, Stuven T, Eichelbaum M. Assessment of the predictive power of genotypes for the *in-vivo* catalytic function of CYP2D6 in a German population. *Pharmacogenetics* 1998;8:15–26.

Hermann M, Ruschitzka F. Cardiovascular risk of cyclooxygenase-2 inhibitors and traditional non-steroidal anti-inflammatory drugs. *Ann Med* 2007;39(1):18–27.

Ingelman-Sundberg M. Duplication, multiduplication, and amplification of genes encoding drug-metabolizing enzymes: evolutionary, toxicological, and clinical pharmacological aspects. *Drug Metab Rev* 1999;31:449–459.

Johansson I, Lundqvist E, Bertilsson L, Dahl ML, Sjoqvist F, Ingelman-Sundberg M. Inherited amplification of an active gene in the cytochrome P450 CYP2D locus as a cause of ultrarapid metabolism of debrisoquine. *Proc Natl Acad Sci USA* 1993;90:11825–11829.

Mardis ER. The impact of next-generation sequencing technology on genetics. *Trends Genet* 2008;24:133–141.

Meyer T, Sartipy P, Blind F, Leisgen C, Guenther E. New cell models and assays in cardiac safety profiling. *Expert Opin Drug Metab Toxicol* 2007;3:507–517.

Petricoin EF, Belluco C, Araujo RP, Liotta LA. The blood peptidome: a higher dimension of information content for cancer biomarker discovery. *Nat Rev Cancer* 2006;6: 961–967.

Sachse C, Brockmoller J, Bauer S, Roots I. Cytochrome P450 2D6 variants in a Caucasian population: allele frequencies and phenotypic consequences. *Am J Hum Genet* 1997;60:284–295.

Solomon SD. Cyclooxygenase-2 inhibitors and cardiovascular risk. *Curr Opin Cardiol* 2006;21:613–617.

Solus JF, Arietta BJ, Harris JR, Sexton DP, Steward JQ, McMunn C, Ihrie P, Mehall JM, Edwards TL, Dawson EP. Genetic variation in eleven phase I drug metabolism genes in an ethnically diverse population. *Pharmacogenomics* 2004;5:895–931.

van Ravenzwaay B, Cunha GC, Leibold E, Looser R, Mellert W, Prokoudine A, Walk T, Wiemer J. The use of metabolomics for the discovery of new biomarkers of effect. *Toxicol Lett* 2007;172:21–28.

Wagner JA. Strategic approach to fit-for-purpose biomarkers in drug development. *Annu Rev Pharmacol Toxicol* 2008;48:631–651.

Wallace KB, Herman E, Holt GD, MacGregor JT, Metz AL, Murphy E, Rosenblum IY, Sistare FD, York MJ. Serum troponins as biomarkers of drug-induced cardiac toxicity. *Toxicol Pathol* 2004;32:106–121.

Wood AJ. A proposal for radical changes in the drug-approval process. *N Engl J Med* 2006;355:618–623.

Woodcock J. The prospects for "personalized medicine" in drug development and drug therapy. *Clin Pharmacol Ther* 2007;81:164–169.

5

THE ROLE OF DRUG METABOLISM IN DRUG DISCOVERY

TONIKA BOHNERT AND LIANG-SHANG GAN

5.1 INTRODUCTION

Drug discovery is an extremely challenging, convoluted, and expensive process, which demands a very close knit interaction of a variety of multidisciplinary teams. Success of such an intricate and integrated alliance of highly trained, multifaceted professionals is ultimately governed by the development of a *safe* and *efficacious* new drug or to improve upon the safety and efficacy of existing, competitive drugs, already in the market. Central to the development of a safe and efficacious drug is the role of drug metabolism. The science of drug metabolism focuses primarily on evaluating and understanding the dynamic interplay of absorption, distribution, metabolism, and excretion, most commonly referred to as ADME properties of the drug or the new chemical entity (NCE). Collectively, these ADME properties ultimately govern the overall disposition of an NCE. High intrinsic permeability across the intestinal epithelium is one of the primary players responsible for a good absorption of a compound. The recent advances in the area of uptake and efflux transporters (also considered as Phase 0 and 3 metabolism, respectively) have significantly improved the understanding of human absorption *in vivo* and the vital role transporters play in tissue distribution of drugs and hence the overall pharmacological response of the NCE. Drug distribution illustrates the passage of the drug from the systemic circulation to the target organ/target sites to elicit the desired pharmacodynamic response. Depending on whether the drug is available as free fraction, plasma and tissue protein bound,

Enzyme Inhibition in Drug Discovery and Development, edited by Lu and Li
Copyright © 2010 John Wiley & Sons, Inc.

partitioned into red blood cells, or lipid bound, it can have remarkably different outcomes to the overall pharmacology and toxicology profile. Drug metabolism generally refers to the science of biotransformation of an NCE. As simple as it sounds, metabolism not only entails the basic concept of the rate and extent of enzymatic breakdown of the NCE, but also involves (a) detailed understanding of major enzymes involved in the metabolism pathways, (b) their relative contribution *in vivo*, (c) potential for drug–drug interactions (DDI) with the NCE and existing drugs to be commonly used in combination, (d) formation of reactive metabolic intermediate of the NCE, and (e) all of these parameters in context to their clinical significance. Lastly, excretion refers to the irreversible removal of the NCE and all its metabolites from the body, with liver and kidneys being the primary organs of drug excretion. Drug metabolizing enzymes and transporters (drug disposition proteins [DDPs]) play a very important role in excretion processes as liver and kidneys are the primary organs of drug removal. Drug excretion and biotransformation together are the key regulators of total drug clearance from the body, which, in concert with drug absorption, is a determinant of good oral exposure. Good oral exposure, along with the optimum drug distribution properties, ultimately regulates the safety and efficacy of an NCE. The objective of this chapter is to (a) highlight the intricate interplay of several ADME processes (such as metabolic stability, formation of reactive intermediates, and inhibition and induction of drug disposition enzymes) that play a crucial role in the biotransformation of an NCE and the potential DDI risk and (b) emphasize the need to evaluate these processes at various stages of drug discovery and development.

5.2 DRUG METABOLISM

Exposure and duration of action are two important factors that ultimately guide the successful development of a safe and efficacious drug. During the entire course of drug discovery and development, optimizing pharmacokinetic (PK) properties of an NCE is a primary goal, in addition to improving its potency and selectivity toward the target receptor. Poor metabolic stability is the culprit for causing poor PK properties [high clearance, low exposure, short half-life, and low oral bioavailability (BA)] in several cases. Metabolic stability portrays an NCE's ability to survive enzymatic modification or degradation once it enters the body. It is often desired to have a metabolically stable compound (except in cases of prodrugs—discussed later) as the successful drug candidate. Optimum metabolic stability also ensures minimal variations in interindividual responses for exposure and duration of action. Determination of metabolic stability, also commonly referred to as *in vitro* intrinsic clearance (Cl_{int}) by DMPK scientists, is one of the key *in vitro* ADME parameters that is evaluated and optimized very early on [in lead identification (LI) or lead optimization (LO) stages] during drug discovery process in the pharmaceutical industry (Brown et al., 2007; Baranczewski et al., 2006b; Jolivette and Ekins, 2007; Masimirembwa et al., 2003; Nassar et al., 2004b; Riley et al., 2002; Smith and van der Waterbeemd, 1999; Yengi et al., 2007). Optimization of metabolic stability or Cl_{int} (intrinsic clearance) is a highly iterative process that involves close alliance between medicinal chemists and DMPK scientists to understand the intricate structure–stability relationship (SSR) of individual chemical series. It requires a thorough understanding of the chemistry of metabolism or biotransformation to successfully modulate the existing functional groups of a chemical series and transform a less desirable, metabolically unstable chemotype into one that has favorable Cl_{int}. Metabolic stability *in vitro* is also believed to imitate hepatic clearance *in vivo*, which in turn is a major contributor to total body clearance of majority of xenobiotics. High metabolic stability and hence low Cl_{int} are often desirable. In most cases, low Cl_{int} also leads to favorable secondary PK properties such as good bioavailability (Stoner et al., 2004) and long half-life which result in good exposure and desired duration of action, respectively. Compounds that possess low Cl_{int} are also less susceptible to DDI.

5.2.1 Drug Metabolizing Enzymes (DMEs)

There are a large number of drug metabolizing enzymes that are responsible for metabolism and bioactivation of xenobiotics and endogenous substrates (Table 5.1) (Guengerich, 2006a; Ioannides, 2002; Lynch and Price, 2007; Parkinson, 2001; Rendic and Di Carlo, 1997; Williams et al., 2005; You and Morris, 2007). Cytochrome P450s (CYPs) represent the major class of drug metabolizing enzymes and are responsible for metabolism of 70–80% of the drugs on market. In addition to CYPs, flavin-containing monooxygenase (FMO), monoamine oxidase (MAO), molybdenum-containing oxidoreductases [aldehyde oxidase, (AO) and xanthine oxidase (XO)], and esterases are important contributors to Phase I metabolism; and enzymes like UGT, sulfotransferase (SULT), *N*-acetyl transferase (NAT), and glutathione *S*-transferase (GST) are important contributors toward Phase II conjugation reactions. In addition to these primary oxidative, reductive, and conjugative enzymes, there are important specialized deconjugative enzymes, such as γ-glutamyl transpeptidase, β-glucuronidase, and sulfatase, that are involved in further metabolism of conjugative Phase II metabolites. Each of these enzymes is also classified into families (40% or more sequence homology) and subfamilies (55% or more sequence homology) or isoforms, based on similarities in their amino acid sequence. A substrate's physicochemical properties sometimes govern which enzyme will preferentially metabolize an NCE. For example, both CYPs and AO facilitate oxidation of carbon atoms. However, CYPs prefer carbon atoms with high electron density—in contrast to AO, which preferentially metabolizes carbon atoms with low electron density. For example, electron-rich ring system naphthalene is oxidized by CYPs (primarily 1A2 and 3A4), while incorporation of additional nitrogen atoms into naphthalene

TABLE 5.1. Major Human Drug Disposition Enzymes and Their Tissue Distribution (Ioannides, 2002; You and Morris, 2007)

Enzyme	Isoforms	Tissue Expression	Reaction Catalyzed	Phase 1 or 2
Cytochrome P450	1, 2, 3, 4	Liver, GI, kidney, brain, heart, lung, spleen, skin	Oxidation, reduction, hydrolysis, dehydrogenation, dealkylation	1
Flavin mono-oxygenase	1, 2, 3, 4, 5, 6	Liver, GI, kidney, brain, lung, skin	Oxidation	1
Monoamine oxidase	A, B	Liver, GI, kidney, brain, stomach, blood	Oxidative deamination	1
Aldehyde oxidase		Liver, GI, kidney, brain, lung	Oxidation, reduction	1
Xanthine oxidase		Liver, GI, kidney, brain, heart, lung, spleen, skeletal muscle	Oxidation, reduction	1
Prostaglandin synthase	1, 2	Kidney, lung, GI, sex glands, heart, brain	Oxidation	1
Lipoxygenases		Liver, GI, kidney, brain, lung, skin	Oxidation	1
UDP-glucuronosyltransferase	1, 2	Liver, esophagus, GI, kidney	Conjugation with glucuronic acid	2
Glutathione *S*-transferase	A, M, P, S, T, O, K, Z	Liver, GI, kidney, brain, heart, lung, spleen, testis, skeletal muscle	Conjugation with GSH	2
Sulfotransferase	1, 2, 4	Liver, GI, kidney, brain, lung,	Conjugation with sulfate	2
N-Acetyl-transferase	1,2	Liver, GI, blood epithelium	Conjugation with acetyl moiety	2
PGP	MDR1/ ABCB1, MDR2/ ABCB2	GI, kidney, bile duct, BBB, liver	Transport of wide variety of substrates	0/3

TABLE 5.1. ***Continued***

Enzyme	Isoforms	Tissue Expression	Reaction Catalyzed	Phase 1 or 2
ABCC/MRP	1, 2, 3, 4, 5, 6, 8, 9, 10, 11, 12	Liver, GI, kidney, heart, lung, brain, macrophages, spleen, testis, adrenal/cerebral cortex	Transport of wide variety of substrates	0/3
BCRP	ABCG1/2/3/4/5/8	Liver, GI, kidney, lung, stomach, pancreas, CNS, placenta	Transport of wide variety of substrates	0/3
OCT	1, 2, 3	Liver, kidney, heart, lung, brain, spinal chord, pancreas	Transport of low-MW hydrophilic cation	0/3
OAT	1, 2, 3, 4, URAT1/RST	Kidney, liver, brain, eyes, adrenal gland	Transport of organic anions	0/3
OATP	1, 2, 3, 4, 5, 6	Liver, GI, kidney, heart, lung, brain, skin, eye, skeletal muscle, testis	Transport of relatively large steroid- or peptide-like substrates	0/3
Peptide transporter	PEPT1, 2 PHT 1, 2	Liver, GI, kidney, lung, brain, pancreas	Transport of peptide-like substrates	0/3
BSEP		Liver	Transport of bile salts	0/3
CNT/ENT	1, 2, 3/1, 2, 3, 4	Liver, GI, kidney, heart, lung, brain, pancreas, spleen, testis, skeletal muscle, prostate, lymph node, ovary, bone marrow	Transport of nucleoside and nucleoside analogs	0/3
MCT/SMCT	1, 2, 3, 4/1, 2	Liver, GI, kidney, heart, lung, brain, pancreas, spleen, testis, skeletal muscle	Transport of monocarboxylate compounds	0/3

(thus decreasing electron density of the original aromatic ring system) yield compounds (phthalazine and pteridine) that are primarily metabolized by AO (Parkinson, 2001).

Metabolism—or, more strictly, biotransformation of xenobiotics—is traditionally classified as Phase I and Phase II metabolism (Ioannides, 2002; Parkinson, 2001). Phase I metabolism primarily entails oxidation (hydroxylation, epoxide formation, heteroatom oxidation, and dealkylation), reduction, or hydrolysis of lipophilic xenobiotics, which results in their conversion to more hydrophilic metabolites. Phase I metabolism often introduces new functional groups (-OH, -NH_2, -SH, -COOH) or unmasks these functional groups, already present in the parent drug. Phase II metabolism involves conjugation of the metabolites formed from Phase I oxidation or sometimes direct conjugation of parent xenobiotic itself (when parent drug structure contains -OH, -NH_2, -SH, -COOH) with endogenous substrates such as uridine diphosphate glucuronic acid (UDPGA), 3′-phosphoadenosine-5′-phosphosulphate (PAPS), glutathione, or amino acids. For example, the imidazobenzodiazepine compound, midazolam (MDZ), often used in treatment of insomnia, is hydroxylated (an oxidation reaction) to yield 1′-OH and 4-OH MDZ. The hydroxylated metabolites can further undergo conjugation with UDPGA into hydrophilic glucuronide conjugates, which undergo renal excretion and are cleared from the body. Recent evidence of direct glucuronidation of the parent, MDZ, has also been reported *in vitro* (Klieber et al., 2008). Unlike MDZ, where a new -OH functional group was introduced due to oxidative enzymatic reaction, the analgesic drug phenacetin, undergoes a Phase I deethylation first to unmask the existing -OH group within the molecule, which further undergoes Phase II conjugation with either glucuronide or sulfate metabolites (Hinson, 1983). A list of examples of Phase I & II biotransformation pathways are listed in Table 5.1. Ultimate result of Phase I & II metabolism is conversion of a lipophilic parent xenobiotic into a hydrophilic metabolite, which can be easily cleared from the body and hence detoxify body of the xenobiotic. There are instances however when Phase II biotransformation has been attributed to cause toxicity as in case of NAT-mediated toxicity of arylamines (Hein et al., 2000) and GSH-mediated toxicity of reactive metabolic intermediates (discussed in detailed in later section). In addition to these well studied and commonly observed biotransformation, there are several examples of complex or unusual reactions (Guengerich, 2001; Isin and Guengerich, 2006a), mediated by CYP that include ipso substitution (replacement of aromatic halide by hydroxyl group), oxidative/reductive ring coupling, contraction and expansion, aryl migration (Fig. 5.1).

5.2.2 Drug Metabolism Studies

Designing correct metabolism studies *in vitro* to best predict *in vivo* scenario is quite challenging; and as simple as an *in vitro* metabolic stability assay setup might appear, there are several considerations that have to be carefully thought

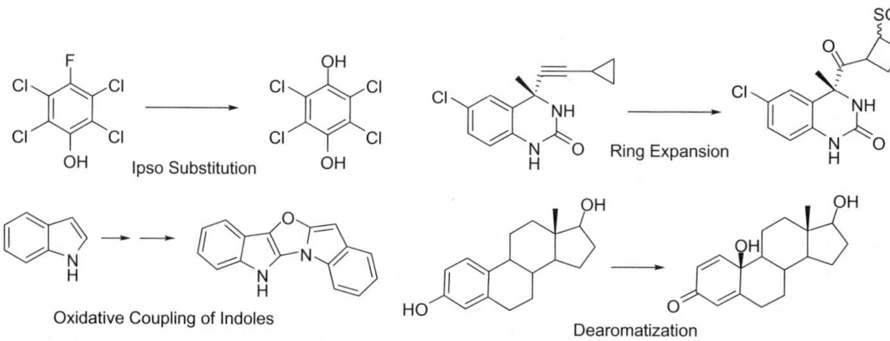

Figure 5.1. Some uncommon reactions catalyzed by CYP (Guengerich, 2001; Isin and Guengerich, 2006a).

out before simply measuring a compound's metabolic stability and using that data to gauge properties *in vivo*. This is because factors that govern metabolism of an NCE, other than its physicochemical properties, as briefly alluded to above, are numerous and in some cases complex to predict. Some major determinants of metabolism are: a substrate's stereochemistry; substrate concentration that the enzyme is exposed to; presence of multiple metabolic pathways that may be mediated by a single or multiple enzymes; route of administration; tissue where metabolism occurs; dose; gender; age; and species in which metabolism is studied (Lin and Lu, 1997). These are illustrated by a few examples here.

Stereochemistry. Differential metabolism of enantiomers of the oral anticoagulant warfarin is well documented in humans. *S*-Warfarin is metabolized by CYP2C9 to 7-hydroxywarfarin (used as 2C9 marker reaction). *R*-warfarin, on the other hand, is metabolized by CYP1A2 to 6- and 8-hydroxywarfarin and by CYP3A4 to the 10-hydroxymetabolite (Kaminsky and Zhang, 1997). The triazole-containing antifungal agent, itraconazole, administered as a racemic mixture of four isomers, also exhibits stereoselective metabolism *in vivo* and *in vitro*, where only two out of the four isomers are differentially metabolized by CYP3A4 (Kunze et al., 2006). The proton pump inhibitor omeprazole, used to treat peptic and duodenal ulcers, also demonstrates stereoselective metabolism: In human liver microsomes, the *R*-omeprazole is hydroxylated approximately 10 times faster than its *S*-enantiomer by CYP2C19, but the *S*-omeprazole is preferentially *O*-demethylated over its R-isomer by the same enzyme CYP2C19. Also, *S*-omeprazole preferentially undergoes CYP3A4-mediated sulfoxidation over the *R*-enantiomer (Abelo et al., 2000). The serotonin 5-HT_4 receptor agonist cisapride, prescribed to gastroesophageal reflux patients, exhibits preferential *N*-dealkylation and parahydroxylation in (+)-cisapride and a predominant orthohydroxylation in (−)-cisapride, all mediated by CYP3A4 (Desta et al., 2001). Another well-known example is that of the anticonvulsant mephenytoin, a marker substrate for CYP2C19, where the

R-enantiomer isomer is 4′-hydroxylated two- to sixfold faster in dog, rat, and rabbit liver microsomes but the *S*-enantiomer is preferentially 4′-hydroxylated in monkey and human liver microsomes (Yasumori et al., 1993). The (–)-enantiomer of synthetic analgesic tramadol was *O*-demethylated at a much higher rate than (+)-enantiomer in rat liver microsomes and was subject to inhibition by quinine and quinidine (Liu et al., 2003).

Substrate Concentration. Diazepam, a benzodiazepine commonly prescribed for anxiety, alcohol dependency, and insomnia, undergoes CYP3A4-mediated 3-hydroxylation *in vitro*; but at low substrate concentration, both *in vitro* and *in vivo*, CYP2C19-mediated *N*-demethylation is the major pathway (Andersson et al., 1994). Another example is that of the proton pump inhibitor lansoprazole, used to treat gastrointestinal ulcers and acid reflux. Lansoprazole is metabolized by CYP3A4 to 5-hydroxylansoprazole at high concentrations, encountered in *in vitro* studies; but at pharmacological low substrate concentration, CYP2C19 is the major contributor to the formation of this metabolite (Pearce et al., 1996). Molinate, a thiocarbamate herbicide that is attributed to cause testicular toxicity, undergoes hydroxylation at low doses but sulfoxidation at higher doses (Jewell and Miller, 1999). Several examples of dose-dependent metabolism have been cited in the literature (Jewell and Miller, 1999; Higashikawa et al., 1999; Kharasch et al., 1999; Powis, 1983). One of the common reasons for dose-dependent metabolism of an NCE is saturation of enzyme or transporter responsible for its clearance via either metabolism or active transport (carrier-mediated) respectively. In the examples cited above, two major CYP isoforms are involved in the metabolism of the substrate: (1) a high-affinity CYP, responsible for metabolism at low substrate concentration, and (2) a low-affinity CYP, responsible for metabolism at high substrate concentration. With increasing substrate concentration, relative contribution by the high-affinity CYP decreases while that by the low-affinity CYP increases, and at high substrate concentration the high-affinity CYP is saturated and the low-affinity CYP is the primary contributor to the metabolism of the substrate, resulting in the difference of metabolism pathways at different substrate concentration.

Multiple Metabolic Pathways. Most compounds are metabolized by multiple enzymes to yield either the same metabolite or different metabolites. Metabolism via multiple pathways/enzymes is highly preferred for NCEs to minimize risks associated with drug–drug interactions (discussed later in the chapter). The tricyclic antidepressant imipramine is metabolized by a high-affinity CYP1A2-mediated pathway and a low-affinity CYP3A4-mediated pathway to give the major *N*-desmethyl metabolite and is also metabolized by CYP2D6 to give a minor hydroxylated metabolite (Ohmori et al., 1993). The steroid hormone testosterone, commonly used as CYP3A4 marker substrate, undergoes CYP2C9, 2C19, and 3A4-mediated metabolism to yield 1β-, 2α-, 2β-, 6β-, 11β-, 15β-, and 16β-hydroxylated metabolites (Yamazaki and

Shimada, 1993). Caffeine undergoes CYP2E1-mediated N1- and N7-demethylation, CYP1A2-mediated N3-demethylation, and CYP3A4-mediated 8-hydroxylation (Tassaneeyakul et al., 1994). The cough suppressant and CYP2D6 activity marker substrate dextromethorphan undergoes CYP2D6-mediated *O*-demethylation (minor contribution from CYP2C9, CYP2C19) plus CYP2E1 and 3A3/4 mediated *N*-dealkylation (minor contribution from CYP2C9, CYP2C19, and CYP2D6) (Aming and Haining, 2001). The non-benzodiazepine hypnotic zolpidem forms three hydroxylated metabolites, mediated by the CYPs 3A4, 2C9, 1A2, 2D6, and 2C19 in that order (Von Moltke et al., 1999). The endogenous substrate estrone is metabolized at multiple sites to its (a) 2- and 4-hydroxy metabolites by CYP 1A1, 1A2, and 1B1 and (b) 16α-hydroxy metabolite by CYPs 1A1, 2C19, and 3A5 (Cribb et al., 2006). The NSAID acetaminophen undergoes both glucuronidation and sulfation at the same site in humans, with glucuronidation being the predominant contributor (Kane et al., 1995). Benzodiazepine drug flunitrazepam is demethylated by CYP2C19 and is hydroxylated at the 3-position by CYP3A4 (Kilicarslan et al., 2001).

Tissue Specificity. Although liver is the primary organ for metabolism of most xenobiotics, drug metabolizing enzymes are ubiquitous and are found in most organs such as small intestine, kidney, lung, heart, brain, and spleen. Therefore, it is not uncommon for involvement of certain tissue specific metabolizing enzymes toward biotransformation of a xenobiotic. Tissue preferential metabolism can also result in tissue-specific bioactivation of xenobiotics. CYP1A1 is essentially an extrahepatic enzyme in human, which is found in intestine, lung, skin, lymphocytes (Parkinson, 2001) but not in the liver and is predominantly involved in metabolism of aromatic hydrocarbons like benzo[a]pyrene, while 3-methylindole (skatole), is a lung-specific toxin due to its bioactivation by lung-specific CYP2F1 (Thornton-Manning et al., 1996). Commonly used industrial and laboratory solvent benzene causes severe hemopoietic disorders, including bone marrow cancer. The reason for this is believed to be the initial CYP2E1-mediated hepatic metabolism of benzene to an epoxide, which is then rapidly transported via blood into bone marrow, rich in peroxidases (Raj et al., 2001). Peroxidases are believed to convert the benzene-epoxide into toxic semiquinones and hydroquinones, resulting in bone-marrow-specific toxicities. The synthetic heroin 1-methyl-4-phenyl-1,2-3,6-tetrahydropyridine (MPTP) crosses the blood–brain barrier and is metabolized by brain MAO-B to produce the pyridinium ion, which by virtue of its positive charge, cannot permeate out of the brain and causes neurotoxicity (Yoshihara et al., 2000). A commonly prescribed NSAID, acetamoniphen, is metabolized via CYP2E1 in the liver, into a reactive iminoquinone metabolite, which is attributed to severe hepatotoxicity (James et al., 2003). The hypersensitive reaction of phenytoin is attributed to its preferential bioactivation in the skin, due to abundance of CYP2C18 in the skin (poor expression in liver) (Kinobe et al., 2005).

Route of Administration. Upon PO dosing, bisphenol, a dietary additive, was found to be eliminated almost exclusively via monoglucuronide conjugate. Upon IP or SC administration, the majority of the dose was recovered as parent drug, resulting in much higher OBA (oral bioavailability) of bisphenol upon IP or SC administration compared to the PO route (Pottenger et al., 2000) due to overcoming the first pass effect via PO route. A route-dependent pulmonary first pass of biphenylacetic esters in rats has also been observed (Dickinson and Taylor, 1998). Toluene 2,4-diisocyanate (2,4-TDI) demonstrates route-dependent carcinogenicity because 2,4-TDI is primarily metabolized to the carcinogenic compound toluene 2,4-diamine upon PO administration but forms noncarcinogenic conjugated metabolites via inhalation exposure (Timchalk et al., 1994). It is common to see poor exposure of NCEs when administered via the PO route due to the first pass effect, which refers to metabolism of the compound in the gut lumen and liver (metabolism or biliary excretion) before the NCE reaches systemic circulation.

Species Specificity. The aromatic amine-containing carcinogen, 2-amino-3,8-dimethylimidazol[4,5-f]quinoxaline (MeIQx), is differentially metabolized to two genotoxic *N*-hydroxylated metabolites in humans and rodents by CYP1A2, but not in cynomologous monkeys since monkeys are deficient in CYP1A2 (Edwards et al., 1994; Frantz et al., 1995). As a result, monkeys are not suitable tox species to assess human carcinogenicity mediated by MeIQx. Coumarin, a common food and fragrance constituent, exhibits metabolism-based species-specific toxicity. In humans, it is primarily metabolized by the high-affinity CYP2A6 to the nontoxic 7-hydroxylated metabolite, which is eliminated by further glucuronidation and sulfation conjugation reactions. In mice and rats, on the other hand, coumarin primarily undergoes 3,4-epoxide formation, mediated by both CYP1A2 and 2E1 (Born et al., 1997). This epoxide rearranges to a phenylacetaldehyde, which is a potent hepatotoxin in these rodents. The dopamine D3-receptor antagonist SB277011 exhibited high OBA in rats but was projected to have very poor OBA in humans. This was attributed to species difference in metabolism of the compound by aldehyde oxidase (AO) whereby rats had a much lower rate of AO-mediated metabolism of SB277011 compared to humans in liver homogenates (Austin et al., 2001). Species difference in metabolism and bioactivation of felbamate has been attributed to a lack of toxicity of felbamate in animal models, but a species-selective toxicity in humans. Factors such as increase in CYP-mediated oxidation, poor esterase activity, higher aldehyde dehydrogenase activity, and higher UGT-mediated glucuronide formation of the reactive atropaldehyde metabolite of felbamate in rats contribute to lack of toxicity in this species compared to humans (Dieckhaus et al., 2000).

Gender Specificity. It is not uncommon to find gender-specific differences in activity and pathway of metabolism and therefore even in drug response (Franconi et al., 2007; Harris et al., 1995; Kuperman et al., 2001). This effect, however, is more pronounced in animals than in humans. For example, CYP3A4-mediated metabolism of testosterone and dexamethasone to the 6β-hydroxytestosterone (Bogaards et al., 2000) and 6-hydroxydexamethasone (Tomlinson et al., 1997), respectively, is considerable higher in male rats than in female rats. (–)-Verbenone is a bicyclic ketone, used in herbal tea, spices, and perfume, and limonene is a hydrocarbon used in cosmetics and flavoring. In male rats, both were found to be metabolized by male-specific isoform CYP2C11 to the respective hydroxylated metabolites. In contrast, in female rats the female-specific isoform CYP2C12 was unable to generate these metabolites (Miyazawa et al., 2003). The (–)-enantiomer of the analgesic drug tramadol undergoes substantially faster conversion to its desmethyl metabolite and the secondary glucuronide metabolite in female RLM than in male RLM (Jin and Liu, 2004). In general, in rodents, males have been seen to have higher activity of both Phase I and Phase II drug metabolizing enzymes than females (Guo et al., 1993).

Age Specificity. Activities of drug metabolizing enzymes, especially Phase I enzymes, have been demonstrated to diminish with age in rodents (Guo et al., 1993; Stohs et al., 1980), and this difference has been also seen to be gender specific. For most drug-metabolizing enzymes, there is no definite pattern of increasing or decreasing metabolizing activity: Some enzymes increase in content with age since birth (CYP2E1 and 2D6 in humans), while some others decrease with age (CYP1A2 in humans) (Hunt et al., 1992; Wauthier et al., 2007). Some, like CYP3A4 in humans, have been suggested to increase in childhood, peak at adolescence, and decrease in adults (Cotreau et al., 2005).

Since liver is considered to be the primary site of metabolism for the majority of the xenobiotics and endogenous compounds, subcellular fractions, intact cells, and tissue slices derived from liver are powerful *in vitro* tools, routinely used in the pharmaceutical industry to assess the metabolism of NCEs. Subcellular fractions (liver microsomes, microsomes with cDNA-expressed metabolic enzymes, S9 homogenates) from a variety of rodent and nonrodent animal species, along with human, are easily commercially available, easy to handle, and amenable to long-term storage. Liver microsomes are the most commonly used *in vitro* system and although they contain enzymes located in the endoplasmic reticulum only, such as CYP, FMO, UGT, and esterases, microsomes are viewed as a very good system for assessing oxidative potential of NCEs. cDNA-expressing microsomes have the advantage of expressing high concentration of one particular enzyme/isoform and are often used when assessing the DDI potential of NCEs. S9 homogenates, although they have the advantage over microsomes of containing both oxidative and conjugative enzymes, have not attained as much popularity as the microsomes due to relatively lower enzyme activity levels. Predictability of *in vivo* clearance from the subcellular-fraction-derived-*in vitro* Cl_{int} is often questionable because of the need to fortify the subcellular fraction with supraphysiological proportions of cofactors (e.g., NADPH, NADPH reductase, $MgCl_2$, cytochrome b5). In spite of the mixed success of *in vivo* Cl_{int} predictions from subcellular-metabolism-derived data (discussed further in Section 5.2.4), it is well accepted that in the early stages of drug discovery, the subcellular fractions play an extremely important role in rank-ordering compounds. Intact cells (cryopreserved and fresh hepatocytes) and tissue slices have certain distinct advantages over subcellular fractions in that they possess intact cellular morphology and are therefore believed to mimic the *in vivo* situation more closely than subcellular fractions. However, it must be remembered that cryopreserved hepatocytes mimic compound uptake through the cell membrane as in an *in vivo* situation; but, due to internalization of efflux transporters, compound efflux is not captured. They also do not need any added cofactors to facilitate metabolism, and metabolism studies can be carried out over much more extended time periods than with the subcellular fractions. However, handling and storage of intact cells and especially tissue slices are also a lot more complicated than subcellular fractions, which are by far the most commonly used *in vitro* tools to evaluate metabolism of NCEs in drug discovery stages. Several excellent reviews have been published recently to compare the advantages and disadvantages of these various *in vitro* models and overview of experimental conditions required (Balani et al., 2005; Blanchard et al., 2004; Brown et al., 2007; De Graaf et al., 2002; Fisher et al., 2002; Hewitt et al., 2007a; Lam and Benet, 2004; Ponsoda et al., 2004; Reddy et al., 2005; Gomez-Lechon et al., 2007; Soars et al., 2007a).

5.2.3 Estimation of Metabolic Stability: Cl_{int} and Cl_{hep}

Metabolic stability of compounds in various stages of drug discovery is routinely assessed in readily available subcellular fractions (Lin et al., 2003), such as microsomes, S9 homogenates, and cryopreserved hepatocytes in various animal species (rodents and human the least). Although the ultimate goal is the successful development of a human drug, metabolic stability screens are performed in other species, especially rodents, because animal models are routinely used to serve as the disease or efficacy models for proof of concept and as a PK model to predict human exposures. Assessment of metabolic stability of the NCEs in animal subcellular fractions serves to predict whether the NCEs will be stable enough to maintain a desired exposure in the animal PD and efficacy studies and for IVIVC studies discussed later. Therefore, metabolic stability assessment in subcellular fractions from human and the disease model animals, with attention to gender (if any effect), is routinely performed by the discovery DMPK groups. Metabolic stability at the early discovery stages is commonly expressed as either percent parent compound remaining after incubation for a certain period of time (up to 30 min for microsomes and up to 2–4 h for hepatocytes) or more scientifically as intrinsic Cl (Cl_{int}). Cl_{int} represents the maximum capacity of the liver to metabolize a compound, in absence of limitations of hepatic blood flow, plasma protein binding of the compound, and permeability/uptake into hepatocytes. Another term related to Cl_{int} is the extraction ratio (E) (Fig. 5.2, Eq. 5.4). As the name suggests, extraction ratio is a dimensionless number, which signifies the fraction of a compound removed by liver, during compound transit through the liver. E values of <0.25, 0.25–0.75, and >0.75 represent low, moderate, and high extraction compounds, respectively.

It is highly advisable to translate the percent parent remaining to Cl_{int} and scale the resulting Cl_{int} further to predict total body Cl (Cl_{hep}). Methods to scale Cl_{int} to Cl_{hep} are discussed later on in this section and in Fig. 5.2, Eqs. 3 and 5 (Barter et al., 2007; Fagerholm, 2007a,b; Hakooz et al., 2006; Houston and Galetin, 2003; Lin, 1998; Obach, 2001; Riley et al., 2005). The reason why percent parent remaining is not very informative is because firstly the percent parent reported is highly dependent on the incubation conditions with a particular subcellular fraction. It has been well-documented that extrinsic variables (Hermann et al., 2006) such as incubation time, microsomal protein concentration/hepatocyte count (amount of subcellular fractions), nature and strength of buffer, incubation temperature, substrate concentration, substrate solubility, percent and nature of organic solvents used to prepare substrate stock solutions, and speed of shaking during incubation interval can all significantly affect the percent parent remaining reported. Hence to minimize the inter-lab and inter-day assay variability resulting from the factors mentioned above, it is highly recommended to report metabolic stability in terms of Cl_{int}, which is essentially an assay independent parameter. Commonly used methods to calculate Cl_{int} involves determination of either (a) *in vitro* half-life ($t_{1/2}$) or (b) calculating the enzyme kinetic parameters, namely, maximum velocity of the enzymatic reaction (V_{max}) and substrate concentration yielding the half of V_{max} (K_m).

Cl_{int}–$T_{1/2}$ Method. This is a simple model-independent method, where percent parent remaining over time is first determined and then the log percent remaining versus time is plotted. The negative slope of the resulting linear regression ($-k$) is used to obtain *in vitro* $t_{1/2}$, where $t_{1/2} = 0.693/k$ (Mohutsky et al., 2006) (Fig. 5.2). The *in vitro* $t_{1/2}$ is further converted to Cl_{int} using Eq. (1) in Fig. 5.2. To illustrate the importance of calculating Cl_{int}, let's consider a compound that is 50% remaining at 30 min in mouse, rat, beagle, cynomologous (cyno), and human microsomes. Apparently, it would be tempting to conclude that it possesses the same stability across all species; however, upon converting this 50% remaining to Cl_{int} via Eq. 1, this corresponds to a Cl_{int} between 40 and 180 mL/min/kg across species.

Cl_{int}–V_{max}/K_m Method. Cl_{int} can also be obtained as a ratio of the enzyme kinetic parameters, V_{max} and K_m, of the pathway of interest (Fig. 5.2). However, this method involves quantitation of the metabolite of interest resulting from a particular biotransformation pathway. In most cases of early discovery, knowledge of metabolites, or the number of metabolites, and analytical

CALCULATION OF CLint

$$Cl_{int} = \frac{0.693}{\text{in vitro } T_{1/2}} * \frac{\text{mL incubation}}{\text{mg microsomes}} * \frac{\text{45 mg microsomes}}{\text{gm liver}} * \frac{\text{X gm liver}}{\text{kg B.W.}} \quad (1)$$

X = 20 -Human; 25- Dog , Pig; 30- Cyno; 45-Guinea Pig, Rat; 87.5-Mouse

$$Cl_{int} = \frac{V_{max}\ [S]}{K_m + [S]} \quad \text{at S << Km, simplifies to} \quad Cl_{int} = \frac{V_{max}}{K_m}$$

involvement of one enzyme (2)

$$Cl_{int} = \frac{V_{max1}\ [S]}{K_{m1} + [S]} + \frac{V_{max1}\ [S]}{K_{m1} + [S]} \quad \text{at S << Km, simplifies to} \quad Cl_{int} = \frac{V_{max1}}{K_{m1}} + \frac{V_{max2}}{K_{m2}}$$

involvement of two enzymes

CALCULATION OF CLhep via Well-Stirred Model

$$Cl_{hep} = \frac{Q_h\ Cl_{int}}{Q_h + Cl_{int}}$$

no protein binding

$$Cl_{hep} = \frac{Q_h\ f_{u,blood}\ (Cl_{int}/f_{u,microsome})}{Q_h + f_{u,blood}\ (Cl_{int}/f_{u,microsome})}$$

blood and microsome protein binding

$$Cl_{hep} = \frac{Q_h \left(\frac{1}{B/P}\right) (f_{u,serum}/f_{u,microsome})\ Cl_{int}}{Q_h + \left(\frac{1}{B/P}\right) (f_{u,serum}/f_{u,microsome})\ Cl_{int}} \quad (3)$$

plasma and microsome protein binding and blood:plasma partitioning

EQUATION FOR EXTRACTION RATIO

$$E = \frac{f_{u,plasma}\ Cl_{int}}{Q_h + f_{u,plasma}\ Cl_{int}} \qquad E = \frac{f_{u,plasma}\ Cl_{influx}\ Cl_{int}}{Q_h Cl_{efflux} + Cl_{int}(Q_h + f_{u,plasma}\ Cl_{influx})} \quad (4)$$

incorporating transporters

CALCULATION OF CLhep INCORPORATING TRANSPORTERS

$$Cl_{hep} = \frac{Q_h\ f_{u,plasma}\ Cl_{influx}\ (Cl_{int,sec} + Cl_{int,met})}{Q_h\ (Cl_{efflux} + Cl_{int,sec} + Cl_{int,met}) + f_{u,plasma}\ Cl_{influx}\ (Cl_{int,sec} + Cl_{int,met})} \quad (5)$$

Cl_{int} = intrinsic clearance
Cl_{hep} = hepatic clearance
[S] = Substrate concentration
V_{max} = maximum reaction velocity
K_m = substrate concentration that results in half of maximum velocity
Q_h = hepatic blood flow
$f_{u,blood/serum/plasma}$ = fraction unbound in blood/serum/plasma
B/P = blood:plasma partition ratio
Cl_{influx} = intrinsic clearance of influx
$Cl_{int,sec}$ = intrinsic clearance of biliary secretion
$Cl_{int,met}$ = intrinsic clearance of metabolism
Cl_{efflux} = intrinsic clearance of efflux

Figure 5.2. Equations used to calculate Cl_{int}, Cl_{hep}, and E (Fagerholm, 2007; Mohutsky et al., 2006; Naritomi et al., 2001; Wilkinson, 1987; Wilkinson and Shand, 1975; Pang et al., 2007).

methodologies to quantify these metabolites are not available. So, the simple $t_{1/2}$ method is the most common method for Cl_{int} determination at early stages of discovery.

Conversion of percent parent remaining to Cl_{int} is still insufficient to unambiguously evaluate an NCE as a potential high or low clearance compound, unless the Cl_{int} is scaled to its total body or systemic clearance (Cl_{hep}) (Fagerholm, 2007b; Naritomi et al., 2001). Cl_{int} can be further converted to Cl_{hep} via several different methods, such as the well-stirred or venous equilibrium model, sinusoidal perfusion or parallel tube model, and physiological-based dispersion model (Wilkinson, 1987; Wilkinson and Shand, 1975) (Fig. 5.2). Due to the complex nature of the dispersion model, the first two models, especially the well-stirred model, are more commonly used in a discovery setting to obtain the predicted Cl_{hep}. In most cases, when the NCE in question is a low clearance compound ($E < 0.25$), Cl_{hep} predicted from either well-stirred or parallel tube models will not be much different. However, for a high clearance compound ($E > 0.75$), Cl_{hep} predicted from the parallel tube model yields a higher value than that from a well-stirred model. Hence it is prudent to take into consideration the nature of the model with high clearance compounds. Once the predicted Cl_{hep} is calculated, the values can be expressed as percent hepatic blood flow to classify compounds as low, moderate, and high Cl based on whether the Cl_{hep} is <25%, between 25% and 75%, or >75% hepatic blood flow of a particular species. With the recent surge of knowledge in transporters, the traditional Cl_{hep} equation has been modified to incorporate the transporter contribution to total hepatic clearance (Fig. 5.2) (Pang et al., 2007).

5.2.4 *In Vitro* and *In Vivo* Correlation (IVIVC)

In the early stages of discovery, since it is not possible to get any actual human data, the results obtained from human subcellular fractions are extremely valuable in predicting the human PK parameters and assessing the potential of an NCE as a successful human drug candidate. But before a significant amount of effort is dedicated to optimize and modulate DMPK properties of a new chemotype, it is very important to evaluate the ability of the *in vitro* ADME assays to predict the *in vivo* situation—that is, establish *in vitro–in vivo* correlations (IVIVC) early on in discovery. There are several reports of successful IVIVC and of unsuccessful IVIVC in the literature (Dunne et al., 1997; Ito et al., 1998a; Iwatsubo et al., 1997; Nagilla et al., 2006; Obach, 2000a; Rostami-Hodjegan and Tucker, 2007; Venkatakrishnan et al., 2001). The primary reasons why IVIVC fail are some underlying assumptions that are made in predicting Cl_{hep} from Cl_{int}, namely:

1. Metabolism is the major route of Cl of the NCE. In cases where NCE is subjected to high efflux clearance (Gardner et al., 1995), *in vitro* metabolic stability data in microsomes or hepatocytes do not accurately predict the overall *in vivo* body clearance.

2. Liver is the primary organ responsible for the Cl of the NCE with insignificant contribution from renal or biliary excretion pathways. Comparison of *in vitro* versus *in vivo* data would normally suggest additional studies. For example, the role of biliary excretion can be evaluated in the event that $Q_h > \mathrm{Cl}_{\mathrm{plasma}} >$ predicted *in vitro* $\mathrm{Cl}_{\mathrm{hep}}$. When $\mathrm{Cl}_{\mathrm{plasma}} > Q_h$ (and $\mathrm{Cl}_{\mathrm{renal}}$ is low), then extrahepatic metabolism should be assessed, along with NCE stability in plasma and its blood:plasma partition ratio.
3. CYP-mediated oxidative metabolism is the key metabolism pathway, and contributions by other phase I and phase II enzymes are not prevalent. This is only true for $\mathrm{Cl}_{\mathrm{hep}}$ prediction from microsomes since hepatocytes constitute the complete cellular enzyme systems, both phase I and phase II.
4. $\mathrm{Cl}_{\mathrm{int}}$ is similar in microsomes and hepatocytes *in vivo*. Several compounds demonstrate uptake rate-limited clearance in hepatocytes; this means that at low physiological NCE concentration, rate of uptake into hepatocytes is the rate-limiting step compared to rate of metabolism. As a result of this, $\mathrm{Cl}_{\mathrm{int}}$ in microsomes is much higher than hepatocytes at lower NCE concentration (Hallifax and Houston, 2006; Lu et al., 2006a; Parker and Houston, 2008; Soars et al., 2007b).
5. Enzyme expression and activities in isolated subcellular fractions are representative of those in the physiological milleu.
6. In the case of IVIVC using microsomes, the assumption is that no uptake transporter is involved in drug uptake *in vivo* and accumulation into the hepatocytes.
7. In the case of IVIVC using hepatocytes, the assumption is that efflux transport does not impact the overall metabolism of NCE since efflux transporters get internalized and hence are nonfunctional in cryopreserved hepatocytes.
8. *In vitro* conditions used to determine metabolic stability represent physiological conditions where NCE (substrate) concentrations are much lower than the K_m of the metabolizing enzymes responsible for its clearance. This is a very basic, yet extremely important, assumption since under these conditions, it is assumed that rate of metabolism is linear with incubation time and enzyme concentration and Michaelis–Menten (MM) first-order kinetics is conserved. In the early stages of drug discovery where knowledge of K_m of the NCEs is commonly nonexistent, and highly sensitive analytical methods for NCE quantitation are not generally developed, caution must be used in assuming that MM kinetics are followed.
9. Secondary metabolism pathways, which are not captured in a static *in vitro* system, are nonexistent. For example, it is highly possible, especially with microsomes, that the metabolites that are initially formed via oxidative metabolism are either (a) further oxidized by non-CYP-mediated conjugation pathways or (b) cleared by alternate pathways.

10. Nonspecific binding to microsomes and hepatocytes (Austin et al., 2001; Hallifax and Houston, 2007; Kalvaas et al., 2001; McLure et al., 2000; Obach, 1996, 1997, 1999; Siebert et al., 2004; Venkatakrishnan et al., 2000). It is generally agreed upon that depending on the physicochemical properties of NCE, nonspecific binding to subcellular fractions will be affected. For example, basic and neutral compounds are more highly bound to microsomes than acidic compounds, which are more highly bound to plasma proteins. Numerous excellent publications in this field highlight the need to assess nonspecific binding of compounds to microsomes or hepatocytes for an accurate projection of *in vivo* Cl. However, caution must be exercised in interpretation of binding data since issues of whether fraction unbound in microsomes (fu_{mic}) is reflective of nonspecific binding *in vivo*, or whether compounds bind to different binding compartments in hepatocytes vs. microsomes, or whether to measure fu_{plasma} over fu_{blood} have all been dealt with skepticism, and a lot of these approaches are still under investigation. It is thus advisable to investigate the appropriateness of nonspecific binding—hepatocyte versus microsome binding, plasma versus blood binding—during IVIVC studies and establish the best combination of these binding parameters that is important in predicting *in vivo* Cl.
11. Atypical kinetics has been extensively studied with the CYP in recent years and has been attributed to the promiscuity of the CYP enzymes to accommodate multiple substrate molecules within the active site (Houston and Kenworthy, 2000; Isin and Guengrich, 2006b; Kramer and Tracy, 2008; Tracy, 2006). Its manifestation is seen either via simple kinetic plots, where the reaction rate versus substrate graphs are not the usual hyperbolic or via Eadie–Hofstee plots where the reaction rate versus reaction rate/substrate ratio plots deviate from the usual linear relationship. Depending on the data fit of these plots, they can be classified as homotropic (due to autoactivation, biphasic kinetics, or substrate inhibition) or heterotropic cooperativity (activation) (Fig. 5.3). Failure to recognize atypical kinetics results in inaccurate estimation of *in vitro* Cl, which is translated into inaccurate prediction of *in vivo* Cl.
12. Blood/plasma partition ratio. Due to ease of biochemical analysis, in most cases, plasma is analyzed with the assumption that due to rapid equilibration of compounds between plasma and blood, concentration of the NCE is same in plasma and in blood, resulting in $Cl_{plasma} = Cl_{blood}$. However, in cases when compound is extensively bound to plasma as in acidic compounds (i.e., $C_{plasma} > C_{blood}$, and therefore $AUC_{plasma} > AUC_{blood}$, which leads to $Cl_{plasma} < Cl_{blood}$), calculation of compound clearance in plasma leads to underestimation of *in vivo* Cl. Conversely, when compound is extensively bound to red blood cells, especially for basic compounds, Cl_{plasma} overestimates *in vivo* Cl by the same logic as above. Also, in case of plasma instability of an NCE, Cl_{plasma} will overestimate *in vivo* Cl. It is thus highly advisable for drug

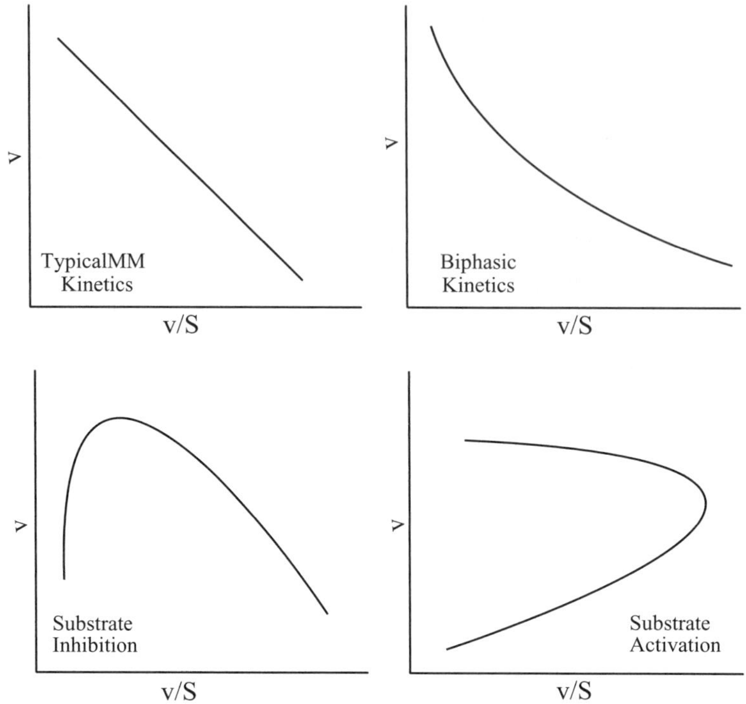

Figure 5.3. Eadie–Hoftsee plots for atypical enzyme kinetics (Houston and Kenworthy, 2000; Isin and Guengrich, 2006; Kramer and Tracy, 2008; Tracy, 2006).

discovery teams to ensure early on that the proper compartment is sampled and analyzed and chemical stability of the NCE in that compartment is assessed for accurate *in vivo* Cl calculation. Blood/plasma partition ratio and plasma stability of NCE should be assessed as an important part of IVIVC. Due to some of these primary reasons, *in vitro* data may not reflect the *in vivo* situation; and in these cases, compound optimization based on *in vitro* data will, most likely, not lead to the development of a successful drug candidate. It is a good practice to establish IVIVC early on for each chemical series and interpret the data accordingly.

Accurate projection of human PK parameters (Cl and V_{dss}) and efficacious dose is extremely valuable in discovery and ensures that NCE with poor PK properties or safety concerns (due to high dose requirement) are not selected as a potential drug candidate for development (Fagerholm, 2007; De Buck et al., 2007; Feng et al., 2000; Ito and Houston, 2005; Obach et al., 1997; Shiran et al., 2006; Sinha et al., 2008). Human PK projection can be performed via some common methods (Fig. 5.4): (a) Allometric scaling of *in vivo* PK parameters (Mahmood, 2005; Mahmood et al., 2003) (namely, Cl and V_{dss}), obtained

Simple Allometry:

$$Cl = a\ BW^{b}$$

$$V_{dss} = c\ BW^{d}$$

BW = Body Weight
MLP= Maximum Life Span
a = coefficient (log a is the y-intercept) of CL
b= exponent (b is slope of log-log plot ofCL and BW) of CL
c = coefficient (log a is the y-intercept) of V_{dss}
d= exponent (b is slope of log-log plot ofV_{dss} and BW) of V_{dss}

ROE used for Cl prediction:
if $0.55 < b < 0.7$ then Simple allometry best (above equation)
if $0.71 < b < 0.99$ then $Cl \times MLP = a\ BW^{b}$
if b > or = 1, then $Cl \times \text{brain weight} = a\ BW^{b}$

Elementary Dedrick Plot

$$Y = \frac{\text{Concentration}}{(\text{Dose/BW})} \qquad X = \frac{\text{Time}}{BW^{1-b}}$$

Complex Dedrick Plot

$$Y = \frac{\text{Concentration}}{(\text{Dose}/BW^{d})} \qquad X = \frac{\text{Time}}{BW^{d-b}}$$

Dienetichrons

$$Y = \frac{\text{Concentration}}{(\text{Dose}/BW^{d})} \qquad X = \frac{\text{Time}}{MLP \times BW^{1-b}}$$

CALCULATION OF OBA in HUMAN

$$f_a \times f_g = \frac{F_{PO,\ animal}}{\left(1 - \frac{Cl_{hep,\ animal}}{Q_{h,animal}}\right)}$$

Cl_{hep}= hepatic clearance
Q_h= hepatic blood flow
f_a =fraction absorbed
f_g =fraction escaping gut first pass
F_{PO} =Bioavailability

$$F_{PO,\ human} \quad f_a \times f_g \times = \left(1 - \frac{Cl_{hep,\ human}}{Q_{h,\ human}}\right)$$

Figure 5.4. Equations used in human PK projection (calculation of CL, V_{dss}, and OBA) (De Buck et al., 2007; Feng et al., 2000; Ito and Houston, 2005; Obach et al., 1997; Shiran et al., 2006; Sinha et al., 2008; Mahmood, 2005, 2006; Mahmood et al., 2003; Boxenbaum and Ronfeld, 1983; Abdel-Rahman and Kauffman, 2004; Huang et al., 2008).

in preclinical animal species via several methods based on simple allometry, rule of exponents (ROE), unbound Cl/F, or unbound fraction-corrected intercept (FCIM) (Mahmood, 2006); (b) *in vivo* Cl projection from *in vitro* Cl_{int} data discussed above—*in vitro* Cl_{int} (either $t_{1/2}$ or enzyme kinetic method described above) obtained in multiple animal species is scaled to Cl_{hep} in respective species, which is then compared to plasma Cl, observed in each of these animal species; if the correlation of the predicted Cl_{hep} and Cl_{plasma} is within two- to threefold, then Cl_{human} is predicted using *in vitro* human Cl_{int} data; (c) species-invariant time methods such as equivalent time, kallynochrons (elementary Dedrick plot), apolysichrons, dienetichrons, or syndesichrons—each of which are some modifications of the elementary Dedrick plot method (Boxenbaum and Ronfeld, 1983); and (d) C_{ss}-MRT method. For human OBA estimation, the fraction of dose absorbed in portal vein ($F_g \times F_a$) of each preclinical animal species can be calculated, the average of which can be used to estimate human $F_g \times F_a$. This estimate can be combined with the predicted human Cl, obtained by methods outlined above, to yield the predicted human OBA. Human OBA value can further be used to estimate the predicted human exposure at a desired dose (first-in-human as well as efficacious dose) (Mahmood, 2005). It is higly advisable to use multiple scaling methods to obtain a range of values for each of the predicted PK parameters since quite often, estimates obtained via different methods can be 5- to 10-fold different, in which case it is always desirable to use the best-case and worst-case scenario predictions of PK profile. With the predicted human PK parameters, a thorough understanding of the PK–PD relationship (Abdel-Rahman and Kauffman, 2004) (in multiple animal species—if data are available), and considerations such as role of species difference in plasma protein binding of NCE in the efficacious–exposure relationship, the efficacious human dose can also be predicted (Mahmood, 2005; Huang et al., 2008).

5.3 METABOLITE IDENTIFICATION AND REACTIVE INTERMEDIATES

In the previous section we discussed in detail how improving metabolic stability of NCEs also results in improved plasma clearance for NCEs that are primary cleared by metabolism. To successfully modulate metabolism of an NCE, it is very crucial to know which site(s) of the NCE is prone to metabolism and what types of metabolites are formed from the parent compound. This knowledge helps medicinal chemists strategically block these reactive sites, also called metabolic soft spots and design better, lower metabolic-clearance analogues. Knowledge of metabolites is very crucial in understanding the various biotransformation pathways of an NCE and the enzymes that are involved in those (Baranczewski et al., 2006a,b). This information, in turn, can aid in reducing the risk of DDI, which could be precipitated due to the modulation of those enzymes that are involved in the NCE biotransformation. Structure elucidation of metabolites may also indicate formation of reactive intermediates, which have the potential to bind to biomolecules and cause immune-mediated toxicity. Knowledge of formation of reactive metabolites in early discovery stages is very helpful in minimizing the risk associated with reactive intermediates (metabolites). Metabolites themselves could be toxic, resulting in unexpected toxicity during drug development (Tang, 2007). In several cases, metabolites, due to structural similarity with their parent, are also pharmacologically active, which results in a disconnect between PK–PD data in most cases. Metabolite characterization *in vitro* is performed in human and multiple animals' subcellular fractions to guide species selection for safety assessment studies in preclinical development—the species that most closely resembles human with respect to metabolite profile is selected as the preferred species for toxicity evaluations. When comparative studies of NCE metabolite formation with human and animal subcellular fractions indicate possible formation of unique human metabolite(s), the human-specific metabolite(s) has to be separately assessed in animal studies, especially if it is an active or toxic metabolite. When unique human metabolite(s) is detected in subcellular fractions, proper precautions must also be exercised to address the first clinical trials. Metabolite identification is therefore very important in drug discovery for NCE progression and should be incorporated very early on to eliminate structural liabilities in the later stages of drug development.

5.3.1 Metabolite Identification

In the early stages of LO, metabolite identification (met ID) is routinely performed in liver subcellular fractions, cDNA-expressed enzymes, or cryopreserved hepatocytes. Although it is highly preferable to maintain low NCE concentration (1 μM) in incubation mixture to mimic physiologically relevant conditions, analytical difficulties in detection of metabolites that are formed in low quantities force most metabolite identification reactions to be performed around 10 μM or higher (depending on the extent of metabolism of the NCE). At supraphysiological incubation conditions of most met ID studies, the biotransformation pathways can sometimes be different from an *in vivo* situation. For example, in the case of CYP2D6-mediated metabolism of dextromethorphan, dextrorphan is the major metabolite both *in vitro* and *in vivo*. However, in the presence of low dextromethorphan concentration or high enzyme activity (in cDNA expressed enzyme), dextrorphan is further metabolized to secondary metabolites (Van et al., 2009), highlighting the importance of incubation conditions in determination of biotransformation pathways. Phase II conjugation of phenol is found to be dose-dependent, with the high-affinity (low-capacity) pathway of sulfate ester formation being predominant at low phenol concentration but low-affinity (high-capacity) glucuronide formed at higher phenol concentration (Timbrell, 1982). In the majority of cases, this difference is mostly quantitative rather than qualitative; in other words, even at artificially higher NCE concentrations, commonly used for met ID studies, the profile of metabolites would not alter significantly (in most cases), although their relative amounts and the enzymes involved in their formation may vary significantly. Primary and secondary amines are capable of forming cabamoyl glucuronide via reaction with dissolved CO_2 and UDPGA *in vivo*. This is illustrated by tertiary amine-containing compounds Org3770 and Org5222, which undergo demethylation and the resulting secondary amine further reacts with dissolved CO_2 and UDPGA to form the respective carbamoyl glucuronide (Fig. 5.5). The β-blocker amosulol, which contains a primary amine, undergoes direct conjugation to form its carbamoyl glucuronide (Delbressine et al., 1990; Schaefer, 2006; Suzuki and Kamimura, 2007). However, under usual *in vitro* met ID assay conditions, where there is no dissolved CO_2, carbamoyl glucuronides cannot form. But if the structure of NCE suggests the possibility of formation of such metabolites, *in vitro* met ID should be conducted in the presence of dissolved CO_2.

Very sensitive analytical LC/MS/MS- or LC/NMR-based methods have revolutionized metabolite identification, quantitation, and characterization (Castro-Perez, 2007; Chen et al., 2007; Clarke and Haselden, 2008; Gao et al.,

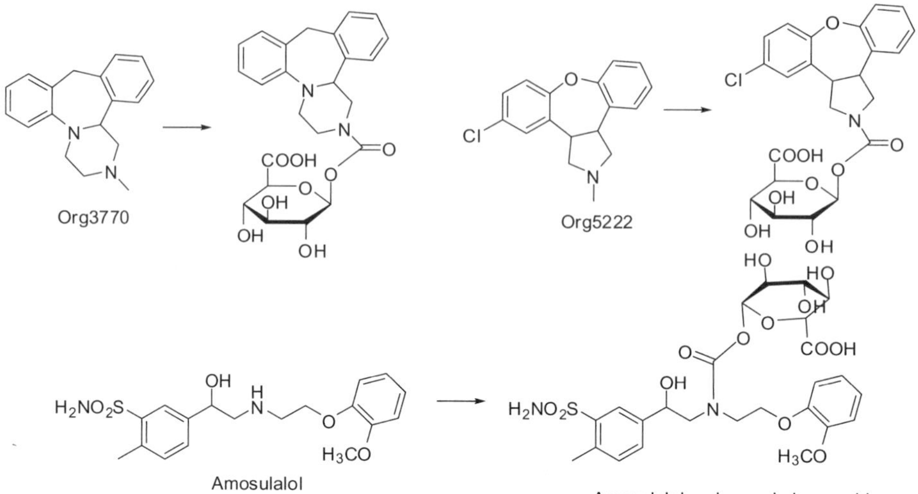

Figure 5.5. Structures of known carbamoyl glucuronides (Delbressine et al., 1990; Schaefer, 2006; Suzuki and Kamimura, 2007).

2007; Liu and Hop, 2005; Nassar and Lee, 2007; Prakash et al., 2007; Staack and Hopfgartner, 2007; Yan et al., 2008). Significant improvement of analyzers, such as the ion trap, quadrupole linear ion trap, and quadrupole time-of-flight, over the triple quadrupole analyzer has enabled biotransformation chemists to detect metabolites present in very tiny quantities. From a physiologically relevant point of view, it is not very practical to characterize all possible metabolites of an NCE, regardless of how tiny their concentrations may be; however, in cases where there is a possibility of formation of a toxic or active metabolite, this information can be very useful. Advanced instruments (Qtrap, QTOF, LTQ-FT, LTQ-OrbiTrap) that allow accurate mass determination and LC/NMR (1-D and 2D-NMR) that allow unambiguous elucidation of the site of metabolism, met ID has become an integral part of most small molecule drug discovery. It must be remembered that met ID with cold NCE is qualitative (best semi-quantitative) since it is assumed that the ionization efficiency of the parent and the metabolites are all similar in the mass spectrometer (MS). In reality, due to structural differences between the parent and metabolites, ionization efficiencies can be substantially different. Due to this reason, relative intensity of parent/metabolites in the MS is not a good indication of their absolute abundance and relative ratios. For example, it was observed that the response of the ring-opened metabolite of ziprasidone is almost 50% lower than that of its parent (Hop et al., 2005). Met ID of tetrahydroquinoline antimalarial compound PB-93 demonstrated that its major metabolite, resulting from *N*-dealkylation of its imidazole ring, has poor ionization efficiency compared to parent and could not even be detected by MS (van Voorhis et al., 2007). Similar dramatic decrease in ionization efficiency of an *N*-dealkylated metabolite has also been reported elsewhere (Hop et al., 2005). Difference in metabolite ionization efficiency was also reported for the anthracycline drug, doxorubicin (Arnold et al., 2004). Few examples just described emphasize the fact that met ID with cold NCE is limited by its qualitative nature. In order to get an accurate estimate of the absolute amounts of metabolite, met ID should be performed either with a radiolabeled NCE or with authentic standards of the metabolites (synthesized separately). Radio-detector response is directly proportional to the amount of radiolabel present, and so intensity of radiochromatograms is reflective of the actual amounts of parent NCE and its metabolites. Authentic standards of metabolites are extremely useful in generating metabolite–calibration curves and eliminate the uncertainties associated with differential ionization efficiencies of parent versus its metabolites. As valuable as radiolabeled NCE or authentic metabolite standards are, their availability is limited to late-stage discovery or development in most cases.

In silico metabolism prediction softwares such as Meteor and Metasite, coupled with several met ID data processing softwares (offered by Agilent, Applied Biosystems, and Thermo Scientific), have made biotransformation prediction for an NCE much easier. However, a strong organic chemistry background, coupled with good expertise in LC/MS/MS methodologies and data interpretation, are the key prerequisites for successful met ID. Met ID *in vivo* is usually done in biological matrices like plasma, urine, bile, and feces. Ion suppression from a variety of endogenous compounds in the biological matrices is a commonly encountered hurdle in *in vivo* met ID and leads to drastic decrease of MS signal intensity of parent and its metabolites. A strong expertise in sample preparation and novel sample introduction into the LC/MS/MS (e.g., UPLC and nano flow techniques) tremendously improves success in detection and identification of *in vivo* metabolites. There are a multitude of very good articles discussing several of these modern analytical strategies involved in met ID and are highly recommended for further reading (Castro-Perez, 2007; Chen et al., 2007; Clarke and Haselden, 2008; Gao et al., 2007; Liu and Hop, 2005; Nassar and Lee, 2007; Prakash et al., 2007; Staack and Hopfgartner, 2007; Yan et al., 2008).

Met ID data should be obtained early on during LO, to enable medicinal chemists to improve an NCE's stability toward metabolism; and there have been numerous reported examples where structural "soft spots" in NCEs were fixed, based on met ID data. The antifungal agent ketoconazole (KTZ) is primarily cleared by hepatic metabolism. Poor metabolic stability, coupled with poor solubility, contributed to poor PK of ketoconazole. Efforts were directed to design a better compound fluconazole, whereby the metabolically labile pyrazole moiety of KTZ was replaced by metabolically inert triazole moieties in fluconazole (Fig. 5.6A) (Smith et al., 1996). Diltiazem is an extensively metabolized benzodiazepine, with *N*-demethylation of its 3° amine linker being the major route of metabolism. The 3° amine was replaced by a 2° cyclic amine (pyrrolidine) that significantly enhanced stability toward

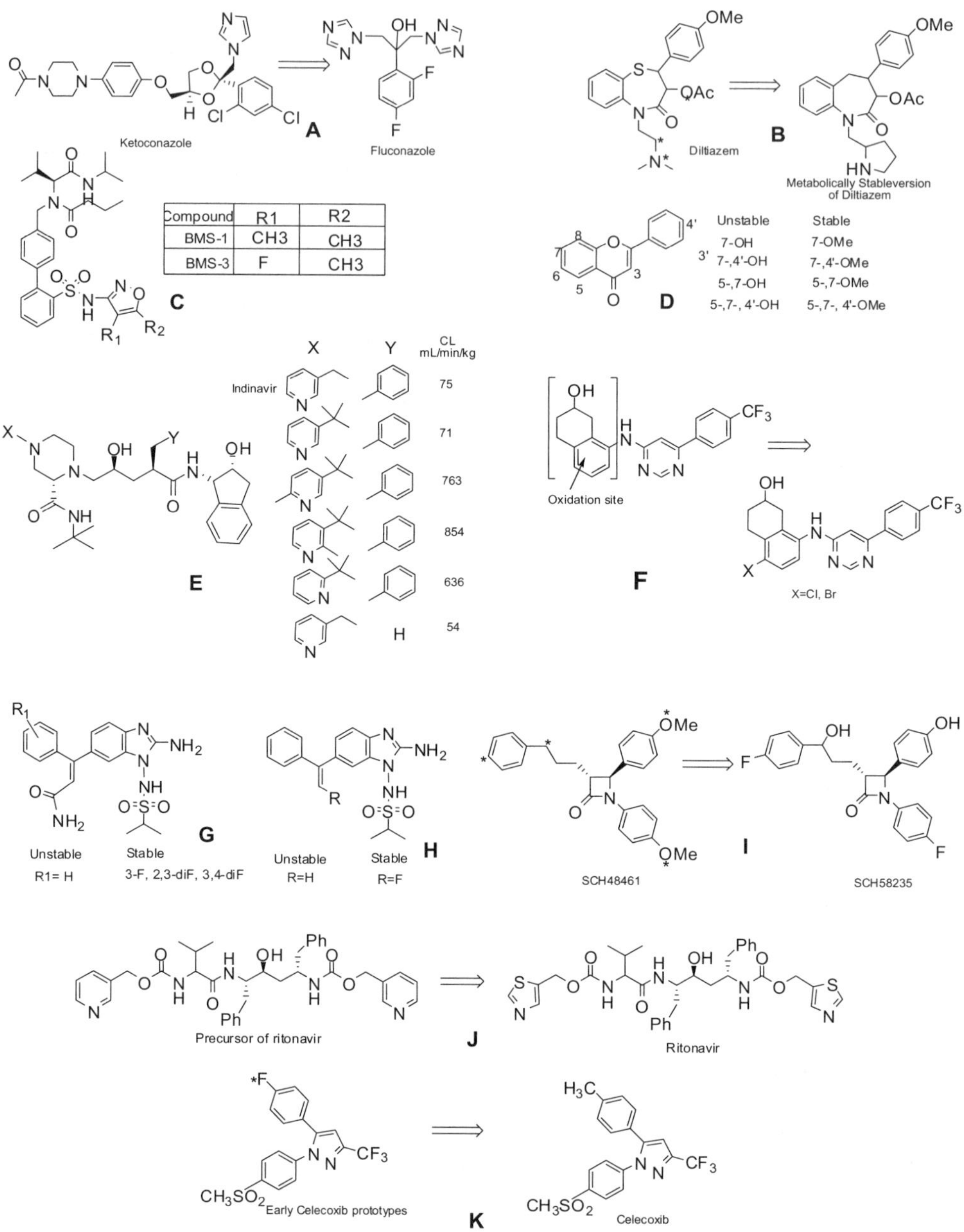

Figure 5.6. Examples of cases of compound structure optimization by improving metabolic stability (Floyd et al., 1992; Zhang et al., 2007; Walle et al., 2007; Campiani et al., 2002; Chiba et al., 2001; Doherty et al., 2008; Stratford et al., 1999; Victor et al., 1997; Rosenblum et al., 1998; Thompson, 2001; Penning et al., 1997; Smith et al., 1996).

metabolism (via decreasing radical stability and affording steric hindrance) (Fig. 5.6B) (Floyd et al., 1992). In the case of the BMS-1 compound, the dimethyl oxazole group was found to be the metabolic soft spot (determined via MS and NMR). Replacement of a Me group by a halogen significantly improved metabolic stability of this compound (also decreased metabolism of the other Me group on the oxazole) (Fig. 5.6C) (Zhang et al., 2007a). Dietary flavonoids have been in use as chemoprotective agents; but due to extensive glucuronidation and sulfation, they exhibit very poor OBA. A strategy to permethylate the polyphenols remarkably improved not only metabolic stability but also GI absorption (Fig. 5.6D) (Walle et al., 2007). A systematic approach in understanding the structure–stability relationship (SSR) has also been demonstrated with the HIV protease inhibitor indinavir, whereby the position of N in the pyridine ring as well as the nature of substitution at another key metabolic site are important determinants of metabolic stability (Fig. 5.6E) (Campiani et al., 2002; Chiba et al., 2001). In the case of the TRPV1 inhibitor compound (backup of a clinical candidate), once the metabolic soft spots were identified, they were blocked by halide substitution, resulting in analogues that possessed much superior metabolic stability and clearance (Fig. 5.6F) (Doherty et al., 2008). Halide substitution to block metabolism on an aromatic ring has been illustrated in the case of benzimidazole antivirals (Fig. 5.6G,H) (Stratford et al., 1999; Victor et al., 1997). In some cases, the most successful blocking of a metabolic soft spot is by synthesizing the metabolite that forms and then using it, because the lead compound-maintaining potency and selectivity is a challenge in these cases. This is well illustrated in the case of cholesterol absorption inhibitor compound SCH48461, which was subjected to hydroxylation of its benzene ring and benzylic position and an *O*-demethylation. The combination of blocking the soft spots and substitution of a –OH at the benzylic position resulted in compound SCH58235, which was resistant to metabolism (Fig. 5.6I) (Rosenblum et al., 1998). Replacement of a metabolically labile functional group by its bioisostere has also been exploited in designing more stable analogues. The HIV protease inhibitor compound, ritonavir, resulted from a strategic replacement of metabolically labile pyridyl groups on its precursors to stable, bioisosteric thiazole groups in ritonavir (Fig. 5.6J) (Thompson, 2001). More often than not, efforts to block metabolic soft spots make other sites more prone to metabolism. In contrast to efforts of improving metabolic stability by blocking soft spots, there may be the need to decrease stability of a compound to avoid an extremely long half-life of the NCE (and avoid toxicity due to NCE's long residence time in the body). Compounds made earlier than the COX2 inhibitor had extremely long half-lives. Replacement of the halogen on a benzene ring by a metabolically labile –Me group in celecoxib resulted in significantly decreased half-life when compared to that of earlier compounds (3.5 h of celecoxib versus 220 h of earlier compounds) (Fig. 5.6K) (Penning et al., 1997). Judicious use of met ID data to guide the design of better analogs is crucial in every stage throughout compound progression. Even in development stages, as more quantitative met ID data becomes available, it is very important to revisit the structural characteristics of an existing series and employ the met ID data to improve upon the existing PK properties.

5.3.2 Prodrug Approaches

Met ID expertise is sometimes harnessed to design prodrugs that are intended to overcome poor PK properties such as poor GI absorption or high first-pass metabolism (hence have low oral BA and duration of action). Prodrugs are inherently inactive compounds that are designed to be converted into their pharmacologically active forms upon administration in the body. It is highly desirable that other than the intended active form of the prodrug, other undesirable biotransformation pathways are minimal. Met ID information is the key to understanding the biotransformation pathways leading to the prodrug activation and formation of other possible metabolites. Esterases, amidases, and phosphatases play a major role in conversion of several prodrugs into their "active" parent drug form (Fig. 5.7) (Stella et al., 2007). Esterases mediate conversion of drugs like enalapril to active enalaprilat, valacyclovir to acyclovir, ceftin to cerufoxime, moexipril to moexiprilat, mycophenolate mofetil to mycophenolic acid, olmesartan medoxomil to olemarsatan, and many more; phosphatases are involved in conversion of clindamycin 2-phosphate to clindamycin, amifostine to WR-1065, fosamprenavir to amprenavir, fosphenytoin to phenytoin, and psilocybin to psilocin; aldehyde oxidase has been involved in conversion of antiviral agent famcyclovir to pencyclovir; codeine undergoes CYP2D6-mediated demethylation to active morphine. Identifying metabolites of prodrugs, other than the active drug, is crucial for successful development of prodrugs. Another strategy involved in the design of prodrugs is their target tissue selective activation. For example, several chemotherapeutic agents are activated in hypoxic tumor cells via redox activation to generate the cytotoxic drug form—for example, tirapazamine (Graham et al., 1997). It is important to have a thorough knowledge of mechanistic biotransformation pathways of the hypoxia-selective prodrugs into their cytotoxic drugs and other possible metabolites (desirable to minimize formation of noncytotoxic metabolites to improve therapeutic efficacy). While dwelling on metabolites, it is worthwhile mentioning/introducing the concept of "antimetabolites." Antimetabolite is a drug (or NCE) that bears very close resemblance to an endogenous essential metabolite and therefore is able to interfere with the cellular pathway(s) in which the endogenous metabolite plays a crucial role (Silverman, 2004). This could lead to severely compromised cellular function such as inhibition of cell growth or multiplication. For example, the anticancer agent methotrexate interferes with folic acid metabolism by binding to dihydrofolate reductase and inhibits DNA synthesis, which in turn inhibits cell replication. Drugs like mercaptopurine, thioguanine, fluorouracil, and gemcitabine, by virtue of their structural similarities to endogenous purines and pyrimidines, prevent these DNA bases from incorporating into the DNA helix and cause cell cycle arrest.

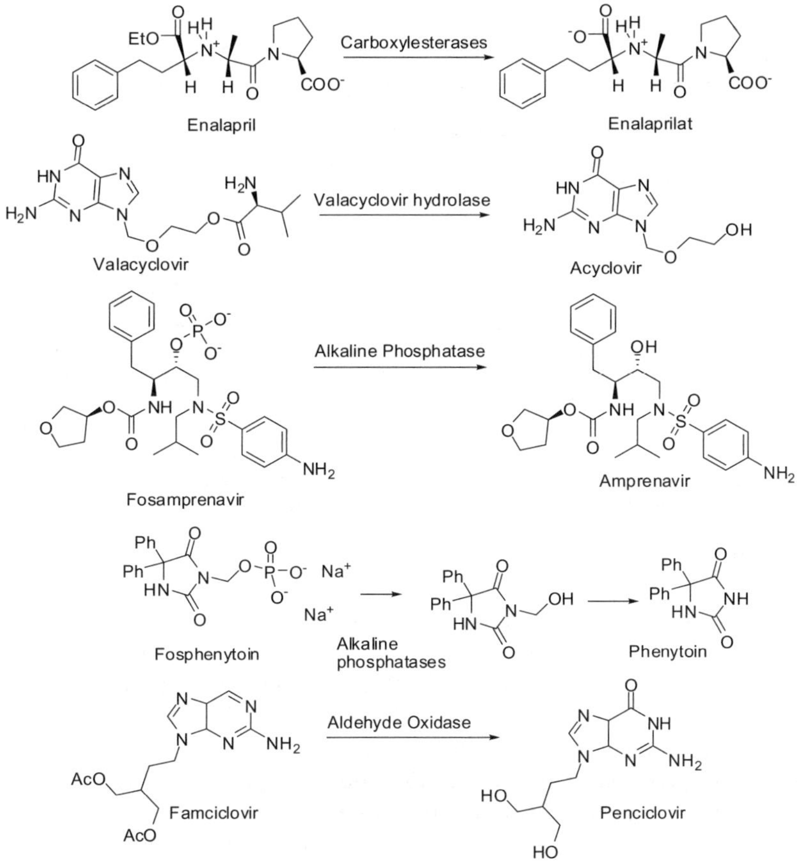

Figure 5.7. Metabolism in conversion of prodrugs into active drugs (Stella et al., 2007).

It is very important to elucidate structure of the endogenous metabolites to be able to target them by designing NCEs that closely mimic them.

5.3.3 Pharmacologically Active Metabolites

Biotransformation of compounds should not be always considered detrimental. In situations where the metabolite formed possesses pharmacological activity toward the same target as the NCE/parent itself, it could increase the therapeutic efficiency of the NCE (Fura, 2006). There may be instances where the metabolite binds to off-target receptors, which could result in either undesirable toxicity or even reversal of parent NCE's pharmacological action. In this section, the discussion will be limited to the concept of pharamacologically beneficial metabolites. The cholesterol-lowering drug atorvastatin is metabolized to two active metabolites, the *ortho-* and *para*-hydroxyatorvastatin (Lennernas, 2003). Proton pump inhibitor omeprazole, which itself does not

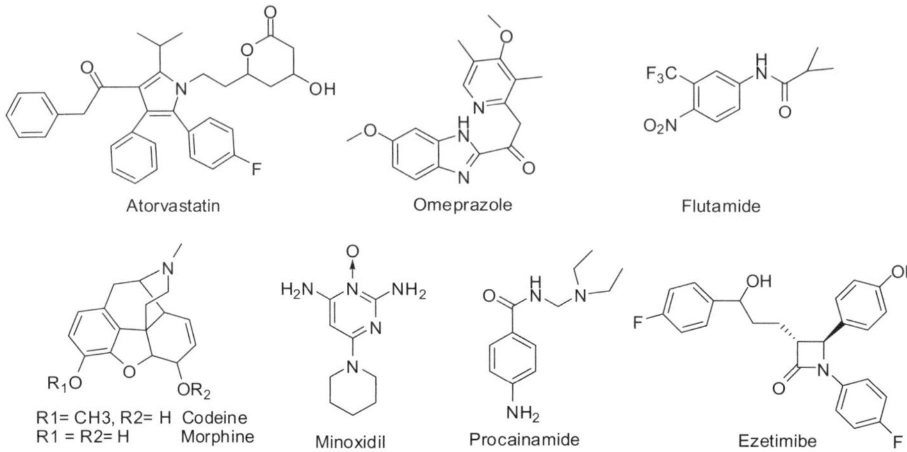

Figure 5.8. Structures of selected drugs that form pharamacologically active metabolites (Fura, 2006; Lennernas, 2003; Olbe et al., 2003; Shet et al., 1997; Joshi, 2008; Lalovic et al., 2006; Milo et al., 2006).

possess any pharmacologic activity, is converted into its reactive sulfenic acid metabolite, which causes inhibition of the H^+, K^+-ATPase enzyme to prevent acid secretion in the parietal cells (Olbe et al., 2003); the prostate cancer drug flutamide is metabolized into hydroxyflutamide, which is more potent than the parent and possesses a longer duration of action (Shet et al., 1997). Phase I metabolism reactions such as oxidation and dealkylation pathways contribute to formation of a major proportion of pharmacologically active metabolites (Fig. 5.8). One reason is that these simple biotransformation pathways result in metabolites that are not significantly different from their parent drug structurally and are therefore able to bind to the same receptor target to mediate the same pharmacological effect. Some drugs that are metabolized into pharmacologically active metabolites via oxidation/hydroxylation are bupropion, propafenone, carbamazepine, thioridazine, tamoxifen, propranolol, and simvastatin, and those metabolized via dealkylation are codeine, imipranime, fluoxetine, amitryptiline, diazepam, phenacetin, and verapamil. Conjugation by Phase II metabolizing enzymes usually do not lead to pharmacologically active metabolites, with a few exceptions: morphine 6-*O*-glucuronide and morphine-6-sulfate (Joshi, 2008), codeine-6-sulfate and codeine-6-acetyl (pain reliever) (Lalovic et al., 2006; Milo et al., 2006), ezetimibe glucuronide (cholesterol absorption inhibitor) (Catapano, 2001), minoxidil sulfate (antihypertension agent) (Anderson et al., 1998), and *N*-acetyl procainamide (antiarrhythmic agent) (Okumura et al., 1997). Knowledge of active metabolites formation is very important in early development, and it is invaluable in understanding their effects on PK–PD, modulation of enzymes and transporters, potency toward the target receptor compared to the parent, their free fraction in plasma, and concentration at the target receptor site (or off-target potency). For example, the active metabolite norfluoextine is a potent CYP2D6 inhibitor and can cause DDI when coadministered with other drugs that are substrates of CYP2D6 (Crewe et al., 1992). If a polymorphic enzyme (such as CYP2D6, 2C9, 2C19, NAT, SULT) is involved in generation of the active metabolite, then the effect of genotype on drug efficacy will have a significant effect and has to be incorporated for efficacious dose calculations and adjustments. Formation of active metabolites results in counterclock hysteresis in the effect versus concentration plot in a PK–PD link model as seen in cases of tramazosin, amitryptiline, imipramine, tesofensine (Lehr et al., 2008), and glibenclamide (Rydberg et al., 1997). The hysteresis can be resolved by inclusion of an effect compartment, which accounts for the "lag" in the pharmacologic effect due to formation of the active metabolite.

5.3.4 Reactive Intermediates

Trapping of reactive intermediates is extremely valuable to mitigate risk associated with the development of an NCE that generates the reactive metabolites. It is implicated that either (a) reactive metabolites bind to crucial proteins and cause direct organ damage or (b) the modified protein elicits a hapten-mediated immune response. A complex cascade of signaling ensues, resulting in unpredictable toxic responses, also called the idiosyncratic toxicity or idiosyncratic drug reaction (IDR). It is called idiosyncratic because the toxicity is not only unprecedented but also occurs in one out of a thousand patients receiving the drug, thus making it virtually impossible to detect earlier at the pre-registration phase. Several IDRs do not show dose-proportionality and the mechanism of idiosyncratic toxicity, in most cases remains elusive. IDR risk is a significant liability and increases cost and risk of development of an NCE. It is therefore crucial to detect reactive metabolites very early on during discovery and strategically remove the toxicophores (the functional group that is responsible for toxicity) to avoid costly drug failures in development or, worse, in post-marketing. Reactive intermediates are envisioned to be electrophiles (either charged species or Michael acceptors) that can be trapped by nucleophilic chemical such as glutathione (GSH) or cyanide (CN), depending on the "hard" (hardly polarizable, e.g., $CH_3C^{+}{=}O$) or "soft" (easily polarizable e.g., $CH_2{=}CH_2{-}CH_2^{+}$) nature of the reactive electrophiles (Gorrod et al., 1990; Nassar and Lopez-Anaya, 2004; Zhang et al., 2000). The majority of electrophiles generated are "soft" and are trapped by the endogenous thiol GSH (Scholz et al., 2005) and *N*-acetyl cysteine, while the "hard" electrophiles are trapped by the hard nucleophiles like cyanide (Argoti et al., 2005), methoxylamine, and semicarbazide. Intermediates such as quinones, quinone methide, quinone imine, iminoquinone methide, aldehyde, epoxides, arene oxides, free radicals, acyl halides, *S*-oxides and *S*-oxidation products, α-,β-unsaturated carbonyl (Michael acceptors), nitro reduction products (e.g., nitroso), isocyanates, and ketenes (Fig. 5.9) can be trapped by either GSH or -CN. GSH adducts can be detected by monitoring neutral loss of 129, 147, or 307 amu

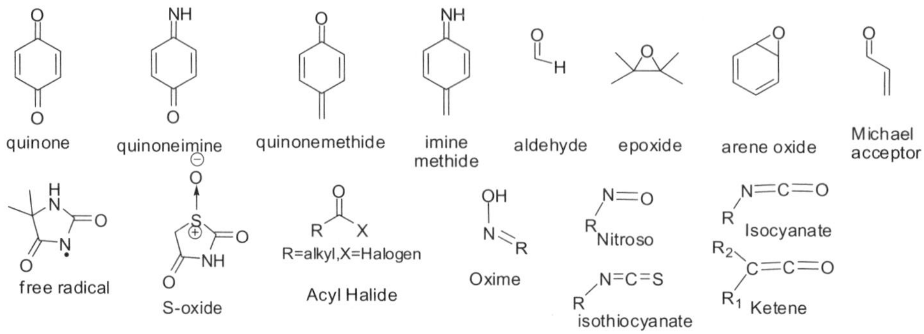

Figure 5.9. Reactive intermediates that are observed in nucleophilic-trapping experiments (Kalgutkar et al., 2005; Kalgutkar and Soglia, 2005; Dalvie et al., 2002).

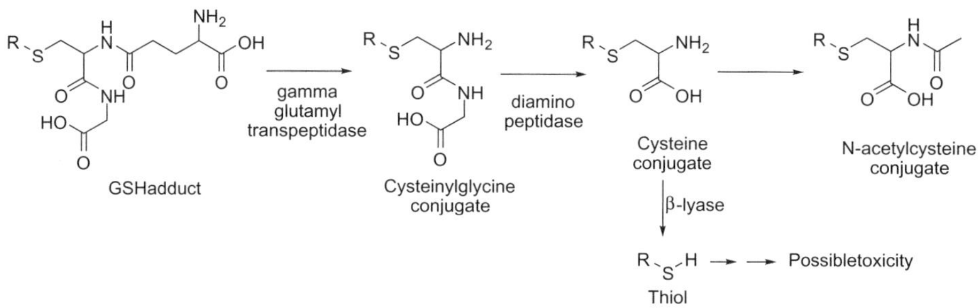

Figure 5.10. GSH adduct formation and subsequent cleavage observed *in vivo* (Scholz et al., 2005; Cooper et al., 2002; Cooper and Pinto, 2006; Dekant et al., 1994).

fragments via CID. A precursor ion scan of *m*/*z* of 272 in negative mode (due to deprotonated γ-glutamyl-dehydroalanyl-glycine fragment) can also be used as a diagnostic to detect GSH-trapped metabolites. A common diagnostic for cyanide-trapped metabolites is to monitor neutral loss of 27 amu fragment (loss of HCN). Use of an equimolar ratio of stable-labeled and unlabeled GSH or –CN is also gaining popularity since this creates an isotopic doublet pattern for neutral losses at 129 (for GSH trap) and 27 (for -CN trap) and rules out interference due to endogenous compounds from biological matrices. Monitoring exact mass neutral loss of 129.0416 or 27.0109 for GSH and CN adducts, respectively, via QTOF, also aids in unambiguously detect the nucleophile-trapped metabolites. GSH conjugates can lose the glutamic acid moiety (mediated by γ-glutamyltranspeptidase) to generate the *S*-cysteinylglycine conjugate, which can undergo acetylation of its free $-NH_2$ group to yield *N*-acetylcysteine conjugate (Fig. 5.10).

Formation of a GSH adduct is normally considered a detoxification pathway, but this may not be necessarily true in several cases where organ toxicity has been attributed to GSH adduct formation (Anders et al., 1992). It has been observed that *in vivo*, the *S*-cysteinylglycine conjugate can be further degraded (via β-lyase) to expose the naked thiol, which can cause toxicity liabilites (Cooper et al., 2002; Cooper and Pinto, 2006; Dekant et al., 1994), demonstrated for haloalkenes (Dekant, 2003; Lock and Schnellmann, 1990), efavirenz (Mutlib et al., 1999), cisplatin (Townsend et al., 2003), and dichloroacetylene (Patel et al., 1994). Acetaminophen is a very well studied compound with detailed characterization of its GSH-trapped adducts reported in literature. *In vitro*, the major metabolites are the GSH, cysteine, and *N*-acetylcysteine conjugates; but *in vivo*, several other peptide conjugates were also detected (Mutlib et al., 2000), illustrating the differences in metabolism *in vivo* versus *in vitro* system. GSH adducts have been detected for a variety of drugs such as acetaminophen, carbamazepine, imipramine, felabamate, remoxipride, clozapine, and many more. Recent advances in the GSH-trapping method have included use of radiolabeled [^{35}S]-GSH (Lim et al., 2008), dansyl-GSH, GSH-ethyl ester, and quartenary ammonium GSH (Soglia et al., 2004, 2006) agents to improve sensitivity and detection simplictity of GSH adducts. Cyanide trapped conjugates have been detected for drugs such as nicotine, ketoconazole, indinavir, mianserin, ciprofloxacin, and rifampin. It is clear that there are several drugs that have had a positive signal in trapping experiments of reactive intermediates by GSH or cyanide. Some of these are also known to cause toxicity (although the mechanism of toxicity remains elusive) while some don't. Consider the case of acetaminophen: This drug causes no heptotoxicity at its usual lower doses but at very high doses causes hepatic necrosis due to formation of a quinine–imine reactive intermediate (which can potentially deplete cellar GSH reserves and form covalent adducts with critical proteins). However, a structurally very similar analog of acetaminophen, 3-hydroxyacetanilide, can bind to GSH and proteins but causes no hepatotoxicity (Rashed et al., 1989). It emphasizes the fact that even though reactive intermediates/metabolites are formed, this does not automatically imply that an NCE will cause toxicity. Conversely, there have been cases where a drug forms reactive intermediates but fails to generate a trapped adduct in GSH trapping studies: A reactive metabolite of valproic acid is formed due to sequential CYP-oxidation and β-oxidation via mitochondrial CoA (Tang and Abbott, 1997) and hence is not trapped via the typical GSH trapping assay. Phenytoin is proposed to form a free radical reactive intermediate (Munns et al., 1997), which does not form a stable GSH adduct. Felbamate undergoes non-CYP-mediated bioactivation into its reactive aldehyde metabolite and hence would also not be detected in typical GSH trapping experimental conditions, conducted in microsomes (Dieckhaus et al., 2000). Also, it should be emphasized here that the trapping methodology is a useful tool to provide mechanistic insight into bioactivation pathways that lead to formation of *moderately* stable reactive intermediates. *In vivo*, nucleophiles such as GSH are ubiquitious in cytosol but reactive intermediates need to possess adequate stability to survive transport from the site of generation to the cytosol to be quenched and eliminated as conjugated moieties. Hence caution must be exercised when designing trapping experiments of reactive metabolites, and a thorough understanding of the chemical structure of the parent NCE and its biotransformation mechanism is extremely important for data interpretation.

GSH and CN trapping are commonly used methods to trap reactive intermediates (Xu et al., 2005). Some of the recently emerging methods to measure "biochemical markers" of toxicity, resulting from reactive intermediate formation, are also gaining popularity (Guengerich and MacDonald, 2007). For example, increase of alanine transaminase level as an indicator of heptotoxicity, induction of glutathione transferase and quinone reductase as indicators of cells stress, and measure of mitochondrial function for cell toxicity have been proposed as methods that could be used for risk assessment toward reactive metabolite-mediated-IDRs. Use of gene chips to explore gene expression patterns to identify individuals susceptible to IDRs has also been proposed. However, as with other existing methods, quantitative risk assessment and its clinical relevance remain a challenging hurdle for all of these methods. Covalent binding assessment has also been extensively used as the gold standard for reactive intermediate screening (Evans et al., 2004; Takakusa et al., 2008): A radiolabeled NCE is incubated with fortified liver microsomes, and the amount of radioactive material covalently bound to the microsomes is indicative of an NCE's propensity toward oxidative bioactivation. An *in vitro* covalent binding value of ≤50 pmol/mg has been accepted as desirable for compound progression. *In vitro* results are corroborated by evaluating the NCE's ability to form covalent adducts *in vivo* in preclinical species. While there are reports of good correlation between covalent binding and GSH-adduct formation (Masubuchi et al., 2007), it is widely accepted that correlation of amount of covalent protein binding *in vitro* to the amount of covalent binding *in vivo* (in preclinical species), and subsequent prediction of organ toxicity (in humans) with respect to its clinical relevance is extremely complicated and should not be attempted. The cutoff values provide a good rank-ordering approach whereby NCEs with the least or no covalent binding are preferentially progressed into development to mitigate the risks associated with reactive intermediate formation. Covalent adducts formation may also provide useful information regarding species difference in the bioactivation of an NCE: A Merck piperidine compound demonstrated high turnover and high covalent binding both *in vitro* (with liver microsomes) and *in vivo* in rhesus monkeys only (Baillie, 2008). Detailed mechanistic studies not only revealed the bioactivation site, but also indicated that the rhesus monkey was not representative of human metabolism and would therefore not be the ideal preclinical species for toxicology studies (Baillie, 2008). One of the most detailed quantitative studies to evaluate covalent adduct formation via metabolic bioactivation of drugs and their ability to cause liver injury was recently reported by Obach et al. (2008a). The authors calculated the Cl_{int} of covalent adduct formation and incorporated (a) the contribution of covalent binding as a fraction of total metabolism and (b) the daily dose of drug and hence the effective daily dose of covalent binding, and (c) competing detoxification pathways for removal of reactive metabolites and observed an acceptable correlation between a set of known hepatotoxic and nonhepatotoxic drugs. The conclusion still remains that *in vitro* covalently binding studies are best used to retrospectively investigate mechanism of toxicity when encountered in a clinical setting and is valuable in identifying imminent bioactivation pathways of NCEs and rank-ordering NCEs in discovery stages. However, efforts to quantitatively link the degree of covalent binding to clinical toxicity should not be attempted prospectively.

5.3.5 Idiosyncratic Drug Reactions (IDRs)

In several cases, reactive metabolites have been implicated to be the cause for idiosyncratic drug reactions (IDRs) (Erve, 2006; Park et al., 2006; Uetrecht, 2003, 2006; Walgren et al., 2005). Several drugs that have been withdrawn from the market and have been implicated to cause toxicity in liver, bone marrow, or skin (usually observed to be the major sites of toxicity) are observed to generate reactive metabolites. Here are a few examples of drugs withdrawn for various reasons: (a) *Hepatotoxicity*: Antidepressant nefadozone (withdrawn in 2004) is bioactivated to a reactive quinoneimine metabolite (Kalgutkar et al., 2005b); the thiophene ring of the diuretic drug tienelic acid is bioactivated into a reactive epoxide and a sulfoxide (Koenigs et al., 1999); NSAIDs, like benoxaprofen and bromfenac, were observed to form reactive acyl glucuronide (Bailey and Dickinson, 2003) (b) *Bone marrow toxicity*: The antiepileptic drug felbamate forms a putative reactive atropaldehyde intermediate, generated from its monocarbamate metabolite (Kapetanovic et al., 1998); antipsychotic drug clozapine is metabolized into a reactive iminium intermediate via peroxidases (Williams et al., 2000); the analgesic drug aminopyrine is metabolized into a highly reactive dication intermediate (Uetrecht et al., 1995). (c) *Skin toxicity*: Anticonvulsant carbamazepine forms reactive quinoneimine and epoxide metabolites (Pirmohamed et al., 1992); and the anticonvulsant phenytoin is bioactivated into a reactive free radical and an arene oxide (Munns et al., 1997); and the bacteriostatic antibiotic sulfamethoxazole undergoes *N*-hydroxylation, which undergoes further oxidation into the reactive nitroso metabolite (Burkhart et al., 2001). In the majority of examples cited above and several others reported in the literature, formation of reactive intermediates has been attributed to covalent binding to critical biomolecules. However, conclusive proof of a quantitative relationship between reactive metabolite formation, degree of covalent binding, and subsequent magnitude of an IDR has been yet to be unambiguously proven. One of the reasons could be due to failure to quantify covalent binding in the organ of toxicity. In addition to failure to quantify covalent binding in organ of toxicity, the following additional factors make toxicity predictions extremely challenging: (a) uncertainty in *in vivo* reactivity of the reactive metabolites (does it bind to macromolecules at the tissue/site of generation or is able to travel throughout the body to several tissues), (b) uncertainty as to which reactive metabolite (in case multiple are formed from the same parent) is responsible for covalent modification of proteins *in vivo*, (c) uncertainty regarding identity of proteins, genes, and cellular pathways that are affected *in vivo*, and (d) inability to trap reactive intermediates that possess high reactivity. Although the mechanism of toxicity remains evasive, it is agreed upon that depending on the structural characteristics of an NCE, it generates chemically reactive intermediates (Ju and Uetrecht, 2002) that can bind covalently to critical proteins. The covalently bound modified protein can elicit a cascade of signaling pathways leading to an immune response. Depending on an individual's genotype, the immune response can potentially cause severe IDRs. Therefore, severity of an IDR depends on an NCE's potential to generate reactive metabolites and an individual's susceptibility to abrogate the potentially harmful cascade of immune response triggered due to formation of the reactive metabolite–biomolecule adducts. It must be borne in mind that metabolite-mediated toxicity can result not only via proposed covalent binding to critical proteins but also due to formation of pharmacologically active metabolites that demonstrate target and off-target potency. In the case of the synthetic analgesic tramadol (opioid and monoaminergic), severe neurological toxicity has been attributed to CYP2D6-mediated conversion into its *O*-desmethyl metabolite, which, in the case of extensive metabolizers, is present in high concentration and demonstrates more potent agonist activity toward the μ-opioid receptor than against tramadol itself (Smith et al., 2009). Cardiotoxicity caused by the active metabolite of the anti-obesity drug fenfluramine has been proposed to off-target the pharmacologic effects of the active metabolite dexfenfuramine (Smith et al., 2009).

5.3.6 Structural Alerts

Structural characterisitics of compounds forming reactive intermediates/metabolites have been extensively reviewed (Kalgutkar et al., 2005a; Kalgutkar and Soglia, 2005; Dalvie et al., 2002). Compounds containing functional groups such as anilines, nitrobenzenes, benzyl/cyclopropyl amines, thiophenes, furans, thiazoles, sulfonylureas, hydrazines, carboxylic acids (those with α-protons), methylenedioxide, alkynes, terminal alkenes, halogenated hydrocarbons, and Michael acceptors (Fig. 5.11) are all recognized to be produce reactive metabolites, most of which can be trapped in *in vitro* assays by nucleophiles like GSH and cyanide. During discovery, when these functional groups are encountered in NCEs, an iterative approach is undertaken by medicinal chemists, pharmacologists, and DMPK scientists to reduce these structural liabilities to avoid potential risk of DDI later in development, while maintaining potency and selectivity of NCEs. The HIV protease inhibitor ritonavir is proposed to undergo CYP3A4 thiazole-ring bioactivation to mediate hepatotoxicity (Koudriakova et al., 1998). In contrast, other protease inhibitors such as indinavir and saquinavir, designed to eliminate thiazole ring-structural liability, were not associated with any heptotoxicity. Antidepressant mianserin is

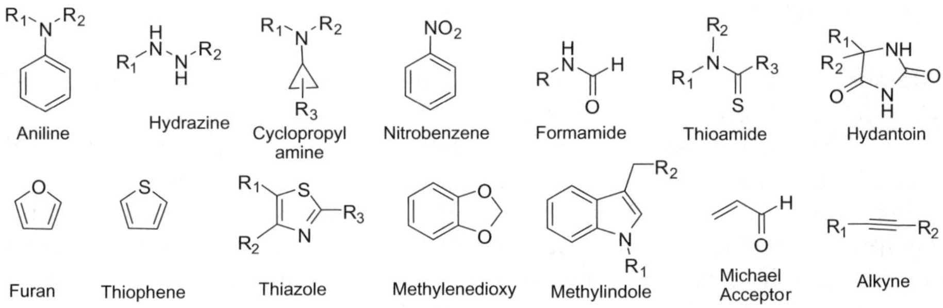

Figure 5.11. Functional groups that lead to generation of reactive intermediates Courtesy of Kalgutkar et al. (2005), Kalgutkar and Soglia (2005), and Dalvie et al., (2002).

known to be bioactivated into an iminium ion that mediates cytotoxicity due to covalent binding to microsomal protein (Roberts et al., 1993). Structurally similar analogues of mianserin, which do not form this iminium ion, resulted in substantial decrease of cytotoxicity. Removal of aniline motifs in prototype drugs such as carbutamide, practolol, and procainamide results in safer therapeutic agents tolbutamide, atenolol, and flecainide, respectively (Kalgutkar and Soglia, 2005). A fluoro-analogue of the anticonvulsant felbamate is shown to prevent reactive aldehyde formation in felbamate due to β-elimination and is therefore hypothesized to reduce risk of hepatotoxicity and aplastic anemia, seen in the parent drug (Thompson et al., 1996). Sometimes it is also possible to preserve the toxicophoric moiety of a parent structure but reroute metabolism via introduction of functional groups that are more prone to metabolism than the toxicophore. This results in preferential metabolism at these alternate sites, other than the original toxicophores, that eliminates or significantly decrease formation of reactive metabolites. Tolcapone, used to treat Parkinson's disease, contains a 3-nitrocatechol moiety that is proposed to undergo bioactivation to a reactive quinoneimine intermediate, which could be the cause of the severe hepatotoxicity of tolcapone (Smith et al., 2003). A structurally similar analogue, entacapone, possesses the 3-nitrocatechol moiety, but nevertheless is primarily cleared via glucuronidation, with no significant bioactivation to the quinoneimine intermediate as tolcapone. As predicted, entacapone also does not cause hepatotoxicity. Leukotriene receptor antagonist compounds CP-80,798 and CP-85,958 both possessed a chromanol structure that underwent ring opening to generate a reactive hydroxyaldehyde intermediate that was proposed to be the culprit for causing toxicity in preclinical species. Structural modifications were made so that metabolism at alternate sites were predominant, resulting in minimizing formation of the reactive aldehyde intermediate. These new analogues, which did not result in formation of the reactive aldehyde metabolite, were found to be nontoxic in preclinical species (Andrews et al., 1995; Chambers et al., 1999).

5.3.7 Metabolites in Safety Testing (MIST)

Metabolites have been implicated in numerous cases of organ toxicity and drug failure. Therefore safety assessment of metabolites is of primary importance during all stages of drug development (Guengerich and MacDonald, 2007; Baillie et al., 2002; Davis-Bruno and Atrakchi, 2006; Guengerich, 2006b; Humphreys and Unger, 2006; Naito et al., 2007; Kumar et al., 2008). The first comprehensive industry guidance to address the role of Metabolites in Safety Testing of a drug was published several years ago, famously called the MIST document, which was followed by a recent FDA guidance addressing the same issue (FDA, 2008). MIST has since become the center stage of many discussions amongst multidisciplinary drug development scientists with the common goal of designing the most informative preclinical and clinical studies to best address safety testing of metabolites. A summary of the MIST highlights is presented here. MIST guidance had originally proposed that human metabolites that represent >25% of drug-related material in circulation after a single dose should be considered for safety evaluation. This 25% criterion was later revised in the most recent guidance, which suggests that circulating metabolites in excess of 10% of parent drug, at steady-state conditions (to account for any metabolite accumulation that might occur after multiple dosing), need further safety evaluation. This criterion was critically reviewed by Obach in several publications (Smith and Obach, 2005, 2006), where some important changes/additional considerations were proposed: (a) Absolute amount of metabolites should be considered compared to the relative amount as suggested by the MIST document. Major justification against using relative metabolite amount is that via the relative amount method, metabolites >10% of low-dose drugs (usually low-dose drugs do not lead to toxicity) would require unnecessary testing while metabolites <10% of high-dose drugs (has more toxicity issues) could escape evaluation (potential risk). (b) In the case of an extensively metabolized parent drug, the parent will represent a very minor amount of all drug-related material in circulation compared to the rest of the metabolites. In this case, it would be necessary to monitor all the metabolites, which may be extremely laborious and expensive. (c) Similarity of the metabolite to its parent's structure is deemed important. Biotransformation can result in significantly different physicochemical properties of metabolites as compared to the parent, resulting in significantly different distribution properties of the metabolite versus the parent. Metabolites are generally more acidic than the parent and hence more bound to plasma proteins as compared to the highly lipophilic parent, which is more highly tissue-bound. So, as per the guidance, quantitating circulating metabolites as percent of parent might be misleading when a major amount of metabolite resides in plasma while a major amount of parent resides in tissues. (d) For a circulating metabolite that results from Phase I metabolism, is structurally similar to its parent, and can possess similar pharmacologic action to the target receptor as its parent, it was proposed that when a metabolite unbound concentration/*in vitro* receptor potency ratio is <0.25 of the ratio in the parent, the metabolite will not cause significant contribution to the overall pharmacology. (e) For circulating metabolites that demonstrate off-target pharmacology/toxicity, an unbound concentration >1 μM warrants its evaluation toward several nontarget receptors. (f) For excreted metabolites that represent the body's exposure to nonselective toxicity via reactive intermediate formation (due to hapten-mediated immunologic IDR or chemical modification of proteins by the reactive intermediates), when amount excreted (generally as conjugated forms like GSH or mercapturic acid-adducts) is >10 mg, this needs to be further evaluated in safety studies. The exceptions where excreted unconjugated, reactive metabolites have been detected are alcofenac-arene epoxide and troglitazone quinine. (g) A recent addition incorporating duration of parent drug administration and age of patients was proposed in the case of metabolites that cause toxicity usually after chronic dosing (Smith et al., 2009). It is clear that before significant resources are invested in safety evaluation of metabolites, a number of factors have to be taken into consideration: The primary ones are the absolute abundance of metabolites, its structural similarity to parent, whether it is circulating or excreted, whether it demonstrates target or off-target pharmacology, and the mechanism of toxicity elicited by the metabolite. In the event that the metabolite of interest in humans does not have adequate coverage in the animal toxicology studies (either not formed in animals or formed in disproportionately high concentration), it is recommended to evaluate its safety following direct administration of the metabolite in toxicology studies. This strategy also has severe limitations since it has been recently discussed that kinetics of administration of preformed metabolites can be substantially different from metabolites generated *in situ* mainly due to differences in absorption, tissue penetrability, clearance rate and mechanism, route of administration, and stability in dosing solutions (Pang, 2009; Prueksaritanont et al., 2006). Moreover, analysis of 24 withdrawn drugs from the market revealed that animals in toxicology studies for all of these drugs had coverage for the corresponding metabolites that precipitated the toxicity event in humans (Smith and Obach, 2006).

Reactive metabolites have been one of the primary reasons, implicated in precipitating IDRs (Uetrecht, 2003), although any concrete correlation between reactive metabolite formation and IDR has not yet been established. The primary reason for this is likely due to lack of understanding of mechanisms that cause IDRs. Nevertheless, reactive metabolites are commonly viewed as one of the possible causes of potentiating IDRs. It is therefore highly recommended that liability of reactive intermediate formation should be minimized early on in discovery. With detailed knowledge of metabolite formation, structural motifs that show propensity of reactive metabolite formation should be avoided or modified to minimize possible risk of IDRs. NCEs with high potency, which are expected to be efficacious at low dose [it is generally believed that drugs with doses <10 mg/day have less propensity for causing IDRs (Uetrecht, 2001)], along with those with high metabolic stability (to avoid metabolism-based DDI), should be developed to keep the absolute exposure of the body to NCE and its metabolites low. These are precautions that may decrease possibility of an IDRs; but since IDRs are not predictable (due to their inherent nature), these strategies do not guarantee elimination of an IDR risk.

5.4 CYP MAPPING AND REACTION PHENOTYPING

About 2 million DDIs are reported every year, resulting in approximately 100,000 fatalities, primarily due to unexpected adverse DDIs postmarketing. Difficulty in accurately predicting DDIs may be the cause of such a high incidence of undesirable side effects of coadministration of drugs. Throughout the course of drug discovery, it is a standard practice to assess NCEs for their ability to be either (1) a victim of a DDI (mediated by another coadministered drug) or (2) a perpetrator (or precipitant) of DDI themselves. It is well established now that the magnitude of a DDI is dependent of the degree of contribution of the clearance pathway that is modulated (inhibited or induced): The higher the fraction of the dose metabolized by the enzyme that is modulated, the greater the magnitude of DDI. It is very favorable for an NCE to have multiple pathways of clearance, so that when one of these is affected, the effect is compensated for by the other pathways. Interindividual differences in drug disposition due to differential expression of polymorphic enzymes is also critical in assessing a DDI risk (Hamberg et al., 2007; Rodrigues and Rushmore, 2002; Williams et al., 2008). For example, CYP2D6 is a well-studied polymorphic enzyme and there is a significant difference in metabolism (and therefore exposure) of some CYP2D6 substrates (antidepressants, antipsychotics, opioids) in the poor metabolizers (PM) and rapid metabolizers (RM). This results in either lack of drug efficacy due to increased clearance of the drug in the RMs or increased toxicity due to accumulation of drug in the PMs. Cases like these, which could warrant a dose adjustment based on the genotype of the individual, further highlight the need to identify major enzymes involved in the clearance of an NCE: Development of an NCE that is primarily metabolized by polymorphic enzymes is a major DDI risk and should be avoided early on during discovery if possible. Hence identifying major pathways that are responsible in the clearance of an NCE is crucial in eliminating DDI risk later on in development. A significant amount of work has been done in the area of the CYP enzyme phenotyping, and therefore most of the discussion in this section will be centralized around these major DMEs.

In early discovery phases, a qualitative idea of the major route of elimination contributing to the total clearance of an NCE is sufficient. Sometimes crude evaluation of clearance pathways may be obtained by coadministration of aminobenzotriazole (ABT), a potent nonspecific inactivator of most CYPs or cyclosporin A (CsA), and a potent nonspecific inhibitor of several major transporters, along with the NCE in a rat PK study. Primary involvement of CYPs or transporters is suggested based on whether ABT or CsA affects the exposure of the NCE. Use of ABT and CsA in *in vitro* experiments is also very common to assess CYP-mediated metabolism contribution or involvement of transporters in clearance of NCEs. To tease out the contribution of FMO- and CYP-mediated metabolism, it is also common to either treat microsomes with Triton X-100 (a detergent that completely inhibits CYP activity but not FMO activity) or incubate microsomes at 50 °C for 2–3 min. FMO is heat-inactivated in contrast to CYP, the catalytic activity of which is only slightly compromised.

During the late LO stage, a little bit more rigorous method called enzyme mapping is implemented. Enzyme mapping is very useful in identifying the enzymes that are involved in the major metabolism pathway of an NCE. In other words, enzyme mapping helps identify the DMEs that are primarily responsible for the overall clearance of an NCE. In early discovery stages, most common method is via the use of recombinant human enzymes. cDNA-expressed major human DMEs (CYP, UGT, FMO, NAT) are easily available commercially. An NCE is simply incubated with each of these recombinant enzymes, and the Cl_{int} (V_{max}/K_m) is determined for each of the enzymes. Care is taken to ensure that reactions are carried out under linear conditions of the enzyme kinetics; that is, NCE disappearance or the metabolite formation (of pathway of interest) is linear with respect to incubation time and protein concentration. The Cl_{int} number gives a rough estimate of relative contribution of each of the enzymes involved in metabolism of the NCE: Higher Cl_{int} of an enzyme suggests higher contribution by that enzyme. A good idea of high- and low-affinity enzymes involved in the metabolism of an NCE can also be obtained from the K_m values generated in a recombinant system: Low K_m for an enzyme suggests high affinity to that enzyme, implying that at low substrate (NCE) concentration, the contribution of that particular enzyme would be more pronounced than one which has a high K_m value. Caveats of this method are as follows: (1) In most of the early stages of discovery, knowledge of metabolite and hence a single major pathway of interest is lacking. It is common to determine the Cl_{int} by rate of parent (NCE) disappearance over time. So when NCE disappearance is monitored over time, it gives no idea about the major biotransformation pathway; instead, it gives just an idea of the overall rate of metabolism of a compound (which, in most discovery settings, is also valuable knowledge). (2) Since the amounts of accessory proteins such as NADPH-cytochrome reductase, cytochrome b5, and heme (in the case of CYP) are different in different c-DNA expression systems, this causes significant lot-to-lot variability of the catalytic activity of the enzymes. (3) Since expression level and specific activities of the enzymes are very different in human liver than what is observed in recombinant enzymes, relative contribution of an enzyme to the total metabolism of the NCE *in vivo* cannot be estimated: The most that these data will suggest is the qualitative involvement of a certain enzyme in metabolism of the NCE (which may be sufficient in early discovery stages). It is also highly possible that in the absence of other DMEs a recombinant system will exhibit NCE turnover, which would otherwise not be observed *in vivo*. To transform Cl_{int} obtained from recombinant enzymes into a physiologically meaningful value, this Cl_{int} must be normalized by applying several scaling factors. However, well-defined scaling factors (such as relative activity factor, RAF, discussed below) are available for CYP isoforms only and cannot be extended to other DMEs. Nonethelesss, in the early stages of discovery, Cl_{int} values from recombinant enzymes do provide useful information about involvement of certain enzymes (or isoforms) in the overall metabolism of an NCE.

When an NCE is moved further into late-stage discovery or early development, a more rigorous version of CYP mapping is performed where a more quantitative estimation of contribution by a CYP isoform to total NCE clearance is evaluated. The integrated approach, also called CYP phenotyping, is usually a combination of a several steps discussed below (Harper and Brassil, 2008; Lu et al., 2003; Zhang et al., 2007b).

Step 1. CYP phenotyping starts with assessment of metabolite formation of an NCE using NADPH-fortified pooled HLM (20–50 donors to reduce intersubject variability) and ruling out non-CYP mediated pathways. The kinetics of major metabolite(s) formation is analyzed under initial rate conditions (NCE depletion is usually ≤15% and metabolite formation is linear with respect to protein concentration and incubation time) over a wide range of NCE concentrations (solubility of NCE permitting). Enzyme kinetic parameters V_{max} and K_m for each major biotransformation pathway are determined via careful analysis of Eadie–Hoftsee plots. In the case where kinetics of a biotransformation pathway is biphasic (curved Eadie–Hoftsee plot seen in this case as compared to straight Eadie–Hoftsee plot), involvement of two enzymes with two different K_m values is suggested. Implication of two K_m values for a single biotransformation pathway is that at low concentrations, physiologically relevant conditions, the low K_m enzyme (and the high-affinity enzyme) predominantly contribute to the biotransformation pathway of interest. To clarify, *N*-demethylation of clomipramine is catalyzed by CYPs 3A4, 2C19, and 1A2 with different K_m values (Nielsen et al., 1996). Hence the result of modulating these different CYP isoforms will have different consequences, depending on the dose of clomipramine. The contribution of CYP isoforms could be altered depending on the biotransformation pathway monitored: CYP2C19 is the high-affinity isoform, while CYP3A4 is the low-affinity isoform, involved in demethylation of the benzodiazepine flunitrazepam; but for the formation of another active 3-hydroxy metabolite of flunitrazepam, CYP3A4 is the high-affinity isoform, while CYP2C19 is the low-affinity isoform (Kilicarslan et al., 2001). It is also possible to have a monophasic Eadie–Hoftsee plot (straight line) due to contribution of two different isoforms toward formation of the same metabolite with similar K_m values. Due to its cardinal role in the CYP phenotyping process, enzyme kinetics determination has to be done carefully especially for low-clearance compounds that are metabolized via multiple pathways. During reaction phenotyping studies, care must also be taken to maintain the NCE concentration as close to physiologically relevant as possible to avoid contribution of low-affinity isoforms at high NCE concentration. This is illustrated in 5-hydroxylation of lansoprazole, where CYP3A4 is the major contributor at high substrate concentrations but CYP2C19 is the major contributor at low substrate concentrations (Pearce et al., 1996).

Step 2. After major biotransformation pathways are identified and their kinetics evaluated, effect of CYP-isoform-specific chemical inhibitors on the particular biotransformation pathway (in HLM) is assessed. This gives a better idea of relative contribution by the CYP isoforms toward the metabolite formation. This is normally done with $[S] \leq K_m$ of the NCE and $[I]/K_i > 10$ of the CYP-specific-inhibitor. The reason for the need to keep $[S] \leq K_m$ of the NCE is because the degree of inhibition by chemical inhibitors is dependent on NCE concentration [S] and mechanism of inhibition (competitive, noncompetitive, uncompetitive, or mechanism-based). However, depending on whether $[S]/K_m < 1$ (when $[S] < K_m$) or $[S]/K_m \approx 1$ (when $[S] \approx K_m$), degree of inhibition would be independent of [S], which mimics more closely an *in vivo* situation. This again highlights the importance of an accurate determination of K_m in Step 1 above. The reason why $[I]/K_i > 10$ is crucial is because in order for almost complete inhibition of a specific isoform, a high inhibitor concentration yielding a high $[I]/K_i$ ratio (normally >10) should be used. Complete inhibition of an isoform is very crucial to determination of relative contribution by that isoform. This is especially true when multiple CYP isoforms are involved in the metabolism of the NCE. Due to the need for high inhibitor concentration for almost complete inhibition of an isoform, the inhibitor used should also be highly specific for that particular CYP isoform, so that relative contribution by the isoform that is being assessed is not miscalculated (for example, ketoconazole is a potent and selective CYP3A4 inhibitor at low concentration, but inhibits several CYP isoforms at high concentrations). If CYP-specific inhibitors are unavailable, then IC50 values of the inhibitor for various CYP isoforms should be determined (using a wide range of [I], solubility permitting); and based on the magnitude of IC50, relative potency of inhibitor toward each CYP isoform can be assessed. It must be borne in mind that absence of selective inhibitors can lead to overestimation of inhibition and hence contribution by a particular CYP isoform. If the inhibitor used is a MBI, then care must be taken to include this into the inhibition study design. In complement to chemical inhibitors, CYP-isoform-specific inhibitory antibodies are also used to evaluate relative contribution of different CYP isoforms toward NCE metabolism. The advantage of antibodies is that since they act as noncompetitive inhibitors, their inhibitory potency is independent of NCE concentration [S]. The drawbacks of antibodies are cross-reactivity with other CYP isoforms (nonselective—for example, antibodies to CYP3A4 also inhibit CYP3A5) and lack of 100% inhibition. To clarify, when extent of inhibition with antibodies is 85%, it is hard to unambiguously conclude whether this is due to involvement of other CYPs (therefore metabolism was not 100% inhibited by a CYP-specific antibody) or merely due to partial inhibition via an suboptimum antibody concentration. Hence it is highly recommended to use various concentration of an antibody to assess its inhibitory potency toward a CYP isoform using an isoform specific substrate. Both chemical and antibody inhibitors gives relative percent contribution of a particular CYP toward the total metabolism of an NCE.

Step 3. Correlation analysis is performed with the specific biotransformation pathway of the NCE of interest and CYP isoform-specific marker substrate metabolism rates or immunoquantified CYP levels to further confirm contribution of a particular CYP isoform. The rate of formation of the metabolite of interest (of the NCE) is studied in a panel of liver microsomes (prepared from at least 10 different human donors) and is correlated with the rate of metabolism of a CYP isoform-specific marker substrate or specific CYP-isoform levels, determined via immunoquantitation. The results are analyzed via statistical analysis of the data, where a correlation coefficient (r) ≥ 0.63 (with Student's t test P ratio ≤ 0.05) is considered significant. When two or more CYP isoforms are involved, it is necessary to use specific inhibitors of the other isoforms to get a successful correlation analysis with one isoform at a time. Activity correlations might also fail due to involvement of polymorphic CYP2B6, 2D6, 2C9, and 2C19 (when levels of CYPs are used for correlation studies) where although levels of CYP (via immunoquantitation) isoforms appear the same, they could have significantly different activities (k_{cat}/K_m).

Step 4. In some cases when polymorphic CYP isoforms are involved, it is advantageous to have genotyped livers to understand the effect of allelic variants versus wild type on the specific biotransformation pathway of an NCE.

Step 5. Confirmation via relative activity factor (RAF) method using cDNA-expressed enzymes and HLM—Cl_{int} (rate of metabolism) values are generated by incubating the NCE with cDNA-expressed enzymes under linear MM-kinetic-conditions. The Cl_{int} values obtained from recombinant enzymes are weighted for relative abundance of each of the isoforms in the liver by a conversion number RAF. Determination of RAF for each CYP isoform is performed by comparing the rate of a CYP-isoform-specific reaction in HLM to the same reaction in recombinant enzyme (Venkatakrishnan et al., 2000; Proctor et al., 2004; Uttamsingh et al., 2005). RAF can be defined in terms of either V_{max} (maximum reaction velocity) (Eq. (1) in Fig. 5.12) (Venkatakrishnan et al., 2000) or Cl_{int} (Eq. (3) in Fig. 5.12) (Uttamsingh et al., 2005). In these assays to determine the rate of metabolism in rCYP and HLM, it is ensured that the concentration of each of the CYP isoform-specific probe substrate is maintained at the maximal velocity conditions (V_{max}), which is different for HLM and the recombinant enzyme. Once the RAF is generated, the relative

Venkatakrishnan et. al RAF method:

$$\text{RAF} = \frac{V_{max,mean} \text{ for isoform-specific reaction in HLM}}{V_{max,mean} \text{ for isoform-specific reaction in cDNA-expressed isoform}} \quad (1)$$

$$f_i\,(\%) = \frac{A_i v_i}{\sum_{i=1}^{n} A_i v_i} \times 100 \quad (2)$$

Uttamsingh et. al RAF method:

$$\text{RAF} = \frac{V_{max}/K_m \text{ for isoform-specific reaction in HLM}}{V_{max}/K_m \text{ for isoform-specific reaction in cDNA-expressed isoform}} \quad (3)$$

$$f_i\,(\%) = \text{RAF} \times (v_i/v_{HLM}) \quad (4)$$

Rodrigues et. al TNR method:

$$\text{NR} = \text{reaction rate} \bullet \text{mean specific content}$$
$$= \text{pmol/min/pmol rCYP}_n \bullet \text{pmol mCYP}_n\text{/mg}$$

$$\%\,\text{TNR} = \frac{\text{pmol/min/pmol rCYP}_n \bullet \text{pmol mCYP}_n\text{/mg}}{\sum \text{pmol/min/pmol rCYP}_n \bullet \text{pmol mCYP}_n\text{/mg}} \times 100 \quad (5)$$

RAF=Relative Activity Factor
V_{max} = Maximum Reaction Velocity
K_m = Substrate concentration that yields half of V_{max}
v_i = velocity of reaction/rate of metabolism for each CYP isoform
A_i = RAF
f_i = relative contribution of each isoform
NR = Normalized Ratio
TNR = Total Normalized Ratio

Figure 5.12. RAF and TNR calculation (Venkatakrishnan et al., 2000; Uttamsingh et al., 2005; Rodrigues, 1999).

contribution of each isoform can be obtained by Eq. (2) in Fig. 5.12 (Venkatakrishnan et al., 2000) or Eq. (4) in Fig. 5.12 (Uttamsingh et al., 2005). Intersystem extrapolation factor (ISEF) has also been proposed as a further modification to RAF (Proctor et al., 2004).

Step 6. Sometimes, a less rigorous method may be used prior to the RAF method. Similar to the RAF method, the data obtained with cDNA is normalized using mean specific content of each of the specific CYP isforms involved (literature reported or immunoquantitated). Drawback of this method is that amounts of CYP isoforms sometimes vary considerably in the microsomal preparations, and so even the normalized rates may not reflect the *in vivo* conditions. The normalized rates are converted to a total normalized rate (TNR) as shown in Eq. (5) in Fig. 5.12 (Rodrigues, 1999). Detailed kinetic

$$\frac{AUC_{PO,\,inhibitor}}{AUC_{PO,\,control}} = \frac{f_{g,inhibitor}}{f_{g,control}} \times \frac{1}{\left(\frac{f_m\ f_{m,CYP}}{Cl_{int,\,control}/Cl_{int,\,inhibitor}}\right) + [1-(f_m\ f_{m,CYP})]}$$

$$Cl_{int,\,control}/Cl_{int,\,inhibitor} = 1 + [I]/K_i \text{ for reversible competitive or non-competitive inhibition}$$

$$= 1 + \frac{k_{inact}}{K_I} \bullet \frac{[I]}{k_{deg}} \text{ for mechanism-based inhibition}$$

AUC = Exposure
f_g = fraction surviving gut first pass
f_m = fraction of total clearance due to all CYP-metabolism
$f_{m,CYP}$ = fraction of total CYP- metabolism catalyzed by the inhibited CYP isoform
[I] = Inhibitor Conc
k_{inact} = Rate constant of CYP inactivation
K_I = Half maximal inactivation rate
K_i = Dissociation constant of enzyme-inhibitor complex

Figure 5.13. Exposure change in the presence of an inhibitor (Bachmann, 2006; Brown et al., 2006; Obach et al., 2006; Venkatakrishnan et al., 2003).

studies of the biotransformation is performed (i.e., K_m, V_{max}, RAF method) only when TNR are ≥20%. TNR is also correlated with percent inhibition, or the correlation coefficient *r*, described in Step 4 above. Due to its limitations, it is advised to use the RAF method over the TNR method.

A combination of Steps 1–6 described above is use to yield an integrated value of f_{mCYP} to assess the contribution of each CYP isoform to the biotransformation pathway of interest (Zhang et al., 2007; Uttamsingh et al., 2005). The magnitude of the product of f_{mCYP} and f_m (fraction of the NCE cleared via metabolism) reflects the propensity of an NCE to be subjected to DDI. The higher the value of $f_{mCYP} \times f_m$, the higher the risk of DDI. Change in exposure of an NCE in the presence and absence of a modulator (inhibitor or inducer) can be simulated by the equation shown in Fig. 5.13 (Bachmann, 2006; Brown et al., 2006; Obach et al., 2006; Venkatakrishnan et al., 2003) (discussed further in Section 5.5). As can be easily seen, the product $f_{mCYP} \times f_m$, plays a crucial role in predicting the change in exposure of an NCE, and therefore the DDI risk. To elucidate, consider the fact that two NCEs, A and B, have the same $f_{mCYP3A4}$ value of 0.9; that is, CYP3A4 contributes to 90% metabolism of both A and B. But in the case of A, renal and biliary clearance play a major role such that the fraction of the total dose of A cleared by CYP-mediated metabolism (f_m) is 0.2. The NCE B is primarily cleared by CYP-mediated metabolism; hence its (f_m) is 0.8. Based on the values of $f_{mCYP} \times f_m$ for compound A ($0.9 \times 0.2 = 0.18$) and that for compound B ($0.9 \times 0.8 = 0.72$), it can be predicted that since compound B has an $f_{mCYP} \times f_m$ value of >0.5, this may pose a DDI risk and warrants further evaluation. Important to note also is that since the change in exposure is also dependent on the change in Cl_{int} (with and without inhibitor), which in turn depends on the $[I]/K_i$ ratio, during simulation to predict the effect of $f_{mCYP} \times f_m$, maximal inhibition (≥95%, i.e., $[I]/K_i \geq 20$) of a CYP isoform is usually assumed. Estimate of f_m, can be most accurately made from a human radiolabel study, or a good estimate can be obtained from a single-dose Phase I human study. In many projects, due to the absence of human data in early stages of development, an estimate of f_m is not available and f_{mCYP} data are used to guide clinical DDI studies via a "rank order" approach discussed in detail in Section 5.5. Essentially the f_{mCYP} values are determined (via a combination of CYP phenotyping methods discussed above) and the CYP with the highest f_{mCYP} is evaluated clinically using CYP-isoform-specific inhibitors. For polymorphic CYP isoforms and CYP2C19, which does not have potent, selective inhibitors, PK of the NCE in genotyped indivduals is recommended.

Unfortunately, assessment of contribution by other enzymes such as FMO, SULT, and UGT (Kaji and Kume, 2005; Sakaguchi et al., 2004) are limited to qualitative evaluation only. The primary reason for this is that for these enzymes, availability of cDNA-expressed enzymes, normalization methods, isoform-specific substrates, and inhibitors (chemical and antibody) is limited. For this reason, identification of these other enzymes to the overall metabolism of an NCE is limited to a "mapping" level at best.

5.5 DRUG–DRUG INTERACTIONS (DDIs)

In this era of polypharmacy, drug–drug interaction (DDI) is a much publicized and highly researched area (Rodrigues, 2008). Well-known severe DDI such as those involving terfenadine, astemizole, cisapride, mibefradil, troglitazone, zomepirac, and benoxaprofen completely warranted the attention that this area has received over the recent years. "Potential risk of DDI" is a phrase that drug development teams highly prefer to stay away from and is a major area of risk assessment during discovery and development stages of any NCE. Depending on the nature of DDI, it can be classified as either pharmacodynamic or pharmacokinetic. A pharmacodynamic DDI (PD-DDI) is where the independent effects of the two drugs produce a change in the nature, magnitude, or duration of the effect expected from either drug alone. The drugs involved independently contribute to the pharmacology due to their separate actions on the same or different target receptor and modulate the overall pharmacodynamic manifestation in the body. PD-DDI can be synergistic (e.g., isoproteronol + epinephrine on β-adrenoreceptors, drugs that prolong QTc and exacerbate ventricular arrythmias such as risperidone + haloperidol) or antagonistic (NSAIDS inhibiting antihypertensive effects of ACE inhibitors, benzodiazepine + theophylline at the GABA receptor) (Oates, 2006). The most common DDIs are pharmacokinetic (PK) in nature where one drug alters the pharmacokinetics of a coadministered drug (i.e., either its absorption, distribution, metabolism, or excretion). Examples are numerous, such as warfarin + cimetidine, cyclosporine + rifampicin, atenolol + ranitidine, and alprazolam + itraconazole (Michalets, 1998). In a pharmacokinetic DDI, one drug may be conceptualized as the victim (i.e., the affected drug), with the other as the perpetrator (i.e., the causative agent). Ultimately, most PK-based DDIs lead to undesirable effects due to (1) modulation of pharmacologic action/PD of the victim drug, resulting in lack of efficacy, and (2) toxicity resulting from either the victim drug itself or formation of a toxic metabolite in undesirable amounts. For the scope of this chapter we will focus on the PK-based DDIs and common mechanistic tools used in the discovery stages of drug development to mitigate risks associate with such DDIs.

5.5.1 Inhibition and Induction

Most of the DDIs result from two basic, yet complex, phenomena when an NCE inhibits or induces the enzymes involved in the clearance or more correctly "disposition" of another coadministered drug (Pelkonen et al., 2008; Zhou, 2008). The DDI risk is more profound if the victim drug has high intrinsic clearance. So, development of an NCE that is either itself a modulator of drug disposition enzymes (DDE) or affected by modulation of the DDEs, especially if the NCE also possesses high Cl_{int}, is not favorable. Therefore, inhibition and induction of drug-metabolizing enzymes is routinely assessed very early on during drug discovery. With the much needed increase in research and knowledge in the transporter area in recent years, the role of transporters in disposition of NCEs has come to the forefront and discovery groups in the pharmaceutical industry are evaluating the role of transporters, along with drug-metabolizing enzymes in the early stages of development (Huang et al., 2008; Ayrton and Morgan, 2008; Balayssac et al., 2005; Chandra and Brouwer, 2004; DeBuske, 2005; Enders et al., 2006; Funk, 2008; Szakacs et al., 2008; Zhang et al., 2008; Zhou et al., 2008).

Inhibition of enzymes can be reversible or irreversible; and no matter what the mechanism, inhibition always results in reduction of intrinsic clearance of the pathway that is inhibited. Reversible inhibition, also known as direct inhibition, can be typically classified as competitive, uncompetitive, noncompetitive, or mixed, with competitive inhibition being the most commonly observed pathway of inhibition. Drug disposition literature is rich in such examples of inhibitors for most major drug disposition enzymes (You and Morris, 2007; Rodrigues, 2008)—for instance, inhibition of CYP1A2 (fluvoxamine), CYP2A6 (tryptamine), CYP2C8 (quercetin), CYP2C9 (sulfaphenazole), CYP2C19 (Omeprazole), CYP2D6(quinidine), CYP2E1 (4-methylpyrazole), CYP 3A4/5 (ketoconazole), UGT (valproic acid), SULT (2,6-dichloro-4-nitrophenol), NAT (methotrexate), PGP (verapamil), MRP (MK571), BCRP (Ko143), OATP1B1(gemfibrozil), and many more. Caution must be exercised during LI and LO stages since more often than not, NCEs act as competitive inhibitors of the enzymes that they are substrates of. So in several instances, NCE, which exhibits metabolic instability or is a transporter substrate, may also be a competitive inhibitor of the corresponding disposition enzymes. Two other types of inhibition are also observed, albeit to a much lower extent—for example, noncompetitive, mixed-type inhibition of CYP2C8 and 2C19 by bisphenol (Niwa et al., 2000), potent noncompetitive inhibition of CYP2D6 by hyperforin (Obach, 2000b), competitive as well as noncompetitive inhibition of UGT by ketoconazole (Takeda et al., 2006).

Inhibitory potential of NCE is routinely determined in early compound optimization stages by a simple, model-independent IC_{50} determination assay (Rodrigues, 2008). IC_{50} determination is usually performed at one substrate concentration ($S \leq K_m$), along with a wide range of inhibitor concentrations to monitor the decrease in reaction velocity with increasing inhibitor concentration. An NCE is classified as potentially high, medium, or low DDI-risk based on whether its IC_{50} value for a particular enzyme is <0.1 µM, 0.1–1 µM, or >1.0 µM, respectively. IC_{50} values should be compared to maximum predicted NCE plasma concentration (C_{max}) at steady state; and only if IC_{50} values are greater than C_{max}, a potential DDI-risk *in vivo* can be anticipated. Drawback from risk prediction based on IC_{50} values is that IC_{50} values are heavily dependent on (a) subcellular fraction used (pooled microsomes, cDNA expressing microsomes, heptocytes), (b) concentrations of the substrate, inhibitor, and enzyme, (c) incubation time, (d) solvent effect, (e) choice of marker substrate used especially for CYP3A4 and 2C9, which show cooperativity, (f) solubility limit of inhibitor and substrate, and (g) extent of protein binding of inhibitor. Failure to optimize several of these parameters also leads to failure of inhibition kinetics to comply with the MM kinetics due to substrate depletion, product inhibition, or allosteric binding at the enzyme active site. This in turn leads to over- or underprediction of the magnitude of the DDI. Sometimes, with all precautions exercised, inhibition kinetics still is complex due to promiscuous nature of the enzyme active site. Care must be used to best address these extrinsic parameters if possible during experimental design and recognize when inhibition kinetics deviate from typical MM kinetics and incorporate these limitations in predicting an *in vivo* DDI, based on *in vitro* data (Weaver, 2001; Yao and Levy, 2002).

A more rigorous assessment of inhibitor potency, usually performed for late-stage discovery compounds, is via determination of K_i, which requires thorough kinetic analysis of a specific biotransformation pathway of interest. Eadie–Hoftsee (or Lineweaver–Burke) plots are generated using multiple substrate and inhibitor concentrations, and the mechanism of inhibition (competitive, noncompetitive, uncompetitive or mixed) is assigned based on model fit of the data, defined by statistical criteria. The K_i value, which, unlike the IC_{50} value, is a more intrinsic number, is substrate-independent, and yields a superior ranking method when compared to IC_{50} determination. Evaluating an NCE's inhibitory potential involves assessment of its $[I]/K_i$ ratio, where [I] represents the mean C_{max} (peak plasma) of inhibitor at steady state, after highest proposed clinical dose and K_i is the dissociation constant of the enzyme–inhibitor complex. An inhibition is considered high, medium, or low risk based on whether its $[I]/K_i$ ratio is >1, 0.1–1, or <0.1, respectively.

In addition to the $[I]/K_i$ ratio of the inhibitor, the contribution of the inhibited pathway to the overall metabolism of an NCE also plays a crucial role in determining the magnitude of a DDI (discussed more in Section 5.4 and later in this section). For a reversible inhibition, which is the most commonly observed inhibition, the change in exposure of the victim drug in the presence of an inhibitor is given by the equation in Fig. 5.13 (Bachmann, 2006; Brown et al., 2006; Obach et al., 2006). In addition to the crucial $[I]/K_i$ ratio, this equation incorporates enzyme interaction at the intestinal site and parallel routes of metabolism. It can be clearly seen that parameters like $f_m \times f_{mCYP}$ and $[I]/K_i$ play a vital role in risk assessment due to potential DDI. F_m represents the fraction of total metabolism that is CYP-mediated while f_{mCYP} is the fraction of total CYP-mediated metabolism of the victim drug, which is due to the inhibited CYP isoform. In situation where an NCE can inhibit multiple CYPs with $[I]/K_i$ ratio >0.1 for each of the CYP-mediated pathways, then a rank order approach is recommended for conduct of clinical DDI (Obach et al., 2005). According to this approach, a clinical DDI study is initiated with the enzyme that is most potently inhibited *in vitro* by an NCE (highest $[I]/K_i$ ratio that is also >0.1) and a selective probe substrate of the enzyme. If the AUC changes are less than twofold, then no further DDI studies are necessary; but if the AUC changes are greater than twofold, then the next most potently inhibited CYP *in vitro* is evaluated in a clinical DDI study, and so on. This approach of predicting clinical DDI from *in vitro* data has retrospectively been proven successful for several drugs such as diltiazem, erythromycin, fluoxetine, and ketoconazole, but also has failed for drugs like cimetidine, troleandomycin, and fluvoxamine. PBPK models, based on parent metabolism and metabolite exposure changes, have also proven valuable for improving quantitative DDI predictions (Pang, 2009; Chien et al., 2006).

5.5.2 Mechanism-Based Inhibition (MBI)

In contrast to reversible inhibition, which does not require bioactivation of the NCE, most of the irreversible inhibition results in conversion of an intrinsically inactive compound into a reactive species. The reactive species binds either irreversibly (covalently bound) or "quasi-irreversibly" (noncovalent, tight intermediate at the catalytically active site) to the enzyme that mediates the bioactivation. This type of irreversible inhibition is often referred to as time-dependent, metabolism- or mechanism-based (MBI) or suicide inhibition. Although used interchangeably, there are some distinct features of each of these types of irreversible inhibition that make them very subtly different from each other. Strictly speaking, time-dependent inhibition (TDI) refers to decrease in rate of the enzyme-catalyzed reaction over time which may result from either (a) slow binding of inhibitor (rare) or (b) enzymatic or nonenzymatic formation of an inhibitory species (Riley et al., 2007). Proper controls must be also be used as well to ensure the viability of the enzymes; for example, CYP2E1 is degraded over time due to preincubation with NADPH.

Irreversibility of TDI has to be established before classifying them as MBI also. Catalytic activation of an NCE by an enzyme is a prerequisite for metabolism- and mechanism-based inactivation, and therefore the potency of both these types of inhibition will increase over time (with preincubation). In addition to being time-dependent, metabolism- and mechanism-based inactivation are also usually inhibitor concentration-dependent and NADPH-dependent. The main difference between metabolism- and mechanism-based inactivation, however, is that in mechanism-based inactivation, the reactive species that is formed, stays within the enzyme active site and immediately inactivates the enzyme. In contrast, in metabolism-based inactivation, the reactive electrophile can leave the enzyme active site and react with nucleophilic biomolecules other than the enzyme that generated the reactive intermediate. For this reason, metabolism-based inactivation potency can be decreased *in vitro*, by addition of external nucleophiles such as GSH, while this will not alter potency of a mechanism-based inactivator. TDI and MBI study design and detailed characteristic properties of these two kinds of inhibition (e.g., irreversibility, saturation kinetics, substrate protection, stoichiometric binding to enzyme) have been reviewed extensively (Hollenberg et al., 2008; Grimm et al., 2009). For the scope of this chapter, we will refer to both metabolism- and mechanism-based inactivation as mechanism-based inactivation (MBI).

Some excellent detailed reviews have been published in this area that emphasize the importance of risk assessment early on in the discovery of NCEs with potential structural liabilities that may be prone to bioactivation and subsequent inactivation of DMEs (Bertelsen et al., 2003; Fontana et al., 2005; Fowler and Zhang, 2008; Maurer et al., 2000; Polasek and Miners, 2007; Venkatakrishnan and Obach, 2007; Von Moltke et al., 2000; Voorman et al., 1998; Yang et al., 2007; Zhao, 2008; Zhou et al., 2004c; Kalgutkar et al., 2007). Acetylene-bearing compound 17-α-ethynylestradiol causes potent MBI of CYP3A4 via heme modification and of CYP 2B1 and 2B6 via covalent binding to the apoprotein due to reactive ketene intermediate (Kent et al., 2006). Tienilic acid is selectively oxidized by 2C9 to an *S*-oxide, which not only binds covalently to 2C9 via Michael addition with an active site amino acid residue, but is also able to escape the enzyme active site and react with microsomal proteins. This latter reaction has been attributed to the immune-mediated hepatotoxicity seen with tienilic acid (Neuberger and Williams, 1989). The hydrazine-containing compound phenelzine inactivates both MAO and CYP via alkylation by a carbon-centered radical intermediate (Kalgutkar et al., 2005). Benzyl isothiocyanate is a potent MBI of several CYP isoforms (1A1, 1A2, 2B1, 2E1) due to formation of an isocyanate intermediate that covalently binds to the amino acid residue at the active site (Kent et al., 2001). Paroxetine, which contains a methylenedioxy group, undergoes bioactivation to a carbene intermediate that results in potent MBI of CYP2D6 (Venkatakrishnan and Obach, 2005). Thiazole-containing NSAID sudoxicam is bioactivated to a thiourea reactive intermediate, which results in severe hepatotoxicity attributed to covalent modification of critical proteins (Obach et al., 2008b).

MBI of CYPs have been extensively investigated, and the presence of functional groups such as aniline, nitrobenzene, hydrazine, benzyl/propargyl/cyclopropyl amine, hydantoin, thioureas, thiazole, furan, thiophene, epoxides, methylene dioxy, methyl indoles, alkyne, isothiocyanate, and terminal alkenes on NCEs warrants immediate and early assessment of inactivation potential of the NCEs to avoid severe DDI liability in late-stage development (Fontana et al., 2005; Kalgutkar et al., 2007). It must be borne in mind that if an NCE does possess a structural alert, it does not automatically imply that it will be a potent inhibitor. Structure of NCE should be carefully studied early in the discovery stages, and its propensity toward metabolic activation should be thoroughly evaluated before classifying it as an MBI. The potency of an MBI should be assessed using the equation in Fig. 5.13 to predict fold change in exposure due to an MBI (Galetin et al., 2006; Obach et al., 2007). It must be recognized that the kinetic parameters (k_{inact}, K_I, k_{deg}) used to characterize an MBI are different from those used for reversible inhibitors (K_i). The parameter k_{inact} is a first-order rate constant for maximal rate of enzyme inactivation, K_I (unlike the reversible inhibitor-enzyme dissociation constant K_i, described above) represents the concentration of the inactivator, which causes half the maximal rate of inactivation, and k_{deg} is the first-order endogenous degradation rate of the enzyme of interest. Distinguishing MBI from simple reversible inhibitor is critical is predicting a clinical DDI, since applying a reversible inhibition model to an MBI may result in significant underprediction of a DDI risk. The major reason is that for reversible inhibition, the degree of inhibition is dependent on the $[I]/K_i$ ratio but in the case of MBI, in addition to this $[I]/K_I$ component, the ratio k_{inact}/k_{deg} is also a very crucial determinant. So, in the case of an MBI, it is possible to have a low $[I]/K_I$ ratio (which, in a reversible inhibition setting, would suggest low DDI risk), but if k_{inact}/k_{deg} ratio is high, this could result in a substantial DDI risk. This is very well demonstrated by correlation studies from Obach's group (Obach et al., 2007): CYP3A4 inactivator fluoxetine, CYP1A2 inactivator zileuton, and CYP 2D6 inactivator paroxetine all have very similar $[I]/K_I$ ratio (approximately 0.02), but only paroxetine, with a very high k_{inact}/k_{deg} ratio, was predicted to yield a significant clinical DDI (the magnitude of DDI predicted was similar to that seen in a clinic).

5.5.3 Nuclear Receptors and Induction

Another mechanism that gives rise to DDIs is when an NCE activates one of the many nuclear receptors and modulates the transcription and hence the expression of DMEs and transporters. Most common cases result in induction of drug disposition enzymes (Fuhr, 2000; Sinz et al., 2008; Xu et al., 2005), where levels of the enzymes are elevated and thus lead to either faster clearance of NCEs or enhanced formation of reactive metabolites. Faster clearance of NCE leads to lower-than-expected exposure, as demonstrated by significant decrease of exposures of several drugs such as midazolam, amitriptyline, cyclosporine, digoxin, indinavir, irinotecan, warfarin, phenprocoumon, alprazolam, dextrometorphane, simvastatin, and ethinylestradiol in the presence of St. John's wort, which has been shown to be a potent inducer of both CYP3A4 and Pgp (Izzo, 2004). Reduced exposure sometimes severely compromises therapeutic efficacy as seen with coadministration of cyclosporine (immunosuppresant and CYP3A4 substrate) and rifampin (potent CYP3A4 inducer commonly used to treat tuberculosis-TB) to organ transplant patients, who were also treated for TB (Chen and Raymond, 2006). CYP2E1 induction by ethanol is attributed to the increased formation of a reactive hepatotoxic quinoneimine metabolite from acetaminophen (Zhao et al., 2003). Induction of Phase I and Phase II DMEs and transporters is regulated by a large family of nuclear receptors, which are activated by a diverse array of xenobiotics as well as endogenous ligands. Some of the most commonly encountered nuclear receptors (also called transcription factors) are aryl hydrocarbon receptor (AhR), pregnane X receptor (PXR), constitutive androstane receptor (CAR), glucocorticoid receptor (GR), peroxisome proliferator-activated receptor-αβγ (PPAR-αβγ), farsenoid receptor (FXR), hepatocyte nuclear factor 4α (HNF-4α), vitamin D receptor (VDR), nuclear factor erythoroid 2 p45-related factor 2 (Nrf2), estrogen receptor (ER), and liver X receptor (LXR) (Matheny et al., 2004; Wang and LeCluyse, 2003). Some of these receptors, like the AhR and CAR, reside in the cytoplasm, while receptors like PXR predominently reside in the nucleus. Upon activation by ligands (via binding at the ligand binding domain except for CAR, which does not require binding of ligand), the ligand–receptor complex forms homodimers or heterodimers in the nucleus [for example, with retinoid X receptor (RXR) in the case of CAR and PXR, and with AhR nuclear translocator (Arnt) in the case of AhR], in the presence of other coactivators. The dimeric complexes then bind to response elements of target gene that yield in the transcription of the corresponding enzyme(s). Unlike enzyme inhibition, which is an immediate effect, induction is a graded effect and more often felt after multiple or chronic dosing of an NCE.

Several features of the mechanism of enzyme induction make data interpretation not so straightforward (Xu et al., 2005; Fuhr, 2000; Sinz et al., 2008; Fahmi et al., 2008; Lin, 2006; Luo et al., 2002, 2004). Just like the DMEs and the transporters, the nuclear receptors demonstrate promiscuity, in that they have broad substrate specificities (due to large substrate binding domain) (Ma and Lu, 2008). As an example, PXR can be activated by several ligands, like phenobarbital, rifampin, paclitaxel, nifedipine, mifepristone, 5β-pregnane-3-20-dione, dexamethasone, ritonavir, and hyperforin, which are of very different chemotypes. 1,7-Phenanthroline, which was viewed as a selective UGT inducer, was found to concurrently increase MRP3 and decrease CYP2C11 and OATP2 expression in rat liver (Wang et al., 2003). Also adding to the complexity is the fact that it is not uncommon to see a DME or transporter be under the regulatory control of multiple nuclear factors (Chen et al., 2005; Lim and Huang, 2008; Puga et al., 2009; Tirona et al., 2003). Following are some cases of regulation of drug disposition enzymes by multilple receptors:

CYP3A4 by PXR, CAR, GR, HNF-4α, FXR, and VDR; CYP2C9 by PXR, CAR, and GR; UGT 1A1 by PXR, CAR, AhR; SULT2A1 by PXR, VDR, and FXR; MRP2 by PXR, CAR, and FXR, OATP1B by HNF-1α and PXR. This also reflects receptor cross-talk or overlap of target specificity, which has been demonstrated in several recent findings. For example, cross-talk such as those of PXR and CAR, FXR, HNF-4α, GR, and LXR and that between CAR and VDR will result in cross-regulation of target genes (of DMEs and transporters), making prediction of DDI mechanism anything but straightforward. Due to regulation of several DMEs and transporters via common receptors, it is advised to monitor a few CYP isoforms that are indicative of activation of the principal nuclear receptors: CYPs 1A2, 2B6, and 3A4 induction are suggested to be indicative of activation of AhR, CAR, and PXR receptors, respectively (Hewitt et al., 2007c). In addition to normally observed agonist activity of xenobiotics on enzyme-expression pathways, there have also been cases where enzyme levels are down-regulated. Examples of negative modulation of enzyme expression have been demonstrated with colchicine, which down-regulates CYPs 2B6, 2C8, 2C9, and 3A4 (alters the GR-CAR/PXR pathway) (Dvorak et al., 2003), with ketoconazole and miconazole, which inhibit the GR-transcription pathway (Duret et al., 2006), with cytokines, which down-regulate expression of CYPs 1A2, 2C, 2E1, and 3A4 (Abdel-Razzak et al, 1993), and with rifampicin, which down-regulates SUL2A1 (Fang et al., 2007).

In vitro models to study enzyme induction have been established that use sandwich culture primary hepatocytes, liver slices, or reporter gene constructs of which the sandwich culture primary hepatocytes model is deemed as the gold standard by DMPK scientists (Hewitt et al., 2007a–c). Readers are advised to refer to these excellent reviews that discuss in detail *in vitro* methodologies and optimization of experimental conditions such as culture format, number of donors, culture medium, confluence number, culture period, and so on, for studying enzyme induction. While experimental details of *in vitro* induction studies is beyond the scope of this chapter, a couple of things are worth mentioning: Having a thorough knowledge of DMEs or transporters that are involved in the disposition of an NCE is an added advantage in designing the best *in vitro* induction model. This is highlighted in a few cases described next (Hewitt et al., 2007c): UGT1A1 induction results obtained for chrysin in primary human hepatocytes and HepG2 cells. While in HepG2 cells, there was a significant induction of UGT1A1, chrysin failed to induce this enzyme in primary human hepatocytes. The reason for this was that due to the presence of high levels of UGTs in primary hepatocytes, chrysin was rapidly glucuronidated but in HepG2 cells, which lack intracellular UGT, high concentration of parent compound chrysin resulted in high enzyme induction. HepG2 cells also have low expression of PXR and CAR, so evaluation of enzyme induction caused by activation of these receptors in HepG2 cell line will result in inaccurate induction-potential prediction. Similarly, Fa2N-4 cells, which are also used to study enzyme induction, have been shown to lack CAR expression, along with uptake transporters OATP1B1 and OATP1B3. This could be a severe limitation in using this cell line for predicting CAR-mediated enzyme inducers or for NCEs that depend on the above-mentioned uptake transporters to get into these cells to activate the induction mechanism. Endpoint measurement in induction studies is usually quantifying the activity or the mRNA levels of the target enzyme(s). However, in certain cases of concurrent inhibition and induction, as demonstrated by troleandomycin and ritonavir (Luo et al., 2002), there may be no change in enzyme activity, and measurement of mRNA levels will be an accurate indicator of the degree of induction. Also to note is that induction can also be tissue-specific as exemplified in a recent study with Wistar rats treated with phenobarbital/β-napthaflavone (PB/NF) or cyclohexanol/albendazole (CH/ABZ). Microsomal fractions were then isolated from esophagus, stomach, duodenum, colon, and liver. Induction of CYP1A1/2 and CYP3A2 isoenzymes was observed in the esophageal, duodenal, and colonic microsomes in rats treated with PB/NF and in hepatic and duodenum microsomes of rats treated with CH/ABZ (Hernandez-Martinez et al., 2007). Benzo(*a*)pyrene-mediated lung toxicity has also been proposed to be due to lung-specific induction of CYP1A1 and 1B1 by the carcinogen (Harrigan et al., 2006). Grapefruit juice has been shown to selectively inhibit CYP3A4 in intestines and not in the liver, when administered at its regular strength (Veronese et al., 2003).

Frequently in early discovery stages, a simple rank-order approach based on the C_{max}/EC_{50} ratio (C_{max} is the maximum plasma concentration of inducer after administration of the highest anticipated clinical dose, and EC_{50} is the concentration of inducer where 50% of maximal induction is observed). A high, medium, or low risk is associated to whether the C_{max}/EC_{50} ratios are >1, 0.1–1, or <0.1, respectively. A more reliable method, called the relative induction score (RIS) method, is also used to better predict clinical DDI based on *in vitro* induction data (Fig. 5.14) (Hewitt et al., 2007c). RIS takes into account the maximum induction response in addition to C_{max} and EC_{50}. However, it was argued that even if two NCEs have similar EC_{50} and E_{max}, the concentration at which induction response is observed may be different, which might affect the DDI predictability. Therefore, the RIS method has also been modified to incorporate an additional NOAEL term (Fig. 5.14), which is the highest concentration at which no induction is observed for an NCE.

$$\frac{E_{max} \cdot C_{max}}{EC_{50} + C_{max}} = \text{Effect}$$

$$\frac{E_{max} \cdot C_{max} \cdot f_{u,plasma}}{EC_{50} + C_{max} \cdot f_{u,plasma}} \times \frac{\text{NOEL}}{C_{max} \cdot f_{u,plasma}} = \text{Effect}$$

E_{max}= maximum induction response
EC_{50}= inducer concentration that results in half maximal response
C_{max}= maximum unbound plasma concentration of inducer
$f_{u,plasma}$= fraction unbound of inducer in plama
NOAEL= No adverse effect level

Figure 5.14. Relative induction score calculation (Hewitt et al., 2007c).

5.5.4 Interplay Between Drug-Metabolizing Enzymes and Transporters

DDI is prevalent in all major therapeutic areas. There are numerous examples in the literature where a DDI may be the result of modulation of a drug-metabolizing enzyme, as exemplified by the DDI between terfenadine–ketoconazole, fluvoxamine–astemizole, cisapride–erythromycin, statins–protease inhibitors, felodipine–nifedipine, codeine–quinidine, β-blockers–ritonavir, and rifampin–cyclosporin/simvastatin/verapamil/midazolam (Levy et al., 2000). Sometimes it may be due to modulation of transporters such as those exemplified by digoxin–quinidine, probenecid–acyclovir/allopurinol/cephalosporins/ciprofloxacin, and talinolol/verapmil (Marchetti et al., 2007a). More often than not, DDIs result from a intricate interplay of modulation of both DMEs and transporters (Pal and Mitra, 2006). Gemfibrozil (lipid-lowering drug) causes six- to eightfold increase of exposures of coadministered drugs like cerivastatin (lipid-lowering drug) and repaglinide (antidiabetic) due to inhibition of not only uptake inhibitor OATP1B1, but also CYP2C8 (Ayrton and Morgan 2008). CYP3A4 and Pgp, which have substantial overlap in substrates and modulators, are widely believed to be coregulated and play a concerted role in drug disposition. Interestingly, it was recently demonstrated in mice that Pgp and CYP3A4 induction is tissue- and inducer-specific and that there is a disconnect between induction of these two enzymes (Matheny et al., 2004) highlighting the complex nature of the interplay of DMEs and transporters. Increase in plasma and CSF concentration of HIV protease inhibitors ritonavir and squinavir, due to coadministration with ketoconazole, has been attributed to potent inhibition of both CYP3A4 and Pgp by ketoconazole (Khaliq et al., 2000). Transporters can modulate the extent of metabolism of an NCE by differential exposure of an NCE to the DME (Lam and Benet, 2004; Lau et al., 2004). This is well demonstrated by the fact that when digoxin (CYP3A4/Pgp substrate) was coadministered with Pgp inhibitor quinidine, digoxin metabolism was enhanced. In contrast, when coadministered with uptake transporter Oatp2-inhibitor rifampicin, metabolism of digoxin was diminished. Concurrent inhibition and induction by drugs can further decrease predictability of DDI from *in vitro* assays if not properly modeled or accounted for (McConn and Zhao, 2004). Coadministration of alprazolam and ritonavir leads to initial inhibition of alprazolam clearance (thereby increasing alprazolam-mediated toxicity) due to CYP3A4 inhibition by ritonavir. However, upon extended coadministration of these two drugs, clearance of alprazolam increased significantly due to induction of CYP3A4 by ritonavir (Greenblatt et al., 2000). St. John's wort has several constituents that are both potent competitive inhibitors and inducers of CYP3A4 (Mannel, 2004). There are extensive data available in the field of quantitative DDI prediction with CYP; but unfortunately, in the area of other DMEs (FMO, UGT, NAT, SULT) or transporters (Pgp, BCRP, MRP, OAT, OATP), DDI predictions are still qualitative at best. The primary reason for this is the lack of DME- or transporter-isoform-specific substrates and inhibitors as well as lack of proper scaling factors to normalize the *in vitro* results with respect to their *in vivo* relevance. Also, in the majority of the *in vitro* DDI studies, the focus is on evaluation of one DME or one transporter only (again due to lack of specific substrates and inhibitors). This does not capture the complex *in vivo* interplay of DMEs and transporters or of multiple DMEs and multiple transporters with each other (You and Morris, 2007). For orally administered drugs, it has been seen that the modulation of DMEs and transporters in intestines and liver are not parallel; for example, even if induction does occur in both intestines and liver together, the levels of DMEs and transporters in each of these sites will not be correlated (Mouly et al., 2002; Zhou et al., 2004a).

5.5.5 Limitations of *In Vitro* DDI Studies

In the discovery stages, DMPK teams rely heavily on *in vitro* tools, and preclinical animal model data for DDI predictions and methodologies and strategies to improve DDI predictions have been extensively published in the literature (Venkatakrishnan et al., 2003; Ito et al., 1998b; Lau et al., 2006; Lu et al., 2006b; Ohno et al, 2007; Sahi et al., 2006; Shou, 2005; Weinkers and Heath, 2005; Youdim et al., 2008). There is a vast amount of knowledge and research in the DDI-prediction area, and examples of successful prediction of occurrence or the magnitude of clinical DDI (Obach et al, 2006; Bachmann and Lewis, 2005; Lu et al., 2008a,b) have been reported. However, more often than not, clinical DDI prediction can be extremely complicated due to a multitude of reasons, some of which are listed as follows (several also discussed in detail earlier in Section 5.2.4). Inhibitors and inducers of DMEs and transporters will be referred to as "modulators" in general:

1. Failure to incorporate high plasma protein binding of NCE (inhibitor) will result in inaccurate assessment of the active concentration of modulators at the pharmacologic-enzyme site.
2. Failure to account for nonspecific binding (NSB) (Grime and Riley, 2006) of modulators to subcellular fractions such as microsomes lead to underestimation of DDI potential. In the case where the NSB is linear—that is, when K_m of a clearance pathway is $<K_d$ (dissociation constant of the bound complex)—a simple correction factor can be used to account for this binding. However, in the case where $K_m \geq K_d$, the enzyme kinetics exhibit sigmoidicity, which can be misinterpreted as auto-inactivation.
3. Additional relevant *in vivo* clearance pathways (both hepatic and extrahepatic) are not captured in *in vitro* experiments (e.g., not accounting for Phase II conjugation reactions when predciting DDIs from microsomes).
4. Shift in major metabolic pathway (and hence the pathway is inhibited) *in vivo* as compared to *in vitro* due to supraphysiological concentrations of modulators in *in vitro* assays.

5. Formation of metabolites that could themselves give rise to potent DDIs, as in the cases of itraconazole (Isoherranen et al. (2004) and fluoxetine (Otton et al., 1993), may not be investigated in an *in vitro* DDI experimental setup.
6. On the contrary to the point just mentioned, formation of inhibitory metabolites (or reactive species) *in vitro*, which would be normally eliminated *in vivo* and hence not pose a DDI risk *in vivo*.
7. Active uptake or efflux in liver *in vivo* (Hallifax and Houston, 2006; Lu et al., 2006; Parker and Houston, 2008; Soars et al., 2007), resulting in higher- or lower-than-predicted concentration of modulators inside the hepatocytes.
8. High partitioning of modulators into liver (Hallifax and Houston, 2007). *In vitro* studies assume instant equilibrium between free drug fraction in plasma and the tissue (enzyme-active site in hepatocytes); but this is not the case in several cases, especially highly basic drugs such as intraconazole, quinine, fluvoxamine, fluoxetine, amitryptiline, diazepam, triazolam, and dextropmethorphan. This could lead to understimation of the DDI magnitude.
9. Atypical kinetics (Isin and Guengrich, 2006; Kramer and Tracy, 2008; Tracy, 2006; Noe et al., 2007; Tracy and Hummel, 2004; Williams et al., 2002). Presence of allosteric multiple binding sites in DMEs like CYP 3A4, 2C8, and 2C9, UGT2B7, SULT1A1, Pgp, OATP1B1, and MRP2 will result in atypical kinetics (Ma and Lu, 2008). The enzyme kinetic parameters generated in these cases (normally assuming typical Michaelis–Menten kinetics) are not accurate, and hence DDI predictions made based on them also do not reflect the *in vivo* situation. To elucidate this, using estrone-3-sulfate (E3S) as a OATP1B1 substrate at its typical concentration, to predict gemfibrozil and statin DDI, would be misleading since it shows substrate-concentration-dependent DDI (Noe et al., 2007). When predicting inhibitory potency of an NCE toward CYP3A4, it is highly advisable to use a couple of different substrates with different chemical structures since CYP3A4 is notorious to exhibit substrate-dependent inhibition due to atypical kinetics.
10. Intestinal efflux, especially for low permeability and low therapeutic index compounds (Murakami and Takano, 2008; Takano et al., 2006).
11. Failure to properly design *in vitro* enzyme inhibition kinetics experiments. This is demonstrated in the case of SKF-525A, which demonstrated shift in inhibition kinetics from reversible competitive inhibition to irreversible MI-complex formation due to preincubation (Franklin and Hathaway, 2008).
12. In the case of MBI, inaccurate determination of k_{deg} value of the modulated enzyme (more details with the MBI inhibition description are discussed above).

13. Effect of f_u in the intestine. Due to the possible presence of mucosal diffusional barrier *in vivo*, unbound modulator concentration in intestine could be higher or lower than unbound portal vein concentration, depending on whether measured at the absorption or postabsorption phase respectively (Thummel et al., 2008).
14. Assumption that the key kinetic parameters such as K_i, k_{cat}, K_I, k_{inact}, k_{deg}, EC_{50}, and E_{max}, which are crucial in DDI assessment, are similar *in vivo* to those determined *in vitro*.
15. In the case of CYP-mediated DDI, accurate determination of f_m (fraction of dose cleared by all CYP) × f_{mCYP} (contribution of each CYP isoform to the total CYP-mediated metabolism) is crucial (discussed in detail in Section 5.4). Knowledge of $f_m \times f_{mCYP}$ is very informative in predicting DDI, since this number also depicts the fraction of total clearance of an NCE that will be affected when the drug disposition enzymes involved in that pathway are modulated (Zhang et al., 2007b). Studies have shown that magnitude of DDI will be greatly diminished if $f_m \times f_{mCYP}$ is <0.5. Unfortunately, *quantitative* reaction phenotyping is not possible for a lot of major DMEs and transporters, due to unavailability of specific substrates and inhibitors for non-CYP enzymes and transporters. This may lead to inaccurate prediction of the magnitude of DDI *in vivo*, from *in vitro* studies, involving non-CYP enzymes and transporters.
16. Route of administration and extraction ratio of the victim drug. For example, in general, modulation of clearance of a victim drug has a more profound effect when the victim drug has a high extraction ratio and is administered via oral route than a low extraction ratio drug, administered via the intravenous route (Bachmann, 2006).
17. When NCE demonstrates concentration-dependent plasma protein binding (PPB) and the unbound concentration (C_u) is significantly higher than protein binding dissociation constant (K_d), then extraction ratio (E_h) will be equal to the unbound fraction of drug and will be independent of Cl_{int}, in contrast to when PPB is linear. This will also have a marked effect in DDI prediction, especially of NCEs with high Cl_{int} (Huang and Oie, 1984).
18. Similar chemical structure but different affinity for disposition enzymes could potentially result in unpredictable DDI for compounds with similar chemical structures. The angiotensin II receptor antagonist telmisartan is shown to be an OATP1B3 substrate (Ishiguro et al., 2006); but valsartan, another drug in the same therapeutic class and with very similar chemical structure, has major contribution by OATP1B1 (Yamashiro et al., 2006) (in addition to OATP1B3) in its uptake. In spite of very close structures of paclitaxel and docetaxel, the enzymes mediating their major route of metabolism are different: CYP2C8 hydroxylates the taxane ring of paclitaxel (Dai et al., 2001), while CYP3A4 catalyzes the oxidation of a *t*-butyl side chain of docetaxel (Cresteil et al., 2002).

$$C_{u,i} \leqslant C_{max} + \frac{k_a \cdot D \cdot F_a}{Q_h}$$

$C_{u,i}$ = Unbound concentration of inhibitior at the inlet of liver after PO administration
C_{max} = Maximum plasma concentration
k_a =absorption rate constant
D = Dose
F_a = fraction absorbed from GI tract into portal vein
Q_h = hepatic blood flow

Figure 5.15. Estimation of inhibitor concentration at the inlet of liver after oral administration (Brown et al., 2005).

$$\frac{AUC_{PO,\ inhibitor}}{AUC_{PO,\ control}} = \frac{Q_{gm} + f_u\ Cl_{int,\ gm}}{\left(Q_{gm} + \frac{f_u \cdot Cl_{int,\ gm}}{[1 + (I_{gm}/K_i)]}\right)} \cdot \left(1 + \frac{I_u}{K_i}\right)$$

I_{gm} = unbound inhibitor concentration in intestine
I_u = unbound inhibitor concentration in liver
Q_{gm} = blood flow to GI mucosa
f_u = fractionun bound of inhibitor in plama
K_i = inhibition constant

Figure 5.16. Exposure change due to inhibition of intestinal metabolism (for orally administered drug in which case this pathway is relevant) (Galetin, 2007).

19. Quantitative models for change in exposure due to modulation of enzymes (and their validation with clinical data) are available for CYPs only; but when disposition of NCE involves non-CYP enzymes and transporters, such quantitative predictions may not be accurate.
20. Uncertainty in quantitative effect of inhibitors toward intestinal clearance (first pass). In most discovery projects, it is common to use the highest predicted unbound inhibitor concentration in systemic circulation to calculate the $[I]/K_i$ ratio described earlier. Although this might be satisfactory in predicting inhibition of hepatic disposition enzymes, for drugs given via the IV route, this would underpredict the magnitude of a DDI of an NCE that is orally administered. The reason for this is that after oral administration, the concentration of inhibitor at hepatic inlet would most likely be much higher than that in systemic circulation, and a fraction of inhibitor absorbed from intestines should also be accounted for (Fig. 5.15) (Brown et al., 2005). Also, due to the significantly higher concentration of drug at the enterocytes than at the hepatocytes, saturation of intestinal DMEs are more likely to be observed compared to saturation of hepatic DMEs (Thummel et al., 2008). Intestinal DME saturation would result in a reduced DDI magnitude than what is predicted by *in vitro* studies. An integrated equation to reflect the contribution of intestinal metabolism to systemic exposure of an orally administered drug is given by equation shown in Fig. 5.16 (Galetin, 2007). Incorporation of intestinal metabolism in prediction of *in vivo* DDI risk is important especially for CYP3A4-metabolized NCEs (Obach et al, 2006).

Additional considerations *in vivo* that are not modeled in *in vitro* DDI assays are as follows:

1. Effect of genetic polymorphism leading to inter-individual variability in exposure (and drug response) *in vivo*, which is not predicted by *in vitro* assays: Polymorphisms (Ioannides, 2002; Rodrigues and Rushmore, 2002; Satoh, 2007; Urquhart et al., 2007) of CYPs 2D6, 2C9, and 2C19 have been extensively demonstrated in clinical studies that have affected exposures of drugs like celecoxib, *S*-warfarin, tolbutamide (CYP2C9 polymorphism), omeprazole, lansoprazole (CYP2C19 polymorphism), haloperidol, propafenone, encanide, and indolealkylamine (CYP2D6 polymorphism). Polymorphism has also been described for other enzymes (Ioannides, 2002) such as FMOs, NATs, GST, UGT1A1, thiopurine-*S*-methyl transferase (TPMT), and transporters (You and Morris, 2007) such as Pgp, ABCC/MRP, ABCG2/BCRP, OCT, OATP, PepT, and NT (nucleoside transporters CNT and ENT); and if polymorphic enzymes are identified to be involved in the clearance of an NCE, it must be taken into consideration with the *in vitro* data for a DDI risk assessment.
2. Species and gender difference: In the literature, there have been numerous examples of species difference in terms of metabolism and transport of drugs that have been attributed to differential regulation of drug-disposition enzymes in different species (a few have been mentioned above in Section 5.2.2). During design of a well-predictive *in vitro* DDI study, it is very important to take these factors into consideration to get the most accurate information from an *in vitro* study. When using data from preclinical animal models in complement with *in vitro* data, species difference in expression and activity of the DDEs should be very carefully evaluated before efforts to predict human DDI from the combined *in vitro* and *in vivo* data are undertaken (Marathe and Rodrigues, 2006). Coadministration of probenecid with famotidine leads to decrease of renal clearance of famotidine in humans but not in rats. It is proposed that human OAT3 is predominantly involved in the uptake of famotidine, while rat OCTs are the major uptake transporter for this drug. Hence coadministration of probenecid, an inhibitor of OAT3, has substantial effect in human renal clearance of famotidine, but not in rats (Tahara et al., 2005, 2006). Species difference in MAO has been demonstrated *in vitro* recently (Ramadan et al., 2007). The Gastric acid suppressor omeprazole is a CYP1A2 inducer in humans but not in rodents (Hewitt et al., 2007b). The barbiturate phenobarbital predominantly induces CYP3A4 in human but CYP 2B in rats (Hewittet et al., 2007b). Species and tissue difference in expression and activity of esterases have been proposed to result in species and tissue differences in the hydrolysis rate of glycovir. In addition, due to gender differences in disposition of drugs (discussed in Section 5.2.2), it is not unexpected to see a gender difference in the magnitude of a DDI, especially in the case of an NCE with narrow therapeutic index.

3. There has been a significant recent growth in the area of use of genetically modified mouse models (Cheung and Gonzalez, 2008; Lin, 2008; Muruganandan and Sinal, 2008) to assess DDI potential and toxicity prediction due to IDRs (discussed in an earlier section). Double transgenic mouse [expressing human CYP3A4/PXR (Ma et al., 2008), CYPs 3A4/2D6 (Felmlee et al., 2008), and OATP1B1 (van de Steeg et al., 2009)], knockout mouse (specific drug disposition enzyme- and transporter-null mouse, SOD2 null mouse (Boelsterli and Hsiao, 2008)], humanized mouse (Gonzalez, 2007; Ma et al., 2007) [mouse expressing human CYPs PXR and PPAR (Gonzalez and Yu, 2006)], and chimeric mouse with humanized liver (Katoh et al., 2006, 2008) have truly revolutionized ways to investigate the mechanism of human-specific biotransformation and toxicity as shown in the case of several drugs. However, quantititative risk prediction of DDI and toxicity in humans from these models is yet to be thoroughly established.

It can be summarized that due to species (and possible gender) difference in expression level, functional activity, and tissue distribution of drug disposition enzymes, extrapolation of data from *in vitro* studies and from preclinical species to predict DDI in humans is very challenging. To compound the difficulty of DDI prediction, a thorough knowledge of known DDIs at the prescriber's level is absent, which further contributes to DDI occurrences in the clinic (Ko et al., 2008).

5.5.6 Herb/Food–Drug Interactions

About 20% of the population use herbal products for either (a) minor symptoms such as the common cold or (b) more serious health conditions such as complementary and alternative medicine (CAM) in combination with conventional chemotherapeutic and HIV treatments (Bent, 2008; Bent and Ko, 2004; Marchetti et al., 2007b; Meijerman et al., 2006). Herbs contain a combination of several chemicals such as flavonoids, steroids, sterols, fatty acids, alkaloids, saponins, tannins, terpenes, and glycosides. Some of the popular herbs are St. John's wort (SJW) (depression), garlic (cholesterol lowering), echinacea (common cold), ginko biloba (dementia), ginger (nausea), ginseng (improve energy and physical/cognitive performance), kava kava (anxiety), vitamin E (antioxidant), quercetin (anemia), β-carotene (antioxidant), milk thistle (helps liver function), and chamomile (sleep disorder/anxiety). SJW is a concurrent inhibitor and inducer of CYP3A4 and Pgp (with induction being the predominant effect in chronic administration) and is a potent inhibitor of other CYP 1A2, 2C9, and 2C19. A multitude of DDIs have been reported with SJW (Zhou et al., 2004b) and several drugs like cyclosporin, digoxin, indinavir, paroxetine, theophylline, warfarin, nefadozone, digoxin, oral contraceptives, irinotecan, and sertraline. Alteration of DMEs and transporters (inhibition of disposition proteins or transcriptional activation of several nuclear receptors such as PXR, CAR, VDR) have been proposed to be the result of lower plasma levels of several anticancer drugs (Meijerman et al., 2006). Herb-drug DDI have been well exemplified in numerous *in vitro* as well as in clinical circumstances (Bent, 2008; Bent and Ko, 2004; Marchetti et al., 2007b; Meijerman et al., 2006; Zhou et al., 2004b): Pgp inhibition by curcumin, green tea extracts, and quercetin; inhibition of BCRP by genistein; inhibition of CYP 2C9, 2C19, 2E1, 3A4/5/7 by garlic; induction of CYPs 1A2, 2B, 3A, and UGT1A1 by flavonoids like chrysin; induction of CYP2B by ginko biloba; induction of hepatic CYP3A4; inhibition of intestinal CYP3A4 by echinacea; and many more. There have also been several clinically significant food–drug DDIs (Wang and Morris, 2007) reported for isoniazid, warfarin, MAO inhibitors, procarbazine, and linezolid. Grapefruit-juice-mediated clinical DDIs have also been well established (Bailey et al., 1998) for the statins, nifedipine, carpamazepine, verapamil, cylcosporin, tacrolimus, midazolam, amiodarone, HIV protease inhibitors, buspirone, and more. The underlying reason is due to inhibition of CYP3A4 (especially intestinal CYP3A4), both via reversible and suicide inactivation mechanisms by furanocoumarins, bergamottin, 6′,7′-dihydroxybergamottin, and furanocoumarin dimers.

5.5.7 Beneficial DDIs

DDIs should not be viewed as solely undesirable, because there have been several cases where the PK of one drug has been modulated by another via a well-planned design to improve the exposure (and hence the efficacy) of the affected drug. Kaletra is a coformulation of lopinavir and ritonavir whereby ritonavir–mediated CYP3A4 inhibition results in higher plasma levels of lopinavir and boosts its anti-HIV protease activity (Cvetkovic and Goa, 2003). Sinamet is a mixture of L-dopa and carbidopa, where carbidopa inhibits decarboxylation of L-dopa outside the CNS and hence significantly increases its efficacy in Parkinson's disease treatments (Yeh et al., 1989). Ketoconazole (KTZ), a potent CYP3A4 inhibitor, is commonly used in combination with cyclosporin A (CsA) to enhance the immunosuppressive properties of the latter as a result of increase in CsA exposure due to KTZ-mediated inhibition of CsA metabolism (Gerntholtz et al., 2004). Primaxin is a combination of antimicrobial drug imipenem and cilastatin (Barza, 1984), where cilastatin inhibits renal dipeptidase and thus prevents loss of imipenem efficacy due to inactivation by this enzyme.

5.5.8 Protein Binding Mediated DDIs

The clinical consequence of protein binding has been debatable over the past several decades (Benet and Hoener, 2002; Christensen et al., 2006; DeVane, 2002; MacKichan, 1984; Mahmood, 2000; McElnay and D'Arcy, 1983; Sellers, 1979; Wilkinson, 1983). If we apply the basic PK principles, protein binding displacement can result in the need to adjust the dosing regimen of an NCE in cases where the unbound concentration of the NCE changes. Theoretically, this may happen with high extraction drugs (also with small V_{dss})

only with narrow therapeutic index (NTI) when (a) they are administered IV, (b) they have a very short PK–PD equilibration time (change in free drug concentration immediately affects PD), and (c) they are administered PO but with nonhepatic route of clearance. No drugs have been found to meet the latter criterion to date. It is worthwhile to mention here that although there have been several reports of clinically relevant DDIs due to protein binding displacement, especially those with warfarin, tolbutamide, and phenytoin (all with NTI) in the past, recent evaluation of these DDIs revealed that inhibition of metabolism of the victim drugs by the corresponding perpetrator drugs, rather than protein binding displacement, is the culprit in these cases (MacKichan, 1989; Rolan, 1994). Nonetheless, when developing NCEs that possess NTI, such as those used in critical treatments such as antiarrhythmia, and anesthetic/pain medication, anticoagulants, anticonvulsants, and antidepressants, it is advisable to evaluate possible protein binding displacements that could lead to DDI. During assessment of DDI due to protein-binding displacement, factors such as species (in which DDI is assessed since species difference in plasma protein binding is well known), gender (women have about 10% lower albumin than men), and disease state (during inflammation, AAG concentration is about fivefold higher, whcih will affect protein binding of basic drugs that are highly bound to AAG) should always be factored in.

5.6 CONCLUSION

Drug discovery is an extremely challenging process, where success is heavily dependent upon very thorough understanding of chemistry and mechanism of biotransformation of NCEs in humans. Due to unavailability of data in humans in early drug discovery stages, DMPK scientists rely heavily on *in vitro* ADME tools, in complement with *in vivo* preclinical animal models, to predict ADME/PK of NCE in humans. In recent times, the availability of various sophisticated *in vitro* ADME tools and *in vivo* preclinical models, has remarkably reduced poor PK-related, late-stage compound attrition. However, it must be realized that *in vitro* assays all have their limitations for example, NCE clearance and its DDI properties *in vitro* are highly dependent on extrinsic parameters (substrate concentration, buffer, incubation time, pH, protein concentration, nature of subcellular fraction, inhibitor concentration, etc., discussed above). Results obtained in preclinical animal models should also be very carefully evaluated because there is significant species and gender difference in biotransformation pathways of NCEs and NCE-mediated inhibition and induction of drug disposition proteins, DDPs (due to differential expression and activity of the DDPs, differential plasma protein binding discussed above), even when the animal model appears to be physiologically relevant to humans. Extreme care must be taken to design *in vitro* and *in vivo* experiments and interpret data obtained from a combination of several of these studies to accurately predict whether an NCE will be a successful human drug candidate. Sometimes, even with the best-designed experiments and prediction tools, extrapolation of animal data to humans may fail due to highly species-specific interplay of DDPs and their effect on the NCE disposition under consideration.

ACRONYMS

AUC	Area under curve
Cl_{hep}	Hepatic clearance
Cl_{int}	Intrinsic clearance
DDE	Drug disposition enzyme
DDI	Drug–drug interactions
DDP	Drug disposition protein
DME	Drug metabolizing enzyme
DMPK	Drug metabolism and pharmacokinetics
GI	Gastrointestinal
GSH	Glutathione
HLM	Human liver microsomes
IDR	Idiosyncratic drug reaction
IP	Intraperitoneal
IV	Intravenous
IVIVC	*In vitro–In vivo* correlation
LC	Liquid chromatography
LI	Lead identification
LO	Lead optimization
MRT	Mean residence time
MS	Mass spectrometry
NCE	New chemical entity
NMR	Nuclear magnetic resonance
NSAID	Nonsteroidal anti-inflammatory drug
OBA	Oral bioavailability
PK	Pharmacokinetics
PD	Pharmacodynamics
PO	Oral
Q_h	Hepatic blood flow
SC	Subcutaneous

REFERENCES

Abdel-Rahman S, Kauffman R. The integration of pharmacokinetics and pharmacodynamics: understanding dose–response. *Annu Rev Pharmacol Toxicol* 2004;44: 111–136.

Abdel-Razzak Z, et al. Cytokines down-regulate expression of major cytochrome P-450 enzymes in adult human hepatocytes in primary culture. *Mol Pharmacol* 1993;44:707–715.

Abelo A, et al. Stereoselective metabolism of omeprazole by human cytochrome P450 enzymes. *Drug Metab Dispos* 2000;28(8):966–972.

Aming Y, Haining RL. *Drug Metab Dispos* 2001;29(11):1514–1520.

Anders M, Dekant W, Vamvakas S. Glutathione-dependent toxicity. *Xenobiotica* 1992;22(9/10):1135–1145.

Anderson R, Kudlacek P, Clemens D. Sulfation of minoxidil by multiple human cytosolic sulfotransferases. *Chem Biol Interact* 1998;109(1–3):53–67.

Andersson T, et al. Diazepam metabolism by human liver microsomes is mediated by both *S*-mephenytoin hydroxylase and CYP3A isoforms. *Br J Clin Pharmacol* 1994;38(2):131–137.

Andrews E, et al. Synthesis and pharmacological profile of two novel heterocyclic chromanols, CP-80,798 and CP-85,958, as potent LTD4 receptor antagonists. *Bioorg Med Chem Lett* 1995;5(13):1365–1370.

Argoti D, et al. Cyanide trapping of iminium ion reactive intermediates followed by detection and structure identification using liquid chromatography-tandem mass spectrometry (LC-MS/MS). *Chem Res Toxicol* 2005;18:1537–1544.

Arnold R, Slack J, Straubinger R. Quantification of Doxorubicin and metabolites in rat plasma and small volume tissue samples by liquid chromatography/electrospray tandem mass spectroscopy. *J Chromatogr B Anal Technol Biomed Life Sci* 2004;808(2):141–152.

Austin N, et al. Pharmacokinetics of the novel, high-affinity and selective dopamine D3 receptor antagonist SB-277011 in rat, dog and monkey: *in vitro*/*in vivo* correlation and the role of aldehyde oxidase. *Xenobiotica* 2001;31(8–9):677–686.

Ayrton A, Morgan P. Role of transport proteins in drug discovery and development: a pharmaceutical perspective. *Xenobiotica* 2008;38(7/8):676–708.

Bachmann K. Inhibition constants, inhibitor concentrations and the prediction of inhibitory drug drug interactions: pitfalls, progress and promise. *Curr Drug Metab* 2006;7:1–14.

Bachmann K, Lewis J. Predicting inhibitory drug–drug interactions and evaluating drug interaction reports using inhibition constants. *Ann Pharmacotherapy* 2005;39: 1064–1072.

Bailey M, Dickinson R. Acyl glucuronide reactivity in perspective: biological consequences. *Chem Biol Interact* 2003;145(2):117–137.

Bailey D, et al. Grapefruit juice–drug interactions. *Br J Clin Pharmacol* 1998; 46(46):101–110.

Baillie T, Metabolism and toxicity of drugs. Two decades of progress in industrial drug metabolism. *Chem Res Toxicol* 2008;21(1):129–137.

Baillie T, et al. Drug metabolites in safety testing. *Toxicol Appl Pharmacol* 2002;182: 188–196.

Balani S, et al. Strategy of utilizing *in vitro* and *in vivo* ADME tools for lead optimization and drug candidate selection. *Curr Topics Med Chem* 2005;5:1033–1038.

Balayssac D, et al. Does inhibition of P-glycoprotein lead to drug–drug interactions? *Toxicol Lett* 2005;156:319–329.

Baranczewski P, et al. Introduction to early *in vitro* identification of metabolites of new chemical entities in drug discovery and development. *Pharmacol Rep* 2006a; 58:341–352.

Baranczewski P, et al. Introduction to *in vitro* estimation of metabolic stability and drug interactions of new chemical entities in drug discovery and development. *Pharmacol Rep* 2006b;58:453–472.

Barter Z, et al. Scaling factors for the extrapolation of *in vivo* metabolic drug clearance from *in vitro* data: reaching a consensus on values of human microsomal protein and hepatocellularity per gram of liver. *Curr Drug Metab* 2007;8:33–45.

Barza M. Imipenem/cilastatin. *Eur J Clin Microbiol* 1984;3(5):453–455.

Benet L, Hoener B-A. Changes in plasma protein binding have little clinical relevance. *Clin Pharmacol Ther* 2002;71(3):115–121.

Bent S. Herbal medicine in the United States: review of efficacy, safety, and regulation: grand rounds at University of California, San Francisco Medical Center. *J Gen Intern Med* 2008;23(6):854–859.

Bent S, Ko R. Commonly used herbal medicines in the United States: a review. *Am J Med* 2004;116:478–485.

Bertelsen K, et al. Apparent mechanism-based inhibition of human CYP2D6 *in vitro* by paroxetine: comparison with fluoxetine and quinidine. *Drug Metab Dispos* 2003;31(3):289–293.

Blanchard N, et al. Impact of serum on clearance predictions obtained from suspensions and primary cultures of rat hepatocytes. *EurJ Pharm Sci* 2004;23: 189–199.

Boelsterli U, Hsiao C-J. The heterozygous Sod2(+/–) mouse: modeling the mitochondrial role in drug toxicity. *Drug Discovery Today* 2008;13(21/22):982–988.

Bogaards J, et al. Determining the best animal model for human cytochrome P450 activities: a comparison of mouse, rat, rabbit, dog, micropig, monkey and man. *Xenobiotica* 2000;30(12):1131–1152.

Born S, et al. Synthesis and reactivity of coumarin 3,4-epoxide. *Drug Metab Dispos* 1997;25(11):1318–1324.

Boxenbaum H, Ronfeld R. Interspecies pharmacokinetic scaling and the Dedrick plots. *Am J Physiol* 1983;245(6):R768–R775.

Brown C, Brown J, Coleman M. How drug metabolism influences treatment outcomes. *Expert Opin Drug Metab Toxicol* 2007;3(6):913–915.

Brown H, et al. Prediction of *in vivo* drug–drug interactions from *in vitro* data: impact of incorporating parallel pathways of drug elimination and inhibitor absorption rate constant. *Br J Clin Pharmacol* 2005;60(5):508–518.

Brown H, et al. Prediction of *in vivo* drug–drug interactions from *in vitro* data: factors affecting prototypic drug–drug interactions involving CYP2C9, CYP2D6 and CYP. *Clin Pharmacokinet* 2006;45(10):1035–1050.

Brown H, Griffin M, Houston J. Evaluation of cryopreserved human hepatocytes as an alternative *in vitro* system to microsomes for the prediction of metabolic clearance. *Drug Metab Dispos* 2007;35(2):293–301.

Burkhart C, et al. Influence of reduced glutathione on the proliferative response of sulfamethoxazole-specific and sulfamethoxazole-metabolite-specific human CD4+ T-cells. *Br J Pharmacol* 2001;132(3):623–630.

Campiani G, et al. Non-nucleoside HIV-1 reverse transcriptase (RT) inhibitors: past, present, and future perspectives. *Curr Pharm Des* 2002;8(8):615–657.

Castro-Perez J. Current and future trends in the application of HPLC-MS to metabolite-identification studies. *Drug Discovery Today* 2007;12(5/6):249–256.

Catapano A. Ezetimibe: a selective inhibitor of cholesterol absorption. *Eur Heart J Suppl* 2001;3:E6–E10.

Chambers R, et al. Discovery of CP-199,330 and CP-199,331: two potent and orally efficacious cysteinyl LT1 receptor antagonists devoid of liver toxicity. *Bioorg Med Chem Lett* 1999;9(18):2773–2778.

Chandra P, Brouwer K. The complexities of hepatic drug transport: current knowledge and emerging concepts. *Pharm Res* 2004;21(5):719–735.

Chen J, Raymond K. Roles of rifampicin in drug–drug interactions: underlying molecular mechanisms involving the nuclear pregnane X receptor. *Ann Clin Microbiol Antimicrob* 2006;5.

Chen Y, et al. The nuclear receptors constitutive androstane receptor and pregnane X receptor cross-talk with hepatic nuclear factor 4alpha to synergistically activate the human CYP2C9 promoter. *J Pharmacol Exp Ther* 2005;314(3):1125–1133.

Chen Y, Monshouwer M, Fitch W. Analytical tools and approaches for metabolite identification in early drug discovery. *Pharm Res* 2007;24(2):248–257.

Cheung C, Gonzalez F. Humanized mouse lines and their application for prediction of human drug metabolism and toxicological risk assessment. *J Pharmacol Exp Ther* 2008;327(2):288–299.

Chiba M, et al. P450 interaction with HIV protease inhibitors: relationship between metabolic stability, inhibitory potency, and P450 binding spec. *Drug Metab Dispos* 2001;29(1):1–3.

Chien J, et al. Stochastic prediction of CYP3A-mediated inhibition of midazolam clearance by ketoconazole. *Drug Metab Dispos* 2006;34(7):1208–1219.

Christensen H, et al. Prediction of plasma protein binding displacement and its implications for quantitative assessment of metabolic drug–drug interactions from *in vitro* data. *J Pharm Sci* 2006;95(12):2778–2787.

Clarke C, Haselden J. Metabolic profiling as a tool for understanding mechanisms of toxicity. *Toxicol Pathol* 2008;36:140–147.

Cooper A, Pinto J. Cysteine *S*-conjugate beta-lyases. *Amino Acids* 2006;30(1):1–15.

Cooper A, Bruschi S, Anders M. Toxic, halogenated cysteine S-conjugates and targeting of mitochondrial enzymes of energy metabolism. *Biochem Pharmacol* 2002;64(4):553–564.

Cotreau M, Von Moltke LL, Greenblatt DJ. The influence of age and sex on the clearance of cytochrome P450 3A substrates. *Clin Pharmacokinet* 2005;44(1):33–60.

Cresteil T, et al. Regioselective metabolism of taxoids by human CYP3A4 and 2C8: structure–activity relationship. *Drug Metab Dispos* 2002;30(4):438–445.

Crewe H, et al. The effect of selective serotonin re-uptake inhibitors on cytochrome P4502D6 (CYP2D6) activity in human liver microsomes. *Br J Clin Pharmacol* 1992;34(3):262–265.

Cribb AE, et al. Role of polymorphic human cytochrome P450 enzymes in estrone oxidation. *Cancer Epidemiol Biomarkers Prev* 2006;15(3):551–558.

Cvetkovic R, Goa K. Lopinavir/ritonavir: a review of its use in the management of HIV infection. *Drugs* 2003;63(8):769–802.

Dai D, et al. Polymorphisms in human CYP2C8 decrease metabolism of the anticancer drug paclitaxel and arachidonic acid. *Pharmacogenetics* 2001;11(7):597–607.

Dalvie D, et al. Biotransformation reactions of five-membered aromatic heterocyclic rings. *Chem Res Toxicol* 2002;15(3):269–299.

Davis-Bruno K, Atrakchi A. A regulatory perspective on issues and approaches in characterizing human metabolites. *Chem Biol Interact* 2006;19:1561–1563.

De Buck S, et al. The prediction of drug metabolism, tissue distribution, and bioavailability of 50 structurally diverse compounds in rat using mechanism-based absorption, distribution, and metabolism prediction tools. *Drug Metab Dispos* 2007; 35(4):649–659.

DeBuske L, The role of P-glycoprotein and organic anion-transporting polypeptides in drug interactions. *Drug Safety* 2005;28(9):789–801.

De Graaf I, et al. Comparison of *in vitro* preparations for semi-quantitative prediction of *in vivo* drug metabolism. *Drug Metab Dispos* 2002;30(10):1129–1136.

Dekant W. Biosynthesis of toxic glutathione conjugates from halogenated alkenes. *Toxicol Lett* 2003;144(1):49–54.

Dekant W, Vamvakas S, Anders M. Formation and fate of nephrotoxic and cytotoxic glutathione S-conjugates: cysteine conjugate beta-lyase pathway. *Adv Pharmacol* 1994;27:115–162.

Delbressine L, et al. On the formation of carbamate glucuronides. *Xenobiotica* 1990;20(1):133–134.

Desta Z, et al. Stereoselective metabolism of cisapride and enantiomer–enantiomer interaction in human cytochrome P450 enzymes: major role of CYP3A. *J Pharmacol Exp Ther* 2001;298(2):508–520.

DeVane C. Clinical significance of drug binding, protein binding, and binding displacement drug interactions. *Psychopharmacol Bull* 2002;36(3):5–21.

Dickinson P, Taylor G. Route dependent pulmonary first-pass metabolism of a series of biphenylacetic acid esters in rats. *Eur J Pharm Sci* 1998;6(1):11–18.

Dieckhaus C, et al. A mechanistic approach to understanding species differences in felbamate bioactivation: relevance to drug-induced idiosyncratic reactions. *Drug Metab Dispos* 2000;28(7):814–822.

Doherty E, et al. 4-Aminopyrimidine tetrahydronaphthols: a series of novel vanilloid receptor-1 antagonists with improved solubility properties. *Bioorg Med Chem Lett* 2008;18:1830–1834.

Dunne A, O'Hara T, Devane J. Level A *in vivo–in vitro* correlation: nonlinear models and statistical methodology. *J Pharm Sci* 1997;86(11):1245–1249.

Duret C, et al. Ketoconazole and miconazole are antagonists of the human glucocorticoid receptor: consequences on the expression and function of the constitutive androstane receptor and the pregnane X receptor. *Mol Pharmacol* 2006;70(1): 329–339.

Dvorak Z, et al. Colchicine down-regulates cytochrome P450 2B6, 2C8, 2C9, and 3A4 in human hepatocytes by affecting their glucocorticoid receptor-mediated regulation. *Mol Pharmacol* 2003;64(1):160–169.

Edwards R, et al. Contribution of CYP1A1 and CYP1A2 to the activation of heterocyclic amines in monkeys and human. *Carcinogenesis* 1994;15(5):829–836.

Enders C, et al. The role of transporters in drug interactions. *Eur J Pharm Sci* 2006;27:501–517.

Erve J. Chemical toxicology: reactive intermediates and their role in pharmacology and toxicology. *Expert Opin Drug Metab Toxicol* 2006;2(6):923–946.

Evans D, et al. Drug–protein adducts: an industry perspective on minimizing the potential for drug bioactivation in drug discovery and development. *Chem Res Toxicol* 2004;17:3–16.

Fagerholm U. Prediction of human pharmacokinetics—evaluation of methods for prediction of hepatic metabolic clearance. *J Pharm Pharmacol* 2007a;59:803–828.

Fagerholm U. Prediction of human pharmacokinetics—improving microsome-based predictions of hepatic metabolic clearance. *J Pharm Pharmacol* 2007b;59:1427–1431.

Fahmi OBS, et al. Prediction of drug–drug interactions from *in vitro* induction data: application of the relative induction score approach using cryopreserved human hepatocytes. *Drug Metab Dispos* 2008;36(9):1971–1974.

Fang H-L, et al. Positive and negative regulation of human hepatic hydroxysteroid sulfotransferase (SULT2A1) gene transcription by rifampicin: roles of hepatocyte nuclear factor 4alpha and pregnane X receptor. *J Pharmacol Exp Ther* 2007; 323(2):586–598.

Felmlee M, et al. Cytochrome P450 expression and regulation in CYP3A4/CYP2D6 double transgenic humanized mice. *Drug Metab Dispos* 2008;36(2):435–441.

Feng M, et al. Allometric pharmacokinetic scaling: towards the prediction of human oral pharmacokinetics. *Pharm Res* 2000;17(4):410–418.

Fisher M, et al. Flavin-containing monooxygenase activity in hepatocytes and microsomes: *in vitro* characterization and *in vivo* scaling of benzydamine clearance. *Drug Metab Dispos* 2002;30(10):1087–1093.

Floyd D, et al. Benzazepinone calcium channel blockers. 2. Structure-activity and drug metabolism studies leading to potent antihypertensive agents. Comparison with benzothiazepinones. *J Med Chem* 1992;35(4):756–772.

Fontana E, Dansette P, Poli S. Cytochrome p450 enzymes mechanism based inhibitors: common sub-structures and reactivity. *Curr Drug Metab* 2005;6:413–454.

Fowler S, Zhang H. *In vitro* evaluation of reversible and irreversible cytochrome P450 inhibition: current status on methodologies and their utility for predicting drug–drug interactions. *AAPS J* 2008;10(2):410–424.

Franconi F, et al. Gender differences in drug responses. *Pharmacol Res* 2007;55: 81–95.

Franklin M, Hathaway L. 2-Diethylaminoethyl-2,2-diphenylvalerate-HCl (SKF525A) revisited: comparative cytochrome P450 inhibition in human liver microsomes by SKF525A, its metabolites, and SKF-acid and SKF-alcohol. *Drug Metab Dispos* 2008;36(12):2539–2546.

Frantz C, et al. Dose–response studies of MeIQx in rat liver and liver DNA at low doses. *Carcinogenesis* 1995;16(2):367–373.

Fuhr U. Induction of drug metabolising enzymes: pharmacokinetic and toxicological consequences in humans. *Clin Pharmacokinet* 2000;38(6):493–504.

Funk C. The role of hepatic transporters in drug elimination. *Expert Opin Drug Metab Toxicol* 2008;4(4):363–379.

Fura A. Role of pharmacologically active metabolites in drug discovery and development. *Drug Discovery Tech* 2006;11(3/4):133–142.

Galetin A. Intestinal first-pass metabolism: bridging the gap between *in vitro* and *in vivo*. *Curr Drug Metab* 2007;8(7):643–644.

Galetin A, et al. Prediction of time-dependent CYP3A4 drug–drug interactions: impact of enzyme degradation, parallel elimination pathways, and intestinal inhibition. *Drug Metab Dispos* 2006;34(1):166–175.

Gao H, et al. Method for rapid metabolite profiling of drug candidates in fresh hepatocytes using liquid chromatography coupled with a hybrid quadrupole linear ion trap. *Rapid Commun in Mass Spec* 2007;21:3683–3693.

Gardner I, et al. Comparison of the disposition of two novel combined thromboxane synthase inhibitors/thromboxane A2 receptor antagonists in the isolated perfused rat liver. *Xenobiotica* 1995;25(2):185–197.

Gerntholtz T, et al. The use of a cyclosporin-ketoconazole combination: making renal transplantation affordable in developing countries. *Eur J Clin Pharmacol* 2004;60(3):143–148.

Gomez-Lechon M, Castell J, Donato M. Hepatocytes—the choice to investigate drug metabolism and toxicity in man: *in vitro* variability as a reflection of *in vivo*. *Chemico-Biol Interactions* 2007;168:30–50.

Gonzalez F. CYP3A4 and pregnane X receptor humanized mice. *J Biochem Mol Toxicol* 2007;21(4):158–162.

Gonzalez F, Yu A-M. Cytochrome P450 expression and regulation in CYP3A4/CYP2D6 double transgenic humanized mice. *Annu Rev Pharmacol Toxicol* 2006;46:41–64.

Gorrod J, Whittlesea C, Lam S. *Biological Reactive Intermediates IV*, Witmer C, editor. 1990, New York: Plenum Press, 657–664.

Graham M, et al. Pharmacokinetics of the hypoxic cell cytotoxic agent tirapazamine and its major bioreductive metabolites in mice and humans: retrospective analysis of a pharmacokinetically guided dose-escalation strategy in a phase I trial. *Cancer Chemother Pharmacol* 1997;40(1):1–10.

Greenblatt D, et al. Alprazolam–ritonavir interaction: implications for product labeling. *Clin Pharmacol Ther* 2000;67(4):335–341.

Grime K, Riley R. The impact of *in vitro* binding on *in vitro–in vivo* extrapolations, projections of metabolic clearance and clinical drug–drug interactions. *Curr Drug Metab* 2006;7:251–264.

Grimm SW, et al. The conduct of *in vitro* studies to address time-dependent inhibition of drug metabolising enzymes: a perspective of the Pharmaceutical Research and Manufacturers of America (PhRMA). *Drug Metab Dispos* 2009;37(7):1355–1370.

Guengerich F. Uncommon P450-catalyzed reactions. *Curr Drug Metab* 2001;2:93–115.

Guengerich F. Cytochrome P450s and other enzymes in drug metabolism and toxicity. *AAPS J* 2006a;8(1):E101–E111.

Guengerich F. Safety assessment of stable drug metabolites. *Chem Res Toxicol* 2006b;19(12):1559–1560.

Guengerich F, MacDonald J. Applying mechanisms of chemical toxicity to predict drug safety. *Chem Res Toxicol* 2007;20:344–369.

Guo Z, et al. Age- and gender-related variations in the activities of drug-metabolizing and antioxidant enzymes in the white-footed mouse (Peromyscus leucopus. *Growth Dev Aging* 1993;57(2):85–100.

Hakooz N, et al. Determination of a human hepatic microsomal scaling factor for predicting *in vivo* drug clearance. *Pharm Res* 2006;23(3):533–539.

Hallifax D, Houston J. Uptake and intracellular binding of lipophilic amine drugs by isolated rat hepatocytes and implications for prediction of *in vivo* metabolic clearance. *Drug Metab Dispos* 2006;34:1829–1836.

Hallifax D, Houston J. Saturable uptake of lipophilic amine drugs into isolated hepatocytes: mechanisms and consequences for quantitative clearance prediction. *Drug Metab Dispos* 2007;35(8):1325–1332.

Hamberg A, et al. A PK-PD model for predicting the impact of age, CYP2C9, and VKORC1 genotype on individualization of warfarin therapy. *Clin Pharmacol Ther* 2007;81(4):529–616.

Harper T, Brassil P. Reaction phenotyping: current industry efforts to identify enzymes responsible for metabolizing drug candidates. *AAPS* J 2008;10(1):200–207.

Harrigan J, et al. Tissue specific induction of cytochrome P450 (CYP) 1A1 and 1B1 in rat liver and lung following *in vitro* (tissue slice) and *in vivo* exposure to benzo(a) pyrene. *Toxicol In Vitro* 2006;20:426–438.

Harris R, Benet L, Schwartz J. Gender effects in pharmacokinetics and pharmacodynamics. *Drugs* 1995;50(2):222–239.

Hein D, et al. Molecular genetics and epidemiology of the NAT1 and NAT2 acetylation polymorphisms. *Cancer Epidemiol Biomarkers Prev* 2000;9:29–42.

Hermann M, et al. Evaluation of microsomal incubation conditions on CYP3A4-mediated metabolism of cyclosporine A by a statistical experimental design. *Curr Drug Metab* 2006;7:265–271.

Hernandez-Martinez N, et al. Tissue-specific induction of the carcinogen-inducible cytochrome P450 isoforms in the gastrointestinal tract. *Environ Toxicol Pharmacol* 2007;24(3):297–303.

Hewitt N, et al. Primary hepatocytes: current understanding of the regulation of metabolic enzymes and transporter proteins, and pharmaceutical practice for the use of hepatocytes in metabolism, enzyme induction, transporter, clearance, and hepatotoxicity studies. *Drug Metab Rev* 2007a;39:159–234.

Hewitt N, de Kanter R, LeCluyse E. Induction of drug metabolizing enzymes: a survey of *in vitro* methodologies and interpretations used in the pharmaceutical industry—do they comply with FDA recommendations? *Chemico-Biol Interactions* 2007b; 168:51–65.

Hewitt N, LeCluyse E, Ferguson S. Induction of hepatic cytochrome P450 enzymes: methods, mechanisms, recommendations, and *in vitro–in vivo* correlations. *Xenobiotica* 2007c;37(10/11):1196–1224.

Higashikawa F, et al. Dose-dependent intestinal and hepatic first-pass metabolism of midazolam, a cytochrome P450 3A substrate with differently modulated enzyme activity in rats. *J Pharm Pharmacol* 1999;51:67–72.

Hinson J. Reactive metabolites of phenacetin and acetaminophen: a review. *Environ Health Persp* 1983;49:71–79.

Hollenberg P, Kent U, Bumpus N. Mechanism-based inactivation of human cytochromes p450s: experimental characterization, reactive intermediates, and clinical implications. *Chem Res Toxicol* 2008;21:189–205.

Hop C, Chen Y, Yu L. Uniformity of ionization response of structurally diverse analytes using a chip-based nanoelectrospray ionization source. *Rapid Commun Mass Spectrom* 2005;19(21):3139–3142.

Houston J, Galetin A. Progress towards prediction of human pharmacokinetic parameters from *in vitro* technologies. *Drug Metab Rev* 2003;35(4):393–415.

Houston J, Kenworthy K. *In vitro–in vivo* scaling of CYP kinetic data not consistent with the classical Michaelis–Menten model. *Drug Metab Dispos* 2000;28(3):246–254.

Humphreys W, Unger S. Safety assessment of drug metabolites: characterization of chemically stable metabolites. *Chem Res Toxicol* 2006;19:1564–1569.

Huang J, Oie S. Hepatic elimination of drugs with concentration-dependent serum protein binding. *J Pharmacokinet Biopharm* 1984;12(1):67–81.

Huang C, et al. Projection of exposure and efficacious dose prior to first-in-human studies: How successful have we been? *Pharm Res* 2008;25(4):713–726.

Hunt C, Westerkam W, Stave G. Effect of age and gender on the activity of human hepatic CYP3A. *Biochem Pharmacol* 1992;44(2):275–283.

Ioannides C, editor. *Enzyme Systems that Metabolise Drugs and Other Xenobiotics*, 1st ed. Vol. 1. New York: John Wiley & Sons, 2002.

Ishiguro N, et al. Predominant contribution of OATP1B3 to the hepatic uptake of telmisartan, an angiotensin II receptor antagonist, in humans. *Drug Metab Dispos* 2006;34(7):1109–1115.

Isin E, Guengerich F. Complex reactions catalyzed by cytochrome P450 enzymes. *Biochim Biophys Acta* 2006a;1770:314–329.

Isin E, Guengerich F. Kinetics and thermodynamics of ligand binding by cytochrome P450 3A4. *J Biol Chem* 2006b;281(14):9127–9136.

Isoherranen N, et al. Role of itraconazole metabolites in CYP3A4 inhibition. *Drug Metab Dispos* 2004;32(10):1121–1131.

Ito K, Houston J. Prediction of human drug clearance from *in vitro* and preclinical data using physiologically based and empirical approaches. *Pharm Res* 2005; 22(1):103–112.

Ito K, et al. Prediction of pharmacokinetic alterations caused by drug–drug interactions: metabolic interaction in the liver. *Annu Rev Pharmacol Toxicol* 1998a;38:461–499.

Ito K, et al. Prediction of pharmacokinetic alterations caused by drug–drug interactions: metabolic interaction in the liver. *Pharmacol Rev* 1998b;50(3):387–411.

Iwatsubo T, et al. Prediction of *in vivo* drug metabolism in the human liver from *in vitro* metabolism data. *Pharmacol Ther* 1997;73(2):147–171.

Izzo A. Drug interactions with St. John's Wort (Hypericum perforatum): a review of the clinical evidence. *Int J Clin Pharmacol Ther* 2004;42(3):139–148.

James L, Mayeux P, Hinson J. Acetaminophen-induced hepatotoxicity. *Drug Metab Dispos* 2003;31(12):1499–1506.

Jewell WT, Miller MG. Comparison of human and rat metabolism of molinate in liver microsomes and slices. *Drug Metab Dispos* 1999;27(7):842–847.

Jin S, Liu H. Gender-related differences in metabolism of the enantiomers of trans tramadol and trans O-demethyltramadol in rat liver microsomes. *Yao Xue Xue Bao* 2004;39(8):581–585.

Jolivette L, Ekins S. Methods for predicting human drug metabolism. *Adv Clin Chem* 2007;43:131–176.

Joshi G. Morphine-6-glucuronide, an active morphine metabolite for the potential treatment of post-operative pain. *Curr Opin Investig Drugs* 2008;9(7):786–799.

Ju C, Uetrecht J. Mechanism of idiosyncratic drug reactions: reactive metabolite formation, protein binding and the regulation of the immune system. *Curr Drug Metab* 2002;3:367–377.

Kaji H, Kume T. Identification of human UDP-glucuronosyltransferase isoform(s) responsible for the glucuronidation of 2-(4-chlorophenyl)- 5-(2-furyl)-4-oxazoleacetic acid (TA-1801A). *Drug Metab Pharmacokinet* 2005;20(3):212–218.

Kalgutkar A, Soglia J. Minimising the potential for metabolic activation in drug discovery. *Expert Opin Drug Metab Toxicol* 2005;1(1):91–142.

Kalgutkar A, et al. A comprehensive listing of bioactivation pathways of organic functional grou. *Curr Drug Metab* 2005a;6:161–225.

Kalgutkar A, et al. Bioactivation of the nontricyclic antidepressant nefazodone to a reactive quinone-imine species in human liver microsomes and recombinant cytochrome P450 3A4. *Drug Metab Dispos* 2005b;33(2):243–253.

Kalgutkar A, Obach R, Maurer T. Mechanism-based inactivation of cytochrome P450 enzymes: chemical mechanisms, structure–activity relationships and relationship to clinical drug–drug interactions and idiosyncratic adverse drug reactions. *Curr Drug Metab* 2007;8(5):407–447.

Kalvaas J, et al. Influence of microsomal concentration on apparent intrinsic clearance: implications for scaling *in vitro*. *Data Drug Metab Dispos* 2001;29(10):1332–1336.

Kaminsky LS, Zhang Z-Y. Human P450 metabolism of warfarin. *Pharmacol Ther* 1997;73(1):67–74.

Kane RE, Li AP, Kaminski DR. Sulfation and glucuronidation of acetaminophen by human hepatocytes cultured on Matrigel and type 1 collagen reproduces conjugation *in vivo*. *Drug Metab Dispos* 1995;23(3):303–307.

Kapetanovic I, et al. Potentially reactive cyclic carbamate metabolite of the antiepileptic drug felbamate produced by human liver tissue *in vitro*. *Drug Metab Dispos* 1998;26(11):1089–1095.

Katoh M, et al. *In vivo* drug metabolism model for human cytochrome P450 enzyme using chimeric mice with humanized liver. *J Pharm Sci* 2006;96(2):428–437.

Katoh M, et al. Chimeric mice with humanized liver. *Toxicology* 2008;245:9–17.

Kent UM, Jushchhyshyn M, Hollenberg P. Mechanism-based inactivators as probes of cytochrome P450 structure and function. *Curr Drug Metab* 2001;2(3):215–243.

Kent U, et al. Identification of 17-alpha-ethynylestradiol-modified active site peptides and glutathione conjugates formed during metabolism and inactivation of P450s 2B1 and 2B6. *Chem Res Toxicol* 2006;19(2):279–287.

Khaliq Y, et al. Effect of ketoconazole on ritonavir and saquinavir concentrations in plasma and cerebrospinal fluid from patients infected with human immunodeficiency virus. *Clin Pharm Ther* 2000;68:637–646.

Kharasch ED, et al. Dose-dependent metabolism of fluoromethyl-2,2-difluoro-1-(trifluoromethyl)vinyl ether (compound A), an anesthetic degradation product, to mercapturic acids and 3,3,3-trifluoro-2-(fluoromethoxy)propanoic acid in rats. *Toxicol Appl Pharmacol* 1999;160:49–59.

Kilicarslan T, et al. Flunitrazepam metabolism by cytochrome P450S 2C19 and 3A4. *Drug Metab Dispos* 2001;29(4):460–465.

Kinobe RT, et al. P450 2C18 catalyzes the metabolic bioactivation of phenytoin. *Chem Res Toxicol* 2005;18(12):1868–1875.

Klieber S, et al. Contribution of the *N*-glucuronidation pathway to the overall *in vitro* metabolic clearance of midazolam in humans. *Drug Metab Dispos* 2008;36(5): 851–862.

Ko Y, et al. Prescribers' knowledge of and sources of information for potential drug–drug interactions: a postal survey of US prescribers. *Drug Safety* 2008;31(6): 525–536.

Koenigs L, et al. Electrospray ionization mass spectrometric analysis of intact cytochrome P450: identification of tienilic acid adducts to P450 2C9. *Biochemistry* 1999;38(8):2312–2319.

Koudriakova T, et al. Metabolism of the human immunodeficiency virus protease inhibitors indinavir and ritonavir by human intestinal microsomes and expressed cytochrome P4503A4/3A5: mechanism-based inactivation of cytochrome P4503A by ritonavir. *Drug Metab Dispos* 1998;26(6):552–561.

Kramer M, Tracy T. Studying cytochrome P450 kinetics in drug metabolism. *Expert Opin Drug Metab Toxicol* 2008;4(5):591–603.

Kumar S, et al. Minimizing metabolic activation during pharmaceutical lead optimization: progress, knowledge gaps and future directions. Curr *Opin Drug Discovery Dev* 2008;11(1):43–52.

Kunze KL, et al. Stereochemical aspects of itraconazole metabolism *in vitro* and *in vivo*. *Drug Metab Dispos* 2006;34(4):583–590.

Kuperman A, et al. Pharmacokinetics and metabolism of a cysteinyl leukotriene-1 receptor antagonist from the heterocyclic chromanol series in rats: *in vitro–in vivo* correlation, gender-related differences, isoform identification, and comparison with metabolism in human hepatic tissue. *Drug Metab Dispos* 2001;29(11):1403–1409.

Lalovic B, et al. Pharmacokinetics and pharmacodynamics of oral oxycodone in healthy human subjects: role of circulating active metabolites. *Clin Pharmacol Ther* 2006;79(5):461–479.

Lam J, Benet L. Hepatic microsome studies are insufficient to characterize *in vivo* hepatic metabolic clearance and metabolic drug–drug interactions: studies of digoxin metabolism in primary rat hepatocytes versus microsomes. *Drug Metab Dispos* 2004;32(11):1311–1316.

Lau Y, et al. *Ex situ* inhibition of hepatic uptake and efflux significantly changes metabolism: hepatic enzyme–transporter interplay. *J Pharmacol Exp Ther* 2004; 308(3):1045–1050.

Lau Y, et al. Multiple transporters affect the disposition of atorvastatin and its two active hydroxy metabolites: application of *in vitro* and *ex situ* systems. *J Pharmacol Exp Ther* 2006;316(2):762–771.

Lehr T, et al. Contribution of the active metabolite M1 to the pharmacological activity of tesofensine *in vivo*: a pharmacokinetic–pharmacodynamic modelling approach. *Br J Pharmacol* 2008;153(1):164–174.

Lennernas H. Clinical pharmacokinetics of atorvastatin. *Clin Pharmacokinet* 2003;42:1141–1160.

Levy R, et al., editors. *Metabolic Drug Interaction*. Baltimore: Lipincott Williams & Wilkins, 2000.

Lim H-K, et al. A generic method to detect electrophilic intermediates using isotopic pattern triggered data-dependent high-resolution accurate mass spectrometry. *Rapid Commun Mass Spec* 2008;22:1295–1311.

Lim Y-P, Huang J-D. Interplay of pregnane X receptor with other nuclear receptors on gene regulation. *Drug Metab Pharmacokinet* 2008;23(1):14–21.

Lin H. CYP induction-mediated drug interactions: *in vitro* assessment and clinical implications. *Pharm Res* 2006;23(6):1089–1116.

Lin J. Applications and limitations of interspecies scaling and *in vitro* extrapolation in pharmacokinetics. *Drug Metab Dispos* 1998;26(12):1202–1212.

Lin J. Applications and limitations of genetically modified mouse models in drug discovery and development. *Curr Drug Metab* 2008;9(5):419–438.

Lin J, Lu A. Role of pharmacokinetics and metabolism in drug discovery and development. *Pharmacol Rev* 1997;49(4):403–449.

Lin J, et al. The role of absorption, distribution, metabolism, excretion and toxicity in drug discovery. *Curr Topics Med Chem* 2003;3:1125–1154.

Liu D, Hop C. Strategies for characterization of drug metabolites using liquid chromatography–tandem mass spectrometry in conjunction with chemical derivatization and on-line H/D exchange approaches. *J Pharm Biomed Analysis* 2005;37:1–18.

Liu H, et al. Stereoselectivity in absorption of trans tramadol in rat intestine. *Yao Xue Xue Bao* 2003;38(12):893–896.

Lock E, Schnellmann R. The effect of haloalkene cysteine conjugates on rat renal glutathione reductase and lipoyl dehydrogenase activities. *Toxicol Appl Pharmacol* 1990;104(1):180–190.

Lu A, Wang R, Lin J. Cytochrome P450 *in vitro* reaction phenotyping: a re-evaluation of approaches used for P450 isoform identification. *Drug Metab Dispos* 2003;31(4):345–350.

Lu C, et al. Comparison of intrinsic clearance in liver microsomes and hepatocytes from rats and humans: evaluation of free fraction and uptake in hepatocytes. *Drug Metab Dispos* 2006a;34(9):1600–1605.

Lu C, et al. A novel model for the prediction of drug–drug interactions in humans based on *in vitro* cytochrome p450 phenotypic data. *Drug Metab Dispos* 2006b; 35(1):79–85.

Lu C, et al. Prediction of pharmacokinetic drug–drug interactions using human hepatocyte suspension in plasma and cytochrome P450 phenotypic data. II. *In vitro–in vivo* correlation with ketoconazole. *Drug Metab Dispos* 2008a;36(7):1255–1260.

Lu C, et al. Prediction of pharmacokinetic drug–drug interactions using human hepatocyte suspension in plasma and cytochrome P450 phenotypic data. III. *In vitro–in vivo* correlation with fluconazole. *Drug Metab Dispos* 2008b;36(7):1261–1267.

Luo G, et al. CYP3A4 induction by drugs: correlation between a pregnane X receptor reporter gene assay and CYP3A4 expression in human hepatocytes. *Drug Metab Dispos* 2002;30(7):795–804.

Luo G, et al. CYP3A4 induction by xenobiotics: biochemistry, experimental methods and impact on drug discovery and development. *Curr Drug Metab* 2004;5:483–505.

Lynch T, Price A. The effect of cytochrome P450 metabolism on drug response, interactions, and adverse effects. *Am Fam Phys* 2007;76:391–396.

Ma Q, Lu A. The challenges of dealing with promiscuous drug-metabolizing enzymes, receptors and transporters. *Curr Drug Metab* 2008;9:374–383.

Ma X, et al. The PREgnane X receptor gene-humanized mouse: a model for investigating drug–drug interactions mediated by cytochromes P450 3A. *Drug Metab Dispos* 2007;35(2):194–200.

Ma X, et al. A double transgenic mouse model expressing human pregnane X receptor and cytochrome P450 3A4. *Drug Metab Dispos* 2008;36(12):2506–2512.

MacKichan J. Pharmacokinetic consequences of drug displacement from blood and tissue proteins. *Clin Pharmacokinet* 1984;9:32–41.

MacKichan J. Protein binding drug displacement interactions fact or fiction? *Clin Pharmacokinet* 1989;16(2):65–73.

Mahmood I. Interspecies scaling: role of protein binding in the prediction of clearance from animals to humans. *J Clin Pharmacol* 2000;40:1439–1446.

Mahmood I, ed. *Interspecies Pharmacokinetic Scaling*. Rockville, MD: Pine House, 2005.

Mahmood I. Prediction of human drug clearance from animal data: application of the rule of exponents and "fu Corrected Intercept Method" (FCIM). *J Pharm Sci* 2006;95(8):1810–1821.

Mahmood I, Green M, Fisher J. Selection of the first-time dose in humans: comparison of different approaches based on interspecies scaling of clearance. *J Clin Pharmacol* 2003;43:692–697.

Mannel M. Drug interactions with St John's wort: mechanisms and clinical implications. *Drug Safety* 2004;27(11):773–797.

Marathe P, Rodrigues AD. *In vivo* animal models for investigating potential CYP3A- and Pgp-mediated drug–drug interactions. *Curr Drug Metab* 2006;7(7):687–704.

Marchetti S, et al. Clinical relevance: drug–drug interaction, pharmacokinetics, pharmacodynamics, and toxicology. In: You G, Morris M, editors. *Drug Transporters: Molecular Characterization and Role in Drug Disposition*, John Wiley & Sons, 2007a, pp. 747–880.

Marchetti S, et al. Concise review: Clinical relevance of drug drug and herb drug interactions mediated by the ABC transporter ABCB1 (MDR1, P-glycoprotein). *The Oncologist* 2007b;12:927–941.

Masimirembwa C, Bredberg U, Andersson T. Metabolic stability for drug discovery and development: pharmacokinetic and biochemical challenges. *Clin Pharmacokinet* 2003;42(6):515–528.

Masubuchi N, Makino C, Murayama N. Prediction of *in vivo* potential for metabolic activation of drugs into chemically reactive intermediate: correlation of *in vitro* and *in vivo* generation of reactive intermediates and *in vitro* glutathione conjugate formation in rats and humans. *Chem Res Toxicol* 2007;20:455–464.

Matheny C, et al. Effect of prototypical inducing agents on P-glycoprotein and CYP3A expression in mouse tissue. *Drug Metab Dispos* 2004;32(9):1008–1014.

Maurer T, Tabrizi-Fard M, Fung H-L. Impact of mechanism-based enzyme inactivation on inhibitor potency: implications for rational drug discovery. *J Pharm Sci* 2000;89(11):1404–1414.

McConn D, Zhao Z. Integrating *in vitro* kinetic data from compounds exhibiting induction, reversible inhibition and mechanism-based inactivation: *in vitro* study design. *Curr Drug Metab* 2004;5:141–146.

McElnay J, D'Arcy P. Protein binding displacement interactions and their clinical importance. *Drugs* 1983;25:495–513.

McLure J, Miners J, Birkett DJ. Nonspecific binding of drugs to human liver microso. *Br J Clin Pharmacol* 2000;49:453–461.

Meijerman I, Beijnen J, Schellens J. Herb–drug interactions in oncology: focus on mechanisms of induction. *The Oncologist* 2006;11:742–752.

Michalets E. Update: clinically significant cytochrome P-450 drug interactions. *Pharmacotherapy* 1998;18(1):84–112.

Milo S, et al. Codeine and 6-acetylcodeine analgesia in mice. *Cell Mol Neurobiol* 2006;26(4–6):1011–1019.

Miyazawa M, Sugie A, Shimada T. Roles of human CYP2A6 and 2B6 and rat CYP2C11 and 2B1 in the 10-hydroxylation of (–)-verbenone by liver microsomes. *Drug Metab Dispos* 2003;31(8):1049–1053.

Mohutsky M, et al. Predictions of the *in vivo* clearance of drugs from rate of loss using human liver microsomes for phase I and phase II biotransformations. *Pharm Res* 2006;23(4):654–662.

Mouly S, et al. Hepatic but not intestinal CYP3A4 displays dose-dependent induction by efavirenz in humans. *Clin Pharmacol Ther* 2002;72(1):1–9.

Munns A, et al. Bioactivation of phenytoin by human cytochrome P450: characterization of the mechanism and targets of covalent adduct formation. *Chem Res Toxicol* 1997;10(9):1049–1058.

Murakami T, Takano M. Intestinal efflux transporters and drug absorption. *Expert Opin Drug Metab Toxicol* 2008;4(7):923–939.

Muruganandan S, Sinal C. Mice as clinically relevant models for the study of cytochrome P450-dependent metabolism. *Nature* 2008;83(6):818–828.

Mutlib A, et al. Identification and characterization of efavirenz metabolites by liquid chromatography/mass spectrometry and high field NMR: species differences in the metabolism of efavirenz. *Drug Metab Dispos* 1999;27(11):1319–1333.

Mutlib A, et al. Disposition of glutathione conjugates in rats by a novel glutamic acid pathway: characterization of unique peptide conjugates by liquid chromatography/mass spectrometry and liquid chromatography/NMR. *J Pharmacol Exp Ther* 2000;294(2):735–745.

Nagilla R, et al. Investigation of the utility of published *in vitro* intrinsic clearance data for prediction of *in vivo* clearance. *J Pharmacol Toxicol Methods* 2006;53:106–116.

Naito S, et al. Current opinion: safety evaluation of drug metabolites in development of pharmaceuticals. *J Toxicol Sci* 2007;32(4):329–341.

Naritomi Y, et al. Prediction of human hepatic clearance from *in vivo* animal experiments and *in vitro* metabolic studies with liver microsomes from animals and humans. *Drug Metab Dispos* 2001;29(10):1316–1324.

Nassar A-EF, Lopez-Anaya A. Strategies for dealing with reactive intermediates in drug discovery and development. *Curr Opin Drug Discovery* 2004;7(1):126–136.

Nassar A-EF, Kamel A, Clairmont C. Improving the decision-making process in the structural modification of drug candidates: enhancing metabolic stability. *Drug Discovery Tech* 2004;9(23):1020–1028.

Nassar A-EF, Lee D. Novel approach to performing metabolite identification in drug metabolism. *J Chromatogr Sci* 2007;45:113–119.

Neuberger J, Williams R. Immune mechanisms in tienilic acid associated hepatotoxicity. *Gut* 1989;30(4):515–519.

Nielsen K, et al. The biotransformation of clomipramine *in vitro*, identification of the cytochrome P450s responsible for the separate metabolic pathways. *J Pharmacol Exp Ther* 1996;277(3):1659–1664.

Niwa T, et al. Inhibition of drug-metabolizing enzyme activity in human hepatic cytochrome P450s by bisphenol A. *Biol Pharm Bull* 2000;23(4):498–501.

Noe J, et al. Substrate-dependent drug–drug interactions between gemfibrozil, fluvastatin and other organic anion-transporting peptide (OATP) substrates on OATP1B1, OATP2B1, and OATP1B3. *Drug Metab Dispos* 2007;35(8):1308–1314.

Oates J. The science of drug therapy. In: Brunton L, Lazo J, Parker K, editors. *Goodman & Gilman's The Pharmacological Basis of Therapeutics*. New York: McGraw-Hill, 2006, pp. 117–136.

Obach R. The importance of nonspecific binding in *in vitro* matrices, its impact on enzyme kinetic studies of drug metabolism reactions, and implications for *in vitro–in vivo* correlations. *Drug Metab Dispos* 1996;24(10):1047–1049.

Obach R. Nonspecific binding to microsomes: impact on scale-up of *in vitro* intrinsic clearance to hepatic clearance as assessed through examination of warfarin, imipramine, and propranolol. *Drug Metab Dispos* 1997;25(12):1359–1369.

Obach R. Prediction of human clearance of twenty-nine drugs from hepatic microsomal intrinsic clearance data: An examination of *in vitro* half-life approach and nonspecific binding to microsomes. *Drug Metab Dispos* 1999;27(11):1350–1359.

Obach R. Metabolism of ezlopitant, a nonpeptidic substance P receptor antagonist, in liver microsomes: enzyme kinetics, cytochrome P450 isoform identity, and *in vitro–in vivo* correl. *Drug Metab Dispos* 2000a;28(9):1069–1076.

Obach R. Inhibition of human cytochrome P450 enzymes by constituents of St. John's wort, an herbal preparation used in the treatment of depression. *J Pharmacol Exp Ther* 2000b;294(1):88–95.

Obach R. The prediction of human clearance from hepatic microsomal metabolism data. *Curr Opin Drug Discovery Dev* 2001;4(1):36–44.

Obach R, et al. The prediction of human pharmacokinetic parameters from preclinical and *in vitro* metabolism data. *J Pharmacol Exp Ther* 1997;283(1):46–58.

Obach R, et al. *In vitro* cytochrome P450 inhibition data and the prediction of drug–drug interactions: qualitative relationships, quantitative predictions, and the rank-order approach. *Clin Pharmacol Ther* 2005;78:582–592.

Obach R, et al. The utility of *in vitro* cytochrome P450 inhibition data in the prediction of drug–drug interactions. *J Pharmacol Exp Ther* 2006;316(1):336–348.

Obach R, Walsky R, Venkatakrishnan K. Mechanism-based inactivation of human cytochrome p450 enzymes and the prediction of drug–drug interactions. *Drug Metab Dispos* 2007;35(2):246–255.

Obach R, et al. *In vitro* metabolism and covalent binding of enol-carboxamide derivatives and anti-inflammatory agents sudoxicam and meloxicam: insights into the hepatotoxicity of sudoxicam. *Chem Res Toxicol* 2008b;21(9):1890–1896.

Obach R, et al. Can *in vitro* metabolism-dependent covalent binding data in liver microsomes distinguish hepatotoxic from nonhepatotoxic drugs? An analysis of 18 drugs with consideration of intrinsic clearance and daily dose. *Chem Res Toxicol* 2008a;21:1814–1822.

Ohmori S, et al. Studies on cytochrome P450 responsible for oxidative metabolism of imipramine in human liver microsomes. *Biol Pharm Bull* 1993;16(6):517–575.

Ohno Y, Hisaka A, Suzuki H. General framework for the quantitative prediction of CYP3A4-mediated oral drug interactions based on the AUC increase by coadministration of standard drugs. *Clin Pharmacokinet* 2007;46(8):681–696.

Okumura K, et al. Genotyping of *N*-acetylation polymorphism and correlation with procainamide metabolism. *Clin Pharmacol Ther* 1997;61(5):509–517.

Olbe L, Carlsson E, Lindberg P. A proton-pump inhibitor expedition: the case histories of omeprazole and esomeprazole. *Nat Rev Drug Discovery* 2003;2(2):132–139.

Otton S, et al. Inhibition by fluoxetine of cytochrome P450 2D6 activity. *Clin Pharmacol Ther* 1993;53(4):401–409.

Pal D, Mitra A. MDR- and CYP3A4-mediated drug–drug interactions. *J Neuroimmune Pharmacol* 2006;1:323–339.

Pang S. Safety testing of metabolites: expectations and outcomes. *Chemico-Biol Interactions* 2009;179(1):45–59.

Pang S, Sun H, Liu S. Interplay of drug transporters and enzymes on hepatic drug disposition. In: You G, Morris M, editors. *Drug Transporters: Molecular Characterization and Role in Drug Disposition*. Hoboken, NJ: John Wiley & Sons, 2007, pp. 709–745.

Park K, et al. Selection of new chemical entities with decreased potential for adverse drug reactions. *Eur J Pharmacol* 2006;549:1–8.

Parker A, Houston J. Rate-limiting steps in hepatic drug clearance: comparison of hepatocellular uptake and metabolism with microsomal metabolism of saquinavir, nelfinavir, and ritonavir. *Drug Metab Dispos* 2008;36(7):1375–1384.

Parkinson A. Biotransformation of Xenobiotics. In: Klaassen CD, editor. *Casarett and Doull's Toxicology*. McGraw-Hill, 2001, pp. 134–224.

Patel N, et al. Glutathione-dependent biosynthesis and bioactivation of *S*-(1,2-dichlorovinyl)glutathione and *S*-(1,2-dichlorovinyl)-L-cysteine, the glutathione and cysteine *S*-conjugates of dichloroacetylene, in rat tissues and subcellular fractions. *Drug Metab Dispos* 1994;22(1):143–147.

Pearce RE, Rodrigues AD, Goldstein JA. Identification of the human P450 enzymes involved in lansoprazole metabolism. *J Pharmacol Exp Ther* 1996;277:805–816.

Pelkonen O, et al. Inhibition and induction of human cytochrome P450 enzymes: current status. *Arch Toxicol* 2008;82(10):667–715.

Penning T, et al. Synthesis and biological evaluation of the 1,5-diarylpyrazole class of cyclooxygenase-2 inhibitors: identification of 4-[5-(4-methylphenyl)-3-(trifluoromethyl)-1*H*-pyrazol-1-yl]benze nesulfonamide (SC-58635, celecoxib). *J Med Chem* 1997;40(9):1347–1365.

Pirmohamed M, et al. An investigation of the formation of cytotoxic, protein-reactive and stable metabolites from carbamazepine *in vitro*. *Biochem Pharmacol* 1992; 43(8):1675–1682.

Polasek T, Miners JO. *In vitro* approaches to investigate mechanism-based inactivation of CYP enzymes. *Expert Opin Drug Metab Toxicol* 2007;3(3):321–329.

Ponsoda X, et al. Drug metabolism by cultured human hepatocytes: how far are we from the *in vivo* reality? *Altern Lab Animals* 2004;32:101–110.

Pottenger L, et al. The relative bioavailability and metabolism of bisphenol A in rats is dependent upon the route of administration. *Toxicol Sci* 2000;54(1):3–18.

Powis G. Dose-dependent metabolism, therapeutic effect, and toxicity of anticancer drugs in man. *Drug Metab Rev* 1983;14(6):1145–1163.

Prakash C, Shaffer C, Nedderman A. Analytical strategies for identifying drug metabolites. *Mass Spec Rev* 2007;26:340–369.

Proctor N, Tucker G, Rostami-Hodjegan A. Predicting drug clearance from recombinantly expressed CYPs: intersystem extrapolation factors. Proctor NJ, Tucker. *Xenobiotica* 2004;34(2):151–178.

Prueksaritanont T, Lin J, Ballie T. Complicating factors in safety testing of drug metabolites: kinetic differences between generated and preformed metabolites. *Toxicol Appl Pharmacol* 2006;217:143–152.

Puga A, Ma C, Marlowe J. The aryl hydrocarbon receptor cross-talks with multiple signal transduction pathways. *Biochem Pharmacol* 2009;77(4):713–722.

Raj H, et al. Chemoprevention of benzene-induced bone marrow and pulmonary genotoxicity. *Teratogenesis Carcinogenesis and Mutagenesis* 2001;21(2):181–187.

Ramadan Z, Wrang M, Tipton K. Species differences in the selective inhibition of monoamine oxidase (1-methyl-2-phenylethyl)hydrazine and its potentiation by cyanide. *Neurochem Res* 2007;32(10):1783–1790.

Rashed M, Streeter A, Nelson S. Investigations of the *N*-hydroxylation of 3′-hydroxyacetanilide, a non-hepatotoxic positional isomer of acetaminophen. *Drug Metab Dispos* 1989;17(4):355–359.

Reddy A, et al. Validation of a semi-automated human hepatocyte assay for the determination and prediction of intrinsic clearance in discovery. *J Pharm Biomed Anal* 2005;37:319–326.

Rendic S, Di Carlo F. Human cytochrome P450 enzymes: a status report summarizing their reactions, substrates, inducers, and inhibitors. *Drug Metab Rev* 1997;29(1/2): 413–580.

Riley R, Martin I, Cooper A. The influence of DMPK as an integrated partner in modern drug discovery. *Curr Drug Metab* 2002;3:527–550.

Riley R, McGinnity D, Austin R. A unified model for predicting human hepatic, metabolic clearance from *in vitro* intrinsic clearance data in hepatocytes and microsomes. *Drug Metab Dispos* 2005;33(9):1304–1311.

Riley R, Grime K, Weaver R. Time-dependent CYP inhibition. *Expert Opin Drug Metab Toxicol* 2007;3(1):51–66.

Roberts P, Kitteringham N, Park B. Elucidation of the structural requirements for the bioactivation of mianserin *in vitro*. *J Pharm Pharmacol* 1993;45(7):663–675.

Rodrigues AD, Integrated cytochrome P450 reaction phenotyping: attempting to bridge the gap between cDNA-expressed cytochromes P450 and native human liver microsomes. *Biochem Pharmacol* 1999;57:465–480.

Rodrigues AD, editor. Drug–Drug Interactions, 2nd edition. *Drugs and Pharmaceutical Sciences*, Vol. 179. New York: Informa Healthcare, 2008.

Rodrigues AD, Rushmore T. Cytochrome P450 pharmacogenetics in drug development: *in vitro* studies and clinical consequences. *Curr Drug Metab* 2002;3:289–309.

Rolan P. Plasma protein binding displacement interactions—why are they still regarded as clinically important? *Br J Clin Pharmacol* 1994;37(2):125–128.

Rosenblum S, et al. Discovery of 1-(4-fluorophenyl)-(3*R*)-[3-(4-fluorophenyl)-(3*S*)-hydroxypropyl]-(4*S*)-(4-hydroxyphenyl)-2-azetidinone (SCH 58235): a designed, potent, orally active inhibitor of cholesterol absorption. *J Med Chem* 1998;41(6): 973–980.

Rostami-Hodjegan A, Tucker G. Simulation and prediction of *in vivo* drug metabolism in human populations from *in vitro*. *Nature Rev Drug Discov* 2007;6(2):140–148.

Rydberg T, et al. Concentration–effect relations of glibenclamide and its active metabolites in man: modelling of pharmacokinetics and pharmacodynamics. *Br J Clin Pharmacol* 1997;43(4):373–381.

Safety Testing of Drug Metabolites, FDA, editor. http://www.fda.gov/CDER/GUIDANCE/6897fnl.pdf. 2008.

Sahi J, et al. Metabolism and transporter-mediated drug–drug interactions of the endothelin—A receptor antagonist CI-1034. *Chemico-Biol Interactions* 2006;159(2): 156–168.

Sakaguchi K, et al. Glucuronidation of carboxylic acid containing compounds by UDP-glucuronosyltransferase isoforms. *Arch Biochem Biophys* 2004;424:219–225.

Satoh T. Genetic polymorphism in drug metabolism and toxicity: linking animal research and risk assessment in man. *AATEX* 2007;14:443–445.

Schaefer W. Reaction of primary and secondary amines to form carbamic acid glucuronides. *Curr Drug Metab* 2006;7/8:873–881.

Scholz K, et al. Rapid detection and identification of *N*-acetyl-L-cysteine thioethers using constant neutral loss and theoretical multiple reaction monitoring combined with enhanced production scans on a linear ion trap mass spectrometer. *J Am Soc Mass Spectrom* 2005;16:1976–1984.

Sellers E. Plasma protein displacement interactions are rarely of clinical significance. *Pharmacology* 1979;18:225–227.

Shet M, et al. Metabolism of the antiandrogenic drug (Flutamide) by human CYP1A2. *Drug Metab Dispos* 1997;25(11):1298–1303.

Shiran M, et al. Prediction of metabolic drug clearance in humans: *in vitro–in vivo* extrapolation vs allometric scaling. *Xenobiotica* 2006;36(7):567–580.

Shou M. Prediction of pharmacokinetics and drug–drug interactions from *in vitro* metabolism data. *Curr Opin Drug Discovery Dev* 2005;8(1):66–77.

Siebert G, et al. Ion-trapping, microsomal binding, and unbound drug distribution in the hepatic retention of basic drugs. *J Pharmacol Exp Ther* 2004;308(1):228–235.

Silverman R, edition. *The Organic Chemistry of Drug Design and Drug Action*, 2nd edition. New York: Elsevier, 2004.

Sinha V, et al. Predicting oral clearance in humans: how close can we get with allometry? *Clin Pharmacokin* 2008;47(1):35–45.

Sinz M, Wallace G, Sahi J. Current industrial practices in assessing CYP450 enzyme induction: preclinical and clinical. *AAPS J* 2008;10(2):391–400.

Smith D, Obach R. Seeing through the mist: abundance versus percentage. Commentary on metabolites in safety testing. *Drug Metab Dispos* 2005;33(10):1409–1417.

Smith D, Obach R. Metabolites and safety: What are the concerns, and how should we address them? *Chem Res Toxicol* 2006;19:1570–1579.

Smith D, van de Waterbeemd H. Pharmacokinetics and metabolism in early drug discovery. *Curr Opin in Chem Biol* 1999;3:373–378.

Smith D, Jones B, Walker D. Design of drugs involving the concepts and theories of drug metabolism and pharmacokinetics. *Med Res Rev* 1996;16(3):243–246.

Smith D, et al. Clearing the MIST (metabolites in safety testing) of time: The impact of duration of administration on drug metabolite toxicity. *Chemico-Biol Interactions* 2009;179(1):60–67.

Smith K, et al. *In vitro* metabolism of tolcapone to reactive intermediates: relevance to tolcapone liver toxicity. *Chem Res Toxicol* 2003;16(2):123–128.

Soars M, et al. The pivotal role of hepatocytes in drug discovery. *Chemico-Biol Interactions* 2007a;168:2–15.

Soars M, et al. Use of hepatocytes to assess the contribution of hepatic uptake to clearance *in vivo*. *Drug Metab Dispos* 2007b;35(6):859–865.

Soglia J, et al. The development of a higher throughput reactive intermediate screening assay incorporating micro-bore liquid chromatography-micro-electrospray ionization-tandem mass spectrometry and glutathione ethyl ester as an *in vitro* conjugating agent. *J Pharm Biomed Anal* 2004;36:105–116.

Soglia J, et al. A semiquantitative method for the determination of reactive metabolite conjugate levels *in vitro* utilizing liquid chromatography–tandem mass spectrometry and novel quaternary ammonium glutathione analogues. *Chem Res Toxicol* 2006;19:480–490.

Staack R, Hopfgartner G. New analytical strategies in studying drug metabolism. *Anal Bioanal Chem* 2007;388:1365–1380.

Stella V, et al. editors. *Prodrugs:Challenges and Rewards*, Part 2. New York: Springer, 2007.

Stohs S, Al-Turk W, Hassing J. *Age* 1980;3(4):88–92.

Stoner C, et al. Integrated oral bioavailability projection using *in vitro* screening data as a selection tool in drug discovery. *Int J Pharmaceutics* 2004;269:241–249.

Stratford RJ, et al. Application of oral bioavailability surrogates in the design of orally active inhibitors of rhinovirus replication. *J Pharm Sci* 1999;88(8):747–753.

Suzuki K, Kamimura H. Pharmacokinetics and metabolism of an alpha,beta-blocker, amosulalol hydrochloride, in mice: biliary excretion of carbamoyl glucuronide. *Biol Pharm Bull* 2007;30(8):1580–1585.

Szakacs G, et al. The role of ABC transporters in drug absorption, distribution, metabolism, excretion and toxicity (ADME-Tox). *Drug Discovery Today* 2008;13(9/10): 379–393.

Tahara H, et al. A species difference in the transport activities of H2 receptor antagonists by rat and human renal organic anion and cation transporters. *J Pharmacol Exp Ther* 2005;315(1):337–345.

Tahara H, et al. Is the monkey an appropriate animal model to examine drug–drug interactions involving renal clearance? Effect of probenecid on the renal elimination of H2 receptor antagonists. *J Pharmacol Exp Ther* 2006;316(3):1187–1194.

Takakusa H, et al. Covalent binding and tissue distribution/retention assessment of drugs associated with idiosyncratic drug toxicity. *Drug Metab Dispos* 2008; 36(9):1770–1779.

Takano M, Yumoto R, Murakami T. Expression and function of efflux drug transporters in the intestine. *Pharmacol Ther* 2006;109(1–2):137–161.

Takeda S, et al. Inhibition of UDP-glucuronosyltransferase 2b7-catalyzed morphine glucuronidation by ketoconazole: dual mechanisms involving a novel noncompetitive mode. *Drug Metab Dispos* 2006;34(8):1277–1282.

Tang W. Drug metabolite profiling and elucidation of drug-induced hepatotoxicity. *Expert Opin Drug Metab Toxicol* 2007;3(3):407–420.

Tang W, Abbott F. A comparative investigation of 2-propyl-4-pentenoic acid (4-ene VPA) and its alpha-fluorinated analogue: phase II metabolism and pharmacokinetics. *Drug Metab Dispos* 1997;25(2):219–227.

Tassaneeyakul W, et al. Caffeine metabolism by human hepatic cytochromes P450: contributions of 1A2, 2E1 and 3A isoforms. *Biochem Pharmacol* 1994;47(10): 1767–1776.

Thompson C, Kinter M, Macdonald T. Synthesis and *in vitro* reactivity of 3-carbamoyl-2-phenylpropionaldehyde and 2-phenylpropenal: putative reactive metabolites of felbamate. *Chem Res Toxicol* 1996;9(8):1225–1229.

Thompson T, Optimization of metabolic stability as a goal of modern drug design. *Med Res Rev* 2001;21(5):412–449.

Thornton-Manning J, et al. Metabolism of 3-methylindole by vaccinia-expressed P450 enzymes: correlation of 3-methyleneindolenine formation and protein-binding. *J Pharmacol Exp Ther* 1996;276(1):21–29.

Thummel KE, Shen D, Isoherranen N. Role of the gut mucosa in metabolically based drug–drug interactions. In: Rodrigues AD, editor. *Drug–Drug Interactions*, Informa Healthcare, 2008, pp. 471–513.

Timbrell J, editor. *Principles of Biochemical Toxicology*. London: Taylor & Francis, 1982.

Timchalk C, Smith F, Bartels M. Route-dependent comparative metabolism of [14C] toluene 2,4-diisocyanate and [14C]toluene 2,4-diamine in Fischer 344 rats. *Toxicol Appl Pharmacol* 1994;124:181–190.

Tirona R, et al. The orphan nuclear receptor HNF4alpha determines PXR- and CAR-mediated xenobiotic induction of CYP3A4. *Nat Med* 2003;9(2):220–224.

Tomlinson E, et al. Dexamethasone metabolism *in vitro*: species differences. *J Steroid Biochem Mol Biol* 1997;62(4):345–352.

Townsend D, et al. High pressure liquid chromatography and mass spectrometry characterization of the nephrotoxic biotransformation products of Cisplatin. *Drug Metab Dispos* 2003;31(6):705–713.

Tracy T, Atypical cytochrome p450 kinetics: implications for drug discovery. *Drugs R D* 2006;7(6):349–363.

Tracy T, Hummel M. Modeling kinetic data from *in vitro* drug metabolism enzyme experiments. *Drug Metab Rev* 2004;36(2):231–242.

Uetrecht J. Prediction of a new drug's potential to cause idiosyncratic reactions. *Curr Opin Drug Discovery Dev* 2001;4(1):55–59.

Uetrecht J. Screening for the potential of a drug candidate to cause idiosyncratic drug reactions. *Drug Discovery Tech* 2003;8(18):832–837.

Uetrecht J. Evaluation of which reactive metabolite, if any, is responsible for a specific idiosyncratic reaction. *Drug Metab Rev* 2006;38:745–753.

Uetrecht J, et al. Oxidation of aminopyrine by hypochlorite to a reactive dication: possible implications for aminopyrine-induced agranulocytosis. *Chem Res Toxicol* 1995;8(2):226–233.

Urquhart B, Tirona R, Kim R. Nuclear receptors and the regulation of drug-metabolizing enzymes and drug transporters: implications for interindividual variability in response to drugs. *J Clin Pharmacol* 2007;47(5):566–578.

Uttamsingh V, et al. Relative contributions of the five major human cytochromes P450, 1A2, 2C9, 2C19, 2D6, and 3A4, to the hepatic metabolism of the proteasome inhibitor bortezomib. *Drug Metab Dispos* 2005;33(11):1723–1728.

Van L, et al. Metabolism of dextrorphan by CYP2D6 in different recombinantly expressed systems and its implications for the *in vitro* assessment of dextromethorphan metabolism. *J Pharm Sci* 2009;98(2):763–771.

van de Steeg E, et al. Methotrexate pharmacokinetics in transgenic mice with liver-specific expression of human OATP1B1 (SLCO1B1). *Drug Metab Dispos* 2009;37(2):277–281.

van Voorhis W, et al. Efficacy, pharmacokinetics, and metabolism of tetrahydroquinoline inhibitors of *Plasmodium falciparum* protein farnesyltransferase. *Antimicrob Agents Chemother* 2007;51(10):3659–3671.

Venkatakrishnan K, Obach R. *In vitro–in vivo* extrapolation of CYP2D6 inactivation by paroxetine: prediction of nonstationary pharmacokinetics and drug interaction magnitude. *Drug Metab Dispos* 2005;33(6):845–854.

Venkatakrishnan K, Obach R. Drug–drug interactions via mechanism-based cytochrome P450 inactivation: points to consider for risk assessment from *in vitro* data and clinical pharmacologic evaluation. *Curr Drug Metab* 2007;8:449–462.

Venkatakrishnan K, et al. Comparison between cytochrome P450 (CYP) content and relative activity approaches to scaling from cDNA-expressed CYPs to human liver microsomes: ratios of accessory proteins as sources of discrepancies between the approaches. *Drug Metab Dispos* 2000;28(12):1493–1504.

Venkatakrishnan K, Von Moltke LL, Greenblatt DJ. Human drug metabolism and the cytochromes P450: application and relevance of *in vitro* models. *J Clin Pharmacol* 2001;41:1149–1179.

Venkatakrishnan K, et al. Drug metabolism and drug interactions: application and clinical value of *in vitro* models. *Curr Drug Metab* 2003;4:423–459.

Veronese M, et al. Exposure-dependent inhibition of intestinal and hepatic CYP3A4 *in vivo* by grapefruit juice. *J Clin Pharmacol* 2003;43(8):831–839.

Victor F, et al. Synthesis, antiviral activity, and biological properties of vinylacetylene analogs of enviroxime. *J Med Chem* 1997;40(10):1511–1518.

Voorman R, et al. Microsomal metabolism of delavirdine: evidence for mechanism-based inactivation of human cytochrome P450 3A. *J Pharmacol Exp Ther* 1998;287(1):381–388.

Von Moltke LL, et al. Zolpidem metabolism *in vitro*: responsible cytochromes, chemical inhibitors, and *in vivo* correlations. *Br J Clin Pharmacol* 1999;48:89–97.

Von Moltke LL, et al. Potent mechanism-based inhibition of human CYP3A *in vitro* by amprenavir and ritonavir: comparison with ketoconazole. *Eur J Clin Pharmacol* 2000;56:259–261.

Walgren J, Mitchell M, Thompson D. Role of metabolism in drug-induced idiosyncratic hepatotoxicity. *Critical Rev Toxicol* 2005;35:325–361.

Walle T, Wen X, Walle U. Improving metabolic stability of cancer chemoprotective polyphenols. *Expert Opin Drug Metab Toxicol* 2007;3(3):379–388.

Wang H, LeCluyse E. Role of orphan nuclear receptors in the regulation of drug-metabolising enzymes. *Clin Pharmacokinet* 2003;42(15):1331–1357.

Wang S, et al. Induction of hepatic phase II drug-metabolizing enzymes by 1,7-phenanthroline in rats is accompanied by induction of MRP3. *Drug Metab Dispos* 2003;31(6):773–775.

Wang X, Morris M. Diet/nutrient interactions with drug transporters. In You G, Morris M, editors. *Drug Transporters: Molecular Characterization and Role in Drug Disposition*, Hoboken, NJ: John Wiley & Sons, 2007, pp. 665–708.

Weaver R. Assessment of drug–drug interactions: concepts and approaches. *Xenobiotica* 2001;31(8/9):499–538.

Wauthier V, Verbeek R, Calderon P. The effect of ageing on cytochrome p450 enzymes: consequences for drug biotransformation in the elderly. *Curr Med Chem* 2007;14:745–757.

Weinkers L, Heath T. Predicting *in vivo* drug interactions from *in vitro* drug discovery data. *Nature Rev* 2005;4:825–833.

Wilkinson G. Plasma and tissue binding considerations in drug disposition. *Drug Metab Rev* 1983;14(3):427–465.

Wilkinson G. Clearance approaches in pharmacology. *Pharmacol Rev* 1987;39(1): 1–47.

Wilkinson G, Shand D. Commentary: a physiological approach to hepatic drug clearance. *Clin Pharmacokinet Ther* 1975;18(4):377–390.

Williams D, et al. Induction of metabolism-dependent and -independent neutrophil apoptosis by clozapine. *Mol Pharmacol* 2000;58(1):207–216.

Williams J, et al. Differential modulation of UDP-glucuronosyltransferase 1A1 (UGT1A1)-catalyzed estradiol-3-glucuronidation by the addition of UGT1A1 substrates and other compounds to human liver microsomes. *Drug Metab Dispos* 2002;30(11):1266–1273.

Williams J, et al. *In vitro* ADME phenotyping in drug discovery: current challenges and future solutions. *Curr Opin in Drug Discovery Dev* 2005;8(1):78–88.

Williams J, et al. PhRMA white paper on ADME pharmacogenomics. *J Clin Pharmacol* 2008;48:849–889.

Xu C, Li Y-T, Kong A-N. Induction of phase I, II and III drug metabolism/transport by xenobiotics. *Arch Pharm Res* 2005;28(3):249–268.

Xu S, et al. Metabolic activation of fluoropyrrolidine dipeptidyl peptidase-IV inhibitors by rat liver microsomes. *Drug Metab Dispos* 2005;33(1):121–130.

Yamashiro W, et al. Involvement of transporters in the hepatic uptake and biliary excretion of valsartan, a selective antagonist of the angiotensin II AT1-receptor, in humans. *Drug Metab Dispos* 2006;34(7):1247–1254.

Yamazaki H, Shimada T. *Arch Biochem Biophys* 1993;346(1):161–169.

Yang J, et al. Theoretical assessment of a new experimental protocol for determining kinetic values describing mechanism (time)-based enzyme inhibition. *Eur J Pharm Sci* 2007;31:232–241.

Yan Z, Caldwell C, Maher N. Unbiased high-throughput screening of reactive metabolites on the linear ion trap mass spectrometer using polarity switch and mass tag triggered data-dependent acquisition. *Anal Chem* 2008;80:6410–6422.

Yao C, Levy R. Inhibition-based metabolic drug–drug interactions: predictions from *in vitro* data. *J Pharm Sci* 2002;91(9):1923–1935.

Yasumori T, et al. Species differences in stereoselective metabolism of mephenytoin by cytochrome P450 (CYP2C and CYP3A). *J Pharmacol Exp Ther* 1993;264(1): 89–94.

Yeh K, et al. Pharmacokinetics and bioavailability of Sinemet CR: a summary of human studies. *Neurology* 1989;39(11(2)):25–38.

Yengi L, Leung L, Kao J. The evolving role of drug metabolism in drug discovery and development. *Pharm Res* 2007;24(5):842–858.

Yoshihara S, Harada K, Ohta S. Metabolism of 1-methyl-4-phenyl-1,2,3,6-tetrahydropyridine (MPTP) in perfused rat liver: involvement of hepatic aldehyde oxidase as a detoxification enzyme. *Drug Metab Dispos* 2000;28(5):538–543.

You G, Morris M, editors. *Drug Transporters: Molecular Characterization and Role in Drug Disposition*. Wiley Series in Drug Discovery and Development, Wang B, series editor. Hoboken, NJ: John Wiley & Sons, 2007.

Youdim K, et al. Application of CYP3A4 *in vitro* data to predict clinical drug–drug interactions; predictions of compounds as objects of interaction. *Br J Clin Pharmacol* 2008;65(5):680–692.

Zhang H, et al. Reduction of site-specific CYP3A-mediated metabolism for dual angiotensin and endothelin receptor antagonists in various *in vitro* systems and in cynomolgus monkeys. *Drug Metab Dispos* 2007a;35(5):795–805.

Zhang H, et al. Cytochrome P450 reaction-phenotyping: an industrial perspective. *Expert Opin Drug Metab Toxicol* 2007b;3(5):667–687.

Zhang K, et al. Metabolism of A dopamine D(4)-selective antagonist in rat, monkey, and humans: formation of A novel mercapturic acid adduct. *Drug Metab Dispos* 2000;28(6):633–642.

Zhang L, et al. A regulatory viewpoint on transporter-based drug interactions. *Xenobiotica* 2008;38(7/8):709–724.

Zhao P, Kalhorn T, Slattery J. *Hepatology* 2003;36(2):326–335.

Zhao P, The use of hepatocytes in evaluating time-dependent inactivation of P450 *in vivo*. *Expert Opin Drug Metab Toxicol* 2008;4(2):151–164.

Zhou C, et al. Tocotrienols activate the steroid and xenobiotic receptor, SXR, and selectively regulate expression of its target genes. *Drug Metab Dispos* 2004a;32(10):1075–1082.

Zhou S, et al. Pharmacokinetic interactions of drugs with St John's wort. *J Psychopharmacol* 2004b;18(2):262–276.

Zhou S, et al. Substrates and inhibitors of human multidrug resistance associated proteins and the implications in drug development. *Curr Med Chem* 2008;15(20): 1981–2039.

Zhou S-F. Drugs behave as substrates, inhibitors and inducers of human cytochrome P450 3A4. *Curr Drug Metab* 2008;9(4):310–322.

Zhou S-F, et al. Therapeutic drugs that behave as mechanism-based inhibitors of cytochrome P450 3A4. *Curr Drug Metab* 2004c;5:415–442.

6

APPLIED PHARMACOKINETICS IN DRUG DISCOVERY AND DEVELOPMENT

Hua Yang, Xingrong Liu, Anjaneya Chimalakonda, Zheng Lu, Cuiping Chen, Frank Lee, and Wen Chyi Shyu

6.1 INTRODUCTION

Drug development attrition due to absorption, distribution, metabolism, and excretion (ADME) reasons has greatly decreased in the past decade. Based on Ismail Kola and John Landis, the drug development attrition rate due to pharmacokinetics (PK) has been decreased from approximately 40% of the total development attrition in 1991 to approximately 10% in 2000 (Kola and Landis, 2004; Prentis et al., 1988). This substantial improvement in reducing drug metabolism and pharmacokinetics (DMPK)-related drug attrition is greatly attributed to the new drug discovery model that has been widely adopted in the pharmaceutical industry in the past two decades. In the new model, drug metabolism and pharmacokinetics are incorporated into the early drug discovery stage. Compounds are evaluated for their potency as well as for their DMPK properties (Roberts, 2003; Alavijeh and Palmer, 2004). Different DMPK evaluation technologies and tools have been developed and implanted in drug discovery to provide valuable DMPK versus structure relationships for chemical structure optimization. The chemical synthesis–DMPK assessment–structure modifications–DMPK assessment cycles have been used to generate compounds that have desirable DMPK properties for clinical development. In the compound structure modification–DMPK assessment

Enzyme Inhibition in Drug Discovery and Development, edited by Lu and Li
Copyright © 2010 John Wiley & Sons, Inc.

cycles, the DMPK efforts have been focused on metabolic stability, cell permeability, clearance (CL), half-life ($t_{1/2}$), oral bioavailability, and drug–drug interaction (DDI) potential. The rapid turnaround time required for DMPK assessment to enable modification of a chemical structure has become one of the key elements in drug optimization (Lin and Lu, 1997; Lin et al., 2003).

There are many different ways to implement DMPK screening in drug discovery. Since a good drug should be efficacious, safe, and with desirable DMPK properties, different strategies have been applied in optimizing all these properties. Some companies only screen compounds for their DMPK properties after the compounds have met the potency and selectivity criteria. Other companies, large-size pharmaceutical companies in particular, characterize DMPK properties on all compounds synthesized. Different DMPK evaluation technologies and assays have been developed in order to support the objectives of drug discovery and development at different stages. In the early discovery stage, rapid testing on many compounds to provide general information on the key chemotypes of the compound is essential. Therefore, *in vitro* and *in vivo* high-throughput assays are important tools to provide adequate information regarding structure–DMPK relationships, as well as to screen or rank compounds. Examples of these assays include *in vitro* metabolic stability assessment using microsomal or S9 preparations, CYP inhibition screening using fluoresce probes, pharmacokinetics screening using cassette dosing, single to limited points pharmacokinetic screening, sampling pooling, and so on. Compared to the traditional or more definitive DMPK study designs, these high-throughput assays are simplified, sometimes automated, and less comprehensive or less rigorous in terms of study design and assay validation. The results may not be definitive in terms of the absolute quantitative values of the DMPK parameters. However, these quick assays offer rich DMPK information with comparative or relative nature to a large number of compounds and enable massive and quick compound screening, structure–activity analysis, issue identification, and compound optimization, and hence they are of great value in early drug discovery (Riley and Grime, 2004; Kansy et al., 2004; Lipinski et al., 1997; Korfmacher et al., 1997; Kuo et al., 1998; Cox et al., 1999; Hop et al., 1998; Berman et al., 1997).

In order to support a discovery program, *in vivo* animal pharmacokinetic information must be obtained. This information, combined with the *in vitro* metabolism data, is crucial to identifying potential human DMPK issues and in exploring the *in vivo–in vitro* correlations to assist the selection of the right screening tools. Discovery program support usually demands quick feedback time, and the available amount of compound material for pharmacokinetic studies is usually limited. In order to utilize the resource optimally, pharmacokinetic studies are usually limited to one species, and they may expand to other species for lead compounds or to investigate specific questions. It is critical in identifying the appropriate animal species for *in vivo* pharmacokinetic studies to provide reliable prediction of drug absorption, distribution, and elimination in humans. Program- and compound-specific information and knowledge, early *in vitro* and *in vivo* DMPK studies, and practical concerns are usually the major determinants in pharmacokinetic animal species selection (Lin and Lu, 1997).

Pharmacokinetics also serves as a "bridging" function in supporting pharmacology, pharmacodynamic, and toxicology studies, as well as formulation development. In discovery or preclinical development programs, DMPK scientists work closely with the pharmacology, toxicology, and formulation functions and contribute to study designs (especially dose, dose regime, and compound exposure assessment) and study results interpretation. Analysis and understanding of PK–pharmacodynamics (PD), PK–efficacy, and PK–toxicology relationships, determination of the efficacy, and determination of toxicity exposure are some of the most important contributions to the discovery or development programs. This information is very important in the internal go/no-go decisions and has great value in guiding the clinical study designs.

There are many common DMPK-related drug development issues, such as poor oral bioavailability, short half-life, drug–drug interactions, metabolites that are related to toxicity, and pharmacokinetic variability. In the past 10–20 years, the pharmaceutical industry has accumulated substantial experience with many successful cases in the projection of human pharmacokinetics based on different preclinical *in vitro* or *in vivo* approaches (Huang et al., 2008; Riviere et al., 1997; Mahmood, 1999, 2007; Sinha et al., 2008). However, the projection of the human efficacious dose, based on the projected human PK and efficacious exposure obtained in preclinical animal models, remains a major challenge. The fundamental question is regarding the quantitative predictability of the animal models, in terms of their mechanistic similarities to the human disease and consequentially the response to drug treatment. As a recent trend in the pharmaceutical industry, new approaches such as the use of mechanism-based PD markers, the establishment of more relevant animal disease models in efficacy assessments, and the incorporation of mechanism-based pharmacokinetics–pharmacodynamics and efficacy analysis are being introduced into drug discovery and development to guide clinical trial design.

The pharmaceutical industry faces considerable challenges, with lower revenue growth, a lower number of new chemical entities approvals, and higher expenses in pharmaceutical R&D. The cost of developing a new drug is spiraling upwards, with estimates of as much as US$1 billion to bring a new medicine to the market (Dimasi et al., 2003). The current status of the pharmaceutical industry demands significant improvement in its efficiency in drug discovery and development, with new thinking, strategies, and new technologies. New technologies, such as computer-aided approaches (*in silico*), use of iRNA technologies, and stem-cell-based test systems, are being widely discussed and explored in the pharmaceutical industry. Since developmental drugs usually fail due to either efficacy or safety reasons, there is a clear need to find human-relevant, predictive, and cost-effective test systems for drug efficacy and safety evaluations (Kapetanovic, 2008; Congreve et al., 2008; Ekins et al., 2007; Sundstrom, 2007).

6.2 OVERVIEW OF DMPK IN DRUG DISCOVERY AND PRECLINICAL DEVELOPMENT

DMPK plays important roles in drug discovery and development by designing optimum DMPK properties, assessing and minimizing the DDI potential of the compounds, supporting the preclinical *in vivo* PK/PD, efficacy, and safety evaluations, profiling the DMPK properties of the development candidates, supporting formulation development, projecting human PK and efficacious dose, and supporting the rational selection of dose and dosing regime for the clinical studies.

The major tasks for preclinical DMPK include initial routine studies to support compound screening and structural modification, and subsequent characterization of the development candidate. In order to effectively support discovery programs in the initial compound screening, the criteria for key assays are high capacity, adequate reliability, appropriate reproducibility, and rapid turnaround time. Once the final candidate(s) is indentified for clinical development, comprehensive DMPK studies are conducted for the development candidate to collect adequate DMPK information to project human PK profiles, to assess the DMPK issues, and to help establish the human dose and dosing regime design. It is important to identify the potential DMPK issues associated with a chemical series or a development candidate and to identify the right tools to assess or manage the issues.

Compounds in different chemical series usually have different DMPK properties and DMPK-related issues. For a given chemical series, the major DMPK-related issues can vary from short elimination half-life, high total body clearance, poor oral bioavailability, or variable plasma drug exposure, to great likelihood of DDI potential. Since drug discovery requires a lot of good science and some luck, it is important to identify the major DMPK issues at the early discovery stage for the given program and manage the issues through compound structural modification with the feedback from the DMPK evaluations. A common approach to identifying the major DMPK issues at the early stage of drug discovery consists of a panel of *in vivo* and *in vitro* studies. Usually a handful of representative compounds from different chemotypes are selected for DMPK evaluation. The DMPK assessment usually includes rat pharmacokinetic studies following intravenous and oral administration with urine collections, *in vitro* metabolic stability or intrinsic clearance studies using microsomal or S9 fractions (usually in rats and humans), CYP inhibition screening (against five major CYP enzymes, 1A2, 2C9, 2C19, 2D6, and 3A4), and cell permeability screening (using PAMPA or Caco-2 cell line). The results from these *in vivo* and *in vitro* studies provide early comprehensive information regarding the major DMPK issues, some initial insights into the human relevance, and potential mechanisms that attribute to the issues. Based on this information, appropriate strategies and tools are identified and applied to further evaluate the compounds and to eliminate or minimize the unwanted DMPK properties.

For example, from rat pharmacokinetic stud, the common DMPK issues are high total body clearance and/or poor oral bioavailability. The primary cause of high total body clearance is rapid metabolism. In this case, rat urinary excretion of the compound is low and the intrinsic clearance in the microsomal or S9 fractions is high. There is a good correlation between the *in vivo* total body clearance and *in vitro* metabolic stability (or intrinsic clearance). It is important to compare the metabolic stability of different compounds, especially in terms of compound ranking, in rat and human microsomal or S9 fractions. Most compounds are metabolically more stable in humans than in rats. A dramatic difference in the *in vitro* metabolic stability between rats and humans or a lack of correlation between the *in vitro* metabolic stability between these two species may suggest different metabolic mechanisms and warrant a follow-up investigation. When there is a good *in vivo* and *in vitro* correlation in clearance between rats and humans, *in vitro* metabolic stability assays, which are usually fast and with high capacities, can be used to screen compounds and to develop a structure–clearance relationship. In this case, one can also rely on rat studies to generate adequate pharmacokinetic information.

It is not uncommon for there to be no correlation between *in vitro* metabolic stability and *in vivo* total body clearance. The *in vivo* system is much more complicated than the *in vivo* system using microsomes or S9 fractions. In addition to liver metabolic intrinsic clearance, other factors may play important roles in the *in vivo* elimination of the compounds. Factors such as plasma protein binding, red blood cell partitioning, extrahepatic metabolism, urinary and biliary excretions can make an *in vitro* prediction fail. In this situation, one might considering using *in vivo* pharmacokinetic studies to screen compounds. *In vivo* studies usually have higher cost and lower capacity and are also associated with a longer turnaround time. Therefore appropriate *in vivo* study strategy is important. Several pharmacokinetic screening methods are widely used, each of which has different advantages and disadvantages. One approach is cassette dosing, in which several compounds are formulated together and dosed to the animals as a mixture (Watt et al., 2000; Brewer and Henion, 1998; Cox et al., 2002; Smith et al., 2007; Korfmacher et al., 2001; Ackermann, 2004). In order to avoid the potential DDI among the mixture of compounds, which may provide misleading results, a cassette dosing approach is usually used at a low dose level, and compounds are selected with minimal potential for DDI based on *in vitro* information. A well-characterized control compound is included in the cassette dosing to help interpret the study results. Cassette dosing can increase the *in vivo* study capacity, minimize the number of animals used, shorten the overall information turnaround time, and provide adequate pharmacokinetic information for compound screening and comparison purposes. The cassette dosing approach can be used in all animal species, such as rat, dog, and monkey. In the event that there is no appropriate *in vitro* screening tool and rat is not the appropriate pharmacokinetic species (i.e., due to significantly different elimination mechanism between rats and humans) for guiding the human PK optimization, cassette dosing is a great cost-effective approach in using large animal species for PK screening. However, an appropriate cassette dosing approach requires significant bioanalytical and formulation support and experience. In addition, there are several other *in vivo* PK screening approaches aimed at lowering the bioanalytical burden (Hopfgartner and Bourgogne, 2003; Atherton et al., 1999; Singh et al., 2005). These approaches include brief PK (only one or limited time points are taken), sample pooling, or single animal PK screening.

Rat is the most commonly used animal species in preclinical pharmacokinetic evaluations. However, one has to keep in mind that the objective of the rat study is to provide reliable pharmacokinetic information to optimize human PK properties of the compounds. It is important to conduct the pharmacokinetic studies in a human-relevant species when significant differences in the elimination pathway between rats and humans are observed or there is no correlation in the *in vitro* metabolic stability between rats and humans. For example, in program A, a panel of *in vitro* and *in vivo* DMPK studies were conducted for six representative compounds, the key results of which are summarized in Table 6.1. The results indicate that the estimated clearance value based on liver microsomal studies is much higher in rats than that in humans. In addition, the estimated clearance value based on liver microsomal studies is similar in humans and in monkeys. Based on the results, a PK study in two monkeys using intravenous administration and cassette dosing was conducted. The results suggest that there is a good correlation between the *in vitro* and *in vivo* clearances in monkey, and high clearance is unlikely a general liability in humans for compounds of this chemotype.

TABLE 6.1. *In Vivo* and *In Vitro* DMPK Parameters for Representative Compounds in Program A

Compound	Rat *in vivo* CL (L/h/kg)	Rat *in vitro* CL (L/h/kg)	Human *in vitro* CL (L/h/kg)	Monkey *in vitro* CL (L/h/kg)
A	3.4	3.7	0.3	0.5
B	3.9	4.2	0.5	0.4
C	1.2	2.2	0.2	0.4
D	5.7	4.1	0.4	0.5
E	2.7	3.5	0.4	0.1
F	1.9	3.1	0.2	0.3

TABLE 6.2. ADME Discovery Assays by Stage—Deliverables

H2L	LO	Late LO and DC
Identify major human relevant DMPK issues	Optimization of PK profiles for development candidate selection	Projected human pharmacokinetics, efficacious dose
Identify relevant *in vitro* or *in vivo* tools for compound screen	ADME profiling and characterize potential drug/drug interactions for development candidate selection	Characterize anticipated human drug–drug interactions
In vivo and *in vitro* correlation		Define exposure and toxicity
		Calculation of preliminary safety
		Analysis PK–PD relationship in efficacy model

Nonclinical drug discovery and development are usually staged based on the different objectives and deliverables at different phases. Drug discovery efforts usually start at the target identification, target validation, and high-throughput screening, where the primary objectives of the program are to (1) identify a potential therapeutic target based on the knowledge in disease etiology and the biological systems associated with it, (2) establish the key biology or biochemistry assays to evaluate the binding affinity and potency of the compounds, and (3) screen library compounds for target affinity to identify attractive chemical series. DMPK support usually starts at the late "hit to lead" stage, where one or more attractive chemical series have been identified and the compound evaluation is started to select the lead chemical series; however, there is a tendency to initiate DMPK-related activities earlier in the drug discovery process than in the past. At the initial stage, the focus of the DMPK effort is to identify the major DMPK issues and *in vivo* or *in vitro* screening tools that could address the DMPK-related issues. As summarized in Tables 6.2 and 6.3, selected compounds are evaluated for *in vitro* metabolic stability (usually in humans and rats), Caco-2 permeability, and rat PK following intravenous and oral doses (Balani et al., 2005). Correlation analysis is commonly conducted to assess the *in vitro* versus *in vivo* and rat versus human relationships. Extremely high *in vivo* clearance (i.e., greater than hepatic blood flow or cardiac output) is not uncommon for the early compounds and is usually an indication of the instability of the compounds in dosing formulation (plasma or blood) or chemical instability. Additional studies are usually warranted to further identify the mechanism and human relevance. When high *in vivo* plasma clearance is associated with high plasma volume of distribution, the partitioning of the testing compounds into the red blood cells should be

TABLE 6.3. ADME Discovery Assays, by Stage

H2L	LO	Late LO and DC
In vivo animal PK in rodents	*In vivo* PK optimization: cassette or discrete PK screen in appropriate species	PK in nonrodent species
Hepatic intrinsic clearance	Use of appropriate *in vitro* assays to screen compounds and address PK liabilities	PK/PD correlation analysis and modeling
Caco-2 permeability	ID major metabolites using *in vitro* assays	Project Human PK and efficacious dose
Studies to resolve exposure issues in pharmacology models	CYP inhibition (competitive and mechanism based)	Determine toxicokinetics in toxicity studies
	ID transporter substrates, RBC partition, and plasma stability	Non-GLP validation of bioanalytical methods for parent drug and method transfer
		Stable isotope synthesis
		Radioisotope synthesis
		Determine CYP and transporter inhibition IC50s
		Transporter mapping
		CYP/UGT mapping
		Plasma protein binding

evaluated. When a compound is highly partitioned into red blood cells, its exposure in whole blood should be monitored and *in vivo* total body blood clearance should be measured. Poor oral bioavailability is also a common DMPK issue. The solubility, cell permeability, and liver microsomal stability information of the compounds should be evaluated together to identify the major contributing factors. Sometimes, additional studies such as portal vein infusion or sampling as well as the use of 1-aminobenzotriazole (ABT, a mechanism-based P450 inhibitor) can be employed to help with the mechanism identification (Balani et al., 2004; Woodcroft et al., 1997). For organ-targeted therapy, such as CNS, liver, or tumor targets, drug exposure in the targeted organ is critical. Screening of compounds for their target organ or tissue exposure could be important. For certain chemical series, target organ or tissue distribution correlates with volume of distribution of the compounds. In this case, volume of distribution might be used as a surrogate for organ or tissue exposure. Identification of the major chemical-series-specific DMPK issues and the related mechanisms is very important in effective drug discovery support. For example, if the major factor contributing to poor oral bioavailability is cell permeability, chemists can then focus their efforts on optimizing the cell permeability of the compounds.

At the lead optimization stage, the major DMPK focus is to screen and optimize compounds (as shown in Table 6.2). The chemical-series-specific ADME tools identified in the hit-to-lead stage are used for this purpose. Metabolite identification is commonly used to identify the major metabolites, and the information is feedback to chemists for structural modification of the compounds. Through lead optimization, a limited number of compounds showing good efficacy and desirable DMPK properties are identified as potential development candidates. Comprehensive DMPK profiling is then conducted to further characterize the compounds and identify potential DMPK liabilities. As shown in Table 6.3, a panel of DMPK studies is conducted to help with the final selection of the development candidate, to enable the projection of human PK and efficacious dose range. A typical discovery DMPK study flow chart is presented in Fig. 6.1 (Balani et al., 2005).

The following sections will focus on the basic PK concepts, basic DMPK strategy in drug discovery, PK data analysis, and human PK and efficacious dose projection.

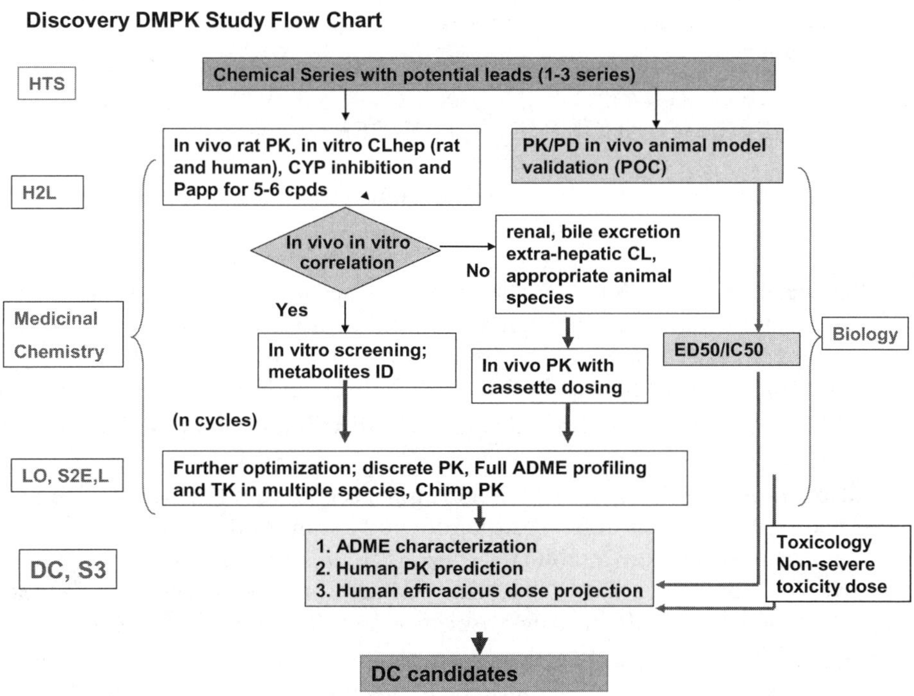

Figure 6.1. A typical discovery DMPK study flow chart.

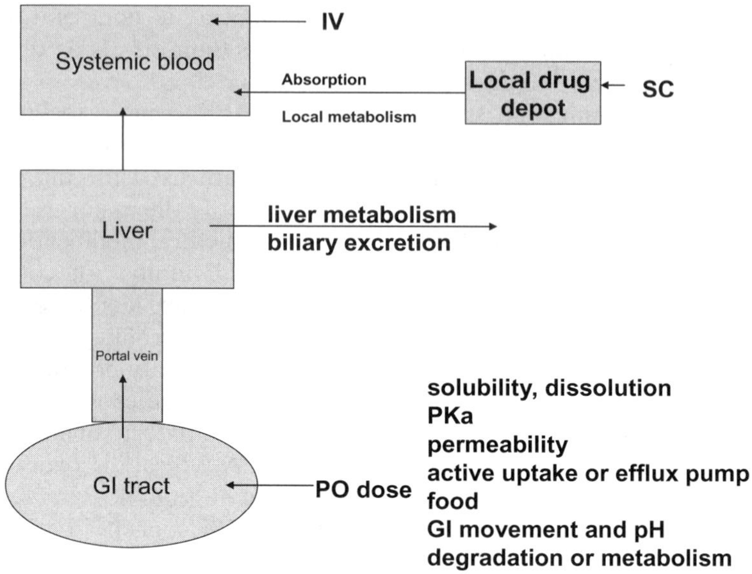

* IV: intravenous administration; SC: subcutaneous administration;

PO: oral administration

Figure 6.2. A brief drug absorption process and some important factors that affect drug absorption.

6.3 BASIC PHARMACOKINETIC CONCEPTS

The commonly used pharmacokinetics concepts are listed here for your reference.

Pharmacokinetics (PK): Describe the relationship between dose and concentrations of a drug in a reference fluid or tissue such as the plasma or blood; what the body does with the drug; and study of the time course of drug absorption, distribution, metabolism, and excretion.

ADME: Abbreviation for absorption, distribution, metabolism, and excretion.

Absorption: A process by which a drug proceeds from the site of administration to blood. It describes the rate and extent of drug entry into body (measured by bioavailability). There is no absorption process for intravenous administration. A brief drug absorption process and some important factors that affect drug absorption are illustrated in Fig. 6.2.

Oral bioavailability (oral F): Measures the extent to which the active ingredient is absorbed and becomes available at the site of action (the bloodstream is used as a surrogate).

Distribution: Following absorption, the compound reaches other tissues or organs throughout the body via blood flow, diffusion, and/or transporter. The extent of drug distribution is determined by volume of distribution. Distribution is a reversible process among all tissues/organs within the body.

Volume of distribution (*V*): A measure of the apparent space in the body throughout which the drug is distributed. Volume of distribution is a proportionality constant between the amount of drug in the body and the concentration measured at certain reference tissue (such as plasma).

$$V = \text{amount of drug in body/concentration at measurement site} \quad (6.1)$$

Major factors that determine drug distribution include delivery of drug to tissue (determine the rate of distribution), cell membrane permeability, binding of drug to plasma and tissue components, and the effects of active transporters (influx and efflux pumps).

Metabolism and excretion: These two processes together are called elimination, which removes drug from the body. Rate of elimination is determined by the sum of all elimination pathways.

Clearance concepts: Clearance (CL) is a proportionality rate constant that relates drug concentration at the reference site (commonly blood or plasma) to the elimination rate (amount/unit of time). Blood CL is equal to the volume of blood completely cleared of drug per unit time (L/h, L/h/kg).

$$\text{CL} = \text{rate of elimination/blood concentration} \quad (6.2)$$

1. CL is the true measure of the ability of organs or the body to remove (or clear) substances from the body.
2. The total body CL (not the $t_{1/2}$) reflects changes in the body's ability to eliminate drugs.
3. CL is a model-independent parameter; it can be estimated with compartmental or noncompartmental pharmacokinetic analysis.
4. CL is an additive parameter; individual organ CL adds up to the total body CL.

$$\text{CL}_{\text{tot}} = \text{CL}_m + \text{CL}_r + \text{CL}_b + \text{CL}_o \quad (6.3)$$

where CL_{tot} is the total body clearance, CL_m is the metabolic clearance, CL_r is the clearance via renal elimination (nonmetabolism), CL_b is the elimination via biliary elimination (nonmetabolism), and CL_o is the clearance via other elimination pathways.

TABLE 6.4. Hepatic Blood Flow in Different Animal Species

	Mouse	Rat	Rabbit	Monkey	Dog	Human
Liver blood flow (mL/min/kg)	86	69	52	44	30	21.4
Liver blood flow (L/h/kg)	5.2	4.1	3.1	2.7	1.8	1.3

Boxenbaum J. *Pharmacokinet Biopharm* 1980;8:165.

Hepatic CL: The volume of plasma (or blood) from which a drug is completely removed by liver in a given time period. It determines the ability of the liver to remove drug from the blood, and is the function of liver metabolic activity and biliary excretion.

$$CL_h = CL_{hm} + CL_b \tag{6.4}$$

where CL_h is the hepatic clearance, CL_{hm} is the hepatic metabolic clearance, and CL_b is the biliary excretion clearance. Drugs can be classified as low, medium, or high clearance based on the hepatic extraction ratio (ε).

$$\varepsilon = CL_h / Q \tag{6.5}$$

where CL_h is the hepatic clearance and Q is the hepatic blood flow. When a drug is mainly eliminated in liver, $\varepsilon = CL/Q$ where CL is the total body clearance. When ε is greater than 0.7, the compound is classified as high clearance; when ε is less than 0.3, the compound is classified as low clearance. Hepatic blood flow in different animal species and humans is summarized in Table 6.4.

Renal CL: The volume of plasma (or blood) from which a drug is completely removed by the kidneys in a given time period is defined as renal clearance. It measures the ability of the kidneys to remove drug from the plasma (or blood), and is the function of kidney filtration, secretion, reabsorption and metabolism.

$$CL_r = CL_f + CL_{se} + CL_{rm} - CL_{rea} \tag{6.6}$$

where CL_r is the renal clearance, CL_f is the glomerular filtration clearance, CL_{se} is the clearance via tubular secretion, CL_{rm} is the clearance via kidney metabolism if there is any, and CL_{rea} is the clearance via reabsorption.

Half-life ($t_{1/2}$): The period of time required for the concentration or amount of drug in the body (usually measured in plasma or blood) to be reduced by one-half. Half-life is a function of clearance and volume of distribution.

$$t_{1/2} = 0.693 * V/CL \tag{6.7}$$

Intravenous bolus dose: A fluid dose of drug administered intravenously and rapidly at one time.

Intravenous infusion: Administration of a drug or other substance as part of a liquid solution injected directly into a vein at a prescribed rate.

Extravascular dose: Any nonvascular (intravenous or intra-arterial) method of drug administration, such as oral, subcutaneous, and so on.

6.4 BASIC DMPK STRATEGY IN DRUG DISCOVERY

DMPK is now considered to be an integral part of drug discovery. The advances in medicinal chemistry (e.g., combinatorial chemistry, multiple parallel synthesis, etc.), along with complementary advances in biological screening technology (HTS, uHTS), have resulted in the necessity of screening many more compounds, more quickly, for DMPK properties. This section will discuss the stages of drug discovery, the typical DMPK studies that are being conducted, and the DMPK strategies used to effectively support drug discovery at each stage.

6.4.1 The Stages of a Typical Drug Discovery Program

Drug discovery for small molecules involves the design and identification of a molecule to treat a human disease. It typically has multiple distinctive stages in the modern pharmaceutical industry. These stages include target identification (TI), high-throughput screening (HTS), hit to lead (HTL), lead optimization (LO), and backup (BU).

In drug discovery, the first step in initiating a research program is to propose and develop a scientific hypothesis that pertains to how perturbation of a specific drug target, such as GPCR receptors, ion channels, enzymes, and so on, can treat a human disease or multiple diseases. The DMPK discipline is to characterize the PK/PD profile of tool compounds to enable biologists to test their hypothesis in animal models though PK/PD data analysis and modeling.

Once the scientific rationale is established based on the preliminary data, the next step, defined as HTS, is to identify chemical leads that can interact with the target. Typically, a large number of compounds are screened using robot in microplates such as 96-well, 384-well, or 1536-well. At this step, DMPK contributions are limited to working with chemists by applying a chemical/physical-based approach, such as Lipinski Rule of Five (Lipinski et al., 1997).

After the compounds that can interact with the target are identified, chemists evaluate the initial hits and select multiple series with diverse chemical structures to design new compounds. The typical length of this stage, which is defined as HTL, is one to two years. The objective of this stage is to demonstrate one or multiple chemical series that show required potency (typically less than 1 μM) and have drug-like properties, such as oral bioavailability. In most organizations, DMPK scientists join the discovery team at this stage and play critical roles in guiding the discovery team to select the right series.

The next stage of drug discovery is LO. This stage requires the most resource-intensive efforts for DMPK scientists while medicinal chemists design and synthesize various compounds in a single or multiple series. DMPK, biology, formulation, and safety then characterize the properties of these compounds in the respective functional areas. Different pieces of information are integrated to provide a holistic picture of the compound. The typical length of this stage is 1–2 years. DMPK scientists work closely with chemists to screen DMPK properties, identify issues, and interpret the data in the context of clinical relevance. DMPK scientists also work closely with the biologists to conduct more resource intensive PK/PD studies in the animal disease models and to further characterize the compounds. Meanwhile, DMPK scientists support safety and formulation studies and the development of an early clinical plan. The end product of this stage is to identify one or multiple compounds for further preclinical development.

Once a drug candidate has been identified for preclinical development, typically the drug discovery team continues to work on the project to design or identify backup compounds. Ideally, the issues or risk factors for the drug candidate(s) have been identified; the goal for the BU program is to design a compound or multiple compound(s) to overcome the issues. For example, if the projected human PK required a twice-a-day dose regimen, the BU program may want to design a compound with a once-a-day dose regimen. Unfortunately, due to the lengthy development time, in most cases it is generally impractical to await validation of the identified issues in humans in order to provide feedback for the discovery scientists to redesign the molecule. Therefore, the typical strategy is to further improve these drug properties that are perceived as posing issues in humans or to simply design a different scaffold to derisk the potential safety issue due to the chemical structure.

The drug discovery process discussed above represents a typical drug discovery program in the pharmaceutical industry with novel targets. However, there are many exceptions to this typical process. Since most pharmaceutical researchers are exploring new territories, new information emerges every day. The new findings from literature, from competitors, or from collaborators often change the discovery path. In order to stay at the frontier of new research, scientists must simultaneously stay focused on their research and be flexible in changing directions. Although this description of drug discovery appears to represent a smooth linear process with one stage finishing before another starts, in reality, the discovery process is seldom smooth and rarely ever linear.

6.4.2 Typical DMPK Studies in Different Stages of Drug Discovery

The advancement of LC-MS/MS technology in the last 30 years has revolutionized the DMPK studies in drug discovery. This analytical technology can rapidly, sensitively, and selectively determine drug concentration in complex biological matrixes. This allows DMPK scientists to run PK studies in preclinical species and have data turnaround in a few days. Another advance is the ready availability of human tissue, such as human liver microsomes and hepatocytes. This allows DMPK to use *in vitro* technique to quickly assess the metabolic stability and screen compounds. The progress of P450, UGT gluconomide transferase, FMO, MAO, and other enzymes has contributed significantly to the use of *in vitro* agents to predict the potential variability of a compound and its DDI potential mediated by the drug metabolism enzymes in humans (Lee et al., 2003). Recently, the understanding of transporters such as the ABC family and OATPs has significantly enhanced our ability to identify the drug transporters and to assess the potential for the DDI due to drug transporters (Giacomini and Sugiyama, 2006).

Optimal DMPK screening requires a careful balance of *in silico* to *in vitro* and *in vivo* screening. Desirable properties include: good solubility; permeability; available parental formulation; high and consistent bioavailability; appropriate concentration at target site; adequate half-life; linear kinetics; small first-pass effect; metabolism; no(auto) induction; no cytochrome P450 inhibition; no interaction with P-glycoprotein (PGP); and a wide therapeutic index.

The following represent the typical studies that are conducted in the DMPK department to advance a drug discovery program.

Target Identification (TI)

- PK profile of tool compounds
- Protein binding of tool compounds
- Exposure of tool compounds in preliminary PK/PD model

High-Throughput Screening (HTS)

- Rule-based library design and compounds selection
- *In silico* prediction

Hit to Lead (HTL)

- *In silico* prediction
- Solubility assay
- *In vitro* metabolic stability assay
- *In vitro* permeability assay
- *In vitro* P450 inhibition assay

- Reactive metabolite
- Rodent PK of the key compounds in the selected series
- Protein binding
- Exposure of tool or lead compounds in preliminary PK/PD studies
- Support of biomarker development

Lead Optimization (LO)

- *In silico* prediction
- Solubility assay
- *In vitro* metabolic stability assay
- *In vitro* permeability assay
- *In vitro* P450 inhibition assay
- *In vitro* P450 induction assay
- *In vitro* transport assay
- Metabolite structure elucidation
- Reaction phenotyping assay
- Reactive metabolite
- Rodent PK of lead compounds
- Nonrodent PK of lead compounds
- Protein binding of lead compounds
- Exposure of lead in PK/PD studies
- Development of PK/PD models to characterize the PK/PD relationship in preclinical species
- Support of safety studies for dose selection and TK analysis
- Support of biomarker development for translational pharmacology
- Simulated human PK, PD, DDI potential, and therapeutic window

6.4.3 DMPK Strategies in Different Stages of Drug Discovery Programs

The purpose of drug discovery is to generate and test scientific hypotheses. Drug discovery itself is a scientific endeavor and exploration. Because of its complexity, it requires skilled and experienced scientists from different disciplines to work together. The progress of science and technology enables us to understand the biological systems better than ever before. But there is far more about the biological systems that we do not know at the present time than we do know. Serendipity or sheer luck also plays an important role in the success of a drug discovery program.

There is no single strategy that can or should be employed by a drug discovery team. The effectiveness of drug discovery is the combination of the strategy, the knowledge of science, and the experience and risk tolerability of the research organization. The following DMPK strategy represents the general strategy that the authors employ.

6.4.3.1 Target Identification. The objective of research at this stage is to generate preliminary data to support the scientific hypothesis that perturbation of a specific biological target may be able to treat a human disease. Proof of the concept is critical for a drug discovery organization, because this will justify the investment in a particular program.

An *in vivo* model is generally required to demonstrate the intended mechanism of action. The *in vivo* proof of concept data will strengthen the hypothesis. If an *in vivo* model is not available due to species-unique differences, alternative models, such as knockout/knock-in animal models or using RNAi approach, might possibly be used to support the proposed hypothesis. The key deliverable for the DMPK scientist is characterization of the PK profile for a tool compound or compounds reported in the scientific community. The information is used to select a dose and dose regimen in the *in vivo* animal model to demonstrate the relationship of dose and the intended pharmacological response in the animal model and to validate the utility and limitation of the animal model. Another deliverable is to determine the plasma concentration or tissue concentration, also called exposure, in the *in vivo* studies in order to develop the concentration–response relationship.

Traditionally, only the dose–response relationship was characterized by pharmacologists. With better understanding of the relationship between pharmacologic response and pharmacokinetic profile, scientists have found that plasma concentration is a better indicator than the dose for representing the drug concentration that interacts with its target. However, for specific tissue/organ targeted mechanisms, tissue/organ concentration could also be important. For example, with CNS targets, the brain tissue concentration-response relationship must be examined thoroughly. Because of the blood–brain barrier, for some drugs, the concentration in plasma does not represent the drug concentration that interacts with the target in the brain. Establishment of the relationship between *in vitro* concentration versus *in vitro* activity and *in vivo* concentration versus *in vivo* activity is the most critical evidence for proof of the concept in preclinical models. The readers are referred to more detailed discussion in a review paper (Liu et al., 2008).

6.4.3.2 High-Throughput Screen (HTS). The objective of research at this stage is to identify scaffolds of molecules as the starting points to design and optimize the structure. This step is mainly an *in vitro* biology-driven activity. Using robots, the capacity of *in vitro* testing has increased tremendously. We used to examine one compound a time by using a single test tube, and now it is possible to examine hundreds or thousands of compounds at a time by using microplates.

As DMPK becomes involved at an ever-earlier stage in the drug discovery process, there is a clear need to be able to screen thousands of hits. In an HTS campaign, all collected compounds or a subset of them may need to be screened rapidly. Because the major objective of this work is to identify compounds that have biological activity, so that medicinal chemists can begin optimizing the starting compounds, typically no DMPK assay is run at this stage. A common approach to identifying compounds with good DMPK properties is to either (a) use *in silico* screening for metabolic stability, permeability, or enzyme inhibition or (b) use a physical/chemical rule approach to design the library or select the library compounds.

6.4.3.3 Targeted Product Profile. First the discovery team needs to work with marketing colleagues to develop a targeted product profile that is believed to bring commercial success by the time the product is lunched. Since it may take more than 6 years from this stage of the project to the final product launch, it is often difficult to project the market situation. The targeted product profile may have to be modified if a competitor introduces a product during this time. For example, product for an indication can be dosed twice a day because there is no drug on the market. If a competitor introduces a drug with the same mechanism, the new product profile may have to be modified to have a once-a-day dose property to be successful in the marketplace. Usually the information regarding the competitor's drug is not accessible at this stage, with the only knowledge being that a competitor is running clinical trials. Sometimes the team has to run head-to-head comparison nonclinical studies between the proprietary compound and a competitor's compound to gain insight into the product differentiations.

6.4.3.4 Hit to Lead (HTL). The objective of research at this stage is to identify one or several chemical series for which the potency, selectivity, and ADME properties can be further improved by modifying the drug structure.

In current drug discovery, the medicinal chemists are often able to generate several viable lead series and rapidly develop these into potent, selective molecules through a directed medicinal chemistry effort. However, it is often more of a challenge to design a compound that is potent and selective and also shows high oral bioavailability, a desirable half-life, and low DDI potential. The emphasis has shifted toward designing molecules possessing not only high affinity at a biological target, but also suitable pharmacokinetic and metabolic characteristics. This has led to a paradigm shift in the integration of DMPK functions into earlier stages of drug discovery.

For most drug discovery projects, the oral route is the preferred dosing route clinically. For a drug to reach its target and manifest activities, a number of barriers must be overcome. In order to gain insight into the clinical human PK for a discovery compound, a variety of *in vitro* tests using human tissue preparation have been developed to project the PK characteristics of the compound in humans. Several strategies have been used to incorporate the data from the *in vitro* assays to make decisions regarding a particular compound and guide the design for new compounds.

One strategy is to use the rule-based decision process. For example, the metabolic stability of compounds obtained in a human liver microsomal assay can be used to classify compounds into low, medium, or high clearance compounds. The classification is typically based on the historical experience observed within the laboratory by comparing a large dataset of compounds with their *in vitro* stability and *in vivo* clearance and oral bioavailability. It should be noted that in the absence of this calibration, caution should be exercised when using these *in vitro* data to make decision for the compounds, because the *in vitro* classification can give false-positive prediction, meaning that high *in vitro* clearance does not necessarily translate into *in vivo* high clearance due to plasma protein binding and blood-plasma partitioning. It is helpful to test the selected compounds *in vivo* in preclinical species, such as in rodents, to verify the *in vitro* prediction.

The *in vitro* data could also be used differently to project the human PK and dose. The *in vitro* metabolic stability data are quantitatively extrapolated into *in vivo* clearance using PK models or using commercially available software such as GastroPlus. This exercise requires skilled DMPK scientists to perform the human PK projection. Similar to the classification method, it is very critical to verify the predictability of this approach. Validation could be performed by extrapolating animal *in vitro* data generated from the same assay to human *in vitro* data to animal *in vivo* PK projection using the same assumption performed in humans.

The typical *in vitro* assays are presented in the previous section. Usually a few compounds from each scaffold that have already demonstrated desirable potency and selectivity are selected for *in vitro* assays. The most important assays are metabolic stability and permeability. Liver tissue preparations such as microsomes, S9 fraction, and hepatocytes are used to determine the metabolic stability. The permeability is determined in epithelial cells such as Caco-2 or MDCK cells. The measurement of apical to basolateral permeability and basolateral to apical permeability provides not only the permeability of the compound across the epithelial membrane in the gastrointestinal track but also the potential involvement of efflux transporters in the testing cell systems.

One additional assay that may be performed is solubility testing. Because solubility is more related to the scaffold and less related to a specific structure, it needs to be determined if the solubility is truly an issue for the whole series or just for a subset of compounds in a series. However, for basic compounds such as most CNS drugs, the solubility is typically high and there is no need to screen their solubility.

One of the major causes of DMPK-related failure in the clinic is the area of drug–drug interactions. Severe drug–drug interactions could produce undue toxicity and/or therapeutic failure. It is now possible to clone and express the multiple major human P450 isoforms (phase I metabolism), along with human uridinediphosphoglucuronyl transferase (phase II), allowing structural series to be screened for their potential to inhibit or induce the metabolism of co-administered drugs that are metabolized by these enzymes. Transporter protein interactions are now also being investigated in a similar way.

At this stage, which DMPK assays should be deployed, how many compounds need to be examined, and the timing for initiating the assays are dependent on the primary issues of the chemotype, the depth and breadth of knowledge about drug research, the target patient population, and available resources. In the authors' views, at the minimum, some solubility, *in vitro* metabolic stability, and *in vitro* fluorescence-probe–based P450 assays should be conducted for one compound in the chemical series that has shown *in vitro* pharmacological potency and selectivity. If the results indicate that it has poor oral bioavailability or severe drug–drug interactions (assuming that severe drug–drug interactions are not acceptable for this indication), several additional compounds should be tested to verify whether the results are due to the specific structure of the compound or whether these issues are a liability for the whole series. In this situation, preliminary metabolite structure elucidation should be conducted to understand the mechanism of the issue and to determine if the liability could be removed during the chemical design (Baranczewski et al., 2006; Prakash et al., 2007).

An alternative strategy is to focus only on the pharmacological potency and selectivity at this stage. Once the compounds showed high potency and selectivity, then their DMPK profile and identify issues can be examined. The drawbacks of this approach is that if the issue is related to the core chemical structure of the scaffold and cannot be readily modified, then it will not be possible, or at least very difficult, to resolve the DMPK-related issues in the LO stage. Sometimes the whole series has to be abandoned simply due to DMPK issues. In addition, the high-throughput capability and relatively low cost of *in vitro* DMPK assays enable early identification of these avoidable issues. Another key activity for DMPK at this stage is to work closely with pharmacologists to develop and/or validate new animal models that are more closely mimic clinical situation and test the lead compounds in these pharmacology models. As discussed above, one of the main contributions from DMPK in the target identification stage is to determine the plasma or tissue drug and sometimes, the concentration of its metabolite(s) in the pharmacological model.

In recent years, progress has been made by DMPK scientists for PK/PD analyses in preclinical studies. The PK/PD analysis allows us to explore the *in vitro* and *in vivo* association for the pharmacology. It helps to validate the animal model and guide the study design. More importantly, if some biomarkers can be used in both animals and humans, the PK/PD model can be used to guide the strategy of early clinical development plan, which includes first-in-human trial and proof-of-concept studies.

6.4.3.5 Lead Optimization (LO). The objective of research at this stage is to identify one or several compounds that meet the targeted profile for preclinical development.

The division between hit-to-lead and LO is sometimes not entirely clear. Generally speaking, more resources are invested in programs at the LO stage, with an expectation that in 1–2 years a final candidate that meets all the target product profile criteria will be identified and advanced for clinical development. At this stage, the knowledge of how to improve the potency–selectivity–acceptable DMPK profile for one series should have been developed, although those attributes may not exhibit in a single molecule. The team should already understand the key issues and develop a plan or screening cascade to design new compounds and screen them. The final goal for the discovery team is to synthesize at least one compound that has the required potency, selectivity, DMPK profile, and physical properties as specified by the project team.

To be able to design an efficient research process, a medicinal chemist, DMPK, and a pharmacologist must work closely together, so that the available resources in each department can be optimized to enable the discovery program to move forward quickly. Below is the typical drug discovery screening program in the pharmaceutical industry. In this hypothetical program, the targeted product profile of the compound includes efficacy, safety, and a once-a-day dosing regimen to treat a disease, with no clinically significant drug–drug interactions.

The research activities are generally as follows:

Tier 1: 100–1000 Compounds/year

1. Chemical synthesis: 1–10 mg
2. *In vitro* biology binding/*in vitro* biology selectivity
3. DMPK assays
   a. *In vitro* metabolic stability
   b. *In vitro* permeability
   c. *In vitro* DDI assay
4. Other issue-driven assays

The key information from the Tier 1 screening is used to determine whether the compound is potent and selective for the intended pharmacological target and if it has the desired DMPK profile.

The Tier 1 assays can be executed in parallel or sequential fashion. Each has its own advantages and disadvantages. The main advantage of parallel screening is a quick data turnaround time, so that all information is ready at the point of decision. The disadvantage of parallel screening is the need to use extensive resources, because a compound could be discarded due to lack of selectivity. The primary advantage of sequential screening is conservation of resources. The disadvantage is that it takes more time to generate the information necessary to make a decision.

To maximize the usage of resources and shorten the critical data turnaround time, the "go/no-go" assays should be performed in parallel fashion, while the less time-sensitive assays could be performed in sequential fashion.

For example, if high metabolic stability is the key issue for poor oral bioavailability, the permeability is generally high, and there are no obvious DDI issues, then only *in vitro* metabolic stability should be run to make the decision on the particular compounds. As a program moves forward, more and more compounds are tested, and knowledge of the chemotypes is accumulated. Developing an *in silico* model for the structure- and biological/DMPK-related properties should be considered. This *in silico* tool may be used to prioritize the compounds for *in vitro* assays, for *in vivo* assays, and for resource-intensive assays. More importantly, the model could guide the medicinal chemists' effort in compound designing for improved properties.

One major decision from Tier 1 screening is to determine whether one or two compounds meet the criteria for the next tier of testing—that is, testing for *in vivo* efficacy and *in vivo* PK profile. Tier 1 screening also allows the team to gain insightful knowledge regarding the chemical structure and property relationship.

Tier 2: 10–100 Compounds/year

1. Chemical synthesis: 10–100 mg
2. Confirm *in vitro* potency
3. Confirm *in vitro* selectivity
4. *In vitro* biology functional assay
5. *In vitro* functional selectivity assay
6. *In vitro* safety assay (e.g., hEGR assay)
7. Preliminary *in vivo* efficacy assay
8. Additional DMPK assay
    a. Protein binding
    b. Reactive metabolite
    c. Time-dependent P450 inhibition
9. Other issue-driven assays
10. *In vivo* rodent PK (IV and PO)
11. Human dose and DDI projection

The main objectives at Tier 2 screening are to confirm Tier 1 screening data for the potency and selectivity for the target, to demonstrate compound's *in vivo* efficacy, and to ensure that the compound has an appropriate *in vivo* PK profile. The *in vivo* animal PK data will confirm whether the projected clearance and oral bioavailability based on the *in vitro* data is consistent with the *in vivo* data. The *in vitro–in vivo* correlation in animals forms the basis for the human PK projection. At this stage an attempt should be made to predict the preliminary human PK profile and DDI potential.

The decisions from the Tier 2 data are used to advance compounds for more detailed evaluation on *in vivo* efficacy and PK profiles in nonrodent species. In addition, *in vitro* safety assays will be initiated to assess safety profiles.

Tie 3: 2–10 Compounds/year

1. Chemical synthesis: 100–1000 mg
2. Confirmative *in vivo* efficacy assay in a dose-related response
3. *In vitro* safety assay (e.g., MNT and Ames)
4. DMPK assay
   a. Blood/plasma ratio
   b. Metabolite identification
   c. Reaction phenotyping
   d. P450 induction
5. Other issue-driven assays
6. *In vivo* nonrodent PK (IV and PO)
7. *In vivo* rat biliary excretion study
8. Refine human dose and DDI projection

The main objective at this stage is to confirm the *in vivo* efficacy in a dose-related response, to demonstrate low risk for *in vivo* genotoxicity, and to further profile the PK in non-rodent PK. These data will provide greater confidence in the *in vitro–in vivo* correlation and human PK and DDI potential assessment.

The decisions from Tier 3 are used to narrow down the number of compounds for further *in vivo* safety evaluation and extensive PK/PD studies, as well as to develop a PK/PD model to assess the therapeutic window and guide development of early clinical plans.

Tier 4: 1–5 Compounds/year

1. Chemical synthesis: 1–10 g
2. Extensive PK/PD studies, such as developing a PK/PD model and time course for the responses
3. DMPK assay
   a. Protein binding
   b. Definitive metabolic stability data
   c. Definitive drug transport data
   d. Definitive drug inhibition data
4. Other issue-driven assays
5. *In vivo* rodent high-dose PK to enable safety study
6. Preliminary *in vivo* safety study
7. Refine human dose, DDI, and therapeutic window projection

The main outcome of Tier 4 screening is the performance of a preliminary assessment for the safety of a compound *in vivo*, typically in rodents for targeted organ toxicity and in dogs or monkeys for cardiovascular toxicity. This information is critical since the outcome of the study could be a go/no-go decision for the compound.

In Tier 4, PK/PD analysis is often performed to further confirm the pharmacological activities of the compound and to develop a quantitative relationship between or mathematic model of the plasma drug concentration and biomarker or efficacy in the animal model. This PK/PD analysis further validates the target with a specific compound to support the program and will also serve as a tool to simulate human pharmacological response, calculate therapeutic and safety window, and guide the subsequent dose range finding studies.

6.4.3.6 Backup (BU). Once a drug candidate is nominated for preclinical development, typically the drug discovery teams will continue working on the program to develop additional compounds as backups for the prototype drug candidate. Because the attrition rate for preclinical and clinical drug development is very high, the main rationale for a BU program is to increase the chance of success by developing multiple compounds for the same biological target.

There are several scenarios for BU programs. In the first scenario, there is a known or projected issue for the drug candidate, such as an unfavorable DMPK property (short half-life, low oral bioavailability, or potential DDI, etc.). In this case, the DMPK strategy for a BU program is identical to the LO strategy. The screening cascade needs to be specifically designed to address that issue and identify a compound or multiple compounds that devoid that issue. In the second scenario, there is no known issue for the current drug candidate. One strategy for a BU program is to identify a structurally dissimilar candidate in case the development of the first candidate has to be terminated due to unexpected toxicity or off-target effects (Greenlee and Desai, 2006). The DMPK strategies discussed above for HTL and LO can be applied in this scenario. The other scenarios in which a BU program is useful include developing a second-generation drug in order to design a single compound to interact with multiple drug targets or to develop a different drug candidate intended for different indications (Greenlee and Desai, 2006).

The primary goal of drug discovery is to design and synthesize a compound with high efficacy and adequate safety, avoiding ADME-related attrition in the clinic while using minimal resources. Although a variety of *in vitro* and *in vivo* assays are available to assess efficacy, safety, and PK properties, the key is to have a research team that can use state-of-the-art technology and integrate all the assays together to design an issue-driven screening cascade. Although drug discovery is full of serendipity, however, the odds of success may favor the fully prepared mind.

6.5 PHARMACOKINETIC DATA ANALYSIS

6.5.1 Introduction

Pharmacokinetics describes the movement of a drug through the body. It captures characteristics of a compound whose concentrations vary over the time course in different biological matrices in our body, like plasma, blood, urine, saliva, cerebrospinal fluid, tissues, and so on. This concentration–time course reflects the biological process of drug absorption, distribution, metabolism, and excretion (ADME). Pharmacokinetic data refer to the concentration data of a drug at different time points, usually in plasma because it is the most conveniently monitored bio-matrix. Basically, classic pharmacokinetic data are a kind of repeated measurement data; the drug concentration is measured across a time scale, providing a continuous type of experimental data.

Based on the features of pharmacokinetics and its data, key questions to be answered include why and how the data should be analyzed prior to the actual data analysis. These are questions of reason and method. Since pharmacokinetic data reflect the drug ADME kinetic process, the ability to extract that instructional information out of the data could greatly enable the development of drug. Specifically, with the information about the features of drug ADME obtained from data analysis, the efficacy and safety of the drug could be further assessed. In terms of methods, since the origin of pharmacokinetics in 1937, scientists in this field have developed many different theories and tools for analyzing pharmacokinetic data. Mathematical equations have been derived to describe and predict a drug concentration–time course; biological meanings can be applied to mathematical parameters that represent drug ADME features. The mathematical equations are further used to predict or simulate pharmacokinetic behavior of the drug under different dosing regimens or different physiological conditions (i.e., disease states, gender, age, etc.). Pharmacokinetic modeling and simulation has become a very powerful tool in drug development.

6.5.2 Pharmacokinetic Models

Based on the anatomy of the human body, the concept of compartments is used to represent the entire body, different body spaces such as blood and tissue, and even specific organs such as the liver and kidneys. Strictly speaking, every pharmacokinetic model is a compartment model with different levels of details within and/or between compartments. Generally speaking, every model used thus far in pharmacokinetic analysis falls into one of the following categories: noncompartmental analysis (NCA), compartmental analysis, or physiological-based pharmacokinetic model (PBPK).

6.5.2.1 Noncompartmental Modeling (Nesch, 1984). Noncompartmenal modeling is the primary pharmacokinetic data analysis method in the pharmaceutical industry because it is possible to quickly obtain several key pharmacokinetic parameters, such as area under the curve (AUC), systemic clearance (CL), volume of distribution at steady state (V_{ss}), and mean residence time (MRT), directly from the concentration data without making any compartmental assumptions or detailed model structures. These key pharmacokinetic parameters describe the basic drug ADME features. In a noncompartmental approach, statistical moment analysis and/or a linear system approach are applied to the pharmacokinetic data. Basically, drug distribution through the body is a stochastic process, and an individual drug molecule passing through the body has its own probability. So considering the drug concentration–time profile *in vivo* as a statistical probability distribution curve, moment analysis is applied to analyze the pharmacokinetic data. Statistically, the expected value (mean) of a random variable is a weighted average of the possible values that it takes, with each value being weighted by thc probability that it assumcs. For a continuous random variable, probability density function $f(t)$, $0 < t < \infty$, the expected value of $f(t)$ is defined as follows:

$$M = \int_0^\infty tf(t)\,dt \tag{6.8}$$

And for the nth moment of a given function $f(t)$, we have

$$n\text{th moment} = \int_0^\infty \frac{t^n}{n!} f(t)\,dt, \quad n = 0, 1, 2 \tag{6.9}$$

It is clear that the zeroth moment (when $n = 0$) is given by

$$m_0 = \int_0^\infty f(t)\,dt \tag{6.10}$$

which is the area under curve (AUC). In pharmacokinetics, we use AUC to indicate the extent of exposure of a drug to the body.

The first moment (when $n = 1$) is expressed as

$$m_1 = \int_0^\infty tf(t)\,dt \tag{6.11}$$

which is the area under the first moment curve (AUMC) and is also the expected value of function $f(t)$ when $f(t)$ is true probability function. Furthermore, the second, third, and fourth moments of the function $f(t)$, when $n = 2, 3, 4$, are the variance, skewness, and kurtosis of the distribution.

Mean residence time (MRT) is the average time that all drug molecules stay in the body. It is also the exit time for drug molecules leaving the body by elimination if the entry time is 0. By comparison, MRT in pharmacokinetics is merely the expected value (mean) of the distribution in statistics, which is described by first moment of the distribution. We can think of a drug concentration–time profile as the probability distribution curve of the MRT of a drug, where the x axis is MRT and the y axis represents probability. In order to obtain a true probability density function for this distribution of MRT, we need the fraction (probability) of drug molecules eliminated at each time point, which is $1 - D_t/D_0$; in IV bolus administration, it is $1 - e^{-kt}$. Then MRT of drug administered by IV bolus is calculated as follows:

$$\text{MRT} = \int_0^\infty tf(t)\,dt = \int_0^\infty t\left(1 - e^{-kt}\right)' dt = \int_0^\infty tke^{-kt} = 1/k \tag{6.12}$$

with the assumption that k is a constant. Using the above method, we can easily calculate MRT for different dosing regimens. Alternatively, we can think of MRT as the weighted average of all possible values that T, which is residence time, can take, each value being weighted by the probability that T assumes in the body.

$$\text{MRT} = \sum_{i=1}^{m}\left(1-\frac{D_t}{D_0}\right)t_i = \frac{1}{D_0}\sum_{i=1}^{m} De_i t_i = \frac{1}{D_0}\sum_{i=1}^{m} t_i k D_0 e^{-kt_i}\Delta t \qquad (6.13)$$

where De_i is the amount of drug eliminated at time t_i in IV bolus administration. When T is a continuous variable, we could rewrite Eq. (6.13) as follows:

$$\text{MRT} = \int_0^{\infty}\frac{tkD_0e^{-kt}dt}{D_0} = \frac{\int_0^{\infty} tD_0e^{-kt}dt}{\frac{D_0}{k}} = \frac{\int_0^{\infty} tC_P(t)dt}{\int_0^{\infty} C_p(t)dt} = \text{AUMC/AUC} \qquad (6.14)$$

The result is still applicable to other routes of dose administration. MRT also represents the time required for 63.2% of drug to be eliminated from the body after IV bolus administration.

We generally use either linear trapezoidal or log-linear trapezoidal methods to calculate AUC and/or AUMC, which can also be used in the calculation of MRT in Eq. (6.14) because sometimes the linear trapezoidal method can underestimate and/or overestimate the area under ascending phase of a drug concentration–time curve. The methods to calculate the AUC and AUMC based on the drug concentration–time curve are summarized in Table 6.5.

Where $\Delta t = t_n - t_{n-1}$ is the time interval between two adjacent time points, C_z is the estimated drug concentration at t_{last}, which is the last time point having measurable drug concentration. λ_z is the slope of the terminal phase of the plasma drug concentration–time profile on a semilog scale, which can obtained from the last few (>2) data points during the terminal phase by linear regression. Because both the linear trapezoidal or log-linear trapezoidal methods

TABLE 6.5. Linear Trapezoidal, Log-Linear Trapezoidal, and Extension Equations for Calculating the AUC and AUMC

Area	Linear Trapezoidal ($0 \sim t_{\text{last}}$)	Log-Linear Trapezoidal ($0 \sim t_{\text{last}}$)	Extension ($t_{\text{last}} \sim \infty$)
AUC	$\left[\frac{C_n + C_{n-1}}{2}\right]\Delta t$	$\frac{C_n - C_{n-1}}{(1/\Delta t)\ln(C_n/C_{n-1})}$	$\frac{C_z}{\lambda_z}$
AUMC	$\left[\frac{t_nC_n + t_{n-1}C_{n-1}}{2}\right]\Delta t$	$\frac{t_nC_n - t_{n-1}C_{n-1}}{(1/\Delta t)\ln(C_n/C_{n-1})} - \frac{C_n - C_{n-1}}{[(1/\Delta t)\ln(C_n/C_{n-1})]^2}$	$\frac{t_zC_z}{\lambda_z} + \frac{C_z}{\lambda_z^2}$

can only calculate $AUC_{0\text{-tlast}}$ and/or $AUMC_{0\text{-tlast}}$, for the area between t_{last} and infinity, we use an extension function. Sometimes the linear trapezoidal rule may be applied for the ascending phase and the log-linear trapezoidal method for the descending phase of the drug concentration–time profile; however, usually the difference between AUC values gained from the linear versus log-linear method is not significant. Other pharmacokinetic parameters, CL and V_{ss}, are defined as follows:

$$\text{CL} = \frac{\text{FD}}{\text{AUC}} \tag{6.15}$$

$$V_{SS} = \text{CL} \cdot \text{MRT} \tag{6.16}$$

Noncompartmental modeling still uses a compartment, called a sampling compartment (body), which is more general and simpler than the compartments used in compartmental modeling. If we have the input function $D(\tau)$, which is the dosing regimen for this compartment, and the output function $C(t)$, which is the drug concentration–time profile, collected at later time, then there should be a probability function $P(t - \tau)$ representing the fraction of drug molecules moving from the input site to the output site during the period $t - \tau$ to link $D(\tau)$ and $C(t)$.

$$C(t) = \int_0^t P(t-\tau)D(\tau)d\tau \tag{6.17}$$

This is called convolution integral. If we let $D(\tau)$ equal the unit bolus dose, the probability function $P(t)$ will be equal to the drug concentration–time course after the unit bolus dose (ubd), which is also called disposition function. Equation (6.17) can be written as

$$C(t) = C(t)_{\text{ubd}} * D(t) \tag{6.18}$$

As long as the system follows the principle of superposition and is time-independent, we can use Eq. (6.18) to analyze pharmacokinetic data. In most instances, when we know the dosing regimen, $D(t)$ and measure the $C(t)$, then it is possible to calculate $C(t)_{\text{ubd}}$, which includes the pharmacokinetic parameters described previously, by deconvolution. The noncompartmental approach only deals with response, that is, the concentration–time curve; in contrast, $C(t)_{\text{ubd}}$, or disposition function, also can be determined with detailed structure by regression method. This latter approach is called compartmental modeling.

6.5.2.2 Compartmental Modeling (Giorgio, 1984; Landaw and DiStefano, 1984; Piotrovskii, 1987). If the noncompartmental model can be seen as having too little structure and the physiological-based pharmacokinetic model as having too much detail, then the classical compartmental model is in the middle. For this reason, it is well-accepted and commonly used in the area of pharmacokinetics. The simplest form of compartmental modeling is called a one-compartment model.

The concept of the sampling compartment in a one-compartment model is similar to the noncompartmental model, but the approach is different. The compartmental approach includes some detailed structures, such as K, which is a first-order elimination rate constant, and V, which is apparent volume of distribution to link dose (D) and concentration (C), where $D = V \cdot C$. In addition, we assume that this compartment is well-stirred, meaning that the drug is homogeneously and instantaneously distributed in this compartment. In the one-compartment model, we use this compartment to represent the whole-body system. However, K is an elimination constant; clinicians are more comfortable with the clearance parameter because of its physiological meanings. Clearance refers to a volume of drug in vascular circulation removed per unit time, which represents the rate at which the drug is eliminated from the body: $CL = K \cdot V$. Clearance (CL) and volume of distribution (V) are defined as primary parameters, whereas the elimination rate constant (K), half-life ($t_{1/2}$, which is time required for 50% of drug eliminated), AUC, and the rest of the parameters are secondary parameters because they are derived from primary parameters.

Besides the one-compartmental model, in the practice of pharmacokinetic data analysis, two-compartment and even three-compartmental models may be reflected by the concentration–time data. If the one-compartment model represents the whole body, the two-compartment model consists of a blood/plasma compartment called the central compartment, which is usually the sampling compartment, and a tissue compartment called the peripheral compartment, which represents the whole group of tissues, such as the liver, skin, fat, and so on. The three-compartment model involves a greater level of detail within the tissue compartment, splitting the tissue compartment as a whole into (a) one tissue compartment to which the drug is more quickly and easily distributed (e.g., liver, kidney, etc.) and (b) another tissue compartment to which the drug is more slowly distributed (e.g., fat, muscle, etc.). If we use an extravascular dosing regimen such as oral drug administration, we can include a so-called depot compartment before our sampling compartment to represent the absorption process. Generally, one- and two-compartment models should be sufficient to describe most pharmacokinetic data. It is best to avoid building overly complex compartmental models (e.g., five or six compartments) for curve fitting unless the data did reflect this, because a model with too many compartments can make biological interpretation of parameter and process difficult and can cause serious problems of model identification.

Although we try to link physiological meanings within the compartmental model with the noncompartmental model, the association is still loose. Often in compartmental analysis, we use a linear mammillary model to describe our system; for this type of model, we assume that all other models are connected with the central compartment and that elimination can only happen in the central compartment.

In a noncompartmental model, one important assumption is superposition, which in pharmacokinetics is considered as dose proportionality. If dose proportionality is not observed over multiple dose levels, the kinetic system is a nonlinear system instead of linear system. The process of nonlinearity could occur during the phase of drug absorption, distribution or elimination. In most instances, nonlinear pharmacokinetics refers to nonlinear elimination. There are two different equations for nonlinear versus linear elimination:

$$\frac{dC}{dt} = -\frac{V_m C}{K_m + C} \tag{6.19}$$

$$\frac{dC}{dt} = -KC \tag{6.20}$$

When C is much lower than K_m, Eq. (6.19) can be reduced to Eq. (6.20); when C is much higher than K_m, the rate of change of concentration is a constant and is independent of concentrations and is saturated. Equation (6.19) is usually applied to the drug absorption, distribution, and elimination phases wherever the nonlinearity occurred.

Table 6.6 lists the pharmacokinetic compartmental models that are built into two different software packages, WinNonlin (version 5.2, Pharsight, Mountain View, CA) and NONMEM (Beal and Sheiner, 1989). Both are widely used for pharmacokinetic and pharmacodynamic data analysis. WinNonlin has 19 compartmental models typically used for PK data analysis; all assume first-order elimination and first-order input for extravascular administration. Exchanges between micro and macro represent different kinds of parameterization for the model. In NONMEM, there are a total of 12 built-in ADVANs for PK/PD analysis and different TRANS representing different parameterizations of the models, as well ADVANs 5 through 9 (not listed) for more general usage and including differential equations, very useful in simultaneous PKPD analysis. In general, the models in these software packages already cover most of the models needed for pharmacokinetic analysis.

There is a standard procedure of pharmacokinetic data analysis with a compartmental model. Because pharmacokinetic data is the drug concentration data at different time points over a period of time, the drug concentration is a response variable, also called a dependent variable; time is a fixed independent variable which is assumed to be no-error. In a compartmental model, a regression analysis is performed between concentration (C) and time (t). Prior to performing the regression analysis, it helps to visualize the relation-

TABLE 6.6. Pharmacokinetic Compartmental Models in WinNonlin and NONMEM

PK Models	WinNonlin	NONMEM
1-Comp	Model 1, IV bolus	ADVAN 1, TRAN1-K
	Model 2, IV infusion	ADVAN 1, TRAN2-CL,V
	Model 15, IV bolus/IV infusion	
1-Comp extravascular	Model 3, first-order input	ADVAN 2, TRAN1-K,K_A
	Model 4, first-order input + lag time	ADVAN 2, TRAN2-CL,V,K_A
	Model 5, first-order input, $k_{10} = k_{01}$	
	Model 6, first-order input + lag time, $k_{10} = k_{01}$	
2-Comp	Model 7, IV bolus, micro	ADVAN 3, TRAN1-K, K12, K21
	Model 8, IV bolus, macro	
	Model 9, IV infusion, micro	ADVAN 3, TRAN3-CL, V, Q, Vss
	Model 10, IV infusion, macro	
	Model 16, IV bolus/IV-infusion, micro	ADVAN 3, TRAN4-CL, V, Q, V2
	Model 17, IV bolus/IV-infusion, macro	ADVAN 3, TRAN5-AOB [A/B], ALPHA, BETA
		ADVAN 3, TRAN6-ALPHA, BETA, K21
2-Comp extravascular	Model 11, first-order input, micro	ADVAN 4, TRAN1-K, K23, K32, KA
	Model 12, first-order input + lag time, micro	ADVAN 4, TRAN2-V2/V3 or K32) + KA
	Model 13, first-order input, macro	
	Model 14, first-order input + lag time, macro	
3-Comp	Model 18, IV bolus, macro	ADVAN 11, TRAN1-
	Model 19, IV infusion, macro	
3-Comp extravascular		ADVAN 12, TRAN1-
1-Comp MM		ADVAN 10, TRAN1-V_{MAX}, K_M

ship between these two variables by plotting drug concentration versus time in a semilog scale. The typical plots of one-compartment and two-compartment models after IV bolus dose are shown in Fig. 6.3.

For a one-compartment model after IV bolus dose, differential Eq. (6.20) is used to describe the system, and Eq. (6.21) is the solution of that differential equation.

$$C = C_0 e^{-kt} = \frac{D_0}{V} e^{-\frac{\mathrm{CL}}{V}t} \tag{6.21}$$

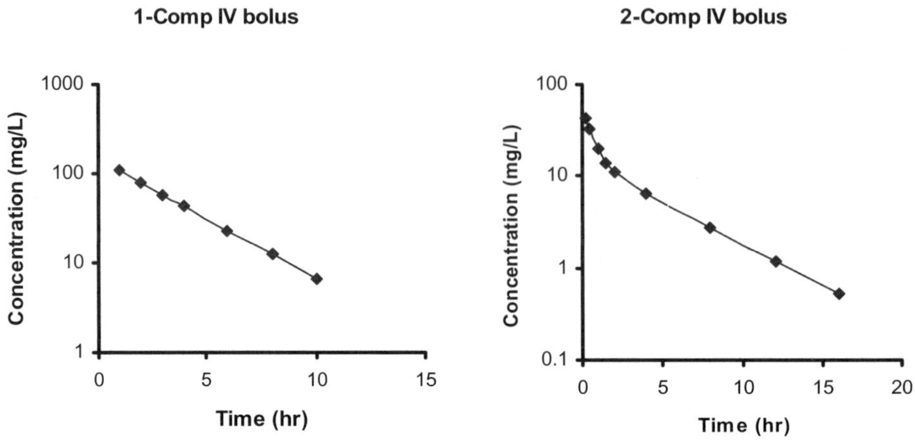

Figure 6.3. Typical plots of one-compartment and two-compartment models after IV bolus dose.

In Eq. (6.21), C is a dependent variable, t is an independent variable, V and CL are called parameters, and D_0, the dose administered, is a constant. Once the plot suggests that the concentration–time relationship appears to be a one-compartment model, an appropriate model from the software could be chosen. For example, Model 1 from WinNonlin could be chosen to perform the regression analysis and derive the values of V and CL, which are interpreted with biological meanings: V represents the extent of drug distribution, and CL represents the rate of drug eliminates from the body. Concentrations at different time points following a new dose (D_0) could be predicted from the values of V and CL. For a two-compartment model after IV bolus, the initial phase represents the distribution process with slope of α and intercept of A, and the terminal phase is called the elimination phase with the slope of β and intercept of B. Typically, the concentration–time profile could be described by the following equation:

$$C = Ae^{-\alpha_t} + Be^{-\beta_t} \tag{6.22}$$

where C, plasma concentration, is a dependent variable, and t, time after dose, is an independent variable.

6.5.2.2.1 Noncompartmental Versus Compartmental Modeling (Cobelli and Toffolo, 1984; Gillespie, 1991). Noncompartmental modeling uses fewer assumptions than compartmental modeling. Dose proportionality (superposition) and time-independent are two important assumptions in noncompartmental modeling, and it is also assumed that the drug concentration–time profile reflects the probability of the drug distributing through the body. The noncompartmental approach does not involve a detailed model structure and can be commonly used for general PK analysis, allowing interpretation of data and parameter estimation without the assignment of a particular model; it is useful in oral bioavailability studies. On the other hand, compartmental modeling consists of a number of homogeneous, well-mixed compartments with similar PK behaviors within the compartment. The exchange between them and with the environment is assumed at a certain order, zero order or first order, and in a certain direction. In comparison with the empiricism of noncompartmental modeling, there may be some rationale for compartmental modeling for a mechanistic study. Currently, the noncompartmental approach is used as an exploratory tool while the model is under construction.

6.5.2.3 Physiologically Based Pharmacokinetic Model (PBPK) (Gibaldi and Perrier, 1982). The physiologically based PK model includes the greatest level of detail within the body: It is still a compartmental model, but each compartment represents an individual organ in the body, and the parameters between or within the compartment have intrinsic biological meanings, such as blood flow, weight, and/or volume of the organ. The technical challenge of this model is the need of a large number of datasets from blood and various tissues, which is especially difficult in human studies. Because almost every compartment in the PBPK approach represents a specific organ in the body, it is not necessary to assume that each compartment is well-stirred when drug distribution is diffusion-limited. Each compartment could be further split into intracellular space, extracellular space, and capillary blood. The partition coefficient of the drug to each tissue/organ is determined based on the concentration of the drug in tissue to blood, with the assumption that the concentration of the drug leaving an organ is equilibrated with the concentration in the organ. Basically, modeling in PBPK analysis is also based on the system of differential equation with the principle of mass balance. PBPK analysis is also one of the methods for PK/PD scaling from animals to humans; however, its application is limited due to the need for extensive information as compared to other methods. The selection of a PK analysis model depends on the objectives of the study and the stage of drug discovery/development.

6.5.3 Parameter Estimation (Draper and Smith, 1998)

In Eq. (6.21), C is a measured concentration which is an observation, $\frac{D_0}{V}e^{-\frac{\mathrm{CL}}{V}t}$ is a model-generated concentration, called predication, which can be generalized as a function $f(t, \theta)$, t is independent variable time in PK data, and θ represent the vector of the parameters V and CL here. In reality, $C-\frac{D_0}{V}e^{-\frac{\mathrm{CL}}{V}t}$ is not zero; therefore, Eq. (6.21) could rewritten as follows:

$$C=\frac{D_0}{V}e^{-\frac{\mathrm{CL}}{V}t}+\varepsilon \tag{6.23}$$

ε is called residual error in Eq. (6.23) and accounts for all sources of error. It includes the difference between observation and prediction, which is not zero, analytical assay error, sample processing error, model misspecification, and unexpected error. There are three unknown parameters, V, CL, and ε, in Eq. (6.23). V and CL are called pharmacokinetic parameters, while ε is called statistical parameter. Because $E(\varepsilon) = 0$ by assumption, the actual statistical parameter is the variance (σ^2) of residual ε. The values of V and CL could be estimated using the classical least squares method, which yields the smallest sum of squares of residuals, ideally 0. Because pharmacokinetic data are repeated measurement data, SSE (sum of squares of residuals) can be expressed as follows:

$$\text{SEE}(V,\text{CL}) = \sum_i^n \left(C_i - \frac{D_0}{V} e^{-\frac{\text{CL}}{V} t_i} \right)^2 \tag{6.24}$$

To find the least squares estimate of V and CL, we need to differentiate Eq. (6.24) with respect to V and CL, respectively, as shown below:

$$\frac{\partial SSE(V,\text{CL})}{\partial V} = \sum_i^n \left(C_i - \frac{D_0}{V} e^{-\frac{\text{CL}}{V} t_i} \right) \frac{\partial \frac{D_0}{V} e^{-\frac{\text{CL}}{V} t_i}}{\partial V} = 0 \tag{6.25}$$

$$\frac{\partial SSE(V,\text{CL})}{\partial \text{CL}} = \sum_i^n \left(C_i - \frac{D_0}{V} e^{-\frac{\text{CL}}{V} t_i} \right) \frac{\partial \frac{D_0}{V} e^{-\frac{\text{CL}}{V} t_i}}{\partial \text{CL}} = 0 \tag{6.26}$$

These two equations need to be solved to obtain the estimate of V and CL. However, practical experience has shown that this is computationally difficult to do because of the nonlinear function $\frac{D_0}{V} e^{-\frac{\text{CL}}{V} t}$. Strictly speaking, nonlinearity in $\frac{D_0}{V} e^{-\frac{\text{CL}}{V} t}$ is not intrinsic, since both sides of Eq. (6.21) have been logarithmically transformed. The relationship between $\ln C \sim t$ is linear; however, for the two-compartmental model, which is described in Eq. (6.22), log transformation would not result in a linear relationship. Here the format of $\frac{D_0}{V} e^{-\frac{\text{CL}}{V} t}$ is kept to show that a nonlinear function is used for nonlinear regression in pharmacokinetic data analysis.

Among the methods of nonlinear regression, the Gauss–Newton method is not only the primary method for managing nonlinear function but also a powerful tool in standard PK software such as WinNonlin and NONMEM.

Basically, the Gauss–Newton method is a method of linearization using the Taylor series, where the nonlinear pharmacokinetic compartmental model is linearlized first, and then linear regression is performed to estimate the values of pharmacokinetic parameters. In linear regression, **Y**, a dependent variable, is the $n \times 1$ vector of the observed concentrations at different time points; **X**, an independent variable, is the $n \times p$ matrix of time; $\boldsymbol{\beta}$ is the $p \times 1$ vector of pharmacokinetic parameters; and ε is the $n \times 1$ vector of residuals, in which n is the number of observations and p is the number of PK parameters. So the following equation represents all of our observations and linear relationship between **Y** and **X**:

$$\mathbf{Y} = \mathbf{X}\boldsymbol{\beta} + \boldsymbol{\varepsilon} \tag{6.27}$$

Following the steps discussed above and computational work because of linearity, $\boldsymbol{\beta}$ (PK parameters), can be solved as follows:

$$\hat{\beta} = (\mathbf{X}^T\mathbf{X})^{-1}\mathbf{X}^T\mathbf{Y} \tag{6.28}$$

Equation (6.29) describes the variance–covariance matrix of $\hat{\beta}$:

$$\mathbf{V}\hat{\beta} = \sigma^2(\mathbf{X}^T\mathbf{X})^{-1} \tag{6.29}$$

Equation (6.30) describes the variance–covariance matrix of $\mathbf{X}\boldsymbol{\beta}$, which is the prediction of **Y**, $\hat{Y}$:

$$\mathbf{V}\hat{Y} = \mathbf{X}(\mathbf{X}^T\mathbf{X})^{-1}\mathbf{X}^T \tag{6.30}$$

Besides the residual ε, the studentized residual is defined as follows:

$$r_i = \frac{\varepsilon_i}{\sqrt{\hat{\sigma}^2(1-h_{ii})}} \tag{6.31}$$

where h_{ii} is the ith diagonal element of **H**, a hat matrix in Eq. (6.30). In order to linearly approximate our pharmacokinetic nonlinear function, we need to select a starting point (t_s C_s), also called an expansion point, on the curve of the nonlinear function; any values of the parameters that result in the linear approximate function and include that starting point can be used as an initial estimate of the parameters (V_0,CL_0). The linear approximate function is polynomial and in pharmacokinetic data analysis, we generally only include the first two terms in the Taylor series, which are zero-order and first-order approximations.

$$\frac{D_0}{V}e^{-\frac{\text{CL}}{V}t_i} = \frac{D_0}{V_0}e^{-\frac{\text{CL}_0}{V_0}t_i} + \frac{\partial \frac{D_0}{V}e^{-\frac{\text{CL}}{V}t_i}}{\partial V}(V - V_0) + \frac{\partial \frac{D_0}{V}e^{-\frac{\text{CL}}{V}t_i}}{\partial \text{CL}}(\text{CL} - \text{CL}_0) + \varepsilon \tag{6.32}$$

If we let **Y** as $n \times 1$ vector of $\frac{D_0}{V}e^{-\frac{\text{CL}}{V}t_i} - \frac{D_0}{V_0}e^{-\frac{\text{CL}_0}{V_0}t_i}$; **β** as $p \times 1$ vector of $(V - V_0, \text{CL} - \text{CL}_0)^T$, **X** as $n \times p$ matrix of $\frac{\partial \frac{D_0}{V}e^{-\frac{\text{CL}}{V}t_i}}{\partial (V, \text{CL})_{V_0, \text{CL}_0}}$, and ε as $n \times 1$ vector of residuals. We will compute Eq. (6.27) in nonlinear regression and then iteratively update the estimates of the parameter until the difference between SSEs from two iterations is very small (<0.0001).

Let's go through an example with the standard pharmacokinetic analysis software WinNonlin. Drug A was administrated to a patient at 50 mg by IV bolus, and the following drug plasma concentrations were measured:

Time (h):	1	2	3	4	6
C (μg/mL):	1.222	0.954	0.864	0.671	0.447

To analyze this data, the first step is to make a plot in a semilog scale (Fig. 6.4). Based on the plot, a one-compartment model appears to fit the data, so the model structure is $C = \frac{D_0}{V}e^{-\frac{\text{CL}}{V}t} = \frac{D_0}{V}e^{-kt}$, because V and K can be approximately determined from the plot; $\frac{D_0}{V}e^{-kt}$ is chosen as structural at this time. $V = \frac{D_0}{C_0}$, where C_0 can be obtained from extrapolation to the Y axis, and K is the slope of line.

After calculation, we can plug the initial estimates of 33 L and 6.6 L/h of V_0 and CL_0 into Eq. (6.32), **X**, a 5×2 matrix here:

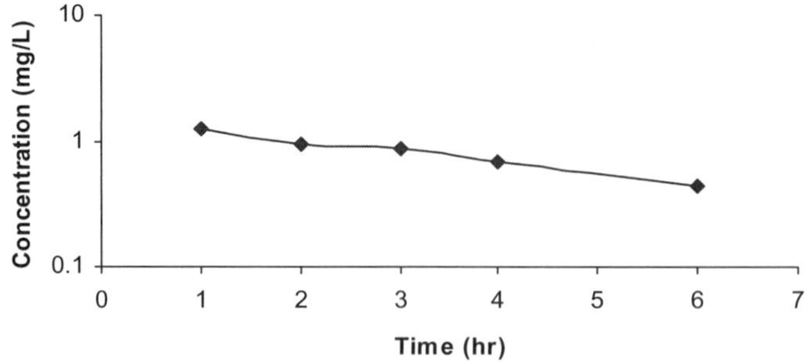

Figure 6.4. Plot in a semilog scale.

$$\mathbf{X} = \begin{bmatrix} \frac{-D_0 e^{-kt_1}}{V^2} & \frac{-D_0 t_1 e^{-kt_1}}{V} \\ \cdot & \cdot \\ \cdot & \cdot \\ \cdot & \cdot \\ \cdot & \cdot \\ \frac{-D_0 e^{-kt_5}}{V^2} & \frac{-D_0 t_5 e^{-kt_5}}{V} \end{bmatrix}$$

in which $D_0 = 50$, $V = V_0 = 33$, $K = K_0 = 0.2$, $t = 1,2,3,4,6$, $\boldsymbol{\beta} = (V - 33, K - 0.2)^T$, and $\mathbf{Y} = \left(1.222 - \frac{50 * e^{-0.2*1}}{33}, \ldots, 0.447 - \frac{50 * e^{-0.2*6}}{33}\right)^T$. By using Eq. (6.28), we have a vector of the estimate of V and K from the first iteration; V_1 and K_1 will then be considered as starting values for the second iteration, with the new values of $\mathbf{X}$, $\boldsymbol{\beta}$, and $\mathbf{Y}$ based on V_1 and K_1. We will continue this cycle until the change in Eq. (6.24) from two iteration is very small, which means that we have already reached the minimal point of SSE. The values of V and K corresponding to this point are our final estimates of the parameters.

The following is typical output from WinNonlin analyzing our pharmacokinetic data after an IV bolus dose.

WinNonlin Compartmental Modeling Analysis. Version 5.2 Build 200701231637.

WinNonlin Compiled Model. 1 compartment IV bolus, no lag time, first-order elimination.

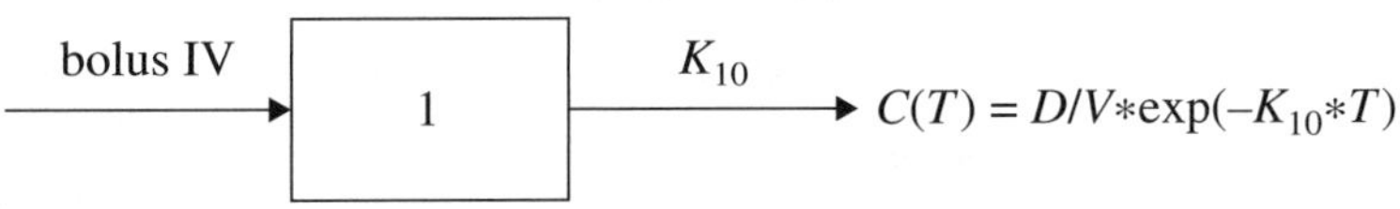

First, Model 1 is chosen from the library of models in WinNonlin (shown in Table 6.6); this model will describe the structure, which is the fixed effect of the data. The following settings indicate that we chose the Gauss–Newton method with modification, which can make process of convergence quicker and more efficient. The process of iteration will stop when the change of SSE between two continuous iterations is less than 0.0001; the maximal number of iterations is here set to be 50.

Settings for Analysis

Input Workbook: [Untitled1]
Input Worksheet: Sheet1
Input Sort Keys: [none]

Gauss–Newton (Levenberg and Hartley) method used
Convergence criteria of 0.0001 used during minimization process
50 maximum iterations allowed during minimization process

Input Data

Time (h)	Concentration (mg/L)
1	1.222
2	0.954
3	0.864
4	0.671
6	0.447

This is the original pharmacokinetic data from IV bolus dose. The following table shows the information from our initial estimate and the lower and upper bound of this estimate.

Output Data

INITIAL PARAMETERS

Parameter	Value	Lower	Upper
V	33	0	330
K_{10}	0.2	0	2

MINIMIZATION PROCESS

Iteration	Weighted SS	V	K_{10}
0	5.38E-03	33	0.2
1	3.98E-03	33.93	0.1955
2	3.98E-03	33.96	0.1951
3	3.98E-03	33.97	0.1951

Basically, the above table lists the results after each iteration, which were described previously with the method of first-order approximation.

FINAL PARAMETERS

Parameter	Units	Estimate	Standard Error	CV%	Univariate CI, Lower Bound	Univariate CI, Upper Bound	Planar CI, Lower Bound	Planar CI, Upper Bound
V	L	33.966	1.23	3.6	30.063	37.869	28.003	39.930
K_{10}	1/h	0.1950	0.014	7.0	0.1514	0.2387	0.1283	0.2617

The above table is the key component in all of this output data, because it gives us the final estimate of our two parameters, V and K_{10}, from the linearization method. The standard error of each parameter was the square root of diagonal elements of the matrix in Eq. (6.29), in which the value of σ^2, called the mean sum of square of residuals (MSE), can be found in the later table of diagnostics, s^2. The last four columns provide information on the lower and upper bound of two kinds of 95% confidence intervals (CIs). Univariate CIs are calculated assuming that the two parameters are not correlated with each other, estimate $\pm\ t_{0.975,\,3}$ * std error. Planar CIs are calculated by taking into account the correlation. As a result, the range of planar CI will be wider than that of univariate CI. To obtain this kind of CI, we need to solve Eq. (6.24).

DOSING

Constant	Value
Number of doses	1
Dose #1	50
Time of dose #1	0

CORRELATION MATRIX

Parameter	V	K_{10}
V	1	
K_{10}	−0.855279	1

If one defines $D = \text{diag}\{V\hat{\beta}\}$ based on Eq. (6.29), a diagonal matrix consisting of diagonal elements of $V\hat{\beta}$, and $D^{1/2}$ is the diagonal matrix in which diagonal elements are the standard error of $\hat{\beta}$, then the correlation matrix **P** is defined as follows:

$$\mathbf{P} = \mathbf{D}^{-1/2}\mathbf{V}\hat{\beta}\mathbf{D}^{-1/2} \tag{6.33}$$

EIGENVALUES

Number	Value
1	26.26
2	8.82E-04

CONDITION NUMBERS

Iteration	Rank	Condition
0	2	4.479587
1	2	4.486198
2	2	4.488673

Eigenvalues here refer to the eigenvalues of matrix $\mathbf{X}^{\mathrm{T}}\mathbf{X}$. A condition number, which is defined as the square root of the ratio of largest to smallest eigenvalues, can indicate the ill condition (large condition number) of the matrix, meaning that the model is overparameterized.

VARIANCE–COVARIANCE MATRIX

Parameter	V	K_{10}
V	1.50311	
K_{10}	−1.44E-02	1.88E-04

The variance–covariance matrix was calculated from Eq. (6.29). The following summary table gives us information about the observations, predications, and residual ε. Note that W represents weight; we applied equal weight to every observation in this example. Besides these values, the eighth column, the standard error of predication, was calculated from Eq. (6.30), and the last column, standard residual, was actually calculated using the method in Eq. (6.31), studentized residual.

SUMMARY TABLE

Time_obs (h)	Conc_obs (mg/L)	Time (h)	Concentration (mg/L)	Predicted (mg/L)	Residual (mg/L)	W	SE $\hat{Y}$	Standard Residual
1	1.222	1	1.2220	1.2112	0.0108	1	0.0307	0.5537
2	0.954	2	0.9540	0.9965	−0.0425	1	0.0189	−1.3679
3	0.864	3	0.8640	0.8199	0.0441	1	0.0175	1.3800
4	0.671	4	0.6710	0.6746	−0.0036	1	0.0205	−0.1208
6	0.447	6	0.4470	0.4567	−0.0097	1	0.0250	−0.3665

DIAGNOSTICS

Function	Item	Value
1	CSS	0.342153
1	WCSS	0.342153
1	SSR	3.98E-03
1	WSSR	3.98E-03
1	S	3.64E-02
1	DF	3
1	CORR_(OBS,PRED)	0.9942
1	WT_CORR_(OBS,PRED)	0.9942
1	AIC	−23.63782
	SBC	−24.41894

The table presenting diagnostics, above, presents the value of the corrected sum of squares (CSS), $\sum_{i=1}^{n}\left(C_i - \bar{C}\right)$; weighted CSS (WCSS) is the same as

CSS because weight is 1 in this example. SSR is the sum of square of residuals,

$$\sum_{i=1}^{n}\left(C_i - \frac{D_0}{V}e^{-kt_i}\right)^2,$$

where V and k are our final estimations. S is the standard error term, calculated by $\sqrt{\text{MSE}}$, that is, $\sqrt{\text{SSR/DF}}$. DF is degree of freedom, $n - p$, which is $5 - 2 = 3$ here. The correlation coefficient is calculated by the square root of coefficient of determination R^2, which is (CSS − SSR)/CSS. AIC and SBC are two commonly used statistical standards for comparing model selection; both AIC and SBC have different terms of penalty for adding more parameters into the model. AIC is formally defined as −2 max log-likelihood +2P; in WinNonlin, AIC is approximately n $\log_e$(SSR) +2P. SBC is defined n $\log_e$(SSR) + $P\log_e n$, where n is the number of observations and p is the number of parameter. The model with the smaller AIC and SBC is considered better.

PARTIAL DERIVATIVES

Function	Time (h)	V	K_{10}
1	1	−0.03562204	−1.21105193
1	2	−0.02930935	−1.99268124
1	3	−0.02411535	−2.45908851
1	4	−0.01984180	−2.69747835
1	6	−0.01343247	−2.73866791

This table is actually the matrix of **X** at the value of the final estimate of the parameters.

SECONDARY PARAMETERS

Parameter	Units	Estimate	Standard Error	CV%
AUC	h*mg/L	7.546648	0.328894	4.36
K10_HL	h	3.553545	0.249506	7.02
Cmax	mg/L	1.472034	0.053080	3.61
CL	L/h	6.625458	0.289035	4.36
AUMC	h * h * mg/L	38.689270	4.298317	11.11
MRT	h	5.126682	0.359961	7.02
V_{ss}	L	33.966618	1.226015	3.61

The last table summarizes the value of secondary parameters that can be calculated from our primary parameters, V and k, by different pharmacokinetic relationships. Generally, if a secondary parameter is a function of one or more primary parameters, $p = f(\theta)$, its standard error can be approximately calculated as follows:

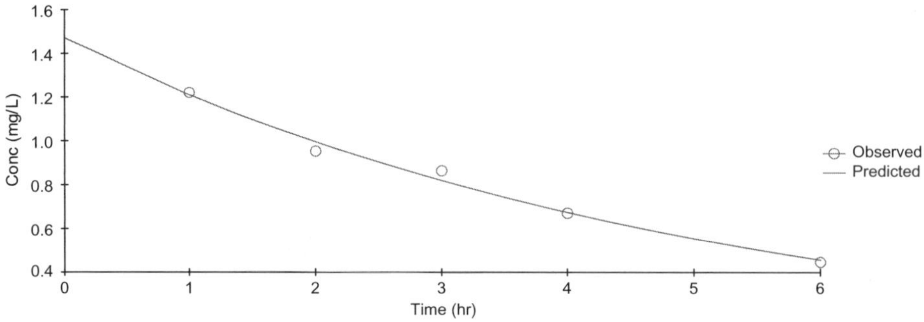

Figure 6.5. Plot of *X* versus observed *Y* and predicted *Y*.

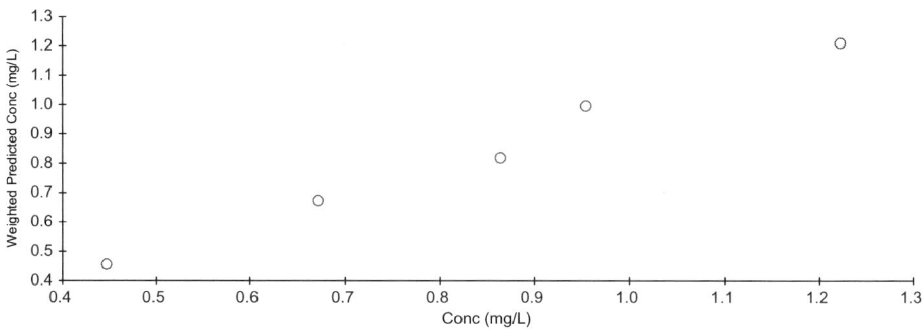

Figure 6.6. Plot of observed *Y* versus weighted predicted *Y*.

$$\sum_{m}^{\#\theta}\sum_{n}^{\#\theta}\frac{\partial f(\theta)}{\partial\theta_m}\frac{\partial f(\theta)}{\partial\theta_n}\operatorname{cov}(\theta_m,\theta_n) \tag{6.34}$$

The plots in Figs. 6.5–6.9 are generated for the aim of diagnosis. We made some assumptions when we performed nonlinear regression, such as no significant error related to model specification and independent variable (x), and $\varepsilon \sim \text{NID}\,(0, \mathbf{I}\sigma^2)$, which means that ε_I have mean 0 and equal variance σ^2 and are uncorrelated. In order to test our assumptions, some key plots, such as the following, were needed.

X Versus Observed Y and Predicted Y. The plot in Fig. 6.5 was made to see if the model's prediction can describe the data of observation well.

Observed Y Versus Weighted Predicted Y. If the model is correct, the plot of predicted *Y* versus observed *Y* (Fig. 6.6) should indicate a straight line.

Weighted Predicted Y Versus Weighted Residual Y. The plot of residuals versus predicted *Y* (Fig. 6.7) could be used to test whether the residuals have

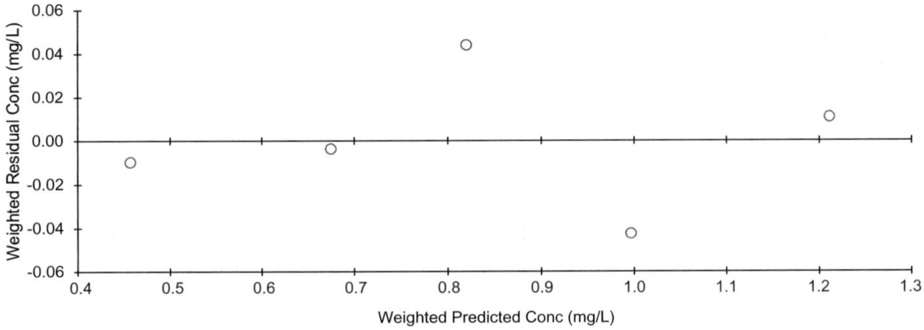

Figure 6.7. Plot of predicted *Y* versus weighted residual *Y*.

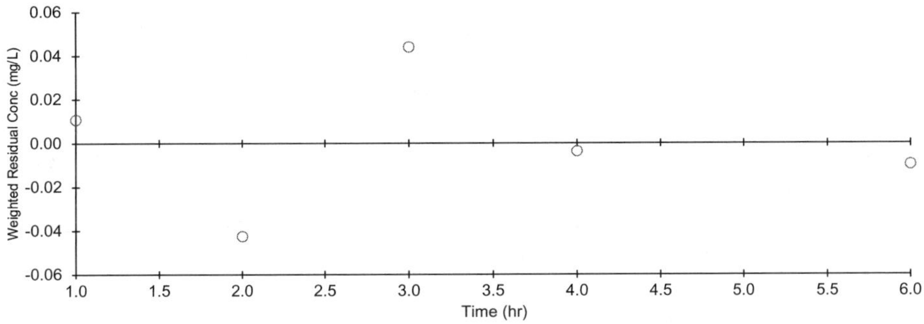

Figure 6.8. Plot of *X* versus residual *Y*.

equal variance. If the pattern in this plot is scattered, it indicates that the model already extracted the describable part of data, leaving only a random part of data there.

X Versus Weighted Residual Y. Residuals versus *X*, an independent variable, should also be scattered (Fig. 6.8).

Partial Derivatives. The plot in Fig. 6.9 can be used to help the selection of optimal sampling time in the experimental design. Partial derivatives here can be considered as a measure of the sensitivity of the model with respect to the change of the model parameters. For example, in order to obtain a reliable estimate the parameter of *k*, the optimal sampling time points should be taken from the earlier time period.

In this example, equal weight is applied to every data point. In PK data analysis, weighting data is another important issue that needs to be considered. In PK data analysis, the need for weighting comes from the fact that we do not have equal confidence in each data point we collected from the experiment (Peck et al., 1984). In the experiment, some data are more trustable, while

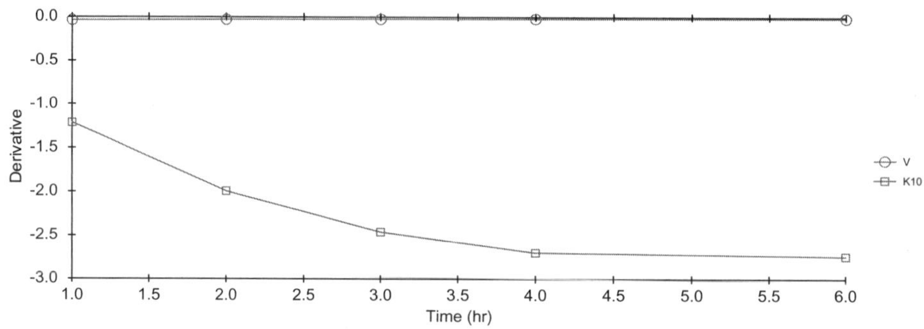

Figure 6.9. Plot of partial derivatives.

some are less trustable. To express the reliability of different data quantitatively, one could choose to use variance. This is a reasonable choice, because data points with less variance after several replications should be more trustable than a data point with very large variance. However, here the variance refers to population variance, not sample variance, so it is not appropriate to use the calculated variance from several replications as the weight in the data analysis. The data with less variance should be more trustable and contribute more weight in the least sum of square than the data with large variance; therefore, the weight is defined as

$$\text{Weight} = \frac{1}{\text{Variance}} \tag{6.35}$$

PK data are considered as random variables, so they are treated as data randomly drawn from a specific probability distribution. It is hard to know the true population distribution, but we can get some idea based on the sample distribution of our PK data. Every distribution can be described by two parameters, expectation and variance. So the variance of that probability distribution is assumed to be the population variance that will be used in Eq. (6.35). Sometimes, PK data can be assumed to be normally distributed, but most of the time it is log-normally distributed. In normally distributed data, variance is constant σ^2 and expectation is μ; for log-normally distributed data, variance and expectation are calculated as follows:

$$\text{Var}(X) = \left(e^{\sigma^2} - 1\right)e^{2\mu+\sigma^2} \tag{6.36}$$

$$E(X) = e^{\mu+\sigma^2/2} \tag{6.37}$$

$\text{Var}(x) = \left(e^{\sigma^2} - 1\right)E^2(x)$ and the coefficient of variation (CV) is $\sqrt{\left(e^{\sigma^2} - 1\right)}$, which is also a constant. So, when we assume that the PK data are log-normally distributed, we are assuming that we have data with a constant CV, not constant variance. This is also the reason that we recommend the weight $1/\hat{y}$ and $1/\hat{y}^2$ when we analyze PK data.

6.5.4 Tools for Pharmacokinetic Data Analysis

There are many tools for PK data analysis. The following is a partial list of these tools:

- Excel
- WinNonlin
- ADAPT II
- Scientist
- Kinetica
- Madonna
- NONMEM
- SAS
- R

Out of all of these software programs, it is likely that many individuals have some experience with Microsoft Excel, since basic spreadsheet operation is common. Excel can generate basic plots for data exploration and model diagnosis in PK data analysis, and it can also be used to summarize the results and data in a table with its friendly graphic user interface (GUI). Usansky et al. of Allergan (Irvine, CA) developed a group of PK functions for Excel. There are eight PK functions that will simplify routine PK calculations in Excel worksheets: C_{max}, t_{max}, ElimRateConstant(k), Half-life($t_{1/2}$), AUC_{0_t}, AUC_{0_inf}, $AUMC_{0\_t}$, and $AUMC_{0\_inf}$. With these add-on PK functions, a preliminary PK data analysis such as noncompartmental analysis, plotting, and summary could be done in Excel.

WinNonlin is a software program that is used extensively in the pharmaceutical industry for PK and PD data analysis. It is a Windows-based software for noncompartmental, compartmental, and PBPK data analysis. Although WinNonlin has a library of common PK and PD models, user-defined models can be designed for special needs.

NONMEM is a software program for population PKPD data analysis, but it also can be used for traditional PK data analysis. It has several subroutines for PK compartmental analysis. It is easier to include information for complex dosing regimens in a dataset for NONMEM analysis. Besides this advantage, NONMEM can also evaluate the relationship between PK parameters and physiological and/or demographic covariates to estimate the population PK parameters.

R is a free statistical software program that now has several packages for PK data analysis, even for population PK analysis. For example, the PK package is designed for basic pharmacokinetics, the PKfit package is a data analysis tool for pharmacokinetics, and the PKtools package is a unified computational interface for population PK. Several other useful packages, such as drc and nlmeODE, can also be used in pharmacokinetic data analysis.

6.6 HUMAN PK AND HUMAN EFFICACIOUS DOSE PROJECTION

6.6.1 Commonly Used Methods in Human Pharmacokinetic Projections

Human PK projections range from projection of PK parameters such as human clearance (CL_H) and volume of distribution at steady state ($V_{SS\ H}$) to projection of the complete human oral concentration–time profile. Therefore, these methodologies can be broadly classified into two major classes:

1. Methodologies used for predicting human pharmacokinetic parameters
2. Methodologies used for predicting human plasma concentration–time profile

6.6.1.1 Methodologies Used for Predicting Human Pharmacokinetic Parameters. The methods commonly used to predict human PK parameters such as human clearance (CL_H), volume of distribution at steady state ($V_{SS\ H}$), and half-life ($t_{1/2}$) will be discussed in this section.

Predicted parameters such as CL_H are in turn used to calculate the area under the curve (AUC) at a certain dose and thus provide estimates of plasma exposure in humans. Commonly used methods include simple allometry or its variants, which rely on various correction factors (Mahmood, 1999, 2005), and the use of *in vitro* systems such as liver microsomes and/or hepatocytes (Obach et al., 1997). In this section we will explore the use of allometry and its associated methodologies.

6.6.1.1.1 Simple Allometric Scaling. The most common method to predict CL_H and $V_{SS\ H}$ is simple allometric scaling. This involves correlation of clearance (CL) or volume of distribution at steady state (V_{SS}) determined in preclinical species to their respective body weights (BW) using a power equation, such as the two shown below (Mahmood, 1999, 2005):

$$\mathrm{CL\ (mL/min)} = a * \mathrm{BW}^b \tag{6.38}$$

$$V_{SS}(\mathrm{L}) = c * \mathrm{BW}^d \tag{6.39}$$

where a and c are intercepts and b and d are the exponents of CL and V_{SS}, respectively. A plot of a representative allometric correlation of CL is provided in Fig. 6.10. To enable visualization of data points that may potentially span across multiple orders of magnitude, a dual logarithmic axis is commonly used (Fig. 6.10, right-hand panel). Additionally, the dual logarithmic axis

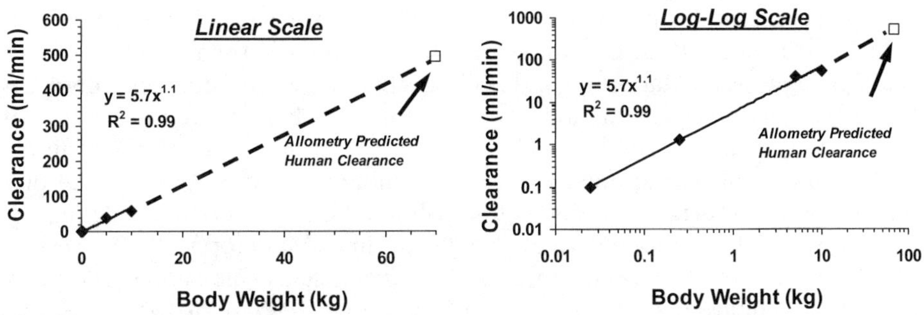

Figure 6.10. Representative plots of clearance of a compound in preclinical species against body weight. The human clearance (CL_H) is predicted using a power equation [Eq. (6.38)]. The left-hand plot presents the body weight in linear axis and the right-hand plot presents the same data in dual logarithmic axis. The solid line is the fit using a power equation, and the broken line is the extrapolation to the CL_H using *a* and *b* from the power equation fit.

linearizes the curve and helps to assess the visual goodness of fit. In practice, the exponents of CL and V_{SS} can span a relatively wide range (~0.2–1.8), although multiple reports in literature propose the concept of a fixed exponent (0.75 and 1 for CL and V_{SS}, respectively).

Although allometry is commonly applied to predict CL_H and $V_{SS\ H}$, it has also been used to scale human half-life (Mahmood, 1999), various compartment model rate constants (Mordenti, 1985), and hepatic extraction ratios, among other parameters (Tannenbaum et al., 1997).

6.6.1.1.2 Simple Allometric Scaling with Correction Factors. Although simple allometric scaling is widely used to predict CL_H and $V_{SS\ H}$, various correction factors have been devised to improve the predictive performance of allometric scaling (Mahmood, 1999, 2005). These correction factors include plasma protein binding, liver microsomal metabolic rates, brain weight (BRW), and maximum life-span potential (MLP), among others. Adding consideration of these factors into simple allometry, although supposed by many to result in more "physiological" predictions, in reality only adds additional empiricism to the process (Mahmood, 1999, 2005). In any case, it has been shown using a relatively large dataset that incorporation of some of these factors results in better prediction of PK parameters, especially CL_H.

Rule of Exponents. The "rule of exponents" proposed by Mahmood and Balian is a widely used method of incorporating correction factors into allometric predictions (Mahmood and Balian, 1996). In this methodology the brain weight (BRW) or maximum life-span potential (MLP) of various species are used as correction factors. The choice of the specific correction factor depends on the value of the exponent of CL [*b* in Eq. (6.38)]. If *b* ranges from 0.55 and 0.70, then this method proposes the use of simple allometry (no correction factors necessary). If *b* ranges from 0.71 and 1.0, then correction with MLP is suggested. If *b* ranges from >1.0 and 1.3, then correction with BRW is suggested (Mahmood and Balian, 1996). However, in cases where *b* is below 0.55 or above 1.3, the method does not provide a correction factor, but only surmises that the predicted CL using simple allometry may be inaccurate. Although the rule of exponents is used commonly, it is important to note that this method is not rigid and that caution should be exercised in predicting CL in cases where the exponent values are borderline (Mahmood, 1999).

Of note is the observation that the CL_H predicted using simple allometry is higher compared to the CL_H predicted using BRW or MLP correction. Therefore, oftentimes in the drug discovery setting when selection of compounds based on CL_H is performed, simple allometry is used instead of allometry with correction factors, because it provides the most conservative (highest possible) value of CL_H.

Plasma Protein Binding and Liver Microsomal/Hepatocyte Scaled Clearance. Introduction of correction factors such as plasma protein binding and/or *in vitro* microsomal or hepatocyte scaled metabolic rate or clearance into the allometric power equation has been proposed. The impact of these correction factors on improving the predictive power of allometry is, however, controversial. For example, Lave et al. (1997) proposed the use of *in vitro* hepatocyte or microsomal scaled clearance values as a correction factor in allometry. They reported that inclusion of *in vitro* results into the correlation could improve the predictive power of allometry compared to other approaches that they evaluated. However, in a reanalysis performed by Mahmood (1998), the authors concluded that the application of the rule of exponents to the dataset used by Lave et al. resulted in similar results. Similarly, incorporation of plasma protein binding into the predictions to estimate unbound clearance (as opposed to total clearance) was proposed. Although Obach et al. (1997) reported a slight improvement in prediction of unbound clearance compared to total clearance, others such as Mahmood (Mahmood, 1999, 2005) and Bjorkman and Redke (2000) reported a lack of improvement in predictive power when protein binding was introduced. In fact, Mahmood states that the introduction of protein binding may result in more error in prediction of human clearance (Mahmood, 2005).

Recently, Tang and Mayersohn (2005) presented a new empirical method using the intercept of the allometric power equation [*a* in Eq. (6.38)] and ratio of protein binding of rats and humans. Implementing their approach on a relatively large dataset (~60 compounds), they demonstrated better predictability of human clearance compared to simple allometry and the rule of exponents (Tang and Mayersohn, 2005). Further use of this methodology with additional datasets will help us gain understanding on this promising technique.

6.6.1.2 Methodologies Used for Predicting Human Plasma Concentration–Time Profile. The prediction of the concentration–time profile becomes

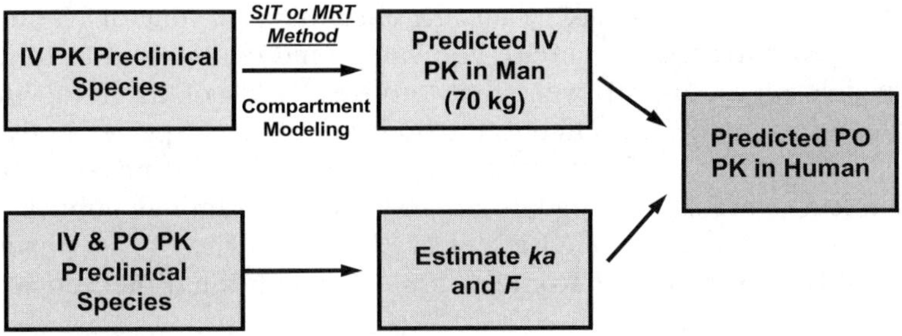

Figure 6.11. Typical scheme used in prediction of a human oral pharmacokinetic profile from preclinical species intravenous and oral pharmacokinetic data.

necessary in discovery programs where the maximum or trough plasma concentrations, as opposed to an AUC estimate, are critical for determining efficacy or toxicity. This is true in case of antiviral drug discovery, wherein the trough concentrations are known to be critical in determining efficacy and development of resistant strains.

Estimation of maximum or trough plasma concentrations is relatively complicated compared to AUC, for which only two predicted parameters are required (bioavailability (F) and CL_H). Prediction of the human PK profile involves two major steps (Fig. 6.11):

- Estimation of the human intravenous (IV) PK profile from preclinical species IV data
- Introduction of an absorption component into the model to estimate the PK profile following extravascular administration

Most methods for predicting the human IV profile are modified forms of the original scaling technique proposed by Dedrick et al. The Dedrick approach, in turn, relies on the principles of allometric scaling.

The central idea introduced initially by Dedrick (Dedrick et al., 1970; Dedrick, 1973) and later expanded by Boxenbaum (Boxenbaum, 1982, 1984; Boxenbaum and Ronfeld, 1983) is the concept of physiological or pharmacokinetic time. The time taken for completing a specific physiological event is called *physiological time*. If the event is a pharmacokinetic process, then it is designated as a *pharmacokinetic time*. This concept is based on observations of life spans of various species in relation to their size (Boxenbaum, 1982, 1984; Boxenbaum and Ronfeld, 1983). Species with a smaller size tend to perform physiological events at a faster pace and thus have a shorter life span compared to animals with a larger size, whose life span is consequently longer. Boxenbaum illustrated this using an example of the potential life spans of a dog and a human, which are 14 and 98 years, respectively. Compared to its potential life span, a dog will "utilize" ~7% of his life in one year, whereas the human will "utilize" ~7% of his life every 7 years (Boxenbaum, 1982, 1984;

Boxenbaum and Ronfeld, 1983). Thus the chronological time of 1 year for the dog and 7 years for the human are equivalent and correspond to time taken for a specific physiological event that "utilizes" ~7% of the total life span. Since these events can be related to the body size of each species, body weight was used to "transform" the time axis in scaling up the plasma concentrations of preclinical species. The readers are referred to seminal publications by Dedrick and Boxenbaum (Dedrick et al., 1970; Dedrick, 1973; Boxenbaum, 1982, 1984; Boxenbaum and Ronfeld, 1983) for additional theoretical details of this technique.

6.6.1.2.1 Species Invariant Time (SIT) Approaches. Multiple types of pharmacokinetic times were introduced to scale the PK profiles of a compound. These are called kallynochrons, apolysichrons, dienetichrons, and syndesichrons (Boxenbaum, 1982, 1984; Boxenbaum and Ronfeld, 1983). A kallynochron is defined as a unit of pharmacokinetic time wherein each species clears the same volume of plasma per unit body weight. Similarly, an apolysichron is defined as a unit of pharmacokinetic time wherein each species clears the same volume of plasma per unit body weight^d, where d is the exponent of volume of distribution. As can be seen here, if the exponent of V_{SS} is unity, then kallynochron and apolysichron are equivalent. Incorporation of additional factors such as MLP and BRW resulted in generation of dienetichrons and syndesichrons, respectively (Boxenbaum, 1982, 1984; Boxenbaum and Ronfeld, 1983).

The data transformation required for prediction of the human IV and subsequently the oral (PO) PK profile will be illustrated below.

In prediction of a human IV PK profile, the first step involves the plotting of the plasma concentrations of the specific drug in various preclinical species (a representative plot is shown in Fig. 6.12).

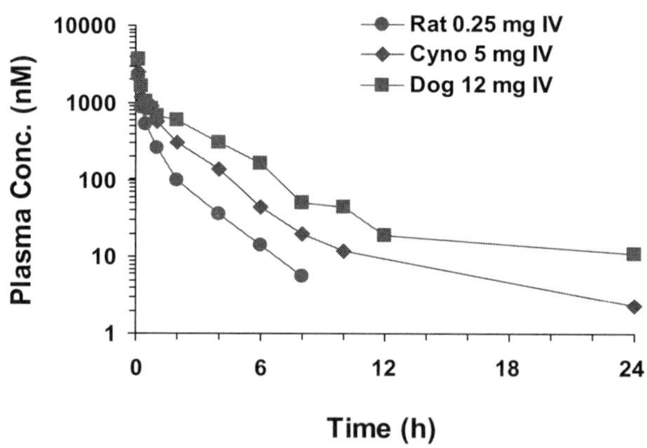

Figure 6.12. Representative plasma concentration time profiles of a potential drug candidate following intravenous administration to rat, dog, and cynomolgus monkey.

Subsequently the X and Y axes are transformed as shown below:

$$X \text{ axis} = \frac{\text{Time}}{\text{BW}^{d-b}} \tag{6.40}$$

$$Y \text{ axis} = \frac{\text{Concentration}}{\frac{\text{Dose}}{\text{BW}^{d}}} \tag{6.41}$$

where b and d are the exponents of clearance and volume of distribution in various preclinical species. First, exponents determined from simple allometry are incorporated along with body weights to transform the preclinical PK data. Second, following this transformation, the "collapsed" concentration–time profiles are fitted using an appropriate compartmental PK model. A modification of this two-stage approach can also be used, wherein both the estimation of the allometric exponents and the subsequent compartmental modeling can be performed using a single-stage approach. A modification of this approach is presented below.

In the provided discussion, based on information about the preclinical PK properties of this drug candidate, a two-compartment model was used for fitting the data. The two-compartment model equation was modified to assist in the transformation of the preclinical PK profiles according to the principles described before. The resulting equation is shown below.

$$\frac{C_p}{\frac{\text{Dose}}{\text{BW}^{d}}} = A * \exp\left(-\alpha * \frac{T}{\text{BW}^{d-b}}\right) + B * \exp\left(-\beta * \frac{T}{\text{BW}^{d-b}}\right) \tag{6.42}$$

where A and B are functions of dose, volume of central compartment, and rate constants, α and β are disposition rate constants, T is the time, and b and d are the exponents of clearance and volume of distribution. The preclinical PK profiles are fitted to Eq. (6.42) to determine A, B, α, β, b, and d simultaneously. The representative observed versus the predicted PK profiles following modeling of the preclinical data using Eq. (6.42) is shown in Fig. 6.13. In this case, the exponent of volume of distribution is >1 (~1.1), and therefore the unit of pharmacokinetic time is presented as apolysichrons. Figure 6.13 illustrates representative plots of observed and the predicted preclinical PK profiles following transformation of preclinical IV PK data using Eq. (6.42). The left-hand panel presents the data in a chronological time format, whereas the right-hand panel presents the data in apolisichron units.

When the PK data are plotted in apolysichrons, the apparent disconnect in the plasma concentration–time profiles in different species is absent and they "collapse" on one another. The human data are also assumed to be similar to the preclinical species in this plot, and so the human IV PK profile is

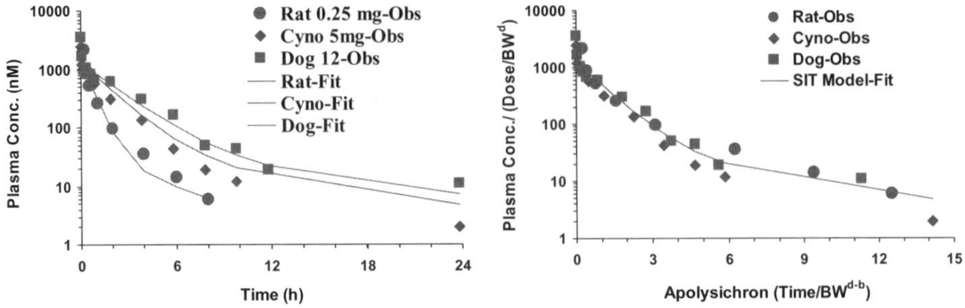

Figure 6.13. Representative plots of observed and the predicted preclinical pharmacokinetic profiles following transformation of preclinical IV PK data using Eq. (6.42). The left-hand panel presents the data in a chronological time format whereas the right-hand panel presents the data in apolisichron units.

back-generated by simulating the profile for a 70-kg individual using the estimated model parameters.

Although the human IV PK profile is useful, given that most drugs are administered by the oral (PO) route, the PK profile following oral administration is desired. In this case, the preclinical species IV or PO PK data are analyzed using appropriate methods (like deconvolution or compartment modeling) to estimate the first-order absorption rate constant (k_a). This, in conjunction with the observed average preclinical species bioavailability, is integrated along with the generated human IV PK profile to simulate the human oral PK profile. In cases where the absorption is complex (not first order), appropriate absorption models should be developed for integration with the IV data, so that an oral profile can be generated. In addition, the bioavailability observed in preclinical species may not be reflective of what is expected in humans, because the formulation can change dramatically (from solution to tablet). Therefore, the availability of preclinical oral data using a clinically relevant formulation would be beneficial in assessing this risk.

For application of other pharmacokinetic time units like dienetichrons and syndesichrons, the time unit in Eq. (6.42) would need to be modified as proposed by Boxenbaum.

6.6.1.2.2 Mean Residence Time (MRT) Normalization Approach. A second methodology for prediction of the human IV PK profile is the relatively recent approach proposed by Wajima et al. (2004). This involves the normalization of the concentration and time axes using the C_{SS} and mean residence time (MRT), respectively. In this methodology, the C_{SS} (not to be mistaken for the concentration at steady state) is calculated by dividing the administered dose by the V_{SS}.

In this approach, the preclinical species plasma concentration and time data are transformed as shown below in Eqs. (6.43) and (6.44).

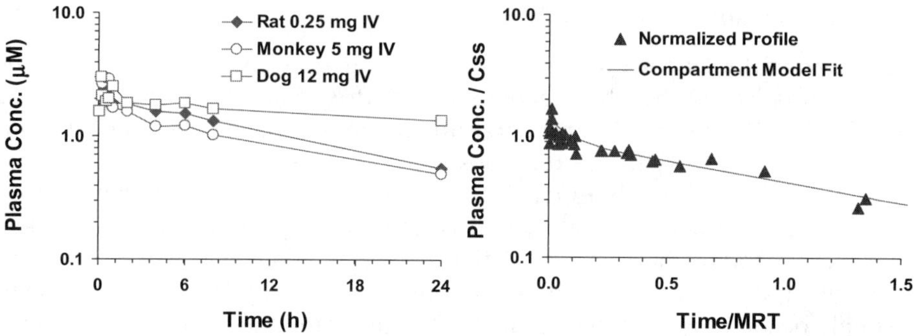

Figure 6.14. A representative plot of preclinical pharmacokinetic data from rat, dog, and cynomolgus monkey (**left-hand panel**) along with their transformation using the MRT normalization approach (**right-hand panel**).

$$X \text{ axis} = \frac{\text{Time}}{\text{MRT}} \tag{6.43}$$

$$Y \text{ axis} = \frac{\text{Concentration}}{\frac{\text{Dose}}{V_{\text{SS}}}} \tag{6.44}$$

These data, when plotted together, "collapse" similar to that observed for the SIT methods (Figure 6.13, right-hand panel). A representative plot of preclinical PK data from rats, dogs, and cynomolgus monkeys (left-hand panel), along with their transformation using the MRT normalization approach, is shown in Fig. 6.14.

Similar to the species invariant time approach, the normalized human concentration–time profile is assumed to be similar to the normalized preclinical species profiles. Therefore, the human IV PK profile is simulated by back-transforming the *X* axis and *Y* axis of the collapsed plots by multiplying them with human MRT and human C_{SS}, respectively. The human MRT and C_{SS} are calculated using the predicted human clearance and V_{SS}. The clearance and V_{ss} can be predicted using either allometry or other *in vitro* or *in silico* methodologies.

Therein lies the major flexibility of this approach, in that a broad spectrum of clearance and/or V_{ss} values, determined using various methodologies, can be incorporated into the modeling process to simulate various scenarios. In contrast, the SIT methodology cannot commonly incorporate *in vitro* or *in silico* scaled clearance and V_{SS} values and is restricted to human clearance and V_{SS} values predicted using the various time transformations. In practice, based on accumulated knowledge in a drug discovery program, if *in vitro* or *in silico* scaled clearance or V_{SS} values are found to be more appropriate, then the MRT normalization approach would be the preferred method over the SIT method.

6.6.2 Projection of Human Efficacious Dose

Projection of the human efficacious dose is not an easy task because it requires good estimation of the pharmacokinetics of drug candidates in humans (as described in Section 6.6.1) and good understanding of the PK–PD relationship. To establish the PK–PD relationship, various animal efficacy models are used, depending on the therapeutic area. One critical assumption here is that the preclinical efficacy models serve as good surrogates of human diseases. However, it should be recognized that different preclinical efficacy models often give quantitatively different (or conflicting) results. When that happens, it simply becomes a question of which preclinical animal models better represent human disease conditions (e.g., rodents versus larger animals). This is not an easy question to answer, and it is beyond the scope of this chapter.

Assuming that the use of these preclinical efficacy models is valid, the main question is, Does *in vivo* efficacy correlate with the overall systemic AUC or a threshold concentration at trough (C_{min})? It is often very difficult to delineate whether AUC or C_{min} drive the efficacy, and the approaches to estimating the human efficacious dose are quite different depending on the endpoint of interest. If the overall exposure (AUC) drives the efficacy, then the projected human dose can be simply estimated by incorporating the overall CL and oral bioavailability.

Projected human efficacious dose = (F * AUC)/CL, where AUC is the drug exposure that produced desirable efficacy in the preclinical animal models, and CL is the projected human CL based on the methods described above. On the other hand, if maintaining a threshold concentration throughout the dosing period is important to sustain efficacy, then the C_{min} (or C_{trough}) needs to be estimated from the preclinical efficacy models or *in vitro* potency derived from human systems. Once C_{min} is established, then the required PK profiles can be constructed using the methods described earlier, such as the species invariant time approach or the mean residence time approach. It should be emphasized that the estimated human efficacious dose is a rough projection based on numerous assumptions. Therefore, the value of using such information should be considered carefully during the decision-making process.

Estimation of the human efficacious dose at the discovery stage can be useful in providing a ballpark figure on the clinically relevant dose with respect to developability. If the projected human efficacious dose exceeds >1000 mg and beyond, then the pharmaceutical development tasks may become unrealistically challenging, and alternative drug candidates should be considered.

6.7 PHARMACOKINETIC (TOXICOKINETIC) STUDIES SUPPORTING SAFETY ASSESSMENT

Toxicokinetics (TK) refers to pharmacokinetics at high doses where toxicity is likely to happen. For a long time, toxicokinetics was used as an evidence that the animals tested in the drug safety study had been exposed to the drug tested; however, the utility of toxicokinetics is now becoming more versatile. Toxicokinetics is a tool used to assess systemic exposure to drugs in preclinical toxicological animal species for the purpose of evaluating drug safety prior to administration into humans.

The objective of preclinical safety studies is to determine a dose or exposure that will produce adverse effects and to determine a dose or exposure that will be safe and well-tolerated. These data are used to explain the observed toxicity in specific tissues or organs in animals, to monitor humans, and to establish safety margins based on drug exposures to ensure that the proposed range of doses in humans will also be safe and well-tolerated. Different toxicokinetic studies are conducted at different stages to support the initiation of the first-in-human phase I study, for chronic drug dosing into phases II and III, and for the inclusion of women of child-bearing potential as well as for pediatrics, if these specific patient population could be benefit from the drug treatment. Toxicokinetics is not only an important part of exposure monitoring in general toxicology studies, but it is also important in developmental and reproductive studies, in genotoxicity studies, in cardiovascular safety monitoring, and in bridging studies when the form of drug changes or when the route of administration of drug changes.

6.7.1 Toxicokinetic Study Design

The Food and Drug Administration (FDA), the Europeans Medicines Agency (EMEA), and the Japan Pharmaceutical Manufacturers Association (JPMA) have adopted guidelines for assessing drug toxicokinetics in preclinical toxicity studies (ICH, 1994). These regulatory documents provide guidance on key elements of toxicokinetic study design, such as dose selection, species selection, route of administration, and toxicokinetic parameters.

6.7.1.1 Study Design. The parameters monitored in toxicokinetics are $AUC_{(0-t)}$, C_{max}, and T_{max}. $AUC_{(0-t)}$ is the area under the concentration curve describing the drug concentration as a function of time up to the last measurable time point (t) and is a measure of drug exposure with respect to time. C_{max} is the maximum concentration of the drug achieved in the plasma following dose administration, and T_{max} is the time at which C_{max} is attained.

Due to the limitation of blood volume that can be collected from the animal, the number of blood samples is often limited to approximately four to six time points. It is impossible to fully characterize all toxicokinetic parameters based on these limited time points. However, intrinsic PK characteristics [e.g., half-life ($T_{1/2}$), volume at steady state (V_{SS}), clearance (CL_S), etc.] are generally obtained from a full PK assessment study that was conducted prior to the toxicokinetic study. Based on the PK parameters, the sample collection time points can be carefully selected in order to capture more accurate toxicokinetic parameters.

6.7.1.2 Species. Generally, drug safety evaluation studies are initiated in two relevant preclinical species—one rodent, the other nonrodent—and in both male and female animals. When animal species selection is under consideration, the species should exhibit species-selective toxicities that are relevant to humans, and the metabolic profile of the animal should be qualitatively similar to that of humans. For developmental and reproductive studies, and genotoxicity studies, rodents are often chosen due to their shorter life spans, allowing evaluation of effects of the drug within a practical time frame.

6.7.1.3 Blood Sampling and Sampling Times. Blood lost affects the well-being of animals, particularly if the animal is to survive for longer-term assessment. In general, the total volume of blood that can be safely sampled for toxicokinetic analysis is less than 10% of total blood volume in the animal. Therefore, blood samples collected for toxicokinetic evaluation should be considered carefully during the study design in order to achieve the study objective(s). In general, time points near the T_{max} should be considered in order to assess C_{max} as accurately as possible (Pai et al., 1996), with one or two samples collected during the distribution phase (for intravenously administered drugs) or absorption phase (for extravascularly administered drugs), as well as two or three samples collected during drug elimination, with the final sampling time at trough, at the end of the dosing interval (i.e., for a QD dosing regimen, this is a 24-h post dose, immediately prior to the next day's dose).

6.7.1.4 Serial Versus Sparse Study Designs. In large animals (e.g., dogs, monkeys, rabbits), multiple blood samples can be collected from each individual animal in a serial sampling design. Drug safety and toxicokinetics of the drug can be evaluated from the same animal. However, for small animals (e.g., mice, rats), due to limited total blood volume constraints, it is challenging to collect multiple blood samples in order to generate a plasma concentration–time profile from an individual animal without compromising the safety evaluation of the animal. Therefore, alternative designs are generally considered for these small animals, such as a sparse sampling design or a satellite group design. In a sparse sampling design, at each dose group, the animal used in the safety evaluation study is assigned into one of the three subgroups with three to four animals per sex per subgroup. Each subgroup is assigned to be bled at different time points, and each animal is generally bled twice. For example, at the lowest dose group, animals in the subgroup A are bled in 0.5 h and 4 h post dose; the subgroup B are bled at 2 h and 8 h post dose; and animals subgroup C are bled at 6 h and 24 h post dose. A complete plasma concentration profile is generated using time points from all animals in each dose level group. In a satellite group design, additional animals (three to four animals per sex per dose) will be included in the drug safety study. These animals will not be included in the safety evaluation, although they will be treated in exactly the same way as the safety evaluation animals, in terms of dosing, handling, feeding, and so on. Similar to PK studies, serial blood samples will be collected from each individual animal and toxicokinetic parameters will be derived.

6.7.1.5 Doses. Dose is selected based on the need for safety evaluation and to provide acceptable exposure margins for human studies (ICH S1C, and S3A Guidelines). Three or four dose levels are generally selected. A good understanding the pharmacology and pharmacokinetics of the drug enables a selection of optimal dose levels for evaluating drug safety in the tested animals. The lowest dose is expected to be the highest pharmacologic clinical dose where no toxicity is expected, based on the pharmacology dose–response study. At the highest dose, it is anticipated that drug-dependent toxicity will be observed but will not be fatal. The mid dose is selected to provide some toxicity and to exhibit dose-related responses.

6.7.1.6 Dose Regimen. The route of administration should be the same as that intended for human use. Frequency of dosing in safety studies is typically once daily. However, if the exposure is limited due to poor absorption (i.e., poor solubility), it is not uncommon that twice-a-day dosing is conducted to enhance the exposure and to provide adequate exposure margin over humans. For the cytotoxic agent, the dose regimen in the safety evaluation study could also mimic the human dose regimen by giving intermittent dosing (e.g., 5 days on cycle, 2 days off cycle) to allow recovery of the animal from the cytotoxic agent.

6.7.1.7 Metabolite Monitoring. FDA and ICH guidelines issued in the mid-1990s described the considerations that must be given to monitor parent drug concentrations in the course of safety assessment (ICH, 1994). These guidelines also acknowledged the role of metabolites and made reference to information that might be considered in various regulatory situations (Center for Drug Evaluation and Research, 2002). In 1999, a PhRMA perspective was issued that described how metabolite monitoring could be conducted and also discussed how metabolite-driven toxicities might be addressed in toxicity testing (Baillie et al., 2002). Considerable open debate followed, which culminated in an FDA-issued guideline (Center for Drug Evaluation and Research, 2008). Because metabolites can have both target (pharmacologically intended) and off-target effects, preclinical safety testing should attempt to ensure that human subjects are not put at unreasonable risk from the metabolites of the drugs they are taking. Ideally, preclinical species will produce a spectrum of metabolites similar to those produced in human subjects, and sufficient exposure margins can be documented for those metabolites which predominate in humans. A particular risk is presented when humans produce large amounts of metabolites not seen to a significant degree in preclinical studies.

6.7.1.7.1 Sex Differences. Exposure differences between male and female animals are often observed in during toxicity testing. These differences can occur through differences in drug metabolism due to differential expression of cytochrome P450 enzymes in male and female animals. In general, CYP2A2, CYP2C11, CYP3A1, CYP3A2 are male-specific and have higher activities in male rats than in female rats; while CYP2A1, CYP2C7 and CYP2C12 are female-specific (Parke and Ioannides, 1996; Waskiewicz et al., 1995). Therefore, for drugs that are substrates of these enzymes, sex-related differences in drug exposures are likely to occur and may contribute to sex-related drug toxicity.

6.7.1.8 Safety Margins. The interpretative use of toxicokinetic information is to derive the relationship between drug exposures and toxicology responses in the drug safety studies using preclinical species. The toxicokinetic data are further used to evaluate a safety margin or exposure multiple for the potential human therapeutic dose that is expected to be safe and well-tolerated in humans. The dose during preclinical safety testing that resulted in no adverse findings is considered as a *no observed adverse effect level* (NOAEL). The safety margin is a quantitative term that describes the risk of toxicity in human at the anticipated therapeutic exposure. It is determined using relevant toxicokinetic parameter (e.g., AUC or C_{max}) at this dose is also characterized, and compared against the anticipated clinical therapeutic exposure.

$$\text{Safety margin} = \frac{\text{AUC}_{\text{NOAEL}}}{\text{AUC}_{\text{Clinical}}} \tag{6.45}$$

or

$$\text{Safety margin} = \frac{C_{\text{max,NOAEL}}}{C_{\text{max,Clinical}}} \tag{6.46}$$

6.7.1.9 Rationale for Selection of Starting Dose. The starting dose for the first in human is based on the result of safety data obtained from the available GLP-compliant animal toxicology studies. Generally, the estimation of the human equivalent dose (HED) is 1/10th of the NOAEL dose. For example, if the NOAEL dose is 30 mg/kg/day (180 mg/m^2/day) for rats based on the findings of toxicity, and the NOAEL dose in the GLP-compliant toxicity study in dogs is 6 mg/kg/day (120 mg/m^2/day), then the estimated HED should be considered based on the more sensitive species. In this case, it is the dog. The NOAEL dose in the dog is 6 mg/kg/day (120 mg/m^2/day), corresponding to 3.2 mg/kg HED. Therefore, a daily oral dose of 25 mg (corresponding to 0.4 mg/kg for a 60-kg subject) will be used as the starting dose. This starting dose may or may not be expected to have minimal pharmacological activity.

The proposed dose escalation will depend upon the toxicology findings and the PK characteristics. If the overall toxicology findings in the animal are modest and the compound exhibits linear pharmacokinetics in the animal, then the next dose could be double or greater. However, if the severity of the toxicology findings increases greatly as the dose increase and the compound exhibits greater than dose-proportional pharmacokinetics, then the dose increment in human should be slow and cautious. Human safety will be evaluated based on the clinical profile (evaluation of adverse events, vital signs, and clinical laboratory parameters). The actual ending dose for dose escalation will be determined by the occurrence of dose-limiting toxicity defined by the investigator.

ACKNOWLEDGMENT

The authors would like to thank Ms. Anastasia Leyden for her kind editorial help.

REFERENCES

Ackermann BL. Results from a bench marking survey on cassette dosing practices in the pharmaceutical industry. *J Am Soc Mass Spectrom* 2004;15:1374–1377.

Alavijeh MS, Palmer AM. The pivotal role of drug metabolism and pharmacokinetics in the discovery and development of new medicines. *IDrugs* 2004;7:755–763.

Atherton JP, Van Noord TJ, Kuo BS. Sample pooling to enhance throughput of brain penetration study. *J Pharm Biomed Anal* 1999;20:39–47.

Baillie TA, Cayen MN, Fouda H, Gerson RJ, Green JD, Grossman SJ, Klunk LJ, LeBlanc B, Perkins DG, Shipley LA. Drug metabolites in safety testing. *Toxicol Appl Pharmacol* 2002;182:188–196.

Balani SK, Li P, Nguyen J, Cardoza K, Zeng H, Mu DX, Wu JT, Gan LS, Lee FW. Effective dosing regimen of 1-aminobenzotriazole for inhibition of antipyrine clearance in guinea pigs and mice using serial sampling. *Drug Metab Dispos* 2004;32:1092–1095.

Balani SK, Miwa GT, Gan LS, Wu JT, Lee FW. Strategy of utilizing *in vitro* and *in vivo* ADME tools for lead optimization and drug candidate selection. *Curr Top Med Chem* 2005;5:1033–1038.

Baranczewski P, Sta czak A, Kautiainen A, Sandin P, Edlund PO. Introduction to early *in vitro* identification of metabolites of new chemical entities in drug discovery and development. *Pharmacol Rep* 2006;58(3):341–352.

Beal SL, Sheiner LB. *NONMEM Users Guide*. San Francisco: NONMEM Project Group, UCSF, 1989.

Berman J, Halm K, Adkison K, Shaffer J. Simultaneous pharmacokinetic screening of a mixture of compounds in the dog using API LC/MS/MS analysis for increased throughput. *J Med Chem* 1997;40:827–829.

Bjorkman S, Redke F. Clearance of fentanyl, alfentanil, methohexitone, thiopentone and ketamine in relation to estimated hepatic blood flow in several animal species: application to prediction of clearance in man. *J Pharm Pharmacol* 2000;52: 1065–1074.

Boxenbaum H. Interspecies scaling, allometry, physiological time, and the ground plan of pharmacokinetics. *J Pharmacokinet Biopharm* 1982;10:201–227.

Boxenbaum H. Interspecies pharmacokinetic scaling and the evolutionary-comparative paradigm. *Drug Metab Rev* 1984;15:1071–1121.

Boxenbaum H, Ronfeld R. Interspecies pharmacokinetic scaling and the Dedrick plots. *Am J Physiol* 1983;245:R768–R775.

Brewer E, Henion J. Atmospheric pressure ionization LC/MS/MS techniques for drug disposition studies. *J Pharm Sci* 1998;87:395–402.

Center for Drug Evaluation and Research UF. Carcinogenicity Study Protocol. Submitted, 2002.

Center for Drug Evaluation and Research UF. *Safety Testing of Drug Metabolites*. 2008.

Cobelli C, Toffolo G. Compartmental vs. noncompartmental modeling for two accessible pools. *Am J Physiol* 1984;247:R488–R496.

Congreve M, Chessari G, Tisi D, Woodhead AJ. Recent developments in fragment-based drug discovery. *J Med Chem* 2008;51:3661–3680.

Cox KA, Dunn-Meynell K, Korfmacher WA. Novel procedure for rapid pharmacokinetic screening of discovery compounds in rats. *Drug Discovery Today* 1999;4: 232–237.

Cox KA, White RE, Korfmacher WA. Rapid determination of pharmacokinetic properties of new chemical entities: *in vivo* approaches. *Comb Chem High Throughput Screen* 2002;5:29–37.

Dedrick R, Bischoff KB, Zaharko DS. Interspecies correlation of plasma concentration history of methotrexate (NSC-740). *Cancer Chemother Rep* 1970;54:95–101.

Dedrick RL. Animal scale-up. *J Pharmacokinet Biopharm* 1973;1:435–461.

Dimasi JA, Hansen RW, Grabowski HG. The price of innovation: new estimates of drug development costs. *Health Econ* 2003;22:151–185.

Draper NR, Smith H. *Applied Regression Analysis*. New York: Wiley, 1998.

Ekins S, Mestres J, Testa B. *In silico* pharmacology for drug discovery: applications to targets and beyond. *Br J Pharmacol* 2007;152:21–37.

Giacomini KM, Sugiyama Y. Membrane transporters and drug response. In: Brunton L, Lazo J, Parker K, editors. *Goodman & Gilman's the Pharmacological Basis of Therapeutics*, 11th ed. New York: McGraw-Hill, 2006, pp. 41–70.

Gibaldi M, Perrier M. *Pharmacokinetics*, 2nd ed. New York: Marcel Dekker, 1982.

Gillespie WR. Noncompartmental versus compartmental modeling in clinical pharmacokinetics. *Clin Pharmacokinet* 1991;20:253–262.

Giorgio S. Relevance, experiences, and trends in the use of compartmental models. *Drug Metab Rev* 1984;15:7–53.

Greenlee WJ, Desai M. Backup programs in drug discovery. *Curr Opin Drug Discovery Dev*, 2006;9(4):412–413.

Hop CE, Wang Z, Chen Q, Kwei G. Plasma-pooling methods to increase throughput for in vivo pharmacokinetic screening. *J Pharm Sci* 1998;87:901–903.

Hopfgartner G, Bourgogne E. Quantitative high-throughput analysis of drugs in biological matrices by mass spectrometry. *Mass Spectrom Rev* 2003;22:195–214.

Huang C, Zheng M, Yang Z, Rodrigues AD, Marathe P. Projection of exposure and efficacious dose prior to first-in-human studies: How successful have we been? *Pharm Res* 2008;25:713–726.

ICH. Note for Guidance on Toxicokinetics: The assessment of systemic exposure in toxicity studies S3A. In: *International Conference on Harmonisation of Technical Requirements for Registration of Pharmaceuticals for Human Use*. 1994.

Kansy M, Avdeef A, Fischer H. Advances in screening for membrane permeability: high-resolution PAMPA for medicinal chemists. *Drug Discovery Today Tech* 2004;1:349–355.

Kapetanovic IM. Computer-aided drug discovery and development (CADDD): *in silico*–chemico–biological approach. *Chem Biol Interact* 2008;171:165–176.

Kola I, Landis J. Can the pharmaceutical industry reduce attrition rates? *Nat Rev Drug Discovery* 2004;3:711–716.

Korfmacher WA, Cox KA, Bryant MS, et al. HPLC-API/MS/MS: a powerful tool for integrating metabolism into the drug discovery process. *Drug Discovery Today* 1997;2:532–537.

Korfmacher WA, Cox KA, Ng KJ. Cassette-accelerated rapid rat screen: a systematic procedure for the dosing and liquid chromatography/atmospheric pressure ionization tandem mass spectrometric analysis of new chemical entities as part of new drug discovery. *Rapid Commun Mass Spectrom* 2001;15:335–340.

Kuo BS, Van Noord T, Feng MR. Wright DS. Sample pooling to expedite bioanalysis and pharmacokinetic research. *J Pharm Biomed Anal* 1998;6:837–846.

Landaw EM, DiStefano JJ. Multiexponential, multicompartmental, and noncompartmental modeling. II Data analysis and statistical considerations. *Am J Physiol* 1984;246:R665–R677.

Lave T, Dupin S, Schmitt C, Chou RC, Jaeck D, Coassolo P. Integration of *in vitro* data into allometric scaling to predict hepatic metabolic clearance in man: application to 10 extensively metabolized drugs. *J Pharm Sci* 1997;86:584–590.

Lee SJ, Obach RS, Fisher MB. *Drug Metabolizing Enzymes: Cytochrome P450 and Other Enzymes in Drug Discovery and Development*. Boca Raton, FL: CRC Press, 2003.

Lin J, Sahakian DC, de Morais SM, Xu JJ, Polzer RJ, Winter SM. The role of absorption, distribution, metabolism, excretion and toxicity in drug discovery. *Curr Top Med Chem* 2003;3:1125–1154.

Lin JH, Lu AY. Role of pharmacokinetics and metabolism in drug discovery and development. *Pharmacol Rev* 1997;49:403–449.

Lipinski CA, Lombardo F, Dominy BW, Feeney PJ. Experimental and computational approaches to estimate solubility and permeability in drug discovery and development settings. *Adv Drug Deliv Rev* 1997;23:3–25.

Liu X, Chen C, Smith BJ. Progress in brain penetration evaluation in drug discovery and development. *J Pharmacol Exp Ther* 2008;325:349–356.

Mahmood I. Integration of *in vitro* data and brain weight in allometric scaling to predict clearance in humans: some suggestions. *J Pharm Sci* 1998;87:527–529; discussion 530.

Mahmood I. Allometric issues in drug development. *J Pharm Sci* 1999;88:1101–1106.

Mahmood I. *Interspecies Pharmacokinetic Scaling: Principles and Application of Allometric Scaling*. Rockville, MD: Pine House Publishers, 2005.

Mahmood I. Application of allometric principles for the prediction of pharmacokinetics in human and veterinary drug development. *Adv Drug Deliv Rev* 2007;59:1177–1192.

Mahmood I, Balian JD. Interspecies scaling: predicting clearance of drugs in humans. Three different approaches. *Xenobiotica* 1996;26:887–895.

Mahmood I, Balian JD. The pharmacokinetic principles behind scaling from preclinical results to phase I protocols. *Clin Pharmacokinet* 1999;36(1):1–11.

Mordenti J. Pharmacokinetic scale-up: accurate prediction of human pharmacokinetic profiles from animal data. *J Pharm Sci* 1985;74:1097–1099.

Nesch EA. Noncompartmental approach in pharmacokinetics using moments. *Drug Metab Rev*. 1984;15:7–53.

Obach RS, Baxter JG, Liston TE, Silber BM, Jones BC, MacIntyre F, Rance DJ, Wastall P. The prediction of human pharmacokinetic parameters from preclinical and *in vitro* metabolism data. *J Pharmacol Exp Ther* 1997;283:46–58.

Pai SM, Fettner SH, Hajian G, Cayen MN, Batra VK. Characterization of AUCs from sparsely sampled populations in toxicology studies. *Pharm Res* 1996;13:1283–1290.

Parke DV, Ioannides C. *Cytochromes P450: Metabolic and Toxicological Aspects*. Boca Raton, FL: CRC Press, 1996.

Peck CC, Sheiner LB, Nichols AI. The Problem of choosing weights in nonlinear regression analysis of pharmacokinetic data. *Drug Metab Rev* 1984;15:133–148.

Piotrovskii VK. Model and model-independent methods of describing pharmacokinetics: the advantages, drawbacks and interrelationship. *Antibiot Med Biotekhnol* 1987;32:492–497.

Prakash C, Shaffer CL, Nedderman A. Analytical strategies for identifying drug metabolites. *Mass Spectrom Rev* 2007;26(3):340–369.

Prentis RA, Lis Y, Walker SR. Pharmaceutical innovation by the seven UK-owned pharmaceutical companies (1964–1985). *Br J Clin Pharmacol* 1988;25:387–396.

Riley RJ, Grime K. Metabolic screening *in vitro*: metabolic stability, CYP inhibition and induction. *Drug Discovery Today Tech* 2004;1:365–372.

Riviere JE, Martin-Jimenez T, Sundlof SF, Craigmill AL. Interspecies allometric analysis of the comparative pharmacokinetics of 44 drugs across veterinary and laboratory animal species. *J Vet Pharmacol Ther* 1997;20:453–463.

Roberts SA. Drug metabolism and pharmacokinetics in drug discovery. *Curr Opin Drug Discovery Dev* 2003;6:66–80.

Singh RP, Singh SK, Gupta RC. A high throughput approach for simultaneous estimation of multiple synthetic trioxane derivatives using sample pooling for pharmacokinetic studies. *J Pharm Biomed Anal* 2005;37:127–133.

Sinha VK, De Buck SS, Fenu LA, Smit JW, Nijsen M, Gilissen RA, Van Peer A, Lavrijsen K, Mackie CE. Predicting oral clearance in humans: How close can we get with allometry? *Clin Pharmacokinet* 2008;47:35–45.

Smith NF, Raynaud FI, Workman P. The application of cassette dosing for pharmacokinetic screening in small-molecule cancer drug discovery. *Mol Cancer Ther* 2007;6:428–440.

Sundstrom LS. Thinking inside the box. *Eur Mol Biol Org* 2007;8:s40–s43.

Tang H, Mayersohn M. A novel model for prediction of human drug clearance by allometric scaling. *Drug Metab Dispos* 2005;33:1297–1303.

Tannenbaum S, Boxenbaum H, Mayersohn M. Allometric analysis of organ extraction ratios. *J Pharm Sci* 1997;86:1319–1320.

Wajima T, Yano Y, Fukumura K, Oguma T. Prediction of human pharmacokinetic profile in animal scale up based on normalizing time course profiles. *J Pharm Sci* 2004;93:1890–1900.

Watt AP, Morrison II, Evans DC. Approaches to higher-throughput pharmacokinetics (HTPK) in drug discovery. *Drug Discovery Today* 2000;5:17–24.

Woodcroft KJ, Webb CD, Yao M, Weedon AC, Bend JR. Metabolism of the cytochrome P450 mechanism-based inhibitor *N*-benzyl-1-aminobenzotriazole to products that covalently bind with protein in guinea pig liver and lung microsomes: comparative study with 1-aminobenzotriazole. *Chem Res Toxicol* 1997;10:589–599.

Waskiewicz MJ, Choudhuri S, Vanderbeck SM, Zhang XJ, Thomas PE. Induction of "male-specific" cytochrome P450 isozymes in female rats by oxandrolone. *Am Soc Pharmacol Exp Ther* 1995;23 (11):1291–1296.

PART II

INHIBITION OF THE DRUG METABOLIZING ENZYMES—THE UNDESIRABLE INHIBITION

7

ENZYME INHIBITION AND INACTIVATION: CYTOCHROME P450 ENZYMES

R. Scott Obach

7.1 INTRODUCTION

Cytochrome P450 enzymes play the predominant role in the clearance of a majority of drugs. Therefore, drugs that can affect the activity of these enzymes can potentially cause pharmacokinetic drug–drug interactions by altering the rate of clearance of other drugs when more than one drug is administered simultaneously. Knowledge of the enzymes of the P450 family, their substrate selectivities, and inhibitors that has developed over the past few decades has formed the basis for approaches that are used to predict drug–drug interactions from *in vitro* data and thus avoid selection of compounds in early drug research that will cause drug–drug interactions. Many reviews, monographs, and even entire books have been written on the topic of cytochrome P450 biochemistry and its importance in drug research, and the reader is referred to these works for greater detail (Ortiz de Montellano, 2005; Lee et al., 2003). The objective of this chapter is to briefly bring together (a) concepts of inhibition of P450 enzymes, (b) methodological considerations for laboratory analysis, and (c) approaches to predict drug interactions from *in vitro* data.

Enzyme Inhibition in Drug Discovery and Development, edited by Lu and Li
Copyright © 2010 John Wiley & Sons, Inc.

7.2 CYTOCHROME P450 CYCLE, UNIFIED MECHANISM, AND INHIBITION MECHANISMS

7.2.1 Catalytic Cycle and Unified Mechanism

In order to understand the role of inhibition of P450 enzymes and drug–drug interactions, some basic elements of these enzymes must be appreciated. There are hundreds of cytochrome P450 enzymes, and humans possess over 50 (Guengerich, 2003). While all of the P450s possess the common feature of a heme with a thiol ligand at their active centers, these enzymes have been divided into families based on their similarities and differences in amino acid sequence. Many of the human enzymes are involved in the metabolism of endogenous compounds, which can be pharmacological targets themselves (e.g., aromatase inhibitors), but for the metabolic clearance of drugs, just three of the P450 families predominate: CYP1, CYP2, and CYP3. Unlike the classic biochemical concept of "one enzyme, one substrate," the drug metabolizing P450 enzymes exhibit a broad array of substrate, reaction, and inhibitor specificities. This broad array of substrates, reactions, and inhibitors contributes to making drug metabolic clearance and drug–drug interactions complex, multivariate phenomena. It requires that a fairly complex experimental approach (*in vitro* and *in vivo*) and analysis is done for understanding the enzymes involved in drug metabolism and prediction of drug–drug interactions for any given drug. These approaches have been the focus of comprehensive reviews and chapters (Zhang et al., 2007; Bjornsson et al., 2003; Madan et al., 2002; Tucker et al., 2001) as well as the focus of government regulatory guidance in the development of new drugs (FDA, 2006).

Despite these complexities, the various P450 enzymes actually utilize a common bioorganic mechanism in their catalytic cycle, and the different reaction outcomes are a function of the chemical substituents residing on the substrates and the orientation(s) of the substrate in the binding pocket which determines which substituent(s) will be oriented toward the active heme center. Understanding the mechanisms of inhibition (described below) requires an understanding of the P450 reaction cycle (Fig. 7.1) which has been developed through many years of research [reviewed in Makris et al. (2005)]. Briefly, the cycle is initiated by the binding of the substrate to the enzyme in its ground-state ferric form, which is followed by a reduction reaction of the ferric to ferrous forms. The source of electrons is NADPH, which transfers the reducing equivalents through NADPH:cytochrome P450 reductase. Binding of O_2 occurs next to form a heme-peroxy species, a second reducing equivalent is introduced (from NADPH:P450 reductase or cytochrome b_5), and the O–O bond undergoes scission resulting in a reactive species hypothesized to be either an oxene (i.e., Fe(V)=O) or radical (Fe(IV)–O$^•$) species. This reactive intermediate is the entity that reacts with various drugs, and the reaction catalyzed (e.g., hydroxylation, heteroatom dealkylation, heteroatom oxygenation, epoxidation, etc.) depends upon the nature of the substrate substituent that is oriented closest to it.

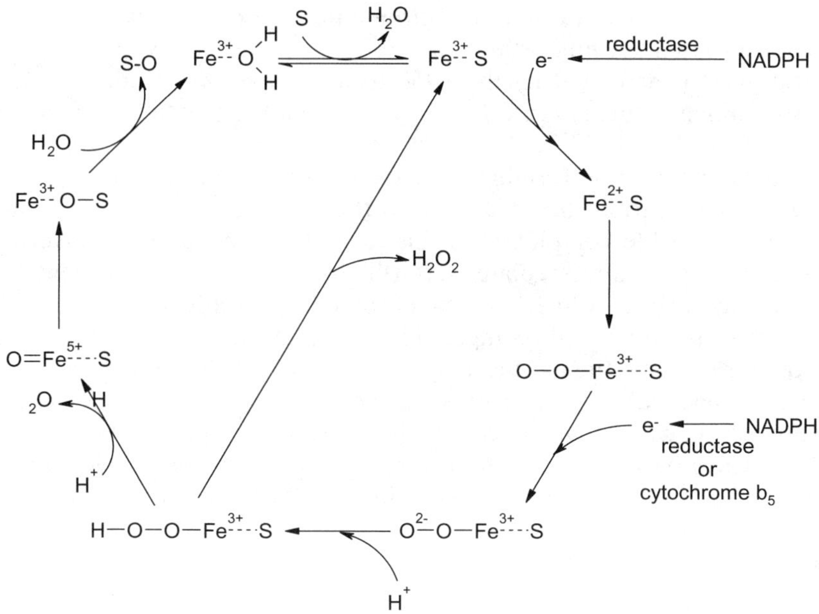

Figure 7.1. Summary of the catalytic cycle of cytochrome P450.

7.2.2 Inhibition Mechanisms

Inhibition and inactivation of P450 enzymes can occur via a few different types of mechanisms. Reversible inhibition can occur when the inhibitor binds to the substrate binding site preventing substrate from binding. Some reversible competitive inhibitors are actually substrates themselves that merely compete for the same binding site, but with one substrate having a greater affinity than the other (e.g., fluxoetine inhibition of CYP2D6), whereas other inhibitors bind in the substrate binding site but are not metabolized by the enzyme (e.g., quinidine inhibition of CYP2D6). Reversible inhibition can also occur if the inhibitor has a chemical substituent that forms a noncovalent ligand interaction with the ferric iron in the heme, such as many imidazole containing inhibitors (e.g., ketoconazole). These different types of binding interactions can be distinguished spectrally, because shifts in the absorbance spectrum of the heme will occur upon ligand binding and are referred to as type I and type II binding spectra. Type I binding results in an absorbance maximum at approximately 390 nm and a minimum at 420 nm, while type II binding results in a maximum at around 430 nm and a minimum at around 400 nm.

While inhibition is reversible and activity of the enzyme is restored when the inhibitor dissociates from the enzyme, inactivation results in permanent disabling of enzyme activity. Inactivators can possess intrinsic chemical reactivity yet possess other portions that permit preferential binding to the enzyme; these are affinity labels. For the purposes of understanding drug interactions and P450s, such inactivators do not have practical relevance and will not be discussed further. For drug interactions, inactivators that are mechanism-based are important. In these cases the inactivator serves as a substrate and is activated by the P450 to a reactive intermediate that irreversibly damages the enzyme by chemically reacting with it. The reaction can occur by forming a covalent bond between the inactivator and the amino acid chain or by forming an adduct with the porphyrin of the heme. A type of inactivation that is unique to the P450 enzymes is the formation of quasi-irreversible complexes. In this case, the inactivator is converted by the enzyme to an intermediate that will form a tight noncovalent ligand with the heme iron, which can be observed spectrally by an increase in absorbance at 456 nm. This complex can be dissociated *in vitro* by the addition of ferricyanide (hence the name "quasi-irreversible"), but *in vivo* this is type of inactivation is irreversible. There are several types of chemical substituents on drugs that can cause mechanism-based inactivation of human P450 enzymes by one of the three mechanisms described above, and some of the better studied ones are listed in Table 7.1 (Kalgutkar et al., 2007; Riley et al., 2007).

TABLE 7.1. Chemical Substituents Known to Cause Mechanism-Based Inactivation of Human Cytochrome P450 Enzymes

Substituent	Inactivating Species	Target	Examples
Amine	Nitroso	Heme Fe (quasi-irreversible)	Erythromycin (CYP3A4)
Alkene	Epoxide	Apoprotein and porphyrin	Secobarbital (CYP2B)
Alkyne	Ketene	Porphyrin and apoprotein	Ethinyl estradiol
Cyclopropylamine	Radical	Aporprotein	Tranylcypramine (CYP1A2)
Methylenedioxy	Carbene	Heme Fe (quasi-irreversible)	Paroxetine (CYP2D6)
Thiophene	Epoxide or *S*-oxide	Apoprotein	Tienilic acid (CYP2C9)
Furan	Unsaturated aldehyde	Apoprotein	Menthofuran (CYP2A6)
Phenol	Quinone	Apoprotein	Raloxifene (CYP3A4)
Benzotriazole	Benzyne	Porphyrin	ABT (several CYPs)
Alkyl imidazole	Imidazomethide	Apoprotein	Furaphylline (CYP1A2)
Acyl glucuronide	Unknown	Unknown	Gemfibrozil (CYP2C8)

7.3 CYTOCHROME P450 ASSAY APPROACHES

7.3.1 Multiplicity of Human Cytochrome P450 Enzymes and Substrate Specificities

The drug-metabolizing human cytochrome P450 enzymes reside primarily in families 1, 2, and 3, although there are a few drugs metabolized by members of family 4. Experimental approaches to measure the inhibition of P450 activities in human liver microsomes are highly dependent on ensuring that the activity assays employed possess selectivity for individual P450 enzymes. Assay selectivity is not critical when the enzymes used are purified or expressed singly from recombinant DNA in heterologous expression systems. Over the years, a good basis of knowledge has been developed regarding the selectivity of various substrates and inhibitors of individual P450 enzymes, and this knowledge forms the basis of the approaches used to determine the P450 enzymes involved in the metabolism of new drugs and the prediction of drug–drug interactions. The most important of the human drug-metabolizing enzymes, along with well-established *in vivo* probe *in vitro* reactions used to explore whether a new compound will inhibit a P450 enzyme are listed in Table 7.2. Note that there is a not a perfect overlap for chemical tools used *in vitro*, and those used in the clinic because some of the best *in vitro* tools are not available for administration to humans.

7.3.2 Cytochrome P450 Assay Approaches: Human Liver Microsomes

The use of human liver microsomes as a source of enzyme activity in the testing of new compounds as inhibitors of P450 enzymes is a common practice. While human liver tissue suitable for preparation of microsomes was rare in the past (i.e., before 1990), over the past 15 years it has become easier to obtain this material by means of ethical nonprofit tissue acquisition networks and commercial suppliers who process the tissue for research. Liver microsomes can be from individual donors or, more frequently, are pooled from multiple individuals to provide an "average human." Generation of inhibition data in liver microsomes from multiple individual donors permits some assessment of variability, however, for prediction of overall drug–drug interactions, the use of pooled liver microsomes is adequate and is resource-sparing. Furthermore, the data gathered using pooled liver microsomes can be entered into computer programs that can model populations to gain a sense of variability.

Liver microsomes provide the advantage that the P450 enzymes are in membranes comprised of a "natural" phospholipid composition and have the appropriate amounts of NADPH:P450 reductase and cytochrome b_5. But it must be kept in mind that the *in vitro* system is far different from an intact *in vivo* situation: oxygen tension is greater *in vitro*, the natural structure of the endoplasmic reticulum has been disrupted into microsomes upon

TABLE 7.2. Frequently Employed *In Vitro* Reactions and Inhibitors for Studying Human Cytochrome P450 Activity

Enzyme	*In Vivo* Activity Probes	*In Vitro* Probe Activities	*In Vivo* Inhibition Probe	*In Vitro* Inhibitors	Comments
CYP1A2	Theophylline Ramelteon Tizanidine	Phenacetin *O*-deethylase Tacrine 1-hydroxylase	Fluvoxamine	Furaphylline	Fluvoxamine is not selective.
CYP2B6	Bupropion Efavirenz	Bupropion hydroxylase	Clopidogrel	Methyl phenethylpiperidine	When used as an *in vivo* probe bupropion metabolites must be monitored.
CYP2C8	Rosiglitazone Repaglinide	Amodiaquine *N*-deethylase Paclitaxel 6α-hydroxylase	Gemfibrozil	Montelukast	Montelukast potency is dependent on protein concentration.
CYP2C9	*S*-Warfarin Tolbutamide	Diclofenac 4′-hydroxylase Tolbutamide Hydroxylase *S*-Warfarin 7-Hydroxylase Flurbiprofen 4′-Hydroxylase	Fluconazole	Sulfaphenazole	CYP2C9 may have different classes of substrates with differing sensitivity to various inhibitors. Fluconazole is not selective.
CYP2C19	Omeprazole *S*-Mephenytoin	*S*-Mephenytoin 4′-hydroxylase Omeprazole 5-hydroxylase	Fluvoxamine	*N*-Benzylnirvanol	Mephenytoin is not a commercially available drug. Fluvoxamine is not selective.
CYP2D6	Dextromethorphan[a] Metoprolol Debrisoquine	Dextromethorphan *O*-demethylase Bufuralol 1′-hydroxylase Debrisoquine 4-hydroxylase	Quindine Paroxetine	Quinidine	
CYP3A	Midazolam Triazolam Nifedipine Felodipine Buspirone	Midazolam 1′-hydroxylase Testosterone 6β-hydroxylase Felodipine dehydrogenase Nifedipine dehydrogenase	Ketoconazole Itraconazole	Ketoconazole Troleandomycin	*In vitro*, CYP3A4 has different substrate classes with differing sensitivities to inhibitors.

homogenization, and natural buffers and other solutes are not present. The extent that these differences between *in vivo* and *in vitro* contribute to confounding *in vitro* to *in vivo* extrapolation and prediction is unknown, and there is really no way to know for certain. An exploration of the effects of buffers used *in vitro* has been done. Endogenous buffering capacity within a hepatocyte is probably a sum of many components such as bicarbonate and many cellular solutes (e.g., amino acids, phosphates, pyrophosphates, etc.). Most investigators utilize phosphate or Tris buffers at concentrations of 25–100 mM to ensure maintaining pH at 7.4 throughout an incubation period. However, it has been shown that different buffers can have different effects on P450 enzymes [e.g., bicarbonate versus others for CYP2D6 activity (Hutzler et al., 2003)]. The effect of Mg^{2+} has also been explored and shown to have an effect on CYP3A activity (Maenpaa et al., 1998). The presence of contaminating fatty acids in liver microsomal preparations may be underlying an effect of added albumin causing an apparent stimulation of some P450 activities by sequestration of inhibitory fatty acids (Rowland et al., 2008; Ishii et al., 2001; Xu et al., 2003; Powis et al., 1977). The use of NADPH or NADPH regeneration systems may impact activity for extended incubation times, with the latter avoiding the buildup of NADP that can inhibit the reductase. These experimental variables may have an effect on absolute activities and cause problems for scaling *in vitro* activity to *in vivo* clearance. However, the measurement of inhibition is a relative measurement (i.e., everything is relative to the control activity); thus the impact of these variables may not be as important for prediction of *in vivo* inhibition. However, this has not been systematically evaluated.

A major complicating factor in the use of human liver microsomes for *in vitro* inhibition assays is that this system possesses a complex mixture of P450 enzymes. This absolutely requires that the activity assays used be selective markers for individual enzymes. In almost all cases, a particular reaction will be catalyzed by more than one enzyme. A key to using a reaction as a selective marker activity is in the selection of an appropriate substrate concentration. A good selective marker activity should be at least 95% attributable to the enzyme of focus at the substrate concentration employed. There are many other practical attributes to marker activities (listed in Table 7.3). A list of well-accepted P450 marker activities is included in Table 7.2; and it should be noted that despite their widespread use and acceptance, many of these suffer from lacking one or more of these desirable attributes.

Sound enzyme kinetic practice is a must for any good inhibition experiment. Preliminary investigation of the linearity of the marker reaction velocity with time and protein concentration is a necessary experiment to establish appropriate conditions. Certainly, consumption of substrate should never exceed 20%, and more ideally this should not exceed 10% and should be as low as possible. For P450 reactions in liver microsomes, reaction rates typically begin to decline after 30 min; this is hypothesized to be due to the buildup of

TABLE 7.3. Characteristics of Good Liver Microsomal Cytochrome P450 Marker Activities

- Substrate, product, and a suitable internal standard are commercially available, inexpensive, stable as solids, and available with sound analytical characterization. In particular, the substrate should not have the product as a contaminant.
- Substrate is readily soluble in water or organic solvents that do not inhibit P450 activities at 1% (v/v).
- The substrate and product are stable in solution.
- The product is readily able to be analyzed (e.g., ionizes on API-MS, has strong chromophore for UV, or is highly fluorescent).
- ≥95% of the activity is catalyzed by one enzyme.
- The activity can be characterized by simple hyperbolic Michaelis–Menten kinetics.
- The activity is high enough that low microsomal protein concentrations can be used.
- Multiple regioisomeric products are either not formed or are readily separated by HPLC without lengthy run times.
- The substrate and product do not undergo appreciable nonspecific binding to microsomes.

reactive oxygen species that arise via nonproductive cycling of the enzyme. In some cases, the marker activity shows some indication of mechanism-based inactivation, as is the case for CYP3A-catalyzed midazolam hydroxylation (Khan et al., 2002), thus necessitating the use of fairly short incubation times. A lack of linearity with protein concentration is most often observed when the substrate nonspecifically binds to microsomal protein or lipid. For the commonly used P450 substrates, this is typically not much of a problem because the rate of reaction is high enough such that low microsomal protein concentrations can be used and the substrates are not excessively lipophilic.

Analytical approaches employed for P450 marker activities in human liver microsomes have almost exclusively included HPLC separation of the product of interest from the substrate (and other products as the case may be), with different detection techniques. Many of the early assays used HPLC-UV (e.g., 6β-hydroxytestosterone, acetaminophen, 7-hydroxywarfarin, etc.) or HPLC with fluorescence (e.g., 1′-hydroxybufuralol) (Fasco et al., 1977; Kronbach et al., 1987; Butler et al., 1989; Purdon and Lehman-McKeeman, 1997). With the increased availability of less expensive, user-friendly HPLC-MS/MS instrumentation, many of the P450 assays have been converted to using this detection approach. In general, HPLC-MS affords a considerable increase in sensitivity that permits the use of lower microsomal protein concentrations, employment of stable isotope-labeled products as internal standards yields an increase in assay precision, and the selectivity of MS detection reduces the need for extensive sample clean-up procedures as a simple filtration step to remove protein can be all that is needed (Walsky and Obach, 2004). Standard aspects of sound analytical method verification are required (i.e., standard curves, quality control samples). A unique quality control sample needed for P450 inhibition assays is the determination that the test inhibitor does not interfere with the assay for the product itself. A less selective detection technique such as HPLC-UV can be subject to interference if the test compound coelutes with the product, and even a selective technique like HPLC-MS can be subject to interference if the test compound coelutes with the product and causes ion suppression. With HPLC-MS/MS, some investigators have devised various "cocktail" approaches in which a set of P450-selective substrates are co-incubated in an attempt to gather the inhibition data for a test compound all at once (Dixit et al., 2007; Tolonen et al., 2007; Smith et al., 2007; Kim et al., 2005). Challenges exist in selecting a reaction time suitable for all activities and in ensuring that all substrates do not interfere with any one of the other assays. The phenomenon of competition for limiting NADPH:P450 reductase and interactions between the P450 enzymes (Backes and Kelley, 2003) themselves can confound this approach.

7.3.3 Cytochrome P450 Assay Approaches: Recombinant Enzymes

Commercially available P450 enzymes that have been cloned and expressed in heterologous expression systems have proven to be a useful tool for *in vitro* inhibition assays. The great advantage offered by such systems is that each individual P450 enzyme is expressed separately; therefore the marker activity employed does not have to be selective. A disadvantage of these systems is that they are further from the actual *in vivo* situation; the amounts of P450, reductase, and cytochrome b_5 do not match what is in liver microsomes, and any effect of P450–P450 interactions is absent.

For testing new compounds as P450 inhibitors, the recombinant P450 systems have permitted the use of fluorogenic substrates that are amenable to plate reader platform assays (Ghosal et al., 2003; Chauret et al., 1999, 2001; Miller et al., 2000). This serves as a great advantage over chromatography-based approaches since the throughput can be increased considerably. The fluorogenic substrates have each been designed such that they will serve as substrates for the human P450 enzymes, and many are based on a coumarin-type core. As with any other enzyme inhibition assay, the assay procedure must be based on a kinetically sound foundation (linearity). The speed and throughput of these assays makes them very appealing for screening in early drug research. However, the correlation of results from fluorogenic assays using recombinant P450 enzymes to those generated using liver microsomes and drug-like substrates does not have a high enough fidelity to merit their use in definitive assessments of P450 inhibition (Cohen et al., 2003). Whether this is due to the recombinant enzyme or the fluorogenic substrate has not been determined. A more recent area of research is the development of luminescent substrates based on the well-known luciferase biochemistry (Cali et al., 2006).

7.3.4 Mechanism-Based Inactivation

Some of the most notorious of the perpetrators of clinical drug–drug interactions are mechanism-based inactivators. Examination of new compounds as potential mechanism-based inactivators utilizes the P450 activity assays described above; however, in this case the test compound must first be preincubated with enzyme and NADPH prior to the addition of the substrate and conduct of the substrate activity assay incubation. There are several experimental approaches that can be used to assess mechanism-based inactivation. The most widely used approach involves preincubation of the test compound with liver microsomes (or other source of P450 enzyme); an aliquot is removed from this preincubation and diluted into a second incubation assay containing the P450 marker substrate and NADPH. When this is done using multiple inactivator concentrations, a plot of the apparent inactivation rate constants measured at each concentration versus the inactivator concentration is constructed and the theoretical maximum inactivation rate constant (k_{inact}) and concentration yielding half-maximal inactivation (K_I) are estimated. Note that for P450 enzymes it is important to correct for any loss of enzyme activity that can occur when the enzyme is incubated with NADPH in the absence of inactivator. (That is, the plot of $k_{inact,app}$ versus [I] may yield a nonzero intercept.) There has been considerable debate in the literature about the best methodological approach to conduct inactivation experiments and whether different approaches yield data that would be interpreted differently (Ghanbari et al., 2006). Co-incubation of test compound and substrate and measuring the product formed at multiple timepoints (i.e., "progress curve"approach) represents another approach to measure mechanism-based inactivation (Fairman et al., 2007). A systematic evaluation of these approaches is warranted for P450 inactivation and prediction of drug interactions.

7.3.5 Complicating Factors

For the P450 enzymes, there are several complicating factors that need to be appreciated when conducting inhibition experiments.

Many P450 reactions do not show straightforward hyperbolic enzyme kinetic behavior. The more complicated the enzyme kinetic behavior, the more challenging it is to determine inhibition kinetics. P450 enzymes can exhibit activation kinetics, substrate inhibition kinetics, and two-site kinetics (Fig. 7.2). Sound characterization of the enzyme kinetics of the marker activity is necessary in order to design and interpret inhibition experiments. Aberrant enzyme kinetic behavior will require careful consideration of substrate concentrations used in inhibition experiments and the number of substrate/inhibitor concentration pairs needed to reliably define the kinetic parameters.

Nonspecific binding of the inhibitor to liver microsomes can counfound the determination of inhibition constants, making them artificially high. While the extent of nonspecific binding of the substrate should already have been taken

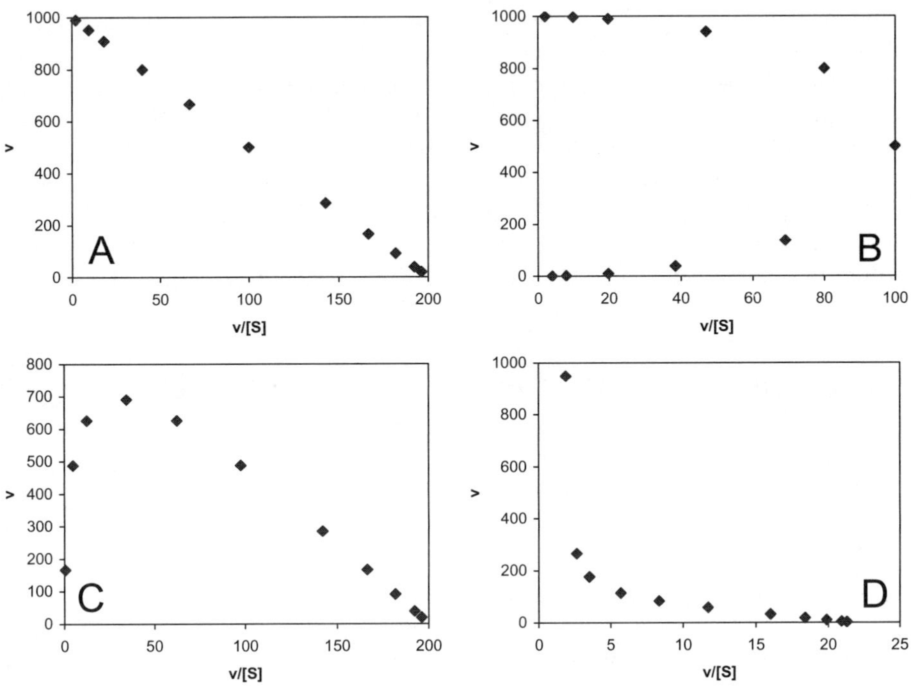

Figure 7.2. Eadie–Hofstee plots showing different typical kinetic behaviors of P450 reactions. **(A)** Normal; **(B)** autoactivation; **(C)** substrate inhibition; **(D)** "two-activity" kinetics.

care of when the assay was developed, and the binding should be low because the substrates used tend to not be highly lipophilic, this may not be the case for the compounds being tested as inhibitors. For some compounds, nonspecific binding can cause meaningful shifts in inhibition constants even with very low microsome concentrations (Margolis and Obach, 2003). The measurement of nonspecific binding can be experimentally measured with equilibrium dialysis or ultrafiltration, but there have been recently described computational approaches to make estimates for the fraction unbound *in vitro* (Gao et al., 2008; Sykes et al., 2006; Hallifax and Houston, 2006; Austin et al., 2002).

Careful investigations of inhibitor and substrate specificities of some P450 enzymes have revealed that there are some instances in which inhibitors will differentially affect the activities of different substrates for the same enzyme. This phenomenon is best characterized for CYP3A (which includes two very closely related enzymes CYP3A4 and 3A5 with overlap in substrate and inhibitor specificities). For this enzyme, three different substrate classes have been proposed (Kenworthy et al., 1999), necessitating the use of more than one activity assay to ensure that a new compound does not inhibit just one class of substrate. The effects of CYP3A inhibitors on midazolam

1′-hydroxylase and testosterone 6β-hydroxylase (or other activities residing in these substrate classes) can be different *in vitro*. Furthermore, the effects on dihydropyridine calcium channel blocker dehydrogenation by CYP3A (i.e., nifedipine or felodipine) represent a third class. The relevance of these *in vitro* substrate classes to *in vivo* drug–drug interactions has yet to be demonstrated. Recent investigations have shown that a similar phenomenon occurs for CYP2C9, and it appears that inhibitors can have different potencies for metabolism of warfarin, flurbiprofen, and a third class of substrates that includes tolbutamde and diclofenac (Nath and Atkins, 2008; Kumar et al., 2006).

7.4 PREDICTING DRUG–DRUG INTERACTIONS FROM *IN VITRO* INHIBITION DATA

7.4.1 Assay Placement Strategy

The myriad approaches to determining the inhibition of P450 enzymes each have their advantages and disadvantages that permit their placement into various stages of drug research. At early stages, when thousands of new compounds are being synthesized in small quantities, a quick assessment of potential P450 inhibition liability can be made through the use of plate-based fluorogenic substrate assays or high-throughput cocktail approaches using HPLC-MS/MS. A single inhibitor concentration is tested. The selection of the concentration needs to be able to be suitable for the widest range of compound and project types, and there is no single concentration that will adequately serve all compounds. In some cases, the anticipated human dose will likely be low (e.g., <1 mg/day) because the potency of the series for the pharmacological target will be high—in these instances the universal inhibitor concentration used for the initial P450 inhibition screen may be too high. For other cases the human dose will likely be high (e.g., >1000 mg/day) because the potency for the target will not be as great (e.g., antibacterials), and in these instances the inhibitor concentration will likely be too low. However, in high-throughput screening, customization of the inhibitor concentration to the need of each chemical series is generally not practical, and these small sacrifices in assay fidelity are made so as to maintain throughput.

Determination of the potential for mechanism-based inactivation in early drug research, in a screening mode, can be challenging. The conduct of mechanism-based inactivation experiments is typically more complex than what is amenable to high throughput. Nevertheless, there may be a need, particularly with some chemical series, to gather *in vitro* inactivation data in order to assist in compound design in efforts to breed out this property. Abbreviated approaches can be employed to merely check if a compound can cause greater inhibition if it is preincubated with the P450 enzyme in the presence of NADPH prior to addition of the P450 marker substrate and conduct of the activity incubation. In such a simple approach, it needs to be assured that the test compound concentration is within a suitable range such that a difference in inhibition between preincubated and non-preincubated can be discerned. Using a test compound concentration such that 25% inhibition would be observed in the absence of preincubation with NADPH is optimal for sensitivity in identifying mechanism-based inactivators in a screening mode where using multiple test compound concentrations is resource-prohibitive (Obach et al., 2007).

As individual compounds become a greater focus of attention, the assay employed to test for P450 inhibition can be a lower-throughput assay that can include a greater number of test compound concentrations. It is most common to run IC_{50} curves at this point in pooled human liver microsomes with P450-selective substrate activity assays. With fewer compounds, greater care and attention can be paid to each to ensure that the data obtained can be used for sound decision-making regarding the continued progression of the compound into the more expensive clinical development phase. At this point, there is more information and greater confidence in the other input parameters, in addition to inhibition potency, needed to make predictions of drug–drug interactions *in vivo*. The need to determine K_i values vs merely determining IC_{50} values (using a substrate concentration equal to K_M) is debatable. The former provide some insight into the kinetic mechanisms, but the latter can also be used to make predictions of drug interactions with the assumption of a competitive inhibition mechanism. For mechanism-based inactivation, determination of the parameters K_I and k_{inact} are necessary to predict the magnitude of drug–drug interactions (Mayhew et al., 2000); and while an abbreviated approach such as an "IC_{50} shift" has been shown to be useful in making predictions of drug interactions for mechanism-based inactivators, such an approach is empirical (Obach et al., 2007).

7.4.2 Prediction of Drug–Drug Interactions

The procedures used to quantitatively predict the magnitude of drug–drug interactions from *in vitro* inhibition and inactivation data have only recently been reasonably well established, although some of the fundamental equations for carrying out this process have been available for decades. Increasing the accuracy of predictions has depended on the incorporation of an increasing number of input parameters.

P450 inhibition potency certainly is a main driver in the potential for a new compound to cause drug interactions. In the past, a value of 1 μM was defined as a cutoff for concern; that is, if a compound showed a K_i or IC_{50} of less than 1 μM, then it was more likely that a drug interaction would be observed (Wrighton et al., 2000). A systematic evaluation of this cutoff showed that it was correct in about four of five instances when an *in vivo* drug interaction was defined as any case where one drug caused a two-fold or greater increase in exposure to a second drug (Obach et al., 2005). The clear limitation of such a simple approach is that the concentration of inhibitor *in vivo* is completely disregarded. Another approach used an empirical cutoff of the ratio between the *in vivo* circulating concentration of inhibitor and the K_i value ($[I]/K_i$) such that if this ratio is greater than 0.1, then a drug interaction was likely and should be explored with a clinical study. The value used for [I] was the total circulating C_{max} value. This approach suffers from being oversimplistic and disregards several other important factors as well as disregards the well-established free-drug hypothesis that states that only unbound drug can exhibit an effect (i.e., that bound drug must first dissociate from nonspecific binding sites before it can interact with the target protein).

Increasing the complexity of the prediction approach and consideration of other important input parameters is necessary to begin to approach reliable and quantitative methods for predicting drug–drug interactions from *in vitro* data. The aforementioned approach of employing a ratio cutoff for $[I]/K_i$ has its origins in the Rowland–Matin equation (Rowland and Matin, 1973), which relates the magnitude of a drug–drug interaction to inhibitor potency, inhibitor concentration, and, very importantly, the fraction of the clearance of the victim drug that is dependent on the affected enzyme:

$$\frac{AUC_i}{AUC} = \frac{1}{\dfrac{f_{CL(CYPX)}}{1+\dfrac{[I]_{in\ vivo}}{K_i}} + (1 - f_{CL(CYPX)})}$$

where AUC_i and AUC are the exposures of the victim drug in the presence and absence of concomitant administration of inhibitor, K_i is the inhibition constant, $[I]_{in\ vivo}$ is the concentration of the inhibitor (see below), and, importantly, $f_{CL(CYPX)}$ is the fraction of the victim drug that is cleared by the affected enzyme (CYPX). The importance of the $f_{CL(CYP)}$ value is illustrated in Fig. 7.3.

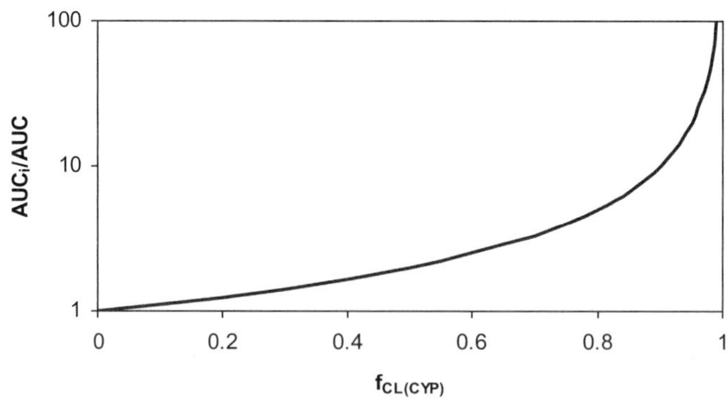

Figure 7.3. Relationship between the maximum magnitude of exposure increase and the fraction of clearance catalyzed by the affected enzyme.

If the value is at 0.5 or below, then even the most potent of inhibitors can only cause a two-fold drug interaction. But as the value increases, especially above 0.8, the potential for large drug interactions becomes possible. Reliable and accurate values for $f_{CL(CYP)}$ are not easy to determine. For some enzymes, observations of clearance differences for a victim drug between wild-type subjects and subjects possessing a null phenotype for a given P450 enzyme (e.g., CYP2D6) can be used to calculate $f_{CL(CYP)}$ (Venkatakrishnan and Obach, 2005).

A very important question arises regarding the value to use for $[I]_{in\ vivo}$. This term truly represents the concentration of inhibitor that is available to bind to the P450 enzyme, which would be a concentration free in solution within the cell (hepatocyte in most cases), which is unmeasurable, especially in humans. Proposals of surrogate concentrations for the intracellular concentration available to bind to P450 enzymes have been made. The aforementioned systemic total C_{max} suffers from the flaws that it ignores the impact of protein binding and its effect on free intracellular concentration, and it also lacks a consideration of the inhibition that can occur during first pass after absorption of orally administered agents. Others have proposed estimates of total liver concentration estimated from tissue partition ratios (e.g., from rats); however, this also represents a total concentration and not a free concentration (von Moltke et al., 1998). Much of what is present in the liver is partitioned into phospholipid membranes, sequestered into subcellular organelles (e.g., lysosomes, fat vesicles) or nonspecifically bound to cellular proteins, and not available to bind to P450. In other approaches, the use of an estimate of the unbound concentration of the inhibitor that is incident to the liver—that is, the free portal vein concentration—has been successfully used in the Rowland–Matin equation to predict the magnitude of drug interactions (Kanamitsu et al., 2000). This concentration is estimated from the dose, absorption rate constant (k_a), and fraction absorbed (F_a) of the inhibitor:

$$[I]_{in\ vivo} = f_u \cdot \left(C_{max} + \frac{\text{Dose} \cdot F_a \cdot k_a}{Q_h} \right)$$

where f_u is the unbound fraction of the inhibitor in blood and Q_h is hepatic blood flow. Use of this concentration has the limitation that the potential for the inhibitor to be actively influxed or effluxed by membrane transporters is not considered.

The impact of inhibition that occurs in the intestine during first-pass absorption is also important for CYP3A4 inhibitors and substrates and needs to be accounted for. In this case, since CYP3A4 is the major P450 enzyme in the intestine, the Rowland–Matin equation can be simplified to remove the $f_{CL(CYPX)}$ term and replace it with unity. However, the impact of the intestine extraction will be a function of the extent to which this tissue extracts the victim drug during first pass (termed $1 - F_g$ in the case of intestine). High-extraction CYP3A-metabolized drugs (e.g., buspirone) will be subject to greater effects at the intestine than low-extraction CYP3A-metabolized drugs (e.g., alprazolam) (Galetin et al., 2007), so it is important to have an estimate of the F_g value. Thus, the effect of a CYP3A4 inhibitor on the intestinal extraction during first pass can be described as

$$\frac{F_{g,\text{inh}}}{F_g} = \frac{1}{F_g + (1 - F_g) \cdot \left(\dfrac{1}{1 + \dfrac{[\text{I}]_g}{K_i}} \right)}$$

where F_g is the fraction of the victim drug that evades extraction by the intestine, and $F_{g,\text{inh}}$ is the same value in the presence of the inhibitor. $[\text{I}]_g$ is the maximum concentration of the inhibitor in the enterocyte during absorption and is estimated as

$$[\text{I}]_g = \frac{\text{Dose} \cdot F_a \cdot k_a}{Q_g}$$

where the terms are the same as described before, with Q_g being the intestinal blood flow (~3.5 mL/min/kg). The effect on intestinal first pass is multiplied with the effect on hepatic metabolism to yield the prediction of the change in AUC caused by an orally administered CYP3A inhibitor on a CYP3A substrate:

$$\frac{\text{AUC}_i}{\text{AUC}} = \frac{1}{F_g + (1 - F_g) \cdot \left[\dfrac{1}{1 + \dfrac{[\text{I}]_g}{K_i}} \right]} \cdot \frac{1}{\dfrac{f_{\text{CL(CYP3A)}}}{1 + \dfrac{[\text{I}]_{in\ vivo}}{K_i}} + (1 - f_{\text{CL(CYP3A)}})}$$

For mechanism-based inactivation, these relationships between magnitude of effect and *in vitro* parameters remain similar in structure; however, the added complexity of the inactivation rate constant (k_{inact}) and *in vivo* enzyme degradation rate constant (k_{deg}) must also now be included. Thus, while drug interactions caused by reversible inhibition are primarily driven by the ratio of inhibitor concentration *in vivo* and inhibition potency, for inactivation this ratio is also still important, but the ratio of inactivation rate and *in vivo* degradation rate of the enzyme must also be considered. If a compound causes a very high rate of inactivation, the concentration needed *in vivo* to cause an effect may not need to be great. If another compound has a low rate of inactivation, the concentrations needed *in vivo* to elicit an effect needs to be high, and the behavior of such a compound begins to resemble that of a reversible inhibitor more than an inactivator.

The equations used to predict the magnitude of a drug interaction caused by an inactivator are still derived from the Rowland–Matin equation, but with inactivation rate and degradation rate constants included:

$$\frac{\mathrm{AUC}_i}{\mathrm{AUC}} = \frac{1}{\left(\dfrac{f_{\mathrm{CL(CYPX)}}}{1+\dfrac{k_{\mathrm{inact}}\cdot[\mathrm{I}]_{in\ vivo}}{k_{\mathrm{deg}}\cdot([\mathrm{I}]_{in\ vivo}+K_{\mathrm{I}})}}\right)+(1-f_{\mathrm{CL(CYPX)}})}$$

where k_{inact} is the maximum inactivation rate constant measured *in vitro*, K_{I} is the inactivator concentration that causes inactivation at half the maximum inactivation rate, and k_{deg} is the *in vivo* rate of degradation of the enzyme in the absence of inactivator. Other parameters are as described before. For CYP3A, interactions including an effect on intestinal extraction are necessary (Wang et al., 2004):

$$\frac{\mathrm{AUC}_i}{\mathrm{AUC}} = \frac{1}{F_g+(1-F_g)\cdot\left[\dfrac{1}{1+\dfrac{k_{\mathrm{inact}}\cdot[\mathrm{I}]_g}{k_{\mathrm{deg(CYP3A,gut)}}\cdot([\mathrm{I}]_g+K_{\mathrm{I}})}}\right]}\cdot\frac{1}{\left(\dfrac{f_{\mathrm{CL(CYP3A)}}}{1+\dfrac{k_{\mathrm{inact}}\cdot[\mathrm{I}]_{in\ vivo}}{k_{\mathrm{deg(CYP3A,hep)}}\cdot([\mathrm{I}]_{in\ vivo}+K_{\mathrm{I}})}}\right)+(1-f_{\mathrm{CL(CYP3A)}})}$$

Values for degradation rate constants for P450 enzymes *in vivo* are controversial (Ghanbari et al., 2006; Yang et al., 2008), since there is no direct method by which such values can be measured in humans. Values used for making predictions of drug interactions for inactivators have come from *in vitro* experiments in human hepatocytes, values from animals, or values estimated by modeling clinical studies using enzyme inducers and inactivators (Obach et al., 2007).

In all of these prediction approaches, the system is considered to be static and single concentrations of inhibitors/inactivators are used (e.g., systemic C_{max}, estimated portal free C_{max}, etc.). However, this is clearly an oversimplification, because both inhibitor and substrate concentrations are dynamic.

Furthermore, these approaches simply target the "average human" and intersubject variability is missed (since several of the parameters in these equations will differ from person to person). Nevertheless, the approaches described using these static, average values perform reasonably well in predicting the average magnitude of drug interactions from *in vitro* data (Obach et al., 2006; Ito et al., 2005; Ito and Houston, 2005). Predicting intersubject variability in drug interactions requires the use of population pharmacokinetic approaches (Inoue et al., 2006) or advanced software (Einolf, 2007).

7.5 CONCLUSIONS

Inhibition and inactivation of human cytochrome P450s represent an important area of drug research. Availability of human reagents (microsomes and expressed P450 enzymes) has led to an understanding of the mechanisms and enzymes involved in drug clearance. Utilizing *in vitro* methods to measure inhibition constants and inactivation parameters can generate data useful in predicting drug–drug interactions. High-throughput methods can permit an early assessment of this potential for newly synthesized compounds and permit the offending structural motifs to be designed out of the molecule. Using these approaches and applying discipline in preventing the further progression of compounds that inhibit or inactivate cytochrome P450 into clinical trials should reduce the occurrence of pharmacokinetic-based drug interactions in clinical practice and improve the overall safety of pharmacotherapy.

REFERENCES

Austin RP, Barton P, Cockroft SL, Wenlock MC, Riley RJ. The influence of nonspecific microsomal binding on apparent intrinsic clearance, and its prediction from physicochemical properties. *Drug Metab Dispos* 2002;30(12):1497–1503.

Backes WL, Kelley RW. Organization of multiple cytochrome P 450s with NADPH-cytochrome P 450 reductase in membranes. *Pharmacol Ther* 2003;98(2):221–233.

Bjornsson TD, Callaghan JT, Einolf HJ, Fischer V, Gan L, Grimm S, Kao J, King SP, Miwa G, Ni L, Kumar G, McLeod J, Obach RS, Roberts S, Roe A, Shah A, Snikeris F, Sullivan JT, Tweedie D, Vega JM, Walsh J, Wrighton SA. The conduct of *in vitro* and *in vivo* drug–drug interaction studies: a Pharmaceutical Research and Manufacturers of America (PhRMA) perspective. *Drug Metab Dispos* 2003;31(7):815–832.

Butler MA, Iwasaki M, Guengerich FP, Kadlubar FF. Human cytochrome P-450PA (P-450IA2), the phenacetin *O*-deethylase, is primarily responsible for the hepatic 3-demethylation of caffeine and *N*-oxidation of carcinogenic arylamines. *Proc Nat Acad Sci USA* 1989;86(20):7696–7700.

Cali JJ, Ma D, Sobol M, Simpson DJ, Frackman S, Good TD, Daily WJ, Liu D. Luminogenic cytochrome P450 assays. *Expert Opin Drug Metab Toxicol* 2006;2(4):629–645.

Chauret N, Tremblay N, Lackman RL, Gauthier JY, Silva JM, Marois J, Yergey JA, Nicoll-Griffith DA. Description of a 96-well plate assay to measure cytochrome P4503A inhibition in human liver microsomes using a selective fluorescent probe. *Anal Biochem* 1999;276(2):215–226.

Chauret N, Dobbs B, Lackman RL, Bateman K, Nicoll-Griffith DA, Stresser DM, Ackermann JM, Turner SD, Miller VP, Crespi CL. The use of 3-[2-(*N,N*-diethyl-*N*-methylammonium)ethyl]-7-methoxy-4-methylcoumarin (AMMC) as a specific CYP2D6 probe in human liver microsomes. *Drug Metab Dispos* 2001;29(9):1196–1200.

Cohen LH, Remley MJ, Raunig D, Vaz ADN. *In vitro* drug interactions of cytochrome P450: An evaluation of fluorogenic to conventional substrates. *Drug Metab Dispos* 2003;31(8):1005–1015.

Dixit V, Hariparsad N, Desai P, Unadkat JD. *In vitro* LC-MS cocktail assays to simultaneously determine human cytochrome P450 activities. *Biopharm Drug Dispos* 2007;28(5):257–262.

Einolf HJ. Comparison of different approaches to predict metabolic drug–drug interactions. *Xenobiotica* 2007;37(10/11):1257–1294.

Fairman DA, Collins C, Chapple S. Progress curve analysis of CYP1A2 inhibition: a more informative approach to the assessment of mechanism-based inactivation? *Drug Metab Dispos Biol Fate Chem* 2007;35(12):2159–2165.

Fasco MJ, Piper LJ, Kaminsky LS. Biochemical applications of a quantitative high-pressure liquid chromatographic assay of warfarin and its metabolites. *J Chromatogr* 1977;131:365–373.

FDA 2006 http://www.fda.gov/cder/drug/drugInteractions/default.htm (accessed December 15, 2007).

Galetin A, Hinton LK, Burt H, Obach RS, Houston JB. Maximal inhibition of intestinal first-pass metabolism as a pragmatic indicator of intestinal contribution to the drug–drug interactions for CYP3A4 cleared drugs. *Curr Drug Metab* 2007;8(7): 685–693.

Gao H, Yao L, Mathieu HW, Zhang Y, Maurer TS, Troutman MD, Scott DO, Ruggeri RB, Lin J. *In silico* modeling of non-specific binding to human liver microsomes. *Drug Metab Dispos* 2008;36(10):2130–2135.

Ghanbari F, Rowland-Yeo K, Bloomer JC, Clarke SE, Lennard MS, Tucker GT, Rostami-Hodjegan A. A critical evaluation of the experimental design of studies of mechanism based enzyme inhibition, with implications for *in vitro–in vivo* extrapolation. *Curr Drug Metab* 2006;7(3):315–334.

Ghosal A, Hapangama N, Yuan Y, Lu X, Horne D, Patrick JE, Zbaida S. Rapid determination of enzyme activities of recombinant human cytochromes P450, human liver microsomes and hepatocytes. *Biopharm Drug Dispos* 2003;24(9):375–384.

Guengerich FP. Cytochromes P450, drugs, and diseases. *Mol Intervent* 2003;3(4): 194–204.

Hallifax D, Houston JB. Binding of drugs to hepatic microsomes: comment and assessment of current prediction methodology with recommendation for improvement. *Drug Metab Dispos* 2006;34(4):724–726.

Hutzler JM, Powers FJ, Wynalda MA, Wienkers LC. Effect of carbonate anion on cytochrome P450 2D6-mediated metabolism *in vitro*: the potential role of multiple oxygenating species. *Arch Biochem Biophys* 2003;417(2):165–175.

Inoue S, Howgate EM, Rowland-Yeo K, Shimada T, Yamazaki H, Tucker GT, Rostami-Hodjegan A. Prediction of *in vivo* drug clearance from *in vitro* data. II: Potential inter-ethnic differences. *Xenobiotica* 2006;36(6):499–513.

Ishii M, Xu BQ, Ding LR, Fischer NE, Inaba T. Interaction of plasma proteins with cytochromes P450 mediated metabolic reactions: inhibition by human serum albumin and γ-globulins of the debrisoquine 4-hydroxylation (CYP2D) in liver microsomes of human, hamster and rat. *Toxicol Lett* 2001;119(3):219–225.

Ito K, Houston JB. Prediction of human drug clearance from *in vitro* and preclinical data using physiologically based and empirical approaches. *Pharm Res* 2005;22(1): 103–112.

Ito K, Hallifax D, Obach RS, Houston JB. Impact of parallel pathways of drug elimination and multiple cytochrome P450 involvement on drug–drug interactions: CYP2D6 paradigm. *Drug Metab Dispos* 2005;33(6):837–844.

Kalgutkar AS, Obach RS, Maurer TS. Mechanism-based inactivation of cytochrome P450 enzymes: chemical mechanisms, structure–activity relationships and relationship to clinical drug–drug interactions and idiosyncratic adverse drug reactions. *Curr Drug Metab* 2007;8(5):407–447.

Kanamitsu S-I, Ito K, Sugiyama Y. Quantitative prediction of *in vivo* drug–drug interactions from *in vitro* data based on physiological pharmacokinetics: use of maximum unbound concentration of inhibitor at the inlet to the liver. *Pharm Res* 2000; 17(3):336–343.

Kenworthy KE, Bloomer JC, Clarke SE, Houston JB. CYP3A4 drug interactions: correlation of 10 *in vitro* probe substrates. *Br J Clin Pharmacol* 1999;48(5):716–727.

Khan KK, He YQ, Domanski TL, Halpert JR. Midazolam oxidation by cytochrome P450 3A4 and active-site mutants: an evaluation of multiple binding sites and of the metabolic pathway that leads to enzyme inactivation. *Mol Pharmacol* 2002; 61(3):495–506.

Kim M-J, Kim H, Cha I-J, Park J-S, Shon J-H, Liu K-H, Shin J-G. High-throughput screening of inhibitory potential of nine cytochrome P450 enzymes *in vitro* using liquid chromatography/tandem mass spectrometry. *Rapid Commun Mass Spectrom* 2005;19(18):2651–2658.

Kronbach T, Mathys D, Gut J, Catin T, Meyer UA. High-performance liquid chromatographic assays for bufuralol 1′-hydroxylase, debrisoquine 4-hydroxylase, and dextromethorphan *O*-demethylase in microsomes and purified cytochrome P-450 isozymes of human liver. *Anal Biochem* 1987;162(1):24–32.

Kumar V, Wahlstrom JL, Rock DA, Warren CJ, Gorman LA, Tracy TS. CYP2C9 inhibition: impact of probe selection and pharmacogenetics on *in vitro* inhibition profiles. *Drug Metab Dispos* 2006;34(12):1966–1975.

Lee JS, Obach RS, Fisher MB. *Drug Metabolizing Enzymes: Cytochrome P450 in Drug Discovery and Development*. 2003. New York: Marcel Dekker.

Madan A, Usuki E, Burton LA, Ogilvie BW, Parkinson A. *In vitro* approaches for studying the inhibition of drug-metabolizing enzymes and identifying the drug-metabolizing enzymes responsible for the metabolism of drugs. In: Rodrigues AD, editor. *Drugs and the Pharmaceutical Sciences*, Vol. 116, *Drug–Drug Interactions*. New York: Marcel Dekker, 2002, pp. 217–294.

Maenpaa J, Hall SD, Ring BJ, Strom SC, Wrighton SA. Human cytochrome P450 3A (CYP3A) mediated midazolam metabolism: the effect of assay conditions and regioselective stimulation by α-naphthoflavone, terfenadine and testosterone. *Pharmacogenetics* 1998;8(2):137–155.

Makris TM, Denisov I, Schlichting I, Sligar SG. Activation of molecular oxygen by cytchrome P450. In: Ortiz de Montellano PR, editor, *Cytochrome P450: Structure, Mechanism, and Biochemistry*, 3rd ed. New York: Kluwer Academic, 2005, pp. 149–182.

Margolis JM, Obach RS. Impact of nonspecific binding to microsomes and phospholipid on the inhibition of cytochrome P4502D6: Implications for relating *in vitro* inhibition data to *in vivo* drug interactions. *Drug Metab Dispos* 2003;31(5): 606–611.

Mayhew BS, Jones DR, Hall SD. An *in vitro* model for predicting *in vivo* inhibition of cytochrome P450 3A4 by metabolic intermediate complex formation. *Drug Metab Dispos* 2000;28(9):1031–1037.

Miller VP, Stresser DM, Blanchard AP, Turner S, Crespi CL. Fluorometric high-throughput screening for inhibitors of cytochrome P450. *Annals of the New York Academy of Sciences*, Vol. 919, *Toxicology for the Next Millennium*. 2000, 26–32.

Nath A, Atkins WM. Principal component analysis of CYP2C9 and CYP3A4 probe substrate/inhibitor panels. *Drug Metab Dispos* 2008;36(11):2151–2155.

Obach RS, Walsky RL, Venkatakrishnan K, Houston JB, Tremaine LM. *In vitro* cytochrome P450 inhibition data and the prediction of drug–drug interactions: qualitative relationships, quantitative predictions, and the rank-order approach. *Clin Pharmacol Ther* 2005;78(6):582–592.

Obach RS, Walsky RL, Venkatakrishnan K, Gaman EA, Houston JB, Tremaine LM. The utility of *in vitro* cytochrome P450 inhibition data in the prediction of drug–drug interactions. *J Pharmacol Exp Ther* 2006;316(1):336–348.

Obach RS, Walsky RL, Venkatakrishnan K. Mechanism-based inactivation of human cytochrome P450 enzymes and the prediction of drug–drug interactions. *Drug Metab Dispos* 2007;35(2):246–255.

O'Donnell CJ, Grime K, Courtney P, Slee D, Riley RJ. The development of a cocktail CYP2B6, CYP2C8, and CYP3A5 inhibition assay and a preliminary assessment of utility in a drug discovery setting. *Drug Metab Dispos* 2007;35(3):381–385.

Ortiz de Montellano PR, editor. *Cytochrome P450: Structure, Mechanism, and Biochemistry*, 3rd ed. New York: Kluwer Academic/Plenum Publishers, 2005.

Powis G, Jansson I, Schenkman JB. The effects of albumin upon the spectral changes and metabolism by the hepatic microsomal fraction. *Arch Biochem Biophys* 1977;179(1):34–42.

Purdon MP, Lehman-McKeeman LD. Improved high-performance liquid chromatographic procedure for the separation and quantification of hydroxytestosterone metabolites. *J Pharmacol Toxicol Methods* 1997;37(2):67–73.

Riley RJ, Grime K, Weaver R. Time-dependent CYP inhibition. *Expert Opin Drug Metab Toxicol* 2007;3(1):51–66.

Rowland A, Elliot DJ, Knights KM, Mackenzie PI, Miners JO. The "albumin effect" and *in vitro–in vivo* extrapolation: sequestration of long-chain unsaturated fatty acids enhances phenytoin hydroxylation by human liver microsomal and recombinant cytochrome P450 2C9. *Drug Metab Dispos* 2008;36(5):870–877.

Rowland M, Matin SB. Kinetics of drug–drug interactions. *J Pharmacokine Biopharm* 1973;1(6):553–567.

Smith D, Sadagopan N, Zientek M, Reddy A, Cohen L. Analytical approaches to determine cytochrome P450 inhibitory potential of new chemical entities in drug discovery. *J Chromatog B: Anal Technol Biomed Life Sci* 2007;850(1–2):455–463.

Sykes MJ, Sorich MJ, Miners JO. Molecular modeling approaches for the prediction of the nonspecific binding of drugs to hepatic microsomes. *J Chem Inf Model* 2006;46(6):2661–2673.

Tolonen A, Petsalo A, Turpeinen M, Uusitalo J, Pelkonen O. *In vitro* interaction cocktail assay for nine major cytochrome P450 enzymes with 13 probe reactions and a single LC/MSMS run: analytical validation and testing with monoclonal anti-CYP antibodies. *J Mass Spectrom* 2007;42(7):960–966.

Tucker GT, Houston JB, Huang SM. Optimizing drug development: strategies to assess drug metabolism/transporter interaction potential—toward a consensus. *Clin Pharmacol Ther* 2001;70(2):103–114.

Venkatakrishnan K, Obach RS. *In vitro–in vivo* extrapolation of CYP2D6 inactivation by paroxetine: Prediction of nonstationary pharmacokinetics and drug interaction magnitude. *Drug Metab Dispos* 2005;33(6):845–852.

von Moltke LL, Greenblatt DJ, Duan SX, Daily JP, Harmatz JS, Shader RI. Inhibition of desipramine hydroxylation (cytochrome P450-2D6) *in vitro* by quinidine and by viral protease inhibitors: relation to drug interactions *in vivo*. *J Pharm Sci* 1998; 87(10):1184–1189.

Walsky RL, Obach RS. Validated assays for human cytochrome P450 activities. *Drug Metab Dispos* 2004;32(6):647–660.

Wang Y-H, Jones DR, Hall SD. Prediction of cytochrome P450 3A inhibition by verapamil enantiomers and their metabolites. *Drug Metab Dispos* 2004;32(2):259–266.

Wrighton SA, Schuetz EG, Thummel KE, Shen DD, Korzekwa KR, Watkins PB. The human CYP3A subfamily: practical considerations. *Drug Metab Rev* 2000;32(3&4): 339–361.

Xu BQ, Ishii M, Ding LR, Fischer NE, Inaba T. Interaction of serum proteins with CYP isoforms in human liver microsomes: inhibitory effects of human and bovine albumin, alpha-globulins, alpha-1-acid glycoproteins and gamma-globulins on CYP2C19 and CYP2D6. *Life Sci* 2003;72(17):1953–1962.

Yang J, Liao M, Shou M, Jamei M, Yeo KR, Tucker GT, Rostami-Hodjegan A. Cytochrome P450 turnover: regulation of synthesis and degradation, methods for determining rates, and implications for the prediction of drug interactions. *Curr Drug Metab* 2008;9(5):384–393.

Zhang H, Davis CD, Sinz MW, Rodrigues AD. Cytochrome P450 reaction-phenotyping: an industrial perspective. *Expert Opin Drug Metab Toxicol* 2007:3(5): 667–687.

8

CYTOCHROME P450 INDUCTION

EDWARD L. LECLUYSE, MICHAEL W. SINZ, NICOLA HEWITT, STEPHEN S. FERGUSON, AND JASMINDER SAHI

8.1 INTRODUCTION

8.1.1 Background

Induction of hepatic drug-metabolizing enzymes has long been recognized in pharmacology and toxicology as a possible consequence of exposure to drugs and other xenobiotics (Conney, 1967). The induction of drug-metabolizing enzymes can have significant consequences on the pharmacokinetics and toxicity of drugs. For example, rifampin can decrease the *in vivo* AUC of coadministered drugs by as much as 70–90%, as in the case of (*R*)-verapamil coadministration with rifampin (Lin 2006). Another classic example of the clinical relevance of CYP induction is organ rejection in patients receiving cyclosporine who were also treated with rifampin for tuberculosis (Hebert et al., 1992). Induction of CYPs can also lead to toxicity by increasing reactive metabolite production or by increasing the CYP1A2-mediated activation of procarcinogens (Guengerich et al., 1990). As such, the efficacy of a drug is reduced if its clearance is increased by enzyme induction. Prior knowledge of potential interactions with co-therapies is needed to guide pharmaceutical companies in the development of a new drug as well as propose post-market labeling. There may be a tendency to link induction effects only to therapeutic drugs; however, a number of cultural- and lifestyle-related agents also induce enzyme activities, such as smoking, alcohol drinking, diet, and nutroceuticals (e.g., St. John's wort).

Enzyme Inhibition in Drug Discovery and Development, edited by Lu and Li
Copyright © 2010 John Wiley & Sons, Inc.

Four of the more important outcomes of hepatic enzyme induction that are of clinical relevance to humans are listed below:

1. Cytochrome P450 induction is one of the key mechanisms for pharmacokinetic drug–drug interactions. Chronic dosing of an inducing agent will increase the enzymatic activity of drug-metabolizing enzymes; and any coadministered drug that utilizes these same enzymes as a major clearance pathway will be more rapidly eliminated, resulting in lower drug concentrations that may be less effective. Drugs that are known to be affected by enzyme inducers include the benzodiazepines, the antiepileptic drug carbamazepine, corticosteroids, and beta-blockers (e.g., propranalol).
2. In some instances, the inducing agent itself is a substrate of the induced enzymes. In this situation, increased enzyme activity means that the drug itself may be metabolized more rapidly in a patient on long-term therapy than in a patient who just started drug therapy. As a result, after being on a drug for a while, a patient may exhibit pharmacokinetic tolerance under the original dosing regimen and subsequently require a dose increase (e.g., carbamazepine).
3. Alternatively, increased enzyme activity means that some drugs and xenobiotics may be bioactivated to a toxic metabolite to a greater extent in patients that have been exposed to enzyme inducers for prolonged periods. The classical example is that of acetaminophen (paracetamol), which is bioactivated to the reactive metabolite *N*-acetyl-*p*-benzoquinoneimine (NAPQI) (Moore et al., 1985). Reactive metabolite formation by inducible CYP and UGT enzymes is often observed during the development of a new drug. Many of these events can lead to increased inactivation of the enzymes and/or formation of other protein adducts.
4. There is a growing awareness of the potential pathophysiological consequences associated with the long-term exposure to inducing agents. These are usually caused by the disruption of cellular homeostasis by overburdening biochemical pathways that regulate levels of physiological substrates. This has led to clinical manifestations of some characteristic pathophysiologies, such as chloracne, hypoporphyria, endocrine disruption, altered inflammatory response, and bone-density loss (Parkinson, 2001; Pascussi and Vilarem, 2005; Zhou et al., 2006; Moreau et al., 2007).

Unlike inhibition, induction generally takes days to result in a clinically relevant effect on drug clearance and may persist for several days beyond ceasing administration of the inducing agent. Upon withdrawal or reduction of blood and tissue concentrations of the inducer, enzyme activities slowly return to normal levels through degradation and turnover of the respective RNA and protein (Parkinson, 2001; Josephy, 2005). Accordingly, the rate and extent of the inductive effect depends on the continued exposure to an inducing agent at sufficiently high concentrations and the biochemical parameters surrounding the regulation and expression of the individual target genes involved.

One of the more important details that is often underappreciated in considering the clinical impact of exposing patients to an inducing agent is that inducers activate multiple genes simultaneously. Thus, an inducer like rifampin can increase the amount of CYP enzymes (such as CYP3A4 and CYP2B6) and phase II enzymes (such as UGT's and GSTs). This is the reason why rifampin and nelfinavir reduce concentrations of UGT substrates like zidovudine, as well as CYP substrates like ethinyl estradiol. Moreover, not all enzyme inducers work alike or cause the same effect. Some drugs are potent inducers, while others are moderate or weak inducers. Rifampin and St. John's wort are two of the most potent inducers. Protease inhibitors and non-nucleotide reverse transcriptase inhibitors such as nelfinavir, efavirenz, and nevirapine are moderate CYP inducers. Many of the anticonvulsive agents, such as phenobarbital and carbamazepine, despite being weak inducers, can cause an observable increase in the clearance of coadministered compounds. As will be discussed later, the relative impact that an inducing agent will have in a clinical setting is a combination of its intrinsic potency and efficacy as well as its corresponding pharmacokinetic profile.

Recognizing the safety and health issues associated with exposure to inducing agents, the US Food and Drug Administration (FDA) has recently placed more emphasis on determining whether a drug candidate has the potential to induce CYP and UGT enzymes and induce drug transporters in humans (FDA, 2006). Although the FDA has always encouraged routine, thorough evaluation of drug interactions *in vitro* whenever feasible and appropriate, the original guidance documents on drug interaction studies published in 1997 contained limited information on the proper conduct of *in vitro* induction studies. The Pharmaceutical Research and Manufacturer's Association (PhRMA) subsequently published a White Paper that addressed many of these issues and acknowledged the importance of appropriate study design (Bjornsson et al., 2003). In 2006, the FDA published a draft guidance on drug interaction studies that addressed some of these limitations and presented details on the principles and conduct of *in vitro* induction studies for IND and/or NDA submissions.

In this chapter, we have discussed the key model systems and study design considerations for evaluating the induction potential of new chemical entities (NCE) in humans. In addition, a major objective of this chapter is to describe the current scientific and regulatory perspectives regarding the molecular mechanisms of enzyme induction and how these factor into the current screening strategies utilized by the pharmaceutical industry.

8.1.2 Mechanisms of CYP Induction

Before embarking on a path to describe the best strategies and methods for determining the potential of a new drug to be an inducer of human clearance

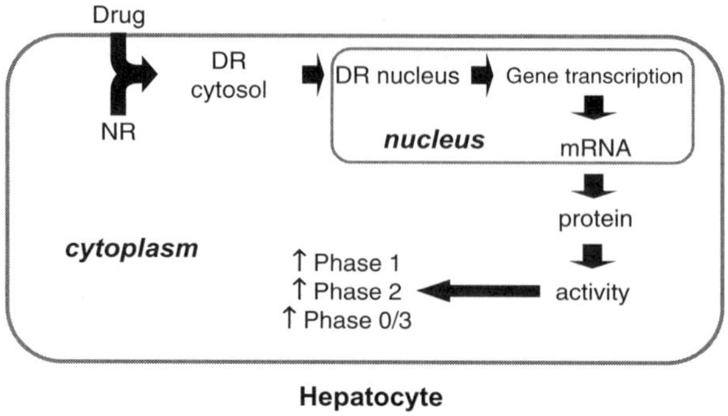

Figure 8.1. General mechanism of enzyme regulation. NR: nuclear receptor; DR: drug receptor complex; Phase 1: oxidation, reduction, and hydrolysis reactions; Phase II: conjugation reactions; Phase 0/3: transporter mediated uptake and efflux.

mechanisms, it is important to understand the molecular mechanisms that are a prerequisite to any clinically significant event. The classical definition of enzyme induction is the biochemical process of initiating or increasing the expression of a gene or the amount of the resulting mRNA or protein product in response to an endogenous or exogenous factor (e.g., hormone, chemical, metabolite, etc.). As is the case of most drugs involved in clinically relevant drug interactions, the inducer molecule either directly or indirectly activates a cytosolic receptor complex, causing it to translocate to the nucleus, of the cell. Once inside the nucleus, the new complex is converted to an active transcription factor through mechanisms that are not well understood, leading to the transcription and translation of target genes that possess the corresponding DNA binding sequences in their promoter regions (Fig. 8.1) (Wang and LeCluyse, 2003). Although increased levels of these enzyme systems can occur by the stabilization of mRNA and/or proteins (e.g., induction of CYP2E1), inductive events that lead to clinically significant changes in the clearance or bioactivation of drugs predominantly occur by a receptor-mediated process that leads to the increased production of an enzyme or other protein at the level of gene transcription.

Drugs causing *de novo* synthesis of enzymes often do so by binding to and activating one or more nuclear receptor that have been identified in recent years to be the primary mediators of the inductive effects of drugs. The principal receptors that have been identified or implicated in nearly all the reported clinically relevant drug interactions are the aryl hydrocarbon receptor, the pregnane X receptor (also known as the steroid and xenobiotic receptor, SXR), and the constitutivc androstane receptor (Fig. 8.2). It is now well known that these nuclear receptors control the expression of the cytochrome P450 enzymes, such as CYP1A (AhR), CYP2B, CYP3A, and CYP2C subfamilies (PXR and CAR), the phase II enzymes such as UGTs, glutathione-*S*-transferases, and sulfotransferases, and transporters including MDR1, MRP2,

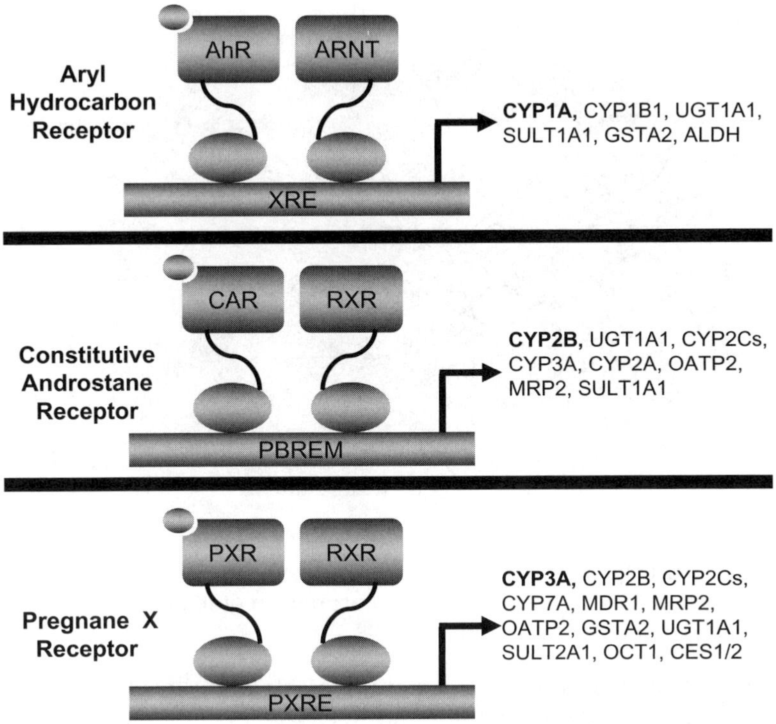

Figure 8.2. Regulation of genes by the nuclear receptors AhR, CAR, and PXR. Upon activation by coactivator peptides and proteins, the heterodimer binds to target xenobiotic response elements located in the proximal and distal P450/transporter gene promoters, resulting in transcription. AhR, aryl hydrocarbon receptor; ARNT, AhR nuclear translocator responsive element; XRE, xenobiotic response element; PXR, pregnane X receptor; PXRE, PXR response element; CAR, constitutive androstane receptor; PBREM, phenobarbital-responsive enhancer module; RXR, retinoid X receptor.

OATP's, and OCT's (Fig. 8.3). It should be kept in mind that not all CYPs are inducible (e.g., CYP2D6 has no known inducer) and not all increases in enzyme levels are due to this mechanism. For example, CYP2E1 induction by isoniazid involves the stabilization of the enzyme itself rather than *de novo* synthesis (Novak and Woodcroft, 2000; Raucy et al., 2004).

AhR and CAR are thought to be predominantly localized in the cytoplasm of hepatocytes, and after receptor activation they are translocated to the nucleus. CAR, in its inactive form, is bound to endogenous steroids; but in the presence of phenobarbital (PB) or "PB-like" inducers, it is dephosphorylated, which initiates its translocation to the nucleus (Squires et al., 2004; Kawana et al., 2003). Notably, CAR does not require ligand binding to be transactivated and is constitutively active in most immortalized cell lines (i.e., up-regulates target gene expression in the absence of an inducing agent);

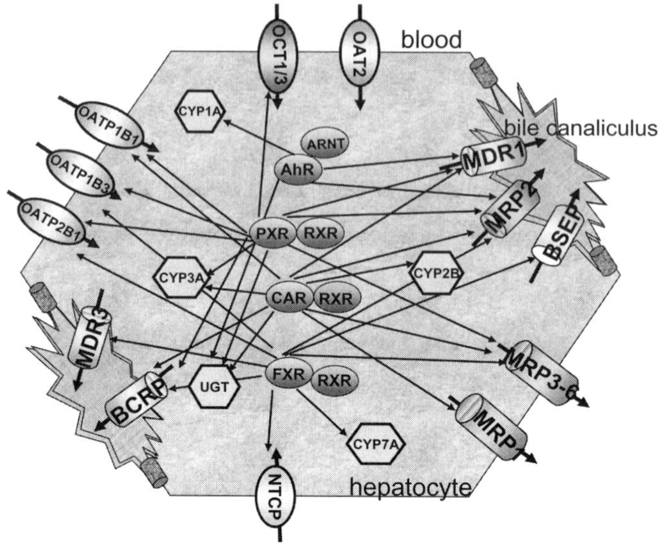

Figure 8.3. Overlapping regulation of CYP and UGT enzymes and transporters by nuclear receptors. Nuclear receptors control expression of multiple drug metabolizing enzymes and xenobiotic transporters.

therefore it is named after this constitutive activation mechanism. PXR has been found in both the cytoplasm and nucleus, depending on the species of origin (Kawana et al., 2003; Squires et al., 2004). Once in the nucleus, these receptors form heterodimers with other factors, such as Arnt (AhR nuclear translocator) for AhR and retinoid X receptor (RXR) for both PXR and CAR. Upon activation by other coactivator peptides and proteins, the heterodimer then binds to the target xenobiotic response elements located in both the proximal and distal P450 gene promoters, resulting in the transcription of the respective CYP isoform (Fig. 8.2) (Lehmann et al., 1998; Goodwin et al., 1999; Sueyoshi et al., 1999; Moore and Kliewer, 2000).

The current practice for investigating induction in preclinical species is the *ex vivo* measurement of CYP activities in livers of animal species. While it is now well-recognized that regulatory mechanisms for induction differ across species, significant induction in preclinical species could trigger early *in vitro* human hepatocyte and/or clinical induction studies. Currently, different *in vitro* methods to analyze CYP enzyme induction are used, all with their advantages and disadvantages. The current gold standard within the industry and the recommended method by regulatory agencies for the prediction of enzyme induction by xenobiotics in humans is the primary human hepatocyte induction model. In this chapter, the *in vitro* methodologies (incubation conditions, relevant CYPs, and positive controls) most commonly utilized to determine the induction potential of an NCE are described. In addition, the relative advantages and disadvantages of each methodology for drug discovery and development purposes are discussed. The major focus of this chapter is on the use of the most widely adopted and accepted model system, namely primary human hepatocyte cultures. The different technologies that are available and currently used to evaluate the potential of a new drug to cause enzyme induction *in vitro* will be discussed. Because the outcome of an *in vitro* assay determines whether further clinical *in vivo* drug–drug interactions are carried out (FDA, 2006), much consideration and discussion has been placed on the analysis and interpretation of the corresponding data. As such, the use of different threshold values and acceptance criteria for relevant endpoints (e.g., fold induction relative to vehicle controls, 40% of the positive response, and EC_{50}/E_{max} values) and the current strategies utilized to predict the induction potential of new drugs are discussed as well.

8.2 METHODS FOR EVALUATING ENZYME INDUCTION

The methodologies utilized for assessing drug–drug interactions caused by enzyme induction change as a drug candidate moves from early discovery to the clinic. Higher-throughput, less definitive methods (e.g., *in silico*) and nuclear receptor-based assays are used early in discovery; but later in discovery and preclinical/clinical development, immortalized cells and primary human hepatocytes are more likely to be used. This section provides an overview of these methodologies and their application in eliminating or predicting drug–drug interactions.

8.2.1 *In Silico*

In silico computational modeling is a high-throughput, cost-effective approach that is typically conducted in collaboration between drug metabolism and medicinal chemistry scientists early in drug discovery to identify and eliminate problematic chemotypes (e.g., CYP inhibitors). Ambiguities in the catalytic cycle and the flexibility of CYP active sites make *in silico* extrapolations of enzyme inhibition challenging (Stjernschantz et al., 2008). It is also complicated to successfully predict induction of CYP enzymes, because this requires combining crystal structure information of the nuclear receptor proteins with the *in silico* structural biology. For this, chemical structures are docked with crystal structures of nuclear receptors (e.g., PXR) and the model built using *in vitro* and *in vivo* data. *In silico* modeling of induction potential is further confounded by the structure of the PXR binding site, because this is large, flexible, and promiscuous, enabling binding to molecules with molecular weights ranging from 300 to greater than 800 daltons (Dickins, 2004).

Moreover, agonists and antagonists bind to several distinct regions of PXR, such as the ligand-binding domain (LBD) and AF-2 region (Ekins et al., 2007). However, merely docking the compound in the protein structure does not always suffice, because the large ligand-binding domain of PXR accommodates ligands of all sizes. Larger ligands (e.g., rifampicin) can alter the protein structure after binding, thereby further increasing volume of the binding cavity (Handschin et al., 2003). The modeling of ligand–receptor complexes is further confounded by multiple binding modes and orientations and the potential for simultaneous binding of additional small molecules. There is evidence to suggest that >30% activation in a PXR reporter gene assay indicates a high risk of CYP3A4 induction (Gao et al., 2007), indicating that transactivation data is a valid measure to guide SAR. The majority of the available *in vitro* PXR data is transactivation data that are expressed relative to a positive control (typically rifampicin) or fold increase over vehicle control and relatively less data related to ligand binding of agonists. The transactivation data are useful in developing and validating models; however, binding data are more directly correlatable to the *in silico* modeling.

There have been successes combining computational and experimental data for different classes of molecules. The PXR crystal structure has been successfully used *in silico* to identify some classes of potential substrates; for example, since the PXR binding site is hydrophobic and contains polar residues, hydrophobic compounds with polar groups have the potential to be PXR ligands (Watkins et al., 2001; Lemaire et al., 2007). By "docking" compounds with the PXR crystal structure, Gao et al. demonstrated that polar groups at the end of an activator reduce PXR activity by destabilizing interactions in the hydrophobic areas of the PXR ligand-binding pocket. This was confirmed by synthesizing analogues incorporating these structural features, using both the transactivation assay and human hepatocytes (Gao et al., 2007). Using structure–activity relationships (SAR) based on these data, chemotypes were made relatively induction-free by introducing polar groups in regions with a higher affinity for hydrophobic groups and making the ligand structure rigid, so as to prevent binding. In addition, *in silico* modeling using an automated docking method (GOLD) successfully "visualized" the ketoconazole binding site to PXR as being in the AF-2 binding pocket and not the ligand-binding pocket (Wang et al., 2007). Xiao et al. (2002) constructed a three-dimensional model of the ligand-binding domain of hCAR that had a ligand-binding cavity the same size as that of PXR, although with less flexibility as it lacked the surface loop. Docking calculations of known hCAR selected ligands correlated well with previously reported receptor binding data, and the hCAR ligands identified by pharmacophore modeling were relatively planar structures with one hydrogen bond acceptor and three hydrophobic features (Xiao et al., 2002).

8.2.2 Reporter Gene Assays

Induction of drug-metabolizing enzymes occurs in the majority of cases via transcriptional gene activation, as is the case with CYP1A1/2, CYP3A4, and UGT. Induction also occurs, albeit less frequently, by stabilization of mRNA or protein—for example, for CYP2E1 (Gonzalez, 2007). For drug-metabolizing enzymes, transcriptional activation is mediated by nuclear receptors that function as transcription factors and include: CAR, PXR, FXR (farnesoid X receptor), PPAR (peroxisome proliferator activated receptor), and HNF-4α (hepatocyte nuclear receptor-4α) (Moore et al., 2002; Tirona and Kim, 2005; Lim and Huang, 2008; Liu et al., 2008). For all these, gene transcription is regulated through binding of the DNA-binding domain of the receptor to specific response elements in the promoter region of target genes. AhR regulates CYP1A enzymes in a similar fashion; however, it is not a member of the nuclear receptor family but belongs to the basic helix–loop–helix PER-ARNT-SIM transcription factor family (Beischlag et al., 2002).

High-throughput *in vitro* receptor binding and transactivation assays that assess the interaction between receptors and potential ligands are employed to evaluate the potential for induction. In particular, PXR-derived nuclear hormone receptor models (ligand binding and transactivation) are widely utilized to predict or evaluate CYP3A4 enzyme induction. To evaluate ligand binding, genetically expressed, isolated receptors are incubated with a test compound and a high-affinity radiolabeled ligand (e.g., SR12813), and the competition of the radiolabeled ligand with a test compound is measured (IC_{50} determination) (Zhu et al., 2004). While simple, effective, and high throughput, this methodology can result in a disproportionate number of false positives and negatives for some chemotypes. Chemical entities that require active transport into the hepatocyte may not have access to the intracellular milieu *in vivo*, but would have direct access to the receptor in this assay, leading to a false-positive or enhanced response. Also, ligand binding may not be followed by the appropriate displacement of corepressors or recruitment of coactivators, as in the case of docetaxel and paclitaxel (Harmsen et al., 2007). For these reasons, the more physiological cell-based transactivation assays are preferable and often provide similar throughput.

The PXR transactivation model requires an immortalized/transformed cell line that is transfected with two vectors: the full-length human PXR and a variation of the CYP3A4 promoter region coupled to a reporter gene (e.g., luciferase). Upon ligand binding to receptor, there is increased expression of luciferase which is proportional to the increased production of luminescent product and induction response (EC_{50} determination) (Sinz et al., 2006). However, this assay can also result in false negatives, because in some cases binding to the PXR receptor does not lead to transactivation or enzyme induction. False negatives or attenuated responses also result in cases where induction is the combined effect of several receptors due to cross-talk between nuclear receptors (e.g., PXR and CAR). For example, CYP3A4 induction can be mediated by both PXR and CAR. Because these test systems are relatively simple, containing single receptors, a full induction response may not be achieved, unlike hepatocytes that contain all relevant and functional nuclear receptors and enzymes (Wang and LeCluyse, 2003).

In general, there appears to be a reasonably good correlation between PXR binding and transactivation assays and PXR transactivation and primary human hepatocyte results (Zhu et al., 2004). However, at this time, the PXR transactivation assay appears to result in less false positives than the binding assay and provides improved correlation to human drug–drug interactions and is the assay of choice in most laboratories.

The methodologies for CAR and AhR-mediated gene transcription are more complex in that they are ligand-activated (e.g., PXR) and also activated by ligand-binding independent mechanisms. AhR can be activated in the cytoplasm upon ligand binding or via protein tyrosine kinases (Backlund and Ingelman-Sundberg, 2005). A transactivation assay, similar to that for PXR, is utilized for AhR, using luminescence to measure the transactivation (Yueh et al., 2005). CAR nuclear translocation can occur by direct binding to the receptor or through a partially elucidated ligand-independent mechanism involving kinases that dephosphorylate CAR; for example, phenobarbital activates CYP2B6 transcription without binding to CAR (Qatanani and Moore, 2005). CAR transactivation assays are relatively difficult to conduct, because cells frequently have high basal CAR activity, due to spontaneous CAR translocation to the nucleus when expressed in cell lines (Chang and Waxman, 2006). In addition, since there are both direct and indirect mechanisms of CAR activation, screening requires a nuclear translocation assay to complement a transactivation assay. The CAR nuclear translocation assay is best conducted in primary hepatocytes, because unlike in cell lines, CAR does not spontaneously translocate in primary hepatocytes. CAR translocation from cytoplasm to nucleus can be monitored by using an expressed CAR protein tagged with fluorescent proteins (e.g., GFP or EYFP) and visualizing this with the aid of confocal microscopy. The ratio of CAR proteins between cytoplasm and nucleus can also be monitored by Western blot analysis (Faucette et al., 2007). A relatively more recent assay utilizes transactivation of a CAR human splice variant, CAR3. Since CAR3 had ~80% lower basal activity compared to wild-type CAR (CAR1) (Auerbach et al., 2005; Faucette et al., 2007), the assay could more readily identify a potential CAR activator.

The hepatocyte nuclear factor 4α (HNF-4α) is a constitutively active transcription factor of the nuclear hormone receptor family that plays a dominant role in the expression and interindividual variations of drug-metabolizing enzymes in human hepatocytes (Kamiyama et al., 2007) and is involved in hepatic xenobiotic metabolism (Nakata et al., 2006). Although HNF-4α is implicated in binding to the regulatory regions of over 1500 genes including GR, SXR, MRP2, CYP1A2, CYP2B6, CYP2C8, CYP2D6, and CYP2E1 (Odom et al., 2004) and has a dose-dependent effect upon mRNA expression levels of CYP2A6, CYP2B6, CYP2C9, CYP2D6, CYP3A4, and CYP3A5 in primary human hepatocytes (Jover et al., 2001), this nuclear receptor is not widely evaluated in drug discovery at this time. This may be because while it is well known that HNF-4α regulates basal expression of multiple liver-specific genes involved in hepatic metabolism and participates in the nuclear receptor responses to xenobiotics through activation of PXR and CAR (Bell and Michalopoulos, 2006; Kamiyama et al., 2007), there is limited information on direct effects of HNF-4α on enzyme regulation. Some indication of this effect is that phenobarbital increases HNF-4α mRNA and protein expression in the nucleus of hepatocytes, independent of CAR and PXR, and this is mediated by cellular phosphatase and kinase (Bell and Michalopoulos, 2006). These data suggest that HNF-4α regulation is likely an integral part of the hepatic response to phenobarbital. Thus far, HNF-4α is studied by using RT-PCR or Western blot analysis, and a binding assay has yet to be reported. Further characterization of this receptor is necessary to determine its overall involvement in regulation of drug-metabolizing enzymes and drug–drug interactions.

No nuclear receptor has been implicated for the regulation and expression of CYP2E1, which appears to be different from the other CYP enzymes. CYP2E1 is regulated by transcriptional activation, and inducers increase mRNA translation and the stability of mRNA and/or protein, resulting in enhanced enzyme activity due to diminished degradation (Gonzalez, 2007). CYP2E1 is induced by DMSO, ethanol, isopentanol, isoniazid, Phenobarbital, and rifampin and also by starvation and diabetes (Koop and Tierney, 1990; Kostrubsky et al., 1995; Madan et al., 2003).

8.2.3 Stem Cells

The advantage of human embryonic (hESC) and adult (hASC) stem cells is the potential to grow indefinitely and provide a continual and readily available source of cells that exhibit and retain hepatocyte-like properties—for example, major xenobiotic transporters, drug metabolizing enzymes, nuclear receptors, and transcription factors. Multiple laboratories are currently working on improving isolation, purification, and culture of hSCs to form specific functional differentiated cell types (Agarwal et al., 2008). Thus far, cell lines developed from hESC have differentiated into hepatocyte–like cells but with limited drug-metabolizing enzymes; for example, the HLC cells exhibit hepatocyte morphology and markers (α-fetoprotein, albumin, HNF-4α) but limited P450 metabolism (Hay et al., 2007). Inducible CYP1A1 and CYP3A4/7 mRNA and immunoreactive protein as well as CYP2C8/9/19 mRNA have been demonstrated in SA002 and SA167 cell lines but with no corresponding catalytic activity (Ek et al., 2007). NeoHep cells were developed from terminally differentiated peripheral blood monocytes. These cells have hepatocyte-like morphology and expression of hepatocyte markers but limited drug metabolism (CYP1A and UGT) and CYP3A induction activities (Ruhnke et al., 2005). Similarly, cell lines derived from hASC that have been characterized are reported to retain limited drug-metabolizing enzyme activity (Schwartz et al., 2002; Herrera et al., 2006; Kazemnejad et al., 2008). Nonetheless, these studies indicate the possibility of hSC-derived hepatocytes being developed for metabolism studies in the future. Thus far, no stem-cell-derived cell lines that develop into mature, differentiated hepatocytes with functional nuclear receptors, drug-metabolizing enzymes, and transporters are available (Sinz et al., 2008b).

8.2.4 Cell Lines

Cell lines (native or transfected) derived from human hepatic tissues are used as an alternative to primary fresh or cryopreserved hepatocytes when specific issues arise (e.g., can a particular chemotype/compound activate PXR?). These hepatocyte cell lines either arise spontaneously from cancerous tissues (HepG2) or are immortalized/transformed in the laboratory, typically either by using the viral oncogene, SV-40 large-T antigen (Fa2N-4) (Mills et al., 2004), or by expression of telomerase reverse transcriptase. Many of these cell lines have altered gene expression and typically little to no endogenous P450 activity. One of the contributing reasons for this is that in hepatoma cell lines there is down-regulation of xenosensors, including PXR and CAR (Pacussi et al., 2005); however, AhR appears to be stable in these cell lines, and often CYP1A enzymes are inducible. The human hepatoma cell line, HepG2, is used for induction studies because some P450 enzymes are inducible to detectable levels even though the expression levels of Phase I and Phase II drug metabolism enzymes are significantly lower than those in human liver and hepatocyte samples (Wilkening et al., 2003). An advantage of HepG2 cells is that these are relatively easy to transfect and demonstrate activation of CAR- and PXR-mediated transcripts (Kodama et al., 2004). In HepG2 cells, CYP1A1 has been induced by 2,3,7,8-tetrachlorodibenzo-*p*-dioxin (TCDD), CYP3A4 mRNA has been induced by *o,p′*-dichloro-diphenyl-trichloroethane (*o,p′*-DDT), and CAR activation of CYP2B6 has been assessed by phenobarbital-type compounds (Bonzo et al., 2005; Medina-Diaz and Elizondo, 2005; Swales et al., 2005). However, due to the low expression of very few drug-metabolizing enzymes, HepG2 cells have limited applications in drug metabolism and induction. The human hepatoma cell line Huh-7 has been utilized to study P450 induction after transfection with PXR to evaluate activation of CYP2B6 and CYP3A4 by known inducers (Wang and LeCluyse, 2003) and after transfection with the CYP2C19 promoter (Arefayene et al., 2003).

The BC2 cell line is derived from a human hepatocarcinoma, is stable for over 2 years of culture, differentiates at confluency, and remains differentiated for several weeks in culture. BC2 cells express CYP1A1/2, CYP2A6, CYP2B6, CYP2C9, CYP2E1, CYP 3A4, GST, and UGT enzymes (Gomez-Lechon et al., 2001). While CYP1A and CYP3A activity were induced to expected levels by 3-MC and dexamethasone, phenobarbital could induce CYP2B6 activity by a maximum of only 1.7-fold. A subpopulation of these cells (ADV-1) demonstrates higher drug-metabolizing enzyme activity than the BC2 parental cell line (O'Connor et al., 2005). Future publications will determine the utility of this cell line to metabolism applications.

HepaRG cells display several hepatocyte-like functions including drug-metabolizing enzymes, nuclear receptors, hepatic drug transporters, and a hepatocyte-like morphology (Aninat et al., 2006; Le Vee et al., 2006). These cells express genes for multiple drug-metabolizing enzymes including CYP1A1, CYP1A2, CYP2B6, CYP2C8, CYP2C9, CYP2E1, and CYP3A4, and the mRNA for these is induced by the prototypical inducers 3-methylcholantrene, rifampicin, and isoniazid (Aninat et al., 2006; Kanebratt and Andersson, 2008b). Activities of phenacetin *O*-dealkylase, bupropion hydroxylase, diclofenac 4′-hydroxylase, and midazolam 1-hydroxylase activities are also induced. However, basal activities are low relative to primary hepatocytes; for example, CYP3A4-mediated basal testosterone hydroxylation activity is about 2 pmol/mg protein/minute (Aninat et al., 2006) relative to 100–500 pmol/mg protein/minute typically observed in primary human hepatocytes. A concern with the HepaRG cells is that this cell line presents as two cell types at confluency: (a) the cholangiocytes that are flattened and have a clear cytoplasm and (b) the hepatocytes that form clusters of granular epithelial cells (Kanebratt and Andersson, 2008a). When cultured for 2–3 weeks in media containing 2% DMSO and 50 μM hydrocortisone hemisuccinate, the hepatocytes make up 50–55% of the total cell population and form structures resembling bile canaliculi (Cerec et al., 2007). Quantitative extrapolations of *in vivo* hepatic induction with this heterogeneous cell line cannot be made with confidence, because the percent response specifically by the hepatocyte-like cells cannot be evaluated without very sensitive analytical methods and the contribution of the cholangiocytes, which make up 50% of the population, is not known.

Fa2N-4 is a nontumorigenic, immortalized hepatic cell line that has been successfully used to assess AhR and PXR-mediated induction (CYP1A & CYP3A4). These cells are easy to maintain in culture and provide consistent data for the first 10–20 passages. At higher passages, the cells start proliferating faster and simultaneously lose the ability to induce (unpublished observations). In the early passages, Fa2N-4 cells demonstrate inducible CYP1A2, CYP3A4, CYP2C9, UGT1A, and MDR1 transcripts and the induction response of CYP1A2, CYP2C9, and CYP3A4 to the prototypical inducers rifampin and beta-naphthoflavone is within the range (albeit the lower level) observed for primary human hepatocytes (Mills et al., 2004; Ripp et al., 2006). Using mRNA analysis, 24 compounds that were previously evaluated for induction in primary human hepatocytes were assessed for induction in Fa2N-4 cells (Ripp et al., 2006). A majority of these showed comparable induction to primary hepatocytes. However, troglitazone, a potent *in vitro* and clinical CYP3A4 inducer (Sahi et al., 2000), did not induce this enzyme in the Fa2N-4 cells. Troglitazone-mediated induction is believed to be a CAR-mediated process, and there is evidence that this nuclear receptor pathway is not functional in Fa2N-4 cells (Hariparsad et al., 2008). Overall, current data indicate that AhR- and PXR-mediated induction correlates well between Fa2N4 cells and primary hepatocytes, while CAR-mediated induction demonstrates significant differences. This is because in the Fa2N4 cells, expression of CAR, sulfotransferase, OATP1B1, and OATP1B3 is significantly lower than cryopreserved primary human hepatocytes, while expression of PXR and AhR is similar (Hariparsad et al., 2008). In this study, the CAR activators CITCO and artemisinin did not induce CYP2B6 in Fa2N4 cells, the PXR/CAR dual activators phenytoin, phenobarbital, and efavirenz caused lower-than-expected induction, and PXR-mediated induction was equivalent (in the case of rifampin, higher than that observed in human hepatocytes). When using this cell line for discovery-stage induction studies, these limitations should be kept in perspective. Another limitation of the Fa2N4 cells is that since enzyme activity is low, it is difficult to differentiate between or rank order the induction potential within a series of compounds based on enzyme activity measurements. Consequently, mRNA is a more appropriate endpoint when using this cell line. It should also be kept in mind that the Fa2N-4 cells are not being capable of metabolizing compounds (unpublished observations) due to the low basal level of drug-metabolizing enzymes. In the case of highly metabolized compounds, this could result in artificially enhanced induction, because there is no metabolic clearance. Conversely, false negatives could result in cases where the metabolite may contribute to, or be responsible for, the induction. The majority of other transformed cell lines used to evaluate induction of drug-metabolizing enzymes are derived from human intestinal or breast tumors. These include the intestinal cell line LS174T, the breast cancer cell line MCF-7, and the colonic tumor cell line, Caco-2. All these cell lines exhibit low expression and activities for limited Phase I and Phase II drug-metabolizing enzymes, and RT-PCR analysis is required to evaluate induction over basal levels, which are in most cases not detectable. These likely do not express all the relevant receptor-mediated pathways; for example, the LZ180 cells express PXR but not CAR and therefore present limited induction response (Gupta et al., 2008). Some cell lines can be cultured in an environment that artificially induces relevant enzymes; for example, long-term treatment of Caco-2 cells with 1,25-dihydroxyvitamin D3 induces CYP3A4 activity to levels comparable to human small intestinal epithelium (Schmiedlin-Ren et al., 2001). While these cell lines can be used for preliminary studies to address specific questions (e.g., does this chemical class carry a CYP1A induction liability?), no available cell line demonstrates the complete set of drug-metabolizing enzymes and associated nuclear receptors at appreciable concentrations; for example, the LS180 and Fa2N-4 cells do not have appreciable concentrations of CAR. Due to the cross-talk between nuclear receptors, using cell lines possessing only some receptor pathways may result in erroneous data/conclusions.

8.2.5 Primary Human Hepatocytes

Primary human hepatocytes have the ability to model metabolism, inhibition, and induction when placed in suspension or culture (Fig. 8.4). Under appropriate culture conditions, human hepatocytes retain all the necessary receptors, regulatory components, drug-metabolizing enzymes, and transporters, as well as the ability to metabolize and induce enzymes, analogous to the *in vivo* liver. As shown in Table 8.1, *in vitro* enzyme induction data utilizing primary human hepatocytes in culture correlates well with clinical observations when pharmacologically relevant concentrations of test article are used (LeCluyse et al.,

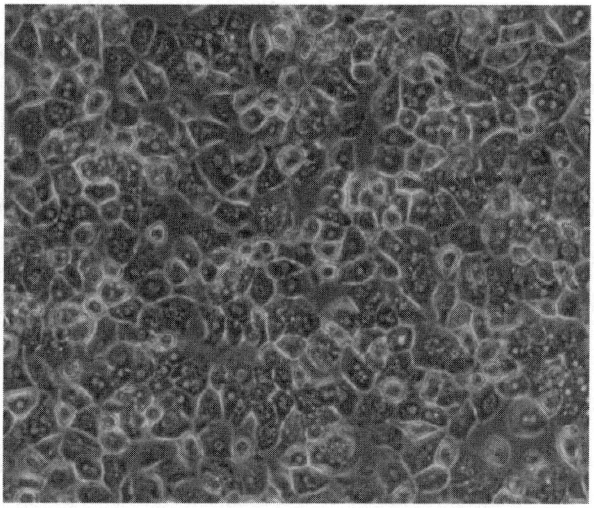

Figure 8.4. Primary human hepatocytes in primary sandwich culture. Hepatocytes from a 41-year-old male donor liver after 3 days in culture. See the insert for color representation of this figure.

TABLE 8.1. Relationship Between *In Vitro* Potency and the Prediction of *In Vivo* Induction Using the EC_{50} and Therapeutic Plasma Concentrations (Hewitt et al., 2007)

Compound	EC_{50} (μM)	C_{max}/C_{ss} (μM)	[I]/EC_{50}	Induction Potential
Rifampin	0.8	14	17.5	Likely
Carbamazepine	0.8	25	28	Likely
Phenobarbital	125	40–180	0.3–1.5	Likely
Phenytoin	25	80	3.2	Likely
Troglitazone	3–5	7	2.3	Likely
Avasimibe	0.2	1–6	5–30	Likely
Rosiglitazone	5–10	0.3–1.2	0.06–0.12	Possible
Simvastatin	0.14	0.024	0.17	Possible
Lovastatin	1–5	0.008	0.008–0.02	Unlikely
Clotrimazole	1–5	Topical	Inhibition	Unlikely
Nifedipine	8	0.008	0.001	Unlikely

2000; Sahi et al., 2000, 2004). Human hepatocytes in primary culture are considered the most appropriate *in vitro* system for assessing induction of drug-metabolizing enzymes (LeCluyse et al., 2005). One of the most widely accepted models is the sandwich culture model where fresh isolated or cryopreserved hepatocytes are plated on collagen, allowed to attach, and then sandwiched on top by an extracellular matrix overlay such as gelled collagen or Matrigel® (LeCluyse et al., 1994, 1999; LeCluyse, 2001). This environment allows for

(a) maintenance of hepatocytes in a more liver-like environment, (b) maintenance of liver-like cell morphology, and (c) formation of functional bile canaliculi (LeCluyse et al., 1994; Kostrubsky et al., 2003; Sahi et al., 2006). This culture system renders possible assessment of the cumulative effects of metabolism, induction, time-dependent inhibition, and metabolite effects, replicating the more steady-state-like effects on CYP activity and gene expression (Sinz et al., 2008a). Hepatocytes can also be cultured as monolayers on a collagen matrix without the overlay and used for metabolism and induction experiments. While the cells are easier to culture and utilize in this environment, they do not exhibit the liver-like morphology or form functional bile canaliculi.

Tissue culture media also influences the morphology, integrity, and cellular function of hepatocytes. Figure 8.5 shows enzymatic activity data generated from cultures of human hepatocytes treated with prototypical inducers for 3 days in three different media, namely, Williams E (WEM), Chee's (MCM), and DMEM/Ham's F-12. Our data demonstrate marked differences in the fold-over-control induction responses over these three routinely used culture media. We used 6β-hydroxytestosterone (6βT) conversion from testosterone as a marker of CYP3A activity. The lower fold-over-control response with WEM appears to arise from an elevated basal CYP3A response in vehicle control (and media-only controls) that is reflected at both the mRNA and functional activity level. This suggests that WEM supports a higher basal CYP3A expression relative to the DMEM/Ham's F-12 and Chee's media. Interestingly, the blunted fold-over-control response with WEM is less profound in microsomal-based studies relative to *in situ* media-based study designs (data not shown). This, coupled with the observation that mRNA responses are less blunted with WEM, suggests that other processes (e.g., testosterone uptake transporters) may also be affected by media compositions. We have also observed profound media formulation effects in cultures of rat hepatocytes where Chee's media was found to elevate basal CYP1A expression while other media types were more effective in modeling CYP1A induction.

Another determinant of cell function is the plating density of hepatocytes (Hamilton et al., 2001). As shown in Fig. 8.6, at high plating densities, >100% confluent, cells in conventional cultures do not spread out. Hepatocytes display hepatocyte-like morphology for 1–2 days, albeit without formation of bile canaliculi, and then begin to detach. When close to confluence (≥90%) at the time of plating, the cells redevelop normal morphology, tight junctions, and bile canaliculi. When plated at less than 50% confluence, hepatocytes appear fibroblast–like after a day in culture and send out philopodial extensions. We also observed that sparsely plated hepatocytes have significantly lower CYP3A4 activity than hepatocytes plated at optimal confluence, although PXR levels remain equivalent (Hamilton et al., 2001); our unpublished observations; Fig. 8.7). Fetal bovine serum commonly used in tissue culture is typically not used long-term when conducting metabolism or toxicology studies. While serum helps to enhance attachment, it contains many components (most

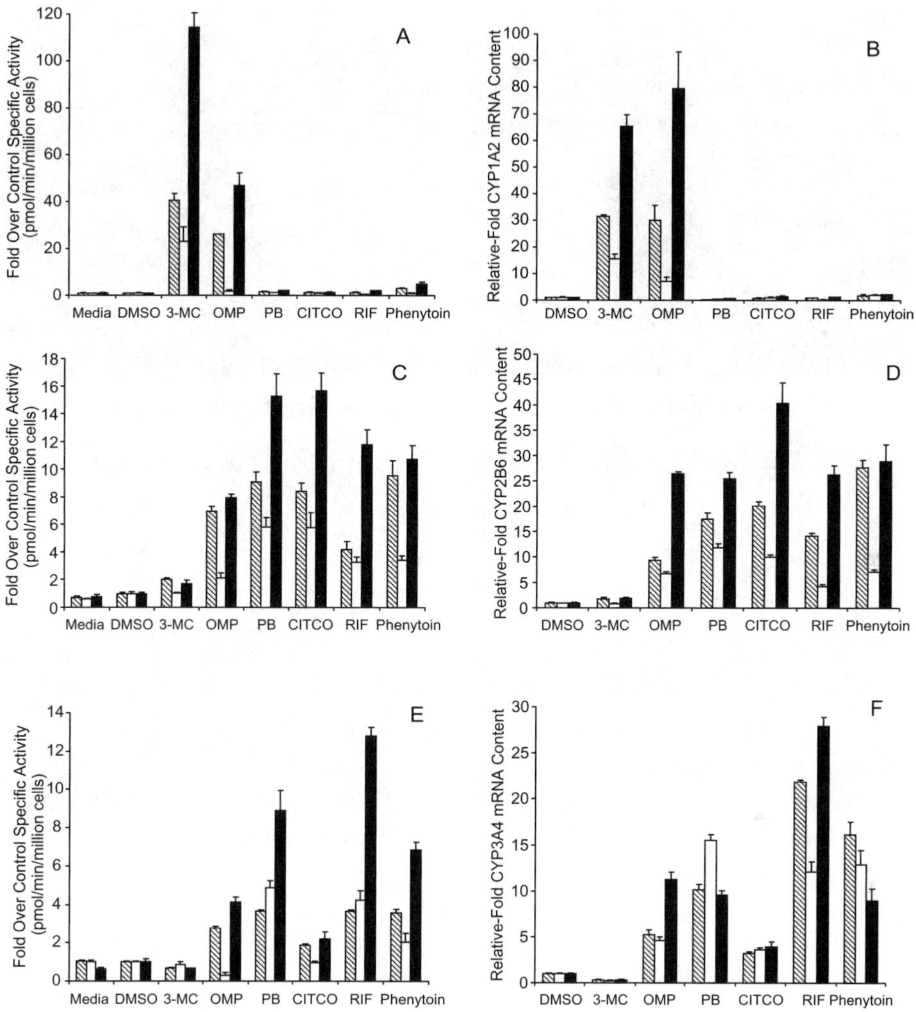

Figure 8.5. Effect of tissue culture media on expression and activity of the major CYP enzymes in primary human hepatocytes. Human hepatocytes in primary sandwich culture were treated with prototypical inducers for 3 days in different media. WEM (dashed), MCM (white), and DMEM/Ham's F-12 (black) with various prototypical inducers including: 3-MC (2 μM), OMP (50 μM), PB (1 mM), CITCO (200 nM), RIF (10 μM), and phenytoin (50 μM). **(A)** Enzymatic activity for CYP1A2 (APAP). **(B)** Relative CYP1A2 mRNA content (TaqMan®). **(C)** Enzymatic activity for CYP2B6 (OHBP). **(D)** Relative CYP2B6 mRNA content (TaqMan®). **(E)** Enzymatic activity for CYP3A (6βT). **(F)** Relative CYP3A4 mRNA content (TaqMan®). Error bars indicate standard errors of triplicate samples.

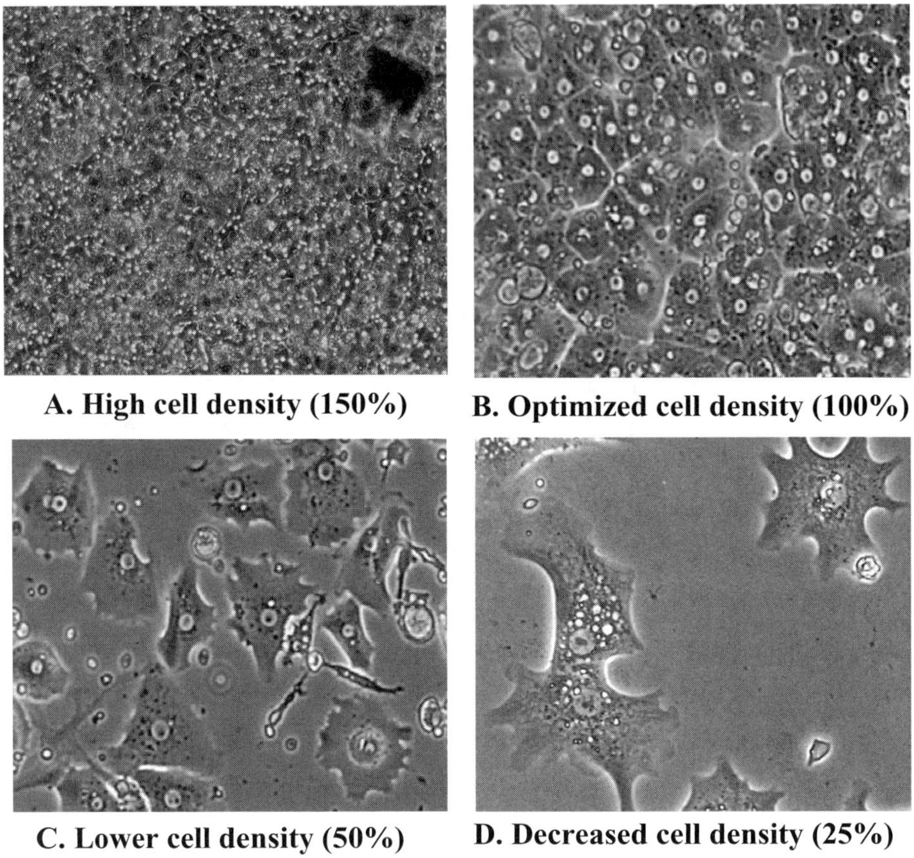

Figure 8.6. Effect of plating cell density on morphology of human hepatocytes in primary culture. **(A)** Too-high plating density. **(B)** Ideal plating density showing normal morphology. **(C)** At 50% confluence, hepatocytes appear fibroblast-like. **(D)** At 25% confluence, hepatocytes form pseudopodia.

of which are undefined) that stimulate cell proliferation and a loss of phenotypic expression patterns. Some serum factors may also inhibit establishment of bile canaliculi in hepatocyte cultures (Terry and Gallin, 1994). As such, it is typically removed after the initial attachment period of 4–6 h (LeCluyse et al., 2005). The additives most commonly used in primary hepatocyte cultures are a glucocorticoid (e.g., dexamethasone), insulin, selenium, and transferring, which aid in the preservation of hepatocyte-specific functions, polygonal hepatocyte morphology, and structural integrity of cytoplasmic membranes, especially bile canaliculi-like structures (LeCluyse et al., 2005).

After 2–3 days in culture, primary hepatocytes can be used for induction experiments and are typically incubated for 2 days with the test compounds when mRNA is the endpoint and for 3 days when enzyme activity is to be

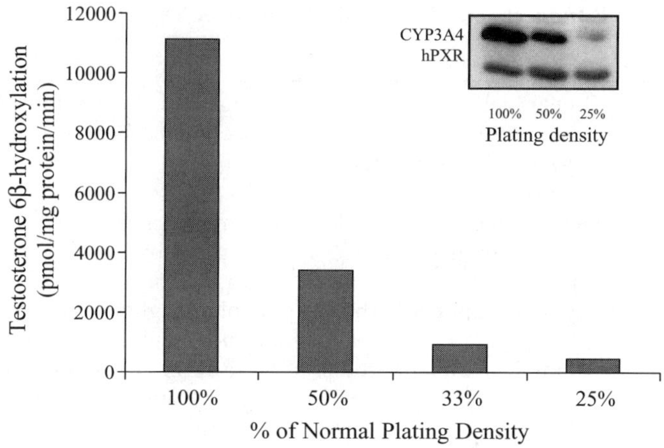

Figure 8.7. Effect of plating density on CYP3A4 protein and activity and PXR in primary human hepatocyte cultures. Hepatocytes have significantly lower CYP3A4 protein and activity when underplated. PXR levels do not appear to be significantly effected by plating density.

assessed. Table 8.2 provides a list of CYP inducers and probe substrates commonly used for these *in vitro* studies (Sinz et al., 2008a, 2008b).

8.2.6 Cryopreserved Hepatocytes

The use of plateable cryopreserved hepatocytes for evaluating *in vitro* induction of P450 enzymes is increasing as these become more accessible. Cryopreservation may moderately change P450 enzyme activities, which were reported to decrease between 3% (CYP1A2) and 22% (CYP2A6) while CYP3A4 activity increased 164% (Li et al., 1999). Since the drug-metabolizing enzymes remain inducible after cryopreservation and due to significant variation in activities of drug-metabolizing enzymes between individual human livers, cryopreserved cells generate results essentially indistinguishable from freshly isolated cells (Schehrer et al., 2000). Expression of mRNA and protein expression, as well as activities of CYP1A2, 2B6, 2C9, 2E1, 3A4, and UGT in cryopreserved hepatocytes, are inducible by prototypical inducers (Reinach et al., 1999; Roymans et al., 2005) as is the RNA expression of CYP3A5, carboxylesterases (CES1 and CES2), and sulfotransferases (CHST1, CHST3, CHST4, CST, SULT2A1, and TPST2) (Nishimura et al., 2004). The advantage that cryopreserved cells offer over freshly isolated cells is that experiments can be planned in advance and are not dependent on availability of fresh primary hepatocytes. Moreover, vials of cells can be stored over several years, making it possible to conduct a series of studies using the same cells, which is a distinct advantage when comparing results within a chemotype. For reasons not yet fully understood, a large percent of

TABLE 8.2. CYP Inducers and Probe Substrates Used *In Vitro*

CYP Enzyme	Inducer	Substrate/Probe
1A2	Omeprazole, β-Naphthoflavone, 3-methylcholantrhrene, Lansoprazole	Phenacetin, theophylline, caffeine, acetaminophen, tacrine
2A6	Dexamethasone, pyrazole	Coumarin, nicotine, butadiene
2B6	Phenobarbital, CITCO, phenytoin	Efavirenz, buproprion, propofol, *S*-mephenytoin
2C8	Rifampicin, phenobarbital	Taxol, repaglinide, rosiglitazone, amodiaquine
2C9	Rifampicin, phenobarbital	*S*-Warfarin, tolbutamide, diclofenac, phenytoin, flurbiprofen
2C19	Rifampicin	*S*-Mephenytoin, fluoxetine, omeprazole, esoprazole, lansoprazole, pantoprazole, citalopram, diazepam, hexobarbital, imipramine, proguanil, propanolol
2D6	None identified	Dextromethorphan, bufuralol, codeine, ethylmorphine, desipramine, atomoxetine, nicotine, debrisoquine
2E1	Ethanol, isoniazid, acetone	Chlorzoxazone, aniline, *p*-nitrophenol, lauric acid, acetaminophen, caffeine, dapsone, enflurane, theophylline
3A4/3A5	Rifampicin, Rifabutin, Rifapentine, Sulfinpyrazone, Carbamazepine, Dexamethasone, Phenytoin, Troleandomycin, Troglitazone, taxol	Midazolam, testosterone, dextromethorphan, triazolam, terfenadine, buspirone, felodipine, lovastatin, eletriptan, sildenafil, simvastatin, triazolam, acetaminophen, carbamezapine, cyclosporin, digitoxin, diazepam, erythromycin, fluoxetine, nifedipine, quinidine, saquinavir, cortisol, terfenadine, verapamil

Compiled from references in Sinz et al. (2008a) and Sinz et al. (2008b).

human hepatocyte preparations lose the ability to attach in culture. While these unattachable cryopreserved cells are useful for metabolism studies, induction cannot be assessed in cell suspension. Due to the limited number of donor hepatocytes that attach after cryopreservation, the commercial supply of attachable cryopreserved human hepatocyte preparations is relatively limited. The USFDA allows for the use of fresh or cryopreserved human hepatocytes for induction assessments.

8.2.7 FDA Draft Guidance on Use of Hepatocytes for Induction Assessments

The 2006 FDA draft guidance on drug–drug interactions states the following: "*A drug that induces a drug-metabolizing enzyme can increase the rate of metabolic clearance of a coadministered drug that is a substrate of the induced pathway. A potential consequence of this type of drug–drug interaction is sub-therapeutic blood concentrations*" (FDA, 2006). This draft guidance provides recommendations of primary hepatocyte-based *in vitro* assays, including substrates and positive controls that can be used in assays measuring catalytic activity of the major drug-metabolizing enzymes CYP1A2 and CYP3A4. It is suggested that freshly isolated and/or attachable cryopreserved hepatocyte cultures from at least three individual donors be treated with the test compound and appropriate vehicle and positive controls for 2–3 days prior to analysis. The recommended positive control activity is greater than twofold increase in enzyme activity of probe substrate. A minimum of three test compound concentrations based on the expected human plasma drug levels are suggested, one of which should be an order of magnitude greater than this concentration. In the absence of knowledge of human plasma levels, concentrations ranging over at least two orders of magnitude should be studied. The FDA draft guidance allows for enzymatic analysis to be conducted in microsomes prepared from the hepatocytes or *in situ*, using media-based assays and probe substrates. The potential for *in vivo* induction should be assessed based on induction activity with the test compound versus induction with the positive control. Since co-induction of CYP2C, CYP2B, and MDR1 occurs along with CYP3A4, negative results for CYP3A4 may eliminate the need to address interactions eliminated by these enzymes and MDR1. Fold increase, percent increase, or EC_{50} values are acceptable; for example, a drug that produces an increase in probe drug enzyme activity that is more than 40% of the positive control will be regarded as an enzyme inducer for CYP3A4.

A 2007 pharmaceutical industry survey indicated that while it was agreed that human hepatocytes in primary culture are the best predictor of *in vivo* induction, the methodologies used have not been standardized across laboratories and there is no consensus on criteria for determining in what cases clinical drug–drug interaction studies are required (Hewitt et al., 2007). A recent position paper on drug interaction evaluations (Huang et al., 2008) agrees with the FDA Draft guidance that since activation of PXR results in co-induction of both CYP2C9 and CYP2C19, a negative *in vitro* result for CYP3A induction eliminates the need for additional induction studies for both CYP3A and CYP2C enzymes. However, this position paper recommends that since CYP2B6 may not be co-induced with CYP3A4, the potential for induction of these enzymes should be evaluated *in vitro* as well.

8.2.8 *In Vivo* Transgenic/Chimeric Models

In vivo animal induction models offer many advantages over the static *in vitro* systems because *in vivo* models incorporate many of the dynamic processes, such as drug absorption, distribution, metabolism, and elimination, all of which occur concurrently. Multiple dynamic processes working in parallel determine the actual test compound concentration over time at the site of action, unlike the static tissue culture systems currently used. Test compounds can be administered over several days at pharmacological doses/exposures equivalent to those expected or observed in human plasma to enable improved assessments of human induction in an *in vivo* setting. However, there are significant interspecies differences in the nuclear receptors as well as drug-metabolizing enzymes. In the case of PXR, the DNA-binding domain is fairly well conserved across species, but there are significant differences in the ligand-binding domain, which contribute to the observed cross-species differences. Some of the newer transgenic/chimeric mouse models have circumvented the LBD dissimilarities between species by incorporating human attributes (genes). Humanized nuclear receptor transgenic mice, knockout mice, and chimeric mice with transplanted human hepatocytes are useful tools for assessing regulation of drug-metabolizing enzymes, and hepatotoxicity of drug candidates. For example, dose-dependent increases in cyp3A mRNA were observed with increasing concentrations of rifampin in hPXR transgenic mice (Xie et al., 2000; Kim et al., 2008). Rifampicin is a potent hPXR agonist but a very poor mouse PXR agonist. In knockout mice, CITCO, phenytoin, and phenobarbital increased CYP2B6 expression in CAR+ but not CAR− mice (Wang et al., 2004). When treated with increasing oral doses of rifampicin, SXR humanized and PXR-knockout (KO) mice demonstrated linear increases in plasma and liver tissue exposure up to 60 mg/kg doses (Kim et al., 2008). Increases in RNA expression and microsomal activity reached maximum responses at 10 mg/kg in SXR humanized mice while at the same dose of rifampicin the induction response was absent in PXR-KO mice. After 3 days of rifampicin treatment (10 mg/kg rifampicin) in SXR humanized mice, triazolam plasma AUC and C_{max} were greatly reduced compared to pretreatment levels (83% and 74%, respectively). These results indicate that the humanized SXR mouse may be used as a quantitative model to predict PXR-mediated CYP3A induction and the resulting pharmacokinetic changes of CYP3A substrates in humans (Kim et al., 2008). The limitation of the transgenic/knockout systems are that only one or two enzymes and/or nuclear receptor are typically "humanized" and the cross-talk between nuclear receptors/enzymes/transporters may be disconnected as the human and mouse systems may or may not be functioning in a similar fashion.

A newer model circumvents some of these disconnects, by using chimeric mice that have livers predominantly populated with human hepatocytes. One such model utilizes SCID mice that are injected with human hepatocytes, and the engraftment is considered successful if over 80% of the hepatocytes in the mouse liver are of human origin (Tateno et al., 2004). CYP1A1/2 and CYP3A4 mRNA, protein content, and enzyme activity was induced in chimeric mice treated with known human inducers and in hepatocytes isolated from these humanized chimeric mice, within the range observed for human hepatocytes (Katoh et al., 2005; Nishimura et al., 2005; Yoshitsugu et al., 2006). A second humanized chimeric mouse model utilizes severely immunodeficient, fumarylacetoacetate hydrolase (Fah)-deficient mice. Mice are pretreated with a urokinase-expressing adenovirus and engrafted (up to 90%) with human hepatocytes. The basal expressions of CYP1A1, CYP1A2, CYP2B6, CYP3A4, CYP3A7, and UGT1A1 in the chimeric mice were in the range expected for adult human livers. Induction of mRNA and enzymatic activity by beta-naphthoflavone (BNF), phenobarbital, and rifampicin was also similar to that of human hepatocytes (Azuma et al., 2007). These humanized models, while technically challenging and not fully validated, could be possible "bridges" between preclinical (*in vivo*) and clinical studies.

8.3 EXPERIMENTAL ENDPOINTS IN CELL-BASED ASSAYS

There is "cross-talk" between receptors involved in the regulation of a large number of phase I and phase II drug-metabolizing enzymes and drug transporters, and it is impractical and superfluous to assess the induction potential of new drugs for the full complement of target genes. A more efficient and effective initial strategy is the use of a single and sensitive target gene for each of the respective nuclear receptors. Most of these principles are reflected in the latest FDA guidance documents and commentaries on drug–drug interactions, which describe for the first time considerations and recommendations for proper study design of *in vitro* induction experiments and interpretation of *in vitro* results (FDA, 2006; Huang et al., 2008). Overall, the current *in vitro* approaches, which are based on our recent knowledge of the mechanisms of enzyme induction, are designed to help determine whether new drugs are likely to be involved in drug–drug interactions at therapeutic doses *in vivo*.

8.3.1 Study Design Considerations

Appropriate design of an induction study employing primary human hepatocyte cultures is essential for the effective evaluation of the induction potential of new drugs. This includes the use of healthy hepatocytes in the most liver-like culture environment, appropriate positive control inducers, relevant concentrations of control inducers that maximize inductive response while minimizing potential cytotoxicity, effective vehicle controls, appropriate time in culture, and the right complement of endpoints to allow effective interpretation of data. The following section describes the most important considerations in the design of an induction study in cultures of human hepatocytes and those which are essential for a relevant interpretation of data and prediction of *in vivo* interactions.

8.3.2 Cell Health and Monolayer Integrity

The most important consideration for achieving optimal outcomes from an *in vitro* induction study is the health and integrity of the hepatocyte cultures. Primary cultures of human hepatocytes are known as the gold standard for assessing the induction potential of new drugs because they are thought to be the most liver-like *in vitro* model capable of sustaining normal levels of metabolism, induction, and transport. However, the importance of maintaining high-quality monolayers for optimal results often goes unappreciated. For example, if large induction responses are observed (e.g., 400-fold CYP3A induction) with positive control inducers, then the common misconception is that these data are "better" for the prediction of clinically relevant induction potential for new chemical entities. Realistically, very large fold changes (>20- to 30-fold) in CYP3A activity are not physiologically relevant in the human liver, and they more likely reflect a problem with the model, such as poor cell health or inappropriate culture conditions. Often these conditions lead to artificially diminished basal CYP3A expression as well as a reduction in many of the other liver-specific endogenous processes (e.g., metabolism, transport, cell–cell signaling) that make a mature hepatocyte culture system such an effective *in vitro* model for predicting responses in human liver (Hamilton et al., 2001).

Additionally, a compromised or stressed culture system produces exaggerated concentration–response profiles and cytotoxicity due to the relatively poor health of these culture models. Alternatively, in the event that a metabolite(s) of a new drug is the relevant inducer or toxicant *in vivo*, the proper responses would not likely be effectively modeled in a compromised primary hepatocyte culture system. Figure 8.6 shows an example of a healthy sandwich culture of human hepatocytes alongside a stressed culture. A simple morphological assessment can help make a decision regarding the suitability of a cell culture monolayer for an induction study. Good cell–cell contacts (e.g., "cobblestone" appearance of monolayers), cytosolic clarity (not granular in appearance), and clearly defined nuclei (single or multiple often observed in health cultures of hepatocytes) are useful markers of cell health (Fig. 8.6B). Recording images of the cell monolayers at each time point is useful for evaluating any trends in cell health that may be related to the effects of new chemical entities at higher concentrations relative to vehicle control cultures. This is especially important in primary cultures relative to immortalized clonal cell lines. Taken together, these considerations point to the need for researchers to evaluate and use robust primary human hepatocyte culture models with the most liver-like features and functions to ensure that the results derived from these model systems produce the most physiologically relevant data for prediction of human induction and hepatotoxicity.

8.3.3 Culture Conditions

In addition to cell health, other considerations, such as the choice of vehicle control, positive control inducers, culture medium (discussed above), and time of incubation, are also important for optimal results. The choice of vehicle control is often dictated by the solubility of the new chemical entity being assessed for induction potential. Common solvents include dimethylsulfoxide (DMSO), methanol, and acetonitrile for most drugs, or simple isotonic buffers (e.g., PBS, HBSS) and cell culture medium for water-soluble compounds. Care should be taken in choosing the concentration of solvent vehicles because they can contribute to an elevated baseline P450 expression and reduce the dynamic range of induction (LeCluyse et al., 2000; Hamilton et al., 2001). Commonly used DMSO concentrations range from 0.1% to 0.2% because they are high enough to maintain chemical solubility in aqueous media yet have minimal impact on basal P450 expression/activity. Methanol is similar to DMSO, however slightly higher concentrations may be an effective solvent depending on the choice of culture medium. Acetonitrile is a less common vehicle for primary hepatocyte-based induction studies; however, 0.1% has minimal impact on basal P450 expression in our laboratory. When initially setting up these types of experiments, it is important to know the effects of the solvent on cell function and health prior to designing the study to ensure that the proper controls are included (or design-appropriate controls within the experiment such as a media-only control).

Table 8.3 provides a list of recommended positive control inducers based on our practical experience and understanding of various nuclear receptor pathways and their role in P450 transcription. The 2006 draft FDA guidance on drug interactions also provides a list of positive control inducers for various P450 targets. There are quite a few similarities and a few notable differences between our recommended list (Table 8.3) and the 2006 Draft Guidance. However, there are a few differences between our recommended list and the 2006 draft guidance. For example, unlike the draft FDA guidelines, we do not recommend dexamethasone as a positive control inducer for CYP2A6 in cultures of human hepatocytes. We have used dexamethasone (along with PB and RIF) as a positive control for CYP2A6 in over 100 preparations of primary human hepatocytes and have not observed it to be an effective inducer in our system. In our hands, phenobarbital at 1 mM concentrations was the most effective inducer of CYP2A6 activity and mRNA followed by rifampin at 10 µM. Our serum-free culture media are supplemented with 50 nM dexamethasone to maintain liver-like expression of other GR-regulated receptor pathways (e.g., CAR and PXR). 50 nM dexamethasone is far below the concentration needed to markedly activate PXR (~10 µM); however, a recent report has shown that CYP2A6 can be regulated indirectly by low concentrations of dexamethasone via the glucocorticoid receptor and increased binding of hepatic nuclear factor 4 receptor (HNF4) to the CYP2A6 promoter (Onica et al., 2008). Therefore, it is likely that the dexamethasone recommendation for CYP2A6 is not compatible with concomitant assessment of CYP2B6 and CYP3A4 induction potential due to the requirement of GR "basal" activation to maintain liver-like expression of CAR and PXR. A second notable difference in our positive control recommendations compared to those listed in the

TABLE 8.3. *In Vitro* Positive Control Inducer Recommendations in Cultures of Primary Human Hepatocytes

Enzyme	Recommended Inducer	Concentration in Primary Human Hepatocytes	Acceptable Inducer	Concentration in Cultures of Primary Human Hepatocytes
CYP1A2	3-MC	2 μM	Omeprazole	50 μM
			β-Napthoflavone	50 μM
CYP2A6	PB	1000 μM	RIF	10 μM
CYP2B6	PB	1000 μM	Phenytoin	50 μM
			CITCO	1 μM
			RIF	10 μM
CYP2C8	RIF	10 μM		
	PB	1000 μM		
CYP2C9	RIF	10 μM		
	PB	1000 μM		
CYP2C19	RIF	10 μM		
CYP2D6	No established PC inducer			
CYP2E1	No established PC inducer			
CYP3A4	RIF	10 μM	PB	1000 μM

recent FDA guidance is that the upper concentration is suggested to result in maximal induction (E_{max}) *in vitro* with primary human hepatocytes cultures. These concentrations are often significantly lower than those required to produce maximal induction for many cell lines (Kanebratt and Andersson, 2008b). The concentrations recommended in Table 8.3 reflect the upper asymptote of the dose–response curves while minimizing the potential for cytotoxicity and cellular stress that can arise and confound data interpretation.

8.3.4 Temporal Effects

Time in culture is also an important consideration for obtaining optimal induction results but also for monolayer and cell integrity. Understanding the time course of diminishing cell integrity can significantly impact the choice of time points to assess induction potential. However, there are general recommendations based on the time required for multistep biological processes (e.g., induction) to manifest. For instance, during the post-plating period (~18–24 h), hepatocytes are refractory to most inducing agents while the cells are attaching to the substratum and cell–cell contacts are being restored. In the case of temporal kinetics of the induction response, DNA is transcribed to mRNA, which is then translated to protein that must be complexed with a heme porphyrin group and inserted into the membrane of the endoplasmic reticulum prior to exhibiting functional enzymatic activity. Each of these steps has multiple components that require both concentration and time to accumulate precursor substrates at sufficient quantities to affect the next downstream process. As a result, short time points (e.g., 1–6 h) are typically too early to observe substantial induction of enzymatic activity. In addition, inductive events that require metabolism of a parent compound to produce active metabolites will require additional time to accumulate significant concentrations of metabolite to illicit the start of the multicomponent induction process. As a result, the general rule for induction assessment is to culture cells for 48–72 h when monitoring enzymatic activity. Longer (>72 h) time points are effective as long as the primary hepatocyte culture system is still of appropriate quality. For some applications (e.g., cytotoxicity assessment), longer-term exposure may be more reflective of a chronic exposure phenomena (e.g., glutathione depletion, proliferation signaling), which may only be important for chronic treatment regimens.

8.3.5 Endpoint Analysis and Data Interpretation

As implied in Fig. 8.1, induction of CYP enzyme expression can be assessed at several different levels, namely mRNA, protein, and enzymatic activity. The choice of endpoint(s) is important because it will affect the initial study design; however, it is often difficult to know what endpoints to choose before knowing the results. Based on the 2006 FDA Draft Guidance for Industry relating to drug–drug interactions, the minimum requirement for assessment of induction potential is CYP1A and CYP3A enzymatic activity in three independent preparations of hepatocytes (or appropriate cell lines if they are suitable to monitor induction of enzymatic activity). However, simply choosing the minimum activity endpoints may be problematic for interpreting the data from enzymatic activity alone, including underestimating or overestimating induction potential. Failure to effectively interpret data from an induction study (e.g., decreases in enzymatic activity with increasing concentrations of an inducer) can lead to indecision regarding the best development strategy and having to repeat the *in vitro* studies, which can cost precious time and resources. In the following section we demonstrate the utility of an effective study design by presenting example induction data using a variety of endpoints and describing ways to interpret the results. We also show how an effective study design anticipates common outcomes based on a mechanistic understanding of induction pathways and provides the prerequisite data to identify compounds that have the potential to be involved in significant inductive effects. In addition, we explain how studies can be designed to de-convolute simultaneous, often opposing, processes that can ultimately determine the ensuing enzymatic activity.

Figure 8.8 contains representative data for CYP1A2 and CYP3A mRNA and enzymatic activity (panels A–D) from primary hepatocyte cultures treated with positive control compounds or three concentrations of a test article (TA).

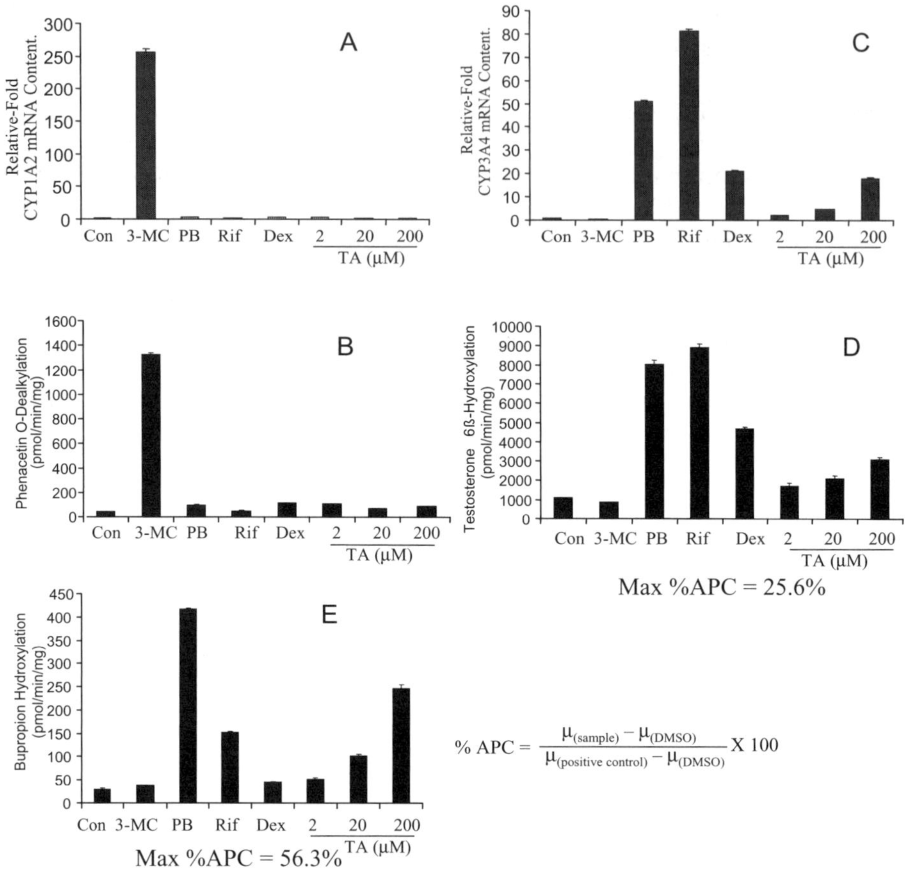

$$\%\ APC = \frac{\mu_{(sample)} - \mu_{(DMSO)}}{\mu_{(positive\ control)} - \mu_{(DMSO)}} X\ 100$$

Figure 8.8. Effect of inducers of CYP mRNA and activity. Examples of no induction (CYP1A2) and small induction (CYP3A4) responses with test article (TA, 2, 20, and 200 μM) relative to positive control responses (3-MC (2 μM), PB (1 mM), RIF (10 μM), and dexamethasone (50 μM)) in cultures of primary human hepatocytes. **(A, C)** CYP1A2 mRNA (TaqMan®) data and enzymatic activity, respectively. **(B, D)** CYP3A4 mRNA (TaqMan®) content and CYP3A enzymatic activity, respectively. **(E)** CYP2B6 enzymatic activity data that demonstrate clear induction with the TA that was not significant in (or predictable from) the CYP3A enzymatic activity data. Equation represents calculation used to determine induction potential. Error bars indicate standard errors of triplicate samples. APC: Adjusted positive control.

This test article was observed to be a noninducer of CYP1A2 activity (phenacetin *O*-deethylation) (Fig. 8.8C) and a weak inducer of CYP3A activity (testosterone 6ß-hydroxylase) (Fig. 8.8D). From these data, one might predict that the potential for this test article to cause clinically relevant induction is minimal. This prediction is supported in the matching mRNA data where CYP1A2 and CYP3A4 do not appear to be significantly induced relative to the corresponding positive controls (i.e., 3-MC and rifampin, respectively) (Fig. 8.8A,B), although CYP3A4 mRNA levels were induced more prominently at the highest concentration of test article than the corresponding CYP3A enzymatic activity. In total, these data suggest low potential for induction with this test article from these data, especially based on the current FDA criterion that an induction response is not considered significant unless it exceeds 40% of the positive control response (FDA, 2006).

However, as stated previously, there are three principal nuclear receptor pathways known to be involved in clinically relevant drug interactions (i.e., AhR, PXR, and CAR); and CYP1A2 and CYP3A do not effectively address all of these, especially CAR-mediated induction (Huang et al., 2008). Figure 8.8E shows the corresponding CYP2B6 (a sensitive CAR target gene) enzymatic activities in hepatocyte cultures treated with this test article. Note the relative percent of adjusted positive control responses for each enzymatic activity at the 200 μM concentration of test article (25% versus 56% for CYP3A and CYP2B6, respectively). These data demonstrate that the CYP3A activity data did not accurately predict the true potential of the compound to induce CYP enzymatic activity. Not only did the initial CYP1A and CYP3A induction results underestimate the CYP2B6 induction response *in vitro*, it is likely that it also has underestimated the induction of CYP2C8 and CYP2C9 as well as other CAR target genes such as UGT1A1 (Smith et al., 2005; Sahi et al., 2009).

Overall, this example illustrates the importance of assessing CYP2B6 induction as a marker for the CAR pathway that continues to emerge as an important mediator of hepatic CYP induction in humans by several drugs (Faucette et al., 2007). Notably, a recent commentary by the FDA acknowledges that assessing CYP1A and CYP3A activities alone may not be sufficient to eliminate completely the possibility of a new drug to induce CYP enzyme activity *in vivo* (FDA, 2006; Huang et al., 2008). The document now recommends the addition of CYP2B6 activity to the screening regimen to account for activators of CAR.

In the example above, monitoring mRNA was a useful, confirmatory tool, but did not provide new insights (relative to the enzymatic activity data) as only induction appeared to be operative with the test article. However, Fig. 8.9 shows two test articles evaluated for CYP3A enzymatic activity where test article 1 (TA1) showed a very small increase at all three concentrations (maximal at 20 μM) while test article 2 (TA2) produced a decrease in enzymatic activity with increasing concentrations. In looking at these activity data, TA1 seems to be a weak inducer while TA2 produces a clear decrease that is not readily explained.

At first glance it is not clear from simply looking at the enzymatic activity alone how to interpret the data or what the cause(s) might be for such an effect. In most cases such as these, there are three possible explanations for the concentration-dependent decrease in enzyme activity: (1) cytotoxicity, (2) enzyme inhibition, and (3) gene suppression. From the cell morphology

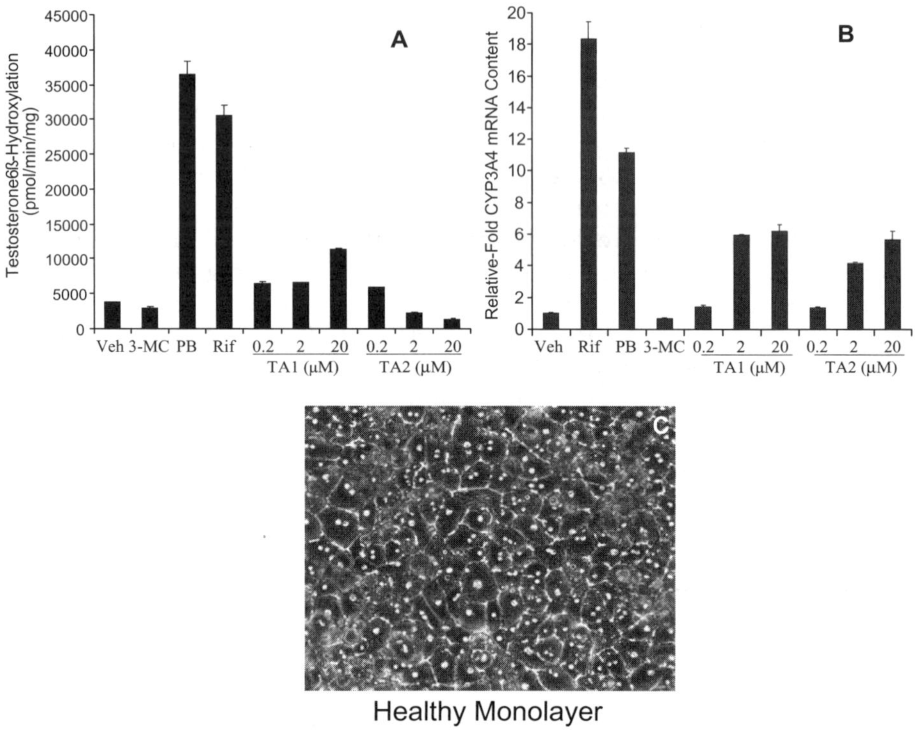

Figure 8.9. CYP3A enzymatic activity data (A), mRNA content (B), and cell morphology (C) from cultures of primary human hepatocytes with two test articles. Data demonstrate an inhibition-like phenotype based on the combination of endpoints indicating that cytotoxicity and suppression were not likely to be operative in these assays. Positive control concentrations were: 3-MC (2 µM), PB (1 mM), and RIF (10 µM) relative to 0.1% DMSO vehicle (Veh) control. Error bars indicate standard errors of triplicate samples. C represents vehicle control hepatocytes; similar morphology was observed in the treated hepatocytes.

assessment (Fig. 8.9C), the integrity of the hepatocyte monolayer does not appear to be compromised by high concentrations of TA2. Assessment of cytotoxicity (e.g., ATP content, mitochondrial function, membrane integrity) can serve as a quantitative measure of cellular stress and identify cultures where cell health has been compromised relative to vehicle control. In the case of TA2, there was no sign of cytotoxicity; therefore this is not likely to be the cause for the decreases in CYP3A enzymatic activity.

When measuring enzymatic activity as the sole endpoint, it is important to understand that changes can arise for multiple reasons that may or may not be related to the amount of enzyme being produced. For example, some mechanism-based inhibitors may go unnoticed in the time frame of routine time-dependent inhibition screening when utilizing a standard microsomal-based *in vitro* assay protocol. In the event that a time-dependent inhibitor is evaluated as an inducer in primary culture of human hepatocytes, it is not unusual to observe a concentration- and time-dependent decrease in enzymatic activity compared to vehicle-control cultures. In fact, primary hepatocyte cultures generally are a more sensitive model system for identifying time-dependent inhibitors due to (1) the longer exposure periods to the test article, (2) the replenishment of test article on a daily basis, and (3) a greater range of concentrations are often tested. Therefore, it is often advantageous to measure an endpoint independent of substrate turnover alone to aid in the interpretation of the results such as those depicted in Fig. 8.9.

A simple, quantitative method for addressing this issue is the inclusion of endpoints for mRNA and/or protein data (e.g., Western immunoblotting, ELISA) in the initial study design. Because mRNA is not sensitive to inhibition phenomena (protein content can sometimes decrease with certain types of mechanism-based inhibitors), yet is decreased with both cytotoxicity and suppression, it is a useful tool to identify compounds that are inhibitors of enzymatic activity in primary cultures of human hepatocytes. In the example illustrated in Fig. 8.9, the corresponding mRNA data demonstrates that TA1 is actually a moderate inducer of CYP3A4 expression (Fig. 8.9B); however, this did not manifest markedly in significant changes in enzymatic activity (Fig. 8.9A). TA1 was later found to be a mechanism-based inhibitor of CYP3A activity in human liver microsomes, which resulted in a blunted induction response relative to changes in the CYP3A4 mRNA content, especially at the 2 μM concentration. At 20 μM there appears to be a balance between induction and inhibition with TA1 under the conditions used. However, other PXR target genes, such as CYP2C enzymes, are likely to be induced by TA1 where inhibition is not operative.

For TA2 in Fig. 8.9, the decrease in enzymatic activity observed was not supported by the mRNA data. In fact, TA2 causes a concentration-dependent increase in CYP3A4 mRNA levels and appears to be a moderate inducer relative to the positive control rifampin. These combined data demonstrate that TA2 is likely an inhibitor of CYP3A enzymatic activity, and based on the net outcome (i.e., predominant loss in enzyme activity), it appears to be a more effective inhibitor than inducer of CYP3A activity. Indeed TA2 was found to be an extremely potent and efficacious mechanism-based inhibitor of CY3A activity in human liver microsomes. Therefore, these data taken together provide valuable clues that demonstrate that TA2 is not cytotoxic at the concentrations examined, but rather is an inhibitor of CYP3A enzymatic activity. In addition, the mRNA results suggest that TA2 has the potential to be involved in the induction of other parallel clearance pathways that are not affected by its inhibitory properties. Without the cell morphology and mRNA data, it would be difficult to prove that the decreases in enzymatic activity were the result of inhibition, even though circumstantially the human liver microsome inhibition data show that TA2 has the ability to inhibit CYP3A in liver microsomes. Therefore, it is important to include an additional endpoint evaluation in the induction study design to avoid inexplicable data when trying

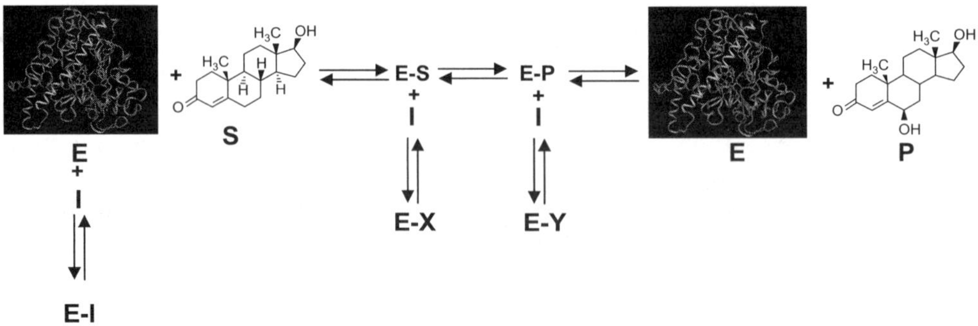

Figure 8.10. Representation of a CYP enzymatic activity reaction. Enzymatic activity is most frequently monitored by the formation of P (product). E is the P450 enzyme (ribbon diagram of enzyme structure), S is the substrate (testosterone), and I represents an inhibitor that can act on E, E-S, or E-P to form inhibitory complexes. The primary point to note is that by only monitoring the production (or depletion) of P, it is not possible to determine if the drug causes decreases in the concentration of P by inhibition of the enzyme or suppressor of the expression of enzyme; or conversely, to determine whether drug is an inducer enzyme expression or activator of catalytic activity for increases in the concentration of P. Enzymatic activity is most frequently monitored by the formation of P (product). E is the P450 enzyme (ribbon diagram of enzyme structure), S is the substrate (testosterone), and I represents an inhibitor that can act on E, E-S, or E-P to form inhibitory complexes. The primary point to note is that by only monitoring the production (or depletion) of P, it is not possible to determine if the drug causes decreases in the concentration of P by inhibition of the enzyme or suppression of enzyme expression; or, conversely, to determine whether drug induces enzyme expression or is an activator of catalytic activity for increases in the concentration of P. See the insert for color representation of this figure.

to use enzymatic activity to rule out the need for human clinical trials. Notably, the current FDA guidance on drug interactions acknowledges the need to evaluate CYP mRNA and enzymatic activities when conducting an *in vitro* induction study of a new drug in the event that time-dependent inhibition is observed (FDA, 2006).

A traditional equation for enzymatic activity is depicted in Fig. 8.10 where the typical endpoint measurement is the formation of product [P]. The increase or decrease in [P] production can arise from multiple factors that may or may not be related to changes in the amount of enzyme [E] being produced. Since induction is defined as an increase in the expression of [E], it is important that an endpoint independent of this biochemical reaction (e.g., mRNA and protein content) be monitored to help deconvolute these data from these dynamic, integrated, whole-cell models.

Figure 8.11 shows another example of a test article that decreases enzymatic activity. Again, based on the enzymatic activity alone, it is not clear why the activity is decreasing. However, the right panel shows the hepatocyte monolayer from these cultures treated with 25 μM TA, and it also shows evidence

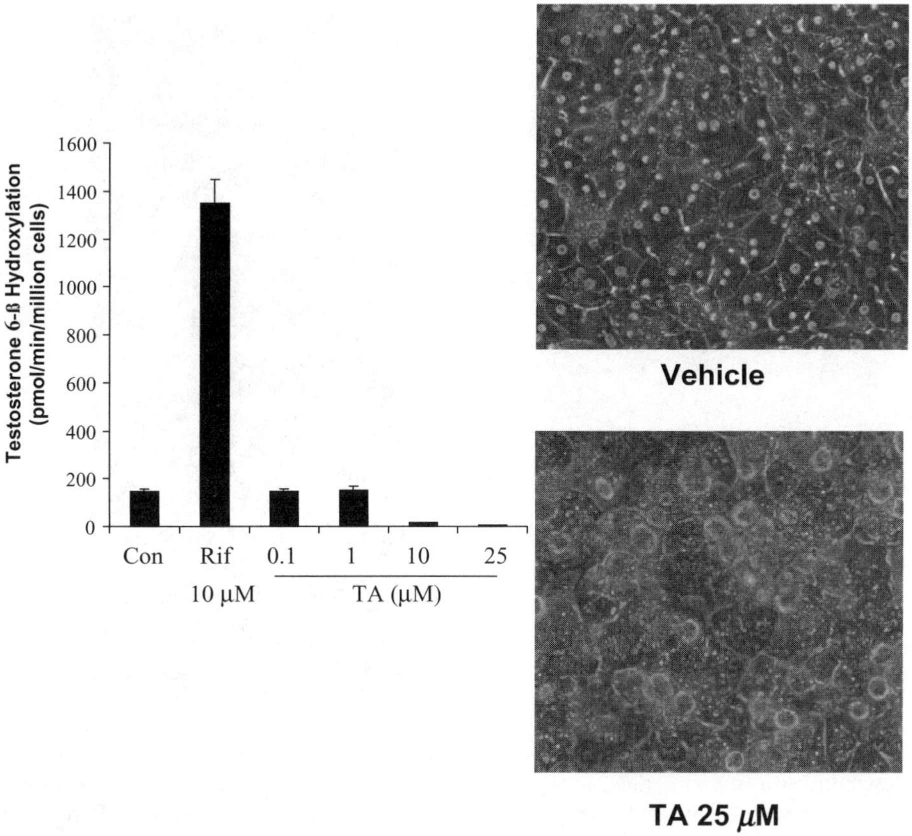

Figure 8.11. Effects of cellular stress on CYP3A enzymatic activity in cultures of primary human hepatocytes. Cell images from vehicle-treated cultures and 25 μM test article (TA) are shown to the right with the resulting 6ßT activity data shown to the left. Error bars indicate standard errors of triplicate samples.

of cell stress and cytotoxicity. It is routinely observed that diminished primary hepatocyte cell health in culture leads to lower cytochrome P450 expression and enzymatic activity. Therefore, assessment of cell health by recording cell morphology images and/or running a cytotoxicity assay at various time points during the treatment period is important for results similar to those observed in Fig. 8.10 and distinguishing this response from inhibition or suppression. It is also important to note that because the effects of cytotoxicity and induction may be simultaneously operative, false-negative or false-positive conclusions regarding the true induction potential of a new drug could be made in the absence of supporting data. As a result, when cytotoxicity is evident, caution should be taken when using these data to make definitive calls regarding the induction potential of an NCE because it is difficult to assess the relative impact of these opposing forces on CYP activity.

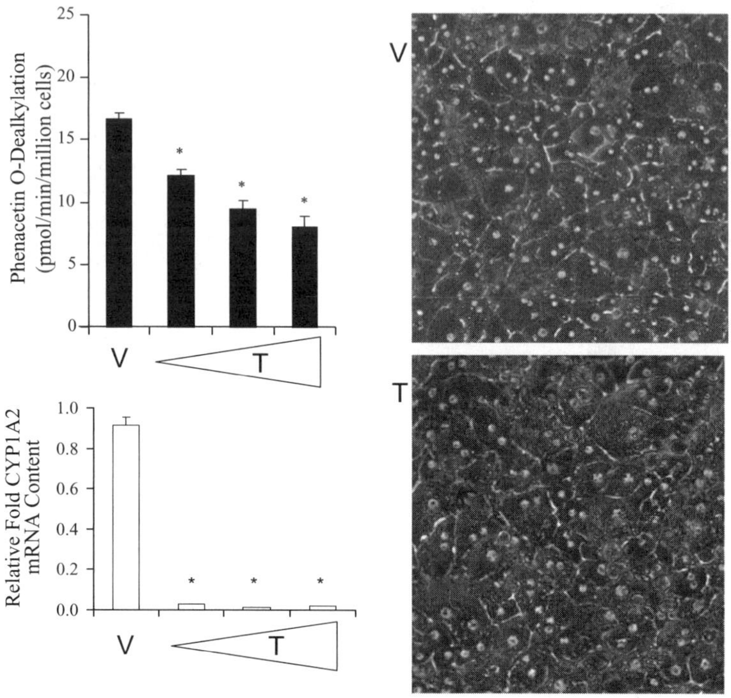

Figure 8.12. Example of suppression in cultures of primary human hepatocytes. Cell morphology data to the right indicate some changes in cell morphology with the highest concentration of test article, but are suggestive of an overt cellular stress phenotype consistent with cytotoxicity when compared with vehicle control cultures. Error bars indicate standard errors of triplicate samples, and an asterisk (*) indicates statistically significant responses (Dunnett's test) at $P < 0.05$). V: vehicle; T: treated. The photomicrograph represents hepatocytes treated with the highest concentration of the compound.

Figure 8.12 shows representative CYP1A2 enzymatic activity and mRNA data for a compound [A]. In this example, the cell health is essentially unperturbed with the highest concentration of compound (Fig. 8.12, image panels), and the membrane integrity was found to be good relative to vehicle control based or lactate dehydrogenase leakage data (data not shown). However, CYP1A2 enzymatic activity clearly decreases with increasing concentration of compound [A]. Initial indications are that compound [A] may be an inhibitor of CYP1A enzymatic activity; however, corresponding mRNA data indicated that CYP1A2 mRNA content is profoundly reduced at all three concentrations of test article. Since the cell morphology and LDH leakage data suggest that this is not cytotoxicity, taken together these data indicate that compound [A] is suppressing CYP1A2 expression and consequentially causing a reduction in CYP1A enzymatic activity.

From this small sample of representative data, we hope to convey the importance of choosing appropriate endpoints to effectively interpret induction data in cultures of primary human hepatocytes. At this point in our understanding of the induction mechanisms, the endpoints for generating the most relevant *in vitro* induction data are CYP1A, CYP2B6, and CYP3A enzymatic activity, as well as an activity-independent endpoint (e.g., mRNA or protein content) to deconvolute inhibition effects from other processes such as suppression and cytotoxicity. We also recommend evaluating effects on the cell monolayers each day prior to dosing to monitor changes in cell integrity and health. In addition, we also recommend a cytotoxicity assay (e.g., ATP, LDH leakage, MTT) to further assess cell health at the conclusion of each study to determine the impact that drug cytotoxicity may have on the experimental outcomes and thus interpretation of the corresponding enzymatic activity data. Therefore, an induction study designed with *in situ* enzymatic activity, mRNA assessment, and cytotoxicity assessment represents a comprehensive study design to completely understand a compounds effect on hepatocyte health, P450 expression, and enzymatic activity. In summary, this section has outlined and briefly discussed the many considerations for designing a comprehensive *in vitro* induction study that provides the type of meaningful data necessary to interpret the results and perform the *in vitro–in vivo* correlations that are described in the next section.

8.4 *IN VITRO–IN VIVO* CORRELATIONS

The methods for calculating CYP induction *in vitro* have evolved over the past 5 years. Initial evaluation involved the calculation of fold increase in enzyme activity relative to the solvent or negative control (fold induction), whereby an induced activity of twofold higher than the vehicle control was considered a positive response. However, the large variation in fold induction between individual CYPs and primary hepatocyte preparations makes it difficult to compare the data from one preparation to the next. In order to take the variation into account, Bjornsson et al. (2003) suggested that a response of greater than 40% of the positive control activity could be used as an alternative, which, in principle, helps eliminate biological variation but may not in fact make the interpretation much clearer (Hewitt et al., 2007). The comparison of the induction response by a drug with that of a positive control (causing a maximal induction response) also reflects the efficacy of the induction response by the drug under investigation. However, the prediction of the induction potential of a compound using *in vitro* models is much more complex than just the calculation of fold increase over control levels or percent induction relative to a positive control, especially if viewed in light of the receptor-mediated pharmacological response that it is. As such, pharmaceutical companies are recognizing the need for better and more predictive strategies to interpret *in vitro* induction data (Smith et al., 2007).

Most inducing agents produce a sigmoidal concentration–response profile that reflects many characteristic aspects related to its potential to be an inducer *in vivo* (Fig. 8.13A). The characteristic features that are reflected in the profile of each inducing agent are: (1) the potency or EC_{50}—effective concentration at half-maximal response; (2) the E_{max} or efficacy—the maximum effect or extent of the induction response; (3) the NOEL—the "no observable effect level" or greatest concentration or amount at which no effect is observed (Fig. 8.13C); and (4) "sigmoidicity"—the relative shape of the concentration–response curve that relates to a number of biochemical and molecular factors, such as cooperativity in receptor binding, ligand metabolism, and biosynthesis of target genes (Kohn and Portier, 1993).

In Fig. 8.13, different induction response curves are compared to illustrate the importance of choosing the most appropriate calculation of the clinical induction risk. Figure 8.13A shows the induction response of drugs A, B, and C, which all have the same maximum response (E_{max} or efficacy) and no-observed-effect level (NOEL) but different EC_{50} values (potencies). In this example, drug A is the most potent, followed by B and then C. This is one method of calculation which is acceptable by the FDA in both the 2004 (FDA, 2004) and 2006 guidance documents; however, the guidance does not specify thresholds for determining the clinical significance of the *in vitro* induction results. In a similar paradigm to that used for categorizing enzyme inhibitors, the likelihood of an *in vivo* interaction can be projected based on the mean steady-state C_{ss} concentration of total drug (bound plus unbound) following administration of the highest clinical dose and, in this case, the *in vitro* potency or EC_{50} of a compound. It should be noted that by using the total C_{ss} concentration, this implies the most conservative prediction of enzyme induction because it often incorporates a higher concentration as compared to using the unbound drug concentration. As with enzyme inhibition, a C_{max}/EC_{50} ratio of greater than 1 would indicate that the compound is likely to cause *in vivo* induction. A C_{max}/EC_{50} ratio of between 0.1 and 1 may indicate a possible induction effect, and a C_{max}/EC_{50} ratio of less than 0.1 indicates that the drug is unlikely to cause CYP induction *in vivo*.

Figure 8.13A illustrates how the C_{max}/EC_{50} ratio can be utilized in a similar fashion as the $[I]/K_i$ ratio to determine the likelihood of a compound to be involved in clinically relevant drug interactions. As the ratio increases, the likelihood of an interaction increases from "remote" to "possible" to "likely." As with inhibitors, an estimated ratio of greater than 0.1 would be considered positive and a follow-up *in vivo* evaluation might be recommended. In the three cases illustrated in Fig. 8.13A, if the C_{max} is less than 1 μM, then the prediction would indicate that none of the drugs are likely to induce this enzyme *in vivo*. However, if the C_{max} is 8 μM, then drug A would be predicted to cause a more significant induction *in vivo* than drugs B and C.

Table 8.1 shows the induction EC_{50} values for CYP3A4 induction for a number of drugs and their *in vivo* C_{max} or steady-state concentrations, and how the C_{max}/EC_{50} ratio relates to the predicted CYP3A4 induction potential. Notable drugs in this table are troglitazone and its successor, rosiglitazone. Troglitazone is known to cause induction *in vivo* (Dimaraki and Jaffe, 2003);

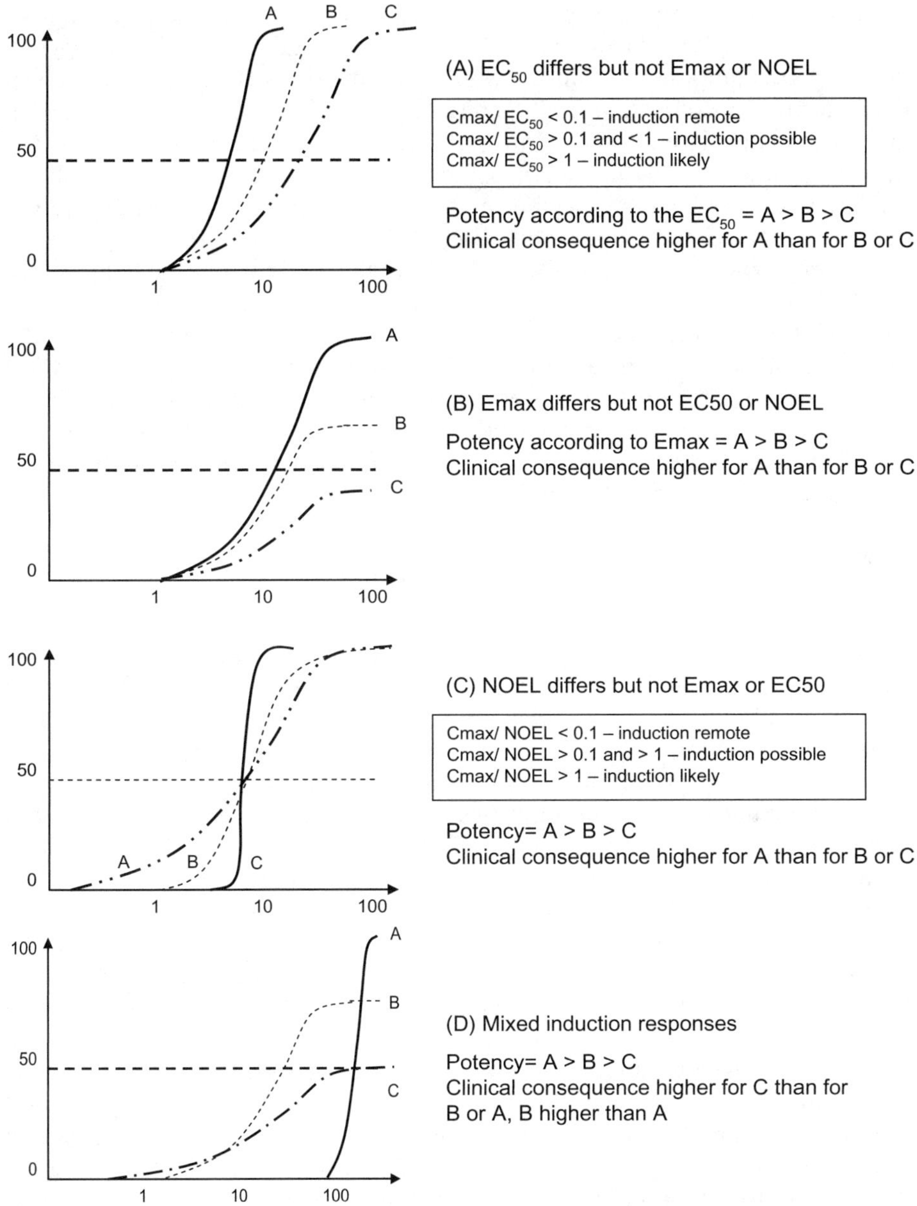

Figure 8.13. Induction response models comparing differences in potency when measured by different parameters. **(A)** EC_{50}, **(B)** E_{max}, **(C)** NOEL, and **(D)** mixed parameters.

however, rosiglitazone has no significant clinical induction effects (Harris et al., 1999) despite it having an EC_{50} similar to troglitazone (which suggests that they are equipotent in causing CYP3A4 induction). Rosiglitazone is given in much smaller doses than troglitazone, consequently, the C_{max} is correspondingly lower and significantly below its EC_{50} value. Unlike the EC_{50} alone, the C_{max}/EC_{50} ratio for troglitazone is sufficiently high to indicate the likelihood for interactions which is reflected in the *in vivo* findings, supporting the use of this calculation for simple estimation of the induction potential.

Although the use of C_{max}/EC_{50} ratio allows for a relatively simple approach to ranking compounds and obtaining an approximation of the likelihood for clinical interactions, it does not take into account all of the complexities of the molecular signaling pathways that mediate an induction response. Inasmuch as induction is a receptor-mediated process, the laws of mass action for drug binding to a receptor are applicable (El-Sankary et al., 2001) and the following equation describes how the net effect of an inducer is affected by the EC_{50}, the E_{max} and drug concentration:

$$\frac{E_{max} \times C_{max}}{EC_{50} + C_{max}} = \text{Effect} \tag{8.1}$$

where E_{max} is the maximum induction response, EC_{50} is the concentration at which the induction response is half the maximum value and C_{max} is the maximum unbound plasma concentration. Using this calculation and a reporter gene HepG2 model, El-Sankary et al. were able to rank compounds in terms of their induction potential *in vivo*. In a more recent report by Ripp et al. (2006) the clinical induction potencies of a number of drugs were compared with those measured in immortalized human hepatocytes. An interesting aspect of this report was the calculation of the induction potency in which the free C_{max}, EC_{50}, and E_{max} were incorporated, using the equation above, to produce a "relative induction score." The authors concluded that the use of an EC_{50} alone was not sufficient to assess the clinical induction risk. Indeed, the correlation of the relative induction score using C_{max}, EC_{50}, and E_{max} with clinical induction data was significantly improved. They found that when the EC_{50} alone is used to determine potency, then mispredictions often occurred. This point is further exemplified in Fig. 8.13B, where all three drugs exhibit the same EC_{50} and NOEL; however, drug A causes a more profound induction than drugs B and C (the E_{max} is over twofold higher than that for drug C). Thus, even though all three drugs caused a maximal induction at 100 μM, drug A would be considered the most potent.

Another factor that should be considered is the lowest concentration at which the induction response occurs (no-observed-effect level, NOEL). In Fig. 8.13C, drug A starts to induce at a very low concentration, whereas drug B and drug C only start to induce at 1 μM and 5 μM, respectively. The significance of the induction response of all three drugs will depend on the C_{max}. Although drug A reached a maximum response at a higher concentration than drug B or C, it has a much lower NOEL. This approach also has the advantage of being more amenable for comparing compounds of limited solubility or those causing cytotoxicity at concentrations that would otherwise be required to complete an entire concentration–response profile. This is similar to the strategy proposed by Kanebratt and Andersson (2008a) using the ratio of the known or predicted drug exposure (*F*) and the concentration at which a drug causes a twofold increase in CYP induction *in vitro* (*F*/2).

Taking the NOEL into account and incorporating it into Eq. (8.1) may give a better estimation of drugs with varying dose response curves:

$$\frac{E_{max} \times C_{max}(\text{fu})}{\text{EC}_{50} + C_{max}(\text{fu})} \times \frac{\text{NOEL}}{C_{max}(\text{fu})} = \text{Induction risk factor} \qquad (8.2)$$

where E_{max} is the maximum induction response, EC_{50} is the concentration at which the induction response is half the maximum value, C_{max} is the maximum unbound plasma concentration and NOEL is the highest concentration at which no induction is evident. In Eq. (8.2), the unbound fraction (fu) of C_{max} is incorporated since only unbound drug is theoretically available to activate the nuclear receptor and cause induction. However, the true effect of protein binding is not known and opinions differ. Some argue that the induction response is unaffected by protein binding because it is a cumulative effect, whereas others have reported a clear correlation between the unbound drug concentration and the induction response (Lin, 2006; Ripp et al., 2006). Another factor that is often overlooked when considering the induction potential of compounds is the difference between plasma and liver drug concentrations, which often can be 10- to 100-fold. In many cases the apparent discrepancies between *in vitro* and predicted *in vivo* induction responses, based on predicted or known plasma levels, can be explained in part by plasma-to-tissue drug concentrations that are >10-fold (McCune et al., 2000).

As suggested before, comparison of induction response curves is more complex than simply determining any one of these parameters alone. Figure 8.13D shows how the efficacy, potency, and NOEL may differ between drugs. Drug A, although having a steep response curve and exhibiting the highest efficacy (E_{max}), would be considered to have a low clinical induction potential, especially if the C_{max} was less than 1 μM. The C_{max}:NOEL ratio would be 100; that is, in order to reach an induction threshold, the plasma concentration would have to be 100 times higher than the normal therapeutic concentration. Drugs B and C have the same EC_{50} values and drug B has a higher E_{max} than drug C, but drug C would be considered to be more of a potential inducer than B because it causes induction at a lower concentration than drug B. As such, approaches that consider all four key parameters (i.e., potency, efficacy, NOEL, and concentration) that relate to the induction potential of a compound in the calculation or interpretation of an induction risk factor gives the best overall predictive power of primary human hepatocyte data.

Whatever method is used to calculate the potential induction risk *in vivo*, once a compound is found to be positive in an *in vitro* hepatocyte induction assay, drug interaction clinical trials are generally warranted, but it is rare that these results would stop the development of a compound. However, attrition sometimes occurs if the compound showed potent induction effects, especially if these were expected and interfered with co-therapies or caused profound autoinduction. Often multiple options are pursued when considering the implications of a positive *in vitro* induction response. Most often additional drug–drug interaction studies are necessary during clinical testing or other pharmacokinetic properties of the compound are assessed to determine the overall induction potential. For example, some of the other properties considered were protein binding, clearance and bioavailability, all of which influence the induction response *in vivo*. Moreover, the induction data are often provided back to the discovery chemists in an attempt to minimize or eliminate the inductive properties of new molecules.

8.5 CONCLUSIONS

There are numerous examples of the effects of CYP inducing drugs which cause a lack of therapeutic efficacy or an increase in toxicity of the drug itself or of coadministered drugs. Although it does not typically cause the demise of a drug, induction is nevertheless an important characteristic to identify and minimize. The CYPs are the main targets of interest for determining the induction potential of new drugs because they metabolize a majority of the drugs currently on the market and marked changes in expression can be observed in the presence of inducing agents. It is also important to acknowledge that the same orphan nuclear receptor pathways regulate enzymes and proteins other than the CYPs and that induction of a CYP may be an indication that other elimination pathways, including phase II enzymes and transporters, also are induced. This is one reason why certain sensitive CYP enzymes are proposed as "sentinel" target genes for each of the key receptor pathways, namely CYP1A2, CYP2B6, and CYP3A4 for AhR, CAR, and PXR, respectively. In this case, one must bear in mind that CYP2B6 is recommended as an endpoint, not because of its overall role in drug metabolism, but rather because of its role as an indicator that other CAR-mediated pathways are induced.

The methods for determining enzyme induction do vary between laboratories, but they mainly adhere to the same principles: primary human hepatocyte culture model; a 2- to 3-day treatment period; and measurement of RNA expression and enzyme activities. Mechanistic studies involving measurement of proteins are encouraged but not generally necessary. Cell lines offer an alternative for drug screening, but the best prediction of *in vivo* induction is universally agreed to be primary human hepatocytes. With the acceptance of cryopreserved hepatocytes as an alternative to fresh cells, access to human hepatocytes for these studies is now much easier. Ultimately, the prediction of the potential of a drug to cause enzyme induction *in vivo* will also depend on the interpretation of the data. Pharmacological-based calculations including the EC_{50}, E_{max}, NOEL, and therapeutic drug concentrations are recommended, especially if these are combined also with other factors such as protein binding. The need for more predictive assessment of enzyme induction *in vitro* still exists; and with more comprehensive studies being conducted to determine the key factors that ultimately determine the *in vivo* outcomes, this goal may soon be achieved.

ACRONYMS AND NOTATION

AhR	Aromatic hydrocarbon receptor
AUC	Area under the plasma concentration–time curve
BSEP	Bile salt export protein
C_{max}	Maxiumum plasma concentration
C_{SS}	Drug concentration at steady state
CAR	Constitutive androstane receptor
CES	Carboxylesterases
CITCO	(6-(4-Chlorophenyl)imidazo[2,1-β][1,3]thiazole-5-carbaldehyde *O*-(3,4-dichlorbenzyl)oxime)
CYP	Cytochrome P450; DDI, Drug–drug interaction
E_{max}	Maximal response (effect)
EC_{50}	Concentration of an agonist that produces 50% of the maximal response
hASC	Human adult stem cells
hESC	Human embryonic stem cells
HNF4α	Hepatocyte nuclear factor 4α
hSC	Human stem cells
IC_{50}	Concentration of drug that produces 50% inhibition
IND	Investigational new drug
LDB	Ligand-binding domain
MDR1	Multidrug resistant protein 1
MRP	Multidrug resistance-associated protein
NDA	New drug application
NCE	New chemical entity
OAT	Organic anion transporter
OATP	Organic anion transporting
OCT	Organic cation transporter
o,p′-DDT	*o,p′*-Dichloro-diphenyl-trichloroethane
PXR	Pregnane X receptor
STs	Sulfo-transferases
SXR	Steroid and xenobiotic receptor
TCDD	2,3,7,8-Tetrachlorodibenzo-*p*-dioxin
UFT	UDP-glucuronosyltransferases

REFERENCES

Agarwal S, Holton KL, Lanza R. Efficient differentiation of functional hepatocytes from human embryonic stem cells. *Stem Cells* 2008;26:1117–1127.

Aninat C, Piton A, Glaise D, Le Charpentier T, Langouet S, Morel F, Guguen-Guillouzo C, Guillouzo A. Expression of cytochromes P450, conjugating enzymes and nuclear receptors in human hepatoma HepaRG cells. *Drug Metab Dispos* 2006;34:75–83.

Arefayene M, Skaar TC, Zhao X, Rae JM, Tanus-Santos JE, Brinkmann U, Brehm I, Salat U, Nguyen A, Desta Z, Flockhart DA. Sequence diversity and functional characterization of the 5′-regulatory region of human CYP2C19. *Pharmacogenetics* 2003;13:199–206.

Auerbach SS, Stoner MA, Su S, Omiecinski CJ. Retinoid X receptor-alpha-dependent transactivation by a naturally occurring structural variant of human constitutive androstane receptor (NR1I3). *Mol Pharmacol* 2005;68:1239–1253.

Azuma H, Paulk N, Ranade A, Dorrell C, Al-Dhalimy M, Ellis E, Strom S, Kay MA, Finegold M, Grompe M. Robust expansion of human hepatocytes in Fah–/–/Rag2–/–/Il2rg–/– mice. *Nat Biotechnol* 2007;25:903–910.

Backlund M, Ingelman-Sundberg M. Regulation of aryl hydrocarbon receptor signal transduction by protein tyrosine kinases. *Cell Signal* 2005;17:39–48.

Beischlag TV, Wang S, Rose DW, Torchia J, Reisz-Porszasz S, Muhammad K, Nelson WE, Probst MR, Rosenfeld MG, Hankinson O. Recruitment of the NCoA/SRC-1/p160 family of transcriptional coactivators by the aryl hydrocarbon receptor/aryl hydrocarbon receptor nuclear translocator complex. *Mol Cell Biol* 2002;22:4319–4333.

Bell AW, Michalopoulos GK. Phenobarbital regulates nuclear expression of HNF-4alpha in mouse and rat hepatocytes independent of CAR and PXR. *Hepatology* 2006;44:186–194.

Bjornsson T, Callaghan J, Einolf H, Fischer V, Gan L, Grimm S, Kao J, King S, Miwa G, Ni L, Kumar G, McLeod J, Obach R, Roberts S, Roe A, Shah A, Snikeris F, Sullivan J, Tweedie D, Vega J, Walsh J, Wrighton S. The conduct of *in vitro* and *in vivo* drug–drug interaction studies: a Pharmaceutical Research and Manufacturers of America (PhRMA) perspective. *Drug Metab Dispos* 2003;31:815–832.

Bonzo J, Chen S, Galijatovic A, Tukey R. Arsenite inhibition of CYP1A1 induction by 2,3,7,8-tetrachlorodibenzo-*p*-dioxin is independent of cell cycle arrest. *Mol Pharmacol* 2005;67:1247–1256.

Cerec V, Glaise D, Garnier D, Morosan S, Turlin B, Drenou B, Gripon P, Kremsdorf D, Guguen-Guillouzo C, Corlu A. Transdifferentiation of hepatocyte-like cells from the human hepatoma HepaRG cell line through bipotent progenitor. *Hepatology* 2007;45:957–967.

Chang TK, Waxman DJ. Synthetic drugs and natural products as modulators of constitutive androstane receptor (CAR) and pregnane X receptor (PXR). *Drug Metab Rev* 2006;38:51–73.

Conney AH. Pharmacological implications of microsomal enzyme induction. *Pharmacol Rev* 1967;19:317–366.

Dickins M. Induction of cytochromes P450. *Curr Top Med Chem* 2004;4:1745–1766.

Dimaraki EV, Jaffe CA. Troglitazone induces CYP3A4 activity leading to falsely abnormal dexamethasone suppression test. *J Clin Endocrinol Metab* 2003;88: 3113–3116.

Ek M, Soderdahl T, Kuppers-Munther B, Edsbagge J, Andersson TB, Bjorquist P, Cotgreave I, Jernstrom B, Ingelman-Sundberg M, Johansson I. Expression of drug metabolizing enzymes in hepatocyte-like cells derived from human embryonic stem cells. *Biochem Pharmacol* 2007;74:496–503.

Ekins S, Chang C, Mani S, Krasowski MD, Reschly EJ, Iyer M, Kholodovych V, Ai N, Welsh WJ, Sinz M, Swaan PW, Patel R, Bachmann K. Human pregnane X receptor antagonists and agonists define molecular requirements for different binding sites. *Mol Pharmacol* 2007;72:592–603.

El-Sankary W, Gibson GG, Ayrton A, Plant N. Use of a reporter gene assay to predict and rank the potency and efficacy of CYP3A4 inducers. *Drug Metab Dispos* 2001;29:1499–1504.

Faucette SR, Zhang TC, Moore R, Sueyoshi T, Omiecinski CJ, LeCluyse EL, Negishi M, Wang H. Relative activation of human pregnane X receptor versus constitutive androstane receptor defines distinct classes of CYP2B6 and CYP3A4 inducers. *J Pharmacol Exp Ther* 2007;320:72–80.

FDA. *Guidance for Industry on Drug Interaction Studies—Study Design, Data Analysis, and Implications for Dosing and Labeling.*

FDA. *Guidance for Industry—Drug Metabolism/Drug Interaction Studies in the Drug Development Process: Studies In Vitro*. Food and Drug Administration Publication, 2006.

Gao YD, Olson SH, Balkovec JM, Zhu Y, Royo I, Yabut J, Evers R, Tan EY, Tang W, Hartley DP, Mosley RT. Attenuating pregnane X receptor (PXR) activation: a molecular modelling approach. *Xenobiotica* 2007;37:124–138.

Gomez-Lechon MJ, Donato T, Jover R, Rodriguez C, Ponsoda X, Glaise D, Castell JV, Guguen-Guillouzo C. Expression and induction of a large set of drug-metabolizing enzymes by the highly differentiated human hepatoma cell line BC2. *Eur J Biochem* 2001;268:1448–1459.

Gonzalez F. Cyp2e1. *Drug Metab Dispos* 2007;35:1–8.

Goodwin B, Hodgson E, Liddle C. The orphan human pregnane X receptor mediates the transcriptional activation of CYP3A4 by rifampicin through a distal enhancer module. *Mol Pharmacol* 1999;56:1329–1339.

Guengerich FP, Shimada T, Iwasaki M, Butler MA, Kadlubar FF. Activation of carcinogens by human liver cytochromes P-450. *Basic Life Sci* 1990;53:381–396.

Gupta A, Mugundu G, Desai PB, Thummel KE, Unadkat JD. Intestinal human colon adenocarcinoma cell line, LS180, is an excellent model to study PXR- but not CAR-mediated CYP3A4 and MDR1 induction: studies with anti-HIV protease inhibitors. *Drug Metab Dispos* 2008;36(6):1172–1180.

Hamilton GA, Jolley SL, Gilbert D, Coon DJ, Barros S, LeCluyse EL. Regulation of cell morphology and cytochrome P450 expression in human hepatocytes by extracellular matrix and cell–cell interactions. *Cell Tissue Res* 2001;306:85–99.

Handschin C, Podvinec M, Meyer UA. *In silico* approaches, and *in vitro* and *in vivo* experiments to predict induction of drug metabolism. *Drug News Perspect* 2003;16:423–434.

Hariparsad N, Carr BA, Evers R, Chu X. Comparison of immortalized Fa2N-4 cells and human hepatocytes as *in vitro* models for cytochrome P450 induction. *Drug Metab Dispos* 2008;36:1046–1055.

Harmsen S, Meijerman I, Beijnen JH, Schellens JH. The role of nuclear receptors in pharmacokinetic drug–drug interactions in oncology. *Cancer Treat Rev* 2007;33: 369–380.

Harris RZ, Inglis AM, Miller AK, Thompson KA, Finnerty D, Patterson S, Jorkasky DK, Freed MI. Rosiglitazone has no clinically significant effect on nifedipine pharmacokinetics. *J Clin Pharmacol* 1999;39:1189–1194.

Hay DC, Zhao D, Ross A, Mandalam R, Lebkowski J, Cui W. Direct differentiation of human embryonic stem cells to hepatocyte-like cells exhibiting functional activities. *Cloning Stem Cells* 2007;9:51–62.

Hebert MF, Roberts JP, Prueksaritanont T, Benet LZ. Bioavailability of cyclosporine with concomitant rifampin administration is markedly less than predicted by hepatic enzyme induction. *Clin Pharmacol Ther* 1992;52:453–457.

Herrera MB, Bruno S, Buttiglieri S, Tetta C, Gatti S, Deregibus MC, Bussolati B, Camussi G. Isolation and characterization of a stem cell population from adult human liver. *Stem Cells* 2006;24:2840–2850.

Hewitt NJ, de Kanter R, LeCluyse E. Induction of drug metabolizing enzymes: a survey of *in vitro* methodologies and interpretations used in the pharmaceutical industry—do they comply with FDA recommendations? *Chem Biol Interact* 2007;168:51–65.

Huang SM, Strong JM, Zhang L, Reynolds KS, Nallani S, Temple R, Abraham S, Habet SA, Baweja RK, Burckart GJ, Chung S, Colangelo P, Frucht D, Green MD, Hepp P, Karnaukhova E, Ko HS, Lee JI, Marroum PJ, Norden JM, Qiu W, Rahman A, Sobel S, Stifano T, Thummel K, Wei XX, Yasuda S, Zheng JH, Zhao H, Lesko LJ. New era in drug interaction evaluation: US Food and Drug Administration update on CYP enzymes, transporters, and the guidance process. *J Clin Pharmacol* 2008;48:662–670.

Josephy P. *Molecular Toxicology*, Josephy P, Mannervik B, editors. New York: Oxford University Press, 2005.

Jover R, Bort R, Gomez-Lechon MJ, Castell JV. Cytochrome P450 regulation by hepatocyte nuclear factor 4 in human hepatocytes: a study using adenovirus-mediated antisense targeting. *Hepatology* 2001;33:668–675.

Kamiyama Y, Matsubara T, Yoshinari K, Nagata K, Kamimura H, Yamazoe Y. Role of human hepatocyte nuclear factor 4alpha in the expression of drug-metabolizing enzymes and transporters in human hepatocytes assessed by use of small interfering RNA. *Drug Metab Pharmacokinet* 2007;22:287–298.

Kanebratt KP, Andersson TB. Evaluation of HepaRG cells as an *in vitro* model for human drug metabolism studies. *Drug Metab Dispos* 2008a;36:1444–1452.

Kanebratt KP, Andersson TB. HepaRG cells as an *in vitro* model for evaluation of cytochrome P450 induction in humans. *Drug Metab Dispos* 2008b;36:137–145.

Katoh M, Watanabe M, Tabata T, Sato Y, Nakajima M, Nishimura M, Naito S, Tateno C, Iwasaki K, Yoshizato K, Yokoi T. *In vivo* induction of human cytochrome P450 3A4 by rifabutin in chimeric mice with humanized liver. *Xenobiotica* 2005;35: 863–875.

Kawana K, Ikuta T, Kobayashi Y, Gotoh O, Takeda K, Kawajiri K. Molecular mechanism of nuclear translocation of an orphan nuclear receptor, SXR. *Mol Pharmacol* 2003;63:524–531.

Kazemnejad S, Allameh A, Seoleimani M, Gharehbaghian A, Mohammadi Y, Amirizadeh N, Esmaeili S. Functional hepatocyte-like cells derived from human bone marrow mesenchymal stem cells on a novel 3-dimensional biocompatible nanofibrous scaffold. *Int J Artif Organs* 2008;31:500–507.

Kim S, Pray D, Zheng M, Morgan D, Pizzano J, Zoeckler M, Chimalakonda A, Sinz M. Quantitative relationship between rifampicin exposure and induction of Cyp3a11 in SXR humanized mice: Extrapolation to human P450 induction potential. *Drug Metab Lett* 2008;2:169–175.

Kodama S, Koike C, Negishi M, Yamamoto Y. Nuclear receptors CAR and PXR cross talk with FOXO1 to regulate genes that encode drug-metabolizing and gluconeogenic enzymes. *Mol Cell Biol* 2004;24:7931–7940.

Kohn M, Portier C. Effects of the mechanism of receptor-mediated gene expression on the shape of the dose-response. *Risk Analysis* 1993;13:565–572.

Koop DR, Tierney DJ. Multiple mechanisms in the regulation of ethanol-inducible cytochrome P450IIE1. *Bioessays* 1990;12:429–435.

Kostrubsky VE, Strom SC, Hanson J, Urda E, Rose K, Burliegh J, Zocharski P, Cai H, Sinclair JF, Sahi J. Evaluation of hepatotoxic potential of drugs by inhibition of bile-acid transport in cultured primary human hepatocytes and intact rats. *Toxicol Sci* 2003;76:220–228.

Kostrubsky VE, Strom SC, Wood SG, Wrighton SA, Sinclair PR, Sinclair JF. Ethanol and isopentanol increase CYP3A and CYP2E in primary cultures of human hepatocytes. *Arch Biochem Biophys* 1995;322:516–520.

Le Vee M, Jigorel E, Glaise D, Gripon P, Guguen-Guillouzo C, Fardel O. Functional expression of sinusoidal and canalicular hepatic drug transporters in the differentiated human hepatoma HepaRG cell line. *Eur J Pharm Sci* 2006;28:109–117.

LeCluyse EL. Human hepatocyte culture systems for the *in vitro* evaluation of cytochrome P450 expression and regulation. *Eur J Pharm Sci* 2001;13:343–368.

LeCluyse EL, Audus KL, Hochman JH. Formation of extensive canalicular networks by rat hepatocytes cultured in collagen-sandwich configuration. *Am J Physiol* 1994;266:C1764–1774.

LeCluyse E, Bullock P, Madan A, Carroll K, Parkinson A. Influence of extracellular matrix overlay and medium formulation on the induction of cytochrome P-450 2B enzymes in primary cultures of rat hepatocytes. *Drug Metab Dispos* 1999;27: 909–915.

LeCluyse E, Madan A, Hamilton G, Carroll K, DeHaan R, Parkinson A. Expression and regulation of cytochrome P450 enzymes in primary cultures of human hepatocytes. *J Biochem Mol Toxicol* 2000;14:177–188.

LeCluyse EL, Alexandre E, Hamilton GA, Viollon-Abadie C, Coon DJ, Jolley S, Richert L. Isolation and culture of primary human hepatocytes. *Methods Mol Biol* 2005;290:207–229.

Lehmann JM, McKee DD, Watson MA, Willson TM, Moore JT, Kliewer SA. The human orphan nuclear receptor PXR is activated by compounds that regulate CYP3A4 gene expression and cause drug interactions. *J Clin Invest* 1998;102: 1016–1023.

Lemaire G, Benod C, Nahoum V, Pillon A, Boussioux AM, Guichou JF, Subra G, Pascussi JM, Bourguet W, Chavanieu A, Balaguer P. Discovery of a highly active ligand of human pregnane x receptor: a case study from pharmacophore modeling and virtual screening to "*in vivo*" biological activity. *Mol Pharmacol* 2007;72:572–581.

Li AP, Lu C, Brent JA, Pham C, Fackett A, Ruegg CE, Silber PM. Cryopreserved human hepatocytes: characterization of drug-metabolizing enzyme activities and applications in higher throughput screening assays for hepatotoxicity, metabolic stability, and drug–drug interaction potential. *Chem Biol Interact* 1999;121:17–35.

Lim YP, Huang JD. Interplay of pregnane X receptor with other nuclear receptors on gene regulation. *Drug Metab Pharmacokinet* 2008;23:14–21.

Lin JH. CYP Induction-Mediated drug interactions: *in vitro* assessment and clinical implications. *Pharm Res* 2006;23:1089–1116.

Liu FJ, Song X, Yang D, Deng R, Yan B. The far and distal enhancers in the CYP3A4 gene co-ordinate the proximal promoter in responding similarly to the pregnane X receptor but differentially to hepatocyte nuclear factor-4alpha. *Biochem J* 2008;409:243–250.

Madan A, Graham RA, Carroll KM, Mudra DR, Burton LA, Krueger LA, Downey AD, Czerwinski M, Forster J, Ribadeneira MD, Gan LS, LeCluyse EL, Zech K, Robertson P, Jr., Koch P, Antonian L, Wagner G, Yu L, Parkinson A. Effects of prototypical microsomal enzyme inducers on cytochrome P450 expression in cultured human hepatocytes. *Drug Metab Dispos* 2003;31:421–431.

McCune JS, Hawke RL, LeCluyse EL, Gillenwater HH, Hamilton G, Ritchie J, Lindley C. *In vivo* and *in vitro* induction of human cytochrome P4503A4 by dexamethasone. *Clin Pharmacol Ther* 2000;68:356–366.

Medina-Diaz IM, Elizondo G. Transcriptional induction of CYP3A4 by *o,p′*-DDT in HepG2 cells. *Toxicol Lett* 2005;157:41–47.

Mills JB, Rose KA, Sadagopan N, Sahi J, de Morais SM. Induction of drug metabolism enzymes and MDR1 using a novel human hepatocyte cell line. *J Pharmacol Exp Ther* 2004;309:303–309.

Moore JT, Kliewer SA. Use of the nuclear receptor PXR to predict drug interactions. *Toxicology* 2000;153:1–10.

Moore LB, Maglich JM, McKee DD, Wisely B, Willson TM, Kliewer SA, Lambert MH, Moore JT. Pregnane X receptor (PXR), constitutive androstane receptor (CAR), and benzoate X receptor (BXR) define three pharmacologically distinct classes of nuclear receptors. *Mol Endocrinol* 2002;16:977–986.

Moore M, Thor H, Moore G, Nelson S, Moldeus P, Orrenius S. The toxicity of acetaminophen and *N*-acetyl-*p*-benzoquinone imine in isolated hepatocytes is associated with thiol depletion and increased cytosolic Ca^{2+}. *J Biol Chem* 1985;260:13035–13040.

Moreau A, Maurel P, Vilarem MJ, Pascussi JM. Constitutive androstane receptor-vitamin D receptor crosstalk: consequence on CYP24 gene expression. *Biochem Biophys Res Commun* 2007;360:76–82.

Nakata K, Tanaka Y, Nakano T, Adachi T, Tanaka H, Kaminuma T, Ishikawa T. Nuclear receptor-mediated transcriptional regulation in Phase I, II, and III xenobiotic metabolizing systems. *Drug Metab Pharmacokinet* 2006;21:437–457.

Nishimura M, Imai T, Morioka Y, Kuribayashi S, Kamataki T, Naito S. Effects of NO-1886 (Ibrolipim), a lipoprotein lipase-promoting agent, on gene induction of cytochrome P450s, carboxylesterases, and sulfotransferases in primary cultures of human hepatocytes. *Drug Metab Pharmacokinet* 2004;19:422–429.

Nishimura M, Yokoi T, Tateno C, Kataoka M, Takahashi E, Horie T, Yoshizato K, Naito S. Induction of human CYP1A2 and CYP3A4 in primary culture of hepatocytes from chimeric mice with humanized liver. *Drug Metab Pharmacokinet* 2005;20:121–126.

Novak RF, Woodcroft KJ. The alcohol-inducible form of cytochrome P450 (CYP 2E1): role in toxicology and regulation of expression. *Arch Pharm Res* 2000;23:267–282.

O'Connor JE, Martinez A, Castell JV, Gomez-Lechon MJ. Multiparametric characterization by flow cytometry of flow-sorted subpopulations of a human hepatoma cell line useful for drug research. *Cytometry A* 2005;63:48–58.

Odom DT, Zizlsperger N, Gordon DB, Bell GW, Rinaldi NJ, Murray HL, Volkert TL, Schreiber J, Rolfe PA, Gifford DK, Fraenkel E, Bell GI, Young RA. Control of pancreas and liver gene expression by HNF transcription factors. *Science* 2004;303:1378–1381.

Onica T, Nichols K, Larin M, Ng L, Maslen A, Dvorak Z, Pascussi JM, Vilarem MJ, Maurel P, Kirby GM. Dexamethasone-mediated up-regulation of human CYP2A6 involves the glucocorticoid receptor and increased binding of hepatic nuclear factor 4 alpha to the proximal promoter. *Mol Pharmacol* 2008;73:451–460.

Pacussi JM, Vilarem MJ. Drug-induced osteomalacia: possible role of PXR, a receptor involved in detoxification. *Med Sci (Paris)* 2005;21:582–583.

Parkinson A. Biotransformation of xenobiotics. In: *Toxicology: The Basic Science of Poisons*. (Klassen C ed), New York: McGraw-Hill, 2001, pp. 133–224.

Qatanani M, Moore DD. CAR, the continuously advancing receptor, in drug metabolism and disease. *Curr Drug Metab* 2005;6:329–339.

Raucy JL, Lasker J, Ozaki K, Zoleta V. Regulation of CYP2E1 by ethanol and palmitic acid and CYP4A11 by clofibrate in primary cultures of human hepatocytes. *Toxicol Sci* 2004;79:233–241.

Reinach B, de Sousa G, Dostert P, Ings R, Gugenheim J, Rahmani R. Comparative effects of rifabutin and rifampicin on cytochromes P450 and UDP-glucuronosyltransferases expression in fresh and cryopreserved human hepatocytes. *Chem Biol Interact* 1999;121:37–48.

Ripp SL, Mills JB, Fahmi OA, Trevena KA, Liras JL, Maurer TS, de Morais SM. Use of immortalized human hepatocytes to predict the magnitude of clinical drug–drug interactions caused by CYP3A4 induction. *Drug Metab Dispos* 2006;34:1742–1748.

Roymans D, Annaert P, Van Houdt J, Weygers A, Noukens J, Sensenhauser C, Silva J, Van Looveren C, Hendrickx J, Mannens G, Meuldermans W. Expression and induction potential of cytochromes p450 in human cryopreserved hepatocytes. *Drug Metab Dispos* 2005;33:1004–1016.

Ruhnke M, Nussler AK, Ungefroren H, Hengstler JG, Kremer B, Hoeckh W, Gottwald T, Heeckt P, Fandrich F. Human monocyte-derived neohepatocytes: a promising alternative to primary human hepatocytes for autologous cell therapy. *Transplantation* 2005;79:1097–1103.

Sahi J, Hamilton G, Sinz M, Barros S, Huang SM, Lesko LJ, LeCluyse EL. Effect of troglitazone on cytochrome P450 enzymes in primary cultures of human and rat hepatocytes. *Xenobiotica* 2000;30:273–284.

Sahi J, Stern RH, Milad MA, Rose KA, Gibson G, Zheng X, Stilgenbauer L, Sadagopan N, Jolley S, Gilbert D, LeCluyse EL. Effects of avasimibe on cytochrome P450 2C9 expression *in vitro* and *in vivo*. *Drug Metab Dispos* 2004;32:1370–1376.

Sahi J, Sinz MW, Campbell S, Mireles R, Zheng X, Rose KA, Raeissi S, Hashim MF, Ye Y, de Morais SM, Black C, Tugnait M, Keller LH. Metabolism and transporter-mediated drug–drug interactions of the endothelin-A receptor antagonist CI-1034. *Chem Biol Interact* 2006;159:156–168.

Sahi J, Shord S, Ferguson S, LeCluyse E. Regulation of cytochrome P450 2C9 expression in primary cultures of human hepatocytes. *J Biochem Mol Toxicol* 2009; 23(1):43–58.

Schehrer L, Regan JD, Westendorf J. UDS induction by an array of standard carcinogens in human and rodent hepatocytes: effect of cryopreservation. *Toxicology* 2000;147:177–191.

Schmiedlin-Ren P, Thummel K, Fisher J, Paine M, Watkins P. Induction of CYP3A4 by 1 alpha,25-dihydroxyvitamin D3 is human cell line-specific and is unlikely to involve pregnane X receptor. *Drug Metab Dispos* 2001;29:1446–1453.

Schwartz RE, Reyes M, Koodie L, Jiang Y, Blackstad M, Lund T, Lenvik T, Johnson S, Hu WS, Verfaillie CM. Multipotent adult progenitor cells from bone marrow differentiate into functional hepatocyte-like cells. *J Clin Invest* 2002;109:1291–1302.

Sinz M, Kim S, Zhu Z, Chen T, Anthony M, Dickinson K, Rodrigues AD. Evaluation of 170 xenobiotics as transactivators of human pregnane X receptor (hPXR) and correlation to known CYP3A4 drug interactions. *Curr Drug Metab* 2006;7:375–388.

Sinz M, Kim S, Ferguson S, LeCluyse E. Evaluating and predicting human cytochrome P450 enzyme induction. In: Nassar A, editor. *Drug Metabolism in Pharmaceuticals: Concepts and Application*. New York: John Wiley & Sons, 2008a.

Sinz M, Wallace G, Sahi J. Current industrial practices in assessing CYP450 enzyme induction: preclinical and clinical. *AAPS J* 2008b;10:391–400.

Smith CM, Faucette SR, Wang H, Lecluyse EL. Modulation of UDP-glucuronosyltransferase 1A1 in primary human hepatocytes by prototypical inducers. *J Biochem Mol Toxicol* 2005;19:96–108.

Smith DA, Dickins M, Fahmi OA, Iwasaki K, Lee C, Obach RS, Padbury G, De Morais SM, Ripp SL, Stevens J, Voorman R, Youdim K. The time to move cytochrome P450 induction into mainstream pharmacology is long overdue. *Drug Metab Dispos* 2007;35:697–698.

Squires EJ, Sueyoshi T, Negishi M. Cytoplasmic localization of pregnane X receptor and ligand-dependent nuclear translocation in mouse liver. *J Biol Chem* 2004;279: 49307–49314.

Stjernschantz E, Vermeulen NP, Oostenbrink C. Computational prediction of drug binding and rationalisation of selectivity towards cytochromes P450. *Expert Opin Drug Metab Toxicol* 2008;4:513–527.

Sueyoshi T, Kawamoto T, Zelko I, Honkakoski P, Negishi M. The repressed nuclear receptor CAR responds to phenobarbital in activating the human CYP2B6 gene. *J Biol Chem* 1999;274:6043–6046.

Swales K, Kakizaki S, Yamamoto Y, Inoue K, Kobayashi K, Negishi M. Novel CAR-mediated mechanism for synergistic activation of two distinct elements within the human cytochrome P450 2B6 gene in HepG2 cells. *J Biol Chem* 2005;280: 3458–3466.

Tateno C, Yoshizane Y, Saito N, Kataoka M, Utoh R, Yamasaki C, Tachibana A, Soeno Y, Asahina K, Hino H, Asahara T, Yokoi T, Furukawa T, Yoshizato K. Near completely humanized liver in mice shows human-type metabolic responses to drugs. *Am J Pathol* 2004;165:901–912.

Terry TL, Gallin WJ. Effects of fetal calf serum and disruption of cadherin function on the formation of bile canaliculi between hepatocytes. *Exp Cell Res* 1994;214: 642–653.

Tirona RG, Kim RB. Nuclear receptors and drug disposition gene regulation. *J Pharm Sci* 2005;94:1169–1186.

Wang H, LeCluyse E. Role of orphan nuclear receptors in the regulation of drug metabolizing enzymes. *Clin Pharmacokinet* 2003;42:1331–1357.

Wang H, Faucette S, Moore R, Sueyoshi T, Negishi M, LeCluyse E. Human constitutive androstane receptor mediates induction of CYP2B6 gene expression by phenytoin. *J Biol Chem* 2004;279:29295–29301.

Wang H, Huang H, Li H, Teotico DG, Sinz M, Baker SD, Staudinger J, Kalpana G, Redinbo MR, Mani S. Activated pregnenolone X-receptor is a target for ketoconazole and its analogs. *Clin Cancer Res* 2007;13:2488–2495.

Watkins RE, Wisely GB, Moore LB, Collins JL, Lambert MH, Williams SP, Willson TM, Kliewer SA, Redinbo MR. The human nuclear xenobiotic receptor PXR: structural determinants of directed promiscuity. *Science* 2001;292:2329–2333.

Wilkening S, Stahl F, Bader A. Comparison of primary human hepatocytes and hepatoma cell line HEPG2 with regard to their biotransformation properties. *Drug Metab Dispos* 2003;31:1035–1042.

Xiao L, Cui X, Madison V, White RE, Cheng KC. Insights from a three-dimensional model into ligand binding to constitutive active receptor. *Drug Metab Dispos* 2002;30:951–956.

Xie W, Barwick JL, Downes M, Blumberg B, Simon CM, Nelson MC, Neuschwander-Tetri BA, Brunt EM, Guzelian PS, Evans RM. Humanized xenobiotic response in mice expressing nuclear receptor SXR. *Nature* 2000;406:435–439.

Yoshitsugu H, Nishimura M, Tateno C, Kataoka M, Takahashi E, Soeno Y, Yoshizato K, Yokoi T, Naito S. Evaluation of human CYP1A2 and CYP3A4 mRNA expression in hepatocytes from chimeric mice with humanized liver. *Drug Metab Pharmacokinet* 2006;21:465–474.

Yueh MF, Kawahara M, Raucy J. Cell-based high-throughput bioassays to assess induction and inhibition of CYP1A enzymes. *Toxicol In Vitro* 2005;19:275–287.

Zhou C, Assem M, Tay J, Watkins P, Blumberg B, Schuetz E, Thummel K. Steroid and xenobiotic receptor and vitamin D receptor crosstalk mediates CYP24 expression and drug-induced osteomalacia. *J Clin Invest* 2006;116:1703–1712.

Zhu Z, Kim S, Chen T, Lin JH, Bell A, Bryson J, Dubaquie Y, Yan N, Yanchunas J, Xie D, Stoffel R, Sinz M, Dickinson K. Correlation of high-throughput pregnane X receptor (PXR) transactivation and binding assays. *J Biomol Screen* 2004;9:533–540.

9

INHIBITION OF DRUG-METABOLIZING ENZYMES IN THE GASTROINTESTINAL TRACT AND ITS INFLUENCE ON THE DRUG–DRUG INTERACTION PREDICTION

ALEKSANDRA GALETIN AND J. BRIAN HOUSTON

9.1 INTRODUCTION

9.1.1 Metabolic Enzymes and Transporters in the Human Small Intestine

Human small intestine expresses a range of phase I and II enzymes (Lown et al., 1994; Madani et al., 1999; Paine et al., 1999; Zhang et al., 1999; Tukey and Strassburg, 2001; Glaeser et al., 2002; Matsumoto et al., 2002; Chen et al., 2003; Wojnowski and Kamdem, 2006; Ritter, 2007; Thummel, 2007): CYP3A accounts for ~80% of total P450s in the gut (range 59–96%), followed by CYP2C9 (~14%). In addition, a number of other P450 enzymes (CYP2C19, CYP2J2, CYP2D6) are expressed but in significantly lower expression levels (<2% of total P450s) (Lin et al., 1999; Paine et al., 2006a). The total amount of CYP3A expressed in the human small intestine (65.7–70.5 nmol) represents approximately 1% of the hepatic estimate (Paine et al., 1997; von Richter et al., 2004; Yang et al., 2004). The zonal expression of the intestinal enzymes differs from the liver and varies along its whole length, with the highest levels of P450 and UGT protein found in the proximal region of the intestine declining distally (Paine et al., 1997; Thummel et al., 1997; Lin et al., 1999; Lapple

Enzyme Inhibition in Drug Discovery and Development, edited by Lu and Li
Copyright © 2010 John Wiley & Sons, Inc.

et al., 2003). This heterogeneous expression of metabolic enzymes is also observed within the small intestinal villi, with the highest levels found in mature enterocyte lining the villus tips (Kolars et al., 1994; Watkins, 1997). A recent study by Cao et al. (2006) indicated a threefold greater expression of UGTs relative to CYP3A4 in the human duodenum, but it is questionable whether this estimate will reflect the UGT:P450 abundance ratio along the whole length of the gut.

The most common method of assessment of intestinal metabolism is based on the comparison of hepatic and intestinal activity expressed per milligram of microsomal protein, resulting in differential catalytic activity in comparison to the liver (Paine et al., 1997; Obach et al., 2001; von Richter et al., 2004; Yang et al., 2004; Galetin and Houston, 2006; Komura and Iwaki, 2007; van de Kerkhof et al., 2007). Great inconsistency across the studies is evident and can be rationalized by differences in enterocyte isolation method (mucosal scraping or enterocyte elution), segment of the intestine used (proximal or the whole length), source of the intestinal tissue (individual or pooled), and interindividual variability in CYP3A4/CYP3A5 expression in both liver and intestine (Kuehl et al., 2001; Lin et al., 2002; Xie et al., 2004). However, comparison of intestinal and hepatic catalytic activities (normalized for the mean population relative abundance of P450 enzymes in specific organs and expressed per pmol P450) results in very good agreement in the metabolic activities between the organs and across the range of different P450s (Yang et al., 2004; Galetin and Houston, 2006).

In addition to metabolic enzymes, the presence of efflux transporters in particular P-glycoprotein (P-gp) on the apical membrane of enterocytes can also modulate intestinal CYP3A first-pass metabolism by decreasing the intracellular concentration of drugs and metabolites via active efflux. This proposed interplay between the two proteins allows recirculation of the drug and in conjunction with the variable abundance of CYP3A4 and P-gp contributes to the interindividual variability in the drug absorption and bioavailability (Lin and Yamazaki, 2003; Kivisto et al., 2004; Wu and Benet, 2005; Benet et al., 2008). A number of studies have shown that the expression of P-gp increases progressively from stomach (Fojo et al., 1987; Mouly and Paine, 2003; Thorn et al., 2005), whereas other reports suggest that the efflux transporters have similar distribution pattern to CYP3A4, with higher expression levels in jejunum in comparison to ileum and colon (Nakamura et al., 2002; Berggren et al., 2007). The abundance of P-gp in the human small intestine has been estimated to be 70–74% relative to CYP3A4 (Paine et al., 2005; Cao et al., 2006). In addition to P-gp, MRP1 and MRP2 transporters are located in the intestine at the basolateral and apical membrane, respectively (Endres et al., 2006) and are primarily involved in the efflux of phase II drug conjugates. In contrast to efflux transporters, uptake transporters and their potential interplay with metabolic enzymes and efflux transporters have received less attention over years. Recent studies have reported the expression of a number of uptake transporters on the apical membrane of enterocytes, including OATP1A2, OATP2B1, OATP1B3 and OATP1B1 (Glaeser et al., 2007); the latter two were previously thought to be liver-specific (Ho and Kim, 2005; Shitara et al., 2006; Niemi, 2007).

9.2 PREDICTION OF METABOLIC DRUG–DRUG INTERACTIONS

CYP3A4 is susceptible to a number of reversible and irreversible metabolic drug–drug interactions (DDI), considering that it is the most abundant human P450 enzyme in both liver and intestine (Galetin et al., 2005, 2006; Obach et al., 2006, 2007; Einolf, 2007; Thummel, 2007; Youdim et al., 2008). In order to progress the quantitative basis of DDI prediction, models have been refined over the years to incorporate various inhibitor concentrations as surrogates for inhibitor concentration at the enzyme active site (Kanamitsu et al., 2000; Houston and Galetin, 2003; Ito et al., 2004; Einolf, 2007) and K_i values obtained under standardized *in vitro* conditions (Galetin et al., 2005; Brown et al., 2006). The contribution of multiple inhibitors (or metabolites) and/or the consequence of multiple inhibition mechanisms has also been addressed (Isoherranen et al., 2004; Rostami-Hodjegan and Tucker, 2004; Galetin et al., 2006; Hinton et al., 2008; Templeton et al., 2008). In contrast to a most commonly used time-averaged value, certain simulation programs (e.g., Simcyp®) can incorporate the time course of the inhibitor concentration (Rostami-Hodjegan and Tucker, 2007), which can be important in the assessment of reversible inhibition DDIs, but less critical for time-dependent inhibition where the reduced state of activity is relatively stable during the dosing interval period of study.

In addition to inhibitor properties, metabolic DDI models have also focused on enzyme properties, e.g., degradation rate constant (Venkatakrishnan and Obach, 2005; Galetin et al., 2006; Yang et al., 2008). Comparable affinities of most of the CYP3A substrates for both CYP3A4 and CYP3A5 (Galetin et al., 2004; Huang et al., 2004), differential expression levels of polymorphic CYP3A5 in the population (Xie et al., 2004; Wojnowski and Kamdem, 2006), and differential potency of both reversible and irreversible inhibitors toward CYP3A5 (Gibbs et al., 1999; McConn et al., 2004; Isoherranen et al., 2008) may represent an additional factor contributing to the interindividual variability in the magnitude of DDI observed. A considerable focus of the DDI prediction model has also been on the characteristics of the victim drugs, namely incorporation of the parallel elimination pathways (Brown et al., 2005; Ito et al., 2005; Galetin et al., 2006; Obach et al., 2006), consideration of *in vitro* CYP3A4 kinetic complexities (Galetin et al., 2005; Houston and Galetin, 2005) and contribution of the intestinal inhibition (Wang et al., 2004; Einolf, 2007; Galetin et al., 2007; Obach et al., 2007). There is an increasing interest in the potential impact of metabolic intestinal interactions, and this has only been recently considered as a part of the prediction strategy (Wang et al., 2004; Galetin et al., 2006; Obach et al., 2006; Einolf, 2007; Galetin, 2007; Obach et al., 2007; Galetin et al., 2008). This chapter will address the ability and limitations of the current approaches to estimate the extent of intestinal DDI, in particularly focusing on the most abundant intestinal P450, CYP3A4.

9.2.1 Drug–Drug Interaction Prediction Models

The most common *in vivo* metric used to assess DDIs is the change in area under the plasma concentration time curve (AUC) of the victim drug following multiple dosing of an inhibitor relative to the control state (Tucker et al., 2001; Bjornsson et al., 2003; Ito et al., 2004; Huang et al., 2007, 2008). DDI prediction models incorporate the contribution of parallel metabolic pathways (defined by fraction of drug metabolized, $fm_{\mathrm{CYP}i}$) and renal clearance of unchanged drug, assuming that the other pathways involved in the metabolism of the substrate are not subject to inhibition (Brown et al., 2005; Galetin et al., 2005; Ito et al., 2005; Chien et al., 2006). Contribution of the intestinal interaction is incorporated into the prediction equation based on a hepatic enzyme interaction as the ratio of the intestinal wall availability in the presence and absence of the inhibitor (F_G' and F_G, respectively). This approach is applicable for both reversible and irreversible inhibition interactions, as shown in Eqs. 9.1 and 9.2, respectively (Rostami-Hodjegan and Tucker, 2004; Wang et al., 2004; Galetin et al., 2006; Obach et al., 2006).

$$\frac{\mathrm{AUC}'}{\mathrm{AUC}}=\frac{F_G'}{F_G}\cdot\frac{1}{\sum_i^n \dfrac{f_{m\mathrm{CYP}i}}{1+\sum_j^m [\mathrm{I}]_j/K_{\mathrm{i},j}}+\left(1-\sum_i^n f_{m\mathrm{CYP}i}\right)} \tag{9.1}$$

where I_j is the estimated unbound inhibitor concentration (either the average systemic plasma concentration after repeated oral administration ($[\mathrm{I}]_{\mathrm{av}}$), or the maximum hepatic input concentration ($[\mathrm{I}]_{\mathrm{in}}$) (Kanamitsu et al., 2000; Ito et al., 2004), $K_{i,j}$, is the particular inhibition constant, $f_{m\mathrm{CYP}i}$ represents the fraction of a substrate drug metabolized by the inhibited pathway via a particular P450 enzyme, $(1-\Sigma f_{m\mathrm{CYP}i})$ represents clearance via other P450 enzymes and/or renal clearance; the terms *i* and *j* indicate the potential to incorporate existence of multiple enzymes and inhibitors, respectively.

The analogous relationship exists for irreversible inactivation, where the model incorporates two parameters to describe *in vitro* inactivation and *in vivo* enzyme degradation rate constant:

$$\frac{\mathrm{AUC}'}{\mathrm{AUC}}=\frac{F_G'}{F_G}\cdot\frac{1}{\sum_i^n \dfrac{f_{m\mathrm{CYP}i}}{\left(1+\sum_j^m \dfrac{k_{\mathrm{inact},j}\times[\mathrm{I}]_j}{k_{\mathrm{deg}}\times\left(K_{\mathrm{I},j}+[\mathrm{I}]_j\right)}\right)}+\left(1-\sum_i^n f_{m\mathrm{CYP}i}\right)} \tag{9.2}$$

where $k_{\mathrm{inact},j}$ is the maximal inactivation rate constant, $K_{I,j}$ is the inhibitor concentration at 50% of $k_{\mathrm{inact},j}$, and k_{deg} is the endogenous degradation rate constant of the enzyme (Wang et al., 2004; Galetin et al., 2006). In addition to inhibition by multiple inhibitors, the model can easily be extended to accommodate the inhibition via different inhibition mechanisms (reversible and time-dependent) (Rostami-Hodjegan and Tucker, 2004; Hinton et al., 2008).

Estimation of $\mathbf{F_G'/F_G}$ ***Ratio.*** The F_G'/F_G ratio can be estimated in three different ways (Galetin et al., 2007), as outlined below; in each case the F_G control values can be obtained from either *in vivo* or *in vitro* data (discussed in detail in Section 9.3).

1. *In vivo* F_G'/F_G ratio—obtained from intravenous (i.v.) and oral data in the presence of an inhibitor (limited availability of such datasets). The inhibitor is assumed not to affect the fraction absorbed (F_a), or the plasma binding of a victim drug and not to alter the hepatic blood flow.
2. Maximal F_G'/F_G ratio—assuming the "worst case" scenario (i.e., maximal inhibition of intestinal CYP3A4 resulting in $F_G'=1$) and therefore the maximal ratio as $1/F_G$.
3. Model predicted F_G'/F_G ratio—from *in vitro* inhibition data using a simple intestinal model. Model predicted F_G ratio is obtained from the decrease in the intestinal intrinsic clearance in the presence of an inhibitor ($CL'_{int,g}$), using the *in vitro*-obtained K_i and the estimated inhibitor concentration in the intestinal wall during absorption phase (I_G) [Eqs. (9.3)–(9.5)] (Rostami-Hodjegan and Tucker, 2004; Obach et al., 2006).

$$\frac{F_G'}{F_G}=\frac{1}{F_G+(1-F_G)\cdot\left(\frac{CL'_{int,g}}{CL_{int,g}}\right)} \tag{9.3}$$

$$\frac{CL'_{int,g}}{CL_{int,g}}=\frac{1}{1+\frac{I_G}{K_i}} \tag{9.4}$$

$$I_G=\frac{Dk_aF_a}{Q_{ent}} \tag{9.5}$$

where $CL'_{int,g}$ and $CL_{int,g}$ are intestinal intrinsic clearance in the presence and absence of the inhibitor, respectively; F_a is the fraction absorbed, k_a is the absorption rate constant, D is dose, and Q_{ent} is the enterocytic blood flow. I_G is estimated from Eq. (9.5) when substrate and inhibitor are coadministered under the assumption that the inhibitor is not subject to extensive first-pass metabolism itself.

Decrease in the intestinal intrinsic clearance of the victim drug in the presence of a time-dependent inhibitor is obtained from corresponding *in vitro*

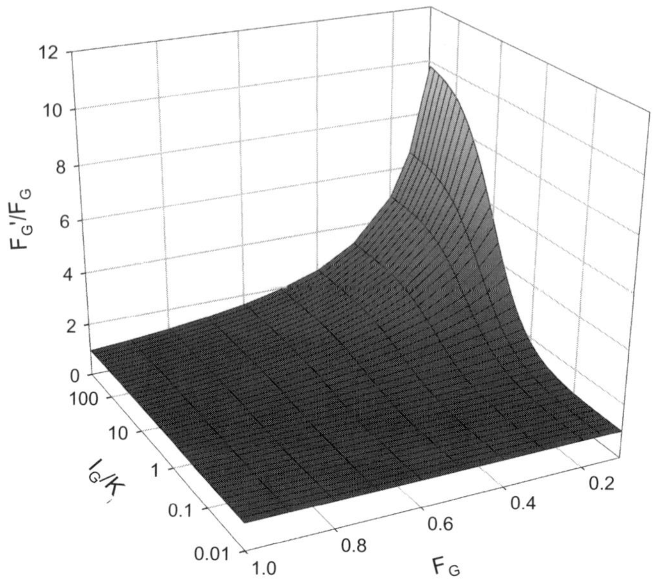

Figure 9.1. Relationship between F'_G/F_G and inhibitor potency (defined by I_G/K_i) for varying degrees of intestinal extraction (F_G). Surface simulated from the relative change in the intestinal intrinsic clearance caused by the reversible inhibitor over an $I_G/K_i = 0.01$–100 and $F_G = 0.1$–1 [Eqs. (9.3) and (9.4)]. The same relationship is applicable for inactivation index ($k_{inact} \cdot I_G/k_{deg} \cdot (K_I + I_G)$) and F'_G/F_G in the case of time-dependent inhibitor [Eqs. (9.3) and (9.6)] (Galetin et al., 2008).

and enzyme parameters (k_{inact}, k_{deg} and K_I) and I_G, as illustrated in Eq. (9.6) (Wang et al., 2004; Obach et al., 2007).

$$\frac{CL'_{int,g}}{CL_{int,g}} = \frac{1}{1 + \dfrac{k_{inact} I_G}{k_{deg}(K_I + I_G)}} \tag{9.6}$$

The potential for a significant interaction in the intestine is determined by the ratio of intestinal concentration of the inhibitor to its estimated potency, that is, I_G/K_i. A relative ratio of >10 in the case of reversible inhibition indicates a risk for interaction in the intestine. Relationship between F'_G/F_G and inhibitor potency (I_G/K_i) and victim drug properties (F_G) is illustrated in Fig. 9.1. Similarly, in the case of time-dependent inhibitors, tight binders ($K_I < 1\,\mu M$) or fast inactivating compounds ($k_{inact} > 0.01\,min^{-1}$) have a potential for interaction in the intestine when $I_G > 1\,\mu M$ (Galetin et al., 2007). Physiological variability in enterocytic blood flow (0.1–0.5 L/min, 2–10% cardiac output) results in a range of I_G after a standard dose (e.g., 8–42 μM ketoconazole after 200 mg dose) (Galetin et al., 2007), contributing potentially to the interindividual differences in the magnitude of inhibition effect in conjunction with other factors (inhibitor dose, intake of food, CYP3A4/3A5 intestinal abundance, and activity). Estimated inhibitor concentrations during the absorption phase for 36 reversible and time-dependent inhibitors in the recent analysis (Galetin et al., 2007) span over three orders of magnitude, with the concentrations of digoxin and azithromycin among the lowest (<0.1 μM) and fluconazole and gemfibrozil among the highest (>100 μM). In addition to inhibitor concentration, changes in splanchic blood flow may also affect the F_G of the victim drug itself. Decreased contact with intestinal CYP3A observed in the case of cirrhotic patients with transjugular intrahepatic portosystemic shunts (Chalasani et al., 2001) results in the faster removal from the vicinity of the enzyme. This can provide the rationale for decreased intestinal extraction and greater F_G observed for midazolam in this group of patients (Rostami-Hodjegan and Tucker, 2002).

9.2.2 Limitations and Accuracy of Different Methods to Estimate the Extent of Intestinal Interaction

In order to improve the predictability of the intestinal inhibition interactions, a comprehensive analysis of various approaches to assess the F_G ratio has been performed recently. An extensive database of 93 reversible and time-dependent inhibition CYP3A4 DDIs (Galetin et al., 2007) was used in this assessment. Evaluation was performed across the range of *in vivo* effects, including a representative number of weak, medium, and potent DDIs (Bjornsson et al., 2003; Huang et al., 2007) in the dataset. The database involved 11 victim drugs with differential importance of intestinal first-pass metabolism; more than half of the interaction studies (27 out of 50) involved victim drugs with an intestinal extraction exceeding 50% (e.g., midazolam, tacrolimus, felodipine, buspirone, atorvastatin, and cyclosporine) and were classified as moderate or potent (increase in AUC of >2 or >5-fold, respectively).

Galetin et al. (2007) have evaluated the advantages and limitations of the maximal F_G ratio ($F_G' = 1$), as a pragmatic indicator of the extent of intestinal inhibition in comparison to the *in vivo* (where available) and model predicted F_G ratios from *in vitro* data. The analysis has indicated a very good agreement between maximal and observed F_G ratios *in vivo* for interactions involving potent inhibitors (regardless of the inhibition mechanism), and substrates predominantly metabolized with no transporter issues (e.g., midazolam, nifedipine). However, the assumption of complete intestinal inhibition ($1/F_G$) overpredicted the F_G ratio observed *in vivo* for dual P-gp-CYP3A4 victim drugs cyclosporine and tacrolimus by 32–223%, respectively; the lack of availability of i.v. data in the presence of an inhibitor prevented a more extensive comparison. In addition, comparison of the maximal F_G and predicted F_G ratios [Eq. 9.3] for 36 inhibitors showed no significant difference between the two approaches for 91% of the studies. This excellent agreement between the two methods supports the application of the $1/F_G$ approach as a pragmatic way to estimate intestinal inhibition interactions; however, this assumption may result in an overprediction of the importance of intestinal inhibition in the interactions with moderate to weak inhibitors (increase in AUC of a victim drug by <2-fold). In addition, the analysis highlighted the sensitivity of the DDI prediction models to the accuracy of the estimated F_G control value, in particular for drugs with a high intestinal first-pass extraction (>75%), due to either extensive intestinal metabolism (e.g., buspirone) or proposed interplay with efflux transporter (e.g., tacrolimus) (Galetin et al., 2007). Different approaches to assess the intestinal extraction from both *in vivo* and *in vitro* data will be discussed in the following section.

9.3 *IN VIVO* AND *IN VITRO* METHODS FOR ESTIMATION OF F_G

Due to ethical reasons, direct measurement of F_G is rarely performed in humans, with the exception of studies in anhepatic patients during liver transplant operations or in patients where the portal blood circulation bypasses the liver (Paine et al., 1996; Thummel et al., 1996). Two approaches to estimate F_G from *in vivo* data, namely, use of i.v. and oral data and interaction studies with grapefruit juice (GFJ), are discussed in more detail in this section.

9.3.1 Assessment of F_G from i.v. and Oral Clinical Data

The estimation of F_G from plasma concentration–time profiles after oral and i.v. administration is based on the assumption of negligible metabolism in enterocytes after i.v. administration and that systemic clearance of a drug after i.v. dose reflects only hepatic elimination. This allows assessment of the intestinal contribution after oral administration by rearranging the following equations [Eqs. (9.7)–(9.9)] (Hall et al., 1999; Galetin et al., 2007).

$$F = F_a F_G F_H \tag{9.7}$$

$$F = F_a(1 - E_G)(1 - E_H) \tag{9.8}$$

$$E_H = \mathrm{CL}_{H,b}/Q_H \tag{9.9}$$

where F represents the overall oral bioavailability, F_a is the fraction of the oral dose absorbed intact across the apical membrane of the epithelial layer, F_G and F_H are fraction of the drug that escapes intestinal and hepatic elimination, respectively, E_G and E_H are the intestinal and hepatic extraction ratios, respectively, $\mathrm{CL}_{H,b}$ is the hepatic blood clearance (determined from net clearance after i.v. administration after correction for renal excretion), and Q_H is the average hepatic blood flow. If plasma clearances are reported, they need to be corrected by the blood:plasma ratio in order to obtain $\mathrm{CL}_{H,b}$ (Yang et al., 2007a). F_aF_G is corrected for F_a when data are available; otherwise complete absorption is assumed ($F_a = 1$). The extent of intestinal extraction of 21 drugs estimated by this method is shown in Table 9.1, ranging from 6% to 86% in

TABLE 9.1. F, F_G and $f_{mCYP3A4}$ Values for 21 CYP3A4 Substrates[a]

CYP3A Substrate	F	F_G	$f_{mCYP3A4}$	References
Tacrolimus	0.14	0.14	0.99	Floren et al. (1997)
Buspirone	0.05	0.21	0.99	Obach et al. (2006)
Atorvastatin	0.14	0.24	0.77	Lennernas (2003)
Cyclosporine	0.22–0.36	0.29–0.68	0.71	Gupta et al. (1989), Hebert et al. (1992), Ducharme et al. (1995), Ku et al. (1998), Lee et al. (2001)
Felodipine	0.14	0.45	0.81	Blychert et al. (1991), Lundahl et al. (1997)
Midazolam	0.24–0.41	0.40–0.79	0.94	Olkkola et al. (1993), Kupferschmidt et al. (1995), Olkkola et al. (1996), Thummel et al. (1996), Gorski et al. (1998), Palkama et al. (1999), Tsunoda et al. (1999), Tateishi et al. (2001), Wang et al. (2001), Lee et al. (2002), Kharasch et al. (2004a, b, 2007), Masica et al. (2004), Saari et al. (2006), Farkas et al. (2007)
Simvastatin	<0.05	0.66	0.99	Obach et al. (2006)
Triazolam	0.55	0.75	0.92	Masica et al. (2004)
Nifedipine	0.41	0.78	0.71	Holtbecker et al. (1996)
Quinidine	0.78	0.90	0.76	Damkier et al. (1999)
Alprazolam	0.84	0.94	0.8	Hirota et al. (2001)
Methadone	0.92	0.78	—	Kharasch et al. (2004a)
Nisoldipine	0.05–0.08	0.11	0.99	Takanaga et al. (2000)
Saquinavir	<0.04	0.18	0.99	Kupferschmidt et al. (1998)
Alfentanil	0.42	0.60	0.87	Kharasch et al. (2004b, 2007)
Rifabutin	0.20	0.21	0.9	Skinner et al. (1989), Skinner and Blaschke (1995)
Sildenafil	0.38	0.54	0.9	Muirhead et al. (2002), Nichols et al. (2002)
Maraviroc	0.23	0.75–0.93	—	Abel et al. (2008)
Trazodone	0.77	0.83	0.35	Greenblatt et al. (1987)
Zolpidem	0.72	0.79	0.6	Patat et al. (1994)
R-Verapamil	0.39	0.66	0.96	
S-Verapamil	0.16	0.52	0.96	Eichelbaum et al. (1979), Fromm et al. (1996)

[a]The F_G values were estimated from i.v. and oral data (Galetin et al., 2008), and $f_{mCYP3A4}$ were obtained from renal excretion data (Brown et al., 2006; Galetin et al., 2006; Houston and Galetin, 2008).

the case of alprazolam and tacrolimus, respectively (Galetin et al., 2008). In all the cases, CYP3A4 contributes predominantly to their elimination ($f_{mCYP3A4} > 0.7$ with the exception of trazodone and zolpidem (Brown et al., 2005; Galetin et al., 2006; Houston and Galetin, 2008). In the case of felodipine and quinidine, the intestinal extraction is lower than hepatic, whereas for certain substrates (e.g., triazolam and tacrolimus) intestinal extraction rivals or even exceeds that of the liver (Galetin et al., 2007). *In vivo* F_G estimates obtained by this method are of limited availability and represent the measure of the net result of the transporter–enzyme interplay.

The i.v./oral approach is based on several assumptions that may lead to inaccuracies in F_G estimates. The assumption that the extent of intestinal metabolism after i.v. administration is negligible may not necessarily be valid as an average 8% extraction ratio after an i.v.-dosed midazolam has been reported in anhepatic patients (up to 26% in one patient) (Paine et al., 1996). Variation in drug-dependent parameters such as $CL_{i.v.}$, F, and blood:plasma ratio contributes to differential F_G estimates. Additional errors may occur due to the use of average Q_H despite its physiological range from 17.1 to 25 mL/min/kg (Kato et al., 2003), which may result in biased E_H and consequently F_G estimates for drugs with $CL_{H,b}$ approaching hepatic blood flow. Alteration of Q_H due to vasodilatation caused by calcium channel antagonists (Bengtsson-Hasselgren et al., 1990; Fromm et al., 1996; Lundahl et al., 1997) could add to the variability in this parameter, resulting in an increased Q_H value (23–36% in case of nifedipine and felodipine, respectively), and this is generally not taken into account when estimating F_G from i.v./oral data. In the absence of accurate estimates of F_a, the indirect method assumes complete absorption (i.e., $F_a = 1$) when estimating E_H, which can contribute to an underestimation of the F_G and subsequently overestimation of the F_G ratio in the DDI prediction, in particular if the latter is obtained assuming maximal intestinal inhibition.

9.3.2 F_G Estimates from Grapefruit Juice Interaction Studies

Since the first grapefruit juice (GFJ) interaction with felodipine (Bailey et al., 1989, 1991) a large number of studies have been reported for drugs from different therapeutic groups (Bailey et al., 1998; Fuhr, 1998; Dresser and Bailey, 2003; Saito et al., 2005; Kirby and Unadkat, 2007; Farkas and Greenblatt, 2008), making the GFJ interaction the most well-characterized food–drug interaction to date. Furanocoumarins have been identified as main contributors to the observed clinical interactions with GFJ (Paine et al., 2006b), causing the irreversible inhibition of intestinal CYP3A4 (Schmiedlin-Ren et al., 1997).

Analysis of a large number of GFJ interaction studies suggested that a single dose of GFJ (on average 175–250 mL used in clinical studies) administered 0–4 h before drug intake can be considered sufficient to produce maximal inhibitory effects. The estimation of F_G from GFJ interaction studies is based on several assumptions: (1) complete inhibition of intestinal CYP3A4 metabolism in the intestine; (2) concomitant administration of GFJ has no effect on the fraction absorbed of the investigated drugs and on hepatic CYP3A4 activity, that is, F_a and F_H remain unchanged in both control and grapefruit juice phase; and (3) the intestine has a negligible contribution to the systemic elimination of the investigated drugs. In this case, the F_G of a CYP3A substrate is estimated from a comparison of the AUC after oral administration in the presence and absence of grapefruit juice, as illustrated in Eq. (9.10) (Gertz et al., 2008a).

$$F_G = \frac{\mathrm{AUC}_{\mathrm{control}}}{\mathrm{AUC}_{\mathrm{GFJ}}} \tag{9.10}$$

The GFJ approach is proposed as a pragmatic method to estimate F_G, because it only requires measurements of AUC in the presence and absence of GFJ, unlike the i.v./oral method. In addition, the wide use and hence abundance of data makes the GFJ approach an attractive alternative to the i.v./oral method. However, this approach makes several assumptions that might lead to biased estimations of intestinal availability, and this has recently been reviewed (Gertz et al., 2008a). The most important assumption in the use of GFJ interaction studies is complete and selective inhibition of the intestinal CYP3A4 enzymes. In the comprehensive analysis of the GFJ interaction data (Gertz et al., 2008a), assessment of this assumption was undertaken for drugs where i.v./oral data in the absence and presence of GFJ were reported. In the case of complete intestinal inhibition, the F_G of the drug in the presence of GFJ should equal 1, and this was confirmed for drugs subject solely to metabolism (e.g., midazolam, alfentanil, and nifedipine). In contrast, GFJ was unable to completely inhibit the intestinal metabolism of cyclosporine and saquinavir, resulting in potentially biased F_G estimate by this method, probably due to transporter involvement.

There is a great inconsistency in the conduct of GFJ clinical studies (different brands and dosage of GFJ), which can result in differential extent of inhibition and consequently differential or erroneous estimates of F_G. Even though Paine et al. (2006) have shown that furanocoumarins represent the main components relevant for the interaction, the effect is most likely to be a combination of various constituents; normalization of active ingredients is therefore highly problematic.

Another potential limitation of the grapefruit juice method is that the intestinal transporter activity is not taken into account. Components of GFJ show only weak inhibition toward the P-gp *in vitro* (Honda et al., 2004) and *in vivo* (Becquemont et al., 2001; Parker et al., 2003), with no significant change in the P-gp protein levels, in contrast to CYP3A4 (Lown et al., 1997; Greenblatt et al., 2003), suggesting that the effect of GFJ on P-gp is unlikely to be as substantial as that on CYP3A4. GFJ has recently been reported to inhibit intestinal uptake transporters (e.g., OATP1A2 and OATP2B1)

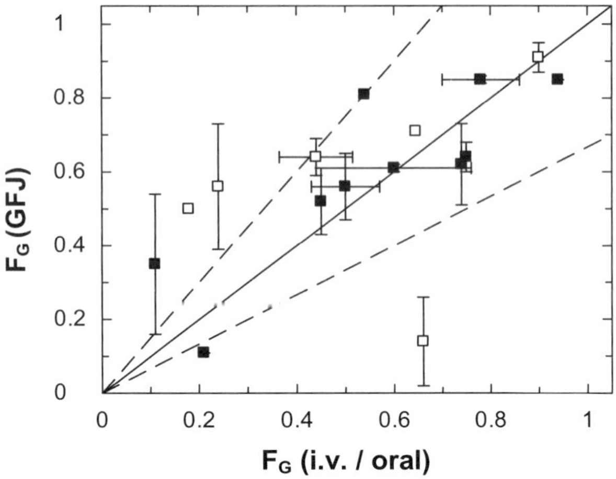

Figure 9.2. Comparison of the F_G for 16 CYP3A4 substrates obtained by i.v./oral and GFJ method (Gertz et al., 2008a). Data represent mean (weighted for the study size) ± sd. ■ represents drugs that are mainly metabolized, and □ represents drugs where either uptake transport or P-gp attributes to the disposition of the drugs. Dashed line represents 1.5-fold difference from the line of unity (full line).

(Glaeser et al., 2007), with no significant effect on the transporter protein expression. Coadministration of GFJ and fexofenadine (substrate for a range of uptake transporters) has shown a differential inhibitory effect of GFJ, depending on the time prior the administration. These findings indicate a potential limitation in the use of this approach to estimate F_G for drugs whose disposition is dependent on both efflux/uptake transporters and metabolic enzymes. This was indeed shown in the direct comparison of F_G estimates from i.v./oral and GFJ data for a range of CYP3A4 substrates (Gertz et al., 2008a), as illustrated in Fig. 9.2. The authors have reported the highest discrepancy between the methods and overestimation of F_G in the case of atorvastatin, cyclosporine, and saquinavir, all substrates for P-gp and OATP transporters (Shitara and Sugiyama, 2006; Lau et al., 2007; Niemi, 2007). In contrast, a very good agreement between the GFJ and i.v./oral F_G estimates was observed for purely metabolised drugs, supporting the interchangeable use of these methods for drugs where transporter mediated uptake/efflux is insignificant.

9.3.3 Interstudy and Interindividual Variability in F_G

Midazolam i.v./oral data are available from several clinical studies, highlighting the complications of interindividual variability in the F_G estimates (Table 9.1). The mean (weighted for population size) F_G estimate for midazolam obtained from 315 individuals was 51.1 ± 9.7%, ranging from 40–79%. Interindividual variability was also observed in the oral bioavailability (29 ± 5.2%) and F_H estimates (57.6 ± 8.8%), as illustrated in Fig. 9.3. The

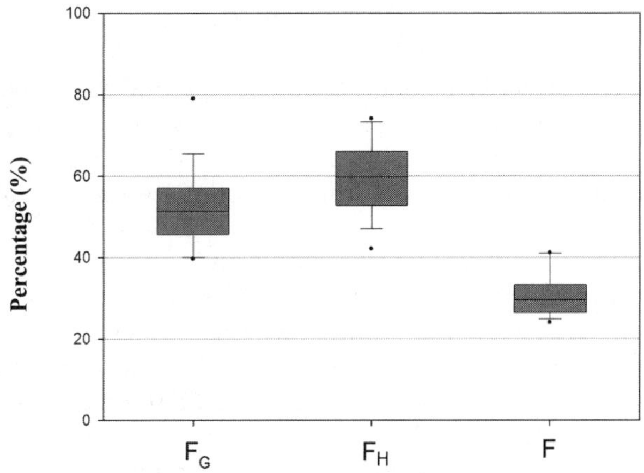

Figure 9.3. F_G, F_H, and F estimates for midazolam obtained from 16 different studies (315 individuals) (Olkkola et al., 1993; Kupferschmidt et al., 1995; Olkkola et al., 1996; Thummel et al., 1996; Gorski et al., 1998; Palkama et al., 1999; Tsunoda et al., 1999; Tateishi et al., 2001; Wang et al., 2001; Lee et al., 2002; Kharasch et al., 2004a, b, 2007; Masica et al., 2004; Saari et al., 2006; Farkas et al., 2007). The box indicates 50%, and the upper and lower whiskers represent, respectively, the highest and lowest value that is not an outlier. The black line represents the median value (Galetin et al., 2008).

variable extent of midazolam F_G values results in a maximal F_G ratio (and hence increase in the hepatic prediction) of up to 2.5-fold. This is in good agreement with the observed 1.8- to 2.7-fold increase in AUC′/AUC ratio after oral administration of midazolam in comparison to i.v. dose in the presence of seven reversible and time-dependent inhibitors (Galetin et al., 2007). In the case of cyclosporine, comparable availability of the GFJ interaction data in both healthy (6 studies, 74 individuals) and kidney transplant patient (8 studies, 64 individuals) population allowed the assessment of the impact of health status on F_G estimates (Gertz et al., 2008a). Significantly lower estimates ($p < 0.01$) were observed in the healthy group compared to stable kidney transplant patients, 0.64 and 0.77, respectively. The analysis has also shown more pronounced interindividual variation in F_G estimates (33% and 27% for the healthy and kidney transplant patients, respectively), in contrast to the interstudy variability (9.2% and 5.7% for the healthy and kidney transplant patients, respectively).

9.3.4 Prediction of F_G from *In Vitro* Data

Modeling of intestinal first-pass requires incorporation of drug absorption, zonal and cellular heterogeneous distribution of metabolic enzyme, and efflux/uptake transporters along the length of the intestine and enterocytic rather than total organ blood flow. A number of physiologically based models have been reported with different levels of complexity and integration of passive permability with the activity of metabolic enzymes and transporters (Ito et al., 1999; Pang, 2003; Tam et al., 2003). Adaptations of the original Compartmental Absorption Transit (CAT) (Yu and Amidon, 1999) and Advanced CAT model (Agoram et al., 2001; Tubic et al., 2006) have now been incorporated in the commercially available softwares (GastroPlus and Simcyp®); an overview of these integrated dynamic models for the prediction of oral drug absorption and metabolism can be found in the recent article by Dokoumetzidis et al. (2007).

In contrast to complex physiologically based models, Yang et al. (2007b) proposed a "minimal" Q_{gut} model. This model overcomes the inadequacy of the well-stirred approach adapted from the corresponding liver model (Rowland et al., 1973; Wilkinson and Shand, 1975) to derive F_G (Lin et al., 1997; Thummel et al., 1997) by combining cellular permeability (including active transport), enterocytic blood flow, intrinsic activity, and spatial heterogeneity of intestinal metabolic enzymes and their relative abundance. A mechanistic model to predict F_G values from *in vitro* metabolism and transporter data is shown in Eq. (9.11) (Yang et al., 2007b).

$$F_G = \frac{Q_{\text{gut}}}{Q_{\text{gut}} + \text{fu}_{\text{gut}} \cdot \text{CL}u_{\text{int},g}} \tag{9.11}$$

$$Q_{\text{gut}} = \frac{\text{CL}_{\text{perm}} \cdot Q_{\text{ent}}}{\text{CL}_{\text{perm}} + Q_{\text{ent}}} \tag{9.12}$$

where fu_{gut} is the fraction unbound in the enterocytes, $\text{CLu}_{\text{int},g}$ is the intestinal clearance, and Q_{gut} is the nominal blood flow in the small intestine. Q_{gut} is a hybrid function of the permeability clearance (CL_{perm}) and enterocytic blood flow (Q_{ent}), as illustrated in the Eq. (9.12). Sensitivity analysis of this model has indicated that intestinal extraction of greater than 50% (i.e., $F_G < 0.5$) is observed for drugs with approximate $\text{CLu}_{\text{int},g} \geq 1\,\mu\text{L/min/pmolCYP3A}$ regardless of Q_{gut} if fu_{gut} is 1. The Q_{gut} model is less sensitive to the drug permeability in comparison to drug clearance (Galetin et al., 2008; Gertz et al., 2008b). Different surrogates can be used for the effective free fraction in the gut, namely $\text{fu}_{\text{gut}} = 1$ or equal to plasma or blood binding; however, an accurate estimate of this parameter is difficult. This represents an issue of concern due to the sensitivity of the model and the predicted F_G on this parameter. The use of plasma or blood binding as fu_{gut} results in a general loss of prediction accuracy in comparison to assumption of no binding ($\text{fu}_{\text{gut}} = 1$) (Yang et al., 2007b).

9.4 IMPACT OF INCORPORATION OF INTESTINAL INHIBITION IN THE PREDICTION STRATEGY

High concentrations of the putative inhibitor achieved during the absorption phase suggest a high potential for DDIs at the level of the gut wall. In addi-

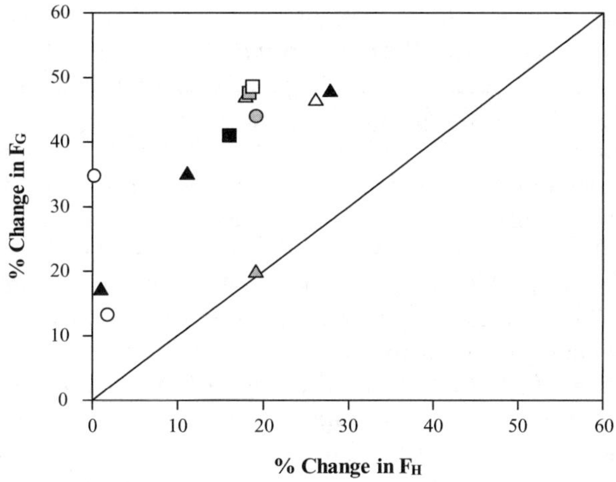

Figure 9.4. The effect of 8 reversible and irreversible inhibitors on intestinal (F_G) and hepatic (F_H) CYP3A (Galetin et al., 2007). Interactions identified according to the inhibitors: ▲ represents ketoconazole (Gomez et al., 1995; Gorski et al., 1998; Tsunoda et al., 1999), △ itraconazole (Olkkola et al., 1996), ▲ fluconazole (Olkkola et al., 1996), □ voriconazole (Saari et al., 2006), ■ clarithromycin (Gorski et al., 1998), ■ erythromycin (Gupta et al., 1989; Olkkola et al., 1993), ○ grapefruit juice (Ducharme et al., 1995; Lundahl et al., 1997), and ● saquinavir (Palkama et al., 1999).

tion, the relatively low blood flow to the intestinal mucosa means that the outflow of the drugs from the enterocyte is not as high as that in the hepatocyte, and hence lower abundance of CYP3A can act severalfold more efficient than in the liver. Therefore, significant intestinal first-pass metabolism may contribute to the interindividual variability in the magnitude of DDI, as seen for a number of CYP3A4 substrates (e.g., midazolam, tacrolimus, and felodipine) (Floren et al., 1997; Lundahl et al., 1997; Quinney et al., 2008).

A number of clinical studies have indicated a more pronounced inhibition (and induction) of intestinal CYP3A in comparison to the liver (Gomez et al., 1995; Palkama et al., 1999; Tsunoda et al., 1999; Saari et al., 2006; Galetin et al., 2007). The effect of eight reversible and irreversible inhibitors (expressed as a percent of change) on the F_G and F_H of a range of CYP3A substrates is shown in Fig. 9.4. A greater increase in F_G (13–49%) in comparison to F_H (0.2–28%) was observed irrespective of the inhibition mechanism. For mutual P-gp and CYP3A4 substrates (e.g., cyclosporine, tacrolimus) the interaction is complex and is a result of interactive nature and differing interrelationship of these two proteins in the intestine and liver (Benet et al., 2004). However, the more pronounced intestinal inhibition cannot entirely be linked to CYP3A4-P-glycoprotein interplay, because the greatest increase in F_G for interactions involving ketoconazole ($n = 3$) was observed with midazolam, a victim drug where the contribution of efflux transporter is minimal.

The incorporation of the intestinal inhibition into the DDI prediction strategy (as the F'_G/F_G ratio) has shown variable success (Wang et al., 2004; Brown et al., 2006; Galetin et al., 2006; Obach et al., 2006; Einolf, 2007). Incorporation of the maximal intestinal inhibition ($1/F_G$) resulted in the successful prediction of time-dependent CYP3A4 DDIs by verapamil (Wang et al., 2004) using unbound average systemic plasma concentration after repeated oral administration. Although successful, this dataset only contained midazolam as a victim drug. Galetin et al. (2006) have also investigated the impact of the maximal intestinal inhibition assumption on the prediction of 28 time-dependent DDIs. However, in this case the dataset included nine victim drugs with a differential extent of intestinal extraction (F_G ranging from 0.21 to 0.98 in the case of buspirone and alprazolam, respectively). The use of the $1/F_G$ approach minimized the number of false negative predictions in this study, but at the same time reduced the number of predictions within twofold of *in vivo* value by 20% (Fig. 9.5). Prediction of interactions with midazolam, triazolam, and nifedipine as victim drugs was overall improved by the incorporation of the intestinal interaction, whereas pronounced overpredictions were observed for the interactions with cyclosporine and buspirone. This is in agreement with findings reported by Brown

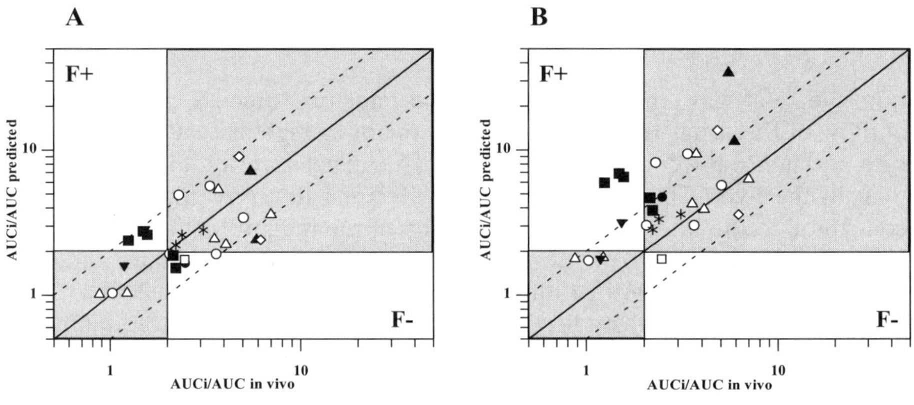

Figure 9.5 Relationship between predicted and observed AUC ratios for time-dependent interactions. **(A)** Predictions of 28 TDI using Eq. (9.2) applying the average unbound plasma concentration of the inhibitor, corresponding fm_{CYP3A4} and CYP3A4 $t_{1/2deg}$ of 3 days and no intestinal interaction. Interactions identified according to the substrates: △ represents midazolam, ○ triazolam, □ alprazolam, ▲ buspirone, ▼ quinidine, ◇ simvastatin, ■ cyclosporine, ● felodipine, and * nifedipine. **(B)** Effect of incorporating the maximal intestinal inhibition ($F'_G = 1$) in the predictions; all other conditions and symbols are as in panel A. The solid line represents line of unity, and dashed lines represent the twofold limit in prediction accuracy. The shaded areas correspond to the true negative and positive time-dependent interactions defined by the twofold increase in the AUC; F+ and F– represent false-positive and false-negative predictions, respectively (Galetin et al., 2006).

et al. (2006) and the overprediction trend observed in the prediction of atorvastatin and tacrolimus reversible CYP3A4-mediated DDIs. Other studies have incorporated intestinal inhibition in the form of the predicted F_G ratio [Eq. (9.3)] rather than assuming maximal inhibition of intestinal enzymes (Obach et al., 2006; Einolf, 2007). Obach et al. (2006) have reported an improvement in the prediction precision and accuracy of a range of reversible CYP3A4 DDIs and also time-dependent DDIs (Obach et al., 2007) using unbound hepatic input concentration. In contrast, use of total inhibitor concentration resulted in the minor impact of intestinal inhibition on the prediction accuracy.

All these studies illustrate that the F_G of a victim drug is an important determinant of the interaction magnitude in addition to the inhibitor properties. A minimal 50% intestinal extraction is indicated as an appropriate cutoff value for a potential significant contribution of the intestinal interaction, irrespective of the potency of the inhibitor and its inhibition mechanism (Galetin et al., 2007). CYP3A substrates with high extent of intestinal first-pass extraction (>75%) are particularly sensitive to any inaccuracy in the initial F_G estimates. The incorporation of the intestinal interaction will correct the prediction based solely on hepatic enzyme interaction by fourfold or more if maximal intestinal inhibition is assumed. This could lead to a potential overprediction of DDIs, as seen in the cases of interactions with cyclosporine, atorvastatin, and buspirone as victim drugs (Brown et al., 2006; Galetin et al., 2006). For the drugs with intestinal extraction ≤50% (e.g., midazolam), the maximal value of the F_G ratio is ≤2 even in the cases of potent inhibition—that is, when the inhibitor decreases $CL_{int,g}$ of a victim drug by >90%. Although this may suggest that the contribution of intestinal inhibition is likely to be relatively minor, for some DDIs this twofold increase may represent a significant improvement in DDI prediction and result in elimination of a false-negative result.

The role of intestinal inhibition cannot be assessed in isolation, but rather in conjunction with other inhibitor- and substrate-related characteristics due to multifactorial nature of the DDI prediction models [Eqs. (9.1) and (9.2)]. The databases used so far to assess the impact of intestinal inhibition differ in some of the parameter estimates that contribute significantly to the DDI prediction success—for example, $f_{m\text{CYP3A4}}$, inhibitor concentrations (use of either single inhibitor concentration or more dynamic concentration–time profile), and enzyme degradation constants (Ito et al., 2004; Wang et al., 2004; Galetin et al., 2005; Brown et al., 2006; Galetin et al., 2006; Obach et al., 2006; Einolf, 2007; Yang et al., 2008). As shown previously, DDI prediction models are particularly sensitive to any inaccuracies in the $f_{m\text{CYP}}$ estimate, especially for victim drugs where the enzyme inhibited contributes more than 80% to the overall elimination (Brown et al., 2005; Galetin et al., 2006). Overestimation of this parameter may lead to significant overpredictions of true positives and increased number of false-positive predictions, regardless of the role of an intestinal interaction.

9.5 CONCLUSIONS

The incorporation of the intestine in DDI prediction models results in the model that represents a "correction" of hepatic prediction by the F_G ratio. The variable success of this prediction strategy raises the question of whether the inclusion of the gut contribution by the relatively simple F_G term describes the role of intestine in an adequate way. Differences in the success of a prediction cannot be associated only with the incorporation of the intestine in the prediction strategy, because other parameters are inconsistent between the datasets (e.g., $f_{mCYP3A4}$) for which the model is particularly sensitive.

The current inconsistency in the use of either hepatic input or average plasma concentration as a surrogate inhibitor concentration at the enzyme (or transporter) active site affects the evaluation of the role of intestinal inhibition. In some instances, a combination of assumptions may result in a comparable prediction outcome; for example, the use of unrealistically high hepatic input concentrations may compensate for ignoring the contribution of the gut, whereas incorporation of the same inhibitor concentration corrected for the plasma binding in conjunction with the intestinal inhibition may also result in a successful prediction. In addition, although plasma inhibitor concentration may be adequate for some compounds, it will underestimate the liver concentration for those drugs where hepatic uptake involves an active process. Comprehensive analysis of different methods to estimate the F_G ratio indicated a very good agreement between predicted estimates and approach based on maximal intestinal inhibition (use of $1/F_G$ in the DDI prediction model). Although attractive, the pragmatic $1/F_G$ approach has limitations and can lead to potential overpredictions in the case of moderate to weak inhibitors or in the interactions involving victim drugs with contribution from both metabolic enzymes and transporters in their disposition. Regardless of the method used to estimate the F_G ratio, several analyses have highlighted the importance of the F_G of the victim drug, in particular for highly extracted victim drugs (i.e., $F_G \leq 0.25$). Accurate assessment of this parameter is essential for the quantitative prediction of human drug clearance and drug–drug interaction potential.

ACKNOWLEDGMENTS

The work was funded by a consortium of pharmaceutical companies (GlaxoSmithKline, Lilly, Novartis, Pfizer and Servier) within the Centre for Applied Pharmacokinetic Research at the University of Manchester.

Financial support for AG's academic position in the School of Pharmacy and Pharmaceutical Sciences, University of Manchester is kindly provided by Pfizer, Sandwich, UK.

REFERENCES

Abel S, Russell D, Whitlock LA, Ridgway CE, Nedderman AN, Walker DK. Assessment of the absorption, metabolism and absolute bioavailability of maraviroc in healthy male subjects. *Br J Clin Pharmacol* 2008;65(Suppl 1):60–67.

Agoram B, Woltosz WS, Bolger MB. Predicting the impact of physiological and biochemical processes on oral drug bioavailability. *Adv Drug Deliv Rev* 2001;50 (Suppl 1):S41–S67.

Bailey DG, Spence JD, Edgar B, Bayliff CD, Arnold JM. Ethanol enhances the hemodynamic effects of felodipine. *Clin Invest Med* 1989;12:357–362.

Bailey DG, Spence JD, Munoz C, Arnold JM. Interaction of citrus juices with felodipine and nifedipine. *Lancet* 1991;337:268–269.

Bailey DG, Malcolm J, Arnold O, Spence JD. Grapefruit juice-drug interactions. *Br J Clin Pharmacol* 1998;46:101–110.

Becquemont L, Verstuyft C, Kerb R, Brinkmann U, Lebot M, Jaillon P, Funck-Brentano C. Effect of grapefruit juice on digoxin pharmacokinetics in humans. *Clin Pharmacol Ther* 2001;70:311–316.

Benet LZ, Cummins CL, Wu CY. Unmasking the dynamic interplay between efflux transporters and metabolic enzymes. *Int J Pharm* 2004;277:3–9.

Benet LZ, Amidon GL, Barends DM, Lennernas H, Polli JE, Shah VP, Stavchansky SA, Yu LX. The use of BDDCS in classifying the permeability of marketed drugs. *Pharm Res* 2008;25:483–488.

Bengtsson-Hasselgren B, Ronn O, Blychert LO, Edgar B, Raner S. Acute effects of felodipine and nifedipine on hepatic and forearm blood flow in healthy men. *Eur J Clin Pharmacol* 1990;38:529–533.

Berggren S, Gall C, Wollnitz N, Ekelund M, Karlbom U, Hoogstraate J, Schrenk D, Lennernas H. Gene and protein expression of P-glycoprotein, MRP1, MRP2, and CYP3A4 in the small and large human intestine. *Mol Pharm* 2007;4:252–257.

Bjornsson TD, Callaghan JT, Einolf HJ, Fischer V, Gan L, Grimm S, Kao J, King SP, Miwa G, Ni L, Kumar G, McLeod J, Obach RS, Roberts S, Roe A, Shah A, Snikeris F, Sullivan JT, Tweedie D, Vega JM, Walsh J, Wrighton SA. The conduct of *in vitro* and *in vivo* drug–drug interaction studies: a Pharmaceutical Research and Manufacturers of America (PhRMA) perspective. *Drug Metab Dispos* 2003;31: 815–832.

Blychert E, Edgar B, Elmfeldt D, Hedner T. A population study of the pharmacokinetics of felodipine. *Br J Clin Pharmacol* 1991;31:15–24.

Brown HS, Ito K, Galetin A, Houston JB. Prediction of *in vivo* drug–drug interactions from *in vitro* data: impact of incorporating parallel pathways of drug elimination and inhibitor absorption rate constant. *Br J Clin Pharmacol* 2005;60:508–518.

Brown HS, Galetin A, Hallifax D, Houston JB. Prediction of *in vivo* drug–drug interactions from *in vitro* data: factors affecting prototypic drug–drug interactions involving CYP2C9, CYP2D6 and CYP3A4. *Clin Pharmacokinet* 2006;45:1035–1050.

Cao X, Gibbs ST, Fang L, Miller HA, Landowski CP, Shin HC, Lennernas H, Zhong Y, Amidon GL, Yu LX, Sun D. Why is it challenging to predict intestinal drug absorption and oral bioavailability in human using rat model. *Pharm Res* 2006;23: 1675–1686.

Chalasani N, Gorski JC, Patel NH, Hall SD, Galinsky RE. Hepatic and intestinal cytochrome P450 3A activity in cirrhosis: effects of transjugular intrahepatic portosystemic shunts. *Hepatology* 2001;34:1103–1108.

Chen G, Zhang D, Jing N, Yin S, Falany CN, Radominska-Pandya A. Human gastrointestinal sulfotransferases: identification and distribution. *Toxicol Appl Pharmacol* 2003;187:186–197.

Chien JY, Lucksiri A, Ernest CS, 2nd, Gorski JC, Wrighton SA, Hall SD. Stochastic prediction of CYP3A-mediated inhibition of midazolam clearance by ketoconazole. *Drug Metab Dispos* 2006;34:1208–1219.

Damkier P, Hansen LL, Brosen K. Effect of fluvoxamine on the pharmacokinetics of quinidine. *Eur J Clin Pharmacol* 1999;55:451–456.

Dokoumetzidis A, Kalantzi L, Fotaki N. Predictive models for oral drug absorption: from *in silico* methods to integrated dynamical models. *Expert Opin Drug Metab Toxicol* 2007;3:491–505.

Dresser GK, Bailey DG. The effects of fruit juices on drug disposition: a new model for drug interactions. *Eur J Clin Invest* 2003;33(Suppl 2):10–16.

Ducharme MP, Warbasse LH, Edwards DJ. Disposition of intravenous and oral cyclosporine after administration with grapefruit juice. *Clin Pharmacol Ther* 1995;57: 485–491.

Eichelbaum M, Ende M, Remberg G, Schomerus M, Dengler HJ. The metabolism of DL-[^{14}C]verapamil in man. *Drug Metab Dispos* 1979;7:145–148.

Einolf HJ. Comparison of different approaches to predict metabolic drug–drug interactions. *Xenobiotica* 2007;37:1257–1294.

Endres CJ, Hsiao P, Chung FS, Unadkat JD. The role of transporters in drug interactions. *Eur J Pharm Sci* 2006;27:501–517.

Farkas D, Greenblatt DJ. Influence of fruit juices on drug disposition: discrepancies between *in vitro* and clinical studies. *Expert Opin Drug Metab Toxicol* 2008;4: 381–393.

Farkas D, Oleson LE, Zhao Y, Harmatz JS, Zinny MA, Court MH, Greenblatt DJ. Pomegranate juice does not impair clearance of oral or intravenous midazolam, a probe for cytochrome P450-3A activity: comparison with grapefruit juice. *J Clin Pharmacol* 2007;47:286–294.

Floren LC, Bekersky I, Benet LZ, Mekki Q, Dressler D, Lee JW, Roberts JP, Hebert MF. Tacrolimus oral bioavailability doubles with coadministration of ketoconazole. *Clin Pharmacol Ther* 1997;62:41–49.

Fojo AT, Ueda K, Slamon DJ, Poplack DG, Gottesman MM, Pastan I. Expression of a multidrug-resistance gene in human tumors and tissues. *Proc Natl Acad Sci* 1987;84:265–269.

Fromm MF, Busse D, Kroemer HK, Eichelbaum M. Differential induction of prehepatic and hepatic metabolism of verapamil by rifampin. *Hepatology* 1996;24: 796–801.

Fuhr U. Drug interactions with grapefruit juice. Extent, probable mechanism and clinical relevance. *Drug Saf* 1998;18:251–272.

Galetin A. Intestinal first-pass metabolism: bridging the gap between *in vitro* and *in vivo*. *Curr Drug Metab* 2007;8:643–644.

Galetin A, Houston JB. Intestinal and hepatic metabolic activity of five cytochrome P450 enzymes: impact on prediction of first-pass metabolism. *J Pharmacol Exp Ther* 2006;318:1220–1229.

Galetin A, Brown C, Hallifax D, Ito K, Houston JB. Utility of recombinant enzyme kinetics in prediction of human clearance: impact of variability, CYP3A5, and CYP2C19 on CYP3A4 probe substrates. *Drug Metab Dispos* 2004;32:1411–1420.

Galetin A, Burt H, Gibbons L, Houston JB. Prediction of time-dependent CYP3A4 drug–drug interactions: impact of enzyme degradation, parallel elimination pathways, and intestinal inhibition. *Drug Metab Dispos* 2006;34:166–175.

Galetin A, Ito K, Hallifax D, Houston JB. CYP3A4 substrate selection and substitution in the prediction of potential drug–drug interactions. *J Pharmacol Exp Ther* 2005;314:180–190.

Galetin A, Hinton LK, Burt H, Obach RS, Houston JB. Maximal inhibition of intestinal first-pass metabolism as a pragmatic indicator of intestinal contribution to the drug–drug interactions for CYP3A4 cleared drugs. *Curr Drug Metab* 2007;8: 685–693.

Galetin A, Gertz M, Houston JB. Potential role of intestinal first-pass metabolism in the prediction of drug–drug interactions. *Expert Opin Drug Metab Toxicol* 2008;4: 909–922.

Gertz M, Davis JD, Harrison A, Houston JB, Galetin A. Grapefruit juice–drug interaction studies as a method to assess the extent of intestinal availability: utility and limitations. *Curr Drug Metab* 2008a;9:785–795.

Gertz M, Fenner KS, Harrison A, Davis J, Houston JB, Galetin A. Impact of permeability on *in vitro* predictions of intestinal availability in the QGut model. *Drug Metab Rev* 2008b;40(Suppl 1):92.

Gibbs MA, Thummel KE, Shen DD, Kunze KL. Inhibition of cytochrome P-450 3A (CYP3A) in human intestinal and liver microsomes: comparison of Ki values and impact of CYP3A5 expression. *Drug Metab Dispos* 1999;27:180–187.

Glaeser H, Drescher S, van der Kuip H, Behrens C, Geick A, Burk O, Dent J, Somogyi A, Von Richter O, Griese EU, Eichelbaum M, Fromm MF. Shed human enterocytes as a tool for the study of expression and function of intestinal drug-metabolizing enzymes and transporters. *Clin Pharmacol Ther* 2002;71:131–140.

Glaeser H, Bailey DG, Dresser GK, Gregor JC, Schwarz UI, McGrath JS, Jolicoeur E, Lee W, Leake BF, Tirona RG, Kim RB. Intestinal drug transporter expression and the impact of grapefruit juice in humans. *Clin Pharmacol Ther* 2007;81: 362–370.

Gomez DY, Wacher VJ, Tomlanovich SJ, Hebert MF, Benet LZ. The effects of ketoconazole on the intestinal metabolism and bioavailability of cyclosporine. *Clin Pharmacol Ther* 1995;58:15–19.

Gorski JC, Jones DR, Haehner-Daniels BD, Hamman MA, O'Mara EM, Jr, Hall SD. The contribution of intestinal and hepatic CYP3A to the interaction between midazolam and clarithromycin. *Clin Pharmacol Ther* 1998;64:133–143.

Greenblatt DJ, Friedman H, Burstein ES, Scavone JM, Blyden GT, Ochs HR, Miller LG, Harmatz JS, Shader RI. Trazodone kinetics: effect of age, gender, and obesity. *Clin Pharmacol Ther* 1987;42:193–200.

Greenblatt DJ, von Moltke LL, Harmatz JS, Chen G, Weemhoff JL, Jen C, Kelley CJ, LeDuc BW, Zinny MA. Time course of recovery of cytochrome p450 3A function after single doses of grapefruit juice. *Clin Pharmacol Ther* 2003;74:121–129.

Gupta SK, Bakran A, Johnson RW, Rowland M. Cyclosporin–erythromycin interaction in renal transplant patients. *Br J Clin Pharmacol* 1989;27:475–481.

Hall SD, Thummel KE, Watkins PB, Lown KS, Benet LZ, Paine MF, Mayo RR, Turgeon DK, Bailey DG, Fontana RJ, Wrighton SA. Molecular and physical mechanisms of first-pass extraction. *Drug Metab Dispos* 1999;27:161–166.

Hebert MF, Roberts JP, Prueksaritanont T, Benet LZ. Bioavailability of cyclosporine with concomitant rifampin administration is markedly less than predicted by hepatic enzyme induction. *Clin Pharmacol Ther* 1992;52:453–457.

Hinton LK, Galetin A, Houston JB. Multiple inhibition mechanisms and prediction of drug-drug interactions: status of metabolism and transporter models as exemplified by gemfibrozil–drug interactions. *Pharm Res* 2008;25:1063–1074.

Hirota N, Ito K, Iwatsubo T, Green CE, Tyson CA, Shimada N, Suzuki H, Sugiyama Y. *In vitro/in vivo* scaling of alprazolam metabolism by CYP3A4 and CYP3A5 in humans. *Biopharm Drug Dispos* 2001;22:53–71.

Ho RH, Kim RB. Transporters and drug therapy: implications for drug disposition and disease. *Clin Pharmacol Ther* 2005;78:260–277.

Holtbecker N, Fromm MF, Kroemer HK, Ohnhaus EE, Heidemann H. The nifedipine–rifampin interaction. Evidence for induction of gut wall metabolism. *Drug Metab Dispos* 1996;24:1121–1123.

Honda Y, Ushigome F, Koyabu N, Morimoto S, Shoyama Y, Uchiumi T, Kuwano M, Ohtani H, Sawada Y. Effects of grapefruit juice and orange juice components on P-glycoprotein- and MRP2-mediated drug efflux. *Br J Pharmacol* 2004;143: 856–864.

Houston JB, Galetin A. Progress towards prediction of human pharmacokinetic parameters from *in vitro* technologies. *Drug Metab Rev* 2003;35:393–415.

Houston JB, Galetin A. Modelling atypical CYP3A4 kinetics: principles and pragmatism. *Arch Biochem Biophys* 2005;433:351–360.

Houston JB, Galetin A. Methods for predicting *in vivo* pharmacokinetics using data from *in vitro* assays. *Curr Drug Metab* 2008;9:940–951.

Huang SM, Temple R, Throckmorton DC, Lesko LJ. Drug interaction studies: study design, data analysis, and implications for dosing and labeling. *Clin Pharmacol Ther* 2007;81:298–304.

Huang SM, Strong JM, Zhang L, Reynolds KS, Nallani S, Temple R, Abraham S, Habet SA, Baweja RK, Burckart GJ, Chung S, Colangelo P, Frucht D, Green MD, Hepp P, Karnaukhova E, Ko HS, Lee JI, Marroum PJ, Norden JM, Qiu W, Rahman A, Sobel S, Stifano T, Thummel K, Wei XX, Yasuda S, Zheng JH, Zhao H, Lesko LJ. New era in drug interaction evaluation: US Food and Drug Administration update on CYP enzymes, transporters, and the guidance process. *J Clin Pharmacol* 2008;48:662–670.

Huang W, Lin YS, McConn DJ, 2nd, Calamia JC, Totah RA, Isoherranen N, Glodowski M, Thummel KE. Evidence of significant contribution from CYP3A5 to hepatic drug metabolism. *Drug Metab Dispos* 2004;32:1434–1445.

Isoherranen N, Kunze KL, Allen KE, Nelson WL, Thummel KE. Role of itraconazole metabolites in CYP3A4 inhibition. *Drug Metab Dispos* 2004;32:1121–1131.

Isoherranen N, Ludington SR, Givens RC, Lamba JK, Pusek SN, Dees EC, Blough DK, Iwanaga K, Hawke RL, Schuetz EG, Watkins PB, Thummel KE, Paine MF. The influence of CYP3A5 expression on the extent of hepatic CYP3A inhibition is substrate-dependent: an *in vitro–in vivo* evaluation. *Drug Metab Dispos* 2008;36: 146–154.

Ito K, Kusuhara H, Sugiyama Y. Effects of intestinal CYP3A4 and P-glycoprotein on oral drug absorption—theoretical approach. *Pharmaceutical Research* 1999;16: 225–231.

Ito K, Brown HS, Houston JB. Database analyses for the prediction of *in vivo* drug–drug interactions from *in vitro* data. *Br J Clin Pharmacol* 2004;57:473–486.

Ito K, Hallifax D, Obach RS, Houston JB. Impact of parallel pathways of drug elimination and multiple cytochrome P450 involvement on drug–drug interactions: CYP2D6 paradigm. *Drug Metab Dispos* 2005;33:837–844.

Kanamitsu S, Ito K, Sugiyama Y. Quantitative prediction of *in vivo* drug–drug interactions from *in vitro* data based on physiological pharmacokinetics: use of maximum unbound concentration of inhibitor at the inlet to the liver. *Pharm Res* 2000;17: 336–343.

Kato M, Chiba K, Hisaka A, Ishigami M, Kayama M, Mizuno N, Nagata Y, Takakuwa S, Tsukamoto Y, Ueda K, Kusuhara H, Ito K, Sugiyama Y. The intestinal first-pass metabolism of substrates of CYP3A4 and P-glycoprotein-quantitative analysis based on information from the literature. *Drug Metab Pharmacokinet* 2003;18: 365–372.

Kharasch ED, Hoffer C, Whittington D, Sheffels P. Role of hepatic and intestinal cytochrome P450 3A and 2B6 in the metabolism, disposition, and miotic effects of methadone. *Clin Pharmacol Ther* 2004a;76:250–269.

Kharasch ED, Walker A, Hoffer C, Sheffels P. Intravenous and oral alfentanil as *in vivo* probes for hepatic and first-pass cytochrome P450 3A activity: noninvasive assessment by use of pupillary meiosis. *Clin Pharmacol Ther* 2004b;76:452–466.

Kharasch ED, Walker A, Isoherranen N, Hoffer C, Sheffels P, Thummel K, Whittington D, Ensign D. Influence of CYP3A5 genotype on the pharmacokinetics and pharmacodynamics of the cytochrome P4503A probes alfentanil and midazolam. *Clin Pharmacol Ther* 2007;82:410–426.

Kirby BJ, Unadkat JD. Grapefruit juice, a glass full of drug interactions? *Clin Pharmacol Ther* 2007;81:631–633.

Kivisto KT, Niemi M, Fromm MF. Functional interaction of intestinal CYP3A4 and P-glycoprotein. *Fundam Clin Pharmacol* 2004;18:621–626.

Kolars JC, Lown KS, Schmiedlin-Ren P, Ghosh M, Fang C, Wrighton SA, Merion RM, Watkins PB. CYP3A gene expression in human gut epithelium. *Pharmacogenetics* 1994;4:247–259.

Komura H, Iwaki M. Species differences in *in vitro* and *in vivo* small intestinal metabolism of CYP3A substrates. *J Pharm Sci* 2007.

Ku YM, Min DI, Flanigan M. Effect of grapefruit juice on the pharmacokinetics of microemulsion cyclosporine and its metabolite in healthy volunteers: does the formulation difference matter? *J Clin Pharmacol* 1998;38:959–965.

Kuehl P, Zhang J, Lin Y, Lamba J, Assem M, Schuetz J, Watkins PB, Daly A, Wrighton SA, Hall SD, Maurel P, Relling M, Brimer C, Yasuda K, Venkataramanan R, Strom S, Thummel K, Boguski MS, Schuetz E. Sequence diversity in CYP3A promoters and characterization of the genetic basis of polymorphic CYP3A5 expression. *Nat Genet* 2001;27:383–391.

Kupferschmidt HH, Ha HR, Ziegler WH, Meier PJ, Krahenbuhl S. Interaction between grapefruit juice and midazolam in humans. *Clin Pharmacol Ther* 1995;58: 20–28.

Kupferschmidt HH, Fattinger KE, Ha HR, Follath F, Krahenbuhl S. Grapefruit juice enhances the bioavailability of the HIV protease inhibitor saquinavir in man. *Br J Clin Pharmacol* 1998;45:355–359.

Lapple F, von Richter O, Fromm MF, Richter T, Thon KP, Wisser H, Griese EU, Eichelbaum M, Kivisto KT. Differential expression and function of CYP2C isoforms in human intestine and liver. *Pharmacogenetics* 2003;13:565–575.

Lau YY, Huang Y, Frassetto L, Benet LZ. effect of OATP1B transporter inhibition on the pharmacokinetics of atorvastatin in healthy volunteers. *Clin Pharmacol Ther* 2007;81:194–204.

Lee JI, Chaves-Gnecco D, Amico JA, Kroboth PD, Wilson JW, Frye RF. Application of semisimultaneous midazolam administration for hepatic and intestinal cytochrome P450 3A phenotyping. *Clin Pharmacol Ther* 2002;72:718–728.

Lee M, Min DI, Ku YM, Flanigan M. Effect of grapefruit juice on pharmacokinetics of microemulsion cyclosporine in African American subjects compared with Caucasian subjects: does ethnic difference matter? *J Clin Pharmacol* 2001;41: 317–323.

Lennernas H. Clinical pharmacokinetics of atorvastatin. *Clin Pharmacokinet* 2003;42: 1141–1160.

Lin JH, Yamazaki M. Role of P-glycoprotein in pharmacokinetics: clinical implications. *Clin Pharmacokinet* 2003;42:59–98.

Lin JH, Chiba M, Baillie TA. *In vivo* assessment of intestinal drug metabolism. *Drug Metab Dispos* 1997;25:1107–1109.

Lin JH, Chiba M, Baillie TA. Is the role of the small intestine in first-pass metabolism overemphasized? *Pharmacol Rev* 1999;51:135–158.

Lin YS, Dowling AL, Quigley SD, Farin FM, Zhang J, Lamba J, Schuetz EG, Thummel KE. Co-regulation of CYP3A4 and CYP3A5 and contribution to hepatic and intestinal midazolam metabolism. *Mol Pharmacol* 2002;62:162–172.

Lown KS, Kolars JC, Thummel KE, Barnett JL, Kunze KL, Wrighton SA, Watkins PB. Interpatient heterogeneity in expression of CYP3A4 and CYP3A5 in small bowel. Lack of prediction by the erythromycin breath test. *Drug Metab Dispos* 1994;22:947–955.

Lown KS, Bailey DG, Fontana RJ, Janardan SK, Adair CH, Fortlage LA, Brown MB, Guo W, Watkins PB. Grapefruit juice increases felodipine oral availability in humans by decreasing intestinal CYP3A protein expression. *J Clin Invest* 1997;99:2545–2553.

Lundahl J, Regardh CG, Edgar B, Johnsson G. Effects of grapefruit juice ingestion—pharmacokinetics and haemodynamics of intravenously and orally administered felodipine in healthy men. *Eur J Clin Pharmacol* 1997;52:139–145.

Madani S, Paine MF, Lewis L, Thummel KE, Shen DD. Comparison of CYP2D6 content and metoprolol oxidation between microsomes isolated from human livers and small intestines. *Pharm Res* 1999;16:1199–1205.

Masica AL, Mayo G, Wilkinson GR. *In vivo* comparisons of constitutive cytochrome P450 3A activity assessed by alprazolam, triazolam, and midazolam. *Clin Pharmacol Ther* 2004;76:341–349.

Matsumoto S, Hirama T, Matsubara T, Nagata K, Yamazoe Y. Involvement of CYP2J2 on the intestinal first-pass metabolism of antihistamine drug, astemizole. *Drug Metab Dispos* 2002;30:1240–1245.

McConn DJ, 2nd, Lin YS, Allen K, Kunze KL, Thummel KE. Differences in the inhibition of cytochromes P450 3A4 and 3A5 by metabolite-inhibitor complex-forming drugs. *Drug Metab Dispos* 2004;32:1083–1091.

Mouly S, Paine MF. P-glycoprotein increases from proximal to distal regions of human small intestine. *Pharm Res* 2003;20:1595–1599.

Muirhead GJ, Rance DJ, Walker DK, Wastall P. Comparative human pharmacokinetics and metabolism of single-dose oral and intravenous sildenafil. *Br J Clin Pharmacol* 2002;53(Suppl 1):13S–20S.

Nakamura T, Sakaeda T, Ohmoto N, Tamura T, Aoyama N, Shirakawa T, Kamigaki T, Nakamura T, Kim KI, Kim SR, Kuroda Y, Matsuo M, Kasuga M, Okumura K. Real-time quantitative polymerase chain reaction for MDR1, MRP1, MRP2, and CYP3A-mRNA levels in Caco-2 cell lines, human duodenal enterocytes, normal colorectal tissues, and colorectal adenocarcinomas. *Drug Metab Dispos* 2002;30: 4–6.

Nichols DJ, Muirhead GJ, Harness JA. Pharmacokinetics of sildenafil after single oral doses in healthy male subjects: absolute bioavailability, food effects and dose proportionality. *Br J Clin Pharmacol* 2002;53(Suppl 1):5S–12S.

Niemi M. Role of OATP transporters in the disposition of drugs. *Pharmacogenomics* 2007;8:787–802.

Obach RS, Zhang QY, Dunbar D, Kaminsky LS. Metabolic characterization of the major human small intestinal cytochrome p450s. *Drug Metab Dispos* 2001;29: 347–352.

Obach RS, Walsky RL, Venkatakrishnan K, Gaman EA, Houston JB, Tremaine LM. The utility of *in vitro* cytochrome P450 inhibition data in the prediction of drug-drug interactions. *J Pharmacol Exp Ther* 2006;316:336–348.

Obach RS, Walsky RL, Venkatakrishnan K. Mechanism-based inactivation of human cytochrome p450 enzymes and the prediction of drug–drug interactions. *Drug Metab Dispos* 2007;35:246–255.

Olkkola KT, Aranko K, Luurila H, Hiller A, Saarnivaara L, Himberg JJ, Neuvonen PJ. A potentially hazardous interaction between erythromycin and midazolam. *Clin Pharmacol Ther* 1993;53:298–305.

Olkkola KT, Ahonen J, Neuvonen PJ. The effects of the systemic antimycotics, itraconazole and fluconazole, on the pharmacokinetics and pharmacodynamics of intravenous and oral midazolam. *Anesth Analg* 1996;82:511–516.

Paine MF, Shen DD, Kunze KL, Perkins JD, Marsh CL, McVicar JP, Barr DM, Gillies BS, Thummel KE. First-pass metabolism of midazolam by the human intestine. *Clin Pharmacol Ther* 1996;60:14–24.

Paine MF, Khalighi M, Fisher JM, Shen DD, Kunze KL, Marsh CL, Perkins JD, Thummel KE. Characterization of interintestinal and intraintestinal variations in human CYP3A-dependent metabolism. *J Pharmacol Exp Ther* 1997;283:1552–1562.

Paine MF, Schmiedlin-Ren P, Watkins PB. Cytochrome P-450 1A1 expression in human small bowel: interindividual variation and inhibition by ketoconazole. *Drug Metab Dispos* 1999;27:360–364.

Paine MF, Criss AB, Watkins PB. Two major grapefruit juice components differ in time to onset of intestinal CYP3A4 inhibition. *J Pharmacol Exp Ther* 2005;312: 1151–1160.

Paine MF, Hart HL, Ludington SS, Haining RL, Rettie AE, Zeldin DC. The human intestinal cytochrome P450 "pie." *Drug Metab Dispos* 2006a;34:880–886.

Paine MF, Widmer WW, Hart HL, Pusek SN, Beavers KL, Criss AB, Brown SS, Thomas BF, Watkins PB. A furanocoumarin-free grapefruit juice establishes furanocoumarins as the mediators of the grapefruit juice–felodipine interaction. *Am J Clin Nutr* 2006b;83:1097–1105.

Palkama VJ, Ahonen J, Neuvonen PJ, Olkkola KT. Effect of saquinavir on the pharmacokinetics and pharmacodynamics of oral and intravenous midazolam. *Clin Pharmacol Ther* 1999;66:33–39.

Pang KS. Modeling of intestinal drug absorption: roles of transporters and metabolic enzymes (for the Gillette Review Series). *Drug Metab Dispos* 2003;31:1507–1519.

Parker RB, Yates CR, Soberman JE, Laizure SC. Effects of grapefruit juice on intestinal P-glycoprotein: evaluation using digoxin in humans. *Pharmacotherapy* 2003; 23:979–987.

Patat A, Trocherie S, Thebault JJ, Rosenzweig P, Dubruc C, Bianchetti G, Court LA, Morselli PL. EEG profile of intravenous zolpidem in healthy volunteers. *Psychopharmacology (Berl)* 1994;114:138–146.

Quinney SK, Haehner BD, Rhoades MB, Lin Z, Gorski JC, Hall SD. Interaction between midazolam and clarithromycin in the elderly. *Br J Clin Pharmacol* 2008;65: 98–109.

Ritter JK. Intestinal UGTs as potential modifiers of pharmacokinetics and biological responses to drugs and xenobiotics. *Expert Opin Drug Metab Toxicol* 2007;3: 93–107.

Rostami-Hodjegan A, Tucker GT. The effects of portal shunts on intestinal cytochrome P450 3A activity. *Hepatology* 35:1549–1550; author reply 2002;1550–1541.

Rostami-Hodjegan A, Tucker G. "In silico" simulations to assess the "*in vivo*" consequences of "*in vitro*" metabolic drug–drug interactions. *Drug Discovery Today: Technologies* 2004;1:441.

Rostami-Hodjegan A, Tucker GT. Simulation and prediction of *in vivo* drug metabolism in human populations from *in vitro* data. *Nat Rev Drug Discovery* 2007;6: 140.

Rowland M, Benet LZ, Graham GG. Clearance concepts in pharmacokinetics. *J Pharmacokinet Biopharm* 1973;1:123–136.

Saari TI, Laine K, Leino K, Valtonen M, Neuvonen PJ, Olkkola KT. Effect of voriconazole on the pharmacokinetics and pharmacodynamics of intravenous and oral midazolam. *Clin Pharmacol Ther* 2006;79:362–370.

Saito M, Hirata-Koizumi M, Matsumoto M, Urano T, Hasegawa R. Undesirable effects of citrus juice on the pharmacokinetics of drugs: focus on recent studies. *Drug Saf* 2005;28:677–694.

Schmiedlin-Ren P, Edwards DJ, Fitzsimmons ME, He K, Lown KS, Woster PM, Rahman A, Thummel KE, Fisher JM, Hollenberg PF, Watkins PB. Mechanisms of enhanced oral availability of CYP3A4 substrates by grapefruit constituents. Decreased enterocyte CYP3A4 concentration and mechanism-based inactivation by furanocoumarins. *Drug Metab Dispos* 1997;25:1228–1233.

Shitara Y, Horie T, Sugiyama Y. Transporters as a determinant of drug clearance and tissue distribution. *Eur J Pharm Sci* 2006;27:425–446.

Shitara Y, Sugiyama Y. Pharmacokinetic and pharmacodynamic alterations of 3-hydroxy-3-methylglutaryl coenzyme A (HMG-CoA) reductase inhibitors: drug-drug interactions and interindividual differences in transporter and metabolic enzyme functions. *Pharmacol Ther* 2006;112:71–105.

Skinner MH, Blaschke TF. Clinical pharmacokinetics of rifabutin. *Clin Pharmacokinet* 1995;28:115–125.

Skinner MH, Hsieh M, Torseth J, Pauloin D, Bhatia G, Harkonen S, Merigan TC, Blaschke TF. Pharmacokinetics of rifabutin. *Antimicrob Agents Chemother* 1989;33: 1237–1241.

Takanaga H, Ohnishi A, Murakami H, Matsuo H, Higuchi S, Urae A, Irie S, Furuie H, Matsukuma K, Kimura M, Kawano K, Orii Y, Tanaka T, Sawada Y. Relationship between time after intake of grapefruit juice and the effect on pharmacokinetics and pharmacodynamics of nisoldipine in healthy subjects. *Clin Pharmacol Ther* 2000;67:201–214.

Tam D, Tirona RG, Pang KS. Segmental intestinal transporters and metabolic enzymes on intestinal drug absorption. *Drug Metab Dispos* 2003;31:373–383.

Tateishi T, Watanabe M, Nakura H, Asoh M, Shirai H, Mizorogi Y, Kobayashi S, Thummel KE, Wilkinson GR. CYP3A activity in European American and Japanese men using midazolam as an *in vivo* probe. *Clin Pharmacol Ther* 2001;69:333–339.

Templeton IE, Thummel KE, Kharasch ED, Kunze KL, Hoffer C, Nelson WL, Isoherranen N. Contribution of itraconazole metabolites to inhibition of CYP3A4 *in vivo*. *Clin Pharmacol Ther* 2008;83:77–85.

Thorn M, Finnstrom N, Lundgren S, Rane A, Loof L. Cytochromes P450 and MDR1 mRNA expression along the human gastrointestinal tract. *Br J Clin Pharmacol* 2005;60:54–60.

Thummel KE. Gut instincts: CYP3A4 and intestinal drug metabolism. *J Clin Invest* 2007;117:3173–3176.

Thummel KE, Kunze KL, Shen DD. Enzyme-catalyzed processes of first-pass hepatic and intestinal drug extraction. *Adv Drug Delivery Rev* 1997;27:99–127.

Thummel KE, O'Shea D, Paine MF, Shen DD, Kunze KL, Perkins JD, Wilkinson GR. Oral first-pass elimination of midazolam involves both gastrointestinal and hepatic CYP3A-mediated metabolism. *Clin Pharmacol Ther* 1996;59:491–502.

Tsunoda SM, Velez RL, von Moltke LL, Greenblatt DJ. Differentiation of intestinal and hepatic cytochrome P450 3A activity with use of midazolam as an *in vivo* probe: effect of ketoconazole. *Clin Pharmacol Ther* 1999;66:461–471.

Tubic M, Wagner D, Spahn-Langguth H, Bolger MB, Langguth P. In silico modeling of non-linear drug absorption for the P-gp substrate talinolol and of consequences for the resulting pharmacodynamic effect. *Pharm Res* 2006;23:1712–1720.

Tucker GT, Houston JB, Huang SM. Optimizing drug development: strategies to assess drug metabolism/transporter interaction potential—toward a consensus. *Clin Pharmacol Ther* 2001;70:103–114.

Tukey RH, Strassburg CP. Genetic multiplicity of the human UDP-glucuronosyltransferases and regulation in the gastrointestinal tract. *Mol Pharmacol* 2001;59:405–414.

van de Kerkhof EG, de Graaf IA, Groothuis GM. *In vitro* methods to study intestinal drug metabolism. *Curr Drug Metab* 2007;8:658–675.

Venkatakrishnan K, Obach RS. *In vitro–in vivo* extrapolation of CYP2D6 inactivation by paroxetine: prediction of nonstationary pharmacokinetics and drug interaction magnitude. *Drug Metab Dispos* 2005;33:845–852.

von Richter O, Burk O, Fromm MF, Thon KP, Eichelbaum M, Kivisto KT. Cytochrome P450 3A4 and P-glycoprotein expression in human small intestinal enterocytes and hepatocytes: a comparative analysis in paired tissue specimens. *Clin Pharmacol Ther* 2004;75:172–183.

Wang YH, Jones DR, Hall SD. Prediction of cytochrome P450 3A inhibition by verapamil enantiomers and their metabolites. *Drug Metab Dispos* 2004;32:259–266.

Wang Z, Gorski JC, Hamman MA, Huang SM, Lesko LJ, Hall SD. The effects of St John's wort (Hypericum perforatum) on human cytochrome P450 activity. *Clin Pharmacol Ther* 2001;70:317–326.

Watkins PB. The barrier function of CYP3A4 and P-glycoprotein in the small bowel. *Adv Drug Delivery Rev* 1997;27:161–170.

Wilkinson GR, Shand DG. Commentary: a physiological approach to hepatic drug clearance. *Clin Pharmacol Ther* 1975;18:377–390.

Wojnowski L, Kamdem LK. Clinical implications of CYP3A polymorphisms. *Expert Opin Drug Metab Toxicol* 2006;2:171–182.

Wu CY, Benet LZ. Predicting drug disposition via application of BCS: transport/absorption/elimination interplay and development of a biopharmaceutics drug disposition classification system. *Pharm Res* 2005;22:11–23.

Xie HG, Wood AJ, Kim RB, Stein CM, Wilkinson GR. Genetic variability in CYP3A5 and its possible consequences. *Pharmacogenomics* 2004;5:243–272.

Yang J, Tucker GT, Rostami-Hodjegan A. Cytochrome P450 3A expression and activity in the human small intestine. *Clin Pharmacol Ther* 2004;76:391.

Yang J, Jamei M, Yeo KR, Rostami-Hodjegan A, Tucker GT. Misuse of the well-stirred model of hepatic drug clearance. *Drug Metab Dispos* 2007a;35:501–502.

Yang J, Jamei M, Yeo KR, Tucker GT, Rostami-Hodjegan A. Prediction of intestinal first-pass drug metabolism. *Curr Drug Metab* 2007b;8:676–684.

Yang J, Liao M, Shou M, Jamei M, Yeo KR, Tucker GT, Rostami-Hodjegan A. Cytochrome p450 turnover: regulation of synthesis and degradation, methods for determining rates, and implications for the prediction of drug interactions. *Curr Drug Metab* 2008;9:384–394.

Youdim KA, Zayed A, Dickins M, Phipps A, Griffiths M, Darekar A, Hyland R, Fahmi O, Hurst S, Plowchalk DR, Cook J, Guo F, Obach RS. Application of CYP3A4 *in vitro* data to predict clinical drug–drug interactions; predictions of compounds as objects of interaction. *Br J Clin Pharmacol* 2008;65:680–692.

Yu LX, Amidon GL. A compartmental absorption and transit model for estimating oral drug absorption. *Int J Pharm* 1999;186:119–125.

Zhang QY, Dunbar D, Ostrowska A, Zeisloft S, Yang J, Kaminsky LS. Characterization of human small intestinal cytochromes P-450. *Drug Metab Dispos* 1999;27: 804–809.

10

ENZYME INHIBITION IN VARIOUS *IN VITRO* SYSTEMS

PING ZHAO

10.1 INTRODUCTION

Drug-metabolizing enzymes (DMEs) in the liver are responsible for the elimination of the majority of medicines. Inhibition of hepatic DMEs constitutes one of the major mechanisms of metabolism-based drug–drug interactions (DDI), where the exposure of the victim drug (substrate, S) can be elevated by the concomitantly administered interacting drug (inhibitor, I) to raise safety concern. Metabolism-based DDI has received broad attention by regulatory agencies and pharmaceutical industries during the development of new chemical entities (NCEs). Recently, the U.S. Food and Drug Administration published draft guidance on drug interactions (http://www.fda.gov/cder/guidance/6695dft.pdf, hereafter as FDA guidance). The gold standard in evaluating the DDI risk is to perform controlled clinical trials by coadministering NCE and P450 substrates or inhibitors (FDA guidance). However, before an NCE can be evaluated in human subject, two basic questions are often asked regarding its potential to cause drug interaction: (i) Does the compound strongly inhibit enzyme(s) responsible for the elimination of commonly used medications or the medications it is expected to be combined with (NCE being inhibitor)? and (ii). Is NCE's clearance predominantly mediated by enzyme to be inhibited (NCE being a victim)?

According to FDA guidance, *in vitro* studies, if appropriately validated and designed, can serve as important first step to determine if the *in vivo* tests are

Enzyme Inhibition in Drug Discovery and Development, edited by Lu and Li
Copyright © 2010 John Wiley & Sons, Inc.

needed. The *in vitro* hepatic systems include perfused liver, liver slices, hepatocytes, S9, cytosol, and microsomal fractions. Accumulating knowledge on the mechanisms and the ongoing optimization of various *in vitro* systems derived from human liver tissues allow pharmaceutical companies to routinely evaluate drug metabolism and DDI with the hope to quantitatively predict the *in vivo* outcomes. The practice of using *in vitro* data to predict *in vivo* situation is often referred to as *in vitro–in vivo* extrapolation (IVIVE), which usually requires information on *in vitro* kinetics and pharmacokinetics of both I and S. Take reversible inhibition, for example, if the victim enzyme is responsible for the elimination of S by means of Michaelis–Menten kinetics and [S] is expected to be much lower than K_m; the rate of metabolism equals product of enzyme's intrinsic clearance (CL'_{int}) and substrate concentration: $CL'_{int} \cdot [S]$. Therefore, the extent of enzyme inhibition can be measured by Eq. (10.1):

$$\frac{CL'_{int}}{CL'_{int,INH}} = 1 + \frac{[I]}{K_i} \tag{10.1}$$

where [I] is the (unbound) concentration of inhibitor, and K_i is the (unbound) inhibition constant. Theoretically, one can use *in vivo* exposure of inhibitor to predict the extent of change in CL'_{int}. Indeed, CL'_{int} of substrate serves as the universal currency because of its relationship to *in vivo* systemic clearance, the key pharmacokinetic parameter dictating drug exposure. For detailed theoretical basis of IVIVE on the inhibition of DME, the readers are recommended to reviews by Ito et al. (1998) and Lin and Lu (1999). Table 10.1 summarizes the equations and some basic assumptions used in conventional IVIVE to predict two types of commonly encountered inhibition: reversible inhibition and time-dependent inactivation.

In this chapter, we will review the utility of *in vitro* hepatic systems in characterizing drug inhibition potential of NCEs.

10.2 PREPARATION OF LIVER SAMPLES

10.2.1 Liver Anatomy

Liver is an extremely complicated organ. For detailed structure of the liver, readers are referred to anatomy and physiology textbooks. A brief overview of liver anatomy is provided in this section to facilitate subsequent review of each *in vitro* hepatic system.

Liver is comprised of several cell types differing in architectures and functions. Communication among different cell types is accomplished through an intricate but highly organized network of blood vessels and bile ducts. Liver lobe is the organ's building block. Lobe is made of single layers of hepatocytes sandwiching sinusoids, whose primary role is to deliver blood from hepatic artery and portal vein to hepatic vein. Accounting for the majority of the cell

TABLE 10.1. Equations Used in Conventional IVIVE of Major Inhibition Mechanisms

Types	Possible Mechanisms	Equations for *In Vitro* Parameters	Equations for IVIVE
Reversible inhibition	Competitive inhibition Noncompetitive inhibition (Segel, 1975; Ito et al., 1998; Lin and Lu, 1999)	$\frac{v}{v_{INH}} = 1 + \frac{[I]_{u,inc}}{K_{i,u}}$ General assumptions: Michaelis–Menten kinetics for S, [S] << K_m	$\frac{CL'_{int}}{CL'_{int,INH}} = 1 + \frac{[I]_{u,in\ vivo}}{K_{i,u}}$
Time-dependent inactivation	Covalent binding of the enzyme Heme complexation Tight binding with slow off-rate (Silverman, 1988)	$\lambda = \frac{k_{inact} \times [I]_{u,inc}}{K_{I,u} + [I]_{u,inc}}$ General assumptions: Michaelis–Menten kinetics for S	$\frac{CL'_{int}}{CL'_{int,INH}} = \frac{K_{deg} + \frac{k_{inact} \times [I]_{u,in\ vivo}}{K_{I,u} + [I]_{u,in\ vivo}}}{K_{deg}}$ General assumptions: steady-state conditions have been reached for both S and I *in vivo*

v and v_{INH}: *in vitro* reaction velocities in the absence and in the presence of [I]; $[I]_{u,inc}$: unbound [I] in incubation; $K_{i,u}$: unbound K_i for reversible inhibition; λ: apparent inactivation rate constant; k_{inact}: apparent maximum inactivation rate constant; $K_{I,u}$: unbound inactivation constant with $[I]_u$ causing 50% maximum inactivaiton; K_{deg}: apparent first-order degradation rate constant of DME; $[I]_{u,in\ vivo}$: unbound [I] *in vivo*, concentration near target enzyme; CL'_{int} and $CL'_{int,INH}$: intrinsic clearances by enzyme to be inhibited/inactivated without and with [I].

population, hepatocytes are responsible for many important biological functions, including the synthesis and disposition of endogenous substances as well as the metabolism and transport of xenobiotics. Hepatocytes exchange nutrients, endogenous substances, and xenobiotics through the plasma membrane facing sinusoids. Side by side, hepatocytes connect with each other through tight junctions. The pocket within tight junctions is called bile canaliculus, which forms the smallest unit for bile collection. The distinct characteristics of sinusoidal membrane and canaliculi membrane of hepatocytes is accentuated by different types of transporters expressed on them. Thus, one needs to appreciate liver heterogeneity before selecting an *in vitro* system to study drug inhibition.

10.2.2 Preparation of Subcellular Fractions

Figure 10.1 illustrates the derivation of different *in vitro* systems from liver. Although theoretically feasible as conducted in animal studies, perfused intact liver is prohibited from both cost and ethical standpoints.

Human liver slices are the most physiologically relevant *in vitro* system. The preparation of the precision-cut liver slices involves two steps. First, liver cylinder cores with several millimeters in diameter are prepared from fresh tissue using a motor-driven coring tool. Second, liver slices are harvested using a semiautomated tissue slicer (de Kanter et al., 2002a,b). Oxygenated (95% oxygen/5% carbon dioxide) medium is used in both steps. The average thickness of liver slices is around 200 μm to allow optimal penetration of test compounds into inner layers of hepatocytes. Liver slices maintain the three-dimensional structure of liver lobule and likely preserve cell–cell interaction.

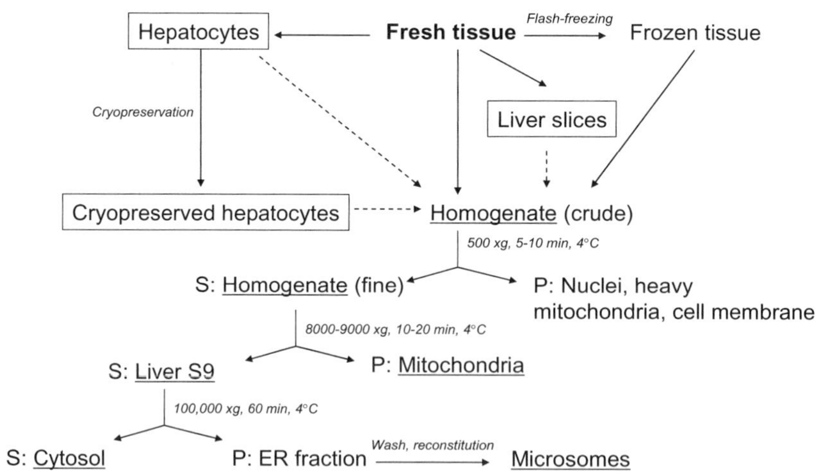

Figure 10.1. Preparation of subcellular fraction from liver tissue.

The next level of tissue preparation is hepatocyte. Depending on how the cells are prepared from fresh tissue, the hepatocyte system can have different final formats. Culture and suspension are now available for both fresh and cryopreserved hepatocytes. For nearly 40 years, collagenase perfusion technique has been adapted as the standard procedure to prepare hepatocytes from fresh tissues (Berry and Friend, 1969). The resulting cells can be plated for longer incubations, or directly used as suspension for shorter incubations. Cryopreservation made it possible to freeze hepatocytes for future use (Li et al., 1999). The resulting cryopreserved hepatocytes can be thawed virtually at any time and hence become attractive "off-the-shelf" reagents like microsomes. Lately, fresh human hepatocytes are becoming "ready-to-use" because several providers have successfully streamlined the process from obtaining fresh tissue to fresh cell preparations as pre-plated or suspension forms. The sandwich-cultured hepatocytes system received its name because a single layer of cells is sandwiched between two collagen layers. This system restores cell–cell interaction and bile canaliculi and has been successfully used to study biliary excretion and drug transport (LeCluyse et al., 1994, Liu et al., 1998, 1999a,b). The interplay between metabolism and transport has also been explored in sandwich-cultured hepatocytes (Hoffmaster et al., 2005; Turncliff et al., 2006).

Liver homogenate and subcellular fractions can be prepared from fresh as well as frozen liver tissues, and in theory from slices and hepatocytes. Homogenate usually requires four to five volumes of homogenizing buffer. The homogeneous preparation is achieved by strong sheer force using a high-speed homogenizer. A "finer" homogenate is centrifuged at low speed (a few hundred $\times g$ force) or filtered to eliminate cell membrane and nuclei components. Centrifugation of homogenate at 8000–9000 $\times g$ results in supernatant known as liver S9 fraction (the number reflecting centrifugation speed of 9000 $\times g$), with pellets containing primarily mitochondria and other "heavier" organelles. Liver S9 fraction can be centrifuged further at ultrahigh speed, normally around 100,000 $\times g$, to precipitate the endothelial reticulum (ER) fraction. The resulting supernatant is cytosol. The ER pellet is usually washed with buffer, recentrifuged, and reconstituted to obtain microsomes. Subcellular fractions and homogenates are normally aliquoted and frozen at below –70 °C before use. Also, time required to obtain each fraction differ markedly from a few minutes for homogenate to more than an hour for microsomes. One expects the significant loss of endogenous substances especially in S9 and microsomes.

Table 10.2 summarizes the physiological relevance of different *in vitro* system to study enzyme inhibition. Liver slices contain the complete metabolism and transport functionalities. Hepatocytes generally preserve major metabolizing enzymes. Hepatocytes in suspension or traditional cultures significantly lose polarity and canaliculi, leading to the alteration of the correct mapping of efflux and uptake transporters. These transporters may play a critical role in the overall disposition of NCE and its metabolites. The

TABLE 10.2. Physiological Relevance of *in Vitro* Human Hepatic Systems for the Study of Metabolism-Based Drug Interactions

In Vitro Systems	Architecture and Cell–cell Interactions	DME	Transporter Functions
Liver slices	Preserved	Complete	Preserved
Hepatocytes			
Suspension	Disrupted; loss of polarity	Complete	Partially preserved; direction unknown
Culture plates (conventional)	Partially preserved; loss of polarity	Complete	Partially preserved; direction unknown
Culture plates (sandwiched)	Restored upon the establishment of canaliculi	Complete; likely down-regulated	Preserved
Homogenate	NA	Complete	NA
S9	NA	Nearly complete	NA
Cytosol	NA	Cytosol specific	NA
Microsomes	NA	ER specific	NA

DME, drug-metabolizing enzyme; ER, endothelial reticulum; NA: not applicable.

metabolism–transport interplay of several compounds has been demonstrated by several groups accentuating the importance of drug transporters (Lam and Benet, 2004; Hoffmaster et al., 2005; Turncliff et al., 2006; Lam et al., 2006). As such, one should be careful interpreting *in vitro* data generated from subcellular systems. Although rarely used, homogenate and S9 possess the advantage of having relatively complete DME machinery. Cytosol is often used when specific liver DMEs are to be studied. Microsomes are the most widely used *in vitro* system due to the high expression of major DMEs (including P450s and UGTs) and the advantage of simplicity and reproducibility.

10.3 *IN VITRO* EXPERIMENTAL DESIGN

Figure 10.2 shows the general experimental procedures to study the inhibition of hepatic DMEs in microsomes. Reversible inhibition experiment aimed at gaining insight in mechanism (i.e., competitive versus noncompetitive) and obtaining K_i values composes a matrix of several inhibitor concentrations and several substrate concentrations (Fig. 10.2A). Definitive time-dependent inactivation experiment aimed at obtaining k_{inact} and K_I values requires an "inactivation incubation" with inactivator for varying times followed by an "activity incubation" with substrate (Fig. 10.2B). During the early stage of drug development, researchers are more interested in IC_{50} values (reversible inhibition)

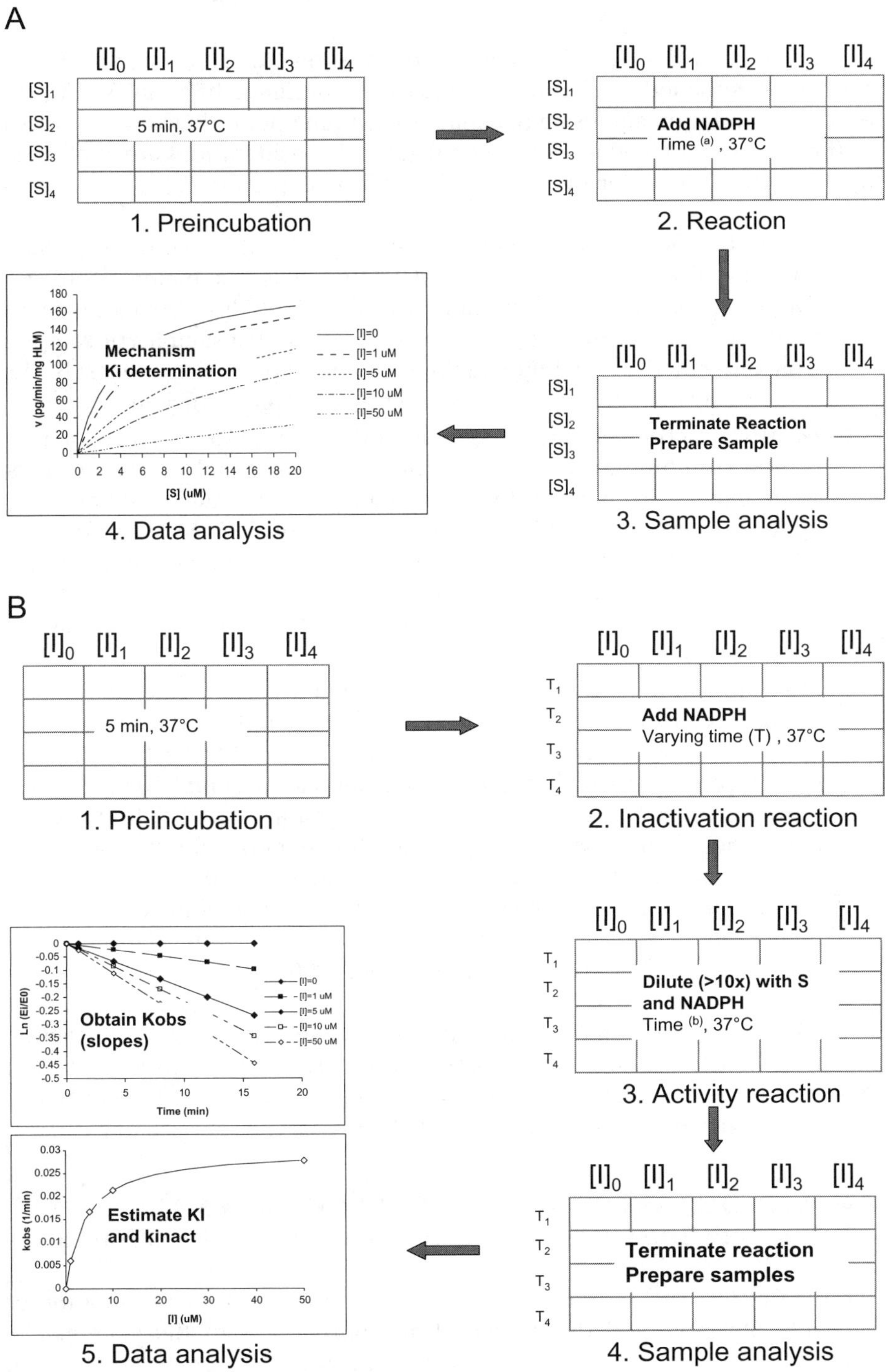

Figure 10.2. General design for drug inhibition experiment *in vitro*. **(A)** Reversible inhibition. **(B)**. Time-dependent inactivation. **(A)** and **(B)**: Incubation time depending on substrate reaction linearity.

or the change in IC_{50} (time-dependent inactivation) to rank-order NCEs. In this case, a simplified version of Fig. 10.2 can be designed by using $S < K_m$ for reversible inhibition and using one inactivation time point (besides time zero) for time-dependent inactivation. Several general considerations are applicable to all *in vitro* systems for the design of inhibition experiment:

a. Given the known interindividual variability of DMEs caused by genetic polymorphisms and environmental factors, it is recommended to use liver samples prepared from multiple donors. In addition, one also needs to decide whether to perform experiment in pooled samples or samples from individual donor. When ranking inhibition potential amongst NCEs is the study objective, a pooled sample is generally sufficient.
b. Relative fractional clearance is an important factor to influence the prediction of inhibition using conventional IVIVE (Table 10.1). For substrate undergoing multiple metabolic pathways, the fraction metabolized by the enzyme to be inhibited (f_m) is the ratio of CL'_{int} by the enzyme to that by all pathways according to the parallel characteristics of clearance concept. As f_m decreases, inhibition becomes less significant. It is important to consider the contribution of noninhibitory pathways when evaluating drug inhibition.
c. Inhibition parameters become equivocal if they are calculated from inhibitor concentrations that do not represent unbound inhibitor concentration at the active site of the enzyme (Table 10.1). Factors influencing I_u include nonspecific binding and metabolic consumption of inhibitor during incubation. It is more critical when parameters are to be generated using hepatocytes, because membrane permeability and transporters are likely to cause concentration gradient across cell membrane. For subcellular systems, the unbound fraction in incubation can be determined both experimentally or by *in silico* prediction (Obach, 1997; Austin et al., 2005; Hellifax and Houston, 2006).

In addition, incubation time, concentration ranges for both S and I, reaction linearity against time and enzyme concentrations, and inhibitory product(s) also need to be considered. For detailed discussions, readers are referred to the FDA guidance and Bjornsson et al. (2003).

10.4 APPLICATION OF *IN VITRO* HEPATIC SYSTEMS IN STUDYING INHIBITION OF DRUG METABOLIZING ENZYMES

Choosing an *in vitro* system to study DDI requires careful consideration of the study objective and the biological appropriateness of the system. The objective of the study certainly dictates the selection process. Although general perception favors the use of intact systems such as hepatocytes, subcellular fractions often provide satisfactory results if the objective is to obtain inhibition parameters, because they represent an environment more closely to purified enzymes. This is especially useful when rank-ordering DDI potential of a series of NCEs of the same chemical class in drug discovery to facilitate the timely decision-making progress. A handful of studies have reported parameters generated using hepatocytes (Yamazaki et al., 2004; Brown et al., 2007; Van et al., 2007). More *in vitro* studies aimed at validating the utility of hepaotcyte parameters for conventional IVIVE are needed. Next, the knowledge of subcellular distribution of the enzyme of interest is critical (Table 10.2). The use of liver microsomes, rather than cytosol, to study P450 inhibition is an obvious example. Equally important, while historically less evaluated, is the impact of concurring mechanisms, such as sequential metabolism and transport of both substrate and inhibitor, on the enzyme to be inhibited. The urge to comprehensively understand the interplay between metabolism and transport is a result of frequent occurrence of mis-prediction by subcellular fractions using conventional IVIVE procedure illustrated in Section 10.1. With the advancement of hepatocytes as a more ready-to-use reagent, it has become popular to study metabolism and drug interactions in hepatocytes (Hewitt et al., 2007). Although it is possible to obtain hepatocyte K_i, the fundamental limitations of conventional IVIVE do not seem to be overcome by directly using hepatocyte parameters. The unique advantage of hepatocytes as an integrated system prompts us to critically assess the alternative utilities other than another system to generate parameters (McGinnity et al., 2005; Van et al., 2007; Paine et al., 2008; Zhao, 2008; Lu et al., 2007, 2008).

In this section, we will discuss the advantage and disadvantage of each *in vitro* system. The flow begins with the simplest system, microsomes, which have been extensively used to study enzyme inhibition (Fig. 10.1).

10.4.1 Subcellular Systems

10.4.1.1 Liver Microsomes. Since P450s are responsible for the majority of metabolism of xenobiotics in humans, microsomes have been the system of choice to study drug metabolism and DDI. Besides its simplicity and reproducibility, the microsomal system has abundant levels of P450s and other major DMEs such as FMO and UGTs. Microsomes are often capable of metabolizing compounds at rates robust enough for researchers to design a variety of inhibition studies. The resulting kinetic parameters can be used in conventional IVIVE (Table 10.1).

As mentioned in Section 10.3, experiments using gender-pooled microsomal sample prepared from 50–60 donors are deemed sufficient if the objective is to rank inhibition potential of a series of compounds. On the other hand, inhibition experiments performed in individual microsomal samples across a panel of donors appear to reveal more interindividual variability of the victim enzyme in response to inhibition, although intensive lab activity and more sophisticated data analysis are required.

Factors influencing microsomal parameters such as nonspecific binding (Obach, 1996, 1999; Tucker et al., 2001; Austin et al., 2002) and metabolic turnover of inhibitors (Zhao et al., 2005; Van et al., 2007) can be corrected to be meaningful for further usage in the equations of IVIVE (Table 10.1). Microsomal unbound fraction ($f_{u,\text{mic}}$) can be obtained in an experimental design similar to plasma protein binding study or from *in silico* predictions (Austin et al., 2002, 2005; Hallifax and Houston, 2006). One approach to correct metabolic consumption of inhibitor is to calculate a time-averaged inhibitor concentration by dividing partial area under the [I] versus incubation time curve by reaction time ($I_{u,\text{ave}}$, Zhao et al., 2005). The abundance of P450 levels and the homogeneous incubation condition grant microsomes the advantage to be applied in the investigation of the effect of parallel pathways on the inhibition of specific CYP isozyme. The value of f_m for substrate and the selectivity of inhibitor can be evaluated from routine CYP phenotyping studies (Bjornsson et al., 2003; FDA guidance). The continuous discovery of selective inhibitors of CYP isoforms makes the chemical inhibition in microsomes the primary approach to obtain f_m for compounds undergoing P450 metabolism. In addition, the challenge on selectivity of chemical inhibitors can be overcome by calculating f_m contributed by each CYP isoform using the approach proposed by Lu et al. (2007).

The disadvantage of microsomal system in studying DDI also stems from its simplicity and homogeneity. When non-CYP enzymes (i.e., phase II such as UGTs) or nonmicrosomal enzymes are involved, relying on f_m generated using CYP phenotyping in microsomes may result in mis-prediction. Theoretically, one can supplement microsomes with uridine phosphate glucuronic acid (UDPGA) in addition to NADPH to identify parallel UGT mediated pathways and to assess its influence on the interpretation of CYP inhibition. However, the utility of this system requires further validation given the known sensitivity of UGT-mediated reaction toward exogenous substances added to microsomes (Fisher et al., 2000).

10.4.1.2 Liver S9. Liver S9 shares microsome's properties of simplicity and homogeneity. The inclusion of enzymes primarily of cytosolic origin overcomes the disadvantage of microsomes. Thus, upon supplement with cofactors, S9 can be a cost-effective substitute of microsomes, although reports on inhibition studies using S9 have been limited. The unpopularity is likely due to lower abundance of ER enzymes in S9 incubations than in microsomes at the same protein concentration (i.e., mg/mL incubation), leading to lower metabolic turnover. Several labs reported the use of S9 to study the inhibition of DMEs, including carbonyl reductase by quercetin (Holleran et al., 2004) and carboxyesterase by tri-bis-(*p*-nitrophenyl phosphate) as well as grapefruit juice contents (Tabata et al., 2004; Li et al., 2007a,b). The mechanism of competitive inhibition of glucuronidation of valpric acid by probenecid was confirmed in rat liver S9 incubations (Ward et al., 2000). These findings suggest that liver S9 can be a cost-effective alternative of liver microsomes, especially when exploration of non-P450 enzymes is of interest.

10.4.1.3 Other Subcellular Fractions. Inhibition studies have been performed for cytosolic-specific enzymes such as *N*-acetyl-transferases (Vrtic et al., 2003; Chen et al., 2000), glutathione-*S*-transferases (Chen et al., 2000), dihydrodiol dehydrogenase (Porter et al., 2000), catechol-*O*-methyl transferase (Nagai et al., 2004), and aldehyde dehydrogenases (Ren and Slattery, 1999). These enzymes may be important in the metabolism of certain drugs. Inhibition kinetics of aldehyde dehydrogenase by carmustine (BCNU) was studied using human liver cytosol and erythrocytes (Ren and Slattery, 1999). The persistent inhibition of aldehyde dehydrogenase-mediated bioactivation of cyclophosphamide by four split doses of BCNU in the BCNU/cyclophosphamide/etoposide regimen was hypothesized to result in lower incidence of cyclophosphamide-induced venoocclusive disease of the liver (Ren and Slattery, 1999). Another important cytosolic enzyme aldehyde oxidase catalyzes both oxidation and reduction reactions (Obach and Walsky, 2005; Obach, 2004). The inhibition of aldehyde oxidase appears to be dictated by the type of reaction. For example, reloxifene inhibits oxidation reaction through noncompetitive mechanism with K_i at subnanomolar values, whereas it is a noncompetitive inhibitor of the reduction reaction with a K_i of 51 nM (Obach, 2004). These results suggest the complexity of these nonmicrosomal enzymes.

Liver homogenate sits one level higher than liver S9, therefore encompassing more complete metabolic machinery; however, the use of homogenates to study enzyme inhibition is rarely reported. Several isoforms of phase I and phase II enzymes residing in mitochondria have been shown to contribute to the metabolism of compounds. For example, CYP2E1 derived from rat liver mitochondria is responsible to approximately 20% of the overall bioactivation of acetaminophen to for the reactive intermediate iminoquinone metabolite (Neve and Ingelman-Sundberg, 2001; Robin et al., 2001); however, the local proximity has significant implication because mitochondrion is likely a primary target of acetaminophen-mediated hepatotoxicity. In order to isolate the contribution from mitochondria, one needs to obtain kinetic parameters from different *in vitro* systems including homogenate.

10.4.2 Intact Systems

10.4.2.1 Hepatocytes. Hepatocyte resembles intact liver more closely than subcellular systems because it contains complete machinery for drug transport and metabolism. In theory, hepatocytes also eliminate the use of cofactors required for subcellular systems. Historically, the use of human hepatocytes has been hindered by cost, labor-intensive cell preparation, and long-term storage. The development of more "ready-to-use" products, including cryopreserved hepatocytes and preplated freshly isolated hepatocytes, have attracted investigators from both pharmaceutical industries and academic laboratories (Li et al., 1999; Lu and Li, 2001; Liu et al., 1999a,b; LeCluyse et al., 1994, 2005). In fact, during a 5-year time period between 2002 and 2007, the number of publications using hepatocytes to study drug metabolism has increased whereas the numbers using microsomes has decreased (Hewitt et al., 2007).

The cell integrity and wider window of incubation time give the hepatocyte unique flexibility to be applied under a variety of experimental designs. Nonspecific binding and uptake transport can be evaluated using techniques such as centrifugation through oil layers or washing the cells with fresh media to obtain intracellular and extracellular drug concentrations. One of the criteria to ensure the quality of hepatocyte system is viability, which is usually tested to last for hours for suspension and days for plated cells. Inhibition experiment aimed at exploring inhibition kinetics described in Section 10.3 can be performed in relatively short period of time to satisfy general requirements such as reaction linearity. On the other hand, hepatocytes can be incubated for hours or even days to reveal competing mechanisms, including concomitant induction and inhibition, the influence of phase II pathways on CYP inhibition potency, the effect of inhibitory metabolites from substrate and/or inhibitor, and the effect of active uptake/efflux transport.

To date, a few studies reported inhibition or inactivation parameters generated using hepatocytes (Yamazaki et al., 2004; Brown et al., 2007; Van et al., 2007). The emphasis has been placed on comparing the unbound parameters between hepatocytes and microsomes or supersomes. More data seem to be required to validate and justify the utility of the hepatocyte parameter to predict drug inhibition using conventional IVIVE (Table 10.2). Take reversible inhibition for example, microsomal unbound K_i is intuitively "cleaner" thus better represents the enzyme of interest, whereas an unambiguous unbound hepatocyte K_i cannot be readily verified until the concurring metabolism and transport events are understood. This should not be confused with the prediction of intrinsic hepatic clearance using hepatocytes as routinely performed in drug discovery, where the metabolism is "globally" assessed in an environment with complete machinery that is not represented in subcellular systems (McGinnity et al., 2004; Lu et al., 2006).

Instead of focusing on hepatocyte inhibition parameters, several groups have attempted to comprehensively evaluate drug inhibition *in vitro* by taking advantage of the physiological resemblance and heterogeneity of hepatocytes. Different from the conventional IVIVE, these new approaches evaluate hepatocyte data and subcellular data together with the aid of additional mathematical treatments including modeling and simulation. In fact, longer incubation time allows hepatocyte experiments to be designed to mimic *in vivo* situation, where multiple mechanisms can be assessed simultaneously (Turncliff et al., 2006; McGinnity et al., 2005; Riley et al., 2005; Zhao et al., 2007; Paine et al., 2008). The concentration–time profiles of both parent drug (either substrate or inhibitor) and their metabolites can be fitted to pharmacokinetic models to reveal the impact of transporters and inhibitory metabolites on the metabolic events, including enzyme inhibition. Fitting the hepatocyte data to a compartmental model, Paine et al. (2008) clearly showed the importance of hepatic uptake with regard to the disposition kinetics of compounds incubated in rat hepatocyte suspension, improved prediction of *in vivo* hepatic clearance in an analogous physiological-based pharmacokinetic model was also presented (Paine et al., 2008). Such mechanistic experimental design can also test the predictability of microsomal parameters (McGinnity et al., 2005; Zhao et al., 2007; Van et al., 2007). Another example of comprehensively using combined data from hepatocytes and microsomes to predict drug inhibition *in vivo* attempt to circumvent the use of *in vitro* K_i, which is known to vary dramatically, depending on how *in vitro* experiments are conducted (Lu et al., 2007, 2008). Reversible CYP inhibition by ketoconazole was titrated in hepatocytes suspended in plasma. Combining microsomal f_m values of substrates and the reported *in vivo* exposure of ketoconazole, the degree of inhibition can be directly extrapolated assuming that the *in vivo* plasma total [I] can be represented by ketoconazole concentration in hepatocytes incubated in plasma. The predicted inhibition appears to agree very well with *in vivo* observations. Theoretically, a sandwich-cultured hepatocyte system should be superior to traditional cell cultures or suspensions (Table 10.2). The utility of this delicate *in vitro* system has focused on drug transport to evaluate hepatic uptake and biliary excretion (Liu et al., 1999a,b). Although the evaluation of enzyme inhibition in sandwich-cultured human hepatocytes using aforementioned mechanistic modeling will be very informative, the effort may have been hampered by the preservation of DMEs after sandwich culturing of the cells (Turncliff et al., 2006).

In summary, besides serving as another *in vitro* system to generate inhibition parameters for conventional IVIVE, hepatocytes can be used in combination with subcellular data and sophisticated modeling and simulation to better predict enzyme inhibition *in vivo*.

10.4.2.2 Liver Slices. Liver slices are unequivocally the closest system to mimic the *in vivo* situation. Besides the advantages and mechanistic experimental design mentioned in the previous section for hepatocytes, liver slices can be used to monitor events other than metabolism and transport, such as pharmacological and toxicological responses. The major drawbacks for liver slices are cost and their labor-intensive preparations. Thus, liver slices appear to be a more appropriate target system where liver enzyme inhibition and its toxicological consequence are evaluated together, rather than being used as a predicting system for conventional IVIVE.

10.5 CONCLUSION

Multiple liver systems have been used to assess the inhibition of DMEs *in vitro*. The selection of one system over another depends on the objective of the study and the composition of DMEs of the system to be selected. Liver microsomes continue to be the system of choice when obtaining inhibition parameters, and ranking inhibition potential are the major objectives. In contrast to conventional IVIVE, which relies primarily on microsomal inhibition parameters, hepatocyte experiments have been designed to comprehensively investigate enzyme inhibition in the presence of competing mechanisms. In addition, alternative approaches taking advantage of the integrity and heterogeneity of hepatocyte systems have emerged to help us better predict drug inhibition *in vivo* from *in vitro* data.

REFERENCES

Austin RP, Barton P, Cockroft SL, Wenlock MC, Riley RJ. The influence of nonspecific microsomal binding on apparent intrinsic clearance and its prediction from physicochemical properties. *Drug Metab Dispos* 2002;30:1497–1503.

Austin RP, Barton P, Mohmed S, Riley RJ. The binding of drugs to hepatocytes and its relationship to physicochemical properties. *Drug Metab Dispos* 2005;33:419–425.

Berry MN, Friend DS. High-yield preparation of isolated rat liver parenchymal cells: a biochemical and fine structural study. *J Cell Biol* 1969;43(3):506–520.

Bjornsson TD, Callaghan JT, Einolf HJ, Fischer V, Gan L, Grimm S, Kao J, King SP, Miwa G, Ni L, Kumar G, McLeod J, Obach RS, Roberts S, Roe A, Shah A, Snikeris F, Sullivan JT, Tweedie D, Vega JM, Walsh J, Wrighton SA. Pharmaceutical Research and Manufacturers of America (PhRMA) Drug Metabolism/Clinical Pharmacology Technical Working Group; FDA Center for Drug Evaluation and Research (CDER). The conduct of in vitro and in vivo drug-drug interaction studies: a Pharmaceutical Research and Manufacturers of America (PhRMA) perspective. *Drug Metab Dispos* 2003;31(7):815–832.

Brown HS, Griffin M, Houston JB. Evaluation of cryopreserved human hepatocytes as an alternative *in vitro* system to microsomes for the prediction of metabolic clearance. *Drug Metab Dispos* 2007;35: 293–301.

Chen TL, Wu CH, Chen TG, Tai YT, Chang HC, Lin CJ. Effects of propofol on functional activities of hepatic and extrahepatic conjugation enzyme systems. *Br J Anaesth* 2000;84(6):771–776.

de Kanter R, Monshouwer M, Meijer DK, Groothuis GM. Precision-cut organ slices as a tool to study toxicity and metabolism of xenobiotics with special reference to non-hepatic tissues. *Curr Drug Metab* 2002a;3(1):39–59.

de Kanter R, De Jager MH, Draaisma AL, Jurva JU, Olinga P, Meijer DK, Groothuis GM. Drug-metabolizing activity of human and rat liver, lung, kidney and intestine slices. *Xenobiotica* 2002b;32(5):349–362.

Fisher MB, Campanale K, Ackermann BL, VandenBranden M, Wrighton SA. *In vitro* glucuronidation using human liver microsomes and the pore-forming peptide alamethicin. *Drug Metab Dispos* 2000;28:560–566.

Hallifax D, Houston JB. Binding of drugs to hepatic microsomes: comment and assessment of current prediction methodology with recommendation for improvement. *Drug Metab Dispos* 2006;34(4):724–726

Hewitt NJ, Lechón MJ, Houston JB, Hallifax D, Brown HS, Maurel P, Kenna JG, Gustavsson L, Lohmann C, Skonberg C, Guillouzo A, Tuschl G, Li AP, LeCluyse E, Groothuis GM, Hengstler JG. Primary hepatocytes: current understanding of the regulation of metabolic enzymes and transporter proteins, and pharmaceutical practice for the use of hepatocytes in metabolism, enzyme induction, transporter, clearance, and hepatotoxicity studies. *Drug Metab Rev* 2007;39(1): 159–234.

Hoffmaster KA, Zamek-Gliszczynski MJ, Pollack GM, Brouwer KL. Multiple transport systems mediate the hepatic uptake and biliary excretion of the metabolically stable opioid peptide [D-penicillamine2,5]enkephalin. *Drug Metab Dispos* 2005; 33(2):287–293.

Holleran JL, Fourcade J, Egorin MJ, Eiseman JL, Parise RA, Musser SM, White KD, Covey JM, Forrest GL, Pan SS. *In vitro* metabolism of the phosphatidylinositol 3-kinase inhibitor, wortmannin, by carbonyl reductase. *Drug Metab Dispos* 2004; 32(5):490–496.

Ito K, Iwatsubo T, Kanamitsu S, Ueda K, Suzuki H, Sugiyama Y. Prediction of pharmacokinetic alterations caused by drug–drug interactions: metabolic interaction in the liver. *Pharmacol Rev* 1998;50(3):387–412.

Lam JL, Benet LZ. Hepatic microsome studies are insufficient to characterize *in vivo* hepatic metabolic clearance and metabolic drug–drug interactions: studies of digoxin metabolism in primary rat hepatocytes versus microsomes. *Drug Metab Dispos* 2004;32(11):1311–1316.

Lam JL, Okochi H, Huang Y, Benet LZ. *In vitro* and *in vivo* correlation of hepatic transporter effects on erythromycin metabolism: characterizing the importance of transporter–enzyme interplay. *Drug Metab Dispos* 2006;34(8):1336–1344.

LeCluyse EL, Audus KL, Hochman JH. Formation of extensive canalicular networks by rat hepatocytes cultured in collagen-sandwich configuration. *Am J Physiol* 1994;266(6 Pt 1):C1764–C1774.

LeCluyse EL, Alexandre E, Hamilton GA, Viollon-Abadie C, Coon DJ, Jolley S, Richert L. Isolation and culture of primary human hepatocytes. *Methods Mol Biol* 2005;290:207–229.

Li AP, Lu C, Brent JA, Pham C, Fackett A, Ruegg CE, Silber PM. Cryopreserved human hepatocytes: characterization of drug-metabolizing enzyme activities and applications in higher throughput screening assays for hepatotoxicity, metabolic stability, and drug–drug interaction potential. *Chem Biol Interact* 1999;121(1): 17–35.

Li P, Callery PS, Gan LS, Balani SK. Esterase inhibition attribute of grapefruit juice leading to a new drug interaction. *Drug Metab Dispos* 2007a;35(7):1023–1031.

Li P, Callery PS, Gan LS, Balani SK. Esterase inhibition by grapefruit juice flavonoids leading to a new drug interaction. *Drug Metab Dispos* 2007b;35(7):1203–1208.

Lin JH, Lu AY. Inhibition and induction of cytochrome P450 and the clinical implications. *Clin Pharmacokinet* 1999;35(5):361–390.

Liu X, Brouwer KL, Gan LS, Brouwer KR, Stieger B, Meier PJ, Audus KL, LeCluyse EL. Partial maintenance of taurocholate uptake by adult rat hepatocytes cultured in a collagen sandwich configuration. *Pharm Res* 1998;15(10):1533–1539.

Liu X, Chism JP, LeCluyse EL, Brouwer KR, Brouwer KL. Correlation of biliary excretion in sandwich-cultured rat hepatocytes and *in vivo* in rats. *Drug Metab Dispos* 1999a;27(6):637–644.

Liu X, LeCluyse EL, Brouwer KR, Lightfoot RM, Lee JI, Brouwer KL. Use of Ca^{2+} modulation to evaluate biliary excretion in sandwich-cultured rat hepatocytes. *J Pharmacol Exp Ther* 1999b;289(3):1592–1599.

Lu C, Li AP. Species comparison in P450 induction: effects of dexamethasone, omeprazole, and rifampin on P450 isoforms 1A and 3A in primary cultured hepatocytes from man, Sprague–Dawley rat, minipig, and beagle dog. *Chem Biol Interact* 2001;16;134(3):271–281.

Lu C, Li P, Gallegos R, Uttamsingh V, Xia CQ, Miwa GT, Balani SK, Gan LS. Comparison of intrinsic clearance in liver microsomes and hepatocytes from rats and humans: evaluation of free fraction and uptake in hepatocytes. *Drug Metab Dispos* 2006;34(9):1600–1605

Lu C, Miwa GT, Prakash SR, Gan LS, Balani SK. A novel model for the prediction of drug–drug interactions in humans based on *in vitro* cytochrome p450 phenotypic data. *Drug Metab Dispos* 2007;35(1):79–85

Lu C, Hatsis P, Berg C, Lee FW, Balani SK. Prediction of pharmacokinetic drug–drug interactions using human hepatocyte suspension in plasma and cytochrome P450 phenotypic data. II. *In vitro–in vivo* correlation with ketoconazole. *Drug Metab Dispos* 2008;36(7):1255–1260

McGinnity DF, Soars MG, Urbanowicz RA, Riley RJ. Evaluation of fresh and cryopreserved hepatocytes as *in vitro* drug metabolism tools for the prediction of metabolic clearance. *Drug Metab Dispos* 2004;32(11):1247–1253.

McGinnity DF, Tucker J, Trigg S, Riley RJ. Prediction of CYP2C9-mediated drug–drug interactions: a comparison using data from recombinant enzymes and human hepatocytes. *Drug Metab Dispos* 2005;33(11):1700–1707.

Nagai M, Conney AH, Zhu BT. Strong inhibitory effects of common tea catechins and bioflavonoids on the *O*-methylation of catechol estrogens catalyzed by human liver cytosolic catechol-*O*-methyltransferase. *Drug Metab Dispos* 2004;32(5):497–504.

Neve EP, Ingelman-Sundberg M. Identification and characterization of a mitochondrial targeting signal in rat cytochrome P450 2E1 (CYP2E1). *J Biol Chem* 2001; 276(14):11317–11322.

Obach RS. The importance of nonspecific binding in *in vitro* matrices, its impact on enzyme kinetic studies of drug metabolism and implications for *in vitro–in vivo* correlations. *Drug Metab Dispos* 1996;24:1047–1049

Obach RS. Nonspecific binding to microsomes: impact on scale-up of *in vitro* intrinsic clearance to hepatic clearance as assessed through examination of warfarin, imipramine, and propranolol. *Drug Metab Dispos* 1997;25(12):1359–1369.

Obach RS. Prediction of human clearance of twenty-nine drugs from hepatic microsomal clearance data: an examination of *in vitro* half-life approach and nonspecific binding to microsomes. *Drug Metab Dispos* 1999;27:1350–1359.

Obach RS. Potent inhibition of human liver aldehyde oxidase by raloxifene. *Drug Metab Dispos* 2004;32(1):89–97.

Obach RS, Walsky RL. Drugs that inhibit oxidation reactions catalyzed by aldehyde oxidase do not inhibit the reductive metabolism of ziprasidone to its major metabolite, *S*-methyldihydroziprasidone: an *in vitro* study. *J Clin Psychopharmacol* 2005; 25(6):605–608.

Paine SW, Parker AJ, Gardiner P, Webborn PJ, Riley RJ. Prediction of the pharmacokinetics of atorvastatin, cerivastatin, and indomethacin using kinetic models applied to isolated rat hepatocytes. *Drug Metab Dispos* 2008;36(7):1365–1374.

Porter SJ, Somogyi AA, White JM. Kinetics and inhibition of the formation of 6beta-naltrexol from naltrexone in human liver cytosol. *Br J Clin Pharmacol* 2000;50(5): 465–471.

Riley RJ, McGinnity DF, Austin RP. A unified model for predicting human hepatic, metabolic clearance from *in vitro* intrinsic clearance data in hepatocytes and microsomes. *Drug Metab Dispos* 2005;33:1304–1311.

Ren S, Slatterly JT. Inhibition of carboxyethylphosphoramide mustard formation from 4-hydroxycyclophosphamide by carmustine. *AAPS PharmSci.* (3) article 14 (http://www.pharmsci.org)

Robin MA, Anandatheerthavarada HK, Fang JK, Cudic M, Otvos L, Avadhani NG. Mitochondrial targeted cytochrome P450 2E1 (P450 MT5) contains an intact N terminus and requires mitochondrial specific electron transfer proteins for activity. *J Biol Chem* 2001;276(27):24680–24689.

Segel, JH. *Enzyme Kinetics*. John Wiley & Sons, New York: 1975, p. 127.

Silverman, RB. *Mechanism-Based Enzyme Inactivation: Chemistry and Enzymology*. Boca Raton, FL: CRC Press.

Tabata T, Katoh M, Tokudome S, Nakajima M, Yokoi T. Identification of the cytosolic carboxylesterase catalyzing the 5'-deoxy-5-fluorocytidine formation from capecitabine in human liver. *Drug Metab Dispos* 2004;32(10):1103–1110.

Tucker GT, Houston JB, Huang S-M. Optimising drug development: strategies to assess drug metabolism/transporter interaction potential—toward a consensus. *Clin Pharmacol Ther* 2001;70:103–114.

Turncliff RZ, Hoffmaster KA, Kalvass JC, Pollack GM, Brouwer KL. Hepatobiliary disposition of a drug/metabolite pair: comprehensive pharmacokinetic modeling in sandwich-cultured rat hepatocytes. *J Pharmacol Exp Ther* 2006;318(2):881–889.

Van LM, Swales J, Hammond C, Wilson C, Hargreaves JA, Rostami-Hodjegan A. Kinetics of the time-dependent inactivation of CYP2D6 in cryopreserved human hepatocytes by methylenedioxymethamphetamine (MDMA). *Eur J Pharm Sci* 2007;31(1):53–61.

Vrtic F, Haefeli WE, Drewe J, Krähenbühl S, Wenk M. Interaction of ibuprofen and probenecid with drug metabolizing enzyme phenotyping procedures using caffeine as the probe drug. *Br J Clin Pharmacol* 2003;55(2):191–198.

Ward ES, Pollack GM, Brouwer KL. Probenecid-associated alterations in valproic acid pharmacokinetics in rats: can *in vivo* disposition of valproate glucuronide be predicted from *in vitro* formation data? *Drug Metab Dispos* 2000;28(12):1433–1439.

Yamazaki S, Toth LN, Black ML, Duncan JN. Comparison of prediction methods for *in vivo* clearance of (*S, S*)-3-[3-(methylsulfonyl)phenyl]-1-propylpiperidine hydrochloride, a dopamine D2 receptor antagonist, in humans. *Drug Metab Dispos* 2004;32(4):398–404.

Zhao P, Kunze KL, Lee CA. Evaluation of time-dependent inactivation of CYP3A in cryopreserved human hepatocytes. *Drug Metab Dispos* 2005;33(6):853–861.

Zhao P, Lee CA, Kunze KL. Sequential metabolism is responsible for diltiazem-induced time-dependent loss of CYP3A. *Drug Metab Dispos* 2007;35(5):704–712.

Zhao P. The use of hepatocytes in evaluating time-dependent inactivation of P450 *in vivo*. *Expert Opin Drug Metab Toxicol* 2008;4(2):151–164. Review.

11

CYTOCHROME P450 DEGRADATION AND ITS CLINICAL RELEVANCE

MINGXIANG LIAO, PING KANG, BERNARD P. MURRAY, AND MARIA ALMIRA CORREIA

11.1 INTRODUCTION

Cytochromes P450 (abbreviated P450s or CYPs) are a very large and diverse superfamily of hemoprotein enzymes present in all organisms, including eukaryotes and prokaryotes. P450s are responsible for the metabolism of a variety of exogenous and endogenous compounds, via multiple mechanisms such as oxidations, reductions, and dehydrogenations (Correia, 2005; Guengerich, 2005). Thus far, more than 7700 distinct P450s have been identified and classified into families and subfamilies; of these, 57 human P450s belonging to P450 1–4 families are believed to be responsible for drug metabolism. More than 85% of the drugs currently on the market are metabolized by P450s (Evans and Relling, 1999).

The P450 content is regulated by various factors including physiological, pathological, genetic, and environmental; indeed, P450 expression is induced by exogenous and endogenous ligands through *de novo* protein synthesis as well as inhibited degradation. Some substrates, such as phenobarbital, induce P450s mainly through increased synthesis, whereas others such as troleandomycin, acetone, and ethanol "induce" other P450s (CYP3A, CYP2E1, and CYP2B1) through protein stabilization. By contrast, some suicide inactivators such as CCl_4, 3,5-dicarbethoxy-2,6-dimethyl-4-ethyl-1,4-dihydropyridine, and

Enzyme Inhibition in Drug Discovery and Development, edited by Lu and Li
Copyright © 2010 John Wiley & Sons, Inc.

6′,7′-dihydroxybergamottin (grapefruit furanocoumarin) accelerate the degradation of certain P450s (Correia et al., 1983; Correia, 1991, 2003; Watkins et al., 1986, 1987; Song et al., 1989). The P450 isoforms exhibit various half-lives ranging from 7 to 20 h (Correia, 1991, 2003; Watkins et al., 1987; Dai and Cederbaum, 1995), and their degradation is proposed to involve different mechanisms.

An increasing amount of evidence indicates that native and inactivated P450s are degraded via two major proteolytic pathways—autophagic-lysosomal and proteasomal degradation—but information regarding the functional impact of P450 degradation on clinical therapeutics is limited. As discussed below, substrate/inactivator-induced P450 stabilization/degradation would alter P450 content and activity, which may lead to drug–drug interactions (DDIs) (Lown et al., 1997; Schmiedlin-Ren et al., 1997, Paine and Oberlies, 2007). Furthermore, mounting evidence documents that covalent binding of metabolites to P450s followed by their proteasomal degradation may lead to idiosyncratic syndromes through antigen presentation engendered during this proteolytic process. Indeed, anti-P450 autoantibodies have been detected in the serum of patients suffering from autoimmune hepatitis or drug-induced hepatitis (Leeder et al., 1996; Bourdi et al., 1996; Uetrecht, 2005, 2007; Eliasson and Kenna, 1996). Furthermore, the exploration of the degradation of various allelic P450 variants may provide some useful information on the influence of P450 polymorphisms on drug action and disease. For instance, CYP1B1 is implicated in chemical carcinogenesis, and thus the variable degradation rates of its polymorphic forms may contribute to differences in cancer susceptibility in individuals (Bandiera et al., 2005). The elucidation of P450 proteolytic processes may provide deep insight into their biochemical and therapeutic significance, thereby assisting researchers in the pharmaceutical industry in new drug discovery and development and also assisting hospital clinicians in improving therapy. This chapter addresses our current concepts about proteolytic pathways of P450 degradation and their potential clinical relevance.

11.2 PROTEIN DEGRADATION

In the 1950s, Christian de Duve discovered the lysosome in the rat liver (de Duve et al., 1953). This not only led to the recognition that protein degradation occurs in this cellular organelle, but also led to the discovery that intracellular proteins are regulated by both synthesis and degradation. Later work in the 1970s showed that lysosomal degradation could not account for all proteolytic processes and led to the discovery of a new proteolytic degradation system in reticulocytes. This marked a new era for mechanistic studies of protein degradation. The proteasome was discovered in 1976 (Goldberg and St. John, 1976), and then, at the beginning of 1980s, ubiquitin was shown to conjugate covalently to protein substrates to tag them for proteolysis

(Ciechanover et al., 1980), a discovery for which the 2004 Nobel Prize in Chemistry was awarded. To date, protein degradation has been widely recognized to play important roles in almost all essential cellular processes, such as signal transduction, cell cycle regulation, antigen presentation, transcriptional regulation, cell differentiation, and quality control (Goldberg, 1991, 2007; Hershko and Ciechanover, 1992, 1998; Shringarpure et al., 2001; Tai and Schuman, 2008). Protein degradation in mammalian cells proceeds largely via two major routes: the autophagic-lysosomal and proteasomal pathways.

11.2.1 Autophagic-Lysosomal Degradation

Before the discovery of the proteasomal degradation system, lysosomes were believed to be the main loci for protein degradation. Lysosomes are organelles containing a wide variety of hydrolases which digest excess or damaged organelles, particles, microbes, and some subcellular components, such as membranes, proteins, lipids, and nucleic acids. Membranes surround the lysosomes, allowing them to maintain an internal acidic pH of ~4.5, for optimal function of the lysosomal proteases. The degradation of organellar/cytoplasmic constituents within lysosomes is termed autophagic-lysosomal degradation (ALD) (Cuervo et al., 2004; Shintani and Klionsky, 2004, Klionsky, 2005, 2007; Mizushima and Klionsky, 2007). ALD plays important roles in various physiological and pathophysiological processes in mammalian cells, such as starvation response, intracellular clearance/quality control, microorganism elimination, cell death, tumor suppression, and antigen presentation of MHC class II molecules (Mizushima, 2005).

In mammalian cells and yeast, ALD occurs via three different processes: microautophagy, chaperone-mediated autophagy, and macroautophagy (Cuervo and Dice, 1998; Dice, 1987; Dice et al., 1990; Kopitz et al., 1990; Seglen, 1987). Microautophagy is relatively nonspecific and occurs when portions of cytoplasm are engulfed directly by the lysosome. In contrast, chaperone-mediated autophagy is limited to proteins and selectively degrades only those which can be recognized and bound by hsc70-containing chaperone–co-chaperone complexes. This substrate–chaperone complex is translocated to the lysosomes on a one-by-one basis. Macroautophagy is the only pathway known to be involved in the degradation of endoplasmic reticulum (ER)-bound proteins among the three identified ALD processes. Macroautophagy consists of several steps: sequestration, transport, degradation, and recycling of the degradation products, and these are illustrated schematically in Fig. 11.1. In this process, the cytoplasmic material is first sequestered by a phagophore or isolation membrane to form an autophagosome. This is then delivered to, and fused with, endosomes; and then it is fused with lysosomes/vacuoles to form autolysosome or autophagolysosome, where its contents are degraded by lysosomal proteases. Once the proteins or peptides have been degraded in the lysosome/vacuole, the resulting amino acids are released into the cytosol for reutilization (Yorimitsu and Klionsky, 2005; Klionsky, 2005).

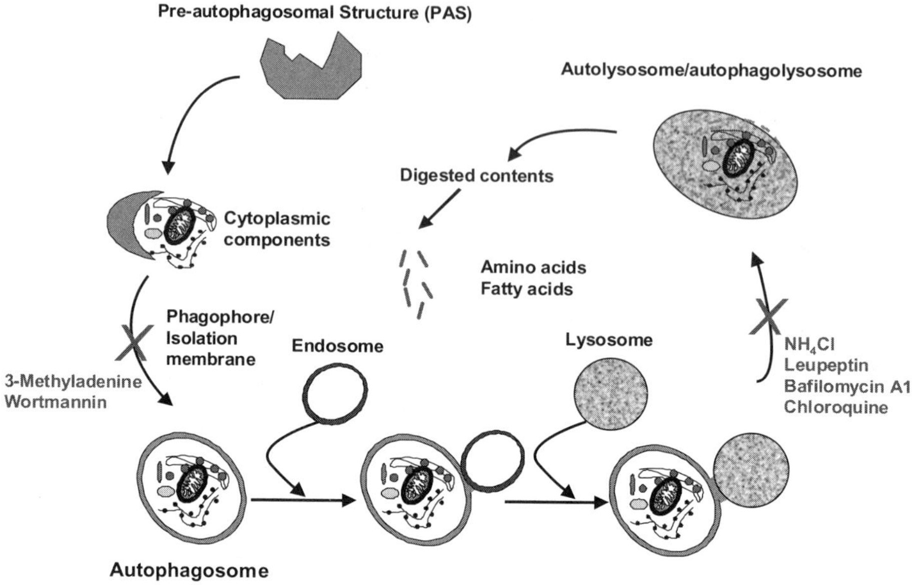

Figure 11.1. The mammalian autophagic-lysosomal degradation (ALD) system: Macroautophagy. Modified from Correia and Liao (2007).

A number of chemicals have been used to inhibit various stages of ALD, such as (a) 3-methyladenine (3-MA) and wortmannin, which block autophagic sequestration, (b) NH_4Cl, which inhibits proteolysis by elevating intralysosomal pH, (c) bafilomycin A1, which strongly inhibits the vacuolar-type Na^+-H^+-ATPase (V-type ATPases), (d) chloroquine, which elevates intralysosomal pH and inhibits several lysosomal degradative processes, and (e) leupeptin, which inhibits lysosomal proteases (Cuervo et al., 2004; Teckman and Perlmutter, 2000; Roccaro et al., 2006; Jin et al., 2008) (Fig. 11.1).

11.2.2 Ubiquitin-Dependent Proteasomal Degradation

The ubiquitin–proteasome system, reportedly responsible for the destruction of greater than 80% of normal and abnormal proteins, is involved in the control of various cellular processes, such as cell cycle, apoptosis, and transcriptional regulation. In addition to its biological importance, ubiquitin-dependent proteasomal degradation (UPD) is also implicated in certain pathological conditions contributing to many diseases such as autoimmune diseases, neurodegenerative disorders, cancer, and viral infections. During the UPD process, proteins are tagged by conjugation with ubiquitin and delivered to the 26S proteasome for degradation (Ciechanover and Ben-Saadon, 2004).

11.2.2.1 Ubiquitin and Ubiquitination. Ubiquitin is a highly evolutionarily conserved 76-residue protein, and as its name implies, it is ubiquitously expressed in all eukaryotic cells. In eukaryotic cells, ubiquitin is covalently bound to proteins through a process called ubiquitination, wherein a single molecule (monoubiquitination) or multiple molecules (polyubiquitination) are attached (Peters et al., 1998). Monoubiquitination can serve as a signal to target membrane proteins to the lysosome for proteolytic turnover, and it may be involved in DNA repair, gene transcriptional regulation, and DNA replication as well (Hicke, 2001). Polyubiquitination is usually, but not always, employed to mark proteins for their degradation by the 26S proteasome, and a chain consisting of at the least four ubiquitin molecules is usually required to target proteins for their proteasomal binding and degradation (Passmore, 2004; Kim et al., 2007). Polyubiquitination may also control the stability, function, and cellular localization of numerous proteins.

The ubiquitination reaction involves three sequential steps catalyzed by three enzymes: E1, E2, and E3 (Peters et al., 1998; Pickart, 2001; Haas and Siepmann, 1997; Hershko and Ciechanover, 1998; Weissman, 2001; Finley, 2002; Thompson et al., 2008). In the initial step, ubiquitin is activated in the presence of ATP by a ubiquitin-activating enzyme (E1) through covalent binding to the cysteine residue of E1 via its C-terminal Gly76. In the second step, the activated ubiquitin is transferred to a cysteine residue of a ubiquitin-conjugating enzyme (E2). Once conjugated to ubiquitin, the E2 molecule binds to a ubiquitin ligase (E3). In the final step, E3 identifies the target substrate and transfers the activated ubiquitin molecule from the cysteine residue of E2 to a lysine ε-NH_2 or an α-NH_2 group of the target protein. Consequently, the first ubiquitin molecule is covalently attached to protein substrate through its C-terminal Gly76. Once one molecule of ubiquitin has been attached to the protein, additional rounds of ubiquitination occur to form a polyubiquitin chain wherein the internal Lys48 of the first ubiquitin is bonded to the C-terminal Gly76 of another ubiquitin molecule. The branched chain may consist of up to 20 ubiquitin molecules. In eukaryotes, only a few E1s exist, with a greater diversity of E2s and an even greater number of E3s. Accordingly, the human UPD system involves only a few E1s, about 50 E2s, and greater than 500 E3s (Peng, 2008). Generally, a given E2 is able to interact with various E3s, whereas E3s can only associate with a limited number of E2s. E3 is responsible for the recognition and binding of targets (Glickman and Ciechanover, 2002) with the structural diversity and binding specificity within the E3 family governing the observed substrate specificity of the proteasome (Risseeuw et al., 2003).

The currently known E3s are grouped into three major classes: the HECT (homologous to E6-AP C-terminus) E3s, the RING-finger E3s, and the U-box (C-terminus of Hsc70 interacting protein) E3s (Weissman, 2001). HECT E3s, which are found in all eukaryotes, contain an N-terminal WW domain (two tryptophan residues, 20–22 residues apart) and a 350-residue C-terminal HECT homology domain. The N-terminal domains of E3s bind substrates and the C-terminal HECT domains directly transfer activated ubiquitin from a given E2 to a conserved HECT cysteine before conjugation to the substrates.

Thus, HECT E3s bind ubiquitin directly. So far, the well-known HECT E3s include Nedd4, E6AP, TOM1, and Rsp5p (Weissman, 2001; Lu et al., 1999; Bernassola et al., 2008). RING-finger E3s contain a 70-amino-acid sequence with eight conserved cysteine and histidine residues that coordinate two zinc atoms to form the small finger, which provides an E2-docking surface. RING-finger E3s may exist as a single subunit with both RING finger-E2 docking and substrate recognition domains on the same polypeptide or as a multisubunit complex including a RING-finger protein for E2 docking, an F-box protein for recognition of phosphorylated substrates, and other adapter subunits. RING-finger E3s mediate substrate ubiquitination by binding to E2-ubiquitin complex and further facilitate the transfer of ubiquitin to the targeted protein. RING-finger E3s catalyze substrate ubiquitination by bringing the E2s in sufficiently close proximity to the protein target rather than by direct thioesterification of the activated ubiquitin as in the case of HECT E3s. Two known subsets of RING-finger E3s include the anaphase-promoting complex (APC) and the Skp1-Cullin-F-box (SCF) protein complex, which contains Skp1 or Skp1-like protein for substrate recognition, the cullin subunit, and an F-box protein for the recognition of phosphorylated substrates. U-box type E3s function as chaperone-dependent ubiquitin ligases, containing an N-terminal tetratricopeptide repeat (TPR) domain that binds molecular chaperones Hsc70 and Hsp90 and a C-terminal U-box which is similar to that of the RING-finger E3 except for its lack of the characteristic zinc-binding motif (Freemont, 2000; Joazeiro and Weissman, 2000; Jackson et al., 2000). The ubiquitin ligase activity of U-box type E3s is similar to that of RING-finger E3s; that is, it facilitates the interaction between E2-bound ubiquitin and their substrates. The U-box type E3s play an active role in the recognition of ubiquitinated, chaperone-associated aberrant proteins for clearance by the proteasome (Murata et al., 2003; Cyr et al., 2002).

11.2.2.2 Proteasome. The key component of UPD is the 26S proteasome (Fig. 11.2). This is a large barrel-like protein complex (<2000 kDa) located in the nucleus and the cytoplasm of all eukaryotes. It is composed of a 20S proteolytic core (~750 kDa) and one or two ATP-dependent assemblies termed the 19S (~700 kDa) regulatory complex, or proteasome activator PA700 (Pickart and Cohen, 2004; Coux et al., 1996; Glickman et al., 1998). The 20S proteasome core consists of four stacked rings (2α and 2β) with outer α-rings flanking the two inner β-rings. Each ring contains seven similar, but functionally distinct subunits. The outer α-rings serve as docking domains for the 19S and regulate protein access into the barrel by controlling the size of its opening. Each β-ring contains three catalytically active sites (β1, β2, and β5) that are located on the interior surface of the rings to ensure that the target protein enters the central pore before it is degraded. These three β-subunits exhibit various proteolytic activities, including chymotrypsin-like, trypsin-like, and peptidyl-glutamyl peptide-hydrolyzing activities. The unique, catalytically active N-terminal threonine residues present in these three β-subunits are

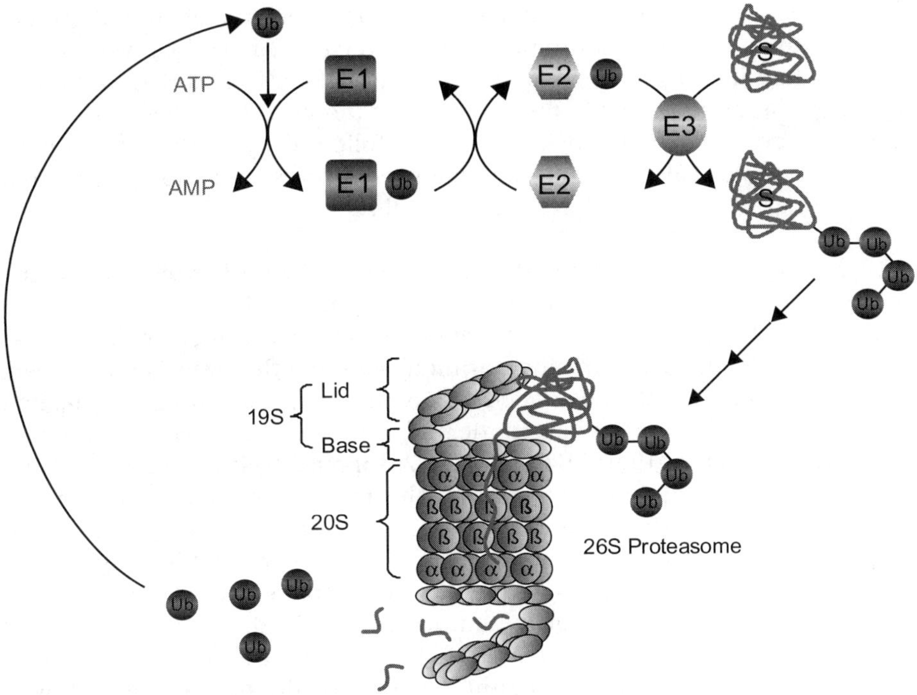

Figure 11.2. The ubiquitin-proteasomal degradation (UPD) system. See the insert for color representation of this figure.

critical in peptide bond hydrolysis. They are also the target for proteasomal inhibitors such as MG-132, MG-262, lactacystin, epoxomycin, and bortezomib (Velcade®) (Rock et al., 1994; Dick et al., 1996; Adams, 2002). The 19S cap at either or both ends of the 20S proteolytic core is composed of an ATPase-containing base and a ubiquitin-recognizing lid hinged together by the Rpn10 subunit. The 19S lid is responsible for recognition of the polyubiquitinated chain and subsequent substrate deubiquitination, which regenerates free ubiquitin for new proteolytic cycles. The 19S base restricts the opening to the proteasomal proteolytic chamber to a mere 13-Å diameter, thus accessible only to unfolded proteins and thereby preventing normally folded proteins from premature proteolytic attack. Recognition of polyubiquitinated proteins by the 19S regulatory particle and their translocation into the 20S core particle requires ATP hydrolysis (Pickart and Cohen, 2004; DeMartino et al., 1994).

Another form of the proteasome is the immunoproteasome, in which the three catalytic β-subunits are replaced by alternate subunits β1i, β2i, and β5i, resulting in the generation of different peptide products. Primarily constitutively expressed in immune cells, immunoproteasomes are also expressed in other cell types, albeit at lower levels, and they can be induced by exposure to interferon gamma, interferon alpha, and cytokines. Immunoproteasomes are intended to be produced and degraded more rapidly than normal proteasomes with an assembly half-time of 21 min versus 82 min, along with a degradation half-life of 27 h compared to the normal 133 h. This rapid assembly and degradation may enable the cell to meet the urgent demands of an efficient and prompt immunological response followed by restoration to their normal physiological status once immunoproteasomes are no longer required (Heink et al., 2005; Yewdell, 2005; Deol et al., 2007).

11.2.2.3 Endoplasmic Reticulum-Associated Degradation. ER is the major locus for protein folding, post-translational modifications, and transport. During the period of ER dysfunction, such as ER stress, protein maturation in the ER may be terminated or inefficient and the resulting misfolded, aberrant proteins are then degraded via an important cellular process, named endoplasmic reticulum-associated degradation (ERAD) (Bonifacino and Lippincott-Schwartz, 1991; Hampton, 2002; Lippincott-Schwartz et al., 1988; Lord et al., 2000; Raasi and Wolf, 2007). This process involves three steps: (i) ubiquitination of misfolded or damaged target ER proteins, (ii) extraction/translocation of substrates from the ER to the cytoplasm, and (iii) subsequent degradation of target proteins by the 26S proteasome. ERAD is responsible not only for the destruction of the unfolded, misfolded, or aberrant luminal and membrane-bound ER proteins, but also for the regulated degradation of ER proteins, such as the rate-limiting enzyme of the mevalonate pathway, HMG-CoA reductase (3-hydroxy-3-methylglutaryl-coenzyme A reductase), and for the maintenance of physiological levels of ER proteins such as P450s (Hampton and Rine, 1994; Ward et al., 1995; Correia and Liao, 2007).

Because of the high evolutionary conservation of ERAD in all eukaryotic cells, our current understanding of mammalian ERAD was mainly derived from the analyses of protein degradation pathways in the yeast, *Saccharomyces cerevisiae* (Hampton et al., 1996; Hampton, 2002; Bays et al., 2001a,b; Biederer et al., 1997; Raasi and Wolf, 2007; Wolf, 2004). Analyses of the ERAD of model yeast proteins, such as Hmg2p and CPY*, led to the identification of cellular ERAD genes, such as HRD (HMGR degradation) and DER (degradation in ER) (Hampton et al., 1996; Hampton, 2002; Bays et al., 2001a; Schafer and Wolf, 2005). This HRD/DER machinery is responsible for the ERAD of integral and lumenal ER proteins. This machinery involves the ER-associated ubiquitin-conjugating enzymes (E2s) including Ubc6p and Ubc7p; the ER-associated ubiquitin-ligase (E3), Hrd1p/Der3p (an integral ER-membrane protein with a C-terminal RING-H2 motif); and the E3 cofactor, Hrd3p (a C-terminally ER-membrane-anchored protein). The HRD/DER machinery also includes Hrd2p/Rpn1 (a 19S cap subunit and thus an indispensable component of the 26S proteasome), Cdc48p (an AAA ATPase homologous to p97 or VCP in mammalian cells), and Npl4p/Hrd4p and Ufd1p. The Cdc48p–Ufd1p–Npl4p complex is apparently involved in the recognition of polyubiquitinated luminal and integral ER proteins, the dislocation of these proteins from the ER, and the subsequent delivery to the 26S proteasome.

Hrd1p/Der3p targets soluble, luminal, and membrane-bound ER proteins, whereas another ER-associated ubiquitin-ligase, Doa10p, mainly targets the integral ER-membrane proteins with large cytosolic domains and soluble cytosolic/nuclear proteins (Hochstrasser et al., 1995; Swanson et al., 2001; Ravid et al., 2006). The DOA10 machinery requires two ubiquitin-conjugating enzymes, Ubc6 and Ubc7/Cue1p, as well as the Cdc48p-Ufd1p-Npl4p complex (Meyer et al., 2000; Dai and Li, 2001; Bazirgan and Hampton, 2005; Huyer et al., 2004). Another HRD/DER-independent ERAD pathway also exists in *S. cerevisiae*, which involves an Rsp5p/Ubc4p/Ubc5p or Rsp5p/Ubc7p complex (Haynes et al., 2002; Arnason et al., 2005; Belgareh-Touzé et al., 2008).

11.2.2.4 Signals for Differential Proteolytic Degradation. Although the precise signals for the differential degradation of proteins remain to be identified, intrinsic molecular structural determinants ("*degrons*") or a variety of post-translational modifications may target the proteins for their diverse degradation. *Degrons* are specific primary sequences of amino acid residues in proteins that mediate the degradation of target proteins (Varshavsky, 1991; Ravid and Hochstrasser, 2008); and when incorporated into heterogeneous proteins, the *degron* usually steers the fusion proteins to be degraded via the same proteolytic pathway as that of the original *degron*-containing protein (Hochstrasser and Varshavsky, 1990; Glotzer et al., 1991; King et al., 1996). Some well-known *degrons* include: KFERQ, NPXY, YXXØ, [DE]XXXL[LI], DXXLL, and acidic clusters as sorting signals for ALD system; lysine or N-terminal residues for ubiquitin conjugation; proline, glutamic acid, serine, and threonine/PAGE sequences for kinases/calpains and other Ca^{2+}-dependent proteases; and Deg-1 sequence, destruction boxes, degradation motifs, KK motifs, PEST sequences, Pro-rich domains, and some N-terminal and C-terminal residues for proteasomal degradation (Hochstrasser and Varshavsky, 1990; Glotzer et al., 1991; King et al., 1996; Bachmair and Varshavsky, 1989; Varshavsky, 1996; Dice et al., 1990; Sokolik and Cohen, 1992; Bonifacino and Traub, 2003). In ERAD, *degrons* not only exist as discrete modular units, but also are "distributed" along the entire primary sequence or tertiary structure of the protein (Doolman et al., 2004; Gardner and Hampton, 1999).

In addition to primary sequences, some post-translational modifications also serve as important signals for protein degradation. These include ubiquitination, phosphorylation, deamidation, glycosylation, glutathione-protein-mixed disulfide formation, and histidine, cysteine, or methionine oxidations (Dice, 1987; Dunlop et al., 2002). It is clear that ubiquitination is a major post-translational modification required for the highly selective degradation of particular proteins. Thus, monoubiquitination targets proteins for lysosomal destruction (Hicke, 2001), while polyubiquitination (>4 ubiquitin molecules) primarily marks proteins for their destruction by the 26S proteasome (Peters et al., 1998; Pickart, 2001; Kim et al., 2007). In the late 1990s, phosphorylation was proposed to contribute to the disposal of some proteins via UPD. In this process, only phosphorylated proteins would be recognized by specific E3-ligases as substrates. One well-known example is the activation of the transcription factor NF-κB through the degradation of its inhibitor, IκBα. Upon stimulation, two serine residues (Ser32 and Ser36) of IκBα are first phosphorylated, and this targets IκBα for UPD. With the release from inhibition by IκBα, the NF-κB is activated and freed to enter the cellular nucleus, wherein it plays a pivotal role in the regulation of various cellular processes through binding to the corresponding DNA-binding elements (Whiteside et al., 1995; Miyamoto et al., 1994; Chen et al., 1995).

11.2.3 Cytochrome P450 Degradation

The accelerated/decelerated degradation of P450s can alter their content, thereby influencing their function. This has the potential to trigger serious complications related to P450-mediated drug metabolism, such as drug–drug interactions (DDIs). Some substrate/inactivators could contribute to such physiological and pathological processes by inducing or inhibiting P450 degradation. Elucidation of the mechanisms whereby P450 degradation affects drug metabolism would enhance our understanding and the predictability of potential DDIs or idiosyncratic syndromes. Such mechanistic elucidation of P450 degradation, however, requires a fundamental understanding of their cellular biology.

The high evolutionary conservation of the eukaryotic proteolytic pathways has permitted the use of various models, including *in vitro* reconstitution systems, yeast, mammalian cell lines, hepatocytes, and intact animals for P450 degradation analyses. Several P450 proteolytic mechanisms have been proposed, and a vast amount of evidence indicates that the two major proteolytic pathways: UPD and ALD are involved in the degradation of native and inactivated P450s. Other mechanisms such as ubiquitin-independent 20S proteasomal degradation have also been reported (Huan et al., 2004; Yang and Cederbaum, 1997; Roberts, 1997; Correia, 2003; Correia and Liao, 2007). The degradation of native hepatic CYP2B1, CYP1A1, and substrate/inhibitor-ligated CYP2E1 is shown to occur mainly via ALD, while the turnover of CYP3A and substrate-free CYP2E1 is reported to occur via different pathways. By contrast, most of suicidally inactivated P450s, including CYP2B1, CYP2C11, CYP2E1, and CYP3A, are targeted to UPD (Correia and Liao, 2007).

The basis for the differential P450 degradation remains to be elucidated. As discussed above, it is conceivable that *degrons* in each P450 primary sequence and/or post-translational modifications are required for such heterogeneity (Aguiar et al., 2005). Not much is really known about P450 *degrons*, but the focus to date has been toward the C-terminal region of these proteins. A CYP2E1 variant with nine of its C-terminal residues deleted is rapidly degraded (Huan and Koop, 1999), while the incorporation of the CYP3A4 C-terminal (CT) heptapeptide onto the CYP2B1 C-terminus switched its degradation from predominantly ALD to UPD (Liao et al., 2006).

Compared to the paucity of our current knowledge about P450 *degrons*, increasing evidence documents the role of posttranslational modifications in accelerating P450 degradation (Correia et al., 2005). P450 ubiquitination is an unequivocal modification that commits both native and inactivated P450 enzymes to degradation. Other modifications, such as heme pyrrole N-modification, oxidation, and phosphorylation, are also reported to contribute to P450 turnover. P450 heme modification caused by oxidative uncoupling or inactivation includes irreversible binding of heme mono- and dipyrrolic fragments to the P450 protein. Such a post-translational modification is reported to result in the rapid degradation of heme-modified CYP3A enzymes (*in vivo*), cumene-hydroperoxide (CuOOH) inactivated CYP3A4 protein (*in vitro*), and heme-modified CYP2 families (CYP2B1, CYP2C11, and CYP2E1) via UPD (He et al., 1998; Correia et al., 1987, 1989, 1992; Dai and Cederbaum, 1995; Roberts, 1997; Wang et al., 1999; Korsmeyer et al., 1999; Tierney et al., 1992.).

Phosphorylation is another post-translational modification that can also mark P450 enzymes for degradation. For instance, the phosphorylation of CYP3A1 and CYP2E1 by cAMP-dependent serine kinase is thought to result in their ER-protease-mediated degradation (Eliasson et al., 1992, 1994; Zhukov et al., 1993). On the other hand, the phosphorylation of CYP2B1 and 2C11 by protein kinase C (PKC) is thought to cause their catalytic inactivation and protein denaturation (Jansson et al., 1990; Correia, 2003). Very recently, a direct correlation between P450 phosphorylation, ubiquitination, and degradation was described and three CYP3A4 phosphorylation sites (S478, T264, and S420) were identified. Phosphorylation of these residues significantly enhanced CYP3A4 ubiquitination leading to subsequent CYP3A4 proteasomal degradation (Wang et al., 2009).

11.3 INDIVIDUAL P450 TURNOVER

11.3.1 CYP2B1

Much insight into the mechanisms of CYP2B1 degradation has been gained from studying a variety of different systems, such as chemical-treated or untreated animals, hepatocytes, transfected cell lines, subcellular fractions, recombinant CYP2B1, and *S. cerevisiae* (Bolender and Weibel, 1973; Yamamoto et al., 1985; Correia and Liao, 2007). Masaki et al. (1987) first reported that CYP2B1 accumulates in autolysosomes isolated from leupeptin-treated rat livers and free of other organellar contaminants. Leupeptin blocks lysosomal cathepsins B, H, and L, thereby preventing CYP2B1 proteolytic degradation (Masaki et al., 1987). Similar results were obtained with acetone-regulated stabilization of CYP2B1 in rat liver (Ronis and Ingelman-Sundberg, 1989; Ronis et al., 1991). CYP2B1 was also found to be degraded in lysosomes in leupeptin-treated or untreated rats (Ueno et al., 1991). The major protease responsible for CYP2B1 ALD was subsequently identified as lysosomal cathepsin D (Tsuji and Akasaki, 1994). Studies in *S. cerevisiae* strains genetically deficient in specific components involved in ubiquitination, proteasomal, or lysosomal degradation also showed that native CYP2B1 heterologously expressed in *S. cerevisiae* is degraded via the ALD pathway (Liao et al., 2005).

However, ALD may not be the only pathway reportedly involved in CYP2B1 turnover. The degradation of CYP2B1 expressed in tetracycline (Tc)-inducible HeLa cell was unaffected by lysosomal inhibitors, but blocked by proteasomal inhibitors, suggesting that CYP2B1 may be a target for proteasomal degradation rather than ALD in this system (Huan et al., 2004). Thus, native CYP2B1 may be degraded differently in model systems, rat liver, *S. cerevisiae* and HeLa cells. Another *in vitro* study with hepatic microsomal fractions revealed that CYP2B1 proteolysis is an Mg^{2+}-ATP-independent process (Eliasson et al., 1992). By contrast, CuOOH-inactivated, heme-modified CYP2B1 undergoes ubiquitin-dependent 26S proteasomal degradation *in vitro*, while the native protein did not appear to be a substrate of this proteolytic pathway (Korsmeyer et al., 1999). Recently, studies in hepatocytes and Hela cells transfected with Myc-tagged CYP2B1, confirmed that UPD is also responsible for IL-1- or NO-donor-induced CYP2B1 turnover (Lee et al., 2008). This is consistent with the report that no native microsomal CYP2B1/2B2 from ethanol-treated rats is degraded after incubation with an ATP-supplemented cytosolic system (Roberts et al., 1995).

The above findings reveal that native CYP2B1 primarily undergoes lysosomal degradation both in *in vivo* and *in vitro* systems, whereas the turnover of CYP2B1 after suicide inactivation or NO-donor inactivation is via UPD.

11.3.2 CYP2E1

CYP2E1 represents 7% of total P450 content in human liver microsomes and is involved in the metabolism of organic solvents and important drugs, such as halothane. Several *in vivo* studies showed that native CYP2E1 exhibits biphasic turnover, consisting of (a) a rapid phase with a half-life of 7 h and (b) a slow phase with a half-life of 38 h (Barmada et al., 1995; Roberts et al., 1995; Song et al., 1989). However, when complexed with substrates, such as acetone and ethanol, the rapid phase of degradation is eliminated (Roberts et al., 1995; Song et al., 1989). Such substrate-induced stabilization of CYP2E1 is proposed to be due to the inhibition of reactive oxygen radical generation because, in the absence of substrates, CYP2E1 exhibits a high oxidase activity resulting in oxidative uncoupling during unproductive cycling (Goasduff and Cederbaum, 2000). The hypothesis is that the reactive oxygen radicals generated lead to the oxidation and/or heme modification of the protein, which in turn trigger the rapid phase of CYP2E1 proteolytic degradation. This hypothesis was supported by findings in Fao cells cultured under serum withdrawal, in which CYP2E1 was significantly stabilized by treatment with diphenyleneiodonium, which inhibits NADPH-cytochrome P450 reductase and thus CYP2E1-dependent hydroxylase activities. The above results suggest that CYP2E1 catalytic cycling in the absence of its substrates is responsible for its rapid degradation (Zhukov and Ingelman-Sundberg, 1999).

There is no consensus for the identity of the proteolytic pathway responsible for the degradation of native CYP2E1. Several distinct degradation pathways have been proposed, including cAMP-serine kinase-dependent MgATP-activated ER serine-protease-mediated CYP2E1 degradation (Eliasson et al., 1992; Zhukov et al., 1993), UPD of ethanol-free CYP2E1 (Bardag-Gorce et al., 2002; Morishima et al., 2005), ubiquitin-independent 20S proteasomal degradation of CYP2E1 (Roberts, 1997; Huan et al., 2004; Yang and Cederbaum, 1996), and ALD (Ronis and Ingelman-Sundber, 1989; Ronis et al., 1991; Roberts et al., 1995; Bardag-Gorce et al., 2002).

cAMP-serine kinase-dependent phosphorylation is associated with CYP2E1 degradation. In cultured rat hepatocytes, the rapid degradation of CYP2E1 is enhanced by glucagon or 8-bromoadenosine 3′,5′-cyclic monophosphate which activates cAMP-dependent pathways. Such CYP2E1 degradation is not influenced by lysosomal inhibitors, but involves its phosphorylation at Ser129, and this phosphorylation and the subsequent degradation can be blocked by the ligands, ethanol and imidazole (Eliasson et al., 1990, 1992). Two ER serine proteinases with molecular weight of 32,000 Da were identified to be responsible for this rapid degradation of CYP2E1 (Zhukov et al., 1993). However, mutation of Ser129 failed to impair CYP2E1 degradation, thereby revealing that additional phosphorylation sites may exist and contribute to the degradation of CYP2E1 expressed in COS 7 cells (Freeman and Wolf, 1994).

Acetone-induced CYP2E1 is degraded at the least partially by the ALD pathway in the intact rat liver and cultured rat hepatocytes (Ronis and Ingelman-Sundberg, 1989; Ronis et al., 1991). The fraction of CYP2E1 protein with the longer half-life, stabilized by substrate complexation, is most likely the fraction subject primarily to ALD disposal (Eliasson et al., 1992; Song et al., 1989) and upon withdrawal of its substrates, CYP2E1 is susceptible to UPD (Goasduff and Cederbaum, 2000; Bardag-Gorce et al., 2002). For example, ethanol stabilizes CYP2E1 protein by preventing its catalytic uncoupling and thus reducing its flux via the UPD pathway. In addition to suppressing autocatalytic inactivation, it has been proposed that ligand-mediated stabilization of CYP2E1 is due to blocking of a ubiquitination site within the CYP2E1 substrate-binding cavity (Bardag-Gorce et al., 2002; Banerjee et al., 2000). Identification of the precise CYP2E1 residues ubiquitinated should resolve this controversy. Interestingly, in cultured Fao hepatoma cells, CYP2E1 apparently relies partially on lysosomes for its degradation, and this requires CYP2E1 transport to the lysosomes (Zhukov and Ingelman-Sundberg, 1997).

CYP2E1 is also reported to undergo ubiquitin-*independent* 20S proteasomal degradation in rat hepatocytes and transfected HepG2 cells expressing human CYP2E1 (Roberts, 1997; Yang and Cederbaum, 1997; Huan et al., 2004). In these studies, proteasomal inhibitors, including lactacystin, ALLN

(*N*-acetyl-Leu-Leu-Norleucinal), and PSI (Czb-Ile-Glu(OtBu)-Ala-Leucinal), protected CYP2E1 from degradation. Additionally, Goasduff and Cederbaum (2000) reported that in an *in vitro* reconstituted system, human liver microsomal CYP2E1 incurs an ATP-dependent degradation, which is inhibited by geldanamycin, an inhibitor of the chaperone, Hsp90, and such inhibition could be reversed by the addition of Hsp90. Similar findings were obtained after treatment of transfected human skin fibroblasts cells expressing CYP2E1 with another hsp90 inhibitor, radicicol (Morishima et al., 2005). In that study, CYP2E1 was reportedly ubiquitilated by the hsp70-dependent ubiquitin-ligase, CHIP, in a cell-free system and in HEK cells coexpressing CYP2E1 and CHIP (Morishima et al., 2005). By contrast, the hsp90 inhibitors, geldanamycin, herbimycin, and radicicol, failed to influence CYP2E1 turnover in tetracycline (Tc)-inducible HeLa cells or in FR-8a2 cells expressing CYP2E1 (Huan et al., 2004). Ethanol-inducible CYP2E1 is apparently also a target for proteasomal degradation *in vivo*. Treatment of ethanol-treated rats with the proteasome inhibitor bortezomib resulted in inhibited proteasomal chymotrypsin-like activity with consequently induced CYP2E1 protein level. After ethanol withdrawal, both the proteasomal chymotrypsin-like activity and liver CYP2E1 content returned to basal levels (Bardag-Gorce et al., 2002).

From the above examples, it is clear that the degradation of CYP2E1 can involve the proteasome, but the role of ubiquitination in native CYP2E1 degradation is somewhat controversial. Some investigators attribute the failure to detect ubiquitinated CYP2E1 to low abundance of ubiquitinated protein and/or the existence of deubiquitinating hydrolases in the liver cytosol (Haas et al., 1985; Wilkinson, 1997). However, additional evidence indicates that CYP2E1 is indeed ubiquitinated in a rabbit reticulocyte lysate system. Two putative ubiquitination sites in CYP2E1, Lys317, and Lys324 residues in a predicted CYP2E1 cytosolic domain were also identified through molecular modeling (Banerjee et al., 2000).

Interestingly, CYP2E1 inactivated by CCl_4 or 3-aminotriazole is ubiquitinated and rapidly degraded presumably by the 26S proteasome in mouse liver (Tierney et al., 1992), but ubiquitination and degradation are not observed when CYP2E1 is inactivated by aminobenzotriazole (ABT) (Huan and Koop, 1999).

11.3.3 CYP3A

P450s of the CYP3A subfamily are major enzymes which account for >30% of total human hepatic P450 content and responsible for the metabolism of at least 50% of marketed drugs. Different proteolytic mechanisms have been reported for the degradation of CYP3A enzymes.

The cAMP-dependent phosphorylation of native CYP3A1/3A23 on its Ser393 has been proposed to cause its turnover by MgATP-dependent serine proteases at the ER (Eliasson et al., 1994). Another pathway proposed for CYP3A23 degradation was high-molecular mass (HMM)-dependent, ubiquitin/26S proteasome-independent degradation (Zangar et al., 2002). But later this HMM CYP3A species was found to be a crosslinking artifact due to the aging of the microsomal preparations (Kimzey et al., 2003). The native human CYP3A4 was degraded by an ATP-/ubiquitin-dependent 26S proteasome pathway in a liver cytosolic system (Fraction II) (Korsmeyer et al., 1999). This was confirmed by studies of native rat liver CYP3A2/3A23 degradation in cultured hepatocytes (Faouzi et al., 2007) or suspensions of primary hepatocytes (Wang et al., 1999) treated with a number of proteasomal or lysosomal inhibitors. This was also documented by degradation analyses of native human CYP3A4 heterologously expressed in *S. cerevisiae* strains deficient in specific UPD or ALD components and their corresponding isogenic wt strains (Murray and Correia, 2001; Liao et al., 2006). Subsequent studies in *in vivo* and/or *in vitro* reconstituted systems revealed that CYP3A enzymes are ubiquitinated by the Ubc7/gp78 ubiqutin–ligase complex and recruited by the p97-Npl4-Ufd1 complex before their degradation by the 26S proteasome (Wang et al., 1999; Murray et al., 2002; Liao et al., 2006; Correia, 2003; Correia and Liao, 2007; Faouzi et al., 2007; Pabarcus et al., 2009).

On the other hand, the predominant proteolytic pathway for suicidally inactivated CYP3A is well recognized to be UPD, as documented in intact rat livers, suspensions or cultures of rat hepatocytes and transfected cell lines (Correia et al., 1992; Wang et al., 1999; Correia, 2003). This is also supported by *in vitro* findings of the degradation of CuOOH-inactivated, heme-modified CYP3A4 (Korsmeyer et al., 1999). However, in HepG2 cells, CYP3A4 degradation induced by high doses of acetaminophen is proposed to occur predominately via a pepstatin A-inhibitable lysosomal cathepsin D-mediated process, but also partially via proteasomal pathway (Zhang et al., 2004). These observations attest to diverse proteolytic pathways responsible for inactivated CYP3A degradation.

Much effort has been made to identify the cellular participants involved in CYP3A UPD. The ER-associated E2, Ubc7p, plus its ER-membrane anchor, Cue1p, but not Ubc6p, were shown to play a role in the degradation of heterologously expressed CYP3A4 in *S. cerevisiae* (Murray and Correia, 2001; Liao et al., 2006). On the other hand, deletion of the three ubiquitin ligases examined (the two canonical ER-anchored proteins Hrd1p/Hrd3p and Doa10p, along with Rsp5p) did not affect CYP3A4 turnover in this system. The findings from *in vitro* reconstituted systems also revealed that Ubc7/gp78 ubiquitin ligase complex is responsible for CYP3A4 ubiquitination before its 26S proteasomal degradation (Pabarcus et al., 2009). This conclusion is somewhat unexpected, because gp78 is ~30% related to the mammalian homolog of Hrd1p, *HRD1* (Doolman et al., 2004). The latter has been convincingly excluded in CYP3A4 UPD in yeast (Murray and Correia, 2001; Liao et al., 2006). Although another mammalian hepatic E3 CHIP is reported to ubiquitinate CYP2E1 and CYP2B4 (Morishima et al., 2005), its role in ubiquitinating CYP3A4 is unclear. No clear *S. cerevisiae* homolog of CHIP exists, so genetic CHIP deletion analyses of CYP3A4 degradation are not feasible, and thus the

E3 ubiquitin ligase involved in CYP3A4 UPD in *S. cerevisiae* remains to be identified. In *S. cerevisiae*, CYP3A4 UPD was also dependent on the AAA ATPase Cdc48p–Ufd1–Npl4p complex (Liao et al., 2006). The AAA ATPase Cdc48p–Ufd1–Npl4p complex is required in (a) the translocation and/or extraction of both integral and luminal ubiquitinated proteins from the ER membrane into the cytosol and (b) subsequent delivery to the 26S proteasome. A possible role for p97/Cdc48p complex in the CYP3A ER extraction and delivery for proteasomal degradation is also suggested in primary cultured rat hepatocytes, after *in situ* protein crosslinking. p97 is detected cross-linked in CYP3A-immunoprecipitates by both immunoblotting and proteomic analyses (Faouzi et al., 2007).

Little is known about the precise signals that trigger CYP3A degradation. Although CYP3A4 C-terminal (CT)-heptapeptide incorporation into CYP2B1 switches its degradation in *S. cerevisiae* from predominantly ALD into UPD, the deletion of its CT domain is insufficient to alter CYP3A4 UPD. These findings exclude a role of CYP3A4 CT heptapeptide as a proteasomal *degron* for its degradation and infer the existence of additional CYP3A4 UPD *degrons*. Such possible determinants exist in either single or multiple domains, or they are even distributed throughout its structure (Liao et al., 2006).

The potential role of post-translational modifications in CYP3A turnover has been examined by several groups. Although a role for ubiquitination has been established in CYP3A degradation, the association of its phosphorylation, ubiquitination, and degradation is unclear. As discussed above, phosphorylation of native CYP3A1/23 by cAMP-dependent kinase results in a marked degradation of the enzyme (Eliasson et al., 1994), and native CYP3A4 is also phosphorylated in a system catalyzed by rat liver cytosolic Fraction II (Korsmeyer et al., 1999; Wang et al., 2001). This phosphorylation was greatly enhanced when CYP3A4 was inactivated by CuOOH, and followed by its ubiquitination and 26S proteasomal degradation. PKC and PKA were identified as the major FII kinases responsible for CYP3A4 phosphorylation (Wang et al., 2001). Two CYP3A4 target sites, T264 and S420, phosphorylated *in vitro* by PKC were identified by HPLC-peptide mapping and LC-MS/MS analyses (Wang et al., 2001). Recent studies identified a third site, S478, phosphorylated by rat liver cytosolic kinases. Further studies of CYP3A4 ubiquitination and degradation *in S. cerevisiae* and in reconstituted systems indicated that PKA/PKC-mediated phosphorylation enhances CYP3A4 degradation by enhancing its ubiquitination, suggesting a direct link between P450 phosphorylation, ubiquitination, and degradation (Wang et al., 2009).

11.4 CLINICAL RELEVANCE OF P450 DEGRADATION

11.4.1 P450 Turnover and Drug Interactions

Pharmacokinetic drug interactions occur when the absorption or disposition of one agent (the "victim" or "object") is affected by treatment with another agent (the "perpetrator" or "precipitant"). It is also common to consider interactions with environmental factors (such as diet or herbal medications) and pharmacogenetic variability as forms of drug interaction. The turnover of P450s can influence both the severity and the time course for some classes of pharmacokinetic drug interactions, and the key factor that determines the level of influence is how the time course for the interaction of the perpetrator with the target P450 compares with that of the turnover of the enzyme. For example, drug interactions due to rapidly reversible enzyme inhibition are unlikely to be influenced by protein turnover because even very potent inhibitors dissociate from the enzyme with a very short half-life (small fractions of a second), which is many orders of magnitude shorter than the half-life for P450 protein turnover (hours or days). The classes of drug interaction influenced by P450 protein turnover are discussed below.

11.4.1.1 Effects of Disease States. It has been known for some time that disease states can influence the pharmacokinetics of P450 substrates (Morgan et al., 2008), the most common observation being a reduction in clearance of those substrates. In many cases this is due to increased degradation and/or decreased synthesis of the P450 proteins and the resulting effects can be very dramatic when the substrates have a low therapeutic index (e.g., Kraemer et al., 1982). Decreased synthesis can be due to suppression by inflammatory mediators of induction pathways controlled by xenobiotic receptors (Moreau et al., 2008). Studies with animals have also shown decreased expression of cytochromes P450 after bacterial or viral infection, or after the administration of model inflammatory mediators, such as bacterial lipopolysaccharide or poly rI:rC, and an increase in the rate of P450 degradation has been observed (e.g., Gooderham and Mannering, 1986). The increased P450 degradation is hypothesized in some cases to be due to inactivation of the enzyme by nitric oxide generated by nitric oxide synthase, NOS2, induced by a variety of cytokines (Aitken et al., 2008). Inflammatory cytokines (such as interferon alpha) and immunomodulating agents are now finding clinical use in antiviral and cancer chemotherapy. Drug interactions with these cytokines have been reported (Israel et al., 1993; Islam et al., 2002); but these are not widely appreciated, likely due to the confounding effects of the underlying diseases.

11.4.1.2 Enzyme Inhibition. While disease states can lead to drug interactions due to relatively nonspecific increases in P450 degradation, enzyme inhibition affects a single P450 protein (or a relatively small subsets of proteins). As outlined above, for P450 protein degradation to affect an inhibitory drug interaction, the duration of the inhibitory effect must be of the same order as that of the half-life of the protein, so simple reversible inhibition of a P450 enzyme is relatively insensitive to the rate of turnover of the enzyme.

Long-term inhibition of an enzyme requires tight ("quasi-irreversible") or covalent interaction with the inhibitor or destruction of prosthetic groups

(Correia and Ortiz de Montellano, 2005; Hollenberg et al., 2008). In many cases this is achieved by the inhibitor being a substrate for the enzyme and being converted to the inactivating form. This has led to the process being termed suicide inhibition, mechanism-based inhibition, or time-dependent inhibition (Silverman, 1988). In some cases, covalent binding of metabolites to the P450 apoprotein can be the cause of idiosyncratic reactions (as described in the section below). The biggest impact of P450 protein turnover on the response to enzyme inactivation is upon the time course for recovery of activity after the inhibitor has left the system (see below).

In addition to inhibitory drug interactions being caused by other drugs, a number of dietary and herbal compounds are also known to display this property. Grapefruit juice, shown nearly 20 years ago to be the culprit of an interaction with felodipine, is probably the most famous example and it contains a variety of furanocoumarin derivatives that are potent mechanism-based inhibitors of human intestinal CYP3A activity (Bailey et al., 1998), as well as acutely inhibiting some drug transporters. Clinical studies have shown that the inhibition of intestinal CYP3A activity is associated with increased protein degradation because there is rapid and profound loss of the protein after consumption of a single glass of grapefruit juice (Schmiedlin-Ren et al., 1997; Lown et al., 1997). This is in contrast to the results with another mechanism-based inhibitor, diltiazem, which inactivates intestinal CYP3A but does not accelerate its degradation appreciably (Pinto et al., 2005). This suggests that there are inhibitor-dependent qualitative differences in the sequelae following enzyme inactivation. Since, under normal conditions, hepatic CYP3A activity is relatively unaffected by grapefruit juice, interaction studies with this agent are now performed to determine the relative contributions of intestinal and hepatic CYP3A to first-pass metabolism of drugs (Kharasch et al., 2004; Gertz et al., 2008). Other dietary and herbal components that have been shown to inactivate human P450 enzymes include: psoralens (Koenigs and Traeger, 1998) and phenethylisothiocyanate (Nakajima et al., 2001), found in a variety of vegetables; resveratrol, which is found in red wine (Chang et al., 2001); and methylenedioxyphenyl alkaloids, found in goldenseal herbal extract (Chatterjee and Franklin, 2003).

11.4.1.3 Induction. Induction is a phenomenon wherein the amount of active protein affecting a drug's absorption and/or disposition is increased, usually leading to reduced exposure and/or increased elimination of the victim drug. Since the amount of active protein present is dependent upon the turnover of that protein, protein turnover can have a strong influence on the rate and extent of the induction effect.

Induction of P450s can be achieved by increasing their rates of synthesis and/or decreasing their rates of degradation. The most-studied mechanism is increased transcription through activation of xenobiotic receptors such as the aryl hydrocarbon receptor (AhR), the pregnane X receptor (PXR), and the constitutive androstane receptor (CAR) (Wang and LeCluyse, 2003), but mRNA stabilization and post-translational events can also influence the rates of synthesis. An increase in the content of some P450s, notably CYP1A2, CYP2E1, and CYP3A, can also be achieved through decreases in their rates of degradation through ligand stabilization (e.g., Steward et al., 1985; Eliasson et al., 1994; Chien et al., 1997; Watkins et al., 1986) as discussed above. As with enzyme inhibition, environmental constituents can also act as inducers—for example, activation of AhR by tobacco smoke (e.g., Thum et al., 2006) and charcoal-grilled meat, as well as activation of PXR by herbal preparations such as St. John's wort (Zhou and Lai, 2008).

Protein degradation also plays a more fundamental role in the CYP1A induction process: Ligand-activated AhR is degraded by the proteasome machinery to attenuate the signal (Ma, 2007), and similar mechanisms may also be in place for PXR (Masuyama et al., 2002). Proteasomal protein degradation also plays a role in the response to antioxidants through the Nrf2/KEAP1 pathway (Zhang, 2006) that regulates murine CYP2A5 (Abu-Bakar et al., 2007).

11.4.1.4 Pharmacogenetics. From the above it is clear that P450 drug interactions can be influenced by protein degradation whether the precipitating event is another drug, a disease state, or a dietary or other environmental factor. Protein degradation can also affect the magnitude of pharmacogenetic drug interactions because some P450 allelic variants have been shown to encode proteins with reduced stability. For example, Hichiya et al. (2005) demonstrated that introduction of the R186G mutation into CYP2C8, a change found in the naturally occurring CYP2C8*8 allele, resulted in poor protein expression in COS-1 cells and that this could be overcome by treatment with the proteasome inhibitor, MG-132. A similar finding was reported for the N453S mutation in CYP1B1*4 (Bandiera et al., 2005). Allelic variants of CYP1A2, CYP2A6, CYP2B6, CYP2C9, CYP2C19, CYP2D6, CYP3A4, and CYP3A5 leading to reduced stability of expressed protein have also all been reported (summarized at www.cypalleles.ki.se).

11.4.1.5 Protein Degradation Kinetics and Drug Interactions. The examples above stress the importance of the relative rate of protein turnover on the potential for drug interactions. Here we discuss how the rate affects the magnitude and time course of the effect.

Proteins are turned over continuously, and under steady-state conditions the rate of degradation is balanced by the rate of synthesis of new protein. The kinetics of this process have been studied for many years using model proteins (e.g., Berlin and Schimke, 1965; Steinberg et al., 1975; Watson et al., 1981), and the general finding is that the rate of protein degradation largely follows first-order kinetics, so the loss depends upon the amount of protein present. In contrast, the rate of protein synthesis obeys zero-order kinetics and so is not directly dependent upon the amount of protein present. Thus the basal turnover of enzyme (E) with time (t) can be expressed as follows, with k_{synth} and k_{deg} being the rate constants for synthesis and degradation, respectively.

$$d[\mathrm{E}]/dt = k_{synth} - k_{deg}[\mathrm{E}]$$

Under equilibrium conditions, the apparent protein half-life ($T_{½}$) is then

$$T_{1/2} = \ln 2/k_{deg}$$

The actual rate of protein synthesis reflects contributions from both constitutive and induced synthesis, and the balance of the contributions of the two is enzyme-dependent. For example, CYP2D6 is largely regulated by the constitutively expressed transcription factor, hepatic nuclear factor 4α (HNF4α), and is thus unaffected by inducers (Cairns et al., 1996). In contrast, CYP1A1 is expressed in the lung in only tiny amounts but can be increased enormously by activation of AhR by cigarette smoke (Thum et al., 2006). CYP3A4 lies between these extremes because it shows substantial constitutive expression, likely mediated by HNF4α, but it is also strongly inducible through activation of PXR or CAR (Tirona et al., 2003). Induction effectively increases the rate of synthesis while inactivation can be considered to increase the rate of degradation; and in either situation if a change is persistent enough, a new steady-state level of protein expression is achieved.

As discussed above, both the rates and routes of P450 degradation can also be influenced by inducers and enzyme inactivators. For some enzymes, such as CYP2E1, there are contributions from the slower ALD and faster UPD routes and the extent of the latter can be increased by enzyme inactivation, triggering ERAD. No clear feedback pathways for regulation of drug metabolizing P450s have been identified, so inhibition of the enzymes does not result in a compensatory increase in the rates of their synthesis.

Extending the model described above, the turnover of the protein in the presence of an inducer or an inhibitor can be expressed as follows:

$$d[\mathrm{E}]/dt = (k_{synth} + k_{induc}) - (k_{deg} + k_{MBI})[\mathrm{E}]$$

where k_{induc} is the increase in the rate of synthesis due to induction and k_{MBI} is the increase in rate of degradation (or inactivation) due to mechanism-based inhibition. However, the same principle can be used to calculate turnover changes triggered by an inhibitor (or more formally a repressor) that decreases the rate of synthesis (k_{synth} is lower) or an inducer that acts through protein stabilization (k_{deg} is lower) *in vitro* systems capable of determining such rate changes are not widely available. The approach summarized in the above equation has been used successfully to model pharmacokinetic data without making assumptions about the underlying mechanism(s) for the change in rates (e.g., Pitlick et al., 1976). However, for prediction of the quantitative and temporal effects of the interacting agents, it should be possible to determine k_{induc} and/or k_{MBI} in *in vitro* systems.

Attempts have been made to predict k_{induc}, the increase in synthesis rate, from *in vitro* induction data using a simple $E_{\max}$ model (e.g., Kato et al., 2005);

$$k_{\text{induc},t+t_{\text{lag}}} = \frac{k_{\max}[\text{C}]_t}{\text{EC}_{50} + [\text{C}]_t}$$

The concentration of inducer at time = t, $[\text{C}]_t$ affects the induced rate of synthesis, k_{induc}, after a lag time, t_{lag}. The lag time represents the delay from, for example, occupation of the xenobiotic receptor to the point when freshly synthesized enzyme is present. The two model parameters are $k_{\max}$, the theoretical maximal rate of induced synthesis, and EC_{50}, the concentration of inducer resulting in half-maximal activation in the *in vitro* system. Note that although there are good correlations between responses in *in vitro* systems with different endpoints (e.g., increases in mRNA, protein, enzyme activity, or reporter gene expression), the dynamic ranges of the various systems can differ considerably. Although t_{lag} is likely inversely related to the rate of turnover (more rapidly cycling proteins showing a shorter delay), this parameter has not been explored in detail.

Similarly, the increase in inactivation rate caused by a mechanism-based inhibitor, k_{MBI}, can be calculated from *in vitro* data as follows (Silverman, 1988):

$$k_{\text{MBI}} = \frac{k_{\text{inact}}[\text{C}]_t}{K_I + [\text{C}]_t}$$

where k_{inact} and K_I are the inactivation kinetic parameters and $[\text{C}]_t$ is the inhibitor concentration. It should be noted that not all agents act purely as inducers or inhibitors. A well-known example is the HIV protease inhibitor, ritonavir, which is both a potent mechanism-based inhibitor of human CYP3A enzymes (Ernest et al., 2005) and a significant inducer in the clinic (Foisy et al., 2008). Another striking example in a preclinical species is the simultaneous induction and inactivation of rat CYP2B1 by secobarbital (He et al., 1996).

The drug concentrations used in the above calculations, as well as the affinity constants for induction and inactivation (EC_{50} and K_I), may need to be corrected for *in vitro* or *in vivo* protein binding, depending upon the *in vitro* test system used and the disposition of the compound *in vivo*. In some cases, Hill slopes need to be applied to the concentration and affinity constants, such as has been seen with AhR-mediated induction (Broccardo et al., 2004). For the relevant concentration, $[\text{C}]_t$, *in vivo*, many researchers have used a single value, such as the average plasma concentration at steady state while others have used pharmacokinetic–pharmacodynamic (PKPD) modeling to link the pharmacokinetics of the agent with the turnover of the enzyme. This can be especially complex when the clearance of the agent depends upon the enzyme it is affecting.

The kinetics of protein turnover strongly affect both the magnitude and the time course of the response to inducers and inactivators. For instance, for mechanism-based inhibition a common approach used is the one described by Mayhew et al. (2000). The effects of inhibition on the area under the plasma concentration–time curve (AUC) of a victim drug, cleared by the inhibited enzyme, can be estimated by assuming that clearance of that drug is proportional to the amount of active enzyme present. Then

$$\frac{\text{AUC}_{\text{inhibited}}}{\text{AUC}_{\text{normal}}} \approx \frac{k_{\text{deg}} + k_{\text{MBI}}}{k_{\text{deg}}}$$

So an enzyme that is turned over relatively slowly ($k_{\text{MBI}} >> k_{\text{deg}}$) will respond more strongly to an inactivator. The rate of return of the enzyme level to normal after the inactivator is removed depends upon k_{synth}. Similarly, the response to an inducer is dependent upon the basal rate of synthesis:

$$\frac{\text{AUC}_{\text{induced}}}{\text{AUC}_{\text{normal}}} \approx \frac{k_{\text{synth}}}{k_{\text{synth}} + k_{\text{induc}}}$$

Although their values are critical to the prediction of responses to inducers and mechanism-based inhibitors, the rates of turnover of P450s, and particularly the human enzymes, are the subject of debate with wide-ranging values for protein half-life being reported (Correia, 2003; Yang et al., 2008). Some of this variability is likely due to the confounding influence of background exposure to varying levels of inducing agents in the environment. This makes it difficult to separate k_{synth} from (k_{synth} + k_{induc}). The ultimate upper limit on protein half-life is constrained by the turnover of the cell or organelle in which the protein is located. In relatively short-lived continuously cycling cells, such as small intestinal epithelium or skin keratinocytes, the lifetime of P450 protein depends upon the transit time of the cells. For example, in short-lived (transit time ~24 h) mature small intestinal villus tip enterocytes, where the majority of intestinal P450 is located, it is likely that the levels of CYP3A do not have time to recover to baseline after the cells are no longer exposed to an inhibitor or inducer before the cells are shed into the intestinal lumen. In longer-lived "conditional renewal" populations, such as the liver and kidney, the lifetime of a P450 protein may be limited more by the continuous turnover of the ER in which the proteins are located (Omura et al., 1967; Parkinson et al., 1983; Shiraki and Guengerich, 1984). These values only represent the maximum limit to the lifetime of the proteins because it is clear that the components of the ER are turned over asynchronously by varying mechanisms.

To illustrate the quantitative importance of protein turnover, Fig. 11.3 shows the simulated effects of an inducer (Fig. 11.3A) and a mechanism-

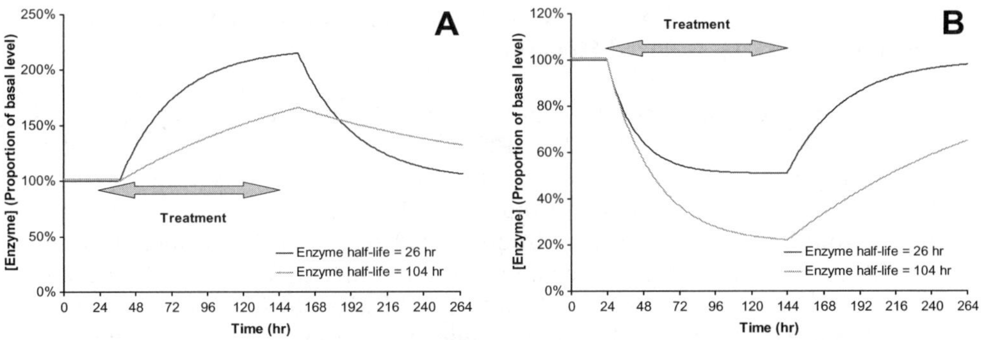

Figure 11.3. The effect of protein degradation half-life on drug interactions. **(A)** Simulation of the effect of treatment with an inducer, bosentan, upon hepatic cytochrome P450 levels. Regimen consists of 24 h of pretreatment, 5 days of dosing, and then 5 days of recovery. Parameters used were: $C_{average}$ 4 μM, EC_{50} 19.9 μM, k_{max} 600% of basal synthesis rate, and a lag time of 12 h. **(B)** Simulation of the effect of treatment with a mechanism-based cytochrome P450 inhibitor, sitaxsentan, upon hepatic enzyme levels. Regimen consists of 24 h of pretreatment, 5 days of dosing, and then 5 days of recovery. Parameters used were: $C_{average}$ 1.8 μM, k_{inact} 0.1 min^{-1}, K_I 250 μM.

based inhibitor (Fig. 11.3B) on the levels of target enzymes during the treatment and recovery phases. Despite their differing effects, both compounds are of the same therapeutic class (endothelin receptor antagonists) and both are used in chronic treatment of pulmonary arterial hypertension. Bosentan is a PXR activator (van Giersbergen et al., 2002) and chronic treatment leads to a reduction in the plasma concentration of coadministered sildenafil due to a two-fold increase in clearance (Paul et al., 2005). Sitaxsentan contains a methylenedioxyphenyl moiety and inactivates human CYP2C9, resulting in a need for dose-adjustment (80% reduction) of coadministered warfarin and a requirement for regular liver function monitoring (Kabunga and Coghlan, 2008). In each case, two different half-lives were used for turnover of the target enzyme, those being 26 h (reported for CYP3A4) and 104 h (reported for CYP2C9).

It is clear that the more rapidly cycling enzyme responds more quickly, in both onset and recovery, to inducers and mechanism-based inhibitors. For this enzyme, the magnitude of induction is higher but the nadir for the enzyme level is not as low when exposed to a mechanism-based inhibitor. The more slowly cycling enzyme responds sluggishly to induction and de-induction but is more strongly affected by the mechanism-based inhibitor.

These two examples graphically highlight the importance of the rates of turnover of human P450s on the magnitude and time course of drug interactions. The degradation of P450s can thus affect a drug's pharmacokinetics and hence its pharmacodynamics through their response to other drugs, diet, and disease and through the manifestation of pharmacogenetic variability.

11.4.2 P450 Degradation and Anti-P450 Autoantibodies in Autoimmune Hepatitis and Drug-Induced Hepatitis

P450 degradation may also have pathological relevance in autoimmune hepatitis and drug-induced hepatitis because aberrant processing of the protein can lead to the generation of autoantibodies. Indeed, there are serum autoantibodies that specifically recognize individual P450 enzymes in some patients with autoimmune hepatitis and drug-induced hepatitis. These include anti-CYP2D6 antibodies in type II chronic autoimmune hepatitis (AIH-2) (Kerkar et al., 2003; Imaoka et al., 2005), anti-CYP11A1, anti-CYP17A, and anti-CYP21 antibodies in both autoimmune polyendocrino-pathycandidiasis-ectodermal dystrophy and Addison's disease (Liiv et al., 2002; Uibo et al., 1994). Anti-P450 antibodies have also been found in idiosyncratic drug reactions, such as halothane hepatitis (anti-CYP2E1), tienilic acid hepatitis (anti-CYP2C9), dihydralazine hepatitis (anti-CYP1A2), and hypersensitivity reactions to the aromatic anticonvulsants phenytoin, phenobarbital, and carbamazepine (anti-rat CYP3A) (Mizutani et al., 2005; Liu and Kaplowitz, 2002; Leeder et al., 1996). In drug-induced hepatitis, reactive drug metabolites have been shown to covalently modify the particular P450 enzyme that bioactivates the drug, resulting in the formation of P450-drug adducts such as CYP2E1-trifluoroacetate (Bourdi et al., 1996), CYP1A2-dihydralazine (Masubuchi et al., 1999), CYP2C9-tienilic acid (Koenigs et al., 1999), and CYP3A4-carbamazepine (Kang et al., 2008).

It is generally believed that the generation of autoantibodies against P450s involves the lysosomal/proteasomal degradation of native or modified P450s to produce intracellular antigenic peptides of defined length and sequence. A working hypothesis linking P450 degradation and immune responses is shown in Fig. 11.4. In hepatocytes, native (e.g., CYP2D6 in AIH-2 hepatitis) or drug-modified P450 (e.g., CYP2C9 in tienilic acid hepatitis) proteins are degraded to peptides which can be further degraded by peptidases to free amino acids. A few epitopic peptides that manage to avoid the proteolytic processes wind up on major histocompatibility complex (MHC) class I molecules on the surface of hepatocytes. The peptide–MHC pair can then be recognized by specific T-cell receptors on CD8 T cells, which in turn results in cytotoxicity to hepatocytes. It has been demonstrated in *ex vivo* systems that CYP2D6-specific CD8 T-cell immune responses are vigorous in AIH-2 and correlate with disease activity, implicating their direct role in disease pathogenesis (Ma et al., 2006; Longhi et al., 2007). Hepatocytes normally do not express MHC class II molecules, but a change can be triggered by viral or autoimmune hepatitis. Degraded P450 peptides are presented on MHC class II molecules to CD4 T helper cells. Alternatively, after hepatocyte death, P450 proteins can be taken up by macrophage and degraded by lysosomes. The resulting antigenic peptides then bind to MHC class II molecules and are recognized by helper T cells. After presentation, both B and T cells may be activated and undergo clonal expansion. B cells

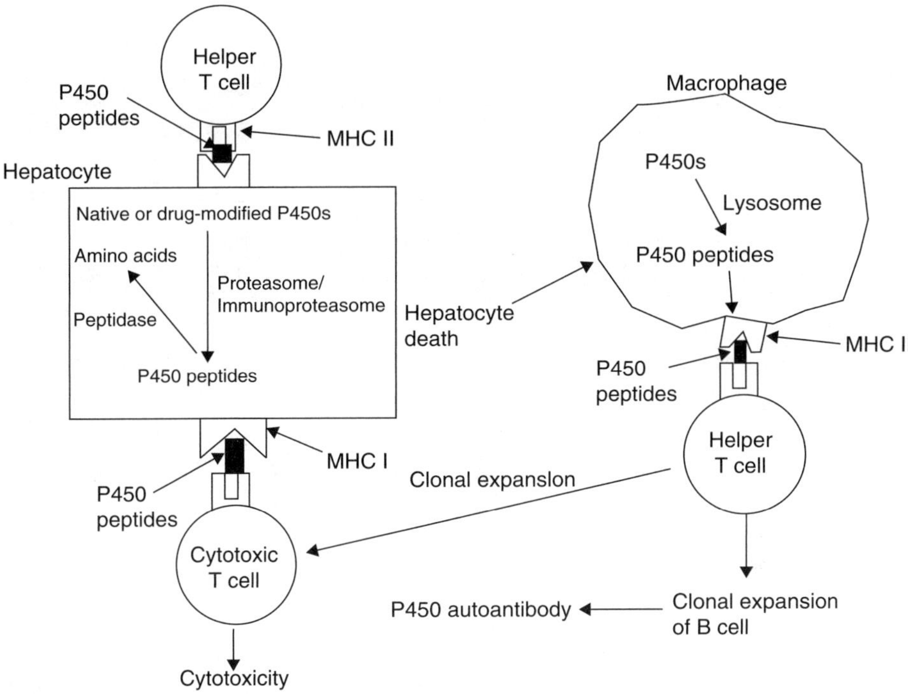

Figure 11.4. Generation of anti-P450 autoantibodies in autoimmune hepatitis or drug-induced hepatitis. Modified from Pessayre (1993).

mature into plasma cells that eventually secrete an autoantibody directed against a P450 protein.

AIH type 2 is characterized by the presence of anti-liver/kidney microsomal autoantibodies type 1 (LKM-1), for which CYP2D6 has been identified as the major antigen. Epitopes of anti-CYP2D6 autoantibodies have been identified by epitope mapping and are immunoreactive with sera of patients with autoimmune hepatitis. The most frequently targeted epitopes are residues 193–212 and 257–269, followed by 321–351, 373–389, and 410–429 (Manns et al., 1991; Yamamoto et al., 1993; Kerkar et al., 2003). A recent study has also revealed conformational epitopes residues 316–327 (Ma et al., 2002). Although B-cell, helper T-cell and CD8 T-cell responses are associated with liver damage and may play a combined role in AIH-2 pathogenesis, it remains uncertain whether CYP2D6 becomes an autoantigen through a mechanism of molecular mimicry or as a consequence of antigenic alteration (Ma et al., 2006; Longhi et al., 2007). Susceptibility to type 2 AIH is most likely determined by a complex combination of environmental and genetic factors, which include allelic polymorphisms in HLA class II genes.

After chronic treatment with the antihypertensive drug dihydralazine, a large number of patients with hepatitis showed anti-CYP1A2 antibodies.

TABLE 11.1. Some Key P450 Degradation: Substrate Ligand, Protein Modification, Proteolytic Pathways, and Test Systems

P450	Substrate/ Inactivator	Protein Modification	Half-Life, $t_{1/2}$ (hours)	Turnover Pathways	Test System	Reference
1A1				Lysosomal (?)	Hepatocytes	Roberts (1997)
1A2				Proteasomal (20S?)	Hepatocytes	Roberts (1997)
1B1		Ubiquitination/No phosphorylation	1.6–4.8	Proteasomal	Transfected COS-1 cells	Bandiera et al. (2005)
2B1		None	>37	Lysosomal	Rat	Masaki et al. (1987)
	Acetone	None	>37	Lysosomal	Rat	Ronis et al. (1991)
			21	ATP-independent	Rat hepatocytes, rat liver microsomes	Eliasson et al. (1992)
				Lysosomal (Cathepsin D)	Purified CYP2B1	Tsuji and Akasaki (1994)
	CuOOH	Heme[a]/ Ubiquitination		Proteasomal (26S)	Purified P450s	Korsmeyer et al. (1999)
		No ubiquitination	8.7	Proteasomal	Tc-Hela cells	Huan et al. (2004)
				Lysosomal	*S. cerevisiae*	Liao et al. (2005)
	IL-1/NO donors	Ubiquitination		Proteasomal	Hepatocytes, transfeced Hela cells, rat liver microsomes	Lee et al. (2008)
2C11				Lysosomal	*S. cerevisiae*	Murray et al. (2002)
	DDEP	Heme[a]/ Ubiquitination		Proteasomal (26S)		Song and Correia[b]
2E1	Acetone		>37	?	Rat	Song et al. (1989)
			7–37	?	Rat	Song et al. (1989)
	Glucagon or 8-Br-cAMP	Phosphorylation/ Ser129			Rat hepatocytes	Eliasson et al. (1990)
	Acetone	None	>37	Lysosomal	Rat hepatocytes	Ronis et al. (1991)
	CCl_4, imidazole	Phosphorylation/ Ser129	9	ER Serine proteases	Rat hepatocytes, rat liver microsomes	Eliasson et al. (1992)

CCl_4, aminotriazole	Heme[a]/ Ubiquitination	3	Proteasomal (26S?)	Mice	Tierney et al. (1992)
	PO_4/Ser129	7	ER Serine proteases	Fao cells	Zhukov et al. (1993)
CCl_4	Ubiquitination		Proteasomal (26S)	Rat liver microsomes	Roberts et al. (1995)
CCl_4	CCl_4 binding		Proteasomal (26S or 20S?)	MVh2E1-9 (HepG2 subline)	Dai and Cederbaum (1995)
Ethanol	None		Lysosomal	Rat	Roberts et al. (1995)
		2.5–6	Proteasomal (26S or 20S?)	MVh2E1-9 (HepG2 subline)	Yang and Cederbaum (1996, 1997)
Serum depletion		7–26	Lysosomal	Fao cells	Zhukov et al. (1997)
None/CCl_4			Proteasomal (26S or 20S?)	Transfected HepG2 cells	Yang and Cederbaum (1996, 1997)
			Proteasomal (20S)	Rat hepatocytes	Roberts (1997)
CCl_4			Proteasomal (20S)	Rat	Roberts (1997)
Diphenylene iodonium		4–26		Fao cells	Zhukov and Ingelman-Sundberg (1999)
NADPH inactivation	Oxidation		Proteasomal (20S)	Human liver microsomes, HepG2	Goasduff and Cederbaum (1999)
			Proteasomal (20S?)	Transfected Hela cells	Huan and Koop (1999)
			Proteasomal (20S, Hsp90)	Human liver microsomes and cytosol	Goasduff and Cederbaum (2000)
			Proteasomal (26S)	Rat	Bardag-Gorce et al. (2002)
	No ubiquitination	3.8	Proteasomal	Tc-Hela cells, CHO	Huan et al. (2004)

TABLE 11.1. *Continued*

P450	Substrate/ Inactivator	Protein Modification	Half-Life, $t_{1/2}$ (hours)	Turnover Pathways	Test System	Reference
	CCl_4	Ubiquitination (CHIP-dependent)		Proteasomal (26S)	GM2E1 cells, transfected HEK293T, Recombinant purified P450s	Morishima et al. (2005)
3A		Phosphorylation/ Ser393	0.3	ER serine proteases	Rat hepatocytes, rat, purified P450	Eliasson et al. (1994)
	DDEP			Proteasomal (20S)	Rat	Roberts (1997)
				Proteasomal (20S)	Rat hepatocytes	Roberts (1997)
	GFJ (FCs?)				Healthy volunteers	Lown et al. (1997)
	GFJ (FCs?)				Healthy volunteers	Schmiedlin-Ren et al. (1997)
	DDEP	Heme[a]/ Ubiquitination	3	Proteasomal (26S)	Hepatocytes	Correia et al. (1992); Wang et al. (1999)
	CuOOH	Heme[a]/		Proteasomal (26S)	Recombinant P450s	Korsmeyer et al. (1999)
	acetaminophen	Ubiquitination		Lysosomal	Transfected HepG2 cells, Human liver microsomes and cytosol	Zhang et al. (2004)
		Ubiquitination		Proteasomal (26S)	*S. cerevisiae*	Murray and Correia (2001), Liao et al. (2006)
		Ubiquitination	6	Proteasomal	Rat hepatocytes	Faouzi et al. (2007)
		Ubiquitination/ UBC7/gp78		26S Proteasomal	*In vitro* reconstituted system	Pabarcus et al. (2009)
		Phosphorylation/ Ser478		Proteasomal	*S. cerevisiae*, recombinant P450s	Wang et al (2009)
4A				Proteasomal (20S)	Rat hepatocytes	Roberts (1997)

[a]Heme modification.
[b]Song J-Z and Correia MA, unpublished observations.
Source: Modified from Correia (2003).

An epitope-mapping study showed that a CYP1A2 amino acid 335–471 domain serves as a conformational epitope that is immunoreactive with 100% sera of patients with dihydralazine-induced hepatitis (Belloc et al., 1997). A recent report has shown that CYP1A was present on the plasma membrane and was further increased by dihydralazine administration in rat hepatocytes (Robin et al., 2000). This observation suggests that the anti-CYP1A2 autoantibodies may participate in the immunologic destruction of hepatocytes.

In 60% of patients suffering from severe hepatitis after administration of tienilic acid, a specific antibody directed against CYP2C9 was detected. Studies indicate that native epitopes, as well as tienilic acid-modified epitopes, are recognized by anti-CYP2C9 autoantibodies. Three regions of CYP2C9 (amino acids 314–322, 345–356, and 439–455) form a major conformational autoantibody binding site which is immunoreactive with sera of patients with tienilic acid-induced hepatitis (Lecoeur et al., 1996). Studies in a rat model have shown that CYP2C11-tienilic acid adducts were present on the plasma membrane of hepatocytes, indicating that antibody-dependent cytotoxicity and complement-related mechanisms associated with anti-CYP2C9 antibodies may be involved in the disease process (Robin et al., 1996).

Sera from 70% of patients with halothane-induced hepatitis contains autoantibodies that recognize rat CYP2E1. Two distinct conformational epitopes on the CYP2E1 surface containing the G-helix and an area formed by juxtaposition of the J′ and K″ helices, respectively, are immunoreactive with sera of patients with halothane- and alcohol-induced hepatitis (Vidali et al., 2004). Eliasson and Kenna (1996) have shown that native CYP2E1 and trifluoroacetate (a halothane metabolite) adducts of CYP2E1 were present on the cell surface of hepatocytes and could act as cell surface target antigens that are recognized by the antibodies of patients and potentially by other immune effector mechanisms.

Curiously, a subset of patients with anticonvulsant hypersensitivity has circulating antibodies that recognize members of the rat CYP3A but not the related human CYP3A proteins (Leeder et al., 1996). Less intense reactivity has been observed with rat CYP2C11, whereas CYP2C6 and CYP3A2 are minor antigens in some patients (Leeder et al., 1992). Amino acids 355–367 of the CYP3A1 protein were identified as the epitope recognized by antibodies in the serum of anticovulsant hypersensitive patients (Leeder et al., 1996).

Despite their usefulness for diagnosis and research, the pathogenetic role of the anti-P450 autoantibodies is still unclear. It is yet to be documented whether autoantibodies observed in autoimmune or drug-induced hepatitis themselves contribute to liver damage. Although in some cases cell surface targeting of intact P450 protein may play a role in developing autoantibodies, it is clear that the process of P450 degradation, either chemically or pathologically induced, is important in the production of antigenic peptides capable of engendering an immune response. A better understanding of P450 degradation may lead to the selection of therapeutic agents with less liability of causing hepatitis in the clinic.

11.5 CONCLUSIONS

P450s are important enzymes responsible for the metabolism of 85% marketed drugs and a wide range of endogenous substrates. Their expression reflects a balance between synthesis and degradation, and the pathways for their turnover are gradually being elucidated. Over the past several decades, the two major protein degradation pathways (lysosomal and proteasomal) and their roles in the regulation of various cellular processes, from cell cycle to quality control, have drawn enormous attention. P450s are substrates for both of these pathways, and their relative contributions are enzyme-dependent and are affected by numerous factors including the state of the protein (native or inactivated). These protein degradation pathways are no longer of strictly academic interest, because the first drug that targets proteasomes, bortezomib (Velcade®), has been marketed for anticancer therapies (Russo et al., 2007). As discussed here, P450 degradation is involved in DDIs, drug-induced hepatitis, and hypersensitivity syndromes. Therefore, there is a critical need to understand P450 degradation mechanisms and their pathological and physiological relevance.

REFERENCES

Abu-Bakar A, Lämsä V, Arpiainen S, Moore MR, Lang MA, Hakkola J. Regulation of CYP2A5 gene by the transcription factor nuclear factor (erythroid-derived 2)-like 2. *Drug Metab Dispos* 2007;35:787–794.

Adams J. Proteasome inhibition: a novel approach to cancer therapy. *Trends Mol Med* 2002;8(4 Suppl):S49–S54.

Aguiar M, Masse R, Gibbs BF. Regulation of cytochrome P450 by posttranslational modification. *Drug Metab Rev* 2005;37:379–404.

Aitken AE, Lee CM, Morgan ET. Roles of nitric oxide in inflammatory downregulation of human cytochromes P450. *Free Radic Biol Med* 2008;44:1161–1168.

Arnason TG, Pisclevich MG, Dash MD, Davies GF, Harkness TA. Novel interaction between Apc5p and Rsp5p in an intracellular signaling pathway in *Saccharomyces cerevisiae*. *Eukaryot Cell* 2005;4:134–146.

Bachmair A, Varshavsky A. The degradation signal in a short-lived protein. *Cell* 1989;56:1019–1032.

Bailey DG, Malcolm J, Arnold O, Spence JD. Grapefruit juice–drug interactions. *Br J Clin Pharmacol* 1998;46:101–110.

Bandiera S, Weidlich S, Harth V, Broede P, Ko Y, Friedberg T. Proteasomal degradation of human CYP1B1: effect of the Asn453Ser polymorphism on the post-translational regulation of CYP1B1 expression. *Mol Pharmacol* 2005;67:435–443.

Banerjee A, Kocarek TA, Novak RF. Identification of a ubiquitination target/substrate-interaction domain of cytochrome P-450 (CYP) 2E1. *Drug Metab Dispos* 2000;28:118–124.

Bardag-Gorce F, Li J, French BA, French SW. Ethanol withdrawal induced CYP2e1 degradation *in vivo* blocked by proteasome inhibitor PS-341. *Free Radic Biol Med* 2002;32:17–21.

Barmada S, Kienle E, Koop DR. Rabbit P450 2E1 expressed in CHO-K1 cells has a short half-life. *Biochem Biophys Res Commun* 1995;206:601–607.

Bays NW, Gardner RG, Seelig LP, Joazeiro CA, Hampton RY. Hrd1p/Der3p is a membrane-anchored ubiquitin ligase required for ER-associated degradation. *Nat Cell Biol* 2001a;3:24–29.

Bays NW, Wilhovsky SK, Goradia A, Hodgkiss-Harlow K, Hampton RY. HRD4/NPL4 is required for the proteasomal processing of ubiquitinated ER proteins. *Mol Biol Cell* 2001b;12:4114–4128.

Bazirgan OA, Hampton RY. Cdc48-Ufd2-Rad23: the road less ubiquitinated? *Nat Cell Biol* 2005;7:207–209.

Belgareh-Touzé N, Léon S, Erpapazoglou Z, Stawiecka-Mirota M, Urban-Grimal D, Haguenauer-Tsapis R. Versatile role of the yeast ubiquitin ligase Rsp5p in intracellular trafficking. *Biochem Soc Trans* 2008;36:791–796.

Belloc C, Gauffre A, André C, Beaune PH. Epitope mapping of human CYP1A2 in dihydralazine-induced autoimmune hepatitis. *Pharmacogenetics* 1997;7:181–186.

Berlin CM, Schimke RT. Influence of turnover rates on the responses of enzymes to cortisone. *Mol Pharmacol* 1965;1:149–156.

Bernassola F, Karin M, Ciechanover A, Melino G. The HECT family of E3 ubiquitin ligases: multiple players in cancer development. *Cancer Cell* 2008;14:10–21.

Biederer T, Volkwein C, Sommer T. Role of Cue1p in ubiquitination and degradation at the ER surface. *Science* 1997;278:1806–1809.

Bolender R, Weibel E. A morphometric study of the removal of phenobarbital induced membranes from hepatocytes after cessation of treatment. *J Cell Biol* 1973;56:746–761.

Bonifacino JS, Lippincott-Schwartz J. Degradation of proteins within the endoplasmic reticulum. *Curr Opin Cell Biol* 1991;3:592–600.

Bonifacino JS, Traub LM. Signals for sorting of transmembrane proteins to endosomes and lysosomes. *Annu Rev Biochem* 2003;72:395–447.

Bourdi M, Chen W, Peter RM, Martin JL, Buters JT, Nelson SD, Pohl LR. Human cytochrome P450 2E1 is a major autoantigen associated with halothane hepatitis. *Chem Res Toxicol* 1996;9:1159–1166.

Broccardo CJ, Billings RE, Chubb LS, Andersen ME, Hanneman WH. Single cell analysis of switch-like induction of CYP1A1 in liver cell lines. *Toxicol Sci* 2004;78: 287–294.

Cairns W, Smith CA, McLaren AW, Wolf CR. Characterization of the human cytochrome P4502D6 promoter: A potential role for antagonistic interactions between members of the nuclear receptor family. *J Biol Chem* 1996;271:25269–25276.

Chang TK, Chen J, Lee WB. Differential inhibition and inactivation of human CYP1 enzymes by trans-resveratrol: evidence for mechanism-based inactivation of CYP1A2. *J Pharmacol Exp Ther* 2001;299:874–882.

Chatterjee P, Franklin MR. Human cytochrome P450 inhibition and metabolic-intermediate complex formation by goldenseal extract and its methylenedioxyphenyl components. *Drug Metab Dispos* 2003;31:1391–1397.

Chen Z, Hagler J, Palombella VJ, Melandri F, Scherer D. Signal-induced site-specific phosphorylation targets I kappa B alpha to the ubiquitin-proteasome pathway. *Genes Dev* 1995;9:1586–1597.

Chien JY, Thummel KE, Slattery JT. Pharmacokinetic consequences of induction of CYP2E1 by ligand stabilization. *Drug Metab Dispos* 1997;25:1165–1175.

Ciechanover A, Ben-Saadon R. N-terminal ubiquitination: more protein substrates join in. *Trends Cell Biol* 2004;14:103–106.

Ciechanover A, Heller H, Elias S, Haas AL, Hershko A. ATP dependent conjugation of reticulocyte proteins with the polypeptide required for protein degradation. *Proc Natl Acad Sci USA* 1980;77:1365–1368.

Correia MA. Cytochrome P450 turnover. *Methods Enzymol* 1991;206:315–325.

Correia MA. Hepatic cytochrome P450 degradation: mechanistic diversity of the cellular sanitation brigade. *Drug Metab Rev* 2003;35:107–143.

Correia MA. Human and rat liver cytochromes P450: functional markers, diagnostic inhibitor probes and parameters frequently used in P450 studies. In: Ortiz de Montellano P, editor. *Cytochrome P450: Structure, Mechanism and Biochemistry*. New York: Kluwer–Academic/Plenum Press, 2005, pp. 619–657.

Correia MA, Liao M. Cellular proteolytic systems in P450 degradation: evolutionary conservation from Saccharomyces cerevisiae to mammalian liver. *Expert Opin Drug Metab Toxicol* 2007;3:33–49.

Correia MA, Ortiz de Montellano PR. Inhibition of cytochrome P450 enzymes. In: Ortiz de Montellano, PR, editor. *Cytochrome P-450. Structure, Mechanism, and Biochemistry*. New York: Kluwer–Plenum Press, 2005, pp. 247–322.

Correia MA, Litman DA, McColl KEL, Schmid R, Thompson GC. Effect of cimetidine on phenobarbitone-induced changes in hepatic cytochrome P450 and *d*-aminolaevulinic acid synthase. *Proc Br Pharmacol* 1983;C24.

Correia MA, Decker C, Sugiyama K, Caldera P, Bornheim L, Wrighton SA, Rettie AE, Trager WF. Degradation of rat hepatic cytochrome P-450 heme by 3,5-dicarbethoxy-2,6-dimethyl-4-ethyl-1,4-dihydropyridine to irreversibly bound protein adducts. *Arch Biochem Biophys* 1987;258:436–451.

Correia MA, Sugiyama K, Yao KQ. Degradation of rat hepatic cytochrome P-450p. *Drug Metab Rev* 1989;20:615–628.

Correia MA, Davoll SH, Wrighton SA, Thomas PE. Degradation of rat liver cytochromes P-450 3A after their inactivation by 3,5-dicarbethoxy-2,6-dimethyl-4-ethyl-1,4-dihydropyridine: Characterization of the proteolytic system. *Arch Biochem Biophys* 1992;297:228–238.

Correia MA, Sadeghi S, Mundo-Paredes E. Cytochrome P450 ubiquitination: branding for the proteolytic slaughter? *Annu Rev Pharmacol Toxicol* 2005;45:439–464.

Coux O, Tanaka K, Goldberg AL. Structure and functions of the 20S and 26S proteasomes. *Annu Rev Biochem Biochem* 1996;65:801–847.

Cuervo AM, Dice JF. Lysosomes, a meeting point of proteins, chaperones, and proteases. *J Mol Med* 1998;76:6–12.

Cuervo AM, Stefanis L, Fredenburg R, Lansbury PT, Sulzer D. Impaired degradation of mutant alpha-synuclein by chaperone-mediated autophagy. *Science* 2004;305: 1292–1295.

Cyr DM, Höhfeld J, Patterson C. Protein quality control: U-box-containing E3 ubiquitin ligases join the fold. *Trends Biochem Sci* 2002;27:368–375.

Dai RM, Li CC. Valosin-containing protein is a multi-ubiquitin chain-targeting factor required in ubiquitin–proteasome degradation. *Nat Cell Biol* 2001;3:740–744.

Dai Y, Cederbaum A. Inactivation and degradation of human cytochrome P4502E1 by CCl4 in a transfected HepG2 cell line. *J Pharmacol Exp Ther* 1995;275:1614–1622.

de Duve C, Gianetto R, Appelmans F, Wattiaux R. Enzymic content of the mitochondria fraction. *Nature (London)* 1953;172:1143–1144.

DeMartino GN, Moomaw CR, Zagnitko OP, Proske RJ, Chu-Ping M, Afendis SJ, Swaffield JC, Slaughter CA. PA700, an ATP-dependent activator of the 20 S proteasome, is an ATPase containing multiple members of a nucleotide-binding protein family. *J Biol Chem* 1994;269:20878–20884.

Deol P, Zaiss DM, Monaco JJ, Sijts AJ. Rates of processing determine the immunogenicity of immunoproteasome-generated epitopes. *J Immunol* 2007;178:7557–7562.

Dice JF. Molecular determinants of protein half-lives in eukaryotic cells. *FASEB J* 1987;1:349–357.

Dice JF, Terlecky SR, Chiang HL, Olson TS, Isenman LD, Short-Russell SR, Freundlieb S, Terlecky LJ. A selective pathway for degradation of cytosolic proteins by lysosomes. *Semin Cell Biol* 1990;1:449–455.

Dick LR, Cruikshank AA, Grenier L, Melandri FD, Nunes SL, Stein RL. Mechanistic studies on the inactivation of the proteasome by lactacystin: a central role for clasto-lactacystin beta-lactone. *J Biol Chem* 1996;271:7273–7276.

Doolman R, Leichner GS, Avner R, Roitelman J. Ubiquitin is conjugated by membrane ubiquitin ligase to three sites, including the N terminus, in transmembrane region of mammalian 3-hydroxy-3-methylglutaryl coenzyme A reductase: implications for sterol-regulated enzyme degradation. *J Biol Chem* 2004;279:38184–38193.

Dunlop RA, Rodgers KJ, Dean RT. Recent developments in the intracellular degradation of oxidized proteins. *Free Radic Biol Med* 2002;33:894–906.

Eliasson E, Kenna JG. Cytochrome P450 2E1 is a cell surface autoantigen in halothane hepatitis. *Mol Pharmacol* 1996;50:573–582.

Eliasson E, Johansson I, Ingelman-Sundberg M. Substrate-, hormone-, and cAMP-regulated cytochrome P450 degradation. *Proc Natl Acad Sci USA* 1990;87:3225–3229.

Eliasson E, Mkrtchian S, Ingelman-Sundberg M. Hormone- and substrate-regulated intracellular degradation of cytochrome P450 (2E1) involving MgATP-activated rapid proteolysis in the endoplasmic reticulum membranes. *J Biol Chem* 1992;267:15765–15769.

Eliasson E, Mkrtchian S, Halpert JR, Ingelman-Sundberg M. Substrate-regulated, cAMP-dependent phosphorylation, denaturation, and degradation of glucocorticoid-inducible rat liver cytochrome P450 3A1. *J Biol Chem* 1994;269:18378–18383.

Ernest CS 2nd, Hall SD, Jones DR. Mechanism-based inactivation of CYP3A by HIV protease inhibitors. *J Pharmacol Exptl Ther* 2005;312:583–591.

Evans WE, Relling MV. Pharmacogenomics: translating functional genomics into rational therapeutics. *Science* 1999;286:487–491.

Faouzi S, Medzihradszky KF, Hefner C, Maher JJ, Correia MA. Characterization of the physiological turnover of native and inactivated cytochromes P450 3A in cultured rat hepatocytes: a role for the cytosolic AAA ATPase P97? *Biochemistry* 2007;46:7793–7803.

Finley D. Ubiquitin chained and crosslinked. *Nat Cell Biol* 2002;4:E121–E123.

Foisy MM, Yakiwchuk EM, Hughes CA. Induction effects of ritonavir: implications for drug interactions. *Ann Pharmacother* 2008;42:1048–1059.

Freeman JE, Wolf CR. Evidence against a role for serine 129 in determining murine cytochrome P450 Cyp2E-1 protein levels. *Biochemistry* 1994;33:13963–13966.

Freemont PS. RING for destruction? *Curr Biol* 2000;10:R84–R87.

Gardner RG, Hampton RY. A "distributed degron" allows regulated entry into the ER degradation pathway. *EMBO J* 1999;18:5994–6004.

Gertz M, Davis JD, Harrison A, Houston JB, Galetin A. Grapefruit juice-drug interaction studies as a method to assess the extent of intestinal availability: utility and limitations. *Curr Drug Metab* 2008;9:785–795.

Glickman MH, Ciechanover A. The ubiquitin–proteasome pathway: Destruction for the sake of construction. *Physiol Rev* 2002;82:373–428.

Glickman MH, Rubin DM, Fried VA, Finley D. The regulatory particle of the *Saccharomyces cerevisiae* proteasome. *Mol Cell Biol* 1998;18:3149–3162.

Glotzer M, Murray AW, Kirschner MW. Cyclin is degraded by the ubiquitin pathway. *Nature* 1991;349:132–138.

Goasduff T, Cederbaum AI. NADPH-dependent microsomal electron transfer increases degradation of CYP2E1 by the proteasome complex: role of reactive oxygen species. *Arch Biochem Biophys* 1999;370:258–270.

Goasduff T, Cederbaum AI. CYP2E1 degradation by *in vitro* reconstituted systems: role of the molecular chaperone hsp90. *Arch Biochem Biophys* 2000;379:321–330.

Goldberg AL, St. John AC. Intracellular protein degradation in mammalian and bacterial cells: Part 2. *Annu Rev Biochem* 1976;45:747–803.

Goldberg AL. ATP-dependent proteases in prokaryotic and eukaryotic cells. *Semin Cell Biol* 1991;1:423–432.

Goldberg AL. Functions of the proteasome: from protein degradation and immune surveillance to cancer therapy. *Biochem Soc Trans* 2007;35:12–17.

Gooderham NJ, Mannering GJ. Depression of cytochrome P-450 and alterations of protein metabolism in mice treated with the interferon inducer polyriboinosinic acid : polyribocytidylic acid. *Arch Biochem Biophys* 1986;250:418–425.

Guengerich FP. Human cytochrome P450 enzymes. In: Ortiz de Montellano P, editor. *Cytochrome P450: Structure, Mechanism and Biochemistry*. New York: Kluwer-Academic/Plenum Press, 2005, pp. 377–530.

Haas AL, Siepmann TJ. Pathways of ubiquitin conjugation. *FASEB J* 1997;11:1257–1268.

Haas AL, Murphy KE, Bright PM. The inactivation of ubiquitin accounts for the inability to demonstrate ATP, ubiquitin-dependent proteolysis in liver extracts. *J Biol Chem* 1985;260:4694–4703.

Hampton RY. ER-associated degradation in protein quality control and cellular regulation. *Curr Opin Cell Biol* 2002;14:476–482.

Hampton RY, Rine J. Regulated degradation of HMG-CoA reductase, an integral membrane protein of the endoplasmic reticulum, in yeast. *J Cell Biol* 1994;125: 299–312.

Hampton RY, Gardner RG, Rine J. Role of 26S proteasome and HRD genes in the degradation of 3-hydroxy-3-methylglutaryl-CoA reductase, an integral endoplasmic reticulum membrane protein. *Mol Biol Cell* 1996;7:2029–2044.

Haynes CM, Caldwell S, Cooper AA. An HRD/DERindependent ER quality control mechanism involves Rsp5p-dependent ubiquitination and ER–Golgi transport. *J Cell Biol* 2002;158:91–101.

He K, He YA, Szklarz GD, Halpert JR, Correia MA. Secobarbital-mediated inactivation of cytochrome P450 2B1 and its active site mutants: partitioning between heme and protein alkylation and epoxidation. *J Biol Chem* 1996;271: 25864–25872.

He K, Bornheim LM, Falick AM, Maltby D, Yin H, Correia MA. Identification of the heme-modified peptides from cumene hydroperoxide-inactivated cytochrome P450 3A4. *Biochemistry* 1998;37:17448–17457.

Heink S, Ludwig D, Kloetzel P, Kruger E. IFNγ-induced immune adaptation of the proteasome system is an accelerated and transient response. *Proc Natl Acad Sci USA* 2005;102:9241–9246.

Hershko A, Ciechanover A. The ubiquitin system for protein degradation. *Annu Rev Biochem* 1992;61:761–807.

Hershko A, Ciechanover A. The ubiquitin system. *Annu Rev Biochem* 1998;67: 425–479.

Hichiya H, Tanaka-Kagawa T, Soyama A, Jinno H, Koyano S, Katori N, Matsushima E, Uchiyama S, Tokunaga H, Kimura H, Minami N, Katoh M, Sugai K, Goto Y, Tamura T, Yamamoto N, Ohe Y, Kunitoh H, Nokihara H, Yoshida T, Minami H, Saijo N, Ando M, Ozawa S, Saito Y, Sawada J. Functional characterization of five novel CYP2C8 variants, G171S, R186X, R186G, K247R and K383N found in a Japanese population. *Drug Metab Dispos* 2005;233:630–636.

Hicke L. Protein regulation by monoubiquitin. *Nat Rev Mol Cell Biol* 2001;2: 195–201.

Hochstrasser M, Varshavsky A. *In vivo* degradation of a transcriptional regulator: the yeast alpha 2 repressor. *Cell* 1990;61:697–708.

Hochstrasser M, Papa FR, Chen P, Swaminathan S, Johnson P, Stillman L, Americk AY, Li SY, et al. The DOA pathway: studies on the functions and mechanisms of ubiquitin-dependent protein degradation in the yeast *Saccharomyces cerevisiae*. *Cold Spring Harb Symp Quant Biol* 1995;60:503–513.

Hollenberg PF, Kent UM, Bumpus NN. Mechanism-based inactivation of human cytochromes P450s: experimental characterization, reactive intermediates, and clinical implications. *Chem Res Toxicol* 2008;21:189–205.

Huan JY, Koop DR. Tightly regulated and inducible expression of rabbit CYP2E1 using a tetracycline-controlled expression system. *Drug Metab Dispos* 1999;27: 549–554.

Huan JY, Streicher JM, Bleyle LA, Koop DR. Proteasome-dependent degradation of cytochromes P450 2E1 and 2B1 expressed in tetracycline-regulated HeLa cells. *Toxicol Appl Pharmacol* 2004;199:332–343.

Huyer G, Piluek WF, Fansler Z, Kreft SG, Hochstrasser M, Brodsky JL, Michaelis S. Distinct machinery is required in Saccharomyces cerevisiae for the endoplasmic reticulum-associated degradation of a multispanning membrane protein and a soluble luminal protein. *J Biol Chem* 2004;279:38369–38378.

Imaoka S, Obata N, Hiroi T, Osada-Oka M, Hara R, Nishiguchi S, Funae Y. A new epitope of CYP2D6 recognized by liver kidney microsomal autoantibody from japanese patients with autoimmune hepatitis. *Biol Pharm Bull* 2005;28:2240–2243.

Islam M, Frye RF, Richards TJ, Sbeitan I, Donnelly SS, Glue P, Agarwala SS, Kirkwood JM. Differential effect of IFNα-2b on the cytochrome P450 enzyme system: a potential basis of IFN toxicity and its modulation by other drugs. *Clin Cancer Res* 2002;8:2480–2487.

Israel BC, Blouin RA, McIntyre W, Shedlofsky SI. Effects of interferon-α monotherapy on hepatic drug metabolism in cancer patients. *Br J Clin Pharmacol* 1993;36:229–235.

Jackson PK, Eldridge AG, Freed E, Furstenthal L, Hsu JY, Kaiser BK, Reimann JD, et al. The lore of the RINGs: substrate recognition and catalysis by ubiquitin ligases. *Trends Cell Biol* 2000;10:429–439.

Jansson I, Curti M, Epstein PM, Peterson JA, Schenkman JB. Relationship between phosphorylation and cytochrome P450 destruction. *Arch Biochem Biophys* 1990;283:285–292.

Jin S, Yi F, Zhang F, Poklis JL, Li PL. Lysosomal targeting and trafficking of acid sphingomyelinase to lipid raft platforms in coronary endothelial cells. *Arterioscler Thromb Vasc Biol* 2008;28:2056–2062.

Joazeiro CA, Weissman AM. RING finger proteins: mediators of ubiquitin ligase activity. *Cell* 2000;102:549–552.

Kabunga P, Coghlan G. Endothelin receptor antagonism: role in the treatment of pulmonary arterial hypertension related to scleroderma. *Drugs* 2008;68:1635–1645.

Kang P, Liao M, Wester MR, Leeder JS, Pearce RE, Correia MA. CYP3A4-Mediated carbamazepine (CBZ) metabolism: formation of a covalent CBZ-CYP3A4 adduct and alteration of the enzyme kinetic profile. *Drug Metab Dispos* 2008;36:490–499.

Kato M, Chiba K, Horikawa M, Sugiyama Y. The quantitative prediction of *in vivo* enzyme-induction caused by drug exposure from *in vitro* information on human hepatocytes. *Drug Metab Pharmacokinet* 2005;20:236–243.

Kerkar N, Choudhuri K, Ma Y, Mahmoud A, Bogdanos DP, Muratori L, Bianchi F, Williams R, Mieli-Vergani G, Vergani D. Cytochrome P4502D6$_{193-212}$: a new immunodominant epitope and target of virus/self cross-reactivity in liver kidney microsomal autoantibody type 1-positive liver disease. *J Immunol* 2003;170:1481–1489.

Kharasch ED, Whittington D, Hoffer C. Influence of hepatic and intestinal cytochrome P4503A activity on the acute disposition and effects of oral transmucosal fentanyl citrate. *Anesthesiology* 2004;101:729–737.

Kim HT, Kim KP, Kisselev AF, Lledias F, Scaglione KM, Skowyra D, Gygi SP, Goldberg AL. Certain E2–E3 pairs synthesize nondegradable forked ubiquitin chains containing all possible isopeptide linkages. *J Biol Chem* 2007;282:17375–17386.

Kimzey AL, Weitz KK, Guengerich FP, Zangar RC. Hydroperoxy-10,12-octadecadienoic acid stimulates cytochrome P450 3A protein aggregation by a mechanism that is inhibited by substrate. *Biochemistry* 2003;42:12691–12699.

King RW, Glotzer M, Kirschner MW. Mutagenic analysis of the destruction signal of mitotic cyclins and structural characterization of ubiquitinated intermediates. *Mol Biol Cell* 1996;7:1343–1357.

Klionsky DJ. The molecular machinery of autophagy: unanswered questions. *J Cell Sci* 2005;118:7–18.

Klionsky DJ. Autophagy: from phenomenology to molecular understanding in less than a decade. *Nat Rev Mol Cell Biol* 2007;8:931–937.

Koenigs LL, Trager WF. Mechanism-based inactivation of P450 2A6 by furanocoumarins. *Biochemistry* 1998;37:10047–10061.

Koenigs LL, Peter RM, Hunter AP, Haining RL, Rettie AE, Friedberg T, Pritchard MP, Shou M, Rushmore TH, Trager WF. Electrospray ionization mass spectrometric analysis of intact cytochrome P450: identification of tienilic acid adducts to P450 2C9. *Biochemistry* 1999;38:2312–2319.

Kopitz J, Kisen GO, Gordon PB, Bohley P, Seglen PO. Nonselective autophagy of cytosolic enzymes by isolated rat hepatocytes. *J Cell Biol* 1990;111:941–953.

Korsmeyer KK, Davoll S, Figueiredo-Pereira ME, Correia MA. Proteolytic degradation of heme-modified hepatic cytochromes P450: a role for phosphorylation, ubiquitination and the 26S proteasome? *Arch Biochem Biophys* 1999;365:31–44.

Kraemer MJ, Furukawa CT, Koup JR, Shapiro GG, Pierson WE, Bierman CW. Altered theophylline clearance during an influenza B outbreak. *Pediatrics* 1982;69:476–480.

Lecoeur S, André C, Beaune PH. Tienilic acid-induced autoimmune hepatitis: anti-liver and -kidney microsomal type 2 autoantibodies recognize a three-site conformational epitope on cytochrome P4502C9. *Mol Pharmacol* 1996;50:326–333.

Lee C, Kim B, Li L, Morgan ET. Nitric oxide-dependent proteasomal degradation of cytochrome P450 2B proteins. *J Biol Chem* 2008;283:889–898.

Leeder JS, Riley RJ, Cook VA, Spielberg SP. Human anti-cytochrome P450 antibodies in aromatic anticonvulsant-induced hypersensitivity. *J Pharmacol Exp Ther* 1992;263:360–367.

Leeder JS, Gaedigk A, Lu X, Cook VA. Epitope mapping studies with human anti-cytochrome P450 3A antibodies. *Mol Pharmacol* 1996;49:234–243.

Liao M, Faouzi S, Karyakin A, Correia MA. Endoplasmic reticulum-associated degradation of cytochrome P450 CYP3A4 in *Saccharomyces cerevisiae*: Further characterization of cellular participants and structural determinants. *Mol Pharmacol* 2006;69:1897–1904.

Liao M, Zgoda VG, Murray BP, Correia MA. Vacuolar degradation of rat liver CYP2B1 in *Saccharomyces cerevisiae*: Further validation of the yeast model and structural implications for the degradation of mammalian endoplasmic reticulum P450 proteins. *Mol Pharmacol* 2005;67:1460–1469.

Liiv I, Teesalu K, Peterson P, Clemente MG, Perheentupa J, Uibo R. Epitope mapping of cytochrome P450 cholesterol side-chain cleavage enzyme by sera from patients with autoimmune polyglandular syndrome type 1. *Eur J Endocrinol* 2002;146: 113–119.

Lippincott-Schwartz J, Bonifacino JS, Yuan LC, Klausner RD. Degradation from the endoplasmic reticulum: disposing of newly synthesized proteins. *Cell* 1988;54: 209–220.

Liu ZX, Kaplowitz N. Immune-mediated drug-induced liver disease. *Clin Liver Dis* 2002;6:755–774.

Longhi MS, Hussain MJ, Bogdanos DP, Quaglia A, Mieli-Vergani G, Ma Y, Vergani D. Cytochrome P450IID6-specific CD8 T cell immune responses mirror disease activity in autoimmune hepatitis type 2. *Hepatology* 2007;46:472–484.

Lord J, Davey J, Frigerio L, Roberts L. Endoplasmic reticulum-associated protein degradation. *Semin Cell Dev Biol* 2000;11:159–164.

Lown KS, Bailey DG, Fontana RJ, Janardan SK, Adair CH, Fortlage LA, Brown MB, Guo W, Watkins PB. Grapefruit juice increases felodipine oral availability in humans by decreasing intestinal CYP3A protein expression. *J Clin Invest* 1997;99: 2545–2553.

Lu P, Zhou X, Shen M, Lu K. Function of WW domains as phosphoserine- or phosphothreonine-binding modules. *Science* 1999;283:1325–1328.

Ma Q. Aryl hydrocarbon receptor degradation-promoting factor (ADPF) and the control of the xenobiotic response. *Mol Intervent* 2007;7:133–137.

Ma Y, Thomas MG, Okamoto M, Bogdanos DP, Nagl S, Kerkar N, Lopes AR, Muratori L, Lenzi M, Bianchi FB, Mieli-Vergani G, Vergani D. Key residues of a major cytochrome P4502D6 epitope are located on the surface of the molecule. *J Immunol* 2002;169:277–285.

Ma Y, Bogdanos DP, Hussain MJ, Underhill J, Bansal S, Longhi MS, Cheeseman P, Mieli-Vergani G, Vergani D. Polyclonal T-cell responses to cytochrome P450IID6 are associated with disease activity in autoimmune hepatitis type 2. *Gastroenterology* 2006;130:868–882.

Manns MP, Griffin KJ, Sullivan KF, Johnson EF. LKM-1 autoantibodies recognize a short linear sequence in P450IID6, a cytochrome P-450 monooxygenase. *J Clin Invest* 1991;88:1370–1378.

Masaki R, Yamamoto A, Tashiro Y. Cytochrome P450 and NADPH-cytochrome P450 reductase are degraded in the autolysosomes in rat liver. *J Cell Biol* 1987;104: 1207–1215.

Masubuchi Y, Horie T. Mechanism-based inactivation of cytochrome P450s 1A2 and 3A4 by dihydralazine in human liver microsomes. *Chem Res Toxicol* 1999;12: 1028–1032.

Masuyama H, Inoshita H, Hiramatsu Y, Kudo T. Ligands have various potential effects on the degradation of pregnane X receptor by proteasome. *Endocrinology* 2002;143:55–61.

Mayhew BS, Jones DR, Hall SD. An *in vitro* model for predicting *in vivo* inhibition of cytochrome P450 3A4 by metabolic intermediate complex formation. *Drug Metab Dispos* 2000;28:1031–1037.

Meyer HH, Shorter JG, Seemann J, Pappin D, Warren G. A complex of mammalian ufd1 and npl4 links the AAA-ATPase, p97, to ubiquitin and nuclear transport pathways. *EMBO J* 2000;19:2181–2192.

Miyamoto S, Maki M, Schmitt MJ, Hatanaka M, Verma IM. Tumor necrosis factor alpha-induced phosphorylation of I kappa B alpha is a signal for its degradation but not dissociation from NF-kappa B. *Proc Natl Acad Sci USA* 1994;91:12740–12744.

Mizushima N. The pleiotropic role of autophagy: from protein metabolism to bactericide. *Cell Death Differ* 2005;12(Suppl 2):1535–1541.

Mizushima N, Klionsky DJ. Protein turnover via autophagy: implications for metabolism. *Annu Rev Nutr* 2007;27:19–40.

Mizutani T, Shinoda M, Tanaka Y, Kuno T, Hattori A, Usui T, Kuno N, Osaka T. Autoantibodies against CYP2D6 and other drug-metabolizing enzymes in autoimmune hepatitis type 2. *Drug Metab Rev* 2005;37:235–252.

Moreau A, Vilarem MJ, Maurel P, Pascussi JM. Xenoreceptors CAR and PXR activation and consequences on lipid metabolism, glucose homeostasis, and inflammatory response. *Mol Pharm* 2008;5:35–41.

Morgan ET, Goralski KB, Piquette-Mille, M, Renton KW, Robertson GR, Chaluvadi MR, Charles KA, Clarke SJ, Kacevska M, Liddle C, Richardson TA, Sharma R, Sinal CJ. Regulation of drug-metabolizing enzymes and transporters in infection, inflammation, and cancer. *Drug Metab Dispos* 2008;36:205–216.

Morishima Y, Peng HM, Lin HL, Hollenberg PF, Sunahara RK, Osawa Y, Pratt WB. Regulation of cytochrome P450 2E1 by heat shock protein 90-dependent stabilization and CHIP-dependent proteasomal degradation. *Biochemistry* 2005;44:16333–16340.

Murata S, Chiba T, Tanaka K. CHIP: a quality-control E3 ligase collaborating with molecular chaperones. *Int J Biochem Cell Biol* 2003;35:572–578.

Murray BP, Correia MA. Ubiquitin-dependent 26S proteasomal pathway: A role in the degradation of the native human liver CYP3A4 expressed in *Saccharomyces cerevisiae*? *Arch Biochem Biophys* 2001;393:106–116.

Murray BP, Zgoda VG, Correia MA. Native CYP2C11: heterologous expression in *Saccharomyces cerevisiae* reveals a role for vacuolar proteases rather than the ubiquitin-26S proteasome system in the degradation of this endoplasmic reticulum enzyme. *Mol Pharmacol* 2002;61:1146–1153.

Nakajima M, Yoshida R, Shimada N, Yamazaki H, Yokoi T. Inhibition and inactivation of human cytochrome P450 isoforms by phenethyl isothiocyanate. *Drug Metab Dispos* 2001;29:1110–1113.

Omura T, Siekevitz P, Palade GE. Turnover of constituents of the endoplasmic reticulum membranes of rat hepatocytes. *J Biol Chem* 1967;242:2389–2239.

Pabarcus MK, Hoe N, Sadeghi S, Patterson C, Wiertz E, Correia MA. CYP3A4 ubiquitination by gp78 (the tumor autocrine motility factor receptor, AMFR) and CHIP E3 ligases. *Arch Biochem Biophys* 2009;483:66–74.

Paine MF, Oberlies NH. Clinical relevance of the small intestine as an organ of drug elimination: drug–fruit juice interactions. *Expert Opin Drug Metab Toxicol* 2007;3:67–80.

Parkinson A, Thomas PE, Ryan DE, Levin W. Differential time course of induction of rat liver microsomal cytochrome P-450 isozymes and epoxide hydrolase by Aroclor 1254. *Arch Biochem Biophys* 1983;225:203–215.

Passmore LA. The anaphase-promoting complex (APC): the sum of its parts? *Biochem Soc Trans* 2004;32:724–727.

Paul GA, Gibbs JS, Boobis AR, Abbas A, Wilkins MR. Bosentan decreases the plasma concentration of sildenafil when coprescribed in pulmonary hypertension. *Br J Clin Pharmacol* 2005;60:107–112.

Peng J. Evaluation of proteomic strategies for analyzing ubiquitinated proteins. *BMB Rep* 2008;41:177–183.

Pessayre D. Toxic and immune mechanisms leading to acute and subacute drug induced liver injury. In: Miguet JP, Dhumeaux D, editors. *Progress in Hepatology*. Paris: John Libbey Eurotext, 1993, pp. 23–39.

Peters JM, Harris JR, Finley D, editors. *Ubiquitin and the Biology of the Cell*. New York: Plenum Press, 1998.

Pickart CM. Mechanisms underlying ubiquitination. *Annu Rev Biochem* 2001;70: 503–533.

Pickart CM, Cohen RE. Proteasomes and their kin: proteases in the machine age. *Nat Rev Mol Cell Biol* 2004;5:177–187.

Pinto AG, Horlander J, Chalasani N, Hamman M, Asghar A, Kolwankar D, Hall SD. Diltiazem inhibits human intestinal cytochrome P450 3A (CYP3A) activity *in vivo* without altering the expression of intestinal mRNA or protein. *Br J Clin Pharmacol* 2005;59:440–446.

Pitlick WH, Levy RH, Tropin AS, Green JR. Pharmacokinetic model to describe self-induced decreases in steady-state concentrations of carbamazepine. *J Pharm Sci* 1976;65:462–463.

Raasi S, Wolf DH. Ubiquitin receptors and ERAD: a network of pathways to the proteasome. *Semin Cell Dev Biol* 2007;18:780–791.

Ravid T, Hochstrasser M. Diversity of degradation signals in the ubiquitin–proteasome system. *Nat Rev Mol Cell Biol* 2008;9:679–690.

Ravid T, Kreft SG, Hochstrasser M. Membrane and soluble substrates of the Doa10 ubiquitin ligase are degraded by distinct pathways. *EMBO J* 2006;25: 533–543.

Risseeuw EP, Daskalchuk TE, Banks TW, Liu E, Cotelesage J, Hellmann H, Estelle M, Somers DE, Crosby WL. Protein interaction analysis of SCF ubiquitin E3 ligase subunits from Arabidopsis. *Plant J* 2003;34:753–767.

Roberts BJ. Evidence of proteasome-mediated cytochrome P450 degradation. *J Biol Chem* 1997;272:9771–9778.

Roberts BJ, Song BJ, Soh Y, Park SS, Shoaf SE. Ethanol induces CYP2E1 by protein stabilization. *J Biol Chem* 1995;270:29632–29635.

Robin MA, Maratrat M, Le Roy M, Le Breton FP, Bonierbale E, Dansette P, Ballet F, Mansuy D, Pessayre D. Antigenic targets in tienilic acid hepatitis. Both cytochrome P450 2C11 and 2C11-tienilic acid adducts are transported to the plasma membrane of rat hepatocytes and recognized by human sera. *J Clin Invest* 1996;98:1471–1480.

Robin MA, Descatoire V, Le Roy M, Berson A, Lebreton FP, Maratrat M, Ballet F, Loeper J, Pessayre D. Vesicular transport of newly synthesized cytochromes P4501A to the outside of rat hepatocyte plasma membranes. *J Pharmacol Exp Ther* 2000;294:1063–1069.

Roccaro AM, Hideshima T, Richardson PG, Russo D, Ribatti D, Vacca A, Dammacco F, Anderson KC. Bortezomib as an antitumor agent. *Curr Pharm Biotechnol* 2006;7:441–448.

Rock KL, Gramm C, Rothstein L, Clark K, Stein R, Dick L, Hwang D, Goldberg AL. Inhibitors of the proteasome block the degradation of most cell proteins and the generation of peptides presented on MHC class I molecules. *Cell* 1994;78:761–771.

Ronis MJ, Ingelman-Sundberg M. Acetone-dependent regulation of cytochrome P450j (IIE1) and P450b (IIB1) in rat liver. *Xenobiotica* 1989;19:1161–1165.

Ronis MJ, Johansson I, Hultenby K, Lagercrantz J, Glaumann H, Ingelman-Sundberg M. Acetone-regulated synthesis and degradation of cytochrome P450E1 and cytochrome P4502B1 in rat liver. *Eur J Biochem* 1991;198:383–389.

Russo A, Fratto ME, Bazan V, Schiró V, Agnese V, Cicero G, Vincenzi B, Tonini G, Santini D. Targeting apoptosis in solid tumors: the role of bortezomib from preclinical to clinical evidence. *Expert Opin Ther Targets* 2007;11:1571–1586.

Schafer A, Wolf DH. Endoplasmic reticulum-associated protein quality control and degradation: screen for ERAD mutants after ethylmethane sulfonate mutagenesis. *Methods Mol Biol* 2005;301:283–288.

Schmiedlin-Ren P, Edwards DJ, Fitzsimmons ME, He K, Lown KS, Woster PM, Rahman A, Thummel KE, Fisher JM, Hollenberg PF, Watkins PB. Mechanisms of enhanced oral availability of CYP3A4 substrates by grapefruit constituents. Decreased enterocyte CYP3A4 concentration and mechanism-based inactivation by furanocoumarins. *Drug Metab Dispos* 1997;25:1228–1233.

Seglen PO. Regulation of autophagic protein degradation in isolated liver cells. In: Glaumann H, Ballard FJ, editors. *Lysosomes: Their Role in Protein Breakdown.* London: Academic Press, 1987, pp. 371–414.

Shintani T, Klionsky DJ. Autophagy in health and disease: a double-edged sword. *Science* 2004;306:990–995.

Shiraki H, Guengerich FP. Turnover of membrane proteins: kinetics of induction and degradation of seven forms of rat liver microsomal cytochrome P-450, NADPH-cytochrome P-450 reductase and epoxide hydrolase. *Arch Biochem Biophys* 1984;235:86–96.

Shringarpure R, Grune T, Davies KJA. Protein oxidation and 20S proteasome dependent proteolysis in mammalian cells. *Cell Mol Life Sci* 2001;58:1442–1450.

Silverman RB. *Mechanism-Based Enzyme Inactivation: Chemistry and Enzymology*, Vol. 1. Boca Raton, FL: CRC Press, 1988, pp. 3–30.

Sokolik CW, Cohen RE. Ubiquitin conjugation to cytochromes c. Structure of the yeast iso-1 conjugate and possible recognition determinants. *J Biol Chem* 1992;267:1067–1071.

Song BJ, Veech RL, Park SS, Gelboin HV, Gonzalez FJ. Induction of rat hepatic *N*-nitrosodimethylamine demethylase by acetone is due to protein stabilization. *J Biol Chem* 1989;264:3568–3572.

Steinberg RA, Levinson BB, Tomkins GM. Kinetics of steroid induction and deinduction of tyrosine aminotransferase synthesis in cultured hepatoma cells. *Proc Natl Acad Sci USA* 1975;72:2007–2011.

Steward AR, Wrighton SA, Pasco DS, Fagan JB, Li D, Guzelian PS. Synthesis and degradation of 3-methylcholanthrene-inducible cytochromes P-450 and their mRNAs in primary monolayer cultures of adult rat hepatocytes. *Arch Biochem Biophys* 1985;241:494–508.

Swanson R, Locher M, Hochstrasser M. A conserved ubiquitin ligase of the nuclear envelope/endoplasmic reticulum that functions in both ER-associated and Matalpha2 repressor degradation. *Genes Dev* 2001;15:2660–2674.

Tai HC, Schuman EM. Ubiquitin, the proteasome and protein degradation in neuronal function and dysfunction. *Nat Rev Neurosci* 2008;9:826–838.

Teckman JH, Perlmutter DH. Retention of mutant alpha(1)-antitrypsin Z in endoplasmic reticulum is associated with an autophagic response. *Am J Physiol Gastrointest Liver Physiol* 2000;279:G961–G974.

Thompson SJ, Loftus LT, Ashley MD, Meller R. Ubiquitin-proteasome system as a modulator of cell fate. *Curr Opin Pharmacol* 2008;8:90–95.

Thum T, Erpenbeck VJ, Moeller J, Hohlfeld JM, Krug N, Borlak J. Expression of xenobiotic metabolizing enzymes in different lung compartments of smokers and nonsmokers. *Environ Health Perspect* 2006;114:1655–1661.

Tierney DJ, Haas AL, Koop DR. Degradation of cytochrome P450 2E1: Selective loss after labilization of the enzyme. *Arch Biochem Biophys* 1992;29:9–16.

Tirona RG, Lee W, Leake BF, Lan LB, Cline CB, Lamba V, Parviz F, Duncan SA, Inoue Y, Gonzalez FJ, Schuetz EG, Kim RB. The orphan nuclear receptor HNF4alpha determines PXR- and CAR-mediated xenobiotic induction of CYP3A4. *Nat Med* 2003;9:220–224.

Tsuji H, Akasaki K. Identification and characterization of lysosomal enzymes involved in the proteolysis of phenobarbital-inducible cytochrome P450. *Biol Pharm Bull* 1994;17:568–571.

Ueno T, Muno D, Kominami E. Membrane markers of endoplasmic reticulum preserved in autophagic vacuolar membranes isolated from leupeptin-administered rat liver. *J Biol Chem* 1991;266:18995–18999.

Uetrecht J. Current trends in drug-induced autoimmunity. *Autoimmun Rev* 2005;4: 309–314.

Uetrecht J. Idiosyncratic drug reactions: current understanding. *Annu Rev Pharmacol Toxicol* 2007;47:513–539.

Uibo R, Aavik E, Peterson P, Perheentupa J, Aranko S, Pelkonen R, Krohn KJ. Autoantibodies to cytochrome P450 enzymes P450scc, P450c17, and P450c21 in autoimmune polyglandular disease types I and II and in isolated Addison's disease. *J Clin Endocrinol Metab* 1994;78:23–28.

van Giersbergen PL, Gnerre C, Treiber A, Dingemanse J, Meyer UA. Bosentan, a dual endothelin receptor antagonist activates the pregnane X nuclear receptor. *Eur J Pharmacol* 2002;450:115–121.

Varshavsky A. Naming a targeting signal. *Cell* 1991;64:13–15.

Varshavsky A. The N-end rule: functions, mysteries, uses. *Proc Natl Acad Sci USA* 1996;93:12142–12149.

Vidali M, Hidestrand M, Eliasson E, Mottaran E, Reale E, Rolla R, Occhino G, Albano E, Ingelman-Sundberg M. Use of molecular simulation for mapping conformational CYP2E1 epitopes. *J Biol Chem* 2004;279:50949–50955.

Wang H, LeCluyse EL. Role of orphan nuclear receptors in the regulation of drug-metabolising enzymes. *Clin Pharmacokinet* 2003;42:1331–1357.

Wang HF, Figueiredo Pereira ME, Correia MA. Cytochrome P450 3A[a] degradation in isolated rat hepatocytes: 26S proteasome inhibitors as probes. *Arch Biochem Biophys* 1999;365:45–53.

Wang X, Medzihradszky KF, Maltby D, Correia MA. Phosphorylation of native and heme-modified CYP3A4 by protein kinase C: a mass spectrometric characterization of the phosphorylated peptides. *Biochemistry* 2001;40:11318–11326.

Wang Y, Liao M, Hoe N, Acharya P, Deng C, Krutchinsky AN, Correia MA. A role for protein phosphorylation in cytochrome P450 3A4 ubiquitin-dependent proteasomal degradation. *J Biol Chem* 2009;284:5671–5684.

Ward CL, Omura S, Kopito RR. Degradation of CFTR by the ubiquitinproteasome pathway. *Cell* 1995;83:121–127.

Watkins PB, Wrighton SA, Schuetz EG, Maurel P, Guzelian PS. Macrolide antibiotics inhibit the degradation of the glucocorticoid-responsive cytochrome P-450p in rat hepatocytes *in vivo* and in primary monolayer culture. *J Biol Chem* 1986;261:6264–6271.

Watkins PB, Wrighton SA, Schuetz EG, Molowa DT, Guzelian PS. Identification of glucocorticoid-inducible cytochromes P-450 in the intestinal mucosa of rats and man. *J Clin Invest* 1987;80:1029–1036.

Watson G, Davey RA, Labarca C, Paigen K. Genetic determination of kinetic parameters in beta-glucuronidase induction by androgen. *J Biol Chem* 1981;256:3005–3011.

Weissman A. Themes and variations on ubiquitylation. *Nat Rev Mol Cell Biol* 2001;2:169–178.

Whiteside S, Ernst M, LeBail O, Laurent-Winter C, Rice N, Israel A. N- and C-terminal sequences control degradation of MAD3/I kappa B alpha in response to inducers of NF-kappa B activity. *Mol Cell Biol* 1995;15:5339–5345.

Wilkinson KD. Regulation of ubiquitin-dependent processes by deubiquitinating enzymes. *FASEB J* 1997;11:1245–1256.

Wolf DH. From lysosome to proteasome: the power of yeast in the dissection of proteinase function in cellular regulation and waste disposal. *Cell Mol Life Sci* 2004;61:1601–1604.

Yamamoto A, Masaki R, Tashiro Y. Is cytochrome P-450 transported from the endoplasmic reticulum to the Golgi apparatus in rat hepatocytes? *J Cell Biol* 1985;101:1733–1740.

Yamamoto AM, Cresteil D, Boniface O, Clerc FF, Alvarez F. Identification and analysis of cytochrome P450IID6 antigenic sites recognized by anti-liver–kidney microsome type-1 antibodies (LKM1). *Eur J Immunol* 1993;23:1105–1111.

Yang J, Liao M, Shou M, Jamei M, Yeo KR, Tucker GT, Rostami-Hodjegan A. Cytochrome P450 turnover: regulation of synthesis and degradation, methods for determining rates, and implications for the prediction of drug interactions. *Curr Drug Metab* 2008;9:384–394.

Yang MX, Cederbaum AI. Role of the proteasome complex in degradation of human CYP2E1 in transfected HepG2 cells. *Biochem Biophys Res Commun* 1996;226:711–716.

Yang MX, Cederbaum AI. Characterization of cytochrome P4502E1 turnover in transfected HepG2 cells expressing human CYP2E1. *Arch Biochem Biophys* 1997;341:25–33.

Yewdell JW. Immunoproteasomes: regulating the regulator. *Proc Natl Acad Sci USA* 2005;102:9089–9090.

Yorimitsu T, Klionsky DJ. Autophagy: molecular machinery for self-eating. *Cell Death Differ* 2005;12(Suppl 2):1542–1552.

Zangar RC, Kimzey AL, Okita JR, Wunschel DS, Edwards RJ, Kim H, Okita R. Cytochrome P450 3A conjugation to ubiquitin in a process distinct from classical ubiquitination pathway. *Mol Pharmacol* 2002;61:892–904.

Zhang DD. Mechanistic studies of the Nrf2-Keap1 pathway. *Drug Metab Rev* 2006;38:769–789.

Zhang QX, Melnikov Z, Feierman DE. Characterization of the acetaminophen-induced degradation of cytochrome P450–3A4 and the proteolytic pathway. *Basic Clin Pharmacol Toxicol* 2004;94:191–200.

Zhou SF, Lai X. An update on clinical drug interactions with the herbal antidepressant St John's wort. *Curr Drug Metab* 2008;9:394–409.

Zhukov A, Ingelman-Sundberg M. Selective fast degradation of cytochrome P450 2E1 in serum-deprived hepatoma cells by a mechanism sensitive to inhibitors of vesicular transport. *Eur J Biochem* 1997;247:37–43.

Zhukov A, Ingelman-Sundberg M. Relationship between cytochrome P450 catalytic cycling and stability: fast degradation of ethanol-inducible cytochrome P450 2E1 (CYP2E1) in hepatoma cells is abolished by inactivation of its electron donor NADPH-cytochrome P450 reductase. *Biochem J* 1999;340:453–458.

Zhukov A, Werlinder V, Ingelman-Sundberg M. Purification and characterization of two membrane bound serine proteinases from rat liver microsomes active in degradation of cytochrome P450. *Biochem Biophys Res Commun* 1993;197:221–228.

12

COMPLEXITIES OF WORKING WITH UDP-GLUCURONOSYLTRANSFERASES (UGTs): FOCUS ON ENZYME INHIBITION

MICHAEL B. FISHER

12.1 INTRODUCTION

The discovery and development of a new chemical entity (NCE) involves significant investment in money and time. Therefore, it is important to understand and try to minimize or eliminate factors that contribute to the attrition of NCEs in development. There have been several high-profile drug withdrawals due to adverse events precipitated by drug–drug interactions (Ajayi et al., 2000). The most high-profile inhibitory drug interaction occurred when the cytochrome P4503A (CYP3A) inhibitor ketoconazole was administered along with terfenadine. Normally, terfenadine is converted almost completely by CYP3A to the active metabolite fexofenadine; inhibiting this pathway resulted in patients getting significant exposure to the parent, terfenadine, which resulted in prolongation of the cardiac Q-T interval. Indeed, since the ketoconazole–terfenadine interaction was elucidated [reviewed in Kivistö et al. (1994)], there has been an increased focus on especially minimizing the potential for inhibitory drug–drug interactions, both in research and development. Early assessment of the drug interaction potential of molecules in discovery is crucial to finding chemical space lacking inhibition potential

Enzyme Inhibition in Drug Discovery and Development, edited by Lu and Li
Copyright © 2010 John Wiley & Sons, Inc.

toward the CYPs, and this assessment continues through the development of NCEs, both *in vitro* and in clinical studies. However, a recent retrospective examination of new drug interactions in Japan from 2000 to 2003 (Yoshida et al., 2006) indicated that metabolic interactions were the most common (45%), with pharmacological interactions less common (27%), suggesting that drug interactions are still a liability in NCEs.

In recent years the reagents and methods to measure P450 metabolism have been widely available and have been universally applied to measure CYP inhibition (Bachmann et al., 2003). Compounds in research will be tested at one or several concentrations *in vitro* in incubations to assess effects on the activity of CYP1A2, 2C9, 2C19, 2D6, and 3A and sometimes CYP2B6, 2C8, and 2E1. Human liver microsomes (HLM), recombinant enzymes, and even hepatocytes have been applied, and each system has advantages and disadvantages that would need to be considered before applying one to a given situation.

These *in vitro* CYP inhibition data have been applied over the years using various models to predict *in vivo* drug–drug interactions (Venkatakrishnan et al., 2003). There exist common aspects to the various models. Firstly, the inhibitor potency (K_i, IC_{50}) toward the target must be known; secondly, the *in vivo* inhibitor concentration (I) must be surmised. It is the ratio of I/K_i that drives the observed drug–drug interaction. Additionally, the fraction of the victim drug's clearance pathway that this inhibitable reaction represents (fraction metabolized, f_m) also determines the magnitude of its maximal inhibition (Obach et al., 2006). An appreciation of these variables as they apply to enzyme inhibition is imperative in understanding the drug interaction potential of any drug.

An understanding of the enzymology and cell biology is obviously very useful in understanding the caveats, or complexities, associated with studying P450s *in vitro*. For example, it has been appreciated for at least a decade that complexities, such as atypical kinetics, heterotropic activation, and substrate dependence to inhibition profiles, can often be observed with CYP3A4 and 2C9 (at least) due to multiple binding regions in their active sites (Atkins, 2006). An understanding of these complexities has led to the derivation of new and better ways to deal with complex enzyme kinetics. Additionally, the presence of the heme prosthetic group has allowed a further understanding of some of these intricate ligand-binding events through spectroscopic measurements (Roberts et al., 2005). Potent inhibition often involving the heme has also added complexity to unmasking and predicting drug interactions with CYPs (Hutzler et al., 2006). Decades of research with P450 enzymology and biochemistry has helped elucidate and predict drug interaction potential of this enzyme system.

A recent survey of drugs on the market demonstrated that while metabolism via the P450s was most frequently observed, the next most common pathway was conjugative metabolism by the UDP-glucuronosyl transferases (Williams et al., 2004). The UGTs are a superfamily of enzymes that add a glucuronic acid moiety to a nucleophilic atom in a molecule. Once thought to occur only subsequent (hence the nomenclature "phase 2 metabolism" referring to conjugation reactions) to the addition or unmasking of a nucleophilic handle, such as aromatic hydroxylation to form a phenol, copious examples of direct conjugation are known to occur (Kiang et al., 2005).

As with the P450s, an understanding of the enzymology and cell biology is also very useful in understanding the caveats and complexities associated with studying UGTs *in vitro*, as will be discussed. However, an advanced understanding of the biochemistry and enzymology of this enzyme system has lagged compared with the P450s. There are several reasons for this, not the least of which is the more recent commercial availability of recombinant UGT isoforms, the lack of a well-characterized battery of selective probe substrates and inhibitors, and the lack of a spectrally sensitive prosthetic group like in P450s. This chapter will provide a perspective on clinical drug interactions and observations and expectations from *in vitro* UGT inhibition data. It will then try to examine the known biochemical complexities associated with UGTs and how these observations helped to refine our methods to quantitatively assess UGT inhibition *in vitro*.

12.2 INTRODUCTION TO UGT INHIBITION

Since the first *in vitro* glucuronidations and the isolation and identification of the vital cofactor UDPGA were performed by Geoffrey Dutton in the first half of the 20th century (Dutton 1997, and references therein), our understanding of this enzyme system has been slowly advancing. The UGTs are a superfamily of membrane-bound enzymes that catalyze the addition of a glucuronic acid moiety from UDPGA onto nucleophilic atoms, such as oxygen, nitrogen, sulfur, or even carbon, in endo- and xenobiotics. The UGTs are expressed largely in the liver, but some forms have a very tissue-specific expression pattern. For example, UGT1A7, 1A8, and 1A10 appear to be extrahepatic isoforms, showing only gastrointestinal expression (Tukey and Strassburg, 2000; Miners et al., 2006). Unlike P450s, where the active site is thought to be cytosolic, the UGT active site resides in the lumen of the endoplasmic reticulum (Guéraud and Paris, 1998) (see Fig. 12.1). According to a recent review, 17 human UGT proteins have been identified to date: UGT1A1, 1A3, 1A4, 1A5, 1A6, 1A7, 1A8, 1A10, 2A1, 2B4, 2B7, 2B10, 2B11, 2B15, 2B17, and 2B28 (Miners et al., 2006). Currently, the ready availability of *in vitro* systems such as HLM, recombinant enzymes, and hepatocytes has allowed the kinetic characterization of glucuronide formation and inhibition. Some complexities unique to UGTs relative to P450s have likely caused a lag in their entry into mainstream research.

General observations with regards to UGTs are that they are low affinity enzymes in general, and that no really potent inhibitors are known. Williams et al. (2004) took a weight of evidence approach and determined that since

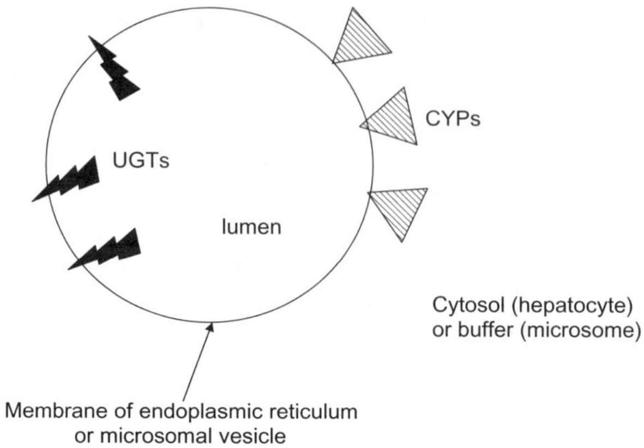

Figure 12.1. The different subcellular localization of cytochrome P450s and UDP-glucuropnosyltransferases.

substrate affinities are typically low (high K_m), inhibitor affinities are typically low (high K_i or IC_{50}), glucuronidation is usually not a major clearance pathway (unlike oxidation by P450s), and since multiple UGTs are usually capable of performing the catalysis, the expectation is that clinically relevant drug interactions involving inhibition of glucuronidation will be rare. This was realized in that manuscript in the few clinical studies examined where AUC increases for the aglycone are typically <2×. While this is a reasonable conclusion given the available data at the time, recent data discussed later in the chapter identify artifacts or unique situations that suggest that liver microsomes, the most common matrix for examining interactions, may not be the most appropriate system to examine without incorporating some recent learnings. Kinetic differences due at least in part to a buffer system, fatty acids liberated during microsomal incubation acting as UGT inhibitors, and kinetic differences due to heterodimerization are a few of the phenomena recently shown to affect ligand affinities in microsomal or recombinant systems.

12.3 INHIBITION OF UGT

12.3.1 Bilirubin

UGTs can metabolize endogenous compounds as well as xenobiotics. For example, UGT2B15 and 2B17 are able to glucuronidate endogenous steroids with high affinity (Bélanger et al., 2003). However, probably the most important potential interaction involves metabolism of an endogenous compound, bilirubin, by UGT1A1. Bilirubin is a heme degradation product and is an important antioxidant at low concentrations. Higher concentrations can lead to jaundice and kernicterus (bilirubin encephalopathy) (Fevery, 2008). While this is a known phenotype with UGT1A genetic mutations (Tukey and

Strassburg, 2000) due to an inability to sufficiently glucuronidate bilirubin, several drugs have been known to inhibit UGT1A1 and cause elevated bilirubin. In fact, the elevated bilirubin can be misinterpreted as a biomarker of hepatotoxicity instead of UGT1A1 inhibition (Zhang et al., 2005). Some of the HIV protease inhibitors, especially indinavir and atazanavir, have been associated with elevated bilirubin levels *in vivo*. A recent study attempted to link this observation with the propensity to inhibit UGT1A1 for six protease inhibitors. Interestingly, the unbound C_{max}/K_i was >0.1 for the two compounds associated with clinical hyperbilirubinemia, while it was <0.1 for the four compounds without this observation (Zhang et al., 2005). This preliminary observation suggests that *in vitro* data may be of some value in rank ordering compounds with regard to their propensity to inhibit UGT1A1 and cause hyperbilirubinemia.

Trubetskoy et al. (2007) describe a high-throughput 384-well assay for UGT1A1 inhibition that may have some value in research for assessing risk of UGT1A1 inhibition. The assay uses recombinant enzyme and the fluorescent substrate 7-hydroxy-6-methoxycoumarin (7-HMC). Glucuronide conjugation of 7-HMC by UGT1A1 reduces the fluorescence, and thus inhibiting UGT1A1 would result in enhanced fluorescence relative to control. Formatting the assay in recombinant enzyme eliminates the need to use a selective probe. However, as discussed later, there may be some disadvantages to using recombinant enzyme as the *in vitro* system to study UGTs, but using something like HLM or hepatocytes would require the use of a selective probe to study the isoform of interest. If available, one could envision a lower-throughput UGT1A1 inhibition assay in a nonfluorescent format that utilized a selective probe (Table 12.1) in HLM, with the probe metabolite detected using liquid chromatography–mass spectrometry.

As stated in Miners et al. (2006), only a subset of known human UGT isoforms are likely to be important in hepatic drug metabolism. Table 12.1 shows one attempt to assemble a list of selective probes.

TABLE 12.1. Potential Probes for Studying Major Human Drug Metabolizing UGTs

UGT Isoform	Selective Probe	Selective Inhibitor
1A1	Bilirubin, β-estradiol (3-gluc.), etoposide	Ketoconazole, atazanavir
1A3	Hexafluoro-1α,25-dihydroxyvitamin D(3)	
1A4	Trifluoperazine	Hecogenin
1A6	Serotonin	
1A9	Propofol, sulfinpyrazone	
2B7	Zidovudine (AZT)	Mefanamic acid, flurbiprofen
2B15	(*S*)-Oxazepam, dihydrotestosterone	

Source: Table created in part using Miners et al. (2006) and Court (2005).

12.4 CLINICAL DRUG–DRUG INTERACTIONS

As mentioned above, when examining inhibition of glucuronidation to date, few drug interactions manifest clinically as a >2× AUC for the victim drug. A relatively recent review demonstrates that a few significant drug interactions can be attributed to inhibition at the UGT active site (Kiang et al., 2005). This is not meant to be an exhaustive list of clinical drug interactions involving UGT, but is meant to highlight some that are clinically significant and some that have warranted further study, both *in vitro* and *in vivo*.

Diflunisal and indomethacin are both anti-inflammatory drugs that are excreted at least in part as glucuronides (Hardman and Limbird, 1996). Van Hecken et al. (1989) assessed a potential interaction between diflunisal (inhibitor) and indomethacin (victim). Co-medication with diflunisal caused a 72% decrease in urinary indomethacin glucuronide, 50% decrease in indomethacin clearance, and a 2–3× increase in the AUC of indomethacin. Conversely, the oxidative metabolites and their downstream products were unaffected, strongly suggesting selective inhibition of glucuronidation. Interestingly, the authors speculate that the observed decrease in volume of distribution could result from decreased enterohepatic recirculation of indomethacin glucuronide during diflunisal cotreatment. This is an interesting speculation that points out how glucuronidation is unique relative to P450-mediated oxidation. Glucuronidation is a reversible process that can have a downstream effect on a drug's pharmacokinetics, such as enterohepatic recirculation (Roberts et al., 2002), such that perturbation of its metabolism can affect the *in vivo* equilibrium and influence more than just clearance.

To try and understand the *in vivo* drug–drug interaction, an *in vitro* investigation (Mano et al., 2006) was performed using human liver and intestinal microsomes. Indomethacin was found to have a 10-fold lower K_m value in intestine relative to liver (17 versus 210 μM). Diflunisal was also shown to more potently inhibit indomethacin glucuronidation in intestine relative to liver (IC_{50} 15–49 μM versus 100–230 μM). The authors speculated that a different UGT isoform was involved in intestine, and this was an explanation for the divergent behavior of substrate and inhibitor. They also concluded that intestinal concentrations of diflunisal likely exceed the IC_{50}, and so inhibition of intestinal glucuronidation of indomethacin is a possible cause of the interaction.

Mycophenolic acid (MPA), the active component of the prodrug mycophenolate mofetil used in the prevention of transplant rejection, is extensively glucuronidated in humans. MPA is often given in combination with cyclosprin or tacrolimus (Hardman and Limbird, 1996). In 1997, Zucker et al. (1997) reported an increase in the exposure of mycophenolic acid (MPA) after dosing mycophenolate mofetil (MMF) and tacrolimus. Patients on tacrolimus co-medication had a 1.6-fold increase in the AUC for MPA relative to the control group who received cyclosporine and MMF. There was also a 2.3× higher MPA C_{min} and 40% lower MPA-glucuronide exposure in plasma with tacrolimus. This increased exposure manifested as an enhancement in observed efficacy. Interestingly, the MPA pharmacokinetics with coadministered tacrolimus could be mimicked in the cyclosporine group by increasing the cyclosporine dose, suggesting that cyclosporine may be a less potent inhibitor than tacrolimus and that this lower potency can be overcome by increasing the dose of cyclosporine.

An *in vitro* investigation of the basis for this interaction was performed by Zucker et al. (1999). A partially purified UGT from human kidney was used to demonstrate that while both agents are inhibitors of MPA glucuronidation (~2 and 0.033 µM for cyclosporin and tacrolimus, respectively), clinically relevant plasma concentrations of tacrolimus (~0.012 to 0.03 µM) reach its *in vitro* K_i value, whereas cyclosporin plasma levels (~0.166 to 0.830 µM) do not. Thus, in this instance the *in vitro* data adequately align with the clinical observations.

12.5 AZT AS AN EXAMPLE TO UNDERSTAND UGT INHIBITION *IN VITRO* AND *IN VIVO*

An ideal tool compound to study UGT drug–drug interactions and their *in vivo* predictability from *in vitro* data would be a drug cleared almost completely through hepatic metabolism by a single UGT isoform. This would simplify the consideration of fraction metabolized and parallel elimination pathways and their ability to attenuate the effect of a pathway inhibitor. 3′-Azido-3′-deoxythymidine (AZT, zidovudine) has been reported to be a selective probe substrate for UGT2B7, and its major clearance pathway is through hepatic metabolism (Court et al., 2003). Thus, drug interactions involving AZT as the victim drug could likely be more straightforward to understand than interactions involving multiple UGTs. In 1998, Trapnell et al. (1998) tried to use *in vitro* data to try and understand in vivo drug interaction results for four perpetrator drugs (atovaquone, fluconazole, methadone, and valproic acid) toward AZT glucuronidation. Curiously, the rate of AZT glucuronidation in HLM was substantially increased in the presence of 2% bovine serum albumin (BSA), a protein typically used to model plasma protein binding. This increase is the opposite of what would occur if substrate binding to albumin were taking place; if BSA decreased the available AZT for metabolism, a decrease in metabolism would have been the result. Therefore, something more complex than a free fraction effect is occurring. This will be discussed in some detail later, but this may be the first mention of this albumin effect for UGT2B7. All four compounds (atovaquone, fluconazole, methadone, and valproic acid) were inhibitory to AZTG formation to varying degrees, with IC_{50} values of >100, 50, 8, and 100 µg/mL. Compared with reported clinical plasma concentrations, the authors conclude that only fluconazole and valproic acid are likely to cause clinically significant interactions with AZT.

Clinical drug interaction studies examining the effects of atovaquone, fluconazole, methadone, and valproic acid on AZT metabolism and pharmacokinetics have been reported [Trapnell et al. (1998) and references therein]. Interestingly, atovaquone caused a 30% decrease in AZTG/AZT AUC ratio and a 30% increase in AZT AUC. Fluconazole caused a 48% decrease in oral formation CL of AZTG, a 34% decrease in urinary AZTG/AZT, and a 74%, 84%, and 128% increase in AZT AUC, C_{max}, and $T_{1/2}$. Methadone caused a 52% increase in the oral AZT AUC, an 18% increase in bioavailability, and a 22% decrease in clearance. The intravenous AZT AUC increased 30% upon coadministration of methadone, and after AZT oral administration it revealed a 42% average increase in excretion of unchanged AZT. Valproic acid caused a 22% decrease in AZTG AUC, along with a near doubling and halving, respectively, of the AUC and oral clearance of AZT. Methadone caused a 30% decrease in AZT intravenous clearance and a 42% increase in the excretion of unchanged AZT. Therefore, since a significant alteration in AZT pharmacokinetics was seen for all perpetrators, it appears that the *in vitro* approach utilized actually underestimated the drug interaction potential of co-medications toward AZT. While the prediction of UGT drug interactions is still developing relative to predicting drug interactions with P450s, it is becoming apparent that additional factors need to be accounted for with UGTs to more fully exploit the available *in vitro* systems.

12.5.1 Complexities in Studying UGTs *In Vitro*

There are several characteristics that make UGTs a fairly complex enzyme system to study *in vitro* and extrapolate to *in vivo*. As described below, the subcellular localization, biochemistry, and genetics of UGTs are unique and differ from P450s. These need to be considered and accounted for in the design, execution, and extrapolation of *in vitro* data.

12.5.1.1 Latency. Probably the earliest complexity observed with regard to UGTs is their "latency" in liver microsome preparations. Historically, it was observed that treatment of liver microsomes with detergents, for example, appeared to activate enzyme activity manyfold. Two hypotheses were provided to explain this observation (Guéraud and Paris, 1998). Firstly, the luminal localization of the UGTs provided a membrane barrier to the UGT active site. Detergent served to permeabilize the membrane, and it allowed substrate and cofactor to access the active site more readily. Secondly, a conformational change hypothesis proposed that the detergent inducing a conformational change to a more active conformer was the explanation for the "activation." However, with the identification of the pore-forming peptide alamethicin as a convenient reagent for circumventing latency (Fisher et al., 2000), the barrier hypothesis appeared to be more fully supported. Additionally, the observation that some forms of UGT are very susceptible to detergent inhibition (Kurkela et al., 2003) makes this treatment much less practical than alamethicin treatment to circumvent latency.

12.5.1.2 Different* In Vitro *Systems. Multiple *in vitro* systems are candidates for predicting drug–drug interactions. For CYPs, HLM or recombinant P450 isoforms are the simplest and most convenient, with hepatocytes being a more complex system. With the relatively recent commercial availability of expressed individual UGT isoforms, the same statement could be made regarding UGTs. However, the elucidation of some considerable effects on enzyme activity and enzyme kinetics of hetero-oligomerization between different UGT isoforms, and also between UGT and CYP isoforms, cautions against the sole use of recombinant enzymes to study glucuronidation (*vide supra*). Additionally, significant lack of *in vitro–in vivo* correlations for the clearance via glucuronidation of some drugs as measured in liver microsomes argues against the analogous application of only non-cell-based systems to studying glucuronidation without a more complete elucidation of the cause of the *in vitro–in vivo* disconnect.

Firstly, in the 1980s, Mistry and Houston (1987) observed a 10- to 30-fold underprediction of clearance in the rat using rat liver microsomes for drugs primarily glucuronidated *in vivo*. It was more than 10 years later that a handful of publications reported the same phenomenon in HLM, while also indicating that human hepatocytes better predict *in vivo* clearance (Boase and Miners, 2002; Soars et al., 2002; Engtrakul et al., 2005). Follow-up studies (Engtrakul et al., 2005) using AZT glucuronidation as a model and ibuprofen as an inhibitor showed that substrate and inhibitor both were 5- to 10-fold more potent in human hepatocytes compared with HLM. V_{max}, when normalized for microsomal content, was also significantly altered. Kinetic deficiencies in HLM were partially abrogated through the use of carbonate buffer systems, a more physiologically relevant buffer system. Clearance predictions for AZT were significantly more accurate when using hepatocytes. The above differences between liver microsomes and hepatocytes really suggest that caution should be exercised when using liver microsomes to generate quantitative data for glucuronidation reactions.

12.5.1.3 Albumin Effect. As mentioned above, the general observation of high K_m values for glucuronidated substrates has led to the generalization that extrapolation from *in vitro* data from HLM may be unsuccessful. Interestingly, it was recently reported that the addition of fatty acid free albumin to HLM and recombinant UGT incubations reduced the K_m for AZT by nearly an order of magnitude, and this is due to the albumin sequestering UGT2B7-inhibitory fatty acids produced during microsomal incubation (Rowland et al., 2007, 2008a). The K_m reported in this system was comparable to what was measured as mentioned above in human hepatocytes (Engtrakul et al., 2005). The phenomenon appears to be similar to the albumin effect documented for CYP2C9 (Rowland et al., 2008b). While this is a relatively new finding, it is very significant and has been used to examine some previously poorly predicted interactions *in vitro* (*vide infra*).

12.5.1.4 UDP Production. A byproduct of glucuronidation is the release of UDP from UDPGA. 1-Naphthol is a probe substrate and has been used as an inhibitor of some UGT isoforms. In HLM the rapid glucuronidation of this aglycone by UGT1A6 liberates significant amounts of UDP (Fujiwara et al., 2008). Curiously, UDP has been shown to be a somewhat potent inhibitor of UGTs, and consistent with this, 1-naphthol inhibited other UGTs (UGT1A1, 1A4, 1A9) when examined in HLM only. Inhibition of the other *recombinant* UGT forms was dependent on the addition of recombinant UGT1A6, presumably to generate the inhibitory concentrations of UDP. UDP was shown to *noncompetitively* inhibit the metabolism of estradiol, imipramine, and propofol by UGT1A1, 1A4, and 1A9, respectively. The inhibition type for serotonin metabolism by UGT1A6 was a mixed type of competitive and noncompetitive. In contrast, UDP competitively altered the K_m for UDPGA for all: UGT1A1 (K_i 7 μM), UGT1A4 (K_i 47 μM), UGT1A6 (K_i 259 μM), and UGT1A9 (K_i 143 μM). This might exist as a type of regulatory feedback inhibition, but so far this has not been characterized. This mechanism of UGT inhibition could obviously complicate the quantitative and even qualitative inhibition result, depending on the *in vitro* system used. It is tempting to speculate that the latency observed with *in vitro* glucuronidation is due to UDP production and its subsequent enzyme inhibition and that permeabilization of the microsomal membrane allows sufficient diffusion of the UDP out of the lumen of the microsomal vesicle to restore full activity.

12.5.1.5 Oligomerization. Three recent reviews have summarized the significant amount of research that has been performed to understand the oligomerization state of the UGT enzymes (Finel and Kurkela, 2008). The reader is directed to these reviews and references therein. A few observations will be made here to emphasize the complexities to which the oligomerization state appears to contribute.

Meech and Mackenzie (1997) reported results from experiments to explore the role of dimerization of UGTs in 1997. Rat UGT2B1 was mutated into two different inactive mutants toward testosterone. When the two mutants were coexpressed, testosterone glucuronidation was restored. The data suggest that the two inactive forms can interact somehow to form an active enzyme complex with restored glucuronidation activity. This work strongly suggested that UGTs form functional dimers/oligomers.

More recently, Kurkela et al. (2007) examined the interaction between human UGT isoforms. UGT1A6 is known to selectively glucuronidate serotonin, while UGT2B7 is inactive toward this substrate. When these two isoforms were coexpressed and the amount of UGT1A6 incubated was kept the same as in the single expression experiment, the activity of UGT1A6 toward serotonin was nearly five times higher than UGT1A6 alone. This suggests that an activating interaction was occurring between these two isoforms. These same authors also studied the effect of coexpression with UGT1A4 on UGT activity. The Y485D mutation in UGT1A6 results in a 10- and 50-fold increase in the K_m toward UDPGA and 1-naphthol, respectively. Coexpression of this mutant with UGT1A4 restored the K_m values at or near the wild-type values. Coexpresion with UGT1A4 was shown to increase the enzyme-normalized activities of several other wild-type UGTs. Therefore, UGT1A4 is able to interact with other UGTs and somehow enhance activity. Given the number of possible hetero-oligomer partners available in the UGT superfamily [which may have just increased with the recent identification of alternative splicing that can occur at exon 5 to generate a negative modulator of the wild-type enzyme (Girard et al., 2007)] strongly suggests that we've just scratched the surface of this work.

Clearly, oligomerization can change the catalytic activity and kinetic parameters of UGTs. These differences would likely affect any UGT inhibition observed. How does this affect the use of recombinant single isoforms? Actually, the case to use single expressed UGT isoforms is further cautioned with consideration of UGT/CYP interactions. Takeda et al. (2005) showed that coexpression of UGT2B7 with CYP3A4, but not CYP1A2 or CYP2C9, increased the K_m for morphine-3-glucuronide formation by 10-fold. The V_{max} was not affected. So what is the evidence that these members of two different enzyme superfamilies actually physically interact? This same laboratory showed that CYP3A4 was co-immunoprecipitated using anti-UGT antibody. The antibody was shown to not recognize CYP3A4. Similar results were seen with UGT2B1 and CYP3A2 in rat, suggesting that a consistent interaction occurs in rodents (Ishii et al., 2007). Interestingly, this work with UGT2B7 and CYP3A4 was independently confirmed when Fremont et al. (2005) showed that UGT2B7 co-immunoprecipitated with anti-CYP3A4 antibody. Much more work needs to be done to understand the functional consequences of UGT–UGT and UGT–CYP interactions, but it is clear that differences in kinetic parameters can occur in systems more complex than a single isoform. For example, Operaña and Tukey (2007) recently showed homo- and hetero-oligomerization to varying extents for all of the UGT1A isoforms in a whole-cell system using fluorescently tagged UGT isoforms by fluorescence resonance energy transfer.

The reader is encouraged to use some caution when interpreting data generated from recombinant UGTs versus liver microsomes. For example, the *N*-glucuronidation of (*S*)-ketotifen shows a high-affinity and a low-affinity component in HLMs (Breyer-Pfaff et al., 2000). UGT1A4 is often involved in *N*-glucuronidation. Indeed, upon screening a battery of recombinant UGTs, UGT1A4 was involved. However, enzyme kinetics showed that UGT1A4 was the low-affinity enzyme. No other isoform was found to correspond to the high-affinity component. It is tempting to speculate that a hetero-oligomer is the high-affinity component.

12.6 PUTTING IT ALL TOGETHER: CORRECT PREDICTION OF FLUCONAZOLE–ZIDOVUDINE AND VALPROIC ACID–LAMOTRIGINE INTERACTIONS

These two drug interactions will be used to exemplify how our understanding of, and tools to apply to, UGTs have improved in the last couple of years. Valproic acid (VPA) and lamotrigine are both antiepileptic drugs that are significantly glucuronidated *in vivo* (Hardman and Limbird, 1996). For the VPA–lamotrigine interaction, it has been reported that VPA co-treatment will increase the exposure of lamotrigine in a dose-dependent fashion (Morris et al., 2000). However, neither molecule is considered to have much affinity for UGTs. In 2006 (Rowland et al., 2006), using alamethicin-treated microsomes, it was demonstrated that the K_m (or S50) of lamotrigine N2-glucuronidation by UGT2B7 in HLM went from 1869 μM to 255 μM with the inclusion of 2% BSA, more than a seven-fold decrease. Interestingly, the *in vitro* K_i of VPA for lamotrigine N2-glucuronidation in HLM also decreased with BSA, going from 2465 μM to 387 μM. Clearly, the inclusion of BSA increases the affinity of both ligands, and it was demonstrated that albumin is sequestering inhibitory fatty acids that were released from the microsomal membrane (Rowland et al., 2007). The authors then attempted to extrapolate all these data to predict a fold change in AUC for lamotrigine that would be expected with VPA co-medication. When the data without albumin were used, no significant interaction was predicted. However, when the data generated with albumin present were used, the authors conclude that a 1.9- to 2.3-fold change in AUC would be expected. *In vivo*, a 2.6-fold change was observed. The addition of albumin helped to make this interaction predictable.

Zidovudine (AZT), as mentioned above, is a selective probe for UGT2B7 and is cleared almost completely through glucuronidation (Hardman and Limbird, 1996). It is known that co-medicating patients with fluconazole causes a 1.92-fold increase on the AUC for AZT (Uchaipichat et al., 2006). *In vitro* data generated using alamethicin in recombinant UGT2B7 and in HLM indicate the K_m for AZT-glucuronidation is 500–900 μM. However, with the addition of 2% BSA, the K_m values decreased 85–90%, to 70–90 μM. Interestingly, BSA also caused a decrease in the K_i of fluconazole for AZT-glucuronidation, going from 500–1100 μM to 70–140 μM. Only the data generated with BSA present predicted an *in vivo* interaction, with fold AUC changes ranging from 41% to 217% of the *in vivo* interaction.

While an examination of the literature suggests that most substrates for UGTs are low-affinity substrates, the vast majority of these data were likely generated with no knowledge of some of the complexities mentioned above. The use of alamethicin and the inclusion of BSA, for example, can dramatically affect the apparent affinity. Compounds like AZT can go from showing a K_m approaching mM without BSA, to K_m values approaching low μM with BSA. More importantly, extrapolation of the data can possibly be more clinically relevant and useful to the scientist.

12.7 CONCLUSIONS

Inhibition of an endogenous process, such as the glucuronidation of bilirubin, can have toxic consequences via hyperbilirubinemia. Inhibition of the glucuronidation of xenobiotics has shown a generally low potency of inhibition and some lack of *in vitro–in vivo* correlation. One conclusion from this observation could be that drug interactions are not that important for this enzyme system. However, a recent appreciation of several *in vitro* phenomena has been shown to significantly affect the substrate and inhibitor kinetics. Indeed, factors such as inhibitory fatty acids being liberated during microsomal incubations can affect the kinetics of substrates and inhibitors, and the inclusion of albumin to sequester these fatty acids can improve the *in vitro* kinetics and *in vitro–in vivo* correlations. However, for highly bound substrates and/or inhibitors, the addition of albumin could complicate data interpretation. The production of the inhibitory byproduct UDP, complex buffer system effects on UGTs, and homo- and heterodimerization of UGTs and UGT-CYPs are only recently appreciated phenomena that likely lead to complex *in vitro* artifacts. Further work is required to establish a standard set of *in vitro* conditions to best study UGTs *in vitro*.

REFERENCES

Ajayi FO, Sun H, Perry J. Adverse drug reactions: a review of relevant factors. *J Clin Pharmacol* 2000;40:1093–1101.

Atkins WM. Current views on the fundamental mechanisms of cytochrome P450 allosterism. *Expert Opin Drug Metab Toxicol* 2006;2:573–579.

Bachmann KA, Ring BJ, Wrighton SA. Drug–drug interactions and the cytochromes P450. In: Lee JS, Obach RS, Fisher MB, editors. *Drug Metabolizing Enzymes: Cytochrome P450 and Other Enzymes in Drug Discovery and Development.* New York: Marcel Dekker, 2003, pp. 311–336.

Bélanger A, Pelletier G, Labrie F, Barbier O, Chouinard S. Inactivation of androgens by UDP-glucuronosyltransferase enzymes in humans. *Trends Endocrinol Metab* 2003;14:473–479.

Boase S, Miners JO. *In vitro–in vivo* correlations for drugs eliminated by glucuronidation: investigations with the model substrate zidovudine. *Br J Clin Pharmacol* 2002;54:493–503.

Breyer-Pfaff U, Mey U, Green MD, Tephly TR. Comparative *N*-glucuronidation kinetics of ketotifen and amitriptyline by expressed human UDP-glucuronosyltransferases and liver microsomes. *Drug Metab Dispos* 2000;28: 869–872.

Court MH. Isoform-selective probe substrates for *in vitro* studies of human UDP-glucuronosyltransferases. *Methods Enzymol* 2005;400:104–116.

Court MH, Krishnaswamy S, Hao Q, Duan SX, Patten CJ, Von Moltke LL, Greenblatt DJ. Evaluation of 3′-azido-3′-deoxythymidine, morphine, and codeine as probe substrates for UDP-glucuronosyltransferase 2B7 (UGT2B7) in human liver microsomes: specificity and influence of the UGT2B7*2 polymorphism. *Drug Metab Dispos* 2003;31:1125–1133.

Dutton GJ. Raising the colors: personal reflections on the glucuronidation revolution 1950–1970. *Drug Metab Rev* 1997;29:997–1024.

Engtrakul JJ, Foti RS, Strelevitz TJ, Fisher MB. Altered AZT (3′-azido-3′-deoxythymidine) glucuronidation kinetics in liver microsomes as an explanation for underprediction of *in vivo* clearance: comparison to hepatocytes and effect of incubation environment. *Drug Metab Dispos* 2005;33:1621–1627.

Fevery J. Bilirubin in clinical practice: a review. *Liver Int* 2008;28:592–605.

Finel M, Kurkela M. The UDP-glucuronosyltransferases as oligomeric enzymes. *Curr Drug Metab* 2008;9:70–76.

Fisher MB, Campanale K, Ackermann BL, VandenBranden M, Wrighton SA. *In vitro* glucuronidation using human liver microsomes and the pore-forming peptide alamethicin. *Drug Metab Dispos* 2000;28:560–566.

Fremont JJ, Wang RW, King CD. Coimmunoprecipitation of UDP-glucuronosyltransferase isoforms and cytochrome P450 3A4. *Mol Pharmacol* 2005;67:260–262.

Fujiwara R, Nakajima M, Yamanaka H, Katoh M, Yokoi T. Product inhibition of UDP-glucuronosyltransferase (UGT) enzymes by UDP obfuscates the inhibitory effects of UGT substrates. *Drug Metab Dispos* 2008;36:361–367.

Girard H, Lévesque E, Bellemare J, Journalt K, Caillier B, Guillemette C. Genetic diversity at the UGT1 locus is amplified by a novel 3′ alternative splicing mechanism leading to nine additional UGT1A proteins that act as regulators of glucuronidation activity. *Pharmacogenet Genomics* 2007;17:1077–1089.

Guéraud F, Paris A. Glucuronidation: a dual control. *Gen Pharmacol* 1998;31: 683–688.

Hardman JG, Limbird LE. *Goodman and Gilman's The Pharmacological Basis of Therapeutics*, 9th ed. New York: McGraw-Hill, 1996.

Hutzler JM, Melton RJ, Rumsey JM, Schnute ME, Locuson CW, Wienkers LC. Inhibition of cytochrome P450 3A4 by a pyrimidineimidazole: evidence for complex heme interactions. *Chem Res Toxicol* 2006;19:1650–1659.

Ishii Y, Iwanaga M, Nishimura Y, Takeda S, Ikushiro S-I, Nagata K, Yamazoe Y, Mackenzie PI, Yamada H. Protein-protein interactions between rat hepatic cytochromes P450 (P450s) and UDP-glucuronosyltransferases (UGTs): Evidence for the functionally active UGT in P450-UGT complex. *Drug Metab Pharmacokinet* 2007;22:367–376.

Kiang TKL, Ensom MHH, Chang TKH. UDP-glucuronosyltransferases and clinical drug–drug interactions. *Pharmacol Ther* 2005;106:97–132.

Kivistö KT, Neuvonen PJ, Klotz U. Inhibition of terfenadine metabolism: pharmacokinetic and pharmacodynamic consequences. *Clin Pharmacokinet* 1994;27: 1–5.

Kurkela M, García-Horsman JA, Luukkanen L, Mörsky S, Taskinen J, Baumann M, Kostiainen R, Hirvonen J, Finel M. Expression and characterization of recombinant human UDP-glucuronosyltransferases (UGTs). UGT1A9 is more resistant to detergent inhibition than other UGTs and was purified as an active dimeric enzyme. *J Biol Chem* 2003;278:3536–3544.

Kurkela M, Patana AS, Mackenzie PI, Court MH, Tate CG, Hirvonen J, Goldman A, Finel M. Interactions with other human UDP-glucuronosyltransferases attenuate the consequences of the Y485D mutation on the activity and substrate affinity of UGT1A6. *Pharmacogenet Genomics* 2007;17:115–126.

Mano Y, Usui T, Kamimura H. *In vitro* drug interaction between diflunisal and indomethacin via glucuronidation in humans. *Biopharm Drug Dispos* 2006;27:267–273.

Miners JO, Knights KM, Houston JB, Mackenzie PI. *In vitro–in vivo* correlation for drugs and other compounds eliminated by glucuronidation in humans: pitfalls and promises. *Biochem Pharmacol* 2006;71:1531–1539.

Meech R, Mackenzie PI. UDP-glucuronosyltransferase, the role of the amino terminus in dimerization. *J Biol Chem* 1997;272:26913–26917.

Mistry M, Houston JB. Glucuronidation *in vitro* and *in vivo*. Comparison of intestinal and hepatic conjugation of morphine, naloxone, and buprenorphine. *Drug Metab Dispos* 1987;15:710–717.

Morris RG, Black AB, Lam E, Westley IS. Clinical study of lamotrigine and valproic acid in patients with epilepsy: using a drug interaction to advantage? *Ther Drug Monit* 2000;22:656–660.

Obach RS, Walsky RL, Venkatakrishnan K, Gaman EA, Houston JB, Tremaine LM. The utility of *in vitro* cytochrome P450 inhibition data in the prediction of drug–drug interactions. *J Pharmacol Exp Ther* 2006;316:336–348.

Operaña TN, Tukey RH. Oligomerization of the UDP-glucuronosyltransferase 1A proteins: homo- and heterodimerization analysis by fluorescence resonance energy transfer and co-immunoprecipitation. *J Biol Chem* 2007;282:4821–4829.

Roberts MS, Magnusson BM, Burczynski FJ, Weiss M. Enterohepatic circulation: physiological, pharmacokinetic and clinical implications. *Clin Pharmacokinet* 2002;41:751–790.

Roberts AG, Campbell AP, Atkins WM. The thermodynamic landscape of testosterone binding to cytochrome P450 3A4: ligand binding and spin state equilibria. *Biochemistry* 2005;44:1353–1366.

Rowland A, Elliot DJ, Williams JA, Mackenzie PI, Dickinson RG, Miners JO. *In vitro* characterization of lamotrigine N2-glucuronidation and the lamotrigine–valproic acid interaction. *Drug Metab Dispos* 2006;34:1055–1062.

Rowland A, Gaganis P, Elliot DJ, Mackenzie PI, Knights KM, Miners JO. Binding of inhibitory fatty acids is responsible for the enhancement of UDP-glucuronosyltransferase 2B7 activity by albumin: implications for *in vitro–in vivo* extrapolation. *J Pharmacol Exp Ther* 2007;321:137–147.

Rowland A, Knights KM, Mackenzie PI, Miners JO. The "albumin effect" and drug glucuronidation: bovine serum albumin and fatty acid-free human serum albumin enhance the glucuronidation of UDP-glucuronosyltransferase (UGT) 1A9 substrates but not UGT1A1 and UGT1A6 activities. *Drug Metab Dispos* 2008a;36: 1056–1062.

Rowland A, Elliot DJ, Knights KM, Mackenzie PI, Miners JO. The "albumin effect" and *in vitro–in vivo* extrapolation: sequestration of long-chain unsaturated fatty acids enhances phenytoin hydroxylation by human liver microsomal and recombinant cytochrome P450 2C9. *Drug Metab Dispos* 2008b;36:870–877.

Soars MG, Burchell B, Riley RJ. *In vitro* analysis of human drug glucuronidation and prediction of *in vivo* metabolic clearance. *J Pharmacol Exp Ther* 2002;301: 382–390.

Takeda S, Ishii Y, Iwanaga M, Mackenzie PI, Nagata K, Yamazoe Y, Oguri K, Yamada H. Modulation of UDP-glucuronosyltransferase function by cytochrome P450: evidence for the alteration of UGT2B7-catalyzed glucuronidation of morphine by CYP3A4. *Mol Pharmacol* 2005;67:665–672.

Trapnell CB, Klecker RW, Jamis-Dow C, Collins JM. Glucuronidation of 3′-azido-3′-deoxythymidine (Zidovudine) by human liver microsomes: relevance to clinical pharmacokinetic interactions with atovaquone, fluconazole, methadone, and valproic acid. *Antimicrob Agents Chemother* 1998;42:1592–1596.

Trubetskoy OV, Finel M, Kurkela M, Fitzgerald M, Peters NR, Hoffman FM, Trubetskoy VS. High throughput screening assay for UDP-glucuronosyltransferase 1A1 glucuronidation profiling. *Assay Drug Dev Technol* 2007;5:343–354.

Tukey RH, Strassburg CP. Human UDP-glucuronosyltransferases: metabolism, expression, and disease. *Annu Rev Pharmacol Toxicol* 2000;40:581–616.

Uchaipichat V, Winner LK, Mackenzie PI, Elliot DJ, Williams JA, Miners JO. Quantitative prediction of *in vivo* inhibitory interactions involving glucuronidated drugs from *in vitro* data: the effect of fluconazole on zidovudine glucuronidation. *Br J Clin Pharmacol* 2006;61:427–439.

Van Hecken A, Verbesselt R, Tjandra-Maga TB, De Schepper PJ. Pharmacokinetic interaction between indomethacin and diflunisal. *Eur J Clin Pharmacol* 1989;36: 507–512.

Venkatakrishnan K, von Moltke LL, Obach RS, Greenblatt DJ. Drug metabolism and drug interactions: application and clinical value of *in vitro* models. *Curr Drug Metab* 2003;4:423–459.

Williams JA, Hyland R, Jones BC, Smith DA, Hurst S, Goosen TC, Peterkin V, Koup JR, Ball SE. Drug-drug interactions for UDP-glucuronosyltransferase substrates: a pharmacokinetic explanation for typically observed low exposure (AUCi/AUC) ratios. *Drug Metab Dispos* 2004;32:1201–1208.

Yoshida N, Yamada A, Mimura Y, Kawakami J, Adachi I. Trends in new drug interactions for pharmaceutical products in Japan. *Pharmacoepidemiol Drug Saf* 2006;15:421–427.

Zhang D, Chando TJ, Everett DW, Patten CJ, Dehal SS, Humphries WG. *Drug Metab Dispos* 2005;33:1729–1739.

Zucker K, Rosen A, Tsaroucha A, de Faria L, Roth D, Ciancio G, Esquenazi V, Burke G, Tzakis A, Miller J. Unexpected augmentation of mycophenolic acid pharmacokinetics in renal transplant patients receiving tacrolimus and mycophenolate mofetil in combination therapy, and analogous *in vitro* findings. *Transplant Immunol* 1997;5:225–232.

Zucker K, Tsaroucha A, Olson L, Esquenazi V, Tzakis A, Miller J. Evidence that tacrolimus augments the bioavailability of mycophenolate mofetil through the inhibition of mycophenolic acid glucuronidation. *Ther Drug Monit* 1999;21:35–43.

13

EVALUATION OF INHIBITORS OF DRUG METABOLISM IN HUMAN HEPATOCYTES

Albert P. Li and Chuang Lu

13.1 INTRODUCTION

One major challenge in the selection of drug candidates for clinical trials is that, due to species–species differences in drug properties, human-specific drug effects cannot be detected using nonhuman animal experimental systems. The high rate of clinical trial failures has been attributed to this species–species difference (DiMasi et al., 2003). One of the reasons for species–species differences in drug properties is the occurrence of species-specific xenobiotic metabolism pathways. Species differences in P450-dependent monooxygenases, a major group of enzymes responsible for drug metabolism, are well established (Guengerich, 2006) (Table 13.1).

In vitro experimental systems with human-specific properties represent an attractive tool for the assessment of human-specific drug properties. *In vitro* experimental systems derived from the human liver, namely, human hepatocytes and human liver tissue fractions, are now used routinely for the assessment of human drug metabolism. The combined use of human *in vitro* hepatic systems and relevant nonhuman animal models is believed to be responsible for the reduction in the contribution of pharmacokinetics as a major factor in human clinical trial failures from approximately 40% in 1991 to approximately 10% in 2000 (Kola and Landis, 2004).

Enzyme Inhibition in Drug Discovery and Development, edited by Lu and Li
Copyright © 2010 John Wiley & Sons, Inc.

TABLE 13.1. Predominant P450 Isoforms in Various Animal Species[a]

	P450 Isoforms				
P450 Family	Human	Mouse	Rat	Dog	Monkey
CYP1A	1A1/2	1A1/2	1A1/2	1A1/2	1A1/2
CYP2A	2A6	2A5	2A5	2A13/25	2A23/24
CYP2B	2B6	2B9/10	2B1	2B11	2B17
CYP2C	2C8/9/19	2C29/37/38/40/44	2C6/7/11	2C21/41	2C20/43
CYP2D	2D6	2D22	2D1	2D15	2D17
CYP2E	2E1	2E1	2E1	2E1	2E1
CYP3A	3A4/5	3A11/13	3A1/2	3A12/26	3A8

[a]The P450-dependent monooxygenases are the major xenobiotic metabolizing enzymes in the liver. The table here illustrates one of the scientific bases for species-species differences in drug properties. Species specific P450 isoforms may lead to different affinities and rates of metabolism of xenobiotics, leading to species differences in metabolic fate and/or toxicity. For the five animal species shown here, species differences in P450 isoforms are present for all P450 families except for CYP1A and CYP2E.

For metabolism and drug–drug interaction studies, the human-based *in vitro* systems include cell-free systems such as liver homogenates, post-mitochondrial supernatants (S-9 or S-10), and purified liver microsomes. Liver microsomes, due to their ease of use and relatively low cost, are an experimental system of choice for the evaluation of metabolic stability, metabolite profiling, metabolite identification, and P450 inhibition studies. Genetically engineered microsomes (cDNA-expressed microsomes) with only one specific P450 isoform are used for the evaluation of isoform-specific properties (Li, 2001, 2004a).

There are limitations to the use of cell-free systems. Specifically, the use of liver microsomes excludes experimentation with enzymes present in the plasma membranes, mitochondria, and cytosol, which may play important roles in the metabolism of the drugs being studied. Furthermore, the lack of an intact plasma membrane and the associated uptake transporters precludes the evaluation of selective distribution of drugs between the intracellular and extracellular compartments.

Intact hepatocytes may represent a physiologically more relevant experimental system than cell-free systems. The parenchymal cells of the liver, commonly known as hepatocytes, contain the majority of, if not all, hepatic xenobiotic biotransformation enzymes. Hepatocytes isolation techniques were developed in the 1970s, and the isolated hepatocytes were proposed to be used as a relevant experimental system for the evaluation of drug properties (e.g., Fry, 1982), a view that continues to be held by the scientific community (e.g., Li, 2007; Gomez-Lechon et al., 2007). The use of hepatocytes in the evaluation of drug metabolism, drug–drug interaction potential, and drug toxicity is now routine practice in both academic and industrial laboratories (Li, 2005; Lu et al., 2007).

In this chapter, the use of hepatocytes in the evaluation of enzyme inhibition will be reviewed. Our discussion will be limited to the application of human hepatocytes, although the principles described here are also applicable to hepatocytes from laboratory animals.

13.1.1 Human Hepatocytes for the Evaluation of Human Drug Properties

Hepatocytes isolated from human livers represent a valuable experimental system in drug development. For decades, efforts in drug development have been handicapped by the inability in the accurate prediction of human *in vivo* drug properties in preclinical studies with laboratory animals. This human–nonhuman animal differences in drug-properties are mainly a result of species differences, especially in drug-metabolizing enzyme activities. As illustrated in Table 13.1, the isoforms of the major family of drug-metabolizing enzymes, the P450-dependent monooxygenases, are different between laboratory animals and humans. The different isoforms may lead to differences in rates of metabolism and formation of different metabolites from a chemical entity, resulting in species–species differences in metabolic fate and toxicity. A clear example of species differences in metabolism is the formation of 7-hydroxycoumarin in humans but not in rodents (Easterbrook et al., 2001).

Hepatocytes isolated from human livers would retain human-specific hepatic metabolism activities and therefore represent a valuable preclinical experimental system for the early assessment of human-specific drug properties.

13.1.2 Isolation of Human Hepatocytes

The general procedures for the isolation of hepatocytes from all animal species, including humans, are essentially similar, involving firstly the perfusion of the liver or liver fragment with an isotonic, divalent ion-free buffer containing the calcium chelator EGTA to remove blood, dissolve clots, and loosen cell–cell junctions. This is followed by perfusion with a collagenase-containing isotonic buffer with the divalent ions calcium and magnesium, which are required for collagenase activity. The collagenase serves as an enzyme to dissociate the hepatocytes from the liver parenchyma into single cell suspension. Our laboratory represents one of the first to isolate and cryopreserve human hepatocytes, and a detailed procedure for human hepatocyte isolation based on the original method of Berry and Friend (1969) was previously reported by our laboratory (Li et al., 1992) for human liver fragments. The procedures have now been modified for large-scale hepatocyte isolation from the whole human liver (Li, 2007).

Procurement of human livers for research as well as human hepatocyte isolation are activities that are not commonly available to most laboratories, and they had represented the major hindrance to research with human hepatocytes when this experimental system was initially introduced to the scientific community. This major hindrance to the use of human hepatocytes is now circumvented by the cryopreservation of the hepatocytes.

13.1.3 Cryopreservation of Human Hepatocytes

Our laboratory was one of the first to report successful cryopreservation of human hepatocytes (Loretz et al., 1989), as well as the first to show (a) similar drug-metabolizing enzymes between cryopreserved and freshly isolated human hepatocytes and (b) the development of assays for metabolic stability, drug–drug interactions, and cytotoxicity using cryopreserved human hepatocytes (Li et al., 1999a,b). The similarity between freshly isolated and cryopreserved human hepatocytes in drug-metabolizing enzyme activities is now generally accepted by the scientific community (Li et al., 1999a,b; Li, 2007; Hewitt et al., 2007; Jouin et al., 2006).

Until recently, cryopreserved hepatocytes in general would lose their ability to be cultured as attached, monolayer cultures, presumably due to the unavoidable membrane damage occurring during the cryopreservation and subsequent thawing processes. It has been projected in the past that one out of 10–20 human hepatocyte isolations would lead to "plateable" cryopreserved hepatocytes. A focused research effort was initiated in our laboratory in 2005 to overcome this deficiency in hepatocyte cryopreservation. Our research resulted in the development of highly optimized hepatocyte isolation, cryopreservation, and recovery procedures (Li, 2007). In our laboratory, approximately 50% of the isolations would lead to "plateable" cryopreserved hepatocytes (Li, 2007).

Besides the retention of high viability and plateability, human hepatocytes after cryopreservation have been shown to retain human drug-metabolizing enzyme activities, including the activities of P450 isoforms, UDP-dependent glucuronosyl transferase activity (UGT), and sulfotransferase activity (ST) (Li et al., 1999a,b). The originally proposed applications of cryopreserved hepatocytes in drug metabolism studies (Li et al., 1999a,b) have been generally accepted by the scientific community at large (Jouin et al., 2006; Brown et al., 2007). Plateable cryopreserved human hepatocytes can also be used for enzyme induction studies (Kafert-Kasting et al., 2006; Hewitt et al., 2007). Cryopreserved human hepatocytes are found to retain uptake transporters such as Na^+-taurocholate cotransporting polypeptide (NTCP), organic anion transporting polypeptide (OATP), and organic cation transporter (OCT) (Shitara et al., 2003; Maeda et al., 2006). Besides the retention of uptake transporter activities, cryopreserved human hepatocytes were found, upon multiple days of culture, to form functional bile canaliculi and have been applied toward the evaluation of efflux transporter activities (Bi et al., 2006; Li et al., 2008).

Human hepatocyte cryopreservation is an enabling technology for the use of human hepatocytes. The advantages of cryopreserved hepatocytes over freshly isolated cells include long-term storage, ease of experimental scheduling, choice of precharacterized lots for experimentation, and repeat experimentations with hepatocytes from the same donors. Thus, the U.S. FDA has listed in the latest guidance document (FDA, 2006) that cryopreserved human hepatocytes are an acceptable experimental system for the generation of drug metabolism and drug–drug interaction data to support IND and NDA submissions.

13.1.4 Prepooled Cryopreserved Human Hepatocytes from Multiple Donors

Cryopreserved human hepatocytes were first available commercially in the mid-1990s. Since then, this experimental system is widely accepted by the pharmaceutical industry in drug development studies. Until the early 2000s, cryopreserved hepatocytes from individual donors were used. Recognizing that for the evaluation of "general" drug properties such as metabolic stability and metabolite profiling, one would like to have results representing the average of multiple individuals, practitioners of the field would pool cryopreserved hepatocytes from multiple donors for their studies. This is akin to the use of liver microsomes that are prepared from multiple individuals.

To eliminate the need to thaw hepatocytes from multiple donors, it is now discovered that cryopreserved human hepatocytes can be thawed, pooled, and recryopreserved without significant changes in viability or drug-metabolizing enzyme activities. These prepooled human hepatocytes (usually pooled from five male and five female donors) are now available commercially for experimentation. The viability, cell morphology, and enzyme activities of one of the lots are illustrated in Fig. 13.1. The ability to prepare prepooled human hepatocytes represent another major advance in hepatocyte technology, allowing this experimental system to be used to replace liver microsomes in studies that the use of hepatocytes represent a more relevant approach.

13.1.5 Applications of Human Hepatocytes in Drug Development

The following are the current routine applications of human hepatocytes in drug development. Cryopreserved human hepatocytes, especially prepooled human hepatocytes, are recommended to be used for these assays. The general scientific principles of the *in vitro* screening methodologies have been previously reviewed (Li, 2001, 2004a, 2007), the specific procedures for the assays are described here.

1. *Metabolic Stability Screening.* An important "drug-like" property for new chemical entities (NCE) is appropriate metabolic stability to allow a practical frequency of drug administration. In the past, liver microsomes were used routinely for metabolic stability screening. However, as liver microsomes contain mainly enzymes such as the P450 isoforms for Phase I oxidation, the assay would only yield metabolic stability toward microsomal oxidative enzyme metabolism, while in humans *in*

HuP58 -Human -10 Donors Pool (5 Male; 5 Female)

INVENTORY		CELL BIOLOGY (CHRM Method)			METABOLISM PROFILE pmol/1 x 10^6 cells/min							
LOT#	VIALS	Initial viability (%)	YIELD (Viable cells per vial)	Viability after 2 hrs at 37 deg. C (%)	ECOD	7-HCG/S	CYP 1A2	CYP 2C8	CYP 2C9	CYP 2C19	CYP 2D6	CYP 3A4
HuP58	469	81%	5.8 Million	71%	21	33	38	34	83	82	16	567

DONOR INFORMATION		
GENDER	RACE	AGE
M	C	39
M	C	41
M	C	56
M	C	59
M	C	20
F	C	18
F	C	17
F	C	26
F	C	53
F	C	17

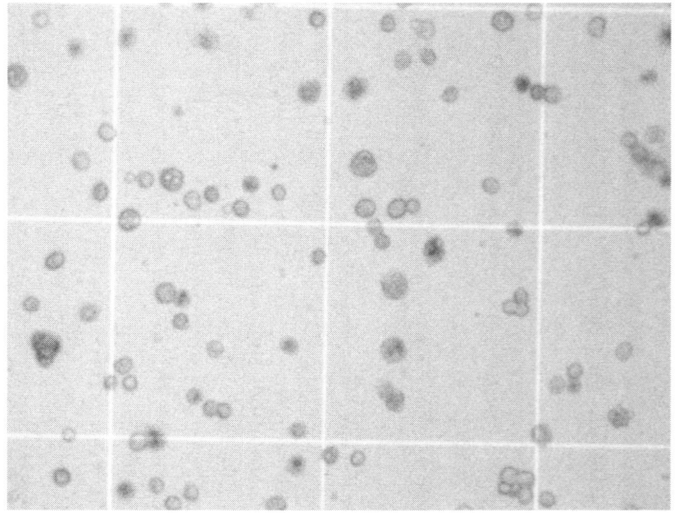

Figure 13.1. Viability, drug metabolic enzyme activities, and cell morphology of thawed prepooled cryopreserved human hepatocytes. For the preparation of prepooled cryopreserved human hepatocytes, cryopreserved cells from 10 donors (five male, five female) were thawed, pooled, and recryopreserved. The figure here shows the properties of the cells thawed after the second cryopreservation. The results show that although the cells have been cryopreserved and thawed twice, they retain their typical morphology, high viability, and normal drug-metabolizing enzyme activities. The age and race of the donors are also shown. Viability was determined by trypan blue exclusion. ECOD, 7-ethoxycoumarin-*O*-deethylase activity; 7-HCG/

vivo, the chemicals studied may be cleared via nonmicrosomal enzyme pathways such as conjugating enzyme pathways. Intact hepatocytes therefore represent a more relevant experimental system for metabolic stability evaluation than liver microsomes (Lavé et al., 1997; Li, 2001, 2004a, 2007; Jouin et al., 2006).

2. *Metabolite Profiling and Species Comparison.* The identification of metabolites formed from the parent drug (metabolite profiling) is important to drug development, because it allows the design of chemical structure to improve metabolic stability or to decrease cytotoxicity (see below). Metabolite identification is also important for the determination of the key drug metabolizing-enzyme pathways (e.g., oxidation or conjugation) as part of the program to understand drug–drug interaction potential. Lastly, metabolite profiling allows the selection of laboratory animal species most relevant to humans for *in vivo* experimentation. An animal species that forms metabolites found in humans would be more relevant than one with metabolites different from those formed in humans. This species comparison is routinely performed using *in vitro* systems such as hepatocytes (e.g., from human, rat, mouse, guinea pig, dog, monkey) (Lee et al., 1994; Li, 2001, 2004a,b; Zhang et al., 2007).
3. *Drug–Drug Interaction Evaluation.* A major adverse drug property with fatal outcome is drug–drug interaction. A drug may inhibit the metabolic clearance of a co-administered drug, leading to toxicity due to high systemic exposure to the affected drug (inhibitory drug–drug interactions). Conversely, a drug may enhance the metabolic clearance of a coadministered drug, leading to inefficacy due to lower than optimal systemic exposure (inductive drug–drug interactions). Inhibitory drug–drug interactions are caused by the inhibition of drug-metabolizing enzyme activities. Inductive drug–drug interactions are caused by the induction of drug-metabolizing enzyme activities. Both types of drug–drug interactions can be evaluated with human hepatocytes (Li, 2001, 2004a,b)

13.1.6 Evaluation of the Inhibition of Drug-Metabolizing Enzymes in Human Hepatocytes

In this review, emphasis is placed on the use of human hepatocytes for enzyme inhibition studies. The use of human hepatocytes for enzyme induction studies have been reviewed recently (Hewitt et al., 2007). As discussed earlier, intact hepatocytes represent an ideal experimental system for the evaluation of drug metabolism. For the same reasons, they also represent an ideal experimental system for the evaluation of the inhibition and induction of drug-metabolizing enzymes, either in the context of drug–drug interactions or in toxicity evaluation. The advantages of the use of intact hepatocytes to evaluate the inhibitory potential of a chemical on drug-metabolizing enzymes are as follows (Li, 2007):

1. *Intact Plasma Membranes.* Chemicals that are not permeable to the plasma membranes will not be available to intracellular enzymes. Intact hepatocytes have intact plasma membranes and therefore will model differential distribution between extracellular and intracellular environments as occur *in vivo*. Cell-free systems such as liver microsomes do not have a biological barrier between the chemicals to be evaluated and the drug-metabolizing enzymes and may lead to nonphysiological interactions between the nonpermeable chemicals and intracellular enzymes.
2. *Active Uptake Transporters.* Besides the intact plasma membranes, intact hepatocytes have active uptake transporters that can actively accumulate a chemical, leading to a higher intracellular than extracellular concentrations. This differential distribution may lead to a higher inhibitory potential than would be expected from the plasma (extracellular) concentrations. Furthermore, drug–drug interactions may also occur at the site of the active transporters that can be studied with intact hepatocytes.
3. *Complete Drug-Metabolizing Enzyme Pathways.* The effects of an enzyme inhibitor are determined by its concentration at the active site of the affected enzyme. Two of the determinants of intracellular enzyme-site concentrations, as discussed earlier, are its permeability and, if it is a substrate, uptake transporter activity. A third determinant is its metabolism by intracellular enzymes. For instance, an inhibitor may be rapidly conjugated by glucuronosyl transferases or sulfotransferases to metabolites that have lower or no inhibitory activities. In some cases, these metabolites may demonstrate higher inhibition potential. Metabolism by these or other cytosolic enzymes cannot be effectively modeled with liver microsomes.
4. *Physiological Enzyme and Cofactor Concentrations.* One major advantage of intact hepatocytes is that enzymes and cofactors are present in physiological concentrations. Furthermore, the multiple enzyme systems are present uninterrupted unlike other cell-free systems that are prepared by homogenization of the liver.

13.2 P450 INHIBITORS ON DRUG METABOLIZING ENZYME ACTIVITIES IN HUMAN HEPATOCYTES

Evaluation of P450 inhibition can be performed readily using cryopreserved human hepatocytes prepooled from multiple donors. The P450 isoform-specific substrates used routinely for the inhibitory drug–drug interaction assay and isoform-specific inhibitors that can be used as positive controls for the assay are shown in Table 13.2. The procedures for hepatocyte P450 inhibition assay is as shown below:

1. Add 125 µL of HMM containing 4× concentration of the drug to be evaluated into each well of a 24-well plate.

TABLE 13.2. Isoform-Selective Substrates and Their Respective Metabolites and Inhibitors that Can Be Used as Positive Controls for the Cryopreserved Human Hepatocyte P450 Inhibition Study[a]

P450 Isoform	Substrate	Metabolite	Positive Control
1A2	Phenacetin	Acetaminophen	Furafylline
CYP2A6	Coumarin	7-OH Coumarin	8-Methoxypsoralen
CYP2B6	Buproprion	Hydroxybuproprion	Triethylenethiophosphoramide (ThioTEPA)
CYP2C8	Paclitaxel	6a-OH Paclitzxel	Quercetin
CYP2C9	Tolbutamide	4-OH Tolbutamide	Sulfaphenazole
CYP2C19	*S*-Mephenytoin	4-OH Mephenytoin	Ticlopidine hydrochloride
CYP2D6	Dextromethorphan	Dextrophan	Quinidine
CYP2E1	Chlorzoxazone	6a-OH Chlozoxazone	Diethyldithiocarbamate
CYP3A4	Testosterone	6b-Hydroxytestosterone	Ketoconazole

[a]It is recommended that prepooled cryopreserved human hepatocytes are used for this assay, akin to the use of liver microsomes pooled from multiple donors. The intact hepatocytes provide intact cell properties including an intact plasma membrane with uptake transporter activities as well as complete, uninterrupted enzyme pathways at physiological cofactors that are not present in liver microsomes.

2. Add 125 μL of HMM containing 4× concentration of the drug-metabolizing enzyme substrate into the same well.
3. Add 250 μL of HMM containing 250,000 human hepatocytes.
4. Incubate for 30 min at 37 °C.
5. Add 1 mL of ACN to terminate reaction.
6. Centrifuge to remove cellular macromolecules.
7. LC/MS or HPLC quantification of metabolites.

13.2.1 Towards a More Physiological Model: Human Hepatocyte/Whole Plasma System for P450 Inhibition

A major challenge of *in vitro* drug–drug interaction studies is to predict the extent of clinical drug–drug interactions. A common approach is to use the ratio of the inhibitor's physiological concentration [I] over the inhibitor's inhibition constant K_i([I]/K_i) (Soars et al., 2003; Ito et al., 2004; Blanchard et al., 2004; Cook et al., 2004; Obach et al., 2006; Bachmann, 2006; Galetin et al., 2006). This simplistic approach suffers the drawback of the difficulties of obtaining accurate [I] and K_i values.

Theoretically, K_i is an absolute value dependent only on the inhibitor–enzyme affinity. However, there is substantial difficulty in the accurate determination of K_i values. The determined K_i value, or so-called apparent K_i value, is known to be influenced by the following factors: (1) *Ubiquitous protein binding*: For inhibitors that have high protein-binding potential, the amount of inhibitor available for interaction with the enzyme active site can be reduced due to nonspecific binding (ubiquitous binding). What has been observed is that for inhibitors that are highly protein-bound, the higher the liver microsomal concentration is used, the higher would be the apparent K_i values. (2) *Enzyme substrates*: Different substrates often generate different K_i values. (3) *Enzyme system*: Different systems, for instance, recombinant enzymes and human liver microsomes, could generate different K_i values. Thus, the K_i values for ketoconazole in the literature vary from 0.015 to 8 μM (Thummel and Wilkinson, 1998), a range of 500-fold. The K_i values for ketoconazole listed in the recently published FDA Drug–Drug Interaction Draft Guidance (2006) also carry a 50-fold difference range from 0.0037 to 0.18 μM.

In addition to the difficulty in determining a reliable value of K_i, the determination of physiological inhibitor concentration [I] is also extremely difficult. While plasma [I] can be experimentally measured, what is needed is not plasma [I], but [I] at the active site of the enzymes to be studied. As discussed earlier, intracellular hepatic concentration of an inhibitor is affected by its binding to plasma proteins, active transporter activities (if it is a uptake transporter substrate), and intracellular metabolism by hepatic enzymes. While various methods have been employed to estimate [I] (Ito et al., 2004; Blanchard et al., 2004; Cook et al., 2004; Obach et al., 2006; Bachmann, 2006), not a single approach could be applied to different classes of drugs.

Because of the difficulties of accurately measuring both [I] and K_i, the use of the ratio of these two parameters, [I] / K_i, to predict drug–drug interaction is often found to yield results that are not quantitatively representative of the *in vivo* situation. To overcome this challenge, a novel experimental model, the hepatocytes suspension/whole plasma system, has been developed (Lu et al., 2006). Because this model encompasses the key *in vivo* parameters—(a) plasma proteins to account for protein binding, (b) intact hepatocytes to allow partitioning across the intact plasma membrane, and (c) uptake transporter and complete drug-metabolizing enzyme pathways—the observed results should represent *in vivo* inhibitory potential. The human hepatocyte–plasma system accurately predicted drug–drug interaction effects of ketoconazole (Lu et al., 2008a) and fluconazole (Lu et al., 2008b) and is believed to be applicable to most inhibitors. One drawback of the assay is that hepatocytes in suspension are known to lack efflux transporters. Efflux pump substrates that can be pumped out of hepatocytes, leading to reduced hepatocyte concentration and hence reduced potency, therefore cannot be effectively studied.

The predicted drug–drug interaction for ketoconazole and fluconazole as compared to actual clinical findings are shown in Tables 13.3 and 13.4, which were modified from Lu et al. (2008a,b).

13.2.2 Inhibition of Drug-Metabolizing Enzymes in Hepatotoxicity Screening

Hepatotoxicity is a major manifestation of drug toxicity, because the liver usually would receive the highest bolus concentration of an ingested drug.

TABLE 13.3. Drug–Ketoconazole DDI Prediction Using the Hepatocytes in Plasma Model[a]

Compound	Fold of AUC Change		f_{renal}
	Predicted	Observed	
Theophylline	1.13–1.16	1.11	0.18
Desipramine	1.00	1.02	0.70
Midazolam	16.7	17.0	<0.01
Tolbutamide	1.43	1.77	0.001
Omeprazole	1.74–2.74	1.36–2.05	<0.01
Loratadine	2.48	3.47	Negligible
Cyclosporine	3.45	4.39	<0.01
Alprazolam	2.75–4.91	3.98	0.20
Budesonide	5.11	5.39	Negligible
Buprenorphine	5.17	2.30	Negligible
Docetaxel	3.48	2.22	0.02
Loratadine	1.92	3.47	Negligible
Methylprednisolone	2.19	2.36	0.05
Sirolimus	8.66	10.9	Negligible
Tacrolimus	1.95	2.39	<1

[a]The fold AUC change was calculated based on data obtained *in vitro* using human hepatocytes suspended in whole human plasma as described in Lu et al. (2008). The results illustrate the accuracy of this experimental system in the prediction of *in vivo* effects, which is attributed to the presence of whole plasma for the modeling of plasma factors and intact human hepatocytes for the modeling of intact cell properties.

TABLE 13.4. Drug–Fluconazole DDI Prediction Using the Hepatocytes in Plasma Model[a]

Compound	Fold of AUC Change		Prediction Error (%)	f_{renal}
	Predicted	Observed		
Theophylline	1.00	1.19	−16.0	0.18
Tolbutamide	2.44	2.09	16.5	0.001
Omeperazole	2.61	6.29	−58.5	<0.01
S-Wafarin	2.76	2.84–4.31	−22.9	<0.02
Phenytoin	2.19	1.75	25.1	0.02
Midazolam	2.89	3.60	−19.7	<0.01
Sirolimus	3.02	4.70	−35.7	Negligible
Cyclosporine	1.98	1.84	7.6	<0.01
Tacrolimus	1.57	1.19	31.8	<0.01

[a]The fold AUC change were calculated based on data obtained in vitro using human hepatocytes suspended in whole human plasma as described in Lu et al. (2008b). The results illustrate the accuracy of this experimental system in the prediction of *in vivo* effects, which is attributed to the presence of whole plasma for the modeling of plasma factors and intact human hepatocytes for the modeling of intact cell properties.

Furthermore, the hepatocytes, being the cells responsible for drug metabolism, are the first cells to be affected by reactive or toxic metabolites. Isolated hepatocytes therefore represent a physiologically relevant experimental model for the evaluation of hepatotoxicity. *In vitro* hepatocyte cytotoxicity measurements have been found to be effective in the delineation of hepatotoxic and help to find less hepatotoxic structures (Li, 2007).

While drug metabolism is generally believed to be an important parameter of drug toxicity, it is often difficult to ascertain the role of drug metabolism in the observed adverse drug effects. One approach is to identify the metabolites for experimental evaluation of the toxicity of the metabolites. This approach, however, has the following drawbacks:

1. *Expense.* Extensive resources are needed for the identification and subsequent purification or manufacturing of the metabolites for testing.
2. *Relevance.* Testing of the observed metabolites may or may not reflect the toxic mechanisms *in vivo*. Firstly, the final metabolites may or may not represent the toxic metabolite: The toxic species may be a highly reactive, relatively unstable metabolite or intermediate. Secondly, due to reactivity and membrane permeability, metabolites formed *in situ* may react with different cellular targets if added to extracellular media.

Building on the successes in using hepatocytes in drug metabolism, drug–drug interactions, and hepatotoxicity studies, two assays have been recently developed with plateable cryopreserved human hepatocytes for the evaluation of the role of metabolism in xenobiotic toxicity: The Metabolic Comparative Cytotoxicity Assay (MCCA) and the Cytotoxic Metabolic Pathway Identification Assay (CMPIA) (Li, 2009).

In the MCCA, the cytotoxicity of a drug is evaluated in the metabolically incompetent cell line [e.g., Chinese hamster ovary (CHO) cells] and in the metabolically competent primary human hepatocytes, in both the absence and presence of a potent cytochrome P450 (CYP) inhibitor, 1-aminobenzotriazole (ABT). ABT has been reported to cause autocatalytic inactivation of P450, leading to a nonspecific, mechanism-based inhibition of multiple human P450 isoforms including CYP1A2, CYP2B6, CYP2C9, CYP2C19, CYP2D6, CYP2E1, and CYP3A4. In the MCCA, a chemical that is transformed by metabolism to cytotoxic metabolites would be more cytotoxic in the metabolically competent cells than in the metabolically incompetent cells. Furthermore, if P450-dependent oxidative metabolism is involved in the metabolic activation, ABT would attenuate the cytotoxicity of this chemical in the metabolically competent cells.

The CMPIA is proposed to be performed when a toxicant is found to require xenobiotic metabolism in order to be cytotoxic in the MCCA. In this assay, the cytotoxicity of the toxicant is evaluated in human hepatocytes, in the presence and absence of isoform-selective P450 inhibitors, to define which P450 isoforms are involved in metabolic activation. If a specific metabolic pathway is involved in metabolic activation, a selective inhibitor of this metabolic pathway is expected to attenuate the cytotoxicity of the chemical in question.

The MCCA was developed in our laboratory for the definition of the general role of hepatic metabolism, especially P450-dependent metabolism in cytotoxicity, and the CMPIA for the evaluation of the role of specific P450 isoforms in metabolic activation. Because our major interest is to define human drug toxicity, human hepatocytes—and, more specifically, plateable cryopreserved human hepatocytes—were used in these assays. Human hepatocytes are known to retain human-specific drug metabolism and therefore can be used for the evaluation for the evaluation of the role of human drug metabolism on drug toxicity. As a control, a metabolically incompetent cell, the CHO cell, is used for comparison with the metabolically competent human hepatocytes in these assays to further define metabolism-related cytotoxic effects.

The application of MCCA and CMPIA to evaluate the role of drug metabolism in the toxicity of a toxicant has been illustrated with aflatoxin B1 (AFB_1), a hepatotoxicant and hepatocarcinogen that is known to require P450-dependent drug metabolizing enzyme activities for its toxicity (Figs. 13.2A–13.2D; Li, 2009). In the MCCA, AFB_1 was found to be more cytotoxic in the metabolically competent human hepatocytes than in the metabolically incompetent CHO cells, therefore confirming the known requirement of hepatic metabolism for its hepatotoxicity (Figs. 13.2A and 13.2D). Furthermore, the nonspecific P450 inhibitor, ABT, was found to be effective in attenuating AFB_1 cytotoxicity, thereby confirming that P450-dependent mixed function monooxygenase activity is involved in metabolic activation (Fig. 13.2A). In the CMPIA, ketoconazole and diethyldithiocarbamate, but not furafylline, sulfaphenazole, nor quinidine, were found to be effective in attenuating AFB_1 cytotoxicity in human hepatocytes (Figs. 13.2B and 13.2C). Ketoconazole was found to be as effective as the nonspecific inhibitor, ABT, therefore suggesting that the pathways inhibited by ketoconazole (mainly CYP3A4) are key pathways for metabolic activation. As diethyldithiocarbamate is known to inhibit CYP2A6 and CYP2E1, the results suggest that one or both of these isoforms may also be involved in AFB1 activation. The lack of effects of furafylline, sulfaphenazole, and quinidine on AFB1 cytotoxicity suggest that CYP1A2, CYP2C9, and CYP2D6 are not key pathways for AFB1 activation. The negative results are important to aid in ruling out key isoforms for metabolic activation, because isoform-selective inhibitors, while selective for certain isoforms, are known to have effects on multiple pathways. The lack of effects by furafylline, sulfaphenazole, and quinidine combined with the attenuating effects of ketoconazole and diethyldithiocarbamate suggest that CYP3A4 (the major P450 isoform inhibited by ketoconazole) and CYP2E1 and/or CYP2A6 (the major P450 isoforms inhibited by diethyldithiocarbamate) are the major isoforms involved in the metabolic activation of AFB1. Our observation, that CYP3A4 is involved in the metabolic activation of AFB1 to

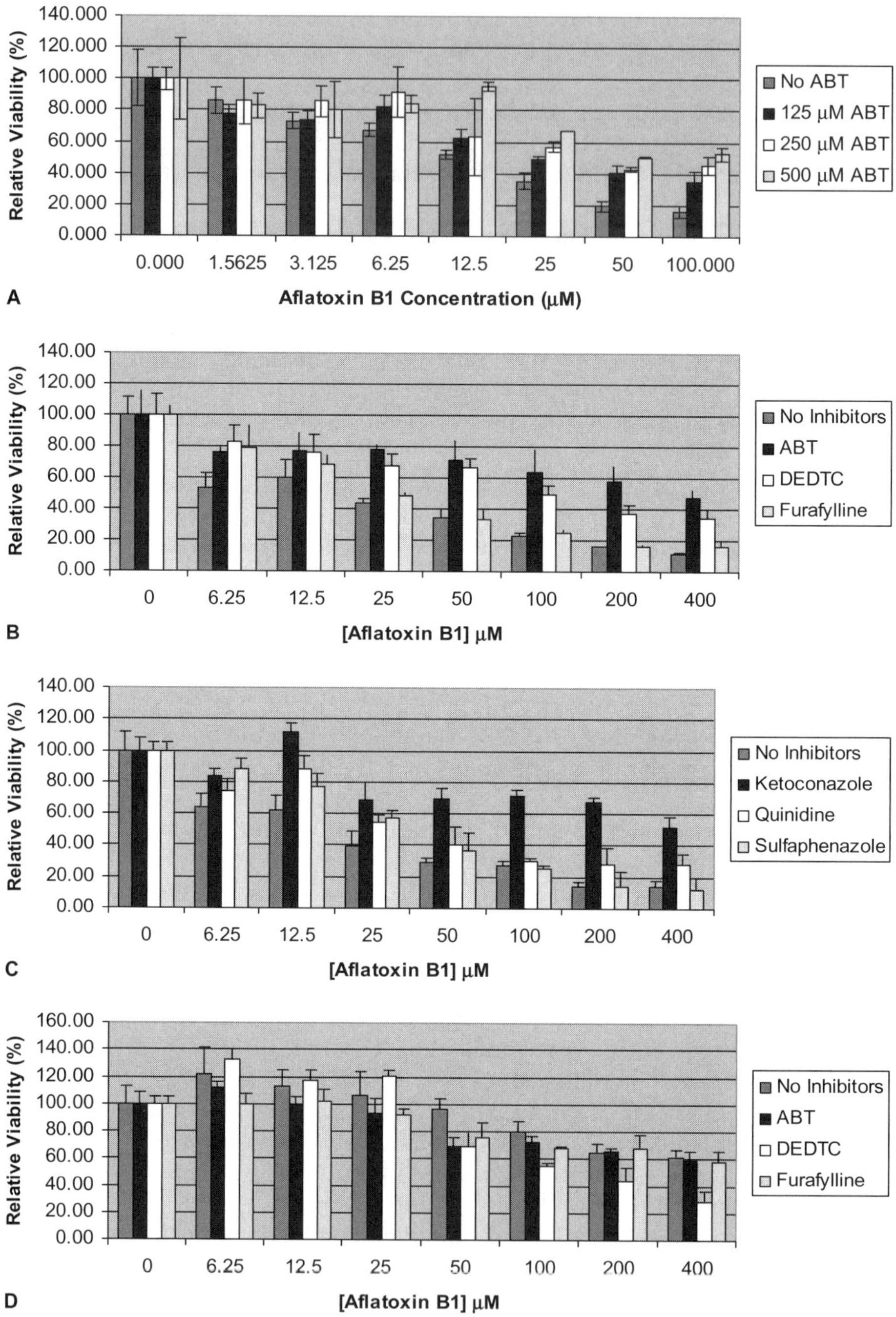

Figure 13.2. **(A)** Effect of ABT on aflatoxin B1 cytotoxicity in human hepatocytes. **(B)** Effect of ABT, EDDTC, and furafylline on aflatoxin B1 cytotoxicity in human hepatocytes. **(C)** Effect of ketoconazole, quinidine, and sulfaphenozole on aflatoxin B1 cytotoxicity in human hepatocytes. **(D)** Effect of ABT, EDDTC, and furafylline on aflatoxin B1 cytotoxicity in CHO cells.

cytotoxic metabolite, is consistent with the results obtained by others based on the quantification of toxic metabolites.

The results suggest that the MCCA and CMPIA are useful assays to estimate the role of xenobiotic metabolism, especially metabolism by P450 isoforms, in drug toxicity. Because P450-related pathways may not be the only pathways involved in metabolic activation, we are also developing approaches to evaluate non-P450 metabolic pathways such as alcohol dehydrogenase, esterase, monoamine oxidase, and flavin-dependent monooxygenases in drug toxicity in the MCCA and CMPIA. We believe that the MCCA and CMPIA can be used to identify drugs that would require metabolism to be toxic. These assays can be followed with analytical chemistry studies in metabolite identification for a definitive elucidation of the key pathways involved in the generation of toxic metabolites. Knowledge of the key pathways may allow the identification of human subpopulations that, due to genetic and environmental conditions, would be more susceptible to the toxicity of the drugs in question.

13.3 DISCUSSION

While this chapter deals with enzyme inhibition studies in human hepatocytes, the underlining theme is that one should always critically evaluate the tools used for drug development. To be able to predict accurately human *in vivo* effects, one needs to consider the key parameters and critically asks if they are adequately modeled in the chosen experimental system. The intact cell properties such as intact plasma membrane, active transporters, and complete, uninterrupted drug metabolizing enzyme pathways are properties in human hepatocytes that are not present in liver microsomes. Using intact human hepatocytes suspended in whole plasma is another step forward in terms of physiological relevance.

The big picture is that the earlier one can determine human-specific drug properties, the more likely is one to successfully develop a nontoxic and efficacious drug. An ideal drug candidate is one that is readily absorbed, has an acceptable plasma half-life to accommodate a convenient drug administration schedule, and has high efficacy, minimum toxicity, and minimum drug–drug interaction potential. Successful selection of drug candidates with these desired properties would greatly enhance the efficiency of drug development.

Because of species–species differences in drug properties, results with laboratory animals are not always predictive of human drug properties. The use of human-based *in vitro* experimental systems during preclinical trials allows the early assessment of human-specific drug properties. The reduction of pharmacokinetics as a contributing factor in clinical trial failures of drug candidates is attributed to the application of the *in vitro* hepatic system in the definition of human drug metabolism.

The successes of the hepatic *in vitro* systems underscore the importance of a science-based approach to the prediction of human drug properties—a key step toward the enhancement of the efficiency of drug development. The scientific understanding of the basis of species–species differences in drug-metabolizing enzymes has argued strongly for the deficiency of nonhuman laboratory animal models and the relevance of human-based metabolically competent experimental systems. In past two to three decades, the persistence of the pioneers of the field in the painstaking characterization of the *in vitro* systems and the generation of supportive data has led to the endorsement of the U.S. FDA, which ultimately led to the universal acceptance of the use of *in vitro* approaches to define human drug metabolism and drug-drug interaction potential. The positive experience with *in vitro* hepatic systems has paved the way for an important goal of alternative experimental systems, namely, the reduction, refinement, and replacement of the use of nonhuman laboratory animals in the assessment of human drug toxicity. Human hepatocyte technologies—isolation, cryopreservation, prepooling, and applications in drug metabolism, enzyme inhibition, and enzyme induction—have proven to be critical scientific advancements that, when applied intelligently, will contribute to the ultimate success in the preclinical prediction of human drug properties.

REFERENCES

Bachmann KA. Inhibition constants, inhibitor concentrations and the prediction of inhibitory drug drug interaction: pitfalls, progress and promise. *Curr Drug Metab* 2006;7:1–14.

Berry MN, Friend DS. High-yield preparation of isolated rat liver parenchymal cells—a biochemical and fine structural study. *J Cell Biol* 1969;43:506–520.

Bi YA, Kazolias D, Duignan DB. Use of cryopreserved human hepatocytes in sandwich culture to measure hepatobiliary transport. *Drug Metab Dispos* 2006;34(9): 1658–1665.

Blanchard N, Richert L, Coassolo P, Lave T. Qualitative and quantitative assessment of drug–drug interaction potential in man, based on K_i, IC50 and inhibitor concentration. *Curr Drug Metab* 2004;5:147–156.

Brown HS, Griffin M, Houston JB. Evaluation of cryopreserved human hepatocytes as an alternative in vitro system to microsomes for the prediction of metabolic clearance. *Drug Metab Dispos* 2007;35(2):293–301.

Cook CS, Berry LM, Burton E. Prediction of *in vivo* drug interaction with eplerenone in man from *in vitro* metabolic inhibition data. *Xenobiotica* 2004;34:215–228.

DiMasi JA, Hansen RW, Grabowski HG. The price of innovation: new estimates of drug development costs. *J Health Econ* 2003;22(2):151–185.

Easterbrook J, Fackett D, Li AP. A comparison of aroclor 1254-induced and uninduced rat liver microsomes to human liver microsomes in phenytoin *O*-deethylation, coumarin 7-hydroxylation, tolbutamide 4-hydroxylation, *S*-mephenytoin 4′-hydroxylation, chloroxazone 6-hydroxylation and testosterone 6beta-hydroxylation. *Chem Biol Interact* 2001;134(3):243–249.

FDA. Drug interaction studies-study design, data analysis, and implications for dosing and labeling. http://www.fda.gov/cder/guidance/2635fnl.pdf. 2006.

Fry JR. The metabolism of drugs by isolated hepatocytes. *Q Rev Drug Metab Drug Interact* 1982;4(2–3):99–122.

Galetin A, Burt H, Gibbons L, Houston JB. Prediction of time-dependent CYP3A4 drug–drug interactions: impact of enzyme degradation, parallel elimination pathways, and intestinal inhibition. *Drug Metab Dispos* 2006;34:166–175.

Gomez-Lechon MJ, Castell JV, Donato MT. Hepatocytes—the choice to investigate drug metabolism and toxicity in man: *in vitro* variability as a reflection of *in vivo*. *Chem Biol Interact* 2007;168(1):30–50.

Guengerich FP. Cytochrome P450s and other enzymes in drug metabolism and toxicity. *AAPS J* 2006;8(1):E101–E11.

Hewitt NJ, Lechon MJ, Houston JB, et al. Primary hepatocytes: current understanding of the regulation of metabolic enzymes and transporter proteins, and pharmaceutical practice for the use of hepatocytes in metabolism, enzyme induction, transporter, clearance, and hepatotoxicity studies. *Drug Metab Rev* 2007;39(1): 159–234.

Ito K, Brown H, Houston JB. Database analyses for prediction of *in vivo* drug–drug interactions from *in vitro* data. *Br J Clin Pharmacol* 2004;57:473–486.

Jouin D, Blanchard N, Alexandre E, et al. Cryopreserved human hepatocytes in suspension are a convenient high throughput tool for the prediction of metabolic clearance. *Eur J Pharm Biopharm* 2006;63(3):347–355.

Kafert-Kasting S, Alexandrova K, Barthold M, et al. Enzyme induction in cryopreserved human hepatocyte cultures. *Toxicology* 2006;220(2–3):117–125.

Kola I, Landis J. Can the pharmaceutical industry reduce attrition rates? *Nat Rev Drug Discov* 2004;3:711–715.

Lavé T, Dupin S, Schmitt C, et al. The use of human hepatocytes to select compounds based on their expected hepatic extraction ratios in humans. *Pharm Res* 1997;14(2): 152–155.

Lee K, Vandenberghe Y, Herin M, et al. Comparative metabolism of SC-42867 and SC-51089, two PGE2 antagonists, in rat and human hepatocyte cultures. *Xenobiotica* 1994;24(1):25–36.

Li AP. Screening for human ADME/Tox drug properties in drug discovery. *Drug Discov Today* 2001;6(7):357–366.

Li AP. *In vitro* approaches to evaluate ADMET drug properties. *Curr Top Med Chem* 2004a;4(7):701–706.

Li AP. An integrated, multidisciplinary approach for drug safety assessment. *Drug Discov Today* 2004b;9(16):687–693.

Li AP. Cell culture tool and method. U.S. Patent 20050101010, 2005.

Li AP. Human hepatocytes: isolation, cryopreservation and applications in drug development. *Chem Biol Interact* 2007;168(1):16–29.

Li AP. Metabolism comparative cytotoxicity assay (MCCA) and cytotoxic metabolic pathway Identification assay (CMPIA) with cryopreserved human hepatocytes for the evaluation of metabolism-based cytotoxicity *in vitro*: proof of concept study with aflatoxin B1. *Chem Biol Interact* 2009;179(1):4–8.

Li AP, Roque MA, Beck DJ, Kaminski DL. Isolation and culturing of hepatocytes from human liver. *J Tiss Culture Methods* 1992;14:139–146.

Li AP, Gorycki PD, Hengstler JG, et al. Present status of the application of cryopreserved hepatocytes in the evaluation of xenobiotics: consensus of an international expert panel. *Chem Biol Interact* 1999a;121(1):117–123.

Li AP, Lu C, Brent JA, et al. Cryopreserved human hepatocytes: characterization of drug-metabolizing enzyme activities and applications in higher throughput screening assays for hepatotoxicity, metabolic stability, and drug–drug interaction potential. *Chem Biol Interact* 1999b;121(1):17–35.

Li M, Yuan H, Li N, Song G, Zheng Y, Baratta M, Hua F, Thurston A, Wang J, Lai Y. *Eur J Pharm Sci* 2008;35:114–126.

Loretz LJ, Li AP, Flye MW, Wilson AG. Optimization of cryopreservation procedures for rat and human hepatocytes. *Xenobiotica* 1989;19(5):489–498.

Lu C, Li P, Gallegos R, Uttamsingh V, Xia CQ, Miwa G, Balani S, Gan LS. Comparison of intrinsic clearance in liver microsomes and hepatocytes from rats and humans, evaluation of free fraction and uptake in hepatocytes. *Drug Metab Dispos* 2006;34:1600–1650.

Lu C, Miwa GT, Prakash SR, Gan, LS, Balani SK. A novel model for the prediction of drug–drug interactions in humans based on *in vitro* CYP phenotyping. *Drug Metab Dispos* 2007;35:79–85.

Lu C, Hatsis P, Berg C, Lee FW, Balani SK. Prediction of pharmacokinetic drug-drug interactions using human hepatocyte suspension in plasma and cytochrome P450 phenotypic data. II. *In vitro–in vivo* correlation with ketoconazole. *Drug Metab Dispos* 2008a;36:1255–1260.

Lu C, Berg C, Prakash SR, Lee FW, Balani SK. Prediction of pharmacokinetic drug–drug interactions using human hepatocytes suspension in plasma and cytochrome P450 phenotypic data. III. *In vitro–in vivo* correlation with fluconazole. *Drug Metab Dispos* 2008b;36:1261–1266.

Maeda K, Kambara M, Tian Y, Hofmann AF, Sugiyama Y. Uptake of ursodeoxycholate and its conjugates by human hepatocytes: role of Na^+-taurocholate cotransporting polypeptide (NTCP), organic anion transporting polypeptide (OATP)1B1 (OATP-C), and OATP1B3 (OATP8). *Mol Pharm* 2006;3:70–77.

Obach RS, Walsky RL, Venkatakrishnan K, Gaman EA, Houston JB, Tremaine LM. The utility of *in vitro* cytochrome P450 inhibitioon data in the prediction of drug–drug interactions. *J Pharmacol Exp Ther* 2006;316:336–348.

Osowaski CL, Dix SP, Lin LS, Mullins RE, Geller RB, Wingard JR. Evaluation of the drug interaction between intravenous high-dose fluconazole and cyclosporine or tacrolimus in bone marrow transplant patients. *Transplantation* 1996;61:1268–1272.

Shitara Y, Li AP, Kato Y, Lu C, Ito K, Itoh T, Sugiyama Y. Function of uptake transporters for taurocholate and estradiol 17beta-D-glucuronide in cryopreserved human hepatocytes. *Drug Metab Pharmacokinet* 2003;18:33–41.

Soars MG, Gelboin HV, Krausz KW, Riley RJ. A comparison of relative abundance, activity factor and inhibitory monoclonal antibody approaches in the characterization of human CYP enzymology. *Br J Clin Pharmacol* 2003;55:175–181.

Thummel KE, Wilkinson GR. *In vitro* and *in vivo* drug interactions involing human CYP3A4. *Annu Rev Pharmacol Toxicol* 1998;38:389–430.

Zhang D, Wang L, Raghavan N, et al. Comparative metabolism of radiolabeled muraglitazar in animals and humans by quantitative and qualitative metabolite profiling. *Drug Metab Dispos* 2007;35(1):150–167.

14

GRAPEFRUIT JUICE AND ITS CONSTITUENTS AS NEW ESTERASE INHIBITORS

SURESH K. BALANI

14.1 INTRODUCTION

The major emphasis on grapefruit juice (GFJ) in the literature has been tied to its inhibition of CYP3A4/5 (hereafter referred to as CYP3A) (Bailey et al., 1989, 1991; Saito et al., 2005; Schmiedlin-Ren et al., 1997; He et al., 1998; Zhou et al., 2004) and OATP (Dresser et al., 2002), with sparse data on the inhibition of Pgp (De Castro et al., 2007). Extensive literature exists on the inhibition of CYP3A *in vitro* as well as *in vivo*, in animals and in humans (Huang et al., 2004). The irreversible inhibition of CYP3A by GFJ has largely been attributed to the gut, with loss of enteric CYP3A protein, but not mRNA (Lown et al., 1997). The potency of the CYP3A inactivation could be similar to that observed with troleandomycin. As is well established, the inhibition of CYP3A is due to the mechanism-based effect of the constituent furanocoumarins bergamottin and 6′,7′-dihydroxybergamottin (DHB) and DHB's dimers. Notably, Greenblatt et al. (2003) have utilized the mechanism-based effect in humans to assess the rate of resynthesis of CYP3A in human gut. They showed that CYP3A activity was restored back to the baseline level in 3 days after stopping the daily GFJ dose. This has important implications in setting inclusion/exclusion criteria for clinical protocols.

The focus of this chapter is on a new finding that GFJ also inhibits esterase activity. Thus, this, expands on the GFJ–drug interaction concerns. Esterases,

Enzyme Inhibition in Drug Discovery and Development, edited by Lu and Li
Copyright © 2010 John Wiley & Sons, Inc.

broadly, are present in the blood, liver, intestine, muscles, kidney, brain, and other tissues and fluids, with the liver typically showing the highest activity. In the tissues, they are localized in the microsomes, cytosol, and lysosomes. In the past, strawberry and banana extracts have shown inhibitory effects on esterase activity (Van Gelder et al., 1999, 2000), but GFJ has only recently been shown to inhibit esterase activity *in vitro* in animal and human tissues and *in vivo* in animals (Li et al., 2007a,b). The finding has strong implications of interactions with ester prodrugs such as enalapril, lovastatin, irinotecan (CPT11), and capecitabine, as discussed below, and the illicit drugs heroin and cocaine. The type, distribution, and biochemistry of esterases (e.g., paraoxonase, carboxylesterase, acetylcholinesterase, and cholinesterase) involved in the activation of ester prodrugs have been reviewed by Liederer et al. (2006). Most esterases have broad, overlapping substrate selectivity. Species differences have also been noted, with rodents generally showing higher esterase activity than higher species. Enalapril is metabolized only by esterases, and hence it is an ideal probe substrate to investigate esterase-mediated GFJ–drug interactions. Because lovastatin is a substrate for both CYP3A and esterases, in humans the oral exposure to lovastatin has been reported to increase by as high as 15-fold with coadministration of GFJ. With the potential adverse effects including myopathy, this increase can be of concern for this statin. Because GFJ inhibits multiple enzymes, the combined effects of inhibition by GFJ could be large on the pharmacokinetics of some compounds. The antitumor activity of CPT11 was shown to be directly related to the carboxylesterase activity that leads to the active product SN38; so the higher the carboxylesterase activity, the higher the antitumor activity observed in nude mice with HT-29 xenograft tumors (Morishita et al., 2005). The level of esterase activity in that experiment was inversely proportional to the dose of the esterase inhibitor bis-(*p*-nitrophenylphosphate) (BNPP). In other words, the antitumor activity of CPT11 was reduced by BNPP in a dose-dependent manner. The anticancer prodrug capecitabine undergoes a three-step activation process *in vivo* to become 5-fluorouracil. One of the steps, conversion of the prodrug to 5′-deoxy-5-fluorocytidine, is catalyzed by carboxylesterases, and the next two steps are catalyzed by cytidine deaminase and thymidine phosphorylase, specifically inside colorectal tumors (Quinney et al., 2005; Verweij, 1999; Miwa et al., 1998). Inhibition of prodrug activation thus would lead to a reduced clinical response. Loperamide is often required to treat chemotherapy-associated diarrhea (e.g., with the use of CPT11), and loperamide is known to competitively inhibit carboxylesterase 2 ($K_i = 1.5\,\mu M$). Thus, the pharmaceutical companies and clinicians must be adequately informed of such drug–drug interaction potentials. Apart from these interactions, it is also important to be aware of the potential for increased toxicity, if any, due to the prodrug itself in the circulation, and also of the need for any potential prodrug-related exposure coverage in animals during the toxicity testing phase. On the other hand, the potential beneficial effects of this interaction also need to be recognized.

This chapter dwells on GFJ and its constituent-mediated inhibition of, specifically, carboxylesterase activity, while providing background information on CYP3A, Pgp, and OATP inhibitory effects. For more details on the types, distribution, and regulation of carboxylesterases, readers are referred to some of the literature reviews (Satoh and Hosokawa, 1998; Satoh et al., 2002; Xie et al., 2002).

14.2 GRAPEFRUIT JUICE CONSTITUENTS

GFJ is a milieu of chemicals, including flavonoids, fruity esters, psoralen derivatives, sugars, polysaccharides, organic acids, vitamins, and minerals. The structures of some of the key flavonoid ingredients discussed in this chapter are shown in Figs. 14.1 and 14.2. Several volatile components have also been identified and quantitated. These include alcohols, ketones, aldehydes, esters, and others (Shaw et al., 2000). These constituents vary widely by GFJ type, which, along with the amount of juice consumed (a typical serving in clinical interaction studies is 8 oz), could result in differential inhibitory activities toward enzymes and transporters (Ross et al., 2000). De Castro et al. (2006) looked at the concentrations of four important constituents in GFJ: naringin (the primary bitter component, as well as a predominant flavonoid in GFJ), naringenin, bergamottin, and 6′,7′-dihydroxybergamottin (DHB). They found extensive variability in their composition, with concentrations of naringin

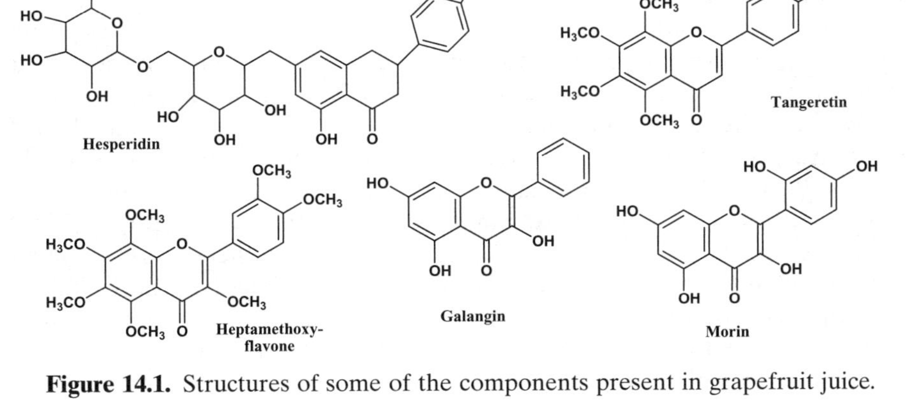

Figure 14.1. Structures of some of the components present in grapefruit juice.

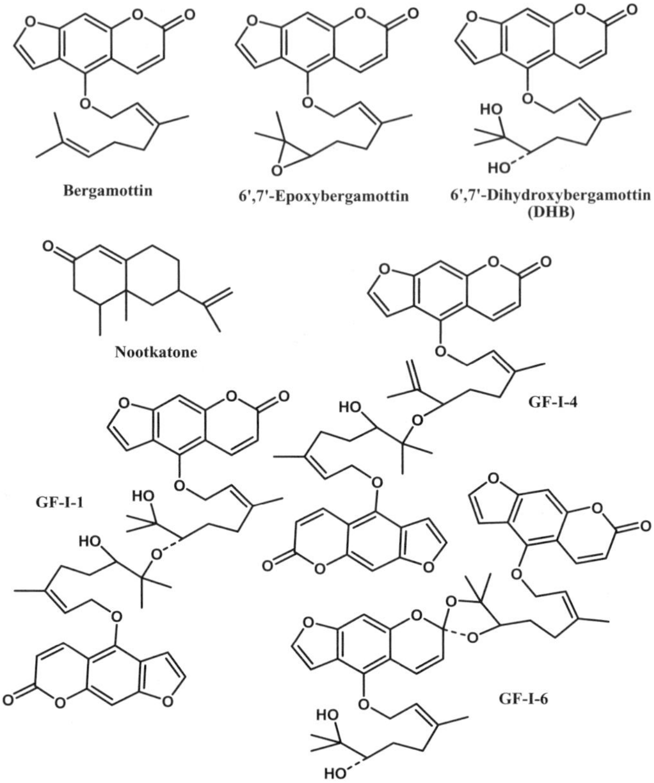

Figure 14.2. Structures of furanocoumarins and nootkatone present in grapefruit juice.

ranging from 174 to 1492 μM, bergamottin from 1 to 37 μM, DHB from 0.22 to 52 μM, and naringenin not detectable. The concentrations of all constituents in white grapefruit were generally greater than those in red grapefruit. Although naringenin was not detected in grapefruit juice (Kuhnau, 1976), oral administration of grapefruit juice resulted in renal excretion of naringenin glucuronide conjugates demonstrating *in vivo* formation of this active species. Also, the microflora of the gut probably hydrolyzed the naringin in GFJ to naringenin (Fuhr and Kummert, 1995). The 6′,7′-epoxybergamottin concentration in white grapefruit juice and its concentrate were 7.4 and 0.1 μM, respectively (Schmiedlin-Ren et al., 1997). The concentration of some of the methoxyflavonoids—nobelitin, tangeretin, sinensetin, and heptamethoxyflavone—was found to be approximately 1.5 μM (Manthey and Buslig, 2005). These methoxyflavonoids are known to inhibit Pgp. Other flavonoid glycosides identified in most GFJs are hesperidin, neohesperidine, didymin, and poncirin. The aglycon quercetin was detected only in certain brands. The total flavonoid concentration ranged from 19 to 84 mg/100 mL. The flavonoids with highest concentrations, next to naringin, were hesperidin and narirutin (Ross et al., 2000). The results of other studies quantitating the flavonoids are covered elsewhere (Gattuso et al., 2007; Paine et al., 2006; Ranganna et al., 1983). Additionally, dimeric 6′,7′-dihydroxybergamottins (GF-I-1, GF-I-4, and GF-I-6) and a sesquiterpene nootkatone have been isolated from GFJ (Paine et al., 2006; Tassaneeyakul et al., 2000; Guo et al., 2000). The dimeric furanocoumarins, however, have received unusually little attention, even though they are more potent inhibitors of CYP3A than bergamottin and DHB. The GF-I-1, -4, and -6 concentrations in a variety of GFJs ranged from 0.1 to 8.7 μM. Upon centrifugation, while DHB was found primarily in the centrifugate portion, its epoxide and dimeric forms and bergamottin were associated more with the particulate part of GFJ. Nootkatone has some inhibitory activity toward CYP2A6 and CYP2C19. The concentration of the dimeric DHBs GFJ-I-1 and GFJ-I-4 in GFJ was claimed to be 20–200 times lower than that of bergamottin or DHB.

14.3 ATTRIBUTES OF GFJ

14.3.1 Effect on CYPs

The CYP3A inhibitory effect of GFJ has been demonstrated to be a result of the furanocoumarins *in vivo* in humans as shown by comparison of the effect of regular GFJ with the one from which the furanocoumarins had been extracted out, through resins treatment, on the pharmacokinetics of felodipine, a good substrate of CYP3A (Paine et al., 2006). The median area under the plasma concentration–time curve (AUC) of felodipine was approximately two-fold greater in the healthy volunteers treated with regular GFJ than in those treated with furanocoumarin-free GFJ or orange juice, with no apparent differences in the T_{max} or $t_{1/2}$. The effect has been considered to be largely due to CYP3A inhibition at the gut rather than hepatic level. The GFJ consituents that have received the least recognition are the dimers of DHB, particularly GFJ-I-1 and GFJ-I-4, that are actually more potent inactivators of CYP3A than are bergamottin or DHB, although they are present in low concentrations. Another dimer, GFJ-I-6, was a weaker inhibitor compared to the two other dimers. This dimeric epoxide analog of DHB could not be tested for its inhibitory potency, because of its instability, but is also expected to be an active inhibitor. In liver microsomes, the inhibitor concentrations producing the half-maximum rate of inactivation (K_I) of CYP3A-mediated nifedipine oxidation for bergamottin, DHB, GFJ-I-1, and GFJ-I-4 (isolated and purified from GFJ) were 40, 5.56, 0.31, and 0.13 μM, respectively, without much difference in the inactivation rate constant (k_{inact}, 0.05–0.08 min^{-1}). Even though the dimers' concentrations are 20–200 times lower than that of DHB, the potency with which they inhibited CYP3A, as assessed in terms of their K_I values, was approximately 20–40 times stronger. Thus, these dimers are also expected to play a respectable role in drug–GFJ interactions *in vivo* in humans. Other components kaempferol and limonin were shown to be weaker in inhibitory potency toward CYP3A by two orders of magnitude. Bergamottin, DHB, GFJ-I-1, and GFJ-I-4 besides being inhibitors of CYP3A, showed decent inhibition of CYP1A2, CYP2C9, CYP2C19, and CYP2D6 (IC_{50} in the range of 0.14–5 μM) (Tassaneeyakul et al., 2000). DHB inhibited CYP1A2 activity with similar potency as CYP3A. CYP2E1 was the least sensitive to the inhibitory effect of furanocoumarin components. The sesquiterpene nootkatone showed strong inhibition of CYP2A6 and CYP2C19 (K_i = 0.8 and 0.5 μM, respectively) (Tassaneeyakul et al., 2000). Thus, coadministration of drugs with GFJ can be expected to have the potential for varying degree of interaction with multiple CYPs.

Bergamottin in GFJ has also been shown to be a substrate and inhibitor of CYP2B6 in an *in vitro* system (Lin et al., 2005). The inactivation was demonstrated to be mechanism-based and irreversible, with covalent binding to the CYP apoprotein and destruction of heme. In fact, it was a more potent inactivator of CYP2B6 than of CYP3A. Bergamottin is metabolized to 6′,7′-epoxy- and 6′,7′-dihydroxybergamottins, which are also potent CYP3A inactivators (He et al., 1998). Bergamottin is readily absorbed from the gut (Kane and Lipsky, 2000). Thus, drugs that are substantially cleared by CYP2B6, which is present in the liver, gut, kidney, and brain, may interact with GFJ. Examples of such drugs are efavirenz, bupropion, and cyclophosphamide. Human *in vivo* data on such interactions are yet to be generated.

14.3.2 Effect on Transporters

GFJ components also have been shown to have differential effects on the oral absorption of drugs that are primarily affected by Pgp. De Castro et al. (2008) and Uno et al. (2006) showed that, in rats, exposure to talinolol, which is known to be a good substrate of Pgp and is not metabolized, is increased by bergamottin, naringenin, and naringin by approximately 80%. The underlying reason for the interaction was indeed found to be Pgp inhibition by these and some other components, and of course by GFJ itself, as demonstrated in a Caco-2 transport study (de Castro et al., 2007). 6′,7′-Epoxybergamottin, 6′,7′-dihydroxybergamottin, aglycone naringenin, and glycoside naringin inhibited the talinolol transport with IC_{50} values of 0.7, 34, 236, and 2409 μM, respectively. Bergamottin did not show much inhibitory activity at concentrations up to 10 μM. These concentrations are in the range that is normally observed in GFJ. Likewise, many polymethoxylated flavones, like tangeretin, nobiletin, 3,5,6,7,8,3,4′-heptamethoxyflavone, and sinensetin, were also potent inhibitors of the basolateral-to-apical transport of talinolol in the Caco-2 model, with IC_{50} values of <5 μM (Mertens-Talcott et al., 2007). Since the concentration of these polymethoxylated flavones is generally small in GFJ, their true interaction potential has yet to be assessed *in vivo*.

In contrast to increases in exposure upon coadministration of drugs with GFJ, fexofendine C_{max} actually decreased by 30% (Banfield et al., 2002).

Fexofenadine is known not to undergo much metabolism, but is a substrate of Pgp and OATP. Glaeser et al. (2007) studied the human intestinal expression of OATP by taking duodenal biopsies and measured the effect of GFJ on OATP, Pgp, and CYP3A expression in relation to the oral dose pharmacokinetics of fexofenadine. The intestine was found to express several OATPs (OATP 1A2, 1B1, 1B3, and 2B1); and efflux transporters, along with CYP enzymes. OATP1B1 and 1B3 were previously thought to be expressed only in the liver, but the Glaeser et al. study showed expression in the gut as well. Both OATP1A2 and Pgp were found to be colocalized and expressed on the apical side of enterocytes. Because OATP1A2 is predominantly responsible for the uptake and Pgp for the efflux of fexofenadine, as shown by *in vitro* data, GFJ ingestion 2 hours before or concomitant with administration of fexofenadine led to reduced plasma exposure in humans, without affecting the expression of OATP1A2 or Pgp; but in contrast, CYP3A expression was reduced. Inhibition of both OATP1A2 and Pgp was inferred from these results, without their down-regulation.

14.4 *IN VITRO* ESTERASE INHIBITION BY GFJ

14.4.1 Purified Porcine Esterase

In vitro studies using commercially available purified porcine esterase and human hepatic and intestinal S9 have shown that the hydrolysis of esterase substrates *p*-nitrophenylacetate (PNPA), lovastatin, and enalapril is inhibited in a GFJ concentration-dependent manner (Li et al., 2007a) (Fig. 14.3). These

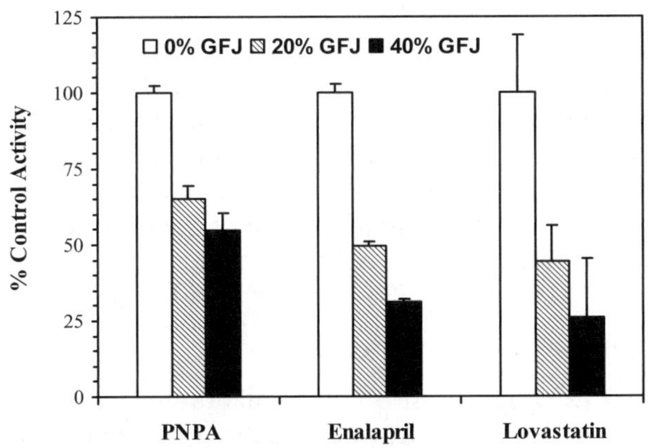

Figure 14.3. Inhibitory activity of grapefruit juice on purified porcine esterase activity using *p*-nitrophenylacetate as a substrate. Reproduced with permission from Li et al. (2007a). Copyright 2007 by the American Society for the Pharmacology and Experimental Therapeutics.

studies provide proof of concept that GFJ does have esterase inhibition properties.

14.4.2 Esterase Activity in Gut Lumen

A matrix that has not received wide attention in the past is the gut lumen. The gut lumen is made of various digestive juices and enzyme-rich fluids secreted into the mouth, stomach, and duodenum. Examples include saliva (lysozymes, α-amylases), pepsin, chymotrypsin, proteases, lipases, carbohydrases, nucleases, carboxypeptidases, pancreatic amylases, steapsin, lactases, bile, and so on. Thus, it has a high mix of various enzymes that have hydrolase-type activities, and hence it can have the potential to hydrolyze prodrugs vastly before they could even enter the gut membrane. The lumen from rat intestine, collected from control rats, was shown to carry substantial esterase activity, and the activity toward enalapril and lovastatin was substantially inhibited in the presence of diluted GFJ (frozen, Minute Maid brand) (Li et al., 2007a). Lovastatin hydrolysis was reduced to 65% and 48% of the control by 20% and 40% of 1:3 diluted GFJ at pH 7, and to 34% and 26% of the control by 20% and 40% of 1:3 diluted GFJ at pH 3.5 (pH of GFJ), respectively. The extent of GFJ reduction of hydrolysis was similar for enalapril and lovastatin and is depicted in Fig. 14.4. *In vivo*, using regular-strength, GFJ would be expected to result in even greater degrees of inhibition for these compounds. Likewise, hydrolytic activities in human liver and gut S9 fractions were affected by GFJ. The overall effect of GFJ could thus be large in stabilizing prodrugs in the gut lumen and making them more available for absorption in the enterocytes, with the net effect being higher bioavailability.

There are obvious challenges in obtaining gut lumen from humans. Balani et al. (1997) and Stoeckel et al. (1998) have noninvasively collected juices from the guts of healthy volunteers. They used a nasogastric tube guided by fluoroscopy to reach into the duodenum; and, using cholecystokinenin octapeptide (CCK8) to induce gallbladder contractions, they collected the fluids, with or without an investigational drug onboard. These studies show the possibility of conducting experiments in the human gut lumen similar to those done in rat gut lumen with GFJ or its constituents. Gut lumen also contains microflora, at increasing concentrations along the length of the GI tract, which could also contribute to the esterase activity in that matrix.

14.4.3 Effect on Gut and Liver Enzymes

Van Gelder et al. (2002) have shown in an elaborate study the use of discrete and mixed fruity esters and a strawberry extract in the inhibition of esterase activity toward the HIV drug tenofovir disoproxil fumarate (tenofovir DF), a bis-ester prodrug of an acyclic nucleoside phosphonate drug. The carboxylic ester is hydrolyzed by carboxylesterases to tenofovir monoester. Van Gelder et al. used rat intestinal preparations, such as homogenate and *in situ* rat

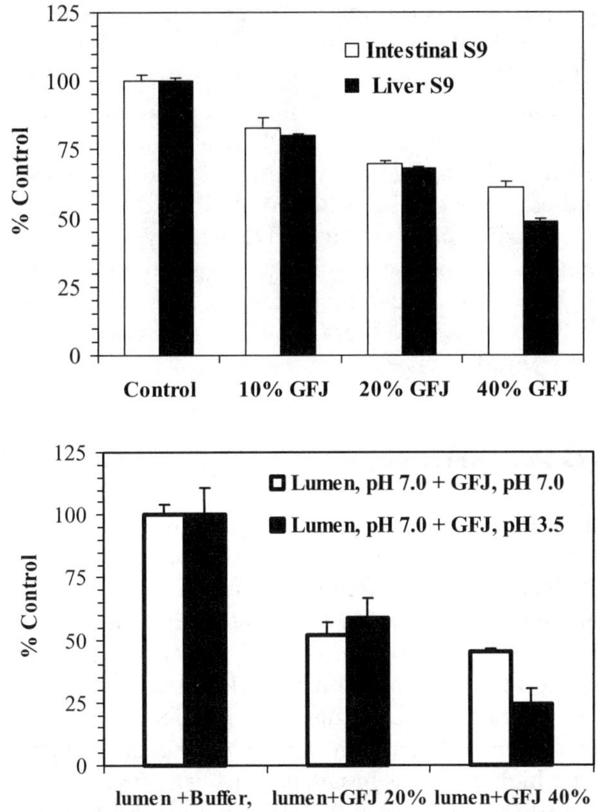

Figure 14.4. Inhibition of esterase activity by grapefruit juice in rat intestinal and liver S9 fractions and intestinal lumen using enalapril as a substrate. Reproduced with permission from Li et al. (2007a). Copyright 2007 by the American Society for the Pharmacology and Experimental Therapeutics.

intestinal preparations, to study esterase inhibition. The strawberry extract (1% volume : volume) almost completely inhibited the hydrolysis of tenofovir DF in the intestinal homogenate. Since strawberry extract is a complex mixture of chemicals, they next focused on some of the fruity esters. Among the 30 different esters tested at a high concentration of 1 mM, about 20% (e.g., phenyl benzoate, propyl paraben, phenethyl isobutyrate, etc.) showed ≥90% inhibition of the hydrolysis of tenofovir DF. Notably, aspirin, an ester, also inhibited the hydrolysis of tenofovir DF by approximately 35%. Not all esters tested had an effect on the hydrolysis. Combinations of some of the esters were also found to have substantial inhibitory effects on the hydrolysis. To assess the effect of esters and strawberry juice on Pgp-mediated transport of tenofovir DF, Caco-2 studies were performed, which showed that the esters also inhibited Pgp-mediated cyclosporin A transport. Thus, these inhibitors can serve a dual purpose as esterase and Pgp inhibitors to boost the prodrug stability and its net absorptive permeability, thereby increasing the bioavailability of tenofovir DF in the whole system. An *in situ* study involving rat intestinal coperfusion with strawberry extract or an ester mixture of propyl paraben, octyl acetate, and ethyl caprylate showed increases in tenofovir equivalents in mesenteric plasma. Although, based on *in vitro* homogenate and Caco-2 studies, these results could be attributed to both the esterase and Pgp inhibition, coperfusion of tenofovir DF with cyclosporin A did not show any effect on net permeability of tenofovir, suggesting that the hydrolysis may be rapid inside the mucosa, leading to less availability for Pgp to efflux the compound out. Thus, *in vitro* effects should be considered carefully to make inferences for potential *in vivo* effects.

14.5 *IN VIVO* INHIBITION OF ESTERASE BY GFJ

The interaction studies of various drugs with GFJ have shown varied responses to drugs' exposures *in vivo*. Most studies have attributed GFJ-mediated changes to CYP3A inhibition, depending on the extent to which they are metabolized or cleared by CYP3A. The classic examples are those of felodipine (Bailey et al., 1989, 1991) and cyclosporine (Ducharme et al., 1995), for which the severity of the interaction is tied to the therapeutic range of the drug. As also mentioned earlier, drugs that are cleared by CYP3A and esterase have shown a greater impact on the drug's exposure change, like the case of lovastatin in which the oral bioavailability was increased to greater than 15-fold upon coadministration with GFJ in humans. For lovastatin, also a substrate of carboxylesterase, which is known to hydrolyze it to the active hydroxyacid drug, the contribution of esterase inhibition cannot be ignored. The definitive involvement in lovastatin metabolism of esterase inhibition by GFJ has been demonstrated in rats (Li et al., 2007a). Unlike CYP3A inhibition at the gut level, esterase inhibition could occur at the gut, hepatic, and other tissue levels, as shown in the above *in vitro* section. Mechanistic studies to elucidate the role of esterases in the bioavailability for two ester prodrugs lovastatin and enalapril are shown below.

14.5.1 Involvement of Esterases

Bis-(*p*-nitrophenylphosphate) (BNPP) is a nonspecific esterase inhibitor (Walker and Mackness, 1983). To show that esterases do play a role in the bioavailability of enalapril and lovastatin (10 mg/kg, PO) in the rat model, the animals were dosed orally with these prodrugs without (control, water) or with BNPP at 25 mg/kg PO (Li et al., 2007a). The blood samples (0.3 mL) were collected into heparin tubes containing 3 μL of 200 mM phenylmethyl sulfonyl fluoride (PMSF) and 5 μL of acetic acid (6:4 dilution with water), to ensure that additional esterase activity was arrested. BNPP cotreatment increased the C_{max} and AUC of enalaprilat, the active drug, relative to the control level by 56% and 57%, and of lovastatin acid by 235% and 141%, respectively. The exposure increase for enalaprilat is mostly due to esterase inhibition, while that of lovastatin acid is mainly due to both esterase and CYP inhibition.

14.5.2 Dose-Dependent Effects of GFJ

As a next step, coadministrations of 1:3, 1:2, and 1:0 diluted GFJ concentrate (Minute Maid brand, frozen concentrate) were tested against the water control (pH 3.5, the pH of GFJ) (Li et al., 2007a). The dose-dependent effects on the exposure to the active products from enalapril and lovastatin were quite interesting and, initially, a bit puzzling, as shown in Fig. 14.5. Relative to the water control, with 1:3, 1:2, and concentrated GFJ coadministration the C_{max}

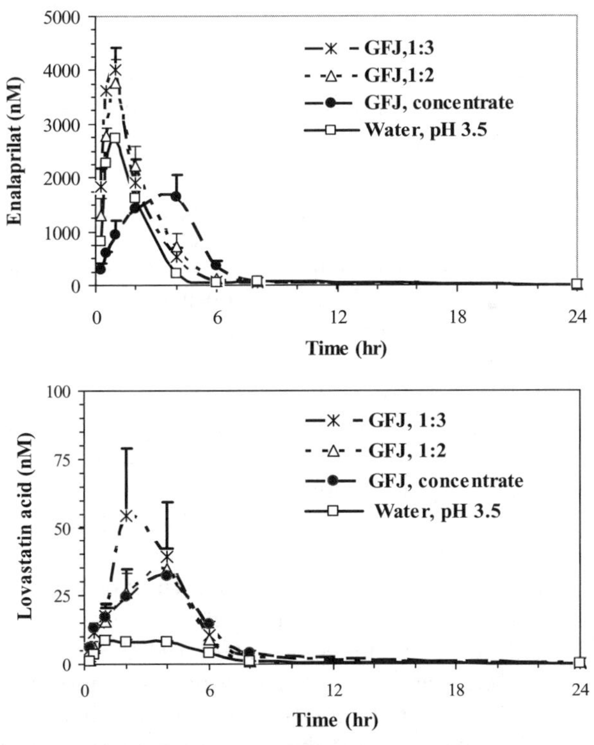

Figure 14.5. Plasma concentration–time profiles of the active products enalaprilat (**top panel**) and lovastatin acid (**bottom panel**) following oral administration of the prodrug enalapril and lovastatin at 10 mg/kg to Sprague–Dawley rats, administered with either water (pH 3.5) or grapefruit juice [1:3, 1:2 diluted (pH 3.5), or concentrate]. Reproduced with permission from Li et al. (2007a). Copyright 2007 by the American Society for the Pharmacology and Experimental Therapeutics.

increased by 311%, 135% and 157%, respectively, while the AUC increased by 279%, 157% and 170%, respectively, for lovastatin acid; and the C_{max} increased by 60% and 50% with 1:3 and 1:2 diluted GFJ, respectively, whereas it decreased by 43% when the GFJ concentrate was used, and the AUC increased by 65%, 70%, and 16%, respectively, with increasing GFJ strength for enalaprilat (Li et al., 2007a). The puzzling part of the relatively decreased enhancement of exposure with increased GFJ strength could be explained by differential trapping and/or binding of the prodrug to the juice components, including pulp. This was indeed found to be the case when the extent of lovastatin was measured after mixing at a fixed concentration into the different strengths of GFJs and centrifuging to produce pulp-free fractions. The concentration of the prodrug in the centrifugate was 88%, 74%, and 66% of the buffer control at 12.5%, 25%, and 100% GFJ (1:3 diluted), respectively. With the use of GFJ concentrate, one can expect that only a limited amount of the prodrug will remain free (available for absorption), and hence the observation of substantially reduced enhancement in exposure *in vivo* at the highest GFJ strength. The reduced enhancement of the exposure was also in line with the fact that the absorption was delayed, with the shift of T_{max} from 1h to 4h at the highest GFJ concentration. A possibility of delayed gastric emptying, when high-strength GFJ was administered, and its effects on the pharmacokinetics are yet to be explored.

14.5.3 Esterase Versus CYP3A Inhibitory Effects

The results of the case study of enalaprilat exposure enhancement upon coadministration of enalapril with GFJ were evidently due to an esterase inhibitory effect. For the case of the high-exposure enhancement of lovastatin acid upon coadministration of lovastatin with GFJ, the relative contributions of the two inhibitory attributes were discerned using a creative approach in which the effects due to competitive and mechanism-based processes were segregated at the gut mucosal level. Irreversible CYP3A inactivation in the gut was established by pretreatment of portal vein-cannulated rats with GFJ at 15 and 2h, to achieve maximal CYP3A inactivation, before administering lovastatin without (Group B) or with (Group C) another dose of GFJ (Li et al., 2007a). The later GFJ coadministration helped to discern exposure increases due to esterase inhibition in the gut of the pretreated animals. Pretreatment with water (pH 3.5) and cotreatment with water (pH 3.5) served as a control (Group A). Portal blood was sampled, and the authors measured the concentrations of (a) CYP3A-mediated oxidative products 6′β-hydroxylovastatin and the corresponding hydroxyacid, (b) the esterase-mediated product lovastatin acid, and (c) lovastatin. GFJ pretreatment (CYP3A-inactivated) led to a 49% increase in the AUC of lovastatin acid compared with the control animals with no GFJ pretreatment (Fig. 14.6). This was in line with the reduction in the oxidative metabolite's AUC to 12% of the control level, demonstrating that gut CYP3A activity was inactivated after GFJ pretreatment. Group C animals,

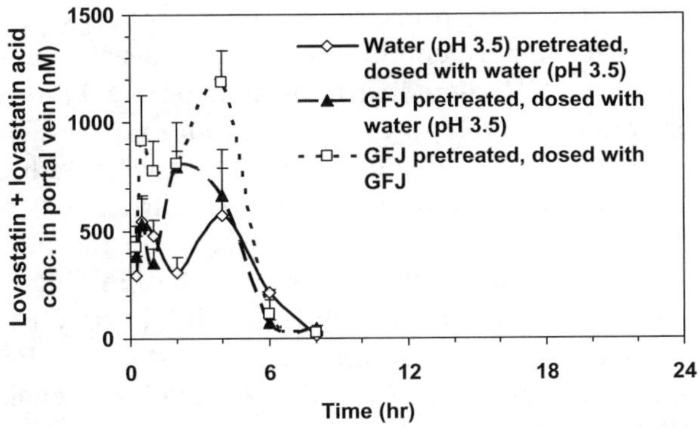

Figure 14.6. Portal plasma concentration of lovastatin acid + lovastatin following oral administration of lovastatin at 10 mg/kg with water (pH 3.5) or 1:3 diluted grapefruit juice (pH 3.5) to portal vein-cannulated Sprague–Dawley rats pretreated at 15 h and 2 h with water (pH 3.5) or 1:3 diluted grapefruit juice (pH 3.5). Reproduced with permission from Li et al. (2007a). Copyright 2007 by the American Society for the Pharmacology and Experimental Therapeutics.

in which CYP3A activity had been inactivated, showed that, upon coadministration of lovastatin with GFJ, the AUC of lovastatin acid was further increased by 46%, compared to the animals with GFJ pretreatment and coadministration of lovastatin with water (Group B). This additional increase in the AUC of the active drug lovastatin acid was considered to be due to the effect of esterase inhibition in the gut lumen and enterocytes by GFJ, leading to enhanced absorption of the prodrug. A similar trend was obtained when the combined increases of lovastatin and lovastatin acid were considered. The pharmacokinetic changes noted were due solely to the effects of GFJ at the gut level and not at the hepatic level, since the rats pretreated with GFJ at 0.5 h before an IV dose of lovastatin or enalapril did not show any significant changes in the pharmacokinetics of either the prodrugs or their active components. These data unequivocally supported the premise of major contributions from both CYP3A and esterase in the increased oral bioavailability of lovastatin acid as a result of prodrug stabilization by GFJ in the rat gut. Thus, on the basis of the *in vitro* and/or *in vivo* data in rats and humans, GFJ can potentially cause drug interactions for ester prodrugs in humans as well.

14.6 EFFECTS ON GUT PERMEABILITY

As described in the previous section, GFJ stabilizes prodrugs in the gut lumen; and with more of the intact, less polar prodrug available, more of it would be expected to permeate across the gut membrane. This was substantiated using Caco-2 cell cultures—with little, if any, CYP3A activity—and studying apical-to-basolateral (A-to-B) transport of these prodrugs (Li et al., 2007a). The data showed that the permeability of lovastatin was increased by 40% in the presence of 6.25% GFJ, demonstrating that permeability was enhanced by the inhibitory effect of GFJ on the esterase and not on CYP3A, because the Caco-2 system was nearly devoid of CYP3A activity. At higher GFJ concentrations, there was entrapment of drug in GFJ particulates, thereby lowering the free fraction, with a less dramatic increase in the permeability. The increase in the A-to-B permeability was not a result of inhibition of Pgp activity by GFJ, shown by the fact that a Pgp/BCRP inhibitor, GF120918, had no effect on the permeability of lovastatin. This suggested that Pgp contributed little to lovastatin permeability in the Caco-2 model. For enalapril, which is absorbed mainly by the passive diffusion process, the permeability increased with GFJ concentrations up to 25%, and thereafter it started diminishing for the reasons stated above.

Discrete ester mixtures (e.g., propyl *p*-hydroxybenzoate, octyl acetate, and ethyl caprylate), as mentioned earlier, have also been shown to stabilize the antiviral prodrug tenofovir disoproxil fumerate (DF) and increase its permeability across Caco-2 cell membranes. This was demonstrated to be a product of both esterase and Pgp inhibition (Van Gelder et al., 2002). Considering that GFJ is a milieu of fruity esters and other chemicals, the inhibitory effects of GFJ could also result from a combination of several different mechanisms, by different chemical classes. GFJ and orange juice are known to inhibit Pgp (Takanaga et al., 2000a, 1998; Spahn-Langguth and Langguth, 2001). This also means that the utility of selective ester mixtures are potentially viable for use in humans to enhance the exposure of active components, delivered as ester prodrugs. On the other hand, one must be cognizant of the potential adverse consequences of the higher exposure to prodrugs and their active components. Reviews of the carboxylesterases and prodrug design to increase oral absorption are described elsewhere (Beaumont et al., 2003; Calogeropoulou et al., 2003; Hatfield et al., 2008; Hosokawa, 2008).

14.7 EFFECTS AT THE HEPATIC LEVEL

The CYP3A-based drug-GFJ interaction occurs with the drugs that are cleared largely by gut CYP3A. Thus, in acute studies, one can use GFJ to selectively knock out gut CYP3A activity allowing an assessment of possible hepatic CYP3A contributions (Hall et al., 1999). Though the effects of GFJ are largely seen at the gut level after consumption of a standard volume, higher volumes (triple dose) have been shown to have an effect on CYP3A activity at the hepatic level (Veronese et al., 2003). Interestingly, there is also a report on the inductive effects of repeated administration of GFJ on CYP3A activity in rat livers. Mohri et al. (2000) studied the acute and chronic effects of GFJ ingestion (2 mL) on gut and hepatic CYP3A levels in rats. The pharmacokinetics of nifedipine were not affected by GFJ after IV administration of nifedipine, compared to control (saline administration after IV administration of nifedipine). After intraduodenal (ID) administration of nifedipine, however, its AUC was increased 1.6-fold. When GFJ was administered BID for 10 consecutive days, the plasma clearance after IV or ID administration of nifedipine was substantially higher relative to that after a single GFJ administration. Microsomes prepared from gut mucosa from these animals still showed lower CYP3A activity. In hepatic microsomes, however, there was significant ($p < 0.005$) induction of CYP3A activity (1.43 ± 0.17 nmol/mg/min versus 1.00 ± 0.06 nmol/mg/min) and also an increase in CYP3A contents, concluding that the lower exposure of nifedipine after chronic GFJ administration led to CYP3A induction at the hepatic level. The GFJ ingredient responsible for the inductive effect is not known. Whether this is just a species-specific effect or will ever be observed in humans is also unknown at this time, because most studies in humans involving subchronic GFJ administration (Lown et al., 1997; Brunner et al., 1998) have shown only exposure increases for the drugs tested as a result of down-regulation of CYP3A in the intestine. Unfortunately, the field of esterase inhibition by GFJ is relatively new and much work has yet to be done.

14.8 GFJ FLAVONOIDS AS ESTERASE INHIBITORS

Among the well-known chemical inhibitors of esterases are diethyl *p*-nitrophenyl phosphate (BNPP) and some aromatic diones (Wadkins et al., 2004, 2005). Some of these compounds are rather toxic for their clinical use. Esterase inhibition by fruity esters in strawberry juice and banana extracts has been documented in the literature (Van Gelder et al., 2002).

In search of the active components that were responsible for the esterase-inhibiting properties of GFJ, Li et al. (2007b) tested 10 flavonoids and furanocoumarins and found that morin, galangin, kaempferol, quercetin, and naringenin inhibited the esterase activity of purified porcine esterase, as well as in human and rat liver microsomes, with IC_{50} values in human liver microsomes of ≥30 μM, using PNPA as a model substrate. Hesperidin, naringin, bergamottin, DHB, and bergapten did not show much activity, with IC_{50} values of >100 μM in human liver microsomes.

In a Caco-2 cell membrane assay, the A-to-B permeability of lovastatin, which was not affected by the Pgp/BCRP inhibitor GF120918, was enhanced >60% by the addition of kaempferol or naringenin. In this model, lovastatin was shown not to be a substrate of Pgp. This enhancement effect was probably a result of the stabilization of the ester prodrug, as shown by the reduction of hydrolysis in rat hepatic S9 to 55% and 72% of the control by kaempferol and naringenin, respectively. The corresponding drop produced by the esterase inhibitors BNPP and phenylmethyl sulfonyl fluoride (PMSF) was to 54% and 24%, respectively, of the control activity. Similar results were obtained for the ester prodrug enalapril when incubated with kaempferol, naringenin, BNPP,

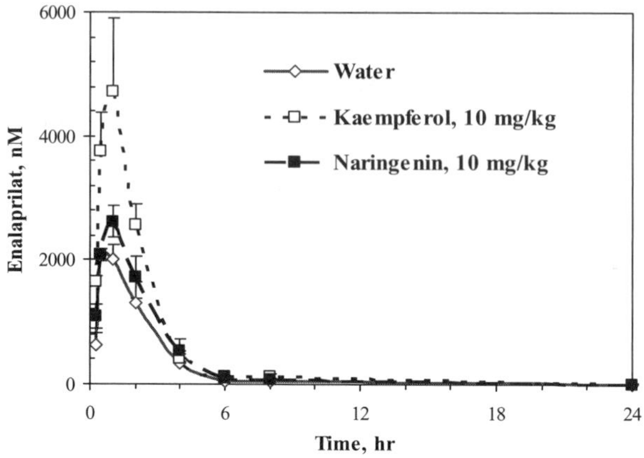

Figure 14.7. Plasma concentration–time profiles of enalaprilat following 10mg/kg oral administration of enalapril with water, or 10mg/kg PO of naringenin or kaempferol, to Sprague–Dawley rats.

or PMSF in rat liver microsomes, leading to reduction of hydrolysis to 29%, 66%, 19%, or 1%, respectively, of the control. Thus, in these systems, kaempferol appeared to be a stronger inhibitor than naringenin.

The ultimate effect of esterase inhibition was then documented *in vivo* in rats for lovastatin and enalapril, given orally at 10mg/kg, in the presence of kaempferol or naringenin, also at 10mg/kg. The AUC enhancement of the active component, lovastatin acid, resulting from kaempferol and naringenin was 246% and 288%, respectively. These effects were similar to those obtained with BNPP coadministration. Likewise, the AUC of enalaprilat upon coadministration of enalapril with kaempferol or naringenin was increased by 109% or 38%, respectively. The plasma concentration–time plots for these are shown in Fig. 14.7.

Since enalapril is metabolized only by carboxylesterase, these data provide strong evidence that these GFJ components have the potential to be used *in vivo* in other species, including humans, to enhance the stability of carboxylester prodrugs. The safety of these flavonoids administered alone on a chronic basis is yet to be determined. But given that GFJ has high concentrations of these flavonoids and that life-threatening toxicities, other than those due to GFJ–drug interactions, have not been reported in humans, the potential for severe toxicity with these flavonoids is expected to be low. Because the situation with lovastatin differs, in the sense that both CYP3A and carboxylesterase metabolize this prodrug, additional studies, similar to those described earlier (Li et al., 2007a) using GFJ itself, were conducted using portal vein-cannulated rats. Oral coadministration of kaempferol with lovastatin, followed by portal plasma collection and analysis for lovastatin, lovastatin acid, and the CYP3A product 6′β-hydroxylovastatin, revealed AUC increases of

113%, 154%, and 208%, respectively. The ratio of 6′β-hydroxylovastatin to lovastatin with and without kaempferol provided a measure of the extent of CYP3A activity alteration due to kaempferol cotreatment. The ratio remained low, at 0.06 to 0.08, demonstrating a weak CYP3A inhibitory effect of kaempferol on lovastatin metabolism. This is consistent with the literature reports that kaempferol is a weak inhibitor of CYP3A with an IC_{50} of >100 μM (Ho et al., 2001). Thus the exposure enhancement of lovastatin/lovastatin acid in rats upon coadministration with kaempferol is largely a result of its esterase inhibitory effect. This is also consistent with the poor absorption of flavonoids due to their high polarity (Schmiedlin-Ren et al., 1997; He et al., 1998).

14.9 UTILITIES OF GFJ APPROACHES

A pharmacokinetic modeling approach was used by Takanaga et al. (2000b) incorporating *in vitro* CYP3A inhibition data for the effect of GFJ on felodipine to predict the interaction *in vivo* in humans. On the basis of different schedules of GFJ ingestion, along with simulation of pharmacokinetics, they estimated a CYP3A turnover $t_{1/2}$ of approximately 8 h. These results show that, to avoid GFJ interaction based on CYP3A inhibition, GFJ ingestion must be at least 2–3 days apart from dosing of a CYP3A substrate drug. This was also consistent with the findings from a clinical study in which the recovery of CYP3A activity, as measured by the pharmacokinetics of midazolam (6 mg PO), was assessed after a single, standard intake of GFJ (300 mL) in humans. The AUC of midazolam increased by 1.65-fold when given 2 h after the GFJ intake. Subsequent administration of midazolam at 26, 50, and 74 h post-GFJ ingestion showed that exposure to midazolam returned to baseline by 74 h, with midazolam-to-control AUC ratios of 1.29, 1.29, and 1.06, respectively (Greenblatt et al., 2003). From these results, Greenblatt et al. estimated that the CYP3A recovery $t_{1/2}$ was 23 h. Overall, the study showed that the gut CYP3A returned to the baseline levels within 3 days. Takanaga et al. (2000c) also noted similar findings. For a recent review on the controversial values of CYP turnover rates, readers are referenced to an article by Yang et al. (2008). It is plausible that there will be a large degree of intersubject variability in drug exposure after inactivation of CYP3A in the gut by GFJ, depending on the frequency, GFJ dosing time relative to drug dosing time, quantity of GFJ intake, and the extent of gut CYP3A expression in the subjects, such that the higher the CYP3A expression, the higher the AUC increase of the drug. As opposed to the time-dependent nature of CYP3A inhibition, esterase and transporter inhibition at the gut level will happen competitively. Thus, the time gap between GFJ intake and an ester drug, which is not metabolized by CYP3A administration, does not have to be as long to avoid esterase- or transporter-mediated interactions: A gap of 4–6 h generally would suffice for most drugs that are absorbed relatively rapidly. Glaeser et al. (2007) demonstrated that a lapse of 4 h (i.e., administering GFJ at least 4 h postdose) was sufficient to eliminate the inhibitory effect on OATP and Pgp for the pharmacokinetics of fexofenadine. However, clinical studies are needed to establish that this is true for esterase effects.

14.10 CLINICAL EFFECTS OF GFJ

Apart from the drug exposure-boosting properties of GFJ, its health-related benefits have been widely recognized throughout the health industry. GFJ is known to have antioxidant, antiseptic, cardiotonic, anti-inflammatory, anticarcinogenic, antimutagenic, and hypocholesterlemic activities. The anti-inflammatory properties of the citrus flavonoids are considered to be a result of their activities against the proinflammatory mediators, like arachidonic acid derivatives, prostaglandins, and thromboxane (Benavente-García et al., 2008). Published literature also shows benefits in its inhibition of atherogenesis (Cerda et al., 1994). These results appear to be related to ingestion of the pectin in GFJ, as shown in studies in microswines in which less plaque and coronary artery narrowing were observed. Some studies in humans have shown reduction in plasma cholesterol (Cerda et al., 1988). GFJ and flavonoids are shown to help in inhibition of breast cancer growth *in vitro* and *in vivo* in rats (So et al., 1996; Guthrie et al., 1998). Effects on multicancer pathways of flavonoids have also been described (Benavente-García et al., 2007). In glioma cells, kaempferol is known to elevate intracellular oxidative stress to induce apoptosis or suppress survivin in tumor necrosis factor-related apoptosis-inducing ligand-mediated apoptosis (Sharma et al., 2007; Siegelin et al., 2008). The fact that some of these flavonoids are efflux pump inhibitors raises the possibility of their use in combination with other cytotoxic agents that are substrates of these pumps. Their combined use raises the possibility of increased concentrations of these agents in cancer cells, with a subsequent increase in efficacy. In cell cultures, various flavonoids increased the rate of apoptosis, inhibited cell proliferation, and inhibited angiogenesis. However, data on intervention studies in humans have been few.

In some epidemiological studies, GFJ intake has been associated with the formation of renal stones (Curhan et al., 1996, 1998), probably via an indirect mechanism, such as via the binding of GFJ flavonoids to calcium and their presence in high concentrations in the kidney (Ameer, 1998).

14.11 FUTURE SCOPE

The field of esterase inhibition by GFJ and its constituents is relatively new, and much work in animals and humans is yet to be done. With scores of chemicals identified in GFJ, additional potential effects of GFJ and its constituents on various other enzymes and transporters need to be explored. The hope is that the recent findings discussed here will stimulate more interest in this field and increase both drug developers' and regulatory agencies' awareness of GFJ in the development of safer drugs. Since flavonoids act as esterase inhibitors, they may be of utility as bioavailability-boosting agents for various prodrugs *in vivo* in humans.

ACKNOWLEDGMENTS

The author gratefully acknowledges the work of Drs. Ping Li and Liang-Shang Gan of Biogen Idec, Inc., and Dr. Patrick S. Callery of West Virginia University, included in this manuscript. The author is thankful for the excellent editorial assistance of Ms. Alexis Khalil of NCDS Management, Strategy and Operations.

REFERENCES

Ameer B. Grapefruit juice and kidney stones. *Ann Intern Med* 1998;129:913.

Bailey DG, Spence JD, Edgar B, Bayliff CD, Arnold JM. Ethanol enhances the hemodynamic effects of felodipine. *Clin Invest Med* 1989;12:357–362.

Bailey DG, Spence JD, Munoz C, Arnold JM. Interaction of citrus juices with felodipine and nifedipine. *Lancet* 1991;337:268–269.

Balani SK, Xu X, Pratha V, Koss MA, Amin RD, Dufresne C, Miller RR, Arison BH, Doss GA, Chiba M, Freeman A, Holland SD, Schwartz JI, Lasseter KC, Gertz BJ, Isenberg JI, Rogers JD, Lin JH, Baillie TA. Metabolic profiles of Montelukast Sodium (Singulair), a potent cysteinyl leukotriene$_1$ receptor antagonist, in human plasma and bile. *Drug Metab Dispos* 1997;25:1282–1287.

Banfield C, Gupta S, Marino M, Lim J, Affrime M. Grapefruit juice reduces the oral bioavailability of fexofenadine but not desloratadine. *Clin Pharmacokinet* 2002;41: 311–318.

Beaumont K, Webster R, Gardner I, Dack K. Design of ester prodrugs to enhance oral absorption of poorly permeable compounds: challenges to the discovery scientist. *Curr Drug Metab* 2003;4:461–485.

Benavente-García O, Castillo J, Alcaraz M, Vicente V, Del Río JA, Ortuño A. Beneficial action of Citrus flavonoids on multiple cancer-related biological pathways. *Curr Cancer Drug Targets* 2007;7:795–809.

Benavente-García O, Castillo J. Beneficial action of citrus flavonoids on multiple cancer-related biological pathways. Update on uses and properties of citrus flavonoids: new findings in anticancer, cardiovascular, and anti-inflammatory activity. *J Agric Food Chem* 2008;56:6185–6205.

Brunner LJ, Munar MY, Vallian J, Wolfson M, Stennett DJ, Meyer MM, Bennett WM. Interaction between cyclosporine and grapefruit juice requires long-term ingestion in stable renal transplant recipients. *Pharmacotherapy* 1998;18:23–29.

Calogeropoulou T, Detsi A, Lekkas E, Koufaki M. Strategies in the design of prodrugs of anti-HIV agents. *Curr Top Med Chem* 2003;3:1467–1495.

Cerda JJ, Robbins FL, Burgin CW, Baumgartner TG, Rice RW. The effects of grapefruit pectin on patients at risk for coronary heart disease without altering diet or lifestyle. *Clin Cardiol* 1988;11:589–594.

Cerda JJ, Normann SJ, Sullivan MP, Burgin CW, Robbins FL, Vathada S, Leelachaikul P. Inhibition of atherosclerosis by dietary pectin in microswine with sustained hypercholesterolemia. *Circulation* 1994;89:1247–1253.

Curhan GC, Willett WC, Rimm EB, Spiegelman D, Stampfer MJ. Prospective study of beverage use and the risk of kidney stones. *Am J Epidemiol* 1996;143:240–247.

Curhan GC, Willett WC, Speizer FE, Stampfer MJ. Beverage use and risk for kidney stones in women. *Ann Intern Med* 1998;128:534–540.

De Castro WV, Mertens-Talcott S, Rubner A, Butterweck V, Derendorf HJ. Variation of flavonoids and furanocoumarins in grapefruit juices: a potential source of variability in grapefruit juice–drug interaction studies. *Agric Food Chem* 2006;11: 249–255.

De Castro WV, Mertens-Talcott S, Derendorf H, Butterweck V. Grapefruit juice–drug interactions: Grapefruit juice and its components inhibit P-glycoprotein (ABCB1) mediated transport of talinolol in Caco-2 cells. *J Pharm Sci* 2007;96:2808–2817.

De Castro WV, Mertens-Talcott S, Derendorf H, Butterweck V. Effect of grapefruit juice, naringin, naringenin, and bergamottin on the intestinal carrier-mediated transport of talinolol in rats. *J Agric Food Chem* 2008;56:4840–4845.

Dresser GK, Bailey DG, Leake BF, Schwarz UI, Dawson PA, Freeman DJ, Kim RB. Fruit juices inhibit organic anion transporting polypeptide-mediated drug uptake to decrease the oral availability of fexofenadine. *Clin Pharmacol Ther* 2002;71: 11–20.

Ducharme MP, Warbasse LH, Edwards DJ. Disposition of intravenous and oral cyclosporine after administration with grapefruit juice. *Clin Pharmacol Ther* 1995;57: 485–491.

Fuhr U, Kummert AL. The fate of naringin in humans: a key to grapefruit juice–drug interactions? *Clin Pharmacol Ther* 1995;58:365–373.

Gattuso G, Barreca D, Gargiulli C, Leuzzi U, Caristi C. Flavonoid composition of citrus juices. *Molecule* 2007;12:1641–1673.

Glaeser H, Bailey DG, Dresser GK, Gregor JC, Schwarz UI, McGrath JS, Jolicoeur E, Lee W, Leake BF, Tirona RG, Kim RB. Intestinal drug transporter expression and the impact of grapefruit juice in humans. *Clin Pharmacol Ther* 2007;81: 362–370.

Greenblatt DJ, von Moltke LL, Harmatz JS, Chen G, Weemhoff JL, Jen C, Kelley CJ, LeDuc BW, Zinny MA. Time course of recovery of cytochrome p450 3A function after single doses of grapefruit juice. *Clin Pharmacol Ther* 2003;74:121–129.

Guo LQ, Fukuda K, Ohta T, Yamazoe Y. Role of furanocoumarin derivatives on grapefruit juice-mediated inhibition of human CYP3A activity. *Drug Metab Dispos* 2000;28:766–771.

Guthrie N, Carol KK. Inhibition of mammary cancer by citrus flavonoids. *Adv Exp Med Biol* 1998;439:227–236.

Hall SD, Thummel KE, Watkins PB, Lown KS, Benet LZ, Paine MF, Mayo RR, Turgeon DK, Bailey DG, Fontana RJ, Wrighton SA, Fontana RJ, Wrighton SA. Molecular and physical mechanisms of first-pass extraction. *Drug Metab Dispos* 1999;27:161–166.

Hatfield JM, Wierdl M, Wadkins RM, Potter PM. Modifications of human carboxylesterase for improved prodrug activation. *Expert Opin Drug Metab Toxicol* 2008;4:1153–1165.

He K, Iyer KR, Hayes RN, Sinz MW, Woolf TF, Hollenberg PF. Inactivation of cytochrome P450 3A4 by bergamottin, a component of grapefruit juice. *Chem Res Toxicol* 1998;11:252–259.

Ho PC, Saville DJ, Wanwimolruk S. Inhibition of human CYP3A4 activity by grapefruit flavonoids, furanocoumarins and related compounds. *J Pharm Pharm Sci* 2001;4:217–227.

Hosokawa M. Structure and catalytic properties of carboxylesterase isozymes involved in metabolic activation of prodrugs. *Molecules* 2008;13:412–431.

Huang SH, Hall SD, Watkins P, Love LA, Serabjit-Singh C, Betz JM, Hoffman FA, Honig P, Coates PM, Bull J, Chen ST, Kearns GL, Murray MD. Drug interactions with herbal prodrugs and grapefruit juice: a conference report. *Clin Pharmacol Ther* 2004;75:1–12.

Kane GC, Lipsky JJ. Drug–grapefruit juice interaction. *Mayo Clin Proc* 2000;75:933–942.

Kuhnau J. The flavonoids: a class of semi-essential food components; their role in human nutrition. *World Rev Nutr Diet* 1976;24:117–191.

Li P, Callery PS, Gan L-S, Balani SK. Esterase inhibition by grapefruit juice leading to a new drug interaction. *Drug Metab Dispos* 2007a;35:1023–1031.

Li P, Callery PS, Gan L-S, Balani SK. Esterase inhibition by grapefruit juice flavonoids leading to a new drug interaction. *Drug Metab Dispos* 2007b;35:1203–1208.

Liederer BM, Borchardt RT. Enzymes involved in the bioconversion of ester-based prodrugs. *J Pharm Sci* 2006;95:1177–1195.

Lin H-L, Kent UM, Hollenberg PF. The grapefruit juice effect is not limited to cytochrome P450 (P450) 3A4: evidence for bergamottin-dependent inactivation, heme destruction, and covalent binding to protein in P450s 2B6 and 3A5. *J Pharm Exp Ther* 2005;313:154–164.

Lown KS, Bailey DG, Fontana RJ, Janardan SK, Adair CH, Fortlage LA, Brown MB, Guo W, Watkins PB. Grapefruit juice increases felodipine oral availability in humans by decreasing intestinal CYP3A protein expression. *J Clin Invest* 1997;99:2545–2553.

Manthey JA, Buslig BS. Distribution of furanocoumarins in grapefruit juice fractions. *J Agric Food Chem* 2005;53:5158–5163.

Mertens-Talcott SU, De Castro WV, Manthey JA, Derendorf H, Butterweck V. Polymethoxylated flavones and other phenolic derivates from citrus in their inhibitory effects on P-glycoprotein-mediated transport of talinolol in Caco-2 cells. *J Agric Food Chem* 2007;55:2563–2568.

Miwa M, Ura M, Nishida M, Sawada N, Ishikawa T, Mori K, Shimma N, Umeda I, Ishitsuka H. Design of a novel oral fluoropyrimidine carbamate, capecitabine, which generates 5-fluorouracil selectively in tumours by enzymes concentrated in human liver and cancer tissue. *Eur J Cancer* 1998;34:1274–1281.

Mohri K, Uesawa Y, Sagawa K-I. Effects of long-term grapefruit juice ingestion on nifedipine pharmacokinetics: induction of rat hepatic P-450 by grapefruit juice. *Drug Metab Dispos* 2000;28:482–486.

Morishita Y, Fujii M, Kasakura Y, Takayama T. Effect of carboxylesterase inhibition on anti-tumour effects of irinotecan. *J Int Med Res* 2005;33:84–89.

Paine MF, Widmer WW, Hart HL, Pusek SN, Beavers KL, Criss AB, Brown SS, Thomas BF, Watkins PB. A furanocoumarin-free grapefruit juice establishes furanocoumarins as the mediators of the grapefruit juice-felodipine interaction. *Am J Clin Nutr* 2006;83:1097–2105.

Quinney SK, Sanghani SP, Davis WI, Hurley TD, Sun Z, Murry DJ, Bosron WF. Hydrolysis of Capecitabine to 5′-deoxy-5-fluorocytidine by human carboxylesterases and inhibition by Loperamide. *J Pharm Exp Ther* 2005;313:1011–1016.

Ranganna S, Govindarajan VS, Ramana KV. Citrus fruits—varieties, chemistry, technology, and quality evaluation. Part II. Chemistry, technology, and quality evaluation. A Chemistry. *Crit Rev Food Sci Nutr* 1983;18:313–386.

Ross SA, Ziska DS, Zhao K, El Sohly MA. Variance of common flavonoids by brand of grapefruit juice. *Fitoterapia* 2000;71:154–161.

Saito M, Hirata-Koizumi M, Matsumoto M, Urano T, Hasegawa R. Undesirable effects of citrus juice on the pharmacokinetics of drugs: focus on recent studies. *Drug Saf* 2005;28:677–694.

Satoh T, Hosokawa M. The mammalian carboxylesterases: from molecules to functions. *Annu Rev Pharmacol Toxicol* 1998;38:257–288.

Satoh T, Taylor P, Bosron WF, Sanghani SP, Hosokawa M, La Du BN. Current progress on esterases: from molecular structure to function. *Drug Metab Dispos* 2002;30:488–493.

Schmiedlin-Ren P, Edwards DJ, Fitzsimmons ME, He K, Lown KS, Woster PM, Rahman A, Thummel KE, Fisher JM, Hollenberg PF, Watkins PB. Mechanisms of enhanced oral availability of CYP3A4 substrates by grapefruit constituents. Decreased enterocyte CYP3A4 concentration and mechanism-based inactivation by furanocoumarins. *Drug Metab Dispos* 1997;25:1228–1233.

Sharma V, Joseph C, Ghosh S, Agarwa Al, Mishra MK, Sen E. Kaempferol induces apoptosis in glioblastoma cells through oxidative stress. *Mol Cancer Ther* 2007;6: 2544–2553.

Shaw PE, Moshonas MG, Heran CJ, Goodner KL. Volatile constituents in fresh and processed juices from grapefruit and new grapefruit hybrids. *J Agric Food Chem* 2000;48:2425–2429.

Siegelin MD, Reuss DE, Habel A, Herold-Mende C, von Deimling A. The flavonoid kaempferol sensitizes human glioma cells to TRAIL-mediated apoptosis by proteasomal degradation of surviving. *Mol Cancer Ther* 2008;7:3566–3574.

So FV, Guthrie N, Chambers AF, Moussa M, Carol KK. Inhibition of human cancer cell proliferation and delay of mammary cell tumorigenesis by flavonoids and citrus juices. *Nutr Cancer* 1996;26:167–181.

Spahn-Langguth H, Langguth P. Grapefruit juice enhances intestinal absorption of the P-glycoprotein substrate talinolol. *Eur J Pharm Sci* 2001;12:361–367.

Stoeckel K, Hofheinz W, Laneury JP, Duchene P, Shedlofsky S, Blouin RA. Stability of cephalosporin prodrug esters in human intestinal juice: Implications for oral bioavailability. *Antimicrob Agents Chem* 1998;42:2602–2606.

Takanaga H, Ohnishi A, Matsuo H, Sawada Y. Inhibition of vinblastine efflux mediated by *p*-glycoprotein by grapefruit juice components in Caco-2 cells. *Biol Pharm Bull* 1998;21:1062–1066.

Takanaga H, Ohnishi A, Yamada S, Matsuo H, Morimoto S, Shoyama Y, Ohtani H, Sawada Y. Polymethoxylated flavones in orange juice are inhibitors of *p*-glycoprotein but not cytochrome P450 3A4. *J Pharmacol Exp Ther* 2000a;293:230–236.

Takanaga H, Ohnishi A, Matsuo H, Murakami H, Sata H, Kuroda K, Urae A, Higuchi S, Sawada Y. Pharmacokinetic analysis of felodipine–grapefruit juice interaction based on an irreversible enzyme inhibition model. *Br J Clin Pharmacol* 2000b;49: 49–58.

Takanaga H, Ohnishi A, Murakami H, Matsuo H, Higuchi S, Urae A, Irie S, Furuie H, Matsukuma K, Kimura M, Kawano K, Orii Y, Tanaka T, Sawada Y. Relationship between time after intake of grapefruit juice and the effect on pharmacokinetics and pharmacodynamics of nisoldipine in healthy subjects. *Clin Pharmacol Ther* 2000c;67:201–214.

Tassaneeyakul W, Guo LQ, Fukuda K, Ohta T, Yamazoe Y. Inhibition selectivity of grapefruit juice components on human cytochromes P450. *Arch Biochem Biophys* 2000;378:356–363.

Uno T, Yasui-Furukori N. Effect of grapefruit juice in relation to human pharmacokinetic study. *Curr Clin Pharmacol* 2006;1:157–161.

Van Gelder J, Annaert P, Naesens L, De Clercq E, Van den Mooter G, Kinget R, Augustijns P. Inhibition of intestinal metabolism of the antiviral ester prodrug bis(POC)-PMPA by nature—identical fruit extracts as a strategy to enhance its oral absorption: an *in vitro* study. *Pharm Res* 1999;16:1035–1040.

Van Gelder J, Deferme S, Annaert P, Naesens L, De Clercq E, Van den Mooter G, Kinget R, Augustijns P. Increased absorption of the antiviral ester prodrug tenofovir disoproxil in rat ileum by inhibiting its intestinal metabolism. *Drug Metab Dispos* 2000;28:1394–1396.

Van Gelder J, Deferme S, Annaert P, Naesens L, De Clercq E, Van den Mooter G, Kinget R, Augustijns P. Intestinal absorption enhancement of the ester prodrug tenofovir disoproxil fumerate through modulation of the biochemical barrier by defined ester mixtures. *Drug Metab Dispos* 2002;30:924–930.

Veronese ML, Gillen LP, Burke JP, Dorval EP, Hauck WW, Pequignot E, Waldman SA, Greenberg HE. Exposure-dependent inhibition of intestinal and hepatic CYP3A4 *in vivo* by grapefruit juice. *J Clin Pharmacol* 2003;43:831–839.

Verweij J. Rational design of new tumoractivated cytotoxic agents. *Oncology* 1999;57: 9–15.

Wadkins RM, Hyatt JL, Yoon KJ, Morton CL, Lee RE, Damodaran K, Beroza P, Danks MK, Potter PM. Discovery of novel selective inhibitors of human intestinal carboxyl esterase for the amelioration of irinotecan-induced diarrhea: synthesis, quantitative structure–activity relationship analysis, and biological activity. *Mol Pharmacol* 2004;65:1336–1343.

Wadkins RM, Hyatt JL, Wei X, Yoon KJ, Wierdl M, Edwards CC, Morton CL, Obenauer JC, Damodaran K, Beroza P, Danks MK, Potter PM. Identification and characterization of novel benzil (diphenylethane-1,2-dione) analogues as inhibitors of mammalian carboxylesterases. *J Med Chem* 2005;48:2906–2915.

Walker CH, Mackness MI. Esterase: problems of identification and classification. *Biochem Pharmacol* 1983;32:3265–3269.

Xie M, Yang D, Liu L, Xue B, Yan B. Human and rodent carboxylesterases: immunorelatedness, overlapping substrate specificity, differential sensitivity to serine enzyme inhibitors, and tumor-related expression. *Drug Metab Dispos* 2002;30: 541–547.

Yang J, Liao M, Shou M, Jamei M, Yeo KR, Tucker GT, Rostami-Hodjegan A. Cytochrome P450 turnover: regulation of synthesis and degradation, methods for determining rates, and implications for the prediction of drug interactions. *Curr Drug Metab* 2008;9:384–393.

Zhou S, Lim LY, Chowbay B. Herbal modulation of P-glycoprotein. *Drug Metab Rev* 2004;36:57–104.

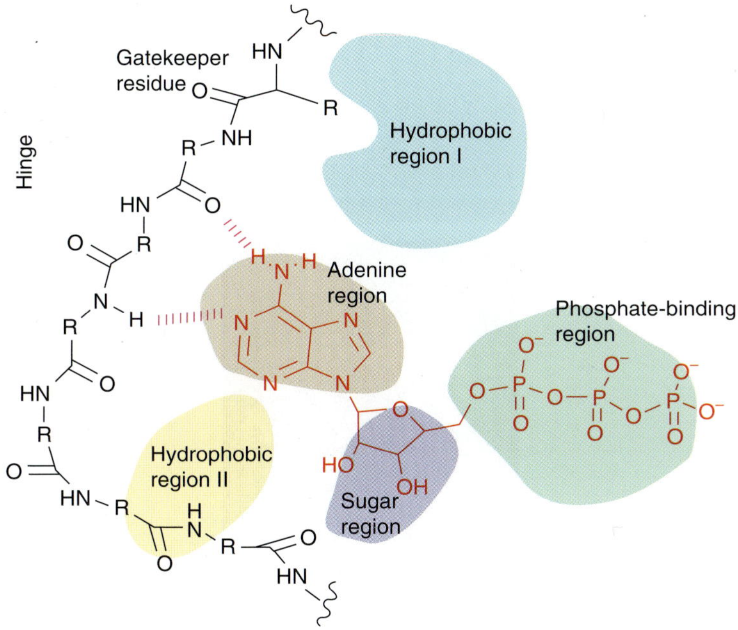

Figure 2.2. ATP binding in the active site, illustrating the important hydrogen bonds made between the backbone of the protein (the hinge region) and its substrate ATP. See pages 17–18 for text discussion.

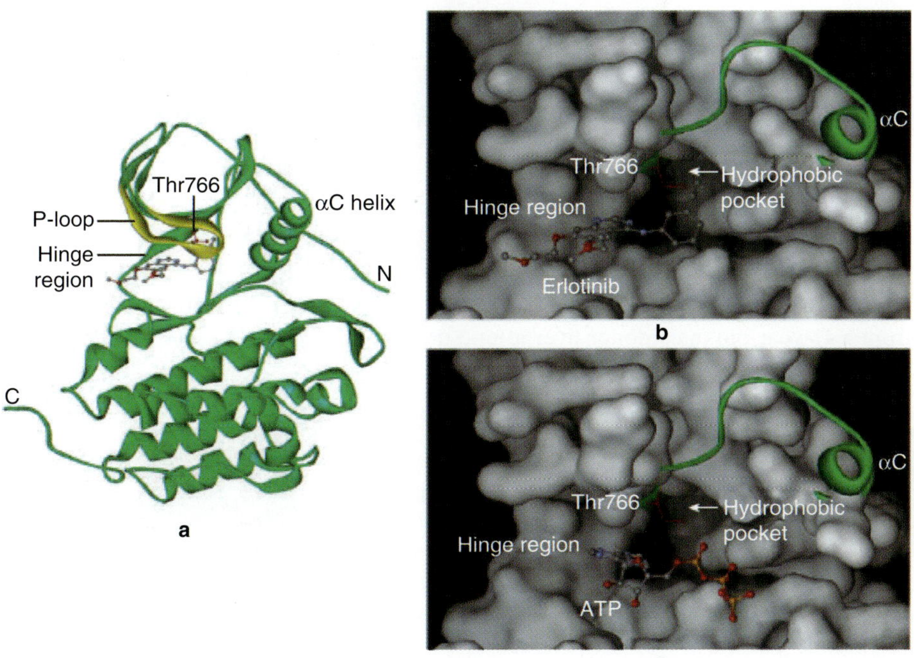

Figure 2.4. Erlotinib bound in the ATP binding pocket. See pages 18–19 for text discussion.

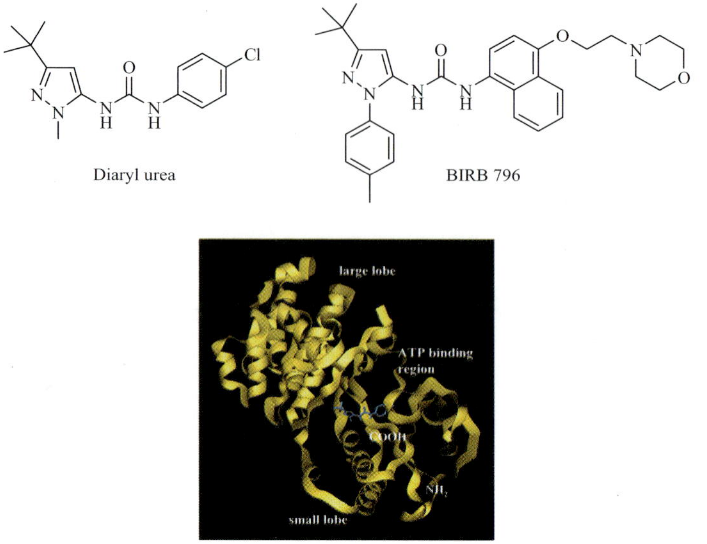

Figure 2.11. Diaryl urea and BIRB 796 are allosteric inhibitors of p38 MAPK. Also shown is the crystal structure of human p38 MAP kinase and diaryl urea (blue). See page 22 for text discussion.

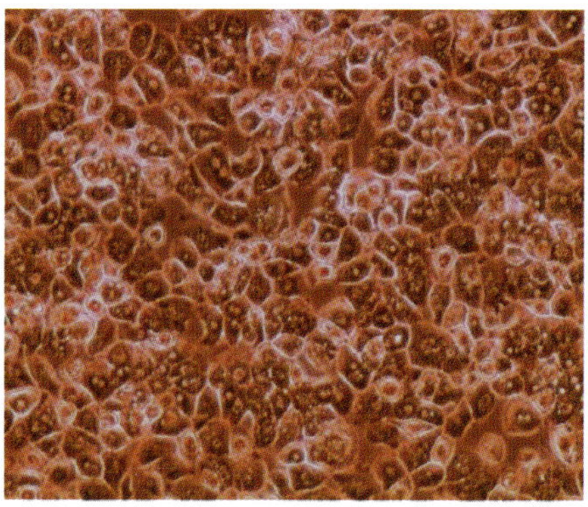

Figure 8.4. Primary human hepatocytes in primary sandwich culture. Hepatocytes from a 41-year-old male donor liver after 3 days in culture. See pages 278–279 for text discussion.

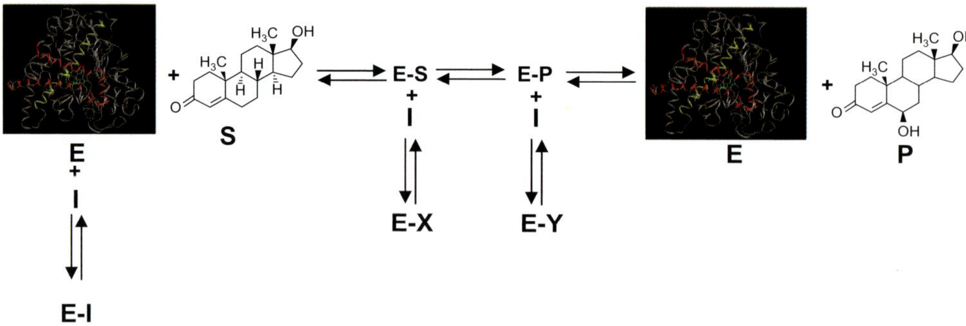

Figure 8.10. Representation of a CYP enzymatic activity reaction. Enzymatic activity is most frequently monitored by the formation of P (product). See page 296 for text discussion.

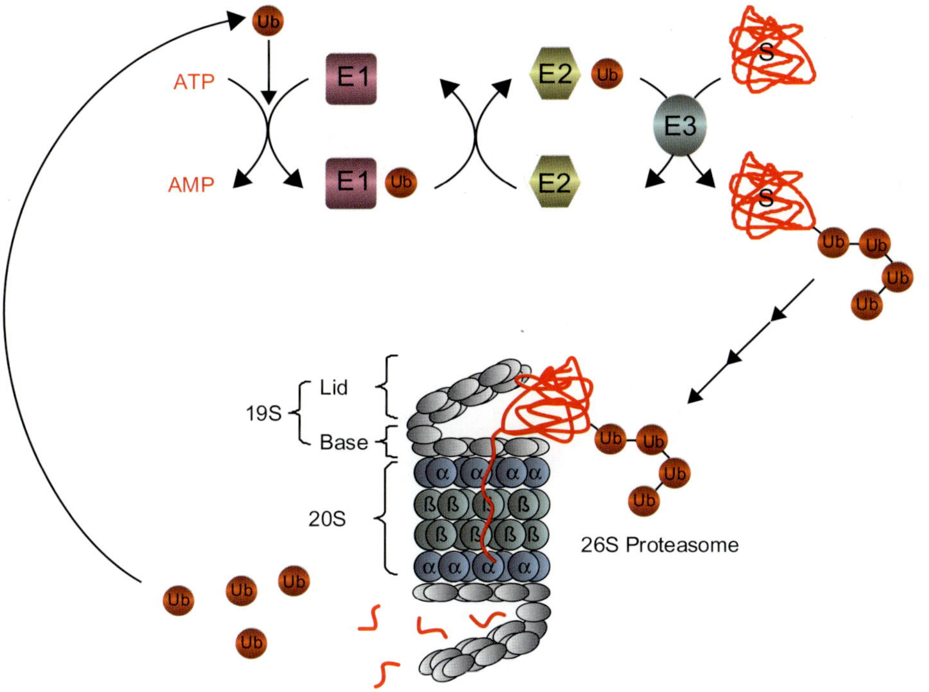

Figure 11.2. The ubiquitin-proteasomal degradation (UPD) system. See page 369 for text discussion.

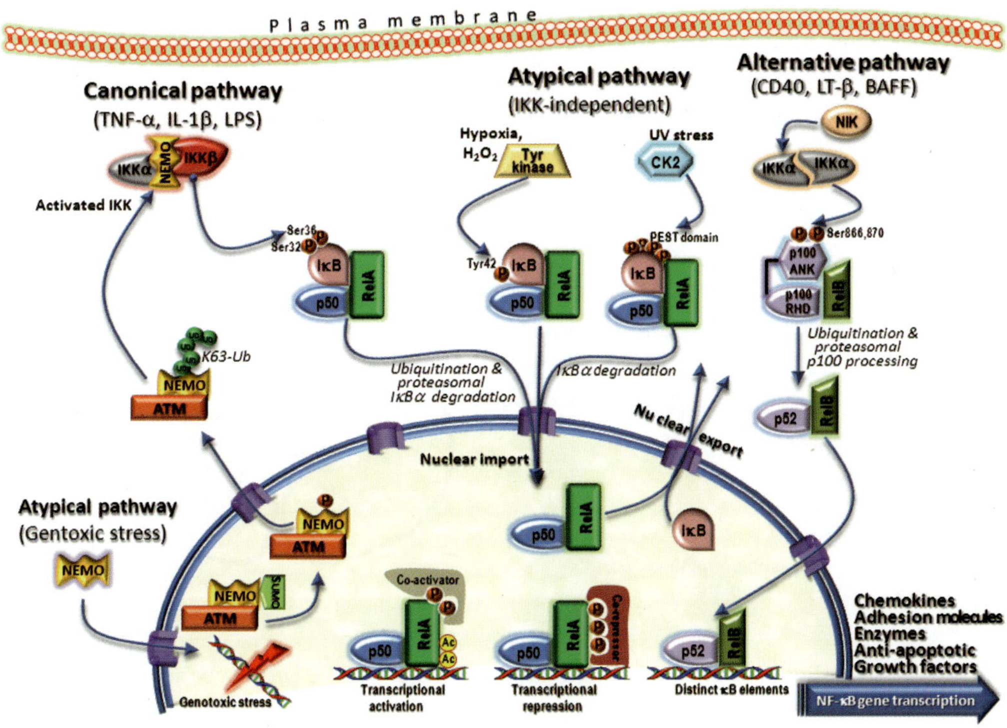

Figure 19.2. NF-κB signaling pathways. Schematic diagram illustrating the three described activation pathways of NF-κB: canonical, alternative, and atypical. See pages 575–576 for text discussion.

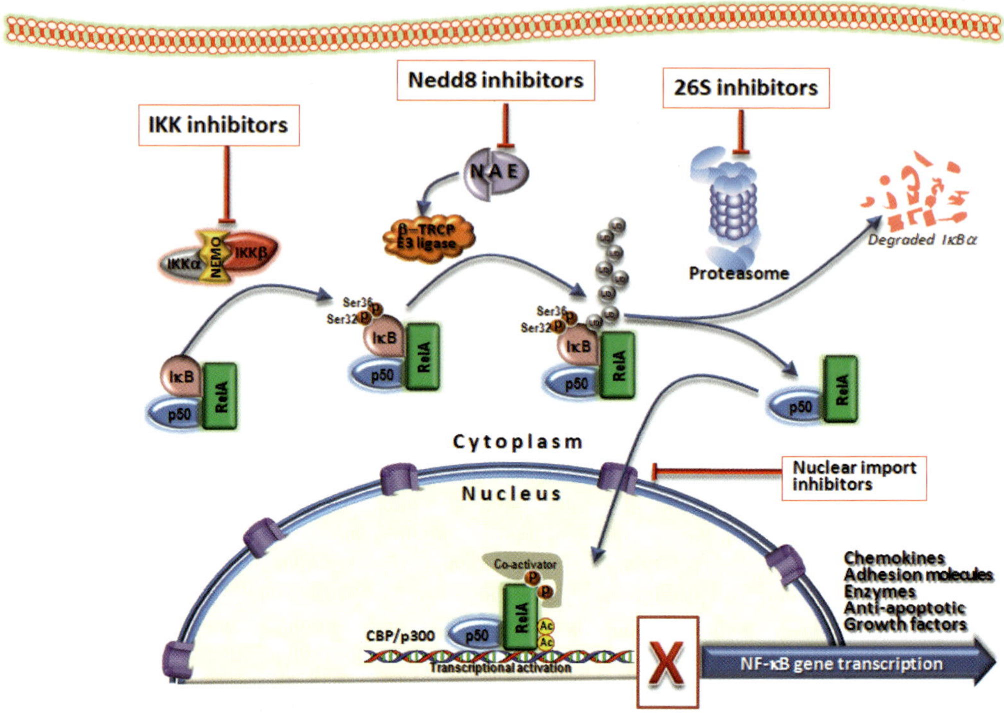

Figure 19.3. Targeting the NF-κB pathway. Exemplified strategies used by various groups in developing therapeutics against NF-κB. See pages 597–598 for text discussion.

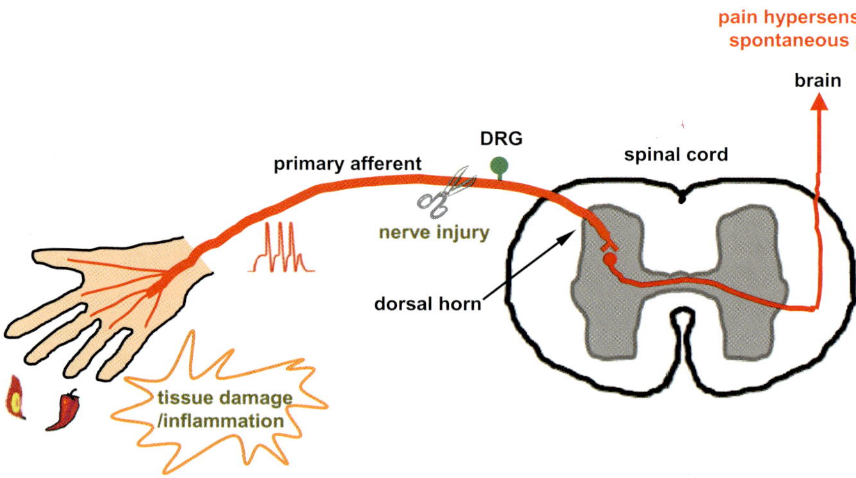

Figure 21.2. Peripheral inflammation and nerve injury cause chronic pain. See pages 680–681 for text discussion.

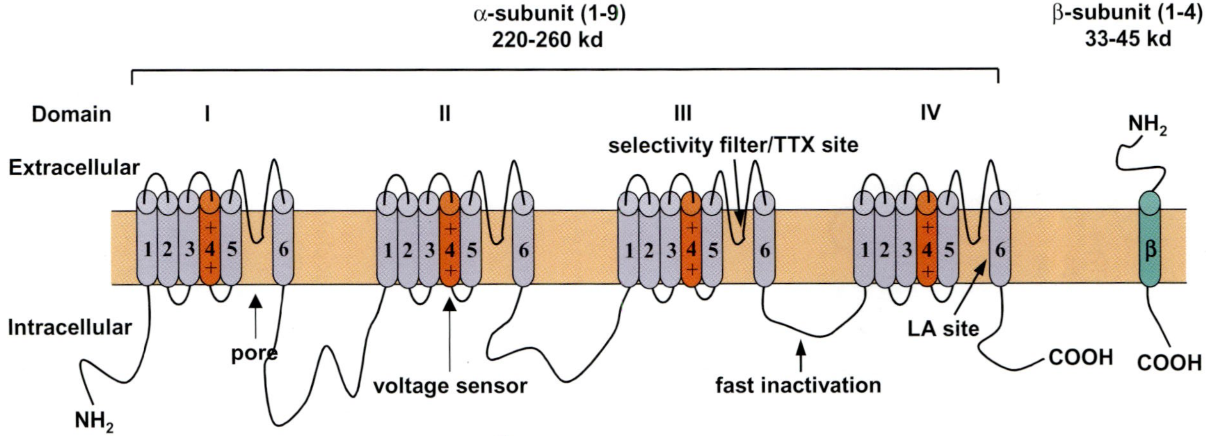

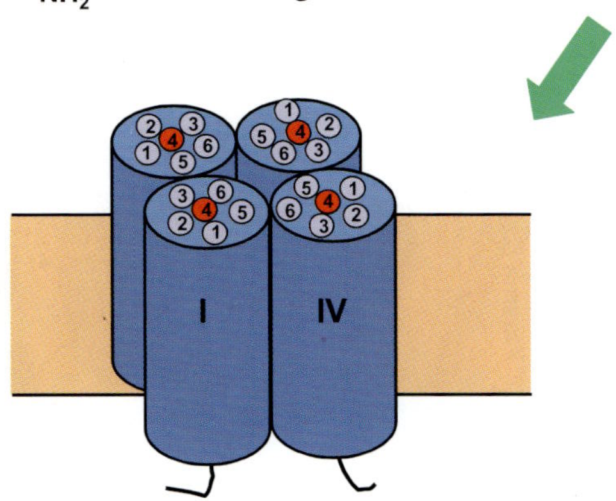

channel	gene	TTX sensitivity	primary localization	implicated in pain
$Na_v1.1$	SCN1A	S	CNS	
$Na_v1.2$	SCN2A	S	CNS	
$Na_v1.3$	SCN3A	S	PNS (post injury)/CNS	Yes
$Na_v1.4$	SCN4A	S	skeletal muscle	
$Na_v1.5$	SCN5A	R	heart	
$Na_v1.6$	SCN8A	S	PNS	
$Na_v1.7$	SCN9A	S	PNS	Yes
$Na_v1.8$	SCN10A	R	PNS	Yes
$Na_v1.9$	SCN11A	R	PNS	Yes

S: Sensitive; R: Resistant; TTX: Tetrodotoxin
C(P)NS: Central(Peripheral) Nervous System

Figure 21.3. Voltage-gated sodium channels: structure, classification, primary localization, and role in pain. Secondary structure of Na_v channel α and β subunits (**top**) and assembly into a pore-forming sodium channel by the four homologous domains (I–IV) with a pseudo-fourfold symmetry (**lower left**). Four of the nine functional Na_v channels are implicated in chronic pain (**lower right**). See pages 681–682 for text discussion.

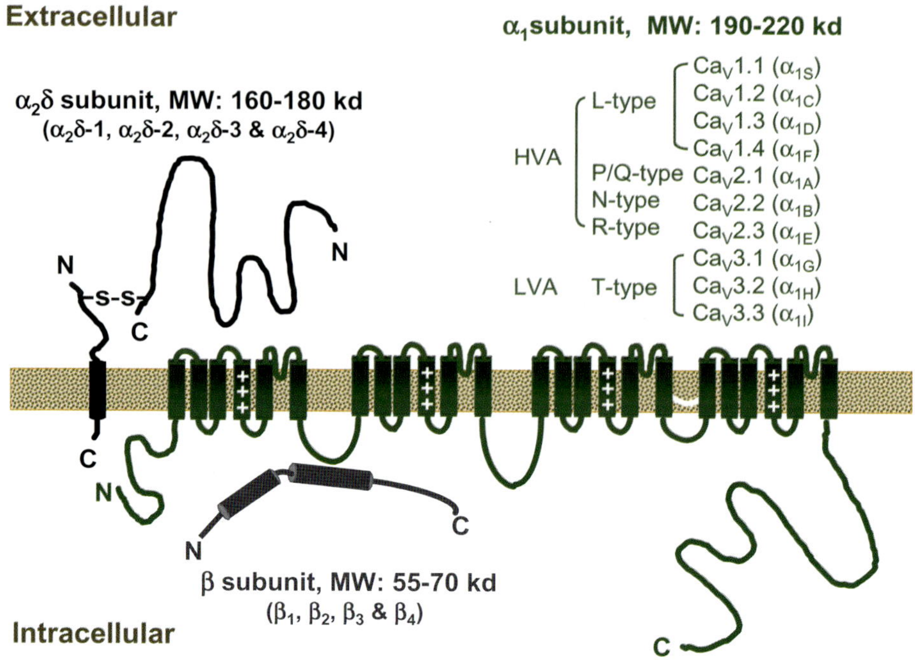

Figure 21.4. Voltage-gated calcium channels—structure and classification. See pages 690–691 for text discussion.

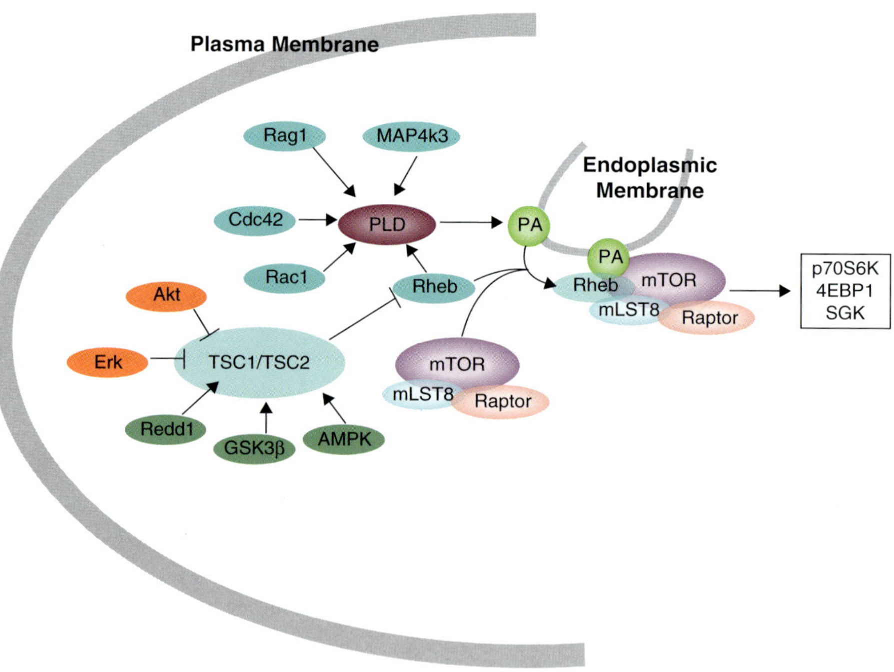

Figure 22.1. mTORC1 activation. See page 725 for text discussion.

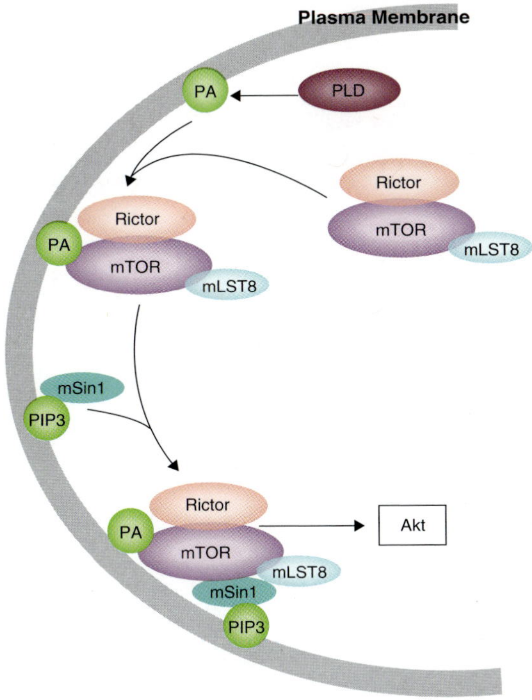

Figure 22.2. mTORC2 activation. See page 728 for text discussion.

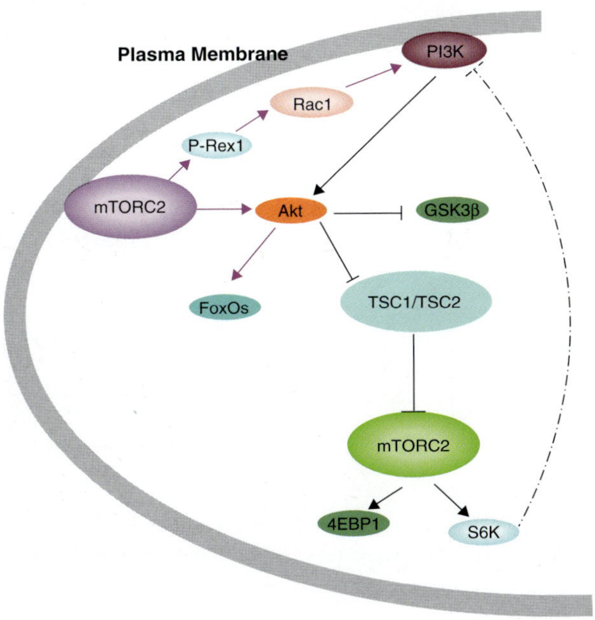

Figure 22.3. Cross-talk between mTORC2 and mTORC1. See page 731 for text discussion.

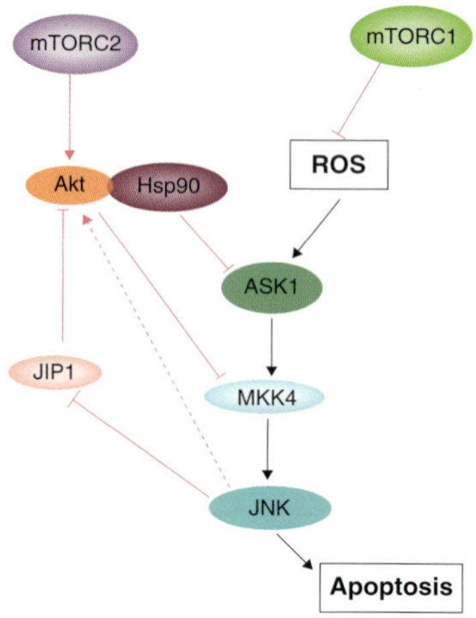

Figure 22.4. Feedback loop of Akt to inhibit JNK activation. See page 735 for text discussion.

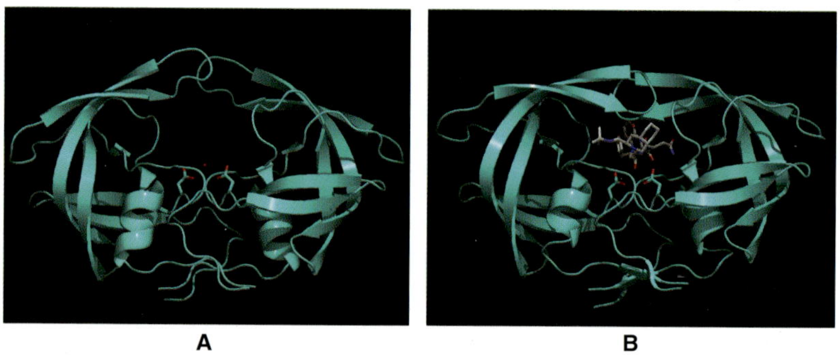

Figure 23.2. A ribbon diagram representation of HIV PR 3-D structure. See pages 751–752 for text discussion.

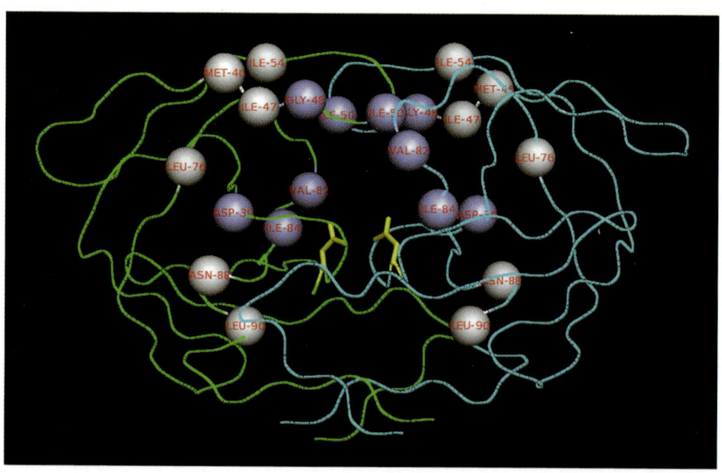

Figure 23.6. A ribbon diagram of HIV PR showing the position of major PI resistant mutations highlighted in bold in Table 23.1. See pages 768–769 for text discussion.

15

TRANSPORTER–XENOBIOTIC INTERACTIONS: AN IMPORTANT ASPECT OF DRUG DEVELOPMENT STUDIES

GANG LUO, RICHARD RIDGEWELL, AND THOMAS GUENTHNER

15.1 INTRODUCTION

Transporters are constitutive transmembrane proteins that actively move small molecules across biological membranes. Those small molecules, known as transporter substrates, include endogenous compounds, xenobiotics, and their metabolites. Transporters can work in both directions, moving substrates into cells (known as influx transport or uptake) or out of cells (known as efflux transport).

Transporters can move substrates against transcellular concentration gradients, either by utilizing energy directly from hydrolysis of adenosine triphosphate (ATP) or by coupling to a secondary active transport system that extracts energy from an electrochemical gradient. The former are known as ATP-binding cassette (ABC) transporters, while the latter are known as secondary active transporters, also called solute carrier transporters (SLC). Facilitative transporters, a specific subtype of SLC transporter, allow solutes to flow downhill with their electrochemical gradient.

Human ABC transporters are members of a large ancient superfamily that has seven currently recognized families (ABCA to ABCG) and 53 members. Readers are referred to the cited website for nomenclature as well as other

Enzyme Inhibition in Drug Discovery and Development, edited by Lu and Li
Copyright © 2010 John Wiley & Sons, Inc.

detailed and updated information (http://www.genenames.org/genefamily/abc.php). Notably, many transporters have multiple aliases (Dean et al., 2001). For example, the most important ABC transporters ABCB1, ABCB11, ABCC1 and 2, and ABCG2 are also known as P-glycoprotein (P-gp), breast cancer resistance protein (BCRP), multidrug resistance-associated protein (MRP) 1 and 2, and bile salt export pump (BSEP), respectively. ABC transporters generally exist as a dimer, consisting of two half-molecules, each containing one ATP-binding domain (Allikmets et al., 1996; Choudhuri and Klaassen, 2006; Hollenstein et al., 2007a,b; Sharom, 2008). However, some ABC transporters, such as BCRP and adrenoleukodystrophy protein (ALDP, encoded by ABCD1), exist only as a single half-molecule monomer (Mao and Unadkat, 2005; Zaher et al., 2006; El-Sheikh et al., 2007; Robey et al., 2007; Velamakanni et al., 2007; Koshiba et al., 2008; Zhou et al., 2008). Unlike cytochrome P450 (CYP) superfamily members, ABC superfamily members share little overall amino acid homology. The greatest degree of homology is seen in the ATP-binding domain, which typically contains characteristic conserved residue motifs (called Walker A and B) separated by nonconserved 90–120 amino acid residues (Allikmets et al., 1996). In terms of their function, all currently known ABC transporters are unidirectional efflux transporters, pumping their substrates out of cells only.

The human SLC superfamily comprises 46 families and approximately 360 members. Readers are referred to the cited website for nomenclature as well as other detailed and updated information (http://www.bioparadigms.org/slc/menu.asp). Again, most of the important SLC members have aliases. The overall sequence homology is also low among SLC families; family members are considered related if they share as much as 20% homology (Zhang et al., 2002; Hediger et al., 2004; Srimaroeng et al., 2008). Like ABC transporters, SLC transporters are related in function, but not in terms of evolution. Unlike the ABC transporters, SLC transporters are smaller (monomers, compared to dimeric ABC transporters) and operate bidirectionally, moving their substrates both into and out of cells.

Because many endogenous compounds are substrates of transporters, and almost all transporters significantly affect the intercellular disposition of endogenous compounds, transporters play many critical roles in human physiology and pathology (Enomoto et al., 2003; Xia et al., 2007; Choi and Song, 2008; Zhou et al., 2008). Transporter dysfunctions are known to cause many diseases, as shown in Table 15.1. For example, defects in 14 ABC transporters are associated with 13 genetic diseases, including cystic fibrosis, Stargardt disease, age-related macular degeneration, adrenoleukodystrophy, Tangier disease, Dubin–Johnson syndrome, and progressive familial intrahepatic cholestasis (Ambudkar et al., 2000; Gottesman and Ambudkar, 2001). Overexpression of urate transporter 1 (URAT1) in male humans may be the major reason that gout (hyperuricemia) occurs more often in men than in women, and more often in humans than in other species (Enomoto et al., 2002, 2003; Enomoto and Endou, 2005).

TABLE 15.1. Inherited Diseases Caused by Genetic Deficiency of Transporters[a]

Disease	Symptom	Defected Transporter	Reference
Adrenoleukodystrophy	Lethal peroxisomal neurodegradation	ALDP (ABCD1)	Takahashi et al. (2007)
Byler disease	Cholestasis, jaundice, and failure to thrive	P-type ATPase	Alissa et al. (2008)
Byler syndrome	Cholestasis, jaundice, and failure to thrive	BSEP (ABCB11)	Alissa et al. (2008)
Congenital hyperinsulinism	Congenital hyperinsulinism	SUR (ABCC8)	Hussain et al. (2008)
Cystic fibrosis	Cystic fibrosis (CF) and congenital bilateral aplasia	CFTR (ABCC7)	Lewis-Jones et al. (2000)
Dubin–Johnson syndrome	Mild jaundice until puberty or adulthood	MRP2 (ABCC2)	Jedlitschky et al. (2006)
MDR3 deficiency	Cholestasis, jaundice, and failure to thrive	MDR3 (ABCB4)	Alissa et al. (2008)
Surfactant deficiency	Fatal respiration disease	ABCA3	Hamvas et al. (2007)
Stargardt's disease	Juvenile macular degeneration	ABCR (ABCA4)	Westerfeld and Mukai (2008)
Tangier disease	Very low levels of high-density lipoprotein, along with an enlarged liver and spleen	ABCA1	Nofer and Remaley (2005)
Wilson disease	Copper accumulation in the liver and brain	ATP7B	Forbes and Cox (1998)

[a]Progressive familial intrahepatic cholestasis includes Byler disease, Byler syndrome, and MDR3 deficiency. Fourteen ABC transporters have been associated with 13 genetic diseases, including cystic fibrosis, Stargardt disease, age-related macular degeneration, adrenoleukodystrophy, Tangier disease, Dubin–Johnson syndrome, and progressive familial intrahepatic cholestasis (Lewis-Jones et al., 2000; Gottesman and Ambudkar, 2001; Nofer and Remaley, 2005; Jedlitschky et al., 2006; Hamvas et al., 2007; Takahashi et al., 2007; Alissa et al., 2008; Hussain et al., 2008; Westerfeld and Mukai, 2008). Wilson disease is associated with ATP7B mutations (Forbes and Cox, 1998).

Transporters also play a critical role in drug disposition. In addition to endogenous compounds, many xenobiotics and their metabolites are substrates of transporters (Ho et al., 2000; Tirona et al., 2003; Mao and Unadkat, 2005; Nozawa et al., 2005; Tahara et al., 2005; Zaher et al., 2006; Behravan and Piquette-Miller, 2007; Luo et al., 2007b; Xia et al., 2007a; Choi and Song, 2008; El-Sheikh et al., 2008; Funk, 2008; Srimaroeng et al., 2008; Tian et al., 2008; Zhou et al., 2008). Furthermore, transporters are highly expressed in intestine, liver, and kidney (Launay-Vacher et al., 2006; Xia et al., 2007a; Terada and Inui, 2007; El-Sheikh et al., 2008; Funk, 2008), which are critical organs for determining xenobiotic disposition in the body. Specifically, transporters have been documented to be major determinants of drug and toxicant absorption, distribution, and elimination.

In this chapter, we will discuss the importance of transporters to drug development research. We will cover aspects of transporter tissue expression and functions, substrates, and inhibitors. In particular, the importance of drug–transporter interactions in the drug discovery and development process will be discussed, including common transporter study models and, in particular, partial or complete deactivation of transporters by chemical inhibition or genetic manipulation, including knockout.

15.2 TISSUE EXPRESSION AND FUNCTIONS

Expression of transporters is practically ubiquitous and can be seen in all tissues, cells, and species. Species differences in drug transporter expression are not extensive in terms of variations in substrate specificity, but are more often seen in terms of overall transporter activity. For example, rats generally demonstrate high levels of biliary transport, compared to humans, of many structurally varied endogenous compounds (e.g., taurocholate), xenobiotics (e.g., methotrexate, pravastatin, and temocaprilat), and metabolites (2,4-dinitrophenyl-*S*-gluthione and estradiol-17β-glucuronide) (Ishizuka et al., 1999; Sasaki et al., 2004). This species difference is likely attributable to a much higher high expression/activity of multiple transporters in the canalicular membrane in rats than in other species (including dogs, monkeys, and humans) as demonstrated by Li et al. (2009) for MRP2/Mrp2.

The expression of transporters can also be induced by many xenobiotics, including therapeutic drugs (e.g., rifampin) and natural products (e.g., St. John's wort extract) (Westphal et al., 2000b; Gurley et al., 2008). It should be noted that induction of drug transporters and metabolic enzymes are mediated by a number of common signal transduction pathways (Geick et al., 2001; Kauffmann et al., 2002; Luo et al., 2002; Sahi et al., 2003; Cerveny et al., 2007). Therefore, co-induction of drug-metabolizing and drug-transporting proteins (e.g., CYP3A4 and P-gp) is frequently observed after exposure to a single xenobiotic compound. The potential for such co-induction must be taken into consideration when assessing or predicting potential drug–drug interactions.

The highest levels of transporter expression are normally seen in tissues and organs that are critical to life and reproduction (such as brain, testis, and placenta), or that play significant roles in drug metabolism and disposition (including intestine, liver, and kidney). In addition, ABC transporters are often highly expressed in tumor cells, exerting a negative effect on patient survival by protecting tumor cells from the cytotoxicity of anticancer drugs (conferring anticancer drug resistance). The expression patterns of transporters in tissues and organs usually reflect the critical role that transporters play in maintaining cell growth and viability. For example, transporters play an important role in the protection of brain, testis, and fetus from the harmful effects of environmental toxins and constitute a critical component of the blood–brain barrier (BBB) and blood–placenta barrier. P-gp is highly expressed in brain endothelial cells and forms an important component of the blood–brain barrier. Laboratory studies with ivermectin demonstrate this protective role. Ivermectin, a neurotoxin that produces central nervous system depression and ataxia, very poorly crosses the BBB of wild-type animals. However, hypersensitivity was found to ivermectin in mice in which the expression of the Mdr1a (also called Mdr3) P-gp gene was disrupted (Schinkel et al., 1994), as well as in certain dogs of the Collie breed whose Mdr1 gene is naturally mutated (Roulet et al., 2003).

In addition, transporters contribute significantly to the maintenance of overall organ function and homeostasis, such as regulation of the enterohepatic circulation of bile acids (Alrefai and Gill, 2007; Pellicoro and Faber, 2007; Kosters and Karpen, 2008). Table 15.1 lists currently known examples of inherited diseases caused by genetically determined deficiencies of ABC superfamily transporter expression and function (Forbes and Cox, 1998; Lewis-Jones et al., 2000; Gottesman and Ambudkar, 2001; Nofer and Remaley, 2005; Jedlitschky et al., 2006; Hamvas et al., 2007; Takahashi et al., 2007; Alissa et al., 2008; Hussain et al., 2008; Westerfeld and Mukai, 2008). Transporters are also important mediators and determinants of the pharmacokinetics of therapeutic drugs affecting multiple aspects of absorption, distribution, metabolism, and elimination. Currently, oral administration is the most common route of drug treatment, requiring absorption from the stomach and small intestine. Notably, P-gp and Peptide transporter 1 (PEPT1) are highly expressed in the small intestine. The former may limit the absorption of certain drugs such as digoxin and taxol by retrograde efflux transport (Westphal et al., 2000a; Breedveld et al., 2006), while the latter can promote the absorption of certain drugs containing a peptide-like moiety, such as valacyclovir, through influx transport (MacDougall and Guglielmo, 2004; Li et al., 2008).

The liver is the most important organ in drug metabolism and contributes significantly to drug clearance through both metabolism and biliary clearance. In liver, many transporters (including both SLC and ABC transporters) are expressed on the sinusoidal membrane and facilitate drug movement in both directions, in and out of liver cells. In addition, the aforementioned multiple ABC transporters are expressed on the canalicular membrane. These transporters are responsible for the biliary excretion of drugs (e.g., methotrexate) and drug metabolites (e.g., estradiol-17β-glucuronide) (Funk, 2008). In addition, transporter activity in the kidney is a major determinant of the renal clearance and disposition of many drugs (Somogyi, 1996; Launay-Vacher et al., 2006). While drug metabolism activity in the kidney is low compared to that of the liver, due to low levels of expression of CYPs in kidney cells, the kidney determines overall plasma clearance by actively secreting or reabsorbing drugs and drug metabolites through transporters expressed on the renal tubular membrane. Although transporters affect drug disposition at multiple points, the impact is often much more significant on distribution to a specific organ than on the overall systemic circulation levels. For example, after intravenous administration of taxol (10 mg/kg) to mice, the brain exposure (AUC_{0-8h}) was approximately 12-fold higher in Mdr1a knockout mice than in wild-type mice, while the plasma exposure (AUC_{0-8h}) was only 50% higher (Kemper et al., 2003). Similarly, the brain and plasma drug levels at 4 h after an intravenous dose of digoxin (1 mg/kg) were 35-fold and 1.9-fold, respectively, higher in Mdr1a knockout mice than in wild-type mice (Schinkel et al., 1995). Likewise, P-gp plays a major role in the efflux of fexofenadine in the small intestine and BBB, but only a limited role in its biliary excretion (Tahara et al., 2005). Xenobiotic-metabolizing enzymes normally play the major role in hepatic clearance of most drugs. However, the primary route of elimination of sulfasalazine in mice is through hepatic clearance by biliary excretion, mediated exclusively by BCRP. This provides an exceptional example of an individual transporter species that can predominantly determine the rate of clearance of a drug (Zaher et al., 2006).

An important clinical phenomena connected in a large part to transporter activity is that of anticancer drug resistance in tumor cells. Efflux transporters, including P-gp, MRP1, and BCRP, are highly expressed on the plasma membranes of many resistant tumor cells, and this overexpression can be induced even further in the presence of anticancer drugs. These transporters actively pump anticancer drugs such as taxol and mitoxantrone out of tumor cancer cells. Therefore, efflux transporters can protect tumor cells from cytotoxicity, the intended drug effect (Gottesman and Ambudkar, 2001). Finally, the level of transporter expression may provide a pharmacokinetic basis for species and gender differences, as exemplified by URAT1. Uric acid, the final product of human purine metabolism, is poorly soluble at pH 7.4 and exists in extracellular fluid predominantly as monosodium urate. Human cells do not contain uricase, the enzyme that catalyzes the conversion of urate to allantoin, which is more water-soluble and therefore easily eliminated from the body by renal excretion. Furthermore, in humans, and particularly in men, URAT1 is highly expressed on the apical membrane of renal proximal tubule epithelial cells and mediates the reabsorption of most of the filtered urate back into the peritubular circulation, contributing to systemic hyperuricemia and gout. The high levels of this urate-specific transporter in men may explain the inordinately high rate of occurrence of gout in male humans (Enomoto et al., 2002, 2003; Enomoto and Endou, 2005).

The tissue expression and function of transporters results in functional collaboration not only among different transporters, but between transporters and metabolic enzymes as well. There are hundreds of known transporters in humans, with both overlapping and complementary substrate specificities and activities. We currently view these transporters as great "team players," although in many casess a single transporter often plays a predominant role. Collaboration among transporters can be seen from their tissue expression patterns, membrane topography and spatial juxtaposition, overlap in substrate specificity (as well as in response to xenobiotic chemical inducers), and cooperative versatility. For example, enterohepatic circulation of bile acids is critical for nutrition and maintenance of homeostasis. Bile acids synthesized in hepatocytes are excreted into bile by BSEP expressed on the canalicular membrane, reabsorbed from the intestine into the systemic circulation with the help of an apical sodium-dependent bile acid transporter expressed in ileum (IBAT), and finally taken up the into liver by the sodium-dependent taurocholate cotransporting polypeptide (NTCP) expressed on the sinusoidal membrane of hepatocytes (Kosters and Karpen, 2008). Similarly, there are many known cases of transporters and metabolic enzymes that share both substrate and inducer specificity and also play their own cooperative and complementary roles in the overall physiological process of xenobiotic disposition. For example, in the case of phase I metabolism, dozens or hundreds of endogenous chemicals and xenobiotics are common substrates and/or inducers of both P-gp and CYP3A4 (Zhou, 2008; Zhou et al., 2008). In the case of phase II metabolism, many conjugates generated by Phase II metabolic enzymes are subsequently excreted into the bile by MRP2, or pumped into the bloodstream by other MRP family members expressed on hepatocyte sinusoidal membranes (Chu et al., 2004; Matsushima et al., 2005).

15.3 SUBSTRATES AND INHIBITORS

Table 15.2 lists compounds that have been useful as model substrates and inhibitors for transporters that play critical roles in drug disposition (Xia et al., 2005, 2007b; Zhou et al., 2008). Because of the versatility and broad substrate specificity of the transporters, inclusion in this chapter of an exhaustive list of all substrates and inhibitors for each transporter is not practical. Instead, it is more important to discuss the characteristics of transporter substrates and inhibitors that have potential uses in the drug development process, concentrating primarily on their substrate spectra, structural requirements, and species differences.

In general, the substrate requirements of most transporters are quite broad and nonspecific. This is easy to rationalize in light of the generalization that transporter superfamilies are functionally similar to members of individual drug metabolizing enzyme superfamilies, in that a limited number of related isoforms with similar catalytic properties have the capacity to dispose of a theoretically unlimited number of chemically diverse xenobiotics. For example, substrates for MRP2 include structurally diverse endogenous compounds (e.g., leukotriene C4), xenobiotics (e.g., methotrexate), and Phase II conjugates of lipophilic substances (e.g., bilirubin glucuronides, estrone 3-sulfate, and *S*-glutathionyl 2,4-dinitrobenzene) (Xia et al., 2005, 2007b; Zhou et al., 2008). The substrate binding affinities and transport capacities of the transporters themselves are largely substrate-dependent. Human MRP2 has an extremely high affinity for its specific endogenous substrates, including

TABLE 15.2. Basic Information for Critical Transporters in Humans

Gene	Aliases	Tissue Expression	Substrate	Inhibitor
ABCB1	P-gp, MDR1	Intestine, liver, kidney, brain, testes, placenta	Digoxin	Elacridar, LY335979
ABCB11	BSEP	Liver	Taurocholate	Bosentan
ABCC1	MRP1	Intestine, liver, kidney, brain	Adefovir	
ABCC2	MRP2	Liver, intestine, kidney, brain	Leukotriene C4	MK-571
ABCG2	BCRP	Liver, intestine, breast, placenta	Mitoxantrone	Ko134, elacridar
SLC10A1	NCTP	Liver, pancreas	Taurocholate	
SLC10A2	IBAT, ISBT	Intestine, kidney, biliary tract	Taurocholate	
SLC15A1	PEPT1	Intestine, kidney	Glycylsarcosine	
SLC15A2	PEPT2	Kidney	Glycylsarcosine	
SLCO1B1	OATP-C	Liver	Rifampin	Cyclosporin
SLCO2B1	OATP-1B3	Liver	Digoxin	
SLCO2B1	OATP-B	Intestine, liver, kidney, brain	Pravastatin	
SLC22A1	OCT-1	Liver	Acyclovir	Ritonavir
SLC22A4	OCTN1	Kidney, skeleton muscle, placenta	Carnitine	
SLC22A5	OCTN2	Kidney, skeleton muscle, placenta, intestine, liver	Carnitine	Grepafloxacin
SLC22A6	OAT1	Kidney, brain	*p*-Aminohippurate	Probenecid
SLC22A7	OAT2	Liver, kidney	Zidovudine	

bilirubin glucuronides and leukotriene C4, with a K_m of approximately 1 μM, but it has a quite low binding affinity to some xenobiotic substrates such as methotrexate (K_m ~480 μM) and *p*-aminohippurate (K_m ~1 mM) (El-Sheikh et al., 2007; Zhou et al., 2008). BCRP is active in the transport of a number of glucuronide, glutathione, and sulfate conjugates. However, BCRP shows some degree of substrate selectivity, because its K_m for sulfate conjugates such as estrone-3-sulfate is approximately 10 μM, while its K_m's for glucuronide and glutathione conjugates are much higher (Chen et al., 2003; Imai et al., 2003; Suzuki et al., 2003; Mao and Unadkat, 2005). The same substrate can have very different binding affinities to different members of the same transporter family. For instance, PEPT1, which is predominantly expressed in intestine, has in general a low binding affinity (high K_m, up to 30 mM) but a high catalytic capacity for most of its substrates (Brandsch et al., 2004). On the other hand, PEPT2, which is mainly expressed in kidney, exhibits a higher binding affinity and lower catalytic capacity for most of its substrates (Inui et al., 2000). Differences in substrate binding affinities to important transporters should help us understand, predict, and even prevent transporter-based drug–drug interactions. For example, both methotrexate and the glucuronide conjugate of DPC 333 are substrates of MRP2, with differing binding affinities; methotrexate has a relatively high K_m, while the K_m for DPC 333 glucuronide is much lower. As a consequence of this interaction, metabolism of DPC 333 produces a glucuronide conjugate that potently inhibits the biliary excretion of methotrexate mediated by canalicular MRP2. This interaction decreases the overall elimination of methotrexate *in vivo* with the potential for higher drug toxicity (Luo et al., 2007b).

The structural diversity of most transporter substrates and inhibitors is so broad that it is usually difficult to specifically define their structural requirements. However, some general structural features required by specific transporters for their substrates have been revealed. For example, most P-gp substrates contain a nitrogen group, aromatic moieties, planar domains, and molecular size larger than 300 Da, are positively charged at physiological pH, and are amphiphilic and lipophilic (Xia et al., 2007b). Substrates for PEPT1 generally contain a peptide or peptide-like moiety and include dipeptides, tripeptides, or peptidomimetic drugs, but not tetrapeptides (Terada et al., 2000; Vig et al., 2006). Notably, not all dipeptides are PEPT1 substrates. Its binding affinity to dipeptides is influenced by charge, hydrophobicity, molecular size, and side-chain flexibility. It should be noted that the current understanding of the key features for substrates and inhibitors is still not at a highly advanced state, as demonstrated by the continued inability of studies using experimental pharmacophores to accurately predict transporter substrate preferences *in vivo* (Chang et al., 2006). Furthermore, substrate and inhibitor specificities overlap to a great degree among many transporters, and overlap between many transporters and xenobiotic-metabolizing enzymes is also quite common. Examples are pravastatin, which is a substrate of P-gp, MRP family members, BCRP and organic anion transporters (OATs) (Matsushima et al., 2005), and GF120918, which is a potent inhibitor of both P-gp and BCRP (Xia et al., 2005). Finally, species differences exist not only in transporter expression levels, as discussed, but also in their substrate binding affinities. For example, the transcellular transport across monolayers of cells transfected with MDR1 from human, monkey, canine, rat (Mdr3 and Mdr1), and mouse (Mdr1a and Mdr1b) transfected cells demonstrated up to a 16.5-fold difference in apparent K_m for diltiazem (Katoh et al., 2006).

15.4 EXPERIMENTAL MODELS FOR STUDYING TRANSPORTERS IN DRUG DEVELOPMENT

The study of the effects of transporters on the disposition of drug candidates—and, conversely, the study of the effects of drug candidates on transporter activity—should be considered as a critical component of any drug development process, as part of the routine ADME (absorption, distribution, metabolism, and elimination) assessment. Drug–transporter interactions are investigated to determine the roles of transporters in the distribution, elimination, and toxicity of drug candidates, to determine whether a compound of interest is a substrate and/or inhibitor of a specific transporter, to examine transporter-based drug–drug interactions, and often to contribute to an understanding of the mechanism of an observed toxicity or a drug–drug interaction. These studies usually employ a combination of *in vivo*, *in situ*, and *in vitro* models. The selection of appropriate experimental models will obviously be dependent on the purpose of the study, the availability and costs of transporter models, and the required study timeline.

15.4.1 *In Vivo* Models

In vivo studies in animals as well as in human subjects are generally superior to *in vitro* or *in situ* models in (a) providing direct evidence that a transporter is important in overall drug disposition and toxicity and (b) investigating potential drug–drug interactions in an intact multiorgan system. More importantly, results obtained *in vivo* should be more immediately relevant and applicable to the clinical pharmacology of the candidate molecule. However, caution is necessary when interpreting and applying data obtained in whole animal studies, because effects on disposition that appear to be linked to a single transporter may, in fact, involve multiple transporters and secondary mechanisms, and pharmacokinetically important species differences may also exist and cannot be ignored. For example, a single compound, namely methotrexate, is a known substrate of multiple transporters including reduced folate carrier 1, MRP2 and MRP3, OATs, and folate binding protein 1 (Saito et al., 1996; Masuda et al., 1997; Uwai et al., 1998; Hirohashi et al., 1999; Kool et al., 1999; Borst et al., 2000; Uwai et al., 2000; Van Aubel et al., 2000; Birn et al., 2005) as well as metabolic enzymes (Rajgopal et al., 2001), and a single drug such as ritonavir can inhibit (and even induce) many transporters and enzymes (Luo et al., 2002, 2003). Species comparison studies are often conducted using both normal (wild-type) and animals that are transporter-deficient, as a result of gene knockout, naturally genetic deficiency, or chemical knockout (see below). In humans, special attention should be paid to genetically determined deficiencies in important transporters (listed in Table 15.1). Such patients may be hypersensitive to certain drugs compared to normal subjects.

15.4.2 *In Situ* Models

Isolated organ perfusion provides a test system that closely mimics whole-body physiology, with certain practical advantages. In nondrug-development studies, this model has been frequently used to demonstrate the role of transporters in drug disposition and to examine the mechanisms of observed drug–drug interactions. Isolated organ perfusion permits close control of experimental conditions, the effects of the organ of interest are observed directly, and data interpretation is usually relatively straightforward. Furthermore, expensive reagents can be used relatively sparingly and efficiently. The most often utilized perfused organs are liver, kidney, intestine, and brain, usually from rats and mice. As is the case for *in vivo* whole-animal studies, an experimental comparison between perfused organs from wild-type and specific transporter-deficient animals is often extremely instructive (Luo et al., 2007b; Tian et al., 2008). Such transporter deficiency could be a result of gene knockout, natural genetic deficiency, or chemical inhibition. However, the use of isolated organ perfusion in transporter studies has limitations as well. Experimental procedures and preparations are usually relatively complex and technically difficult, throughput is normally relatively low, and training requirements are often demanding. In addition, the organ donors are usually rats and mice because of cost considerations, although other animals such as dog and monkey could theoretically be used in exceptional circumstances. In addition, data interpretation and extrapolation may be complicated by species differences in expression levels and substrate specificity of transporters and coexisting xenobiotic-metabolizing enzymes.

15.4.3 *In Vitro* Models

For addressing specific mechanism-focused questions, *in vitro* transporter models are often more direct and less complicated than whole-animal or whole-organ models. Standard *in vitro* models include cell-based assays, using freshly isolated intact cells or established cell lines, isolated membrane fraction-based assays (including ATPase assays, membrane vesicular uptake, and radio-ligand binding assays), and assays utilizing oocytes and yeast.

Established cell lines have been extensively used in transporter studies. The most commonly used cell lines include immortalized human colorectal carcinoma-derived cell line (Caco-2), MDR human KB carcinoma (KB-V-1) cells, immortalized Madine–Darby canine kidney (MDCK) cells, Chinese hamster ovary (CHO) cells, human embryonic kidney (HEK) 293 cells, porcine proximal tubular epithelial (LLK-PK1) cells, and MCF-7 cells, derived from human metastatic breast cancer. Caco-2 cells, which in 21-day cultures naturally express high levels of P-gp on the apical membrane, are used directly as monolayers in transwells to determine the polarized permeability ratio of basolateral to apical and apical to basolateral transport, usually in the absence and presence of a selective P-gp inhibitor. P-gp has been thoroughly studied in cultured cell models and has provided the most consistent data among different laboratories. Studies of other related transporters, including MRP2, BCRP, and PEPT1, are sparser with varying results. Human epidermoid carcinoma KB cells (KB-3-1) express little (if any) P-gp and are sensitive to chemotherapeutic agents such as doxorubicin, daunorubicin, mitoxantrone, vinblastine, paclitaxel, and colchicine. However, KB-V-1 cells, derived from the parental KB-3-1 cells in the presence of vinblastine at a low concentration, express a much higher level of P-gp (Galmarini, 2002). KB-V-1 cells are currently used as a standard drug development model to determine the pharmacokinetic contribution of P-gp to the cytotoxicity of candidate anticancer drugs in the absence and presence of a P-gp inhibitor (Galmarini, 2002). MDCK cells, which spontaneously express P-gp at a very low level, are often used as host cells for transfection with transporter genes of interest. While MDCK cells are probably the most commonly used transfected cell in transport studies, other mammalian cell lines including CHO, HEK293, LLC-PK1, and MCF-7 also provide useful models (Xia et al., 2007a). After MDCK cells, Caco-2 is probably the most frequently used transfection host cell line. Important differences between Caco-2 and MDCK cells are that MDCK cells do not require 21 days of culture before they are used; and because MDCK cells are naturally P-gp nonexpressing, they can be transfected with a sham vector and used as a direct control. As a practical point for transport studies, Caco-2 cell monolayers usually have tighter junctions than MDCK cells.

The main utility of transfected cells as an experimental model is that interactions with a single transporter of interest in the disposition of test compound can be studied in relative isolation. Therefore, this test system is excellent for mechanism-based pharmacokinetic determinations. Both transiently and stably transfected cells are useful experimental models. Although approximately one month is required to establish stable transfectants and select for expressing colonies (usually by antibiotics Geneticin, also known as G418), stably transfected cell lines eventually provide much higher throughput and experimental reproducibility.

Transfected cells are good model systems for studying SLC-dependent drug transport. For example, CHO cells transfected with hOAT1 showed efficient uptake of adeforvia and cidofovir and exhibited a 100-fold higher intracellular accumulation of drug and/or metabolites than did nontransfected CHO cells. Consequently, adefovir and cidofovir were approximately 400-fold more cytotoxic to the hOAT1 transfected CHO cells compared to nontransfected cells. On the other hand, the prodrug form of cidofovir, which is a poor substrate for hOAT1, produced greatly reduced nephrotoxicity compared to the active drug *in vivo*, and it only marginally increased cytotoxicity *in vitro* to hOAT1 transfected CHO cells (Ho et al., 2000). Another example of metabolism-dependent uptake by a specific transporter is provided by a study where HEK293 cells stably transfected with organic anion transporter polypeptide (OATP) 1B1 were able to accumulate 7-ethyl-10-hydroxycamptothecin (SN-38), the active metabolite of irinotecan, but not the parent compound or its glucuronide conjugate metabolite. Because OATP1 is predominantly expressed on the sinusoidal membrane of hepatocytes, this study concludes that OATP1B1 is important for the specific hepatic uptake of SN-38 (Nozawa et al., 2005).

Because the liver has a central role in drug disposition, the active transport of test compounds by liver cells is often measured using freshly isolated hepatocytes. However, biliary excretion cannot be directly examined using suspended hepatocytes because of the loss of the canalicular network and the internalization of transporters expressed on canalicular membranes during the isolation and preparation of hepatocytes. In the last 10 years, the B-Clear hepatocyte model has been developed to overcome this problem. In this model, primary cultures of freshly isolated hepatocytes are cultured in a sandwich format permitting a canalicular network to be formed on the hepatocyte monolayer. A compound that is a substrate for hepatocyte transporters can be taken up and excreted into the bile pocket. Because the canalicular network disappears when the surrounding buffer is made calcium-free, the difference between the total uptake of test compound by hepatocytes in calcium-containing buffer and in calcium-free buffer represents the biliary excretion of test compound. Currently, this model has been validated in hepatocytes from rats, dogs, monkeys, and humans. Most results obtained so far using the B-Clear model are consistent with those seen in *in vivo* studies (Liu et al., 1999a,b). The B-Clear model allows comparison of transport-dependent hepatocyte uptake between different species (notably, between human and rat or dog or monkey hepatocytes). Also the biliary excretion of multiple interactive compounds can be tested *in vitro* in a single experiment.

Another useful category of *in vitro* models for the measurement of drug–transporter interaction is the isolated membrane-based transporter assay. Currently useful models include the ATPase assay, membrane vesicle uptake, and radioligand-binding assays. The ATPase assay provides rapid and high throughput for screening substrates and/or inhibitors of ABC transporters. An ABC transporter contains one or two ATP binding sites and hydrolyzes ATP when a substrate binds to and activates it. The ABC transporter-associated ATPase is vanadate-sensitive and magnesium-dependent. Based on this principle, a simple and efficient ATPase assay has been devised, which measures the amount of inorganic phosphate liberated in reactions containing an ABC transporter preparation and a test compound, in the presence and absence of vanadate. For example, verapamil, a substrate of P-gp, stimulates magnesium-dependent and vanadate-sensitive membrane ATPase activity in preparations containing P-gp enriched cell membranes (Sarkadi et al., 1992). ABC transporter-enriched cell membrane preparations isolated from baculovirus-infected insect cells or from stably transfected cell lines are commercially available from BD Gentest (Woburn, Massachusetts, U.S.) and Solvo Biotechnology (Budapest, Hungary). However, this method is only valid for ABC transporters because SLC transporters do not have ATPase activity.

Because the isolation of membrane vesicles from intact cells requires some degree of specialized skill and training, the use of membrane vesicles for rapid high-throughput screening of drug candidate–transporter activity has not been extensive. However, isolated membrane vesicles provide an excellent model for the study of molecular aspects of ABC transporter-mediated efflux. For instance, rat liver canalicular membrane vesicles (CMV) have been used to demonstrate the active uptake of daunomycin, daunorubicin, and vinblastine, whose biliary excretion is extensive (Kamimoto et al., 1989; Sinicrope et al., 1992; Bohme et al., 1994; Kwon et al., 1996) and is likely primarily mediated by P-gp. Plasma membrane vesicles purified from KB-V-1 cells have been used to study P-gp-dependent efflux of vinblastine from cancer cells (Horio et al., 1988). Brush border membrane vesicles (BBMV) prepared from rat intestine were used to show expression of P-gp on the apical membrane, but not on the basolateral membrane (Hsing et al., 1992) and were used to demonstrate and differences in P-gp expression in ileum, jejunum, and duodenum (Makhey et al., 1998). Renal BBMV were used to examine the effect of P-gp on its substrate disposition in the kidney (Dutt et al., 1994).

Radiolabeled ligand-binding assays, including photoaffinity assay and scintillation proximity assay (SPA), are reliable and well-characterized methods for the determination of binding affinity of a substrate to transporters of interest. Photoaffinity assays have been used to characterize molecular features of the binding sites of ABC transporters including P-gp (using [^{125}I]-iodoary-azidoprazosin) (Limtrakul et al., 2007), MRP2 (using [^{3}H]-leukotriene C4) (Jedlitschky and Keppler, 2002), and BCRP (using [^{125}I]-iodoary-azidoprazosin) (Limtrakul et al., 2007). Photoaffinity assays are time-consuming and labor-intensive and are therefore suitable for transporter characterization rather than high-throughput screening of drug–transporter interactions.

The SPA assay provides an excellent system for high-throughput screening of interactions between transporters and their substrates and inhibitors. There are two different SPA assays. In the first type of SPA assay, cell membrane fractions containing a transporter of interest are preincubated and covalently bound to microscopic beads which contain a scintillant that can be stimulated to emit light when a radiolabeled transporter substrate ligand is added. The beads produce measurable light in proportion to the binding of ligand to the transporter. Displacement of the standard radioligand by a test compound produces a proportional decrease in light emission. SPA has been successfully used to screen compounds that interact with P-gp modulators, using [^{3}H]-vinblastine as a standard probe substrate in P-gp-enriched cell membrane fractions prepared from KB-V-1 cells. However, the use of SPA to study interactions with other transporter species has not been developed yet. A second type of SPA assay has been developed to measure interactions of small molecules with uptake transporters. Cells are transfected with an uptake transporter and are grown in microplate wells that are impregnated with scintillant. Transporter-dependent cellular uptake of a standard probe radioligand that is incubated with cells brings it into proximity with the scintillant in the plate base and produces measurable light emissions. The interactions of test molecules with these transporters will decrease uptake of the standard probe and proportionally decrease light emissions. This method saves the time required to isolate and prepare transporter-enriched cell membrane fractions (Bonge et al., 2000; Lohmann et al., 2007).

Xenopus laevis oocytes provide an *in vitro* model for transporter study that may be useful in some special circumstances. These oocytes are approximately

1.0 mm in diameter and are precursors to mature eggs of the African clawed frog. Defolliculated oocytes are micro-injected with complementary RNA (cRNA) that is synthesized from the cDNA of a transporter gene of interest and cultured for 3 days so that the transporter encoded by cRNA is expressed and functions. For instance, *Xenopus* oocytes transfected with OATP-C and its three common variants (polymorphisms) were able to take up SN-38, pravastatin, estrone-3-sulfate, and estradiol-17-glucuronide differently (Nozawa et al., 2005). This model provides a useful method for measuring uptake of a test compound mediated by SLC transporters, but has not been demonstrated to be useful for measuring efflux mediated by ABC transporters. In addition its limited usefulness, there are technical difficulties inherent in this assay. Variations among experiments are large, partially due to seasonal variations in oocyte properties.

Finally, both efflux and uptake transporters have been functionally expressed in yeast (Ruetz et al., 1993; Osato et al., 2003). Yeasts that are manipulated to express mammalian transporters can be used to measure transporter-dependent drug uptake, either by measuring the content of the test compound in centrifugal filtrates of transfected yeast incubated with the test compound or by measuring cytotoxicity, expressed by yeast growth in the presence of the compound, as a reflection of cellular uptake. Also, transporter proteins can be produced in large quantities by transfected yeast, which can serve as a useful source of membrane vesicles. The advantages of yeast-based systems include simple genetic manipulation, a low level of background expression in controls, and high levels of production of transporter proteins. However, not every mammalian transporter can be expressed in yeast expression systems because of the presence of codons at the N-terminus of mammalian transporters that are not expressible in yeast and also because of occasional inaccurate post-translational modifications of mammalian proteins in yeast hosts (Sawamiphak et al., 2005).

15.5 EXPERIMENTAL MANIPULATION OF TRANSPORTER ACTIVITY

Design of optimally informative studies using the model transport systems discussed above usually requires some form of experimental manipulation of transporter activity. Such manipulations are required to distinguish (a) affinities of drug candidates for multiple transport systems and (b) the contributions of individual transporters to overall drug transport and disposition. Studies of metabolism of drug candidates have typically employed up-regulation (induction or transfection) or down-regulation (inhibition or blockade of expression) of individual metabolic enzymes to establish their contributions to the overall metabolism of the proposed drug. In a similar fashion, transporter studies have, in the past, relied on induction or inhibition of individual transporters to determine specific transport interactions. More recently, knockout animals, which are bred to lack expression of one or more constitutive transporter proteins, have found increasing usefulness in the demonstration or verification of specific individual transporter activity toward a drug or drug candidate. The use of induction, inhibition, and knockout of individual transporter species will be discussed in light of their applicability to drug development studies.

15.5.1 Induction of Transporter Activity

Chemical induction of individual transporter species can be useful in the demonstration of overall transporter effects, but does not offer satisfactory precision as a means of examining a single transporter species in isolation. Many transporters do not respond to induction by any known inducer compound; and those that do are usually not induced in isolation, but rather as a consequence of activation of broad-scale chemical induction pathways. As a result, expression of multiple transporters, as well as multiple drug-metabolizing enzymes, can be simultaneously induced, complicating interpretation of experimental results. For example, the PXR ligands rifampin and pregnenolone 16α-carbonitrile induce human MRP2 (Kauffmann et al., 2002) and rodent Mrp2 (Johnson and Klaassen, 2002), respectively. However, interpretation of results obtained with animals pretreated by PXR ligands may be greatly complicated by the fact that these compounds simultaneously greatly increase hepatic clearance of many therapeutic drugs by inducing the expression of highly nonspecific monooxygenase CYP3A4 (Luo et al., 2007a). The complex pattern of chemical induction of Mrp transporters provides an example of extreme complexity within a single transporter superfamily. The toxicity of acetaminophen is partially mitigated by the fact that it induces several members of the Mrp superfamily in liver, which decreases its net concentration in hepatocytes (Aleksunes et al., 2008). However, its induction of Mrp3 and Mrp4 appears to be mediated by NFE-Related Factor 2 (Nrf2), which also mediates the induction of other cytoprotective enzymes such as NADPH quinone oxidoreductase and glutamylcysteine ligase, while the induction of Mrp1 and Mrp2 in liver by acetaminophen appears to be Nrf2-independent (Aleksunes et al., 2008). Inducers that stimulate multifactorial enzyme induction via constitutive androstane receptor (CAR), such as phenobarbital and trans-stilbene oxide, also induce Mrp2, Mrp3, Mrp4, and Mrp7, but not Mrp5 and Mrp6, in male mice (Slitt et al., 2006). Aromatic hydrocarbon receptor (AhR) ligands, such as TCDD, induce another combination of Mrp family members, namely, Mrp2, Mrp3, Mrp5, and Mrp6 (Maher et al., 2005). The lesson for drug development studies is that an overall picture of drug-transport interaction can be identified by using chemical induction as a means of experimentally altering transporter activity only on a fairly global level. However, a simplistic analysis of the interaction of a compound of interest with a specific transporter, or even with a smaller subset of transporters, may not be possible using simple chemical induction as a means of altering transporter activity. A more precise analysis may be possible by using down-regulation, rather than up-regulation of individual transporter species as an experimental approach.

15.5.2 Inhibition of Transport Activity

The addition of a chemical inhibitor of transporter activity to an experimental model offers a simple method of down-regulating transporter activity. However, as is the case with using chemical induction to manipulate transporter activity, a simple picture of global transport effects is often possible, but a detailed analysis of the interaction of drug candidates with individual transporter species is probably not obtainable. Historically, inhibition of transport proteins has been a useful therapeutic modality. In fact, the clinical introduction of probenecid, probably the first example of an effective transport inhibitor drug, predates the actual discovery of transport proteins and therefore predates a mechanistic understanding of that drug's effect as well. Probenecid was initially used to enable more efficient use of penicillin by inhibiting its renal clearance (Somogyi, 1996). Penicillin was a "wonder drug" for prevention and cure of infection among wounded soldiers and the spread of infectious disease among civilians due to destruction of public hygiene infrastructure during the World War II. However, its supply was a problem because the drug was difficult and expensive to manufacture at that time, and even large doses were extremely therapeutically inefficient, due to the drug's extremely rapid and efficient renal clearance, which involves active secretion through renal transporters resulting in a high clearance. Concomitant administration of probenecid with penicillin resulted in more efficient dosing of penicillin, by inhibiting its active secretion by renal tubular transporters, thus permitting the same clinical outcome with much smaller doses (Somogyi, 1996). Subsequently, probenecid also became a very useful drug for the treatment of gout or hyperuricemia, which it does by inhibiting the organic anion transporter (OAT)-mediated reabsorption of uric acid in kidneys and thus reducing uric acid levels in plasma (Silverman et al., 2008).

As discussed above, inhibition by one drug of transporter-dependent movement of a second drug is a major source of drug–drug interactions. The study of these interactions has provided a large database of compounds that can act as inhibitors of transport activity, and many of these compounds have been used to experimentally manipulate overall and individual transporter activity. A recent review of transporter-related drug–drug interactions (Xia et al., 2007a) contains a 12-page table that exhaustively describes such potential inhibitors of transport activity in drug development studies. However, as is the case with using chemical inducers to modulate transport activity, chemical inducers are often very nonspecific, toward both multiple types of transport proteins but also toward other metabolic enzymes as well. For example, most transporters are broadly substrate-specific for hydrophobic or amphipathic anionic and cationic small molecules. These same types of compounds are good substrates for nonspecific phase 1 and 2 enzymes such as CYP3A4. The complications that arise when a compound used to inhibit a specific transporter in fact inhibits multiple transporters (such as probenecid or cyclosporine A does), or inhibits one or more metabolic enzymes as well (such as erythromycin or verapamil does), will usually preclude the use of this strategy to specifically identify or characterize individual transporter interactions. Nevertheless, broad-scale transporter inhibition by compounds such as verapamil, which inhibits most Mdr and Mrp transporters, can often give a good picture of the overall transporter-related disposition of a drug candidate.

15.5.3 Knockout of Transporter Genes

A more selective and informative strategy for examining the contribution of individual transporters to the disposition of a candidate molecule involves selective repression of transporter gene expression, either by employing naturally deficient genetic mutants or through the development of knockout animal models. Selective chemical repression of transporter gene expression is not a common experimental approach, although moderate inhibition of expression of the adenosine transporter Ent1 at the blood–brain barrier is seen following pretreatment of mice with ethanol (Choi et al., 2004). A more selectively directed chemical method for knocking down transporter activity might involve the use of siRNA technology to target expression of specific or closely related transporter genes (Yu, 2007). However, a more fully developed experimental approach, with dozens of existing genetically altered strains available, is the knockout mouse model, where expression of specific transporter subtypes or individual gene products has been deleted in a targeted fashion. A recent review by Klaassen and Lu (2008) lists 29 currently available rodent transporter knockout models and two currently available naturally transporter-deficient mutant strains. Table 15.3 is an adaptation of a table from that review. To date, the most extensive number of studies using transporter knockout to establish and characterize drug interactions with specific transporter species has been directed toward members of the ABC superfamily (Mdr and Mrp family members), with a more limited number of knockout models for members of the SLC superfamily, including several Oat and organic cation transporters (Octs) (Klaassen and Lu, 2008).

The first animal model in which a specific target gene was knocked out was the Mdr1-null mouse, developed in the mid-1990s. Knockout of Mdr1a and 1b genes produced mice whose intestinal and blood–brain barriers lost the ability to protect against a number of cytotoxic xenobiotics (Schinkel et al., 1995, 1997). The same knockout strains were used to demonstrate a loss of protection at the placental barrier against xenobiotic uptake in pregnant Mdr1-null mice (Smit et al., 1999). In addition to Mdr1 null mice, knockout models for other members of the ABC transporter superfamily have also been subsequently developed. Mrp2- and Mrp3-null mice have been used to establish the central role of these transporters in biliary uptake and excretion of sulfate, glucuronide, and glutathione conjugated metabolites and in the

TABLE 15.3. Available Transporter-Deficient Rodent Models

Gene Product *(Gene Symbol)*	Rodent Model	Gene Product *(Gene Symbol)*	Rodent Model	Gene Product *(Gene Symbol)*	Rodent Model
Abca1 *(Abca1)*	KO	Mdr1a *(Abca1)*	MM	Oat1 *(Slc22a6)*	KO
Abcg5/g8 *(Abcg5/g8)*	KO	Mdr2 *(Abcb4)*	KO	Oat3 *(Slc22a8)*	KO
Asbt *(Slc10a2)*	KO	Mrp1 *(Abcc1)*	KO	Oatp1a1 *(Slco1a1)*	KO
Atp7b *(Atp7b)*	KO	Mrp2 *(Abcc2)*	KO	Oatp1a4 *(Slco1a4)*	KO
Atp8b1 *(Atp8b1)*	KO	Mrp2 *(Abcc2)*	MR	Oatp1b2 *(Slco1b2)*	KO
Bcrp *(Abcg2)*	KO	Mrp3 *(Abcc3)*	KO	Oct1 *(Slc22a1)*	KO
Bsep *(Abcb11)*	KO	Mrp4 *(Abcc4)*	KO	Oct2 *(Slc22a2)*	KO
Dmt1 *(Slc11a2)*	MR	Mrp5 *(Abcc5)*	KO	Octn2 *(Slc22a5)*	KO
Ent1 *(Slc29a1)*	KO	Mrp6 *(Abcc6)*	KO	Pept2 *(Slc15a2)*	KO
Mct8 *(Slc16a2)*	KO	Npc1L1 *(Npc1l1)*	KO	Znt1 *(Slc30a1)*	KO
Mdr1 *(Abcb1)*	KO	Npt2 *(Slc34a1)*	KO	Znt3 *(Slc30a3)*	KO

Abbreviations: KO, knockout mouse; MM, mutant mouse; MR, mutant rat.
References are: Umbenhauer et al. (1997) for Mdr1a mutant mouse, Muller et al. (1996) for Mrp2 mutant TR$^-$ rat, Ito et al. (1997) for Mrp2 mutant Eisai hyperbilirubinemic rat, Knopfel et al. (2005) for divalent metal transporter1 (Dmt1) mutant Belgrade rat, and Klaassen and Lu (2008) for all others.
Source: Adapted from Klaassen and Lu (2008).

reuptake of bile acids *in vivo* (Belinsky et al., 2005; Manautou et al., 2005; Zelcer et al., 2005; Lagas et al., 2006; Vlaming et al., 2006). Although Mrp4 is expressed primarily in the kidney, it also plays a role in hepatic transport of bile acids, because livers from Mrp4-null mice showed marked accumulation of bile acids after bile duct ligation (Mennone et al., 2006). The role of Mrp6 in transporting small peptides in hepatocyte membranes has been investigated in Mrp6-knockout mice (Klement et al., 2005). The targeted knockout of Abcg family members has also been used to examine the role of these transporters in transport of xenobiotics at the intestinal and blood–brain barriers. Abcg5- and Abcg8-null mice show elevated blood and brain levels of dietary phytosterols (Jansen et al., 2006). Bcrp (Abcg2)-null mice are highly sensitive to phototoxicity caused by accumulation of breakdown products of dietary chlorophyll (Jonker et al., 2002). In addition, this transporter is highly expressed in rodent kidney, and Bcrp-null mice exhibit decreased urinary clearance of a number of xenobiotic sulfate metabolites (Mizuno et al., 2004). Finally, Abca1 knockout mice have been used to study (a) the role of this transporter in intestinal and hepatic absorption of cholesterol and (b) the potential effects of drugs upon this specific uptake pathway. Abca1-null mice show a marked decrease in the efflux of cholesterol from hepatocytes into blood, along with an overall decreased blood level of circulating cholesterol and HDL lipoproteins (Drobnik et al., 2001; Timmins et al., 2005).

Knockout of a number of Slc superfamily transporters has been used to generate models for the study of the role of these proteins in renal secretion and clearance of xenobiotics. Oct1 and Oct2 are found on the basolateral proximal tubule membrane where they transport cations into the tubular fluid. Knockout of either Oct1 or Oct2 alone has a minimal effect on cation clearance, but simultaneous knockout of both completely abolished tetraethylammonium secretion (Jonker et al., 2003). Oct1 is also expressed on hepatocyte basolateral membranes, and Oct1-null mice show decreased hepatic uptake of cationic compounds, including some anticancer drugs and toxins (Jonker et al., 2001). Knockout animals deficient in Slc superfamily anion transporters Oat1 and Oat3 show lower renal uptake and clearance for a variety of anionic xenobiotic compounds (Sweet et al., 2002; Eraly et al., 2006). In addition, impaired renal uptake of furosemide in Oat1-null animals produces a decreased diuretic response to this drug (Eraly et al., 2006). Oat family transporters are also present in hepatocyte membranes and at the blood–brain barrier; and knockout of Oatp1b2 allows mice to resist phalloidin-induced hepatotoxicity (Klaassen and Lu, 2008), while Oat3-null mice show a marked decrease in uptake of fluorescein at the choroid plexus compared to intact animals (Sweet et al., 2002). Knockout of Pept2, another Slc transporter that is highly expressed at the choroid plexus, results in a markedly impaired transport of dipeptides across the blood–brain barrier (Shen et al., 2003; Ocheltree et al., 2005).

In addition to mouse gene knockout models, some preclinical animals are naturally genetically deficient in certain specific transporters and therefore can provide useful experimental models. So far, Mdr1a deficiency has been discovered in a subpopulation of the CF-1 mouse strain (Umbenhauer et al., 1997) and in certain dogs of the collie breed (Roulet et al., 2003), Mrp2 deficiency is seen in Wistar-derived TR^- rats (Muller et al., 1996) and in Sprague–Dawley-derived Eisai hyperbilirubinemic rats (Ito et al., 1997), and Divalent metal transporter1 (Dmt1) deficiency is seen in Belgrade rats (Knopfel et al., 2005).

Although these genetically modified strains have been primarily used to examine basic questions of transport specificity and physiology, they should also provide valuable resources for screening and development of new drug candidates. The ability to screen drug bioavailability, disposition, clearance, and dose-related therapeutic and toxic effects in animals deficient in specific or multiple transport proteins will allow the efficient detection of both problematic and beneficial drug–transporter interactions very early on in the drug development process.

15.6 SPECIFIC EXAMPLES OF TRANSPORTER STUDIES IN DRUG DISCOVERY AND DEVELOPMENT

Over the last decade, drug development programs have increasingly incorporated studies on the effects of physiologically important transporters on the pharmacokinetics of new drug candidates and, conversely, the effects of candidate drugs on transporter activities. The reasons for incorporation of studies of drug–transporter interactions into the drug development process can be roughly grouped into four categories (although certainly other additional reasons for studying drug–transporter interactions may exist): (1) These studies will help explain and predict those tissue- and organ-specific effects of drugs, either toxic or therapeutic, that are attributable to tissue-specific transport. (2) The propensity of drug candidates to provoke drug–drug interactions with existing therapeutic drugs may be explained or predicted by their common interaction with important transporters. (3) Characterization of drug–transporter interactions may suggest structural alterations in the drug molecule that will increase its bioavailability. (4) In some cases, transporters themselves may be primary therapeutic targets for new drugs, as revealed by drug–transporter interaction studies during the development process. Specific examples of these types of studies are presented below.

15.6.1 Transporter Contribution to Tissue Specificity

15.6.1.1 Prediction and Explanation of Tissue-Specific Toxicity. Statins inhibit hepatic 3-hydroxy-3-methylglutaryl-coenzyme A (HMG-CoA) reductase, which lowers the rate of synthesis of cholesterol and reduces total cholesterol in the body. All known statins share the undesirable side effect of muscle toxicity (rhabdomyolysis). In fact, because of its association with multiple cases of lethal rhabdomyolysis, Baycol (cerivastatin sodium) was withdrawn from the market. The development of other statins (e.g., pravastatin) provided similar products with greatly reduced muscle toxicity. Pravastatin has been shown to be a good substrate for OATPs that are highly expressed on the liver sinusoidal membrane, thereby mediating the uptake of pravastatin into hepatocytes (Funk, 2008). Therefore, pravastatin distributes preferentially to the liver (its therapeutic target organ), but poorly to muscle (its major potential site of toxicity). In addition, pravastatin has been found to be poorly metabolized by CYPs *in vivo*; instead, the nonmetabolized drug is transported directly into the bile by MRP2, which is highly expressed on the canalicular membrane. This promotes enterohepatic circulation of the active drug, and it allows maintenance of effective concentrations in the liver and reduction of doses required for treatment. Such observations show the impact of drug–transporter interactions on potential strategies for screening and selecting HMG-CoA reductase inhibitor candidates.

Cidofovir and adefovir, two potent antiviral agents, provoke a significant tissue-specific nephrotoxicity in humans. Transporter studies (Ho et al., 2000) clearly demonstrated that both cidofovir and adefovir are substrates of human OAT1, which is highly expressed on the basolateral membrane of the renal proximal tubule. This transporter generates an intracellular accumulation of cidofovir and adefovir, leading to nephrotoxicity. Adefovir has a higher binding affinity to human OAT1 than to rat OAT1, and therefore a higher dose is required to induce nephrotoxicity in rats. Such transporter-mediated tissue-specific toxicity may be predicted early on in the drug development process by properly designed screening for drug–transporter interactions (Ho et al., 2000).

15.6.1.2 Desirable Pharmacokinetic Effects of Drug–Transporter Interactions at Target Tissues. An important pharmacokinetic consideration for drug candidates developed for treatment of human immunodeficiency virus (HIV) infections is the problem of virus "sheltering" in the central nervous system (CNS), where it is protected by the blood–brain barrier (Huisman et al., 2001; Park and Sinko, 2005), making total eradication of the virus in the body potentially more difficult. For example, saquinavir, an HIV protease inhibitor, does not penetrate the blood–brain barrier and is therefore free of CNS toxicity. However, it also lacks ability to target residual virus sheltered in the CNS. Saquinavir is a good substrate for P-gp, which is a highly expressed component of the BBB. Substrates for P-gp are kept at low concentrations in the brain due to active efflux transport. On the other hand, sustiva (efavirenz) is a non-nucleotide reverse transcriptase inhibitor, which is not a substrate of P-gp, and therefore it is able to reach higher concentrations in the brain to eradicate the remaining virus in central nervous system. On the negative side, efavirenz has a greater potential for provoking CNS toxicity (Marzolini et al., 2001). Tissue specificity for similar drug candidates should be predictable on the basis of properly designed transporter studies during drug development.

It has been well documented that rifampin induces multiple human CYPs and drug transporters (Westphal et al., 2000b; Luo et al., 2002, 2007a), and individual differences in the rifampin-mediated CYP/transporter induction ratio are very significant to its overall pharmacokinetics (Luo et al., 2002, 2003). However, the molecular mechanisms responsible for this high degree of interindividual variation have not been fully elucidated. Tirona et al. (2003) demonstrated that the uptake of rifampin into liver, the target organ for hepatic CYP and transporter induction, is mediated by OATP-C. Furthermore, the activity of OATP-C-mediated hepatic uptake of rifampin was very different in subjects having naturally occurring allelic variants of OATP-C (Tirona et al., 2003). These findings may not fully explain the individual difference in CYP/transporter induction by rifampin, but certainly further our understanding.

15.6.2 Potential Drug–Drug Interactions Due to Transporter Induction or Inhibition

Fexofenadine, an antihistamine drug used to treat hay fever and similar allergic symptoms, is a poor substrate for human CYP enzymes but a good substrate for P-gp and OATP (Miura et al., 2007). In clinical studies, fexofenadine showed diminished oral bioavailability in subjects who were pretreated with rifampin (Hamman et al., 2001), but showed increased bioavailability when itraconazole was coadministered (Shimizu et al., 2006). The decreased bioavailability was attributed to induction of P-gp by rifampin, which resulted in a retrograde efflux that lowered the net gastrointestinal uptake and absorption of fexofenadine. Likewise, the increased bioavailability of fexofenadine in the presence of itraconazole was likely due to inhibition of P-gp by itraconazole. Such potential transporter-based drug–drug interactions may be predictable by transporter studies early in the drug development process.

15.6.3 Recognition of Drug–Transporter Interactions to Improve Oral Bioavailability

The oral route is the most convenient, economic, and safe route of drug administration. However, poor oral bioavailability is a problem that can adversely affect the potential clinical usefulness of a drug candidate. In fact, many drug candidates have failed because of their poor absorption from the stomach and small intestine. Careful characterization of the properties of important intestinal transporters has provided an effective approach to overcoming low oral bioavailability in some cases. Essentially, this approach advocates the oral administration of a prodrug that will provide a better substrate for transporters expressed in the small intestine, including PEPT1 and IBAT. In particular, PEPT1 is expressed highly in the small intestine, but at low levels in the kidney, and shows a high capacity and low affinity for peptide-bond containing substrates (Brandsch et al., 2004). Once a prodrug is absorbed, the active drug can be metabolically generated by the first pass through the liver. Valacyclovir, a prodrug form of acyclovir, is one such successful example. The efficacy of oral acyclovir is limited because of its poor oral bioavailability; however, valacyclovir (the L-valyl ester of acyclovir), is well absorbed in the small intestine through uptake transport mediated by PEPT1 (MacDougall and Guglielmo, 2004; Li et al., 2008). In healthy human volunteers, the oral bioavailability of acyclovir after oral administration of valacyclovir was 54%, in comparison to the 10–20% bioavailability of acyclovir itself (MacDougall and Guglielmo, 2004; Li et al., 2008).

15.6.4 Transporters as Potential Novel Drug Targets

Because transporters play such an important role in physiology and drug disposition *in vivo*, it is rational to consider transporters as potential targets for drugs developed for treatment of certain diseases. For instance, Tangier disease is associated with a genetic mutation of ABCA1 (Nofer and Remaley, 2005; Bennett et al., 2006). Patients with this mutation are characterized by very low levels of high-density lipoprotein in plasma accompanied by accumulation of cholesterol in tissue macrophages. This results in prevalent atherosclerosis and an enlarged liver and spleen. ABCA1 is known to transport excess cholesterol from macrophages into the high-density lipoprotein (HDL) metabolic pathway. Because of this, ABCA1 has been proposed as a drug target for the reduction of high total-body cholesterol levels. Diseases resulting from high cholesterol may be amenable to prevention by drugs designed to induce ABCA1 through activation of the hepatic X receptor signal transduction pathway (Bennett et al., 2006).

15.7 SUMMARY

The key roles of transporter proteins in determining the overall disposition of drugs and toxic xenobiotics is attracting increased awareness among academic and industrial pharmacological scientists. Drug–transporter interactions will affect multiple aspects of drug disposition, including absorption and bioavailability, tissue-specific distribution and uptake, hepatic uptake and metabolism, and renal and hepatic elimination and clearance. The possibility of numerous potential interactions between drug candidates and the multiple transporters present *in vivo* is an aspect of drug disposition that must be assessed early on in the drug development process. Such assessment may provide an early warning of undesirable toxic effects or drug–drug interactions that would not otherwise be indicated by pharmacodynamic studies. Fortunately, recent advances in characterization of transport proteins, along with technological developments of models for the efficient determination of drug–transporter interactions, allow the assessment of the potential for these interactions, early on in the drug development process. In particular, the development of genetic knockout models for important transport proteins should permit a continued development of cost- and time-efficient assessment of the interaction of drug candidates with specific individual transporters, even against a background of often redundant transport proteins with overlapping substrate selectivities.

ACRONYMS AND ABBREVIATIONS

ABC	Adenosine triphosphate (ATP)-binding cassette
BCRP	Breast cancer resistance protein
BBB	Blood–brain barrier
BBMV	Brush border membrane vesicle
BSEP	Bile salt export pump
CHO	Chinese hamster ovary
CMV	Canalicular membrane vesicle
HEK	Human embryonic kidney
IBAT	Ileal sodium-dependent bile acid transporter
LLC-PK	Porcine-derived proximal tubular epithelial
MDR	Multidrug resistance
MRP	Multidrug resistance-associated protein
NCTP	Sodium taurocholate cotransporting polypeptide

OAT	Organic anion transporter
OCT	Organic cation transporter
OCTN	Novel organic cation transporter
PEPT	Peptide transporter
P-gp	P-glycoprotein
SLC	Solute carrier
SLCO	Solute carrier organic anion transporter
SN-38	7-Ethyl-10-hydroxycamptothecin
SPA	Scintillation proximity assay
SUR	Sulfonylurea receptor
URAT	Urate transporter

REFERENCES

Aleksunes LM, Slitt AL, Maher JM, Augustine LM, Goedken MJ, Chan JY, Cherrington NJ, Klaassen CD, Manautou JE. Induction of Mrp3 and Mrp4 transporters during acetaminophen hepatotoxicity is dependent on Nrf2. *Toxicol Appl Pharmacol* 2008;226(1):74–83.

Alissa FT, Jaffe R, Shneider BL. Update on progressive familial intrahepatic cholestasis. *J Pediatr Gastroenterol Nutr* 2008;46(3):241–252.

Allikmets R, Gerrard B, Hutchinson A, Dean M. Characterization of the human ABC superfamily: isolation and mapping of 21 new genes using the expressed sequence tags database. *Hum Mol Genet* 1996;5(10):1649–1655.

Alrefai WA, Gill RK. Bile acid transporters: structure, function, regulation and pathophysiological implications. *Pharm Res* 2007;24(10):1803–1823.

Ambudkar SV, Rosen BP, Gottesman MM. Workshop on ABC Transporters and Human Diseases. *Drug Resist Update* 2000;3(1):51–54.

Behravan J, Piquette-Miller M. Drug transport across the placenta, role of the ABC drug efflux transporters. *Expert Opin Drug Metab Toxicol* 2007;3(6):819–830.

Belinsky MG, Dawson PA, Shchaveleva I, Bain LJ, Wang R, Ling V, Chen ZS, Grinberg A, Westphal H, Klein-Szanto A, Lerro A, Kruh GD. Analysis of the *in vivo* functions of Mrp3. *Mol Pharmacol* 2005;68(1):160–168.

Bennett DJ, Cooke AJ, Edwards AS. Non-steroidal LXR agonists; an emerging therapeutic strategy for the treatment of atherosclerosis. *Recent Patents Cardiovasc Drug Discovery* 2006;1(1):21–46.

Birn H, Spiegelstein O, Christensen EI, Finnell RH. Renal tubular reabsorption of folate mediated by folate binding protein 1. *J Am Soc Nephrol* 2005;16(3): 608–615.

Bohme M, Jedlitschky G, Leier I, Buchler M, Keppler D. ATP-dependent export pumps and their inhibition by cyclosporins. *Adv Enzyme Regul* 1994;34:371–380.

Bonge H, Hallen S, Fryklund J, Sjostrom JE. Cytostar-T scintillating microplate assay for measurement of sodium-dependent bile acid uptake in transfected HEK-293 cells. *Anal Biochem* 2000;282(1):94–101.

Borst P, Evers R, Kool M, Wijnholds J. A family of drug transporters: the multidrug resistance-associated proteins. *J Natl Cancer Inst* 2000;92(16):1295–1302.

Brandsch M, Knutter I, Leibach FH. The intestinal H^+/peptide symporter PEPT1: structure–affinity relationships. *Eur J Pharm Sci* 2004;21(1):53–60.

Breedveld P, Beijnen JH, Schellens JH. Use of P-glycoprotein and BCRP inhibitors to improve oral bioavailability and CNS penetration of anticancer drugs. *Trends Pharmacol Sci* 2006;27(1):17–24.

Cerveny L, Svecova L, Anzenbacherova E, Vrzal R, Staud F, Dvorak Z, Ulrichova J, Anzenbacher P, Pavek P. Valproic acid induces CYP3A4 and MDR1 gene expression by activation of constitutive androstane receptor and pregnane X receptor pathways. *Drug Metab Dispos* 2007;35(7):1032–1041.

Chang C, Ekins S, Bahadduri P, Swaan PW. Pharmacophore-based discovery of ligands for drug transporters. *Adv Drug Deliv Rev* 2006;58(12–13):1431–1450.

Chen ZS, Robey RW, Belinsky MG, Shchaveleva I, Ren XQ, Sugimoto Y, Ross DD, Bates SE, Kruh GD. Transport of methotrexate, methotrexate polyglutamates, and 17beta-estradiol 17-(beta-D-glucuronide) by ABCG2: effects of acquired mutations at R482 on methotrexate transport. *Cancer Res* 2003;63(14):4048–4054.

Choi DS, Cascini MG, Mailliard W, Young H, Paredes P, McMahon T, Diamond I, Bonci A, Messing RO. The type 1 equilibrative nucleoside transporter regulates ethanol intoxication and preference. *Nat Neurosci* 2004;7(8):855–861.

Choi MK, Song IS. Organic cation transporters and their pharmacokinetic and pharmacodynamic consequences. *Drug Metab Pharmacokinet* 2008;23(4):243–253.

Choudhuri S, Klaassen CD. Structure, function, expression, genomic organization, and single nucleotide polymorphisms of human ABCB1 (MDR1), ABCC (MRP), and ABCG2 (BCRP) efflux transporters. *Int J Toxicol* 2006;25(4):231–259.

Chu XY, Huskey SE, Braun MP, Sarkadi B, Evans DC, Evers R. Transport of ethinylestradiol glucuronide and ethinylestradiol sulfate by the multidrug resistance proteins MRP1, MRP2, and MRP3. *J Pharmacol Exp Ther* 2004;309(1):156–164.

Dean M, Hamon Y, Chimini G. The human ATP-binding cassette (ABC) transporter superfamily. *J Lipid Res* 2001;42(7):1007–1017.

Drobnik W, Lindenthal B, Lieser B, Ritter M, Christiansen Weber T, Liebisch G, Giesa U, Igel M, Borsukova H, Buchler C, Fung-Leung WP, Von Bergmann K, Schmitz G. ATP-binding cassette transporter A1 (ABCA1) affects total body sterol metabolism. *Gastroenterology* 2001;120(5):1203–1211.

Dutt A, Heath LA, Nelson JA. P-glycoprotein and organic cation secretion by the mammalian kidney. *J Pharmacol Exp Ther* 1994;269(3):1254–1260.

El-Sheikh AA, van den Heuvel JJ, Koenderink JB, Russel FG. Interaction of nonsteroidal anti-inflammatory drugs with multidrug resistance protein (MRP) 2/ABCC2- and MRP4/ABCC4-mediated methotrexate transport. *J Pharmacol Exp Ther* 2007;320(1):229–235.

El-Sheikh AA, Masereeuw R, Russel FG. Mechanisms of renal anionic drug transport. *Eur J Pharmacol* 2008;585(2–3):245–255.

Enomoto A, Endou H. Roles of organic anion transporters (OATs) and a urate transporter (URAT1) in the pathophysiology of human disease. *Clin Exp Nephrol* 2005;9(3):195–205.

Enomoto A, Kimura H, Chairoungdua A, Shigeta Y, Jutabha P, Cha SH, Hosoyamada M, Takeda M, Sekine T, Igarashi T, Matsuo H, Kikuchi Y, Oda T, Ichida K, Hosoya T, Shimokata K, Niwa T, Kanai Y, Endou H. Molecular identification of a renal urate anion exchanger that regulates blood urate levels. *Nature* 2002;417(6887): 447–452.

Enomoto A, Niwa T, Kanai Y, Endou H. Urate transporter and renal hypouricemia. *Rinsho Byori* 2003;51(9):892–897.

Eraly SA, Vallon V, Vaughn DA, Gangoiti JA, Richter K, Nagle M, Monte JC, Rieg T, Truong DM, Long JM, Barshop BA, Kaler G, Nigam SK. Decreased renal organic anion secretion and plasma accumulation of endogenous organic anions in OAT1 knock-out mice. *J Biol Chem* 2006;281(8):5072–5083.

Forbes JR, Cox DW. Functional characterization of missense mutations in ATP7B: Wilson disease mutation or normal variant? *Am J Hum Genet* 1998;63(6):1663–1674.

Funk C. The role of hepatic transporters in drug elimination. *Expert Opin Drug Metab Toxicol* 2008;4(4):363–379.

Galmarini CM. P-glycoprotein expression by cancer cells affects cell cytotoxicity and cell-cycle perturbations induced by six chemotherapeutic drugs. *J Exp Ther Oncol* 2002;2(3):146–152.

Geick A, Eichelbaum M, Burk O. Nuclear receptor response elements mediate induction of intestinal MDR1 by rifampin. *J Biol Chem* 2001;276(18):14581–14587.

Gottesman MM, Ambudkar SV. Overview: ABC transporters and human disease. *J Bioenerg Biomembr* 2001;33(6):453–458.

Gurley BJ, Swain A, Williams DK, Barone G, Battu SK. Gauging the clinical significance of P-glycoprotein-mediated herb–drug interactions: comparative effects of St. John's wort, Echinacea, clarithromycin, and rifampin on digoxin pharmacokinetics. *Mol Nutr Food Res* 2008;52(7):772–779.

Hamman MA, Bruce MA, Haehner-Daniels BD, Hall SD. The effect of rifampin administration on the disposition of fexofenadine. *Clin Pharmacol Ther* 2001;69(3): 114–121.

Hamvas A, Cole FS, Nogee LM. Genetic disorders of surfactant proteins. *Neonatology* 2007;91(4):311–317.

Hediger MA, Romero MF, Peng JB, Rolfs A, Takanaga H, Bruford EA. The ABCs of solute carriers: physiological, pathological and therapeutic implications of human membrane transport proteinsIntroduction. *Pflugers Arch* 2004;447(5): 465–468.

Hirohashi T, Suzuki H, Sugiyama Y. Characterization of the transport properties of cloned rat multidrug resistance-associated protein 3 (MRP3). *J Biol Chem* 1999;274(21):15181–15185.

Ho ES, Lin DC, Mendel DB, Cihlar T. Cytotoxicity of antiviral nucleotides adefovir and cidofovir is induced by the expression of human renal organic anion transporter 1. *J Am Soc Nephrol* 2000;11(3):383–393.

Hollenstein K, Dawson RJ, Locher KP. Structure and mechanism of ABC transporter proteins. *Curr Opin Struct Biol* 2007a;17(4):412–418.

Hollenstein K, Frei DC, Locher KP. Structure of an ABC transporter in complex with its binding protein. *Nature* 2007b;446(7132):213–216.

Horio M, Gottesman MM, Pastan I. ATP-dependent transport of vinblastine in vesicles from human multidrug-resistant cells. *Proc Natl Acad Sci USA* 1988;85(10): 3580–3584.

Hsing S, Gatmaitan Z, Arias IM. The function of Gp170, the multidrug-resistance gene product, in the brush border of rat intestinal mucosa. *Gastroenterology* 1992;102(3): 879–885.

Huisman MT, Smit JW, Wiltshire HR, Hoetelmans RM, Beijnen JH, Schinkel AH. P-glycoprotein limits oral availability, brain, and fetal penetration of saquinavir even with high doses of ritonavir. *Mol Pharmacol* 2001;59(4):806–813.

Hussain K, Flanagan SE, Smith VV, Ashworth M, Day M, Pierro A, Ellard S. An ABCC8 gene mutation and mosaic uniparental isodisomy resulting in atypical diffuse congenital hyperinsulinism. *Diabetes* 2008;57(1):259–263.

Imai Y, Asada S, Tsukahara S, Ishikawa E, Tsuruo T, Sugimoto Y. Breast cancer resistance protein exports sulfated estrogens but not free estrogens. *Mol Pharmacol* 2003;64(3):610–618.

Inui K, Terada T, Masuda S, Saito H. Physiological and pharmacological implications of peptide transporters, PEPT1 and PEPT2. *Nephrol Dial Transplant* 2000;15(Suppl 6):11–13.

Ishizuka H, Konno K, Shiina T, Naganuma H, Nishimura K, Ito K, Suzuki H, Sugiyama Y. Species differences in the transport activity for organic anions across the bile canalicular membrane. *J Pharmacol Exp Ther* 1999;290(3):1324–1330.

Ito K, Suzuki H, Hirohashi T, Kume K, Shimizu T, Sugiyama Y. Molecular cloning of canalicular multispecific organic anion transporter defective in EHBR. *Am J Physiol* 1997;272(1 Pt 1):G16–G22.

Jansen PJ, Lutjohann D, Abildayeva K, Vanmierlo T, Plosch T, Plat J, von Bergmann K, Groen AK, Ramaekers FC, Kuipers F, Mulder M. Dietary plant sterols accumulate in the brain. *Biochim Biophys Acta* 2006;1761(4):445–453.

Jedlitschky G, Hoffmann U, Kroemer HK. Structure and function of the MRP2 (ABCC2) protein and its role in drug disposition. *Expert Opin Drug Metab Toxicol* 2006;2(3):351–366.

Jedlitschky G, Keppler D. Transport of leukotriene C4 and structurally related conjugates. *Vitam Horm* 2002;64:153–184.

Johnson DR, Klaassen CD. Regulation of rat multidrug resistance protein 2 by classes of prototypical microsomal enzyme inducers that activate distinct transcription pathways. *Toxicol Sci* 2002;67(2):182–189.

Jonker JW, Wagenaar E, Mol CA, Buitelaar M, Koepsell H, Smit JW, Schinkel AH. Reduced hepatic uptake and intestinal excretion of organic cations in mice with a targeted disruption of the organic cation transporter 1 (Oct1 [Slc22a1]) gene. *Mol Cell Biol* 2001;21(16):5471–5477.

Jonker JW, Buitelaar M, Wagenaar E, Van Der Valk MA, Scheffer GL, Scheper RJ, Plosch T, Kuipers F, Elferink RP, Rosing H, Beijnen JH, Schinkel AH. The breast cancer resistance protein protects against a major chlorophyll-derived dietary phototoxin and protoporphyria. *Proc Natl Acad Sci USA* 2002;99(24):15649–15654.

Jonker JW, Wagenaar E, Van Eijl S, Schinkel AH. Deficiency in the organic cation transporters 1 and 2 (Oct1/Oct2 [Slc22a1/Slc22a2]) in mice abolishes renal secretion of organic cations. *Mol Cell Biol* 2003;23(21):7902–7908.

Kamimoto Y, Gatmaitan Z, Hsu J, Arias IM. The function of Gp170, the multidrug resistance gene product, in rat liver canalicular membrane vesicles. *J Biol Chem* 1989;264(20):11693–11698.

Katoh M, Suzuyama N, Takeuchi T, Yoshitomi S, Asahi S, Yokoi T. Kinetic analyses for species differences in P-glycoprotein-mediated drug transport. *J Pharm Sci* 2006;95(12):2673–2683.

Kauffmann HM, Pfannschmidt S, Zoller H, Benz A, Vorderstemann B, Webster JI, Schrenk D. Influence of redox-active compounds and PXR-activators on human MRP1 and MRP2 gene expression. *Toxicology* 2002;171(2–3):137–146.

Kemper EM, van Zandbergen AE, Cleypool C, Mos HA, Boogerd W, Beijnen JH, van Tellingen O. Increased penetration of paclitaxel into the brain by inhibition of P-Glycoprotein. *Clin Cancer Res* 2003;9(7):2849–2855.

Klaassen CD, Lu H. Xenobiotic transporters: ascribing function from gene knockout and mutation studies. *Toxicol Sci* 2008;101(2):186–196.

Klement JF, Matsuzaki Y, Jiang QJ, Terlizzi J, Choi HY, Fujimoto N, Li K, Pulkkinen L, Birk DE, Sundberg JP, Uitto J. Targeted ablation of the abcc6 gene results in ectopic mineralization of connective tissues. *Mol Cell Biol* 2005;25(18):8299–8310.

Knopfel M, Zhao L, Garrick MD. Transport of divalent transition-metal ions is lost in small-intestinal tissue of b/b Belgrade rats. *Biochemistry* 2005;44(9):3454–3465.

Kool M, van der Linden M, de Haas M, Scheffer GL, de Vree JM, Smith AJ, Jansen G, Peters GJ, Ponne N, Scheper RJ, Elferink RP, Baas F, Borst P. MRP3, an organic anion transporter able to transport anti-cancer drugs. *Proc Natl Acad Sci USA* 1999;96(12):6914–6919.

Koshiba S, An R, Saito H, Wakabayashi K, Tamura A, Ishikawa T. Human ABC transporters ABCG2 (BCRP) and ABCG4. *Xenobiotica* 2008;38(7–8):863–888.

Kosters A, Karpen SJ. Bile acid transporters in health and disease. *Xenobiotica* 2008;38(7–8):1043–1071.

Kwon Y, Kamath AV, Morris ME. Inhibitors of P-glycoprotein-mediated daunomycin transport in rat liver canalicular membrane vesicles. *J Pharm Sci* 1996;85(9):935–939.

Lagas JS, Vlaming ML, van Tellingen O, Wagenaar E, Jansen RS, Rosing H, Beijnen JH, Schinkel AH. Multidrug resistance protein 2 is an important determinant of paclitaxel pharmacokinetics. *Clin Cancer Res* 2006;12(20 Pt 1):6125–6132.

Launay-Vacher V, Izzedine H, Karie S, Hulot JS, Baumelou A, Deray G. Renal tubular drug transporters. *Nephron Physiol* 2006;103(3):97–106.

Lewis-Jones DI, Gazvani MR, Mountford R. Cystic fibrosis in infertility: screening before assisted reproduction: opinion. *Hum Reprod* 2000;15(11):2415–2417.

Li F, Maag H, Alfredson T. Prodrugs of nucleoside analogues for improved oral absorption and tissue targeting. *J Pharm Sci* 2008;97(3):1109–1134.

Li N, Zhang Y, Hua F, Lai Y. Absolute difference of hepatobiliary transporter multidrug resistance-associated protein (MRP2/Mrp2) in liver tissues and isolated hepatocytes from rat, dog, monkey, and human. *Drug Metab Dispos* 2009;37(1):66–73.

Limtrakul P, Chearwae W, Shukla S, Phisalphong C, Ambudkar SV. Modulation of function of three ABC drug transporters, P-glycoprotein (ABCB1), mitoxantrone resistance protein (ABCG2) and multidrug resistance protein 1 (ABCC1) by tetrahydrocurcumin, a major metabolite of curcumin. *Mol Cell Biochem* 2007;296(1–2): 85–95.

Liu X, Chism JP, LeCluyse EL, Brouwer KR, Brouwer KL. Correlation of biliary excretion in sandwich-cultured rat hepatocytes and *in vivo* in rats. *Drug Metab Dispos* 1999a;27(6):637–644.

Liu X, LeCluyse EL, Brouwer KR, Gan LS, Lemasters JJ, Stieger B, Meier PJ, Brouwer KL. Biliary excretion in primary rat hepatocytes cultured in a collagen-sandwich configuration. *Am J Physiol* 1999b;277(1 Pt 1):G12–G21.

Lohmann C, Gelius B, Danielsson J, Skoging-Nyberg U, Hollnack E, Dudley A, Wahlberg J, Hoogstraate J, Gustavsson L. Scintillation proximity assay for measuring uptake by the human drug transporters hOCT1, hOAT3, and hOATP1B1. *Anal Biochem* 2007;366(2):117–125.

Luo G, Cunningham M, Kim S, Burn T, Lin J, Sinz M, Hamilton G, Rizzo C, Jolley S, Gilbert D, Downey A, Mudra D, Graham R, Carroll K, Xie J, Madan A, Parkinson A, Christ D, Selling B, LeCluyse E, Gan LS. CYP3A4 induction by drugs: correlation between a pregnane X receptor reporter gene assay and CYP3A4 expression in human hepatocytes. *Drug Metab Dispos* 2002;30(7):795–804.

Luo G, Lin J, Fiske WD, Dai R, Yang TJ, Kim S, Sinz M, LeCluyse E, Solon E, Brennan JM, Benedek IH, Jolley S, Gilbert D, Wang L, Lee FW, Gan LS. Concurrent induction and mechanism-based inactivation of CYP3A4 by an L-valinamide derivative. *Drug Metab Dispos* 2003;31(9):1170–1175.

Luo G, Gan L, Guenthner T. Testing drug candidates for CYP3A4 induction. In: Zhang D, Zhu M, Humphreys W, editors. *Drug Metabolism in Drug Design and Development*. Hoboken, NJ: John Wiley & Sons, 2007a, pp. 545–571.

Luo G, Garner CE, Xiong H, Hu H, Richards LE, Brouwer KL, Duan J, Decicco CP, Maduskuie T, Shen H, Lee FW, Gan LS. Effect of DPC 333 [(2R)-2-{(3R)-3-amino-3-[4-(2-methylquinolin-4-ylmethoxy)phenyl]-2-oxopyrrolidin-1-yl}-*N*-hydroxy-4-methylpentanamide], a human tumor necrosis factor alpha-converting enzyme inhibitor, on the disposition of methotrexate: a transporter-based drug–drug interaction case study. *Drug Metab Dispos* 2007b;35(6):835–840.

MacDougall C, Guglielmo BJ. Pharmacokinetics of valaciclovir. *J Antimicrob Chemother* 2004;53(6):899–901.

Maher JM, Cheng X, Slitt AL, Dieter MZ, Klaassen CD. Induction of the multidrug resistance-associated protein family of transporters by chemical activators of receptor-mediated pathways in mouse liver. *Drug Metab Dispos* 2005;33(7):956–962.

Makhey VD, Guo A, Norris DA, Hu P, Yan J, Sinko PJ. Characterization of the regional intestinal kinetics of drug efflux in rat and human intestine and in Caco-2 cells. *Pharm Res* 1998;15(8):1160–1167.

Manautou JE, de Waart DR, Kunne C, Zelcer N, Goedken M, Borst P, Elferink RO. Altered disposition of acetaminophen in mice with a disruption of the Mrp3 gene. *Hepatology* 2005;42(5):1091–1098.

Mao Q, Unadkat JD. Role of the breast cancer resistance protein (ABCG2) in drug transport. *AAPS J* 2005;7(1):E118–E133.

Marzolini C, Telenti A, Decosterd LA, Greub G, Biollaz J, Buclin T. Efavirenz plasma levels can predict treatment failure and central nervous system side effects in HIV-1-infected patients. *AIDS* 2001;15(1):71–75.

Masuda M, I'Izuka Y, Yamazaki M, Nishigaki R, Kato Y, Ni'inuma K, Suzuki H, Sugiyama Y. Methotrexate is excreted into the bile by canalicular multispecific organic anion transporter in rats. *Cancer Res* 1997;57(16):3506–3510.

Matsushima S, Maeda K, Kondo C, Hirano M, Sasaki M, Suzuki H, Sugiyama Y. Identification of the hepatic efflux transporters of organic anions using double-transfected Madin-Darby canine kidney II cells expressing human organic anion-transporting polypeptide 1B1 (OATP1B1)/multidrug resistance-associated protein 2, OATP1B1/multidrug resistance 1, and OATP1B1/breast cancer resistance protein. *J Pharmacol Exp Ther* 2005;314(3):1059–1067.

Mennone A, Soroka CJ, Cai SY, Harry K, Adachi M, Hagey L, Schuetz JD, Boyer JL. Mrp4-/- mice have an impaired cytoprotective response in obstructive cholestasis. *Hepatology* 2006;43(5):1013–1021.

Miura M, Uno T, Tateishi T, Suzuki T. Pharmacokinetics of fexofenadine enantiomers in healthy subjects. *Chirality* 2007;19(3):223–227.

Mizuno N, Suzuki M, Kusuhara H, Suzuki H, Takeuchi K, Niwa T, Jonker JW, Sugiyama Y. Impaired renal excretion of 6-hydroxy-5,7-dimethyl-2-methylamino-4-(3-pyridylmethyl) benzothiazole (E3040) sulfate in breast cancer resistance protein (BCRP1/ABCG2) knockout mice. *Drug Metab Dispos* 2004;32(9):898–901.

Muller M, Roelofsen H, Jansen PL. Secretion of organic anions by hepatocytes: involvement of homologues of the multidrug resistance protein. *Semin Liver Dis* 1996;16(2):211–220.

Nofer JR, Remaley AT. Tangier disease: still more questions than answers. *Cell Mol Life Sci* 2005;62(19–20):2150–2160.

Nozawa T, Minami H, Sugiura S, Tsuji A, Tamai I. Role of organic anion transporter OATP1B1 (OATP-C) in hepatic uptake of irinotecan and its active metabolite, 7-ethyl-10-hydroxycamptothecin: *in vitro* evidence and effect of single nucleotide polymorphisms. *Drug Metab Dispos* 2005;33(3):434–439.

Ocheltree SM, Shen H, Hu Y, Keep RF, Smith DE. Role and relevance of peptide transporter 2 (PEPT2) in the kidney and choroid plexus: *in vivo* studies with glycylsarcosine in wild-type and PEPT2 knockout mice. *J Pharmacol Exp Ther* 2005;315(1):240–247.

Osato DH, Huang CC, Kawamoto M, Johns SJ, Stryke D, Wang J, Ferrin TE, Herskowitz I, Giacomini KM. Functional characterization in yeast of genetic variants in the human equilibrative nucleoside transporter, ENT1. *Pharmacogenetics* 2003;13(5):297–301.

Park S, Sinko PJ. P-glycoprotein and mutlidrug resistance-associated proteins limit the brain uptake of saquinavir in mice. *J Pharmacol Exp Ther* 2005;312(3):1249–1256.

Pellicoro A, Faber KN. Review article: The function and regulation of proteins involved in bile salt biosynthesis and transport. *Aliment Pharmacol Ther* 2007;26(Suppl 2):149–160.

Rajgopal A, Sierra EE, Zhao R, Goldman ID. Expression of the reduced folate carrier SLC19A1 in IEC-6 cells results in two distinct transport activities. *Am J Physiol Cell Physiol* 2001;281(5):C1579–C1586.

Robey RW, Polgar O, Deeken J, To KW, Bates SE. ABCG2: determining its relevance in clinical drug resistance. *Cancer Metastasis Rev* 2007;26(1):39–57.

Roulet A, Puel O, Gesta S, Lepage JF, Drag M, Soll M, Alvinerie M, Pineau T. MDR1-deficient genotype in collie dogs hypersensitive to the P-glycoprotein substrate ivermectin. *Eur J Pharmacol* 2003;460(2–3):85–91.

Ruetz S, Raymond M, Gros P. Functional expression of P-glycoprotein encoded by the mouse mdr3 gene in yeast cells. *Proc Natl Acad Sci USA* 1993;90(24):11588–11592.

Sahi J, Milad MA, Zheng X, Rose KA, Wang H, Stilgenbauer L, Gilbert D, Jolley S, Stern RH, LeCluyse EL. Avasimibe induces CYP3A4 and multiple drug resistance protein 1 gene expression through activation of the pregnane X receptor. *J Pharmacol Exp Ther* 2003;306(3):1027–1034.

Saito H, Masuda S, Inui K. Cloning and functional characterization of a novel rat organic anion transporter mediating basolateral uptake of methotrexate in the kidney. *J Biol Chem* 1996;271(34):20719–20725.

Sarkadi B, Price EM, Boucher RC, Germann UA, Scarborough GA. Expression of the human multidrug resistance cDNA in insect cells generates a high activity drug-stimulated membrane ATPase. *J Biol Chem* 1992;267(7):4854–4858.

Sasaki M, Suzuki H, Aoki J, Ito K, Meier PJ, Sugiyama Y. Prediction of *in vivo* biliary clearance from the *in vitro* transcellular transport of organic anions across a double-transfected Madin–Darby canine kidney II monolayer expressing both rat organic anion transporting polypeptide 4 and multidrug resistance associated protein 2. *Mol Pharmacol* 2004;66(3):450–459.

Sawamiphak S, Sophasan S, Endou H, Boonchird C. Functional expression of the rat organic anion transporter 1 (rOAT1) in *Saccharomyces cerevisiae*. *Biochim Biophys Acta* 2005;1720(1–2):44–51.

Schinkel AH, Smit JJ, van Tellingen O, Beijnen JH, Wagenaar E, van Deemter L, Mol CA, van der Valk MA, Robanus-Maandag EC, te Riele HP, et al. Disruption of the mouse mdr1a P-glycoprotein gene leads to a deficiency in the blood-brain barrier and to increased sensitivity to drugs. *Cell* 1994;77(4):491–502.

Schinkel AH, Wagenaar E, van Deemter L, Mol CA, Borst P. Absence of the mdr1a P-Glycoprotein in mice affects tissue distribution and pharmacokinetics of dexamethasone, digoxin, and cyclosporin A. *J Clin Invest* 1995;96(4):1698–1705.

Schinkel AH, Mayer U, Wagenaar E, Mol CA, van Deemter L, Smit JJ, van der Valk MA, Voordouw AC, Spits H, van Tellingen O, Zijlmans JM, Fibbe WE, Borst P. Normal viability and altered pharmacokinetics in mice lacking mdr1-type (drug-transporting) P-glycoproteins. *Proc Natl Acad Sci USA* 1997;94(8):4028–4033.

Sharom FJ. ABC multidrug transporters: structure, function and role in chemoresistance. *Pharmacogenomics* 2008;9(1):105–127.

Shen H, Smith DE, Keep RF, Xiang J, Brosius FC, 3rd. Targeted disruption of the PEPT2 gene markedly reduces dipeptide uptake in choroid plexus. *J Biol Chem* 2003;278(7):4786–4791.

Shimizu M, Uno T, Sugawara K, Tateishi T. Effects of single and multiple doses of itraconazole on the pharmacokinetics of fexofenadine, a substrate of P-glycoprotein. *Br J Clin Pharmacol* 2006;62(3):372–376.

Silverman W, Locovei S, Dahl G. Probenecid, a gout remedy, inhibits pannexin 1 channels. *Am J Physiol Cell Physiol* 2008;295(3):C761–C767.

Sinicrope FA, Dudeja PK, Bissonnette BM, Safa AR, Brasitus TA. Modulation of P-glycoprotein-mediated drug transport by alterations in lipid fluidity of rat liver canalicular membrane vesicles. *J Biol Chem* 1992;267(35):24995–25002.

Slitt AL, Cherrington NJ, Dieter MZ, Aleksunes LM, Scheffer GL, Huang W, Moore DD, Klaassen CD. trans-Stilbene oxide induces expression of genes involved in metabolism and transport in mouse liver via CAR and Nrf2 transcription factors. *Mol Pharmacol* 2006;69(5):1554–1563.

Smit JW, Huisman MT, van Tellingen O, Wiltshire HR, Schinkel AH. Absence or pharmacological blocking of placental P-glycoprotein profoundly increases fetal drug exposure. *J Clin Invest* 1999;104(10):1441–1447.

Somogyi A. Renal transport of drugs: specificity and molecular mechanisms. *Clin Exp Pharmacol Physiol* 1996;23(10–11):986–989.

Srimaroeng C, Perry JL, Pritchard JB. Physiology, structure, and regulation of the cloned organic anion transporters. *Xenobiotica* 2008;38(7–8):889–935.

Suzuki M, Suzuki H, Sugimoto Y, Sugiyama Y. ABCG2 transports sulfated conjugates of steroids and xenobiotics. *J Biol Chem* 2003;278(25):22644–22649.

Sweet DH, Miller DS, Pritchard JB, Fujiwara Y, Beier DR, Nigam SK. Impaired organic anion transport in kidney and choroid plexus of organic anion transporter 3 (Oat3 (Slc22a8)) knockout mice. *J Biol Chem* 2002;277(30):26934–26943.

Tahara H, Kusuhara H, Fuse E, Sugiyama Y. P-glycoprotein plays a major role in the efflux of fexofenadine in the small intestine and blood–brain barrier, but only a limited role in its biliary excretion. *Drug Metab Dispos* 2005;33(7):963–968.

Takahashi N, Morita M, Maeda T, Harayama Y, Shimozawa N, Suzuki Y, Furuya H, Sato R, Kashiwayama Y, Imanaka T. Adrenoleukodystrophy: subcellular localization and degradation of adrenoleukodystrophy protein (ALDP/ABCD1) with naturally occurring missense mutations. *J Neurochem* 2007;101(6):1632–1643.

Terada T, Inui K. Gene expression and regulation of drug transporters in the intestine and kidney. *Biochem Pharmacol* 2007;73(3):440–449.

Terada T, Sawada K, Irie M, Saito H, Hashimoto Y, Inui K. Structural requirements for determining the substrate affinity of peptide transporters PEPT1 and PEPT2. *Pflugers Arch* 2000;440(5):679–684.

Tian X, Swift B, Zamek-Gliszczynski MJ, Belinsky MG, Kruh GD, Brouwer KL. Impact of basolateral multidrug resistance-associated protein (Mrp) 3 and Mrp4 on the hepatobiliary disposition of fexofenadine in perfused mouse livers. *Drug Metab Dispos* 2008;36(5):911–915.

Tian X, Zamek-Gliszczynski MJ, Li J, Bridges AS, Nezasa K, Patel NJ, Raub TJ, Brouwer KL. Multidrug resistance-associated protein 2 is primarily responsible for the biliary excretion of fexofenadine in mice. *Drug Metab Dispos* 2008;36(1):61–64.

Timmins JM, Lee JY, Boudyguina E, Kluckman KD, Brunham LR, Mulya A, Gebre AK, Coutinho JM, Colvin PL, Smith TL, Hayden MR, Maeda N, Parks JS. Targeted inactivation of hepatic Abca1 causes profound hypoalphalipoproteinemia and kidney hypercatabolism of apoA-I. *J Clin Invest* 2005;115(5):1333–1342.

Tirona RG, Leake BF, Wolkoff AW, Kim RB. Human organic anion transporting polypeptide-C (SLC21A6) is a major determinant of rifampin-mediated pregnane X receptor activation. *J Pharmacol Exp Ther* 2003;304(1):223–228.

Umbenhauer DR, Lankas GR, Pippert TR, Wise LD, Cartwright ME, Hall SJ, Beare CM. Identification of a P-glycoprotein-deficient subpopulation in the CF-1 mouse strain using a restriction fragment length polymorphism. *Toxicol Appl Pharmacol* 1997;146(1):88–94.

Uwai Y, Okuda M, Takami K, Hashimoto Y, Inui K. Functional characterization of the rat multispecific organic anion transporter OAT1 mediating basolateral uptake of anionic drugs in the kidney. *FEBS Lett* 1998;438(3):321–324.

Uwai Y, Saito H, Inui K. Interaction between methotrexate and nonsteroidal anti-inflammatory drugs in organic anion transporter. *Eur J Pharmacol* 2000;409(1):31–36.

Van Aubel RA, Masereeuw R, Russel FG. Molecular pharmacology of renal organic anion transporters. *Am J Physiol Renal Physiol* 2000;279(2):F216–F232.

Velamakanni S, Wei SL, Janvilisri T, van Veen HW. ABCG transporters: structure, substrate specificities and physiological roles: a brief overview. *J Bioenerg Biomembr* 2007;39(5–6):465–471.

Vig BS, Stouch TR, Timoszyk JK, Quan Y, Wall DA, Smith RL, Faria TN. Human PEPT1 pharmacophore distinguishes between dipeptide transport and binding. *J Med Chem* 2006;49(12):3636–3644.

Vlaming ML, Mohrmann K, Wagenaar E, de Waart DR, Elferink RP, Lagas JS, van Tellingen O, Vainchtein LD, Rosing H, Beijnen JH, Schellens JH, Schinkel AH. Carcinogen and anticancer drug transport by Mrp2 *in vivo*: studies using Mrp2 (Abcc2) knockout mice. *J Pharmacol Exp Ther* 2006;318(1):319–327.

Westerfeld C, Mukai S. Stargardt's disease and the ABCR gene. *Semin Ophthalmol* 2008;23(1):59–65.

Westphal K, Weinbrenner A, Giessmann T, Stuhr M, Franke G, Zschiesche M, Oertel R, Terhaag B, Kroemer HK, Siegmund W. Oral bioavailability of digoxin is enhanced by talinolol: evidence for involvement of intestinal P-glycoprotein. *Clin Pharmacol Ther* 2000a;68(1):6–12.

Westphal K, Weinbrenner A, Zschiesche M, Franke G, Knoke M, Oertel R, Fritz P, von Richter O, Warzok R, Hachenberg T, Kauffmann HM, Schrenk D, Terhaag B, Kroemer HK, Siegmund W. Induction of P-glycoprotein by rifampin increases intestinal secretion of talinolol in human beings: a new type of drug/drug interaction. *Clin Pharmacol Ther* 2000b;68(4):345–355.

Xia CQ, Milton MN, Gan LS. Evaluation of drug–transporter interactions using *in vitro* and *in vivo* models. *Curr Drug Metab* 2007a;8(4):341–363.

Xia C, Yang J, Bbalani S. Drug transporters in drug disposition, drug interactions, and drug resistance. In: Zhang D, Zhu M, Humphreys W, editors. *Drug Metabolism in Drug Design and Development*. Hoboken, NJ: John Wiley & Sons, 2007b, pp. 137–202.

Xia CQ, Yang JJ, Gan LS. Breast cancer resistance protein in pharmacokinetics and drug–drug interactions. *Expert Opin Drug Metab Toxicol* 2005;1(4):595–611.

Yu AM. Small interfering RNA in drug metabolism and transport. *Curr Drug Metab* 2007;8(7):700–708.

Zaher H, Khan AA, Palandra J, Brayman TG, Yu L, Ware JA. Breast cancer resistance protein (Bcrp/abcg2) is a major determinant of sulfasalazine absorption and elimination in the mouse. *Mol Pharm* 2006;3(1):55–61.

Zelcer N, van de Wetering K, Hillebrand M, Sarton E, Kuil A, Wielinga PR, Tephly T, Dahan A, Beijnen JH, Borst P. Mice lacking multidrug resistance protein 3 show altered morphine pharmacokinetics and morphine-6-glucuronide antinociception. *Proc Natl Acad Sci USA* 2005;102(20):7274–7279.

Zhang EY, Knipp GT, Ekins S, Swaan PW. Structural biology and function of solute transporters: implications for identifying and designing substrates. *Drug Metab Rev* 2002;34(4):709–750.

Zhou SF. Drugs behave as substrates, inhibitors and inducers of human cytochrome P450 3A4. *Curr Drug Metab* 2008;9(4):310–322.

Zhou SF, Wang LL, Di YM, Xue CC, Duan W, Li CG, Li Y. Substrates and inhibitors of human multidrug resistance associated proteins and the implications in drug development. *Curr Med Chem* 2008;15(20):1981–2039.

16

POLYMORPHISMS OF DRUG TRANSPORTERS AND THEIR CLINICAL IMPLICATIONS

CINDY Q. XIA AND JOHNNY J. YANG

16.1 INTRODUCTION

The role of transporters in drug disposition was gradually recognized over the last three decades (Juliano and Ling, 1976), much after the emergence of knowledge on CYPs. We are currently following in the footsteps of CYPs knowledge-base development and are steadily increasing our understanding of transporter functions, types, locations, and their roles in a qualitative fashion. Classically, transporters are proteins that translocate endogenous compounds (such as bile acids, lipids, sugars, amino acids, steroids, hormones, and electrolytes) and xenobiotics (such as drugs and toxins) across biological membrane to maintain cellular physiological solute concentrations and fluid balance as well as to provide a mechanism of detoxification for any potentially harmful foreign substances in the cells. Transporter proteins are divided into the adenosine triphosphate (ATP)-binding cassette (ABC) transporter superfamily and the solute carrier (SLC) family of proteins. SLC transporters act by facilitating the uptake of their substrates into the cells. This family of transporters contains 46 subfamilies and 360 transporters including sodium-bile acid cotransporters (NTCP, SLC10 family), proton oligopeptide cotransporters (PEPT, SLC15 family), organic anion transporting polypeptides (OATP, SLC21 family), organic cation/anion/zwitterion transporters (OCT/OAT, SLC22 family), and nucleoside transporters (NT, SLC29 family). SLC

Enzyme Inhibition in Drug Discovery and Development, edited by Lu and Li
Copyright © 2010 John Wiley & Sons, Inc.

transporters are divided into facilitative transporter and active transporter classes. Facilitative transporters are not coupled to any energy source and passively facilitate the diffusion of molecules across the membrane down their concentration gradients allowing a rapid equilibrium across the membrane. The active SLC transporters use an energy source that is provided by an ion-exchanger which causes pH alteration in the microenvironment of the cell surface, or is indirectly coupled to Na^{+}/K^{+} ATPase which can create an intracellular negative membrane potential due to the imbalance in charge movement. ABC efflux membrane transporters consist of transmembrane domains (TMDs) and nucleotide binding domains (NBDs). They are directly coupled to ATPase activity and hydrolyze ATP to derive energy for pumping substrates across the cell membrane. The full efflux transporters, such as P-glycoprotein (P-gp, also known as multidrug resistance 1 (MDR1) protein) and multidrug resistance protein (MRP), possess two NBDs in one polypeptide chain. The half-transporters, such as breast cancer resistance protein (BCRP, also known as mitoxantrone resistance protein [MXR], ABCG2, ABCP), contain only one NBD (Borst and Elferink, 2002). The half-transporters function as a dimer or tetramer bridged by specific linkages. Among 49 human genes in seven subfamilies of ABC transporters, P-gp in ABCB family, MRP1 and MRP2 in ABCC family, and BCRP in ABCG family are the major ABC transporters to confer resistance in the tumor cells and to efflux xenobiotics (such as drugs or toxins) out of normal tissues. Uptake (SLC family) and efflux (ABC family) transporters interact dynamically to mediate the accumulation and translocation of drugs or endogenous substrates into the cells.

Transporter proteins affect drug absorption in small intestine and drug elimination in liver and/or kidney by governing drug substance in and out of the intestinal enterocytes, hepatocytes, or renal tubular cells (Xia et al., 2007, 2008). Transporters can also limit or facilitate penetration of drugs into brain, placenta, tumor, T cells, and so on. The inhibition of transporters or lack of transporter functions can alter the exposure of drugs into the tissues and potentially result in either lack of efficacy or increased toxicity. A classic example of that is provided by studies on antiparasitic agent avermectin, which caused neurotoxicity in CF-1 mice deficient in P-gp (Lankas et al., 1997). The roles of transporters in drug disposition have been evaluated by using transporter knockout or deficient animals or by using transporter inhibitors in both animals and humans.

Polymorphisms of transporters have been drawn great attention due to their potential impact on interindividual differences in pharmacokinetics of a drug and subsequent pharmacological and toxicological effects. Many *in vitro* and clinical studies have demonstrated that some polymorphisms are associated with a change in expression and function of transporters and pharmacokinetics of drugs. However, compared to well-studied polymorphisms of metabolic enzymes, the clinical significance of the generic polymorphisms in transporters is not well understood. Tissue distributions, functions, and evaluation models of drug transporters can be found in Chapter 15 in this book. The present chapter will focus on our current understanding of studies on genetic polymorphisms of the most important transporters including P-gp, BCRP, MRP, OATP, OAT, OCT, PEPT, and their implications in interindividual variations of drug exposure. This should be considered as an updated understanding of this topic, but not an exclusive review. Some of the past reviews for this area are provided in the references for extended reading.

Polymorphic genetic variations have been reported in human MDR1 (P-gp), MRP1, MRP2, BCRP, OATP (OATP1B1, OATP1B3 and OATP2B1), OAT1, OCT1, and OCT2 (Beketic-Oreskovic et al., 1995; Sparreboom et al., 2003; Ho and Kim, 2005; Kerb, 2006). Generally, the role of single nucleotide polymorphisms (SNPs) in drug disposition is confirmed *in vitro* by measuring the efflux or uptake activities of specific substrates in the cells or membranes expressing recombinant protein, and it is confirmed *in vivo* by measuring the expression of mRNA or protein in tissue samples, by assessing the intracellular accumulation of substrates, by evaluating the pharmacokinetic alterations of drug substrates, or by associating clinical outcomes from drug substrates (Kerb, 2006).

16.2 GENETIC POLYMORPHISMS OF MDR1 (P-gp) AND THEIR IMPLICATIONS

16.2.1 Genetic Variability

The MDR1 gene (ABCB1) was the first human ABC transporter gene cloned and characterized in drug-resistance cancer cells (Chen et al., 1986), and its product P-gp has been best characterized among ABC transporters. The MDR1 gene, which is composed of 28 exons, locates on chromosome 7q21.1. Kioka et al. (1989) first reported the polymorphisms of the MDR1 gene from *in vitro* studies. Two amino acid substitutions, Gly185Val and Ala893Ser, were detected in P-gp isolated from colchicine-selected drug-resistant normal human adrenal gland cell cultures. The Gly185Val in P-glycoprotein from colchicine-resistant cells occurred during selection of cells in colchicine treatment. Cells transfected with the MDR1 cDNA carrying Val185 acquire increased resistance to colchicine compared to other drugs. The other amino acid substitution Ala893Ser was suggested to reflect genetic polymorphism. Subsequently, Hoffmeyer et al. (2000) were the first to systemically screen the MDR1 gene for the presence of single-nucleotide polymorphisms (SNP) in 2000. In their report, the original MDR1 gene was defined as the wild type; and all 28 exons, including the core promoter region and exon–intron boundaries, were amplified by PCR. By sequencing the MDR1 gene from 185 Caucasian individuals, 15 polymorphisms have been revealed, which include eight in exons and seven in introns. This report had a tremendous impact because a polymorphism, C3435T in exon 26, which caused no amino acid change, was associated with altered P-gp expression in human duodenum and thereby intestinal absorption of digoxin, a prototypical MDR1 substrate. Individuals with the homozygous T allele (TT genotype) had a twofold decreased P-gp protein level ($P = 0.056$) in duodenum biopsies as compared to those with the wild-type alleles (CC genotype). Consequently, the exposure of digoxin was significantly higher in T allele after oral administration. Later, various investigators have reported that the T allele is associated with increased P-gp protein expression or has no clearly distinguishable effect (Sparreboom et al., 2003; Marzolini et al., 2004; Sakaeda, 2005; Kerb, 2006).

To date, genetic polymorphisms of human MDR1 has been extensively investigated. There are more than 40 SNPs and 64 haplotypes of MDR1 that have been identified. Broad attentions have been focused on the silent mutation C3435T in exon 26 and nonsynonymous variants in exon 21 (Mickley et al., 1998), G2677T (Ala893Ser), and G2677A (Ala893Thr) (Sakaeda, 2005; Kerb, 2006). Potential functional consequence of MDR1 polymorphisms can be assumed from their location within MDR1 gene in relation to the domain structure of P-gp. Both polymorphisms, G2677T/A and G2995A (in exon 24, Ala999Thr), are located in the second transmembrane domain, and the exon G2995A variant is closer to the ABC domain. Since certain serine residues in P-glycoprotein are subject to phosphorylation by protein kinase C, G2677T (Alz893Ser) may result in altered protein function (Tang et al., 2002). The A61G mutation (Asn21Asp) results in a net charge change (basic to acidic) close to the N-terminus of P-glycoprotein, which appears to be of minor functional importance (Ambudkar, 1999). The protein alteration Phe103Leu in exon 5 is located next to the second transmembrane domain on the extracellular side of P-glycoprotein and closes to glycosylation sites of P-glycoprotein. A protein structural alteration may happen due to the change from a large aromatic to a large lipophilic residue. The nonsynonymous G1199A SNP in exon 11 (Ser400Asn) results in a significant size change dependent on pH and isoelectric environment of the residue, which possibly result in a charge change in the protein. This SNP is located on the cytoplasmatic side just preceding the first ATP binding domain (Ambudkar, 1999). The C3435T SNP at a wobble position in exon 26 does not alter its encoded Ile amino acid (Ile1145Ile) and is therefore of apparent silent nature. Although this polymorphism was associated with altered P-glycoprotein expression and function as mentioned above, the molecular basis of this observation is still poorly understood.

16.2.2 Interethnic Variability

There are marked differences in genotype and allele frequencies among African, Caucasian, and Asian populations. The frequencies depend on the origin, even in the same ethnic group living in the same country (Ostrovsky et al., 2004). Interestingly, C3435T was found to be closely linked with C1236 (locates in exon 12) and G2677T (Ieiri et al., 2004a). Over 90% of Japanese, 62% of European American, and 80% of Caucasian German individuals have C3435T and G2677T SNPS simultaneously (Ieiri et al., 2004a). The allelic frequency distributions of SNPs in MDR1 gene have been reported in different racial populations [see reviews: Sparreboom et al. (2003), Sakaeda (2005), Kerb (2006)]. C3435T has been detected in all ethnic populations, albeit with variation in frequencies. The frequency of the C3435T allele has been shown as 43–54% in Caucasians, 34–63% in Asians, and 73–90% in Africans. The incidence of C/T and C/C 3435 genotypes in the African is much higher than those in other racial populations. For G2677T/A, Caucasians (57%) and Japanese (43%) share a similar frequency G2677 allele, while there is a trend of lower frequency in the Indian population (34%) and higher frequency in African American (87%). Additionally, the frequency of Caucasians homozygous for 1236TT is about one-third of the value in Japanese (13.3% versus 37.5%). Moreover, several mutations are only identified in single ethnic populations. For instance, the nonsynonymous SNP at position 3421 was only found in Ghanaians and African Americans (2.4% and 4.3%, respectively), but not in Caucasians. An A61G variant (Asn21Asp) was observed only in Caucasians. Although ethnic difference in P-gp activities have not widely compared, inter-ethnic differences in the distribution of the MDR1 variants are a possible cause of the inter-ethnic differences in the pharmacokinetics of P-gp substrate drugs.

16.2.3 Impact of MDR1 Polymorphism on P-gp Tissue Expression

Hoffmeyer et al. (2000) first reported that C3435T was associated with a lower level of MDR1 in the duodenum and resulted in a high plasma concentration of digoxin. Since then, clinical investigations on MDR1-genotype-related MDR1 activity have been performed primarily focusing on C3435T. However, there seems to be no consensus in the reports on the association of C3435T with MDR1 expression. The association pattern might differ among tissues and be altered depending on pathological conditions or ethnicity, even in the same tissues. In contrast to the observations by Hoffmeyer et al. (2000), Nakamura et al. (2002) reported nonsignificantly increased MDR1 mRNA levels in Japanese subjects with the 3435TT in comparison to the CT and CC groups. Tanabe et al. (2001) found that the P-glycoprotein expression in 100 human placentas was in relation to genotype at position 3435 with the trend of CC > CT > TT, although not significantly. The same group also observed a nonsignificant trend for P-glycoprotein expression in relation to the polymorphism at position 2677 (GG > G/A,T > A,T/A,T). They demonstrated significantly higher P-glycoprotein levels in human placentas from patients with TT genotype in exon 1b (T-129C) in comparison to the group with CT genotype (Tanabe et al., 2001).

To date, there is no rational explanation for the correlation between C3435T and MDR1 expression. The connection might reflect linkage disequilibium between C3435 and a SNP in the promoter region and/or in the exon–intron boundaries that are important for mRNA splicing.

16.2.4 Impact of MDR1 Polymorphisms on Pharmacokinetic

The SNPs of MDR1 was demonstrated to be associated with the changes in P-gp expression and function, as well as subsequent alteration in drug disposition. However, much of the clinical data have been contradictory or inconclusive (Sparreboom et al., 2003). Discordant clinical data have been noted in C3435T for P-gp drug substrates, including digoxin, fexofenadine, cyclosporine, and tacrolium. Similar conflicting findings have also been observed for the exon 21 (G2677T/A) polymorphisms (Sparreboom et al., 2003; Ieiri et al., 2004b; Marzolini et al., 2004, 2005; Sakaeda, 2005; Kerb, 2006). Several clinical studies demonstrated that P-gp polymorphisms have no effect on the plasma concentrations of its substrates, such as squinavir (La Porte et al., 2007), gabapentin (Kang et al., 2007), talinolol (Bernsdorf et al., 2006), lopinavir, efavirenz, loperamide, dicloxacillin, and docetaxel (Sparreboom et al., 2003; Ieiri et al., 2004b; Marzolini et al., 2004; Marzolini et al., 2005; Sakaeda, 2005; Kerb, 2006).

Influence of the ABCB1 polymorphisms 2677G>T/A and 3435C>T on placental P-gp expression and function was evaluated by using dually perfused human placental and well-established P-gp substrate saquinavir. The results indicate that the variant allele 3435T was associated with significantly higher placental P-gp expression than the wild-type alleles. Although the ABCB1 polymorphism 3435C>T altered the expression levels of P-gp in the human placenta, this did not have any consequences on P-gp-mediated placental transfer of saquinavir (Rahi et al., 2008). The P-gp function at the blood–brain barrier, evaluated by integration plot analysis with the first 3-min data using ^{11}C-verapamil as a probe, was not significantly different between the haplotypes of MDR1 genes (1236TT, 2677TT, 3435TT versus 1236CC, 2677GG, 3435CC) (Takano et al., 2006).

Because of conflicting results of the functional significance of MDR1 exon 26 C3435T SNP on the disposition of digoxin in different ethnic groups, Chowbay et al. (2005) performed a meta-analysis on published data investigating the influence of C3435T SNP on the pharmacokinetics of digoxin and the expression of MDR1. Meta-analysis was performed on data from published studies investigating the influence of MDR1 C3435T SNP on digoxin pharmacokinetics, as well as MDR1 expression in Caucasian and Japanese populations. The following outcomes were included: exposures to digoxin measured by area under the concentration–time curve (AUC) and maximum concentration (C_{max}), the mean intestinal MDR1 mRNA expression, and P-gp expression in the absence of digoxin administration. The meta-analysis results of available studies indicate that the synonymous MDR1 C3435T SNP does not affect the pharmacokinetics of digoxin and the expression of MDR1 mRNA. Future studies should focus on the impact of MDR1 haplotypes on the pharmacokinetics of MDR1 substrates rather than the C3435T SNP alone (Chowbay et al., 2005).

Although it is not uncommon to see controversial conclusions from published clinical observations about P-gp polymorphism in drug disposition, even using the same probe drug and in the same racial group, there are several possible reasons for such inconsistent data including different experimental conditions (e.g., probe drug used, dose level, single dose versus repeat dose), small sample sizes, sample selection, or heterogeneity in ethnical diverse populations. Many transporter probe substrates are also substrates for drug-metabolizing enzymes or other transporters. For example, cyclosporine and tacrolimus may be complicated by the involvement of CYP3A metabolism. Digoxin and fexofenadine are also substrates of OATPs. Thus, it is possible that other metabolism and multiple transport mechanisms apart from P-gp contribute to the variable drug disposition.

16.2.5 Impact of MDR1 Polymorphisms on Drug Response and Disease Risk

Besides the direct effect of genetic polymorphism on the drug disposition of P-gp substrates, the association between genetic variant and clinical outcomes remains largely unexplored. The effects of P-gp polymorphisms on drug response are summarized in Table 16.1. C3435T polymorphism was analyzed in 116 patients with allogenic kidney graft treated with cyclosporin A and 144 randomly selected healthy individuals. The prevalence of MDR1 gene genotypes 3435CC, 3435CT, and 3435TT were also compared in patients after allogenic kidney graft with both acute and chronic graft rejection and control groups. The results of the study demonstrated that the allelic frequency and MDR1 genotype distribution were similar in all evaluated groups, indicating that MDR1 gene polymorphism was not a predisposing factor for terminal kidney failure leading to renal transplantation. Therefore, evaluation of C3435T polymorphism of MDR1 gene will probably not be useful for characterization of groups of patients at increased risk of acute and chronic kidney graft rejection (Kotrych et al., 2007).

HIV patients with the MDR1 TT genotype 6 months after starting treatment had a greater rise in CD4-cell count (257 cells/μL) than did patients with the CT (165 cells/μL) and CC (121 cells/μL) genotype ($p = 0.0048$), and they also had the best recovery of naive CD4-cells. These data suggest that P-glycoprotein has an important role in admittance of antiretroviral drugs to restricted compartments *in vivo*, and the polymorphism MDR1 3435 C/T predicts immune recovery after initiation of antiretroviral treatment (Fellay et al., 2002). MDR1 variant C3435T showed association with lower P-gp expression in acute myeloid leukemia (AML) and childhood ALL patients but not in adult ALL patients (Illmer et al., 2002; Jamroziak et al., 2005).

The physiological importance of P-gp in the GI tract has come from the description of the mdr1 knock-out mice model, which develops a spontaneous colitis in a specific pathogen-free environment. A meta-analysis of the available findings obtained with two-SNPs polymorphisms (C3435T and G2677T/A) in inflammation bowel disease (IBD) demonstrated a significant association of 3435T allele and 3435TT genotype with ulcerative colitis, but no association

TABLE 16.1. Effects of P-gp Polymorphisms on Drug Resposne

Drug Response	Populations	Reference
Efficacy of HIV Drug Therapy		
3435T allele associated with better immune recovery after HAART[a]	Caucasian	Fellay et al. (2002)
3435C allele associated with failure of vial and immune response to HAART	Caucasian	Brumme et al. (2003)
C3435T not associated with immune response to HAART	Caucasian	Winzer et al. (2005)
1236T with better immune recovery after HIV protease inhibitor treatment, no effect of G2677GT and C3435CT	Asian	Zhu et al. (2004)
3435.Cn heterozygot children had more rapid virologic responses to HAART	Caucasian	Saitoh et al. (2005)
3435T, 2677T and 2677T/13435T not associated with viral/immune response to HAART	Caucasian	Winzer et al. (2005)
3435T. 2677T and 2677Tfi435T 1101 associated with viral/ immune response to HAART	Caucasian	Verstuyft et al. (2005)
3435TT with better viral response and fewer viral resistance in efavirenz patients	Caucasian and African-American	Haas et al. (2005)
Efficacy of AML Treatment		
1236C/2677G/3435C associated with lower P-gp expression and decreased survival	Caucasian	Illmer et al. (2002)
Efficacy of ALL Treatment		
C3435T not associated with prognostic effects	Caucasian	Jamroziak et al. (2005)
Efficacy of Atorvastatin Treatment		
3435T and 2677C/3435T with increased response	Asian	Kajinami et al. (2004)
Radiotherapy Response		
Patients with 2677G-3435C haplotype had a significant better response to radiotherapy than those with the other haplotypes		Wang et al. (2007)
Efficacy of Antiepileptic and Antipsychiatric Therapy		
3435C associated with drug-resistance in epilepsy	Caucasian	Siddiqui et al. (2003)
C3435T not associated with drug-resistance in epilepsy	Caucasian	Tan et al. (2004)

TABLE 16.1. ***Continued***

Drug Response	Populations	Reference
No association with C3435T	Caucasian	Sills et al. (2005)
G2677T and C3435T are not associated, with therapeutic response to paroxetine in patients with major depressive disorder		Mihaljevic-Peles et al. (2008)
C3435T but not G2677 is associated, with some therapeutic response to bromperidol in schizophrenic patients		Yasui-Furukori et al. (2006)
Steroid Dose/Weaning		
3435T and 2677T associated with success of steroid weaning in pediatric heart transplantation patients	Caucasian and African-American	Zheng et al. (2002)
3435T—and not 2677T—associated with success of steroid weaning in pediatric heart transplantation patients	Caucasian and African-American	Zheng et al. (2004)
Toxicity/Safety		
3435TT associated with cyclosporin nephrotoxicity	Caucasian	Hauser et al. (2005)
3435TT associated with nortriptyline-induced postural hypotension	Caucasian	Roberts et al. (2002)
2677T positive predictor of tacroliinus-induced neurotoxicity	Asian	Yamauchi et al. (2002)
3435T and 2677T associated with lower risk for steroid-induced osteonecrosis in kidney transplantation patients	Asian	Asano et al. (2003)

[a]HAART, high active antiretroviral therapy.

with Crohn's disease (Annese et al., 2006). The MDR1 exon 26 C3435T allele has also been suggested to have a protective role for patients with parkinsonism and especially among those with a history of exposure to pesticides. This assumes that pesticides in general are substrates of P-gp and that CNS exposure to such agents may cause neurodegenerative disease (Furuno et al., 2002; Drozdzik et al., 2003). Recent findings also suggested that the C3435T allele was significantly greater in subjects with drug-resistant seizure disorders (Siddiqui et al., 2003). Although C3435T MDR1 polymorphism is not an important genetic risk factor for rheumatoid arthritis (RA) susceptibility, the risk of having an active form of RA resistance to therapy with disease-modifying antirheumatic drugs (DMARDs) in patients with C3435C and C3435T genotype was 2.89-fold higher than in homozygous T3435T subjects (Pawlik et al., 2004; Drozdzik et al., 2006).

16.3 GENETIC POLYMORPHISMS OF MRP AND THEIR IMPLICATIONS

MRPs are organic anion pumps belonging to the C subfamily of ABC transporter families. MRP1, MRP2, and MRP3 have similar substrates, including glutathione, glucuronate, and sulfate conjugates of the drugs and unconjugated drugs, such as methotrexate, vinblatine, irinotecan, daunorubicin, pravastatine, and HIV inhibitors (Xia et al., 2007, 2008). The MRP1 and MRP2 genes locate in chromosome 16p13.1 and 10q24, respectively.

ABCC genes have been screened in Japanese, Chinese, and Caucasians (Ito et al., 2001; Saito et al., 2002; Lang et al., 2004; Niemi et al., 2004; Wang et al., 2004). Although a high number of rare mutations, which lead to amino acid changes, has been reported in literature and database (Pharmcogenetics and Pharmacogenomics Knowledge Base, www.pharmgkb.org), there are few striking common nonsynonymous. Nonsynonymous SNPs that occur with a frequency of clearly more than 1% have only been reported for MRP2 (ABCC2): G1249A (Val471Ile; 14% in African Americans, 13% in Asians, and 24% Caucasians), C2943G (Phe981Leu; 4% in Caucasians), and G4544A (Cysl515Tyr; 2% in Caucasians), as well as for ABCC3, C202T (His68Tyr; 2% in Caucasians), and G3890A (Arg1297His; 5% in Caucasian). The contribution of common nonsynonymous SNPs to the genetic variability of expression and function of MRP transporters is limited. Little success has been attained to characterize the functional effects of MRP1, MRP2, and MRP3 variants by *in vitro* assays and clinical trials. The more common ABCC2 nonsynonymous variants Val471Ile and Ala1450Thr did not differ from wild type regarding their *in vitro* transport activity for various substrates (Hirouchi et al., 2004). Niemi et al. (2004) screened the ABCC2 gene and failed to find a correlation between ABCC2 SNPs with pravastatin plasma levels, while Lang et al. (2004) systematically screened the entire ABCC3 gene for genetic variants and found a statistical significant correlation of a polymorphism in the promoter region with mRNA expression in hepatocytes, which apparently affected the binding of nuclear factors as demonstrated by electrophoretic mobility shift assay.

16.4 GENETIC POLYMORPHISMS OF BCRP AND THEIR IMPLICATIONS

BCRP (ABCG2, formerly known as MXR/ABCP) gene locates in chromosome 4q22. BCRP has been systemically screened for single nucleotide polymorphism (SNP) in 90 different ethnic populations. More than 40 non-synonymous and synonymous SNPs have been revealed in the promoter as well as in both the exon and intron sequences (Honjo et al., 2002; Lepper et al., 2005). The two most frequent naturally occurred SNPs G34A and C421A have been identified in humans (Zamber et al., 2003). G34A variant in exon 2 resulting in a Val12Met amino acid change has been associated with a low BCRP protein expression and an altered efflux function in cancer cells (Imai et al., 2002; Zamber et al., 2003; Mizuarai et al., 2004). All Mexican-Indians screened possess at least one variant allele, while the frequency in Caucasians was only 4.7%. Recent studies suggested that nasopharyngeal carcinoma patients who were wild type for the G34A showed a trend toward lower systemic exposure of irinotecan compared with patients with one or two variant alleles (Table 16.2) (Zhou et al., 2005).

The mutation C421A at exon 5 with an amino acid change of Gln to Lys at codon 141 has been reported with (a) a reduced BCRP protein level but not mRNA level in PA317 cells and (b) a decreased BCRP efflux function in transfected LLC-PK and HEK283 cells (Imai et al., 2002; Mizuarai et al., 2004; Kobayashi et al., 2005). Although C421A did not influence interindividual variation in the expression of BCRP mRNA and protein in human intestines (Zamber et al., 2003), it significantly affected the pharmacokinetics of diflomotecan. In five cancer patients heterozygous for this allele, plasma diflomotecan levels after intravenous administration were about threefold higher than those in 15 patients with wild-type alleles (Table 16.3) (Sparreboom et al., 2004). These findings provide the first evidence linking variant BCRP C421A allele to altered BCRP substrate drug exposure and suggest that BCRP genotype might contribute to interindividual variability. The same group recently also demonstrated that the oral bioavailability of topotecan in two cancer patients heterozygous for C421A allele increased 1.3-fold compared with 10 wild-type patients, due to the increased oral absorption (Table 16.2) (Sparreboom et al., 2005). C421A allele also showed increased oral absorption of sulfasalazine in Japanese healthy volunteers (Yamasaki et al., 2008) and rosuvastatin in Chinese healthy volunteers (Zhang et al., 2006) (Table 16.2). The AUC_{0-48hr} of sulfasalazine increased about 3.5-fold in AA^{421} homozygous than in CC^{421} homozygous (Yamasaki et al., 2008). In contrast, no significant changes in irinotecan pharmacokinetics were observed in relation to the BCRP C421A genotype in European Caucasians (De Jong et al., 2004) (Table 16.2). There were no significant differences in nitrofurantoin plasma exposure (AUC_{0-72h}) and urinary elimination among the ABCG2 CC^{421}, CA^{421}, and AA^{421} genotypic cohorts from 36 prescreened Chinese healthy subjects (12 subjects per genotype). The lack of the contribution of this genetic variant to irinotecan and nitrofurantoin disposition might be obscured by a functional role of some other polymorphic proteins. C421A allele frequency varies highly among different populations (Imai et al., 2002; Backstrom et al., 2003; Zamber et al., 2003). C421A allele appears to be very common in Japanese and Chinese populations, with allele frequency between 26% and 35%. In both European

TABLE 16.2. Effects of ABCG2 Polymorphism on Pharmacokinetics of Drug Substrate

Drug	Population	Parameter	Effect	Reference
Irinotecan	Cancer patients (G34A) (GG/GA+AA) (11/10)	C_{max}	Significantly higher in A allele	Zhou et al. (2005)
Diflomotecan	Cancer patients (C421A) (CC/CA/AA) (15/5/0)	AUC, C_{max} (i.v.)	Significantly higher in CA	Sparreboom et al. (2004)
		AUC, C_{max} (p.o.)	No significant difference	Sparreboom et al. (2004)
Irinotecan	European Caucasians cancer patients (C421A) (CC/CA/AA) (68/14/2)	AUC, C_{max}	No significant difference	De Jong et al. (2004)
Nitrofurantoin	Chinese healthy volunteer (C421A) (CC/CA/AA) (12/12/12)	AUC (p.o.), urinary elimination	No significant difference	Adkison et al. (2008)
Sulfasalazine	Japanese healthy volunteer (C421A) (CC/CA/AA) (12/12/12)	AUC (p.o.)	Increased in 1.9-fold in CA and 3.5-fold in AA	Yamasaki et al. (2008)
Rosuvastatin	Chinese healthy volunteer (C421A) (CC/CA/AA) (12/12/12)	AUC, C_{max} (p.o.)	Increased in CA + AA	Zhang et al. (2006)
Topotecan	Cancer patients (C421A) (CC/CA/AA) (10/2/0)	AUC (i.v.)	No significant difference	Sparreboom et al. (2005)
		AUC (p.o.)	Increased in CA but not significant ($P = 0.16$)	
		Bioavailability	Significantly higher in CA	Sparreboom et al. (2005)

AUC, area under the time-concentration curve; AA, homozygous variant; GA or CA, heterozygous variant; GG or CC, wild type; C_{max}, The maximal drug plasma concentration; i.v., intravenous injection; p.o., oral administration.

TABLE 16.3. Effects of SLCO1B1 (OATP1B1, OATP-C) Polymorphism on Pharmacokinetics of Drug Substrate

Drug	Allele (Haplotype): Population (Frequency)[a]	Parameter	Effect	Reference
Pravastatin	*1b (130Asp): AS (54%), CA (13%)	AUC_{0-6hr} Urine elimination	Decrease[b] Increased	Mwinyi et al. (2004)
	*15 (130Asp/174Ala): AS (10%), CA (12%)	AUC	Increased	Nishizato et al. (2003), Niemi et al. (2004)
	*17 -1187A/130Asp/174Ala: CA(1%)	AUC_{0-12hr}	Increased	Niemi et al. (2004)
	*18 (130Asp/155Thr): CA (9%)	AUC_{0-12hr}	No change	Niemi et al. (2004)
	*19 (643Phe): CA (1%)	AUC_{0-12hr}	No change	Niemi et al. (2004)
	*20/*21: CA (7%)	AUC_{0-12hr}	No change	Niemi et al. (2004)
	*5 (174Ala): AS (1%), CA (1%)	AUC_{0-12hr}	Increased	Mwinyi et al. (2004)
	*15 (130Asp/174Ala): AS (10%), CA (12%)	Myopathy	Higher	Morimoto et al. (2004)
Statins	*5 (174Ala): AS (1%), CA (1%)	Cholesterol level	Decreased	
Fluvastatin	*14 (130Asp/667Gly): AS (0%), CA (rare)	Cholesterol level	Decreased	Couvert et al. (2008)
	*5 (174Ala): AS (1%), CA (1%)	AUC_{0-12hr} and C_{max}	Increased	Niemi et al. (2006b)
Simvastatin acid	*5 (174Ala): AS (1%), CA (1%)	AUC_{0-12hr} and Cmax	Increased	Pasanen et al. (2006)
Simvastatin	rs4149056 C variant	Myopathy	Higher	Link et al. (2008)
Pitavastatin	*5 (174Ala): AS (1%), CA (1%)	AUC	Increased	Chung et al. (2005)
Fexofenadine	*5 (174Ala): AS (1%), CA (1%)	AUC	Increased	Niemi et al. (2005)
Repaglinide	*5 (174Ala): AS (1%), CA (1%)	AUC	Increased	Kalliokoski et al. (2008a)
Nateglinide	*5 (174Ala): AS (1%), CA (1%)	AUC	No change	Kalliokoski et al. (2008a)
Rosiglitazone	*5 (174Ala): AS (1%), CA (1%)	AUC	No change	Kalliokoski et al. (2008b)
Pioglitazone	*5 (174Ala): AS (1%), CA (1%)	AUC	No change	Kalliokoski et al. (2008b)
Rosuvastatin	*5 (174Ala): AS (1%), CA (1%)	Drug exposure	Increased	Lee et al. (2005a)
	*15 (130Asp/174Ala): AS (10%), CA (12%)	AUC and C_{max}	Increased	Choi et al. (2007)

[a]AS, Asians; CA, Caucasians.
[b]Statically not significant.

and American Caucasian, the allele frequency is approximately 10–12%. The low frequency (<1%) is observed in African Americans and Africans from North or South of Sahara.

The R482T or R482G mutant of BCRP, where the wild-type arginine on the amino acid position 482 is replaced by threonine or glycine, alters the substrate specificity (Allen et al., 2002; Chen et al., 2003). Interestingly, these mutations were only found in drug-resistant human tumor cell lines (Honjo et al., 2001; Robey et al., 2001), but not in human individuals (Honjo et al., 2002; Zamber et al., 2003).

16.5 GENETIC POLYMORPHISMS OF OATP (SLC 21) AND THEIR IMPLICATIONS

Genetic variants of uptake transporter have been predominantly investigated for OATPs (gene symbol SLCO). OATPs mediate the uptake of anionic, neutral, and cationic compounds. The major OATPs includes OATP1A2 (OATP-A, SLCO1A2), OATP1B1 (OATP-C, SLCO1B1), OATP1B3, (OATP-8, SLCO1B3), and OATP2B1 (OATP-B, SLCO2B1).

16.5.1 Polymorphisms of OATP1A2 (SCLO1A2)

OATP1A2 has been reported to be the only transporter in the brain capillary endothelium and may play a critical role in the CNS penetration of many drugs and hormones across the blood–brain barrier (BBB) (Gao et al., 2000). Therefore, genetic variability of OATP1A2 could have an important impact on the efficacy and central nerve system (CNS) toxicity of various drugs. OAP1A2 mRNA has also been detected in liver and kidney. Its substrates include endogenous compounds such as bile acids, steroids, hormones, and their conjugates, thyroid hormones as well as drugs such as 3-hydroxy-3-methylglutaryl–coenzyme A (HMG-CoA) reductase inhibitors, fexofenadine, ouabain, and peptides (Xia et al., 2008). In 2001, several SLCO1A2 mutations were identified in regulatory regions which can affect the gene expression from 48 Japanese individuals. Later, six nonsynonymous polymorphisms within the coding region of SLCO1A2, T38C (Ile3Thr), A516C (Glu172Asp), G559A (Ala187Thr), A382T (Asn128Tyr), A404T (Asn187Thr), and C2003G (Thr668Ser) have been detected in 96 individuals from various ethnic backgrounds (Lee et al., 2005b). The allelic frequencies of these variants appeared to be ethnicity-dependent, and some of these genetic variants were associated with markedly reduced uptake transporter activity. T38C and A516C polymorphisms were more common in European-Americans (11.1% and 5.3%, respectively) than in Hispanic- and African-Americans, whereas these SNPs were not observed with Chinese-Americans. G559A variation was observed only in Hispanic-Americans (0.5%), whereas A382T was found only in African-Americans (1.0%). C2003G variation was observed in African- and

Hispanic-Americans with varying allelic frequencies of 3.7% and 1.0%, respectively. None of the six nonsynonymous SNPs tested were observed in Chinese-Americans. *In vitro* functional assessment revealed that the A516C and A404T variants had markedly reduced capacity for mediating the cellular uptake of OATP1A2 substrates, estrone 3-sulfate, and two δ-opioid receptor agonists, deltorphin II and [D-penicillamine2,5]-enkephalin. On the other hand, the G559A and C2003G variants appeared to have substrate-dependent changes in transport activity. Cell surface biotinylation and immunofluorescence confocal microscopy suggested that altered plasma membrane expression of the transporter may contribute to reduced transport activity associated with the A516C, A404T, and C2003G variants. The A404T (N135I) variant also showed a shift in the apparent molecular size suggesting alterations in glycosylation status (Lee et al., 2005b). The contribution of SLCO1A2 genetic variants to the interindividual variability in drug disposition and CNS entry needs to be further investigated in clinic.

16.5.2 Polymorphisms of OATP1B1 (SLCO1B1)

OATP1B1, OATP1B3, and OATP2B1 are mainly expressed in sinusoidal (basolateral) membrane of hepatocytes. OATP2B1 mRNA has been also detected in intestine, pancreas, lung, ovary, testes, heart, and spleen. They have been demonstrated to extract their substrates from the blood into the liver. OATP1B1 and OATP1B3 share the similar substrates as for OATP1A2, albeit with difference in affinity. OATP2B1 exhibits restricted substrate specificity and only transport cholyltaurine and pravastin at acidic pH; thus its importance in the drug hepatic elimination is not clear yet (Marzolini et al., 2005). To date, 44 polymorphism of SLCO1B1 (encoding OATP1B1) have been identified in coding regions, including 17 nonsynonymous (change of an amino acid), 4 conservative (no change of amino acid), 20 intronic, and 3 in the promoter regions. Of the nonsynonymous variant, seven were common: A388G (Asnl30Asp, in OATP1B1*1b), and T521C (Va1174Ala, OATP1B1*5) occurred in African-Americans (74% and 1%), Asians (63% and 16%) and Caucasians (40% and 14%) while A452G (Asnl51Ser, OATP1B1*16) was detected specifically in Asians (3.8%), and C463A (Prol55Thr, OATP1B1*4) (8%) and A1929C (Leu643Phe, OATP1B1*19) (9%) were specific for Caucasians and G1463C (Gly488Ala, OAP1B1*9) (9%) and A2000G (Glu667Gly, OATP1B1*11) (34%) was found only in African-Americans (Tirona et al., 2001; Michalski et al., 2002; Nozawa et al., 2002; Niemi et al., 2004; Sakaeda, 2005). *In vitro* experiments with cultured cells expressing wild type and mutated OATP1B1 revealed that several variants exhibited markedly reduced uptake of estrone sulfate and estradiol 17β-D-glucoronide (Tirona et al., 2001). Alterations in transport activity were specifically associated with SNPs that introduced amino acid changes within the transmembrane-spanning domains and extracellular loop 5. The common amino acid variants with altered transporter function were Va1174Ala (*5) Gly488Ala (*9). Interestingly, the variant A1964G (Asp655Gly, *10) reduced the transport activity of estrone sulfate but not that of 17β-D-glucuronide, suggesting possible substrate-dependent polymorphisms. Cell surface trafficking defects proved to be responsible for altered transport function of many of these SLCO1B1 variants (Tirona et al., 2001).

16.5.2.1 Impact of OATP1B1 Polymorphisms on Pharmacokinetics. Nishizato et al. (2003) provided the first evidence in human that SLCO1B1 variants were associated with altered pharmacokinetics of pravastatin. Subjects with the OATP1B1*15 allele (130Asp174Ala) had a reduced total and nonrenal clearance and increase plasma concentrations of pravastatin as compared with those with the SLCO1B1 allele (130Asp). These findings suggest much of the loss of function defect associated with the SLCO1B1 haplotype is related to the Val174Ala mutation. Subsequently, several groups demonstrated that SLCO1B1 variant haplotypes had significant effects on the disposition, efficacy, and toxicity of HMG-CoA reductase inhibitors. The T521C (Val174Ala) genetic polymorphism of SLCO1B1 considerably increases the plasma concentration of simvastatin acid and moderately increases those of pravastatin but seems to have no significant effect on fluvastatin. Effects of SLCO1B1 polymorphism on the response of drug substrates in humans are summarized in Table 16.3. After a single oral dose of 40 mg pravastatin, the $AUC_{(0-6h)}$ of pravastatin from the OATP1B1*1b (A388G) group was more than 60% lower than those derived from carriers of the wild-type OATP1B1*1a haplotype, although this difference failed to reach statistical significance. However, the amount of pravastatin excreted into the urine from time 0 h to 12 h was significantly diminished in the OATP1B1*1b haplotype group compared with *1a wild-type control subjects. Whereas the AUC_{0-6h} of pravastatin from OATP1B1*5 (T521C) was more than twofold higher than those from the OATP1B1*1b (A388G) group. It seems that *5 expression delayed the hepatocellular uptake of pravastatin, while *1b expression accelerated OATP1B1-dependent uptake of the drug (Mwinyi et al., 2004). In subjects with the G-11187A or T521C genotype, the mean AUC_{0-12h} of pravastatin was 98% (P = 0.0061) or 106% (P = 0.0034) higher, respectively, compared to subjects with the reference genotype. These results were substantiated by haplotype analysis. In heterozygous carriers of *15B (containing the A388G and T521C variants) and *17 (containing the G-11187A, A388G and T521C variants), the mean AUC_{0-12h} of pravastatin was 93% (P = 0.024) and 130% (P = 0.0053) higher, respectively, compared to noncarriers. These results suggest that haplotypes are more informative in predicting the OATP1B1 phenotype than single SNPs (Niemi et al., 2004). In summary, individuals carrying the *SLCO1B1* T521C (V174A) variant, as found in SLCO1B1*5, *15, *16 and *17 haplotypes, are associated with higher pravastatin exposure, a result consistent with decreased *in vitro* transport function.

Systemic exposure to rosuvastatin had been observed to be approximately twofold higher in Japanese subjects living in Japan compared with white subjects in Western Europe or the United States. Lee et al. (2005a) observed that plasma exposure to rosuvastatin and its metabolites was significantly higher in Chinese, Malay, and Asian-Indian subjects compared with white subjects living in the same environment. While Lee et al. found an association between the homozygous *SLCO1B1* 521CC genotype and increased rosuvastatin levels in white subjects, heterozygosity at this allele did not confer differences in drug concentrations within each ethnic group. It is worth noting that the genetic polymorphism in *SLCO1B1* at the 521 position did not account for the clear population differences in rosuvastatin exposures among white subjects and the Asian groups. These results may not be surprising, given the lack of profound differences in the allelic frequencies of the 521C variant among white (0.222), Chinese (0.086), Malay (0.129), and Asian-Indian (0.071) subjects. It seems that the pharmacogenetics of other rosuvastatin disposition pathways may better explain the ethnic differences in pharmacokinetics because no other known *SLCO1B1* polymorphism is more predominant in Asians and white subjects yet.

16.5.2.2 Impact of OATP1B1 Polymorphisms on the Treatment Outcome. Notably, the factors that may determine statin's therapeutic response are as important as those that affect statins' clearance from the systemic circulation because the liver is both the site of the drug target and the major eliminating organ. Hence, interindividual differences in the expression and function of hepatocyte drug transporters that take up (i.e., basolateral uptake transporters) or take away (i.e., biliary or basolateral efflux transporters) statin to and from its intracellular target, HMG-CoA reductase, would be expected to translate into variability in hepatocellular drug levels and the resulting antihyperlipidemic effects. In support of this notion, a small retrospective study by Tachibana-Iimori et al. indicated that individuals with *SLCO1B1* 521C alleles had attenuated lipid-lowering effects by statins (pravastatin, atorvastatin, and simvastatin) in comparison with those homozygous for 521T, consistent with *in vitro* evidence for decreased hepatocyte uptake transport function of this variant. Because individuals with *SLCO1B1* 521C alleles have higher drug levels of pravastatin than those with 521T, the concentration–response relationship is likely to differ among individuals depending on the *SLCO1B1* genotype (Tachibana-Iimori et al., 2004).

In rare cases, statins can cause muscle pain or weakness in association with elevated creatine kinase levels (i.e., myopathy); occasionally, this leads to muscle breakdown and myoglobin release (i.e., rhabdomyolysis), with a risk of renal failure and death. The mechanisms by which statins cause myopathy remain unknown but appear to be related to statin concentrations in the blood. Morimoto et al. have found that the frequency of OATP-C*15 is significantly higher in patients who experienced myopathy after receiving pravastatin or atorvastatin than in patients without myopathy. However, there were two patients who experienced pravastatin-induced myopathy despite the fact that they did not possess OATP-C*15 or other known mutations of OATP1B1 that have been reported to decrease the function of OATP1B1 (Morimoto et al., 2004). Recently, The ongoing Study of the Effectiveness of Additional Reductions in Cholesterol and Homocysteine (SEARCH) collaboration group has carried out a genome wide association study using approximately 300,000 markers (and additional fine-mapping) in 85 subjects with definite or incipient myopathy and 90 controls, all of whom were taking 80 mg of simvastatin daily as part of a trial involving 12,000 participants. A single strong correlation of myopathy with the rs4363657 SNP located within SLCO1B1 on chromosome 12 was yielded from 300,000 markers. The noncoding rs4363657 SNP was in nearly complete linkage disequilibrium with the nonsynonymous rs4149056 SNP ($r^2 = 0.97$), which has been linked to statin disposition. The prevalence of the rs4149056 C allele in the population was 15%. Among participants taking 80 mg of simvastatin daily, rs4149056 CC homozygotes had an 18% cumulative risk, with myopathy occurring primarily during the first year, whereas the CT genotype was associated with a cumulative risk of about 3%. In contrast, the cumulative risk of myopathy was only 0.6% among TT homozygotes. Overall, more than 60% of these myopathy cases could be attributed to the rs4149056 C variant in *SLCO1B1*. No SNPs in any other region were clearly associated with myopathy. This genome-wide study from the SERACH group has identified common genetic variants in *SLCO1B1* that are associated with substantial alterations in the risk of simvastatin-induced myopathy. These findings are likely to apply to other statins because myopathy is a class effect, and *SLCO1B1* polymorphisms affect the blood levels of several statins. Moreover, these variants may be relevant to the effects of other classes of drugs transported by OATP1B1. Consequently, the genotyping of *SLCO1B1* polymorphisms may be useful in the future for tailoring both the statin dose and safety monitoring in order to obtain the benefits of statin therapy more safely and effectively.

Tirona et al. (2003) demonstrated that various SNPs in SLCO1B1 markedly reduced the hepatocellular uptake of rifampin, suggesting that carriers of these functionally deficient OATP-C variants may exhibit reduced capacity for rifampin-mediated induction of hepatic drug-metabolizing enzymes and transporters. This finding also indicates that transporters, metabolizing enzymes, and regulatory factors should be viewed and evaluated as an integrated system when analyzing the genetic of drug response. Later, Niemi et al. (2006a) demonstrated that SLCO1B1 polymorphisms did not affect the extent of induction of hepatic CYP3A4 by rifampicin in humans, probably because other uptake transporters, such as OATP1B3, can compensate for reduced uptake of rifampicin by OATP1B1.

16.5.3 Polymorphisms of OATP1B3 (SLCO1B3) and OATP2B1 (SLCO2B1)

Compared to OATP1B1, OATP1B3 (SLC1B3) and OATP2B1 (SLCO2B1) are less intensively studied. OATP1B3 is the only OATP family member able to transport digoxin and cholecystokinin, thus it would be of interest to study the effect of genetic variants of OATP1B3 on the transport of its specific substrates *in vivo*. SLCOIB3 was genetically characterized with three nonsynonymous SNPs: 112Ala and 233Ile, which were the major variants in Caucasians with frequency of 78% and 71%, respectively, and 522Cys, which occurred in 2% of Caucasians. While protein from the two frequent variants showed equal localization and uptake of bromosulfophthalein and bile acids *in vitro*, OATPIB3 522Cys protein was retained intracellularly and abolished bromosulfophthalein uptake completely (Letschert et al., 2004).

SLCO2B1 gene (encoding OATP2B1) has two nonsynonymous SNPs: assigned names as OATP-B*2 (C1175T, T392I) and OATP-B*3 (C1457T, S486F). Nozawa et al. (2002) described the presence of a nonsynonymous SNP in SLCO2B1 at codon 486 (S486F, OATP-B*3) that reduces the uptake rate of estrone sulfate in cells expressing the variant transporter. The functional importance of this variant to the *in vivo* disposition of substrate drugs remains to be defined.

16.6 GENETIC POLYMORPHISMS OF OAT AND OCT (SLC22) AND THEIR IMPLICATIONS

OAT1 (gene symbol SLC22A6), OAT2 (gene symbol SLC22A7), and OAT3 (gene symbol SLC22A8) mainly locate in kidney proximal tubular cells and have important roles in renal clearance and drug disposition of organic anions. Among these three OATs, OAT1 and OAT3 are polymorphic. Race-specific SNPs of SLC22A6 (product OAT1) and SLC22A8 (product OAT3) have been identified. Among 12 OAT1 coding region variants, G1361A (Arg454Gln) resulted in the function loss for the uptake of methotrexate, *p*-aminohippurate, and ochratoxin in transfected cells. However, there was no significant difference in the renal clearance of adefovir observed between the subjects with OAT1 454Gln variant and reference transporter OAT1 454Arg. These data indicate that the coding region of OAT1 has low genetic and functional diversity and the coding region variants of OAT1 may not contribute substantially to inter-individual differences in renal elimination of xenobiotics (Zair et al., 2008). For OAT3, more than 12 nonsynonymous SNPs have been identified in the coding regions. C1166T (Ala389Val) showed little difference in pravastatin clearance when compared with wild-type transporter in the clinic studies (Nishizato et al., 2003). Other SNPs exhibited altered transporter functions *in vitro*, but the clinical relevant genetic variants have not been well established (Zair et al., 2008).

OCT1 (gene symbol SLC22A1) and OCT2 (gene symbol SLC22A2) have an extensive overlap in their substrate specificity, although they show distinctive distribution in tissue distributions (Xia et al., 2008). OCT1 is primarily found in sinusoidal membrane of hepatocytes and to a less extent in intestinal cells, while OCT2 and its splice variant OCT2-A are in mainly found in kidney proximal tubular cells (Xia et al., 2008).

Genotyping studies have identified over 200 SNPs in the SLC22A1 genes, and most of the variants have been functionally characterized with *in vitro* studies. The SNPs and the results of their functions have been published in the PharmaGKB database (Kerb, 2006). Of the SNPs isolated, C41T (S14F), identified in African-Americans, is associated with increased uptake of toxin 1-methyl 1–4 phenylpyridinium (MPP) *in vitro* (Shu et al., 2003). C848T (Pro283Leu) and C859G (Arg287Gly), which are the SNPs expressed in Asian subjects, showed reduced uptake of MPP and tetraethylammonium bromide (TEA) with a reduction in K_m and V_{max} *in vitro* (Itoda et al., 2002; Sakata et al., 2004). The OCT1 protein variants Ser14Phe (C41T), Arg61Cys (C181T), Ser189Leu (C566T), Gly401Ser (G1203A), 420del (1260delATG), and Gly465Arg (G1393A) exhibited reduced or lost uptake of metformin (an OCT substrate and widely used in the treatment of Type 2 diabetes) in stable transfected HEK293 cells, as a result of lowered V_{max} when compared with the reference proteins (Shu et al., 2008). Among these seven nonsynonymous, 420del and Arg61Cys are common polymorphisms of OCT1, with allele frequencies of 19% and 7.2%, respectively, in individuals of European descent. Recently, the same research group (Shu et al., 2007) has demonstrated that metformin has higher AUC, higher C_{max}, and lower oral volume of distribution (V/F) in the individuals carrying a reduced function OCT1 allele (Arg61Cys, Gly401Ser, 420del, or Gly465Arg). Clinical and Oct1 (–/–) knockout mice studies demonstrated that OCT1 expression may associate with metfomin efficacy (Shu et al., 2008). OCT1 mRNA levels tend to be lower in Met408Val (A1222G) variant, although not statistically significant. In healthy volunteers, metformin treatment resulted in significantly elevated plasma glucose concentrations and prolonged plasma insulin level in subjects expressing M408V (V1222G), although the basal plasma glucose levels were similar in the OCT variant and reference subjects. However, in Japanese patients the polymorphism of OCT1 (including Met408Val) and OCT2 have little contribution to the clinical efficacy of metformin (Shikata et al., 2007).

In contrast to those in the SLC22A1 (OCT1) gene, it appears that the number of nonsynonymous variants in the SLC22A2 (encoding OCT2) gene and their allelic frequencies were lower than in other known drug transporter genes such as MDR1, MRP1, MRP2, and OATP-C. The observations are consistent with the finding of a lower frequency of nonsynonymous variants in ethnically diverse genomic DNA samples (Leabman et al., 2002). Recent population-genetic analysis has demonstrated that selection has acted against amino acid changes in OCT2 (Leabman et al., 2002), suggesting that OCT2 is relatively intolerant of nonsynonymous changes. This selection may be due to a necessary role of OCT2 in the renal elimination of endogenous amines or xenobiotics, including environmental toxins, neurotoxic amines, and therapeutic drugs. A number of race-specific SLC22A2 SNPs have been identified from genetic variant screening studies. Of the SNPs, C596T (Thr199Ile) and

C602T (Thr201Met) are in Korean and Japanese subjects, A493G (Met165Val), G495A (Met165Ile), and C1198T (Arg400Cys) are in African-American subjects, and A1294C (Lys432Gln) is present in both African-American and Mexican-American populations (Leabman et al., 2002; Kerb, 2006; Zair et al., 2008). In general, the less frequent nonsynonymous variants resulted in more significant and deleterious functional changes. However, the G808T (Ala270Ser) variant was reported to exhibit subtle functional differences from the reference form of OCT2 (Leabman et al., 2002). A clinical study in Chinese healthy volunteers showed that the G808T (Ala270Ser) polymorphism is associated with a reduced metformin renal or tubular clearance (Wang et al., 2008). The renal tubular clearance of metformin averaged 8.78 ± 1.75, 7.68 ± 0.672, and 6.32 ± 0.954 ml/min/kg for participants with GG (n = 6), GT (n = 5), and TT (n = 4) genotypes, respectively (P = 0.037, one-way analysis of variance). Moreover, the inhibition of metformin renal tubular secretion by cimetidine also appeared to be dependent on this mutation. In the presence of cimetidine, metformin clearance was decreased in all participants, but the decrease was significantly lower in the TT group than in the GG group (18.7 versus 48.2%, P = 0.029). Although the altered transporter activity and substrate selectivity of other SLC22A2 SNPs have been demonstrated *in vitro*, the clinical relevant remains to be further investigated.

16.7 GENETIC POLYMORPHISMS OF PEPT (SLC15) AND THEIR IMPLICATIONS

PEPT1 is predominantly localized at the apical (luminal) membrane of absorptive enterocytes of small intestines and mainly responsible for the active absorption of di- and tripeptides as well as hydrophilic drugs such as β-lactam antibiotics and angiotensin-converting enzyme inhibitors (Xia et al., 2008). To date, 40 coding polymorphisms and over 100 haplotypes have been identified in the SLC15A1 gene (encoding PEPT1). Although the allelic frequencies of many PEPT1 SNPs are low, race-specific SNPs have been observed. Although PEPT1 is an attractive target for improving oral drug delivery, functional characterization and protein expression of genetic variants still remain *in vitro* studies and no clinical relevant variant has been identified yet.

16.8 SUMMARY

Drug transporters play important roles in drug disposition, efficacy, and toxicity. Several polymorphic variants of ABC and SLC transporters, which have been summarized in this chapter, may alter protein expression and function in humans, such as BCRP C421A SNP and OATP1B1 T521C SNP, which have the potential to be the most clinically relevant in terms of *in vitro* versus *in vivo* data correlation and altered transporter activities. Recently, advances in high-throughput DNA sequencing technologies have provided a lot of information on occurrence and frequency of drug transporter polymorphisms. Transfected cells have been served as a great tool to evaluate the expression, localization, and function of transporter genetic variants. However, the effects of genetic variants in drug transporter genes associated with its phenotypical consequence are still controversial, because contradictory results have been reported. Most published studies have experienced small sample sizes in relation to the allele and genotype frequency of the studied variant, and they have also encountered subjects and probe drugs with confounding factors.

Transporters may interplay with drug metabolism enzymes and are regulated by several nuclear receptors. A probe drug is usually a substrate for multiple transporters and metabolism enzymes. To evaluate the genetic component of drug transporters, a more integrative approach, considering several genes with involvement of functional units and pathways, is necessary. Given the presence of linkage disequilibrium that exists for many SNP investigated so far, studies on the effects of haplotypes rather than of SNPs have been increased. Other factors such as lifestyle, comedication, and comortality must also be considered in line with a patient's genetic makeup. With more knowledge and advanced technologies, we will continue increasing our appreciation of understanding the underlying biological mechanisms on interindividual variability in drug disposition and efficacy and lowering the drug-associated risk by tailoring the doses based on patient genetic profiling.

REFERENCES

Adkison KK, Vaidya SS, Lee DY, Koo SH, Li L, Mehta AA, Gross AS, Polli JW, Lou Y, Lee EJ. The ABCG2 C421A polymorphism does not affect oral nitrofurantoin pharmacokinetics in healthy Chinese male subjects. *Br J Clin Pharmacol* 2008;66: 233–239.

Allen JD, Jackson SC, Schinkel AH. A mutation hot spot in the Bcrp1 (Abcg2) multidrug transporter in mouse cell lines selected for doxorubicin resistance. *Cancer Res* 2002;62:2294–2299.

Ambudkar SV. DS, Hrycyna CA, Ramachandra M, Pstan I, Gottesman MM. Biochemical, cellular and pharcological aspects of the multidrug transporter. *Annu Rev Pharmacol Toxicol* 1999;39:361–398.

Annese V, Valvano MR, Palmieri O, Latiano A, Bossa F, Andriulli A. Multidrug resistance 1 gene in inflammatory bowel disease: a meta-analysis. *World J Gastroenterol* 2006;12:3636–3644.

Asano T, Takahashi KA, Fujioka M, Inoue S, Okamoto M, Sugioka N, Nishino H, Tanaka T, Hirota Y, Kubo T. ABCB1 C3435T and G2677T/A polymorphism decreased the risk for steroid-induced osteonecrosis of the femoral head after kidney transplantation. *Pharmacogenetics* 2003;13:675–682.

Backstrom G, Taipalensuu J, Melhus H, Brandstrom H, Svensson A-C, Artursson P, Kindmark A. Genetic variation in the ATP-binding Cassette Transporter gene ABCG2 (BCRP) in a Swedish population. *Eur J Pharm Sci* 2003;18:359–364.

Beketic-Oreskovic L, Duran GE, Chen G, Dumontet C, Sikic BI. Decreased mutation rate for cellular resistance to doxorubicin and suppression of mdr1 gene activation by the cyclosporin PSC 833. *J Nat Cancer Inst* 1995;87:1593–1602.

Bernsdorf A, Giessmann T, Modess C, Wegner D, Igelbrink S, Hecker U, Haenisch S, Cascorbi I, Terhaag B, Siegmund W. Simvastatin does not influence the intestinal P-glycoprotein and MPR2, and the disposition of talinolol after chronic medication in healthy subjects genotyped for the ABCB1, ABCC2 and SLCO1B1 polymorphisms. *Br J Clin Pharm* 2006;61:440–450.

Borst P, Elferink RO. Mammalian ABC transporters in health and disease. *Annu Rev Biochem* 2002;71:537–592.

Brumme ZL, Dong WWY, Chan KJ, Hogg RS, Montaner JSG, O'Shaughnessy MV and Harrigan PR. Influence of polymorphisms within the CX3CR1 and MDR-1 genes on initial antiretroviral therapy response. *AIDS (London, U. K.) FIELD Full Journal Title:AIDS (London, United Kingdom)* 2003;17:201–208.

Chen CJ, Chin JE, Ueda K, Clark DP, Pastan I, Gottesman MM, Roninson IB. Internal duplication and homology with bacterial transport proteins in the mdr1 (P-glycoprotein) gene from multidrug-resistant human cells. *Cell* 1986;47:381–389.

Chen Z-S, Robey RW, Belinsky MG, Shchaveleva I, Ren X-Q, Sugimoto Y, Ross DD, Bates SE, Kruh GD. Transport of methotrexate, methotrexate polyglutamates, and 17b-estradiol 17-(b-D-glucuronide) by ABCG2: effects of acquired mutations at R482 on methotrexate transport. *Cancer Res* 2003;63:4048–4054.

Choi JH, Lee MG, Cho JY, Lee JE, Kim KH, Park K. Influence of OATP1B1 genotype on the pharmacokinetics of rosuvastatin in Koreans. *Clin Pharmacol Ther* 2007;83:251–257.

Chowbay B, Li H, David M, Cheung YB, Lee EJ. Meta-analysis of the influence of MDR1 C3435T polymorphism on digoxin pharmacokinetics and MDR1 gene expression. *Br J Clin Pharmacol* 2005;60:159–171.

Chung J-Y, Cho J-Y, Yu K-S, Kim J-R, Oh D-S, Jung H-R, Lim K-S, Moon K-H, Shin S-G, Jang I-J. Effect of OATP1B1 (SLCO1B1) variant alleles on the pharmacokinetics of pitavastatin in healthy volunteers. *Clin Pharmacol Ther (NY)* 2005;78: 342–350.

Couvert P, Giral P, Dejager S, Gu J, Huby T, Chapman MJ, Bruckert E, Carrie A. Association between a frequent allele of the gene encoding OATP1B1 and enhanced LDL-lowering response to fluvastatin therapy. *Pharmacogenomics* 2008;9:1217–1227.

De Jong FA, Marsh S, Mathijssen RHJ, King C, Verweij J, Sparreboom A, McLeod HL. ABCG2 pharmacogenetics: ethnic differences in allele frequency and assessment of influence on Irinotecan disposition. *Clin Cancer Res* 2004;10:5889–5894.

Drozdzik M, Bialecka M, Mysliwiec K, Honczarenko K, Stankiewicz J, Sych Z. Polymorphism in the P-glycoprotein drug transporter MDR1 gene: a possible link between environmental and genetic factors in Parkinson's disease. *Pharmacogenetics* 2003;13:259–263.

Drozdzik M, Rudas T, Pawlik A, Kurzawski M, Czerny B, Gornik W, Herczynska M. The effect of 3435C>T MDR1 gene polymorphism on rheumatoid arthritis treatment with disease-modifying antirheumatic drugs. *Eur J Clin Pharmacol* 2006; 62:933–937.

Fellay J, Marzolini C, Meaden ER, Back DJ, Buclin T, Chave J-P, Decosterd LA, Furrer H, Opravil M, Pantaleo G, Retelska D, Ruiz L, Schinkel AH, Vernazza P,

Eap CB, Telenti A. Response to antiretroviral treatment in HIV-1-infected individuals with allelic variants of the multidrug resistance transporter 1: a pharmacogenetics study. *Lancet* 2002;359:30–36.

Furuno T, Landi M-T, Ceroni M, Caporaso N, Bernucci I, Nappi G, Martignoni E, Schaeffeler E, Eichelbaum M, Schwab M, Zanger UM. Expression polymorphism of the blood–brain barrier component P-glycoprotein (MDR1) in relation to Parkinson's disease. *Pharmacogenetics* 2002;12:529–534.

Gao B, Hagenbuch B, Kullak-Ublick GA, Benke D, Aguzzi A, Meier PJ. Organic anion-transporting polypeptides mediate transport of opioid peptides across blood–brain barrier. *J Pharmacol Exp Ther* 2000;294:73–79.

Haas DW, Smeaton LM, Shafer RW, Robbins GK, Morse GD, Labbe L, Wilkinson GR, Clifford DB, D'Aquila RT, De Gruttola V, Pollard RB, Merigan TC, Hirsch MS, George AL, Jr, Donahue JP, Kim RB. Pharmacogenetics of long-term responses to antiretroviral regimens containing efavirenz and/or nelfinavir: an adult AIDS clinical trials group study. *J Infect Dis* 2005;192:1931–1942.

Hauser IA, Schaeffeler E, Gauer S, Scheuermann EH, Wegner B, Gossmann J, Ackermann H, Seidl C, Hocher B, Zanger UM, Geiger H, Eichelbaum M, Schwab M. ABCB1 genotype of the donor but not of the recipient is a major risk factor for Cyclosporine-related nephrotoxicity after renal transplantation. *J Am Soc Nephrol* 2005;16:1501–1511.

Hirouchi M, Suzuki H, Itoda M, Ozawa S, Sawada J, Ieiri I, Ohtsubo K, Sugiyama Y. Characterization of the cellular localization, expression level, and function of SNP variants of MRP2/ABCC2. *Pharm Res* 2004;21:742–748.

Ho RH, Kim RB. Transporters and drug therapy: implications for drug disposition and disease. *Clin Pharmacol Ther* 2005;78:260–277.

Hoffmeyer S, Burk O, Von Richter O, Arnold HP, Brockmoller J, Johne A, Cascorbi I, Gerloff T, Roots I, Eichelbaum M, Brinkmann U. Functional polymorphisms of the human multidrug-resistance gene: multiple sequence variations and correlation of one allele with P-glycoprotein expression and activity *in vivo*. *Proc Natl Acad Sci USA* 2000;97:3473–3478.

Honjo Y, Hrycyna CA, Yan Q-W, Medina-Perez WY, Robey RW, Van de Laar A, Litman T, Dean M, Bates SE. Acquired mutations in the MXR/BCRP/ABCP gene alter substrate specificity in MXR/BCRP/ABCP-overexpressing cells. *Cancer Res* 2001;61:6635–6639.

Honjo Y, Morisaki K, Huff LM, Robey RW, Hung J, Dean M, Bates SE. Single-nucleotide polymorphism (SNP) analysis in the ABC half-transporter ABCG2 (MXR/BCRP/ABCP1). *Cancer Biol Ther* 2002;1:696–702.

Ieiri I, Suzuki H, Kimura M, Takane H, Nishizato Y, Irie S, Urae A, Kawabata K, Higuchi S, Otsubo K, Sugiyama Y. Influence of common variants in the pharmacokinetic genes (OATP-C, UGT1A1, and MRP2) on serum bilirubin levels in healthy subjects. *Hepatol Res* 2004a;30:91–95.

Ieiri I, Takane H, Otsubo K. The MDR1 (ABCB1) gene polymorphism and its clinical implications. *Clin Pharmacokinet* 2004b;43:553–576.

Illmer T, Schuler US, Thiede C, Schwarz UI, Kim RB, Gotthard S, Freund D, Schakel U, Ehninger G, Schaich M. MDR1 gene polymorphisms affect therapy outcome in acute myeloid leukemia patients. *Cancer Res* 2002;62:4955–4962.

Imai Y, Nakane M, Kage K, Tsukahara S, Ishikawa E, Tsuruo T, Miki Y, Sugimoto Y. C421A polymorphism in the human breast cancer resistance protein gene is associated with low expression of Q141K protein and low-level drug resistance. *Mol Cancer Ther* 2002;1:611–616.

Ito S, Ieiri I, Tanabe M, Suzuki A, Higuchi S, Otsubo K. Polymorphism of the ABC transporter genes, MDR1, MRP1 and MRP2/cMOAT, in healthy Japanese subjects. *Pharmacogenetics* 2001;11:175–184.

Itoda M, Saito Y, Soyama A, Saeki M, Murayama N, Ishida S, Sai K, Nagano M, Suzuki H, Sugiyama Y, Ozawa S, Sawada Ji J. Polymorphisms in the ABCC2 (cMOAT/MRP2) gene found in 72 established cell lines derived from Japanese individuals: an association between single nucleotide polymorphisms in the 5′-untranslated region and exon 28. *Drug Metab Dispos* 2002;30:363–364.

Jamroziak K, Balcerczak E, Cebula B, Kowalczyk M, Panczyk M, Janus A, Smolewski P, Mirowski M, Robak T. Multi-drug transporter MDR1 gene polymorphism and prognosis in adult acute lymphoblastic leukemia. *Pharmacol Rep* 2005;57:882–888.

Juliano RL, Ling V. A surface glycoprotein modulating drug permeability in Chinese hamster ovary cell mutants. *Biochim Biophys Acta Biomembranes* 1976;455: 152–162.

Kajinami K, Brousseau ME, Ordovas JM, Schaefer EJ. Polymorphisms in the multidrug resistance-1 (MDR1) gene influence the response to atorvastatin treatment in a gender-specific manner. *Am J Cardiol* 2004;93:1046–1050.

Kalliokoski A, Neuvonen M, Neuvonen PJ, Niemi M. Different effects of SLCO1B1 polymorphism on the pharmacokinetics and pharmacodynamics of repaglinide and nateglinide. *J Clin Pharmacol* 2008a;48:311–321.

Kalliokoski A, Neuvonen M, Neuvonen PJ, Niemi M. No significant effect of SLCO1B1 polymorphism on the pharmacokinetics of rosiglitazone and pioglitazone. *Br J Clin Pharmacol* 2008b;65:78–86.

Kang H-A, Cho H-Y, Lee Y-B. The effect of MDR1 G2677T/A polymorphism on pharmacokinetics of gabapentin in healthy Korean subjects. *Arch Pharmacol Res* 2007;30:96–101.

Kerb R. Implications of genetic polymorphisms in drug transporters for pharmacotherapy. *Cancer Lett (Amsterdam)* 2006;234:4–33.

Kioka N, Tsubota J, Kakehi Y, Komano T, Gottesman MM, Pastan I, Ueda K. P-glycoprotein gene (MDR1) cDNA from human adrenal: normal P-glycoprotein carries gly185 with an altered pattern of multidrug resistance. *Biochem Biophys Res Commun* 1989;162:224–231.

Kobayashi D, Ieiri I, Hirota T, Takane H, Maegawa S, Kigawa J, Suzuki H, Nanba E, Oshimura M, Terakawa N, Otsubo K, Mine K, Sugiyama Y. Functional assessment of ABCG2 (BCRP) gene polymorphisms to protein expression in human placenta. *Drug Metab Dispos* 2005;33:94–101.

Kotrych K, Sulikowski T, Domanski L, Bialecka M, Drozdzik M. Polymorphism in the P-glycoprotein drug transporter MDR1 gene in renal transplant patients treated with cyclosporin A in a Polish population. *Pharmacol Rep* 2007;59:199–205.

La Porte CJ, Li Y, Beique L, Foster BC, Chauhan B, Garber GE, Cameron DW, van Heeswijk RP. The effect of ABCB1 polymorphism on the pharmacokinetics of saquinavir alone and in combination with ritonavir. *Clini Pharmacol Ther* 2007; 82:389–395.

Lang T, Hitzl M, Burk O, Mornhinweg E, Keil A, Kerb R, Klein K, Zanger UM, Eichelbaum M, Fromm MF. Genetic polymorphisms in the multidrug resistance-associated protein 3 (ABCC3, MRP3) gene and relationship to its mRNA and protein expression in human liver. *Pharmacogenetics* 2004;14:155–164.

Lankas GR, Cartwright ME, Umbenhauer D. P-glycoprotein deficiency in a subpopulation of CF-1 mice enhances avermectin-induced neurotoxicity. *Toxicol Appl Pharmacol* 1997;143:357–365.

Leabman MK, Huang CC, Kawamoto M, Johns SJ, Stryke D, Ferrin TE, DeYoung J, Taylor T, Clark AG, Herskowitz I, Giacomini KM. Polymorphisms in a human kidney xenobiotic transporter, OCT2, exhibit altered function. *Pharmacogenetics* 2002;12:395–405.

Lee E, Ryan S, Birmingham B, Zalikowski J, March R, Ambrose H. Rosuvastatin pharmacokinetics and pharmacogenetics in white and Asian subjects residing in the same environment. *Clin Pharmacol Ther* 2005a;78:330–341.

Lee W, Glaeser H, Smith LH, Roberts RL, Moeckel GW, Gervasini G, Leake BF, Kim RB. Polymorphisms in human organic anion-transporting polypeptide 1A2 (OATP1A2): implications for altered drug disposition and central nervous system drug entry. *J Biol Chem* 2005b;280:9610–9617.

Lepper ER, Nooter K, Verweij J, Acharya MR, Figg WD, Sparreboom A. Mechanisms of resistance to anticancer drugs: the role of the polymorphic ABC transporters ABCB1 and ABCG2. *Pharmacogenomics* 2005;6:115–138.

Letschert K, Keppler D, Koenig J. Mutations in the SLCO1B3 gene affecting the substrate specificity of the hepatocellular uptake transporter OATP1B3 (OATP8). *Pharmacogenetics* 2004;14:441–452.

Link E, Parish S, Armitage J, Bowman L, Heath S, Matsuda F, Gut I, Lathrop M, Collins R, Meade T, Sleight P, Collins R, Barton J, Bray C, Wincott E, Clarke R, Graham I, Simpson D, Warlow C, Wilcken D, Tobert J, Musliner T, Wilhelmsen L, Doll R, Fox KM, Hill C, Sandercock P, Peto R. SLCO1B1 variants and statin-induced myopathy—a genomewide study. *N Engl J Med* 2008;359:789–799.

Marzolini C, Paus E, Buclin T, Kim RB. Polymorphisms in human MDR1 (P-glycoprotein): recent advances and clinical relevance. *Clin Pharmacol Ther (St Louis)* 2004;75:13–33.

Marzolini C, Tirona RG, Kim RB. Pharmacogenetics of drug transporters. *Drugs Pharm Sci* 2005;156:109–155.

Michalski C, Cui Y, Nies AT, Nuessler AK, Neuhaus P, Zanger UM, Klein K, Eichelbaum M, Keppler D, Konig J. A naturally occurring mutation in the SLC21A6 gene causing impaired membrane localization of the hepatocyte uptake transporter. *J Biol Chem* 2002;277:43058–43063.

Mickley LA, Lee J-S, Weng Z, Zhan Z, Alvarez M, Wilson W, Bates SE, Fojo T. Genetic polymorphism in MDR-1: a tool for examining allelic expression in normal cells, unselected and drug-selected cell lines, and human tumors. *Blood* 1998;91: 1749–1756.

Mihaljevic-Peles A, Bozina N, Sagud M, Rojnic Kuzman M, Lovric M. MDR1 gene polymorphism: therapeutic response to paroxetine among patients with major depression. *Prog Neuro-Psychopharmacol Biol Psychiatry* 2008;32:1439–1444.

Mizuarai S, Aozasa N, Kotani H. Single nucleotide polymorphisms result in impaired membrane localization and reduced ATPase activity in multidrug transporter ABCG2. *Int J Cancer* 2004;109:238–246.

Morimoto K, Oishi T, Ueda S, Ueda M, Hosokawa M, Chiba K. A novel variant allele of OATP-C (SLCO1B1) found in a Japanese patient with pravastatin-induced myopathy. *Drug Metab Pharmacokinet* 2004;19:453–455.

Mwinyi J, Johne A, Bauer S, Roots I, Gerloff T. Evidence for inverse effects of OATP-C (SLC21A6) 5 and 1b haplotypes on pravastatin kinetics. *Clin Pharmacol Ther (St Louis)* 2004;75:415–421.

Nakamura T, Sakaeda T, Horinouchi M, Tamura T, Aoyama N, Shirakawa T, Matsuo M, Kasuga M, Okumura K. Effect of the mutation (C3435T) at exon 26 of the MDR1 gene on expression level of MDR1 messenger ribonucleic acid in duodenal enterocytes of healthy Japanese subjects. *Clin Pharmacol Ther (St Louis)* 2002; 71:297–303.

Niemi M, Schaeffeler E, Lang T, Fromm MF, Neuvonen M, Kyrklund C, Backman JT, Kerb R, Schwab M, Neuvonen PJ, Eichelbaum M, Kivistoe KT. High plasma pravastatin concentrations are associated with single nucleotide polymorphisms and haplotypes of organic anion transporting polypeptide-C (OATP-C, SLCO1B1). *Pharmacogenetics* 2004;14:429–440.

Niemi M, Kivistoe KT, Hofmann U, Schwab M, Eichelbaum M, Fromm MF. Letter to the editors: Fexofenadine pharmacokinetics are associated with a polymorphism of the SLCO1B1 gene (encoding OATP1B1). *Br J Clin Pharmacol* 2005;59:602–604.

Niemi M, Kivistoe KT, Diczfalusy U, Bodin K, Bertilsson L, Fromm MF, Eichelbaum M. Effect of SLCO1B1 polymorphism on induction of CYP3A4 by rifampicin. *Pharmacogenet Genomics* 2006a;16:565–568.

Niemi M, Pasanen MK, Neuvonen PJ. SLCO1B1 polymorphism and sex affect the pharmacokinetics of pravastatin but not fluvastatin. *Clin Pharmacol Ther (NY)* 2006b;80:356–366.

Nishizato Y, Ieiri I, Suzuki H, Kimura M, Kawabata K, Hirota T, Takane H, Irie S, Kusuhara H, Urasaki Y, Urae A, Higuchi S, Otsubo K, Sugiyama Y. Polymorphisms of OATP-C (SLC21A6) and OAT3 (SLC22A8) genes: consequences for pravastatin pharmacokinetics. *Clin Pharmacol Ther* 2003;73:554–565.

Nozawa T, Nakajima M, Tamai I, Noda K, Nezu J-I, Sai Y, Tsuji A, Yokoi T. Genetic polymorphisms of human organic anion transporters OATP-C (SLC21A6) and OATP-B (SLC21A9): allele frequencies in the Japanese population and functional analysis. *J Pharmacol Exp Ther* 2002;302:804–813.

Ostrovsky O, Nagler A, Korostishevsky M, Gazit E, Galski H. Genotype and allele frequencies of C3435T polymorphism of the MDR1 gene in various Jewish populations of Israel. *Ther Drug Monit* 2004;26:679–684.

Pasanen MK, Neuvonen M, Neuvonen PJ, Niemi M. SLCO1B1 polymorphism markedly affects the pharmacokinetics of simvastatin acid. *Pharmacogenet Genomics* 2006;16:873–879.

Pawlik A, Wrzesniewska J, Fiedorowicz-Fabrycy I, Gawronska-Szklarz B. The MDR1 3425 polymorphism in patients with rheumatoid arthritis. *Int J Clin Pharmacol Ther* 2004;42:496–503.

Rahi M, Heikkinen T, Hakkola J, Hakala K, Wallerman O, Wadelius M, Wadelius C, Laine K. Influence of adenosine triphosphate and ABCB1 (MDR1) genotype on the P-glycoprotein-dependent transfer of saquinavir in the dually perfused human placenta. *Hum Exp Toxicol* 2008;27:65–71.

Roberts RL, Joyce PR, Mulder RT, Begg EJ, Kennedy MA. A common P-glycoprotein polymorphism is associated with nortriptyline-induced postural hypotension in patients treated for major depression. *Pharmacogenomics J* 2002;2:191–196.

Robey RW, Honjo Y, van de Laar A, Miyake K, Regis JT, Litman T, Bates SE. A functional assay for detection of the mitoxantrone resistance protein, MXR (ABCG2). *Biochim Biophys Acta* 2001;1512:171–182.

Saito S, Iida A, Sekine A, Miura Y, Ogawa C, Kawauchi S, Higuchi S, Nakamura Y. Identification of 779 genetic variations in eight genes encoding members of the ATP-binding cassette, subfamily C (ABCC/MRP/CFTR). *J Hum Genet* 2002;47:147–171.

Saitoh A, Singh KK, Powell CA, Fenton T, Fletcher CV, Brundage R, Starr S, Spector SA. An MDR1-3435 variant is associated with higher plasma nelfinavir levels and more rapid virologic response in HIV-1 infected children. *AIDS (London)* 2005;19:371–380.

Sakaeda T. MDR1 genotype-related pharmacokinetics: fact or fiction? *Drug Metab Pharmacokinet* 2005;20:391–414.

Sakata T, Anzai N, Shin HJ, Noshiro R, Hirata T, Yokoyama H, Kanai Y, Endou H. Novel single nucleotide polymorphisms of organic cation transporter 1 (SLC22A1) affecting transport functions. *Biochem Biophys Res Commun* 2004;313:789–793.

Shikata E, Yamamoto R, Takane H, Shigemasa C, Ikeda T, Otsubo K, Ieiri I. Human organic cation transporter (OCT1 and OCT2) gene polymorphisms and therapeutic effects of metformin. *J Hum Genet* 2007;52:117–122.

Shu Y, Leabman MK, Feng B, Mangravite LM, Huang CC, Stryke D, Kawamoto M, Johns SJ, DeYoung J, Carlson E, Ferrin TE, Herskowitz I, Giacomini KM, Benet LZ, Brett CM, Burchard EG, Castro R, de la Cruz M, Edwards RH, Gitschier J, Glatt CE, Ho C, Kroetz DL, Lin ET, Reus VI, Sadee W, Salazar M, Schaefer C, Sheiner LB, Tran C, Urban TJ, Vulpe C, Wright EM. Evolutionary conservation predicts function of variants of the human organic cation transporter, OCT1. *Proc Natl Acad Sci USA* 2003;100:5902–5907.

Shu Y, Sheardown SA, Brown C, Owen RP, Zhang S, Castro RA, Ianculescu AG, Yue L, Lo JC, Burchard EG, Brett CM, Giacomini KM. Effect of genetic variation in the organic cation transporter 1 (OCT1) on metformin action. *J Clin Invest* 2007;117:1422–1431.

Shu Y, Brown C, Castro RA, Shi RJ, Lin ET, Owen RP, Sheardown SA, Yue L, Burchard EG, Brett CM, Giacomini KM. Effect of genetic variation in the organic cation transporter 1, OCT1, on metformin pharmacokinetics. *Clin Pharmacol Ther (NY)* 2008;83:273–280.

Siddiqui A, Kerb R, Weale M, Brinkmann U, Smith A, Goldstein DB, Wood NW, Sisodiya SM. Association of multidrug resistance in epilepsy with a polymorphism in the drug-transporter gene ABCB1. *N Engl J Med* 2003;348:1442–1448.

Sills GJ, Mohanraj R, Butler E, McCrindle S, Collier L, Wilson EA, Brodie MJ. Lack of association between the C3435T polymorphism in the human multidrug resistance (MDR1) gene and response to antiepileptic drug treatment. *Epilepsia* 2005;46:643–647.

Sparreboom A, Danesi R, Ando Y, Chan J, Figg WD. Pharmacogenomics of ABC transporters and its role in cancer chemotherapy. *Drug Resistance Updates* 2003;6:71–84.

Sparreboom A, Gelderblom H, Marsh S, Ahluwalia R, Obach R, Principe P, Twelves C, Verweij J, McLeod HL. Diflomotecan pharmacokinetics in relation to ABCG2 421C>A genotype. *Clin Pharmacol Ther (St Louis)* 2004;76:38–44.

Sparreboom A, Loos WJ, Burger H, Sissung TM, Verweij J, Figg WD, Nooter K, Gelderblom H. Effect of ABCG2 genotype on the oral bioavailability of topotecan. *Cancer Biol Ther* 2005;4:e7–e10.

Tachibana-Iimori R, Tabara Y, Kusuhara H, Kohara K, Kawamoto R, Nakura J. Effect of genetic polymorphism of OATP-C (SLCO1B1) on lipid-lowering response to HMG-CoA reductase inhibitors. *Drug Metab Pharmacokinet* 2004;19:375–380.

Takano A, Kusuhara H, Suhara T, Ieiri I, Morimoto T, Lee YJ, Maeda J, Ikoma Y, Ito H, Suzuki K, Sugiyama Y. Evaluation of in vivo P-glycoprotein function at the blood–brain barrier among MDR1 gene polymorphisms by using 11C-verapamil. *J Nuclear Med* 2006;47:1427–1433.

Tan NCK, Heron SE, Scheffer IE, Pelekanos JT, McMahon JM, Vears DF, Mulley JC, Berkovic SF. Failure to confirm association of a polymorphism in ABCB1 with multidrug-resistant epilepsy. *Neurology* 2004;63:1090–1092.

Tanabe M, Ieiri I, Nagata N, Inoue K, Ito S, Kanamori Y, Takahashi M, Kurata Y, Kigawa J, Higuchi S, Terakawa N, Otsubo K. Expression of P-glycoprotein in human placenta: relation to genetic polymorphism of the multidrug resistance (MDR)-1 gene. *J Pharmacol Exp Ther* 2001;297:1137–1143.

Tang K, Ngoi SM, Gwee PC, Chua JM, Lee EJ, Chong SS, Lee CG. Distinct haplotype profiles and strong linkage disequilibrium at the MDR1 multidrug transporter gene locus in three ethnic Asian populations [see comment]. *Pharmacogenetics* 2002;12: 437–450.

Tirona RG, Leake BF, Merino G, Kim RB. Polymorphisms in OATP-C. Identification of multiple allelic variants associated with altered transport activity among European- and African-Americans. *J Biol Chem* 2001;276:35669–35675.

Tirona RG, Leake BF, Wolkoff AW, Kim RB. Human organic anion transporting polypeptide-C (SLC21A6) is a major determinant of rifampin-mediated pregnane X receptor activation. *J Pharmacol Exp Ther* 2003;304:223–228.

Verstuyft C, Marcellin F, Morand-Joubert L, Launay O, Brendel K, Mentre F, Peytavin G, Gerard L, Becquemont L, Aboulker JP. Absence of association between MDR1 genetic polymorphisms, indinavir pharmacokinetics and response to highly active antiretroviral therapy. *AIDS (Hagerstown, MD)* 2005;19:2127–2131.

Wang H, Hao B, Zhou K, Chen X, Wu S, Zhou G, Zhu Y, He F. Linkage disequilibrium and haplotype architecture for two ABC transporter genes (ABCC1 and ABCG2) in Chinese population: implications for pharmacogenomic association studies. *Ann Hum Genet* 2004;68:563–573 [erratum appears in *Ann Hum Genet.* 2005;69(Pt 4):499].

Wang Z-J, Yin OQP, Tomlinson B, Chow MSS. OCT2 polymorphisms and *in-vivo* renal functional consequence: studies with metformin and cimetidine. *Pharmacogenet Genomics* 2008;18:637–645.

Wang Z-Y, Chen L-H, Fan Q, Yan W-P, Chen Y-Q, Li Q-S, Yuan Y-W. Relationship between radiosensitivity of nasopharyngeal carcinoma and MDR1 gene polymorphism. *Nanfang Yike Daxue Xuebao* 2007;27:580–583.

Winzer R, Langmann P, Zilly M, Tollmann F, Schubert J, Klinker H, Weissbrich B. No influence of the P-glycoprotein polymorphisms MDR1 G2677T/A and C3435T on the virological and immunological response in treatment of naive HIV-positive patients. *Ann Clin Microbiol Antimicrob* 2005;4:3–9.

Xia CQ, Milton MN, Gan L-S. Evaluation of drug-transporter interactions using *in vitro* and *in vivo* models. *Curr Drug Metab* 2007;8:341–363.

Xia CQ, Yang JJ, Balani SK. Drug transporters in drug disposition, drug interactions, and drug resistance. In *Drug Metabolism in Drug Design and Development*. Zhang D, Zhu M, Humphreys WG, eds. Wiley-Interscience, 2008, pp. 137–203.

Yamasaki Y, Ieiri I, Kusuhara H, Sasaki T, Kimura M, Tabuchi H, Ando Y, Irie S, Ware J, Nakai Y, Higuchi S, Sugiyama Y. Pharmacogenetic characterization of sulfasalazine disposition based on NAT2 and ABCG2 (BCRP) gene polymorphisms in humans. *Clin Pharmacol Ther* 2008;84:95–103.

Yamauchi A, Ieiri I, Kataoka Y, Tanabe M, Nishizaki T, Oishi R, Higuchi S, Otsubo K, Sugimachi K. Neurotoxicity induced by tacrolimus after liver transplantation: relation to genetic polymorphisms of the ABCB1 (MDR1) gene. *Transplantation* 2002;74:571–573.

Yasui-Furukori N, Saito M, Nakagami T, Kaneda A, Tateishi T, Kaneko S. Association between multidrug resistance 1 (MDR1) gene polymorphisms and therapeutic response to bromperidol in schizophrenic patients: a preliminary study. *Prog Neuro-Psychopharmacol Biol Psychiatry* 2006;30:286–291.

Zair ZM, Eloranta JJ, Stieger B, Kullak-Ublick GA. Pharmacogenetics of OATP (SLC21/SLCO), OAT and OCT (SLC22) and PEPT (SLC15) transporters in the intestine, liver and kidney. *Pharmacogenomics* 2008;9:597–624.

Zamber CP, Lamba JK, Yasuda K, Farnum J, Thummel K, Schuetz JD, Schuetz EG. Natural allelic variants of breast cancer resistance protein (BCRP) and their relationship to BCRP expression in human intestine. *Pharmacogenetics* 2003;13: 19–28.

Zhang W, Yu B-N, He Y-J, Fan L, Li Q, Liu Z-Q, Wang A, Liu Y-L, Tan Z-R, Fen J, Huang Y-F, Zhou H-H. Role of BCRP 421C>A polymorphism on rosuvastatin pharmacokinetics in healthy Chinese males. *Clin Chim Acta* 2006;373:99–103.

Zheng H, Webber S, Zeevi A, Schuetz E, Zhang J, Lamba J, Bowman P, Burckart GJ. The MDR1 polymorphisms at exons 21 and 26 predict steroid weaning in pediatric heart transplant patients. *Hum Immunol* 2002;63:765–770.

Zheng HX, Webber SA, Zeevi A, Schuetz E, Zhang J, Lamba J, Boyle GJ, Wilson JW, Burckart GJ. The impact of pharmacogenomic factors on steroid dependency in pediatric heart transplant patients using logistic regression analysis. *Pediatr Transplant* 2004;8:551–557.

Zhou Q, Sparreboom A, Tan E-H, Cheung Y-B, Lee A, Poon D, Lee EJD, Chowbay B. Pharmacogenetic profiling across the irinotecan pathway in Asian patients with cancer. *Br J Clin Pharmacol* 2005;59:415–424.

Zhu D, Taguchi-Nakamura H, Goto M, Odawara T, Nakamura T, Yamada H, Kotaki H, Sugiura W, Iwamoto A, Kitamura Y. Influence of single-nucleotide polymorphisms in the multidrug resistance-1 gene on the cellular export of nelfinavir and its clinical implication for highly active antiretroviral therapy. *Antiviral Ther* 2004;9:929–935.

17

CLINICAL DRUG INTERACTIONS DUE TO METABOLIC INHIBITION: PREDICTION, ASSESSMENT, AND INTERPRETATION

LISA L. VON MOLTKE AND DAVID J. GREENBLATT

Drug interactions are an important consideration in clinical practice, especially among populations where chronic therapy with multiple medications is common. As most countries continue to see growth in their elderly demographic sectors, and more diseases become amenable to chronic management, the issue of interactions will likely remain an important consideration for scientists and clinicians involved in drug development, the regulatory sector, and patient care.

17.1 CLASSIFICATION OF DRUG INTERACTIONS

Interactions are generally considered to fall into three broad categories: physiochemical, pharmacodynamic, or pharmacokinetic. For an intravenously administered drug, physiochemical interactions may in theory occur at any time before the drug is administered. In the case of an oral medication, this kind of interaction can occur any time prior to absorption, including the time that the drug is within the gastrointestinal tract. Because of the customary cautions and controls surrounding the mixing of intravenous medications without prior knowledge that the combination is compatible, the physiochemi-

Enzyme Inhibition in Drug Discovery and Development, edited by Lu and Li
Copyright © 2010 John Wiley & Sons, Inc.

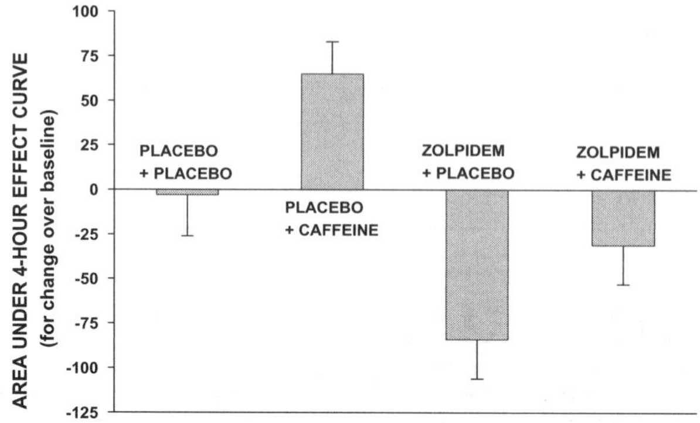

Figure 17.1. Mean (±SE, n = 12) area under the 4-h pharmacodynamic effect area (AUCE) for a test of manual tapping speed, done at multiple time points following concurrent single doses of zolpidem 7.5 mg, or matching placebo, together with caffeine 500 mg, or matching placebo, administered to healthy volunteers in a crossover study. The test reflects speed of motor function. The pharmacokinetic component of the study showed no drug interaction—zolpidem and caffeine did not alter each other's kinetic properties. However, a pharmacodynamic interaction was evident. Placebo plus placebo produced an AUCE no different from zero. Placebo plus caffeine improved tapping speed, consistent with a stimulant effect, whereas zolpidem plus placebo depressed tapping speed. The combination of zolpidem and caffeine also produced depressed tapping speed, but less so than with zolpidem alone (zolpidem plus placebo). See Cysneiros et al. (2007).

cal interactions of intravenous medications would be expected to be less common than interactions that occur following oral administration, and can result from chelation (such as occurs with certain cations and tetracyclines), changes in gastrointestinal motility, and changes in pH.

Pharmacodynamic interactions are important, but relatively difficult to predict with the kind of quanitative estimates that now accompany many kinetic interaction predictions. Interactions of this type can be additive or antagonistic. Examples include the increased sedation and respiratory depression that benzodiapepines and opioids produce, each working via different molecular receptors. Antagonism is evident in the partial reversal of the sedating effects of benzodiazepines by caffeine (Cysneiros et al., 2007) (Fig. 17.1). In these types of interactions, neither drug affects the clearance or bioavailabilty of the other, such that the altered effects are not due to a change in systemic exposure to either drug.

Drug interactions that do occur as the result of changes in systemic exposure are classified as pharmacokinetic, and result when the bioavailability and/or clearance of one compound is changed by another, such that the plasma concentrations of one of the compounds (which can be referred to as the "victim") are altered to some measurable extent by the other (which can be termed the "perpetrator"). The mechanisms underlying most of these kinetic interactions involve the perturbation of enzyme or transporter activity through either inhibition or induction of activity (Farkas et al., 2008; Greenblatt and von Moltke, 2005, 2008). The clinical result can be advantageous and desirable, adverse and undesirable, or of no clinical consequence whatsoever. Interactions that occur as the result of inhibition have received relatively more attention, since this situation often results in exaggerated drug action or overt toxicity, and can occur quite quickly. In contrast, induction is a slower process and often results in suboptimal exposure to the victim drug or loss of efficacy, which is not necessarily dramatic or clearly evident. Drugs which are "prodrugs" can have differing scenarios.

17.2 PREDICTABILITY OF DRUG INTERACTIONS

Despite many limitations and caveats, the use of *in vitro* tools and approaches have significantly improved the ability to anticipate and predict pharmacokinetic interactions involving the cytochrome P450 (CYP) system, glucuronyl transferses (UGTs), and transporters such as ABCB1 (P-glycoprotein). This has been of tremendous importance to the efforts to focus finite clinical resources for description of interactions likely to be of interest and clinical significance. In essence, new molecular entities are evaluated to determine whether they are candidates to be "victim" drugs (whose own kinetics can be altered to a meaningful extent by other medications) or "perpetrator" drugs (that can alter other compounds' pharmacokinetics appreciably), or both.

In vitro models, based on human liver microsomal preparations or recombinant CYPs or UGTs, can be used to generate values of an inhibition constant (K_i) or 50% inhibitory concentration (IC_{50}) for a specific candidate perpetrator drug suspected of being an inhibitor of a particular metabolic pathway (Volak et al., 2008; Venkatakrishnan et al., 2001, 2003). The predictive scaling paradigm compares the anticipated or measured *in vivo* exposure to the perpetrator ([I]) to the *in vitro* K_i or IC_{50}. Based on current regulatory guidelines (Huang et al., 2007), a ratio ($[I])/K_i$) of less than 0.1 is reasonable evidence that an *in vivo* drug interaction is unlikely, whereas a ratio exceeding 10.0 indicates a high probability of an *in vivo* interaction. Ratios in the intermediate range ($0.1 < [I]/K_i < 10.0$) only indicate that an interaction is "possible" and that it can be delineated only with a clinical drug interaction study. A number of authors have pointed to reasons for uncertainties in predictive scaling (Greenblatt et al., 2008; von Moltke et al., 1998; Ito et al., 1998, 2004; Lin, 2000, 2006; Bachmann, 2006; Obach et al., 2006; Brown et al., 2006). Although extreme values of $[I]/K_i$ (<0.1 or >10.0) generally are predictively correct, there are cases of "false-negative" predictions even at these extremes. An example is that involving bupropion as perpetrator and desipramine as victim. The *in vivo* interaction (inhibition of desipramine clearance by bupropion) is large and clinically important, even though $[I]/K_i$ is less than 0.1 (Reese et al., 2008).

17.3 CONSEQUENCES OF DRUG INTERACTIONS

17.3.1 Clinical Implications of Being a Victim Drug

When a drug in being examined as a potential victim of an interaction, two key pieces of information help inform clinical relevance. The first is the relative contributions(s) of the enzymes(s) and/or transporters that determine the victim drug's baseline pharmacokinetic exposure. A drug whose clearance is fully dependent on a single enzyme or subfamily (such as CYP3A) may have a greater chance of a clinically meaningful interaction when combined with a potent CYP3A inhibitor such as ketoconazole, as opposed to a drug that is dependent on multiple enzymes for its clearance (Venkatakrishnan et al., 2001, 2003). Additionally, it will likely be important to know whether the enzymes or transporters play a role in bioavailability (whether there is significant presystemic extraction or enteric efflux transport, for example) as well as systemic clearance.

The other important determinant of clinical relevance is the character of the dose-response or concentration–response relationship of the victim drug, which in turn determines the width of its therapeutic window. As noted above, alteration of victim profiles do not become clinically important unless systemic plasma levels are pushed into ranges of toxicity, or dropped below those which are needed for efficacy (Fig. 17.2). In many drug development paradigms where the safety profile is established in normal healthy volunteers very early, there may be some understanding of this relationship such that a prediction of altered clearance based on *in vitro* data can have some validity. But even in these cases, the therapeutic window may itself be altered in more fragile patient populations. When safety is first assessed in patients as opposed to healthy volunteers, questions may arise as to whether to allow these patients to continue therapy with known inhibitors or inducers that may be necessary for their medical conditions. This evaluation is further complicated when little or nothing is actually known about the therapeutic window of the proposed victim in humans, in which case an evaluation of possible clinical relevance is difficult, and concomitant administration with potent inhibitors is of substantial concern. The presence of potent inducers, many of which are pleiotropic and induce multiple enzymes and tranpsorters, in a Phase I setting may result in an underestimation of exposure and clinical effect at a given dose. The importance of this could be magnified in settings where the cohorts exposed to a particular dose in a Phase I escalation scheme contain small numbers (as with some adaptive design protocols), or in settings where large numbers of individuals are on inducing medications (such as patients with central nervous system lesions on enzyme-inducing antiepileptic drugs).

17.3.2 Case Examples

17.3.2.1 Terfenadine and Cardiac Arrhythmias. Once the therapeutic window of the victim drug is known, the routes of clearances have been

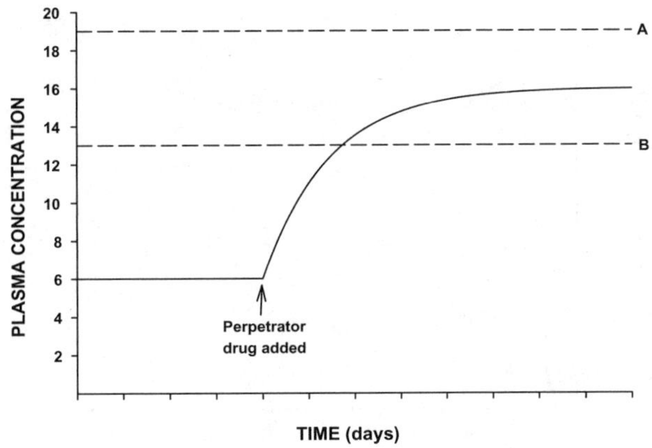

Figure 17.2. Hypothetical effect of addition of an inhibitory perpetrator drug to ongoing treatment with constant doses of a victim drug. Plasma concentrations of the victim are shown by the solid line. Levels are constant at 6 units, until the inhibitory perpetrator is added (*arrow*). Due to impairment of clearance, plasma levels of the victim increase until the new steady-state concentration (13 units) is reached. The clinical importance of the interaction depends on the concentration–response relationship for the victim drug. If the lower limit of the toxic plasma concentration range is 19 units (dashed line labeled A), the pharmacokinetic interaction might increase the therapeutic effect of the victim drug, but does not produce adverse effects. However, if the lower border of toxicity is 13 units (dashed line labeled B), the interaction results in clinical toxicity.

established, and the ability of known prototype inhibitors identified as potentially important have been evaluated in clinical studies, the impact of the fact that the drug can be "victimized" by particular inhibitors will depend on (a) the toxicity under conditions of inhibitions and (b) the therapeutic use of the drug. In a case such as terfenadine, one of the first nonsedating antihistamines marketed in the United States, the fact that a potent inhibitor such as ketoconazole could produce a life-threatening cardiac arrhythmia (Torsades de Pointes), and the drug was indicated for a relatively mild condition (allergic rhinitis) with other therapeutic options available, resulted in a public health risk-benefit assessment that dictated the drug be removed from the market (Monahan et al., 1990; Woosley et al., 1993; Honig et al., 1993; von Moltke et al., 1994a).

17.3.2.2 Context-Dependent Clinical Relevance. For drugs that have clinically monitorable pharmacodynamic effects under conditions of metabolic inhibition (such as increased sedation with inhibition of CYP3A-dependent benzodiazepines), the result of being a victim may have relevance in certain clinical situations but less so in others. An example is the interaction of the viral protease inhibitor ritonavir as perpetrator with the benzodiazepine derivative midazolam as victim. Midazolam is metabolized by hepatic and enteric

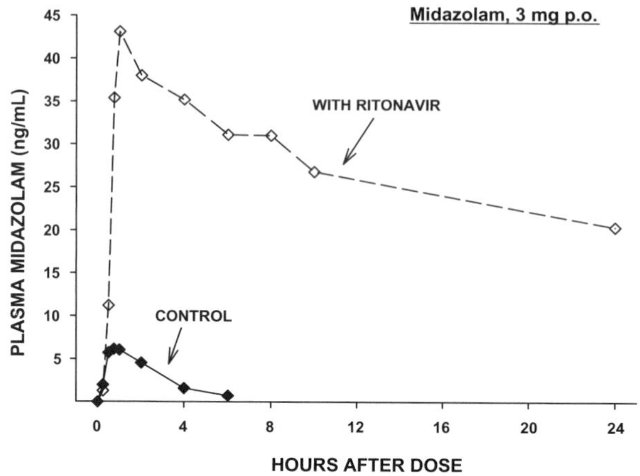

Figure 17.3. Plasma midazolam concentrations following a 3-mg oral dose of midazolam given to a healthy volunteer subject on two occasions: in the control condition without cotreatment, and during coadministration of ritonavir, three doses of 100 mg over a 30-h period.

CYP3A isoforms, with low net systemic availability (approximately 30%) after oral dosage (Tsunoda et al., 1999; Greenblatt et al., 1984). Coadministration of midazolam with ritonavir causes greatly increased systemic exposure (increased area under the plasma concentration curve, AUC) to midazolam, evident as prolongation of half-life and particularly reduced presystemic extraction after oral dosage (Knox et al., 2008). If midazolam is intended as an outpatient oral sedative-hypnotic, the interaction with ritonavir could be dangerous—the maximum intensity of midazolam-induced sedation would be increased, and the duration of sedation also prolonged (Fig. 17.3). However, if midazolam is intended as an intravenous sedative prior to general anesthesia, the interaction with ritonavir might be of minimal clinical importance. Since ritonavir changes the clearance but not the volume of distribution of midazolam (Knox et al., 2008), the initial sedative effect of midazolam might not be detectably altered. In the context of general anesthesia and surgery, ventilation is in any case monitored and controlled. Any change in the intensity or duration of midazolam sedation would be easily managed.

17.3.2.3 Managing Versus Avoiding Drug Interactions. For "victim drugs" used to treat life-threatening diseases, especially with limited alternative options, the overall risk–benefit analysis will yield a greater willingness to accept and manage a potential drug–drug interaction rather than avoid a model drug combination. An example is the use of cyclosporine and azole antifungals (such as ketoconazole, itraconazole, or voriconazole) in transplant patients. It may be that this drug combination is the only reasonable option for a transplant patient with complicating fungal infection. Although these azole antifungals impair CYP3A metabolic activity and increase systemic exposure to cyclosporine (Gomez et al., 1995; Venkatakrishnan et al., 2000), clinicians may choose to accept the interaction and manage it by monitoring cyclosporine concentrations, watching for signs of toxicity and reducing cyclosporine dosage if necessary (Albengres and Tillement, 1992; Sobh et al., 1995). Discontinuing either cyclosporine or the azole antifungal would avoid the potential drug interaction, but raise the unacceptably greater hazard of transplant rejection or untreated disseminated mycosis.

17.3.3 Clinical Implications of Being a Perpetrator Drug

The key clinical consideration for an inhibitory perpetrator drug is the extent to which systemic exposure to the inhibitor at clinically relevant concentration leads to impaired clearance of the victim drug. This question can ultimately be answered only through a clinical pharmacokinetic study in which clearance of the victim drug is determined in the control condition (without the inhibitory perpetrator), and again during exposure to relevant concentrations of the inhibitor. If the specific study design involves single-dose administration of the victim drug on two occasions in a crossover paradigm, systemic exposure to the victim in the control condition (AUC_0) is compared to that during coadministration of the inhibitor (AUC_I). The relative increase in AUC (AUC_I/AUC_0 ratio) is equivalent to the reciprocal of the relative clearance decrement caused by the inhibitory perpetrator.

The problem faced by pharmaceutical sponsors in the process of drug development is that clinical drug interaction studies are costly, time-consuming, and involve a low but finite risk to study participants. Unavoidable limitations in development time and resources make it inevitable that not every needed clinical study can actually be done. Some (or most) studies that are done will prove to be negative, and some clinically important drug interactions may be delayed in identification due to the logistics of study scheduling.

As discussed above, the use of *in vitro* models early on in the course of development may allow identification of these interactions that are probable, possible, or unlikely. The hope is that this information will allow more effective and efficient channeling of development resources to those drug interaction studies that are likely to be positive, and/or those that will yield information of substantial clinical importance. Still, it must be recognized that the *in vitro–in vivo* scaling paradigms continue to have major limitations, despite ongoing refinement of technical procedures and theoretical scaling models. The hope of a direct and reliable *in vitro–in vivo* link (Fig. 17.4) has not been achieved. Explanations for flawed predictability are described in detail previously (Greenblatt et al., 2008), and fall into the following three categories: (1) biased or inaccurate determination of *in vitro* K_i or IC_{50} attributable to technical issues; (2) uncertainty in identifying *in vivo* "inhibitor exposure" ([I]) consistent with actual concentrations available to the enzyme;

In vitro ⟶ **In vitro**

$$1 + \frac{[I]}{K_i} = \frac{AUC_I}{AUC_O}$$

Figure 17.4. Paradigm for predictive scaling of *in vitro* data on metabolic inhibition to clinical drug interaction studies. See text for explanation of symbols.

and (3) physiologic factors modulating *in vivo* disposition of perpetrator or victim drugs that are not reflected *in vitro*.

17.3.4 Case Examples

17.3.4.1 Underprediction. The antidepressant drug fluvoxamine is known to be a potent inhibitor of drug disposition mediated by the CYP1A2 enzyme (Rasmussen et al., 1995, 1998; Greenblatt et al., 1999; Jeppesen et al., 1996a; Becquemont et al., 1996). Victim drugs whose clearance is known to be significantly inhibited by coadministration of fluvoxamine include: caffeine, theophylline, tacrine, tizandidine, olanzapine, clozapine, duloxetine, and ramelteon (Jeppesen et al., 1996b; Christensen et al., 2002; Culm-Merdek et al., 2005; Yao et al., 2001; Orlando et al., 2006; Rasmussen et al., 1997; Becquemont et al., 1997; Granfors et al., 2004; Wang et al., 2004; Lobo et al., 2008). However, a consistent finding in the literature is that values of *in vitro* K_i for fluvoxamine versus specific victim substrates, together with measured systemic plasma concentrations of fluvoxamine ([I]), yield a large underestimation of the actual clinical drug interaction (AUC_I/AUC_O), by a factor of 10-fold or more (Yao et al., 2001; Culm-Merdek et al., 2005). An example of this is the fluvoxamine–ramelteon interaction (Lee and Doddapaneni, 2005). Using the paradigm in Fig. 17.4, the predicted AUC_I/AUC_O for ramelteon is approximately 3.0, whereas the actual value observed *in vivo* is 190 (Fig. 17.5). The reasons for the gross underprediction are not established. One possible explanation is that enzyme-available intrahepatic concentrations of fluvoxamine greatly exceed systemic plasma concentration. This has been demonstrated for a number of lipophilic drugs (Yamano et al., 1999a,b, 2000, 2001; von Moltke et al., 1994b, 1998; Parker and Houston, 2008), and specifically for fluvoxamine in experimental models (von Moltke et al., 1995).

17.3.4.2 Selective Enteric Inhibition. A number of naturally occurring furanocoumarin derivatives found in grapefruit juice are irreversible (mechanism-based) inhibitors of the CYP3A enzyme subfamily (Farkas and Greenblatt, 2008; Mertens-Talcott et al., 2006). The most important of these furanocoumarins is 6′7′-dihydroxybergamottin (DHB). The IC_{50} for DHB as an *in vitro* inhibitor of CYP3A falls in the nanomolar range (Greenblatt et al., 2003). However, it appears that DHB and other furanocoumarins are ordinarily consumed or inactivated in the gastrointestinal tract, and they will not reach the portal or systemic circulation in clinically important amounts. This

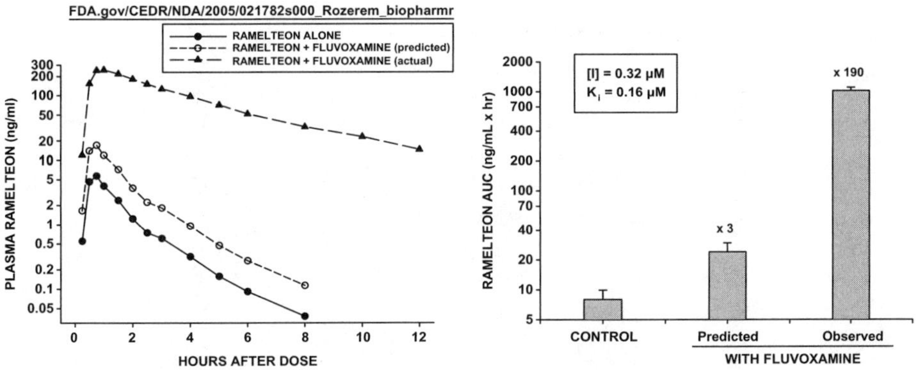

Figure 17.5. Left: Mean plasma concentrations of ramelteon, a substrate for metabolism by CYP1A2, administered to healthy volunteers in the control condition (ramelteon alone: closed circles) and during coadministration of fluvoxamine (closed triangles, ramelteon plus fluvoxamine, actual). The open circles represent the plasma concentrations of ramelteon predicted to occur from the relationship in Fig. 17.4 based on *in vitro* data, together with systemic plasma concentrations of the inhibitory perpetrator, fluvoxamine. **Right**: Mean (±SE) total AUC for ramelteon in the control condition, and the predicted and actual AUC during coadministration of fluvoxamine. The scaling paradigm, using an *in vitro* K_i of 0.16 μM for fluvoxamine versus ramelteon together with measured systemic plasma fluvoxamine concentrations ([I]) of approximately 0.33 μM, predicts a three-fold increase in ramelteon AUC, but the observed increase in the clinical drug interaction study was 190-fold.

is generally the case with nutrients and natural substances (Markowitz et al., 2008). As a consequence, metabolic inhibition is limited to enteric CYP3A, and clinical drug interactions with grapefruit juice will be observed only when the victim drug is given orally and normally undergoes presystemic extraction by enteric CYP3A. Inhibitory interactions will not occur when the victim drug is given intravenously. This has been clearly demonstrated in the case of midazolam, cyclosporine, nifedipine, and nicardipine (Kupferschmidt et al., 1995; et al., 1995; Rashid et al., 1995; Uno et al., 2000) (Fig. 17.6).

17.3.4.3 Boosting or Augmentation. Often we implicitly assume that inhibitory drug interactions are unwanted or hazardous. However, inhibitory interactions are sometimes deliberately produced for therapeutic benefit, in which case the interaction phenomenon is termed "boosting" or "augmentation." The viral protease inhibitor ritonavir now is more commonly used for its boosting properties (attributed to strong inhibition of CYP3A and P-glycoprotein) than for its antiretroviral properties as such. A number of other protease inhibitors (lopinavir, atazanvir, saquinavir) are coadministered with low-dose ritonavir (the perpetrator) to increase systemic exposure, reduce dosing requirements, and extend the interval between doses for the victim antiretroviral (Flexner, 2000; Sahali et al., 2008). Lopinavir, in fact, is available only as fixed-dose combination with ritonavir (Kaletra) (Magnum

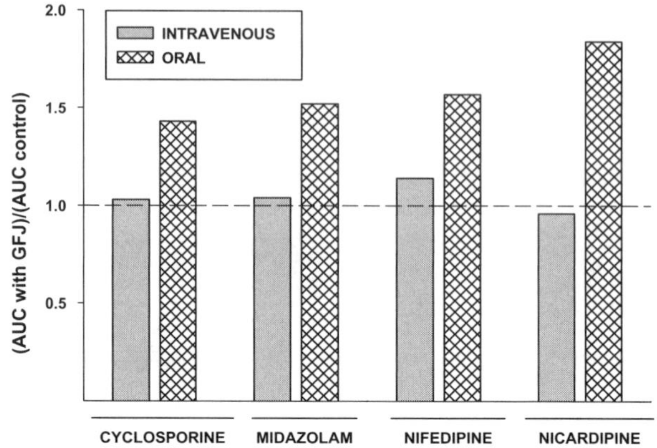

Figure 17.6. Effect of grapefruit juice (GFJ) coadministration on systemic exposure (total AUC) to four different victim drugs metabolized by CYP3A enzymes. Each bar is the ratio of the mean AUC during GFJ coadministration divided by the mean AUC in the control condition, based on studies in healthy volunteers. The effect of GFJ depends on the route of administration of the victim drug. When given intravenously, GFJ has no effect—the AUC ratios are not significantly different from 1.0 (dashed line). However, GFJ significantly increases the AUC ratio when the victim is given orally. This indicates the selective inhibition by GFJ of enteric but not hepatic CYP3A.

and Graham, 2001; Sham et al., 1998). It is of interest that doses of ritonavir required for maximal boosting are very low—far below those needed for antiretroviral activity (Mathias et al., 2009). Pharmacokinetic boosting is also used in other clinical situations, such as reduction of cyclosporine dosage requirements by coadministration of ketoconazole (Albengres and Tillement, 1992; Sobh et al., 1995).

17.4 COMMENT

The evaluation of pharmacokinetic drug interactions requires the integration of scientific and clinical information. The two core questions are:

1. How big is the interaction? The question is mainly a pharmacokinetic one, which can be answered in quantitative terms as the AUC_I/AUC_0 ratio for the victim drug, either after single doses or at steady state.
2. Is the interaction of clinical importance? This question is more difficult, and requires clinical insight rather than simply prespecified statistical procedures. Factors to consider include: the magnitude of the interaction; the concentration–response relationship for the victim drug; and the underlying health and vulnerability of the affected patient population. The question does not necessarily have a "yes/no" answer, but rather an assignment of graded risk.

In the context of the prevalence of polypharmacy in contemporary clinical practice, clinically important drug interactions actually are relatively uncommon (Bergk et al., 2004). Drug interactions requiring avoidance of a drug combination, or even dosage adjustment of victim or perpetrator, are quite unusual relative to the number of possible drug interactions.

REFERENCES

Albengres E, Tillement JP. Cyclosporin and ketoconazole, drug interaction or therapeutic association? *Int J Clin Pharmacol Ther Toxicol* 1992;30:555–570.

Bachmann KA. Inhibition constants, inhibitor concentrations and the prediction of inhibitory drug drug interactions: pitfalls, progress and promise. *Curr Drug Metab* 2006;7:1–14.

Becquemont L, Le Bot MA, Riche C, Beaune P. Influence of fluvoxamine on tacrine metabolism *in vitro*: potential implication for the hepatotoxicity *in vivo*. *Fundam Clin Pharmacol* 1996;10:156–157.

Becquemont L, Ragueneau I, Le Bot MA, Riche C, Funck-Bretano C, Jaillon P. Influence of the CYP1A2 inhibitor fluvoxamine on tacrine pharmacokinetics in humans. *Clin Pharmacol Ther* 1997;61:619–627.

Bergk V, Gasse C, Rothenbacher D, Loew M, Brenner H, Haefeli WE. Drug interactions in primary care: impact of a new algorithm on risk determination. *Clin Pharmacol Ther* 2004;76:85–96.

Brown HS, Galetin A, Hallifax D, Houston JB. Prediction of *in vivo* drug–drug interactions from *in vitro* data: factors affecting prototypic drug-drug interactions involving CYP2C9, CYP2D6 and CYP3A4. *Clin Pharmacokinet* 2006;45:1035–1050.

Christensen M, Tybring G, Mihara K, Yasui-Furokori N, Carrillo JA, Ramos SI, Andersson K, Dahl ML, Bertilsson L. Low daily 10-mg and 20-mg doses of fluvoxamine inhibit the metabolism of both caffeine (cytochrome P4501A2) and omeprazole (cytochrome P4502C19). *Clin Pharmacol Ther* 2002;71:141–152.

Culm-Merdek KE, von Moltke LL, Harmatz JS, Greenblatt DJ. Fluvoxamine impairs single-dose caffeine clearance without altering caffeine pharmacodynamics. *Br J Clin Pharmacol* 2005;60:486–493.

Cysneiros RM, Farkas D, Harmatz JS, von Moltke LL, Greenblatt DJ. Pharmacokinetic and pharmacodynamic interactions between zolpidem and caffeine. *Clin Pharmacol Ther* 2007;82:54–62.

Ducharme MP, Warbasse LH, Edwards DJ. Disposition of intravenous and oral cyclosporine after administration with grapefruit juice. *Clin Pharmacol Ther* 1995; 57:485–491.

Farkas D, Greenblatt DJ. Influence of fruit juices on drug disposition: discrepancies between *in vitro* and clinical studies. *Expert Opin Drug Metab Toxicol* 2008;4: 381–393.

Farkas D, Shader RI, von Moltke LL, Greenblatt DJ. Mechanisms and consequences of drug–drug interactions. In: Gad SC, editor. *Preclinical Development Handbook: ADME and Biopharmaceutical Properties*. Philadelphia: Wiley-Interscience, 2008, pp. 879–917.

Flexner C. Dual protease inhibitor therapy in HIV-infected patients: pharmacologic rationale and clinical benefits. *Annu Rev Pharmacol Toxicol* 2000;40:649–674.

Gomez DY, Wacher VJ, Tomlanovich SJ, Hebert MF, Benet LZ. The effects of ketoconazole on the intestinal metabolism and bioavailability of cyclosporine. *Clin Pharmacol Ther* 1995;58:15–19.

Granfors MT, Backman JT, Neuvonen M, Ahonen J, Neuvonen PJ. Fluvoxamine drastically increases concentrations and effects of tizanidine: a potentially hazardous interaction. *Clin Pharmacol Ther* 2004;75:331–341.

Greenblatt DJ, von Moltke LL. Pharmacokinetics and drug interactions. In: Sadock BJ, Sadock VA, editors. *Comprehensive Textbook of Psychiatry*, 8th ed. Philadelphia: Lippincott Williams & Williams, 2005, pp. 2699–2706.

Greenblatt DJ, von Moltke LL. Drug–drug interactions: Clinical perspectives. In: Rodrigues AD, editor. *Drug–Drug Interactions*, 2nd ed. New York, Informa Healthcare, 2008, pp. 643–664.

Greenblatt DJ, Abernethy DR, Locniskar A, Harmatz JS, Limjuco RA, Shader RI. Effect of age, gender, and obesity on midazolam kinetics. *Anesthesiology* 1984;61: 27–35.

Greenblatt DJ, von Moltke LL, Harmatz JS, Shader RI. Human cytochromes and some newer antidepressants: kinetics, metabolism, and drug interactions. *J Clin Psychopharmacol* 1999;19(Suppl 1):23S–35S.

Greenblatt DJ, von Moltke LL, Harmatz JS, Chen G, Weemhoff JL, Jen C, Kelley CJ, LeDuc BW, Zinny MA. Time-course of recovery of cytochrome P450 3A function after single doses of grapefruit juice. *Clin Pharmacol Ther* 2003;74: 121–129.

Greenblatt DJ, He P, von Moltke LL, Court MH. The CYP3 family. In: Ioannides C, editor. *Cytochrome P450: Role in the Metabolism and Toxicology of Drugs and Other Xenobiotics*. Cambridge, UK: Royal Society of Chemistry, 2008, pp. 354–383.

Honig PK, Wortham DC, Zamani K, Conner DP, Mullin JC, Cantilena LR. Terfenadine–ketoconazole interaction: pharmacokinetic and electrocardiographic consequences. *JAMA* 1993;269:1513–1518.

Huang SM, Temple R, Throckmorton DC, Lesko LJ. Drug interaction studies: study design, data analysis, and implications for dosing and labeling. *Clin Pharmacol Ther* 2007;81:298–304.

Ito K, Iwatsubo T, Kanamitsu S, Ueda K, Suzuki H, Sugiyama Y. Prediction of pharmacokinetic alterations caused by drug–drug interactions: metabolic interaction in the liver. *Pharmacol Rev* 1998;50:387–412.

Ito K, Brown HS, Houston JB. Database analyses for the prediction of *in vivo* drug–drug interactions from *in vitro* data. *Br J Clin Pharmacol* 2004;57:473–486.

Jeppesen U, Gram LF, Vistisen K, Loft S, Poulsen HE, Brøsen K. Dose-dependent inhibition of CYP1A2, CYP2C19, and CYP2D6 by citalopram, fluoxetine, fluvoxamine and paroxetine. *Eur J Clin Pharmacol* 1996a;51:73–78.

Jeppesen U, Loft S, Poulsen HE, Brøsen K. A fluvoxamine–caffeine interaction study. *Pharmacogenetics* 1996b;6:213–222.

Knox TA, Oleson L, von Moltke LL, Kaufman RA, Wanke CA, Greenblatt DJ. Ritonavir greatly impairs CYP3A activity in HIV infection with chronic viral hepatitis. *JAIDS* 2008;49:358–368.

Kupferschmidt HHT, Ha HR, Ziegler WH, Meier PJ, Krähenbühl S. Interaction between grapefruit juice and midazolam in humans. *Clin Pharmacol Ther* 1995; 58:20–28.

Lee D, Doddapaneni S. Clinical pharmacology and biopharmaceutics review. Application #21-782: Ramelteon (TAK-375). FDA.gov/CEDR/FOI/NDA/2005/ 021782s000_Rozerem_Biopharmr.pdf 2005.

Lin JH. Sense and nonsense in the prediction of drug–drug interactions. *Curr Drug Metab* 2000;1:305–331.

Lin JH. CYP induction-mediated drug interactions: *in vitro* assessment and clinical implications. *Pharm Res* 2006;23:1089–1116.

Lobo ED, Bergstrom RF, Reddy S, Quinlan T, Chappell J, Hong Q, Ring B, Knadler MP. *In vitro* and *in vivo* evaluations of cytochrome P450 1A2 interactions with duloxetine. *Clin Pharmacokinet* 2008;47:191–202.

Mangum EM, Graham KK. Lopinavir–ritonavir: a new protease inhibitor. *Pharmacotherapy* 2001;21:1352–1363.

Markowitz JS, von Moltke LL, Donovan JL. Predicting interactions between conventional medications and botanical products on the basis of *in vitro* investigations. *Mol Nutr Food Res* 2008;52:747–754.

Mathias AA, West S, Hui J, Kearney BP. Dose–response of ritonavir on hepatic CYP3A activity and elvitegravir oral exposure. *Clin Pharmacol Ther* 2009;85: 64–70.

Mertens-Talcott SU, Zadezensky I, De Castro WV, Derendorf H, Butterweck V. Grapefruit-drug interactions: can interactions with drugs be avoided? *J Clin Pharmacol* 2006;46:1390–1416.

Monahan BP, Ferguson CL, Killeavy ES, Lloyd BK, Troy J, Cantilena LR. Torsades de pointes occurring in association with terfenadine use. *JAMA* 1990;264:2788–2790.

Obach RS, Walsky RL, Venkatakrishnan K, Gaman EA, Houston JB, Tremaine LM. The utility of *in vitro* cytochrome P450 inhibition data in the prediction of drug-drug interactions. *J Pharmacol Exp Ther* 2006;316:336–348.

Orlando R, Padrini R, Perazzi M, De Martin S, Piccoli P, Palatini P. Liver dysfunction markedly decreases the inhibition of cytochrome P450 1A2-mediated theophylline metabolism by fluvoxamine. *Clin Pharmacol Ther* 2006;79:489–499.

Parker AJ, Houston JB. Rate-limiting steps in hepatic drug clearance: comparison of hepatocellular uptake and metabolism with microsomal metabolism of saquinavir, nelfinavir, and ritonavir. *Drug Metab Dispos* 2008;36:1375–1384.

Rashid TJ, Martin U, Clarke H, Waller DG, Renwick AG, George CF. Factors affecting the absolute bioavailability of nifedipine. *Br J Clin Pharmacol* 1995;40:51–58.

Rasmussen BB, Mäenpää J, Pelkonen O, Loft S, Poulsen HE, Lykkesfeldt J, Brøsen K. Selective serotonin reuptake inhibitors and theophylline metabolism in human liver microsomes: potent inhibition by fluvoxamine. *Br J Clin Pharmacol* 1995;39: 151–159.

Rasmussen BB, Jeppesen U, Gaist D, Brosen K. Griseofulvin and fluvoxamine interactions with the metabolism of theophylline. *Ther Drug Monit* 1997;19:56–62.

Rasmussen BB, Nielsen TL, Brosen K. Fluvoxamine is a potent inhibitor of the metabolism of caffeine *in vitro*. *Pharmacol Toxicol* 1998;83:240–245.

Reese MJ, Wurm RM, Muir KT, Generaux GT, St John-Williams L, McConn DJ. An in vitro mechanistic study to elucidate the desipramine/bupropion clinical drug-drug interaction. *Drug Metab Dispos* 2008;36:1198–1201.

Sahali S, Chaix ML, Delfraissy JF, Ghosn J. Ritonavir-boosted protease inhibitor monotherapy for the treatment of HIV-1 infection. *AIDS Rev* 2008;10:4–14.

Sham HL, Kempf DJ, Molla A, Marsh KC, Kumar GN, Chen CM, Kati W, Stewart K, Lal R, Hsu A, Betebenner D, Korneyeva M, Vasavanonda S, McDonald E, Saldivar A, Wideburg N, Chen X, Niu P, Park C, Jayanti V, Grabowski B, Granneman GR, Sun E, Japour AJ, Norbeck DW, et al. ABT-378, a highly potent inhibitor of the human immunodeficiency virus protease. *Antimicrobial Agents Chemother* 1998;42:3218–3224.

Sobh M, El-Agroudy A, Moustafa F, Harras F, El-Bedewy M, Ghoneim M. Coadministration of ketoconazole to cyclosporin-treated kidney transplant recipients: a prospective randomized study. *Am J Nephrol* 1995;15:493–499.

Tsunoda SM, Velez RL, von Moltke LL, Greenblatt DJ. Differentiation of intestinal and hepatic cytochrome P450 3A activity with use of midazolam as an *in vivo* probe: effect of ketoconazole. *Clin Pharmacol Ther* 1999;66:461–471.

Uno T, Ohkubo T, Sugawara K, Higashiyama A, Motomura S, Ishizaki T. Effects of grapefruit juice on the stereoselective disposition of nicardipine in humans: evidence for dominant presystemic elimination at the gut site. *Eur J Clin Pharmacol* 2000;56:643–649.

Venkatakrishnan K, von Moltke LL, Greenblatt DJ. Effects of the antifungal agents on oxidative drug metabolism in humans: clinical relevance. *Clin Pharmacokinet* 2000;38:111–180.

Venkatakrishnan K, von Moltke LL, Greenblatt DJ. Human drug metabolism and the cytochromes P450: application and relevance of in vitro models. *J Clin Pharmacol* 2001;41:1149–1179.

Venkatakrishnan K, von Moltke LL, Obach RS, Greenblatt DJ. Drug metabolism and drug interactions: application and clinical value of *in vitro* models. *Curr Drug Metab* 2003;4:423–459.

Volak LP, Greenblatt DJ, von Moltke LL. In vitro approaches to anticipating clinical drug interactions. In: Li AP, editor. *Drug–Drug Interactions in Pharmaceutical Development*. Hoboken, NJ: John Wiley & Sons, 2008, p. 75–93.

von Moltke LL, Greenblatt DJ, Duan SX, Harmatz JS, Shader RI. In vitro prediction of the terfenadine-ketoconazole pharmacokinetic interaction. *J Clin Pharmacol* 1994a;34:1222–1227.

von Moltke LL, Greenblatt DJ, Cotreau-Bibbo MM, Duan SX, Harmatz JS, Shader RI. Inhibition of desipramine hydroxylation *in vitro* by serotonin-reuptake-inhibitor antidepressants, and by quinidine and ketoconazole: a model system to predict drug interactions *in vivo*. *J Pharmacol Exp Ther* 1994b;268:1278–1283.

von Moltke LL, Greenblatt DJ, Court MH, Duan SX, Harmatz JS, Shader RI. Inhibition of alprazolam and desipramine hydroxylation *in vitro* by paroxetine and fluvoxamine: comparison with other selective serotonin reuptake inhibitor antidepressants. *J Clin Psychopharmacol* 1995;15:125–131.

von Moltke LL, Greenblatt DJ, Schmider J, Wright CE, Harmatz JS, Shader RI. *In vitro* approaches to predicting drug interactions *in vivo*. *Biochem Pharmacol* 1998;55:113–122.

Wang CY, Zhang ZJ, Li WB, Zhai YM, Cai ZJ, Weng YZ, Zhu RH, Zhao JP, Zhou HH. The differential effects of steady-state fluvoxamine on the pharmacokinetics of olanzapine and clozapine in healthy volunteers. *J Clin Pharmacol* 2004;44: 785–792.

Woosley RL, Chen Y, Freiman JP, Gillis RA. Mechanism of the cardiotoxic actions of terfenadine. *JAMA* 1993;269:1532–1536.

Yamano K, Yamamoto K, Kotaki H, Takedomi S, Matsuo H, Sawada Y, Iga T. Correlation between *in vivo* and *in vitro* hepatic uptake of metabolic inhibitors of cytochrome P-450 in rats. *Drug Metab Dispos* 1999a;27:1225–1231.

Yamano K, Yamamoto K, Kotaki H, Sawada Y, Iga T. Quantitative prediction of metabolic inhibition of midazolam by itraconazole and ketoconazole in rats: implication of concentrative uptake of inhibitors into liver. *Drug Metab Dispos* 1999b;27:395–402.

Yamano K, Yamamoto K, Kotaki H, Takedomi S, Matsuo H, Sawada Y, Iga T. Quantitative prediction of metabolic inhibition of midazolam by erythromycin, diltiazem, and verapamil in rats: implication of concentrative uptake of inhibitors into liver. *J Pharmacol Exp Ther* 2000;292:1118–1126.

Yamano K, Yamamoto K, Katashima M, Kotaki H, Takedomi S, Matsuo H, Ohtani H, Sawada Y, Iga T. Prediction of midazolam-CYP3A inhibitors interaction in the human liver from *in vivo*/*in vitro* absorption, distribution, and metabolism data. *Drug Metab Dispos* 2001;29:443–452.

Yao C, Kunze KL, Kharasch ED, Wang Y, Trager WF, Ragueneau I, Levy RH. Fluvoxamine–theophylline interaction: gap between *in vitro* and *in vivo* inhibition constants toward cytochrome P4501A2. *Clin Pharmacol Ther* 2001;70:415–424.

18

PREDICTING INTERINDIVIDUAL VARIABILITY OF METABOLIC DRUG–DRUG INTERACTIONS: IDENTIFYING THE CAUSES AND ACCOUNTING FOR THEM USING A SYSTEMS APPROACH

Amin Rostami-Hodjegan

18.1 INTRODUCTION

As discussed in previous chapters, metabolic drug–drug interactions (M-DDI) may result in restricted use, withdrawal, or nonapproval by regulatory agencies. However, many of these M-DDIs, particularly during the clinical use, are not related to observations in an "average" person, but rather they are seen in a subgroup of patients. Therefore, understanding the theoretically conceivable extreme effects in a certain subgroup of patients following coadministration of given doses of interacting drugs is desirable. In fact, this has been advocated by some of the scientists working in regulatory agencies, as well as those in industry, following high-profile cases of drug withdrawals from the market due to M-DDI (Krayenbuhl et al., 1999).

Full understanding of the issue of variability in M-DDI requires the knowledge of the parameters influencing the exposure to a "perpetrator" drug and the proportional metabolism by the pathway of interest for a "victim" drug. This knowledge can be built as a progressive process during drug development

Enzyme Inhibition in Drug Discovery and Development, edited by Lu and Li
Copyright © 2010 John Wiley & Sons, Inc.

(Jamei et al., 2009a). Accordingly, the available information for interindividual variability in each system parameter (human body) and trial design can be incorporated into simulation of various populations (as opposed to single average individual).

Earlier identification of the covariates for M-DDI (prior to conduct of any clinical studies) is desirable and can avoid suboptimal study designs. Ignoring these covariates in the design of M-DDI studies during so-called Phase II, or inadequate planning for collection of relevant information during Phase III studies, may lead to delays in approval or nonapproval by regulatory bodies. More importantly, observation of unexpected M-DDI during clinical use, which may happen only in subgroups of patients, may lead to withdrawal that has significant implications regarding the cost. Thus, accounting for variability can prevent false conclusions regarding the absence of significant M-DDI when certain groups who were not examined in the early studies remain susceptible to such effects.

The use of physiologically based pharmacokinetics (PBPK) helps to build a framework that captures and integrates mechanisms (/sources) of variations in absorption, distribution, metabolism, and elimination (ADME) processes (Jamei et al., 2009a). However, separation of information on human body from those of the drug and study conditions (/design) is an essential step in identifying the sources of interindividual variability and this may require sophisticated software and databases (Jamei et al., 2009b).

This chapter summarizes the kinetic basis of M-DDI which is described in detail in previous chapters, and it outlines the components that lead to interindividual variability in M-DDI. The complementary nature of the chapter means that readers are frequently referred to information provided in previous chapters.

18.2 PHARMACOKINETIC BUILDING BLOCKS DETERMINING M-DDI

A number of previous chapters have described the most common mathematical formalism in predicting M-DDI related to "inhibition" of metabolism. These indicated that, under "static" (non-time-variant) concentrations for the "perpetrator," M-DDI can be estimated using the following equation [after Rowland and Matin (1973), with expansions introduced by Rostami-Hodjegan and Tucker (2004) for multiple interactions and generalization to consider all mechanisms of inhibition instead of restricting the model to competitive inhibition]:

$$\frac{\mathrm{AUC(inhibited)}}{\mathrm{AUC}} = \frac{1}{\sum_{j=1}^{n} \frac{\mathrm{fm}_j}{\text{Fold reduction in } \mathrm{CL}u_{\mathrm{int},j}} + \left(1 - \sum_{j=1}^{n} \mathrm{m}_j\right)} \tag{18.1}$$

where fm_j is the fraction of substrate clearance mediated by the inhibited metabolic pathway j and $CLu_{int,j}$ is the intrinsic metabolic clearance of substrate down pathway j. The fold reduction in this value will be defined later as it varies by the type of inhibition.

In the case of induction, the fold change in activity of the induced enzyme can be incorporated in a way similar to that of inhibition; however, the overall impact on AUC is still dependent on the fractional metabolism by the pathway (fm):

$$\frac{\text{AUC(induced)}}{\text{AUC}} = \frac{1}{\sum_{j=1}^{n}(\text{fm}_j \times \text{Fold induction in CL}u_{\text{int},j}) + \left(1 - \sum_{j=1}^{n}\text{fm}_j\right)} \tag{18.2}$$

Fold induction in CLu_{int} can be defined based on the concentration of the inducer at the site of effect, [I]; the maximal increase in the level of induced enzyme, E_{max}; and $IndC_{50}$ (the concentration of inducer associated with half-maximal induction):

$$\text{Fold induction in CL}u_{\text{int},j} = 1 + \left(\frac{E_{\max} \times [I]}{\text{IndC}_{50} + [I]}\right) \tag{18.3}$$

E_{max} is measured as a fold of the uninduced value when the enzyme is exposed to a high concentration of inducer.

It is important to realize that the derivation of Eqs. (18.1) and (18.2) relies on a number of assumptions that are often ignored. Details of these can be found in a review by Rostami-Hodjegan and Tucker (2004). The equation applies only to orally administered drugs undergoing hepatic metabolism and it ignores, amongst many other factors, the possibility of inhibition of gut "first-pass" metabolism (see Chapter 9 in this book) and time-variant inhibitor concentration (see later sections). Nonetheless, despite its relative simplicity, the equation captures the two main elements that define the level of any M-DDI, namely: (i) fractional metabolism (fm) and (ii) potency of interacting drug (fold reduction or induction in CLu_{int}).

18.2.1 Sensitivity of M-DDI to Fractional Metabolism

It is evident from the Eq. (18.1) that in the presence of a "strong inhibitor" the uninhibited pathways, defined by 1 – fm, determine the level of interaction:

$$\text{Fold reduction} \to \infty \Rightarrow \frac{\text{AUC(inhibited)}}{\text{AUC(uninhibited)}} = \frac{1}{1 - \text{fm}} \tag{18.4}$$

We will later discuss how fm value varies within populations, but typical examples include the following cases (Rostami-Hodjegan and Tucker, 2004):

- Genetic control of noninhibited metabolic pathways
- Renal impairment influencing 1 – fm
- Variable rates of maturation of CYP enzymes in the liver
- Disparity between maturation of renal function and metabolic routes

The two latter cases can lead to an age-varying value of 1 – fm and the consequent differences in M-DDIs in neonates compared to adults, even under similar exposure to an inhibitor; which of course are difficult to study experimentally.

18.2.2 Sensitivity of M-DDI to Potency of Inhibitors

The fold inhibition can be defined according to the type of inhibition that occurs (Rostami-Hodjegan and Tucker, 2004). Thus, for multiple ("p") competitive inhibitors acting via same mechanism to inhibit enzyme "j":

$$\text{Fold reduction in CL}u_{\text{int},j} = 1 + \sum_{k=1}^{p} \frac{[\text{I}_k]}{Ki_k} \tag{18.5}$$

where $[\text{I}_k]$ is the concentration of inhibitor k at the enzyme site, and Ki_k is the inhibition constant for inhibitor k obtained from *in vitro* studies after accounting for nonspecific binding. The same equation applies for multiple noncompetitive inhibitors acting at the same enzyme site. However, if the mechanisms of inhibition by multiple inhibitors are different (independent), the fold reduction in clearance would be greater:

$$\text{Fold reduction in CL}u_{\text{int},j} = \prod_{k=1}^{p} \left(1 + \frac{[\text{I}_k]}{Ki_k}\right) \tag{18.6}$$

For the case of mechanism-based (suicidal) inactivation of enzymes:

$$\text{Fold reduction in CLu}_{\text{int},j} = \frac{k_{\text{deg}}}{k_{\text{deg}} + \dfrac{[\text{I}] \times k_{\text{inact}}}{[\text{I}] + K_{\text{I}}}} \tag{18.7}$$

where k_{deg} is the natural degradation rate constant for the enzyme, k_{inact} is the maximum degradation rate constant in the presence of a very high concentrations of inhibitor, and K_{I} is the concentration of inhibitor associated with half-maximal inactivation.

Under the assumption that the inhibited metabolic pathway is the only elimination pathway (i.e., fm = 1), the effect of competitive inhibition can be described by the following equation:

$$[1-\text{fm}] \to 0 \Rightarrow \frac{\text{AUC(inhibited)}}{\text{AUC(uninhibited)}} = \begin{cases} \text{Competitive:} & 1+\dfrac{[\text{I}]}{K_i} \\ \text{Mechanism-based:} & \dfrac{k_{\text{deg}}}{k_{\text{deg}} + \dfrac{[\text{I}] \times k_{\text{inact}}}{[\text{I}] + K_{\text{I}}}} \end{cases} \tag{18.8}$$

This equation has been used for "semiquantitative" predictions of M-DDI as described in previous chapters. It is evident from this equation that the level of interaction, under the assumption of 1 – fm approaching 0, is independent of the variability in the circulating plasma concentrations of the "victim" drug. However, the equation clearly shows the sensitivity to [I] and k_{deg} and the fact that interindividual variability in these parameters may propagate to M-DDI. K_i and K_{I} are considered to be inherent characteristics of the inhibitor binding to the enzyme, and hence no interindividual variability is usually assumed for them; this may not be true if in certain genotypes, mutations at an active site of enzyme influence the affinity of binding for the inhibitors.

Figure 18.1 captures the range of sensitivities for inhibitory M-DDI defined by Eq. (18.1). It shows that the variation in fm will have considerable impact when dealing with M-DDI caused by strong inhibitors, while the outcome is not sensitive to variation of [I] under this condition.

18.2.3 Sensitivity of M-DDI to Inhibition of Gut "First-Pass" Metabolism

The importance of gut-wall metabolism is explained in Chapter 9 in this book. The amendment to Eq. (18.1), when there is a significant contribution by gut-wall metabolism to bioavailability, is described by Eq. (9.1) in Chapter 9, and as a reminder it is shown below:

$$\frac{\text{AUC(inhibited)}}{\text{AUC(uninhibited)}} = \frac{F'_G}{F_G} \times \frac{1}{\sum_{j=1}^{n} \text{fm}_j \times \text{Fold change in CL}u_{\text{int},j} + \left(1 - \sum_{j=1}^{n} \text{fm}_j\right)} \tag{18.9}$$

where F_G and F'_G describe the fraction of drug that escapes first-pass gut metabolism in the absence and presence of any interacting drug, respectively. The mechanistic models for F_G such as "segmental and segregated intestinal transport and metabolism" described by Pang and colleagues (Pang, 2003; Tam et al., 2003) have not been used widely since many of the parameter values for the model cannot be generated from the current *in vitro* studies. Variations of these models are now implemented in commercially available software (Jamei et al., 2009b) which may facilitate a wider use. Less

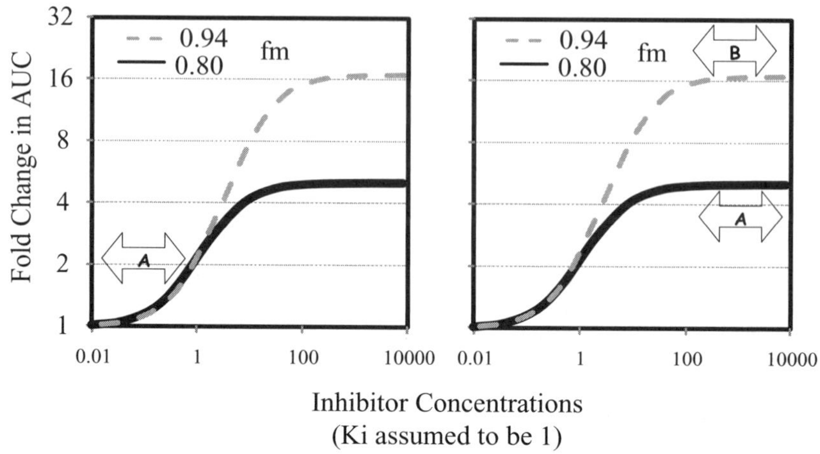

Figure 18.1. Concentration–effect relationship for M-DDI by an inhibitor (measured as the fold change in AUC for "victim" drug). **Left Panel:** Arrow A shows that a change in the [I] (alongside X axis) for weak or moderate inhibitors will lead to a changes in AUC, and hence the variability in [I] is associated with the level of variability in the interaction. This happens regardless of fm (i.e., compare solid and broken lines which are superimposable at low ranges of [I]). **Right panel:** Arrows A shows that for strong inhibitors, any variation in circulating concentrations of ([I]) will not lead to any appreciable effect on M-DDI level; hence interindividual variability in [I] does not propagate to inhibitory effect. However, change in fm (or its interindividual variability) is associated with variation in the inhibitory response as indicated by arrow B and the big gap between broken and solid lines at the top end of the curves.

mechanistic, more pragmatic, models have gained more popularity as they could link some of the currently available *in vitro* data to clinical observations. Readers are referred to a recent report by Yang et al. (2007), as an example, where the F_G was described using an operational Q_{Gut} model. The model retains the form of the "well-stirred" model, but the flow term (Q_{Gut}) is a hybrid of both permeability through the enterocyte membrane and villous blood flow.

$$F_G = \frac{Q_{\mathrm{Gut}}}{Q_{\mathrm{Gut}} + \mathrm{f}u_{\mathrm{G}} \cdot \sum_{j=1}^{n} \mathrm{CL}u_{\mathrm{int},G_j}} \tag{18.10}$$

Thus, expanding Q_{Gut} in terms of its fundamental parameters leads to

$$Q_{\mathrm{Gut}} = \frac{Q_{\mathrm{villi}} \cdot \mathrm{CL}_{\mathrm{perm}}}{Q_{\mathrm{villi}} + \mathrm{CL}_{\mathrm{perm}}} \tag{18.11}$$

where $\mathrm{CL}_{\mathrm{perm}}$ is a clearance term defining permeability through the enterocyte and Q_{villi} is actual villous blood flow. Interindividual variability in intrinsic metabolic clearance ($\mathrm{CL}u_{\mathrm{int\ Gut}}$) and its inhibition will impact the outcome of

M-DDI. However, two compounds with exactly the same intrinsic clearance (or same degree of inhibition on clearance at the level of gut wall) may exhibit quite different extents of first-pass gut metabolism if the exposure to enzymes during transit through the gut wall varies between individuals. The degree of exposure depends on the effects of uptake and efflux transporters, passive membrane permeability, and enterocytic blood flow.

In the presence of an inhibitor that only alters intrinsic gut metabolic clearance, Eq. (18.10) can be rewritten to include an estimate of the concentration of inhibitor in the enterocyte ($[I]_{gut}$) and its inhibitory constant for affected pathway j:

$$F_G' = \frac{Q_{Gut}}{Q_{Gut} + fu_{Gut} \cdot \sum_{j=1}^{n} \frac{CLu_{int,G_j}}{\text{Fold inhibition of pathway J}}} \tag{18.12}$$

Fold inhibition can be defined depending on the type of inhibition:

$$\text{Fold inhibition of } CLu_{int,Gut} = \begin{cases} \text{Competitive:} & 1 + \dfrac{[I]_{Gut}}{Ki} \\ \text{Mechanism-based:} & \dfrac{k_{deg}}{k_{deg} + \dfrac{[I]_{Gut} \times k_{inact}}{[I] + K_I}} \end{cases} \tag{18.13}$$

When an inhibitor is not coadministered with a substrate, the unbound concentration of inhibitor in the systemic circulation can be used as an estimate of $[I]_{gut}$. However, when inhibitor and substrate are coadministered, the value of $[I]_{gut}$ may be estimated from the prehepatic absorption rate of inhibitor knowing enterocytic blood flow (Q_{ent}) according to Eq. (18.14):

$$[I]_{gut} = \frac{f_a \times k_a(I) \times \text{Dose}(I)}{Q_{ent}} \tag{18.14}$$

where f_a is the fraction of the inhibitor dose that is absorbed into the gut wall and k_a(I) and Dose(I) are the absorption rate constant and dose of inhibitor, respectively. The above equation assumes that the inhibitor is not subject to major first-pass gut metabolism itself.

Operating concentrations for induction of gut metabolism would be similar to those described above; however, Eq. (18.15) is used instead of (18.12) to account for the increase (rather than decrease) in the activity (see Almond et al. (2009) for further details):

$$F_G' = \frac{Q_{Gut}}{Q_{Gut} + fu_{Gut} \cdot \sum_{j=1}^{n} \left(CLu_{int,G_j} \times \left[1 + \left(\frac{E_{max} \times [I]_{Gut}}{IndC_{50} + [I]_{Gut}} \right) \right] \right)} \tag{18.15}$$

Equations (18.12) through (18.15) highlight the nonlinear relationship between the inhibitor concentrations (/dose) and the extent of effect on F_G. Thus, the interindividual variations related to $[I]_{Gut}$ (see Section 18.3.2) may not necessarily lead to variable inhibition of intestinal first-pass metabolism, and the propagation of variability in $[I]_{Gut}$ will depend on the nature of "victim" drug. If the "victim" drug has a high F_G to start with, variability in $[I]_{Gut}$ becomes irrelevant for any moderate inhibitory effects. Similarly, if the "victim" drug has high F_G, variation in $[I]_{Gut}$ is not going to cause variations in the level of M-DDI involving moderate induction of gut first-pass. Variability in $[I]_{Gut}$ (and the dose of "perpetrator" drug) will have a proportional influence on the inhibition and induction of "victim" drugs with low F_G.

18.3 CURRENT KNOWLEDGE OF ADME VARIABILITY RELEVANT TO M-DDI

Examples of incorporating variability into predictions of mDDI using first principles are given in this section with extended examples from an earlier report (Rostami-Hodjegan and Tucker, 2004).

18.3.1 Sources of Variability in Fractional Metabolism

If we consider that variation in fm depends on relative variability of metabolic routes, and that many of these routes are influenced by genetic polymorphism, then it is intuitive that, even at a fixed level of a given inhibitor (fixed [I] and $[I]_{Gut}$), a large variability in the extent of M-DDI is likely. The aspects related to genetic variability in noninhibited pathways has been reviewed by Collins et al. (2006) in relation to approved drugs on the market. However, as shown by Eq. (18.2), the notion of fm determining the variability in M-DDI is only true when dealing with strong inhibitors (i.e., M-DDI is more sensitive to fm than to [I]). A hypothetical example was considered by Rostami-Hodjegan and Tucker (2004) to show the interindividual variability of M-DDI by a strong inhibitor for a "victim" compound that is metabolized 50% by CYP1A2, 30% by CYP2D6 (an enzyme under genetic control), and 20% by renal excretion. As shown in Fig. 18.2, while complete inhibition of CYP1A2 in a typical healthy adult may not increase the AUC more than twofold, the same level of inhibition in an elderly patient with a compromised renal function or in an individual who lacks functional CYP2D6 for genetic reasons may cause up to 2.7- and 3.5-fold increases, respectively. The extent of inhibition in a poor CYP2D6 metabolizer with renal failure would be even greater (up to 11-fold assuming 20% normal renal function).

In the case of strong mechanism-based (suicidal) inactivation (MBI) (Ghanbari et al., 2006) and induction (Shou et al., 2008), the fold change in clearance depends on the natural degradation rate (k_{deg}) of the enzyme and it changes with the time parameter. The k_{deg} is an intrinsic parameter of the

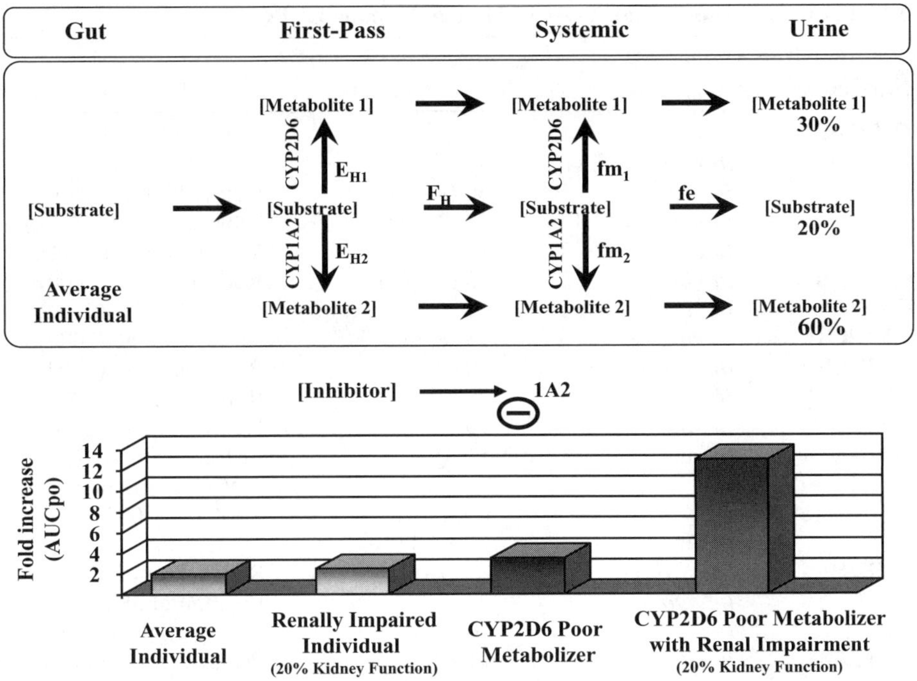

Figure 18.2. Amplification of inhibition effect on a particular enzyme (CYP1A2) by compromised level of activity in uninhibited routes (1-fm). The use of strong inhibitors will not significantly raise the plasma concentrations unless the parallel pathways (escape routes) are compromised. In the example given above, uninhibited routes, namely renal excretion and CYP2D6 activity, might be lower in some patients due to renal impairment or poor metabolizer genotype. These covariates will be associated with higher exposure to the "victim" drug under any inhibition of the major pathway (CYP1A2 in this case). However, the highest susceptibility is seen in (rare) poor metabolizers of CYP2D6 whose kidney function is also impaired. Studying such a specific group during Phase II is fraught with problems; nonetheless, it is valuable to know the extent of theoretically conceivable M-DDI in such a group.

system (human body), rather than a drug-dependent parameter (Yang et al., 2008). Thus, differential equations incorporating the dynamics of enzyme synthesis and degradation (Riley et al., 2007) are necessary. Interindividual variability in the degradation could be substantial (Yang et al., 2008) as shown by the range of values in Table 18.1.

18.3.2 Sources of Variability in Operating Inhibitor Concentrations [I]

The circulating concentration of the inhibitor (and hence concentration at the active site) is subject to variability due to variation in the administered dose

TABLE 18.1. Turnover Half-Lives of Human Hepatic CYPs

Enzyme	Method of Assessment	*N*	$t_{1/2}$ (h)[a]	Reference
CYP1A2	*In vitro*	1	51	Diaz et al. (1990)
	In vitro	NC[b]	43[c]	Maurel (1996)
	In vitro	5	36 (8–58)[d]	Renwick et al. (2000)
	In vivo	12	39 (27–54)	Faber and Fuhr (2004)
	In vivo	7	105	Magnusson et al. (2008)
CYP2A6	*In vitro*	2	26 (19–37)	Renwick et al. (2000)
CYP2B6	*In vitro*	1	32	Renwick et al. (2000)
CYP2C8	*In vitro*	5	23 (8–41)	Renwick et al. (2000)
CYP2C9	*In vitro*	5	104	Renwick et al. (2000)
CYP2C19	*In vitro*	3	26 (7–50)	Renwick et al. (2000)
CYP2D6	*In vitro*	4	70	Renwick et al. (2000)
	In vivo	13	51	Liston et al. (2002) Venkatakrishnan and Obach (2005)
	In vivo	15	46.6 ± 11.8	O'Mathuna et al. (2008)
CYP2E1	*In vitro*	5	27 (7–40)	Renwick et al. (2000)
	In vivo	6	60	Berthou et al. (1995)
	In vivo	11	50 ± 19	Emery et al. (1999)
CYP3A4	*In vitro*	1	44	Pichard et al. (1992)
	In vitro	NC	26[c]	Maurel (1996)
	In vitro	4	79	Renwick et al. (2000)
	In vivo	15	72[c]	Tran et al. (1999)
	In vivo	6	96 ± 38 (53–154)	Pitlick et al. (1976)
	In vivo	7	72 (20–146)	Lai et al. (1978)
	In vivo	3	(85–806)	Warren et al. (1980)
	In vivo	8	(36–50)	Fromm et al. (1996)
	In vivo	13	10[c] (2–158)	Boddy et al. (1995)
	In vivo	35	94 (62–205)	Rostami-Hodjegan et al. (1999)
	In vivo	7	70	Magnusson et al. (2008)
	In vivo	16	85 ± 61	Hsu et al. (1997)
	In vivo	6	140 (48–284)	von Bahr et al. (1998)
CYP3A5	*In vitro*	3	36 (15–70)	Renwick et al. (2000)

[a] $t_{1/2} = \ln(2)/k_{deg}$.
[b] NC, not clear.
[c] Estimated from reported data.
[d] Values in parentheses are ranges.
[e] ± indicates SD.

Source: Updated from Yang et al. (2008) and Ghanbari et al. (2006).

of the "perpetrator" or issues related to variable compliance. However, even under the administration of an equal dose to a group of subjects, [I] can vary substantially due to interindividual variation in all ADME processes that govern pharmacokinetics. Many of these variations can be simulated, and a detailed account is given for metabolism (Rostami-Hodjegan and Tucker, 2007) as well as other aspects (Jamei et al., 2009c). Here we summarize such sources of variability:

18.3.2.1 Variability in [I] Due to Interindividual Difference in Oral Absorption. Absorption of inhibitor is determined by its release from the formulation and permeation to enterocytes (fa), proportional escape from gut wall (F_G) and hepatic metabolism (F_H). All of these are subject to interindividual variation. For instance, variations in gastrointestinal (GI) motility and pH can lead to variable extent and rate of absorption, and hence variable concentration time profiles after the administration of the same dose in different individuals. Variable [I] (/and $[I]_{Gut}$) may propagate to the M-DDI according to Eqs. (18.3), (18.7), and (18.12). However, it should be noted that reported large variation in motility and residence time in the intestine [e.g., by Yu and co-workers (Yu and Amidon, 1998)], as well as variations in pH (Dressman and Fleisher, 1986; Evans et al., 1988; Dressman et al., 1990; Russell et al., 1993), may or may not translate into pharmacokinetic differences depending on drug and formulation characteristics (see Jamei et al. (2009c) for further details of issues related to variability in absorption).

F_G for any inhibitor would be sensitive to the abundance of drug-metabolizing enzymes. The abundance of enzymes in the gut wall is influenced by genetics and diet. There are various enzymes in the gut as described by Galetin and Houston in Chapter 9; however, CYP3A (Kolars et al., 1992; Watkins, 1992; Paine et al., 1997, 2006; Zhang et al., 1999) and UGTs (Ritter, 2007) are probably the most influential enzymes affecting gut wall metabolism, and hence their variability can affect the F_G of inhibitor profile (as well as F_G of the "victim" drug).

Active secretion of the drug from the gut wall into the gut lumen by the multidrug efflux pump, P-glycoprotein (P-gp), as well as influx and efflux by other transporters, is subject to interindividual variations that affect transporter abundance and/or activity. However, knowledge in this area is sparse.

18.3.2.2 Variability in [I] due to Interindividual Differences in Elimination. The sources of variability in intrinsic metabolic clearance that determine fm of the victim drug can equally affect the F_H of the "perpetrator" drug and hence cause variability in [I]. Induction and inhibition of enzymes by environmental substances or toxins contributes to interindividual differences in drug metabolism as much as the genetic makeup of the individual (Lin et al., 1997).

If renal excretion of the inhibitor (/inducer) was the major route of its elimination, individual variations in urine flow, urine pH, and plasma protein binding can lead to interindividual differences in [I]. The extent of sensitivity to these parameters depends on the nature of the compound and the mechanisms involved in its renal elimination (e.g., glumerular filtration, active secretion, passive reabsorption).

Recently, more attention has been paid to the genetic differences in transporters responsible for active renal secretion of drugs [e.g., OATP 1B1 (Yamashiro et al., 2006)] as well as active uptake to hepatocytes (Watanabe et al., 2009). Obviously, these may influence the circulating levels of [I] with potential impacts on interindividual variability of M-DDI.

18.3.3 Problems Associated with Identifying Covariates of M-DDI Using Clinical Data

Compliance (/adherence) to dosage regimens is an important aspect of clinical studies that if not monitored (or if the inhibitor or inducer plasma concentrations were unknown) could lead to difficulty in interpreting the dose–effect relationship (Vrijens and Goetghebeur, 1999) and should be considered in interpretation of any results from clinical data.

An argument that has often been used against paying attention to variability in fold inhibition (or induction) is the inverse relationship between the CL of the "victim" drug and the level of fold induction. It is argued that when fm is high in an individual due to high baseline activity of the inhibited enzyme, the concentration of the drug would be low to start with; under this hypothesis, high fold inhibition would not be of concern since it suppresses the initial high CL to bring it at a level similar to that of average individuals and raises the concentration of the "victim" drug from very low baseline levels to something in line with an average patient. This hypothesis is only valid if the prescribed dose is considered to be the same in all individuals and does not co-vary with their clearance. There are considerable data to show that prescribers, even in absence of monitoring drug concentrations, adjust the dose in line with the clearance of the drug, by using marker of activity. Hence the majority of patients receiving the "victim" drug under a realistic clinical setting will have similar concentrations in plasma at baseline (which is achieved via variable dose). A typical example of this would be the correlation between the genotype of CYP2C9 and the prescribed level of maintenance dose even when the prescribers were unaware of the patient genotype (Fig. 18.3); if these patients are exposed to an inhibitor, fold inhibition will raise the concentration to a much higher level in individuals who are at high doses [because of high CL]). Nonetheless, it is not always possible to link the level of M-DDI to the level of therapeutic dose of a "victim" drug, particularly when the reason for high fm is compromised status of noninhibited routes in some subjects. Figure 18.4 schematically describes such complexity in analyzing the covariate effects in M-DDI studies.

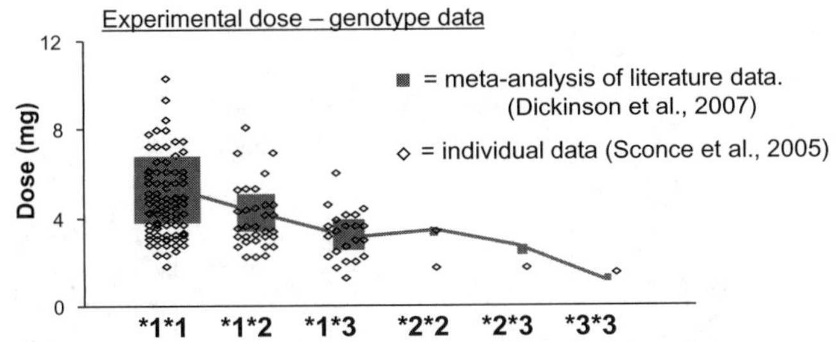

Figure 18.3. Meta-analysis of data for warfarin maintenance dose in relation to *CYP2C9* genotype. The prescribers were not aware of the patients' genotype and genotyping took place retrospectively. Drawn from the data collected and analyzed by Dickinson et al. (2007) from the literature reports (Aithal et al., 1999; Margaglione et al., 2000; Taube et al., 2000; Loebstein et al., 2001; Higashi et al., 2002; Kamali et al., 2004; Siguret et al., 2004).

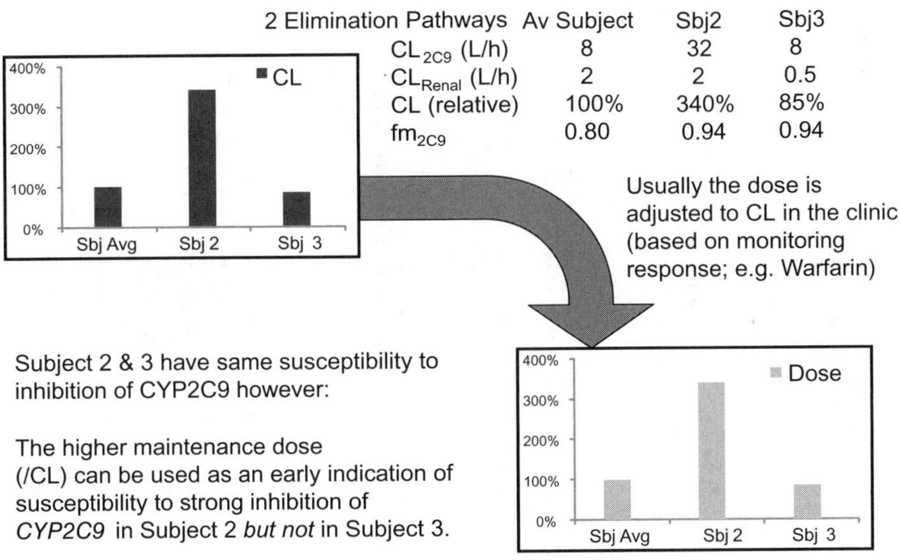

Figure 18.4. Schematic representation of reasons to exercise caution in taking initial high clearance in an individual as an indication for safety of high "fold-interaction." Some assume that the concentration of the "victim" drug will rise only to the level of concentrations in average people after M-DDI; the latter notion ignores the natural dose adjustment that takes place in a clinical setting based on markers of drug effect.

18.4 TOOLS FOR INCORPORATING INTERINDIVIDUAL VARIABILITY IN M-DDI

To our knowledge, currently the only automated tool with the capability of assessing interindividual variability in M-DDI is the Simcyp® Population-based ADME Simulator (Jamei et al., 2009b). The Simulator

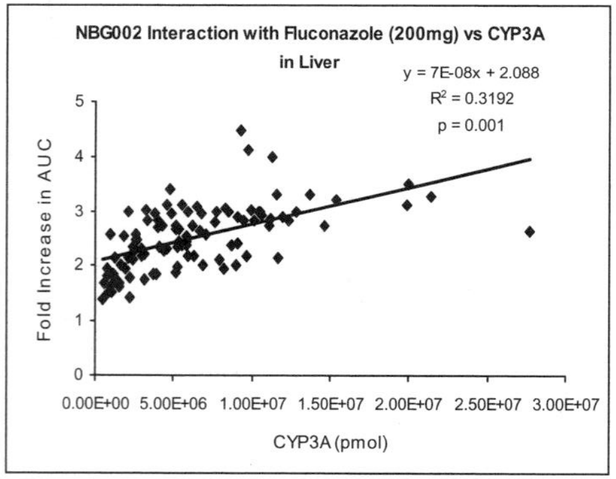
NBG002 Interaction with Fluconazole (200mg) vs CYP3A in Liver
y = 7E-08x + 2.088
R^2 = 0.3192
p = 0.001
Fold Increase in AUC
0
1
2
3
4
5
0.00E+00
5.00E+06
1.00E+07
1.50E+07
2.00E+07
2.50E+07
3.00E+07
CYP3A (pmol)

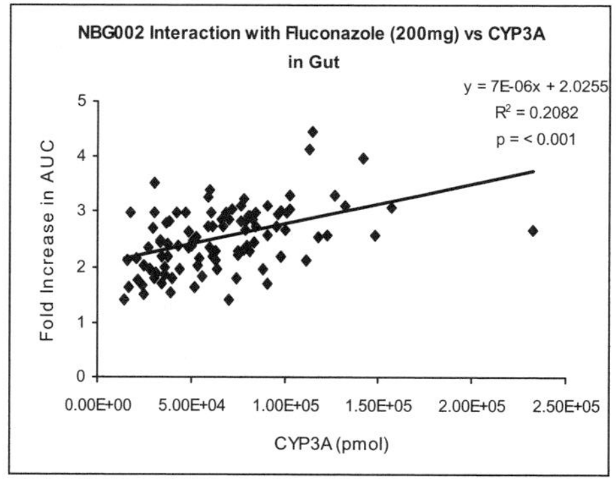
NBG002 Interaction with Fluconazole (200mg) vs CYP3A in Gut
y = 7E-06x + 2.0255
R^2 = 0.2082
p = < 0.001
Fold Increase in AUC
0
1
2
3
4
5
0.00E+00
5.00E+04
1.00E+05
1.50E+05
2.00E+05
2.50E+05
CYP3A (pmol)

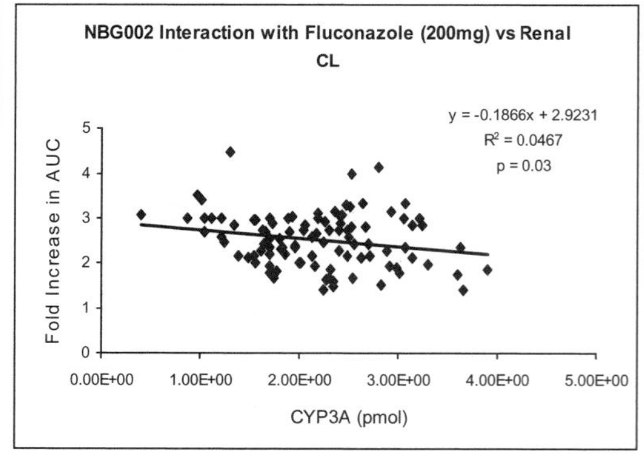
NBG002 Interaction with Fluconazole (200mg) vs Renal CL
y = -0.1866x + 2.9231
R^2 = 0.0467
p = 0.03
Fold Increase in AUC
0
1
2
3
4
5
0.00E+00
1.00E+00
2.00E+00
3.00E+00
4.00E+00
5.00E+00
CYP3A (pmol)

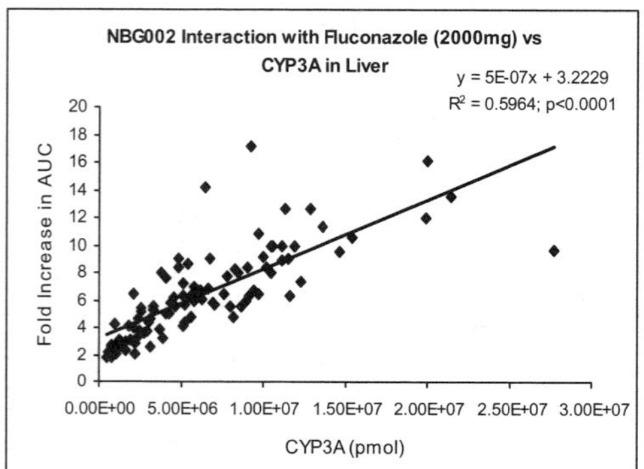
NBG002 Interaction with Fluconazole (2000mg) vs CYP3A in Liver
y = 5E-07x + 3.2229
R^2 = 0.5964; p<0.0001
Fold Increase in AUC
0
2
4
6
8
10
12
14
16
18
20
0.00E+00
5.00E+06
1.00E+07
1.50E+07
2.00E+07
2.50E+07
3.00E+07
CYP3A (pmol)

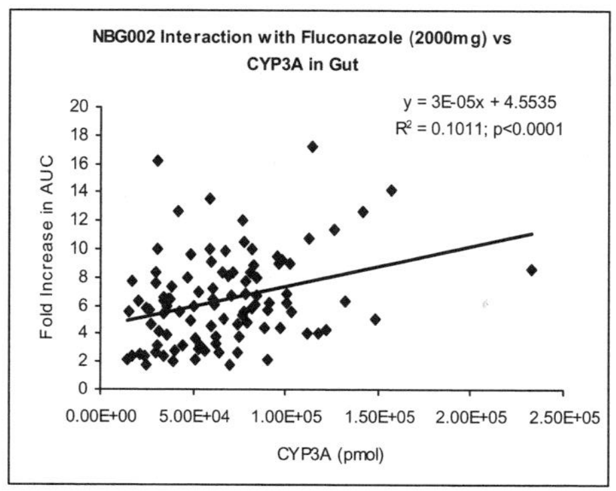
NBG002 Interaction with Fluconazole (2000mg) vs CYP3A in Gut
y = 3E-05x + 4.5535
R^2 = 0.1011; p<0.0001
Fold Increase in AUC
0
2
4
6
8
10
12
14
16
18
20
0.00E+00
5.00E+04
1.00E+05
1.50E+05
2.00E+05
2.50E+05
CYP3A (pmol)

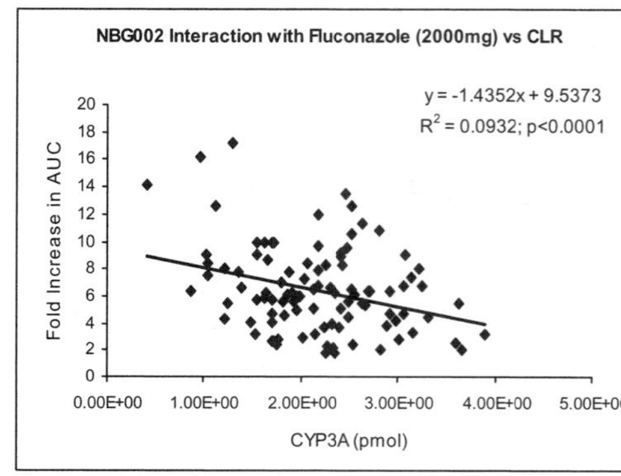
NBG002 Interaction with Fluconazole (2000mg) vs CLR
y = -1.4352x + 9.5373
R^2 = 0.0932; p<0.0001
Fold Increase in AUC
0
2
4
6
8
10
12
14
16
18
20
0.00E+00
1.00E+00
2.00E+00
3.00E+00
4.00E+00
5.00E+00
CYP3A (pmol)

contains large databases of information relevant to ADME on healthy and disease populations. The "bottom-up" nature of the platform, which uses mechanistic modeling and simulation of the processes of oral absorption, tissue distribution, metabolism, and excretion of drugs, assists the user with a more realistic assessment of theoretically conceivable extremes in M-DDI. The models are particularly useful to M-DDI involving multiple drug interactions, combined effects of parent drug and metabolite, and time- and dose-dependent phenomena such as auto-induction and auto-inhibition. In many cases, M-DDIs are primarily the result of inhibition of "first-pass" metabolism because the concentration of inhibitor is highest during its absorption. This is demonstrated by simulations showing the large impact that staggering the times of administration of inhibitor and substrate can have on the degree of an interaction (Yang et al., 2003). Predictions of M-DDI with the commonly used Eq. (18.1) do not account for different levels of inhibition during "first-pass" and systemic exposure. An account of various strategies in using qualitative, semiquantitative, and quantitative predictions of M-DDI was provided recently by Einolf (2007). More recent examples are provided by Fahmi and co-workers (2009). Readers are also referred to two review articles which outline the position by the experts from leading pharmaceutical companies in the USA (i.e., PhARMA reports by Grimm et al. (2009) and Chu et al. (2009)).

As an example, the capability of the currently available *in silico* techniques regarding covariate analysis is shown in a simulation study that was carried out and described in Fig. 18.5. The results highlight a fundamental concept that was described at the beginning of this chapter, namely, "covariates of M-DDI depend on the strength of the inhibitor at a given dose." Another example of the use of program for assessment of variations due to administration time and study design is given by Zhao et al. (2009).

Figure 18.5. Investigating covariates by creating *virtaal patients*. Fold increase in AUC resulting from an interaction between a hypothetical CYP3A4 substrate (NBG002) and orally adminstered fluconazole is simulated in a Caucasian population at two dose levels for the inhibitor using Simcyp ADME Simulator. The covariation of M-DDI with the hepatic and gut expression levels of CYP3A4 (according to R^2 values for the regression lines) and the renal function of individuals was investigated in the virtual population created by Simcyp (V 7). The graphs show that CYP3A4 in the gut and liver is a determinant of the level of interaction ("regression to mean effect") particularly for weak inhibitors (/doses), which do not abolish metabolic activity completely. However, the remaining elimination pathway (renal CL in this case) also plays a significant role, and this is largely dependent on whether a major part of the metabolic activity disappears due to inhibition or not. As the inhibitor dose is increased from 200 to 2000 mg, renal function is a stronger covariate for increased AUC. Therefore, the level of inhibition, especially in the presence of a strong inhibitor (or larger dose), is determined by the function of the noninhibited routes (i.e., 1-fm). Thus, it should be noted that covariates determining the magnitude of drug–drug interactions are dose-dependent. A degree of inhibition of CYP3A4 that is not important in normal subjects may produce a pronounced effect in renally impaired subjects and elderly patients (or CYP2D6 poor metabolizers if this enzyme is involved, etc.).

18.5 CONCLUSIONS

The speed and cost of drug development requires optimization to reduce the ever-rising cost/benefit of the process of drug development. The majority of the procedures involving assessment of covariates for M-DDI are still being built based on observed data ("top-down") rather than taking advantage of broader knowledge of the system in which the M-DDI happens (i.e., human body) under the so-called "bottom-up" approach. Although the identification and quantification of covariates is now considered in drug development, determining covariates using traditional small parallel studies or using large-scale sparse sampling coupled with population-based pharmacokinetic studies is not straightforward and is not possible all the time. *In vitro–in vivo* extrapolation (IVIVE) has become possible now because of advances in the understanding of the extrapolation factors (physical chemistry, biology, physiology, and genetics) and the ability to "integrate" such information using mechanistic models of the human body (Rostami-Hodjegan and Tucker, 2007) combined with the power of computers (Jamei et al., 2009b). These efforts are in parallel to other scientific activities under the umbrella of "Systems Biology" and are embraced increasingly by the pharmaceutical industry [e.g., see Lalonde et al. (2007)]. Population-based simulations of M-DDI considering the full profiles of interacting drugs (accounting for the mutual effect), and additional complexities of inhibitory metabolites (simultaneous inhibition and induction) to find the most appropriate study designs and assess the M-DDI in rare conditions (which clinically exist but difficult to design as part of any study) are a small part of a big effort in accelerating the drug development while keeping it as safe as ever, if not safer.

REFERENCES

Aithal GP, Day CP, Kesteven PJ, Daly AK. Association of polymorphisms in the cytochrome P450 CYP2C9 with warfarin dose requirement and risk of bleeding complications. *Lancet* 1999;353:717–719.

Almond LM, Yang J, Jamei M, Tucker GT, Rostami-Hodjegan A. Towards a quantitative framework for the prediction of DDI's arising from Cytochrome P450 induction. *Curr Drug Metab* 2009;10:420–432.

Berthou F, Goasduff T, Lucas D, Dreano Y, Le Bot MH, Menez JF. Interaction between two probes used for phenotyping cytochromes P4501A2 (caffeine) and P4502E1 (chlorzoxazone) in humans. *Pharmacogenetics* 1995;5:72–79.

Boddy AV, Cole M, Pearson AD, Idle JR. The kinetics of the auto-induction of ifosfamide metabolism during continuous infusion. *Cancer Chemother Pharmacol* 1995;36:53–60.

Chu V, Einolf HJ, Evers R, Kumar G, Moore D, Ripp S, Silva J, Sinha V, Sinz M, Skerjanec A. In vitro and in vivo induction of cytochrome p450: a survey of the current practices and recommendations: a Pharmaceutical Research and Manufacturers of America perspective. *Drug Metab Dispos* 2009;37:1339–1547.

Collins C, Levy R, Ragueneau-Majlessi I, Hachad H. Prediction of maximum exposure in poor metabolizers following inhibition of nonpolymorphic pathways. *Curr Drug Metab* 2006;7:295–299.

Diaz D, Fabre I, Daujat M, Saint Aubert B, Bories P, Michel H, Maurel P. Omeprazole is an aryl hydrocarbon-like inducer of human hepatic cytochrome P450. *Gastroenterology* 1990;99:737–747.

Dickinson GL, Lennard MS, Tucker GT, Rostami-Hodjegan A. The use of mechanistic DM-PK-PD modelling to assess the power of pharmacogenetic studies—CYP2C9 and warfarin as an example. *Br J Clin Pharmacol* 2007;64:14–26

Dressman JB, Fleisher D. Mixing-tank model for predicting dissolution rate control or oral absorption. *J Pharm Sci* 1986;75:109–116.

Dressman JB, Berardi RR, Dermentzoglou LC, Russell TL, Schmaltz SP, Barnett JL, Jarvenpaa KM. Upper gastrointestinal (GI) pH in young, healthy men and women. *Pharm Res* 1990;7:756–761.

Einolf HJ. Comparison of different approaches to predict metabolic drug–drug interactions. *Xenobiotica* 2007;37:1257–1294.

Emery MG, Jubert C, Thummel KE, Kharasch ED. Duration of cytochrome P-450 2E1 (CYP2E1) inhibition and estimation of functional CYP2E1 enzyme half-life after single-dose disulfiram administration in humans. *J Pharmacol Exp Ther* 1999;291:213–219.

Evans DF, Pye G, Bramley R, Clark AG, Dyson TJ, Hardcastle JD. Measurement of gastrointestinal pH profiles in normal ambulant human subjects. *Gut* 1988;29:1035–1041.

Faber MS, Fuhr U. Time response of cytochrome P450 1A2 activity on cessation of heavy smoking. *Clin Pharmacol Ther* 2004;76:178–184.

Fahmi OA, Hurst S, Plowchalk D, Cook J, Guo F, Youdim K, Dickins M, Phipps A, Darekar A, Hyland R, Obach RS. Comparison of different algorithms for predicting clinical drug-drug interactions, based on the use of CYP3A4 in vitro data: predictions of compounds as precipitants of interaction. *Drug Metab Dispos* 2009;37: 1658–1666.

Fromm MF, Busse D, Kroemer HK, Eichelbaum M. Differential induction of prehepatic and hepatic metabolism of verapamil by rifampin. *Hepatology* 1996;24:796–801.

Ghanbari F, Rowland-Yeo K, Bloomer JC, Clarke SE, Lennard MS, Tucker GT, Rostami-Hodjegan A. A critical evaluation of the experimental design of studies of mechanism based enzyme inhibition, with implications for *in vitro–in vivo* extrapolation. *Curr Drug Metab* 2006;7:315–334.

Grimm SW, Einolf HJ, Hall SD, He K, Lim HK, Ling KH, Lu C, Nomeir AA, Seibert E, Skordos KW, Tonn GR, Van Horn R, Wang RW, Wong YN, Yang TJ, Obach RS. The conduct of in vitro studies to address time-dependent inhibition of drug-metabolizing enzymes: a perspective of the pharmaceutical research and manufacturers of America. *Drug Metab Dispos* 2009;37:1355–1370.

Higashi MK, Veenstra DL, Kondo LM, Wittkowsky AK, Srinouanprachanh SL, Farin FM, Rettie AE. Association between CYP2C9 genetic variants and anticoagulation-related outcomes during warfarin therapy. *Jama* 2002;287:1690–1698.

Hsu A, Granneman GR, Witt G, Locke C, Denissen J, Molla A, Valdes J, Smith J, Erdman K, Lyons N, Niu P, Decourt JP, Fourtillan JB, Girault J, Leonard JM.

Multiple-dose pharmacokinetics of ritonavir in human immunodeficiency virus-infected subjects. *Antimicrob Agents Chemother* 1997;41:898–905.

Jamei M, Dickinson GL, Rostami-Hodjegan A. A framework for assessing inter-individual variability in pharmacokinetics using virtual human populations and integrating general knowledge of physical chemistry, biology, anatomy, physiology and genetics: a tale of "bottom-up" vs "top-down" recognition of covariates. *Drug Metab Pharmacokin* 2009a;24:53–75.

Jamei M, Marciniak S, Feng K, Barnett A, Tucker GT, Rostami-Hodjegan A. The Simcyp® population-based ADME simulator. *Expert Opin Drug Metab Toxicol* 2009b;5:211–223.

Jamei M, Turner D, Yang J, Neuhoff S, Polak S, Rostami-Hodjegan A, Tucker GT. Population-based mechanistic prediction of oral drug absorption. *AAPS Journal* 2009c;11:225–237.

Kamali F, Khan TI, King BP, Frearson R, Kesteven P, Wood P, Daly AK, Wynne H. Contribution of age, body size, and CYP2C9 genotype to anticoagulant response to warfarin. *Clin Pharmacol Ther* 2004;75:204–212.

Kolars JC, Schmiedlin-Ren P, Schuetz JD, Fang C, Watkins PB. Identification of rifampin-inducible P450IIIA4 (CYP3A4) in human small bowel enterocytes. *J Clin Invest* 1992;90:1871–1878.

Krayenbuhl JC, Vozeh S, Kondo-Oestreicher M, Dayer P. Drug–drug interactions of new active substances: mibefradil example. *Eur J Clin Pharmacol* 1999;55:559–565.

Lai AA, Levy RH, Cutler RE. Time-course of interaction between carbamazepine and clonazepam in normal man. *Clin Pharmacol Ther* 1978;24:316–323.

Lalonde RL, Kowalski KG, Hutmacher MM, Ewy W, Nichols DJ, Milligan PA, Corrigan BW, Lockwood PA, Marshall SA, Benincosa LJ, Tensfeldt TG, Parivar K, Amantea M, Glue P, Koide H, Miller R. Model-based drug development. *Clin Pharmacol Ther* 2007;82:21–32.

Lin JH, Chiba M, Baillie TA. *In vivo* assessment of intestinal drug metabolism. *Drug Metab Dispos* 1997;25:1107–1109.

Liston HL, DeVane CL, Boulton DW, Risch SC, Markowitz JS, Goldman J. Differential time course of cytochrome P450 2D6 enzyme inhibition by fluoxetine, sertraline, and paroxetine in healthy volunteers. *J Clin Psychopharmacol* 2002;22:169–173.

Loebstein R, Yonath H, Peleg D, Almog S, Rotenberg M, Lubetsky A, Roitelman J, Harats D, Halkin H, Ezra D. Interindividual variability in sensitivity to warfarin—nature or nurture? *Clin Pharmacol Ther* 2001;70:159–164.

Magnusson MO, Dahl ML, Cederberg J, Karlsson MO, Sandstrom R. Pharmacodynamics of carbamazepine-mediated induction of CYP3A4, CYP1A2, and Pgp as assessed by probe substrates midazolam, caffeine, and digoxin. *Clin Pharmacol Ther* 2008;84:52–62.

Margaglione M, Colaizzo D, D'Andrea G, Brancaccio V, Ciampa A, Grandone E, Di Minno G. Genetic modulation of oral anticoagulation with warfarin. *Thromb Haemost* 2000;84:775–778.

Maurel P. The use of adult human hepatocytes in primary culture and other *in vitro* systems to investigate drug metabolism in man. *Adv Drug Deliv Rev* 1996;22:105–132.

O'Mathuna B, Farre M, Rostami-Hodjegan A, Yang J, Cuyas E, Torrens M, Pardo R, Abanades S, Maluf S, Tucker GT, de la Torre R. The consequences of 3,4-methylenedioxymethamphetamine induced CYP2D6 inhibition in humans. *J Clin Psychopharmacol* 2008;28:523–529.

Paine MF, Hart HL, Ludington SS, Haining RL, Rettie AE, Zeldin DC. The human intestinal cytochrome P450 "pie". *Drug Metab Dispos* 2006;34:880–886.

Paine MF, Khalighi M, Fisher JM, Shen DD, Kunze KL, Marsh CL, Perkins JD, Thummel KE. Characterization of interintestinal and intraintestinal variations in human CYP3A-dependent metabolism. *J Pharmacol Exp Ther* 1997;283:1552–1562.

Pang KS. Modeling of intestinal drug absorption: roles of transporters and metabolic enzymes (for the Gillette Review Series). *Drug Metab Dispos* 2003;31:1507–1519.

Pichard L, Fabre I, Daujat M, Domergue J, Joyeux H, Maurel P. Effect of corticosteroids on the expression of cytochromes P450 and on cyclosporin A oxidase activity in primary cultures of human hepatocytes. *Mol Pharmacol* 1992;41:1047–1055.

Pitlick WH, Levy RH, Tropin AS, Green JR. Pharmacokinetic model to describe self-induced decreases in steady-state concentrations of carbamazepine. *J Pharm Sci* 1976;65:462–463.

Renwick AB, Watts PS, Edwards RJ, Barton PT, Guyonnet I, Price RJ, Tredger JM, Pelkonen O, Boobis AR, Lake BG. Differential maintenance of cytochrome P450 enzymes in cultured precision-cut human liver slices. *Drug Metab Dispos* 2000;28:1202–1209.

Riley RJ, Grime K, Weaver R. Time-dependent CYP inhibition. *Expert Opin Drug Metab Toxicol* 2007;3:51–66.

Ritter JK. Intestinal UGTs as potential modifiers of pharmacokinetics and biological responses to drugs and xenobiotics. *Expert Opin Drug Metab Toxicol* 2007;3:93–107.

Rostami-Hodjegan A, Tucker GT. "*In Silico*" simulations to assess the "*in vivo*" consequences of "*in vitro*" metabolic drug–drug interactions. *Drug Discovery Today: Technologies* 2004;1:441–448.

Rostami-Hodjegan A, Tucker GT. Simulation and prediction of in vivo drug metabolism in human populations from *in vitro* data. *Nat Rev Drug Discov* 2007;6:140–148.

Rostami-Hodjegan A, Wolff K, Hay AW, Raistrick D, Calvert R, Tucker GT. Population pharmacokinetics of methadone in opiate users: characterization of time-dependent changes. *Br J Clin Pharmacol* 1999;48:43–52.

Rowland M, Matin SB. Kinetics of drug–drug interactions. *J Pharmacokinet Biopharm* 1973;1:553–567.

Russell TL, Berardi RR, Barnett JL, Dermentzoglou LC, Jarvenpaa KM, Schmaltz SP, Dressman JB. Upper gastrointestinal pH in seventy-nine healthy, elderly, North American men and women. *Pharm Res* 1993;10:187–196.

Sconce EA, Khan TI, Wynne HA, Avery P, Monkhouse L, King BP, Wood P, Kesteven P, Daly AK, Kamali F. The impact of CYP2C9 and VKORC1 genetic polymorphism and patient characteristics upon warfarin dose requirements: proposal for a new dosing regimen. *Blood* 2005;106:2329–2333.

Shou M, Hayashi M, Pan Y, Xu Y, Morrissey K, Xu L, Skiles GL. Modeling, prediction, and *in vitro–in vivo* correlation of CYP3A4 induction. *Drug Metab Dispos* 2008;36:2355–2370.

Siguret V, Gouin I, Golmard JL, Geoffroy S, Andreux JP, Pautas E. [Cytochrome P450 2C9 polymorphisms (CYP2C9) and warfarin maintenance dose in elderly patients]. *Rev Med Interne* 2004;25:271–274.

Tam D, Tirona RG, Pang KS. Segmental intestinal transporters and metabolic enzymes on intestinal drug absorption. *Drug Metab Dispos* 2003;31:373–383.

Taube J, Halsall D, Baglin T. Influence of cytochrome P-450 CYP2C9 polymorphisms on warfarin sensitivity and risk of over-anticoagulation in patients on long-term treatment. *Blood* 2000;96:1816–1819.

Tran JQ, Kovacs SJ, McIntosh TS, Davis HM, Martin DE. Morning spot and 24-hour urinary 6 beta-hydroxycortisol to cortisol ratios: intraindividual variability and correlation under basal conditions and conditions of CYP 3A4 induction. *J Clin Pharmacol* 1999;39:487–494.

Venkatakrishnan K, Obach RS. *In vitro–in vivo* extrapolation of CYP2D6 inactivation by paroxetine: prediction of nonstationary pharmacokinetics and drug interaction magnitude. *Drug Metab Dispos* 2005;33:845–852.

von Bahr C, Steiner E, Koike Y, Gabrielsson J. Time course of enzyme induction in humans: effect of pentobarbital on nortriptyline metabolism. *Clin Pharmacol Ther* 1998;64:18–26.

Vrijens B, Goetghebeur E. The impact of compliance in pharmacokinetic studies. *Stat Methods Med Res* 1999;8:247–262.

Warren JW Jr, Benmaman JD, Wannamaker BB, Levy RH. Kinetics of a carbamazepine–ethosuximide interaction. *Clin Pharmacol Ther* 1980;28:646–651.

Watanabe T, Kusuhara H, Maeda K, Shitara Y, Sugiyama Y. Physiologically based pharmacokinetic modeling to predict transporter-mediated clearance and distribution of pravastatin in humans. *J Pharmacol Exp Ther* 2009;328:652–662.

Watkins PB. Drug metabolism by cytochromes P450 in the liver and small bowel. *Gastroenterol Clin North Am* 1992;21:511–526.

Yamashiro W, Maeda K, Hirouchi M, Adachi Y, Hu Z, Sugiyama Y. Involvement of transporters in the hepatic uptake and biliary excretion of valsartan, a selective antagonist of the angiotensin II AT1-receptor, in humans. *Drug Metab Dispos* 2006;34:1247–1254.

Yang JS, Kjellsson M, Rostami-Hodjegan A, Tucker GT. The effects of dose staggering on metabolic drug–drug interactions. *Eur J Pharm Sci* 2003;20:223–232.

Yang J, Jamei M, Yeo KR, Tucker GT, Rostami-Hodjegan A. Prediction of intestinal first-pass drug metabolism. *Curr Drug Metab* 2007;8:676–684.

Yang J, Liao M, Shou M, Jamei M, Yeo KR, Tucker GT, Rostami-Hodjegan A. Cytochrome p450 turnover: regulation of synthesis and degradation, methods for determining rates, and implications for the prediction of drug interactions. *Curr Drug Metab* 2008;9:384–394.

Yu LX, Amidon GL. Characterization of small intestinal transit time distribution in humans. *Int J Pharm* 1998;171:157–163.

Zhang QY, Dunbar D, Ostrowska A, Zeisloft S, Yang J, Kaminsky LS. Characterization of human small intestinal cytochromes P-450. *Drug Metab Dispos* 1999;27: 804–809.

Zhao P, Ragueneau-Majlessi I, Zhang L, Strong JM, Reynolds KS, Lev RH, Thummel KE, Huang SM. Quantitative evaluation of pharmacokinetic inhibition of CYP3A substrates by ketoconazole—a simulation study. *J Clin Pharmacol* 2009;49:351–359.

PART III

INHIBITION OF THE DRUG TARGET ENZYMES—THE DESIRABLE INHIBITION

19

NF-κB: MECHANISM, TUMOR BIOLOGY, AND INHIBITORS

LENNY DANG

19.1 NF-κB

Nuclear factor kappa B (NF-κB) is an evolutionary conserved family of transcription factors involved in responses to a vast array of external stress stimuli. Intense research in this field, as evident by the volume of scientific papers published on this subject over the last 20 years, has yielded a great deal of understanding the basic mechanism underlying this important signaling pathway. NF-κB has emerged to be one of the most-studied biological pathways in eukaryotes. Yet, as much as we appreciate the importance of NF-κB in the pathogenesis of a plethora of diseases ranging from inflammation to cancer, diabetes, pain, and neurological disorders, among many others, the diversity of tissue-specific responses and the complexity of the *in vivo* dynamic network of this pathway are only beginning to unravel with the use of classic genetic techniques in combination with genomic and pharmacological tools that recently have become more readily available. There is great promise in targeting this pathway with so much compelling evidences, yet further studies in both normal physiology and disease states will still be needed in order to develop safe and effective therapeutics for a myriad of diseases.

This review is intended to introduce readers to basic mechanistic biology, associated pathology, and exemplary inhibitors of NF-κB pathway. But given the importance and the breadth of this research field, this overview cannot possibly be sufficiently comprehensive to cover all disease areas that are

Enzyme Inhibition in Drug Discovery and Development, edited by Lu and Li
Copyright © 2010 John Wiley & Sons, Inc.

relevant to the discovery and development of therapeutics for various NF-κB-driven diseases. For instance, NF-κB in the context of inflammation and autoimmune diseases in and of itself is a vast area of research deserving a much more extensive review than this space allows. As such, a number of excellent review articles are available for more in-depth understanding of NF-κB mechanism in a variety of disease settings. In particular, review series authored or edited by Thomas D. Gilmore over the years are highly recommended readings.

With that consideration, the discussion in this chapter will focus on current knowledge of the basic regulation of NF-κB and its role in hemaetological malignancies. Toward the end of the chapter, we will discuss specific approaches to inhibit NF-κB with the aim toward disease management.

19.2 THE NF-κB PROTEIN FAMILY

In eukaryotes, the generically known NF-κB actually composes five members of the Rel/NF-κB proteins: p65 (RelA), RelB, c-Rel, p50 (NF-κB1), and p52 (NF-κB2) (Fig. 19.1). This is characterized by a 300 amino-acid N-terminal sequence known as the Rel-homology domain (RHD) (Hayden and Ghosh, 2004; Gilmore, 2006, reviews). The RDH sequence mediates dimerization, inhibitory kappa-B binding interaction, nuclear localization, and DNA binding. These five members can be further subdivided into two groups based on their sequence C-terminal to their RDH: the "NF-κB" proteins (p50/p105 and p52/p100) and the "Rel" proteins (RelA, RelB, and c-Rel). The Rel protein family members, but not the NF-κB proteins, possess a potent transactivation domain (TA) C-terminal to the RHD which is required for competent gene transcription. On the other hand, the NF-κB(1/2) family proteins are synthesized as precursors and are distinguished by possessing an I kappa-B like inhibitory domain characterized by multiple ankyrin-like repeats. Consequently, members of this subgroup become activated when they are partially processed proteolytically by the proteasome to the transcriptionally competent forms of p50 and p52 (Palombella et al., 1994; Sears et al., 1998). Without processing, however, these NF-κB precursors in some instances can indeed act as inhibitors when combined with Rel family members. These NF-κB and Rel family members can associate as homo- or heterodimers, depending on context. The prototypical dimer most often studied is the p50/RelA heterodimer; however, other combinations are also frequently observed. For example, p50/p50 or p52/p52 homodimers can form to negatively regulate gene expression (Nolan et al., 1993). These different NF-κB dimers bind to kappa-B sites within the promoter/enhancer region of the target genes and regulate gene transcription positively or negatively via recruitment of coactivators and corepressors (Lenardo and Baltimore, 1989). This ability to dimerize in a combinatorial manner and be activated under a variety of internal and external stimuli allows NF-κB to orchestrate unique sets of gene transcription programs

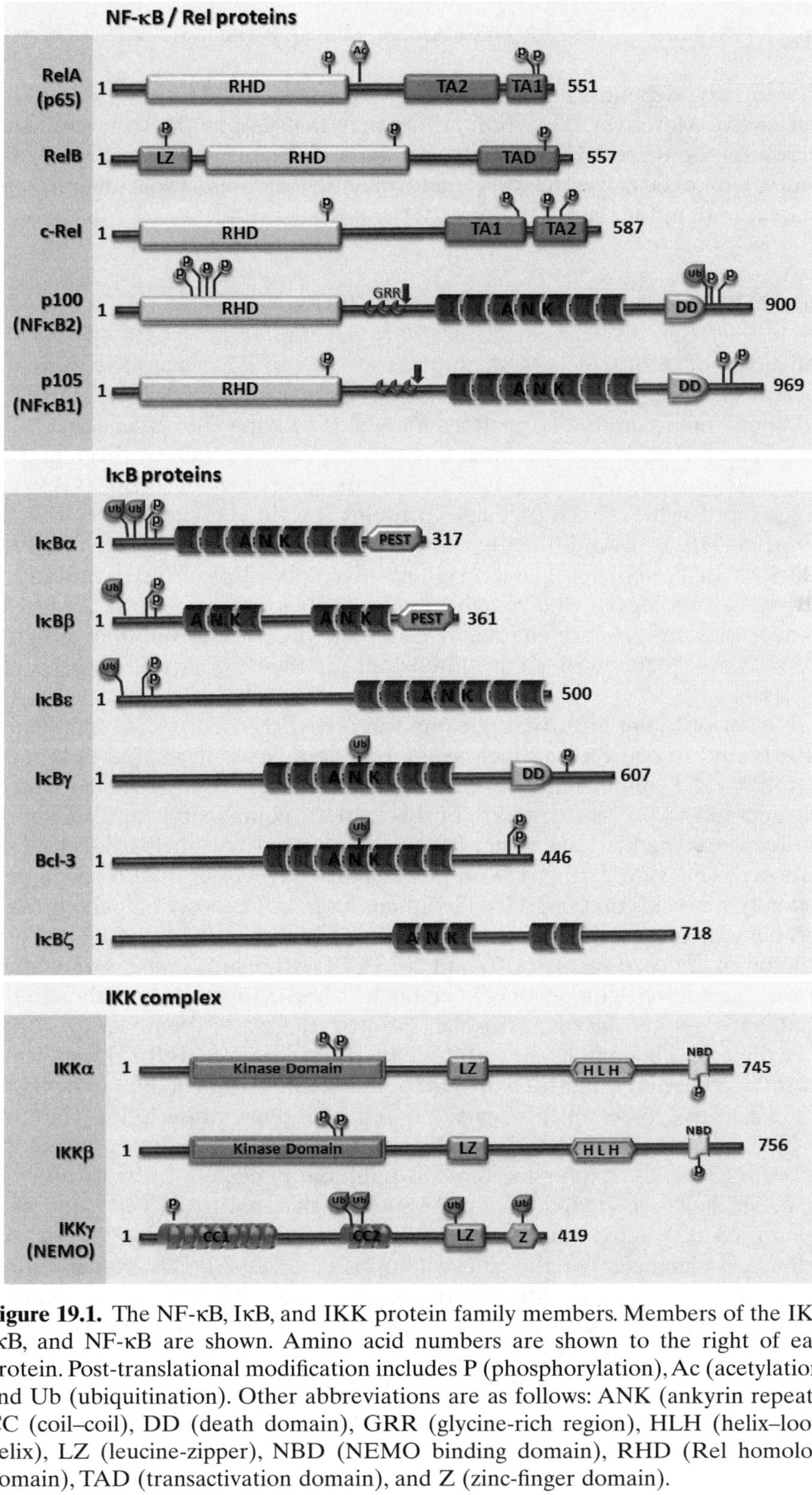

Figure 19.1. The NF-κB, IκB, and IKK protein family members. Members of the IKK, IκB, and NF-κB are shown. Amino acid numbers are shown to the right of each protein. Post-translational modification includes P (phosphorylation), Ac (acetylation), and Ub (ubiquitination). Other abbreviations are as follows: ANK (ankyrin repeats), CC (coil–coil), DD (death domain), GRR (glycine-rich region), HLH (helix–loop–helix), LZ (leucine-zipper), NBD (NEMO binding domain), RHD (Rel homology domain), TAD (transactivation domain), and Z (zinc-finger domain).

differentially, depending on the context of other cellular processes (Xiao et al., 2004). Moreover, it is worth noting here that despite the structural similarities among these NF-κB family members and their ability to bind to the same cognate DNA sequence, genetic deletion studies reveal distinct and nonoverlapping functions (Perkins, 2007).

19.3 MECHANISM OF NF-κB ACTIVATION

Activation of NF-κB occurs at multiple levels, but it is clear that the main mechanism of regulation occurs through the association of tightly controlled and highly labile inhibitory proteins known as I kappa Bs. Under most basal conditions, NF-κB dimers are maintained in an inactive state by this family of I kappa B proteins. The typical IκB family members include IκBα, IκBβ, and IκBε, as well as the NF-κB precursor proteins described earlier, p100 (NFκB1) and p105 (NFκB2). Additionally, there are two atypical IκB proteins that are inducible, IκBζ and Bcl-3, and an alternative spliced of NFκB1 resulting in IκBγ whose biological role remains unclear (Hayden and Ghosh, 2008). All of these IκB proteins are characterized by the presence of multiple ankyrin-like repeat sequences, which are important for their interaction with NF-κB dimers.

In most cells, the prototypical canonical NF-κB is the trimeric complex of IκBα bound to p50-RelA, which as a latent complex is incapable of binding to DNA and importantly preferentially partitioned to the cytoplasm. Examination the crystal structure of this trimeric complex reveals two important features. The first one is that IκBα masks one of two nuclear localization sequences (NLS), only the one on p65 at the RHD, but not the NLS on p50. Secondly, it reveals that IκBα itself contains a surface-exposed signal sequence for nuclear export (NES) (Phelps et al., 2000; Johnson et al., 1999). The combination of exposed NLS of p50 and NES of IκBα results in a constant shuttling of the trimeric complex between nucleus and cytoplasm, even though the basal steady-state favors cytosolic NF-κB. Regulated removal of IκBα from the trimeric complex via proteolytic degradation of IκBα dramatically alters the dynamic equilibrium in favor of the accumulation of unbound NF-κB in the nucleus, allowing for DNA binding and gene transcription (Perkins, 2007). Interestingly, one of the early response gene transcription events by NF-κB is in fact the expression of its own inhibitory protein IκBα, creating an important negative feedback loop. As such, in the absence of IκBα, termination of NF-κB activation following canonical stimulation by TNF-α, for instance, is significantly delayed (Pasparakis et al., 2006). This example illustrates the importance of not only the intensity and duration of the stimulus signal, but also the kinetics of feedback loop that determine the degree of NF-κB pathway activation. Moreover, it has been suggested that isoforms of typical IκB proteins (α, β, ε) participate differentially in controlling the amplitude, duration, and oscillation frequency of NF-κB activation resulting in distinct gene transcription programs (Hoffmann et al., 2002). Consistently, while genome substitution of IκBβ knock-in can mostly replace IκBα functionally, gene knockout studies suggest that each IκB protein has unique functions even in the same pathway (Tergaonkar et al., 2005).

19.4 NF-κB SIGNALING PATHWAYS

Depending on the type of stimulus and cellular context, there are several distinct pathways through which NF-κB can be activated, often referred to as canonical, alternative, and atypical (Fig. 19.2). The canonical pathway involves the degradation of IκBα regulated via signal-induced phosphorylation by the IκBα kinase complex (IKK) consisting of IKKα, IKKβ, and NEMO. The noncanonical or alternative pathway involves the proteasome processing of p100, which is regulated via phosphorylation not by the IKK complex that contains IKKβ, but rather by the IKKα homodimer. And finally, the atypical pathway modulates NF-κB in manners entirely independent of IKK phosphorylation.

In the canonical pathway, external stimuli such as TNF-α, IL-1β, and LPS promote the rapid degradation of IκBα. This process is regulated by the phosphorylation of IκBα at two conserved residues at the N-terminus, Ser^{32} and Ser^{36}, by the IKK complex (Ghosh and Baltimore, 1990; Mercurio et al., 1997). Phosphorylation on these two serines creates a conserved binding motif (DS*GXXS*) for β-transducin repeat-containing protein (β-TrCP), a receptor subunit of a large Skip1/Cullin1/F-box (SCF) E3 ligase complex which in coordination with the E2 UbcH5 mediates poly-ubiquitination of consecutive lysines 21 and 22 on IκBα (Campbell et al., 2004). This K48-linked polyubiquitinated IκBα is then targeted to the 26S proteasome for proteolytic degradation while sparing p65 and p50, exposing the previously masked NLS on p65. With two NLS exposed on both p65 and p50, combined with the loss of NES on IκBα, unbound NF-κB is then preferentially partitioned from the cytoplasm to the nucleus for DNA-binding. This process is highly efficient and rapid. In fact, immunofluorescent imaging techniques now readily available commercially using anti-p65 antibodies are often used to assay for NF-κB nucleus translocation as a convenient and highly quantitative measurement of canonical pathway activation in cells. Moreover, in a demonstration of how the NLS and the process of nuclear translocation serve to tightly regulate pathway activation, a cell-permeable polypeptide was designed based on the NLS of p50. This peptide, known as SN50, but not a scrambled-sequence control peptide, inhibited NF-κB transcription in a dose-dependent manner in endothelial or monocytic cells stimulated with either LPS or TNFα (Lin et al., 1995). In fact, SN50 is commonly used as an NF-κB inhibitor research tool in a variety of cellular biology experiments.

In the alternative or noncanonical pathway, instead of utilizing proteasomal degradation, activation of NF-κB proceeds through proteasomal processing of the precursor form of p100 to 52, an interesting mechanism where the 26S

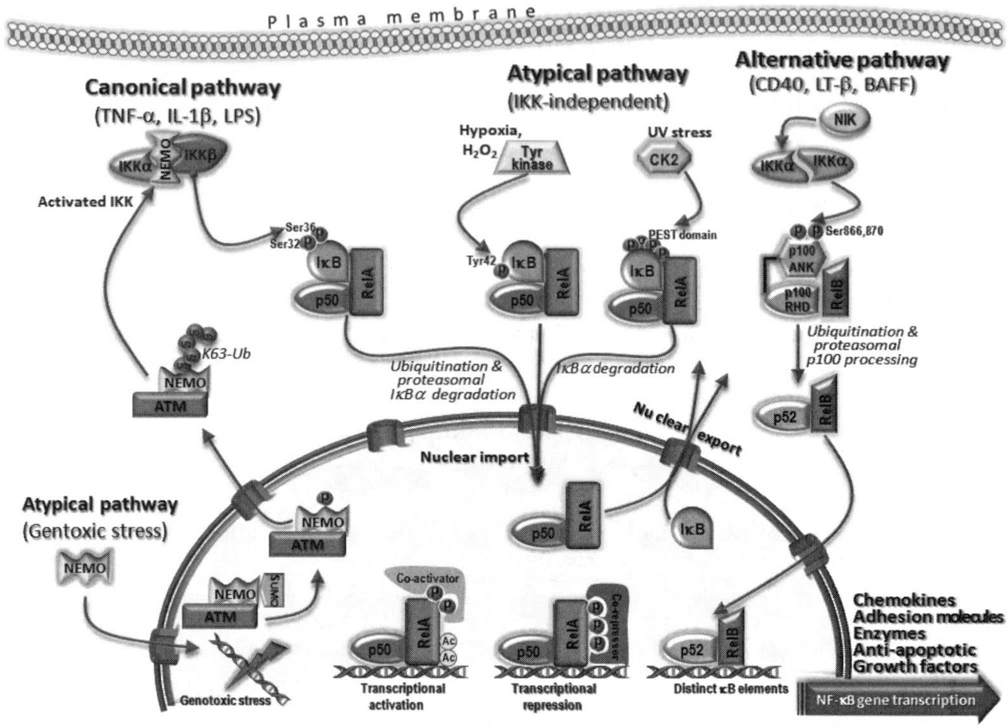

Figure 19.2. NF-κB signaling pathways. Schematic diagram illustrating the three described activation pathways of NF-κB: canonical, alternative, and atypical. Canonical NF-κB pathway mediates TNFα, IL-1β, or LPS responses leading to the activation of the IKK complex, ubiquitination, and proteasomal degradation of IκBα, which then allows p50/RelA to accumulate in the nucleus for transcription. Alternative NF-κB signals from LT-β, CD40, or BAFF, which lead to activation of the IKKα homodimer, results in proteasomal processing of p100 to p52 to form the transcriptionally competent p52/RelB complex. An atypical NF-κB pathway is involved during genotoxic, UV, or hypoxia stress responses, which in many cases results in IKK-independent activation of NF-κB. See the insert for color representation of this figure.

proteasome partially proteolyzes p100 only up to the glycine-rich region of p100, resulting in releasing active p52 (Palombella et al., 1994). This proteolysis thereby allows p52 to form a transcriptionally competent complex with RelB in the nucleus. This p52/RelB heterodimer complex binds to distinct κB elements and drives specific transcriptional responses in part due to the fact that RelB contains an efficient transactivator domain (TAD) (Perkins, 2007). Stimuli such as lymphotoxin β, CD40L, and LPS activate this pathway via NF-κB-inducing kinase (NIK) phosphorylation and activation of IKKα, leading to specific phosphorylation of the C-terminus of p100 on conserved Ser^{866} and Ser^{870} that signals for its proteolytic processing. In gene deletion studies, mouse embryonic fibroblasts (MEFs) from *Ikk-alpha*$^{-/-}$ mice showed normal NF-κB activation in response to TNFα or IL-1β that stimulate the canonical pathway, suggesting that IKKα is dispensable for canonical signaling. However, MEFs from these mice were defective in NF-κB responses to stimulation via B-cell activating factor of the TNF family (BAFF) or lymphotoxin (LT-β), members of the TNF-α superfamily that are important in B-cell functions (Senftleben et al., 2001). Such selective defect in NF-κB signaling delineates the alternative NF-κB pathway from the canonical pathway and defines the distinct physiological functions of IKKα from IKKβ. Given the difference between the two mechanisms of NF-κB activation, one via proteasomal degradation and one via proteasomal processing resulting in different NF-κB dimers that drive transcription, the canonical pathway and the alternative pathway each regulates distinct sets of gene expression.

In the atypical pathway, NF-κB can be activated independently of IKK by inducers such as UV irradiation, oxidation, hypoxia, and chemotherapeutic stress. For example, c-Src or Syk dependent phosphorylation of IκBα on tyrosine 42 can activate NF-κB following hypoxia or glucose stress, respectively, leading to the dissociation but not degradation of IκBα. This mechanism of IκBα dissociation has been suggested to be mediated by phosphoinositide-3 kinase (PI-3K) (Singh et al., 1996; Campbell et al., 2004; Fan et al., 2003; Yang et al., 2008). Under basal conditions, IκBα protein can also be destabilized by constitutive casein kinase II (CK2) dependent phosphorylation of a cluster of serine/threonine residues at its C-terminal PEST domain (Beauparlant, 1996). However, under UV irradiation stimulation, p38 MAP kinase activates CK2 in a stress-responsive manner leading to the same IκBα PEST domain phosphorylation (Neumann and Naumann, 2007). And finally, there is yet another unique twist of NF-κB activation mechanism following DNA damage via the less well studied "nuclear-to-cytoplasmic" signaling. In this case, nuclear NEMO is first sumoylated by a ubiquitin-like protein modifier called SUMO, a process that is enhanced by p53-inducible death domain-containing protein (PIDD) and receptor-interacting protein 1 (RIP1). Sumoylated NEMO is then phosphorylated by ataxia telangiectasia-mutated (ATM) kinase, resulting in the mono-ubiquitination of NEMO which signals for cytoplasmic translocation where NEMO activates IKK complex (Wu et al., 2006). This last example serves to illustrate that even at the most fundamental level of regulation, the NF-κB pathway integrates a complex signaling network in response to a remarkably wide range of cellular conditions.

19.5 REGULATION OF THE IKK COMPLEX

Despite the many different possible ways to activate NF-κB in response to a plethora of inducers, it is generally accepted that the majority of these signals converge onto IκB kinase complexes. Understanding how IKK is regulated is central to the understanding of NF-κB pathway. Therefore, considerable efforts were directed at isolating and identifying the bona fide IκB kinase.

Using classical protein chemistry techniques, the IκB kinase complex was first purified from HeLa cell extracts as a high-molecular-weight protein of approximately 700–900 kDa which was capable of specifically phosphorylating Ser^{32} and Ser^{36}, the two residues critical for signal-induced degradation of IκBα (Chen et al., 1995). Subsequently, the identities of the kinases within similarly purified complexes were unambiguously revealed by several independent groups to be IKKα (IKK1) and IKKβ (IKK2). Moreover, following biochemical analysis of the 700- to 900-kDa complex, as well as complementation studies using a cell line lacking responsiveness to NF-κB stimuli, it was clear that there was an additional scaffolding protein essential for proper canonical NF-κB activation, identified as NEMO (IKKγ) (Mercurio et al., 1997; DiDonato et al., 1997; Woronicz et al., 1997; Rothwarf et al., 1998). Indeed, recombinant expression techniques in yeast recapitulated the native assembly of the IKK complex and biochemically confirmed that the core component of the canonical IKK complex composed of IKKα, IKKβ, and NEMO (Miller and Zandi, 2001). IKKα and IKKβ both associate with NEMO through their C-terminal stretch of six residues known as the NEMO-binding domain (NBD). It is worth noting that IKKβ has higher affinity to NEMO than does IKKα, possibly explaining the fact that IKKα can exist as a homodimer in the alternative pathway while IKKβ must be associated with NEMO as a complex for catalytic activity. Indeed, an 11-mer polypeptide from the NBD sequence that interferes with NEMO-IKKβ assembly could completely inhibit IKK complex kinase activity and subsequently IκBα degradation, NF-κB DNA-binding, and κB gene transcription (Choi et al., 2003). While the exact stoichiometry of the endogenous IKK complex has not been conclusively determined due to limited purity and variability from the conventional purification schemes, it is thought that the complex is composed of a dimer of trimers (Miller and Zandi, 2001; Karin and Ben-Neriah, 2000).

The identification of the catalytic components of the IKK complex led to the identification of two other related kinases, IKKε (IKKi) and TANK-binding kinase 1 (TBK1), of which the four together comprise the family of IKK proteins (Peters and Maniatis, 2001). The IKK family is distinguished by three important architectural features: a kinase domain located at the N-terminus, a helix–loop–helix (HLH); and leucine-zipper (LZ) motifs at the C-terminus. Mutagenesis of these regions suggests that dimerization of IKKα/β, which is required for catalytic activity, is mediated by the LZ domains (Mercurio et al., 1997). The kinase domain of either IKKα or IKKβ is conserved among other serine/threonine kinases, such as the presence of the activation loop with virtually identical sequence between IKKα and IKKβ, whose phosphorylation confers an active conformation change necessary for substrate binding. Mutating Ser^{177} and Ser^{181} on IKKβ "activation-loop" renders the kinase inactive (similarly for IKKα on $Ser^{176/180}$). Moreover, computational modeling shows general conservation in the ATP binding pocket, particularly at Lys^{44} where the MgATP-coordination occurs. Lys^{44} is often mutated to generate kinase-defective IKK proteins. Overexpression studies using such Lys[44] mutants, which can incorporate into and dimerize with wild-type endogenous IKK complex, indicates that phosphorylation of IκBα by the IKK complex is indeed the rate-limiting step in the activation of canonical NF-κB (Zandi et al., 1997; Woronicz et al., 1997; Karin and Ben-Neriah, 2000).

Surmised from the knowledge of how MAP family kinases are generally activated and from mutational studies, it is clear that phosphorylation on the activation loop of the catalytic subunits is required to activate the enzymatic activity of the IKK complex. However, it is less clear exactly how and which kinase(s) are the bona fide activators of IKK enzymes. Purified directly from unstimulated HeLa cell extracts, IKK complex is inactive against GST-tagged IκBα or short polypeptide spanning the Ser[32/36] of IκBα, unless it is pre-incubated in the presence of MgATP with either recombinant NF-κB inducing kinase (NIK) or with constitutively activated kinase domain of mitogen-activated protein kinase kinase 1 (MEKK1) (unpublished results; Lee et al., 1997), suggesting that phosphorylation by an upstream IKK kinase is required to activate IKK. In addition to NIK and MEKK1, several groups have proposed other candidates for the IKK-K, including Akt/PKB, MEKK2, MEKK3, Cot/TPL2, and TGF-β activated kinase 1 (TAK1), TANK binding kinase 1 (TBK1, also called NAK) (Karin and Ben-Neriah, 2000). Many of these studies, however, rely mainly on forced overexpression of target proteins which in some cases may exaggerate promiscuous interactions that otherwise may not be observed under normal physiological conditions. With that being said, the relevance of each of these kinases is likely to be cellular context-dependent.

Regardless of the identity of IKK kinases, however, recent works suggest that the key to enzymatic activation of the IKK complex lies the combination of regulated protein–protein interaction via K63 linked ubiquitination and the ability of NEMO to modulate oligomerization, behaving as an assembler of higher-order protein complexes (Poyet et al., 2000; Hayden and Ghosh, 2008). This line of thought is supported by analysis of how signal is transduced from the receptors via TNF-receptor-associated factor (TRAF), whose family members include TRAF2 and TRAF6, which serve to mediate NF-κB activation from TNF-R and from IL-1R/Toll-like receptors, respectively. TRAF proteins are RING-domain-containing E3 ligases that mediate K63-linked ubiquitination in conjunction with E2 Ubc13/Uev1A that direct their substrates for specific protein–protein interaction rather than for proteasomal protein degradation (Deng et al., 2000). In response to cytokine signals, TRAF proteins oligomerize and mediate K63-linked ubiquitination of RIP1 by TRAF2 (or in case of TRAF6 by auto-ubiquitination). This K63-linked ubiquitination on RIP or TRAF6 allows for specific binding of TAB2 or TAB3, which are K63-polyubiquitin receptor modules that serve as regulatory subunits in the TAK1 kinase complex (Kanayama et al., 2004). The recruitment through the TAB activates TAK1 that is now primed to phosphorylate the activation loop on IKK. But in addition to RIP1 and TRAF6, NEMO is also ubiquitinated as it is recruited to the TRAF/RIP complex by another E3 ligase called inhibitor of apoptosis protein (cIAP), a component of the TNF receptor complex (Tang et al., 2003). In this setting, NEMO coordinates the association with RIP1 with its C-terminus and the association with IKK with its N-terminus to expose the IKK activation loop for TAK1 to phosphorylate, leading to enzymatic activation of IKK. The exposed activation loop of IKK upon assembly into the signalsome has also been proposed to be trans-autophophorylated by IKK itself, because there is equally compelling evidence for IKK autophosphorylation as there is for TAK1 as an IKK-K (Hayden and Ghosh, 2008).

As an integrator of signals, the mechanism for NF-κB signal suppression and termination is equally complex as the activation process. But what is clear is that NF-κB activation is transient from most stimuli; and interestingly, often the stronger the stimuli the more transient the responses (Karin and Ben Neriah, 2000). In addition to transcriptional regulation of IκBα as a negative feedback loop to dampen NK-κB signaling, aside from constitutive cytoplasmic PP2A phosphatase activity, regulation of the IKK activation itself appears to possess two distinct negative-feedback mechanisms: deubiquitination and deactivating phosphorylation. First, underscoring the importance of K63 linked ubiquitination in the process of IKK activation, both the A20 ubiquitin-editing and CYLD de-ubiquitinating enzymes that are capable of removing the K63-linked polyubiquitin chains conjugated on RIP or TRAF6 protein serve as potent negative regulators of NF-κB. For example, when the proteins are overexpressed in cells, A20 and CYLD can diminish TAK1 activation and abolish subsequent IKK/NF-κB activation (Chen et al., 2006). Second, phosphorylation on Ser^{740} at the NEMO binding domain (NBD) and on Ser^{68} of NEMO results in the disruption of NEMO-IKK interaction and thereby terminates enzymatic activity of IKK complex. It has been generally thought that this negative-feedback phosphorylation is mediated by IKK itself (Palkowitsch et al., 2008). However, a recent report suggests that a Ser^{740} kinase purified from HeLa cell extract capable of specifically phosphorylate NBD of IKK is in fact polo-like kinase 1 (PLK1), a mitotic cell-cycle kinase (Higashimoto et al., 2008).

19.6 IKK GENE KNOCKOUTS

At first approximation, IKKα and IKKβ appear to share overlapping function because both exist in the same enzyme complex *in vivo*; and when expressed individually as recombinant enzymes, they both can recognize the same substrate motif on IκB proteins with relatively high specificity, although with varying efficiency from a kinetics standpoint. Despite their general biochemical similarities, however, the divergent physiological roles of IKKα and IKKβ only came to light with IKK-deficient mouse studies. Indeed, extensive studies using targeted gene deletions, conditional knockouts, and transgenic models, combined with critical analyses of naturally occurring mutations in diseases, have provided profound insights into the physiological functions of IKK proteins (Pasparakis et al., 2006; Courtois and Gilmore, 2006).

Ikk-alpha$^{-/-}$ knockout mice survive embryonic development but die prenatally mainly as a consequence to an unexpected and severe skin barrier defect that results in dehydration (Hu et al., 1999). These mice superficially exhibit developmental abnormalities that are characterized by truncated limbs, ears, snouts, and tails. But upon closer examination, these IKKα-deficient mice contained normal skeletal embryonic development features, but were unable to emerge into its normal size due to the constrain of a severe thickening skin defect caused by the combination of hyperproliferation and underdifferentiation of keratinocytes (Hu et al., 1999). The thickened epidermis of these IKKα-deficient mice measured about 5- to 10-fold thicker when compared to their wild-type litter mates (Karin and Ben-Neriah, 2000). As such, IKKα appears to play a critical role in skeletal morphogenesis during development by regulating karatinocyte differentiation and formation of the epidermis. Moreover, by extension of its role in morphogenesis from embryonic development into adulthood, and given that *Ikk-alpha*-deficient keratinocytes lack terminal differentiation markers and are unable to stop proliferation, IKKα protein appears to also act as a tumor suppressor in the skin by virtue of its function as a potent cofactor of the transforming-growth factor beta (TFG-β) signaling pathway (Descargues et al., 2008). However, it is important to note that while its tumor suppressor function is dependent on the intact protein, its kinase activity may be dispensable in this role.

More relevant to the understanding of NF-κB signaling, a second key observation from *Ikk-alpha* knockout studies showed that IKKα is dispensable for NF-κB signaling from pro-inflammatory stimuli like TNFα, IL-1β, or LPS. However, NF-κB DNA-binding activity was reduced about twofold in MEFs lacking IKKα, suggesting that IKKα participates in some role modulating NF-κB. Indeed, IKKα is found to be necessary for lymphoid organogenesis because it mediates NF-κB signaling from CD40, BAFF, and LT-β stimulation (Bonizzi et al., 2004). More interestingly, detailed analyses of IKKα-deficient mice reveals that IKKα plays a critical role in negatively regulating macrophage NF-κB activation and contributes to inflammation resolution in a complementary fashion with the pro-inflammatory nature of IKKβ (Lawrence et al., 2005).

Unlike the surprised results generated from the *Ikk-alpha* knockout mice, phenotypes of mice with *Ikk-beta* gene deletion were less surprising and essentially consistent with genetic deletion of p65/RelA or double deletion of p65 and p50 (Karin and Ben-Neriah, 2000). Constitutive *Ikk*-beta$^{-/-}$ knockout mice died *in utero* at embryonic day E12.5–E14.5 from excessive loss of hepatocytes due to massive apoptosis; and in a similar fashion, *relA/p65*$^{-/-}$ mice also died at E14.5 and p65/p50 double knockout mice died at E12.5 (Beg et al., 1995). Apoptosis appears to be modulated by TNFα, since the inactivation of TNFR1 by additionally deleting *TNFR1* in *Ikk-beta*$^{-/-}$ mice indeed rescued their hepatocytes from apoptosis and embryonic death (Li et al., 1999; Tanaka et al.,

1999). These physiological observations were confirmed at the cellular level, because MEFs from *Ikk-beta*$^{-/-}$ mice characteristically undergo apoptosis in response to TNF-α challenge at much higher sensitivity compared to normal MEFs, presumably due to a loss of anti-apoptotic signals normally provided by NF-κB activation via IKKβ (Geisler et al., 2007; Li et al., 1999; Tanaka et al., 1999). And not unexpectedly, *Ikk-gamma* knockout is embryonic lethal due to hepatocyte apoptosis as well (Beraza et al., 2007); suggesting that NEMO is essential for IKKβ function and share functions with IKKβ at least with respect to TNFα/NF-κB signaling axis.

Relevant to therapeutic considerations, these results beg the question of whether complete inhibition of IKK complex in adult hepatocytes would lead to the same hypersensitivity to apoptosis. To answer this question, hepatocyte-targeted conditional knockout studies provided some clues as to what may happen upon NF-κB drug intervention. First, it is informative to compare the severity of TNFα-induced apoptosis between knockouts of RelA/p65, IKKγ, and IKKβ of hepatocyte in adult animals. For instance, disruption of RelA/p65 rendered adult mice sensitive to lethal liver injury upon TNFα administration, as reported in many studies. Primary RelA/p65-deficient hepatocytes showed no NF-κB activation and committed rapid apoptosis after TNFα treatment. Similarly, hepatocyte-specific deletion of IKKγ results in near complete inhibition of NF-κB activation and these MEFs are highly sensitive to TNFα-induced apoptosis (Luedde et al., 2007). In contrast, hepatocytes deficient for IKKβ in this particular study displayed some residual NF-κB activity and consequently were susceptible to only moderate apoptosis in response to TNFα (Geisler et al., 2007). More recently, studies using tissue-specific targeted gene transfer of dominant-negative kinase-dead IKKβ-DN as a way to suppress endogenous IKKβ activity specifically in mouse hepatocytes concluded that *in vivo* status and levels of hepatic TNF-α and NF-κB determined the cellular fate between necrosis and apoptosis under endotoxemia condition (Dajani et al., 2007). What these studies suggest is that in post-embryonic development some residual NF-κB is required for proper hepatic tissue regeneration where TNFα is present at a relatively high level. What this could mean to drug development is that determination of appropriate dose and schedule to preserve a minimum level of hepatic NF-κB activity will be paramount in order to maximize therapeutic window.

A second key observation from the constitutive knockout experiment was that embryonic fibroblasts from these mice clearly show impaired activation of NF-κB in response to TNFα and IL-1β stimulation. Importantly, cells derived from *ikk-beta+/–* heterozygotes as a result of loss of copy on one allele showed a corresponding twofold reduction in IKKβ expression, correlated to an approximately half of the kinase activity and a 70–80% decrease in NF-κB DNA binding and transcription activity in response to TNFα or IL-1β. This tight correlation between loss of IKKβ expression and NF-κB responsiveness demonstrates that despite their biochemical similarities, IKKα clearly cannot compensate for IKKβ functions at least in MEFs; furthermore, it solidifies

IKKβ catalytic subunit within the IKK complex as the main signal transducer of pro-inflammatory activation of NF-κB.

19.7 NF-κB IN CANCER

The first clue to linking NF-κB to cancer was the realization upon identification and cloning of the *c-rel* gene as a homolog of the *v-rel* oncogene from the avian Rev-T retrovirus. And consistent with its acutely oncogenic nature, expression of v-Rel potently promotes cellular oncogenic transformation that leads to fatal development of B-cell lymphomas and leukemias in birds (Gilmore, 1992). Since that initial discovery, more compelling evidence has come to light connecting NF-κB to various human cancer types. As an example, in various subtypes of human B-cell lymphomas, and to a lesser extent some subtypes of T-cell lymphomas, REL gene amplifications indeed have been detected at significant frequency (Courtois and Gilmore, 2006). From a broad perspective, NF-κB participates in multiple aspects of oncogenesis, growth, and progression by virtue of its transcriptional control of important factors mediating cell survival, growth, angiogenesis, anti-apoptosis, and tissue invasiveness, among other functions. Indeed, it appears that the NF-κB pathway is intricately involved in key mechanisms that are the "hallmarks of cancer." In a seminal paper published in 2000, Hanahan and Weinberg outlined the fundamental phenotypic attributes of cancer and cancer-associated cells that distinguish them from normal cells under healthy homeostasis; these attributes are referred to as the six "hallmarks of cancer": self-sufficiency from external growth signals, insensitivity to anti-growth signals, evasion of apoptosis, limitless replicative potential, sustained angiogenesis, and capacity to invade tissue and metastasis to distant sites (Hanahan and Weinberg, 2000). It is worth noting that as relevant as the role of NF-κB is in the intrinsic oncogenic signaling in specific tumor types, these principles are equally applicable to the host environment changes that are necessary to support invasion and progression of tumors, where NF-κB plays a particularly important role.

19.8 PROMOTION OF GROWTH

NF-κB contributes to tumorigenesis by providing a direct stimulus toward proliferation via the transcription of proto-oncogenes such as c-Myc and cyclin D1 (Kessler et al., 1992; Guttridge et al., 1999), as well as many growth factors such as IL-2, GM-CSF, and IL-6. In particular, NF-κB was shown to promote cell growth in embryonic fibroblasts in correlation with its transcriptional regulation of cyclin D1. In diploid fibroblasts, NF-κB was shown to be required to induce cyclin D1 expression and retinoblastoma protein (pRb) hyperphosphorylation to promote G_1-to-S progression, demonstrating cyclin D1 as a major target gene of NF-κB and establishing the role of NF-κB in early phase of cell cycle to regulate cell growth (Guttridge et al., 1999). In addition, IKKα links mitogenic signaling to cyclin D1 transcription via Wnt/β-catenin/Tcf pathway by direct phosphorylation at the N-terminus of β-catenin (Albanese et al., 2003). In Burkitt's lymphoma, the *c-myc* gene is translocated to one of the immunoglobulin genes resulting in the dysregulation of *c-myc* transcription. In this setting, translocated c-Myc expression driven by immunoglobulin enhancers requires activated NF-κB binding to the enhancer region. The consequence of deregulated c-Myc expression has been linked to increased expression of cyclin A and cyclin E (Jansen-Durr et al., 1993), concomitant with decreases in $p27^{kip1}$ levels and interferes with the function of the CDK inhibitor $p27^{kip1}$ (Vlach et al., 1996), all of which are consistent with the thought that c-Myc promotes S-phase entry.

In some tumors, constitutive expression of proinflammatory cytokines correlated with growth. For example, head and neck squamous cell carcinoma (HNSCC) constitutively expresses IL-1α, IL-6, IL-8, and granulocyte-macrophage colony-stimulating factor (GM-CSF) via activation of NF-κB. Overexpression of IκBα-SR reduced tumor burden in a mouse xenograft model system, although cells *in vitro* were not specifically sensitive to canonical NF-κB inhibition, suggesting the importance of these cytokines in driving tumor survival and growth *in vivo* (Duffey et al., 1999).

Finally, it is interesting to note that many carcinogens (including nicotine and carcinogenic metals) that are known to initiate tumorgenesis in lung cancer and in head and neck squamous cell carcinomas (HNSCC) are also potent activators of NF-κB in promoting cell proliferation and survival (Van Waes, 2007; Baldwin, 2001). Nicotine has also been shown to activate reactive oxygen species (ROS) and to activate NF-κB in a dose-dependent manner (Barr et al., 2007).

19.9 EVASION OF APOPTOSIS

Apoptosis is a critical process in the maintenance of tissue homeostasis by eliminating unnecessary or harmful cells. Dysregulation of apoptotic programmed cell death is recognized as one of the hallmarks of cancer, allowing for uncontrolled growth and progression of tumors. Failure in the timely apoptosis contributes to tumorgenesis by creating a permissive environment for genetic instability resulting in genetic mutations and conferring resistance to chemotherapeutics. As an advantage to survival, many solid and hematological tumors are frequently associated with high expression of known anti-apoptotic factors such as Bcl-2, or the loss of pro-apoptotic factors such as the tumor suppressor p53 (Basseres and Baldwin, 2006). As some examples, lymphomas that are driven by Epstein–Barr virus (EBV), the cause of mononucleosis, override cell apoptosis by expressing the viral protein BHRF1, a homolog Bcl-2 (Dawson et al., 1998). Human T-lymphocyte virus I (HTLV) gene Tax activates NF-κB to hijack survival and proliferation machinery contributing the pathogenesis of lymphomas, T-cell leukemias, and nasopharyngeal carcinomas (Van Waes, 2007). Similarly, human papilloma virus (HPV), implicated in causing cervical cancer, can activate NF-κB and produces a protein call E6 that binds and inactivates the apoptosis promoter p53. Malignant melanoma cells avoid apoptosis by down-regulating the expression of the gene encoding apoptotic protease activated factor-1 (Apaf-1), a component of the apoptosome that interacts with Cytochrome c released from mitochondria to initiate the caspase cascade.

It is well established that there are two main pathways of apoptosis, the extrinsic pathway involving death receptors and the intrinsic mitochondria pathway (Elmore, 2007). Although NF-κB has been shown to mediate both anti- and pro-apoptotic signaling depending on cellular context, overwhelming evidence strongly supports its role in anti-apoptosis. NF-κB regulates a variety of factors that are known to suppress apoptosis, including inhibitor of apoptosis proteins 1 and 2 (c-IAP1 and c-IAP2), X-linked IAP (XIAP), A1/Bfl-1, Bcl-x (BCL-2 antiapoptosis family members), cellular Fas-associated death domain-like IL-1β converting enzyme (FLICE) inhibitor protein (c-FLIP), TRAF1, and TRAF2 (Kucharczak et al., 2003; Wang et al., 1999). As examples of anti-apoptotic mechanisms, IAPs block apoptosis induced either via extrinsic death receptor activation or via intrinsic mitochondrial pathways by directly binding to and inhibiting effector caspases-3, -6, and -7. c-FLIP blocks apoptosis for instance by acting as a decoy protein to bind to TNFR and competes with caspase-8 binding with TNFR due to its similarity to caspase-8 but without any catalytic activity. There are many other similar examples in which these NF-κB factors exert their anti-apoptotic effects.

In context of cancer transformation, for example, NF-κB is required to suppress apoptosis for efficient Ras transformation (Mayo et al., 1997). The *ras* proto-oncogene is frequently mutated in human tumors, and its mechanism of cellular transformation involves the stimulation of key transcription factors, namely c-Myc, c-Jun, Ets, and NF-κB. As expression of mutant H-Ras leads to transformation of immortalized mouse fibroblasts, NF-κB activation is absolutely required to keep p53-independent apoptosis in check, since the coexpression the genetic inhibitor IκBα-superrepressor (IκB-SR) leads to spontaneous apoptosis in these cells (Basseres and Baldwin, 2006; Mayo et al., 1997). Consistent with the thought that NF-κB activation provides survival mechanisms, many tumor cell types are sensitive to inhibition of NF-κB. In some Hodgkin Reed/Sternberg cells, for example, where IκBα is defective resulting in constitutive NF-κB activation, forced expression of IκBα leads to apoptosis (Cabannes et al., 1999). Notably, inhibition of NF-κB in H/R-S cells concomitantly leads to loss of key anti-apoptotic effectors such as c-IAPs, TRAF, and A1/Bfl-1 (Hinz et al., 2001). Similarly, diffuse large B-cell lymphoma or multiple myeloma cells that exhibit high constitutive NF-κB are also sensitive to commit apoptosis upon inhibition of NF-κB by either the IκBα–SR or small-molecules against IKKβ (Annunziata et al., 2007; Lam et al., 2005; Hideshima et al., 2002). In addition to trancriptionally regulating anti-apoptotic proteins via NF-κB activation, IKKβ can play a direct role in regulating apoptosis. In breast cancer cells, IKKβ acts in part as an oncogene by directly phosphorylates and promotes the degradation of forkhead O-box 3a (FOXO3a) transcription factor, whose functions include inhibition of cell-cycle and the promotion of apoptosis (Hu et al., 2004).

19.10 INVASION AND METASTASIS

Cancer metastasis is a complex process involving a series of steps that include invasion, angiogenesis, and trafficking of cancer cells through blood vessels, extravasations, organ-specific homing, and growth at distal sites. As part of such disease progression, cancer-associated angiogenesis is critical for continued tumor growth under increasingly nutrient-constrained environment as the tumor increases both in size burden and in reach of distal tissue sites. Angiogenesis itself is a process with a series of coordinated steps to build new blood vessels to support this growth. One of the earliest events in angiogenesis involves the degradation of vascular basement membrane and the remodeling of extra cellular matrix (ECM). In this setting, the role NF-κB plays is well documented. NF-κB activation leads to expression key enzymes involved in this early step, namely matrix metalloprotease -2, -3, and -9 (MMP-2, -3,-9) and urokinase-type and tissue-type plasminogen activators (uPA and tPA) (Tabruyn and Griffioen, 2007). Not surprisingly, highly metastatic breast cancer cell lines frequently have high expression of MMPs. But in addition to matrix MMPs, uPA, tPA, and inflammatory cytokines, other chemokines regulated by NF-κB play important roles in invasiveness and angiogenesis. As an example, the chemokine stromal-derived factor-1α (SDF-1α) and its receptor CXCR4 are thought to be equally important in motility, homing, and proliferation of cancer cells at specific metastatic sites. Indeed, NF-κB transcriptionally up-regulates CXCR4 in metastatic breast cancer cells, and inhibition of NF-κB by overexpresing IκBα–SR leads to loss of SDF-1α mediated migration *in vitro* (Helbig et al., 2003). On the other hand, it should also be pointed out that on the balance NF-κB can also modulate anti-angiogenesis by blocking endothelial cell migration by activating inhibitors of MMP-1 and plasminogen activator inhibitors (PAI-1, -2) in response to TNFα and reactive oxygen species (ROS) (Tabruyn and Griffioen, 2007).

Endothelial cells utilize integrins as the principal adhesion receptors to interact with the extracellular environment. Integrins serve to modulate endothelial cell migration, proliferation, and survival. In glioblastoma multiforme (GBM), treatment of PMA activates NF-κB and leads to an increase in mRNA of β3 and αV integrin subunits (Ritchie et al., 2000). Moreover, interaction of αV-β3 integrin with the ECM activates NF-κB via activation of the IKK complex and degradation of IκBα. In rat aortic endothelial cells (RAECs), ligation of αV-β3 integrin by its ligand osteopontin triggers a prosurvival signal that can be inhibited by forced expression of a dominant-negative IKKβ–DN (Rice et al., 2006).

In addition to formation of new blood vessels to support growth and maintenance of tumor burden, inflammatory cytokines from either tumor cells or infiltrating tumor associated macrophages (TAM) provide a rich array of stimuli for progression, angiogenesis, and invasion. Endothelial angiogenic factors such as vascular endothelial growth factor (VEGF) can be produced by a variety of cells. But in particular to TAMs, macrophage VEGF expression is up-regulated by a hypoxia-induced mitogenic factor through the activation of the NF-κB pathway (Kiriakidis et al., 2003). And in an *in vitro* model system, coculturing of macrophages with ovarian or breast cancer cell lines led to tumor-cell-specific activation of NF-κB activity with correspondent-enhanced invasiveness that was inhibitable by blocking NF-κB with TNFα neutralizing Abs, an NF-κB inhibitor, RNAi to RelA, or by overexpression of IκBα (Hagemann et al., 2006). Even more interestingly, when NF-κB is genetically inhibited specifically in TAMs, these macrophages revert back to a "classical" phenotype in a process to which Hagemann et al. (2008) refer as "re-educating" as they then become cytotoxic to tumor cells *in vitro* and *in vivo*. These studies suggest that inhibiting NF-κB in the tumor microenvironment (specifically TAMs) to suppress invasion and metastasis may be of great therapeutic benefit.

19.11 CHEMOTHERAPY RESISTANCE

Chemo-induced activation of NF-κB is a well-established mechanism for suppression of the apoptotic potential of the chemotherapy through induction of antiapoptotic gene expression (Tapia et al., 2007). Several studies have shown that topoisomerase I inhibitor irinotecan (CPT-11) activates NF-κB in most colorectal cancer cell lines. An adjuvant approach to promote chemosensitivity by adenovirus-mediated transfer of IκBα–SR (as a genetic inhibitor of NF-κB) demonstrated effective chemosensitization of CPT-11 *in vitro* in human colon cancer cell lines via increased induction of apoptosis, as well as in a tumor xenograft model, resulting in a significantly enhanced tumor response to CPT-11 (Cusack et al., 2000). In a similar finding, suppression of the canonical NF-κB pathway via siRNA silencing of p65 expression in the HCT-116 cell line did not impact cell viability on its own; however, depletion of p65-sensitized cells contributed to the cytotoxic effects of CPT-11 by severalfold. Knockdown of p65 by siRNA decreased expression of two cellular inhibitors of apoptosis that are known to be regulated by NF-κB, c-IAP1, and c-IAP2 and subsequently increased the ability of CPT-11 to activate caspase-3. Importantly, the reduction in viability translated into decreased colony formation in a longer-term survival assay, as well as increased tumoricidal responses to CPT-11 treatment in a mouse xenograft model (Guo et al., 2004). Other studies confirmed the importance of NF-κB in chemo-induced resistance in a variety of cancer cells. Constitutive activation of NF-κB in a subset of breast cancers leads to overexpression of c-IAP2 and manganese superoxide dismutase (Mn-SOD) and resistance to paclitaxel cytotoxic effects, which can be reversed by expression of the genetic inhibitor IκB-SR or by parthenolide (a natural product inhibitor of NF-κB DNA binding) leading to increased paclitaxel-sensitivity (Patel et al., 2000). A recent study identified sonic hedgehog (Shh) as an NF-κB target gene that confers TRAIL resistance to apoptosis, and this Shh knockdown prevents NF-κB-stimulated proliferation; conversely, the addition of Shh rescues the proliferation defect via NF-κB inhibition in a pancreatic cancer model (Kasperczyk et al., 2009). Several other studies noted similar resistance to chemotherapeutics via the up-regulation of NF-κB—for instance, in non-small-cell lung carcinomas (Jones et al., 2000), in pancreatic cancer cells under treatment with etoposide or doxorrubicin (Arlt et al., 2001), in gliomas cells under treatment with carboplatin or SN-38 (Weaver et al., 2003), in prostate cancer cells with paclitaxel (Flynn et al., 2003), in human stomach cancer cell lines with 5-fluorouracil (5-FU) (Uetsuka et al., 2003), in colon cancer (Voboril et al., 2004), in gastric cancer (Camp et al., 2004), in ovarian carcinoma cells with anthracyclines (Salvatore et al., 2005), and in breast cancer cells (Montagut et al., 2006). Finally, in chronic myelogeneous leukemia (CML) where the disease is characterized by the Philadelphia chromosome resulting in the constitutively active BCR-Abl-kinase, imatinib is an effective treatment for BCR-Abl positive leukemia. However, many CML patients eventually relapse and develop resistance to imatinib. A common mechanism of resistance is mutation of the gatekeeper residue where imatinib binds in the active site of BCR-Abl. But another mechanism for resistance is the up-regulation of NF-κB. Indeed, the IKKβ inhibitor PS-1145 not only was shown to inhibit the proliferation of CML cell lines, but also overcame imatinib resistance and acted synergistically with imatinib against resistant cell lines and cells from resistant patients with increased apoptosis and inhibition of proliferation and colony growth (Cilloni et al., 2006). These results, among many others, make a compelling case for the development of NF-κB pathway inhibitors as adjuvant therapeutics in combination with the current armament of standard DNA-damage and anti-metabolite agents for treatment of a variety of malignancies.

19.12 NF-κB IN HEMATOLOGICAL TUMORS

While NF-κB plays many essential roles in various aspects of tumorgenesis and tumor progression of a variety of solid cancer types, the most compelling rationale for NF-κB therapeutic intervention remains in the areas of hematologic malignancies. In particular, B lymphocytes depend on IKK/NF-κB pathway for survival and maintenance. For instance, conditional knockout of either NEMO or IKKβ in adult B cells by using B lineage-specific disruption led to the disappearance of mature B lymphocytes (Pasparakis et al., 2002).

Indeed, specific pharmacological inhibition of NF-κB in adult mice led to rapid lymphocytes depletion in a TNFR1-dependent manner. In this study, treatment of the selective IKKβ inhibitor MLN120B *in vivo* led to disappearance of thymocytes and bone marrow B cells at various stages of development, but interestingly leaving granulocytes population well intact (Nagashima et al., 2006). These results suggest a special role NF-κB plays in the survival and maintenance specifically for lymphocytes with which can readily be adapted as an advantage for tumorgenesis and proliferation. Some examples of NF-κB-driven hemaetological malignancies are discussed below.

19.13 HODGKIN'S DISEASE

Hodgkin's disease (HD), also known as Hodgkin's lymphoma, is a cancer of the lymphatic system named after the British physician Thomas Hodgkin, who in 1832 first described the disease where he noted enlarged lymph nodes without obvious signs of inflammation. Histologically, HD is characterized by the presence of large, multi- or binucleated malignant Hodgkin/Reed–Sternberg (H/RS) cells derived from clonal expansion of germinal-center B cells, surrounded by reactive immune cells. In most cases, H/RS cells lack the expression of B-cell receptor (BCR) that normal B cells require for survival and maintenance (Jost and Ruland, 2007). As such, many of these H/RS cells bypass BCR maintenance signals and become autonomous via altering intrinsic signaling pathways for survival. Indeed, high constitutive NF-κB activity is a defining characteristic of H/RS cell lines as well as of primary cells derived from HD patients (Scheidereit, 2006; Bargou et al., 1997; Hinz et al., 2001).

Constitutively activated NF-κB in HD can arise from several distinct mechanisms. A commonly shared mechanism of oncogenic alterations is the constitutive activation of IKK via modulating upstream TRAFs from autonomously activated receptors such as CD30, receptor activator of nuclear factor κB (RANK), and CD40. For example, self-oligomerization of CD30 receptor recruits TRAF2 and TRAF5 to their cytoplasmic tails and propagates activating signals to both the canonical and alternative NF-κB pathways to directly contribute to survival of H/RS cells (Jost and Ruland, 2007). RANK, a member of the TNF-receptor superfamily, is highly expressed in H/RS cells (Fiumara et al., 2004). In Hodgkin's disease, overexpressed RANK constitutively activates canonical NF-κB pathway via TRAF molecules because its ligand RANK-L is also frequently overexpressed by H/RS and other bystander cells leading to autonomous stimulation. High expression of CD40, also another member of the TNF-receptor superfamily, is another feature of malignant H/RS cells. Via CD40, NF-κB is constitutively activated in H/RS cells because its cognate ligand CD40L (CD154) is also highly expressed on activated CD4+ T cells and other surrounding cells, including eosinophils, neutrophils, dendritic cells, and even subsets of activated normal B lymphocytes (Carbone, 2005; Pham et al., 2005; O'Grady, 1994). This activated cellular background constantly provides the necessary NF-κB stimulus to malignant H/RS cell for its survival. Finally, HD is often associated with Epstein–Barr virus (EBV) involvement at an estimated frequency of up to 50% of all HD patients (Braun et al., 2006a,b). The degree of EBV involvement in HD cases may vary with subtypes, and the correlative incidence of EBV-positive HD may be age-related (Carbone, 2005). Nonetheless, EBV is thought to contribute to pathogenesis of HD wherein EBV-positive cells NF-κB pathway is hijacked and aberrantly activated through expression of latent membrane protein 1 (LMP1), a virally encoded gene whose intracytoplasmic tail mimics the action of ligated CD40 receptor (Deacon et al., 1993; Braun et al., 2006a,b).

In most EBV-negative and some EBV-positive HD cases, NF-κB can also be constitutively activated downstream of IKK from the various inactivating mutations of inhibitory IκB proteins. Complete loss of IκBα has been found in HD patient biopsies where loss-of-function mutations in one allele of IκBα gene (i.e., insertions, non-sense, or deletion) are frequently combined with deletion of the second allele of IκBα. More commonly, mutations in the ankyrin repeats region of IκBα were often detected, leading to its inability to suppress NF-κB (Emmerich et al., 1999; Jungnickel et al., 2000). In addition, frameshift mutations in IκBε generating pre-terminal stop codon have also been identified in some H/RS cells (Emmerich et al., 2003).

19.14 NON-HODGKIN'S LYMPHOMA

Non-Hodgkin's lymphoma (NHL) is a collection of clinically and histologically diverse set of malignancies of the immune cells. NHL can be divided into aggressive (fast-growing) and indolent (slow-growing) types and can be either B-cell or T-cell non-Hodgkin lymphoma. But generally speaking, NHL is prevalently a set disease of B-cell origin, which, among many, includes diffuse large B-cell lymphoma (DLBCL), MALT lymphoma, Burkitt's lymphoma, follicular lymphoma, immunoblastic large-cell lymphoma, precursor B-lymphoblastic lymphoma, and mantle-cell lymphoma.

19.15 DIFFUSE-LARGE B-CELL LYMPHOMA

Among NHL, the most common subtype is diffuse large B-cell lymphoma (DLBCL) because it accounts for about 40% of all NHL cases and represents the majority of the aggressive types. DLBCL is characterized by lymphoid cells whose nuclei are approximately twice the size of that of a small lymphocyte, and these malignant cells displace the normal lymph node architecture in a diffuse pattern. Most recently, the use of gene expression profiling using cDNA microarrays led to the discovery of biologically and clinically distinct subtypes of DLBCL. On the basis of correlative analysis of gene expression patterns and clinical outcomes, it is now possible to classify DLBCL into three subgroups that arise from distinct genetic pathways: germinal-center B-cell-like (GCB-DLBCL), activated B-cell-like (ABC-DLBCL), and Type 3 DLBCL lymphoma (Alizadeh et al., 2000; Rosenwald et al., 2002; Wright et al., 2003; Lenz et al., 2008). This cell-of-origin gene-expression-based classification is important not only for rationalizing current management of DLBCL, but also for identifying key target pathways for future novel drug development. Profile analysis of the poorest-prognosis ABC-DLBCL subgroup, distinguishing from the better-prognosis GCB-DLBCL, revealed a differentially elevated NF-κB gene signature that includes cancer relevant NF-κB target genes like BCL-2, cFLIP, and cyclin D2. Importantly, this gene signature was found with concomitantly activated IKK complex leading to persistent NF-κB transcription (Davis et al., 2001). The mechanism of upstream oncogenic activation of NF-κB in this setting is a subject still under investigation, but as one example, recent findings uncover CARD11 (an oligomeric cytoplasmic scaffold protein required for NF-κB signal transduction) as a bona fide oncogene that is often missense mutated at its coil–coil domain in ABC-DLBCL leading to constitutive IKK activation (Lenz et al., 2008). Inhibition of NF-κB by expression of IκB-SR or use of IKK inhibitor (β-carboline derivatives PS-1145 or MLX105) was selectively toxic to ABC-DLBCL cells but not to GCB-DLBCL cells. Phenotypically, blocking NF-κB in ABC-DLBCL cells arrested cells in G1 phase followed by apoptosis. Importantly, to demonstrate that ABC-DLBCL cells are addicted to NF-κB pathway for survival, RelA (p65) was introduced in an estrogen receptor inducible manner in ABC cells under treatment with IKK inhibitor. Since RelA is downstream from IKK and its overexpression functionally bypasses IKK, its induced expression indeed was shown to fully rescue the toxic effect of IKK inhibitor on ABC cells, suggesting the sole dependence on NF-κB pathway by ABC-DLBCL for growth and survival (Davis et al., 2001). Taken together, ABC-DLBCL represents one of the most compelling clinical opportunities to develop IKK or NF-κB drug as single-agent therapy or in combination with existing chemotherapeutics for hematological malignancies.

19.16 MALT LYMPHOMA

MALT lymphoma is a relatively rare form of lymphoma accounting for about 8% of NHL cases, and this type belongs to a group of neoplastic disease known as marginal zone B-cell lymphoma (Bertoni and Zucca, 2006). While most NHL develops in the lymph nodes (i.e., nodal lymphoma), MALT lymphoma develops from the mucosa-associated lymphoid tissue, which is any lymphatic tissue found in other parts of the body such as the stomach, thyroid gland, lungs, salivary glands, or eyes, and is therefore commonly referred to as extranodal lymphoma. Primary MALT lymphoma cases arising from gastric mucosa are commonly associated with *Helicobacter pylori* infection, in which the disease can be simply treated with antibiotic therapy. But MALT lymphoma can also develop at other sites, especially in the context of chronic inflammation. Moreover, many of these cases harbor genetic lesions that constitute advanced diseases that are unresponsive to antibiotics. At least three distinct recurrent chromosomal translocations have been implicated in the pathogenesis of MALT lymphoma, all of which interestingly converge onto NF-κB via the activation of IKK complex despite their seemingly disparate mutation events.

The first, t(1;14)(p22;q32), results in the transfer of the entire BCL10 gene to chromosome 14 where BCL10, being juxtaposed next to the promoter of Ig heavy-chain gene, is inappropriately overexpressed by the Ig enhancer. The second, t(14;18)(q32;q21), in a similar manner to the first, also is a translocation event that juxtaposes the *MALT1* gene (MALT lymphoma translocation 1) next to the promoter region of the Ig heavy-chain genes also resulting in MALT1 overexpression via the Ig enhancer. The third, t(11;18)(q21;q21), juxtaposes a fragment of the cIAP2 gene on chromosome 11 next to the MALT1 gene on chromosome 18 resulting in the synthesis of a novel fusion protein, cIAP2-MALT1 (Lucas et al., 2001). Among these events, one common observation is the net strong expression of BCL10, supporting the close interaction between BCL10 and MALT1 in the activation of IKK. In normal lymphocytes, BCL10 is thought to oligomerize with MALT1 in response to antigen leading to the K63-linked ubiquitination of NEMO (IKKγ) and activation of IKK complex (Zhou et al., 2005). But in MALT lymphoma, BCL10/MALT1-mediated ubiquitination of NEMO is constitutive, since the translocation mutations uncouple the cascade from the homeostasis control of ligand–receptor signaling (Sagaert et al., 2006; Bertoni and Zucca, 2006). In the case of the cIAP2-MALT1 fusion, autonomous NF-κB activation also occurs through the deregulation of NEMO ubiquitination. It is now known that in normal lymphocytes, cIAP2 functions as the ubiquitin ligase (E3) for BCL10 which targets BCL10 for proteasome-mediated degradation, which normally inhibits antigen-receptor-mediated cytokine production. However, in MALT lymphoma, cIAP2-MALT1 fusion protein lacks E3 activity. As a consequence, BCL10 protein is stabilized to interact with MALT1 to activate IKK.

The MALT lymphoma example illustrates the central role of NF-κB in the relationship between the pathogenesis of chronic inflammation and cancer exemplified by other diseases like colitis-associated colorectal cancer and hepatitis-associated hepatocarcinoma, among others (Greten et al., 2004). It also poignantly illustrates how genetic mutations can lead to aberrant NF-κB pathway activation that drives lymphomagenesis. This latter point is further exemplified by elegant genetic investigative work in the pathogenesis of multiple myeloma.

19.17 MULTIPLE MYELOMA

Multiple myeloma (MM) is a progressive disease characterized by the proliferation of malignant plasma cells and subsequent overabundant production of monoclonal paraprotein. Plasma cells reside in the bone marrow compartment as they become differentiated from plasmablasts that derive earlier from post-germinal-center B cells that have undergone somatic hypermutation, antigen selection, and IgG switching (Cheng et al., 2007). Notably, MM is a clonal disease where MM cells have strong dependence on the bone marrow microenvironment for survival and growth (Hideshima et al., 2002). An intriguing feature of MM is that the antibody-producing cells (i.e., plasma cells) are malignant and thus may cause unusual manifestations. While MM remains an incurable disease, it has also now become a highly treatable disease with the advent of bortezomib, a potent reversible proteasome inhibitor now approved for use in front-line therapy, as well as second-generation derivatives of thalidomide when used in combination with dexamethasone (Barlogie et al., 2008). The mechanisms by which these drugs act in MM setting are complex and not well understood. However, it has been demonstrated at least *in vitro* that both of these agents can block NF-κB transcription in cells. In particular, bortezomib is a potent inhibitor of both canonical and alternative NF-κB by virtue of its potent biochemical activity against degradation of IκBα or processing of p100, respectively. In fact, bortezomib has nanomolar potency when tested in a κB-driven luciferase reporter assay (unpublished).

While bortezomib's considerable efficacy in MM may be attributed to other additional mechanisms of action such as the unfolded protein stress responses, its ability to potently inhibit NF-κB at least in part affirms the NF-κB hypothesis in MM pathogenesis. Indeed, both transformed MM cell lines and freshly drawn bone-marrow aspirates of MM patients exhibit a high level of constitutive NF-κB nuclear expression (Ni et al., 2001). Consequently, it is thought that NF-κB plays a dual role in driving MM progression, an intrinsic dependence on the pathway for survival, and an extrinsic dependence on bone marrow microenvironment for growth via cell-adhesion induced secretion of growth factors such as IL-6, BAFF, and APRIL from bone marrow stroma cells (BMSCs). Treatment with the IKKβ inhibitor PS-1145 on RMPI8226 or MM.1S cells when cultured alone could only suppress intrinsic cell viability only modestly, even though the inhibitor could completely block TNFα-induced NF-κB activation as measured by expression of intracellular adhesion molecule ICAM-1. However, when MM cells were cocultured with BMSCs, PS-1145 was able to inhibit both (a) the IL-6 secreted from BMSCs triggered by MM cell adhesion and (b) the proliferation of MM cells adherent to BMSCs (Hideshima et al., 2002). Such results highlight the NF-κB signaling interplay between cancer cells and their microenvironment. Recently, treatment of IKKβ inhibitor AS602868 on a panel of 14 MM cell lines and primary cells from 13 patients showed a dose-dependent induction of apoptosis and inhibition of cell cycle progression, suggesting the role of NF-κB as the oncogenic growth driver in this setting. More relevant to clinical settings, combining AS602868 with suboptimal doses of melphalan or VELCADE® showed an additive effect in growth inhibition of these MM cells (Jourdan et al., 2007).

Other studies further solidify and expand our understanding of the biological consequences of constitutive NF-κB in MM pathogenesis (Mitsiades et al., 2002a,b; Hideshima et al., 2006; Jourdan et al., 2007). But it was only until recently that the molecular basis underpinning constitutive oncogenic NF-κB activation in MM been revealed. Works from two independent labs showed that many MM cell lines and primary tumor cells harbor diverse genetic mutations that cause NF-κB activation necessary for their survival (Annunziata et al., 2007; Keats et al., 2007). Through the use of various genomic and molecular techniques, genetic and epigenetic mutations were identified to be either loss-of-function of negative regulators or gain-of-function of positive regulators of NF-κB.

Genes that are positive regulators of NF-κB having activating mutations in MM include NIK, CD40, LTβR, TACI, NF-κB1 (p50/p105), and NF-κB2 (p52/p100). Expression of NIK normally directly activates the alternative NF-κB pathway. However, interestingly, it appears that either amplification or overexpressed NIK in MM activates IKKβ and canonical pathway because expression of IκB-SR or treatment with the IKKβ-specific inhibitor MLN120B caused cell death, but knockdown of IKKα using silencing RNAi did not (Annunziata et al., 2007). In other cases with high NF-κB, truncation of NF-κB2 to generate constitutively active p52 appears to be a mechanism to bypass altogether upstream signal-induced controls. In MM cases where CD40, lymphotoxin β receptor, or TACI was overexpressed, the proposed mechanism is that these receptors could lead to efficient activation of NF-κB either by autonomous oligomerization or by increased sensitivity to their cognate ligands produced by the BM microenvironment. [Transmembrane activator and calcium modulator and cyclophilin ligand interactor (TACI) is a lymphocyte specific member of the TNF-receptor superfamily].

Negative regulators of NF-κB signaling with inactivating mutations found include TRAF3, CYLD, and BIRC2 (cIAP1)/BIRC3 (cIAP2). Homozygous deletion of CYLD found in many primary MM tumors was associated with high NF-κB signature expression (Annunziata et al., 2007). CYLD serves to de-ubiquitinate K63-linked chains from activated TRAF2, TRAF6, IKKγ, or BLC3; hence, the loss of CYLD expression positively up-regulates NF-κB at several different levels. Another common recurrent mutation among these was TRAF3 inactivation. TRAF3 functions as a RING-finger containing E3 ligase-mediating degradation of NIK and thus acts as an essential negative regulator of the NIK-IKK signaling axis (He et al., 2007). Reintroducing TRAF3 expression in TRAF3-deficient MM cells reduced NF-κB activity in both canonical and alternative pathways and was toxic to these cells (Annunziata et al., 2007). Indeed, a recent study confirmed that receptor-mediated inhibition of TRAF3 intricately links activation of alternative pathway to activation of the canonical NF-κB pathway (Zarnegar et al., 2008).

The MM example illustrates two important themes. For one, NF-κB activity within the host microenvironment is as important in promoting survival as intrinsic activity of the cancer cell. Second, canonical and alternative NF-κB pathways are intimately linked at the IKK complex node, since diverse mutations in both pathways appear to converge onto IKKβ to drive NF-κB transcription programs for survival. It is also worth noting that some aspects of NF-κB signaling revealed in the MM example above may be unique to MM, because cancer cells often adapt and evolve to gain growth and survival advantage by rewiring essential pathway networks upon which it eventually becomes dependent. This is the concept of oncogene addiction (Weinstein, 2002).

19.18 MYELOID MALIGNANCIES

19.18.1 Myelodysplastic Syndrome

Myelodysplastic syndrome (MDS) is a collection of hematopoietic stem cell disorders in which the bone marrow does not produce enough blood cells. While MDS itself is not considered cancerous, late-stage MDS can progress and transform to acute myeloid leukemia (AML), which usually does not respond well to chemotherapy. Therefore, this group of disease is classified based on a continuum of cases from those without blast cells to those of high proportion of blast cells, and to those with transformed blasts in AML. MDS therefore is sometimes referred to as pre-leukemia. Interestingly, both low-risk and high-risk MDS lead to cytopenia, however, the loss of blood production in each case is caused by different disease mechanisms. Low-risk MDS is characterized by excessive apoptosis of progenitors, leading to inefficient hemaetopoiesis and cytopenia. But high-risk MDS is characterized by a progressive increase in blast cells infiltration with reduced apoptotic potential that leads to cytopenias through marrow failure (Braun et al., 2006).

While activation of NF-κB is not generally observed in early-stage MDS patients, it is found mainly in high-risk MDS, especially those who carry cytogenetic alterations (Cilloni et al., 2007). A recent analysis of tissues from MDS patients showed a correlation between NF-κB activation to the risk of disease progression toward AML development. This correlation was significant not only in cross-patient population analysis, but also longitudinal follow-up with individual patients (Braun et al., 2006a,b). Such finding suggests the requirement of activated NF-κB for the progressive suppression of apoptosis and increased survival and proliferation of MDS clonal population as the disease progresses from early MDS toward AML transformation. The same study furthermore demonstrated the specificity of activated NF-κB in high-risk MDS but not in low-risk MDS. Analysis of the same samples using fluorescent *in situ* hybridization (FISH) of common MDS-associated mutations showed that activated NF-κB was restricted to only MDS cells carrying cytogenetic alterations, rather than an attribute of general stress responses from surrounding stroma cells (Braun et al., 2006a,b). In a panel of MDS and AML cell lines, siRNA knockdown of p65 (or the IKK complex components) or a cell-permeable IKKγ/NEMO-antagonistic peptide could block constitutive NF-κB activation and induced apoptotic cell death (Carvalho et al., 2007). In a similar finding, inhibition of NF-κB activation in P39 cell line using either the IKK inhibitor BAY11–7082 or bortezomib, also resulted in the apoptotic cell death (Braun et al., 2006a,b). These studies strongly suggest that therapeutic inhibition of NF-κB has the potential to specifically eliminate blasts that pathologically interfere with hematopoiesis in high-risk MDS.

19.18.2 Acute Myeloid Leukemia

Acute myeloid leukemia (AML) is a cancer of stem or progenitor cells. Clonal leukemic stem cells (LSC) may carry a heterogeneous set of cytogenetic alterations and molecular mutations to promote their proliferation, evasion of apoptosis, and defects in differentiation (Jordan, 2002; Haferlach, 2008). LSCs provide the seeds for the generation of leukemic blasts which can proliferate in an uncontrolled manner initially in the bone marrow leading to loss of platelets and red blood cells, but can then quickly spread to lymph nodes, liver, spleen, brain, and testis. As an acute leukemia, fatal outcome is likely if left untreated. AML is treated initially with chemotherapy aimed at inducing a remission, and eligible patients may go on to receive hematopoietic stem cell transplantation. Although rapidly cycling leukemic blasts are susceptible to chemotherapy-induced cell death, primitive LSCs are quiescent cells that possess the self-renewal capacity of stem cells but are cycling slowly and thus tend to be chemoresistant leading to disease relapses (Cilloni et al., 2007). Importantly, current chemotherapeutics do not effectively distinguish LSCs from hematopoeitic stem cells that are important to regenerate various healthy blood cells. Ultimately, successful therapeutics will target unique vulnerabilities in LSCs that are distinct from normal hemaetopoietic stem cells. One of the distinguishing features of LSCs is the constitutive activation of NF-κB. Indeed, constitutive NF-κB DNA-binding activity was observed in 16 of 22 (73%) investigated AML cases and was associated with resistance to apoptosis (Birkenkamp et al., 2004). Enrichment of AML leukemic cells ($CD34^+/CD38^-/CD123^+$) showed a consistent high level of nuclear NF-κB which was confirmed by positive expression of κB-driven genes, which is in contrast to unstimulated $CD34^+$ progenitor cells where nuclear NF-κB level remained low. Moreover, treating these cells with proteasome inhibitor MG-132 blocked NF-κB nuclear activity and induced cell death, whereas normal $CD34^+/CD38^-$ cells showed little, if any, effect (Guzman et al, 2001).

Such tantalizing results suggest the link between NF-κB activation and cell proliferation in AML. However, the degree of dependence on NF-κB for disease progression is not well understood, nor is the underlying molecular mechanism by which NF-κB is activated. In one study, NF-κB activation levels correlated with blast counts in the peripheral blood (Frelin et al., 2005). But this was contrasted to another study where 14 of 30 AML cases showed activated NF-κB but that the levels of activation were not directly correlated with blast counts (Bueso-Ramos et al., 2004). Perhaps more sensitive assays and larger cohort studies would help to establish a more clear correlation. What seems to be evident, however, is that constitutive NF-κB in AML occurs via disregulation of IKK kinase activity. Two independent studies involving 35 and 18 patients both showed increased IKK activity in bone marrow LSCs and AML peripheral blood blasts, respectively (Baumgartner et al., 2002; Frelin et al., 2005). Pharmacological inhibition of IKKβ by the anilino-pyrimidine AS602868 induced apoptosis of primary AML cells and sensitized AML blasts to neoplastic drugs (Frelin et al., 2005). What is interesting was the finding that AS602868 also contributed to *in vitro* efficacy by additionally inhibiting Fms-like tyrosine kinase 3 (FLT3), which is also frequently overexpressed or mutated in AML patients (Griessinger et al., 2007). In this study, results from transfecting either Ba/F3 murine pre-B cells or the MV4–11 human AML line with various known FLT3 mutants (ITD, D835V, D835Y) strongly suggested that oncogenic stimulation of FLT3 leads to activation of NF-κB. In another similar finding, suppression of NF-κB activity by the SN-50 peptide in AML cells resulted in enhanced chemotherapy-induced apoptosis (Birkenkamp et al., 2004). And in further exploring the mechanism by which NF-κB may be activated in AML, this study suggested that constitutive NF-κB activity was mediated by the PI3-K/Akt-mediated pathway and that activation of Ras may lead to NF-κB activation in AML cells.

The importance of NF-κB in AML is now well established. Nonetheless, heterogeneity of cancer would dictate that there are likely subpopulations of AML patients who harbor stronger dependence on NF-κB for disease progression than others and that any NF-κB-based therapeutics would be most likely to be successful when used in combination with cytotoxic chemotherapeutics or with inhibitors of other growth pathways such as RTK, PI-3K, and MAPK signaling networks.

19.19 THERAPEUTIC APPROACHES TO NF-κB

With a deeper understanding of NF-κB mechanistic biology and better appreciation of its contribution to pathogenesis of diseases ranging from cancer to other debilitating immune diseases like rheumatoid arthritis, the clinical research community has steadily uncovered the mechanism of action for a number of old drugs that act partly through their NF-κB inhibitory activities, all with the hope to leverage the full utility of these drugs in various disease settings. And given the attractiveness of the NF-κB pathway as a target, there are a growing number of drug discovery efforts to develop novel inhibitors and translational approaches to blocking NF-κB. See Figure 19.3.

19.19.1 NSAIDs and ImiDs

Nonsteroidal anti-inflammatory drugs (NSAIDS), such as aspirin and sulindac, exert their anti-inflammatory effects largely through blocking prostaglandin

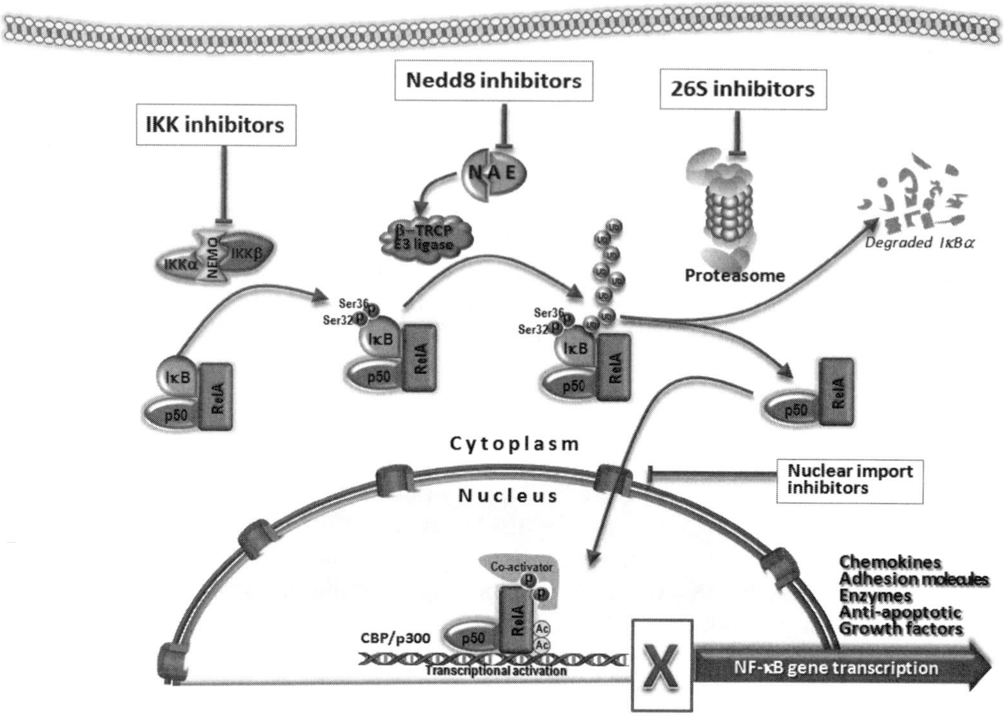

Figure 19.3. Targeting the NF-κB pathway. Exemplified strategies used by various groups in developing therapeutics against NF-κB. This schematic illustrates various points of intervention, including 26S proteasome, neddylation, IKK, and specific nuclear import inhibitors. See the insert for color representation of this figure.

synthesis by inhibiting cyclooxygenase-2 (COX-2). But many NSAIDS also mediate their effects through non-COX2 mechanisms (Olivier et al., 2006). Aspirin and sodium salycilate were demonstrated to inhibit NF-κB by blocking the degradation of IκBα (Kopp and Ghosh, 1994). Follow-up studies into the mechanism of action of aspirin and sulindac against NF-κB showed direct inhibition of IKKβ in a manner which prevents ATP binding (Yin et al., 1998; Yamamoto et al., 1999). Aspirin was also tested *in vivo* as an NF-κB inhibitor in several studies. For example, in a model of angiotensin-II induced vascular wall inflammation using rats harboring human renin and angiotensinogen genes, high doses of aspirin were shown to inhibit NF-κB and AP-1 activation and inflammation in heart and kidney, reduced mortality, cardiac hypertrophy, fibrosis, and albuminuria independent of blood pressure (Muller et al., 2001). Aspirin was also demonstrated to have efficacy in blocking influenza virus propagation *in vivo* via its NF-κB inhibiting activity (Mazur et al., 2007).

Sulfasalazine is a derivative of 5-aminosalicylic acid (5-ASA), used primarily as an anti-inflammatory agent in the treatment of inflammatory bowel disease such as ulcerative colitis, as well as for rheumatoid arthritis. Sulfasalazine is actually a prodrug, because it is not active in its ingested form. It is broken down by bacteria in the colon into two products, 5-ASA and sulfapyridine. According to its label, following oral administration, approximately 33% of the sulfasalazine is absorbed, all of the sulfapyridine is absorbed, and about 33% of the 5-ASA is absorbed. There is controversy as to which of these two products are responsible for the anti-inflammatory activity (Olivier et al., 2006). Nonetheless, sulfasalazine has been shown to inhibit NF-κB via direct inhibition of both IKKα and IKKβ (Weber et al., 2000). Treatment of adipose tissue and skeletal muscle with sulfasalazine significantly inhibited the release of IL-6, IL-8, and TNF-α and p65 DNA-binding activity, with concomitant increase in expression of insulin receptor β, which suggests that regulating aberrant cytokine release via interfering NF-κB by sulfasalazine can thereby alleviate insulin resistance in type-2 diabetes mellitus (Lappas et al., 2005). Sulfasalazine can also suppress drug resistance and invasiveness of lung adenocarcinoma cells expressing AXL partly through its ability to inhibit NF-κB (Lay et al., 2007).

Thalidomide and its derivatives are known collectively as immunomodulatory drugs (IMiDs). They have been shown to be active in multiple myeloma when combined with dexamethasone, but also as single agents in NHL and MDS. IMiDs *in vivo* mechanism of action is poorly understood. It is thought that IMiDs activate T helper 1 response with an increase of IL-2 and IFN-γ secretion, leading to stimulation of natural killer (NK) cell response against tumor cells (Davies et al., 2001; Zhu et al., 2008). But IMiDs have also been shown to act partly through the inhibition of NF-κB, although the mechanism by which it regulates NF-κB is not clear. In MM cells, IMiDs down-regulate NF-κB and suppress the expression of cIAP-2 and FLICE inhibitory protein, together with activation of caspase-8 resulting in enhanced MM cell sensitivity to Fas-induced apoptosis (Mitsiades et al., 2002). In another study, thalidomide inhibited IL-1β-induced NF-κB transcriptional activation and IL-8 production in Caco-2 colon cancer cells, where thalidomide was shown to suppress IκBα degradation and NIK-induced NF-κB transcriptional activation (Jin et al., 2002).

19.19.2 Proteasome Inhibitors

Protein degradation is as essential as protein synthesis to living cell. Proteins are degraded by two major processes: lysosomal degradation and proteasomal degradation. The latter primarily degrades endogenous proteins that are covalently tagged with K48-linked polyubiquitin chains. The proteasome itself is a large multiprotein complex inside virtually all living cells, mainly residing in the cytoplasm, the nucleus, and smooth endoplasmic reticulum. Mammalian 26S proteasome usually consists of a 19S (or in some cases 11S/PA28) regulatory particle encapping a 20S catalytic core particle consisting of four concentric stacked rings. The two inner rings are composed of seven catalytic β subunits, while the two outer rings are composed of seven noncatalytic α subunits that serve as docking domains for the regulatory particles with the allosteric N-termini of each α subunit acting as a gating mechanism against unregulated entry of polypeptides. These four rings together create a narrow cylinder forming a hollow aperture through which unfolded proteins being degraded are threaded for proteolytic processing to lengths of seven to eight aminoacids for further metabolic recycling. The interior central chamber of the core particle formed by the β subunits contains three distinct active sites, namely β5, β2, and β1, each having distinct substrate specificities that are enzymatically characterized as chymotrypsin-like, trypsin-like, and peptidyl-glutamyl peptide-hydrolyzing (PHGH) activities (Adams et al., 2002; Voges et al., 1999; Orlowski and Wilk, 2000). In addition to seven constitutively synthesized β subunits, three inducible β subunits, namely β1i, β2i, and β5i (also known as LMP2, MECL, and LMP7 respectively), are synthesized in response to interferon (IFN-γ) that are incorporated into the 20S catalytic core particle during biogenesis and maturation of the proteasome complex (Griffin et al., 1998). The substitution of these immunoproteasome βi subunits for the constitutive β subunits alters the substrate specificity of the catalytic core, creating an interesting drug development opportunity for selective targeting the tissue-restricted immunoproteasome for autoimmune diseases.

26S proteasome-mediated protein degradation is essential to the metabolic and regulatory processes of cellular physiology. In broad terms, 26S proteasome functions to remove abnormal and misfolded proteins, serves as an essential component in cellular stress responses by degrading key regulatory proteins, controls the timely transitions between cell cycle stages, regulates signal-induced cell fate and differentiation, and lastly participates in the immune system by processing antigenic peptides presented by the major histocompatibility complex (MHC) class I molecules. Although the 26S proteasome serves as an effector to a vast array of cellular processes under homeostasis, early observations that blocking the proteasome could induce apoptosis in a broad range of cell lines suggest that it could be modulated as a cancer target. Indeed, inhibition by cell-permeable peptidyl aldehyde-based or lactacystin-based proteasome inhibitors with relatively high specificity toward the β5 chymotrypsin-like activity could readily induce apoptosis in a variety of transformed cell lines and primary cancer cells *in vitro* (Orlowski, 1999). This may certainly be explained by the fact that apoptosis cascade modulators like IAPs, Bik, Bim, Bad, and Bid, as well as pro-apoptotic proteins like p53, $p21^{Cip1}$, $p27^{Kip1}$, and IκBα, are all regulated by proteasomal degradation (Nikrad et al., 2005; Olivier et al., 2006). But most intriguingly, the relationship between proteasome and programmed cell death seems in many cases to depend on cell type and cellular context. In one example, in freshly isolated MM cells from bone marrow of patients or in dexamethasone-resistant transformed MM cell lines, complete inhibition of cell proliferation was observed at bortezomib concentrations of less than 100 nM, compared to EC50 values of greater than 5–10 μM against normal bone marrow stromal cells (Hideshima et al., 2001). In another instance, limited inhibition of proteasome appears to protect cultured primary thymocytes or neurons from apoptosis, suggesting that quiescent and nonproliferating cells are not as susceptible to cytotoxic effects of proteasome inhibitors as those that are undergoing rapid cell division, especially in cases of cancerous lymphocytes and monocytes such as CLL-derived cells (Masdehors et al., 1999; Orlowski, 1999). Moreover, it has been postulated that proteasome inhibition in cancer cells promotes apoptosis by renormalizing the overall balance from antiapoptotic signals back toward proapoptotic signals. Such preferential induction of programmed cell death toward transformed cells *in vitro* was the impetus to further investigate proteasome inhibitors *in vivo*, where these inhibitors were indeed shown to be efficacious in a Burkitt's lymphoma xenograft model (Orlowski et al., 1998).

Given that many lymphoma and leukemia cells often have constitutive NF-κB and are hypersensitive to proteasome inhibitors, one of the early hypotheses for the mechanism-of-action of proteasome inhibitors was that they transduce their effects through modulating the NF-κB pathway in hematologic malignant cells. Indeed, in our laboratory as in many others, proteasome inhibitors including the early prototype aldehyde-based MG-132 is still commonly used as a potent reference inhibitor of NF-κB in cell-based assays owing to its ability to effectively stabilize phospho-IκBα expression. In particular, as discussed earlier, multiple myeloma and non-Hodgkin's lymphoma have an especially strong therapeutic rationale for such potent NF-κB blockers demonstrated in preclinical settings. With compelling preclinical results of these early peptidyl aldehyde-based inhibitors, a series of boronic acids were optimized for potency, selectivity, and biodistribution that ultimately resulted in the development molecule PS-341 (bortezomib, VELCADE®) (Adams, 2002).

Bortezomib was the first proteasome inhibitor to be extensively studied in murine models of cancer and inflammation, and to have rapidly progressed through clinical trials in cancer patients due to its positive clinical activity in MM and NHL. Bortezomib showed selective and reversible inhibition with K_i of 0.62 nM against the 20S proteasome and about 1400- to 14,000 fold-lower K_i values measured against a panel of 11 known proteases. Bortezomib demonstrated anti-proliferation activity against a broad range of carcinoma cell lines when it was tested in a National Cancer Institute panel of 60 cell lines (Adams et al., 1999). In animal model studies, bortezomib inhibited myeloma tumor growth and prolonged survival in a RPMI-8226 murine xenograft model (LeBlanc et al., 2002). Bioavailability appears to be low, regardless of administration route and species. Pharmacokinetic studies showed rapid plasma clearance and tissue distribution after subcutaneous injection, with an initial half-life $t_{1/2\alpha}$ of 0.2–0.5 h, a slower terminal elimination half-life $t_{1/2\beta}$ after first dose (1.45–2.0 mg/m^2) of 9–15 h, and a large volume of distribution of >500 L (Adams, 2002). A number of Phase I studies demonstrated bortezomib to be well tolerated at various dosing schedules (Orlowski and Kuhn, 2008). Early positive clinical activities were seen in non-small-cell lung carcinoma and in prostate carcinoma (Aghajanian et al., 2002; Papandreou et al., 2004). However, the most impressive clinical responses were observed in multiple myeloma and in mantle-cell and follicular non-Hodgkin's lymphoma (Orlowski et al., 2002). A multicenter, open-label Phase II study indeed confirmed bortezomib activity where it was found to produce durable responses with meaningful survival benefits in patients with recurrent and/or refractory multiple myeloma. In a population of heavily pretreated MM patients, bortezomib produced an overall response rate of 35% and a median duration of response of 12 months, including a number of patients who experienced complete response with undetectable myeloma protein (Richardson et al., 2003). As a side note, common adverse events associated with bortezomib include thrombocytopenia, neutropenia, fatigue, and peripheral neuropathy.

Building on positive results from these early trials, a confirmatory randomized Phase III trial comparing bortezomib to dexamethasone showed that bortezomib gave better response quality and higher overall response rate of 43% versus 18%, longer median time to progression at 6.2 versus 3.5 months, and better overall survival rate of 29.3 versus 23.7 months, respectively (Richardson et al., 2005, 2007). More recently, a large multicenter, randomized front-line Phase III study coded as VISTA was conducted in previously untreated MM patients with VELCADE® in combination with melphalan and prednisone (VcMP) compared to melphalan and prednisone (MP) alone. The VcMP arm demonstrated a 30% complete response rate compared to 4% with MP; and more importantly, a significant survival benefit with the VcMP arm showing overall survival with a 39% reduction in risk of death (hazard ratio = 0.61; $p = 0.008$) with a follow-up of 16.3 months (San Miguel et al., 2008). Together, these and other similar clinical studies led to the approval of VELCADE® for treatment of multiple myeloma patients in more than 80 countries worldwide to date.

In addition to its activity in MM, bortezomib is also quite active in a number of other B-cell malignancies where NF-κB is known to contribute to the pathogenesis. Several Phase II studies demonstrated significant clinical activity in mantle cell, follicular, and marginal zone lymphomas, as well as in Waldenstrom's macroglobulinemia (Goy et al., 2005; O'Connor et al., 2005; Treon et al., 2007). In mantle cell lymphoma (MCL) cell lines *in vitro,* bortezomib induces G1 cell cycle arrest followed by induction of apoptosis. Cell cycle arrest was associated with reduced expression of cyclin D1, which is a molecular genetic marker of MCL. Programmed cell death was associated with downregulation of the antiapoptotic factors Bcl-xL and bfl/A1 with concomitant mitochondrial cytochrome c release and activation of caspase-3 (Ludwig et al., 2005). A follow-up pivotal trial in relapsed MCL showed overall response rate of 33%, including 8% complete remission with a median duration of response of 9.2 months and time to progression of 6.2 months, leading to FDA approval of VELCADE® for second-line mantle cell lymphoma. Currently, there are more than 200 ongoing VELCADE® clinical trials worldwide.

Given the clinical benefits already demonstrated by bortezomib, a number of other proteasome inhibitors are under development. Two examples currently in clinical trials are NPI-0052 (salinosporamide) and PR-171 (carfilzomib). However, unlike bortezomib, whose interaction mode can be described in enzymatic terms as reversible, slow, tight binding to the proteasome, NPI-0052 and carfilzomib are covalently irreversible inhibitors. Preclinical studies have shown that these inhibitors can partially overcome bortezomib resistance *in vitro* and have enhanced *in vivo* efficacy compared to bortezomib in a number of mouse xenograft models, including MM and chronic lymphocytic leukemia, suggesting they may have broader antitumor activities (Orlowski and Kuhn, 2008). Clinical studies are still in progress for these inhibitors, but early reports of clinical responses for carfilzomib, including some bortezomib-refractory MM and Waldenstrom's macroglobulinemia patients, are indeed promising (Stewart et al., 2007).

19.19.3 IKK Inhibitors

Since the cloning of the IKK complex, considerable efforts have been focused on the development of specific IKKβ inhibitors mainly to target the canonical NF-κB signaling pathway for inflammatory diseases. Aspirin and early salicylate-derived compounds were rather weak IKKβ inhibitors, but nonetheless were useful in the demonstration of their anti-inflammatory properties both *in vitro* and *in vivo* (Yin et al., 1998; Yuan et al., 2001). Since then, a number of more potent, orally active, specific IKKβ inhibitors have been published. Some examples are TPCA1, BMS-345541, Bayer Compound A, and PS-1145 and its derivative MLN120b.

At Millennium Pharmaceuticals, in collaboration with scientists from Hoechst (Aventis), a series of β-carbolines were synthesized as inhibitors of IKKβ (Castro et al., 2003). The prototype β-carboline PS-1145 is sufficiently potent and selective against IKK kinase complex with K_i value of 90 nM in an ATP-competitive manner, while essentially inactive against a panel of 14 known kinases. This compound demonstrated specific NF-κB pathway inhibition and cellular efficacy in a broad range of cancer cells. For example, PS-1145 was shown to block the protective effect of IL-6 against dexamethasone-induced apoptosis in a number of MM cell lines (Hideshima et al., 2002). PS-1145 also induced apoptosis in prostate cancer cells and significantly sensitized these cells to apoptosis through TNFα-mediated cytotoxicity, which correlated with down-regulation of NF-κB-regulated gene expression including IL-6, cyclin D1, cyclin D2, IAP-1, and IAP-2. Moreover, PS-1145 inhibited the invasion activity of highly invasive PC3-S cells in invasion chamber assay (Yemelyanov et al., 2006). Furthermore, docetaxel-induced NF-κB activation and IL-6 expression were effectively inhibited by treatment of PS-1145, leading to synergistic docetaxel cytotoxic effects on PC-3 and DU-145 androgen-independent prostate cancer cells (Domingo-Domenech et al., 2006). Similar results were observed with pancreatic cancer cells, where treatment of

PS-1145, treatment of bortezomib, or siRNA knockdown of p65 led to sensitization of Panc-1 and HS776T cells to TRAIL-induced apoptosis (Khanbolooki et al., 2006). And finally, PS-1145 and a derivative β-carboline were shown to be specifically cytotoxic to a subsets of diffuse-large B-cell lymphoma (DLBCL) and primary mediastinal B-cell lymphoma (PMBL) cells. This cytotoxic effect from PS-1145 was fully dependent on canonical NF-κB signaling because overexpression of an inducible p65 alone could rescue these cells from apoptosis (Lam et al., 2005).

Given the selective *in vitro* inhibition of NF-κB by PS-1145, a number of studies were conducted to assess its activity *in vivo*. Although the compound has low aqueous solubility when dosed as a suspension in methylcellulose, PS-1145 exhibits excellent oral bioavailability in rodent at 100% and has quick absorption with T_{max} at <0.5 h and a half-life $t_{1/2\alpha}$ of 1.5–2 h. In an acute LPS challenge model, this compound showed consistent inhibition of 60% of total serum TNFα release at 50 mg/kg when dosed 1 h prior to LPS challenge. In a murine graft-versus-host disease (GVHD) model, PS-1145 protected mice from lethality of GVHD, did not compromise donor engraftment, and effected marked reduction in the levels of serum cytokines that are normally increased during GVHD (Vodanovic-Jankovic et al., 2006). In a pancreatic tumor model, where Panc-1 tumor sizes were monitored dynamically and noninvasively by chemiluminescence, 50 mg/kg PS-1145 treatment alone was not active in inhibiting tumor growth. However, PS-1145 overcame TRAIL-induced resistance *in vivo* and was efficacious in combination with 10 mg/kg TRAIL resulting in Panc-1 tumor regressions (Khanbolooki et al., 2006).

Further optimization of the β-carbolines series identified another molecule that has excellent selectivity, good potency, and excellent pharmacokinetics profile, namely MLN120b [*N*-(6-chloro-7-methoxy-9*H*-beta-carbolin-8-yl)-2-methyl-nicotinamide]. MLN120b is an ATP-competitive inhibitor and is highly selective for IKKβ with K_i value of 45 nM and is essentially inactive against a large panel of kinases. *In vitro*, this compound effectively inhibits TNFα-stimulated NF-κB translocation into the nucleus, blocks TNFα- or IL-1β-induced RANTES and MCP-1 production in human fibroblast-like synoviocytes (HFLS), and also potently inhibits LPS- or peptidoglycan-induced cytokines production in human cord-blood-derived mast cells. In addition, in human chondrocytes MLN120b inhibits IL-1β-induced matrix metalloproteinase (MMP) production, and blocks IL-1β-induced prostaglandin E2 production (Wen et al., 2006). All of this suggests that MLN120b would be effective in rheumatoid arthritis where these cellular processes are manifested in the inflammation and destruction of arthritic joints. Indeed, oral administration of MLN120b suppressed both clinical and histopathologic manifestations of disease in a murine model of antibody-induced arthritis, where *in vivo* imaging demonstrated that NF-κB activity in inflamed arthritic paws was inhibited by MLN120b, resulting in significant suppression of inflammatory genes transcription that is dependent on NF-κB (Izmailova et al., 2007). Similarly, in a rat adjuvant-induced arthritis model, MLN120b inhibited paw swelling in a dose-dependent manner (median effective dosage 12 mg/kg BID) and offered significant protection against arthritis-induced weight loss as well as cartilage and bone erosion (Schopf et al., 2006).

Scientists at GlaxoSmithKline synthesized [5-(*p*-fluorophenyl)-2-ureido] thiophene-3-carboxamide (TPCA-1) as a potent ATP-competitive inhibitor with IC50 values of 20 nM against IKKβ and 400 nM against IKKα. TPCA-1 appears to be at least 500-fold selective against a panel of other known kinases, as well as COX-1 and COX-2. In human monocytes, TPCA-1 inhibits the LPS-stimulated production of TNF-α, IL-6, and IL-8 in a concentration-dependent manner with IC_{50} values ranging from 170 to 320 nM. In a murine Type II collagen-induced arthritis (CIA) model where NF-κB activation is known to correlate closely with appearance of disease, prophylactic or therapeutic intraperitoneal administration of TPCA-1 limited the onset and severity of disease in a manner comparable to a maximally effective dose of etanercept (Podolin et al., 2005). In a rat model of airway inflammation, oral administration of TPCA-1 effectively blocked NF-κB nuclear translocation in the lung tissues following antigen challenge, which was associated with a reduction in expression of inflammatory cytokines, airway eosinophilia, and late asthmatic reaction (Birrell et al., 2005).

At Bristol-Myers Squibb, 4(2′-aminoethyl)amino-1,8-dimethylimidazo(1,2-a)quinoxaline (BMS-345541) was shown to have IC_{50} values of 300 nM against IKKβ and 4000 nM against IKKα and to be relatively selective against a panel of 15 known kinases. Interestingly, kinetic analyses revealed nonlinear Michaelis kinetics with competitive inhibition with respect to IκBα peptide substrate and noncompetitive inhibition with respect to ATP. BMS-345541 inhibited LPS-stimulated TNFα, IL-1β, IL-8, and IL-6 in THP-1 cells with IC_{50} values at 1–5 μM *in vitro*; and oral administration of the drug blocked serum cytokine production *in vivo* following LPS challenge in a dose-dependent manner (Burke et al., 2003). This molecule was also shown to be efficacious in a CIA chronic model of inflammation when dosed prophylactically or therapeutically, and it was shown to have activity in a dextran sulfate sodium (DSS)-induced colitis mouse model (McIntyre et al., 2003; MacMaster et al., 2003). In malignant melanoma disease model, consistent with the thought that aberrant IKK/NF-κB activation leads to constitutive expression of angiogenic CXCL8 and CXCL1 that promote tumorigenesis in melanoma, treatment of BMS-345541 blocked cellular secretion of CXCL1, induced mitochondrial-mediated apoptosis *in vitro*, and inhibited tumor growth in xenograft mice bearing SK-MEL-5, A375, or Hs 294T tumors (Yang et al., 2006).

At Bayer, scientists optimized a series of hydroxyphenyl pyridines and derived the "Compound A" (2-amino-6-[2-(cyclopropylmethoxy)-6-hydroxyphenyl]-4-piperidin-4-yl nicotinonitrile), which showed broad anti-inflammatory activity *in vitro* and *in vivo* (Murata et al., 2004). This compound was shown to be selective for and potent against IKKβ with K_i of 2–4 nM (K_i = 135 nM versus IKKα; IC_{50} > 10 μM versus other kinases) and exhibited competitive inhibition with respect to ATP and noncompetitive against the

IκBα peptide substrate. Pharmacokinetic studies showed 36–69% oral bioavailability in rodents with half-life $t_{1/2\alpha}$ of 1.3–2.1 h and large volumes of distribution of 5.3–6.9 L/kg. In animal studies, "Compound A" demonstrated efficacy in several chronic and acute disease models, including allergen-induced airway inflammation and hyperreactivity model, OVA-induced lung inflammation, and arachidonic-acid-induced and phorbol-ester (PMA)-induced ear edema rat models (Ziegelbauer et al., 2004).

Other non-small molecule approaches to blocking IKK include utilizing NEMO-binding domain (NBD) peptide to disrupt IKK complex formation. In murine colitis models, NBD peptide was shown to ameliorate colonic inflammatory injury through down-regulation of proinflammatory cytokines, suggesting that IKK inhibitors could be effective in the treatment of inflammatory bowel diseases (Shibata et al., 2007). In another study, NEMO-binding domain peptide was shown to block osteoclastogenesis and bone erosion in inflammatory arthritis (Dai et al., 2004a). These studies further validate IKK as potent regulator of cytokine-induced inflammatory arthritis and autoimmune diseases.

As an interesting note, the www.NF-κB.org website created by Thomas D. Gilmore curates known NF-κB inhibitors and pathway modulators that have been described in the literature at a count of more than 750 different compounds, although many of these are natural products that are not fully characterized mechanistically or pharmacologically. This website is an excellent encyclopedic reference for published NF-κB inhibitors and all things NF-κB.

19.19.4 Neddylation Inhibitor

Another interesting approach to blocking NF-κB is by targeting the ubiquination machinery that controls the signal-induced degradation of IκB's and processing of NF-κB proteins that are mediated by β-TRCP containing SCF E3 ligase. This class of E3 ligases is generically formed by four subunits: Skp1, Cul-1, Rbx1, and a variable F-box protein such as β-TrCP that provides substrate specificity; hence they are termed SCF E3 ligases. These SCF E3 ligases are tightly regulated by the Nedd8 pathway. Nedd8 is a ubiquitin-like protein modifier that is conjugated exclusively on Cullin proteins in the SCF complex. Following the cloning of the E1 activating enzyme for Nedd8 (APPBP1-UBA3, or NAE) by scientists at Millennium, molecular characterization of NAE biology quickly revealed that neddylation is absolutely essential for proper SCF E3 ligase functions and SCF-mediated ubiquitination of regulatory proteins like IκBα, p105, or p27 (Read et al., 2000; Podust et al., 2000; Amir et al., 2002; Ohh et al., 2002). Indeed, targeting NAE with small-molecule inhibitors resulted in potent suppression of pathways controlled by SCF-mediated ubiquitination such as NF-κB. Medicinal chemistry optimization of a series adenosine derivatives resulted in the development molecule MLN4924, a potent and selective inhibitor of NAE. In ABC-like DLBCL cells, MLN4924 induces marked stabilization of phospho-IκBα and inhibits p65 nuclear translocation and NF-κB gene transcription. Inhibition of NF-κB signaling in OCI-Ly10 cells results in a G1 cell cycle arrest and an acute induction of apoptosis, a phenotype similar to that observed from treatment with IKKβ inhibitors like MLN120b. Interestingly, in contrast to NF-κB-dependent cells, however, in GCB-like OCI-Ly19 cells MLN4924 instead induces accumulation of several other SCF-ligase substrates including Cdt1, leading to an accumulation of cells with increased DNA content (>4N) followed by a DNA damage response and induction of cell death (Soucy et al., 2009). This mechanism of action in GCB-like cells is observed in other tumor cell lines that are not dependent on activated NF-κB signaling for survival. In a xenograft ABC-like OCI-Ly10 tumor model, a single dose of MLN4924 *in vivo* resulted in inhibition neddylated-cullin proteins, marked elevation of phospho-IκBα levels, and an induction of apoptosis via observation of cleaved caspase. In therapeutic treatment mode, daily or intermittent dosing of MLN4924 below maximum tolerated doses resulted in significant tumor regression in the OCI-Ly10 model, affirming the addiction of these tumors to NF-κB signaling *in vivo* (Milhollen et al., 2008). Currently, MLN4924 is in early clinical trials for both solid and hematological cancers.

This example aptly illustrates the concept of oncogene addiction and context inhibition in cancer, in that even though a neddylation inhibitor like MLN4924 can simultaneously block several different signaling pathways, the mechanism of cell death is dependent on the context of genetic susceptibility of individual cancer cell types. Therefore, neddylation inhibitors, or proteasome inhibitors for that matter, can indeed uncover NF-κB susceptibility in cancer cells where NF-κB pathway is overactivated.

19.20 FUTURE DIRECTIONS

Great strides have been made over the past two decades in advancing our understanding of NF-κB biology through the synergy between academic research and industry drug discovery engines (see Fig. 19.4). To be sure, there is much more to learn about this pleiotropic pathway in order to develop an effective therapy in the many disease settings involving NF-κB. The most specific way to surgically inhibit canonical NF-κB seems to be the IKKβ inhibitors, as evident by the efforts concentrated thus far on discovering selective inhibitors against IKKβ. While there has been tremendous progress in this field, blocking IKKβ is not without concerns. The most critical concern for potent IKKβ blockers is likely to be liver toxicity given the precarious balance of apoptotic fate for hepatocytes in the presence of TNFα under loss of IKKβ activity, as seen in animal knockouts. However, since NF-κB is such a potent immune modulator, sustained or extensive inhibition of IKKβ may not be necessary, depending on the degree of NF-κB dependence in a particular application. Short-term or intermittent dosing could be considered for many

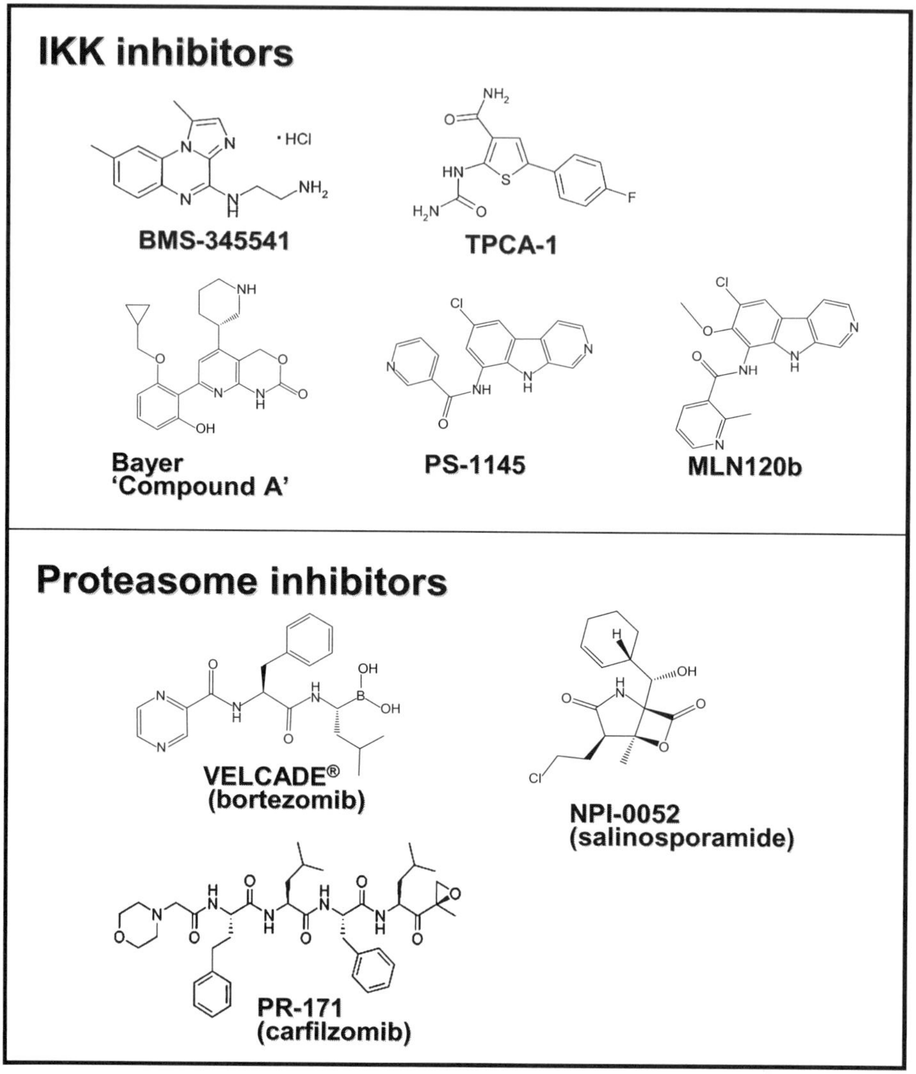

Figure 19.4. Chemical structures of published NF-κB inhibitors.

of these applications. Moreover, given the new appreciation for the alternative NF-κB pathway contribution to the pathogenesis of some cancers, targeting IKKα may be another effective way to block NF-κB. So far, the major concern with regard to blocking IKKα may of the potential long-term consequences of skin defects including possible carcinoma development. But such concern may be less relevant for short-term therapy. Other IKK family members, such as IKKε or TBK1, should also be considered for some solid tumor types including subsets of breast, lung, and prostate. No potent and selective inhibitors of these kinases have been reported so far, but targeting IKKε should be less pleiotropic in their effects when compared to IKKα or IKKβ, given its inducible nature and restricted tissue expression.

Other ways to uncover NF-κB susceptibility in certain disease settings such as targeting the proteasome or the ubiquitination machinery appear to be extremely promising, given the success of VELCADE® in hematological malignancies and highly encouraging early results thus far with the neddylation inhibitor discovered recently. Among these targets to be considered should include the regulatory component of the 26S proteasome, the ubiquitin E1, and the β-TrCP E3 ligase and its associated E2, although the difficulty to achieve protein–protein disruption of E3's by drug-like small molecules is not underestimated here. Other difficulties in developing such pleiotropic inhibitors certainly lie in the management of optimal dosing and schedules in order to maximize their therapeutic index. But in diseases where NF-κB is strongly activated, such multitasking inhibitors have proven to be highly effective and well-tolerated so far in preclinical as well as early-clinical testing.

All of these targets mentioned above are highly promising approaches to modulate NF-κB with the goal to advance disease treatments. However, advancement in two areas could help accelerate this endeavor. The first is the development of more sensitive and robust pharmacodynamic markers in order to manage optimal dosing and scheduling of these inhibitors. The second is the development of more robust disease model systems to better recapitulate NF-κB signaling network *in vivo*. Having such tremendous potential to treat a broad range of diseases with equally tremendous challenges facing the efforts to develop safe and effective NF-κB drugs, especially following the fruitful two decades of past, the next decade of NF-κB research promises to be an even more exciting one for scientists and clinicians in this field.

REFERENCES

Adams J. Proteasome inhibition: a novel approach to cancer therapy. *Trends Mol Med* 2002;8:S49–S54.

Adams J, Palombella VJ, Sausville EA, Johnson J, Destree A, Lazarus DD, et al. Proteasome inhibitors: a novel class of potent and effective anti-tumor agents. *Cancer Res* 1999;59:2615–2622.

Aghajanian C, Soignet S, Dizon DS, Pien CS, Adams J, Elliott PJ, Sabbatini P, Miller V, Hensley ML, Pezzulli S, Canales C, Daud A, Spriggs DR. A phase I trial of the novel proteasome inhibitor PS341 in advanced solid tumor malignancies. *Clin Cancer Res* 2002;8:2505.

Albanese C, Wu K, D'Amico M, Jarrett C, Joyce D, Hughes J, Hulit J, Sakamaki T, Fu M, Ben-Ze'ev A, Bromberg JF, Lamberti C, Verma U, Gaynor RB, Byers SW, Pestell RG. IKKα regulates mitogenic signaling through transcriptional induction of cyclin D1 via Tcf. *Mol Biol Cell* 2003;14(2):585–599.

Alizadeh AA, et al. Distinct types of diffuse large B-cell lymphoma identified by gene expression profiling. *Nature* 2000;403:503–511.

Amir RE, Iwai K, Ciechanover A. The NEDD8 pathway is essential for SCF(beta-TrCP)-mediated ubiquitination and processing of the NF-kappa B precursor p105. *J Biol Chem* 2002;277(26):23253–23259.

Annunziata CM, Davis RE, Demchenko Y, Bellamy W, Gabrea A, Zhan F, Lenz G, Hanamura I, Wright G, Xiao W, Dave S, Hurt EM, Tan B, Zhao H, Stephens O, Santra M, Williams DR, Dang L, Barlogie B, Shaughnessy JD Jr, Kuehl WM, Staudt LM. Frequent engagement of the classical and alternative NF-kappaB pathways by diverse genetic abnormalities in multiple myeloma. *Cancer Cell* 2007;12(2): 115–130.

Arlt A, Vorndamm J, Breitenbroich M, Fölsch UR, Kalthoff H, Schmidt WE, Schäfer H. Inhibition of NF-kappaB sensitizes human pancreatic carcinoma cells to apoptosis induced by etoposide (VP16) or doxorubicin. *Oncogene* 2001;20(7):859–868.

Bargou RC, Emmerich F, Krappmann D, Bommert K, Mapara MY, Arnold W, Royer HD, Grinstein E, Greiner A, Scheidereit C, Dorken B. Constitutive nuclear NF-κB/RelA activation is required for proliferation and survival of Hodgkin's disease tumor cells. *J Clin Invest* 1997;100:2961–2969.

Barlogie B, Shaughnessy JD Jr, Crowley J. Duration of survival in patients with myeloma treated with thalidomide. *N Engl J Med* 2008;359(2):210–212.

Barr J, Sharma CS, Sarkar S, Wise K, Dong L, Periyakaruppan A, Ramesh GT. Nicotine induces oxidative stress and activates nuclear transcription factor kappa B in rat mesencephalic cells. *Mol Cell Biochem* 2007;297(1–2):93–99.

Basseres DS, Baldwin AS. NF-κB and IκB kinase pathways in oncogenic initiation and progression. *Oncogene* 2006;25:6817–6830.

Baumgartner B, Weber M, Quirling M, Fischer C, Page S, Adam M, Von Schilling C, Waterhouse C, Schmid C, Neumeier D, Brand K. Increased IKK activity is associated with activated NF-κB in acute myeloid blasts. *Leukemia* 2002;16:2062–2071.

Beg A, Sha W, Bronson R, Ghosh S, Baltimore D. Embryonic lethality and liver degeneration in mice lacking the RelA component of NF-κB. *Nature* 1995;376: 167–170.

Beraza N, Lüdde T, Assmus U, Roskams T, Vander Borght S, Trautwein C. Hepatocyte-specific IKK gamma/NEMO expression determines the degree of liver injury. *Gastroenterology* 2007;132(7):2504–2517.

Bertoni F, Zucca E. Delving deeper into MALT lymphoma biology. *J Clin Invest* 2006;116(1):22.

Birkenkamp KU, Geugien M, Schepers H, Westra J, Lemmink HH, Vellenga E. Constitutive NF-κB DNA-binding activity in AML is frequently mediated by a Ras/PI3-K/PKB-dependent pathway. *Leukemia* 2004;18(1):103–112.

Birrell MA, Hardaker E, Wong S, McCluskie K, Catley M, De Alba J, Newton R, Haj-Yahia S, Pun KT, Watts CJ, Shaw RJ, Savage TJ, Belvisi MG. Ikappa-B kinase-2 inhibitor blocks inflammation in human airway smooth muscle and a rat model of asthma. *Am J Respir Crit Care Med* 2005;172(8):962–971.

Bonizzi G, Bebien M, Otero DC, Johnson-Vroom KE, Cao Y, Vu D, Jegga AG, Aronow BJ, Ghosh G, Rickert RC, Karin M. Activation of IKKalpha target genes depends on recognition of specific kappaB binding sites by RelB:p52 dimers. *EMBO J* 2004;23(21):4202–4210.

Braun T, Carvalho G, Coquelle A, Vozenin MC, Leppelley P, Hirsch F, Kiladjian JJ, Ribrag V, Fenaux P, Kroemer G. NF-κB constitutes a potential therapeutic target in high-risk myelodysplastic syndromes. *Blood* 2006a;107:1154–1165.

Braun T, Carvalho G, Fabre C, Grosjean J, Fenaux P, Kroemer G. Targeting NF-κB in hematologic malignacies. *Cell Death Diff* 2006b;13:748–758.

Bueso-Ramos CE, Rocha FC, Shi-shodia S, Medeiros LJ, Kantarjian HM, Vadhan-Raj S, et al. Expression of constitutively active NF-κB/RelA transcription factor in blasts of acute myeloid leukaemia. *Hum Pathol* 2004;35:246–253.

Burke JR, Pattoli MA, Gregor KR, Brassil PJ, MacMaster JF, McIntyre KW, Yang X, Iotzova VS, Clarke W, Strnad J, Qiu Y, Zusi FC. *J Biol Chem* 2003;278(3): 1450–1456.

Cabannes E, Khan G, Aillet F, Jarrett RF, Hay RT. Mutations in the IκBa gene in Hodgkin's disease suggest a tumour suppressor role for IkappaBalpha. *Oncogene* 1999;18(20):3063–3070.

Camp ER, Li J, Minnich DJ, Brank A, Moldawer LL, MacKay SL, Hochwald SN. Inducible nuclear factor-kappaB activation contributes to chemotherapy resistance in gastric cancer. *J Am Coll Surg* 2004;199(2):249–258.

Campbell K, Rocha S, Perkins N. Active repression of antiapoptotic gene expression by RelA/p65 NF-κB. *Mol Cell* 2004;13:853–865.

Carbone A. KSHV/HHV-8 associated Kaposi's sarcoma in lymph nodes concurrent with Epstein–Barr virus associated Hodgkin lymphoma. *J Clin Pathol* 2005;58; 626–628.

Carvalho G, Fabre C, Braun T, Grosjean J, Ades L, Agou F, Tasdemir E, Boehrer S, Israel A, Véron M, Fenaux P, Kroemer G. Inhibition of NEMO, the regulatory subunit of the IKK complex, induces apoptosis in high-risk MDS. *Oncogene* 2007;26:(16)2299–2307.

Castro AC, Dang LC, Soucy F, Grenier L, Mazdiyasni H, Hottelet M, Parent L, Pien C, Palombella V, Adams J. Novel IKK inhibitors: beta-carbolines. *Bioorg Med Chem Lett* 2003;13(14):2419–2422.

Chen F, Castranova V. Nuclear factor-kB, an unappreciated tumor suppressor. *Cancer Res* 2007;67(23):11093–11098.

Chen F, Bhatia D, Chang Q, Castranova V. Finding NEMO by K63-linked polyubiquitin chain. *Cell Death Diff* 2006;13:1835–1838.

Chen Z, Hagler J, Palombella VJ, Melandri F, Scherer D, Ballard D, Maniatis T. Signal-induced site-specific phosphorylation targets I kappa B alpha to the ubiquitin–proteasome pathway. *Genes Dev* 1995;9(13):1586–1597.

Cheng WJ, Glebov O, Bergsagel PL, Kuehl WM. Genetic events in the pathogenesis of multiple myeloma. *Best Pract Res Clin Haematol* 2007;20(4):571–596.

Choi M, Rolle C, Wellner M, Cardoso MC, Scheidereit C, Luft FC, Kettritz R. Inhibition of NF-κB by a TAT-NEMO–binding domain peptide accelerates constitutive apoptosis and abrogates LPS-delayed neutrophil apoptosis. *Blood* 2003;102(6):2259–2267.

Cilloni D, Messa F, Arruga F, Defilippi I, Morotti A, Messa E, et al. The NF-κB pathway blockade by the IKK inhibitor PS1145 can overcome imatinib resistance. *Leukemia* 2006;20:61–67.

Cilloni D, Martinelli G, Messa F, Baccarani M, Saglio G. Nuclear factor kB as a target for new drug development in myeloid malignancies. *Haematologica* 2007;92(9): 1224–1229.

Courtois G, Gilmore TD. Mutations in the NF-κB signaling pathway: implications for human disease. *Oncogene* 2006;25:6831–6843.

Coussens L, Werb Z. Inflammation and cancer. *Nature* 2002;420:860–867.

Cusack JC, Liu R, Baldwin AS. Inducible chemoresistance to 7-ethyl-10-[4-(1-piperidino)-1-piperidino]-carbonyloxycamptothe cin (CPT-11) in colorectal cancer cells and a xenograft model is overcome by inhibition of nuclear factor-kappaB activation. *Cancer Res* 2000;60(9):2323–2330.

Dai S, Hirayama T, Abbas S, Abu-Amer Y. The IKK inhibitor, NEMO-binding domain peptide, blocks osteoclastogenesis and bone erosion in inflammatory arthritis. *J Biol Chem* 2004a;279(36):37219–37222.

Dajani R, Sanlioglu S, Zhang Y, Li Q, Monick MM, Lazartigues E, Eggleston T, Davisson RL, Hunninghake GW, Engelhardt JF. Pleiotropic functions of TNF-alpha determine distinct IKKbeta-dependent hepatocellular fates in response to LPS. *Am J Physiol Gastrointest Liver Physiol* 2007;292(1):G242–G252.

Davies FE, Raje N, Hideshima T, Lentzsch S, Young G, Tai YT, Lin B, Podar K, Gupta D, Chauhan D, Treon SP, Richardson PG, Schlossman RL, Morgan GJ, Muller GW, Stirling DI, Anderson KC. Thalidomide and immunomodulatory derivatives augment natural killer cell cytotoxicity in multiple myeloma. *Blood* 2001;98(1): 210–216.

Davis RE, et al. Constitutive NF-κB activity is required for survival of activated B-like diffuse large B-cell lymphoma cells. *J Exp Med* 2001;194:1861–1874.

Dawson CW, Dawson J, Jones R, Ward K, Young LS. Functional differences between BHRF1, the Epstein-Barr virus-encoded Bcl-2 homologue, and Bcl-2 in human epithelial cells. *J Virol* 1998;72(11):9016–9024.

Deacon EM, Pallesen G, Niedobitek G, Crocker J, Brooks L, Rickinson AB, Young LS. Epstein–Barr virus and Hodgkin's disease: transcriptional analysis of virus latency in the malignant cells. *J Exp Med* 1993;177(2):339–349.

Deng L, Wang C, Spencer E, Yang L, Braun A, You J, Slaughter C, Pickart C, Chen ZJ. Activation of the IkappaB kinase complex by TRAF6 requires a dimeric ubiquitin-conjugating enzyme complex and a unique polyubiquitin chain. *Cell* 2000;103(2):351–361.

Descargues P, Sil AK, Karin M. IKK α critical regulator of epidermal differentiation and a suppressor of skin cancer. *The EMBO J* 2008;27:2639–2647.

Domingo-Domenech J, Oliva C, Rovira A, Codony-Servat J, Bosch M, Filella X, Montagut C, Tapia M, Campás C, Dang L, Rolfe M, Ross J, Gascon P, Albanell J, Mellado B. Interleukin 6, a nuclear factor-kappaB target, predicts resistance to docetaxel in hormone-independent prostate cancer and nuclear factor-kappaB inhibition by PS-1145 enhances docetaxel antitumor activity. *Clin Cancer Res* 2006;12(18):5578–5586.

Duffey DC, Chen Z, Dong G, Ondrey FG, Wolf JS, Brown K, Siebenlist U, Van Waes C. Expression of a dominant-negative mutant inhibitor-kappaBalpha of nuclear factor-kappaB in human head and neck squamous cell carcinoma inhibits survival, proinflammatory cytokine expression, and tumor growth *in vivo*. *Cancer Res* 1999;59(14):3468–3474.

Elmore S. Apoptosis: a review of programmed cell death. *Toxicol Pathol* 2007;35(4):495–516.

Emmerich F, Meiser M, Hummel M, Demel G, Foss HD, Jundt F, Mathas S, Krappmann D, Scheidereit C, Stein H, Dörken B. Overexpression of I kappa B alpha without inhibition of NF-κB activity and mutations in the I kappa B alpha gene in Reed–Sternberg cells. *Blood* 1999;94(9):3129–3134.

Emmerich F, Theurich S, Hummel M, Haeffker A, Vry MS, Döhner K, Bommert K, Stein H, Dörken B. Inactivating I kappa B epsilon mutations in Hodgkin/Reed–Sternberg cells. *J Pathol* 2003;(3):413–420.

Fan C, Li Q, Ross D, Engelhardt JF. Tyrosine phosphorylation of I kappa B alpha activates NF kappa B through a redox-regulated and c-Src-dependent mechanism following hypoxia/reoxygenation. *J Biol Chem* 2003;278(3):2072–2080.

Fiumara P, Snell V, Li Y, Mukhopadhyay A, Younes M, Gillenwater AM, Cabanillas F, Aggarwal BB, Younes A. Functional expression of receptor activator of nuclear factor kappaB in Hodgkin disease cell lines. *Blood* 2004;98(9):2784–2790.

Flynn V Jr, Ramanitharan A, Moparty K, Davis R, Sikka S, Agrawal KC, Abdel-Mageed AB. Adenovirus-mediated inhibition of NF-kappaB confers chemo-sensitization and apoptosis in prostate cancer cells. *Int J Oncol* 2003;23(2):317–323.

Frelin C, Imbert V, Griessinger E, Peyron AC, Rochet N, Philip P, Dageville C, Sirvent A, Hummelsberger M, Berard E, Dreano M, Sirvent N, Peyron JF. Targeting NF-kB activation via pharmacologic inhibition of IKK2-induced apoptosis of human acute myeloid leukemia cells. *Blood* 2005;105:804–811.

Gatto S, Scappini B, Pham L, Onida F, Milella M, Ball G, et al. The proteasome inhibitor PS-341 inhibits growth and induces apoptosis in Bcr-Abl positive cell lines sensitive and resistant to imatinib mesylate. *Haematologica* 2003;88:853–863.

Geisler F, Algül H, Paxian S, Schmid RM. Genetic inactivation of RelA/p65 sensitizes adult mouse hepatocytes to TNF-induced apoptosis *in vivo* and *in vitro*. *Gastroenterology* 2007;132(7):2489–2503.

Ghosh S, Baltimore D. Activation *in vitro* of NF-kappa B by phosphorylation of its inhibitor I kappa B. *Nature* 1990;344(6267):678–682.

Gilmore TD. Role of rel family genes in normal and malignant lymphoid cell growth. *Cancer Surv* 1992;15:69–87. Review.

Gilmore TD. Introduction to NF-κB: players, pathways, perspectives. *Oncogene* 2006;25:6680–6684.

Goy A, Younes A, McLaughlin P, et al. Phase II study of proteasome inhibitor bortezomib in relapsed or refractory B-cell non-Hodgkin's lymphoma. *J Clin Oncol* 2005;23:667.

Greten FR, Eckmann L, Greten TF, Park JM, Li ZW, Egan LJ, Kagnoff M, Karin M. IKK links inflammation and tumorigenesis in a mouse model of colitis-associated cancer. *Cell* 2004;118:285–296.

Greten FR, Arkan MC, Bollrath J, Hsu LC, Goode J, Miething C, Goktuna SI, Neuenhahn M, Fierer J, Paxian S, Van Rooijen N, Xu Y, O'Cain T, Jaffee BB, Busch DH, Duyster J, Schmid RM, Eckmann L, Karin M. NF-κB is a negative regulator of IL-1β secretion as revealed by genetic and pharmacological inhibition of IKKβ. *Cell* 2007;130:918–931.

Griessinger E, Imbert V, Lagadec P, Gonthier N, Dubreuil P, Romanelli A, Dreano M, Peyron JF. AS602868, a dual inhibitor of IKK2 and FLT3 to target AML cells. *Leukemia* 2007;21(5):877–885.

Griffin TA, Nandi D, Cruz M, Fehling HJ, Kaer LV, Monaco JJ, Colbert RA. Immunoproteasome Assembly: cooperative incorporation of interferon gamma (IFN-gamma)-inducible subunits. *J Exp Med* 1998;187(1):97–104.

Guo J, Verma U, Gaynor R, Frenkel E, Becerra C. Enhanced chemosensitivity to Irinotecan by RNA interference-mediated down-regulation of the nuclear factor-{kappa}B p65 subunit. *Clin Can Res* 2004;10:3333–3341.

Guttridge DC, Albanese C, Reuther JY, Pestell RG, Baldwin AS Jr. NF-kappaB controls cell growth and differentiation through transcriptional regulation of cyclin D1. *Mol Cell Biol* 1999;19(8):5785–5799.

Guzman M, Neering S, Upchurch D, Grimes B, Howard DS, Rizzieri DA, Luger SM, Jordan C. NF-κB is constitutively activated in primitive human acute myelogenous leukaemia cells. *Blood* 2001;98:2301–2307.

Haferlach T. Molecular genetic pathways as therapeutic targets in acute myeloid leukemia. *Hematol Am Soc Hematol Educ Program* 2008;2008:400–411.

Hagemann T, Wilson J, Burke F, Kulbe H, Li NF, Plüddemann A, Charles K, Gordon S, Balkwill FR. Ovarian cancer cells polarize macrophages toward a tumor-associated phenotype. *J Immunol* 2006;176(8):5023–5032.

Hagemann T, Lawrence T, McNeish I, Charles KA, Kulbe H, Thompson RG, Robinson SC, Balkwill FR. "Re-educating" tumor-associated macrophages by targeting NF-kappaB. *J Exp Med* 2008;205(6):1261–1268.

Hanahan D, Weinberg R. The hallmarks of cancer. *Cell* 2000;100:57–70.

Hayden MS, Ghosh S. Signaling to NF-κB. *Genes Dev* 2004;18:2195–2124.

Hayden MS, Ghosh S. Shared principles in NF-κB signaling. *Cell* 2008;132:344–362.

He JQ, Saha SK, Kang JR, Zarnegar B, Cheng G. Specificity of TRAF3 in its negative regulation of the noncanonical NF-kappa B pathway. *J Biol Chem* 2007;282(6):3688.

Helbig G, Christopherson KW, Bhat-Nakshatri P, Kumar S, Kishimoto H, Miller KD, Broxmeyer HE, Nakshatri H. NF-kappaB promotes breast cancer cell migration and metastasis by inducing the expression of the chemokine receptor CXCR4. *J Biol Chem* 2003;278(24):21631–21638.

Hideshima T, Richardson P, Chauhan D, et al. The proteasome inhibitor PS-341 inhibits growth, induces apoptosis, and overcomes drug resistance in human multiple myeloma cells. *Cancer Res* 2001;61:3071–3076.

Hideshima T, Chauhan D, Richardson P, Mitsiades C, Mitsiades N, Hayashi T, Munshi N, Dang L, Castro A, Palombella V, Adams J, Anderson KC. NF-κB as a therapeutic target in multiple myeloma. *J Biol Chem* 2002;277(19):16639–16647.

Hideshima T, Neri P, Tassone P, Yasui H, Ishitsuka K, Raje N, Chauhan D, Podar K, Mitsiades C, Dang L, Munshi N, Richardson P, Schenkein D, Anderson KC. MLN120B, a novel IkappaB kinase beta inhibitor, blocks multiple myeloma cell growth *in vitro* and *in vivo*. *Clin Cancer Res* 2006;12(19):5887–5894.

Higashimoto T, Chan N, Lee YK, Zandi E. Regulation of IKK by phosphorylation of gamma binding domain of IκB kinase beta by Polo-like kinase 1. *J Biol Chem* 2008;283(51):35354–35367.

Hinz M, Loser P, Mathas S, Krappmann D, Dorken B, Scheidereit C. Constitutive NF-κB maintains high expression of a characteristic gene network, including CD40 and CD86, and a set of anti-apoptotic genes in Hodgkin/Reed–Sternberg cells. *Blood* 2001;97:2798–2807.

Hoffmann A, Levchenko A, Scott ML, Baltimore D. The IkappaB-NF-kappaB signaling module: temporal control and selective gene activation. *Science* 2002;298(5596): 1241–1245.

Hu MC, Lee DF, Xia W, Golfman LS, Ou-Yang F, Yang JY, Zou Y, Bao S, Hanada N, Saso H, Kobayashi R, Hung MC. IKK promotes tumorigenesis through inhibition of forkhead Foxo3a. *Cell* 2004;117:225–237.

Hu Y, Baud V, Delhase M, Zhang P, Deerinck T, Ellisman M, Johnson R, Karin M. Abnormal morphogenesis but intact IKK activation in mice lacking the IKKalpha subunit of IkappaB kinase. *Science* 1999;284(5412):316–320.

Izmailova E, Paz N, Alencar H, Chun M, Schopf L, Hepperle M, Lane JH, Harriman G, Xu Y, Ocain T, Weissleder R, Mahmood U, Healy AM, Jaffee B. Use of molecular imaging to quantify response to IKK-2 inhibitor treatment in murine arthritis. *Arthritis Rheum* 2007;56(1):117–128.

Jansen-Durr P, Meichle A, Steiner P, Pagano M, Finke D, Botz J, Wessbecher J, Draetta G, Eilers M. Differential modulation of cyclin gene expression by MYC. *Proc Natl Acad Sci USA* 1993;90:3685–3689.

Jin SH, Kim TI, Han DS, Shin SK, Kim WH. Thalidomide suppresses the IL-1β-induced NF-κB signaling pathway in colon cancer cells. *Ann N Y Acad Sci* 2002;973:414–418.

Johnson C, Van Antwerp D, Hope TJ. An N-terminal nuclear export signal is required for the nucleocytoplasmic shuttling of IkappaBalpha. *EMBO J* 1999;18(23): 6682–6693.

Jones DR, Broad RM, Madrid LV, Baldwin AS Jr, Mayo MW. Inhibition of NF-kappaB sensitizes non-small cell lung cancer cells to chemotherapy-induced apoptosis. *Ann Thorac Surg* 2000;70(3):930–936.

Jordan C. Unique molecular and cellular features of acute myelogenous leukaemia stem cells. *Leukemia* 2002;16:559–562.

Jost PJ, Ruland J. Aberrant NF-κB signaling in lymphoma: mechanisms, consequences, and therapeutic implications. *Blood* 2007;109(7):2700.

Jourdan M, Moreaux J, De Vos J, Hose D, Mahtouk K, Abouladze M, Robert N, Baudard M, Rème T, Romanelli A, Goldschmidt H, Rossi JF, Dreano M, Klein B. Targeting NF-κB pathway with an IKK2 inhibitor induces inhibition of multiple myeloma cell growth. *Br J Haematol* 2007;138(2):160–168.

Jungnickel B, Staratschek-Jox A, Bräuninger A, Spieker T, Wolf J, Diehl V, Hansmann ML, Rajewsky K, Küppers R. Clonal deleterious mutations in the IkappaBalpha gene in the malignant cells in Hodgkin's lymphoma. *J Exp Med* 2000;191(2): 395–402.

Kanayama A, Seth RB, Sun L, Ea CK, Hong M, Shaito A, Chiu YH, Deng L, Chen ZJ. TAB2 and TAB3 activate the NF-kappaB pathway through binding to polyubiquitin chains. *Mol Cell* 2004;15(4):535–548.

Karin M, Ben-Neriah Y. Phosphorylation meets Ubiquitination: The control of NF-κB activity. *Annu Rev Immun* 2000;18:621–663.

Kasperczyk H, Baumann B, Debatin KM, Fulda S. Characterization of sonic hedgehog as a novel NF-{kappa}B target gene that promotes NF-{kappa}B-mediated apoptosis resistance and tumor growth *in vivo*. *FASEB J* 2009;23:21–33.

Keats JJ, Fonseca R, Chesi M, Schop R, Baker A, Chng WJ, Van Wier S, Tiedemann R, Shi CX, Sebag M, Braggio E, Henry T, Zhu YX, Fogle H, Price-Troska T, Ahmann G, Mancini C, Brents LA, Kumar S, Greipp P, Dispenzieri A, Bryant B, Mulligan G, Bruhn L, Barrett M, Valdez R, Trent J, Stewart AK, Carpten J, Bergsagel PL. Promiscuous mutations activate the noncanonical NF-kappaB pathway in multiple myeloma. *Cancer Cell* 2007;12(2):131–144.

Kessler DJ, Duyao MP, Spicer DB, Sonenshein GE. NF-kappa B-like factors mediate interleukin 1 induction of c-myc gene transcription in fibroblasts. *J Exp Med* 1992;176(3):787–792.

Khanbolooki S, Nawrocki ST, ArumugamT, Andtbacka R, Pino MS, Kurzrock R, Logsdon CD, Abbruzzese JL, McConkey DJ. Nuclear factor-kappaB maintains TRAIL resistance in human pancreatic cancer cells. *Mol Cancer Ther* 2006;5(9): 2251–2260.

Kiriakidis S, Andreakos E, Monaco C, Foxwell B, Feldmann M, Paleolog E. VEGF expression in human macrophages is NF-kappaB-dependent: studies using adenoviruses expressing the endogenous NF-kappaB inhibitor IkappaBalpha and a kinase-defective form of the IkappaB kinase 2. *J Cell Sci* 2003;116(Pt 4):665–674.

Kopp E, Ghosh S. Inhibition of NF-kappa B by sodium salicylate and aspirin. *Science* 1994;265(5174):956–959.

Kucharczak J, Simmons MJ, Fan Y, Gélinas C. To be, or not to be: NF-kappaB is the answer—role of Rel/NF-kappaB in the regulation of apoptosis. *Oncogene* 2003; 22(56):8961–8982.

Lam LT, Davis RE, Pierce J, Hepperle M, Xu Y, Hottelet M, Nong Y, Wen D, Adams J, Dang L, Staudt LM. Small molecule inhibitors of IKK are selectively toxic for subgroups of diffuse large B-cell lymphoma defined by gene expression profiling. *Clin Cancer Res* 2005;11:28–40.

Lappas M, Yee K, Permezel M, Rice GE. Sulfasalazine and BAY 11–7082 interfere with the nuclear factor-kappa B and I kappa B kinase pathway to regulate the release of proinflammatory cytokines from human adipose tissue and skeletal muscle *in vitro*. *Endocrinology* 2005;146(3):1491–1497.

Lawrence T, Bebien M, Liu GY, Nizet V, Karin M. IKK limits macrophage NF-κB activation and contributes to the resolution of inflammation. *Nature* 2005;434: 1138–1143.

Lay JD, Hong CC, Huang JS, Yang YY, Pao CY, Liu CH, Lai YP, Lai GM, Cheng AL, Su IJ, Chuang SE. Sulfasalazine suppresses drug resistance and invasiveness of lung adenocarcinoma cells expressing AXL. *Cancer Res* 2007;67(8):3878–3887.

LeBlanc R, Catley LP, Hideshima T, et al. Proteasome inhibitor PS-341 inhibits human myeloma cell growth *in vivo* and prolongs survival in a murine model. *Cancer Res* 2002;62:4996–5000.

Lee FS, Hagler J, Chen ZJ, Maniatis T. Activation of the IkappaB alpha kinase complex by MEKK1, a kinase of the JNK pathway. *Cell* 1997;88(2):213–222.

Lenardo MJ, Baltimore D. NF-kappa B: a pleiotropic mediator of inducible and tissue-specific gene control. *Cell* 1989;58(2):227–229.

Lenz G, Davis, RE, Ngo VN, Lam L, George TC, Wright GW, Dave SS, Zhao H, XuW, Rosenwald A, Ott G, Muller-Hermelink HK, Gascoyne RD, Connors JM, Rimsza LM, Campo E, Jaffe ES, Delabie J, Smeland EB, Fisher RI, Chan WC, Staudt LM. Oncogenic CARD11 mutations in human diffuse large B cell lymphoma. *Science* 2008;319(5870):1676–1679.

Li ZW, Chu W, Hu Y, Delhase M, Deerinck T, Ellisman M, Johnson R, Karin M. The IKKβ subunit of IκB kinase (IKK) is essential for NF-κB activaton and prevention of apoptosis. *J Expt Med* 1999;189:1839–1845.

Lin YZ, Yao SY, Veach RA, Torgerson TR, Hawiger J. Inhibition of nuclear translocation of transcription factor NF-kappa B by a synthetic peptide containing a cell membrane-permeable motif and nuclear localization sequence. *J Biol Chem* 1995;270(24):14255–14258.

Lucas PC, Yonezumi M, Inohara N, McAllister-Lucas LM, Abazeed ME, Chen FF, Yamaoka S, Seto M, Nunez G. Bcl10 and MALT1, independent targets of chromosomal translocation in malt lymphoma, cooperate in a novel NF-kappa B signaling pathway. *J Biol Chem* 2001;276(22):19012–19019.

Ludwig H, Khayat D, Giaccone G, Facon T. Proteasome inhibition and its clinical prospects in the treatment of hematologic and solid malignancies. *Cancer* 2005;104(9):1794–1807.

Luedde T, Beraza N, Kotsikoris V, van Loo G, Nenci A, De Vos R, Roskams T, Trautwein C, Pasparakis M. Deletion of NEMO/IKKgamma in liver parenchymal cells causes steatohepatitis and hepatocellular carcinoma. *Cancer Cell* 2007;11(2): 119–132.

MacMaster JF, Dambach DM, Lee DB, Berry KK, Qiu Y, Zusi FC, Burke JR. An inhibitor of IkappaB kinase, BMS-345541, blocks endothelial cell adhesion molecule expression and reduces the severity of dextran sulfate sodium-induced colitis in mice. *Inflamm Res* 2003;52(12):508–511.

Masdehors P, Omura S, Merle-Beral H, Mentz F, Cosset JM, Dumont J. Increased sensitivity of CLL-derrived lymphocytes to apoptotic death activation by the proteasome-specific inhibitor lactacystin. *Br J Haematol* 1999;105:752–757.

Mayo MW, Wang CY, Cogswell PC, Rogers-Graham KS, Lowe SW, Der CJ, Baldwin AS Jr. Requirement of NF-kappaB activation to suppress p53-independent apoptosis induced by oncogenic Ras. *Science* 1997;278(5344):1812–1815.

Mazur I, Wurzer WJ, Ehrhardt C, Pleschka S, Puthavathana P, Silberzahn T, Wolff T, Planz O, Ludwig S. Acetylsalicylic acid (ASA) blocks influenza virus propagation via its NF-kappaB-inhibiting activity. *Cell Microbiol* 2007;9(7):1683–1694.

McIntyre KW, Shuster DJ, Gillooly KM, Dambach DM, Pattoli MA, Lu P, Zhou XD, Qiu Y, Zusi FC, Burke JR. A highly selective inhibitor of IκB kinase, BMS-345541, blocks both joint inflammation and destruction in collagen-induced arthritis in mice. *Arthritis Rheum* 2003;48:2652–2659.

Mercurio F, Zhu H, Murray BW, Shevchenko A, Bennett BL, Li J, Young DB, Barbosa M, Mann M, Manning A, Rao A. IKK-1 and IKK-2: cytokine-activated IkappaB kinases essential for NF-kappaB activation. *Science* 1997;278(5339): 860–866.

Milhollen MA, et al. ASH abstract, 2008.

Miller BS, Zandi E. Complete reconstitution of human IkappaB kinase (IKK) complex in yeast. Assessment of its stoichiometry and the role of IKKgamma on the complex activity in the absence of stimulation. *J Biol Chem* 2001;276(39):36320–36326.

Mitsiades N, Mitsiades CS, Poulaki V, Chauhan D, Fanourakis G, Gu X, Bailey C, Joseph M, Liberman TA, Treon SP, Munshi NC, Richardson PG, Hideshima T, Anderson KC. Molecular sequelae of proteasome inhibition in human multiple myeloma cells. *Proc Natl Acad Sci USA* 2002a;99:14374–14379.

Mitsiades N, Mitsiades C, Poulaki V, Chauhan D, Richardson P, Hideshima T, Munshi N, Treon S, Anderson KC. Apoptotic signaling induced by immunomodulatory thalidomide analogs in human multiple myeloma cells: therapeutic implications. *Blood* 2002b;99:4525–4530.

Montagut C, Tusquets I, Ferrer B, Corominas JM, Bellosillo B, Campas C, Suarez M, Fabregat X, Campo E, Gascon P, Serrano S, Fernandez PL, Rovira A, Albanell J. Activation of nuclear factor-kappa B is linked to resistance to neoadjuvant chemotherapy in breast cancer patients. *Endocr Relat Cancer* 2006;13(2):607–616.

Muller DN, Heissmeyer V, Dechend R, Hampich F, Park JK, Fiebeler A, Shagdarsuren E, Theuer J, Elger M, Pilz B, Breu V, Schroer K, Ganten D, Dietz R, Haller H, Scheidereit C, Luft FC. Aspirin inhibits NF-kappaB and protects from angiotensin II-induced organ damage. *FASEB J* 2001;15(10):1822–1824.

Munzert G, Kirchner D, Ottmann O, Bergmann L, Schmid R. Constitutive NF-κB/Rel activation in Philadelphia chromosome positive acute lymphoblastic leukemia. *Leuk Lymphoma* 2004;45:1181–1184.

Murata T, et al. Synthesis and structure-activity relationships of novel IKK-beta inhibitors. Part 3: Orally active anti-inflammatory agents. *Bioorg Med Chem Lett* 2004;14(15):4013–4017.

Nagashima K, Sasseville VG, Wen D, Bielecki A, Yang H, Simpson C, Grant E, Hepperle M, Harriman G, Jaffee B, Ocain T, Xu Y, Fraser CC. Rapid TNFR1-dependent lymphocyte depletion in vivo with a selective chemical inhibitor of IKKβ. *Blood* 2006;107(11):4266–4273.

Nencioni A, Grunebach F, Patrone F, Ballestrero A, Brossart P. Proteasome inhibitors: antitumor effects and beyond. *Leukemia* 2007;21:30–36.

Neumann M, Naumann M. Beyond IκBs: alternative regulation of NF-κB activity. *FASEB J* 2007;21:2642.

Ni H, Ergin M, Huang Q, Qin JZ, Amin HM, Martinez RL, Saeed S, Barton K, Alkan S. Analysis of expression of nuclear factor kappa B (NF-kappa B) in multiple myeloma: downregulation of NF-κB induces apoptosis. *Br J Haematol* 2001; 115(2):279–286.

Nikrad M, Johnson T, Puthalalath H, Coultas L, Adams J, Kraft AS. The proteasome inhibitor bortezomib sensitizes cells to killing by death receptor ligand TRAIL via BH3-only proteins Bik and Bim. *Mol Cancer Ther* 2005;4(3):443–449.

Nolan GP, Fujita T, Bhatia K, Huppi C, Liou HC, Scott ML, Baltimore D. The bcl-3 proto-oncogene encodes a nuclear I kappa B-like molecule that preferentially interacts with NF-kappa B p50 and p52 in a phosphorylation-dependent manner. *Mol Cell Biol* 1993;13(6):3557–3566.

O'Connor OA, Wright J, Moskowitz C, et al. Phase II clinical experiencewith the novel proteasome inhibitor bortezomib in patients with indolent non-Hodgkin's lymphoma and mantle cell lymphoma. *J Clin Oncol* 2005;23:676.

Ohh M, Kim WY, Moslehi JJ, Chen Y, Chau V, Read MA, Kaelin WG Jr. An intact NEDD8 pathway is required for Cullin-dependent ubiquitylation in mammalian cells. *EMBO Rep* 2002;3(2):177–182.

Olivier S, Robe P, Bours V. Can NF-kappa B be a target for novel and efficient anti-cancer agents? *Biochem Pharmacol* 2006;72(9):1054–1068.

Orlowski M, Wilk S. Catalytic activities of the 20 S proteasome, a multicatalytic proteinase complex. *Arch Biochem Biophys* 2000;383(1):1–16.

Orlowski RZ. The role of the ubiquitin–proteasome pathway in apoptosis. *Cell Death Differ* 1999;6(4):303–313.

Orlowski RZ, Eswara JR, Lafond-Walker A, Grever MR, Orlowski M, Dang CV. Tumor growth inhibition induced in a murine model of human Burkitt's lymphoma by a proteasome inhibitor. *Cancer Res* 1998;58(19):4342–4348.

Orlowski RZ, Stinchcombe TE, Mitchell BS, Shea TC, Baldwin AS, Stahl S, et al. Phase I trial of the proteasome inhibitor PS-341 in patients with refractory hematologic malignancies. *J Clin Oncol* 2002;20:4420–4427.

Orlowski RZ, Kuhn DJ. Proteasome inhibitors in cancer therapy: lessons from the first decade. *Clin Cancer Res* 2008:14(6):1649–1657.

Palkowitsch L, Leidner J, Ghosh S, Marienfeld RB. Phosphorylation of serine 68 in the IkappaB kinase (IKK)-binding domain of NEMO interferes with the structure of the IKK complex and tumor necrosis factor-alpha-induced NF-kappaB activity. *J Biol Chem* 2008;283(1):76–86.

Palombella VJ, Rando OJ, Goldberg AL, Maniatis T. The ubiquitin-proteasome pathway is required for processing the NF-kappa B1 precursor protein and the activation of NF-kappa B. *Cell* 1994;78(5):773–785.

Papandreou CN, Daliani DD, Nix D, et al. Phase I trial of the proteasome inhibitor bortezomib in patients with advanced solid tumors with observations in androgen-independent prostate cancer. *J Clin Oncol* 2004;22:2108.

Pasparakis M, Schmidt-Supprian M, Rajewsky K. IκB kinase signaling is essential for maintenance of mature B cells. *J Exp Med* 2002;196(6):743–752.

Pasparakis M, Luedde T, Schmidt-Supprian M. Dissection of the NF-kappaB signalling cascade in transgenic and knockout mice. *Cell Death Differ* 2006;13(5):861–872.

Patel NM, Nozaki S, Shortle NH, Bhat-Nakshatri P, Newton TR, Rice S, Gelfanov V, Boswell SH, Goulet RJ Jr, Sledge GW Jr, Nakshatri H. Paclitaxel sensitivity of breast cancer cells with constitutively active NF-kappaB is enhanced by IkappaBalpha super-repressor and parthenolide. *Oncogene* 2000;19(36):4159–4169.

Perkins ND. Integrating cell-signalling pathways with NF-κB and IKK function. *Nat Rev Mol Cell Biol* 2007;8:49–62.

Peters RT, Maniatis T. A new family of IKK-related kinases may function as IκB kinase kinases. *Biochim Biophys Acta* 2001;1471:M57–M62.

Pham LV, Tamayo AT, Yoshimura LC, Lin-Lee YC, Ford RJ. Constitutive NF-κB and NFAT activation in aggressive B-cell lymphomas synergistically activates the CD154 gene and maintains lymphoma cell survival. *Blood* 2005;106(12):3940–3947.

Phelps CB, Sengchanthalangsy LL, Huxford T, Ghosh G. Mechanism of I kappa B alpha binding to NF-kappa B dimers. *J Biol Chem* 2000;275(38):29840–29846.

Podolin PL, Callahan JF, Bolognese BJ, Li YH, Carlson K, Davis TG, Mellor GW, Evans C, Roshak AK. Attenuation of murine collagen-induced arthritis by a novel, potent, selective small molecule inhibitor of IκB kinase 2, TPCA-1 (2-[(aminocarbonyl)amino]-5-(4-fluorophenyl)-3-thiophenecarboxamide), occurs via reduction of proinflammatory cytokines and antigen-induced T cell proliferation. *J Pharmacol Exp Ther* 2005;312:373–381.

Podust VN, Brownell JE, Gladysheva TB, Luo RS, Wang C, Coggins MB, Pierce JW, Lightcap ES, Chau V. A Nedd8 conjugation pathway is essential for proteolytic targeting of p27Kip1 by ubiquitination. *Proc Natl Acad Sci USA* 2000;97(9): 4579–4584.

Poyet JL, Srinivasula SM, Lin JH, Fernandes-Alnemri T, Yamaoka S, Tsichlis PN, Alnemri ES. Activation of the IκB kinases by RIP via IKKγ/NEMO-mediated oligomerization. *J Biol Chem* 2000;275(48):37966–37977.

Read MA, Brownell JE, Gladysheva TB, Hottelet M, Parent LA, Coggins MB, Pierce JW, Podust VN, Luo RS, Chau V, Palombella VJ. Nedd8 modification of cul-1 activates SCF(beta(TrCP))-dependent ubiquitination of IkappaBalpha. *Mol Cell Biol* 2000;20(7):2326–2333.

Rice J, Courter DL, Giachelli CM, Scatena M. Molecular mediators of alphavbeta3-induced endothelial cell survival. *J Vasc Res* 2006;43(5):422–436.

Richardson PG, Barlogie B, Berenson J, Singhal S, Jagannath S, Irwin D, et al. A phase 2 study of bortezomib in relapsed, refractory myeloma. *N Engl J Med* 2003;348: 2609–2617.

Richardson PG, Sonneveld P, Schuster MW, et al. Bortezomib or high-dose dexamethasone for relapsed multiple myeloma. *N Engl J Med* 2005;352: 2487.

Richardson PG, Sonneveld P, Schuster M, et al. Extended follow-up of a phase 3 trial in relapsedmultiple myeloma: final time-to-event results of the APEX trial. *Blood* 2007;110:3557.

Ritchie CK, Giordano A, Khalili K. Integrin involvement in glioblastoma multiforme: possible regulation by NF-kappaB. *J Cell Physiol* 2000;184(2):214–221.

Rosenwald A, et al. The use of molecular profiling to predict survival after chemotherapy for diffuse large-B-cell lymphoma. *N Engl J Med* 2002;346(25):1937–1947.

Rothwarf DM, Zandi E, Natoli G, Karin M. IKK-gamma is an essential regulatory subunit of the IkappaB kinase complex. *Nature* 1998;395(6699):297–300.

Saccani S, Marazzi I, Beg AA, Natoli G. Degradation of promoter-bound p65/RelA is essential for the prompt termination of the NF-κB response. *J Exp Med* 2004;200:107–113.

Sagaert X, Laurent M, Baens M, Wlodarska I, De Wolf-Peeters C. MALT1 and BCL10 aberrations in MALT lymphomas and their effect on the expression of BCL10 in the tumour cells. *Mod Pathol* 2006;19(2):225–232.

Salvatore C, Camarda G, Maggi CA, Goso C, Manzini S, Binaschi M. NF-kappaB activation contributes to anthracycline resistance pathway in human ovarian carcinoma cell line A2780. *Int J Oncol* 2005;27(3):799–806.

San Miguel JF, et al., and Richardson PG for the VISTA Trial Investigators. Bortezomib plus melphalan and prednisone for initial treatment of multiple myeloma. *N Engl J Med* 2008;359(9):906–917.

Scheidereit C. IκB kinase complexes: gateways to NF-κB activation and transcription. *Oncogene* 2006;25:6685–6705.

Schopf L, Savinainen A, Anderson K, Kujawa J, DuPont M, Silva M, Siebert E, Chandra S, Morgan J, Gangurde P, Wen D, Lane J, Xu Y, Hepperle M, Harriman G, Ocain T, Jaffee B. IKKbeta inhibition protects against bone and cartilage destruction in a rat model of rheumatoid arthritis. *Arthritis Rheum* 2006; 54(10):3163–3173.

Sears C, Olesen J, Rubin D, Finley D, Maniatis T. NF-κB p105 processing via the ubiquitin-proteasome pathway. *J Biol Chem* 1998;273(3):1409–1419.

Senftleben U, Cao Y, Xiao G, Greten FR, Krahn G, Bonizzi G, et al. Activation of IKKα of a second, evolutionary conserved NF-κB signalling pathway. *Science* 2001;293:1495–1499.

Shibata W, Maeda S, Hikiba Y, Yanai A, Ohmae T, Sakamoto K, Nakagawa H, Ogura K, Omata M. The IkappaB kinase (IKK) inhibitor, NEMO-binding domain peptide, blocks inflammatory injury in murine colitis. *J Immunol* 2007;179(5):2681–2685.

Singh S, Darnay BG, Aggarwal BB. Site-specific tyrosine phosphorylation of IkappaBalpha negatively regulates its inducible phosphorylation and degradation. *J Biol Chem* 1996;271(49):31049–31054.

Soucy TA, Smith PG, Milhollen MA, Berger AJ, Gavin JM, Adhikari S, Brownall JE, Burke KE, Cardin DP, Critchley S, Cullis CA, Doucette A, Garnsey JJ, Gaulin JL, Gershman RE, Lubinsky AR, McDonald A, Mizutani H, Narayanan U, Olhava EJ, Peluso S, Rezaei M, Sintchak MD, Talreja T, Thomas MP, Traore T, Vyskocil S, Wheatherhead GS, Yu J, Zhang J, Dick LR, Claiborne GF, Rolfe M, Bolen JB, Langston SP. An inhibitor of NEDD8-activating enzyme as a new approach to treat cancer. *Nature* 2009;458(7239):732–736.

Stewart KA, O'Connor OA, Alsina M, Trudel S, Urquilla PR, Vallone MK, Molineaux CJ, Goy A, Orlowski RZ. Phase I evaluation of carfilzomib (PR-171) in hematological malignancies: Responses in multiple myeloma and Waldenstrom's macroglobulinemia at well-tolerated doses. *J Clin Oncol* 2007;25(18S):8003.

Sun B, Karin M. NF-κB signaling, liver disease and hepatoprotective agents. *Oncogene* 2008;27:6228–6244.

Tabruyn SP, Griffioen AW. A new role for NF-κB in angiogenesis inhibition. *Cell Death Diff* 2007;14:1393–1397.

Tanaka M, Fuentes ME, Yamaguchi K, Durnin MH, Dalrymple SA, Hardy KL, Goeddel DV. Embryonic lethality, liver degeneration, and impaired NF-κB activation in IKKβ-deficient mice. *Immunity* 1999;10:421–429.

Tang ED, Wang CY, Xiong Y, Guan KL. A role for NEMO/IKKγ ubiquitination in the activation of the IκB kinase complex by TNFα. *J Biol Chem* 2003;278(39): 37297–37305.

Tapia MA, González-Navarrete I, Dalmases A, Bosch M, Rodriguez-Fanjul V, Rolfe M, Ross JS, Mezquita J, Mezquita C, Bachs O, Gascón P, Rojo F, Perona R, Rovira A, Albanell J. Inhibition of the canonical IKK/NF kappa B pathway sensitizes human cancer cells to doxorubicin. *Cell Cycle* 2007;6(18):2284–2292.

Tergaonkar V, Correa RG, Ikawa M, Verma IM. Distinct roles of IkappaB proteins in regulating constitutive NF-kappaB activity. *Nat Cell Biol* 2005;7(9): 921–923.

Treon SP, Hunter ZR, Matous J, et al. Multicenter clinical trial of bortezomib in relapsed/refractory Waldenstrom'smacroglobulinemia: results of WMCTG Trial 03-248. *Clin Cancer Res* 2007;13:3320.

Uetsuka H, Haisa M, Kimura M, Gunduz M, Kaneda Y, Ohkawa T, Takaoka M, Murata T, Nobuhisa T, Yamatsuji T, Matsuoka J, Tanaka N, Naomoto Y. Inhibition of inducible NF-kappaB activity reduces chemoresistance to 5-fluorouracil in human stomach cancer cell line. *Exp Cell Res* 2003;289(1):27–35.

Van Waes C. Nuclear factor κB in development, prevention, and therapy of cancer. *Clin Cancer Res* 2007;13(4):1076.

Vlach J, Hennecke S, Alevizopoulos K, Conti D, Amati B. Growth arrest by the cyclin-dependent kinase inhibitor p27Kip1 is abrogated by c-Myc. *EMBO J* 1996;15: 6595–6604.

Voboril R, Hochwald SN, Li J, Brank A, Weberova J, Wessels F, Moldawer LL, Camp ER, MacKay SL. Inhibition of NF-kappa B augments sensitivity to 5-fluorouracil/ folinic acid in colon cancer. *J Surg Res* 2004;120(2):178–188.

Vodanovic-Jankovic S, Hari P, Jacobs P, Komorowski R, Drobyski WR. NF-kappaB as a target for the prevention of graft-versus-host disease: comparative efficacy of bortezomib and PS-1145. *Blood* 2006;107(2):827–834.

Voges D, Zwickl P, Baumeister W. The 26S proteasome: a molecular machine designed for controlled proteolysis. *Annu Rev Biochem* 1999;68:1015–1068.

Wang CY, Cusack JC, Liu R, Baldwin AS. Control of inducible chemoresistance: enhanced anti-tumor therapy through increased apoptosis by inhibition of NF-kB. *Nat Med* 1999;5:412–417.

Weaver KD, Yeyeodu S, Cusack JC Jr, Baldwin AS Jr, Ewend MG. Potentiation of chemotherapeutic agents following antagonism of nuclear factor kappa B in human gliomas. *J Neurooncol* 2003;61(3):187–196.

Weber CK, Liptay S, Wirth T, Adler G, Schmid RM. Suppression of NF-kappaB activity by sulfasalazine is mediated by direct inhibition of IkappaB kinases alpha and beta. *Gastroenterology* 2000;119(5):1209–1218.

Wen D, Nong Y, Morgan JG, Gangurde P, Bielecki A, Dasilva J, Keaveney M, Cheng H, Fraser C, Schopf L, Hepperle M, Harriman G, Jaffee BD, Ocain TD, Xu Y. A selective small molecule IkappaB Kinase beta inhibitor blocks nuclear factor kappaB-mediated inflammatory responses in human fibroblast-like synoviocytes, chondrocytes, and mast cells. *J Pharmacol Exp Ther* 2006;317(3):989–1001.

Weinstein IB. Addiction to oncogenes—the Achilles heal of cancer. *Science* 2002; 297(5578):63–64.

Woronicz JD, Gao X, Cao Z, Rothe M, Goeddel DV. IκB kinase-beta: NF-κB activation and complex formation with IκB kinase-alpha and NIK. *Science* 1997;278(5339): 866–869.

Wright G, Tan B, Rosenwald A, Hurt E, Weistner A, Staudt LM. A gene expression-based method to diagnose clinically distinct subgroups of diffuse large B cell lymphoma. *PNAS* 2003;100(17):9991–9996.

Wu CJ, Conze DB, Li T, Srinivasula SM, Ashwell JD. Sensing of Lys 63-linked polyubiquitination by NEMO is a key event in NF-kappaB activation. *Nat Cell Biol* 2006;8(4):398–406.

Xiao G, Fong A, Sun SC. Induction of p100 processing by NF-κB-inducing kinase involves docking IKK to p100 and IKK-mediated phosphorylation. *J Biol Chem* 2004;279:30099–30105.

Yamamoto Y, Gaynor RB. Therapeutic potential of inhibition of the NF-κB pathway in treatment of inflammation and cancer. *J Clin Invest* 2001;107:134–142.

Yamamoto Y, Yin MJ, Lin KM, Gaynor RB. Sulindac inhibits activation of the NF-kappaB pathway. *J Biol Chem* 1999;274(38):27307–27314.

Yang J, Amiri KI, Burke JR, Schmid JA, Richmond A. BMS-345541 targets inhibitor of kappaB kinase and induces apoptosis in melanoma: involvement of nuclear factor kappaB and mitochondria pathways. *Clin Cancer Res* 2006;12(3 Pt 1):950–960.

Yang M, Huang J, Pan HZ, Jin J. Triptolide overcomes dexamethasone resistance and enhanced PS-341-induced apoptosis via PI3k/Akt/NF-kappaB pathways in human multiple myeloma cells. *Int J Mol Med* 2008;22(4):489–496.

Yemelyanov A, Gasparian A, Lindholm P, Dang L, Pierce J W, Kisseljov F, Karseladze A, Budunova I. Effects of IKK inhibitor PS1145 on NF-kappaB function, proliferation, apoptosis and invasion activity in prostate carcinoma cells. *Oncogene* 2006;25(3):387–398.

Yin MJ, Yamamoto Y, Gaynor RB. The anti-inflammatory agents aspirin and salicylate inhibit the activity of IκB kinase-beta. *Nature* 1998;396:77–80.

Yuan M, Konstantopoulos N, Lee J, Hansen L, Li ZW, Karin M, Shoelson SE. Reversal of obesity- and diet-induced insulin resistance with salicylates or targeted disruption of Ikk-beta. *Science* 2001;293(5535):1673–1677.

Zandi E, Rothwarf DM, Delhase M, Hayakawa M, Karin M. The IkappaB kinase complex (IKK) contains two kinase subunits, IKKalpha and IKKbeta, necessary for IkappaB phosphorylation and NF-kappaB activation. *Cell* 1997;91(2):243–252.

Zarnegar B, Yamazaki S, He JQ, Cheng G. Control of canonical NF-kappaB activation through the NIK-IKK complex pathway. *Proc Natl Acad Sci USA* 2008;105(9): 3503–3508.

Zhou H, Du M, Dixit VM. Constitutive NF-κB activation by the t(11;18)(q21;q21) product in MALT lymphoma is linked to deregulated ubiquitin ligase activity. *Cancer Cell* 2005;7:425–431.

Zhu D, Corral LG, Fleming YW, Stein B. Immunomodulatory drugs Revlimid (lenalidomide) and CC-4047 induce apoptosis of both hematological and solid tumor cells through NK cell activation. *Cancer Immunol Immunother* 2008;57(12):1849–1859.

Ziegelbauer K, Gantner F, Lukacs NW, Berlin A, Fuchikami A, Ohta T, Kawagoe J, Takahashi K, Yada-Hashimoto N, Seino-Noda H, Sakata M, Motoyama T, Kurachi H, Testa JR, Tasaka K, Murata Y. *Clin Cancer Res* 2004;10(22):7645–7654.

20

G-PROTEIN-COUPLED RECEPTORS AS DRUG TARGETS

WENYAN MIAO AND LIJUN WU

20.1 INTRODUCTION

G-protein-coupled receptors (GPCR) form the largest gene family in mammals, comprising ~2% of the human genome. They are cell surface proteins with seven transmembrane (TM) domains; therefore they are also referred to as the 7TM receptors or heptahelical receptors. GPCRs trigger intracellular signaling in response to a plethora of extracellular stimuli such as light, neurotransmitters, hormones, peptides and proteins, amino acids, biogenic amines, nucleotides and nucleosides, lipids, and ions.

GPCRs are expressed on many cell types and play vital roles in generating cellular responses in reaction to changes in their extracellular environment. These receptors are highly selective in ligand recognition, allowing them to generate diverse responses to different physiological settings. Not surprisingly, GPCRs have been implicated in numerous diseases including inflammatory, cardiovascular, metabolic, neurological, psychiatric, infectious diseases, and cancer. They have served as successful targets for drug development, and over 30% of current marketed drugs target GPCRs, with over $50 billion annual sales worldwide.

Enzyme Inhibition in Drug Discovery and Development, edited by Lu and Li
Copyright © 2010 John Wiley & Sons, Inc.

20.2 GPCR CLASSES AND STRUCTURES

There are over 800 GPCR genes in the human genome. They are divided into three major classes based primarily on their physiological ligand and sequence homology (Kristiansen, 2004; Kolakowski, 1994):

The Class A rhodopsin family has a conserved DRY (Asp-Arg-Tyr) motif at the intracellular boundary of the third transmembrane domain. This is the largest GPCR family with ~700 members, about 400 of which encode sensory GPCRs, including olfactory, taste, and pheromone receptors (Gaillard et al., 2004).

The Class B secretin family consists of receptors to hormones and neuropeptides. This is a family with ~50 members, more than half of which are orphan receptors with ligands yet to be identified. The N-terminal portion of these receptors is ~120 residues long and functions as the hormone-binding domain.

The Class C glutamate receptor family has 17 members. These receptors share a unique feature of exceptionally long N-terminus. Examples of receptors in this family include metabotropic glutamate receptors and $GABA_B$ receptors.

It is inherently difficult to obtain high-resolution structural information for membrane proteins. In fact, less than 1% of the 30,000 structures deposited in the public database represent membrane proteins. All GPCRs share the common structural signature of seven transmembrane helices connected with alternating intracellular and extracellular loops, with extracellular N-terminus and intracellular C-terminus. The large percentage of hydrophobic regions makes it difficult to obtain receptor protein with sufficient quantity and purity for structural studies. In addition, recombinant GPCR expression levels are often low because high levels of exogenous GPCRs on the cell surface often causes stress to the host cells. Furthermore, detergent is almost always required when purifying GPCRs. In conjunction with the inherent structure flexibility of the extracellular and intracellular loops and protein instability in the presence of detergent, it has been difficult to study GPCR structure using X-ray crystallography.

In 2000 the first crystal structure of a mammalian GPCR, bovine rhodopsin, was solved (Palczewski et al., 2000). This study revealed that the seven transmembrane helices were arranged in a counterclockwise fashion, with various amount of kinks and bends within each helix. There is an eighth helix in the cytoplasmic tail that runs parallel to the plasma membrane. The first crystal structure for a human GPCR was solved in 2007 for a β2-adrenergic receptor bound by an inverse agonist and an antigen-binding fragment (Fab) of an antibody to its third intracellular loop (Rasmussen et al., 2007). The overall assembly of the transmembrane helices is similar to that of the bovine rhodopsin, with differences in the arrangement of the cytoplasmic ends of these helices. Specifically, the β2-adrenergic receptor has a more "open" structure for G-protein interaction, which was thought to account for its constitutive basal activity in the absence of ligand.

The extracellular domains play critical roles in ligand recognition and thus are highly variable among GPCRs. They are often glycosylated differently under different conditions in different cell types. The glycosylation can affect receptor expression, ligand binding, and receptor signaling (Wheatley and Hawtin, 1999; Duvernay et al., 2005). Evidently, the heterogeneity of receptor glycosylation contributes to added difficulty in obtaining structure information for the extracellular domains, an invaluable tool for GPCR drug discovery.

Due to the limited structure information, computational modeling has been widely used to facilitate rational drug design for GPCRs. Pharmacophore models can be built based on the structure of available receptor agonists and antagonists. These models have been used with reasonable success to design "focused" compound libraries to a GPCR target in order to discover new chemical scaffolds and reduce the cost and time in drug screening (Holenz et al., 2005). Additionally, homology models can be established with information from known GPCR crystal structures and target receptor mutagenesis studies (Deupi et al., 2007). Virtue screens are often conducted with these models to predict the potential binders, agonists, or antagonists in the compound libraries.

20.3 GPCR SIGNALING

Upon agonist binding, GPCRs undergo conformational changes and engage intracellular heterotrimeric G proteins for downstream signaling. Ligand may bind to various sites on the receptor, including the N-terminus, the extracellular loops, and the transmembrane region, or any combinations of the above. Small-molecule ligands such as biogenic amines and lipids tend to bind the transmembrane regions of the receptors, whereas large ligands such as proteins and peptides usually bind to the N-terminus and the extracellular loops.

G proteins respond to the receptor activation signal by binding to the intracellular loops of the receptor (Bourne, 1997). Trimeric G proteins consist of three subunits: α, β, and γ. α is the largest subunit, ranging from 40 kD to 46 kD. The β and γ subunits are 35–36 kD and 8–9 kD, respectively. In their quiescent state, G proteins are bound by GDP on the α subunits. Upon GPCR activation, G proteins are recruited to the receptors, triggering GTP/GDP exchange. The GTP-bound α subunit then dissociates from $\beta\gamma$, and the $\beta\gamma$ subunits remain tightly bound and function as a single unit. The free α and $\beta\gamma$ transduce the activation signal to their respective downstream effectors. The α subunit has intrinsic GTPase activity that hydrolyzes the bound GTP into GDP. The GDP-bound α subunit ceases its signaling and reunites with $\beta\gamma$ to their original trimeric quiescent state ready for reactivation (Fig. 20.1).

So far, 20 Gα, 6 Gβ, and 12 Gγ have been identified in mammals. The large array of G proteins with different subunit combinations gives them the capacity to generate specific and diverse signal for a large number of GPCRs

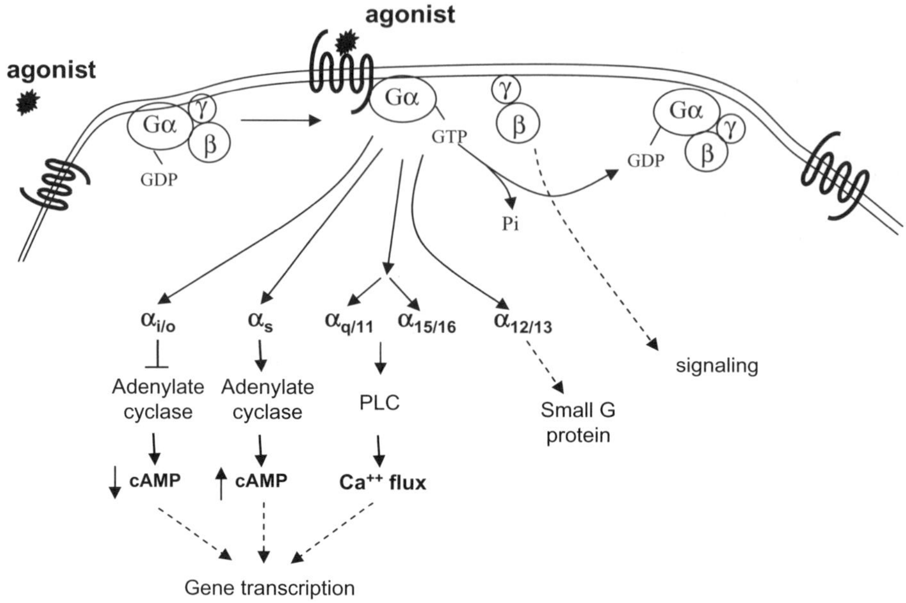

Figure 20.1. GPCR signaling pathways.

(Neves et al., 2002). G proteins can be categorized into four classes based primarily on the sequences and signaling of their G_α subunit:

1. $G_{i/o}$, which inhibits adenylyl cyclase, causes reduction of intracellular cAMP levels. Subtypes in this class include G_i (G_{i1}, G_{i2} and G_{i3}), G_o and G_z.
2. G_s, which activates adenylyl cyclase, leads to increase of intracellular cAMP levels.
3. $G_{q/11}$, which activates phospholipase Cβ (PLCβ). PLCβ breaks down plasma membrane phosphatidylinositol 4,5-bisphosphate (PIP_2) into diacylglyceride (DAG) and inositol 1,4,5-triphosphate (IP3). DAG activates protein kinase C and IP3 induces calcium flux from ER into the cytoplasm. There are five members in the $G_{q/11}$ family: G_q, G_{11}, G_{14}, G_{15}, and G_{16}.
4. $G_{12/13}$, whose downstream signaling remains elusive. There are reports showing that $G_{12/13}$ are involved in the activation of small G proteins Rho, Ras, Na^+/H^+ exchanger, nitric acid synthase and cytoskeleton rearrangement (Dhanasekaran and Dermott, 1996; Kitamura et al., 1996). This family has two members, G_{12} and G_{13}.

Two toxins, pertussis toxin and cholera toxin, have been useful to distinguish the G_i versus G_s action. Both toxins catalyze the ADP-ribosylation reaction to the α subunit, with distinct specificity. Pertussis toxin specifically modifies $G_{\alpha i}$. The ADP-ribosylated $G_{\alpha i}$ is unable to release GDP and therefore loses its ability to bind to the GPCRs. As a result, pertussis toxin inhibits the G_i signal. In comparison, cholera toxin can lock $G_{\alpha s}$ in the GTP-bound form through ADP-ribosylation, resulting in constitutive G_s signaling that leads to higher intracellular cAMP levels.

It is important to keep in mind that although a particular GPCR preferentially signals through one class of G proteins, they may often couple to other family of G proteins as well. This promiscuity in G-protein coupling allows a receptor to generate a wider range of responses based on the nature of the ligands and the state of the receptor, such as post-translational phosphorylation and palmitoylation (Daaka et al., 1997; Okamoto et al., 1997). Therefore, cellular context can add complexity to GPCR signaling based on the availability and concentration of different classes of G proteins and various downstream effectors.

Given its essential role in regulating a wide range of cellular activities in response to extracellular stimuli, GPCR signaling is fine-tuned by elaborate mechanisms to ensure not only fast and specific turn-on but also prompt turn-off, even in the continued presence of ligand. This is achieved by rapid receptor desensitization. A group of kinases called GPCR kinases (GRKs) are recruited and turned on by activated receptors to phosphorylate the serine/threonine residues of the C-terminal tail and the intracellular loops of the receptors. The phosphorylated receptors have high affinity for intracellular adaptor protein β-arrestins. β-arrestins turn off the receptor signal by sterically preventing further receptor/G-protein interaction. The receptors are then recruited into the clathrin-coated pits and undergo endocytosis. Subsequently, receptors in the intracellular vesicles either recycle back to the plasma membrane for another round of signaling (resensitization) or undergo degradation in the lysosomes for long-term signal down-regulation. It has been shown that the relative cellular levels of GRKs and β-arrestins determine the fate of receptor toward resensitization or degradation (Menard et al., 1997; Premont and Gainetdinov, 2007).

Seven GRKs (GRK1-7) and four β-arrestins have been cloned so far. GRKs have specificity for their target receptors, and their expression level varies in different tissues and cell types. Their subcellular localization and activity are tightly regulated by G-protein subunits, lipids, membrane anchoring proteins, and calcium-sensing proteins (Penela et al., 2003). Additionally, as scaffolding proteins, β-arrestins not only orchestrate receptor internalization but also mediate G-protein-independent receptor signaling by bringing receptors into close proximity of signaling molecules. Recent evidence revealed the role of β-arrestins in the activation of MAPK cascade, Akt and PI3 kinases and small G-protein RhoA (DeWire et al., 2007).

The GTP hydrolysis by the G_α subunit is enhanced 100- to 1000-fold by GTPase activating proteins (GAP) or regulators of G protein signaling (RGS). Over 20 mammalian RGS proteins have been identified. Some (RGS3 and RGS5) have ubiquitous expression, whereas others are more restricted to certain cell types or cell differentiation and activation stages. RGS proteins have remarkable selectivity toward their target G_α proteins (Xie and Palmer, 2007). In addition, recent evidence suggests that RGS proteins are also capable of generating their own signal, functioning as antagonists to receptor/G-protein interaction, as scaffolding proteins for G-protein signaling, and engaging their own downstream effectors (Bansal et al., 2007).

Another element that further adds to the complexity of GPCR signaling is receptor dimerization and oligomerization. Increasing literature information revealed the ability of GPCRs to form homo- and hetero-oligomers (Milligan et al., 2003). Compared to their monomeric counterparts, oligomerized receptors may have different expression, G-protein coupling, signaling, and pharmacological properties (Franco et al., 2007; Milligan et al., 2006). For example, two CNS GPCRs, the orexin-1 receptor and the CB1 receptor are capable of forming heterodimers. Antagonists for one receptor reduced signaling of the other receptor (Ellis et al., 2006). As a result, drugs to one partner may affect the pharmacology of the other partner in the dimer. This is an important aspect to consider when we study the heterogeneity of GPCR signaling, drug–drug interaction and the off-target effect of GPCR modulators.

GPCR signaling is still a fast growing area after decades of extensive research. New assay technologies are shedding light on many aspects of GPCR signaling that either complement or challenge the traditional dogmas. As the name implies, the vast majority of GPCRs exert their functions by activating G proteins. However, there are recent examples of 7-transmembrane domain receptors that function independently of G proteins (Sun et al., 2007a). Sun et al. (2007b) observed that β_2-adrenergic receptor switched from G-protein-dependent to independent signaling when high concentration of ligand is present. Also, there exist many phosphorylation-independent GPCR desensitization mechanisms (Ferguson, 2007).

20.4 GPCR PHARMACOLOGY

GPCRs are believed to constitutively toggle among conformations with different levels of activity. An agonist stabilizes the active conformation and tips the balance of the receptor population toward the active state. A full agonist can activate a receptor to its highest potential in the test system, whereas a partial agonist can only partially turn on the receptor. One explanation for the action of full and partial agonists is their differential ability to induce the optimal receptor conformation required to maximally engage the signaling molecules in a given assay system.

In the absence of agonists, when the population of receptors with active conformation is high enough to generate signal above the detection threshold of a certain assay system, constitutive activity is observed. Agents stabilizing the inactive conformation will shift the balance of the receptor population toward the inactive conformation and suppress the constitutive activity. These agents are called inverse agonists.

An antagonist blocks the action of an agonist. It may occupy the agonist binding site to prevent agonist/receptor interaction, in which case the antagonist and agonist are mutually exclusive. This type of antagonists is referred as orthosteric antagonists. Alternatively, an antagonist and an agonist may not share the same binding site; instead they exist in a ternary complex with the receptor. These antagonists are called allosteric antagonists. Another term often used is "neutral antagonist." It refers to an agent that by itself does not elicit any response through the receptor but only demonstrates activity by interfering with the agonist action. Like the agonist, an antagonist can be a full antagonist, which completely depresses the agonist induced activity; or it can be a partial antagonist that only suppresses a portion of the agonist activity.

In recent years, more and more attention has been directed toward the allosteric receptor modulators (Raddatz et al., 2007; Milligan and Smith, 2007). Occupancy of the allosteric sites by these modulators changes the receptor conformation and the changes are transmitted to the ligand binding sites. In other words, allosteric modulators can turn a receptor into a "new" receptor with different properties to its agonists, antagonists, and downstream signaling molecules.

The effect of a given allosteric modulator is influenced by the receptor, the agonist, and the assay system. For example, the neuromuscular blocker alcuronium can increase the affinity of muscarinic antagonist *N*-methylscopolamine specifically for M_2 and M_4 but not other subtypes of muscarinic receptors (Jakubik et al., 1995). In addition, an alcuronium analogue eburnamonine can modulate the affinity of several muscarinic agonists toward M_2 differently: it caused a 25-fold and 7-fold increase of affinity for agonists pilocarpine and oxotremorine, respectively, while having no effect on some other agonists (Jakubik et al., 1997). A more remarkable example is the HIV drug AK602. Chemokine receptor CCR5 is one of the two receptors that mediate HIV viral entry into the immune cells. AK602 has potent anti-HIV activity by blocking the binding between the viral protein gp120 and CCR5. CCR5 has three ligands: CCL3 (MIP1α), CCL4 (MIP1β), and CCL5 (RANTES). As an allosteric inhibitor, AK602 blocks CCL3 binding to CCR5 but allows the signaling of CCL4 and CCL5 toward CCR5 (Maeda et al., 2004). Therefore allosteric modulators can exert not only receptor specific but also agonist-dependent effect.

As a result, it is possible to target the allosteric sites to inhibit the nondesirable while preserving the desirable agonist activity. In practice, however, it is important to keep in mind that allosterism is contingent upon multiple elements such as cellular background, signal readout, assay format, and so on. A proper assay system is vital to uncover allosteric modulators. Most importantly, the activity of allosteric modulators needs to be assessed in multiple biologically relevant systems to evaluate their potential impact *in vivo*.

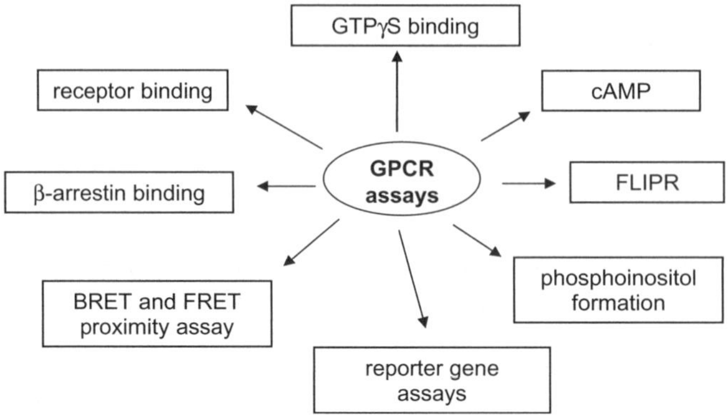

Figure 20.2. Common GPCR assays.

20.5 GPCR ASSAYS FOR DRUG DISCOVERY

GPCRs have a long history as drug targets with well-studied signaling pathways. As a result, a plethora of bioassays have been developed as powerful tools to support GPCR drug discovery (Fig. 20.2).

Careful considerations should be given before one initiates assay development for a GPCR target. It is important to establish the assays in ways that best mimic the receptor's physiological setting, so that data from the assays are more likely to predict the physiological effect of the compounds. One of the most important variables to consider is the assay cell line. Cellular background not only affects receptor expression and post-translational modification, but also dictates whether appropriate intracellular signaling machinery is present to preserve the physiological pharmacology and signaling. In this section, we will discuss the most widely used GPCR assay formats.

A receptor binding assay is used to study the pharmacology of compound–receptor interaction. The most widely used format is the radioligand binding assay. This assay measures the specific binding of radioisotope-labeled ligand to the receptor and the ability of compounds to displace this binding. Traditional radioligand binding assays require wash steps to remove the unbound ligand. To increase the throughput, homogeneous assays that do not require any wash steps have been developed, such as the scintillation proximity assays (SPA). Radioligand binding assay is a simple assay format with straightforward data interpretation. As a result, it is one of the most important workhorses in GPCR drug discovery.

However, many labs consider radioactivity an undesirable feature, especially for high-throughput screening. Therefore, various formats of nonradioactive binding assays have been developed. Ligand can be labeled with a fluorescence tag that can be detected directly, or a tag that allows amplification of the signal through a conjugated enzymatic reaction, such as a horseradish peroxidase (HRP) tag.

In some cases, a compound is directly labeled with a radioisotope in order to study the direct interaction between the compound and the receptor, rather than the ability of the compound to compete off the labeled ligand. This is especially useful to study the agonistic compounds or functional allosteric compounds that do not inhibit ligand binding.

Cell-based functional assays are indispensable for compound activity characterization. There has been an explosion of new assay technologies in the last 10–15 years with major improvements to make the assays more simple, robust, reliable, and amenable for automation. Generally, assays measuring signaling events proximal to the receptor are likely to have less false positives but also less signal amplification. Conceivably, assays examining the distal functions of the receptor are generally more robust due to signal amplification, but likely with more false positives. Usually a combination of several assays is employed to meet the needs of different stages of compound characterization.

Guanine nucleotide exchange assays measure the GTP/GDP exchange triggered by receptor activation. Traditionally, the assay is done with ^{35}S-labeled GTPγS. GTPγS can bind to G_α upon G-protein activation. GTPγS is nonhydrolyzable, therefore locking G_α in this GTPγS bound form, which can be readily quantified. Recently a nonradioactive format was developed by PerkinElmer using time-resolved fluorescence (TRF), in which europium-labeled GTP was used in place of [^{35}S]GTPγS. The advantage of the guanine nucleotide exchange assay is that it requires no knowledge of the G-protein coupling property of the target receptor. However, since the event is in very close proximity to the receptor, it has very little signal amplification. A robust guanine nucleotide exchange assay requires abundant receptors and G proteins and reasonably fast guanine nucleotide exchange rate. Thus for some receptors it is difficult to achieve a satisfactory assay window with this format.

Recruitment of β-arrestin to the receptor is another signaling event proximal to receptor activation. In recent years, two assay platforms have been developed to measure this event. Taking advantage of the fact that β-arrestins co-internalize with the receptors when they bind GPCR with high affinity, Cellomics developed a high content imaging system to measure the β-arrestin clustering (http://www.cellomics.com/). In this system, β-arrestins are tagged with a fluorescence tag such as a green fluorescent protein (GFP). In quiescent cells, β-arrestins distribute evenly within the cytosol, resulting in diffused fluorescence. Co-internalizing with the activated receptor leads to β-arrestin redistribution into punctate aggregates within the cells. The images are captured by a high-throughput High Content Reader and analyzed by image analysis softwares. This platform does not require any modification of the receptor; however, it is not homogeneous and requires complicated algorithms to accurately interpret the imaging data.

In addition, DiscoveRx developed the PathHunter assay to measure the β-arrestin/GPCR interaction using the Enzyme Fragment Complementation (EFC) Technology (http://www.discoverx.com/pathhunter_overview.php). In this assay platform, β-arrestin and the target GPCR are each fused to one of the two inactive β-galactosidase fragments. When β-arrestin comes to close proximity of the receptor, the active enzyme is reconstituted and generates chemiluminescent signal with a β-galactosidase substrate. This assay is in a simple homogeneous format; however, it requires tagging the target receptor with a β-galactosidase enzyme fragment. It is important to ensure that the tag does not affect the receptor function.

Many assays are developed to measure the activity of second messengers as the result of receptor activation. As described above in GPCR signaling, both G_i and G_s affect adenylyl cyclase activity, which can be quantified by the change of intracellular cAMP levels. It is straightforward to measure the G_s-induced increase of cAMP. However, to measure G_i activation, one needs to first use an adenylyl cyclase activator such as forskolin to increase the intracellular cAMP, then measure the reduction of cAMP levels as a result of G_i activation. Traditional cAMP assays are done with ^{125}I-labeled cAMP as the tracer. Over the years the radioactive assay has been replaced by fluorescence or luminescence-based assays using various new technologies, such as non-homogeneous dissociation-enhanced lanthanide fluorescence immunoassay (DELFIA), homogeneous time-resolved fluorescence resonance energy transfer (HTRF), fluorescence polarization, AlphaScreen™ chemiluminescence, and enzyme complementation. All these assays employ the same principal as the radioactive assays, in that cellular cAMP compete with a labeled cAMP tracer for cAMP antibody binding. A detailed comparison of these assay formats was described by Gabriel et al. (2003). In their studies, the highest sensitivity was observed in the DELFIA assay. Among the homogeneous assays, the HTRF and AlphaScreen™ assays showed the biggest dynamic range and highest sensitivity.

The G_q activation can be gauged by two products in the pathway: the inositol phosphate and the cytosolic calcium. As mentioned previously, G_q activation leads to PLCβ activation, which generates IP3. IP3 is rapidly converted to IP2 and then IP1 within the cells. As a result, assays have been developed to measure the IP1 accumulation. Early assay formats measure cellular IP1 in the competition binding with a radioactive IP1 tracer to an IP1 antibody. A nonradioactive HTRF assay called IP-One was developed by Cisbio (http://www.htrf.com/products/gpcr/ipone/). In this format the TRF signal is generated between a europium-IP1 antibody and the IP1 acceptor. Free IP1 from the cell lysates competes with the IP1 acceptor and results in the reduction of the TRF signal. In addition, intracellular calcium flux is another hallmark of G_q activation. This calcium flux can be accurately captured by various fluorescent calcium dyes and Flexstation or a high-throughput real-time fluorescence imaging plate reader (FLIPR) developed by Molecular Devices (http://www.moleculardevices.com/).

Importantly, by coexpressing a promiscuous G protein with a target GPCR, calcium flux and IP-One assays can be used as universal GPCR assays regardless of a receptor's natural G-protein coupling property. Most GPCRs can couple to the promiscuous G proteins, namely $G_{\alpha 15}$ and $G_{\alpha 16}$, and generate signal in the G_q pathway. This strategy enables a single assay platform for rapid high-throughput screening of different classes of GPCRs. Another important utility of this universal assay system is to screen for natural or surrogate ligand for orphan GPCRs without the knowledge of their preferred G proteins. For the receptors that do not couple well to $G_{\alpha 15/16}$, an alternative strategy is to use the chimeric G_α proteins. These chimeric G_α have the backbone of $G_{\alpha q}$ or $G_{\alpha 16}$, with the last five to nine amino acids replaced by the corresponding sequences of $G_{\alpha i}$ or $G_{\alpha s}$. For some G_i- or G_s-coupled receptors, these engineered G proteins better channel receptor signaling into the G_q pathway (Milligan and Rees, 1999). One caution to keep in mind is that promiscuous G proteins and the chimeric G proteins may alter the pharmacology of the receptors.

Another productive assay format widely used for GPCR drug discovery is the homogeneous reporter gene assays. In these assays, a reporter gene is placed under transcriptional regulation of a responsive element for transcription factors such as CRE (responding to cAMP), AP1, or NFAT (responding to calcium). The reporter proteins are synthesized when these second messengers are generated as the result of GPCR activation. Commonly used reporter proteins include luciferase, various fluorescent proteins, β-galactosidase, and β-lactamase. These reporter assays are generally simple, robust, and easily adaptable to automation. However, a longer incubation period is often needed to accumulate enough reporter proteins to generate sufficient signal. Having long assay duration in combination with measuring distal signaling events to receptor activation, the reporter assays have a higher probability of generating false positives.

In recent years, assays using biosensors have been developed to investigate real-time GPCR interaction with other receptors or proteins. These assays usually use bioluminescence resonance energy transfer (BRET) or fluorescence resonance energy transfer (FRET) technology (Pfleger and Eidne, 2005). Both technologies measure the energy transfer between a donor and an acceptor protein when they come to close proximity to each other. A GPCR and its binding partner are each fused to one of the donor/acceptor proteins, and binding events can be detected as they are happening in live cells. A few energy donor/acceptor pairs have been established for these assays, such as (a) Renilla luciferase and green or yellow fluorescent protein (GFP or YFP) for BRET assays and (b) cyan fluorescent protein (CFP) and YFP for FRET assays. People have successfully used these assays to detect real-time interaction between receptors and G proteins, β-arrestins, or GRKs. New donor/acceptor protein pairs are constantly being explored to increase the robustness of the assays in order for them to be amenable for high-throughput screens.

20.6 CHALLENGES AND OPPORTUNITIES FOR GPCR DRUG DISCOVERY

GPCRs have an excellent track record as productive drug targets; nevertheless, this target class faces challenges in drug discovery. The significant structure homology in the GPCR transmembrane domains makes the off-target toxicity for small-molecule inhibitors a liability. There are many examples of GPCR inhibitors with toxicity issues due to their cross-reactivity with other GPCRs. Therefore, improving the compound selectivity has been one of the most important goals in GPCR drug discovery.

Recent years have seen an explosion of knowledge toward receptor allosteric modulation, GPCR homo- and heterodimerization and oligomerization, and so on. We now realize that GPCR signaling is a more complex system than we previously understood. This presents both challenges and opportunities for future GPCR drug discovery. The allosteric modulators may provide us with the much needed drug-target specificity and specific inhibition of undesired and preservation of the desired target signaling. GPCR oligomerization necessitates potential GPCR drugs to be evaluated in a more complex system, so that we can better predict their efficacy *in vivo* and understand their side effects and their interaction with other drugs. Meanwhile, GPCR oligomerization offers a new receptor modulation mechanism that we may be able to exploit in order to achieve specific therapeutic goals.

In addition, given the long history of GPCR drug discovery, most of the "low hanging fruit" have been picked. Nevertheless, many opportunities still lie ahead in the GPCR field. For example, about 100 nonsensory GPCRs remain "orphan" with their ligands remain unidentified. Though challenging, pairing these orphan receptors with their physiological ligands may unearth fruitful targets for GPCR drug discovery (Chung et al., 2008; Levoye et al., 2006).

20.7 GPCRs AS DRUG TARGETS IN INFLAMMATION

Immune cells express a wide range of GPCRs, including receptors for chemokines, lipids, complement anaphylatoxins, various neuropeptides, nucleotides, and biogenic amines. Their expression is often cell-type-specific and changes depending on the stages of cell development, differentiation, activation, and cell cycle. This receptor expression regulation along with differential intracellular signaling machinery in each cell type enables GPCRs to generate intricate responses to complex stimuli in the immune system. A diverse array of host- or pathogen-derived molecules serve as ligands for these GPCRs. They modulate the immune system by regulating immune cell migration, survival, differentiation, and activation. As a result, one can envisage that aberrant GPCR expression and signaling can lead to a variety of immune disorders.

20.7.1 Chemokine Receptors

Chemokine receptors are a class of over 20 G_i-coupled GPCRs that direct cell migration, induce leukocyte activation and differentiation, and promote cell proliferation. They are crucial participants in multiple inflammatory and autoimmune diseases, HIV entry into the cells, and cancer metastasis (Viola and Luster, 2008; O'Hayre et al., 2008).

Chemokine receptors are activated by about 50 chemotactic cytokines (chemokines). These are small proteins with 60–130 amino acids. Although chemokines have varied sequence homology, they share similar tertiary structures. The binding specificity is largely conveyed through the flexible N-terminal tail and the more rigid N-loop by their interaction with the N-terminus and extracellular loops of the chemokine receptors (Goncharova and Tarakanov, 2008). Most chemokines are secreted molecules except for CX_3CL1 and CXCL16, which are membrane-bound with a mucin-rich transmembrane domain.

Chemokines are classified into four subclasses based on the sequential arrangement of the first two of the four conserved cysteines: C-C, C-X-C (X being a non-cysteine amino acid), C, and CX_3C (Prieschl et al., 1995). They can also be divided into constitutive and inducible chemokines based on their expression regulation. The constitutive chemokines are usually expressed at a constant level at certain sites within the body. These chemokines are responsible for directing homeostatic leukocyte trafficking for immune surveillance and tissue/organ development. In contrast, the inducible chemokines are usually up-regulated under inflammatory conditions and participate in recruiting inflammatory cells to the disease tissues. The local chemokine concentration gradient provides a directional migration signal for leukocytes. Chemokine receptor activation up-regulates integrins on the cell surface, thereby promoting cell rolling and adhesion to the endothelial cells and finally leading to leukocyte extravasation into the inflammatory sites. As a result, chemokine and chemokine receptors participate in a wide range of processes, such as innate and adaptive immune responses, organogenesis, hematopoiesis and tumor growth, and metastasis.

There is considerable promiscuity among chemokines and their receptors. Many chemokines activate multiple receptors and many chemokine receptors can be activated by multiple chemokines (Allen et al., 2007) (Table 20.1). This apparent functional redundancy adds complexity in exploiting these receptors as drug targets. However, there is growing evidence suggesting that the distinct *in vivo* spatial and temporal expression of the apparently overlapping chemokines and chemokine receptors fine-tunes the system so that it can respond appropriately under complex immunological circumstances. In addition, both chemokines and chemokine receptors may form homo- and heterodimers and oligomers, adding further convolution to the system.

Given the importance of the chemokine system in the immune disease pathology, significant effort has been made to generate therapeutics to modulate the activity of this system (Viola and Luster, 2008). Specifically, various modified forms of chemokines, antibodies to chemokines or chemokine receptors, and small-molecule antagonists to chemokine receptors have been tested preclinically and clinically for their disease-modifying activity. Below we will

TABLE 20.1. Chemokine and Chemokine Receptor Families for (a) CC Chemokines and (b) CXC, CX_3C, and C Chemokines[a]

	CCR1	CCR2	CCR3	CCR4	CCR5	CCR6	CCR7	CCR8	CCR9	CCR10
CCL1 (I-309)								■		
CCL2 (MCP-1)		■								
CCL3 (MIP-1α)	■				■					
CCL4 (MIP-1β)					■					
CCL5 (RANTES)	■		■		■					
CCL7 (MCP-3)	■	■	■							
CCL8 (MCP-2)		■			■					
CCL11 (eotaxin)			■							
CCL13 (MCP-4)	■	■								
CCL14 (HCC-1)	■									
CCL15 (HCC-2)	■		■							
CCL16 (HCC-4)	■		■							
CCL17 (TARC)				■						
CCL18 (PARC)										
CCL19 (MIP-3β)							■			
CCL20 (MIP-3α)						■				
CCL21 (SLC)							■			
CCL22 (MDC)				■						
CCL23 (MPIF-1)	■									
CCL24 (eotaxin-2)			■							
CCL25 (TECK)									■	
CCL26 (eotaxin-3)			■							
CCL27 (CTACK)										■
CCL28 (MEC)										■

describe several inflammatory chemokine receptors that have been studied in the clinic.

CCR1 is induced primarily on activated T cells, monocytes and macrophages under inflammatory conditions. Being a key receptor for a panel of inflammatory chemokines including CCL3 (MIP-1α), CCL5 (RANTES), and CCL7 (MCP-3), a number of CCR1 antagonists have been developed as potential therapeutics for treating various immune disorders (Gladue et al., 2004; Pease and Horuk, 2005). Pfizer's CP-481715 was efficacious in reducing infiltrating macrophages and other CCR1-positive cells in the synovium of rheumatoid arthritis (RA) patients in a phase Ib trial. However, this compound failed the phase II RA trial because of the lack of efficacy. Currently,

TABLE 20.1. *Continued*

	CXCR1	CXCR2	CXCR3	CXCR4	CXCR5	CXCR6	XCR1	CX_3CR1
CXCL1 (GROα)		■						
CXCL2 (GROβ)		■						
CXCL3 (GROγ)		■						
CXCL4 (PF4)								
CXCL5 (ENA-78)		■						
CXCL6 (GCP-2)	■	■						
CXCL7 (NAP-2)	■	■						
CXCL8 (IL-8)	■	■						
CXCL9 (Mig)			■					
CXCL10 (IP-10)			■					
CXCL11 (I-TAC)			■					
CXCL12 (SDF-1)				■				
CXCL13 (BCA-1)					■			
CXCL14 (BRAK)								
CXCL16						■		
XCL1 (lymphotactin)							■	
XCL2 (SCM-1β)							■	
CX_3CL1 (fractalkine)								■

[a]Significant redundancy exists where a receptor may have multiple ligand, and a ligand may be able to activate multiple receptors. The ligand–receptor relationship is represented by the shaded boxes.

several other CCR1 antagonists from multiple companies are being tested in clinical trials for RA and multiple sclerosis (MS).

CCR2 is primarily expressed on monocytes and macrophages and directs their tissue infiltration in response to inflammatory chemokines CCL2 (MCP-1), CCL7 (MCP-3) and CCL13 (MCP-4). Due to the critical roles of monocytes and macrophages in disease pathogenesis, CCR2 and its ligand have been implicated preclinically in animal models for multiple diseases such as RA, MS, atherosclerosis, organ transplantation, myocardial infarction, vascular remodeling, and so on (Quinones et al., 2005; Mahad and Ransohoff, 2003; Charo and Peters, 2003). As a result, considerable effort has been put into testing the clinical benefit of inhibiting CCR2 for many diseases. Both CCR2 small-molecule antagonists and a humanized monoclonal anti-CCR2 antibody are under clinical evaluation for indications including RA, MS, atherosclerosis, and COPD.

CCR3 is expressed mainly on eosinophils but also on a variety of other inflammatory cells such as T_H2 cells, basophils, mast cells during airway inflammation. It responds mostly to CCL11 (eotaxin), CCL24 (eotaxin-2), and CCL26 (eotaxin-3). CRTH2 is a lipid receptor primarily expressed on the T_H2 cells responding to prostaglandin D2 (PGD2). CCR3 and CRTH2 are responsible for mobilizing the eosinophils and T_H2 cells trafficking to the lung during airway inflammation and allergic reaction (Pease, 2006; Pettipher, 2008). Several small molecule antagonists against CRTH2 or CCR3 are being developed in the clinic for asthma and allergy. In addition, CCR3 antisense oligonucleotides and a monoclonal anti-eotaxin-1 antibody are also under clinical evaluation for the same indications.

CCR9 is up-regulated on gut-homing T cells and plasmacytoid dendritic cells during intestinal inflammation (Johansson-Lindbom and Agace, 2007). ChemoCentryx and GSK have a CCR9 antagonist in phase II/III clinical trials for Crohn's disease and ulcerative colitis. Being a selective antagonist to CCR9 which is almost exclusively expressed in the gut, this compound may have fewer side effects than the existing IBD drugs that cause general immune suppression.

CXCR1 and CXCR2 are primarily expressed on neutrophils and mediate migration of these cells in response to pro-inflammatory chemokines CXCL8 (IL-8), CXCL5 (ENA-78), CXCL1-3 (GROα, β and γ), and others. Pharmacopeia and Schering-Plough are developing a CXCR1/CXCR2 dual inhibitor in a phase II trial for chronic obstructive pulmonary disease (COPD), because neutrophils have been implicated in the pathogenesis of this disease (Magno and Di Stefano, 2007).

CCR5 is another inducible chemokine receptor up-regulated on activated T cells, monocytes and macrophages. It responds to a panel of inflammatory chemokines and induces inflammatory cell migration into the disease tissues (Turner et al., 2007). AstraZeneca is developing a CCR5 antagonist AZD-5672 in phase II trials for RA.

In addition, HIV viruses use CCR5 or CXCR4 as co-receptors for entering into the immune cells. Chemically modified forms of CCL5 (RANTES, a CCR5 ligand) and small-molecule antagonists to the receptor have shown efficacy in blocking the HIV entry (Copeland, 2006). Selzentry, a CCR5 blocker developed by Pfizer, has been approved for AIDS treatment. A CXCR4 antagonist from Genzyme, Plerixafor, was tested in a phase Ib/IIa trial for HIV. It dose dependently eliminated CXCR4-dependent viral entry in 9 out of 19 patients, but the trial was discontinued due to what was believed to be non-mechanism-based adverse effects. Currently plerixafor is in phase III clinical trials for non-Hodgkin's lymphoma and multiple myeloma due to its ability to mobilize white blood cells after stem-cell transplantation in these cancer patients. This compound is also being investigated preclinically for the treatment of rheumatoid arthritis. Many companies have compounds targeting CCR5 or CXCR4 in various stages of AIDS clinical trials. In addition, a CXCR4/CXCR5 dual antagonist is under evaluation in clinical trials for multiple indications, including phase II trials for viral infection, HIV/AIDS, RA, wound healing, cancer, cachexia, anorexia, and phase I trials for type II diabetes. This highlights the great potential of the CCR5 and CXCR4 antagonist as therapeutics for many diseases.

20.7.2 *N*-Formyl Peptide Receptors

Formyl peptide receptor (FPR) and FPR-like receptor (FPRL1) are highly expressed on neutrophils and monocytes. They mediate the chemotactic activity of *N*-Formyl peptides. *N*-Formyl peptides are derived either exogenously from bacteria or endogenously from mitochondria. It has been long established that FPR and FPRL1 are involved in attracting phagocytes to sites of bacterial infection and host tissue damage. Therefore they are part of the innate host defense system (Panaro and Mitolo, 1999).

In addition to pathogen-derived FPR ligand, more and more host endogenous FPR ligand are being discovered, including the serum amyloid A, annexin 1, lipid metabolite (LXA4), β amyloid ($A\beta_{42}$), and prion peptide (aa 106–126) (Le et al., 2002), implicating the involvement of the formyl peptide receptors in inflammation, neurodegenerative diseases, and infection. Functionally, in addition to inducing chemotaxis, they also induce phagocyte oxidative burst and cytokine production (Sodhi and Biswas, 2002; Arbour et al., 1996). Future research is directed toward understanding how formyl peptide receptors can respond to such a diverse array of agonists. This knowledge may help us craft strategies for generating appropriate FPR receptor antagonists for specific therapeutic indications.

20.7.3 Lipid Receptors

20.7.3.1 Lysophospholipid Receptors. The lysophospholipid receptors are activated by bioactive lysophospholipid mediators lysophosphatidic acid (LPA) and sphingosine-1-phosphate (S1P). Originally, they were referred to as the endothelial differentiation gene (EDG) receptors. Later on, they were named the LPA receptors or the S1P receptors based on the higher-affinity ligand to which they bind.

Currently, there are four identified and validated LPA receptors (LPA_{1-4}), all of which have been detected with various expression levels in the immune tissues including the spleen and the thymus. They are involved in regulating T-cell survival, migration, and IL-2 production. Different LPA receptor subtypes play distinct roles in immune modulation. For example, LPA_2 is responsible for LPA-induced naive T-cell migration and IL-2 production inhibition, whereas LPA_1 is expressed on activated T cells and mediates the opposite effect—that is, inducing IL-2 production and inhibiting cell migration (Zheng et al., 2000, 2001).

There are five members in the S1P receptor family ($S1P_{1-5}$). They play a distinct role in regulating immune cell migration, activation, proliferation, and cytokine secretion (von Wenckstern et al., 2006). The most prominent function of S1P is to regulate lymphocyte homeostatic trafficking between the peripheral blood and the lymphoid organs and tissues. This function is mostly mediated through the $S1P_1$ receptor. $S1P_1R$ is the most widely expressed among all S1P receptor subtypes. It is expressed on T cells, B cells, dendritic cells, macrophages, and mast cells (von Wenckstern et al., 2006). Although $S1P_1R$ mediates migration of all these cell types, it has been most extensively studied for its role in controlling the lymphocyte egress from the lymphoid organs into the circulating blood (Chiba et al., 2006).

In addition to its role in modulating cell migration, S1P also promotes T_H2 polarization during T cell activation by suppressing IL-2, IL-4 and IFNγ and promoting IL-10 production. This effect is mainly mediated through the $S1P_4$ receptor expressed on T cells and dendritic cells (Wang et al., 2005). Finally, S1P induces mast cell degranulation through $S1P_2$ (Jolly et al., 2004).

The importance of the S1P receptors in immune regulation is highlighted by the immunosuppressive drug fingolimod, also known as FTY720 (Zhang et al., 2007). It is a chemical derivative of myriocin, a natural product isolated from a parasitic fungus called *Isaria sinclairii*. This fungus has been used extensively in traditional Chinese medicine as an immune suppressant. Upon phosphorylation by sphingosine kinase *in vivo*, FTY720 becomes a S1P receptor agonist. It is a strong agonist for $S1P_1$ but can also activate $S1P_3$, $S1P_4$, and $S1P_5$. Working through $S1P_1$, FTY720 causes lymphopenia by trapping lymphocytes in the spleen and lymph nodes. FTY720 has shown its promise in treating multiple sclerosis and preventing organ transplantation rejection (Zhang et al., 2007). It is currently sponsored by Novartis in phase III trials for MS (Horga and Montalban, 2008). The phase III trials for renal transplantation rejection were discontinued because FTY720 did not meet the endpoint of the study. Presumably, a selective $S1P_1$ agonist will retain the beneficial effect of FTY720 and steer clear of the side effects invoked by activating other S1P receptors. A selective $S1P_1$ agonist Actelion-2 is being co-developed by Actelion and Roche. It is currently in phase I trial for the treatment of RA, psoriasis and MS.

20.7.3.2 Leukotriene Receptors. Leukotrienes comprise a family of products of the 5-lipoxygenase pathway of arachidonic acid metabolism. Leukotriene A_4 (LTA_4) can be converted to leukotriene B_4 (LTB_4) and cysteinyl leukotrienes LTC_4, LTD_4, and LTE_4.

Leukotrienes act by binding to several GPCRs expressed on neutrophils, macrophages, eisonophils, T lymphocytes, and mast cells. Type 1 cysteinyl leokotriene receptor (CysLT1) recognizes cysteinyl leokotrienes in a descending order of affinity of $LTD_4 > LTC_4 = LTE_4$, whereas CysLT2 receptor recognizes cysteinyl leokotrienes in a descending order of affinity of LTC_4 =

LTD_4 > LTE_4. CysLT1 mediates sustained brochoconstriction, mucus secretion, and edema in the airways, whereas CysLT2 contributes to inflammation, vascular permeability, and tissue fibrosis (Kim and Luster, 2007; Massoumi and Sjolander, 2007).

Two receptors have been identified for LTB_4. BLT1 is the high-affinity receptor for LTB_4 that mediates most, if not all, of its potent chemoattractant and proinflammatory action. BLT2 is a lower-affinity receptor for LTB_4 and other lipooxygenase products. Little is known about the physiological function of BLT2.

By binding to their specific cognate receptors, leukotrienes promote the accumulation and function of virtually all subgroups of leukocytes at sites of inflammation. Such responses are important in the pathogenesis of diseases including asthma, cardiovascular diseases, and cancer (Nicosia et al., 2001; Back and Hansson, 2006).

Several anti-leukotriene agents have been on the market, including montelukast, zafirlukast, and pranlukast. All these drugs are selective CysLT1 antagonists. Several BLT-1 antagonists are undergoing clinical trials for the treatment of a wide range of immune diseases such as RA, cystic fibrosis, and COPD. In addition, several LTD_4 antagonists are under clinical evaluation for the treatment of asthma by inhibiting bronchoconstriction.

20.7.3.3 Prostaglandin E2 (PGE_2) Receptors. PGE_2 is a cyclooxygenase (COX) metabolite of arachidonic acid. It is a potent inducer for acute inflammatory responses including pain and swelling at the inflammatory sites. The importance of PGE_2 in rheumatoid arthritis and osteoarthritis pathogenesis is highlighted by the COX-2 inhibitors, which are highly efficacious in treating these diseases (Akaogi et al., 2006; Baraf, 2007).

There are four receptors (EP1-4) for PGE_2. Each receptor mediates different biological activities of PGE_2. For example, EP3 is responsible for PGE_2-induced fever and pain, whereas EP4 mediates the inflammation related bone resorption (Kobayashi and Narumiya, 2002). Pfizer has an EP4 antagonist in phase I trial for osteoarthritis. Amgen is developing AMG-009, a prostaglandin receptor antagonist for asthma.

20.7.3.4 Platelet-Activating Factor (PAF) Receptor. PAF is a potent proinflammatory phospholipid with diverse functions, such as wound healing, angiogenesis, and pathogen clearance (Stafforini et al., 2003). It is produced by a wide variety of cells in the immune system, including platelets, neutrophils, eosinophils, monocytes, and macrophages. PAF not only mediates cell migration, but also induces inflammatory molecule secretion. Pathologically, PAF and its receptor have been implicated in multiple immune disorders such as acute inflammation, asthma, anaphylactic shock, transplantation rejection, skin inflammation and allergy, thrombosis, and so on (Braquet et al., 1987).

A PAF antagonist is being tested in phase I trials for ulcerative colitis and Crohn's disease.

20.7.4 Complement Receptors

C3a and C5a are two anaphylatoxins generated by the complement cascade during infection or tissue injury (Rambach et al., 2008). The pro-inflammatory responses induced by C3a and C5a are mediated by C3aR and C5aR, respectively. These receptors are widely expressed on neutrophils, phagocytes, eosinophils, mast cells, bronchial epithelial cells, vascular endothelial cells, and smooth muscle cells. They participate in multiple cellular responses including cell migration, degranulation, and activation.

C5a has been implicated in multiple immune disorders such as rheumatoid arthritis, inflammatory bowel diseases, sepsis, systemic lupus erythematosus, asthma, and so on. It is up-regulated at inflammatory sites, serving not only as a potent chemoattractant for neutrophils and monocytes but also as an activator for these cells (Guo and Ward, 2005). C5a has been shown to enhance the monocyte and granulocyte phagocytic activity by boosting other complement receptor expression and augmenting oxidative burst (Mollnes et al., 2002). In addition, it has been implicated in modulating macrophage cytokine production such as interleukin-1 (IL-1) and tumor necrosis factor (TNF) (Cavaillon et al., 1990). Multiple C5a inhibitors are being evaluated in clinical trials for RA and psoriasis.

20.7.5 Protease-activated Receptors (PARs)

PARs have a unique activation mechanism: A fragment of the N-terminus of the receptor is cleaved by extracellular serine proteases and the remaining N-terminus serves as a tethered ligand by binding intramolecularly to the receptor.

There are four members in the PAR family (PAR1–4). PAR1, PAR3, and PAR4 are activated by thrombin, whereas PAR2 is activated by a panel of trypsin-like serine proteases, including trypsin, tryptases, mast cell granzymes, neutrophil proteinase 3, and factor VIIa and Xa. Among the PAR family members, PAR2 has demonstrated the most association with the inflammatory responses by being a sensor to tissue injuries (Bunnett, 2006). PAR2 is widely expressed on various cells and tissues in the airway, GI tract, joint, and CNS. It is involved in regulating the epithelium contractility, leukocyte adhesion, neutrophil tissue infiltration, and cytokine secretion. Studies with PAR2 agonists, PAR2 antagonists, and the PAR2 knockout animals point out contradictory roles that PAR2 plays in inflammatory processes (Coelho et al., 2003). It is pro-inflammatory in arthritis, inflammatory bowel diseases, allergic airway, and skin inflammation (Ferrell et al., 2003; Mall et al., 2002; Ebeling et al., 2007; Seeliger et al., 2003), whereas it is protective in colitis, asthma, and ischemia (Fiorucci et al., 2001; Milia et al., 2002; Cocks et al., 1999).

Further studies with potent and selective PAR2 agonists and antagonists will aid better understanding of the role PAR2 plays under various inflammatory conditions.

20.7.6 Other Peptide Receptors

20.7.6.1 Bradykinin Receptors. Two bradykinin receptors (BK1 and BK2) regulate vascular functions and promote local inflammatory responses such as redness, swelling, and pain. They also modulate blood flow and bronchoconstriction. Therefore, antagonizing these receptors may provide therapeutic opportunities for arthritis, sepsis, asthma, allergic rhinitis, and diabetes (Sharma et al., 2003). Sanofi-Aventis is investigating a BK1 antagonist for the treatment of inflammation and pain. This compound inhibited paw edema induced by bradykinin and ear edema induced by capsaicin in animal models. It is currently in phase II clinical trial for chronic pain. In addition, a BK2 antagonist is being developed in phase II trial for trauma head injury.

20.7.6.2 Corticotropin Releasing Factor and Urocortin Receptors. Corticotropin releasing factor (CRF) and urocortins are neuropeptides that mediate stress-related responses through their receptors CRF1 and CRF2. CRF activates CRF1, whereas urocortins activate both CRF1 and CRF2. CRF and urocortins can be synthesized and released by mast cells in the skin in response to IgE receptor crosslinking. In addition, both CRF receptors are up-regulated on the skin mast cells following stress. As a result, the interplay between the CRF peptides and their receptors leads to exacerbation of the inflammatory skin conditions such as psoriasis and atopic dermatitis. Additionally, antagonists to CRF1 and agonists to CRF2 may be used in treating GI inflammation (Gravanis and Margioris, 2005).

20.7.6.3 Endothelin Receptors. The three endothelins (ET1-3) are primarily known to regulate contractility in the vascular system. They exert their activity through two GPCRs, namely, ET_AR and ET_BR. ET-1 is produced by cardiovascular endothelial cells and stimulates vasoconstriction and smooth muscle proliferation (Levin, 1995). It can also be produced by macrophages upon bacterial or LPS challenge and is associated with macrophage functions such as (a) pro-inflammatory cytokine secretion and (b) superoxide and matrix metalloprotease production (Aung et al., 2006). Antagonists to the endothelin receptors have shown anti-inflammatory activity in various preclinical animal models including colitis, asthma, and allergy (Nett et al., 2006).

20.7.6.4 Melanocortin Receptors. The melanocortin receptors (MC1R through MC5R) are activated by four melanocortin peptides—α-, β-, and γ-melanocortin and adrenocorticotropic hormone (ACTH). These peptides are derived from a single peptide called POMC (pro-opiomelanocortin) through post-translational protease cleavage. The expression and maturation of the melanocortin peptides are enhanced by inflammatory cytokines and interferons. The primary melanocortin receptors expressed on leukocytes are MC1R and MC3R. Notably, these receptors are anti-inflammatory rather than pro-inflammatory. Upon activation by melanocortins, they suppress the NF-κB-induced transcription of pro-inflammatory cytokines and chemokines, as well as inhibiting inflammatory cell migration by reducing adhesion molecule expression on these cells. Accordingly, the melanocortin peptides demonstrated a protective role in multiple animal models for human diseases including rheumatoid arthritis, gouty arthritis, inflammatory bowel diseases, sepsis, allergy, organ transplatation, and autoimmune uveoretinitis (Catania et al., 2004).

20.7.6.5 Tachykinin Receptors. The tachykinin family of neuropeptides consists of at least nine members that are widely expressed in the central and periphery nervous system. Two members of this family, substance P and neurokinin A, play important roles in inducing neurogenic inflammation by influencing the vascular functions, including vasodilation and constriction, mucous secretion, and vascular permeability. In addition, substance P has been shown to regulate the production of pro-inflammatory cytokines, lipids, and reactive oxygen species by macrophages (Ho et al., 1996; Murris-Espin et al., 1995). As a result, these neuropeptides have been implicated in the airway, gastrointestinal, and joint immune disorders, such as obstructive airway diseases, chronic ulcerative colitis, and arthritis (Mazzone, 2004; Rijnierse et al., 2006; Keeble and Brain, 2004).

Neurokinins exert their functions through three receptors NK1-3. Several NK_2 receptor antagonists are being evaluated in the clinic for asthma and irritable bowel syndrome (IBS). In addition, a NK_3 antagonist is being developed in phase II trials for IBS and COPD. It is also in phase I trial for respiratory infection.

20.7.7 Other Nonpeptide Receptors

20.7.7.1 Histamine Receptors. Histamine is largely released from mast cells and basophils and plays a vital role in allergic inflammation, such as vasodilation and bronchoconstriction, leukocyte migration, cytokine production, and T-cell differentiation and activation (Thurmond et al., 2008). Increased level of histamine has been detected in the disease tissues of allergy, asthma, MS, and psoriasis. It has been demonstrated that elevated histamine level in asthmatic bronchoalveolar lavage fluid is responsible for airway hypersensitivity and obstruction (Casale et al., 1987; Jarjour et al., 1991). There are four histamine receptors, H_{1-4}. H_1, H_2, and H_4, expressed on many inflammatory cells, including endothelial cells, smooth muscle cells, eosinophils, basophils, mast cells, dendritic cells, monocytes, and T cells. In contrast, H_3 is mainly expressed in the nervous system and inhibits of excitatory neurotransmission in the airway (Ichinose and Barnes, 1989).

H_1 has a well-established role in allergy induction. Many nonsedative H_1 inhibitors are being tested in clinical trials for allergy, hives, and contact dermatitis. In recent years, H_4 has emerged as a promising target for various immune indications. It has more restricted expression on the hematopoietic cells and is capable of mediating chemotaxis of these leukocytes. H_4 may work synergistically with H_1 in mediating the histamine-induced pro-inflammatory responses. It is conceivable that H_4 antagonists are under rigorous investigation as potential anti-inflammatory therapeutics. Finally, based on the evidence that H_3 activation inhibits bronchoconstriction (O'Connor et al., 1993), a H_3 agonist is being evaluated in phase II clinical trial for asthma.

20.7.7.2 Muscarinic Receptors. Muscarinic receptors (M_{1-5}) respond to acetylcholine, the primary parasympathetic neurotransmitter in the airway. Acetylcholine induces smooth muscle contraction and mucous secretion, mainly by activating M_3. In addition to the nerve cells, other cells involved in airway inflammation were recently found to produce acetylcholine as well. These include epithelial and endothelial cells, smooth muscle cells, and leukocyte infiltrates such as lymphocytes, macrophages, neutrophils, eosinophils, and mast cells. These immune cells also express the muscarinic receptors, enabling acetylcholine to act in an autocrine fashion to induce cell proliferation, activation, and cytokine production (Gosens et al., 2006). In addition, non-neuronal origin of acetylcholine has been implicated in airway tissue remodeling during chronic airway inflammation in COPD and asthma (Gosens et al., 2004).

Several M_3 antagonists are under clinical evaluation for the treatment of COPD and asthma either as stand-alone drugs or in combination with β2-adrenoceptor agonists.

20.7.7.3 Adenosine Receptors. Adenosine is an important extracellular signal molecule for multiple systems within the body. It induces sedation through the central nervous system and has an antihypertensive effect in the periphery nervous system and an antidiuretic effect in the kidney.

There are four subtypes of adenosine receptors—A_1, A_{2A}, A_{2B}, and A_3—all of which can be detected in the immune system. They differ in their affinity to adenosine as well as in their intracellular signaling pathways. A_1, A_{2A}, and A_3 are high-affinity adenosine receptors, whereas A_{2B} is a low-affinity receptor. A_1 and A_3 are G_i-coupled, whereas A_{2A} and A_{2B} are G_s-coupled. Given above, it is not surprising that adenosine plays a complex role in the immune system—it is released under various inflammatory conditions and can be pro- and anti-inflammatory depending on its local concentrations, cells, and receptor subtypes involved.

It has been established that the A_1 receptor activation alleviates inflammatory and neuropathic pain (Sawynok, 1998). However, its role in asthma pathogenesis is more convoluted. The A_1 receptor is at least partially responsible for bronchoconstriction in asthma by acting on the airway smooth muscle (Livingston et al., 2004). On the other hand, the A_1 receptor was also shown to mediate the anti-inflammatory effect of adenosine in the lung (Sun et al., 2005). More in-depth studies are ongoing to understand the role of A_1 in modulating pulmonary inflammation.

The A_{2A} receptor activation dampens inflammatory signal during tissue damage (Ohta and Sitkovsky, 2001). Since A_{2A} activation also causes vasodilation and hypotension, thc A_{2A} agonists may have better utility as local agents rather than systematic drugs. Preclinical studies have demonstrated the therapeutic potential of the A_{2A} agonists in treating multiple immune diseases including asthma, COPD, and sepsis (Mohsenin and Blackburn, 2006; Fozard et al., 2002), while the A_{2A} antagonists have shown efficacy in a mouse inflammatory pain model (Bilkei-Gorzo et al., 2008). An A_{2A} receptor agonist in the form of inhaled dry powder is currently in phase II clinical trials for asthma and COPD. Another A_{2A} receptor agonist is in phase I trial for acute inflammation such as sickle cell crisis.

The A_{2B} receptor on airway mast cells participates in allergen-induced mast cell degranulation, contributing to airway inflammation and bronchial hypersensitivity in asthmatic patients (Sun et al., 2006; Fozard and McCarthy, 2002). Thus A_{2B} antagonists may be beneficial in treating the airway inflammatory diseases. Several A_{2B} antagonists are being tested in clinical trials for asthma, cardiopulmonary diseases, and IBS.

The A_3 receptor is the most enigmatic adenosine receptor subtype. It has demonstrated both pro- and anti-inflammatory activity. For airway diseases such as asthma and allergy, A_3 plays a pro-inflammatory role. It is up-regulated during airway inflammation and its activation leads to lung mast cell degranulation, increased eosinophil lung infiltration, and prolonged inflammatory cell survival. A_3 antagonists are being explored preclinically and clinically as anti-asthmatic and anti-allergic therapeutics (Gessi et al., 2008).

On the other hand, literature also pointed out the anti-inflammatory role for A_3. In both human and murine macrophages, A_3 inhibits LPS-induced TNF-α production. As expected, A_3 agonists were effective in preclinical models for sepsis, colitis, and reperfusion lung injury (Gessi et al., 2008). Moreover, exciting preclinical and clinical data have proved the efficacy of A_3 agonists in treating rheumatoid arthritis (Silverman et al., 2008). CF-101 is an oral A_3 receptor agonist that suppresses TNF-α. It is under development in phase II trial for RA and psoriasis. It is also being investigated as a potential therapeutics for IBD.

20.7.7.4 Purinergic Receptors. Purinergic receptors respond to extracellular nucleotides. These receptors are classified into the ionotropic P2X receptors and the metabotropic P2Y receptors. The seven members of the P2X family ($P2X_{1-7}$) are all ATP-gated nonselective cation channels (North, 2002). These receptors are beyond the scope of this chapter and thus will not be discussed here. The P2Y family consists of eight GPCRs: $P2Y_1$, $P2Y_2$, $P2Y_4$, $P2Y_6$, $P2Y_{11}$, $P2Y_{12}$, $P2Y_{13}$, and $P2Y_{14}$. Depending on the subtype, their high-affinity ligand repertoire includes ATP and ADP ($P2Y_1$, $P2Y_{11}$, $P2Y_{12}$, and $P2Y_{13}$), UTP and UDP ($P2Y_2$, $P2Y_4$, and $P2Y_6$), and UDP-glucose ($P2Y_{14}$). The P2Y receptors are widely expressed on leukocytes, and often multiple subtypes are present on the same cells and in the same tissues. These receptors have subtype-dependent G-protein coupling and subsequently distinct intracellular signaling cascades. They are also likely to form homo- and heterodimers and oligomers under certain conditions, further adding complexity of their biological functions.

Nucleotides are important extracellular signaling molecules. ATP is one of the most important danger signals that sense local tissue damage and cell activation. For instance, normally the pericellular ATP concentration is in the nanomolar range, whereas intracellular ATP is in the millimolar range. However, extracellular ATP can dramatically increase during cell activation and cell death at the inflammatory sites.

Not surprisingly, the P2Y purinoceptors participate in a host of immune responses such as cell migration, differentiation, survival, and activation. For example, $P2Y_2$ and $P2Y_4$ likely mediate the ATP- and UTP-induced mucous secretion in the airway (Burnstock, 2004; Cressman et al., 1999). $P2Y_{11}$ is selectively activated by ATP and participates in dendritic cell maturation (Wilkin et al., 2001). $P2Y_{13}$ has been proposed to play a role in regulating the immune responses due to its predominant expression in the spleen, lymph nodes, bone marrow, and brain (Communi et al., 2001).

Among the P2Y receptors, $P2Y_6$ sparked the most interest in the inflammatory disease arena. It is expressed in the spleen and thymus and on blood leukocytes (Communi et al., 1996). The highest affinity ligand for $P2Y_6$ is UDP. In the innate immune system, $P2Y_6$ is involved in the IL-8 production by monocytes upon UDP and LPS challenge (Warny et al., 2001). In the adaptive immune system, $P2Y_6$ is up-regulated on activated circulating T cells but not resting T cells. It is also highly expressed on the bowel-infiltrating T cells in the diseased but not the healthy tissues (Somers et al., 1998). Given this, $P2Y_6$ may participate in causing tissue damage in IBD. Indeed, $P2Y_6$ was recently incriminated in stress-associated intestinal inflammation (Grbic et al., 2008).

20.7.7.5 Adrenergic Receptors. The adrenergic receptors mediate the "fight-or-flight" response to the neurotransmitters epinephrine and norepinephrine in the sympathetic nervous system. They are further classified into the α- and β-adrenoceptor subtypes, with each subtype controlling different aspect of the sympathetic responses. For example, the α-adrenoceptor activation causes vasoconstriction and intestinal relaxation, whereas the β-adrenoceptor activation leads to bronchial dilation and vasodilation.

The adrenergic receptors also take part in various immune responses. In IBS, a high level of α2-adrenergic receptor was detected on the monocyte infiltrate, contributing to local intestinal inflammation (Blandizzi, 2007). In addition, aberrant β-adrenoceptor function has been implicated in airway hypersensivity in bronchial asthma (Townley, 2007). Furthermore, β-adrenoceptors on dendritic cells have been shown to mediate the inhibitory effect of norepinerphrine on dendritic cell migration and activation during pathogen challenge, demonstrating the role of β-adrenoceptors in the innate immunity (Maestroni et al., 2005).

Currently, marked β-agonists are short-acting and mostly used as rescue drugs. Multiple long-lasting β2-adrenoceptor agonists in the inhaled form, either stand-alone or in combination with corticosteroid, are in various phases of clinical trials for the treatment of COPD and asthma. Nadolol, an established antihypertensive drug that is an antagonist for the β1 receptor and an inverse agonist for the β2 receptor, is being tested as an antiasthmatic drug for patients who are nonresponsive to β-agonist treatment. In addition, multiple β3-adrenoceptor agonists are being investigated in the clinic for the treatment of IBS.

20.7.7.6 Cannabinoid Receptor CB2. The cannabinoids are anti-inflammatory agents that modulate lymphocyte differentiation, lymphocyte and macrophage migration, and cytokine production (Klein et al., 2003; Raborn et al., 2008). The cannabinoid receptors are CB1 and CB2. CB1 is mainly expressed in the brain, whereas CB2 is expressed primarily on cells with the hematopoietic origin, including B cells, NK cells, monocytes, neutrophils, and T cells. The levels of CB2 on lymphocytes and macrophages are regulated based on the stages of cell differentiation and activation. Studies with CB2 knockout mice have indicated that CB2 activation could be beneficial in treating multiple disorders such as multiple sclerosis, allergic dermatitis, atherosclerosis, osteoporosis, and pain (Buckley, 2008). CB2 activation has also been implicated in suppressing intestinal, pulmonary, and arthritic inflammation (Wright et al., 2008; Ashton, 2007).

A selective CB2 agonist Cannabinor has shown efficacy in various animal models for immune disorders including RA, MS, and IBD. Several cannabinoid derivatives are being investigated preclinically and clinically for the treatment of RA, IBD, MS, and asthma.

20.8 GPCRs AS DRUG TARGETS IN OBESITY

Obesity has become a global epidemic in the last two decades. Being overweight increases the risk for numerous diseases such as diabetes, hypertension, dyslipidemia, coronary heart disease, stroke, osteoarthritis, and so on.

Controlling obesity will most likely help resolve numerous public health issues. However, developing efficacious and safe anti-obesity drugs remains a challenging task for the biotech and pharmaceutical industry.

GPCRs have emerged as an important class of molecules involved in energy homeostasis. Most of these GPCRs are located in the hypothalamus in the brain, the control center for food intake and energy expenditure. In this section, several examples will be discussed.

20.8.1 Melanocortin Receptors

As mentioned previously, melanocortins are four peptides derived from the POMC peptide: α-, β-, and γ-melanocortin and ACTH. The melanocortin system is critically linked to feeding and body weight regulation, because POMC deficiency led to morbid obesity (Coll et al., 2004). As predicted, intracerebroventricular (i.c.v.) infusion of α-MSH caused reduced food intake and body weight in rodent (McMinn et al., 2000). Interestingly, two endogenous peptides can antagonize the action of α-MSH: agouti and agouti-related peptide (AGRP). Overexpression of either antagonist peptides causes obesity, while deficiency of these peptides results in increased metabolic rate and longer life span for mice fed with high-fat diet (Stutz et al., 2005; Ilnytska and Argyropoulos, 2008).

Out of the five melanocortin receptors, MC2R specifically interacts with ATCH, while the remaining four receptors are activated by α-MSH. MC4R and MC3R have been linked to body weight regulation. MC4R is mainly expressed in the brain, and its deficiency resulted in severe obesity in both human and mouse (Huszar et al., 1997; Farooqi, 2006). MC3R is expressed both centrally and peripherally. The MC3R knockout mice are obese but not hyperphagic as are the MC4R knockout mice. MC3R is therefore most likely involved in regulating nutrient partitioning and energy expenditure (Butler and Cone, 2002).

Selective MC4R agonists are under clinical evaluation for obesity, type II diabetes, and related metabolic syndrome. In addition, Contrave, a combination of opioid antagonist naltrexone and the dopamine agonist bupropion, is being investigated in a phase III trial for obesity. Contrave can stimulate the α-MSH release by the POMC neurons and can presumably achieve weight loss by boosting MC4R activity.

20.8.2 Melanin Concentrating Hormone Receptors

Melanin concentrating hormone (MCH) is a 19-amino-acid cyclic peptide. It is primarily produced by neurons in the brain and is one of the key hypothalamic orexigenic neuropeptides. Acute i.c.v. injection of MCH peptide increased food intake in rats, although chronic injection led to tolerance to the peptide (Kela et al., 2003; Rossi et al., 1997). As expected, MCH knockout mice are hypophagic with higher metabolic rate and less body fat (Shimada et al., 1998).

There are two MCH receptors, MCH_1 and MCH_2, both of which are primarily expressed in the brain. Most of the drug discovery effort has been directed toward MCH_1 because MCH_2 does not have a functional counterpart in rodent, making it difficult to evaluate the preclinical activities of drug candidates in the rodent obesity models. The MCH_1 knockout mice are leaner and hyperphagic with normal body weight, while the MCH ligand knockout mice had the hypophagic phenotype. In addition, the MCH_1 knockout mice also have remarkably higher locomotor activity and altered metabolic rate (Marsh et al., 2002). Several MCH_1 receptor antagonists are being tested as anti-obesity therapeutics in phase I clinical trials.

20.8.3 Cannabinoid Receptor CB1

In addition to their role in the inflammatory system as discussed above, the endocannabinoids also play a vital role in feeding regulation. While CB2 receptor mediates their anti-inflammatory activities, CB1 is responsible for the appetite and body weight regulation.

CB1 is expressed in the central and peripheral nervous system, as well as in some non-neuronal peripheral tissues. Consistent with the ability of cannabinoids to induce spontaneous compulsive eating, blocking CB1 activity through either genetic knockout or receptor antagonism not only reduces body weight, suppresses appetite, and reduces adiposity, but also improves metabolic parameters including insulin, glucose, and triglyceride levels (Kunos, 2007). Therefore antagonizing CB1 becomes a lucrative strategy in treating obesity and the related metabolic syndromes.

The most advanced CB1 antagonist developed is Rimonabant from Sanofi-Aventis. It is also an inverse agonist for CB1. Rimonabant was approved in Europe in 2006 as an anorectic anti-obesity drug. Two additional CB1 antagonists from AstraZeneca are under clinical evaluation for obesity, despite recent termination of clinical development for several compounds due to safety concerns. In addition, GW Pharmaceuticals is developing Tetrahydrocannabivarin (THCV), a naturally occurring cannabinoid receptor antagonist that inhibits both CB1 and CB2, in a phase I clinical trial for obesity.

20.8.4 Ghrelin Receptor

Ghrelin is a hormone synthesized in the stomach that induces appetite and stimulates gastric emptying (Masuda et al., 2000). Its level is high during fasting and drops after a meal; therefore it is referred as a "hunger" hormone.

As expected, chronic ghrelin administration induces obesity (Wren et al., 2001). Interestingly, the ghrelin level has an inverse relationship with body mass index: It is lower in obese people than in lean people. Consequently, people with weight loss have an increased sense of hunger due to a higher ghrelin level in their blood (Shiiya et al., 2002).

The only obesity treatment that has produced permanent weight loss result is the gastrointestinal bypass surgery. This surgery removes part of the stomach and small intestine for the extremely obese patients. People lose weight by eating less, feeling full more quickly, and absorbing less food. Plasma ghrelin levels in postsurgery patients are dramatically lower, which may at least partly contribute to the long-lasting appetite reduction in these patients (Cummings et al., 2002).

The receptor for ghrelin is the growth hormone secretagogue receptor. It is expressed in the hypothalamus and the arcuate nucleus in the brain. The ghreline receptor is part of the network that links the gut nutriment sensors and the brain appetite control centers to regulate appetite before and after meals. As expected, ghrelin receptor antagonists reduced food intake and body weight and improved metabolic parameters in mice (Asakawa et al., 2003).

20.8.5 Other Peptide Receptors

20.8.5.1 Neuropeptide Y (NPY) Receptors. NPY is distributed in the brain and some of the non-neuronal tissues. It is an orexigenic peptide that regulates feeding based on nutrition status and the presence of other drugs such as nicotine and cannabis. The hypothalamic NPY level is higher in genetically obese animals and it rises with food deprivation. However, it decreases with a long-term high-fat diet (Bina and Cincotta, 2000; Beck et al., 2001; Hansen et al., 2004). Chronic NPY administration has been shown to increase body weight and adiposity in rats (Zarjevski et al., 1993).

There are five NPY receptors, four of which have been associated with appetite regulation: the Y1, Y2, Y4, and Y5 receptors. These receptors are widely distributed in the brain and some peripheral tissues. Considerable effort has been put into understanding the complex roles that these receptors play in energy homeostasis using ligand and receptor knockout animals and receptor agonists and antagonists (Kamiji and Inui, 2007).

Another peptide in the same peptide family as NPY is peptide YY (PYY). PYY is expressed in the small intestine, with high expression in ileum, colon, and rectum (Adrian et al., 1985). It is a gut hormone that is released into the circulation after a meal. Gut hormones are peptides whose expression is modulated by ingested nutrients during and after meals. Their receptors are located in the feeding and satiety center in the brain. Gut hormones function as the feedback mediators from the gastrointestinal tract to influence appetite. The circulating form of PYY is PYY_{3-36}, a high-affinity ligand for the

Y2 receptor. PYY_{3-36} administration reduces food intake in both human and rodent (Halatchev et al., 2004; Batterham et al., 2002). It does so by acting centrally on neurons to induce satiety and peripherally on gut to decrease motility.

Currently, three NPY receptor modulators are under clinical evaluation for obesity, including a dual Y2/Y4 agonist that is a synthetic peptide analogue administered s.c., an intranasal formulation of PYY_{3-36}, and an oral Y5 antagonist.

20.8.5.2 Orexin Receptors. The orexins (orexin-A and -B, also known as hypocretin-1 and -2) are peptides in the nervous system. They are primarily expressed in the brain and function to increase food intake and regulate sleep/arousal cycles. Peripheral orexin was detected in neurons in the gut and it facilitates motility based on food status (Kirchgessner and Liu, 1999). Orexins are part of the brain–gut axis of energy homeostasis control.

There are two orexin receptors, OX_1R and OX_2R, expressed in distinct regions in the CNS. Similar to their ligands, these receptors have also been detected in the peripheral neurons, such as the ones in pituitary, small intestine, and pancreas (Kirchgessner, 2002). The OX_1R antagonists reduces food intake in ob/ob mice (Porter et al., 2001). However, no OX_1R antagonist is in the clinic for obesity, likely due to its liability in disrupting sleep and wakefulness states.

20.8.5.3 Opiate Receptors. The opiate peptides increase short-term food intake. They do so by modulating the reward pathway centrally and regulating gastrointestinal motility peripherally. There are four opiate receptors, μ, δ, κ, and ORL1, out of which μ and κ have been associated with obesity (Zhang et al., 2006; Jarosz, 2007). Elevated level of μ receptor was found in rats susceptible to diet-induced obesity (Barnes et al., 2006). Activation of μ and κ receptors led to over eating, whereas antagonists to these receptors such as naloxone and naltrexone reduce food intake (Yeomans and Gray, 2002). As mentioned above, a combination of opioid antagonist naltrexone and the dopamine agonist bupropion is under clinical evaluation in phase III obesity trials.

20.8.6 Nonpeptide Receptors

20.8.6.1 Serotonin Receptors. Serotonin (5-HT) is believed to have an inhibitory effect on food intake and weight gain by enhancing satiety. 5-HT injected into the paraventricular nucleus decreases food intake in rats (Leibowitz et al., 1989). This anorectic effect of serotonin is mediated mainly through two serotonin receptor subtypes, $5\text{-}HT_{1B}$ and $5\text{-}HT_{2C}$. The $5\text{-}HT_{2C}$

knockout mice had increased appetite and body weight (Tecott et al., 1995), whereas the 5-HT_{1B} knockout mice had higher food and water intake and higher body weight without signs of obesity (Bouwknecht et al., 2001).

The importance of the serotonergic system in body weight control can be highlighted by the anorectic effect of sibutramine, one of the anti-obesity drugs currently on the market. Sibutramine inhibits the reuptake of serotonin, norepinephrine, and dopamine, leading to elevated local concentration of these neurotransmitters. Another compound that has demonstrated anorectic efficacy in humans is fenfluramine. This compound boosts synaptic 5-HT levels by promoting its efflux from the neurons. However, fenfluramine was withdrawn from the market due to side effects including heart valve hypertrophy and primary pulmonary hypertension, most likely due to increased serotonin signal through other 5-HT receptor subtypes expressed in the cardiovascular system. Since 5-HT_{1B} is expressed in the cardiovascular system, 5-HT_{2C} is conceivably a more viable obesity target.

Several 5-HT_{2C} agonists are undergoing clinical evaluation. The most advanced compound is Lorcaserin from Arena. It is an oral drug being tested in phase III trial for obesity. Another 5-HT_{2C} agonist ATHX-105 from Athersys is in phase I trial for obesity. Wyeth also has a 5-HT_{2C} agonist Vabicaserin. It is currently in a phase II trial for schizophrenia and under preclinical evaluation for obesity.

In recent years, selective 5-HT_6 antagonists have demonstrated their potential as highly efficacious and safer therapeutics for obesity treatment (Heal et al., 2008). 5-HT_6 is expressed exclusively in the brain and is involved in the regulation of satiety through the reward system. The 5-HT_6 knockout mice were resistant to diet-induced obesity by reducing food intake; therefore, 5-HT_6 antagonism may be a viable option as appetite suppressant (Heal et al., 2008). Its exclusively central location may reduce the likelihood of peripheral side effects. Studies are ongoing to further unravel the detailed biological pathway linking 5-HT_6 to obesity. Two 5-HT_6 antagonists are being tested as anti-obesity drugs in phase I clinical trials: BVT-74316 from Biovitrum and PRX-07034 from EPIX Pharmaceuticals.

20.8.6.2 β3-Adrenergic Receptor. β3-adrenoceptor is expressed exclusively in the adipose tissues at high levels and mediates lipolysis and thermogenesis. In white adipose tissue, β3-adrenoceptor activation leads to hydrolysis of triglyceride, resulting in less stored fat mass. In brown adipose tissue, β3-adrenoceptor participates in metabolic thermogenesis by stimulating fatty acid oxidation (Galitzky et al., 1993). Therefore agonists to β3-adrenoceptor are postulated to have anti-obesity effect by increasing energy expenditure. Indeed, β3 agonists have demonstrated their efficacy in reducing body fat in mouse, rat, and dog obesity models (Ghorbani et al., 1997; Nagase et al., 1996; Omachi et al., 2007).

Nisshin Kyorin is developing a β3 agonist in phase II trials for obesity and diabetes. Several other β3 agonists are in clinical development for type II diabetes, overactive bladder, and IBS.

20.9 CONCLUSIONS AND PERSPECTIVES

After decades of extensive exploration on GPCR biology, pharmacology, and drug discovery, GPCR research remains a fascinating field today. As one of the most important cell surface molecules, GPCRs play intricate roles in regulating cellular responses to their microenvironment. Therefore, modulating GPCR activity presents itself as an enticing opportunity for therapeutic interventions. Indeed, there is a proven record of GPCR modulators as successful drugs.

We are gaining new knowledge daily on the contribution of GPCRs to the pathogenesis of various diseases. A large effort is being made in an attempt to match orphan receptors with their physiological ligands. More and more new assay technologies are becoming available for GPCR research. A great deal of progress is being made to further understand new GPCR modulation mechanisms such as receptor oligomerization and allosterism. With much attention from scientists in both academia and industry, GPCRs remain strong candidates for future drug development. We believe that new GPCR modulators will continue as prolific therapeutics for disease intervention.

REFERENCES

Adrian TE, Ferri GL, Bacarese-Hamilton AJ, Fuessl HS, Polak JM, Bloom SR. Human distribution and release of a putative new gut hormone, peptide YY. *Gastroenterology* 1985;89(5):1070–1077.

Akaogi J, Nozaki T, Satoh M, Yamada H. Role of PGE2 and EP receptors in the pathogenesis of rheumatoid arthritis and as a novel therapeutic strategy. *Endocr Metab Immune Disord Drug Targets* 2006;6(4):383–394.

Allen SJ, Crown SE, Handel TM. Chemokine: receptor structure, interactions, and antagonism. *Annu Rev Immunol* 2007;25:787–820.

Arbour N, Tremblay P, Oth D. *N*-Formyl-methionyl-leucyl-phenylalanine induces and modulates IL-1 and IL-6 in human PBMC. *Cytokine* 1996;8(6):468–475.

Asakawa A, Inui A, Kaga T, Katsuura G, Fujimiya M, Fujino MA, Kasuga M. Antagonism of ghrelin receptor reduces food intake and body weight gain in mice. *Gut* 2003;52(7):947–952.

Ashton JC. Cannabinoids for the treatment of inflammation. *Curr Opin Investig Drugs* 2007;8(5):373–384.

Aung HT, Schroder K, Himes SR, Brion K, van Zuylen W, Trieu A, Suzuki H, Hayashizaki Y, Hume DA, Sweet MJ, Ravasi T. LPS regulates proinflammatory gene expression in macrophages by altering histone deacetylase expression. *FASEB J* 2006;20(9):1315–1327.

Back M, Hansson GK. Leukotriene receptors in atherosclerosis. *Ann Med* 2006; 38(7):493–502.

Bansal G, Druey KM, Xie Z. R4 RGS proteins: regulation of G-protein signaling and beyond. *Pharmacol Ther* 2007;116(3):473–495.

Baraf HS. Efficacy of the newest COX-2 selective inhibitors in rheumatic disease. *Curr Pharm Des* 2007;13(22):2228–2236.

Barnes MJ, Holmes G, Primeaux SD, York DA, Bray GA. Increased expression of mu opioid receptors in animals susceptible to diet-induced obesity. *Peptides* 2006;27(12):3292–3298.

Batterham RL, Cowley MA, Small CJ, Herzog H, Cohen MA, Dakin CL, Wren AM, Brynes AE, Low MJ, Ghatei MA, Cone RD, Bloom SR. Gut hormone PYY(3-36) physiologically inhibits food intake. *Nature* 2002;418(6898):650–654.

Beck B, Richy S, Dimitrov T, Stricker-Krongrad A. Opposite regulation of hypothalamic orexin and neuropeptide Y receptors and peptide expressions in obese Zucker rats. *Biochem Biophys Res Commun* 2001;286(3):518–523.

Bilkei-Gorzo A, Abo-Salem OM, Hayallah AM, Michel K, Muller CE, Zimmer A. Adenosine receptor subtype-selective antagonists in inflammation and hyperalgesia. *Naunyn Schmiedebergs Arch Pharmacol* 2008;377(1):65–76.

Bina KG, Cincotta AH. Dopaminergic agonists normalize elevated hypothalamic neuropeptide Y and corticotropin-releasing hormone, body weight gain, and hyperglycemia in ob/ob mice. *Neuroendocrinology* 2000;71(1):68–78.

Blandizzi C. Enteric alpha-2 adrenoceptors: pathophysiological implications in functional and inflammatory bowel disorders. *Neurochem Int* 2007;51(5):282–288.

Bourne HR. How receptors talk to trimeric G proteins. *Curr Opin Cell Biol* 1997; 9(2):134–142.

Bouwknecht JA, van der Gugten J, Hijzen TH, Maes RA, Hen R, Olivier B. Male and female 5-HT(1B) receptor knockout mice have higher body weights than wildtypes. *Physiol Behav* 2001;74(4–5):507–516.

Braquet P, Touqui L, Shen TY, Vargaftig BB. Perspectives in platelet-activating factor research. *Pharmacol Rev* 1987;39(2):97–145.

Buckley NE. The peripheral cannabinoid receptor knockout mice: an update. *Br J Pharmacol* 2008;153(2):309–318.

Bunnett NW. Protease-activated receptors: how proteases signal to cells to cause inflammation and pain. *Semin Thromb Hemost* 2006;32(Suppl 1):39–48.

Burnstock G. Introduction: P2 receptors. *Curr Top Med Chem* 2004;4(8):793–803.

Butler AA, Cone RD. The melanocortin receptors: lessons from knockout models. *Neuropeptides* 2002;36(2–3):77–84.

Casale TB, Wood D, Richerson HB, Trapp S, Metzger WJ, Zavala D, Hunninghake GW. Elevated bronchoalveolar lavage fluid histamine levels in allergic asthmatics are associated with methacholine bronchial hyperresponsiveness. *J Clin Invest* 1987;79(4):1197–1203.

Catania A, Gatti S, Colombo G, Lipton JM. Targeting melanocortin receptors as a novel strategy to control inflammation. *Pharmacol Rev* 2004;56(1):1–29.

Cavaillon JM, Fitting C, Haeffner-Cavaillon N. Recombinant C5a enhances interleukin 1 and tumor necrosis factor release by lipopolysaccharide-stimulated monocytes and macrophages. *Eur J Immunol* 1990;20(2):253–257.

Charo IF, Peters W. Chemokine receptor 2 (CCR2) in atherosclerosis, infectious diseases, and regulation of T-cell polarization. *Microcirculation* 2003;10(3–4): 259–264.

Chiba K, Matsuyuki H, Maeda Y, Sugahara K. Role of sphingosine 1-phosphate receptor type 1 in lymphocyte egress from secondary lymphoid tissues and thymus. *Cell Mol Immunol* 2006;3(1):11–19.

Chung S, Funakoshi T, Civelli O. Orphan GPCR research. *Br J Pharmacol* 2008; 153(Suppl 1):S339–S346.

Cocks TM, Fong B, Chow JM, Anderson GP, Frauman AG, Goldie RG, Henry PJ, Carr MJ, Hamilton JR, Moffatt JD. A protective role for protease-activated receptors in the airways. *Nature* 1999;398(6723):156–160.

Coelho AM, Ossovskaya V, Bunnett NW. Proteinase-activated receptor-2: physiological and pathophysiological roles. *Curr Med Chem Cardiovasc Hematol Agents* 2003; 1(1):61–72.

Coll AP, Farooqi IS, Challis BG, Yeo GS, O'Rahilly S. Proopiomelanocortin and energy balance: insights from human and murine genetics. *J Clin Endocrinol Metab* 2004;89(6):2557–2562.

Communi D, Parmentier M, Boeynaems JM. Cloning, functional expression and tissue distribution of the human P2Y6 receptor. *Biochem Biophys Res Commun* 1996;222(2):303–308.

Communi D, Gonzalez NS, Detheux M, Brezillon S, Lannoy V, Parmentier M, Boeynaems JM. Identification of a novel human ADP receptor coupled to G(i). *J Biol Chem* 2001;276(44):41479–41485.

Copeland KF. Inhibition of HIV-1 entry into cells. *Recent Patents Anti-Infect Drug Disc* 2006;1(1):107–112.

Cressman VL, Lazarowski E, Homolya L, Boucher RC, Koller BH, Grubb BR. Effect of loss of P2Y(2) receptor gene expression on nucleotide regulation of murine epithelial Cl(−) transport. *J Biol Chem* 1999;274(37):26461–26468.

Cummings DE, Weigle DS, Frayo RS, Breen PA, Ma MK, Dellinger EP, Purnell JQ. Plasma ghrelin levels after diet-induced weight loss or gastric bypass surgery. *N Engl J Med* 2002;346(21):1623–1630.

Daaka Y, Luttrell LM, Lefkowitz RJ. Switching of the coupling of the beta2-adrenergic receptor to different G proteins by protein kinase A. *Nature* 1997;390(6655): 88–91.

Deupi X, Dolker N, Lopez-Rodriguez ML, Campillo M, Ballesteros JA, Pardo L. Structural models of class a G protein-coupled receptors as a tool for drug design: insights on transmembrane bundle plasticity. *Curr Top Med Chem* 2007;7(10): 991–998.

DeWire SM, Ahn S, Lefkowitz RJ, Shenoy SK. Beta-arrestins and cell signaling. *Annu Rev Physiol* 2007;69:483–510.

Dhanasekaran N, Dermott JM. Signaling by the G12 class of G proteins. *Cell Signal* 1996;8(4):235–245.

Duvernay MT, Filipeanu CM, Wu G. The regulatory mechanisms of export trafficking of G protein-coupled receptors. *Cell Signal* 2005;17(12):1457–1465.

Ebeling C, Lam T, Gordon JR, Hollenberg MD, Vliagoftis H. Proteinase-activated receptor-2 promotes allergic sensitization to an inhaled antigen through a TNF-mediated pathway. *J Immunol* 2007;179(5):2910–2917.

Ellis J, Pediani JD, Canals M, Milasta S, Milligan G. Orexin-1 receptor-cannabinoid CB1 receptor heterodimerization results in both ligand-dependent and -independent coordinated alterations of receptor localization and function. *J Biol Chem* 2006;281(50):38812–38824.

Farooqi IS. Monogenic human obesity syndromes. *Prog Brain Res* 2006;153:119–125.

Ferguson SS. Phosphorylation-independent attenuation of GPCR signalling. *Trends Pharmacol Sci* 2007;28(4):173–179.

Ferrell WR, Lockhart JC, Kelso EB, Dunning L, Plevin R, Meek SE, Smith AJ, Hunter GD, McLean JS, McGarry F, Ramage R, Jiang L, Kanke T, Kawagoe J. Essential role for proteinase-activated receptor-2 in arthritis. *J Clin Invest* 2003;111(1):35–41.

Fiorucci S, Mencarelli A, Palazzetti B, Distrutti E, Vergnolle N, Hollenberg MD, Wallace JL, Morelli A, Cirino G. Proteinase-activated receptor 2 is an anti-inflammatory signal for colonic lamina propria lymphocytes in a mouse model of colitis. *Proc Natl Acad Sci USA* 2001;98(24):13936–13941.

Fozard JR, McCarthy C. Adenosine receptor ligands as potential therapeutics in asthma. *Curr Opin Investig Drugs* 2002;3(1):69–77.

Fozard JR, Ellis KM, Villela Dantas MF, Tigani B, Mazzoni L. Effects of CGS 21680, a selective adenosine A2A receptor agonist, on allergic airways inflammation in the rat. *Eur J Pharmacol* 2002;438(3):183–188.

Franco R, Casado V, Cortes A, Ferrada C, Mallol J, Woods A, Lluis C, Canela EI, Ferre S. Basic concepts in G-protein-coupled receptor homo- and heterodimerization. *Sci World J* 2007;7:48–57.

Gabriel D, Vernier M, Pfeifer MJ, Dasen B, Tenaillon L, Bouhelal R. High throughput screening technologies for direct cyclic AMP measurement. *Assay Drug Dev Technol* 2003;1(2):291–303.

Gaillard I, Rouquier S, Giorgi D. Olfactory receptors. *Cell Mol Life Sci* 2004;61(4):456–469.

Galitzky J, Reverte M, Carpene C, Lafontan M, Berlan M. Beta 3-adrenoceptors in dog adipose tissue: studies on their involvement in the lipomobilizing effect of catecholamines. *J Pharmacol Exp Ther* 1993;266(1):358–366.

Gessi S, Merighi S, Varani K, Leung E, Mac Lennan S, Borea PA. The A3 adenosine receptor: an enigmatic player in cell biology. *Pharmacol Ther* 2008;117(1):123–140.

Ghorbani M, Claus TH, Himms-Hagen J. Hypertrophy of brown adipocytes in brown and white adipose tissues and reversal of diet-induced obesity in rats treated with a beta3-adrenoceptor agonist. *Biochem Pharmacol* 1997;54(1):121–131.

Gladue RP, Zwillich SH, Clucas AT, Brown MF. CCR1 antagonists for the treatment of autoimmune diseases. *Curr Opin Investig Drugs* 2004;5(5):499–504.

Goncharova LB, Tarakanov AO. Why chemokines are cytokines while their receptors are not cytokine ones? *Curr Med Chem* 2008;15(13):1297–1304.

Gosens R, Zaagsma J, Grootte Bromhaar M, Nelemans A, Meurs H. Acetylcholine: a novel regulator of airway smooth muscle remodelling? *Eur J Pharmacol* 2004; 500(1–3):193–201.

Gosens R, Zaagsma J, Meurs H, Halayko AJ. Muscarinic receptor signaling in the pathophysiology of asthma and COPD. *Respir Res* 2006;7:73.

Gravanis A, Margioris AN. The corticotropin-releasing factor (CRF) family of neuropeptides in inflammation: potential therapeutic applications. *Curr Med Chem* 2005;12(13):1503–1512.

Grbic DM, Degagne E, Langlois C, Dupuis AA, Gendron FP. Intestinal inflammation increases the expression of the P2Y6 receptor on epithelial cells and the release of CXC chemokine ligand 8 by UDP. *J Immunol* 2008;180(4):2659–2668.

Guo RF, Ward PA. Role of C5a in inflammatory responses. *Annu Rev Immunol* 2005;23:821–852.

Halatchev IG, Ellacott KL, Fan W, Cone RD. Peptide YY3–36 inhibits food intake in mice through a melanocortin-4 receptor-independent mechanism. *Endocrinology* 2004;145(6):2585–2590.

Hansen MJ, Jovanovska V, Morris MJ. Adaptive responses in hypothalamic neuropeptide Y in the face of prolonged high-fat feeding in the rat. *J Neurochem* 2004;88(4):909–916.

Heal DJ, Smith SL, Fisas A, Codony X, Buschmann H. Selective 5-HT6 receptor ligands: progress in the development of a novel pharmacological approach to the treatment of obesity and related metabolic disorders. *Pharmacol Ther* 2008; 117(2):207–231.

Ho WZ, Kaufman D, Uvaydova M, Douglas SD. Substance P augments interleukin-10 and tumor necrosis factor-alpha release by human cord blood monocytes and macrophages. *J Neuroimmunol* 1996;71(1–2):73–80.

Holenz J, Merce R, Diaz JL, Guitart X, Codony X, Dordal A, Romero G, Torrens A, Mas J, Andaluz B, Hernandez S, Monroy X, Sanchez E, Hernandez E, Perez R, Cubi R, Sanfeliu O, Buschmann H. Medicinal chemistry driven approaches toward novel and selective serotonin 5-HT6 receptor ligands. *J Med Chem* 2005;48(6): 1781–1795.

Horga A, Montalban X. FTY720 (fingolimod) for relapsing multiple sclerosis. *Expert Rev Neurother* 2008;8(5):699–714.

Huszar D, Lynch CA, Fairchild-Huntress V, Dunmore JH, Fang Q, Berkemeier LR, Gu W, Kesterson RA, Boston BA, Cone RD, Smith FJ, Campfield LA, Burn P, Lee F. Targeted disruption of the melanocortin-4 receptor results in obesity in mice. *Cell* 1997;88(1):131–141.

Ichinose M, Barnes PJ. Inhibitory histamine H3-receptors on cholinergic nerves in human airways. *Eur J Pharmacol* 1989;163(2–3):383–386.

Ilnytska O, Argyropoulos G. The role of the agouti-related protein in energy balance regulation. *Cell Mol Life Sci* 2008;65(17):2721–2731.

Jakubik J, Bacakova L, el-Fakahany EE, Tucek S. Subtype selectivity of the positive allosteric action of alcuronium at cloned M1-M5 muscarinic acetylcholine receptors. *J Pharmacol Exp Ther* 1995;274(3):1077–1083.

Jakubik J, Bacakova L, El-Fakahany EE, Tucek S. Positive cooperativity of acetylcholine and other agonists with allosteric ligands on muscarinic acetylcholine receptors. *Mol Pharmacol* 1997;52(1):172–179.

Jarjour NN, Calhoun WJ, Schwartz LB, Busse WW. Elevated bronchoalveolar lavage fluid histamine levels in allergic asthmatics are associated with increased airway obstruction. *Am Rev Respir Dis* 1991;144(1):83–87.

Jarosz PA. The effect of kappa opioid receptor antagonism on energy expenditure in the obese Zucker rat. *Biol Res Nurs* 2007;8(4):294–299.

Johansson-Lindbom B, Agace WW. Generation of gut-homing T cells and their localization to the small intestinal mucosa. *Immunol Rev* 2007;215:226–242.

Jolly PS, Bektas M, Olivera A, Gonzalez-Espinosa C, Proia RL, Rivera J, Milstien S, Spiegel S. Transactivation of sphingosine-1-phosphate receptors by FcepsilonRI triggering is required for normal mast cell degranulation and chemotaxis. *J Exp Med* 2004;199(7):959–970.

Kamiji MM, Inui A. Neuropeptide y receptor selective ligands in the treatment of obesity. *Endocr Rev* 2007;28(6):664–684.

Keeble JE, Brain SD. A role for substance P in arthritis? *Neurosci Lett* 2004; 361(1–3):176–179.

Kela J, Salmi P, Rimondini-Giorgini R, Heilig M, Wahlestedt C. Behavioural analysis of melanin-concentrating hormone in rats: evidence for orexigenic and anxiolytic properties. *Regul Pept* 2003;114(2–3):109–114.

Kim N, Luster AD. Regulation of immune cells by eicosanoid receptors. *Sci World J* 2007;7:1307–1328.

Kirchgessner AL. Orexins in the brain–gut axis. *Endocr Rev* 2002;23(1):1–15.

Kirchgessner AL, Liu M. Orexin synthesis and response in the gut. *Neuron* 1999; 24(4):941–951.

Kitamura K, Singer WD, Star RA, Muallem S, Miller RT. Induction of inducible nitric-oxide synthase by the heterotrimeric G protein Galpha13. *J Biol Chem* 1996;271(13):7412–7415.

Klein TW, Newton C, Larsen K, Lu L, Perkins I, Nong L, Friedman H. The cannabinoid system and immune modulation. *J Leukoc Biol* 2003;74(4):486–496.

Kobayashi T, Narumiya S. Function of prostanoid receptors: studies on knockout mice. *Prostaglandins Other Lipid Mediat* 2002;68–69:557–573.

Kolakowski LF Jr. GCRDb: a G-protein-coupled receptor database. *Receptors Channels* 1994;2(1):1–7.

Kristiansen K. Molecular mechanisms of ligand binding, signaling, and regulation within the superfamily of G-protein-coupled receptors: molecular modeling and mutagenesis approaches to receptor structure and function. *Pharmacol Ther* 2004;103(1):21–80.

Kunos G. Understanding metabolic homeostasis and imbalance: what is the role of the endocannabinoid system? *Am J Med* 2007;120(9 Suppl 1):S18–S24; discussion S.

Le Y, Murphy PM, Wang JM. Formyl-peptide receptors revisited. *Trends Immunol* 2002;23(11):541–548.

Leibowitz SF, Weiss GF, Walsh UA, Viswanath D. Medial hypothalamic serotonin: role in circadian patterns of feeding and macronutrient selection. *Brain Res* 1989;503(1):132–140.

Levin ER. Endothelins. *N Engl J Med* 1995;333(6):356–363.

Levoye A, Dam J, Ayoub MA, Guillaume JL, Jockers R. Do orphan G-protein-coupled receptors have ligand-independent functions? New insights from receptor heterodimers. *EMBO Rep* 2006;7(11):1094–1098.

Livingston M, Heaney LG, Ennis M. Adenosine, inflammation and asthma—a review. *Inflamm Res* 2004;53(5):171–178.

Maeda K, Nakata H, Koh Y, Miyakawa T, Ogata H, Takaoka Y, Shibayama S, Sagawa K, Fukushima D, Moravek J, Koyanagi Y, Mitsuya H. Spirodiketopiperazine-based CCR5 inhibitor which preserves CC-chemokine/CCR5 interactions and exerts potent activity against R5 human immunodeficiency virus type 1 *in vitro*. *J Virol* 2004;78(16):8654–8662.

Maestroni GJ. Adrenergic modulation of dendritic cells function: relevance for the immune homeostasis. *Curr Neurovasc Res* 2005;2(2):169–173.

Mall M, Gonska T, Thomas J, Hirtz S, Schreiber R, Kunzelmann K. Activation of ion secretion via proteinase-activated receptor-2 in human colon. *Am J Physiol Gastrointest Liver Physiol* 2002;282(2):G200–G210.

Magno F, Di Stefano A. Contribution of bronchial biopsies in the evaluation of pathogenesis and progression of COPD. *Monaldi Arch Chest Dis* 2007;67(4): 229–233.

Mahad DJ, Ransohoff RM. The role of MCP-1 (CCL2) and CCR2 in multiple sclerosis and experimental autoimmune encephalomyelitis (EAE). *Semin Immunol* 2003; 15(1):23–32.

Marsh DJ, Weingarth DT, Novi DE, Chen HY, Trumbauer ME, Chen AS, Guan XM, Jiang MM, Feng Y, Camacho RE, Shen Z, Frazier EG, Yu H, Metzger JM, Kuca SJ, Shearman LP, Gopal-Truter S, MacNeil DJ, Strack AM, MacIntyre DE, Van der Ploeg LH, Qian S. Melanin-concentrating hormone 1 receptor-deficient mice are lean, hyperactive, and hyperphagic and have altered metabolism. *Proc Natl Acad Sci USA* 2002;99(5):3240–3245.

Massoumi R, Sjolander A. The role of leukotriene receptor signaling in inflammation and cancer. *Sci World J* 2007;7:1413–1421.

Masuda Y, Tanaka T, Inomata N, Ohnuma N, Tanaka S, Itoh Z, Hosoda H, Kojima M, Kangawa K. Ghrelin stimulates gastric acid secretion and motility in rats. *Biochem Biophys Res Commun* 2000;276(3):905–908.

Mazzone SB. Targeting tachykinins for the treatment of obstructive airways disease. *Treat Respir Med* 2004;3(4):201–216.

McMinn JE, Wilkinson CW, Havel PJ, Woods SC, Schwartz MW. Effect of intracerebroventricular alpha-MSH on food intake, adiposity, c-Fos induction, and neuropeptide expression. *Am J Physiol Regul Integr Comp Physiol* 2000;279(2): R695–R703.

Menard L, Ferguson SS, Zhang J, Lin FT, Lefkowitz RJ, Caron MG, Barak LS. Synergistic regulation of beta2-adrenergic receptor sequestration: intracellular complement of beta-adrenergic receptor kinase and beta-arrestin determine kinetics of internalization. *Mol Pharmacol* 1997;51(5):800–808.

Milia AF, Salis MB, Stacca T, Pinna A, Madeddu P, Trevisani M, Geppetti P, Emanueli C. Protease-activated receptor-2 stimulates angiogenesis and accelerates hemodynamic recovery in a mouse model of hindlimb ischemia. *Circ Res* 2002;91(4): 346–352.

Milligan G, Rees S. Chimaeric G alpha proteins: their potential use in drug discovery. *Trends Pharmacol Sci* 1999;20(3):118–124.

Milligan G, Smith NJ. Allosteric modulation of heterodimeric G-protein-coupled receptors. *Trends Pharmacol Sci* 2007;28(12):615–620.

Milligan G, Ramsay D, Pascal G, Carrillo JJ. GPCR dimerisation. *Life Sci* 2003; 74(2–3):181–188.

Milligan G, Canals M, Pediani JD, Ellis J, Lopez-Gimenez JF. The role of GPCR dimerisation/oligomerisation in receptor signalling. *Ernst Schering Found Symp Proc* 2006(2):145–161.

Mohsenin A, Blackburn MR. Adenosine signaling in asthma and chronic obstructive pulmonary disease. *Curr Opin Pulm Med* 2006;12(1):54–59.

Mollnes TE, Brekke OL, Fung M, Fure H, Christiansen D, Bergseth G, Videm V, Lappegard KT, Kohl J, Lambris JD. Essential role of the C5a receptor in *E. coli*-induced oxidative burst and phagocytosis revealed by a novel lepirudin-based human whole blood model of inflammation. *Blood* 2002;100(5):1869–1877.

Murris-Espin M, Pinelli E, Pipy B, Leophonte P, Didier A. Substance P and alveolar macrophages: effects on oxidative metabolism and eicosanoid production. *Allergy* 1995;50(4):334–339.

Nagase I, Yoshida T, Kumamoto K, Umekawa T, Sakane N, Nikami H, Kawada T, Saito M. Expression of uncoupling protein in skeletal muscle and white fat of obese mice treated with thermogenic beta 3-adrenergic agonist. *J Clin Invest* 1996; 97(12):2898–2904.

Nett PC, Teixeira MM, Candinas D, Barton M. Recent developments on endothelin antagonists as immunomodulatory drugs—from infection to transplantation medicine. *Recent Patents Cardiovasc Drug Discovery* 2006;1(3):265–276.

Neves SR, Ram PT, Iyengar R. G protein pathways. *Science* 2002;296(5573): 1636–1639.

Nicosia S, Capra V, Rovati GE. Leukotrienes as mediators of asthma. *Pulm Pharmacol Ther* 2001;14(1):3–19.

North RA. Molecular physiology of P2X receptors. *Physiol Rev* 2002;82(4):1013–1067.

O'Connor BJ, Lecomte JM, Barnes PJ. Effect of an inhaled histamine H3-receptor agonist on airway responses to sodium metabisulphite in asthma. *Br J Clin Pharmacol* 1993;35(1):55–57.

O'Hayre M, Salanga CL, Handel TM, Allen SJ. Chemokines and cancer: migration, intracellular signalling and intercellular communication in the microenvironment. *Biochem J* 2008;409(3):635–649.

Ohta A, Sitkovsky M. Role of G-protein-coupled adenosine receptors in downregulation of inflammation and protection from tissue damage. *Nature* 2001;414(6866): 916–920.

Okamoto Y, Ninomiya H, Tanioka M, Sakamoto A, Miwa S, Masaki T. Palmitoylation of human endothelin B. Its critical role in G protein coupling and a differential requirement for the cytoplasmic tail by G protein subtypes. *J Biol Chem* 1997; 272(34):21589–21596.

Omachi A, Ishioka K, Uozumi A, Kamikawa A, Toda C, Kimura K, Saito M. Beta3-adrenoceptor agonist AJ-9677 reduces body fat in obese beagles. *Res Vet Sci* 2007;83(1):5–11.

Palczewski K, Kumasaka T, Hori T, Behnke CA, Motoshima H, Fox BA, Le Trong I, Teller DC, Okada T, Stenkamp RE, Yamamoto M, Miyano M. Crystal structure of rhodopsin: A G protein-coupled receptor. *Science* 2000;289(5480):739–745.

Panaro MA, Mitolo V. Cellular responses to FMLP challenging: a mini-review. *Immunopharmacol Immunotoxicol* 1999;21(3):397–419.

Pease JE. Asthma, allergy and chemokines. *Curr Drug Targets* 2006;7(1):3–12.

Pease JE, Horuk R. CCR1 antagonists in clinical development. *Expert Opin Investig Drugs* 2005;14(7):785–796.

Penela P, Ribas C, Mayor F Jr. Mechanisms of regulation of the expression and function of G protein-coupled receptor kinases. *Cell Signal* 2003;15(11):973–981.

Pettipher R. The roles of the prostaglandin D(2) receptors DP(1) and CRTH2 in promoting allergic responses. *Br J Pharmacol* 2008;153(Suppl 1):S191–S199.

Pfleger KD, Eidne KA. Monitoring the formation of dynamic G-protein-coupled receptor–protein complexes in living cells. *Biochem J* 2005;385(Pt 3): 625–637.

Porter RA, Chan WN, Coulton S, Johns A, Hadley MS, Widdowson K, Jerman JC, Brough SJ, Coldwell M, Smart D, Jewitt F, Jeffrey P, Austin N. 1,3-Biarylureas as selective non-peptide antagonists of the orexin-1 receptor. *Bioorg Med Chem Lett* 2001;11(14):1907–1910.

Premont RT, Gainetdinov RR. Physiological roles of G protein-coupled receptor kinases and arrestins. *Annu Rev Physiol* 2007;69:511–534.

Prieschl EE, Kulmburg PA, Baumruker T. The nomenclature of chemokines. *Int Arch Allergy Immunol* 1995;107(4):475–483.

Quinones MP, Estrada CA, Kalkonde Y, Ahuja SK, Kuziel WA, Mack M, Ahuja SS. The complex role of the chemokine receptor CCR2 in collagen-induced arthritis: implications for therapeutic targeting of CCR2 in rheumatoid arthritis. *J Mol Med* 2005;83(9):672–681.

Raborn ES, Marciano-Cabral F, Buckley NE, Martin BR, Cabral GA. The cannabinoid delta-9-tetrahydrocannabinol mediates inhibition of macrophage chemotaxis to RANTES/CCL5: linkage to the CB(2) receptor. *J Neuroimmune Pharmacol* 2008;3(2):117–129.

Raddatz R, Schaffhauser H, Marino MJ. Allosteric approaches to the targeting of G-protein-coupled receptors for novel drug discovery: a critical assessment. *Biochem Pharmacol* 2007;74(3):383–391.

Rambach G, Wurzner R, Speth C. Complement: an efficient sword of innate immunity. *Contrib Microbiol* 2008;15:78–100.

Rasmussen SG, Choi HJ, Rosenbaum DM, Kobilka TS, Thian FS, Edwards PC, Burghammer M, Ratnala VR, Sanishvili R, Fischetti RF, Schertler GF, Weis WI, Kobilka BK. Crystal structure of the human beta2 adrenergic G-protein-coupled receptor. *Nature* 2007;450(7168):383–387.

Rijnierse A, van Zijl KM, Koster AS, Nijkamp FP, Kraneveld AD. Beneficial effect of tachykinin NK1 receptor antagonism in the development of hapten-induced colitis in mice. *Eur J Pharmacol* 2006;548(1–3):150–157.

Rossi M, Choi SJ, O'Shea D, Miyoshi T, Ghatei MA, Bloom SR. Melanin-concentrating hormone acutely stimulates feeding, but chronic administration has no effect on body weight. *Endocrinology* 1997;138(1):351–355.

Sawynok J. Adenosine receptor activation and nociception. *Eur J Pharmacol* 1998; 347(1):1–11.

Seeliger S, Derian CK, Vergnolle N, Bunnett NW, Nawroth R, Schmelz M, Von Der Weid PY, Buddenkotte J, Sunderkotter C, Metze D, Andrade-Gordon P, Harms E, Vestweber D, Luger TA, Steinhoff M. Proinflammatory role of proteinase-activated receptor-2 in humans and mice during cutaneous inflammation i*n vivo*. *FASEB J* 2003;17(13):1871–1885.

Sharma JN, Al-Dhalmawi GS. Bradykinin receptor antagonists: therapeutic implications. *IDrugs* 2003;6(6):581–586.

Shiiya T, Nakazato M, Mizuta M, Date Y, Mondal MS, Tanaka M, Nozoe S, Hosoda H, Kangawa K, Matsukura S. Plasma ghrelin levels in lean and obese humans and the effect of glucose on ghrelin secretion. *J Clin Endocrinol Metab* 2002;87(1): 240–244.

Shimada M, Tritos NA, Lowell BB, Flier JS, Maratos-Flier E. Mice lacking melanin-concentrating hormone are hypophagic and lean. *Nature* 1998;396(6712):670–674.

Silverman MH, Strand V, Markovits D, Nahir M, Reitblat T, Molad Y, Rosner I, Rozenbaum M, Mader R, Adawi M, Caspi D, Tishler M, Langevitz P, Rubinow A, Friedman J, Green L, Tanay A, Ochaion A, Cohen S, Kerns WD, Cohn I, Fishman-Furman S, Farbstein M, Yehuda SB, Fishman P. Clinical evidence for utilization of the A3 adenosine receptor as a target to treat rheumatoid arthritis: data from a phase II clinical trial. *J Rheumatol* 2008;35(1):41–48.

Sodhi A, Biswas SK. fMLP-induced i*n vitro* nitric oxide production and its regulation in murine peritoneal macrophages. *J Leukoc Biol* 2002;71(2):262–270.

Somers GR, Hammet FM, Trute L, Southey MC, Venter DJ. Expression of the P2Y6 purinergic receptor in human T cells infiltrating inflammatory bowel disease. *Lab Invest* 1998;78(11):1375–1383.

Stafforini DM, McIntyre TM, Zimmerman GA, Prescott SM. Platelet-activating factor, a pleiotrophic mediator of physiological and pathological processes. *Crit Rev Clin Lab Sci* 2003;40(6):643–672.

Stutz AM, Morrison CD, Argyropoulos G. The Agouti-related protein and its role in energy homeostasis. *Peptides* 2005;26(10):1771–1781.

Sun CX, Young HW, Molina JG, Volmer JB, Schnermann J, Blackburn MR. A protective role for the A1 adenosine receptor in adenosine-dependent pulmonary injury. *J Clin Invest* 2005;115(1):35–43.

Sun CX, Zhong H, Mohsenin A, Morschl E, Chunn JL, Molina JG, Belardinelli L, Zeng D, Blackburn MR. Role of A2B adenosine receptor signaling in adenosine-dependent pulmonary inflammation and injury. *J Clin Invest* 2006;116(8):2173–2182.

Sun Y, McGarrigle D, Huang XY. When a G protein-coupled receptor does not couple to a G protein. *Mol Biosyst* 2007a;3(12):849–854.

Sun Y, Huang J, Xiang Y, Bastepe M, Juppner H, Kobilka BK, Zhang JJ, Huang XY. Dosage-dependent switch from G protein-coupled to G protein-independent signaling by a GPCR. *EMBO J* 2007b;26(1):53–64.

Tecott LH, Sun LM, Akana SF, Strack AM, Lowenstein DH, Dallman MF, Julius D. Eating disorder and epilepsy in mice lacking 5-HT2c serotonin receptors. *Nature* 1995;374(6522):542–546.

Thurmond RL, Gelfand EW, Dunford PJ. The role of histamine H1 and H4 receptors in allergic inflammation: the search for new antihistamines. *Nat Rev Drug Discovery* 2008;7(1):41–53.

Townley RG. Interleukin 13 and the beta-adrenergic blockade theory of asthma revisited 40 years later. *Ann Allergy Asthma Immunol* 2007;99(3):215–224.

Turner JE, Steinmetz OM, Stahl RA, Panzer U. Targeting of Th1-associated chemokine receptors CXCR3 and CCR5 as therapeutic strategy for inflammatory diseases. *Mini Rev Med Chem* 2007;7(11):1089–1096.

Viola A, Luster AD. Chemokines and their receptors: drug targets in immunity and inflammation. *Annu Rev Pharmacol Toxicol* 2008;48:171–197.

von Wenckstern H, Zimmermann K, Kleuser B. The role of the lysophospholipid sphingosine 1-phosphate in immune cell biology. *Arch Immunol Ther Exp (Warsz)* 2006;54(4):239–251.

Wang W, Graeler MH, Goetzl EJ. Type 4 sphingosine 1-phosphate G protein-coupled receptor (S1P4) transduces S1P effects on T cell proliferation and cytokine secretion without signaling migration. *FASEB J* 2005;19(12):1731–1733.

Warny M, Aboudola S, Robson SC, Sevigny J, Communi D, Soltoff SP, Kelly CP. P2Y(6) nucleotide receptor mediates monocyte interleukin-8 production in response to UDP or lipopolysaccharide. *J Biol Chem* 2001;276(28):26051–26056.

Wheatley M, Hawtin SR. Glycosylation of G-protein-coupled receptors for hormones central to normal reproductive functioning: its occurrence and role. *Hum Reprod Update* 1999;5(4):356–364.

Wilkin F, Duhant X, Bruyns C, Suarez-Huerta N, Boeynaems JM, Robaye B. The P2Y11 receptor mediates the ATP-induced maturation of human monocyte-derived dendritic cells. *J Immunol* 2001;166(12):7172–7177.

Wren AM, Small CJ, Abbott CR, Dhillo WS, Seal LJ, Cohen MA, Batterham RL, Taheri S, Stanley SA, Ghatei MA, Bloom SR. Ghrelin causes hyperphagia and obesity in rats. *Diabetes* 2001;50(11):2540–2547.

Wright KL, Duncan M, Sharkey KA. Cannabinoid CB2 receptors in the gastrointestinal tract: a regulatory system in states of inflammation. *Br J Pharmacol* 2008; 153(2):263–270.

Xie GX, Palmer PP. How regulators of G protein signaling achieve selective regulation. *J Mol Biol* 2007;366(2):349–365.

Yeomans MR, Gray RW. Opioid peptides and the control of human ingestive behaviour. *Neurosci Biobehav Rev* 2002;26(6):713–728.

Zarjevski N, Cusin I, Vettor R, Rohner-Jeanrenaud F, Jeanrenaud B. Chronic intracerebroventricular neuropeptide-Y administration to normal rats mimics hormonal and metabolic changes of obesity. *Endocrinology* 1993;133(4):1753–1758.

Zhang J, Frassetto A, Huang RR, Lao JZ, Pasternak A, Wang SP, Metzger JM, Strack AM, Fong TM, Chen RZ. The mu-opioid receptor subtype is required for the anorectic effect of an opioid receptor antagonist. *Eur J Pharmacol* 2006;545(2–3): 147–152.

Zhang Z, Schluesener HJ. FTY720: a most promising immunosuppressant modulating immune cell functions. *Mini Rev Med Chem* 2007;7(8):845–850.

Zheng Y, Voice JK, Kong Y, Goetzl EJ. Altered expression and functional profile of lysophosphatidic acid receptors in mitogen-activated human blood T lymphocytes. *FASEB J* 2000;14(15):2387–2389.

Zheng Y, Kong Y, Goetzl EJ. Lysophosphatidic acid receptor-selective effects on Jurkat T cell migration through a Matrigel model basement membrane. *J Immunol* 2001;166(4):2317–2322.

21

PHARMACOLOGICAL MODULATION OF ION CHANNELS FOR THE TREATMENT OF CHRONIC PAIN

Yi Liu and Ning Qin

21.1 INTRODUCTION

Ion channels are a diverse class of cell membrane-spanning proteins that regulate the flow of ions across the membrane. They are expressed in all living cells and play important roles in a wide range of biological processes, such as electrical signaling in nerve and muscle, cardiac pacemaking, hormone secretion, and immune cell activation. Consequently, improper functioning of ion channels may have serious deleterious effects and result in disease. This is evidenced by the many channelopathies (diseases caused by defects in ion channel function) that have been identified (Doyle and Stubbs, 1998; Bernard and Shevell, 2008).

Ion channel drug discovery aims to identify drugs that treat diseases with minimal undesirable effects by selectively normalizing or compensating for the function of "diseased" channels while leaving other, uninvolved channels or proteins intact. To this end, ion channels must be druggable targets. A recent analysis estimates that as many as 10% of the 2000–3000 druggable targets in the human genome may be ion channels (Russ and Lampel, 2005). Another study shows that more than 13% of the proteins targeted by currently available drugs are ion channels, placing them second among the gene classes

Enzyme Inhibition in Drug Discovery and Development, edited by Lu and Li
Copyright © 2010 John Wiley & Sons, Inc.

as drug targets (Overington et al., 2006). Moreover, these ion channels encompass diverse gene families—from voltage-gated sodium channels to ligand-gated chloride channels—and are targeted for a variety of diseases (Imming et al., 2006), further demonstrating the general druggability of ion channels as a class.

This chapter concerns those ion channels that are implicated in chronic pain, a group of painful conditions that can severely afflict large populations of people and for which treatment options are limited due to issues of efficacy and/or adverse effects. We begin by giving a general overview of ion channels and introducing methods to study ion channel function and pharmacology, particularly those methods that are commonly employed in ion channel drug discovery. This is followed by a general discussion of the role of ion channels in chronic pain. The rest of the chapter then focuses on several voltage-gated sodium channels ($Na_v1.3$, $Na_v1.7$, $Na_v1.8$, and $Na_v1.9$), a voltage-gated calcium channel ($Ca_v2.2$), and a transient receptor potential (TRP) channel (TRPV1). We review the current understanding of their role in chronic pain and efforts of small-molecule analgesic discovery and development that target these channels. Finally, we end the chapter with a few perspectives.

21.2 ION CHANNELS—AN OVERVIEW

This section gives a general overview of some key elements of ion channels. Readers are advised to consult an excellent book for more in-depth discussions of the subject (Hille, 2001).

21.2.1 Ion Channels Facilitate and Regulate Ion Transport Across Cell Membranes

The lipid membrane that surrounds living cells is impermeable to ions. Specialized transmembrane proteins such as ion channels and transporters are necessary to help ions travel through the membrane. This chapter focuses on ion channels. In contrast to transporters, which use energy to move ions across the membrane against their concentration gradient, ion channels employ a passive transport mechanism by which ions pass through a water-filled pore down their electrochemical gradient without energy expenditure. This ion conduction pathway is formed by a circular arrangement of several (typically 3–5, sometimes identical) protein subunits (e.g., potassium channels) or of several homologous domains of a single protein subunit (e.g., sodium channels). Each subunit has multiple (typically 2–16) membrane-spanning segments. In addition to pore-forming (a.k.a. α) subunits, there are auxililary (e.g., β) subunits that do not form a channel by themselves but interact with α subunits to modulate channel function and expression. Ionic flow through the pore is "gated"—that is, controlled by one or more "gates" that regulate channel opening and closing (Fig. 21.1). In addition to closed states, many

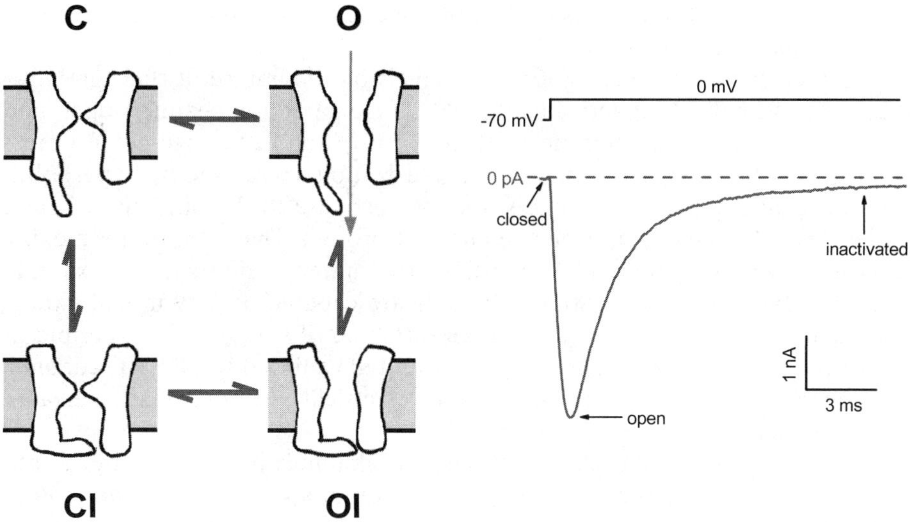

Figure 21.1. Ionic flow is controlled by "gates" in the ion channel protein. **Left panel**: Schematic depiction of ion channel gating. Though oversimplified and hypothetical, this model is sometimes used to approximate the gating of some voltage-gated channels. Permeation is controlled by two sets of gates in this drawing (in the middle and at the bottom of the channel protein, respectively). Ion flow is prevented when either gate is shut. When the upper gate (the activation gate) is shut, the channel is said to be in a closed state (C or CI). When the lower gate (the inactivation gate) is shut, the channel enters a so-called inactivated state (CI or OI). The only conducting state is the open state (O), in which neither gate is in position to block the flow of ions through the pore. **Right panel**: Time course of voltage-gated sodium channel currents recorded from a rat dorsal root ganglion neuron. Channels are opened rapidly in response to a depolarizing voltage step to 0 mV (current shown as downward deflection), followed by inactivation (seen as current decay).

channels can enter other nonconducting states (usually referred to as inactivated or desensitized states) in the prolonged presence of a stimulus. Ion channels can be opened by stimuli as diverse as membrane potential, chemical ligands, temperature, light, and mechanical force.

21.2.2 Ion Channel Classification

Ion channels can be classified in several different ways. One of them is based on their selectivity for physiological, inorganic ions—that is, the ion species that is selectively allowed to pass through the pore. For example, sodium channels only conduct Na^+ ions. Based on this classification, there are also potassium channels, calcium channels, proton channels, chloride channels, and nonselective cation channels (these channels are selective for cations over anions, but do not discriminate among cations, such as Na^+, K^+, and Ca^{2+}).

Specific amino acid sequences in the pore of each type of channel are responsible for their ion selectivity.

Another way of classifying ion channels is by the nature of their gating—that is, the stimulus that activates them. Voltage-gated ion channels open and close in response to membrane potential. For example, voltage-gated potassium, or K_v, channels are opened by depolarization and closed by hyperpolarization. Ligand-gated ion channels open in response to the binding of ligand molecules to specific site(s) on the channel protein. The chloride-permeable $GABA_A$ receptor, opened by the binding of γ-aminobutyric acid, is an example of a ligand-gated channel. Some channels are capable of poly-modal gating; that is, they can be gated by more than one type of stimulus. An example of such a channel is the transient receptor potential vanilloid 1 (TRPV1) channel, which can be activated by depolarization (electrical), noxious heat (thermal), and protons and capsaicin (chemical).

Aside from functional classifications, ion channels are also grouped into families and superfamilies according to the homology of their primary amino acid sequence. For example, the superfamily of voltage-gated ion channels includes three families: voltage-gated sodium (Na_v), potassium (K_v), and calcium channels (Ca_v).

21.2.3 Molecular and Functional Diversity of Ion Channels

One aspect of ion channel diversity is already evident from their classification: They convert a variety of input stimuli into electrical signals carried by different species of ions in a stimulus-dependent manner. Different ion channels often serve vastly different physiological functions in a cell as a result of their differences in ionic selectivity and other functional properties, such as kinetics of channel activation and inactivation. Activation of Na_v channels in a neuron, for instance, causes rapid influx of Na^+, and the resultant membrane depolarization triggers the initiation of action potentials. In contrast, activation of K_v channels in a neuron counters the actions of Na_v channels by repolarizing the membrane (due to the efflux of K^+), which contributes to the termination of action potentials.

The functional diversity of ion channels reflects their diversity at the molecular level. Molecular cloning has unveiled not only a large number of ion channel genes that encode pore-forming α subunits, but also the existence of auxiliary (e.g., β) subunits that associate with α subunits and modulate channel function (Sanguinetti et al., 1996; Nerbonne, 1998). Through alternative splicing of transcripts, ion channel genes can also produce multiple forms of channel subunits that have distinct functional properties and patterns of localization (Nerbonne, 1998; Gray et al., 2007). Because many channels consist of more than one α subunit, variation in functional and physiological properties may also arise from the formation of functionally distinct heteromultimeric channels between different α subunits (Nerbonne, 1998). In addition, subunit proteins can also undergo post-translational modifications, such as phosphorylation by protein kinases, which may regulate and further extend the range of channel functions (Nerbonne, 1998). Thus, electrical signals generated by ion channels can be remarkably diverse and rich in information.

21.2.4 Ion Channel Modulation

Modulation of an ion channel refers to a process in which the function of the channel is modified. Virtually all ion channels are subject to modulation. It is of fundamental importance for the regulation of neuronal electrical activity, including that associated with pain signaling. Many mechanisms have evolved by which channel activity can be modulated, such as those that occur through interaction with other proteins (e.g., G-proteins or Ca^{2+}/calmodulin), phosphorylation by protein kinases, or other post-translational modifications, and those that involve modulation by cations (e.g., Ca^{2+}), metabolites, fatty acids, and other molecules. Of the many potential modulatory mechanisms, channel phosphorylation, which results from a cascade of intracellular events that are often triggered by the binding to specific cell surface receptors of neurotransmitters, hormones, proinflammatory mediators, and so on, is among the most prevalent and of particular interest in pain signaling. Some of the peripheral and central sensitization mechanisms in chronic pain, for example, involve the phosphorylation of various types of ion channels, such as TRPV1, Na_v channels, and *N*-methyl-D-aspartic acid (NMDA) receptors. The most common amino acids to be phosphorylated are serine and threonine. The biophysical manifestation of channel modulation is diverse and can range from changes in the rate of desensitization and threshold for thermal activation to shifts in the voltage dependence and kinetics of channel activation and inactivation.

In ion channel drug discovery, efforts are directed toward identifying chemical or biological molecules that modulate specific ion channel functions in such a way as to treat diseases with minimal undesirable effects. A wide variety of molecules could potentially be useful in this respect, ranging from inorganic molecules and low-molecular-weight organic molecules to small interfering RNA (siRNA) and antisense (AS) oligodeoxynucleotides (ODNs), to toxins, antibodies, and other proteins. The primary emphasis of this chapter is on small organic molecule modulators.

21.3 METHODS FOR STUDYING ION CHANNEL FUNCTION AND PHARMACOLOGY—SCREENING ASSAYS/TECHNOLOGIES

Despite being attractive therapeutic targets, ion channels have until recently been a challenging class of proteins to screen against using target-based, functional, and high-throughput screening (HTS) assays. The past decade has seen great improvement in HTS technologies, particularly for cell-based, functional assays that can be configured to screen a variety of ion channels in a robust and cost-effective manner. The recent advent of automated patch-clamp technologies has also greatly increased the throughput over conventional patch clamp. These technological breakthroughs will undoubtedly help to accelerate the pace of ion channel drug discovery. Here, we describe some of the major screening assays and technology platforms currently used in ion channel drug discovery.

21.3.1 Nonfunctional Methods—Radioligand Binding

This method requires the use of a high-affinity ligand for the channel of interest that is labeled with a radioactive tracer (e.g., ^{3}H). If a (unlabeled) test compound binds to the same site on the channel protein as the labeled ligand, it competes with and decreases the ability of the labeled ligand to bind to the site. The binding affinity of the test compound for the channel can thus be obtained from the reduction of the radioactivity associated with the decreased binding of the labeled ligand.

The binding assay offers some advantages. For example, because of the competitive nature of the assay, novel compounds can be identified that act at a known site (to the extent that the ligand binding site is characterized). In addition, it can be performed with relatively high throughput and at a reasonable cost. There are also some inherent limitations associated with this method. First, a high-affnity ligand must exist and be available in a radiotracer form. Second, it is not a functional assay. As such, the data have relatively low information content: It does not provide information about whether binding to a channel has any effect on channel function, let alone the nature of any effect (agonism, antagonism, etc.). Third, it is prone to false negatives in that it does not report as hits compounds that bind to allosteric sites. This may limit the extent to which hits with structural as well as modulatory diversity can be identified. Lastly, the assay involves radioactivity, which has safety and environmental concerns.

21.3.2 Functional Methods—Membrane Potential

Functional assays exploit the fact that movement of ions across the cell membrane (ion flux) takes place during channel activation. Membrane potential assays can report functional information about the channel by using voltage-sensitive dyes to detect changes in the cell membrane potential, a direct consequence of ion flux.

21.3.2.1 Single-Dye Assays. Oxonol-derived voltage-sensitive fluorescent dyes are negatively charged and move in response to membrane potential, partitioning out of the cell with hyperpolarization and into the cell with depolarization. Thus, depolarization increases and hyperpolarization decreases the intracellular fluorescence upon excitation of the dye. The fast response times and other improved features (e.g., quenching of extracellular fluorescence) afforded by the membrane potential dye kit from MDS Analytical Technologies overcome some of the drawbacks of the earlier dyes, resulting in higher throughput and temporal resolution. Imaging plate readers such as FLIPR$^{TETRA®}$ (MDS Analytical Technologies) and FDSS7000 (Hamamatsu) are well-suited for HTS using this dye kit. This method is applicable to a variety of ion channel types. Due to the indirect nature of this method, nonspecific fluorescent signals such as autofluorescence and interaction of compounds with the dye may result in false positives.

21.3.2.2 Dual-Dye Assays. This approach is based on the principle of fluorescence resonance energy transfer (FRET), whereby an excited donor transfers its energy to a nearby acceptor, which then produces fluorescence. The efficiency of the energy transfer is critically dependent on the distance between the donor and the acceptor. The FRET donor is a coumarin dye that is linked to phospholipids anchored to the outer leaflet of the cell membrane. The FRET acceptor is an oxonol derivative. Upon membrane hyperpolarization, the acceptor moves to the outer leaflet of the membrane near the donor, resulting in FRET. Conversely, FRET is disrupted upon membrane depolarization, resulting in increased donor but decreased acceptor fluorescence. The ratio of fluorescence from the donor and acceptor can thus be used to report membrane potential changes. Several technology platforms are suitable for FRET in HTS formats, including VIPR (Vertex), FLIPR$^{TETRA®}$ and FDSS7000. This method has general applicability, good sensitivity, and temporal resolution (faster kinetics than the single dye method). The ratiometric nature of FRET also helps to reduce artifacts, such as those related to cell number and time-dependent changes in dye concentration. On the other hand, FRET assays are more complex to perform, placing restraints on cost and throughput. False positives can also be an issue. A recently reported method that combines FRET with electrical field stimulation, the E-VIPR (Huang et al., 2006), may help to reduce false positives (at least for voltage-gated channels) with its more physiologically relevant mode of channel stimulation compared with artificial methods such as the use of neurotoxins or high concentrations of extracellular K^+. In addtition, E-VIPR can potentially bridge the gap between patch-clamp and conventional HTS by identifying use-dependent channel modulators.

21.3.3 Functional Methods—Ion Flux

Ion flux assays directly measure the flux of ions that results from channel opening. The major difference among various flux assays lies in the method of detection.

21.3.3.1 Radiolabeled Tracer Flux Assays. These assays follow the movement of radioactive tracer ions down their electrochemical gradient through the channel of interest. Radioactive isotopes of physiological ions, such as

$^{22}Na^+$, $^{45}Ca^{2+}$, and $^{36}Cl^-$, have been used to study sodium (Willow et al., 1984), calcium (Rasmussen et al., 1987), and chloride (Thampy and Barnes, 1984) channels, respectively. Other commonly used tracers include: (for sodium channels) [^{14}C]-guanidinium (Willow et al., 1984), (for potassium channels) $^{86}Rb^+$ (Weir and Weston, 1986), and (for chloride channels) $^{125}I^-$ (Derand et al., 2002). A test compound is typically added prior to channel activation and the cumulative flux over a period of time after channel activation in the presence and absence of the compound is compared to determine its effect. The cumulative flux is determined by measuring the amount of radioactivity in the cell lysate and/or extracellular buffer (with, for example, a scintillation counter). Radiotracer assays are direct, robust, and fairly HTS friendly. A major drawback of this method is the use of hazardous radioactive material. Other limitations include: low temporal resolution, lack of voltage control, relatively low signal-to-noise ratio, and non-real-time measurement.

21.3.3.2 Nonradioactive Tracer Flux Assays. The principle of this method is the same as that for the radioactive tracer flux. The main difference is that tracers used in this method are nonradioactive and detected by atomic absorption spectroscopy (AAS). Most studies have been conducted on potassium or nonselective cation channels using Rb^+ as the tracer (Terstappen, 1999). Use of other ions, such as Li^+ (for sodium channels), has also been reported (Trivedi et al., 2008). The biggest advantage of this method over its radioactive predecessor is the lack of radioactivity. The signal-to-noise ratio is also higher than the radioactive method. It otherwise shares the pros and cons with the radioactive method. A commercial HTS system, ICR12000 (Aurora Biomed), is currently available.

21.3.3.3 Fluorescence-Based Flux Assays. The emission spectrum of an ion-sensitive fluorescent dye shifts upon binding of the ion. Thus, ion-binding-induced changes in fluorescence intensity (often measured near the peak-emission wavelength of the bound dye) can be used as a measure of changes in the intracellular concentration of the free ion. The most widely used are Ca^{2+}-sensitive dyes, such as Fluo-3 and Fluo-4. These dyes have been used successfully in screening against Ca^{2+}-permeable ion channels. Cells expressing these channels are first loaded with a Ca^{2+}-sensitive dye. Channel activation results in Ca^{2+} influx, which causes a significant increase in the intracellular Ca^{2+} concentration, leading to increased Ca^{2+} binding to the dye and fluorescence intensity. If a compound interacts with the channel, it may affect Ca^{2+} influx and alter the fluorescence intensity. Such calcium mobilization assays can be easily configured in HTS formats using FLIPR$^{TETRA®}$ or FDSS7000. This method does not control membrane potential. In addition, false positives/negatives can result from, among other possibilities, autofluorescence, compound-dye interactions, or compound effects on Ca^{2+} release from intracellular calcium stores.

21.3.3.4 Patch-Clamp Electrophysiology. Since its introduction some three decades ago (Hamill et al., 1981), patch-clamp electrophysiology, which directly measures the ionic current (i.e., the rate of ionic flux) flowing through ion channels in real time, has revolutionized ion channel research and remained as the "gold standard" for studying ion channel function and pharmacology. A central feature of this technique involves the formation of a high resistance seal ($>10^9\,\Omega$ or $1\,G\Omega$) between the membrane of a cell and the wall of a glass microelectrode placed against the membrane. The tight seal reduces the background electrical noise to such a low level that it becomes possible to resolve small currents flowing through as few as a single ion channel. Patch clamp allows for accurate control of voltage applied to the cell membrane through the electrode. The high temporal resolution afforded by patch clamp (and unmatched by other techniques) also makes it possible to resolve fast channel events.

There are four basic patch-clamp configurations (Hille, 2001), of which the whole-cell configuration is the most widely used in drug discovery. In this configuration, the recording electrode makes direct contact with the interior of the cell. This provides excellent control of the cell membrane potential and allows the whole-cell current (total current conducted by all the channels of interest in the plasma membrane of the entire cell) to be measured.

Conventional patch clamp, as it is now called, comes with some critical limitations, particularly when it is used in drug discovery. First, only a single cell can be recorded at a time, making its throughput painfully low. Second, the method is technically challenging and labor intensive, requiring the constant attention of a specialist who is "skilled in the art" and further limiting the throughput. As a result, patch clamp has historically been a major bottleneck in ion channel drug discovery, being useful only for a small number of late-stage compounds.

Recent efforts to increase the throughput and ease of use of patch clamp have resulted in several commercially available whole-cell patch-clamp platforms with varying degrees of automation and increased throughput over conventional patch clamp (Dunlop et al., 2008). Three systems in particular, all in planar-array formats that can record multiple cells in parallel, have seen wide applications in drug discovery. IonWorks Quattro (MDS Analytical Technologies) adopts a 384-well format and uses a population patch-clamp strategy (simultaneous recording of 64 cells per well) to achieve high success rate and low variability. It has the highest throughput of all the patch-clamp systems currently available and is capable of primary screening of directed compound libraries. However, its sub-GΩ seal resistance and electronics limitations impose significant restrictions on the scope of its utility (e.g., it is unsuitable for studying certain types or properties of ion channels). PatchXpress (MDS) and the QPatch HT (Sophion) are both capable of achieving GΩ seals and can therefore generate high-quality data comparable to those obtained by conventional patch clamp. The QPatch HT has the additional advantage of higher throughput (48 wells for Qpatch HT vs 16 wells for PatchXpress), microfluidics flow, and continuous operation for several hours without needing operator intervention (which further extends the throughput and efficiency). Neither system has the capacity for primary screening, but both are well-suited for secondary screening, selectivity profiling, and mechanistic studies.

One of the issues common to all three systems is their propensity to cause rightward potency shift for some compounds compared to conventional patch clamp. This will likely be successfully addressed in future automated patch-clamp systems. But the jury is still out as to the feasibility of cost-effective patch-clamp HTS in the foreseeable future. A potentially promising alternative to the "needle-in-a-haystack" HTS approach, however, would be to screen relatively small, but richly ion channel-focused, libraries using current or future (improved) patch clamp technologies.

Importantly, automated patch clamp has in recent years seen increasing integration into the drug discovery workflow and has begun to have significant impact on discovery programs. This trend is likely to continue.

21.3.4 Functional Methods—Label-Free Assays

Cell-based, label-free technologies have recently gained increasing recognition as viable compound screening platforms. Although they have been mostly used for G-protein-coupled receptors (GPCRs) targets, indications are that they may be suitable for screening against ion channels as well. These methods detect ligand-induced, integrated cellular responses (e.g., local mass redistribution caused by protein trafficking and cytoskeletal rearrangement, or changes in cell volume, morphology, adherence, etc.) via a biosensor/transducer (optical, electrical, calorimetric, acoustic, magnetic, etc.) at the cell surface that converts cellular changes into quantifiable signals (Shiau et al., 2008).

Several label-free HTS systems have been developed in recent years based on either impedance (e.g., CellKey™ from MDS) or waveguide (e.g., Epic™ from Corning and BIND™ from SRU Biosystems) technologies (Shiau et al., 2008). These label-free methods offer certain advantages over those that employ labels (e.g., fluorescent dyes). For example, they are minimally invasive and are thus more likely than label-dependent methods to represent endogenous, physiological (rather than artificial) states of the cellular system [in this respect, they are also advantageous compared to whole-cell patch clamp, which, despite itself being a label-free method (albeit of a type beyond the narrow definition used in this section), artificially modifies the native intracellular environment]. This helps to increase the likelihood of obtaining more biologically relevant pharmacology. They are also void of assay artifacts introduced by labels, such as autofluorescence and compound-dye interaction.

On the flipside, these systems generally have low temporal resolution and, because of the integrated nature of the readouts, present a less clear or more complex picture of the underlying mechanism under investigation than the other methods discussed above. In addition, the extent to which these platforms are applicable to studying ion channel pharmacology (particularly in HTS formats) remains to be demonstrated as of this writing.

21.4 CHRONIC PAIN—ROLE OF ION CHANNELS

Pain is an unpleasant sensory experience associated with hurting and soreness. Acute (or nociceptive) pain, produced by the normal, physiological functioning of the nervous system in response to damage or threat of damage to the body, is a natural and necessary alarm/defense mechanism that is essential for survival. In such a scenario, a noxious (i.e., high-threshold) stimulus activates nociceptors (C- and Aδ-fibers) in the periphery, which carry the information to neurons in the spinal cord, which in turn convey the (synaptically integrated) signal to the brain for proper processing and protective action. Acute pain usually ceases after the stimulus is no longer present. In contrast, chronic (or clinical) pain is a persistent and pathologic expression of the nervous system with regard to pain signaling and sensation. It long outlasts the resolution of the initiating event (Watkins and Maier, 2003; Woolf, 2004) and is not known to serve any beneficial function. Patients often experience crippling pain in response to even innocuous (i.e., low-threshold and normally nonpainful) stimuli (such as a light touch to the skin), a condition known as allodynia. Noxious (painful) stimuli produce pain with even higher intensity (hyperalgesia) in chronic pain conditions than normal. In addition, chronic pain may also arise spontaneously without any peripheral stimulus.

Chronic pain can be either inflammatory or neuropathic in nature (Fig. 21.2). Inflammatory pain results from tissue damage and inflammation, which involves the release of inflammatory mediators (e.g., cytokines). Although inflammatory pain can help to minimize further tissue damage and promote healing, it can also become a serious clinical condition, such as that associated with rheumatoid arthritis. Neuropathic pain typically develops with injury to the peripheral or central nervous system due to events such as surgery, cancer, or diabetes. Diabetic neuropathy, postherpetic neuralgia, and pain associated with chemotherapy and spinal cord injury are some examples of neuropathic pain. Although inflammatory pain and neuropathic pain have different causes, it can sometimes be difficult to distinguish between the two in chronic pain conditions in which both mechanisms are involved.

A number of mechanisms can cause chronic pain. Sensitization of peripheral or central neurons following inflammation or injury, which lowers the activation threshold of these neurons and increases their overall responsiveness to input, is a major cause of pain hypersensitivity. Sensitization is brought about (in peripheral neurons) by the release or production of inflammatory mediators (e.g., cytokines, neurotrophins, or prostanoids) and (in central neurons) by synaptic input from the periphery as well as transcriptional regulation. After peripheral inflammation or nerve injury, transcriptional and

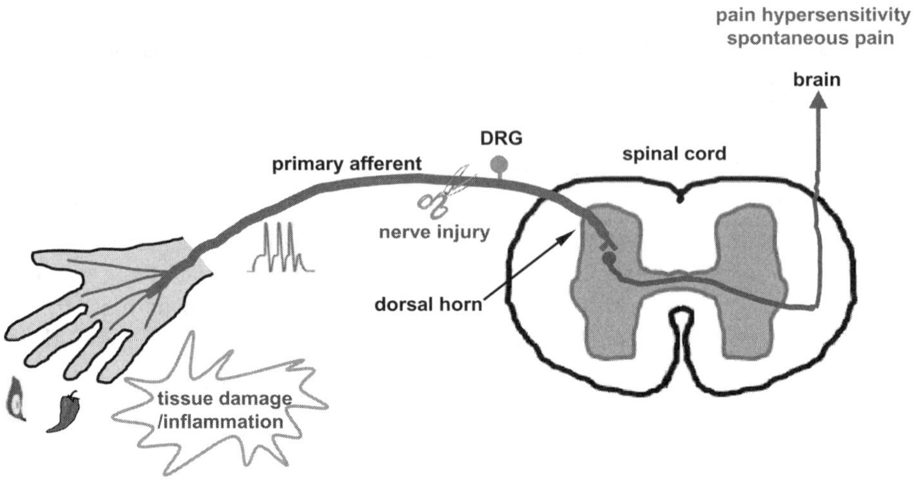

Figure 21.2. Peripheral inflammation and nerve injury cause chronic pain. Peripheral inflammation and nerve injury can result in various changes in the PNS and/or CNS (e.g., peripheral or central sensitization—see text) that cause neuronal hyperexcitability, leading to pain hypersensitivity and spontaneous pain. See the insert for color representation of this figure.

post-transcriptional changes in nociceptors can lead to increased membrane excitability and synaptic transmission. Peripheral nerve injury and inflammation also cause disinhibition (loss of inhibition) in central neurons, resulting in increased excitability and pain. [For review, see Woolf (2004).]

A variety of ion channels are expressed (in normal and/or pain states) in pain pathways and play a central role in pain signaling. In normal nociception, ion channels are intimately involved in the transduction (conversion of peripheral stimuli into electrical signals), conduction (propagation of action potentials from the peripheral to the central terminal of a nociceptor), and transmission (modulation of presynaptic signals and passage of these signals to postsynaptic neurons) of nociceptive signals. In chronic pain, the functional properties (e.g., activation threshold, voltage dependence, functional expression, etc.) of many of these channels are modulated, often by channel phosphorylation, such that pain signaling (intensity and/or duration) becomes greatly enhanced. For example, peripheral inflammation increases the expression of a thermal transducer ion channel, TRPV1, in the peripheral terminal of nociceptors (Ji et al., 2002). Prostaglandin E_2 (PGE_2), whose production is a major element of the inflammatory reaction, decreases the thermal activation threshold of TRPV1 from ~42 °C to ~35 °C (Moriyama et al., 2005). These actions increase the peripheral sensitivity to heat, which may, at least in part, underlie the burning pain experienced by individuals with sunburn (Chizh et al., 2007). Central sensitization increases the surface expression of NMDA receptors in postsynaptic neurons. As a result, the excitability of these neurons in response to presynaptically released glutamate is increased, contributing to allodynia and hyperalgesia (Woolf, 2004). Peripheral inflammation and nerve injury also cause disinhibition in central neurons through glycine and GABA receptor-mediated mechanisms (Zeilhofer, 2008). Other channels, including voltage-gated sodium, potassium, and calcium channels, are also indispensable in pain signaling. In the sections that follow, we focus on some of these channels and discuss in more detail the roles that they play in chronic pain and drugs that target these channels for the treatment of chronic pain.

21.5 PHARMACOLOGICAL MODULATION OF ION CHANNELS FOR THE TREATMENT OF CHRONIC PAIN

In the following sections, we discuss three types of ion channels with respect to their roles in chronic pain and modulation of these channels for the treatment of these painful syndromes.

21.5.1 Voltage-Gated Sodium Channels (Na_v)

Voltage-gated sodium channels are expressed in neurons in nociceptive pathways and play a critical role in the initiation and propagation of sensory nerve action potentials necessary for pain signaling. It is thus not surprising that sodium-channel-blocking drugs have been found effective and used clinically for treating pain syndromes.

21.5.1.1 Basic Molecular, Structural, Functional, and Pharmacological Properties. Voltage-gated sodium channels consist of a large (220–260 kD) α subunit and one or more smaller (33–45 kD) β subunits. The α subunit has four homologous domains (I–IV) that are arranged to form a central conduction pore with a pseudo-fourfold symmetry. Each domain further consists of six transmembrane segments (S1–S6), of which the highly conserved S4 segment is known as the voltage sensor. It contains several positively charged amino acid residues and moves to open and close the channel in response to changes in membrane potential. The S6 segments of the homologous domains are arranged in a square array surrounding the inner pore, whereas the membrane-reentrant pore loops between the S5 and S6 segments line the narrower outer pore and form the ion selectivity filter (responsible for its Na^+ selectivity). The inner portion of the pore contains a highly conserved region that binds local anesthetics (LAs) and a number of antiarrhythmic and anticonvulsant drugs. The linker region between domains III and IV is responsible for the fast inactivation of the channel by plugging the inner pore upon channel activation. [For review, see Catterall (2000a); also see Fig. 21.3]. The α subunit alone can form a functional channel. However, interaction with β subunits (which have a single transmembrane segment) or with other accessory proteins (e.g., p11) results in changes of the function (e.g., channel kinetics and

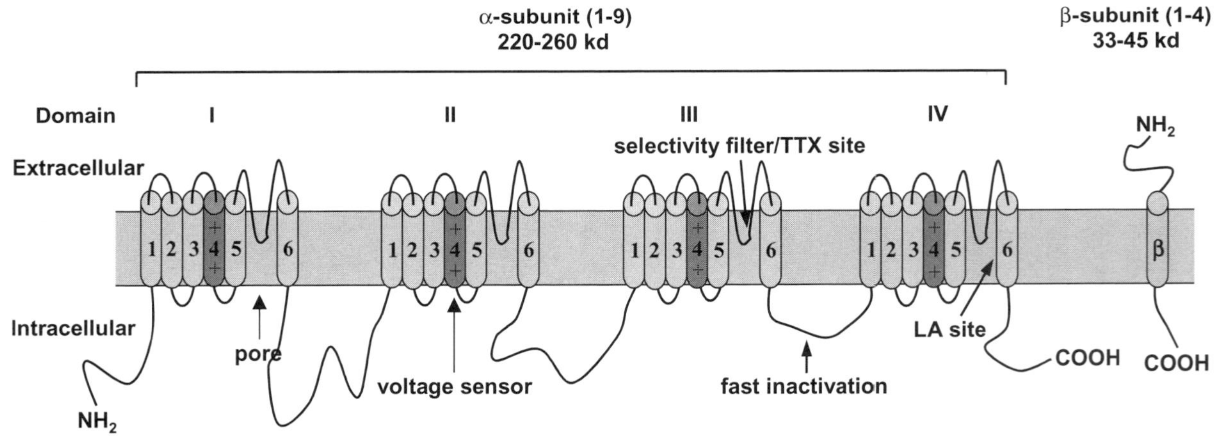

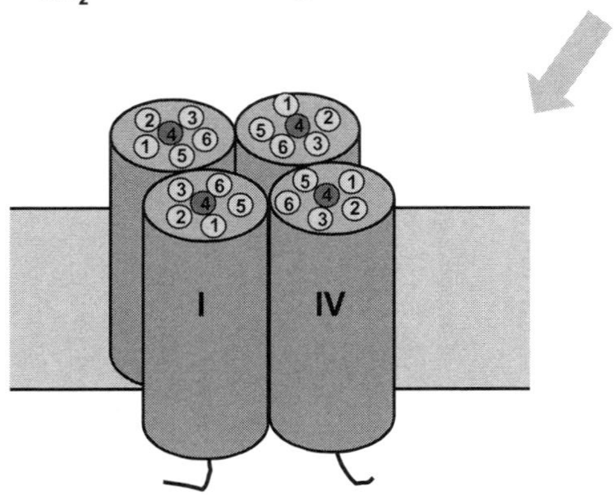

channel	gene	TTX sensitivity	primary localization	implicated in pain
$Na_v1.1$	SCN1A	S	CNS	
$Na_v1.2$	SCN2A	S	CNS	
$Na_v1.3$	SCN3A	S	PNS (post injury)/CNS	Yes
$Na_v1.4$	SCN4A	S	skeletal muscle	
$Na_v1.5$	SCN5A	R	heart	
$Na_v1.6$	SCN8A	S	PNS	
$Na_v1.7$	SCN9A	S	PNS	Yes
$Na_v1.8$	SCN10A	R	PNS	Yes
$Na_v1.9$	SCN11A	R	PNS	Yes

S: Sensitive; R: Resistant; TTX: Tetrodotoxin
C(P)NS: Central(Peripheral) Nervous System

Figure 21.3 Voltage-gated sodium channels: structure, classification, primary localization, and role in pain. Secondary structure of Na_v channel α and β subunits (**top**) and assembly into a pore-forming sodium channel by the four homologous domains (I–IV) with a pseudo-fourfold symmetry (**lower left**). Four of the nine functional Na_v channels are implicated in chronic pain (**lower right**). See the insert for color representation of this figure.

voltage sensitivity) and/or cellular localization of the channel (Tseng et al., 2007). To date, nine functional α subunits, $Na_v1.1$–$Na_v1.9$, and four β subunits, β_1–β_4, have been identified from mammalian species (Goldin et al., 2000; Catterall et al., 2003; Tseng et al., 2007).

Na_v channels reside in one of three classes of conformational states: closed (or resting), open, or inactivated. Transitions from one state to another involve conformational changes of the channel protein. At hyperpolarized membrane potentials (e.g., −100 mV), the channel is in the closed state, preventing Na^+ ions from flowing through the channel. It opens rapidly (<1 ms) upon membrane depolarization and selectively allows Na^+ ions into the cell down its electrochemical gradient. Continued depolarization causes the channel to enter a nonconducting, fast inactivated state within milliseconds. Slower (seconds or longer) inactivation, whose molecular mechanism is less well understood, also occurs after prolonged depolarization. Recovery from inactivation is necessary before the channel can be available again for activation. These functional characteristics help to shape the important physiological roles that Na_v channels play. For example, activation of these channels is responsible for the rapid rising phase of action potentials (Catterall, 2000a). Fast inactivation contributes to the termination of action potentials (McCollum et al., 2003; Ulbricht, 2005) and regulation of action potential firing frequency (Herzog et al., 2003), whereas slow inactivation plays a role in limiting the overall excitability of the cell (Nau and Wang, 2004; Rogawski and Loscher, 2004). Indeed, as we shall see later in this section, many Na_v channel blockers in clinical use exploit the functional significance of Na_v channel inactivation to achieve therapeutic efficacy by stabilizing the channel in the inactivated state.

No fewer than six classes of neurotoxins can modify Na_v channel function through interaction with various sites on the channel (Catterall et al., 2003). Na_v channels can be divided into two groups based on their sensitivity to tetrodotoxin (TTX), a highly Na_v channel-selective, small-molecule neurotoxin from pufferfish. Binding to neurotoxin receptor site 1 at the outer pore (Penzotti et al., 1998), TTX potently (~nM) and tonically (i.e., with no preference for any particular state) blocks six of the nine Na_v isoforms ($Na_v1.1$–$Na_v1.4$, $Na_v1.6$ and $Na_v1.7$). The other three ($Na_v1.5$, $Na_v1.8$, and $Na_v1.9$) are resistant to TTX. Both TTX-sensitive (TTX-S) and TTX-resistant (TTX-R) isoforms are capable of generating action potentials in sensory neurons (Renganathan et al., 2001). Binding of alkaloid toxins such as veratridine and batrachotoxin occurs at a different site (site 2) located in S6 and promotes activation and/or destabilizes inactivation. Many drugs in clinical use, including local anesthetics and a number of anticonvulsants and antiarrythmics, inhibit veratridine and batrachotoxin binding and enhance Na_v channel inactivation by binding to a region in the inner half of S6 that at least partially overlaps with site 2 (Ragsdale et al., 1996; Linford et al., 1998). Preferential binding of these drugs to the inactivated state (e.g., by shifting the voltage dependence of steady-state inactivation in the hyperpolarizing direction, which makes the channel less available for activation than in the absence of a drug) is thought to underlie the phenomenon of use-dependent inhibition, wherein the degree of drug inhibition increases with the increase of frequency of channel activation and inactivation. Use-dependent inhibition is favored over tonic inhibition because drugs that exhibit use dependence are more selective in inhibiting neurons that fire action potentials hyperactively (thought to occur during pain signaling) without much affecting neurons involved in normal, "housekeeping" signaling. These drugs also tend to exhibit voltage-dependent inhibition, blocking the channel more potently at depolarized than hyperpolarized potentials. This is probably largely, if not entirely, secondary to the voltage dependence of inactivation. As with use-dependent inhibition, voltage-dependent inhibition is also desirable in that abnormally hyperactive neurons tend to be depolarized.

21.5.1.2 Role of Na_v Channels in Chronic Pain. A large body of evidence (molecular, biochemical, genetic, pharmacological, etc.) exists that illustrates the critical role of Na_v channels in chronic pain. Multiple subtypes of Na_v channels are expressed in neurons along pain pathways. Their level and/or pattern of expression as well as functional properties may change after nerve injury or tissue inflammation in ways that produce neuronal hyperexcitability, which contributes to spontaneous ectopic activity in chronic pain states. This provides a mechanistic basis for using Na_v channel blockers in the treatment of chronic pain.

Expression of the TTX-R $Na_v1.8$ (a.k.a. SNS or PN3) is restricted almost exclusively to peripheral sensory neurons (Waxman and Wood, 1999; Amaya et al., 2000), making it an attractive target for the development of novel, $Na_v1.8$-selective blockers. $Na_v1.8$ has slow inactivation and fast repriming (recovery from inactivation) kinetics (Gold, 1999) and a high threshold for activation and steady-state inactivation. Aside from being a major contributor to action potentials in nociceptors (Renganathan et al., 2001), the biophysical properties of $Na_v1.8$ make it particularly suited and important for generating high-frequency action potentials (even more so when neurons are depolarized for a prolonged duration). TTX-R sodium currents in peripheral nerves are increased by PGE_2, adenosine, and serotonin, among others, in part through modulation of $Na_v1.8$ (England et al., 1996; Gold, 1999; Cardenas et al., 2001). Expression of $Na_v1.8$ is increased in digital nerve and dorsal root ganglion (DRG) in rat models of inflammatory pain (Tanaka et al., 1998; Coggeshall et al., 2004). Peripheral nerve injury also leads to functional and biochemical changes of $Na_v1.8$ in a way that suggests a potential role of the channel in neuropathic pain (Gold et al., 2003). Furthermore, patients with chronic neurogenic pain or chronic local hyperalgesia and allodynia also have elevated $Na_v1.8$ expression (Coward et al., 2000; Yiangou et al., 2000; Coward et al., 2001).

Studies using animals in which $Na_v1.8$ is selectively knocked out or knocked down provide further insight into the role of the channel and the potential utility of $Na_v1.8$-selective blockers in chronic pain. Although $Na_v1.8$-knockout mice exhibit no change in neuropathic pain behaviors, the development of inflammatory hyperalgesia is delayed in these animals (Kerr et al., 2001; Nassar et al., 2005). Selective knockdown of $Na_v1.8$ by antisense oligodeoxynucleotides or siRNA reverses mechanical allodynia and thermal hyperalgesia after peripheral inflammation and nerve injury (Porreca et al., 1999; Lai et al., 2002; Joshi et al., 2006; Dong et al., 2007). Overall, there is strong evidence in support of an important role for $Na_v1.8$ in chronic pain.

Another TTX-R sodium channel, $Na_v1.9$ (a.k.a. SNS2 or NaN), is also predominantly expressed in nociceptive neurons (Amaya et al., 2000). $Na_v1.9$ exhibits very slow activation and inactivation kinetics (an exception for Na_v channels) and substantial overlap of activation and steady-state inactivation near the resting membrane potential (Cummins et al., 1999; Ostman et al., 2008). These properties enable $Na_v1.9$ to produce a persistent sodium current that amplifies the response of nocicptors to subthreshold stimuli (Herzog et al., 2001). As with $Na_v1.8$, modulation of $Na_v1.9$ by inflammatory mediators also results in increased current amplitude (Rush and Waxman, 2004), helping to maintain inflammation-induced hyperalgesia. Consistent with this, hypersensitivity and spontaneous pain behavior produced by peripheral inflammation (but not by nerve injury) are diminished in $Na_v1.9$-knockout mice (Priest et al., 2005; Amaya et al., 2006).

$Na_v1.3$, which is TTX-sensitive and exhibits rapid recovery from inactivation, is thought to be responsible for the rapidly repriming TTX-S currents in injured DRG neurons and important in maintaining sustained high-frequency firing that underlies neuronal hyperexcitability (Cummins and Waxman, 1997). It is virtually undetectable in normal adult DRG or trigeminal neurons, but is up-regulated in DRG and dorsal horn neurons after inflammation or nerve injury (Cummins and Waxman, 1997; Black et al., 1999; Dib-Hajj et al., 1999; Black et al., 2004). $Na_v1.3$-knockout mice do not show impaired neuropathic pain behavior (Nassar et al., 2006). However, studies using selective knockdown of $Na_v1.3$ by AS ODN indicate that, following peripheral nerve injury, $Na_v1.3$ is a major contributor to the hyperexcitability of dorsal horn neurons and the injury-induced allodynia and hyperalgesia (Hains et al., 2004).

Another TTX-S sodium channel, $Na_v1.7$ (a.k.a. PN1), has received considerable attention in recent years for its participation in pain signaling. $Na_v1.7$ is preferentially expressed in nociceptive and sympathetic neurons (Toledo-Aral et al., 1997). It produces a fast activating and inactivating current. Because it recovers from inactivation only slowly, it is unable to sustain high frequency firing, making it unlikely to have a major role in the depolarizing upstroke of repetitive action potentials (Cummins et al., 1998). However, $Na_v1.7$ also develops closed-state inactivation very slowly. This crucial feature allows it to pass depolarizing currents in response to slow, small depolarizations near the resting potential, suggesting that it may serve as a "threshold" channel to enhance stimulus depolarizations (Cummins et al., 1998).

$Na_v1.7$ is up-regulated in DRG neurons in animal studies of inflammatory pain (Black et al., 2004). Nerve growth factors (NGF) may contribute to this up-regulation since it is known to increase $Na_v1.7$ expression (Toledo-Aral et al., 1995) and inflammatory pain responses evoked by NGF are reduced in mice whose nociceptors lack $Na_v1.7$ (Nassar et al., 2004). Conditional $Na_v1.7$-knockout (Nassar et al., 2004) and knockdown (Yeomans et al., 2005) studies in mice show that inflammation-induced mechanical and/or thermal hyperalgesia are either reduced or abolished in these animals. These preclinical studies support the notion that $Na_v1.7$ plays a key role in inflammatory pain. Although conditional $Na_v1.7$-knockout studies in mice do not implicate $Na_v1.7$ in neuropathic pain (Nassar et al., 2005), clinical studies of human patients suggest otherwise.

A number of gain-of-function mutations in $Na_v1.7$ have been linked to two painful human disorders. One of them, inherited erythromelalgia (IE), is characterized by severe chronic burning pain sensations in the extremities (Dib-Hajj et al., 2007). $Na_v1.7$ mutations found in these patients led to hyperexcitability in DRG neurons (Dib-Hajj et al., 2005; Harty et al., 2006; Rush et al., 2006). The second disorder, with distinct $Na_v1.7$ mutations from those in IE, is an autosomal dominant chronic pain syndrome called paroxysmal extreme pain disorder (PEPD) and is characterized by severe burning rectal, ocular, and submandibular pain sensations (Dib-Hajj et al., 2007). These data indicate that $Na_v1.7$ is sufficient to cause neuropathic pain in humans. There is also compelling evidence that $Na_v1.7$ is necessary for humans to perceive pain, because loss of $Na_v1.7$ function causes an individual to be completely indifferent to pain (Cox et al., 2006; Ahmad et al., 2007; Goldberg et al., 2007). Importantly, individuals lacking functional $Na_v1.7$ appear otherwise fairly normal (no overt motor, cognitive, or cardiac deficits), sparking widespread interest in the prospects of selectively targeting $Na_v1.7$ for the treatment of pain disorders.

21.5.1.3 Inhibition of Na_v Channels for the Treatment of Chronic Pain. The compelling evidence for the involvement of Na_v channels in chronic pain argues for the development of Na_v channel blockers for treating these pain syndromes. Currently available drugs that block Na_v channels were originally developed for treating various indications. They were discovered empirically using traditional, rather than gene target-based, discovery approaches and were only subsequently found to inhibit Na_v channels. A number of these drugs, mostly anticonvulsants, tricyclic antidepressants, and local anesthetics, were later found to be also effective in treating neuropathic pain (see Table 21.1).

Among anticonvulsants, carbamazepine and lamotrigine are both use-dependent Na_v channel blockers and bind preferentially to the inactivated state (Willow et al., 1985; Xie et al., 1995). Carbamazepine is effective for treating trigeminal neuralgia (Wiffen et al., 2005) and PEPD (Fertleman et al., 2006), but not very effective for diabetic neuropathy and certain other

TABLE 21.1. Select Small-Molecule Na_v Channel Blockers for the Treatment of Chronic Pain

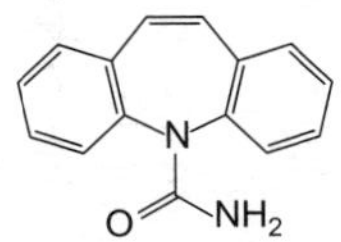

Carbamazepine

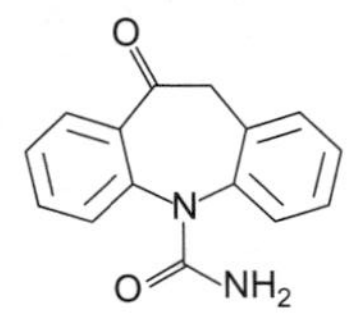

Oxcarbazepine

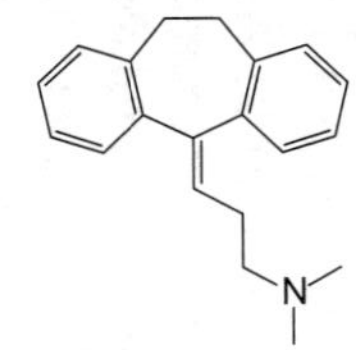

Amitriptyline

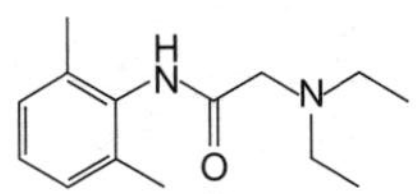

Lidocaine

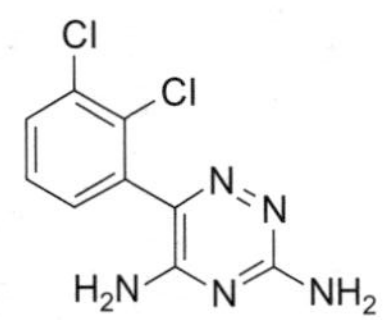

Lamotrigine

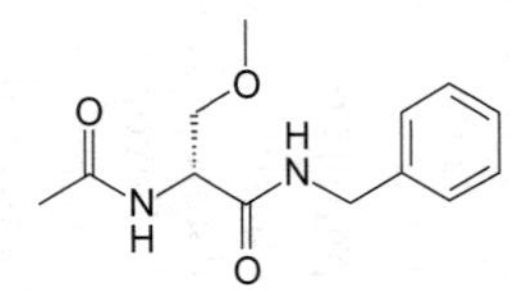

Lacosamide

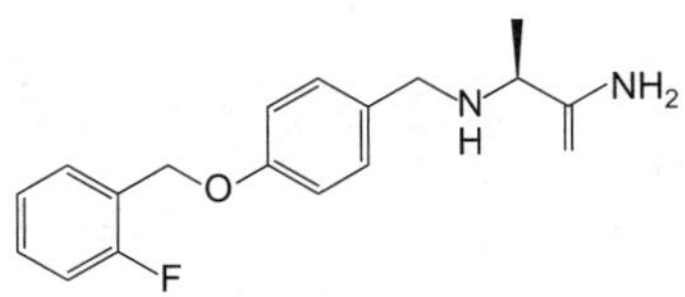

Ralfinamide

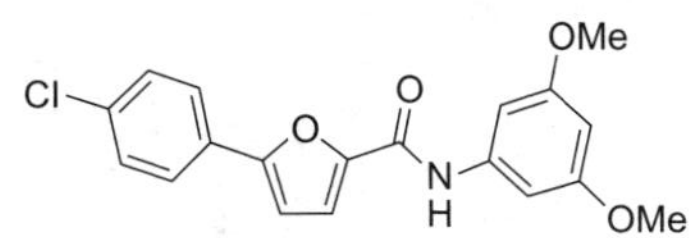

A-803467

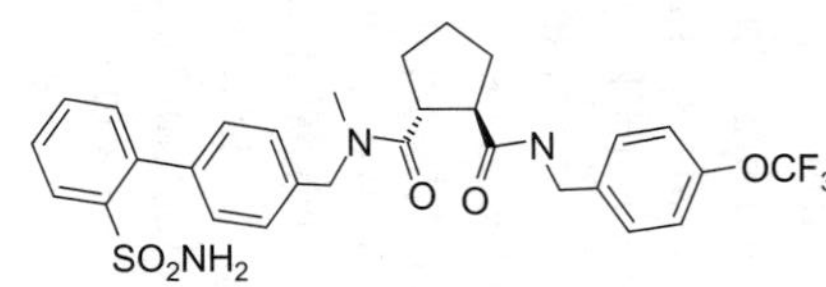

CDA54

painful neuropathic conditions (Wiffen et al., 2005; Eisenberg et al., 2007). Lamotrigine also has efficacy for several types of neuropathic pain, including trigeminal neuralgia, painful diabetic neuropathy, HIV-associated neuropathy, and post-stroke pain (Vestergaard et al., 2001; Eisenberg et al., 2007). But lamotrigine is not effective in treating chemotherapy-induced or spinal cord injury-induced neuropathic pain (Hargus and Patel, 2007). Another anticonvulsant, oxcarbazepine, is an improved derivative of carbamazepine and is thought to derive its anticonvulsant efficacy mainly from inhibition of Na_v channels (Ambrosio et al., 2002). Several clinical studies demonstrate that oxcarbazepine is similar in effectiveness to carbamazepine for treating trigeminal neuralgia and diabetic neuropathy, but has less severe side effects (nausea, dizziness, etc.) than carbamazepine (Hargus and Patel, 2007). Other clinical studies indicate that oxcarbazepine is effective in treating several forms of neuropathic pain (Magenta et al., 2005), including pain associated with multiple sclerosis (Solaro et al., 2007). A new investigational anticonvulsant drug, lacosamide, is in Phase III clinical trials for both epilepsy and neuropathic pain. In a Phase II study, lacosamide is shown to be effective in painful diabetic neuropathy (Rauck et al., 2007). Although it is also an Na_v channel blocker with a higher affinity for inactivated channels, lacosamide is differentiated from the other Na_v channel-blocking drugs by virtue of its preferential interaction with channels in the slow, rather than fast, inactivated state (Errington et al., 2008; Sheets et al., 2008), suggesting that it may be more effective than those drugs in selectively blocking hyperactivity in neurons that are chronically depolarized.

Tricyclic antidepressants have long been used to treat neuropathic pain, sometimes as first-line therapy. Of these, amitriptyline is the most effective and commonly used for the management of neuropathic pain. Clinical studies show that amitriptyline is effective in treating diabetic neuropathy (Turkington, 1980; Max et al., 1992), pain associated with postherpetic neuralgia (Watson et al., 1982; Max et al., 1988), central pain (Leijon and Boivie, 1989), and nerve injury pain (Kalso et al., 1996). Amitriptyline has a higher affinity for the open and inactivated states of Na_v channels (Wang et al., 2004; Leffler et al., 2007) and blocks these channels in a use-dependent manner and at concentrations that are effective for treating neuropathic pain (Song et al., 2000a; Wang et al., 2004; Dick et al., 2007), suggesting that its effect on Na_v channels may be clinically important.

Local anesthetics are very effective state-dependent blockers of Na_v channels, with higher affinities for open/inactivated states (Ragsdale et al., 1996). Clinically, LAs are primarily used to provide local anesthesia for acute pain relief. The most commonly used LA for neuropathic pain treatment is lidocaine, which can be effectively administered either topically via a patch or systemically by intravenous injection. In one study, lidocaine patches (5%) as an add-on therapy alleviate both allodynia and ongoing pain in patients with diverse focal peripheral neuropathic pain syndromes, including postherpetic neuropathy, and compare reasonably well with topically applied capsaicin and systemic administration of gabapentin in separate studies (Meier et al., 2003). Other studies indicate that lidocaine patches (5%) are effective in treating painful diabetic neuropathy (Barbano et al., 2004) and idiopathic distal polyneuropathy (Herrmann et al., 2005) and may be useful for treating lower back pain (Hines et al., 2002) and central neuropathic pain (Hans et al., 2008). Intravenous injection of lidocaine is also shown to be effective in easing neuropathic pain (Ferrante et al., 1996; Tremont-Lukats et al., 2006), including diabetic neuropathy (Bach et al., 1990).

It is important to note that many of the currently available drugs that block Na_v channels also have effects on other channels/proteins. Therefore, it is possible that inhibition of Na_v channels does not fully, or even primarily, account for their efficacy in chronic pain. In addition, these drugs tend to have poor therapeutic windows at least in part because of their nonselective nature in blocking various Na_v isoforms, including those expressed in the peripheral and central nervous systems as well as the cardiovascular system (Catterall, 2000a). As a result, they often have limited clinical utility for treating pain. More recent, largely gene-target based efforts aimed at identifying novel Na_v channel blockers with improved therapeutic index have resulted in analgesic agents with various intriguing properties.

One such molecule is ralfinamide (formerly NW-1029), which, in Phase II clinical trials, is well tolerated (no difference from placebo) and effective in patients with multiple forms of neuropathic pain (http://www.newron.com). A primary mechanism of action of ralfinamide is the use- and frequency-dependent inhibition of both TTX-S and TTX-R Na_v channels, particularly in the inactivated state (Stummann et al., 2005). Interestingly, ralfinamide selectively suppresses firing of action potentials in capsaicin-responsive, nociceptive neurons, but has no such effect on capsaicin-unresponsive neurons (Yamane et al., 2007). This selectivity of ralfinamide may contribute to the effectiveness of the drug in reducing inflammatory and neuropathic pain and to the side effect profile. Ralfinamide is in phase IIb/III trials for neuropathic low back pain.

Another way to achieve selectivity, as intuitive as it is challenging, is to develop novel Na_v channel blockers that are subtype-selective, a property that current Na_v channel-blocking drugs are lacking. One such molecule, A-803467, has recently been reported (Jarvis et al., 2007). A-803467 is a potent blocker of $Na_v1.8$ with higher affinity for the inactivated state and >100-fold selectivity against $Na_v1.2$, $Na_v1.3$, $Na_v1.5$, and $Na_v1.7$. It is the first reported small molecule with selectivity for $Na_v1.8$ over cardiac and other neuronal Na_v channels. A-803467 potently blocks TTX-R sodium currents and spontaneous and evoked action potentials in dissociated rat DRG neurons. It also attenuates the firing in spinal dorsal horn neurons *in vivo*. A-803467 is efficacious in several animal models of inflammatory and neuropathic pain. Despite showing good CNS penetration, A-803467 does not significantly alter motor functions, coordination, or balance, consistent with its $Na_v1.8$-selective profile. The discovery of A-803467 illustrates not only the feasibility of identifying highly subtype-selective Na_v channel blockers, but also the potential of selective $Na_v1.8$ blockers for the treatment of chronic pain with less toxicity.

Yet another strategy to achieve selectivity, as exemplified by CDA54 (Brochu et al., 2006), is to develop compounds that effectively can only block peripheral neuronal sodium channels *in vivo* by virtue of poor CNS penetration. CDA54 blocks multiple subtypes of (preferentially inactivated) Na_v channels with similar potency, including $Na_v1.2$ (mainly expressed in the CNS), $Na_v1.7$, and $Na_v1.8$ (both preferentially expressed in peripheral sensory neurons), and is effective in two animal models of neuropathic pain. However, thanks to its poor CNS penetration (with a brain to plasma concentration ratio of 0.03), CDA54 does not affect motor coordination at a dosage that is effective in pain models. This is in contrast to clinically used Na_v blockers, which accumulate in the CNS and cause impaired motor coordination in animals and CNS side effects in man at all efficacious doses.

21.5.2 Voltage-Gated Calcium Channels (Ca_v)

Voltage-gated calcium channels are expressed in various excitable cells, including neurons in pain pathways. They regulate Ca^{2+} influx across the plasma membrane and play a critical role in coupling excitation to multiple processes, including muscle contraction, neurotransmitter release, gene expression, exocytosis, and Ca^{2+}-dependent enzyme activity.

21.5.2.1 Basic Molecular, Structural, Functional, and Pharmacological Properties. Similar to Na_v channels, Ca_v channels consist of a large (190–250 kD) pore-forming α_1 subunit and two or more of the auxiliary subunits, $\alpha_2\delta$ (160–180 kD), β (50–70 kD), and γ (20–30 kD). The molecular architecture of Ca_v channels also resembles that of Na_v channels in that the four homologous domains (I–IV) of the α_1 subunit, each also consisting of six transmembrane segments (S1–S6, with S4 being the voltage sensor), form the Ca^{2+} conduction pathway. In particular, segments of S5 and S6, along with the membrane-reentrant pore loops between them, line the channel pore. The extracellular pore loops contain amino acid residues that are critical for Ca^{2+} selectivity. Residues in S6 and in the intracellular loop linking domains I and II are implicated in channel inactivation (Stotz et al., 2004). To date, 10 α_1 (Catterall, 2000b), four β ((Birnbaumer et al., 1998), four $\alpha_2\delta$ (Davies et al., 2007), and eight γ subunits (Chen et al., 2007) have been cloned and characterized. Most of the specific functional and pharmacological characteristics reside in the subtype-defining α_1 subunit. The auxiliary subunits play important roles in, among others, channel trafficking and assembly, the modulation of channel function, and pharmacology (Birnbaumer et al., 1998; Klugbauer et al., 2003; Chen et al., 2007). Based on primary sequence homology as well as functional and pharmacological properties, Ca_v channels are divided into three subtypes: $Ca_v1.x$ (x = 1–4, a.k.a. α_{1S}, α_{1C}, α_{1D}, and α_{1F}, the α_1 subunits for L-type

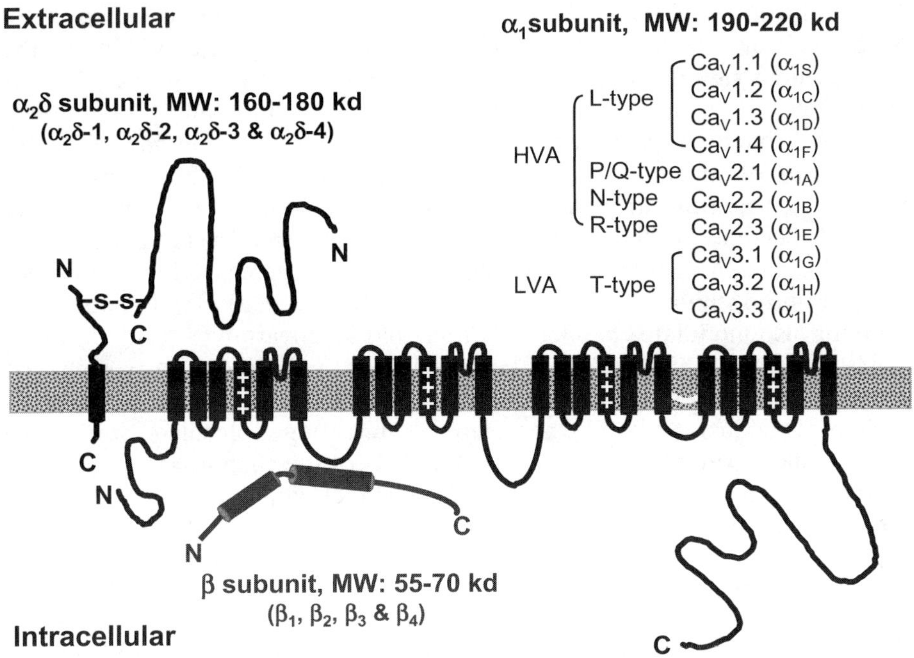

Figure 21.4. Voltage-gated calcium channels—structure and classification. See the insert for color representation of this figure.

channels), $Ca_v2.x$ (x = 1–3, a.k.a. α_{1A}, α_{1B}, and α_{1E}, the α_1 subunits for P/Q-, N-, and R-type channels, respectively) and $Ca_v3.x$ (x = 1–3, a.k.a. α_{1G}, α_{1H}, and α_{1I}, the α_1 subunits for T-type channels). (Also see Fig. 21.4). Ca_v3 channels have a low-voltage threshold for activation and fast rate of inactivation. The other subtypes are activated by more depolarized potentials and also inactivate more slowly. In addition to serving as vehicles of Ca^{2+} influx that triggers various downstream events, Ca_v channels also contribute directly to cell excitability by membrane depolarization that results from the Ca^{2+} entry upon channel activation. Ca_v1 channels can be blocked by various dihydropyridines, phenylalkylamines, and benzothiazepines and potentiated by BAY-K-8644. Ca_v3 channels are sensitive to inhibition by mibefradil and the scorpion toxin kurtoxin, among others. Peptide toxins potently inhibit Ca_v2 channels (e.g., ω-agatoxin IVA for $Ca_v2.1$, ω-conotoxin MVIIA and GVIA for $Ca_v2.2$, and SNX-482 for $Ca_v2.3$), which, in part, helped to validate $Ca_v2.2$ as a target for chronic pain. Small-molecule inhibitors of $Ca_v2.2$ for the treatment of neuropathic pain are also an active area of drug discovery efforts in the pharmaceutical industry. In the remainder of this section, we will focus on this Ca_v channel subtype with respect to its role in pain signaling and its modulation for the treatment of chronic pain. For a general review of Ca_v channels, see Catterall (2000b).

21.5.2.2 Function and Modulation of N-Type Calcium Channels. The N-type calcium channel is a heteromeric protein complex consisting of the pore-forming subunit, $Ca_v2.2$, and auxiliary $\alpha_2\delta$-1 and β_3 subunits (Witcher et al., 1993). It has a high-voltage threshold for activation and preferentially undergoes inactivation via subthreshold, closed states (Patil et al., 1998). This mechanism of inactivation results in rapidly decreasing Ca^{2+} currents in response to repetitive firing of high-frequency action potentials, because fewer channels become available for activation each time they cycle through these states during a firing burst. Functional expression is increased in the presence of $\alpha_2\delta$-1 and β subunits (Williams et al., 1992; Stea et al., 1993). In addition, β subunits also modulate channel activation and inactivation (Stea et al., 1993).

N-type calcium channels can be modulated by various factors, including synaptic proteins, GPCRs, and protein kinases. Binding of synaptic proteins, such as syntaxin-1, SNAP-25, and Rab3A interacting molecule (Rim), to the cytoplasmic linker between domains II and III of the channel (the "synprint" region) is thought to help colocalize the channel with synaptic vesicles and target the channel to the presynaptic terminus (Zamponi, 2003). Syntaxin-1 has also been shown to directly modulate the channel function by right-shifting the voltage dependence of inactivation (Stanley, 2003). Activation of $G_{i/o}$-coupled GPCRs, such as μ-opioid receptors, results in the binding of G-protein $\beta\gamma$ subunits to and voltage-dependent inhibition of N-type calcium channels, which is in part responsible for morphine-mediated analgesia (Seward et al., 1991; Bourinet et al., 1996). GPCRs can also mediate internalization of N-type calcium channels directly (Beedle et al., 2004; Altier et al., 2006) or indirectly (Puckerin et al., 2006; Tombler et al., 2006). Protein kinase C increases N-type calcium channel activity by either reversing G-protein-mediated channel inhibition (Barrett and Rittenhouse, 2000) or recruiting new $Ca_v2.2$ subunits to the plasma membrane (Zhang et al., 2008).

21.5.2.3 Role of N-Type Calcium Channels in Chronic Pain. Both preclinical and clinical studies indicate that N-type calcium channels play an important role in chronic pain. $Ca_v2.2$ is located throughout the presynaptic terminal of peripheral and central neurons, such as those in the superficial dorsal horn of the spinal cord (Gohil et al., 1994; Westenbroek et al., 1998). It is a major player in the regulation of neurotransmitter release and synaptic transmission critical in pain signaling.

Expression of $Ca_v2.2$ and $\alpha_2\delta$-1 is up-regulated in DRG neurons and the dorsal horn in animal models of neuropathic pain and correlates temporally with the development of allodynia (Luo et al., 2001, 2002; Newton et al., 2001; Abe et al., 2002; Cizkova et al., 2002). Mice lacking $Ca_v2.2$ display decreased thermal and mechanical pain responses under inflammatory and nerve injury conditions, suggesting a critical role of the channel in the initiation and/or maintenance of inflammatory and neuropathic pain states (Hatakeyama et al., 2001; Kim et al., 2001; Saegusa et al., 2001). In addition, pain responses in mice lacking the β_3 subunit are also reduced under inflammatory conditions

(Murakami et al., 2002). Blockade of $Ca_v2.2$ by selective peptide inhibitors is effective in several animal models of chronic pain (Chaplan et al., 1994; Malmberg and Yaksh, 1994; Bowersox et al., 1996; Scott et al., 2002). Prialt® (Ziconotide), a synthetic version of ω-conotoxin-MVIIA (found in the venom of the predatory marine snail *Conus magus*) that selectively blocks N-type calcium channels, is effective in human patients with acute and chronic pain (Brose et al., 1997; Atanassoff et al., 2000; Staats et al., 2004), providing clinical validation for the role of N-type calcium channels in chronic pain.

21.5.2.4 Modulation of N-Type Calcium Channels for the Treatment of Chronic Pain. N-type calcium channels are potently blocked by several ω-conopeptides, including GVIA, CVID, and MVIIA. GVIA blocks the channel irreversibly. Although intrathecal administration of GVIA is effective against neuropathic pain in animals (Scott et al., 2002), it may be difficult clinically to achieve safe and stable dosing due to the irreversible nature of the block. As with GVIA, CVID (under the name of AM336 in clinical trials) is also effective in an animal model of neuropathic pain (Scott et al., 2002). In addition, it has a larger therapeutic window than GVIA and MVIIA in animals and may produce less adverse effects in humans.

Ziconotide binds with high affinity to the outer pore/vestibule region of $Ca_v2.2$ (Feng et al., 2001), consistent with a direct pore-blocking mechanism, and potently and tonically inhibits N-type Ca^{2+} currents (Jain, 2000; Feng et al., 2003) as well as neurotransmitter release in cultured neurons (Smith et al., 2002). Ziconotide is efficacious in several animal models of chronic pain (Chaplan et al., 1994; Malmberg and Yaksh, 1994; Yamamoto and Sakashita, 1998; Horvath et al., 2001). Patients that receive epidural ziconotide before, during, and after surgery require significantly less postoperative morphine than patients that receive placebo (Atanassoff et al., 2000). In a Phase III study (Staats et al., 2004) with cancer and AIDS patients who are intolerant of or refractory to opioid treatment, improvement in the Visual Analog Scale of Pain Intensity (VASPI) scores in ziconotide-treated patients (intrathecal administration) is significantly higher than in placebo-treated patients. Ziconotide is also reported to be very effective in a case study of a patient with chronic neuropathic pain (Brose et al., 1997). Unlike morphine, there is no development of tolerance associated with the use of ziconotide (Jain, 2000). Ziconotide intrathecal infusion (Prialt®) has been approved by the US Food and Drug Administration (FDA) for the management of severe chronic pain in patients for whom intrathecal therapy is warranted and who are intolerant of or refractory to other analgesics (also see Table 21.2).

Although ziconotide is effective for the treatment of chronic pain, it requires intrathecal administration and careful titration. A major adverse effect of ziconotide is symptomatic orthostatic hypotension (Jain, 2000). Dizziness, ataxia, sedation, nausea, bradycardia, and other side effects have also been reported (Jain, 2000).

TABLE 21.2. N-Type Calcium Channel Modulators for the Treatment of Chronic Pain

Compound	Trade Name (company)	Structure	Mol. Target	Indication	Status
Ziconotide	Prialt (Elan)	25-residue peptide	$Ca_v2.2$	Severe chronic pain	Launched
Gabapentin	Neurontin (Pfizer)	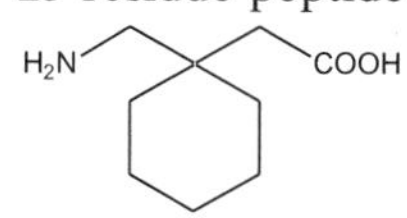	$\alpha_2\delta$-1	Neuropathic pain associated with postherpetic neuralgia	Launched
Pregabalin	Lyrica (Pfizer)	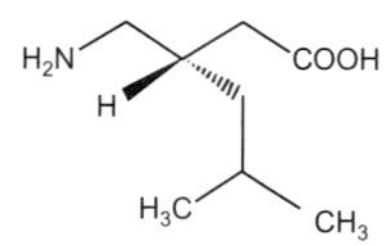	$\alpha_2\delta$-1	Neuropathic pain associated with postherpetic neuralgia/diabetic peripheral neuropathy; fibromyalgia	Launched
NMED-160	(Neuromed/Merck)	Small molecule	$Ca_v2.2$	Neuropathic pain	Phase II[a]

[a]Discontinued.

The anticonvulsants gabapentin (Neurontin®) and its structural analog pregabalin (Lyrica®) appear to have a novel mechanism of action (see below) and are widely used for treating various neuropathic pain conditions. In preclincal studies, both drugs are efficacious in models of chronic pain (Luo et al., 2002; Field et al., 2006). Results from clinical trials demonstrate that gabapentin and pregabalin are effective in treating pain associated with postherpetic neuralgia (Rowbotham et al., 1998; Dworkin et al., 2003), diabetic neuropathy (Backonja et al., 1998; Rosenstock et al., 2004), fibromyalgia (Crofford et al., 2005; Arnold et al., 2007), spinal cord injury (Tzellos et al., 2008), and postsurgical pain (Gilron, 2007). Both drugs are FDA approved for treating postherpetic neuralgia. Pregabalin is additionally approved for the treatment of fibromyalgia (the only drug approved by FDA for treating this condition) and diabetic neuropathy. Sedation, dizziness and ataxia are among some of the common adverse effects of both drugs (Gilron, 2007). Pregabalin is also a Schedule V controlled substance with abuse potential.

Although the mechanism of action of gabapentin and pregabalin may still be a subject of debate (Sills, 2006), it is firmly established that these drugs bind to the $\alpha_2\delta$-1 subunit of Ca_v channels (Gee et al., 1996; Field et al., 2006), which is important in Ca_v channel membrane trafficking and assembly (Klugbauer et al., 2003). It is thought that interaction of these drugs with $\alpha_2\delta$-1 may decrease the cell surface expression of functional $Ca_v2.2$ channels (Hendrich et al., 2008). Importantly, pregabalin loses its analgesic efficacy in mice knocked-in with an $\alpha_2\delta$-1 mutant that reduces the binding affinity of gabapentin and pregabalin (Field et al., 2006), further illustrating the critical role of this auxiliary subunit in mediating the effects of gabapentin and pregabalin. It remains controversial, however, as to whether these drugs directly inhibit neuronal Ca^{2+} currents (Sills, 2006). In addition, it is not clear if/how binding of gabapentin or pregabalin to $\alpha_2\delta$-1 can result in selective targeting of $Ca_v2.2$, because this mechanism of action would be expected to also affect other Ca_v channels.

Target-based efforts to develop selective and orally bioavailable, small-molecule N-type calcium-channel blockers have led to the identification of a number of novel chemotypes of compounds. Incorporation of key structural features of $Ca_v2.2$-selective conotoxins into the rational design of small molecule conotoxin mimetics results in $Ca_v2.2$ blockers with modest (~2–20 μM) potency but relatively high (>20-fold) selectivity against $Ca_v2.1$ (Baell et al., 2004; Schroeder et al., 2004). Several novel scaffolds have been reported to potently block N-type calcium channels (Hu et al., 1999; Song et al., 2000b; Snutch et al., 2001; Seko et al., 2002; Franco et al., 2004; Teodori et al., 2004) and have efficacy in animal models of chronic pain (Seko et al., 2002; Bowen et al., 2004; Teodori et al., 2004). Intriguingly, some of these compounds also exhibit strong use-dependent inhibition of N-type calcium channels (Snutch et al., 2001; Pan et al., 2004), which may be important for increasing the therapeutic window of selective $Ca_v2.2$ blockers for the treatment of chronic pain (Winquist et al., 2005).

21.5.3 TRPV1

The family of TRP channels consists of 28 members that are divided into six subfamilies, one of which is TRPV (vanilloid). TRPV1 is the founding and most extensively studied member of the six known TRPV channels. The cloning of TRPV1 (Caterina et al., 1997) has opened the door to our understanding of the molecular mechanism of sensory transduction of thermal and chemical stimuli. The recognition of its role in transduction/integration of diverse pain stimuli has led to the novel approach of developing selective TRPV1 antagonists to block pain signaling at the sensor, versus the traditional approach of blocking inflammatory responses or other stages such as signal conduction or transmission.

21.5.3.1 Basic Molecular, Structural, Functional, and Pharmacological Properties. TRPV1 is a cation-selective, outwardly rectifying, homotetrameric channel. Each subunit is predicted to have six transmembrane segments (S1–S6) and a molecular weight of ~95 kD. The pore of TRPV1 is in part formed by the loop between S5 and S6 and has a higher permeability for Ca^{2+} over Na^{+}. TRPV1 can be activated by a wide variety of stimuli, including natural products such as vanilloids [e.g., capsaicin, the pungent ingredient in hot chili peppers, and its potent analog, resiniferatoxin (RTX)], endogenous ligands such as endocannabinoids (e.g., anandamide) and other lipids (e.g., *N*-arachidonoyldopamine), eicosanoids [e.g., 15-(*S*)-HPETE], protons and polyamines, synthetic agents such as 2-aminoethoxydiphenyl borate (2-APB) and olvanil, as well as nonchemical stimuli such as membrane depolarization (electrical) and noxious heat (>42 °C; thermal). TRPV1 is sensitized (i.e., the activation threshold is lowered) by proinflammatory mediators such as bradykinin and prostaglandins and is inhibited by phosphatidylinositol 4,5-bisphosphate (PIP_2) and Ca^{2+}/calmodulin. Subthreshold levels of any type of agonist (e.g., protons) also sensitize TRPV1 activation by other types of agonists (e.g., capsaicin). [For a review, see Szallasi et al. (2007); also see Fig. 21.5.]

The polymodal nature of TRPV1 activation has led to the identification of multiple sites/regions involved in channel activation by different stimuli, including capsaicin (linker of S2 and S3, S3/S4) (Jordt and Julius, 2002; Gavva et al., 2004), protons and polyamines (near the outer pore) (Jordt et al., 2000; Ahern et al., 2006; Klionsky et al., 2006) and heat (C-terminus) (Brauchi et al., 2006). In the continued presence of an agonist, TRPV1 undergoes Ca^{2+}-dependent desensitization that is thought to involve phosphatases (e.g., calcineurin) and calmodulin (Tominaga and Tominaga, 2005).

Many TRPV1 antagonists have been identified. These include naturally occurring plant products (e.g., thapsigargin and yohimbine), acylpolyamine toxins from the funnel web spider venom, and endogenous molecules such as adenosine, fatty acids (e.g., eicosapentaenoic acid) and dynorphins. A multitude of synthetic TRPV1 antagonists (e.g., capsazepine, an early capsaicin analog) have been reported to have analgesic properties (more on these

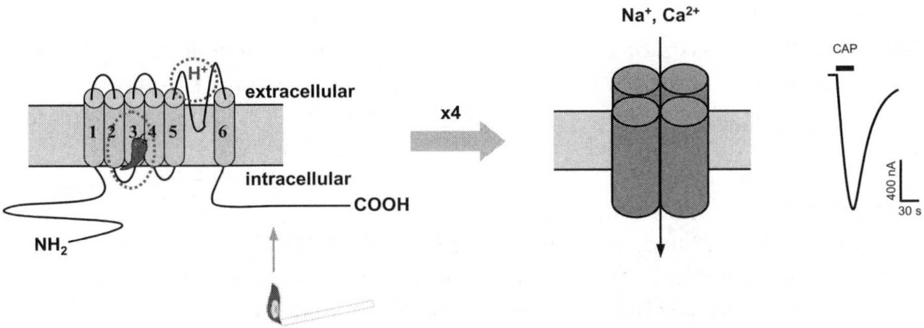

Figure 21.5. TRPV1, a cation-selective channel capable of polymodal gating. TRPV1 is a cation-selective, homotetrameric channel (**middle panel**). Each subunit is predicted to have six transmembrane segments (S1–S6; **left panel**). A wide variety of stimuli can activate TRPV1, including capsaicin, protons, and noxious heat. Left panel shows some of the implicated regions of the channel in these interactions. **Right panel**: Capsaicin-activated currents in *Xenopus* oocytes injected with TRPV1 cRNA. Current trace adapted by permission from Macmillan Publishers Ltd. [*Nature*] (Caterina et al., 1997), copyright 1997.

compounds later). Many of the natural and synthetic small-molecule TRPV1 ligands appear to share at least an overlapping binding site with capsaicin (Immke and Gavva, 2006; Gharat and Szallasi, 2008). Others, such as ruthenium red, a nonselective TRP channel antagonist, and acylpolyamine toxins appear to block the channel pore (Garcia-Martinez et al., 2000; Kitaguchi and Swartz, 2005).

21.5.3.2 Role of TRPV1 in Chronic Pain. Consistent with a role in pain signaling, TRPV1 is highly expressed in peripheral nociceptors, that is, small- to medium-sized sensory neurons (unmyelinated C-fibers and certain thinly myelinated Aδ fibers). After peripheral inflammation and nerve injury, there is an increase in the level of TRPV1 in the peripheral terminal of nociceptors, thereby increasing peripheral sensitivity to heat and other stimuli (Ji et al., 2002; Rashid et al., 2003a,b). Tissue damage or inflammation causes the release of intracellular factors such as protons, ATP, and various proinflammatory mediators (e.g., substance P, calcitonin gene related peptide, bradykinin, NGF and PGE_2), which can either activate TRPV1 directly (Waldmann, 2001), increase its membrane expression (Ji et al., 2002), or otherwise sensitize the channel (e.g., via phosphorylation or PIP_2 hydrolysis) by lowering its threshold of activation (Chuang et al., 2001; Tominaga et al., 2001). In TRPV1-knockout mice, nociceptive responses to painful thermal stimulation are impaired and thermal hypersensitivity in inflammatory models of pain is greatly decreased (Caterina et al., 2000; Davis et al., 2000). Selective knockdown of TRPV1 using siRNA and antisense approaches is also effective in deseasing pain behavior in rat models of neuropathic pain (Christoph et al., 2006, 2007). After initial channel activation, TRPV1 agonists cause channel desensitization and functional antagonism of TRPV1-expressing nocieptors, which alleviates pain-like behaviors in rodents and pain in humans (Szallasi and Blumberg, 1999). TRPV1 is also expressed in higher brain centers, such as hypothalamus and cortex (Acs et al., 1996; Steenland et al., 2006), that are thought to be important for pain processing. Consistent with this, a selective TRPV1 antagonist with good CNS penetration shows a broader spectrum of analgesic activities than one with similar properties except for poor CNS penetration (Cui et al., 2006). Taken together, these results implicate a critical role of TRPV1 in chronic pain as well as the potential utility of selective TRPV1 antagonists and agonists for treating these conditions.

21.5.3.3 Modulation of TRPV1 for the Treatment of Chronic Pain

21.5.3.3.1 TRPV1 Antagonists. The compelling evidence supporting the role of TRPV1 as a molecular integrator of pain signals has led to discovery programs at a number of pharmaceutical companies that aim to develop selective and orally bioavailable small molecule TRPV1 antagonists for the treatment of inflammatory and other chronic pain. HTS and medicinal chemistry efforts have identified many novel chemotypes of potent and selective TRPV1 antagonists with oral bioavailability (Immke and Gavva, 2006; Szallasi et al., 2007). Generally, these compounds completely reverse capsaicin-induced behavioral changes (e.g., eye wiping, flinching, hypothermia, and secondary mechanical hyperalgesia), indicating good on-target effects. They also tend to be at least partially efficacious in various animal models of inflammatory pain, such as thermal and/or mechanical hyperalgesia induced by complete Freund's adjuvant (CFA) or carrageenan. Some compounds also have efficacy in models of noninflammatory pain, including neuropathic pain, sodium monoiodoacetate-induced osteoarthritic pain, bone cancer pain, *p*-benzoquinone-induced writhing, and colorectal distention-induced electromyographic activity. It remains to be seen whether the efficacy in some of these models (e.g., attenuation of mechanical hyperalgesia, which is not observed in TRPV1 knockout mice) may be mediated by potential off-target activity of these compounds; BCTC, for example, is a potent antagonist of not only TRPV1, but also TRPM8, another TRP channel implicated in chronic pain (Colburn et al., 2007). However, this seems unlikely given the structural diversity of the compounds that show efficacy in these models. Overall, results with TRPV1 antagonists not only help to further validate TRPV1 as a target for chronic pain (at least preclinically), but also demonstrate the potential utility of such compounds for treating these syndromes.

Several TRPV1 antagonists have advanced to various stages of clinical development (see Table 21.3). Notably, in Phase I trials with healthy volunteers, SB-705498 (GlaxoSmithKline) significantly reduces capsaicin-evoked flare and acute heat-evoked pain on nonsensitized skin. Furthermore, it reduces heat-evoked pain after ultraviolet B-evoked inflammation. SB-705498

TABLE 21.3. Small-Molecule TRPV1 Antagonists in Clinical Trials for Pain Indications

Compound	Structure	Company	Indications	Clinical Stage
SB-705498	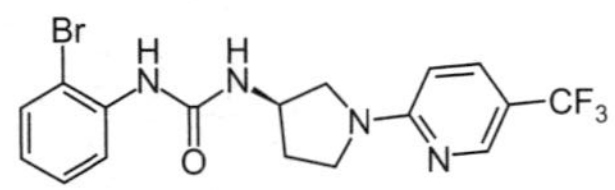	GlaxoSmithKline	Dental pain Migraine Rectal pain	Phase II[a] Phase II[b] Phase II[c]
NGD8243/ MK2295		Neurogen/Merck	Pain	Phase II[d]
GRC 6211		Glenmark/Eli Lilly	Osteoarthritic pain Other pain	Phase II
AZD1386		AstraZeneca	Chronic nociceptive pain	Phase II
AMG 517	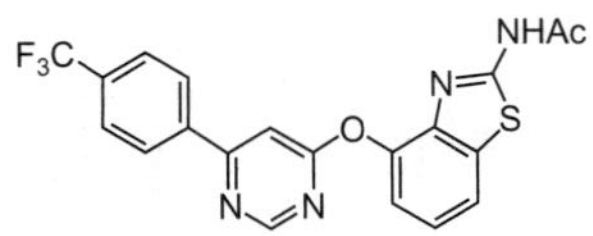	Amgen	Pain	Phase I[e]
AMG 628	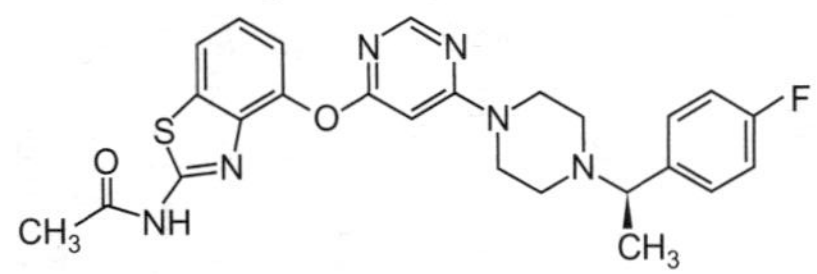	Amgen	Pain	Phase I
ABT-102	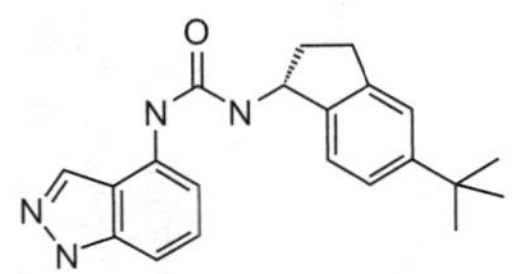	Abbott	Pain	Phase I

[a]Discontinued.

[b]Discontinued.

[c]Discontinued.

[d]Now in phase II for cough.

[e]Discontinued.

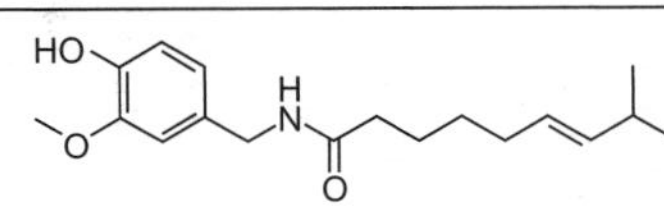

Capsaicin

has completed a Phase II clinical trial for dental pain. NGD8243 (Neurogen/Merck) was in Phase II previously for pain and is currently in Phase II for cough. Also in Phase II trials are GRC 6211 (Glenmark/Eli Lilly) for various pain conditions, including osteoarthritic pain, and AZD1386 (AstraZeneca) for chronic nociceptive pain.

In addtition to expression in pain pathways, TRPV1 is also more broadly expressed in other neuronal and nonneuronal tissues (Caterina, 2003). Consequently, TRPV1 antagonism may also result in other, unintended/undesirable effects, potentially narrowing or even closing the therapeutic window.

One of the potential obstacles facing successful clinical development of TRPV1 antagonists is the tendency for these compounds to cause hyperthermia. Capsaicin has long been known to cause hypothermia in multiple species, including man (Hori, 1984; Szallasi and Blumberg, 1999). However, it was not until recently, with the help of diverse chemotypes of TRPV1 antagonists, that clear evidence began to emerge showing the presence of tonic TRPV1 activity and its involvement in the ongoing regulation of body temperature (Gavva et al., 2007). Indeed, AMG 517, a selective TRPV1 antagonist previously in Phase I clinical trials, causes marked hyperthermia in humans and has been discontinued from further clinical development (Gavva et al., 2008). It appears that the site of action for TRPV1 antagonist-induced hyperthermia is outside the blood–brain barrier (Tamayo et al., 2008), suggesting that it is unlikely for TRPV1 antagonist-induced hyperthermia to be effectively mitigated by restriction to peripheral exposure. Interestingly, a recently identified novel TRPV1 modulator, AMG 8562, has a unique pharmacological profile against TRPV1 (blocks capsaicin, but not heat-induced activation, and potentiates proton-induced activation) and is efficacious in several rodent on-target and pain models without causing hyperthermia (Lehto et al., 2008). These findings raise the possibility of maintaining TRPV1 antagonist efficacy in chronic pain states without causing hyperthermia. However, it remains to be seen whether any off-target activity of AMG 8562 may contribute to its distinct *in vivo* pharmacological profile.

21.5.3.3.2 TRPV1 Agonists. Despite the overwhelming enthusiasm for TRPV1 antagonists in recent years, it was capsaicin, a TRPV1 agonist, that first saw application in pain relief. In fact, topical capsaicin has been used clinically to relieve pain for decades (Mason et al., 2004), well before the identification of its molecular mechanism of action. Although capsaicin causes activation of nociceptors initially, it achieves functional antagonism by taking advantage of the fact that this initial excitation is followed by a lasting refractory period, during which nociceptors are unresponsive to further stimulation, a phenomenon known as capsaicin desensitization. The underlying mechanism of capsaicin desensitization, which requires activation of TRPV1, appears to be complex (Knotkova et al., 2008). Because TRPV1 agonists silence the entire capsaicin-sensitive nerve terminal, it may be argued that, at least in principle, they should be more effective as analgesics than TRPV1 antagonists, which only selectively block TRPV1.

Controlled clinical studies using over-the-counter capsaicin cream/patches or other topical capsaicin formulations (at concentrations ≤ 0.075%) have generally shown moderate to poor efficacy for chronic pain (Mason et al., 2004; Knotkova et al., 2008). This approach may nonetheless be useful as either an adjunct therapy or monotherapy for some patients who are unresponsive to or intolerant of other treatments. Topical application of higher concentrations of capsaicin (5–10%) has been reported to be effective in patients with a number of painful neuropathic conditions (Robbins et al., 1998). NGX-4010, an 8% *trans*-capsaicin patch, significantly reduces pain compared to placebo in Phase III clinical trials for postherpetic neuralgia and painful HIV-associated distal sensory polyneuropathy; it is also in Phase II trials for painful diabetic neuropathy (http://www.neurogesx.com/ngx_4010). The side effects associated with topical capsaicin treatment are generally limited to the initial burning sensations. Topical applications are mostly only helpful for relatively superficial and localized pain that is amenable to topical treatment. In addition, the intense burning sensations upon administration of high doses of capsaicin require pretreatment with local/regional anesthesia.

A potentially more effective therapy involves injection of capsaicin directly to the site of pain (e.g., intra-articularly, perineurally, or within soft tissue). This route of administration leads to prolonged (but reversible) and localized desensitization of nociceptors and provides site-specific pain relief. Adlea™ (formerly ALGRX-4975), an injectable capsaicin formulation, demonstrates statistically significant pain reduction in Phase III clinical trials for knee replacement and bunionectomy surgeries, respectively (http://www.anesiva.com/wt/page/adlea). Other clinical studies also show efficacy of Adlea™ for various pain indications (Knotkova et al., 2008), including joint pain from moderate to severe osteoarthritis, pain following orthopedic surgeries, musculoskeletal pain due to tendonitis of the elbow, and post-trauma neuropathic pain in interdigital neuroma (also see http://www.anesiva.com/wt/page/cl_dataadlea). The systemic exposure of Adlea™ is short ($T_{1/2}$ ~ 1–2 h and undetectable in blood after 24 h) relative to its duration of action (weeks to months with a single injection). Adlea™ is safe and well-tolerated in clinical studies with a side effect profile similar to that of placebo.

Injectable capsaicin therapy is more invasive than topical applications, although this is in part offset by the infrequency of administration due to the long-acting nature of this treatment. In addition, it is typically necessary to precede a capsaicin injection with local/regional anesthesia in order to blunt the capsaicin-induced acute pain. In contrast to topical or injectable capsaicin formulations, which are generally safe, oral administration of capsaicin is not considered viable as a clinical option due to adverse effects on blood pressure, breathing, and other reflex functions (Knotkova et al., 2008).

In addition to capsaicin, RTX, an ultrapotent capsaicin analog, is also a powerful analgesic agent in animal models of neuropathic and bone cancer pain (Brown et al., 2005; Menendez et al., 2006; Kissin et al., 2007) and has more favorable therapeutic profiles than capsaicin (Szallasi and Blumberg, 1999). Currently, RTX clinical trials are being conducted in patients with severe cancer pain (Knotkova et al., 2008).

21.6 CONCLUDING REMARKS

Chronic pain inflicts the lives of millions of people. But safe, effective, and reliable long-term treatments remain a great challenge. Many ion channels are implicated in chronic pain states. Some of the currently available drugs being used for treating chronic pain are known to modulate ion channels. However, the effectiveness of these drugs is limited by undesirable effects and other challenges. Selective targeting (e.g., by use-dependent modulation or otherwise normalization of disease-specific functional, such as gating, changes) of ion channels that are specifically expressed and involved in chronic pain signaling may lead to effective drugs with greater therapeutic windows. Our increasing appreciation of the molecular mechanisms of chronic pain and the significant improvement in ion channel screening technologies, such as the automated patch clamp, have greatly enabled and facilitated target-based ion channel drug discovery in recent years. While challenges remain great, the opportunities and prospects for ion channel analgesic discovery have never been greater.

REFERENCES

Abe M, Kurihara T, Han W, Shinomiya K, Tanabe T. Changes in expression of voltage-dependent ion channel subunits in dorsal root ganglia of rats with radicular injury and pain. *Spine* 2002;27:1517–1524; discussion 1525.

Acs G, Palkovits M, Blumberg PM. Specific binding of [^{3}H]resiniferatoxin by human and rat preoptic area, locus ceruleus, medial hypothalamus, reticular formation and ventral thalamus membrane preparations. *Life Sci* 1996;59:1899–1908.

Ahern GP, Wang X, Miyares RL. Polyamines are potent ligands for the capsaicin receptor TRPV1. *J Biol Chem* 2006;281:8991–8995.

Ahmad S, Dahllund L, Eriksson AB, Hellgren D, Karlsson U, Lund PE, Meijer IA, Meury L, Mills T, Moody A, Morinville A, Morten J, O'Donnell D, Raynoschek C, Salter H, Rouleau GA, Krupp JJ. A stop codon mutation in SCN9A causes lack of pain sensation. *Hum Mol Genet* 2007;16:2114–2121.

Altier C, Khosravani H, Evans RM, Hameed S, Peloquin JB, Vartian BA, Chen L, Beedle AM, Ferguson SS, Mezghrani A, Dubel SJ, Bourinet E, McRory JE, Zamponi GW. ORL1 receptor-mediated internalization of N-type calcium channels. *Nat Neurosci* 2006;9:31–40.

Amaya F, Decosterd I, Samad TA, Plumpton C, Tate S, Mannion RJ, Costigan M, Woolf CJ. Diversity of expression of the sensory neuron-specific TTX-resistant voltage-gated sodium ion channels SNS and SNS2. *Mol Cell Neurosci* 2000; 15:331–342.

Amaya F, Wang H, Costigan M, Allchorne AJ, Hatcher JP, Egerton J, Stean T, Morisset V, Grose D, Gunthorpe MJ, Chessell IP, Tate S, Green PJ, Woolf CJ. The voltage-gated sodium channel nav1.9 is an effector of peripheral inflammatory pain hypersensitivity. *J Neurosci* 2006;26:12852–12860.

Ambrosio AF, Soares-Da-Silva P, Carvalho CM, Carvalho AP. Mechanisms of action of carbamazepine and its derivatives, oxcarbazepine, BIA 2-093, and BIA 2-024. *Neurochem Res* 2002;27:121–130.

Arnold LM, Goldenberg DL, Stanford SB, Lalonde JK, Sandhu HS, Keck PE Jr, Welge JA, Bishop F, Stanford KE, Hess EV, Hudson JI. Gabapentin in the treatment of fibromyalgia: a randomized, double-blind, placebo-controlled, multicenter trial. *Arthritis Rheum* 2007;56:1336–1344.

Atanassoff PG, Hartmannsgruber MW, Thrasher J, Wermeling D, Longton W, Gaeta R, Singh T, Mayo M, McGuire D, Luther RR. Ziconotide, a new N-type calcium channel blocker, administered intrathecally for acute postoperative pain. *Reg Anesth Pain Med* 2000;25:274–278.

Bach FW, Jensen TS, Kastrup J, Stigsby B, Dejgard A. The effect of intravenous lidocaine on nociceptive processing in diabetic neuropathy. *Pain* 1990;40:29–34.

Backonja M, Beydoun A, Edwards KR, Schwartz SL, Fonseca V, Hes M, LaMoreaux L, Garofalo E. Gabapentin for the symptomatic treatment of painful neuropathy in patients with diabetes mellitus: a randomized controlled trial. *JAMA* 1998;280: 1831–1836.

Baell JB, Duggan PJ, Forsyth SA, Lewis RJ, Lok YP, Schroeder CI. Synthesis and biological evaluation of nonpeptide mimetics of omega-conotoxin GVIA. *Bioorg Med Chem* 2004;12:4025–4037.

Barbano RL, Herrmann DN, Hart-Gouleau S, Pennella-Vaughan J, Lodewick PA, Dworkin RH. Effectiveness, tolerability, and impact on quality of life of the 5% lidocaine patch in diabetic polyneuropathy. *Arch Neurol* 2004;61:914–918.

Barrett CF, Rittenhouse AR. Modulation of N-type calcium channel activity by G-proteins and protein kinase C. *J Gen Physiol* 2000;115:277–286.

Beedle AM, McRory JE, Poirot O, Doering CJ, Altier C, Barrere C, Hamid J, Nargeot J, Bourinet E, Zamponi GW. Agonist-independent modulation of N-type calcium channels by ORL1 receptors. *Nat Neurosci* 2004;7:118–125.

Bernard G, Shevell MI. Channelopathies: a review. *Pediatr Neurol* 2008;38:73–85.

Birnbaumer L, Qin N, Olcese R, Tareilus E, Platano D, Costantin J, Stefani E. Structures and functions of calcium channel beta subunits. *J Bioenerg Biomembr* 1998;30:357–375.

Black JA, Cummins TR, Plumpton C, Chen YH, Hormuzdiar W, Clare JJ, Waxman SG. Upregulation of a silent sodium channel after peripheral, but not central, nerve injury in DRG neurons. *J Neurophysiol* 1999;82:2776–2785.

Black JA, Liu S, Tanaka M, Cummins TR, Waxman SG. Changes in the expression of tetrodotoxin-sensitive sodium channels within dorsal root ganglia neurons in inflammatory pain. *Pain* 2004;108:237–247.

Bourinet E, Soong TW, Stea A, Snutch TP. Determinants of the G protein-dependent opioid modulation of neuronal calcium channels. *Proc Natl Acad Sci USA* 1996;93:1486–1491.

Bowen C, Bridson G, Heiser A, Galullo V, Winquist R, Zelle R. Novel blockers of Cav2.2: brain penetration and efficacy in rat models of pain including neuropathic pain. *Neuropathic Pain: Changing Paradigms in Diagnosis and Treatment* 2004;abstract 40.

Bowersox S, Gadbois T, Singh T, Pettus M, Wang Y, Luther R. Selective N-type neuronal voltage-sensitive calcium channel blocker, SNX-111, produces spinal antinociception in rat models of acute, persistent and neuropathic pain. *J Pharmacol Exp Ther* 1996;279:1243–1249.

Brauchi S, Orta G, Salazar M, Rosenmann E, Latorre R. A hot-sensing cold receptor: C-terminal domain determines thermosensation in transient receptor potential channels. *J Neurosci* 2006;26:4835–4840.

Brochu RM, Dick IE, Tarpley JW, McGowan E, Gunner D, Herrington J, Shao PP, Ok D, Li C, Parsons WH, Stump GL, Regan CP, Lynch JJ Jr, Lyons KA, McManus OB, Clark S, Ali Z, Kaczorowski GJ, Martin WJ, Priest BT. Block of peripheral nerve sodium channels selectively inhibits features of neuropathic pain in rats. *Mol Pharmacol* 2006;69:823–832.

Brose WG, Gutlove DP, Luther RR, Bowersox SS, McGuire D. Use of intrathecal SNX-111, a novel, N-type, voltage-sensitive, calcium channel blocker, in the management of intractable brachial plexus avulsion pain. *Clin J Pain* 1997;13:256–259.

Brown DC, Iadarola MJ, Perkowski SZ, Erin H, Shofer F, Laszlo KJ, Olah Z, Mannes AJ. Physiologic and antinociceptive effects of intrathecal resiniferatoxin in a canine bone cancer model. *Anesthesiology* 2005;103:1052–1059.

Cardenas LM, Cardenas CG, Scroggs RS. 5HT increases excitability of nociceptor-like rat dorsal root ganglion neurons via cAMP-coupled TTX-resistant Na(+) channels. *J Neurophysiol* 2001;86:241–248.

Caterina MJ. Vanilloid receptors take a TRP beyond the sensory afferent. *Pain* 2003;105:5–9.

Caterina MJ, Schumacher MA, Tominaga M, Rosen TA, Levine JD, Julius D. The capsaicin receptor: a heat-activated ion channel in the pain pathway. *Nature* 1997;389:816–824.

Caterina MJ, Leffler A, Malmberg AB, Martin WJ, Trafton J, Petersen-Zeitz KR, Koltzenburg M, Basbaum AI, Julius D. Impaired nociception and pain sensation in mice lacking the capsaicin receptor. *Science* 2000;288:306–313.

Catterall WA. From ionic currents to molecular mechanisms: the structure and function of voltage-gated sodium channels. *Neuron* 2000a;26:13–25.

Catterall WA. Structure and regulation of voltage-gated Ca^{2+} channels. *Annu Rev Cell Dev Biol* 2000b;16:521–555.

Catterall WA, Goldin AL, Waxman SG. International Union of Pharmacology. XXXIX. Compendium of voltage-gated ion channels: sodium channels. *Pharmacol Rev* 2003;55:575–578.

Chaplan SR, Pogrel JW, Yaksh TL. Role of voltage-dependent calcium channel subtypes in experimental tactile allodynia. *J Pharmacol Exp Ther* 1994;269:1117–1123.

Chen RS, Deng TC, Garcia T, Sellers ZM, Best PM. Calcium channel gamma subunits: a functionally diverse protein family. *Cell Biochem Biophys* 2007;47:178–186.

Chizh BA, O'Donnell MB, Napolitano A, Wang J, Brooke AC, Aylott MC, Bullman JN, Gray EJ, Lai RY, Williams PM, Appleby JM. The effects of the TRPV1 antagonist SB-705498 on TRPV1 receptor-mediated activity and inflammatory hyperalgesia in humans. *Pain* 2007;132:132–141.

Christoph T, Grunweller A, Mika J, Schafer MK, Wade EJ, Weihe E, Erdmann VA, Frank R, Gillen C, Kurreck J. Silencing of vanilloid receptor TRPV1 by RNAi reduces neuropathic and visceral pain *in vivo*. *Biochem Biophys Res Commun* 2006;350:238–243.

Christoph T, Gillen C, Mika J, Grunweller A, Schafer MK, Schiene K, Frank R, Jostock R, Bahrenberg G, Weihe E, Erdmann VA, Kurreck J. Antinociceptive effect of antisense oligonucleotides against the vanilloid receptor VR1/TRPV1. *Neurochem Int* 2007;50:281–290.

Chuang HH, Prescott ED, Kong H, Shields S, Jordt SE, Basbaum AI, Chao MV, Julius D. Bradykinin and nerve growth factor release the capsaicin receptor from PtdIns(4,5)P2-mediated inhibition. *Nature* 2001;411:957–962.

Cizkova D, Marsala J, Lukacova N, Marsala M, Jergova S, Orendacova J, Yaksh TL. Localization of N-type Ca^{2+} channels in the rat spinal cord following chronic constrictive nerve injury. *Exp Brain Res* 2002;147:456–463.

Coggeshall RE, Tate S, Carlton SM. Differential expression of tetrodotoxin-resistant sodium channels Nav1.8 and Nav1.9 in normal and inflamed rats. *Neurosci Lett* 2004;355:45–48.

Colburn RW, Lubin ML, Stone DJ Jr, Wang Y, Lawrence D, D 'Andrea MR, Brandt MR, Liu Y, Flores CM, Qin N. Attenuated cold sensitivity in TRPM8 null mice. *Neuron* 2007;54:379–386.

Coward K, Plumpton C, Facer P, Birch R, Carlstedt T, Tate S, Bountra C, Anand P. Immunolocalization of SNS/PN3 and NaN/SNS2 sodium channels in human pain states. *Pain* 2000;85:41–50.

Coward K, Jowett A, Plumpton C, Powell A, Birch R, Tate S, Bountra C, Anand P. Sodium channel beta1 and beta2 subunits parallel SNS/PN3 alpha-subunit changes in injured human sensory neurons. *Neuroreport* 2001;12:483–488.

Cox JJ, Reimann F, Nicholas AK, Thornton G, Roberts E, Springell K, Karbani G, Jafri H, Mannan J, Raashid Y, Al-Gazali L, Hamamy H, Valente EM, Gorman S, Williams R, McHale DP, Wood JN, Gribble FM, Woods CG. An SCN9A channelopathy causes congenital inability to experience pain. *Nature* 2006;444:894–898.

Crofford LJ, Rowbotham MC, Mease PJ, Russell IJ, Dworkin RH, Corbin AE, Young JP Jr, LaMoreaux LK, Martin SA, Sharma U. Pregabalin for the treatment of fibromyalgia syndrome: results of a randomized, double-blind, placebo-controlled trial. *Arthritis Rheum* 2005;52:1264–1273.

Cui M, Honore P, Zhong C, Gauvin D, Mikusa J, Hernandez G, Chandran P, Gomtsyan A, Brown B, Bayburt EK, Marsh K, Bianchi B, McDonald H, Niforatos W, Neelands TR, Moreland RB, Decker MW, Lee CH, Sullivan JP, Faltynek CR. TRPV1 receptors in the CNS play a key role in broad-spectrum analgesia of TRPV1 antagonists. *J Neurosci* 2006;26:9385–9393.

Cummins TR, Waxman SG. Downregulation of tetrodotoxin-resistant sodium currents and upregulation of a rapidly repriming tetrodotoxin-sensitive sodium current in small spinal sensory neurons after nerve injury. *J Neurosci* 1997;17:3503–3514.

Cummins TR, Howe JR, Waxman SG. Slow closed-state inactivation: a novel mechanism underlying ramp currents in cells expressing the hNE/PN1 sodium channel. *J Neurosci* 1998;18:9607–9619.

Cummins TR, Dib-Hajj SD, Black JA, Akopian AN, Wood JN, Waxman SG. A novel persistent tetrodotoxin-resistant sodium current in SNS-null and wild-type small primary sensory neurons. *J Neurosci* 1999;19:RC43.

Davies A, Hendrich J, Van Minh AT, Wratten J, Douglas L, Dolphin AC. Functional biology of the alpha(2)delta subunits of voltage-gated calcium channels. *Trends Pharmacol Sci* 2007;28:220–228.

Davis JB, Gray J, Gunthorpe MJ, Hatcher JP, Davey PT, Overend P, Harries MH, Latcham J, Clapham C, Atkinson K, Hughes SA, Rance K, Grau E, Harper AJ, Pugh PL, Rogers DC, Bingham S, Randall A, Sheardown SA. Vanilloid receptor-1 is essential for inflammatory thermal hyperalgesia. *Nature* 2000;405:183–187.

Derand R, Bulteau-Pignoux L, Becq F. The cystic fibrosis mutation G551D alters the non-Michaelis–Menten behavior of the cystic fibrosis transmembrane conductance regulator (CFTR) channel and abolishes the inhibitory Genistein binding site. *J Biol Chem* 2002;277:35999–36004.

Dib-Hajj SD, Fjell J, Cummins TR, Zheng Z, Fried K, LaMotte R, Black JA, Waxman SG. Plasticity of sodium channel expression in DRG neurons in the chronic constriction injury model of neuropathic pain. *Pain* 1999;83:591–600.

Dib-Hajj SD, Rush AM, Cummins TR, Hisama FM, Novella S, Tyrrell L, Marshall L, Waxman SG. Gain-of-function mutation in Nav1.7 in familial erythromelalgia induces bursting of sensory neurons. *Brain* 2005;128:1847–1854.

Dib-Hajj SD, Cummins TR, Black JA, Waxman SG. From genes to pain: Nav1.7 and human pain disorders. *Trends Neurosci* 2007;30:555–563.

Dick IE, Brochu RM, Purohit Y, Kaczorowski GJ, Martin WJ, Priest BT. Sodium channel blockade may contribute to the analgesic efficacy of antidepressants. *J Pain* 2007;8:315–324.

Dong XW, Goregoaker S, Engler H, Zhou X, Mark L, Crona J, Terry R, Hunter J, Priestley T. Small interfering RNA-mediated selective knockdown of Na(V)1.8 tetrodotoxin-resistant sodium channel reverses mechanical allodynia in neuropathic rats. *Neuroscience* 2007;146:812–821.

Doyle JL, Stubbs L. Ataxia, arrhythmia and ion-channel gene defects. *Trends Genet* 1998;14:92–98.

Dunlop J, Bowlby M, Peri R, Vasilyev D, Arias R. High-throughput electrophysiology: an emerging paradigm for ion-channel screening and physiology. *Nat Rev Drug Discovery* 2008;7:358–368.

Dworkin RH, Corbin AE, Young JP Jr, Sharma U, LaMoreaux L, Bockbrader H, Garofalo EA, Poole RM. Pregabalin for the treatment of postherpetic neuralgia: a randomized, placebo-controlled trial. *Neurology* 2003;60:1274–1283.

Eisenberg E, River Y, Shifrin A, Krivoy N. Antiepileptic drugs in the treatment of neuropathic pain. *Drugs* 2007;67:1265–1289.

England S, Bevan S, Docherty RJ. PGE2 modulates the tetrodotoxin-resistant sodium current in neonatal rat dorsal root ganglion neurones via the cyclic AMP-protein kinase A cascade. *J Physiol* 1996;495(Pt 2):429–440.

Errington AC, Stohr T, Heers C, Lees G. The investigational anticonvulsant lacosamide selectively enhances slow inactivation of voltage-gated sodium channels. *Mol Pharmacol* 2008;73:157–169.

Feng ZP, Hamid J, Doering C, Bosey GM, Snutch TP, Zamponi GW. Residue Gly1326 of the N-type calcium channel alpha 1B subunit controls reversibility of omega-conotoxin GVIA and MVIIA block. *J Biol Chem* 2001;276:15728–15735.

Feng ZP, Doering CJ, Winkfein RJ, Beedle AM, Spafford JD, Zamponi GW. Determinants of inhibition of transiently expressed voltage-gated calcium channels by omega-conotoxins GVIA and MVIIA. *J Biol Chem* 2003;278:20171–20178.

Ferrante FM, Paggioli J, Cherukuri S, Arthur GR. The analgesic response to intravenous lidocaine in the treatment of neuropathic pain. *Anesth Analg* 1996;82:91–97.

Fertleman CR, Baker MD, Parker KA, Moffatt S, Elmslie FV, Abrahamsen B, Ostman J, Klugbauer N, Wood JN, Gardiner RM, Rees M. SCN9A mutations in paroxysmal extreme pain disorder: allelic variants underlie distinct channel defects and phenotypes. *Neuron* 2006;52:767–774.

Field MJ, Cox PJ, Stott E, Melrose H, Offord J, Su TZ, Bramwell S, Corradini L, England S, Winks J, Kinloch RA, Hendrich J, Dolphin AC, Webb T, Williams D. Identification of the alpha2-delta-1 subunit of voltage-dependent calcium channels as a molecular target for pain mediating the analgesic actions of pregabalin. *Proc Natl Acad Sci USA* 2006;103:17537–17542.

Franco R, Dong L, Galullo V, Hornsten A, Pan JQ, Sui J, White G, Winquist R, Zelle R. Discovery of novel, orally bioavailable, N-type calcium (Cav2.2) channel blockers for treatment of neuropathic pain. *Neuropathic Pain: Changing Paradigms in Diagnosis and Treatment* 2004;abstract 155.

Garcia-Martinez C, Morenilla-Palao C, Planells-Cases R, Merino JM, Ferrer-Montiel A. Identification of an aspartic residue in the P-loop of the vanilloid receptor that modulates pore properties. *J Biol Chem* 2000;275:32552–32558.

Gavva NR, Klionsky L, Qu Y, Shi L, Tamir R, Edenson S, Zhang TJ, Viswanadhan VN, Toth A, Pearce LV, Vanderah TW, Porreca F, Blumberg PM, Lile J, Sun Y, Wild K, Louis JC, Treanor JJ. Molecular determinants of vanilloid sensitivity in TRPV1. *J Biol Chem* 2004;279:20283–20295.

Gavva NR, Bannon AW, Surapaneni S, Hovland DN Jr, Lehto SG, Gore A, Juan T, Deng H, Han B, Klionsky L, Kuang R, Le A, Tamir R, Wang J, Youngblood B, Zhu D, Norman MH, Magal E, Treanor JJ, Louis JC. The vanilloid receptor TRPV1 is tonically activated *in vivo* and involved in body temperature regulation. *J Neurosci* 2007;27:3366–3374.

Gavva NR, Treanor JJ, Garami A, Fang L, Surapaneni S, Akrami A, Alvarez F, Bak A, Darling M, Gore A, Jang GR, Kesslak JP, Ni L, Norman MH, Palluconi G, Rose MJ, Salfi M, Tan E, Romanovsky AA, Banfield C, Davar G. Pharmacological blockade of the vanilloid receptor TRPV1 elicits marked hyperthermia in humans. *Pain* 2008;136:202–210.

Gee NS, Brown JP, Dissanayake VU, Offord J, Thurlow R, Woodruff GN. The novel anticonvulsant drug, gabapentin (Neurontin), binds to the alpha2delta subunit of a calcium channel. *J Biol Chem* 1996;271:5768–5776.

Gharat LA, Szallasi A. Advances in the design and therapeutic use of capsaicin receptor TRPV1 agonists and antagonists. *Expert Opin Ther Pat* 2008;18:159–209.

Gilron I. Gabapentin and pregabalin for chronic neuropathic and early postsurgical pain: current evidence and future directions. *Curr Opin Anaesthesiol* 2007;20:456–472.

Gohil K, Bell JR, Ramachandran J, Miljanich GP. Neuroanatomical distribution of receptors for a novel voltage-sensitive calcium-channel antagonist, SNX-230 (omega-conopeptide MVIIC). *Brain Res* 1994;653:258–266.

Gold MS. Tetrodotoxin-resistant Na^+ currents and inflammatory hyperalgesia. *Proc Natl Acad Sci USA* 1999;96:7645–7649.

Gold MS, Weinreich D, Kim CS, Wang R, Treanor J, Porreca F, Lai J. Redistribution of Na(V)1.8 in uninjured axons enables neuropathic pain. *J Neurosci* 2003;23: 158–166.

Goldberg YP, MacFarlane J, MacDonald ML, Thompson J, Dube MP, Mattice M, Fraser R, Young C, Hossain S, Pape T, Payne B, Radomski C, Donaldson G, Ives E, Cox J, Younghusband HB, Green R, Duff A, Boltshauser E, Grinspan GA, Dimon JH, Sibley BG, Andria G, Toscano E, Kerdraon J, Bowsher D, Pimstone SN, Samuels ME, Sherrington R, Hayden MR. Loss-of-function mutations in the Nav1.7 gene underlie congenital indifference to pain in multiple human populations. *Clin Genet* 2007;71:311–319.

Goldin AL, Barchi RL, Caldwell JH, Hofmann F, Howe JR, Hunter JC, Kallen RG, Mandel G, Meisler MH, Netter YB, Noda M, Tamkun MM, Waxman SG, Wood JN, Catterall WA. Nomenclature of voltage-gated sodium channels. *Neuron* 2000;28:365–368.

Gray AC, Raingo J, Lipscombe D. Neuronal calcium channels: splicing for optimal performance. *Cell Calcium* 2007;42:409–417.

Hains BC, Saab CY, Klein JP, Craner MJ, Waxman SG. Altered sodium channel expression in second-order spinal sensory neurons contributes to pain after peripheral nerve injury. *J Neurosci* 2004;24:4832–4839.

Hamill OP, Marty A, Neher E, Sakmann B, Sigworth FJ. Improved patch-clamp techniques for high-resolution current recording from cells and cell-free membrane patches. *Pflugers Arch* 1981;391:85–100.

Hans GH, Robert DN, Van Maldeghem KN. Treatment of an acute severe central neuropathic pain syndrome by topical application of lidocaine 5% patch: a case report. *Spinal Cord* 2008;46:311–313.

Hargus NJ, Patel MK. Voltage-gated Na^+ channels in neuropathic pain. *Expert Opin Investig Drugs* 2007;16:635–646.

Harty TP, Dib-Hajj SD, Tyrrell L, Blackman R, Hisama FM, Rose JB, Waxman SG. Na(V)1.7 mutant A863P in erythromelalgia: effects of altered activation and steady-state inactivation on excitability of nociceptive dorsal root ganglion neurons. *J Neurosci* 2006;26:12566–12575.

Hatakeyama S, Wakamori M, Ino M, Miyamoto N, Takahashi E, Yoshinaga T, Sawada K, Imoto K, Tanaka I, Yoshizawa T, Nishizawa Y, Mori Y, Niidome T, Shoji S. Differential nociceptive responses in mice lacking the alpha(1B) subunit of N-type Ca(2+) channels. *Neuroreport* 2001;12:2423–2427.

Hendrich J, Van Minh AT, Heblich F, Nieto-Rostro M, Watschinger K, Striessnig J, Wratten J, Davies A, Dolphin AC. Pharmacological disruption of calcium channel trafficking by the alpha2delta ligand gabapentin. *Proc Natl Acad Sci USA* 2008;105: 3628–3633.

Herrmann DN, Barbano RL, Hart-Gouleau S, Pennella-Vaughan J, Dworkin RH. An open-label study of the lidocaine patch 5% in painful idiopathic sensory polyneuropathy. *Pain Med* 2005;6:379–384.

Herzog RI, Cummins TR, Waxman SG. Persistent TTX-resistant Na^+ current affects resting potential and response to depolarization in simulated spinal sensory neurons. *J Neurophysiol* 2001;86:1351–1364.

Herzog RI, Cummins TR, Ghassemi F, Dib-Hajj SD, Waxman SG. Distinct repriming and closed-state inactivation kinetics of Nav1.6 and Nav1.7 sodium channels in mouse spinal sensory neurons. *J Physiol* 2003;551:741–750.

Hille B. *Ion Channels of Excitable Membranes*, 3rd ed. Sunderland, MA: Sinauer, 2001.

Hines R, Keaney D, Moskowitz MH, Prakken S. Use of lidocaine patch 5% for chronic low back pain: a report of four cases. *Pain Med* 2002;3:361–365.

Hori T. Capsaicin and central control of thermoregulation. *Pharmacol Ther* 1984;26: 389–416.

Horvath G, Brodacz B, Holzer-Petsche U. Role of calcium channels in the spinal transmission of nociceptive information from the mesentery. *Pain* 2001;93:35–41.

Hu LY, Ryder TR, Nikam SS, Millerman E, Szoke BG, Rafferty MF. Synthesis and biological evaluation of substituted 4-(OBz)phenylalanine derivatives as novel N-type calcium channel blockers. *Bioorg Med Chem Lett* 1999;9:1121–1126.

Huang CJ, Harootunian A, Maher MP, Quan C, Raj CD, McCormack K, Numann R, Negulescu PA, Gonzalez JE. Characterization of voltage-gated sodium-channel blockers by electrical stimulation and fluorescence detection of membrane potential. *Nat Biotechnol* 2006;24:439–446.

Imming P, Sinning C, Meyer A. Drugs, their targets and the nature and number of drug targets. *Nat Rev Drug Discovery* 2006;5:821–834.

Immke DC, Gavva NR. The TRPV1 receptor and nociception. *Semin Cell Dev Biol* 2006;17:582–591.

Jain KK. An evaluation of intrathecal ziconotide for the treatment of chronic pain. *Expert Opin Investig Drugs* 2000;9:2403–2410.

Jarvis MF, Honore P, Shieh CC, Chapman M, Joshi S, Zhang XF, Kort M, Carroll W, Marron B, Atkinson R, Thomas J, Liu D, Krambis M, Liu Y, McGaraughty S, Chu K, Roeloffs R, Zhong C, Mikusa JP, Hernandez G, Gauvin D, Wade C, Zhu C, Pai M, Scanio M, Shi L, Drizin I, Gregg R, Matulenko M, Hakeem A, Gross M, Johnson M, Marsh K, Wagoner PK, Sullivan JP, Faltynek CR, Krafte DS. A-803467, a potent and selective Nav1.8 sodium channel blocker, attenuates neuropathic and inflammatory pain in the rat. *Proc Natl Acad Sci USA* 2007;104:8520–8525.

Ji RR, Samad TA, Jin SX, Schmoll R, Woolf CJ. p38 MAPK activation by NGF in primary sensory neurons after inflammation increases TRPV1 levels and maintains heat hyperalgesia. *Neuron* 2002;36:57–68.

Jordt SE, Julius D. Molecular basis for species-specific sensitivity to "hot" chili peppers. *Cell* 2002;108:421–430.

Jordt SE, Tominaga M, Julius D. Acid potentiation of the capsaicin receptor determined by a key extracellular site. *Proc Natl Acad Sci USA* 2000;97:8134–8139.

Joshi SK, Mikusa JP, Hernandez G, Baker S, Shieh CC, Neelands T, Zhang XF, Niforatos W, Kage K, Han P, Krafte D, Faltynek C, Sullivan JP, Jarvis MF, Honore

P. Involvement of the TTX-resistant sodium channel Nav 1.8 in inflammatory and neuropathic, but not post-operative, pain states. *Pain* 2006;123:75–82.

Kalso E, Tasmuth T, Neuvonen PJ. Amitriptyline effectively relieves neuropathic pain following treatment of breast cancer. *Pain* 1996;64:293–302.

Kerr BJ, Souslova V, McMahon SB, Wood JN. A role for the TTX-resistant sodium channel Nav 1.8 in NGF-induced hyperalgesia, but not neuropathic pain. *NeuroReport* 2001;12:3077–3080.

Kim C, Jun K, Lee T, Kim SS, McEnery MW, Chin H, Kim HL, Park JM, Kim DK, Jung SJ, Kim J, Shin HS. Altered nociceptive response in mice deficient in the alpha(1B) subunit of the voltage-dependent calcium channel. *Mol Cell Neurosci* 2001;18:235–245.

Kissin I, Freitas CF, Bradley EL Jr. Perineural resiniferatoxin prevents the development of hyperalgesia produced by loose ligation of the sciatic nerve in rats. *Anesth Analg* 2007;104:1210–1216, table of contents.

Kitaguchi T, Swartz KJ. An inhibitor of TRPV1 channels isolated from funnel Web spider venom. *Biochemistry* 2005;44:15544–15549.

Klionsky L, Tamir R, Holzinger B, Bi X, Talvenheimo J, Kim H, Martin F, Louis JC, Treanor JJ, Gavva NR. A polyclonal antibody to the prepore loop of transient receptor potential vanilloid type 1 blocks channel activation. *J Pharmacol Exp Ther* 2006;319:192–198.

Klugbauer N, Marais E, Hofmann F. Calcium channel alpha2delta subunits: differential expression, function, and drug binding. *J Bioenerg Biomembr* 2003;35:639–647.

Knotkova H, Pappagallo M, Szallasi A. Capsaicin (TRPV1 agonist) therapy for pain relief: farewell or revival? *Clin J Pain* 2008;24:142–154.

Lai J, Gold MS, Kim CS, Bian D, Ossipov MH, Hunter JC, Porreca F. Inhibition of neuropathic pain by decreased expression of the tetrodotoxin-resistant sodium channel, Nav1.8. *Pain* 2002;95:143–152.

Leffler A, Reiprich A, Mohapatra DP, Nau C. Use-dependent block by lidocaine but not amitriptyline is more pronounced in tetrodotoxin (TTX)-Resistant Nav1.8 than in TTX-sensitive Na^+ channels. *J Pharmacol Exp Ther* 2007;320:354–364.

Lehto SG, Tamir R, Deng H, Klionsky L, Kuang R, Le A, Lee D, Louis JC, Magal E, Manning BH, Rubino J, Surapaneni S, Tamayo N, Wang T, Wang J, Wang J, Wang W, Youngblood B, Zhang M, Zhu D, Norman MH, Gavva NR. Antihyperalgesic effects of (*R,E*)-N-(2-hydroxy-2,3-dihydro-1*H*-inden-4-yl)-3-(2-(piperidin-1-yl)-4-(trifluoromethyl)phenyl)-acrylamide (AMG8562), a novel transient receptor potential vanilloid type 1 modulator that does not cause hyperthermia in rats. *J Pharmacol Exp Ther* 2008;326:218–229.

Leijon G, Boivie J. Central post-stroke pain—a controlled trial of amitriptyline and carbamazepine. *Pain* 1989;36:27–36.

Linford NJ, Cantrell AR, Qu Y, Scheuer T, Catterall WA. Interaction of batrachotoxin with the local anesthetic receptor site in transmembrane segment IVS6 of the voltage-gated sodium channel. *Proc Natl Acad Sci USA* 1998;95:13947–13952.

Luo ZD, Chaplan SR, Higuera ES, Sorkin LS, Stauderman KA, Williams ME, Yaksh TL. Upregulation of dorsal root ganglion (alpha)2(delta) calcium channel subunit and its correlation with allodynia in spinal nerve-injured rats. *J Neurosci* 2001;21: 1868–1875.

Luo ZD, Calcutt NA, Higuera ES, Valder CR, Song YH, Svensson CI, Myers RR. Injury type-specific calcium channel alpha 2 delta-1 subunit up-regulation in rat neuropathic pain models correlates with antiallodynic effects of gabapentin. *J Pharmacol Exp Ther* 2002;303:1199–1205.

Magenta P, Arghetti S, Di Palma F, Jann S, Sterlicchio M, Bianconi C, Galimberti V, Osio M, Siciliano G, Cavallotti G, Sterzi R. Oxcarbazepine is effective and safe in the treatment of neuropathic pain: pooled analysis of seven clinical studies. *Neurol Sci* 2005;26:218–226.

Malmberg A, Yaksh T. Voltage-sensitive calcium channels in spinal nociceptive processing: blockade of N- and P-type channels inhibits formalin-induced nociception. *J Neurosci* 1994;14:4882–4890.

Mason L, Moore RA, Derry S, Edwards JE, McQuay HJ. Systematic review of topical capsaicin for the treatment of chronic pain. *BMJ* 2004;328:991.

Max MB, Schafer SC, Culnane M, Smoller B, Dubner R, Gracely RH. Amitriptyline, but not lorazepam, relieves postherpetic neuralgia. *Neurology* 1988;38:1427–1432.

Max MB, Lynch SA, Muir J, Shoaf SE, Smoller B, Dubner R. Effects of desipramine, amitriptyline, and fluoxetine on pain in diabetic neuropathy. *N Engl J Med* 1992;326:1250–1256.

McCollum IJ, Vilin YY, Spackman E, Fujimoto E, Ruben PC. Negatively charged residues adjacent to IFM motif in the DIII-DIV linker of hNa(V)1.4 differentially affect slow inactivation. *FEBS Lett* 2003;552:163–169.

Meier T, Wasner G, Faust M, Kuntzer T, Ochsner F, Hueppe M, Bogousslavsky J, Baron R. Efficacy of lidocaine patch 5% in the treatment of focal peripheral neuropathic pain syndromes: a randomized, double-blind, placebo-controlled study. *Pain* 2003;106:151–158.

Menendez L, Juarez L, Garcia E, Garcia-Suarez O, Hidalgo A, Baamonde A. Analgesic effects of capsazepine and resiniferatoxin on bone cancer pain in mice. *Neurosci Lett* 2006;393:70–73.

Moriyama T, Higashi T, Togashi K, Iida T, Segi E, Sugimoto Y, Tominaga T, Narumiya S, Tominaga M. Sensitization of TRPV1 by EP1 and IP reveals peripheral nociceptive mechanism of prostaglandins. *Mol Pain* 2005;1:3.

Murakami M, Fleischmann B, De Felipe C, Freichel M, Trost C, Ludwig A, Wissenbach U, Schwegler H, Hofmann F, Hescheler J, Flockerzi V, Cavalie A. Pain perception in mice lacking the beta 3 subunit of voltage-activated calcium channels. *J Biol Chem* 2002;277:40342–40351.

Nassar MA, Stirling LC, Forlani G, Baker MD, Matthews EA, Dickenson AH, Wood JN. Nociceptor-specific gene deletion reveals a major role for Nav1.7 (PN1) in acute and inflammatory pain. *Proc Natl Acad Sci USA* 2004;101:12706–12711.

Nassar MA, Levato A, Stirling LC, Wood JN. Neuropathic pain develops normally in mice lacking both Nav1.7 and Nav1.8. *Mol Pain* 2005;1:24.

Nassar MA, Baker MD, Levato A, Ingram R, Mallucci G, McMahon SB, Wood JN. Nerve injury induces robust allodynia and ectopic discharges in Nav1.3 null mutant mice. *Mol Pain* 2006;2:33.

Nau C, Wang GK. Interactions of local anesthetics with voltage-gated Na^+ channels. *J Membr Biol* 2004;201:1–8.

Nerbonne JM. Regulation of voltage-gated K^+ channel expression in the developing mammalian myocardium. *J Neurobiol* 1998;37:37–59.

Newton RA, Bingham S, Case PC, Sanger GJ, Lawson SN. Dorsal root ganglion neurons show increased expression of the calcium channel alpha2delta-1 subunit following partial sciatic nerve injury. *Brain Res Mol Brain Res* 2001;95:1–8.

Ostman JA, Nassar MA, Wood JN, Baker MD. GTP up-regulated persistent Na^+ current and enhanced nociceptor excitability require NaV1.9. *J Physiol* 2008;586: 1077–1087.

Overington JP, Al-Lazikani B, Hopkins AL. How many drug targets are there? *Nat Rev Drug Discovery* 2006;5:993–996.

Pan JQ, White G, Hoger J, Dong L, Franco R, Galullo V, Zelle R, Gribkoff V. Functional characterization of a novel class of Cav2.2 channel blocker. Society for Neuroscience Annual Meeting, abstract 735.7, 2004.

Patil PG, Brody DL, Yue DT. Preferential closed-state inactivation of neuronal calcium channels. *Neuron* 1998;20:1027–1038.

Penzotti JL, Fozzard HA, Lipkind GM, Dudley SC Jr. Differences in saxitoxin and tetrodotoxin binding revealed by mutagenesis of the Na^+ channel outer vestibule. *Biophys J* 1998;75:2647–2657.

Porreca F, Lai J, Bian D, Wegert S, Ossipov MH, Eglen RM, Kassotakis L, Novakovic S, Rabert DK, Sangameswaran L, Hunter JC. A comparison of the potential role of the tetrodotoxin-insensitive sodium channels, PN3/SNS and NaN/SNS2, in rat models of chronic pain. *Proc Natl Acad Sci USA* 1999;96:7640–7644.

Priest BT, Murphy BA, Lindia JA, Diaz C, Abbadie C, Ritter AM, Liberator P, Iyer LM, Kash SF, Kohler MG, Kaczorowski GJ, MacIntyre DE, Martin WJ. Contribution of the tetrodotoxin-resistant voltage-gated sodium channel NaV1.9 to sensory transmission and nociceptive behavior. *Proc Natl Acad Sci USA* 2005;102:9382–9387.

Puckerin A, Liu L, Permaul N, Carman P, Lee J, Diverse-Pierluissi MA. Arrestin is required for agonist-induced trafficking of voltage-dependent calcium channels. *J Biol Chem* 2006;281:31131–31141.

Ragsdale DS, McPhee JC, Scheuer T, Catterall WA. Common molecular determinants of local anesthetic, antiarrhythmic, and anticonvulsant block of voltage-gated Na^+ channels. *Proc Natl Acad Sci USA* 1996;93:9270–9275.

Rashid MH, Inoue M, Bakoshi S, Ueda H. Increased expression of vanilloid receptor 1 on myelinated primary afferent neurons contributes to the antihyperalgesic effect of capsaicin cream in diabetic neuropathic pain in mice. *J Pharmacol Exp Ther* 2003a;306:709–717.

Rashid MH, Inoue M, Kondo S, Kawashima T, Bakoshi S, Ueda H. Novel expression of vanilloid receptor 1 on capsaicin-insensitive fibers accounts for the analgesic effect of capsaicin cream in neuropathic pain. *J Pharmacol Exp Ther* 2003b;304: 940–948.

Rasmussen CA Jr, Sutko JL, Barry WH. Effects of ryanodine and caffeine on contractility, membrane voltage, and calcium exchange in cultured heart cells. *Circ Res* 1987;60:495–504.

Rauck RL, Shaibani A, Biton V, Simpson J, Koch B. Lacosamide in painful diabetic peripheral neuropathy: a phase 2 double-blind placebo-controlled study. *Clin J Pain* 2007;23:150–158.

Renganathan M, Cummins TR, Waxman SG. Contribution of Na(v)1.8 sodium channels to action potential electrogenesis in DRG neurons. *J Neurophysiol* 2001;86:629–640.

Robbins WR, Staats PS, Levine J, Fields HL, Allen RW, Campbell JN, Pappagallo M. Treatment of intractable pain with topical large-dose capsaicin: preliminary report. *Anesth Analg* 1998;86:579–583.

Rogawski MA, Loscher W. The neurobiology of antiepileptic drugs. *Nat Rev Neurosci* 2004;5:553–564.

Rosenstock J, Tuchman M, LaMoreaux L, Sharma U. Pregabalin for the treatment of painful diabetic peripheral neuropathy: a double-blind, placebo-controlled trial. *Pain* 2004;110:628–638.

Rowbotham M, Harden N, Stacey B, Bernstein P, Magnus-Miller L. Gabapentin for the treatment of postherpetic neuralgia: a randomized controlled trial. *JAMA* 1998;280:1837–1842.

Rush AM, Waxman SG. PGE2 increases the tetrodotoxin-resistant Nav1.9 sodium current in mouse DRG neurons via G-proteins. *Brain Res* 2004;1023:264–271.

Rush AM, Dib-Hajj SD, Liu S, Cummins TR, Black JA, Waxman SG. A single sodium channel mutation produces hyper- or hypoexcitability in different types of neurons. *Proc Natl Acad Sci USA* 2006;103:8245–8250.

Russ AP, Lampel S. The druggable genome: an update. *Drug Discovery Today* 2005;10:1607–1610.

Saegusa H, Kurihara T, Zong S, Kazuno A, Matsuda Y, Nonaka T, Han W, Toriyama H, Tanabe T. Suppression of inflammatory and neuropathic pain symptoms in mice lacking the N-type Ca^{2+} channel. *EMBO J* 2001;20:2349–2356.

Sanguinetti MC, Curran ME, Zou A, Shen J, Spector PS, Atkinson DL, Keating MT. Coassembly of K(V)LQT1 and minK (IsK) proteins to form cardiac I(Ks) potassium channel. *Nature* 1996;384:80–83.

Schroeder CI, Smythe ML, Lewis RJ. Development of small molecules that mimic the binding of omega-conotoxins at the N-type voltage-gated calcium channel. *Mol Divers* 2004;8:127–134.

Scott DA, Wright CE, Angus JA. Actions of intrathecal [omega]-conotoxins CVID, GVIA, MVIIA, and morphine in acute and neuropathic pain in the rat. *Eur J Pharmacol* 2002;451:279–286.

Seko T, Kato M, Kohno H, Ono S, Hashimura K, Takenobu Y, Takimizu H, Nakai K, Maegawa H, Katsube N, Toda M. L-Cysteine based N-type calcium channel blockers: structure–activity relationships of the C-terminal lipophilic moiety, and oral analgesic efficacy in rat pain models. *Bioorg Med Chem Lett* 2002;12:2267–2269.

Seward E, Hammond C, Henderson G. Mu-opioid-receptor-mediated inhibition of the N-type calcium-channel current. *Proc Biol Sci* 1991;244:129–135.

Sheets PL, Heers C, Stoehr T, Cummins TR. Differential block of sensory neuronal voltage-gated sodium channels by lacosamide [(2*R*)-2-(acetylamino)-*N*-benzyl-3-methoxypropanamide], lidocaine, and carbamazepine. *J Pharmacol Exp Ther* 2008;326:89–99.

Shiau AK, Massari ME, Ozbal CC. Back to basics: label-free technologies for small molecule screening. *Comb Chem High Throughput Screen* 2008;11:231–237.

Sills GJ. The mechanisms of action of gabapentin and pregabalin. *Curr Opin Pharmacol* 2006;6:108–113.

Smith MT, Cabot PJ, Ross FB, Robertson AD, Lewis RJ. The novel N-type calcium channel blocker, AM336, produces potent dose-dependent antinociception after intrathecal dosing in rats and inhibits substance P release in rat spinal cord slices. *Pain* 2002;96:119–127.

Snutch TP, Feng ZP, Doering C, Cayabyab F, Janke D, Parker DB, Belardetti F, Morimoto B, Vanderah T, Zamponi GW. A New class of N-type calcium channel blocker efficacious in animal models of chronic pain. Society for Neuroscience Annual Meeting, abstract 465.1, 2001.

Solaro C, Restivo D, Mancardi GL, Tanganelli P. Oxcarbazepine for treating paroxysmal painful symptoms in multiple sclerosis: a pilot study. *Neurol Sci* 2007;28: 156–158.

Song JH, Ham SS, Shin YK, Lee CS. Amitriptyline modulation of Na(+) channels in rat dorsal root ganglion neurons. *Eur J Pharmacol* 2000a;401:297–305.

Song Y, Bowersox SS, Connor DT, Dooley DJ, Lotarski SM, Malone T, Miljanich G, Millerman E, Rafferty MF, Rock D, Roth BD, Schmidt J, Stoehr S, Szoke BG, Taylor C, Vartanian M, Wang YX. (S)-4-Methyl-2-(methylamino)pentanoic acid [4, 4-bis(4-fluorophenyl)butyl]amide hydrochloride, a novel calcium channel antagonist, is efficacious in several animal models of pain. *J Med Chem* 2000b;43: 3474–3477.

Staats PS, Yearwood T, Charapata SG, Presley RW, Wallace MS, Byas-Smith M, Fisher R, Bryce DA, Mangieri EA, Luther RR, Mayo M, McGuire D, Ellis D. Intrathecal ziconotide in the treatment of refractory pain in patients with cancer or AIDS: a randomized controlled trial. *JAMA* 2004;291:63–70.

Stanley EF. Syntaxin I modulation of presynaptic calcium channel inactivation revealed by botulinum toxin C1. *Eur J Neurosci* 2003;17:1303–1305.

Stea A, Dubel SJ, Pragnell M, Leonard JP, Campbell KP, Snutch TP. A beta-subunit normalizes the electrophysiological properties of a cloned N-type Ca^{2+} channel alpha 1-subunit. *Neuropharmacology* 1993;32:1103–1116.

Steenland HW, Ko SW, Wu LJ, Zhuo M. Hot receptors in the brain. *Mol Pain* 2006;2:34.

Stotz SC, Jarvis SE, Zamponi GW. Functional roles of cytoplasmic loops and pore lining transmembrane helices in the voltage-dependent inactivation of HVA calcium channels. *J Physiol* 2004;554:263–273.

Stummann TC, Salvati P, Fariello RG, Faravelli L. The anti-nociceptive agent ralfinamide inhibits tetrodotoxin-resistant and tetrodotoxin-sensitive Na^+ currents in dorsal root ganglion neurons. *Eur J Pharmacol* 2005;510:197–208.

Szallasi A, Blumberg PM. Vanilloid (capsaicin) receptors and mechanisms. *Pharmacol Rev* 1999;51:159–212.

Szallasi A, Cortright DN, Blum CA, Eid SR. The vanilloid receptor TRPV1: 10 years from channel cloning to antagonist proof-of-concept. *Nat Rev Drug Discovery* 2007;6:357–372.

Tamayo N, Liao H, Stec MM, Wang X, Chakrabarti P, Retz D, Doherty EM, Surapaneni S, Tamir R, Bannon AW, Gavva NR, Norman MH. Design and synthesis of peripherally restricted transient receptor potential vanilloid 1 (TRPV1) antagonists. *J Med Chem* 2008;51:2744–2757.

Tanaka M, Cummins TR, Ishikawa K, Dib-Hajj SD, Black JA, Waxman SG. SNS Na^+ channel expression increases in dorsal root ganglion neurons in the carrageenan inflammatory pain model. *NeuroReport* 1998;9:967–972.

Teodori E, Baldi E, Dei S, Gualtieri F, Romanelli MN, Scapecchi S, Bellucci C, Ghelardini C, Matucci R. Design, synthesis, and preliminary pharmacological evaluation of 4-aminopiperidine derivatives as N-type calcium channel blockers active on pain and neuropathic pain. *J Med Chem* 2004;47:6070–6081.

Terstappen GC. Functional analysis of native and recombinant ion channels using a high-capacity nonradioactive rubidium efflux assay. *Anal Biochem* 1999;272:149–155.

Thampy KG, Barnes EM Jr. Gamma-aminobutyric acid-gated chloride channels in cultured cerebral neurons. *J Biol Chem* 1984;259:1753–1757.

Toledo-Aral JJ, Brehm P, Halegoua S, Mandel G. A single pulse of nerve growth factor triggers long-term neuronal excitability through sodium channel gene induction. *Neuron* 1995;14:607–611.

Toledo-Aral JJ, Moss BL, He ZJ, Koszowski AG, Whisenand T, Levinson SR, Wolf JJ, Silos-Santiago I, Halegoua S, Mandel G. Identification of PN1, a predominant voltage-dependent sodium channel expressed principally in peripheral neurons. *Proc Natl Acad Sci USA* 1997;94:1527–1532.

Tombler E, Cabanilla NJ, Carman P, Permaul N, Hall JJ, Richman RW, Lee J, Rodriguez J, Felsenfeld DP, Hennigan RF, Diverse-Pierluissi MA. G protein-induced trafficking of voltage-dependent calcium channels. *J Biol Chem* 2006;281: 1827–1839.

Tominaga M, Tominaga T. Structure and function of TRPV1. *Pflugers Arch* 2005;451: 143–150.

Tominaga M, Wada M, Masu M. Potentiation of capsaicin receptor activity by metabotropic ATP receptors as a possible mechanism for ATP-evoked pain and hyperalgesia. *Proc Natl Acad Sci USA* 2001;98:6951–6956.

Tremont-Lukats IW, Hutson PR, Backonja MM. A randomized, double-masked, placebo-controlled pilot trial of extended IV lidocaine infusion for relief of ongoing neuropathic pain. *Clin J Pain* 2006;22:266–271.

Trivedi S, Dekermendjian K, Julien R, Huang J, Lund PE, Krupp J, Kronqvist R, Larsson O, Bostwick R. Cellular HTS assays for pharmacological characterization of Na(V)1.7 modulators. *Assay Drug Dev Technol* 2008;6:167–179.

Tseng TT, McMahon AM, Johnson VT, Mangubat EZ, Zahm RJ, Pacold ME, Jakobsson E. Sodium channel auxiliary subunits. *J Mol Microbiol Biotechnol* 2007;12:249–262.

Turkington RW. Depression masquerading as diabetic neuropathy. *JAMA* 1980;243: 1147–1150.

Tzellos TG, Papazisis G, Amaniti E, Kouvelas D. Efficacy of pregabalin and gabapentin for neuropathic pain in spinal-cord injury: an evidence-based evaluation of the literature. *Eur J Clin Pharmacol* 2008;64:851–858.

Ulbricht W. Sodium channel inactivation: molecular determinants and modulation. *Physiol Rev* 2005;85:1271–1301.

Vestergaard K, Andersen G, Gottrup H, Kristensen BT, Jensen TS. Lamotrigine for central poststroke pain: a randomized controlled trial. *Neurology* 2001;56:184–190.

Waldmann R. Proton-gated cation channels—neuronal acid sensors in the central and peripheral nervous system. *Adv Exp Med Biol* 2001;502:293–304.

Wang GK, Russell C, Wang SY. State-dependent block of voltage-gated Na^+ channels by amitriptyline via the local anesthetic receptor and its implication for neuropathic pain. *Pain* 2004;110:166–174.

Watkins LR, Maier SF. Glia: a novel drug discovery target for clinical pain. *Nat Rev Drug Discovery* 2003;2:973–985.

Watson CP, Evans RJ, Reed K, Merskey H, Goldsmith L, Warsh J. Amitriptyline versus placebo in postherpetic neuralgia. *Neurology* 1982;32:671–673.

Waxman SG, Wood JN. Sodium channels: from mechanisms to medicines? *Brain Res Bull* 1999;50:309–310.

Weir SW, Weston AH. The effects of BRL 34915 and nicorandil on electrical and mechanical activity and on 86Rb efflux in rat blood vessels. *Br J Pharmacol* 1986;88:121–128.

Westenbroek RE, Hoskins L, Catterall WA. Localization of Ca^{2+} Channel subtypes on rat spinal motor neurons, interneurons, and nerve terminals. *J Neurosci* 1998;18: 6319–6330.

Wiffen PJ, McQuay HJ, Moore RA. Carbamazepine for acute and chronic pain. *Cochrane Database System Review*, 2005, CD005451.

Williams ME, Brust PF, Feldman DH, Patthi S, Simerson S, Maroufi A, McCue AF, Velicelebi G, Ellis SB, Harpold MM. Structure and functional expression of an omega-conotoxin-sensitive human N-type calcium channel. *Science* 1992;257:389–395.

Willow M, Kuenzel EA, Catterall WA. Inhibition of voltage-sensitive sodium channels in neuroblastoma cells and synaptosomes by the anticonvulsant drugs diphenylhydantoin and carbamazepine. *Mol Pharmacol* 1984;25:228–234.

Willow M, Gonoi T, Catterall WA. Voltage clamp analysis of the inhibitory actions of diphenylhydantoin and carbamazepine on voltage-sensitive sodium channels in neuroblastoma cells. *Mol Pharmacol* 1985;27:549–558.

Winquist RJ, Pan JQ, Gribkoff VK. Use-dependent blockade of Cav2.2 voltage-gated calcium channels for neuropathic pain. *Biochem Pharmacol* 2005;70:489–499.

Witcher DR, De Waard M, Sakamoto J, Franzini-Armstrong C, Pragnell M, Kahl SD, Campbell KP. Subunit identification and reconstitution of the N-type Ca^{2+} channel complex purified from brain. *Science* 1993;261:486–489.

Woolf CJ. Pain: moving from symptom control toward mechanism-specific pharmacologic management. *Ann Intern Med* 2004;140:441–451.

Xie X, Lancaster B, Peakman T, Garthwaite J. Interaction of the antiepileptic drug lamotrigine with recombinant rat brain type IIA Na^+ channels and with native Na^+ channels in rat hippocampal neurones. *Pflugers Arch* 1995;430:437–446.

Yamamoto T, Sakashita Y. Differential effects of intrathecally administered N- and P-type voltage-sensitive calcium channel blockers upon two models of experimental mononeuropathy in the rat. *Brain Res* 1998;794:329–332.

Yamane H, de Groat WC, Sculptoreanu A. Effects of ralfinamide, a Na^+ channel blocker, on firing properties of nociceptive dorsal root ganglion neurons of adult rats. *Exp Neurol* 2007;208:63–72.

Yeomans DC, Levinson SR, Peters MC, Koszowski AG, Tzabazis AZ, Gilly WF, Wilson SP. Decrease in inflammatory hyperalgesia by herpes vector-mediated knockdown of Nav1.7 sodium channels in primary afferents. *Hum Gene Ther* 2005;16:271–277.

Yiangou Y, Birch R, Sangameswaran L, Eglen R, Anand P. SNS/PN3 and SNS2/NaN sodium channel-like immunoreactivity in human adult and neonate injured sensory nerves. *FEBS Lett* 2000;467:249–252.

Zamponi GW. Regulation of presynaptic calcium channels by synaptic proteins. *J Pharmacol Sci* 2003;92:79–83.

Zeilhofer HU. Loss of glycinergic and GABAergic inhibition in chronic pain—contributions of inflammation and microglia. *Int Immunopharmacol* 2008;8:182–187.

Zhang Y, Helm JS, Senatore A, Spafford JD, Kaczmarek LK, Jonas EA. PKC-induced intracellular trafficking of CaV2 precedes its rapid recruitment to the plasma membrane. *J Neurosci* 2008;28:2601–2612.

22

TARGETING THE mTOR PATHWAY FOR TUMOR THERAPEUTICS

WEI CHEN

The centerpiece of life is survival and growth, and for prehistoric humans it involved balancing the discrepancy between the environment (scarce availability of food) and human development. It has been hypothesized that the dominant genetic pathways in humans evolved in favor of mechanisms to increase caloric intake (Ames, 2006). Once the balance between the environment and human development tilted to high energy intake, exceeding the capacity of our genetic pathways, pathological events arose, like obesity and tumorigenesis (Speakman, 2004; Polednak, 2008). In order to battle diseases caused by this interruption in energy homeostasis, it is important to understand how energy homeostasis is regulated. As one of the major regulators of energy homeostasis, our understanding of the function of mTOR (mammalian target of rapamycin) and its regulation has gradually unfolded since mTOR was identified more than 30 years ago when rapamycin was characterized as an antibiotic (Heitman et al., 1991). In recent years, growing evidence has supported the importance of mTOR for cell growth, proliferation, and survival (Gingras et al., 2001; Schmelze and Hall, 2000; Fingar and Blenis, 2004). It plays a central role in integrating signals from metabolism, energy homeostasis, the cell cycle, and the stress response, which may lead to growth, differentiation, or cell death (Gingras et al., 2001; Schmelze and Hall, 2000; Fingar and Blenis, 2004). This review summarizes recent advances in mTOR biology and its clinical application in oncology, and it focuses on the cross-talk between mTOR pathways with the JNK (Jun N-terminal kinase), Ras/Erk, Wnt, and

Enzyme Inhibition in Drug Discovery and Development, edited by Lu and Li
Copyright © 2010 John Wiley & Sons, Inc.

metabolism regulation pathways. Analysis of mTOR functions related to its physiological and pathological contexts will help us better understand the physiological role and regulation of mTOR, which may have implications for mTOR as a potential target for diseases, particularly for tumor therapeutics.

22.1 mTOR FUNCTIONS AS A KINASE IN TWO DISTINCT COMPLEXES TO PHOSPHORYLATE DIFFERENT SETS OF SUBSTRATES

mTOR, the mammalian target of rapamycin, is a member of phosphatidylinositol 3-kinase-related kinase (PIKK) family, which contains TOR (target of rapamycin), PI3K (phosphatidylinositol triphosphate kinase), ATM (ataxia telangiectasia mutated protein), ATR (ataxia telangiectasia related protein), DNA-PK (DNA-dependent protein kinase), and TRAP (Trp RNA-binding attenuation protein) (Hoekstra, 1997). TOR from mammals and other metazoans is highly conserved and is about 42% identical to yeast TOR in its amino acid sequence (Wullschleger et al., 2006). mTOR is a large protein (~280 kD) consisting of distinct structures like the HEAT repeats domain, FAT domain, FRB domain, kinase domain, NRD domain, and FATC domain (Fingar and Blenis, 2004). The HEAT (Huntington, EF3, a subunit of PP2A, TOR1) repeats and FAT (FKBP12–rapamycin-associated protein, ataxia telangiectasia mutated protein, Trp RNA-binding attenuation protein) domains are involved in protein–protein interaction. FRB (FKBP12 rapamycin binding) domain binds to the FKBP12–rapamycin complex, and NRD (negative regulative domain) is a putative negative regulating domain (Wullschleger et al., 2006). Structure and mutagenesis studies suggest that the FATC (FAT at C-terminal) domain contains a disulfite-bonded loop that may be regulated by cytoplasmic redox (reduction–oxidation) potential, which may play a role in mTOR stability (Dames et al., 2005).

22.1.1 The Subunits of mTOR Complexes Are Shared with Other Pathways

The kinase activity of mTOR requires specific components to form complexes *in vivo* to become functionally dependent on different substrates. The two mTOR complexes that have been identified so far are mTORC1 and mTORC2 in mammalian cells (Wullschleger et al., 2006). mTORC1 contains mTOR, mLST8, Raptor, and Rheb, while mTORC2 includes mTOR, mLST8, Rictor, and mSin1 (Inoki et al., 2005). The components listed here for each complex are required for mTOR kinase activity in cells. Other components also have been shown to associate with either mTORC1 or mTORC2 complexes, which may or may not affect mTOR kinase activity. For example, PRAS40, an Akt substrate, can bind to mTORC1 and inhibit mTORC1 activity once phosphorylated by Akt (Wang et al., 2007). Protor also interacts with mTORC2 with unknown function (Pearce et al., 2007). Both PRR5 and PRR5L interact with mTORC2, but their functions are not very clear (Woo et al., 2007; Thedieck et al., 2007).

The subunit of mTORC2, mSin1 (also named as MIP1), was originally identified as MEKK2 interacting protein, which inhibits JNK kinase activity (Jacinto et al., 2006). mSin1 has a PH (Pleckstrin homology) domain and can bind to PIP3 (phosphatidylinositol triphosphate) and PA (phosphatidic acid), leading to mSin1 translocation to plasma membrane, as evidenced by the fact that the majority of GFP-tagged mSin1 was detected on the plasma membrane (Schroder et al., 2007). It was reported that mSin1 interacts with Ras and affects its activity, but how this interaction affects mTOR activity is not clear (Schroder et al., 2007). There are five different spliced forms of mSin1, and each isoform seems to participate in different biological activities. Further studies are needed to produce a more detailed picture of their functionality. Rictor has been reported to interact strongly with ILK (integrin-linked kinase) and Myo1c to play a role in cell skeleton reorganization, which is independent of mTOR activity (McDonald et al., 2008; Hagan et al., 2008). Rheb is also involved in the MAPK pathway by interacting with B-Raf (Karboniczek et al., 2006).

22.1.2 The List of Identified Substrates for mTOR Complexes Is Growing

The well-studied substrates for mTORC1 are p70S6K, phosphorylated at T386, and 4EBP1, phosphorylated at S42 (Chung et al., 2002; Graves et al., 1995). Phosphorylation at T386 by mTORC1 activates p70S6k, which is involved in protein synthesis through p70S6 in combination with other phosphorylations by PDK1 and SGK (Balendran et al., 1999). Phosphorylation of 4EBP1 at S42 by mTORC1 leads to its dissociation from 4EF, allowing 4EF to participate in translation initiation. It is reported that mTORC1 also phosphorylates NFATc4 at S168/170, which regulates the nucleocytoplasmic location of NFATc4 (Yang et al., 2008). mTORC1 also phosphorylates SGK1 at S422 to control p27 nuclear translocation and regulate the cell cycle (Hong et al., 2008). Another newly identified substrate for mTORC1 is PRAS40, which is phosphorylated by mTORC1 at several sites, leading to increased mTORC1 activity (Oshiro et al., 2007). Most of these mTORC1 substrates contain a TOR signaling motif (TOS motif), which mediates the interaction between substrates with Raptor of mTORC1 (Schalm and Blenis, 2002).

The substrates thus far identified for mTORC2 include Akt, possibly PKCα, and downstream effectors like Rho and Rac1 (Jacinto et al., 2004; Sarbassov et al., 2004, 2005). Residue S473 of Akt is believed to interact with the N-nodule of Akt and affect the substrate binding. So far the only known substrates of Akt requiring phosphorylation at S473 are the FoxO transcription factors (Shiota et al., 2006; Guertin et al., 2006). Knocking out Rictor or mSin1 did not affect the expression level of Akt and its kinase activity toward GSK3β, TSC2, and others (Shiota et al., 2006; Guertin et al., 2006). Two recent studies report that mTORC2 phosphorylates Akt1 at S450, affecting the stability of Akt1 in cells, but this phosphorylation at S450 is delayed compared to phosphorylation at S473 by mTORC2 (Facchinetti et al., 2008; Ikenoue et al., 2008). These data conflict with another report that JNK phosphorylates Akt at S450 under ischemic/reoxygenation condition (Shao et al., 2006). Further studies are needed to reveal the actual mechanism of phosphorylation of Akt S450.

22.2 REGULATION OF mTOR COMPLEXES DIFFER FROM EACH OTHER WITH SOME OVERLAP

Three phosphorylation sites have been identified on mTOR so far: T2446, S2448, and S2481 (Peterson et al., 2000; Holz and Blenis, 2005). T2446 has been proposed to be phosphorylated by AMPK and S2448 has been suggested as an Akt and/or a p70S6K phosphorylation site. No obvious consequence, in terms of either kinase activity or the composition of mTOR complexes, has been observed *in vivo* when these two sites on mTOR are phosphorylated. It is an interesting observation that the phosphorylation at S2448 is associated with mTOR translocation into the nucleus during hypoxia (Li et al., 2007), which may facilitate mTORC1 functioning as a transcription factor during hypoxia. S2481 has been confirmed as an autophosphorylation site on mTOR, while phosphorylation at this site is not sensitive to mitogen treatment, amino acid withdrawal, or rapamycin treatment (Peterson et al., 2000). In our laboratory, dephosphorylation of S2481 on mTOR did not change its kinase activity *in vitro*, although the phosphorylation level of S2481 has been widely used as an indicator of mTOR activity in many published studies (unpublished data).

22.2.1 The Cellular Localization of mTORC1 Is Different from That of mTORC2

The subunit composition of mTORC1 is substantially different from that of mTORC2, and these two kinase complexes have also been shown to function differently from each other. It is interesting to analyze whether their cellular localizations differ, reflecting their differing functions. In live *Saccharomyces cerevisiae* cells, TOR1 and TOR2 have distinct locations: TOR1 is diffusely expressed in the cytoplasm and concentrated near vacuolar membranes, while TOR2 is expressed in the cytoplasm but is predominantly present in dots at the plasma membrane (Sturgill et al., 2008). In mammalian cells, both mTORC1 and mTORC2 are shown to be associated with membranes, with a very small fraction in the nucleus (Hresko et al., 2005; unpublished data). When cells were treated with a PI3K-specific inhibitor that did not inhibit mTORC2, the amount of mTOR associated with the plasma membrane decreased substantially and the phosphorylation of Akt at S473 was blocked, while the amount of mTOR in other fractions of cells was not affected (unpublished data). This result suggests that mTORC2 has to be transferred to the plasma membrane to become activated. Activation of FBXW7 led to partial mTOR degradation accompanied by decreased mTORC1 signaling, while the phosphorylation of Akt1 at S473 was not changed (Mao et al., 2008). This supports the existence of another level of regulation of mTORC2 to prevent the dynamic equilibrium of mTOR between mTORC1 and mTORC2, which maintains mTORC2 complex even mTORC1 is degraded. We hypothesize that mTORC2 has to be associated with plasma membrane to become activated and phosphorylate Akt at S473, which is also bound to plasma membrane through interactions of its PH domain with PIP3.

22.2.2 Activation of mTORC1 Is Regulated Through Multiple Mechanisms

One of mechanism to activate mTORC1 may be through TSC1/TSC2. The TSC1/TSC2 complex was identified in a genetic disease that was found to involve hyperactivity of mTORC1 activity, in which loss-of-function mutations were identified in TSC1/TSC2. These mutations resulted in the loss of inhibition of mTORC1 activity (Gomez, 1991; van Slegtenhorst et al., 1997; European Chromosome 16 Tuberous Sclerosis Consortium, 1993). Later studies indicated that TSC2 actually functions as a GEF to a small GTPase, Rheb, to inhibit the interaction of Rheb with mTORC1 and activate mTORC1 (Castro et al., 2003; Garami et al., 2003; Inoki et al., 2003; Saucedo et al., 2003; Stocker et al., 2003; Tee et al., 2003; Zhang et al., 2003). TSC1 binds to TSC2 and stabilizes TSC2 protein by preventing proteasome degradation mediated through DDB1-Cul4-Roc1 E3 ligase (Nellist et al., 2001; Hu et al., 2008). Therefore, TSC1 and TSC2 form a stable complex to block Rheb activity and mTORC1 activation. The mechanism of how Rheb activates mTORC1 is not well established, although one recent study proposed that Rheb could compete with an mTOC1 endogenous inhibitor, FKBP38. However, this conflicts with the results of another published study (Bai et al., 2007; Wang et al., 2008).

Many signal pathways converge on the TSC1/TSC2 complex to regulate mTORC1 activity. AMPK phosphorylates TSC2 at S1345 to activate TSC2 function, resulting in a decrease in mTORC1 activity when AMPK is activated by a low ratio of ATP to AMP, or by interaction with p53 (Hardie et al., 1998, 2001; Budanov and Karin, 2008). AMPK is also reported to directly phosphorylate Raptor to affect mTORC1 assembly and down-regulate mTORC1 activity (Gwinn et al., 2008). Stress response protein REDD1 also phosphorylates TSC2 to stabilize TSC2 and down-regulate mTORC1 activity (Sofer et al., 2005). Erk activated through the Raf–Mek–Erk pathway phosphorylates TSC2 at S864 to destabilize TSC2 and release its inhibitory effect on Rhomb, activating mTORC1 (Ma et al., 2005). It is well known that PI3K can activate mTORC1 through Akt. But how Akt affects TSC2 is less certain; it has been proposed that Akt affects TSC2 either by directly phosphorylating TSC2 at S939, S981, and T1462 to down-regulate TSC2 or by increasing the ATP-to-

AMP ratio to inhibit AMPK activity, decrease TSC2 activity, and increase mTORC1 activity (Inoki et al., 2002; Hahn-Windgassen et al., 2005). GSK3β is suggested to phosphorylate TSC1 at T357 and T390 to enhance the interaction between TSC1 and TSC2, which leads to the stabilization of TSC2 and inhibition of mTORC1 activity (Mak et al., 2005). GSK3β can be inactivated by Akt phosphorylation or the canonical Wnt pathway, which leads to activation of mTORC1 (Inoki et al., 2006). IKKβ, from the NFκB pathway, also phosphorylates TSC1 at S487 and S511 to break the interaction between TSC1 and TSC2, which leads to the activation of mTORC1. This plays a substantial role in the inflammation response and wound healing (Lee et al., 2007).

Amino acids also stimulate mTORC1 activity, although the mechanism has not been clearly revealed. Vps34, a member of the PI3K kinase family, is required for mTORC1 activation by amino acids (Nobukuni et al., 2005). MAP4K3, a member of MAPKK, was reported to participate in amino acid activation of mTORC1, although the detailed mechanism is unknown (Findlay et al., 2007; Cook and Morley, 2007). One study showed that Rag1, a small GTPase, is also involved in amino acid activation of mTORC1 by facilitating mTORC1 localization to certain endoplasmic compartments where Rab7 exists (Sancak et al., 2008). This raises the question of whether there are more small GTPases besides Rheb and Rag1 that can activate mTORC1 in response to different stimuli to contribute to the control of cell growth and proliferation. It might be an interesting experiment to run a genome-wide search for GTPases that modulate mTORC1 activity.

The growing evidence suggests that mTORC1 is not only temporally regulated, but also spatially controlled. The majority of mTORC1 is localized to the endoplasmic membrane when cells are stained with antibodies against mTORC1 components or prepared as differential fractions after Western blotting with antibodies against mTORC1 subunits (Hresko and Mueckler, 2005; Liu and Zheng, 2007). A small amount of mTORC1 is also detected in the nucleus (Cunningham et al., 2007). mTORC1 is believed to associate with the membrane through Rheb and/or through mTOR interactions with phosphatidic acid (PA) (Liu and Zheng, 2007; Buerger et al., 2006; Fang et al., 2001). mTOR also contains a sequence that directs mTOR to membrane association (Liu and Zheng, 2007). Similar to other members of the Ras superfamily, Rheb has a C-terminal CaaX box that is subject to farnesylation, which has been reported to play a key role in Rheb's subcellular localization and which directs its association with the endoplasmic membrane (Buerger et al., 2006). After brief association with the endoplasmic reticulum, Rheb localizes to highly ordered, distinct structures within the cytoplasm that display characteristics of Golgi membranes where Rab7 also localizes. Failure of Rheb to localize to the endoplasmic membrane impairs its ability to interact with mTOR and activate mTORC1 downstream targets. It has been noted that the interaction between Rheb and mTOR is necessary but not sufficient to activate mTORC1 (Long et al., 2005). Therefore additional proteins must be involved in mTORC1 activation (Avruch et al., 2008).

Phosphatidic acid (PA) is the product of PLD1/2 in cells; it has long been recognized as a secondary messenger (Oude Weernink et al., 2007). PLD has been reported to mediate certain pathways of mitogenic activation of mTORC1 (Fang et al., 2001). Studies have revealed that its product, PA, could directly bind to mTOR at the FBD domain to compete with the FKBP12-rapamycin complex (Chen et al., 2003). The interaction between PA and mTOR may activate mTORC1 activity by directing the mTORC1 complex to localize to certain areas within the cell, since direct treatment with PA did not increase mTORC1 activity in biochemical assay (unpublished data). An unexpected result of a recently published study shows that Rheb binds to PLD1 and stimulates PA production in a manner dependent on its GTPAse acivity (Sun et al., 2008). Knockdown of PLD1 abolished the capability of Rheb to activate mTORC1, which may be the reason Rheb is not sufficient to activate mTORC1 (Sun et al., 2008). PLD1 is found in plasma membranes as well as in perinuclear compartments composed with ER and Golgi and in other dispersed vesicular structures in the cytoplasm (Kim et al., 1999). Upon activation, PLD1 can be transferred into different cellular locations; the consequences of these PLD1 translocations are not well understood.

In summary, it can be hypothesized that Rheb activated by inhibition of TSC2 is transferred to certain endoplasmic membranes, to which PLD1 also locates, to elevate PA production by PLD1. PA then brings mTOR to assembly with Rheb and other components to form mTORC1 at the endoplasmic membrane and activate mTORC1 (Fig. 22.1).

22.2.3 mTORC1 Is Thought to Play a Role as a Transcription Factor

In a variety of studies with different procedures, mTOR has been found in the cell nucleus, which raises the possibility that, in addition to its key role in translation regulation, mTOR could participate in transcription activity. Two recent research reports confirmed this. They found that mTORC1 interacts with the transcription factor YY1 to recruit PCG-1α to specifically transcribe mitochondria genes, which play a key role in cellular energy homeostasis (Cunningham et al., 2007). As discussed later, mTORC1 participates in both transcription and translation to coordinate different physiological activities most efficiently. As another example, mTORC1 is involved in activation of HIF-1α transcription activity (Land and Tee, 2007). HIF-1α utilizes a similar mechanism to interact with CBP/p300 through mTORC1, which activates transcription of genes needed for cell survival and growth under hypoxic conditions.

So far there is no report of mTORC2 functioning as a transcription factor, but our data suggest that it can be detected in the cell nucleus. Whether mTORC2 has a similar function in gene transcription needs further study. Our data suggest that mTORC2 interacts with LRP130, which also interacts with PCG-1α to participate in transcription (Cooper et al., 2006). We are currently characterizing the potential role of mTORC2 in this setting.

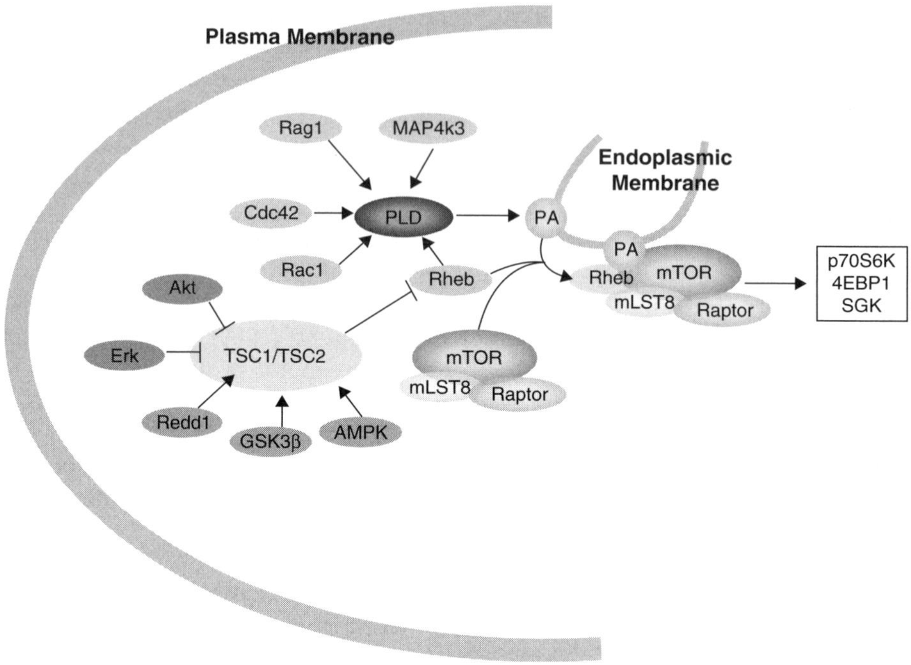

Figure 22.1. mTORC1 activation. See the insert for color representation of this figure.

22.2.4 Multiple Feedback Loops Regulate mTORC1 Activity

Within the PI3K-Akt-mTORC1 axis, there is a feedback loop to maintain the signal pathway response to stimuli in an acute and temporal manner. Insulin activates PI3K by bringing IRS to the insulin receptor where it is phosphorylated by the receptor's kinase activity (Boura-Halfon and Zick, 2008). Phosphorylated IRS serves as a docking site for p85, a subunit of PI3K, and brings PI3K to the membrane to activate it and trigger the signal cascade PI3K-Akt-mTORC1-p70S6K, eventually activating p70S6K. However, activated p70S6K can phosphorylate IRS and lead to degradation of IRS through the proteasome pathway to down-regulate PI3K activity (Wan et al., 2007). Short-term treatment with rapamycin has been shown to lead to increased phosphorylation of Akt1 at S473 because of this feedback loop mechanism, resulting in increased PI3K activity. However, this feedback regulation is a short-term effect. Prolonged treatment with rapamycin in mice and humans has been reported to inhibit mTORC1 activity, actually decreasing the phosphorylation of S473 on Akt (Sarbassov et al., 2006). This prolonged treatment with rapamycin leads to decreased but reversible glucose tolerance and insulin sensitivity (Di Paolo et al., 2006). GSK3β and JNK also play a negative role in IRS stabilization and keep PI3K activity under control (Liberman et al., 2008; Sharfi and Eldar-Finkelman, 2008). GSK3β directly phosphorylates IRS1 at Ser332 while JNK phosphorylates IRS1 at Ser307. On the other hand, p70S6K can directly phosphorylate GSK3β and inactivate GSK3β, which leads to cell growth in TSC2-null cells (Zhang et al., 2006). It is not clear whether this aberrant pathway plays an important role in tumorigenesis.

22.2.5 The Mystery of mTORC2 Regulation Is Still Waiting to Be Solved

There is very limited information about mTORC2 regulation simply because it has only recently been discovered. It is helpful to analyze historical data about regulation of known mTORC2 substrates to understand how mTORC2 may be controlled in cells. So far, Akt is the only protein that has been confirmed to serve as a substrate of mTORC2 kinase (Sarbassov et al., 2005). Three phosphorylation sites on Akt–T308, S450, and S473 have been identified (Manning and Cantley, 2007). T308 localizes to the activation loop of Akt and is required for its kinase activity, and phosphorylation at S450 plays a role in stabilizing newly synthesized Akt (Facchinetti et al., 2008; Ikenoue et al., 2008; Manning and Cantley, 2007). Phosphorylation at S473 is not required for either Akt kinase activity or protein stability but is required for the substrate specificity of Akt (Shiota et al., 2006; Guertin et al., 2006). It has been well established that PDK1 phosphorylates Akt1 at T308 when activated by the PI3K pathway (Manning and Cantley, 2007). But the search for the kinase that phosphorylates S473 has lasted for some time, and several kinases have been proposed, including DNAPK, ILK, ATM/ATR, and PKCs. Only recently, mTORC2 was identified and confirmed to be the kinase that phosphorylates Akt at S473 *in vivo* (Shiota et al., 2006; Guertin et al., 2006). Knockout of either Rictor or mSin1 in mice only affected the phosphorylation of Akt at S473, but did not change Akt protein levels or the intensity of phosphorylation at T308. The phosphorylation of most Akt substrates, like GSK3β, TSC2, and other, was not changed except for that of FoxO transcription factors (Shiota et al., 2006; Guertin et al., 2006). DNAPK may also phosphorylate Akt1 at S473 only when there has been DNA damage, without responding to other mitogenic stimuli (Bozulic et al., 2008).

Since the discovery of mTORC2 as the kinase that phosphorylates the S473 site of Akt1, the list of its direct substrates has not been expanded. Several reports showed that the phosphorylation of PKC and the activity of RhoA and RAC1 are enhanced upon mTORC2 activation (Jacinto et al., 2004; Dada et al., 2008; Waters et al., 2008). However, *in vitro* kinase assays failed to confirm that they serve directly as mTORC2 substrates. The majority of Rictor is associated with endoplasmic membrane, with some localizing to plasma membrane (unpublished data). As discussed above, mSin1 contains a PH domain that binds to PIP3 and PA *in vitro*, while GFP-mSin1 is restricted to the plasma membrane (Schroeder et al., 2007). LPA treatment

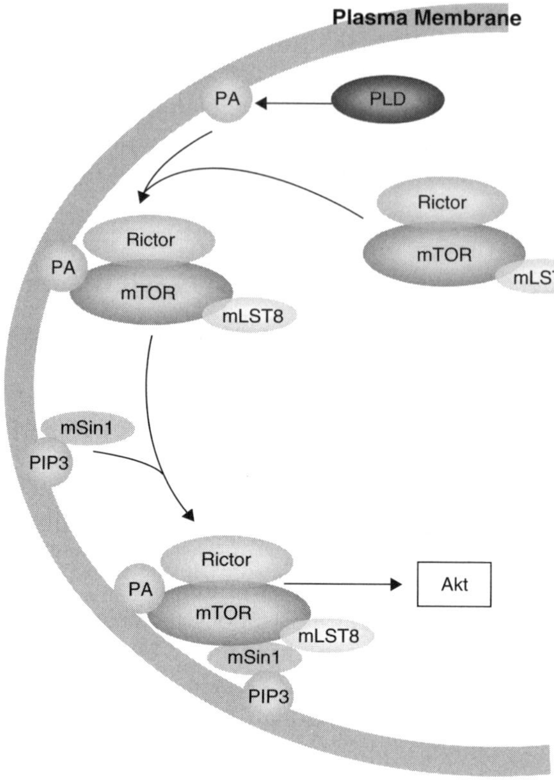

Figure 22.2. mTORC2 activation. See the insert for color representation of this figure.

of cells stimulates PLD1 activity and translocation to the plasma membrane, in which the increased phosphorylation of Akt1 at S473 was observed, and can be inhibited by a PLD-specific inhibitor (Kam and Exton, 2004; Yaghini et al., 2007). However, the activity of purified mTORC2 complex was not affected by PA when tested in an *in vitro* kinase assay (data on file). Therefore, a mechanism for mTORC2 activation may be hypothesized, in which, upon activation, PA generated by PLD1 brings mTOR/Rictor/mLST8 to the plasma membrane to be assembled with mSin1, brought to the plasma membrane by PIP3, to form mTORC2 and be activated (Fig. 22.2). Indeed, PA treatment increased the amount of mTor associated with the plasma membrane (unpublished data), and PLD1 activated by PKC was observed to be transferred to the plasma membrane (Kim et al., 1999), which also leads to the increased phosphorylation of Akt at S473 (Partovian et al., 2008). On the other hand, application of a PI3K-specific inhibitor was able to decrease the amount of mTOR associated with the plasma membrane and reduce the phosphorylation of Akt at S473; this could be due to the dissociation of mSin1 from the plasma membrane (unpublished data).

22.3 IS THE KEY PHYSIOLOGICAL FUNCTION OF mTORC1 AND mTORC2 TO COORDINATE THE CELLULAR RESPONSE TO ENERGY HOMEOSTASIS?

Since rapamycin was discovered as a specific inhibitor for mTORC1, it has been widely used to study the function of the mTORC1 pathway (Heitman et al., 1991). Inhibition of mTORC1 with rapamycin has been found to alter many major processes, including mRNA translation, ribosome biogenesis, nutrient metabolism, cell cycle regulation, and autophagy (Heitman et al., 1991; Gingras et al., 2001). However, the mechanisms through which mTORC1 mediates these processes are not clear, even though two of its substrates, p70S6K1 and 4EBP1, are relatively well understood.

The phosphorylation of p70S6K1 has been widely recognized as an indicator for mTORC1 activation. p70S6K, a member of the AGC family of kinases, becomes activated when phosphorylated by mTORC1 at its hydrophobic C-terminal motif and by PDK1 at its activation loop (Zanchi and Lancha, 2008). There are two members of p70S6K: p70S6K1 and p70S6K2 (Zanchi and Lancha, 2008). A knockout experiment demonstrated that p70S6K1, not P70SK2, controls cell size (Jastrzebski et al., 2007). It was proposed that p70S6K1 controls the translation of an abundant class of mRNA containing a 5′ TOP sequence; however, there have been recent reports against this idea (Jastrzebski et al., 2007). Several substrates of p70S6K1 have been identified, including the ribosomal S6 protein and the translational regulators eEF2 kinase and eIF-4B. Still the mechanism through which p70S6K1 controls cell size has yet to be determined.

The major and earliest identified function of mTORC1 is to control protein synthesis in cells (Heitman et al., 1991). As it is well known that protein synthesis consumes about 30% of the energy of cells (Buttgereit and Brand, 1995), it is necessary to understand how mTORC1 works to influence or regulate cellular energy homeostasis or vice versa. Glucose has been thought to be the major source of energy in cells, supplying it through many steps of oxidization in the cytosol and mitochondria (Xie and Wang, 2000). mTORC1 coupled with TSC1/TSC2 and p70S6K directly regulates glucose transport via regulation of Glut1 translocation to the plasma membrane, where it becomes active, which regulates glucose uptake into cells (Buller et al., 2008). Mitochondria are the organelles that produce the energy supply (ATP) for almost all cellular processes through F_1F_0 ATP synthase, driven by a proton gradient generated by an electron transfer chain reaction that consumes oxygen (Hagen and Vidal-Puig, 2002). A recent report showed that mTORC1 directly controls mitochondrial gene expression through YY1–mTORC1–PCG-1α interaction (Cunningham et al., 2007). Rapamycin treatment of C2C12 cells led to a 32% decrease in mitochondrial DNA expression and a 12% decrease in oxygen consumption (Cunningham et al., 2007). This shows that the cells were able to directly control their energy consumption and maintain ATP levels through mTORC1. On the other hand, mTORC1 is reported to be associated with mitochondria, and oxygen concentration regulates the assembly of mTORC1 on the mitochondrial membrane (Schieke et al., 2006). This observation indicates that mTORC1 can function as an oxygen sensor for mitochondria to coordinate the response of mitochondria and the nucleus and could serve as a signal of nutrition and energy homeostasis.

Treatment of cells with rapamycin also triggers autophagy through stabilization of Atg1, which is constitutively degraded when phosphorylated by mTORC1 (Scott et al., 2004). Similarly, under nutrient restriction, the activity of mTORC1 is down-regulated, resulting in Atg1 stabilization and initiation of the autophagy process (Pattingre et al., 2008). This process is thought to provide necessary metabolites for cell survival during short-term nutrient restriction.

mTORC2 is also involved in glucose metabolism and energy homeostasis either directly or indirectly through its downstream target the FoxO transcription factors. Conditional tissue-specific knockout of Rictor in mice led to decreased glucose uptake in the short-term, while insulin sensitivity in the whole body was not affected (Kumar et al., 2008). The possible mechanism for this may be direct or indirect phosphorylation of AS160, which controls Glut4 transfer from the Golgi to the plasma membrane, by mTORC2 (Kumar et al., 2008). mTORC2 also controls the phosphorylation of Akt at S473, which is required for FoxO phosphorylation by Akt as described above. The phosphorylation of FoxO results in FoxO translocation into the cytosol and degradation through the proteasome pathway (Weidinger et al., 2008). FoxO has been reported to be deeply involved in glucose metabolism and energy homeostasis and other functions like apoptosis, vasculogenesis, oxidative stress resistance, inflammation, tumor suppression, and tissue-specific stem/progenitor cell maintenance (Gross et al., 2008). FoxO increases the level of circulated glucose through gluconeogenesis and glycolysis by regulating G6PC, PGC-1α, and AGRP gene transcription (Salih and Brunet, 2008). Transgenic overexpression of FoxO in mice did not affect overall insulin sensitivity, although *in vitro* experiments showed decreased beta-pancreatic cell proliferation (Buteau et al., 2007; Okamoto et al., 2006). This is consistent with the report that conditional muscle tissue-specific knockout of Rictor in mice does not affect whole-body insulin sensitivity, while it decreases blood glucose uptake (Kumar et al., 2008). Therefore, inhibition of mTORC2 leads to FoxO translocation into the nucleus to trigger transcription of proteins that increase glucose circulation to supply energy from degradation of glycogen. FoxO also plays a role in regulation of angiogenesis by inhibiting vessel formation and migration in endothelial cells, which is similar to the pathological angiogenesis induced by sustained Akt (Potente et al., 2005). Stem/precursor cells and tumorigenic cells are thought to be regulated by FoxO factors that limit their proliferation (Tothova et al., 2007). Transgenic knockout of FoxO1 alone or in combination with deletion of FoxO1 and FoxO4 led to a reduced pool of hematopoietic stem cells as a result of increased ROS and apoptosis, which resulted in reduced quiescence. This was noted to increase with age, but could be reversed with treatment with the antioxidant *N*-acetyl-L-cysteine (Salih and Brunet, 2008).

In summary, mTORC1 and mTORC2 work together to control glucose uptake by cells via regulation of glucose transporters, and at the same time they work to control glucose metabolism by affecting the equilibrium between gluconeogenesis and glycolysis in response to the cell energy supply. Combined with autophagy, mTORC1 and mTORC2 play a key role in supporting tumor cell survival and proliferation under metabolic stress conditions.

22.4 mTORC2 REGULATES ACTIN REORGANIZATION

Another important observation for understanding the function of mTORC2 is that it regulates actin reorganization (Jacinto et al., 2004; Sarbassov et al., 2004). It had been hypothesized that depletion of Rictor in cells, leading to reorganization of actin, occurred via the RhoA/Rac1 pathway (Jacinto et al., 2004). Rac1 GTP exchange factor, P-Rex1, was reported to interact with mTOR and only participate in the mTORC2 signal pathway in order to affect Rac1 activity, since siRNA knockdown of Rictor, but not Raptor, affected Rac1 activity (Hernández-Negrete et al., 2007). P-Rex1 has a PH domain that specifically binds to PIP3, and it has been thought to relay signals to Rac1 from GPR/PI3K when Rac1 is transferred to the plasma membrane (Barber et al., 2007). It has been reported that prolonged treatment with rapamycin decreases Akt phosphorylation at S473 and Rac1 activation, independent of mTORC1, in neurofibromin-deficient cells (Sandsmark et al., 2007). Prolonged treatment with rapamycin leads to Rictor dephosphorylation, translocation into the cytoplasm, and decreased assembly of mTORC2, without affecting the total amount of Rictor protein in cells (Rosner and Hengstschläger, 2008). On the other hand, Rictor also interacts with Myo 1c to affect actin reorganization through Paxillin (Hagan et al., 2008). Therefore, Rictor may regulate actin reorganization through mTORC2-dependent and -independent pathways. The mechanism of Rac1 activation by mTORC2 needs to be further studied (Fig. 22.3).

Simultaneous pharmacological inhibition of mTORC1 and mTORC2 with small molecules in tumor cell lines led to decreased phosphorylation of Akt at both S473 and T308 in most cell lines except MDA-MB-486. MDA-MB-486 has very high PI3K activity, and the phosphorylation of Akt at T308 was not affected by mTOR inhibitors (Chresta et al., 2008). This may suggest that mTORC2 could affect PI3K activity through the Rho family pathway (Wu et al., 2007). Therefore, inhibition of mTORC2 may result in reduced mTORC1 activity through Rho/PI3K/Akt/TSC/Rheb/mTORC1 in cells in which the Rho family plays a substantial role in PI3K activation. But ablation of Rictor did not reduce TSC2 phosphorylation by Akt and mTORC1 activity in a transgenic murine model, which suggests that, at least in the context of the TSC1/TSC2 pathway, mTORC2 does not directly regulate mTORC1 *in vivo* (Shiota

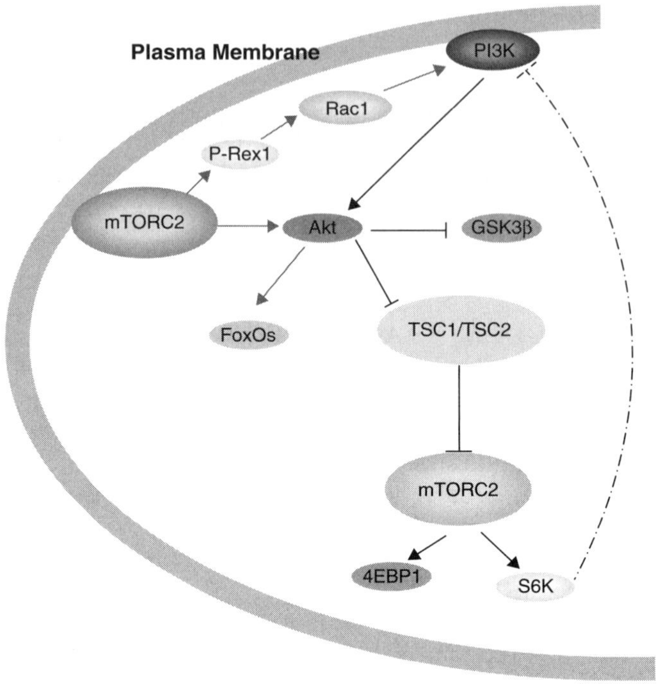

Figure 22.3. Cross-talk between mTORC2 and mTORC1. See the insert for color representation of this figure.

et al., 2006). As discussed above, Rac1 can regulate PLD1, directly activating mTORC1 through the mTORC2/Rac1/PLD1/mTORC1 pathway (Powner and Wakelam, 2002). PKC and another member of Rho family, Cdc42, have also been reported to activate mTORC1 through up-regulation of PLD (Fang et al., 2003). Considering that PLD can be regulated by many other pathways (Maiese et al., 2007), the observation of an mTORC2 pathway disconnected from the PI3K pathway in a transgenic murine model may suggest that tumor cells utilize the mTORC2 pathway to increase the basal level of PI3K signaling. It also has been proposed that the efficacy of an mTOR inhibitor is inversely correlated with K-Ras activity in xenograft models (Chresta et al., 2008). As is well known, Rac1 is required as the effector of Ras, which suggests that constitutively activated Ras can override the mTORC2 regulation of Rac1. This further supports the hypothesis that Rac1 is a downstream target of mTORC2.

22.5 mTOR CAN COORDINATE WITH OTHER PATHWAYS TO REGULATE CELL PROLIFERATION

Cells have to coordinate many external signals to make precise decisions about their fates by sharing components of different pathways to create an information exchange network. For example, GSK3β is controlled by Akt phosphorylation to regulate glucose homeostasis (Maise et al., 2007). Furthermore, GSK3β phosphorylates TSC1 and TSC2 in cooperation with AMPK to activate TSC2, leading to mTORC1 inactivation (Inoki et al., 2006). But this process is inhibited by β-catenin activation, which suggests that β-catenin activation can up-regulate mTORC1 to enhance translation. On the other hand, activated p70S6K can phosphorylate GSK3β to result in inactivation of GSK3β and further enhancement of β-catenin activation (Zhang et al., 2006). This provides a linkage between the mTORC1 and Wnt pathways to coordinate with each other and regulate cell growth. It is reported that activated Rac1 interacts with JNK2 to phosphorylate β-catenin, which results in β-catenin translocation into the nucleus, where it becomes activated (Wu et al., 2008). As discussed above, mTORC2 can activate Rac1 and then lead to β-catenin activation. Activated Akt also directly phosphorylates β-catenin at S522 to result in β-catenin translocation into the nucleus and activation, although it is not clear if the phosphorylation of S473 on Akt is required for this event (Fang et al., 2007). All these studies point to fact that there are tightly controlled feedback loops between the mTORC1/mTORC2 and Wnt pathways.

One of the MAP kinase pathway kinases, ERK, directly phosphorylates TSC2 to inactivate TSC2 and lead to mTORC1 activation (Ma et al., 2005). This plays a special role in ERK-mediated (as opposed to Akt-mediated) mTORC1 activation in tuberous sclerosis and human cancer. ERK is hyperactive in all subependymal giant-cell astrocytomas, which leads one to hypothesize that consistent ERK activation in different tuberous sclerosis-associated tumors is a molecular trigger for the development of these neoplasms (Jozwiak et al., 2008). Although B-Raf has been reported to interact with Rheb and result in decreased dimerization of B-Raf/C-Raf and decreased kinase activity, this process is rapamycin-insensitive, which suggests that Rheb/B-Raf interaction is independent of mTORC1 activity (Karbowniczek et al., 2006).

The hypoxia response plays a key role in tumorigenesis (Wouters and Koritzinsky, 2008). It was reported that when the oxygen concentration becomes very low (hypoxic condition), the overall activity of mTORC1 does not decrease until there is a prolonged low oxygen condition, and it was also reported that there is actually an initial brief (about 60 minutes) increase in activity (Li et al., 2007). Under hypoxic conditions, HIF-1α is stabilized and transferred into the nucleus to activate transcription of a set of genes coding VEGF and other angiogenesis factors (Gardner and Corn, 2008). There is also evidence suggesting that mitochondria-derived ROS might control activation of HIF-1α during hypoxia (Klimova and Chandel, 2008). ROS can activate HIF-1α by increasing its translation through the PI3K-Akt-mTORC1 pathway, in which case an increase of PTEN glutathionylation, due to ROS, leads to PI3K activation. As discussed above, mTORC1 can serve as a transcription factor to bring HIF-1α to CBP/p300, which is required for this transcription activation (Land and Tee, 2007). It has not been determined whether mTORC1 kinase activity is necessary for this process. The translocation of mTORC1 into the nucleus has been reported to be associated with the augmentation of phosphorylation of mTOR at S2481 under hypoxic conditions (Li et al., 2007). Rapamycin treatment only partially reduces the phosphorylation of S2481 on mTOR. It is possible that mTORC1 participates in HIF-1α activation at an early stage of hypoxia and is then down-regulated during prolonged hypoxia to support cell survival.

Treatment with rapamycin in a variety of cell lines leads to cell cycle arrest in G1, although the mechanism is not well understood. It has been reported that cyclin D translation is controlled by mTROC1 activity (Gao et al., 2004). A very recent research article revealed that mTORC1 can phosphorylate SGK at S422, and phosphorylated SGK then phosphorylates p27 at S157 to result in its translocation out of the nucleus, the release of its inhibition of CDKs, and the promotion of mitosis (Hong et al., 2008). The Wnt pathway may also play a role here in directly controlling the transcription of Cyclin D. The Wnt pathway can be activated through the mTORC2/Rac1 pathway (Wu et al., 2008).

22.6 IS JNK SIGNALING PATHWAY RESPONSIBLE FOR CELL APOPTOSIS RESULTING FROM mTORC1/mTORC2 INHIBITION?

Inhibition of mTORC1 leads to mitotic cells arresting in G1, possibly through down-regulation of Cyclin D1 translation (Gao et al., 2004), which may eventually result in cell death through apoptosis and/or other pathways. But for those cells that are not proliferating, the outcome of inhibition of mTORC1/mTORC2 is not well studied. It has been reported that inhibition of mTORC1 can activate the JNK pathway by suppressing protein phosphatase activity and inducing apoptosis (Huang et al., 2004). It is important to elucidate the possible connection between the JNK and mTOR pathways. The c-JNK is a stress-activated member of the MAP kinase family. It includes three JNK genes coding JNK1, JNK2, and JNK3 (Johnson and Lapadat, 2002). JNK1 and JNK2 are ubiquitously expressed, while JNK3 is restricted to the brain, heart, and testes (Raman et al., 2007). JNK activation has been strongly implicated in inflammatory responses, neurodegeneration, and apoptosis (Kyriakis and Avruch, 2001). JNKs can function either in the cytoplasm or in the nucleus, depending on their translocation signals (Raman et al., 2007). However, JNK-induced apoptosis requires JNK translocation into the nucleus, and activation of genes beyond c-Jun and AP-1 (Björkblom et al., 2008). It has been reported that ROS-induced apoptosis in neuronal death goes through the JNK2/FoxO3 pathway in coordination with inhibition of PI3K/Akt pathway (Dávila and Torres-Aleman, 2008). As is well known, tumor cells rely on abnormal aerobic glycolysis to support their proliferation (Ortega et al., 2008). Mitochondrial alterations and dysfunction also are common in cancer cells (Ortega et al., 2008). Inhibition of mTORC1 leads to further decreased mitochondria activity, and it forces tumor cells to increasingly rely on aerobic glycolysis, which also leads to ROS production increases (Hervouet et al., 2007). The increased ROS levels can lead to cell death through either the mitochondria pathway or the JNK pathway (Hervouet et al., 2007).

As discussed above, both mTORC1 and mTORC2 play substantial roles in glucose homeostasis: mTORC1 controls Glut1 traffic and autophagy, and mTORC2 manipulates Glut4- and FoxO-dependent glucose regulation. Inhibition of mTORC1 and mTORC2 could dramatically lower glucose uptake by tumor cells, which may activate the JNK pathway. It has been shown that with depletion of glucose from cell culture media, JNK becomes activated through the ROS/ASK1/MKK4/JNK pathway (Song and Lee, 2003a). Activated JNK phosphorylates Beclin (Atg5), which is required for autophagy (Wei et al., 2008). Also, blockade of mTORC1 leads to autophagy by stabilizing Atg1 and supporting cell survival (Pattingre et al., 2008). Therefore a feedback loop must exist to down-regulate the JNK pathway and prevent cell death due to the constitutively activated JNK pathway.

It has been established that Akt coupling with Hsp90 phosphorylates ASK1 at S83 to block ASK1 function (Zhang et al., 2005). Hsp90 plays an important role in stabilizing newly synthesized Akt (Ikenoue et al., 2008). ASK1 activates MKK4 and leads to JNK activation to promote apoptosis (Kim et al., 2005). Therefore activated Akt can block JNK activation. Inhibition of mTORC1 with rapamycin and its analogues activates ASK1 through decreased phosphorylation at S83 on ASK1 (Huang et al., 2004). This decreased Akt activity may be a result of the decrease in phosphorylation at S473 by mTORC2, which has been reported to be the consequence of prolonged treatment with rapamycin (Sarbassov et al., 2006). Concomitantly, Akt interacts with MKK4 and inhibits its activation of JNK (Song and Lee, 2003b). Therefore, inactivation of Akt can lead to JNK activation through ASK1/MKK4/JNK and release the Akt inhibitory checkpoint. It is interesting to note that JIP1 also interacts with Akt and inhibits its activity, while this inhibition can be released once JIP1 is phosphorylated by JNK via a negative feedback loop under metabolic stress (Song and Lee, 2003b). It is also reported that reactivation of Akt mediated by JNK promotes cardiomyocyte survival after hypoxic injury *in vitro* and *in vivo*, involving phosphorylation of Akt at S450 by JNK (Shao et al., 2006). This indicates that JNK-induced apoptosis under conditions of metabolic stress can be rescued or stopped by activation of Akt through multiple negative feedback loops, which is very important for tumorigenesis (Fig. 22.4). mSin1 was first identified as a component of the MAP kinase pathway; it interacts with MEKK2 and inactivates JNK1 (Schroder et al., 2005). However, it has not been studied whether there is a direct regulatory relationship between the mTORC2 and JNK pathways. And it could be very interesting to find out whether inhibition of mTORC2 could activate the JNK pathway under metabolic stress. In summary, inhibition of mTORC1/mTORC2 could lead to activation of the JNK pathway and sustain it by preventing Akt negative feedback regulation, which results in cell death.

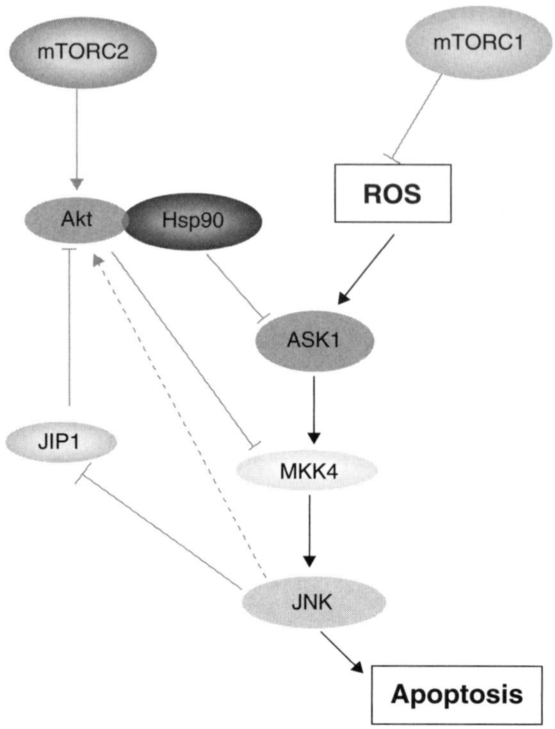

Figure 22.4. Feedback loop of Akt to inhibit JNK activation. See the insert for color representation of this figure.

22.7 TARGETING mTOR IN TUMOR THERAPEUTICS IS PROMISING

mTOR is a validated target for tumor therapy (Kopelovich, 2007). Currently, there are three drugs in clinical trials: temsirolimus (CCI779), everolimus (RAD001), and deforolimus (AP23573) (Gadducci et al., 2008). These compounds are analogues of rapamycin with better pharmacokinetic characteristics. They directly target mTORC1 activity *in vivo*, as confirmed with pharmacodynamic markers like phosphorylation of p70S6K (Gadducci et al., 2008). Temsirolimus has shown improved survival among patients with renal cell carcinoma despite a low response rate in a phase 3 clinical trial of patients with poor prognoses (Rini, 2008). There are also dual inhibitors that target both PI3K and mTOR currently in clinical trials. Compound XL765 from Exelixis is one of these, and is being studied in a phase 1 study of patients with solid tumors (Laird, 2007). This compound has shown 80% to 90% reductions in phosphorylation of pathway components including Akt, 4EBP1, and p70S6K, and a 54% reduction in cell proliferation (as assessed by Ki67 staining) in tumor tissue from a patient with chondrosarcoma (Exelixis Press

Release, 2008). BEZ235 from Novartis also targets both mTOR and PI3K and is being developed as a solid tumor therapeutic; it is currently in phase 1 trials (Schnell et al., 2008).

However, very little is understood about the pharmacological difference between targeting mTOR and PI3K, although these two targets are somewhat located in the same biological pathway. As discussed above, perturbation of PI3K activity leads to changes in both mTORC1 and mTORC2 activities, while mTORC1/mTORC2 can be regulated independently of PI3K. One of the major effectors for PI3K is Akt, which can phosphorylate and regulate many biological pathways. Akt can be activated by phosphorylation at T308 by PI3K/PDK1 without phosphorylation at S473 by mTORC2, which can normally phosphorylate GSK3β, TSC2, and Bcl2 (Guertin et al., 2006). FoxO transcription factors are the only substrate so far that require Akt phosphorylated at both S473 and T308 (Guertin et al., 2006). No systematic examination has been done to check all Akt substrates for the requirement of phosphorylation at S473 by mTORC2. This kind of information will help to understand the difference between PI3K and mTORC2 function. The question about mTORC2 activation also needs to be answered, although some evidence points to PI3K as one of the activators, as discussed above. One of the mTORC1 functions is coordination with other transcription factors to participate in transcription control. The translocation of mTOR into the nucleus from the cytosol is related to its autophosphorylation at S2481 via an unknown mechanism, but independent of PI3K, under hypoxic conditions (Li et al., 2007). This could be another difference to be studied between targeting PI3K and mTOR. Understanding these mechanisms will further elucidate the regulation of mTOR in cell proliferation, survival, and death.

Tumorigenesis has been hypothesized to heavily rely on glycolysis, rather than oxidative phosphorylation, for an energy supply, even in the presence of high oxygen concentrations (Ortega et al., 2008). This can be explained in some tumors as a result of up-regulation of the HIF pathway to increase transcription of genes involved in glycolysis and lactate metabolism, while respiratory chain complex activities and subunit amounts are severely diminished in the mitochondria (Wallace, 2005). However, the transcription of genes involved in the structure and biogenesis of these complexes does not seem to be substantially decreased, which may be attributed to alterations in the respiration chain to increase ROS production (Hervouet et al., 2007). The increased ROS level has been detected in many tumor cells, presumably to change the cell signal transduction pathways (Brandon et al., 2006). ROS can serve as a signaling molecule or cellular toxicants, depending on its spatially and temporally occurring and its level (Liu et al., 2008). High levels of ROS can result in cell apoptosis or senescence through JNK and other pathways. Therefore tumor cells have to develop mechanisms to protect themselves from ROS-induced damage. One of the mechanisms that has been reported is up-regulation of the pentose phosphate pathway by Akt to increase NADPH production and increase Mcl-1 protection against apoptosis, in which Mcl-1 is directly controlled by mTORC1 during translation (Mills et al., 2008). Prolonged high levels of ROS lead to activation of the JNK pathway, which can result in autophagy, necrosis, and apoptosis (Matsuzawa and Ichijo, 2008).

Rapamycin and its analogues for treatment of cancer cells can lead to apoptosis, but the mechanism has not been well elucidated. Inhibition of mTORC1 obviously can lead to cell cycle arrest through down-regulation of cyclin D and SGK/p27 translocation, alteration of Glut4 transference to the plasma membrane to decrease glucose uptake, increases in autophagy to supply nutrition temporarily, and generally decreases in protein translation, including anti-apoptotic Mcl-1. On the other hand, inhibition of mTORC2 leads to alteration of Glut1 translocation to the plasma membrane and decreased glucose uptake, FoxO transcription factors transference into nucleus, which activates glycogenolysis and decreases β-catenin activity to transcribe Cyclin D and other cell cycle genes. Depletion of glucose from cell culture media leads to a dramatic increase in ROS, which results in apoptosis through the ASK/MEKK/JNK pathway. Inhibition of mTORC2 also releases the negative feedback inhibition of JNK by Akt, which could lead to sustained activation of the JNK pathway and eventually lead to cell death through a mechanism like apoptosis, autophagy, and necrosis. It is also possible that prolonged activation of FoxO transcription factors from blockade by mTORC2 results in tumor cell reliance on glycogenolysis to supply energy and eventually leads to depletion of glycogen and cell death. Inhibition of mTORC2 also blocks cell migration, possibly through the Rac1 pathway, to prevent metastasis of the tumor.

22.8 CONCLUSION

Although mTOR is becoming a promising target for tumor therapy, there are still many biological and clinical aspects that need to be clarified. Understanding the difference between inhibition of PI3K, mTORC1, and mTORC2 will provide strong indications for identifying specific patient groups likely to be responsive to each targeted tumor therapy. Further understanding of the outcomes of mTORC1/mTORC2 inhibition will support biomarker identification. Examining the possible relationship between mTORC1 and mTORC2 is especially critical for understanding how cells use mTOR to integrate external and internal signals to coordinate nutrition, energy, and environmental stress with cell growth and proliferation. Because our research has focused on the biological mechanism of mTOR in various models that were specifically selected and manipulated to simplify the studies, it is very important to integrate clinical information into the whole picture in order to gain a better understanding about particular diseases and select specific therapeutic methods for specific groups of patients. As the biology becomes more mature, it is becoming possible to evaluate mTORC1 and mTORC2 as therapeutic oncology targets and to select the right groups of patients that may best respond to these therapies.

ACKNOWLEDGMENTS

I express my great appreciation to Ms. Alexis Khalil and Dr. Chuang Lu for their wonderful advice and careful editing of this manuscript. I am also grateful to Mr. Lenny Dang for his scientific discussion and support of this work.

REFERENCES

Ames BN. Low micronutrient intake may accelerate the degenerative diseases of aging through allocation of scarce micronutrients by triage. *Proc Natl Acad Sci USA* 2006;103(47):17589–17594.

Avruch J, Long X, Ortiz-Vega S, Rapley J, Papageorgiou A, Dai N. Amino Acid Regulation of TOR Complex 1. *Am J Physiol Endocrinol Metab* 2008;296:E592–E602.

Bai X, Ma D, Liu A, Shen X, Wang QJ, Liu Y, Jiang Y. Rheb activates mTOR by antagonizing its endogenous inhibitor, FKBP38. *Science* 2007;318(5852):977–980.

Balendran A, Currie R, Armstrong CG, Avruch J, Alessi DR. Evidence that 3-phosphoinositide-dependent protein kinase-1 mediates phosphorylation of p70 S6 kinase *in vivo* at Thr-412 as well as Thr-252. *J Biol Chem* 1999;274(52):37400–37406.

Barber MA, Donald S, Thelen S, Anderson KE, Thelen M, Welch HC. Membrane translocation of P-Rex1 is mediated by G protein betagamma subunits and phosphoinositide 3-kinase. *J Biol Chem* 2007;282(41):29967–29976.

Björkblom B, Vainio JC, Hongisto V, Herdegen T, Courtney MJ, Coffey ET. All JNKs can kill, but nuclear localization is critical for neuronal death. *J Biol Chem* 2008;283(28):19704–19713.

Boura-Halfon S, Zick Y. Phosphorylation of IRS proteins, insulin action and insulin resistance. *Am J Physiol Endocrinol Metab* 2008;296:E581–E591.

Bozulic L, Surucu B, Hynx D, Hemmings BA. PKBalpha/Akt1 acts downstream of DNA-PK in the DNA double-strand break response and promotes survival. *Mol Cell* 2008;30(2):203–213.

Brandon M, Baldi P, Wallace DC. Mitochondrial mutations in cancer. *Oncogene* 2006;25(34):4647–4662.

Budanov AV, Karin M. p53 target genes sestrin1 and sestrin2 connect genotoxic stress and mTOR signaling. *Cell* 2008;134(3):451–460.

Buerger C, DeVries B, Stambolic V. Localization of Rheb to the endomembrane is critical for its signaling function. *Biochem Biophys Res Commun* 2006;344 (3): 869–880.

Buller CL, Loberg RD, Fan MH, Zhu Q, Park JL, Vesely E, Inoki K, Guan KL, Brosius FC 3rd. A GSK-3/TSC2/mTOR pathway regulates glucose uptake and GLUT1 glucose transporter expression. *Am J Physiol Cell Physiol* 2008;295(3):C836–C843.

Buteau J, Shlien A, Foisy S, Accili D. Metabolic diapause in pancreatic beta-cells expressing a gain-of-function mutant of the forkhead protein Foxo1. *J Biol Chem* 2007;282(1):287–293.

Buttgereit F, Brand MD. A hierarchy of ATP-consuming processes in mammalian cells. *Biochem J* 1995;312(Pt 1):163–167.

Castro AF, Rebhun JF, Clark GJ, Quilliam LA. Rheb binds tuberous sclerosis complex 2 (TSC2) and promotes S6 kinase activation in a rapamycin- and farnesylation-dependent manner. *J Biol Chem* 2003;278:32493–32496.

Chen Y, Zheng Y, Foster DA. Phospholipase D confers rapamycin resistance in human breast cancer cells. *Oncogene* 2003;22(25):3937–3942.

Chresta CM, Brandon D, Critchlow S, Cosulich S, Curtis N, Davenport S, Davies B, Drew L, Duggan HME, Ellston R, Harding T, Jones D, Hickson I, Hummersone M, Malagu K, McCormack R, Pass M, Smith PD. Cellular differentiation of rapamycin, PI3-kinase & TOR-kinase inhibitors. AACR Annual Meeting 2008.

Chung J, Bachelder RE, Lipscomb EA, Shaw LM, Mercurio AM. Integrin (alpha 6 beta 4) regulation of eIF-4E activity and VEGF translation: a survival mechanism for carcinoma cells. *J Cell Biol* 2002;158:165–174.

Cook SJ, Morley SJ. Nutrient-responsive mTOR signalling grows on sterile ground. *Biochem J* 2007;403(1):e1–e3.

Cooper MP, Qu L, Rohas LM, Lin J, Yang W, Erdjument-Bromage H, Tempst P, Spiegelman BM. Defects in energy homeostasis in Leigh syndrome French Canadian variant through PGC-1alpha/LRP130 complex. *Genes Dev* 2006;20(21):2996–3009.

Cunningham JT, Rodgers JT, Arlow DH, Vazquez F, Mootha VK, Puigserver P. mTOR controls mitochondrial oxidative function through a YY1-PGC-1alpha transcriptional complex. *Nature* 2007;450(7170):736–740.

Dada S, Demartines N, Dormond O. mTORC2 regulates PGE2-mediated endothelial cell survival and migration. *Biochem Biophys Res Commun* 2008;372(4):875–879.

Dames SA, Mulet JM, Rathgeb-Szabo K, Hall MN, Grzesiek S. The solution structure of the FATC domain of the protein kinase target of rapamycin suggests a role for redox-dependent structural and cellular stability. *J Biol Chem* 2005;280(21): 20558–20564.

Dávila D, Torres-Aleman I. Neuronal death by oxidative stress involves activation of FOXO3 through a two-arm pathway that activates stress kinases and attenuates insulin-like growth factor I signaling. *Mol Biol Cell* 2008;19(5):2014–2025.

Di Paolo S, Teutonico A, Leogrande D, Capobianco C, Schena PF. Chronic inhibition of mammalian target of rapamycin signaling downregulates insulin receptor substrates 1 and 2 and AKT activation: A crossroad between cancer and diabetes? *J Am Soc Nephrol* 2006;17(8):2236–2244.

European Chromosome 16 Tuberous Sclerosis Consortium. Identification and characterization of the tuberous sclerosis gene on chromosome 16. *Cell* 1993;75:1305–1315.

Exelixis Press Release, May 31, 2008.

Facchinetti V, Ouyang W, Wei H, Soto N, Lazorchak A, Gould C, Lowry C, Newton AC, Mao Y, Miao RQ, Sessa WC, Qin J, Zhang P, Su B, Jacinto E. The mammalian target of rapamycin complex 2 controls folding and stability of Akt and protein kinase C. *EMBO J* 2008;27(14):1932–1943.

Fang D, Hawke D, Zheng Y, Xia Y, Meisenhelder J, Nika H, Mills GB, Kobayashi R, Hunter T, Lu Z. Phosphorylation of beta-catenin by AKT promotes beta-catenin transcriptional activity. *J Biol Chem* 2007;282 (15):11221–11229.

Fang Y, Vilella-Bach M, Bachmann R, Flanigan A, Chen J. Phosphatidic acid-mediated mitogenic activation of mTOR signaling. *Science* 2001;294(5548):1942–1945.

Fang Y, Park IH, Wu AL, Du G, Huang P, Frohman MA, Walker SJ, Brown HA, Chen J. PLD1 regulates mTOR signaling and mediates Cdc42 activation of S6K1. *Curr Biol* 2003;13(23):2037–2044.

Findlay GM, Yan L, Procter J, Mieulet V, Lamb RF. A MAP4 kinase related to Ste20 is a nutrient-sensitive regulator of mTOR signalling. *Biochem J* 2007;403(1):13–20.

Fingar DC, Blenis J. Target of rapamycin (TOR): an integrator of nutrient and growth factor signals and coordinator of cell growth and cell cycle progression. *Oncogene* 2004;23(18):3151–3171.

Gadducci A, Tana R, Cosio S, Fanucchi A, Genazzani AR. Molecular target therapies in endometrial cancer: from the basic research to the clinic. *Gynecol Endocrinol* 2008;24(5):239–249.

Gao N, Flynn DC, Zhang Z, Zhong XS, Walker V, Liu KJ, Shi X, Jiang BH. G1 cell cycle progression and the expression of G1 cyclins are regulated by PI3K/AKT/mTOR/p70S6K1 signaling in human ovarian cancer cells. *Am J Physiol Cell Physiol* 2004;287(2):C281–C291.

Garami A, Zwartkruis FJ, Nobukuni T, Joaquin M, Roccio M, Stocker H, Kozma SC, Hafen E, Bos JL, Thomas G. Insulin activation of Rheb, a mediator of mTOR/S6K/4E-BP signaling, is inhibited by TSC1 and 2. *Mol Cell* 2003;11:1457–1466.

Gardner LB, Corn PG. Hypoxic regulation of mRNA expression. *Cell Cycle* 2008;7(13):1916–1924.

Gingras AC, Raught B, Sonenberg N. Regulation of translation initiation by FRAP/mTOR. *Genes Dev* 2001;15(7):807–826.

Gomez MR. Phenotypes of the tuberous sclerosis complex with a revision of diagnostic criteria. *Ann NY Acad Sci* 1991;615:1–7.

Graves LM, Bornfeldt KE, Argast GM, Krebs EG, Kong X, Lin TA, Lawrence JC Jr. cAMP- and rapamycin-sensitive regulation of the association of eukaryotic initiation factor 4E and the translational regulator PHAS-I in aortic smooth muscle cells. *Proc Natl Acad Sci USA* 1995;92:7222–7226.

Gross DN, van den Heuvel AP, Birnbaum MJ. The role of FoxO in the regulation of metabolism. *Oncogene* 2008;27(16):2320–2336.

Guertin DA, Stevens DM, Thoreen CC, Burds AA, Kalaany NY, Moffat J, Brown M, Fitzgerald KJ, Sabatini DM. Ablation in mice of the mTORC components raptor, rictor, or mLST8 reveals that mTORC2 is required for signaling to Akt-FOXO and PKCalpha, but not S6K1. *Dev Cell* 2006;11(6):859–871.

Gwinn DM, Shackelford DB, Egan DF, Mihaylova MM, Mery A, Vasquez DS, Turk BE, Shaw RJ. AMPK phosphorylation of raptor mediates a metabolic checkpoint. *Mol Cell* 2008;30(2):214–226.

Hagan GN, Lin Y, Magnuson MA, Avruch J, Czech MP. A Rictor–Myo1c complex participates in dynamic cortical actin events in 3T3-L1 adipocytes. *Mol Cell Biol* 2008;28(13):4215–4226.

Hagen T, Vidal-Puig A. Mitochondrial uncoupling proteins in human physiology and disease. *Minerva Med* 2002;93(1):41–57.

Hahn-Windgassen A, Nogueira V, Chen CC, Skeen JE, Sonenberg N, Hay N. Akt activates the mammalian target of rapamycin by regulating cellular ATP level and AMPK activity. *J Biol Chem* 2005;280(37):32081–32089.

Hardie DG, Hawley SA. AMP-activated protein kinase: the energy charge hypothesis revisited. *Bioessays* 2001;23:1112–1119.

Hardie DG, Carling D, Carlson M. The AMP-activated/SNF1 protein kinase subfamily: metabolic sensors of the eukaryotic cell? *Annu Rev Biochem* 1998;67:821–855.

Heitman J, Movva NR, Hall MN. Proline isomerases at the crossroads of protein folding, signal transduction, and immunosuppression. *Science* 1991;253(5022):905–909.

Hernández-Negrete I, Carretero-Ortega J, Rosenfeldt H, Hernández-García R, Calderón-Salinas JV, Reyes-Cruz G, Gutkind JS, Vázquez-Prado J. P-Rex1 links mammalian target of rapamycin signaling to Rac activation and cell migration. *J Biol Chem* 2007;282(32):23708–23715.

Hervouet E, Simonnet H, Godinot C. Mitochondria and reactive oxygen species in renal cancer. *Biochimie* 2007;89(9):1080–1088.

Hoekstra MF. Responses to DNA damage and regulation of cell cycle checkpoints by the ATM protein kinase family. *Curr Opin Genet Dev* 1997;7:170–175.

Holz MK, Blenis J. Identification of S6 kinase 1 as a novel mammalian target of rapamycin (mTOR)-phosphorylating kinase. *J Biol Chem* 2005;280(28):26089–26093.

Hong F, Larrea MD, Doughty C, Kwiatkowski DJ, Squillace R, Slingerland JM. mTOR-raptor binds and activates SGK1 to regulate p27 phosphorylation. *Mol Cell* 2008;30(6):701–711.

Hresko RC, Mueckler M. mTOR.RICTOR is the Ser473 kinase for Akt/protein kinase B in 3T3-L1 adipocytes. *J Biol Chem* 2005;280(49):40406–40416.

Hu J, Zacharek S, He YJ, Lee H, Shumway S, Duronio RJ, Xiong Y. WD40 protein FBW5 promotes ubiquitination of tumor suppressor TSC2 by DDB1-CUL4-ROC1 ligase. *Genes Dev* 2008;22(7):866–871.

Huang S, Shu L, Easton J, Harwood FC, Germain GS, Ichijo H, Houghton PJ. Inhibition of mammalian target of rapamycin activates apoptosis signal-regulating kinase 1 signaling by suppressing protein phosphatase 5 activity. *J Biol Chem* 2004;279(35):36490–36496.

Ikenoue T, Inoki K, Yang Q, Zhou X, Guan KL. Essential function of TORC2 in PKC and Akt turn motif phosphorylation, maturation and signalling. *EMBO J* 2008;27(14):1919–1931.

Inoki K, Li Y, Zhu T, Wu J, Guan KL. TSC2 is phosphorylated and inhibited by Akt and suppresses mTOR signalling. *Nat Cell Biol* 2002;4(9):648–657.

Inoki K, Li Y, Xu T, Guan KL. Rheb GTPase is a direct target of TSC2 GAP activity and regulates mTOR signaling. *Genes Dev* 2003;17:1829–1834.

Inoki K, Ouyang H, Li Y, Guan KL. Signaling by target of rapamycin proteins in cell growth control. *Microbiol Mol Biol Rev* 2005;69(1):79–100.

Inoki K, Ouyang H, Zhu T, Lindvall C, Wang Y, Zhang X, Yang Q, Bennett C, Harada Y, Stankunas K, Wang CY, He X, MacDougald OA, You M, Williams BO, Guan KL. TSC2 integrates Wnt and energy signals via a coordinated phosphorylation by AMPK and GSK3 to regulate cell growth. *Cell* 2006;126(5):955–968.

Jacinto E, Loewith R, Schmidt A, Lin S, Rüegg MA, Hall A, Hall MN. Mammalian TOR complex 2 controls the actin cytoskeleton and is rapamycin insensitive. *Nat Cell Biol* 2004;6(11):1122–1128.

Jacinto E, Facchinetti V, Liu D, Soto N, Wei S, Jung SY, Huang Q, Qin J, Su B. SIN1/MIP1 maintains rictor-mTOR complex integrity and regulates Akt phosphorylation and substrate specificity. *Cell* 2006;127(1):125–137.

Jastrzebski K, Hannan KM, Tchoubrieva EB, Hannan RD, Pearson RB. Coordinate regulation of ribosome biogenesis and function by the ribosomal protein S6 kinase, a key mediator of mTOR function. *Growth Factors* 2007;25(4):209–226.

Johnson GL, Lapadat R. Mitogen-activated protein kinase pathways mediated by ERK, JNK, and p38 protein kinases. *Science* 2002;298(5600):1911–1912.

Jozwiak J, Jozwiak S, Wlodarski P. Possible mechanisms of disease development in tuberous sclerosis. *Lancet Oncol* 2008;9(1):73–79.

Kam Y, Exton JH. Role of phospholipase D1 in the regulation of mTOR activity by lysophosphatidic acid. *FASEB J* 2004;18(2):311–319.

Karbowniczek M, Robertson GP, Henske EP. Rheb inhibits C-raf activity and B-raf/C-raf heterodimerization. *J Biol Chem* 2006;281(35):25447–25456.

Kim SD, Moon CK, Eun SY, Ryu PD, Jo SA. Identification of ASK1, MKK4, JNK, c-Jun, and caspase-3 as a signaling cascade involved in cadmium-induced neuronal cell apoptosis. *Biochem Biophys Res Commun* 2005;328(1):326–334.

Kim Y, Kim JE, Lee SD, Lee TG, Kim JH, Park JB, Han JM, Jang SK, Suh PG, Ryu SH. Phospholipase D1 is located and activated by protein kinase C alpha in the plasma membrane in 3Y1 fibroblast cell. *Biochim Biophys Acta* 1999;1436(3):319–330.

Klimova T, Chandel NS. Mitochondrial complex III regulates hypoxic activation of HIF. *Cell Death Differ* 2008;15(4):660–666.

Kopelovich L, Fay JR, Sigman CC, Crowell JA. The mammalian target of rapamycin pathway as a potential target for cancer chemoprevention. *Cancer Epidemiol Biomarkers Prev* 2007;16(7):1330–1340.

Kumar A, Harris TE, Keller SR, Choi KM, Magnuson MA, Lawrence JC Jr. Muscle-specific deletion of rictor impairs insulin-stimulated glucose transport and enhances basal glycogen synthase activity. *Mol Cell Biol* 2008;28(1):61–70.

Kyriakis JM, Avruch J. Mammalian mitogen-activated protein kinase signal transduction pathways activated by stress and inflammation. *Physiol Rev* 2001;81(2):807–869.

Laird D. 2007 XL765 Targets Tumor Growth, Survival, and Angiogenesis in Preclinical Models by Dual Inhibition of PI3K and mTOR Poster at the AACR-NCI-EOTITC international Conference October 22–26, 2007, San Francisco CA.

Land SC, Tee AR. Hypoxia-inducible factor 1alpha is regulated by the mammalian target of rapamycin (mTOR) via an mTOR signaling motif. *J Biol Chem* 2007;282(28):20534–20543.

Lee DF, Kuo HP, Chen CT, Hsu JM, Chou CK, Wei Y, Sun HL, Li LY, Ping B, Huang WC, He X, Hung JY, Lai CC, Ding Q, Su JL, Yang JY, Sahin AA, Hortobagyi GN, Tsai FJ, Tsai CH, Hung MC. IKK beta suppression of TSC1 links inflammation and tumor angiogenesis via the mTOR pathway. *Cell* 2007;130(3):440–455.

Li W, Petrimpol M, Molle KD, Hall MN, Battegay EJ, Humar R. Hypoxia-induced endothelial proliferation requires both mTORC1 and mTORC2. *Circ Res* 2007;100(1):79–87.

Liberman Z, Plotkin B, Tennenbaum T, Eldar-Finkelman H. Coordinated phosphorylation of insulin receptor substrate-1 by glycogen synthase kinase-3 and protein kinase C betaII in the diabetic fat tissue. *Am J Physiol Endocrinol Metab* 2008;294(6):E1169–E1177.

Liu B, Chen Y, St Clair DK. ROS and p53: a versatile partnership. *Free Radic Biol Med* 2008;44(8):1529–1535.

Liu X, Zheng XF. Endoplasmic reticulum and Golgi localization sequences for mammalian target of rapamycin. *Mol Biol Cell* 2007;18(3):1073–1082.

Long X, Ortiz-Vega S, Lin Y, Avruch J. Rheb binding to mammalian target of rapamycin (mTOR) is regulated by amino acid sufficiency. *J Biol Chem* 2005;280 (25):23433–23436.

Ma L, Chen Z, Erdjument-Bromage H, Tempst P, Pandolfi PP. Phosphorylation and functional inactivation of TSC2 by Erk implications for tuberous sclerosis and cancer pathogenesis. *Cell* 2005;121(2):179–193.

Maiese K, Chong ZZ, Shang YC. Mechanistic insights into diabetes mellitus and oxidative stress. *Curr Med Chem* 2007;14(16):1729–1738.

Mak BC, Kenerson HL, Aicher LD, Barnes EA, Yeung RS. Aberrant beta-catenin signaling in tuberous sclerosis. *Am J Pathol* 2005;167(1):107–116.

Manning BD, Cantley LC. AKT/PKB signaling: navigating downstream. *Cell* 2007;129(7):1261–1274.

Mao JH, Kim IJ, Wu D, Climent J, Kang HC, DelRosario R, Balmain A. FBXW7 targets mTOR for degradation and cooperates with PTEN in tumor suppression. *Science* 2008;321(5895):1499–1502.

Matsuzawa A, Ichijo H. Redox control of cell fate by MAP kinase: physiological roles of ASK1-MAP kinase pathway in stress signaling. *Biochim Biophys Acta* 2008;1780(11):1325–1336.

McDonald PC, Oloumi A, Mills J, Dobreva I, Maidan M, Gray V, Wederell ED, Bally MB, Foster LJ, Dedhar S. Rictor and integrin-linked kinase interact and regulate Akt phosphorylation and cancer cell survival. *Cancer Res* 2008;68(6):1618–1624.

Mills JR, Hippo Y, Robert F, Chen SM, Malina A, Lin CJ, Trojahn U, Wendel HG, Charest A, Bronson RT, Kogan SC, Nadon R, Housman DE, Lowe SW, Pelletier J. mTORC1 promotes survival through translational control of Mcl-1. *Proc Natl Acad Sci USA* 2008;105(31):10853–10858.

Nellist M, Verhaaf B, Goedbloed MA, Reuser AJ, van den Ouweland AM, Halley DJ. TSC2 missense mutations inhibit tuberin phosphorylation and prevent formation of the tuberin-hamartin complex. *Hum Mol Genet* 2001;10:2889–2898.

Nobukuni T, Joaquin M, Roccio M, Dann SG, Kim SY, Gulati P, Byfield MP, Backer JM, Natt F, Bos JL, Zwartkruis FJ, Thomas G. Amino acids mediate mTOR/raptor signaling through activation of class 3 phosphatidylinositol 3OH-kinase. *Proc Natl Acad Sci USA* 2005;102(40):14238–14243.

Okamoto H, Hribal ML, Lin HV, Bennett WR, Ward A, Accili D. Role of the forkhead protein FoxO1 in beta cell compensation to insulin resistance. *J Clin Invest* 2006;116(3):775–782.

Ortega AD, Sánchez-Aragó M, Giner-Sánchez D, Sánchez-Cenizo L, Willers I, Cuezva JM. Glucose avidity of carcinomas. *Cancer Lett* 2008;276:125–135.

Oshiro N, Takahashi R, Yoshino K, Tanimura K, Nakashima A, Eguchi S, Miyamoto T, Hara K, Takehana K, Avruch J, Kikkawa U, Yonezawa K. The proline-rich Akt substrate of 40 kDa (PRAS40) is a physiological substrate of mammalian target of rapamycin complex 1. *J Biol Chem* 2007;282(28):20329–20339.

Oude Weernink PA, López de Jesús M, Schmidt M. Phospholipase D signaling: orchestration by PIP2 and small GTPases. *Naunyn Schmiedebergs Arch Pharmacol* 2007;374(5–6):399–411.

Partovian C, Ju R, Zhuang ZW, Martin KA, Simons M. Syndecan-4 regulates subcellular localization of mTOR Complex2 and Akt activation in a PKCalpha-dependent manner in endothelial cells. *Mol Cell* 2008;32(1):140–149.

Pattingre S, Espert L, Biard-Piechaczyk M, Codogno P. Regulation of macroautophagy by mTOR and Beclin 1 complexes. *Biochimie* 2008;90(2):313–323.

Pearce LR, Huang X, Boudeau J, Pawłowski R, Wullschleger S, Deak M, Ibrahim AF, Gourlay R, Magnuson MA, Alessi DR. Identification of Protor as a novel Rictor-binding component of mTOR complex-2. *Biochem J* 2007;405(3):513–522.

Peterson RT, Beal PA, Comb MJ, Schreiber SL. FKBP12-rapamycin-associated protein (FRAP) autophosphorylates at serine 2481 under translationally repressive conditions. *J Biol Chem* 2000;275(10):7416–7423.

Polednak AP. Estimating the number of U.S. incident cancers attributable to obesity and the impact on temporal trends in incidence rates for obesity-related cancers. *Cancer Detect Prev* 2008;32:190–199.

Potente M, Urbich C, Sasaki K, Hofmann WK, Heeschen C, Aicher A, Kollipara R, DePinho RA, Zeiher AM, Dimmeler S. Involvement of Foxo transcription factors in angiogenesis and postnatal neovascularization. *J Clin Invest* 2005;115(9):2382–2392.

Powner DJ, Wakelam MJ. The regulation of phospholipase D by inositol phospholipids and small GTPases. *FEBS Lett* 2002;531(1):62–64.

Raman M, Chen W, Cobb MH. Differential regulation and properties of MAPKs. *Oncogene* 2007;26(22):3100–3112.

Rini BI. Temsirolimus, an inhibitor of mammalian target of rapamycin. *Clin Cancer Res* 2008;14(5):1286–1290.

Rosner M, Hengstschläger M. Cytoplasmic and nuclear distribution of the protein complexes mTORC1 and mTORC2: rapamycin triggers dephosphorylation and delocalization of the mTORC2 components rictor and sin1. *Hum Mol Genet* 2008;17(19):2934–2948.

Salih DA, Brunet A. FoxO transcription factors in the maintenance of cellular homeostasis during aging. *Curr Opin Cell Biol* 2008;20(2):126–136.

Sancak Y, Peterson TR, Shaul YD, Lindquist RA, Thoreen CC, Bar-Peled L, Sabatini DM. The Rag GTPases bind raptor and mediate amino acid signaling to mTORC1. *Science* 2008;320(5882):1496–1501.

Sandsmark DK, Zhang H, Hegedus B, Pelletier CL, Weber JD, Gutmann DH. Nucleophosmin mediates mammalian target of rapamycin-dependent actin cytoskeleton dynamics and proliferation in neurofibromin-deficient astrocytes. *Cancer Res* 2007;67(10):4790–4799.

Sarbassov DD, Ali SM, Kim DH, Guertin DA, Latek RR, Erdjument-Bromage H, Tempst P, Sabatini DM. Rictor, a novel binding partner of mTOR, defines a rapamycin-insensitive and raptor-independent pathway that regulates the cytoskeleton. *Curr Biol* 2004;14:1296–1302.

Sarbassov DD, Guertin DA, Ali SM, Sabatini DM. Phosphorylation and regulation of Akt/PKB by the rictor-mTOR complex. *Science* 2005;307(5712):1098–1101.

Sarbassov DD, Ali SM, Sengupta S, Sheen JH, Hsu PP, Bagley AF, Markhard AL, Sabatini DM. Prolonged rapamycin treatment inhibits mTORC2 assembly and Akt/PKB. *Mol Cell* 2006;22(2):159–168.

Saucedo LJ, Gao X, Chiarelli DA, Li L, Pan D, Edgar BA. Rheb promotes cell growth as a component of the insulin/TOR signalling network. *Nat Cell Biol* 2003;5:566–571.

Schalm SS, Blenis J. Identification of a conserved motif required for mTOR signaling. *Curr Biol* 2002;12:632–639.

Schieke SM, Phillips D, McCoy JP Jr, Aponte AM, Shen RF, Balaban RS, Finkel T. The mammalian target of rapamycin (mTOR) pathway regulates mitochondrial oxygen consumption and oxidative capacity. *J Biol Chem* 2006;281(37):27643–27652.

Schmelzle T, Hall MN. TOR, a central controller of cell growth. *Cell* 2000;103(2):253–262.

Schnell CR, Stauffer F, Allegrini PR, O'Reilly T, McSheehy PM, Dartois C, Stumm M, Cozens R, Littlewood-Evans A, García-Echeverría C, Maira SM. Effects of the dual phosphatidylinositol 3-kinase/mammalian target of rapamycin inhibitor NVP-BEZ235 on the tumor vasculature: implications for clinical imaging. *Cancer Res* 2008;68(16):6598–6607.

Schroder W, Bushell G, Sculley T. The human stress-activated protein kinase-interacting 1 gene encodes JNK-binding proteins. *Cell Signal* 2005;17(6):761–767.

Schroder WA, Buck M, Cloonan N, Hancock JF, Suhrbier A, Sculley T, Bushell G. Human Sin1 contains Ras-binding and pleckstrin homology domains and suppresses Ras signalling. *Cell Signal* 2007;19(6):1279–1289.

Scott RC, Schuldiner O, Neufeld TP. Role and regulation of starvation-induced autophagy in the *Drosophila* fat body. *Dev Cell* 2004;7(2):167–178.

Shao Z, Bhattacharya K, Hsich E, Park L, Walters B, Germann U, Wang YM, Kyriakis J, Mohanlal R, Kuida K, Namchuk M, Salituro F, Yao YM, Hou WM, Chen X, Aronovitz M, Tsichlis PN, Bhattacharya S, Force T, Kilter H. c-Jun N-terminal kinases mediate reactivation of Akt and cardiomyocyte survival after hypoxic injury *in vitro* and *in vivo*. *Circ Res* 2006;98(1):111–118.

Sharfi H, Eldar-Finkelman H. Sequential phosphorylation of insulin receptor substrate-2 by glycogen synthase kinase-3 and c-Jun NH2-terminal kinase plays a role in hepatic insulin signaling. *Am J Physiol Endocrinol Metab* 2008;294(2):E307–E315.

Shiota C, Woo JT, Lindner J, Shelton KD, Magnuson MA. Multiallelic disruption of the rictor gene in mice reveals that mTOR complex 2 is essential for fetal growth and viability. *Dev Cell* 2006;11(4):583–589.

Sofer A, Lei K, Johannessen CM, Ellisen LW. Regulation of mTOR and cell growth in response to energy stress by REDD1. *Mol Cell Biol* 2005;25(14):5834–5845.

Song JJ, Lee YJ. Effect of glucose concentration on activation of the ASK1-SEK1-JNK1 signal transduction pathway. *J Cell Biochem* 2003a;89(4):653–662.

Song JJ, Lee YJ. Role of the ASK1-SEK1-JNK1-HIPK1 signal in Daxx trafficking and ASK1 oligomerization. *J Biol Chem* 2003b;278(47):47245–47252.

Speakman JR. Obesity: the integrated roles of environment and genetics. *J Nutr* 2004;134(8 Suppl):2090S–2105S.

Stocker H, Radimerski T, Schindelholz B, Wittwer F, Belawat P, Daram P, Breuer S, Thomas G, Hafen E. Rheb is an essential regulator of S6K in controlling cell growth in *Drosophila*. *Nat Cell Biol* 2003;5:559–565.

Sturgill TW, Cohen A, Diefenbacher M, Trautwein M, Martin D, Hall MN. TOR1 and TOR2 have distinct locations in live cells. *Eukaryot Cell* 2008.

Sun Y, Fang Y, Yoon MS, Zhang C, Roccio M, Zwartkruis FJ, Armstrong M, Brown HA, Chen J. Phospholipase D1 is an effector of Rheb in the mTOR pathway. *Proc Natl Acad Sci USA* 2008;105(24):8286–8291.

Tee AR, Manning BD, Roux PP, Cantley LC, Blenis J. Tuberous sclerosis complex gene products, tuberin and hamartin, control mTOR signaling by acting as a GTPase-activating protein complex toward Rheb. *Curr Biol* 2003;13:1259–1268.

Thedieck K, Polak P, Kim ML, Molle KD, Cohen A, Jenö P, Arrieumerlou C, Hall MN. PRAS40 and PRR5-like protein are new mTOR interactors that regulate apoptosis. *PLoS ONE* 2007;2(11):e1217.

Tothova Z, Kollipara R, Huntly BJ, Lee BH, Castrillon DH, Cullen DE, McDowell EP, Lazo-Kallanian S, Williams IR, Sears C, Armstrong SA, Passegué E, DePinho RA, Gilliland DG. FoxOs are critical mediators of hematopoietic stem cell resistance to physiologic oxidative stress. *Cell* 2007;128(2):325–339.

van Slegtenhorst M, de Hoogt R, Hermans C, Nellist M, Janssen B, Verhoef S, Lindhout D, van den Ouweland A, Halley D, Young J, Burley M, Jeremiah S, Woodward K, Nahmias J, Fox M, Ekong R, Osborne J, Wolfe J, Povey S, Snell RG, Cheadle JP, Jones AC, Tachataki M, Ravine D, Kwiatkowski DJ, et al. Identification of the tuberous sclerosis gene TSC1 on chromosome 9q34. *Science* 1997;277:805–808.

Wallace DC. Mitochondria and cancer: Warburg addressed. *Cold Spring Harb Symp Quant Biol* 2005;70:363–374.

Wan X, Harkavy B, Shen N, Grohar P, Helman LJ. Rapamycin induces feedback activation of Akt signaling through an IGF-1R-dependent mechanism. *Oncogene* 2007;26(13):1932–1940.

Wang L, Harris TE, Roth RA, Lawrence JC Jr. PRAS40 regulates mTORC1 kinase activity by functioning as a direct inhibitor of substrate binding. *J Biol Chem* 2007;282(27):20036–20044.

Wang X, Fonseca BD, Tang H, Liu R, Elia A, Clemens MJ, Bommer UA, Proud CG. Re-evaluating the roles of proposed modulators of mammalian target of rapamycin complex 1 (mTORC1) signaling. *J Biol Chem* 2008;283:30482–30492.

Waters JE, Astle MV, Ooms LM, Balamatsias D, Gurung R, Mitchell CA. P-Rex1—a multidomain protein that regulates neurite differentiation. *J Cell Sci* 2008;121(Pt 17):2892–2903.

Wei Y, Sinha S, Levine B. Dual role of JNK1-mediated phosphorylation of Bcl-2 in autophagy and apoptosis regulation. *Autophagy* 2008;4(7):949–951.

Weidinger C, Krause K, Klagge A, Karger S, Fuhrer D. FOXOs: critical conductors of cancer's fate. *Endocr Relat Cancer* 2008;15:917–929.

Woo SY, Kim DH, Jun CB, Kim YM, Haar EV, Lee SI, Hegg JW, Bandhakavi S, Griffin TJ, Kim DH. PRR5, a novel component of mTOR complex 2, regulates platelet-derived growth factor receptor beta expression and signaling. *J Biol Chem* 2007;282(35):25604–25612.

Wouters BG, Koritzinsky M. Hypoxia signalling through mTOR and the unfolded protein response in cancer. *Nat Rev Cancer* 2008;8:851–864.

Wu H, Yan Y, Backer JM. Regulation of class IA PI3Ks. *Biochem Soc Trans* 2007;35(Pt 2):242–244.

Wu X, Tu X, Joeng KS, Hilton MJ, Williams DA, Long F. Rac1 activation controls nuclear localization of beta-catenin during canonical Wnt signaling. *Cell* 2008;133(2):340–353.

Wullschleger S, Loewith R, Hall MN. TOR signaling in growth and metabolism. *Cell* 2006;124(3):471–484.

Xie L, Wang DIC. Energy metabolism and ATP balance in animal cell cultivation using a stoichiometrically based reaction network. *Biotechnol Bioeng* 2000;52(5):591–601.

Yaghini FA, Li F, Malik KU. Expression and mechanism of spleen tyrosine kinase activation by angiotensin II and its implication in protein synthesis in rat vascular smooth muscle cells. *J Biol Chem* 2007;282(23):16878–16890.

Yang TT, Yu RY, Agadir A, Gao GJ, Campos-Gonzalez R, Tournier C, Chow CW. Integration of protein kinases mTOR and extracellular signal-regulated kinase 5 in regulating nucleocytoplasmic localization of NFATc4. *Mol Cell Biol* 2008;28(10):3489–3501.

Zanchi NE, Lancha AH Jr. Mechanical stimuli of skeletal muscle: implications on mTOR/p70s6k and protein synthesis. *Eur J Appl Physiol* 2008;102(3):253–263.

Zhang HH, Lipovsky AI, Dibble CC, Sahin M, Manning BD. S6K1 regulates GSK3 under conditions of mTOR-dependent feedback inhibition of Akt. *Mol Cell* 2006;24(2):185–197.

Zhang R, Luo D, Miao R, Bai L, Ge Q, Sessa WC, Min W. Hsp90-Akt phosphorylates ASK1 and inhibits ASK1-mediated apoptosis. *Oncogene* 2005;24(24):3954–3963.

Zhang Y, Gao X, Saucedo LJ, Ru B, Edgar BA, Pan D. Rheb is a direct target of the tuberous sclerosis tumor suppressor proteins. *Nat Cell Biol* 2003;5:578–581.

23

HIV-1 PROTEASE INHIBITORS AS ANTIRETROVIRAL AGENTS

SERGEI V. GULNIK, ELENA AFONINA, AND MICHAEL EISSENSTAT

23.1 INTRODUCTION

Human immunodeficiency virus (HIV) encodes three enzymes essential for its life cycle and critical for infectivity: HIV protease (HIV PR); RNA-dependent DNA polymerase, or reverse transcriptase (HIV RT); and integrase. All three enzymes have been successfully targeted by small-molecule inhibitors. Historically, nucleoside HIV RT inhibitors were introduced first in 1987, followed by HIV PR inhibitors (PIs) and non-nucleoside HIV RT inhibitors (NNRTIs). The first-in-class integrase inhibitor raltegravir, developed by Merck and approved by the FDA in 2008, is the latest addition to the arsenal of antiretrovirals. The introduction of HIV PIs into clinical practice in the mid-1990s initiated the era of highly active antiretroviral therapy (HAART) and resulted in the dramatic improvement of the management of HIV infection. Further advances were made with newer PIs such as lopinavir, amprenavir, atazanavir and, most recently, tipranavir and darunavir. In addition, boosting with a low dose of ritonavir overcame pharmacokinetic limitations of most drugs in this class (Youle, 2007) and took full advantage of their beneficial properties, such as high potency and high barrier to resistance. This chapter will touch on biochemical and structural properties of HIV PR, the important function of this enzyme in the viral lifecycle, and methods for measuring HIV PR inhibition. It will briefly introduce readers to the history of PI discovery and development, highlight strengths and shortcomings of currently

Enzyme Inhibition in Drug Discovery and Development, edited by Lu and Li
Copyright © 2010 John Wiley & Sons, Inc.

approved HIV PIs, and describe the concept of PK boosting of HIV PIs by ritonavir. Finally, it will discuss the problem of viral resistance and current approaches for the design of HIV PIs with improved resistance profiles.

23.2 HIV PR AS AN ANTIVIRAL TARGET

23.2.1 Role of HIV PR in the Viral Life Cycle

Structural proteins of HIV are translated as a polyprotein precursor, p55 Gag, from a single-spliced mRNA. A minus one ribosomal frame shift at the C-terminus of nucleocapsid protein occurs at a frequency of about 5% and results in the synthesis of p160 Gag-Pol precursor that, in addition to structural proteins, contains viral enzymes. HIV PR is responsible for proper processing of Gag and Gag-Pol to generate mature individual structural proteins and enzymes (Fig. 23.1A). It was shown in early HIV research that mutation of catalytic residues of HIV PR as well as chemical inhibition of the enzyme resulted in the production of immature non-infectious viral particles (Kaplan et al., 1993; Kohl et al., 1988; McQuade et al., 1990; Seelmeier et al., 1988). These studies confirmed the critical role of HIV PR for viral infectivity and provided the rationale for the design of inhibitors of this enzyme as potential antiviral drugs.

23.2.2 Enzyme Structure and Catalytic Mechanism

The determination of the molecular structure of HIV PR by X-ray crystallography in 1989 (Lapatto et al., 1989; Navia et al., 1989; Wlodawer et al., 1989) opened the door for the rational design of inhibitors targeting this enzyme. The field has flourished since then; and more than 150 crystallographic and NMR structures of wild-type (WT) and drug-resistant mutant enzymes, both unliganded and in complex with different inhibitors, have been deposited in the Protein Data Bank (PDB) (http://www.rcsb.org/pdb). Additional structures not deposited in PDB can be found in the National Institute of Standards HIV Structural database (http://xpdb.nist.gov/hivsdb/hivsdb.html). Some of the structures were solved at very high resolution, below 1 Å, allowing the resolution of multiple inhibitor conformations [recently reviewed (Weber et al., 2007)].

HIV PR is a homodimer of two 99-amino-acid monomers related to each other by crystallographic twofold symmetry (Fig. 23.2A). It is a member of the aspartic proteinase family of hydrolases. However, the homodimeric structure is a unique feature of retroviral aspartic proteases, where the architecture of each monomer is similar to that of a single domain of monomeric but bilobal aspartic proteinases of eukaryotes. The secondary structure of the monomer is dominated by beta sheets with only one short helix. The conserved active site triad D25–T26–G27 is positioned in the loop that forms part of the active

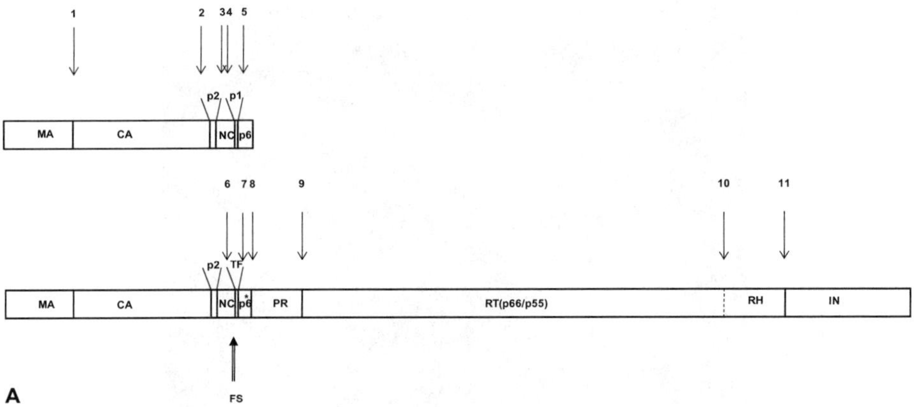

A

cleavage site	P5	P4	P3	P2	P1	P1'	P2'	P3'	P4'	P5'
1	V	S	Q	N	Y	P	I	V	Q	N
2	K	A	R	V	L	A	E	A	M	S
3	S	A	T	A	M	M	Q	R	G	N
4	E	R	Q	A	N	F	L	G	K	I
5	R	P	G	N	F	L	Q	S	R	P
6	E	R	Q	A	N	F	L	R	E	D
7	E	D	L	A	F	L	Q	G	K	A
8	V	S	N	N	F	P	Q	V	T	L
9	C	T	L	N	F	P	I	S	P	I
10	G	A	E	T	F	Y	V	D	G	A
11	I	R	K	V	L	F	L	D	G	I

B

Figure 23.1. (**A**) Structural organization of Gag and Gag-Pol polyproteins. p55 Gag and p160 Gag-Pol polyprotein precursors are translated from single-spliced mRNA. Minus 1 ribosomal frame shift at the C-terminus of NC (indicated by the arrow pointing up, FS) occurs at a frequency of about 5% and results in Gag-Pol synthesis. Abbreviations: MA, matrix protein; CA, capsid protein; NC, nucleocapsid protein; PR, protease; RT, reverse transcriptase; RH, RNAse H domain of RT. The width of the box around each protein is scaled to its size. (**B**) Amino acid sequence around HIV PR cleavage sites indicated in part A spanning the P5–P5′ region.

site. Each monomer provides one of the two catalytic aspartic acids that are almost coplanar and H-bonded to the catalytic water molecule. The active site is covered by two flexible beta hairpin loops termed flaps. The flaps undergo significant movement upon substrate/inhibitor binding (tips of the flap move up to 15 Å) closing on the substrate and locking it in the active site (Fig. 23.2B). The catalytic aspartates face a fairly long and primarily hydrophobic substrate binding cleft with well-defined subsites that can accommodate peptides of seven to eight amino acids in length. This distinctive active site

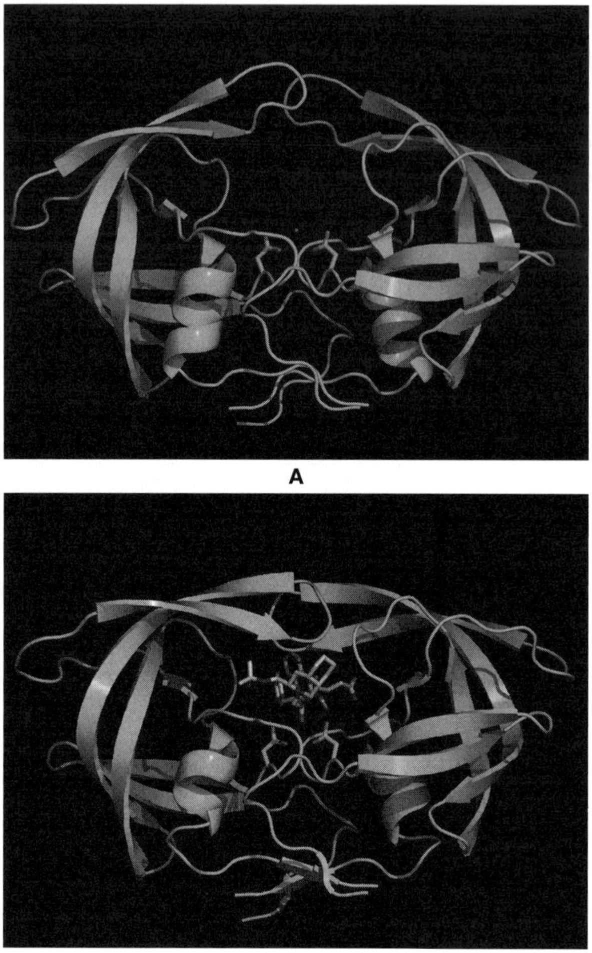

A

B

Figure 23.2. A ribbon diagram representation of HIV PR 3-D structure. The enzyme is a homodimer with a secondary structure dominated by beta-sheets. Substrates and inhibitors bind in the active-site cleft located roughly in the middle of the molecule and interact with active site aspartates, (side chains shown as stick models). (**A**) Unliganded form of the enzyme (PDB code 2PC0). (**B**) HIV PR in complex with saquinavir represented as a stick model colored by atoms (PDB code 2NMW). Note the large conformational shift of the beta-hairpin loops (flaps) located on the top of the molecule, which adopt a closed conformation upon inhibitor binding. The hydroxy group of the inhibitor, present in all clinical HIV PIs, forms H-bonding interactions with the active site aspartates, displacing the catalytic water molecule (shown in part **A** as a red sphere). See the insert for color representation of this figure.

architecture, and especially the prominent and deep binding pockets, allowed the successful design of potent and selective HIV PR inhibitors.

Since each monomer contributes one half of the active site, only dimeric HIV PR is catalytically active. The dimeric structure is stabilized by extensive interactions at the dimer interface, dominated by four stranded antiparallel beta sheets consisting of four N- and four C-terminal PR residues that contribute more than 50% of the interface contacts. The dimer dissociation constant has been determined using different biophysical and biochemical approaches including: urea- and temperature-induced protein unfolding (Grant et al., 1992; Todd et al., 1998), the effect of enzyme concentration on catalytic activity (Cheng et al., 1990; Jordan et al., 1992; Kuzmic, 1993; Louis et al., 1999a; Zhang et al., 1991), binding of fluorescently labeled inhibitors (Darke et al., 1994), and sedimentation equilibrium (Grant et al., 1992; Xie et al., 1999). The wide array of K_d values reported ranged from picomolar to micromolar. It appears that dimer stability depends strongly on the experimental conditions; reduced pH and increased ionic strength favor dimerization. The highest dimerization constant (5.8 μM) was measured at pH 7.0. It was proposed that relatively low dimer stability at neutral pH may serve as a regulatory mechanism that restricts HIV PR dimer formation in infected cells up to the point of viral maturation (Gulnik et al., 2000).

The general base–general acid mechanism of catalysis has been well accepted for aspartic proteinases, including HIV PR. One aspartate of the active site is protonated and serves as a general acid, while the other is negatively charged and plays the role of a general base, which increases the nucleophilicity of the final member of the catalytic triad, a water molecule. Nucleophilic attack of the catalytic water molecule on the carbonyl carbon of the scissile bond results in the formation of metastable tetrahedral gem-diol intermediate, followed by peptide bond breakage. This mechanism is supported by experimental data, including biochemical and structural studies as well as theoretical calculations (Brik and Wong, 2003; Davies, 1990; Dunn, 2002; Hyland et al., 1991a,b; Meek et al., 1994; Northrop, 2001; Polgar et al., 1994). However, the precise mechanistic details—including the proton sharing arrangement between the aspartates and the catalytic water, as well as proton shuttling during catalysis, nucleophilic attack, and cleavage to products—remain controversial (Das et al., 2006; Kovalevsky et al., 2007; Meek et al., 1994; Silva et al., 1996; Tyndall et al., 2008). Recently, the high-resolution crystal structure of a gem-diol tetrahedral intermediate was reported by Weber's group (Kovalevsky et al., 2007), suggesting that at least under these crystallization conditions, its decomposition into products was rate-limiting, in agreement with kinetic studies of the reaction mechanism. This structure, together with *ab initio* calculations, led to the proposal of a modified reaction mechanism in which no hydrogen atoms are transferred from the HIV PR during the catalysis, contrary to what has been generally accepted.

23.2.3 Substrate Specificity and Polyprotein Processing

At least 11 sites within Gag and Gag-Pol polyproteins are recognized and cleaved by HIV PR (Fig. 23.1). Brief examination of the sequences in Fig. 23.1B suggests that the enzyme has rather promiscuous substrate specificity. Interestingly, three out of nine natural cleavage sites have proline in the P1′ position. The ability to cleave peptides with a P1′ proline is rarely observed for other endopeptidases and appears to be a unique feature of the retroviral aspartic proteases. It was proposed that substrates of HIV PR may be broadly grouped into two categories based on the specific amino acids in P1 and P1′ positions: hydrophobic–hydrophobic and aromatic–proline (Griffiths et al., 1992). However, it became clear that even such a broad classification has a number of exceptions in which either small nonpolar or charged amino acids are found in the P1 and P1′ positions. The substrate specificity of HIV PR has been extensively probed using synthetic peptides corresponding to cleavage sites in HIV and other proteins and by introducing systematic substitutions at different subsites [reviewed in Dunn et al. (1994) and Tomasselli and Heinrikson (1994)]. These studies suggest that the enzyme recognizes eight amino acids at positions P4 through P4′. The comprehensive knowledge of subsite preferences uncovered by this research was instrumental in designing autolysis-resistant HIV PR. The Q7K/L33I/L63I triple mutant HIV PR, which has a dramatically reduced autolysis rate, but is biochemically and structurally similar to the wild-type enzyme (Mildner et al., 1994), has been widely used for biochemical and structural studies. Detailed structural analysis of six decapeptides bound to the inactive D25N mutant HIV PR was performed in Schiffer's laboratory (Prabu-Jeyabalan et al., 2002). All peptides were bound in an extended conformation with strict conservation of the main chain but not the side-chain H bonds. This study suggested that HIV PR specificity is governed by the recognition of the asymmetric shape of substrate subsite residues rather than a particular amino acid sequence. Recently, Komorowski and colleagues (Kontijevskis et al., 2007) applied the rough sets approach to the analysis of 374 cleavable and 1251 noncleavable substrates of HIV PR. The predictive model resulting from this study suggests that combination of at least three particular amino acids is required for the productive cleavage of substrates by HIV PR.

Multiple cellular host proteins with important biological functions have been reported to be cleaved by HIV PR *in vitro* (Korant et al., 1998; Snasel and Pichova, 1996). However, the relevance of these observations for HIV pathogenesis *in vivo* remains unclear.

The order of Gag and Gag-Pol polyprotein processing has been studied by adding purified recombinant HIV PR to reticulocyte-expressed polyproteins *in trans* (Pettit et al., 2005; Pettit et al., 1994). These and other studies revealed that proteolytic processing of Gag-Pol polyprotein precursor is a highly orchestrated and tuned process in which timing, order, and extent of the processing is critical for generation of infectious viral particles (Krausslich et al.,

1995; Mervis et al., 1988; Pettit et al., 1994; Wiegers et al., 1998). For example, delaying, but not fully abolishing, cleavage at the CA-p2 site by bevirimat is sufficient for this maturation inhibitor to demonstrate potent antiviral activity in cell culture and efficacy in patients (Zhou et al., 2004). Despite some progress, the mechanism of HIV PR autoactivation and the exact order of Gag-Pol processing by HIV PR embedded in the Gag-Pol precursor remain unclear. It has been shown that maturation occurs concomitant to, or shortly after, viral budding (Kaplan et al., 1994). It appears that HIV maturation does not depend on cellular proteinases and HIV PR is solely responsible for this process. Since each monomer provides one-half of the active site of HIV PR, the dimerization of the protease domain within Gag-Pol polyprotein is a prerequisite for HIV PR activation. Other Gag and Pol proteins including MA, CA, NC, RT, and IN have a tendency to form dimers or oligomers that can influence the strength of the initial polyprotein dimer and the spatial arrangement of the individual proteins in it. The initial events in the full-length Gag-Pol polyprotein precursor autoprocessing were studied using an *in vitro* reticulocyte transcription/translation system. It was shown that the first cleavage occurs at the p2/NC site by what appeared to be an intramolecular reaction, followed by processing at TF-p6* (Pettit et al., 2004). However, in this artificial system the reaction failed to proceed beyond the initial steps for reasons that are not well understood. Studies of truncated model precursors indicated that the intramolecular cleavage at the N-terminus of HIV PR coincided with the appearance of full protease activity and was followed by intermolecular cleavage leading to the liberation of the C-terminus (Louis et al., 1994; Louis et al., 1999a,b; Wondrak et al., 1996). It was proposed that the high dimerization constant of HIV PR having an N-terminal TF fusion peptide may provide an additional mechanism for the regulation of the timing for viral maturation (Louis et al., 2007).

23.2.4 Methods for Evaluating the Potency of HIV PIs

It is important to have a reliable and sensitive assay for monitoring HIV PR activity that can be used for biochemical studies and inhibitor evaluation. Continuous assays using chromogenic or fluorogenic substrates are the most common. Chromogenic substrates usually contain *p*-nitrophenylalanine at the P1′ position. At a pH below 6.5, protonation of the N-terminal nitrophenylalanine generated during cleavage shifts the absorbance maximum from 278 to 272 nm. The product formation can be continuously followed by monitoring the absorbance decrease at 300–310 nm (Dunn et al., 1994). Another technique uses peptides modified by fluorophore and quencher groups. The fluorescence is quenched in substrate by either collision or resonance energy transfer. Cleavage of the peptide leads to an increase in fluorescence due to the separation of fluorophore from quencher. EDANS [5-((2-aminoethyl)amino)naphthalene-1-sulfonic acid] and DABCYL [4-((4-(dimethylamino)phenyl)azo)benzoic acid] (Krafft and Wang, 1994; Matayoshi et al., 1990), as well as tyrosine and *p*-nitrophenylalanine, are among the most commonly utilized fluorophore-quencher pairs (Peranteau et al., 1995). Both chromogenic and fluorogenic substrates of HIV PR are now commercially available, and assays can be performed in high throughput format using plate readers.

All HIV PIs currently approved by the FDA are competitive inhibitors that target the active site of the enzyme. On the enzyme level the potency of inhibitors is usually evaluated by measuring the decrease of the initial rate of substrate cleavage in the presence of different inhibitor concentrations. The IC_{50} values and inhibition constants (K_i) can be obtained from the analysis of such response curves. Examples of different response curves are presented in Fig. 23.3. Several important points that may affect the data analysis and interpretation of the results for HIV PR inhibitors should be emphasized. Rational design and medicinal chemistry efforts have led to the discovery of extremely potent inhibitors. The estimation of inhibition constants for such inhibitors can be challenging, but methods for the analysis of inhibition data for tight binding inhibitors are available. These methods take into account the depletion of inhibitor due to binding to the enzyme and must be used when the inhibition constant is in the range of the enzyme concentration in the assay. The most commonly used equation for analyzing concentration–response data for tight binding inhibitors was introduced by Morrison and colleagues (Williams and Morrison, 1979):

$$\frac{v_i}{v_0} = 1 - \frac{\left(E_t + I_t + K_i \times \left(1 + \frac{S}{K_m}\right)\right) - \sqrt{\left(E_t + I_t + K_i \times \left(1 + \frac{S}{K_m}\right)\right)^2 - 4 \times E_t \times I_t}}{2 \times E_t}$$

where v_i and v_0 are the initial rates of substrate conversion with and without inhibitor, and E_t, I_t, S, and K_m are total enzyme concentration, total inhibitor concentration, substrate concentration, and Michaelis–Menten constant, respectively. Determination of such very low K_i values also requires careful selection of the range of inhibitor concentrations, and active enzyme concentration should be accurately determined in a separate experiment. Increasing the substrate concentration in the assay is also helpful, since it elevates the apparent K_i and increases the curvature in the "elbow" part of the dose–response curve, allowing for more accurate estimation of the inhibition constant (Fig. 23.3C). An excellent discussion on tight binding inhibitors that covers both theoretical and practical aspects of the problem and contains a nice literature review can be found in Copeland (2005). It is not uncommon to see publications where the enzyme concentration in the assay is two to four orders of magnitude higher than the reported K_i. It is important to emphasize that at such high enzyme concentration to K_i ratios it is very challenging (nearly impossible) to get reliable measurements for the inhibition constant, as illustrated in Fig. 23.3B. Therefore, caution should be taken when interpreting low picomolar values of inhibition constants reported for PIs.

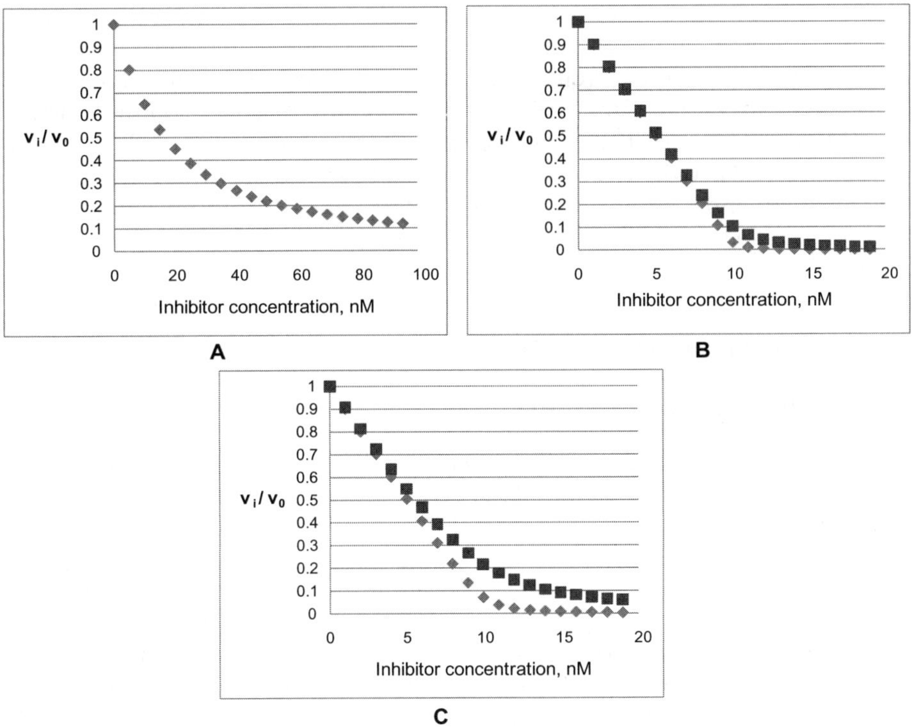

Figure 23.3 Examples of different dose–response curves that can be generated for K_i determination. The curves were created using the program DynaFit (Kuzmic, 1996). Total enzyme concentration, E_t, was fixed at 10 nM. (**A**) Dose–response curve for an inhibitor with K_i = 10 nM. Substrate concentration in this experiment was $S = 0.2K_m$. In this case, K_i can be accurately determined using Morrison's equation. (**B**) Dose–response curves for inhibitors X with K_i = 0.1 nM (squares), and Y with K_i = 0.01 nM (diamonds). The inhibition curves for these two inhibitors are very similar. Substrate concentration in this experiment was $S = 0.2K_m$. Accurate determination of inhibition constants in this case is problematic. (**C**) Dose–response curves for the same inhibitors as in part B, but measured at $S = 5K_m$. Increasing the substrate concentration introduced more curvature into the titration with inhibitor X. In this case, K_i can be determined using a carefully planned experimental set up for inhibitor X, but not for inhibitor Y. It is generally accepted that inhibition constants can be accurately estimated using Morrison's equation when $E_t \leq 100 \times (1 + S/K_m)$ (Copeland, 2005).

It is well documented that the activity and kinetics of HIV PR is highly dependent on experimental conditions such as pH, ionic strength, DMSO concentration, and the presence of common additives such as detergents, PEG and BSA (Gulnik et al., 2000; Porter et al., 2001). The absence of standardized assay conditions makes it difficult to compare inhibition constants obtained in different laboratories.

The most common secondary assays for HIV PR inhibitors involve measuring antiviral activity in cell culture. The relative ease of propagating HIV in cell culture allowed the development of various assays. The inhibition of viral replication can be measured using single or multiple replication cycle assays in CD4 positive T-cell lines or primary macrophages. Protease inhibitors show robust antiviral activity in cell based assays in the nanomolar range (Haubrich, 2004; Kellam and Larder, 1994; Pauwels et al., 1987; Petropoulos et al., 2000; Qari et al., 2002).

23.3 INHIBITORS OF HIV PR

23.3.1 FDA-Approved HIV PIs

Once HIV PR was established as a realistic therapeutic target for inhibition to treat HIV infection in the late 1980s, companies were able to quickly mount synthetic efforts targeting this aspartyl protease. Many companies were able to capitalize on already established programs aimed at inhibiting aspartyl proteases such as renin and were thus familiar with approaches to targeting the catalytic site. Most of the initial approaches involved elaborating a certain peptidomimetic template or core which incorporated a transition state mimetic of the cleavage of the peptide substrate. Elaborating these cores with functionality that made effective interactions with the residues of the enzyme in the catalytic pocket allowed the relatively rapid identification of potent inhibitors. These were relatively large molecules for drugs with molecular weights in the 700 dalton range, and most of the subsequent work in the field was aimed at trying to simplify the molecules to improve bioavailability without giving up the high potency (picomolar to low nanomolar at the enzyme level) required to effectively inhibit the virus at the cellular level (nanomolar). Later efforts targeted the resistance problem observed with all PIs. The structures of the FDA approved HIV PIs are shown in Fig. 23.4.

Saquinavir. Roche won the race to the market with the introduction of saquinavir (Ro31-8959) in 1995. Saquinavir is considered a substrate based inhibitor; that is the peptide substrate amino acids were replaced by amino acid mimetics. Saquinavir uses a hydroxyethylamine core with the amine being incorporated into a decahydroisoquinoline mimic of proline. This moiety at P1′–P2′ provided increased potency and the quinolinecarboxamide at P3 was an effective amino acid mimetic (Roberts et al., 1990). The P1 substituent was a benzyl group, which has been retained in most subsequent PIs. The P2 leucine of the substrate was replaced by asparagine. Saquinavir remains one of the most potent PIs *in vitro* and has a reasonably good profile against mutants (Craig et al., 1991). However, the very low bioavailability of this drug limited its usage despite the 1997 introduction of a moderately improved formulation.

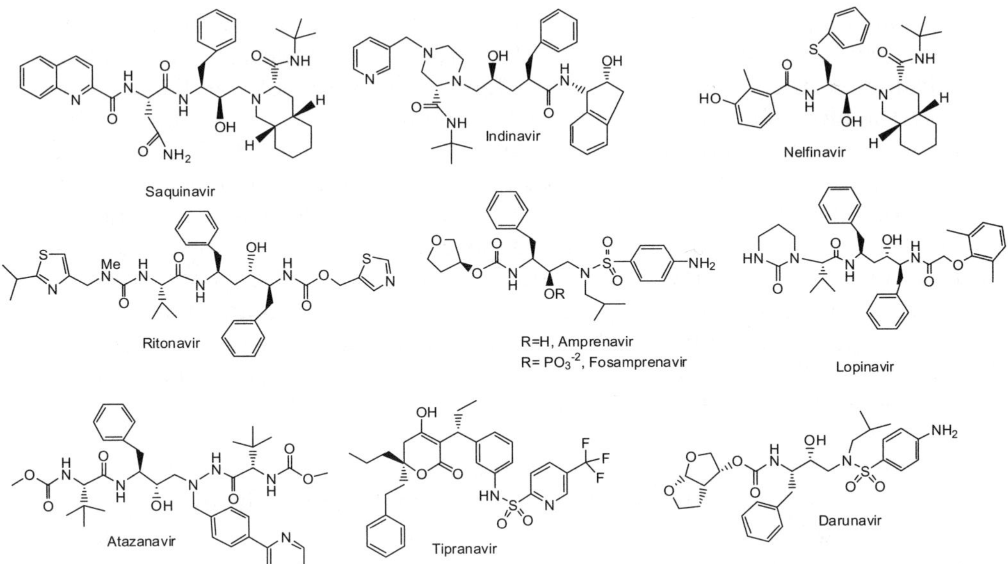

Figure 23.4. Chemical structures of FDA-approved HIV PIs.

Indinavir. Merck screened compounds prepared for its renin program and identified potent protease inhibitors that used a hydroxyethylene as the transition state mimic. An early lead had a benzyl amide at P2′. Constraining the benzyl group into the presumed bioactive conformation provided a more potent indane derivative to which a *cis*-2-hydroxy group was added to provide a further 80x improvement in potency. In order to enhance the solubility of these compounds, a substituted piperazine moiety was introduced to mimic the decahydroisoquinoline of saquinavir. The resulting molecule indinavir (L-735524) had significantly better bioavailability than saquinavir and was much less protein bound (about 60%) (Dorsey et al., 1994; Vacca and Condra, 1997; Vacca et al., 1994). In addition, unlike other HIV PIs that are strongly bound to alpha-acid glycoprotein (AAG) in serum, indinavir is primarily bound to serum albumin. Indinavir captured most of the market when it was introduced in 1996, despite requiring three times per day dosing.

Ritonavir. Abbott researchers used the twofold symmetry of the protease in their design (Erickson et al., 1990). They explored symmetrization of both the C-terminus and N-terminus of hydroxyethylene-based inhibitors—that is, by incorporating the same groups on the left side of the inhibitor that were making interactions with the S residues into the right side of the inhibitor to make interactions with the S′ residues, and vice versa. The N-symmetrization was successful and provided 2,3-dihydroxy-1,4-butanediamine derivatives that were potent PIs. Only one of the hydroxyls was shown to be important for binding to the catalytic Asps and the other was removed (Kempf et al., 1993).

Several compounds were put into clinical trials but generally showed poor bioavailability and high protein binding. The early analogues had pyridine substituents at P3 and P2′. These were found to be susceptible to P450 oxidation. The bioavailability was greatly improved by replacing the pyridines with thiazoles, which resulted in the discovery of ritonavir (ABT-538) (Kempf et al., 1995). Despite being approved at the same time as indinavir and only requiring twice-a-day administration, ritonavir did not achieve as high a market share, presumably due to a wide spectrum of side effects. The excellent bioavailability of ritonavir was found to largely result from its effectiveness as an inhibitor of its own metabolism by Cyp3A4 (Kempf et al., 1997). This property was utilized by Abbott in the development of their second-generation PI lopinavir, ABT-378 (*vide infra*), which was combined with ritonavir to be marketed as Kaletra.

Nelfinavir. Agouron and Lilly collaborated to develop nelfinavir using structure-based design techniques. This compound retained the hydroxyethylamine moiety of saquinavir, but replaced the P3-P2 quinoline–asparagine moiety by a 2-methyl-3-hydroxybenzamide (Kaldor et al., 1997). Structural analysis of the benzamide suggested that the nonplanar binding conformation was favored by the introduction of a 2-methyl substituent (Freskos et al., 1996). Replacing the P1 benzyl by phenythiomethyl gave a 10-fold boost in potency for this compound. Nelfinavir (AG-1343) entered the market in 1997 and soon surpassed indinavir as the market leader.

Amprenavir. Vertex and GSK collaborated in the development of amprenavir (VX-478). Since a general problem with PIs is poor bioavailability, Vertex attempted to address this by reducing the molecular weight of the inhibitor and increasing the solubility. They also used structure-based drug design techniques to identify suitable amino acid mimetics. They used the hydroxyethylamine core of saquinavir, but incorporated the amine into a sulfonamide as originally described by Monsanto-Searle (Vazquez et al., 1995). The major change was the introduction of a tetrahydrofuran moiety at P2, reported earlier by Merck (Ghosh et al., 1993). A *p*-aminobenzensulfonamide was optimum in P2′ (Kim et al., 1995). Amprenavir was introduced in 1999. While smaller and more bioavailable than many other PIs, amprenavir was not very soluble or potent and its formulation led to a high pill burden. In 2003 GSK introduced a phosphate prodrug fosamprenavir (GW-433908) with improved solubility and a reduced pill burden (Furfine et al., 2004).

Lopinavir. In order to have an improved resistance profile, Abbott removed the P3 isopropylthiazolyl group of ritonavir and then introduced a conformational constraint in the resulting urea by incorporating it into a six-membered ring. The P2′ thiazolylmethoxycarbonyl moiety was replaced by a dimethylphenoxyacetyl group that had previously shown good affinity in the Biomega saquinavir analogue, palinavir (Beaulieu et al., 2000; Lamarre et al., 1997).

Lopinavir was about an order of magnitude more potent as an antiviral than ritonavir and was also less affected by the V82A mutation commonly found in ritonavir-resistant viruses (Sham et al., 1998). However, it had very poor bioavailability in animals (and humans). The main route of elimination of lopinavir was via metabolism by CYP3A4. Since ritonavir had previously been identified as a potent 3A4 inhibitor, Abbott investigated the effect of ritonavir on the bioavailability of lopinavir (Kumar et al., 1999). It was found that the blood levels of lopinavir were much higher when dosed in the presence of ritonavir. This led to the development of a fixed-dose combination of the two, named Kaletra. The amount of ritonavir required to boost lopinavir was 100 mg as opposed to the antiviral dose of 600 mg. Kaletra was approved in late 2000 and became the market leader by 2003.

Atazanavir. Ciba-Geigy examined hydroxyethyl hydrazine derivatives as their core (Fassler et al., 1996). In these compounds the N of the hydrazine closer to the hydroxy in the core was substituted with a benzyl derivative to provide a structure that is essentially an aza analog of the Abbott series. The optimum benzyl derivative had a 2-pyridyl attached at the para position. Methoxycarbonyl-*tert*-leucine amides capped the molecule at both termini to give CGP-73547 (Bold et al., 1998). The molecule was purchased by BMS (BMS-232632) and eventually was marketed as atazanavir in 2003. Remarkably for such a large molecule (MW 705 for the free base), atazanavir has good bioavailability as well as good antiviral activity. It has a unique resistance profile; it selects I50L in patients, which sensitizes the virus to all other PIs (Gong et al., 2000). This feature makes atazanavir an excellent choice as a first line PI in PI-containing regimens. Administration with ritonavir boosting allows once-a-day treatment. The major issue in clinical development was hyperbilirubinemia, but this side effect generally appears to be of minor consequence (Zhang et al., 2005). Atazanavir and Kaletra currently have the major share of the PI market.

Tipranavir. Upjohn and Parke-Davis took a different approach to the identification of HIV PIs. Screening of compound libraries against HIV PR provided warfarin as a hit (Fig. 23.5). This pyrone derivative was considered to be an excellent starting point for elaboration since it was nonpeptidic and low molecular weight. Both companies did extensive SAR on substituted pyrones, and the compound had to be built up to a size comparable to other PIs before potency and selectivity comparable to the competitors was achieved (Prasad et al., 1995a,b; Skulnick et al., 1995; Thaisrivongs et al., 1994, 1995, 1996; Tummino et al., 1994; Vara Prasad et al., 1994). An example of one of the Parke-Davis analogues, PD178390, is shown in Fig. 23.5. Upjohn was successful in identifying a clinical candidate. This compound, PNU-140690 (tipranavir), was truly nonpeptidic, lacking any amide bonds. It also lacked the aliphatic hydroxy transition state mimetic found in other PIs, but rather used the enol of the β-dicarbonyl subunit of the pyrone to make interactions with the

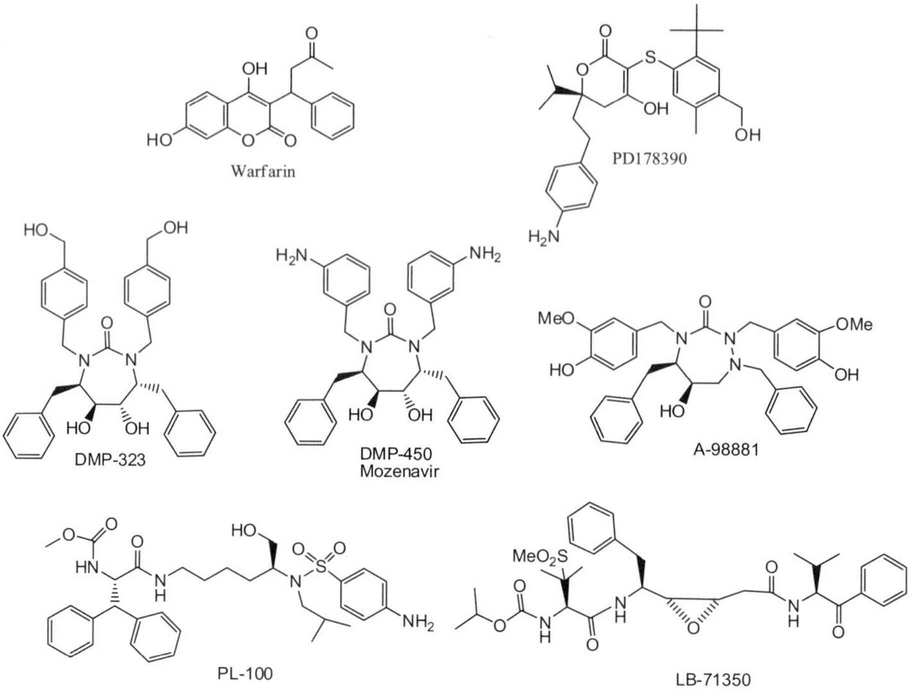

Figure 23.5. Chemical structures of selected experimental HIV PIs described in this chapter.

catalytic aspartates; the other carbonyl interacts directly with Ile50 on the flap, whereas other PIs interact with this residue via a conserved water molecule. Boehringer Ingelheim purchased the rights to tipranavir in 2000, and it was approved in 2005. Tipranavir is effective against resistant mutants, and high blood levels of drug are achieved. Unfortunately, it is a strong CYP 450 inducer, inducing its own metabolism and requiring double the usual boosting dose of ritonavir (200 mg), resulting in multiple drug–drug interactions (King and Acosta, 2006). These problems and the significant incidence of side effects of this drug suggest that its market share will most likely stay restricted to deep salvage regimens.

Darunavir. In the course of elaborating the P2 position of saquinavir analogues, Ghosh replaced the quinaldic–asparaginamide moiety by a fused bicyclic tetrahydrofuran (bis-THF) that could potentially make hydrogen bonding interactions with the main chain Asp 29 and Asp 30 NH's of HIV PR (Ghosh et al., 1994, 1996). Aside from retaining potency, the bis-THF moiety improved aqueous solubility, reduced logP, and reduced MW. The bis-THF was used to replace the THF of amprenavir (Ghosh et al., 1998). The excellent resistance properties of this series were uncovered by NCI researchers (Erickson et al., 1999; Yoshimura et al., 2002). One of the bis-THF-containing analogues was later developed by Tibotec as TMC-114 (darunavir) (De Meyer et al., 2005; Surleraux et al., 2005b). Several groups subsequently reported structural support for the tight bonding of darunavir and some other bis-THF-containing compounds both to wild-type and mutant virus (King et al., 2004b; Kovalevsky et al., 2006a,b; Tie et al., 2004). It was proposed that the high potency and exceptional resistance properties of darunavir derive from its unique binding mode that maximizes hydrogen bonding interactions with conserved main-chain atoms of the enzyme [extensively reviewed recently (Ghosh et al., 2006a,b, 2007, 2008; Mitsuya et al., 2008; Weber et al., 2007)]. A subnanomolar dissociation constant for darunavir with an extremely slow off-rate from WT HIV PR, as estimated by surface plasmon resonance (Dierynck et al., 2007), may contribute to the flat resistance profile of this inhibitor (Gulnik and Eissenstat, 2008). Darunavir requires boosting by ritonavir, and it was initially approved in 2006 for the treatment of PI-experienced patients as salvage therapy. In October 2008 the darunavir–ritonavir combination was approved as a first-line therapy for naive patients.

With the sole exception of tipranavir, the evolution of commercial HIV PIs can be described as incremental structural changes in the initially described leads that led to greater potency, increased activity against mutants, and in some cases improved bioavailability. As will be discussed in Section 23.3.5, the common current practice of using ritonavir to boost blood levels and in particular trough levels of HIV PIs somewhat obviates the struggle to identify compounds with good bioavailability.

23.3.2 Thermodynamics of HIV PI Binding

In general, high binding affinity can be achieved by optimizing either the entropy or enthalpy components of the total Gibbs energy change ($\Delta G = \Delta H - T\Delta S$). Enthalpy and entropy changes originate from different types of interactions between inhibitor and enzyme. Hydrogen bonds and van der Waals interactions primarily contribute to enthalpy changes. On the other hand, favorable entropic changes originate primarily from the burying of hydrophobic groups of the inhibitor in the active site, resulting in an increase in the solvent entropy while minimizing the decrease of conformational freedom of the inhibitor upon binding. Therefore, a combination of enthalpic and entropic optimization is required to achieve high affinity of drug-like molecules. Enthalpic optimization is notoriously the more difficult to achieve. Several factors are responsible for this. First, the hydrogen bond interactions have to be nearly perfect in terms of distance and geometry; otherwise, the potential enthalpic gain will be overcome by the loss of enthalpy due to desolvation of polar groups upon binding. The other issue that makes it difficult to optimize enthalpy-driven affinity is the compensating loss of entropy caused by ordering the enzyme and inhibitor upon binding. The hydrogen bonds should be designed in such a way that they do not interfere with maximal burial from the solvent. In addition, in order to minimize conformational entropy losses upon binding, hydrogen bonds must be targeted to structurally conserved regions of the protein. To increase binding affinity, it is preferable to place hydrogen bond donors or acceptors in a conformationally constrained part of the inhibitor (Lafont et al., 2007).

The early thermodynamic studies by isothermal titration calorimetry of HIV PI interactions with enzyme revealed that binding is associated with positive enthalpy, indicating that binding is an entropically driven process (Hoog et al., 1996; Luque et al., 1998; Xie et al., 1998). While some of the early FDA-approved HIV PIs, including saquinavir, indinavir, and nelfinavir, had this property, binding of others exhibited a negative enthalpy change. However, with the exception of amprenavir and darunavir (Ohtaka and Freire, 2004, 2005; Surleraux et al., 2005b), binding remained mainly entropy-driven. For the latter two inhibitors, the enthalpic contribution to the total binding energy was greater than the entropic. Detailed thermodynamic and structural analysis of the binding of HIV PIs to the enzyme was performed by researchers in Freire's and others' laboratories [see Ohtaka and Freire, (2004, 2005) for excellent reviews]. As will be discussed below, resistance to HIV PIs presents one of the major challenges for their long-term clinical efficacy. Inhibitors that utilize hydrogen bonding with conserved and structurally constrained regions of the active site should be less susceptible to drug resistance mutations. Introduction of flexible groups that interact with the unconserved regions of the protein can also lead to inhibitors with an improved resistance profile, since it may provide entropic compensation for the enthalpic loss caused by mutation (Vega et al., 2004). Interestingly, most of the HIV PIs with improved resistance profiles, such as bis-THF-containing inhibitors (see below), exhibit primarily enthalpically driven binding to the enzyme. For these inhibitors the loss of binding enthalpy caused by mutations is at least partially compensated by favorable entropic gains. On the other hand, tipranavir is an example of an HIV PI with an excellent resistance profile for which binding to the WT HIV PR is primarily entropically driven. This inhibitor maintains high affinity to the mutants by compensating for entropy losses with enthalpy gains (Muzammil et al., 2007).

23.3.3 Molecular Structures of HIV PR-HIV PI Complexes

HIV PR has been a relatively friendly enzyme from a structural biology viewpoint. It can be expressed in high yield as inclusion bodies in *E. coli*, purified under denaturing conditions in quantities sufficient for crystallization, and crystallized in the presence of a variety of common precipitants. The quality of HIV PR crystals and the resulting structural resolution was also improved by use of an autolysis-resistant mutant of the enzyme (Mildner et al., 1994). Consequently, HIV PR is often referred to as a success story in the field of structure-based drug design. The structures of the WT HIV PR with all FDA-approved PIs are available (Clemente et al., 2006; Hong et al., 2000; Kaldor et al., 1997; Kempf et al., 1995; Kim, 1995; Smith et al., 1997; Stoll et al., 2002;

Surleraux et al., 2005b; Tie et al., 2004; Turner et al., 1998). Additionally, many structures with experimental PIs have also been published or deposited in the PDB (http://www.rcsb.org/pdb). The structural aspects of HIV PR–HIV PI interactions as well as the structural basis for resistance to HIV PIs are very well covered in the literature and have been reviewed extensively (Erickson, 1995; Erickson and Burt, 1996; Gulnik et al., 2000; Louis et al., 2007; Weber et al., 2007; Wlodawer and Vondrasek, 1998). All marketed HIV PIs with the exception of tipranavir are peptidomimetics that bind to the active site in an extended conformation similarly to peptide substrates. The central hydroxyl moiety that is present in every inhibitor and mimics the transition-state tetrahedral intermediate makes hydrogen bonding interactions with the conserved active site aspartic acids Asp25 and Asp25′ of the dimeric enzyme. All PIs contain bulky hydrophobic substituents in P1 and P1′ that make favorable van der Waals interactions with the hydrophobic S1 and S1′ pockets of the active site. S2 and S2′ subsites can accommodate more polar residues, while S3 and S3′ are partially exposed to solvent. Carbonyl and NH groups of the peptidomimetic core of PIs make direct and water-mediated hydrogen bonds to the main chain and side chains of HIV PR. Newer-generation PIs have an increased number of direct H-bond interactions with the main chain atoms of the enzyme. This feature increases the enthalpic contribution to the binding energetics of these inhibitors and minimizes the detrimental effect of mutations on inhibitor potency. Tipranavir, the only nonpeptidic marketed PI, has a unique binding mode. It does not bind in the active site in an extended conformation; rather, it makes a direct hydrogen bonding interaction with Ile50 on the flap, while in all other marketed PI-PR complexes this interaction is mediated through a conserved water molecule. In addition, tipranavir makes seven direct H-bonds and fewer water-mediated H-bonds than other inhibitors (Muzammil et al., 2007; Turner et al., 1998).

23.3.4 Viral Resistance to HIV PIs

Similarly to other antiretrovirals, viral resistance to HIV PIs is one of the major roadblocks to their long-term efficacy. Despite expectations that the relatively small size of HIV PR and its dimeric nature would restrict possibilities for mutations, drug resistance mutations have been found in more than one-third of its 99 amino acids. The main factors contributing to rapid generation of resistance in HIV are: relatively low fidelity of HIV reverse transcriptase (one mutation per 10^4–10^5 bases), rapid viral turnover at steady state (approximately 10^9 new virions per day), and relatively high adaptability of biological function of HIV PR to mutations. Therefore, each single mutation is sampled thousands of times daily in each patient. The presence of drug provides a selective advantage for some of the preexisting mutants: They become more fit than wild-type virus and quickly overgrow the wild-type population. There are several ways in which the virus can manifest resistance to HIV PIs. Mutations in the target enzyme, HIV PR, are the most common and obvious possibility. HIV PR mutations can be broadly divided into two classes: active site and nonactive site. Active-site mutations are located in the inhibitor binding cleft and directly interfere with PI binding. Active-site mutations at V32, G48, I50, V82, and I84, and to a lesser extent D30, often result in a significant increase in K_i values (Gulnik et al., 2000; Louis et al., 2007; and references therein). Nonactive-site mutations do not directly contact bound inhibitor, but may interfere with its binding through a long-range protein perturbation. While a single nonactive-site mutation may exhibit a minimal effect on the K_i, accumulation of multiple nonactive-site mutations can significantly affect inhibitor binding. This has been demonstrated using recombinant HIV PR, in which a primary active-site mutation reverted to the WT amino acid (Muzammil et al., 2003; Olsen et al., 1999). To complicate the matter further, biochemical analysis revealed that the effect of additional mutations on resistance is not simply additive and may also be highly sequence-dependent (Clemente et al., 2004). These results underscore the danger of generalizations based on the analysis of single mutants on the WT background. Non-active-site mutations may also compensate for the loss of enzyme activity often associated with active-site mutations. On the enzyme level, such mutations often increase the k_{cat}/K_m value of HIV PR, leading to increases in the enzyme vitality (Gulnik et al., 1995) or mutant processing activity (Tang and Hartsuck, 1995). On the viral level, accumulation of nonactive-site mutations often leads to the restoration of viral fitness (Nijhuis et al., 1999), which may have been significantly reduced by some of the active-site mutations.

Some active-site (V82F, V82F/I84V, V82T/ I84V) as well as nonactive-site (L24I, L90M, F53) mutations have been shown to reduce the dimer stability (Liu et al., 2005, 2006; Xie et al., 1999). Since there is a linkage between inhibitor binding and induced dimer formation, a reduction in dimer stability can lead to a reduction in the apparent binding affinity of the inhibitor.

Mutations in the Gag cleavage site represent another mechanism of viral response to HIV PIs. Such mutations were first reported by Doyon et al. (1996) in the study of the investigational PI, BILA 2186. These cleavage site mutations appear to compensate for the loss of the catalytic efficiency of HIV PR and help to restore viral fitness diminished by mutations in HIV PR (Maguire et al., 2002; Myint et al., 2004; Robinson et al., 2000). These mutations are very common and found both during *in vitro* selection studies and in patients (Shafer and Schapiro, 2008). A recent report suggests that certain cleavage site mutations can result in resistance to a broad spectrum of PIs even in the absence of mutations in HIV PR (Nijhuis et al., 2007).

Many structures of HIV PIs with different drug resistance mutants are available in the literature and the Protein Data Bank (PDB) (Chen et al., 1995; Clemente et al., 2004; Hong et al., 2000; King et al., 2002; Klei et al., 2007; Kozisek et al., 2007; Liu et al., 2005; Mahalingam et al., 2004; Prabu-Jeyabalan et al., 2003) and this subject was recently reviewed by Louis et al. (2007). The structural effect of mutation on inhibitor binding can be explained for active-site mutations based on the specific interaction between the inhibitor and the active-site subsite. Generally speaking, mutations I84V, V82A, and I50V that introduce smaller amino acids result in the loss of van der Waals interactions with inhibitor substituents in contact with these residues. Mutations V32I and G48V introduce larger side chains and force inhibitor rearrangement to avoid unfavorable clashes. The effects of some of these active-site mutations have been successfully modeled. However, it is not uncommon that mutations can cause asymmetric rearrangement of the protein and inhibitor atoms in the two monomers of the enzyme (Baldwin et al., 1995; Tie et al., 2005). In many cases the structural effect of non-active-site mutations is more difficult to rationalize. For example, the M46I mutation that is common to many multi-PI-resistant HIV isolates is located in the flap and does not contribute directly to inhibitor binding. Flap movement is critical for catalysis and inhibitor binding. Structural data, molecular dynamics simulations, and NMR experiments suggest that the flap undergoes large conformational rearrangement upon ligand binding (Collins et al., 1995; Freedberg et al., 2002; Hornak et al., 2006; Ishima et al., 1999; Katoh et al., 2003). Flap mutations may therefore contribute to resistance by interfering with flap dynamics. In addition, some of the non-active-site mutations are located at the interface of domains of the protein that move as rigid bodies with respect to each other (Rose et al., 1998). It was proposed that such mutations may stabilize a more open conformation and contribute to resistance by increasing the off-rate of inhibitors.

It is important to emphasize that although resistance mutations are sampled every day in each patient, mutants are generally less fit than WT virus and may present a replication advantage only in the presence of drug. If the concentration of drug is high enough to fully suppress viral replication, then resistance selection is not possible. Therefore, the main driving force and cause of the emergence of resistant virus is suboptimal exposure to the drug. The selection of drug-resistant HIV has been performed *in vitro* in cell culture in an attempt to predict the emergence of clinical resistance. In these studies, virus was propagated for several passages at increasing drug concentrations, usually starting with approximately IC_{95} levels (Gong et al., 2000; Partaledis et al., 1995; Tisdale et al., 1995). Selection of resistant viruses for the first generation of PIs was achieved with relatively few passages and a two- to threefold increase of selection concentration at each passage, while newer PIs featuring a higher barrier to resistance required a more permissive strategy, with a lower starting selection concentration (IC_{50}) and smaller incremental increases (30–50%) of selection concentrations. Generally, *in vitro* selection studies have been helpful for predicting clinical resistance, although the relative frequencies with which specific mutants and mutant combinations are found in the clinical setting can vary greatly from those found *in vitro*, owing to much different selection pressure, viral dynamics, and virus population sizes.

For most PIs a single mutation is not sufficient to significantly reduce phenotypic susceptibility and the accumulation of multiple mutations may be required to generate a high level of resistance, indicating that PIs as a class are characterized by a higher barrier to resistance than RT or integrase inhibitors (Martinez-Cajas and Wainberg, 2007). The emergence of resistance in patients highly correlates with inadequate drug exposure, which in turn is highly influenced by pharmacokinetic properties of the drug and patient adherence, with severity of adverse effects affecting the latter. In order to overcome these problems, efforts have focused on the discovery of agents with better oral bioavailability and improved potency, resistance, and safety profiles. Newer PIs are characterized by a better pharmacologic profile, lower impact on lipid metabolism, and a higher genetic barrier to resistance. The specific path for viral escape for each HIV PI is usually unique to each individual patient and is determined by such factors as the viral load and the level of residual viral replication, viral polymorphism and background mutations, drug pressure, and so on. Therefore, treatment of different patients with the same PI may often result in the selection of a different constellation of drug-resistant mutations in HIV PR and Gag cleavage sites. Nevertheless, although the specific combination of mutations may be different, each PI usually selects for several characteristic major mutations. Major mutations are defined in general as those selected first in the presence of drug, or those shown at a biochemical or virologic level to result in significant inhibition of activity or replication (Johnson et al., 2008). Figure 23.6 and Table 23.1 illustrate the

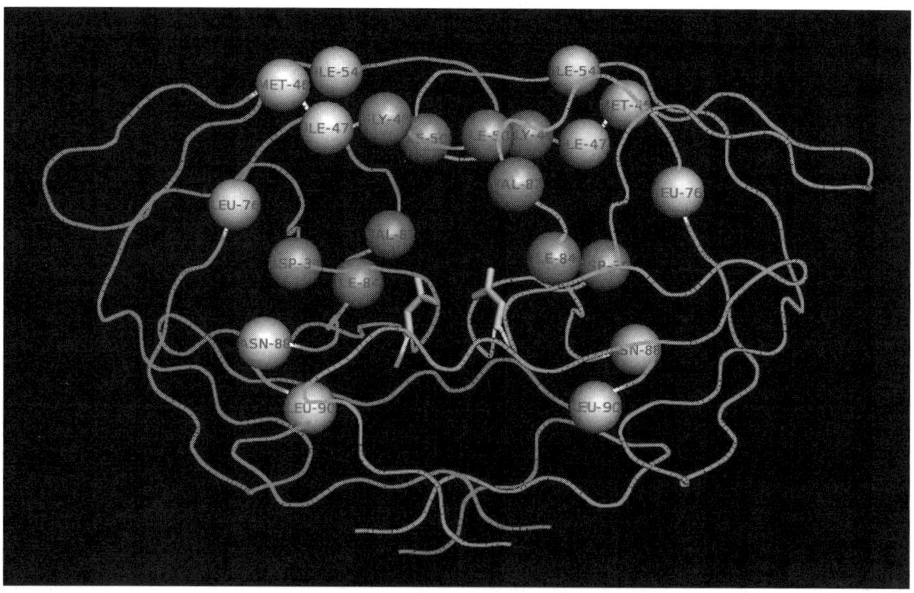

Figure 23.6. A ribbon diagram of HIV PR showing the position of major PI resistant mutations highlighted in bold in Table 23.1. The two HIV PR monomers are shown in green and cyan. The side chains of the active-site aspartates are shown in yellow. Active-site mutations are in blue. Note that most of the major PI-resistant mutations are clustered around the active site of the enzyme or on the flaps. See the insert for color representation of this figure.

TABLE 23.1. HIV-1 Protease Resistance-Associated Mutations (Johnson et al., 2008)

Wt PR aa	L	V	I	G	K	L	N	V	L	E	E	M	K	M	I	G	I	F	I	Q	D	I	L	I	H	A	G	T	L	V	V	N	**I**	I	**N**	L	L	I
	10	11	13	16	20	24	30	32	33	34	35	36	43	46	47	48	50	53	54	58	60	62	63	64	69	71	73	74	76	77	82	83	84	85	88	89	90	93
ATV	I F V C			E	M RI TV	I		I	I F V	Q		I L V		I L		V	**L**	L Y	LV MT A		E	V		L MV		V I T L	C S T A				A T F I		**V**	V	**S**		M	L M
DRV		I						I	F						V		**V**		**M L**								S		**V**				**V**			V		
LPV	F I R V				MR	I		**I**	F					I L	**V A**		V	L	VL A MT S				P			V T	S		V		**A F T S**		V				M	
TPV	V		V		MR				F		G	I	T	L	V				A MV	E					K			P			**L T**	D	**V**				M	
APV	F I R V							I						I L	V		**V**		LV M								S		V		A F T S		**V**				M	
SQV	I R V					I										**V**			VL			V				V T	S			I	A F T S		V				**M**	
NFV	F I						**N**					I		I L												V T				I	A F T S		V		D S		**M**	
IDV	I R V				M R	I		I				I		**I L**					V							V T	S A		V	I	A F T		V				M	

Top row indicates wild-type amino acids in the selected positions. Unique mutations associated with the PI are shaded boxes. Major mutations are in bold.

location of major protease mutations and their association with approved PIs. It is important to note that most of these mutations are associated with multiple PIs and I84V contributes to resistance to all PIs, thus reflecting the high degree of cross-resistance within the whole PI class (Johnson et al., 2008). The cross-resistance between different PIs can be rationalized based on the peptidomimetic character of most of these drugs and their overlapping interactions with the active site of the enzyme. However, some of the inhibitors select for unique mutations. For example, D30N occurs only in patients treated with NFV (Perrin and Mammano, 2003). I50L is associated solely with ATV treatment, while DRV is affected by L76V (De Meyer et al., 2008; Vora et al., 2006). In addition, mutations selected by one inhibitor may increase susceptibility to other PIs. These include the ATV-associated I50L mutation, which sensitizes HIV PR to other PIs (Gianotti et al., 2007); I50V and I54L, which increase tipranavir susceptibility (Elston et al., 2006); N88S, which sensitizes virus to amprenavir (Ziermann et al., 2000); and L76V, which increases susceptibility to atazanavir, saquinavir, and tipranavir (Braun et al., 2007; Vermeiren et al., 2007).

As mentioned earlier, some major active-site HIV PR mutations may impair viral replication potential and viral fitness, and secondary mutations often appear in HIV-infected patients to compensate for these defects. These mutations usually cause low resistance by themselves and are often present as polymorphisms in treatment-naive patients. Distinct amino acids at codons 10, 20, 36, 63, 71, 77, and 93 are recognized as secondary nonactive-site mutations. While these mutations do not cause PI resistance by themselves, when present with major mutations their contribution to resistance and compensation for the decrease of catalytic efficiency sometimes results in a synergistic effect. For example, the naturally occurring polymorphism L63P contributes to resistance by restoration of fitness decreased by major resistance mutations D30N and L90M (Condra et al., 1995; Mammano et al., 2000; Nijhuis et al., 1999; Rose et al., 1996). Compensatory mutations may also occur to overcome fitness loss by improving such Gag functions as polymerization of viral proteins and/or assembly (Doyon et al., 1996; Gatanaga et al., 2002; Maguire et al., 2002; Myint et al., 2004; Zhang et al., 1997).

The availability of several PIs represented a significant breakthrough in the HIV treatment paradigm by allowing sequential therapy. However, this approach may be limited by viral cross-resistance to the PIs. The degree of cross-resistance generally correlates with the level of resistance to the resistance-selecting primary PI and with the total number of mutations that emerge in the HIV PR. The gradual accumulation of resistance mutations and increase of cross-resistance in patients failing PI regimens is well-documented (Yusa and Harada, 2004) and justifies constant monitoring of the viral load and genotypic and/or phenotypic testing of the virus. An early switch from a failing regimen should minimize the potential for high level cross-resistance. A significantly higher frequency of substitutions at positions 10, 54, 71, 82, 84, and 90 is observed in broadly cross-resistant proteases (Hertogs et al., 2000;

Palmer et al., 1999; Shafer et al., 1998). Cross-resistance represents a significant problem in choosing a treatment regimen for salvage therapy. In order to assist clinicians in selecting treatment strategies, a variety of genotypic and phenotypic assays have been developed. Interpretation of genotypic results relies on rules-based algorithms and virtual phenotype where all mutations have a certain score based on their contribution to resistance. The International AIDS Society-USA (IAS-USA) maintains a list of significant resistance-associated mutations in the reverse transcriptase, protease, and envelope genes (see http://www.iasusa.org/resistance_mutations). A database cataloging existing mutations is also available (Stanford database, http://hivdb.stanford.edu/). The Stanford database contains more than 90,000 protease, reverse transcriptase and integrase sequences from more than 80,000 distinct viral isolates obtained from nearly 40,000 individuals. It provides an excellent publicly available online resource for analysis of the correlations between genotypic data and treatment for people from whom sequenced HIV-1 isolates have been obtained (genotype-treatment); correlations between genotype and *in vitro* drug susceptibility (genotype-phenotype); and also correlations between genotype and the clinical response to a new treatment regimen (genotype-outcome).

Phenotyping assays measure the ability of viruses to grow in the presence of different concentrations of antiretroviral drugs. Drug concentrations that inhibit 50% and 90% of viral replication are calculated, and the ratio of the IC_{50} of test and reference viruses is reported as the fold increase in IC_{50} (i.e., fold resistance). Interpretation of phenotyping assay results is based on clinically significant fold increase cutoffs that are now available for the majority of PIs (DHHS, 2008).

23.3.5 Pharmacokinetic Properties of HIV PIs and the Concept of PK Boosting

All HIV PIs are substrates and inhibitors of CYP450 and as such exhibit properties considered undesirable in current drug development, since they can lead to various drug–drug interactions (see Part II of this book). The potent inhibition of CYP450 by ritonavir (RTV) is well documented. It has been shown that RTV inhibits the metabolism of several CYP 3A4 substrates with a nanomolar IC_{50} (Ernest et al., 2005; Kempf et al., 1997; Kumar et al., 1996, 1999; von Moltke et al., 2000). RTV also inhibits CYP 2D6 and 2C9 isoforms to a much lesser extent (Kumar et al., 1996). The exact mechanism of CYP450 inhibition by RTV is not fully understood. Binding of RTV to CYP450 is characterized by a type II difference spectrum, suggesting direct interaction with the ferric heme iron (Kempf et al., 1997). While modification of either of the two thiazole groups maintains the type II spectral shift, the potency of CYP 3A4 inhibition is dramatically reduced, suggesting that the direct interaction with the iron is not the underlying source of the potency of this compound. Several studies have suggested that RTV is a mechanism-based inhibitor of CYP 3A4 based on the observation of reduced IC_{50} values upon preincubation of the CYP with RTV in the presence of NADPH (Ernest et al., 2005; Koudriakova et al., 1998; Kumar et al., 1999; von Moltke et al., 2000). The quasi-irreversible inactivation of CYP 3A4 and 3A5 through the formation of a metabolic intermediate complex (MIC) has been proposed (Ernest et al., 2005), but the nature of the metabolic species involved in MIC formation remains elusive.

The generally hydrophobic nature of the HIV PR active site dictates the hydrophobic character of its inhibitors. This hydrophobicity contributes to the binding affinity of HIV PIs with CYP3A4, and to a much lesser extent of CYP2D6, CYP2C9, and CYP2C19, causing them to be excellent substrates and contributing to their suboptimal pharmacokinetic properties. The short elimination half-life of PIs leads to complicated dosing regimens requiring TID or BID administration in order to keep plasma concentrations above IC_{50} values. It may also contribute to the variability of PI plasma concentrations of HIV infected individuals seen in controlled pharmacokinetic studies (Acosta et al., 2000; Kaletra, prescribing information; Reyataz, prescribing information).

Potent inhibition of the metabolism of several HIV PIs by RTV in liver microsomal preparations as well as dramatic enhancement of the bioavailability of different PIs by RTV in animal studies (Kempf et al., 1997) suggested the possibility of its utilization as a PK booster for other HIV drugs. The ability of RTV to inhibit some efflux transporters, particularly P-glycoprotein (MDR-1) and MRP-1 (Drewe et al., 1999; Olson et al., 2002; Storch et al., 2007), may also contribute to its boosting properties. The clinical enhancement of saquinavir levels by ritonavir was observed in the mid-1990s (Mascolini, 1996). Boosted PI regimens are currently well-accepted and common in clinical practice. They also "boosted" sales of RTV, originally approved by the FDA as an antiretroviral drug in 1996 and not used much anymore as a PI. In modern practice, RTV is predominantly used only as a PK booster at doses that are subtherapeutic for inhibition of viral replication. Moreover, the use of unboosted PIs is gradually declining (or being completely eliminated in the case of PI experienced patients) and the most recent PIs, such as darunavir and tipranavir, were approved only as RTV-boosted regimens. Pharmacokinetic properties of another potent HIV PI, lopinavir, did not support its development as a stand-alone drug. It was developed by Abbott Laboratories coformulated with low-dose RTV (Kaletra, prescribing information). The PK enhancement of HIV PIs by RTV has been the subject of multiple excellent reviews (Becker, 2003; Gallant, 2004; Motwani and Khayr, 2006; Moyle and Back, 2001; Scott, 2005; Walmsley, 2007; Youle, 2007; Zeldin and Petruschke, 2004).

The main pharmacokinetic parameters of nonboosted PIs and the effects of RTV boosting are summarized in Table 23.2. The usual metrics of drug exposure that relate to pharmacologic response are AUC, C_{max}, $t_{1/2}$ and C_{min} (or

TABLE 23.2. Pharmacokinetic Properties of HIV PIs

HIV Protease Inhibitor	SQV HG (Invirase)	SQV SG (Fortovase)	NFV (Viracept)	IDV (Crixivan)	RTV (Norvir)	APV (Agenerase)	FPV (Lexiva)	LPV/RTV (Kaletra)	ATV (Reyataz)	TPV (Aptivus)	DRV (Prezista)
Administered dose	600 mg TID	1200 mg TID	1250 BID	800 mg Q8h	600 mg Q12h	1200 mg BID	1400 mg BID		400 mg QD		
Administered dose with RTV	1000 mg + 100 mg RTV BID	1000 mg + 100 mg RTV BID		800 mg + 100mg RTV BID		1200 mg + 200 mg RTV QD	1400 mg + 200 mg RTV QD	400 mg LPV + 100 mg RTV BID	300 mg + 100 mg RTV QD	500 mg + 200 mg RTV BID	600 mg DRV + 100 mg RTV BID
PK Parameters											
Bioavailability, %	4	NR (331% of HG)	20–80	60–65	NR	NR	NR	NR	60–68	NR	37 (82)
AUC_{0-24}, μg h/ml (w RTV)	2.5 (29.2)	21.7 (38.2)	52.8	18.8 (50.8)	NR	17.7 (28.7)[b]	33 (66)	(92.6)[b]	22.3	427[b]	(61.7)[b]
C_{max}, μg/ml (w RTV)	NR	NR	4	7.73 (12.4)	11.2	6.77 (NC)	4.82 (7.23)	(9.8)	3.55 (5.23)	46.9	(6)
T_{max}, h (w RTV)	NR	NR	NR	0.8	NR	1	1.3	(4)	2 (3)	3	(2.5–4)
$t_{1/2}$, h (w RTV)	1.5–2	NR	3.5–5	1.8	3–5	7.1–10.6	7.7	(5–6)	6.5 (8.6)	6	(15)
C_{min}, μg/ml (w RTV)	0.08 (0.37)	0.22 (0.43)	0.7–2.2[a]	0.154 (1.69)	3.7	0.32 (1.02)	0.35 (1.4)	(5.5)	0.27 (0.86)	21.4	(3.54)

NR, not reported; HG, hard gel capsules; SG, soft gel capsules; NC, No change.
[a]NFV C_{min} is higher in the morning.
[b]AUC_{0-12}.

TABLE 23.3. Inhibition of CYP3A4 by HIV PIs

Inhibitor	CYP 3A4 K_i in Microsomes (μM)	CYP 3A4 K_i in Bactosomes (μM)
RTV	0.017	0.03
APV	0.5	0.11
IDV	0.17–0.9	0.24
NFV	0.31–4.8	0.3
SQV	0.7–4.0	0.76

C_{trough}). For HIV PIs, C_{min} is considered to be the best predictor for virologic response. Plasma C_{min} levels that are above the IC_{90} are usually a good indicator of effective and prolonged viral suppression (DHHS, 2008; Haas et al., 2000; Hoetelmans et al., 1998; Moyle, 2001). "Inhibitory Quotient" (IQ), defined as the ratio of C_{min} to IC_{50}, has proven to be useful in predicting virologic response (Barrios et al., 2004; Gonzalez de Requena et al., 2004; Hoefnagel et al., 2005; Hsu et al., 2003). Different approaches have been taken to calculating IQ values, with the numerator of the formula being C_{min} in the majority of the cases. There is ongoing debate about the denominator: IC_{50} for virus isolated from the patient; protein binding adjusted IC_{50} for WT virus; or virtual phenotype value (virtual IQ).

C_{min} of PIs (as well as other PK parameters) is highly variable between patients, in part reflecting variability in CYP 3A4 levels. RTV increases the C_{min} for all boosted PIs (see Table 23.2), which directly translates into an increased IQ. This is undoubtedly the major benefit of boosting. In most cases, RTV boosting also leads to an increase in the total exposure (AUC) and C_{max}. These effects of RTV boosting can result in decreased daily pill burden and/or frequency of dosing and more convenient dosing by reducing food effects. Simplification of antiviral regimens through RTV boosting is likely to improve patients' adherence to therapy, a critical factor for the successful treatment of HIV where adherence rates of more than 95% are required (Paterson et al., 2000). The beneficial effect of RTV boosting on improved virologic suppression and the decreased development of drug resistance was confirmed in cohort studies and large population-based settings (Lima et al., 2008; Wood et al., 2007). It should be mentioned that other potent 3A4 inhibitors such as ketoconazole that have been examined clinically do not have the same boosting capability as ritonavir (Autar et al., 2007; Sekar et al., 2008). It is thus possible that some other aspect of ritonavir's pharmacological profile makes it particularly effective as a booster. For example, tissue distribution could have a major impact, since the principal metabolizing CYPs are mainly in the gut and the liver.

The benefits associated with RTV boosting come at a cost. There is the potential for an increase in the toxic side effects of "boostees" due to an increase in their exposure: higher incidence of lipid abnormalities and cardiovascular risk and increased risk of drug–drug interactions due to direct CYP and transporter inhibition as well as induction of different metabolic enzymes

(Becker, 2003; Foisy et al., 2008). Ritonavir currently needs to be refrigerated (although Abbott has a new formulation that should solve this problem) and has a spectrum of associated side effects. Despite these issues, the benefits of boosted PI regimens in the management of HIV infection are generally considered to outweigh these costs.

The benefits of PI boosting and the commercial success of RTV have prompted research on new CYP inhibitors as potential boosters, preferably without ritonavir's side effects. Another important consideration in the design efforts was elimination of the PI activity associated with RTV due to concerns that subtherapeutic doses of RTV may select PI class-resistant HIV. Sequoia Pharmaceuticals has described benzofuran-derived P450 inhibitors as PK boosters (Eissenstat and Duan, 2008). They have recently reported that the initial development candidate from this class, SPI-452, has entered Phase 1 clinical studies, and they have suggested the possibility of combining it with their new clinical PI, SPI-256. Recently, Abbott and Gilead have disclosed in patent applications elaboration of dithiazoles related to ritonavir as P450 inhibitors. These inhibitors generally retain the thiazoles at either end of the molecule, but lack the hydroxyl in the core required for protease inhibitory activity (Desai et al., 2008; Flentge et al., 2008; Klein et al., 2008). Gilead has disclosed that one of their compounds, GS-9350, has entered phase 1 clinical trials. In addition, Abbott described amine cores with thiazole-substituted urethanes at both termini (Flentge et al., 2008). While somewhat less potent CYP inhibitors than ritonavir *in vitro*, some of these compounds were excellent boosters of lopinavir in dogs. Pfizer has reported series of sulfonamido pyridines and substituted pyrazoles as PK boosters (Planken et al., 2008a,b). Tibotec has described the clinical boosting properties of a benzoxazole-containing protease inhibitor (Van't Klooster et al., 2006).

23.3.6 PIs' Off-Target Activity

High potency of PIs toward HIV PR translates into high antiviral efficacy. However, interactions of PIs with other (human) cellular targets have been implicated in various side effects, including serious metabolic perturbations and drug–drug interactions.

PIs are predominantly metabolized by CYP3A4, but are also substrates and competitive inhibitors of cell transporters. Each of these properties is known to cause drug–drug interactions in multidrug regimens. Their CYP3A4 inhibitory activity ranges from the micromolar range for most PIs to the low nanomolar inhibition by RTV (Table 23.3) (Granfors et al., 2006). RTV also has some inhibitory activity for other CYP isoforms *in vitro* and is considered an inhibitor of 2D6 *in vivo* (Granfors et al., 2006; Hsu et al., 1998). The potent inhibition of CYP3A4 by RTV is used for pharmacologic enhancement of other PIs (Section 23.3.5). The extent of competitive inhibition of transporters such as Pgp, MRP2, and BCRP varies from relatively weak (low micromolar range) by SQV, NFV, and IDV to potent inhibition by RTV (IC_{50} for Pgp

20–50 nM) (Ford et al., 2003; Gutmann et al., 1999; Janneh et al., 2007; Perloff et al., 2000). Although many clinically significant drug–drug interactions of PIs can be explained solely by their inhibitory activity toward CYPs, sometimes drug interactions, or the lack thereof, do not fit this mechanism. For example, acute administration of RTV significantly increases exposure of CYP3A4 substrates such as alprazolam or midazolam, but its chronic administration has no effect (Culm-Merdek et al., 2006; Greenblatt et al., 1999). Similarly, chronic administration of nelfinavir increases the oral clearance of ethynyl estradiol and zidovudine (Viracept package insert). These data suggest that PIs are inducers as well as inhibitors of metabolic enzymes. CYPs (3A4, 2B6, 2C8, 2C9, and 2C19) as well as other phase I and phase II metabolizing enzymes and drug efflux pumps are regulated by the nuclear receptor PXR. PXR gets activated by direct ligand binding, and it has a promiscuous binding site resembling the substrate cavity of CYP enzymes that can accommodate a variety of different moieties (Orans et al., 2005). Not surprisingly, CYP substrates are very often good ligands of PXR, which results in CYP induction and more efficient clearance of xenobiotics from the body. It has been shown that PIs can induce the expression of CYPs as measured by mRNA and protein levels *in vitro* (Dixit et al., 2007). The net *in vivo* effect of chronic PI administration is thus a combination of CYP inhibition and induction, with the outcome in most cases net inhibition of CYP 3A4, although this may vary between PIs and for other CYPs (Culm-Merdek et al., 2006; Fellay et al., 2005; Fichtenbaum and Gerber, 2002; Vourvahis and Kashuba, 2007; Yeh et al., 2006).

Significant side effects observed upon treatment with PI-based HAART include: fat tissue redistribution with loss of subcutaneous adipose tissue (lipoatrophy), often in combination with fat accumulation in abdominal and dorsocervical regions; hyperlipidemia involving both cholesterol and triglycerides; and insulin resistance. These abnormalities are characterized as HIV-associated lipodystrophy syndrome (HIV-LD). Considerable efforts have been devoted to address the PI-associated HIV-LD syndrome in experimental systems such as animal and cell culture models in hopes that identification of cellular and molecular mechanisms will lead to better understanding of the clinical syndrome. Various investigators have shown that PIs interfere with adipocyte differentiation and adipocyte-specific gene expression. One of the most studied effects of PIs is on expression of sterol regulatory element binding proteins (SREBP), although results are quite complex and controversial (Mallon, 2007). This endoplasmic reticulum (ER) resident membrane-bound protein undergoes activation through specific proteolytic cleavages, release from the membrane, and translocation to the nucleus. Mature activated SREBP up-regulates genes involved in fatty acid and cholesterol biosynthesis (Osborne, 2000). Expression of one of the SREBP isoforms, SREBP-1, is significantly lower in the adipose tissue of patients with HIV-LD (Bastard et al., 2002; Kannisto et al., 2003). Consistent with this observation, exposure of adipocytes to IDV or NFV decreased levels of active SREBP-1 and concomitantly decreased rates of adipocyte differentiation *in vitro* (Caron et al., 2001; Dowell et al., 2000; Miserez et al., 2002). In contrast, RTV increased the level of the mature active form of SREBP-1 both *in vitro* and in mice (Nguyen et al., 2000; Riddle et al., 2001). It has been shown that long-term treatment of mice with LPV/RTV (but not ATV) results in the development of a metabolic syndrome resembling HIV-LD and a 5.5-fold increase in SREBP-1c gene expression in the inguinal depot (Prot et al., 2006). These apparently contradictory results underscore the complexity of the phenomenon. There are several other possible explanations for how PIs may cause lipid abnormalities. HIV PIs have been implicated in the induction of ER stress and the subsequent activation of unfolded protein response (Carr, 2000; Zhou et al., 2005, 2006). In addition, a decrease in expression of peroxisome proliferator-activated receptors (PPAR) was observed upon exposure to PIs (Caron et al., 2001; Lenhard et al., 2000; Mallon, 2007). PPARs are transcription factors that are important for normal function of both differentiating and differentiated adipocytes. Interestingly, certain proteins involved in the modification or activation pathways of SREBP (human site-1 protease, S1P) or PPAR (cytoplasmic retinoid acid binding protein 1, CRABP-1) have sequence homology to the catalytic region of HIV-1 protease (Barbaro, 2006). Any of these mechanisms could contribute to HIV-induced metabolic syndromes.

Another adverse effect associated with PI treatment is insulin resistance, frequently found early in the treatment of HIV-1 infected patients. Several studies have now shown that HIV PIs exert a direct effect on glucose uptake, specifically inhibiting the GLUT4 isoform of glucose transporter, with minimal inhibition of GLUT2 and no inhibition of GLUT1 (Murata et al., 2002; Noor, 2007; Schutt et al., 2000). Therapeutic levels (1–5 μM) of APV, LPV/RTV, and RTV acutely inhibited glucose uptake *in vitro* in primary rat adipocytes and in rats *in vivo*, whereas ATV displayed no inhibition compared to controls (Yan and Hruz, 2005). Recent studies have begun to elucidate the relationship between PI structure and GLUT4-mediated glucose transport. Hertel et al. (2004) screened a panel of short peptides for their ability to inhibit glucose uptake in primary rat adipocytes. Peptides that inhibited GLUT4 contained a highly aromatic core flanked by hydrophobic moieties similar to the phenylalanine derived core structure of most PIs. Thus, inhibition of GLUT4 may represent a direct molecular target of PIs and may provide an off-target design opportunity for future PIs; that is, compounds could be designed to minimize effects on this metabolic pathway. However, the involvement of HIV PIs in insulin resistance may not be solely attributed to inhibition of GLUT4. At least two other possible mechanisms have been recently proposed: inhibition of resistin degradation and inhibition of Akt signaling. Resistin is a hormone that was found to be produced and released from adipose tissue to serve endocrine functions likely involved in insulin resistance. It was called "resistin" because of the observed insulin resistance when it was injected in mice (Degawa-Yamauchi et al., 2003; Gabriely et al., 2002; McTernan et al., 2002; Steppan et al., 2001). Although this field is controversial, there are studies showing that elevated resistin levels correlate with obesity and insulin resistance (Hirosumi et al., 2002; Rajala et al., 2004; Silha et al., 2003). An interesting putative role for HIV PIs in regulation of resistin levels emerged from their inhibition of several cellular aspartic proteases. It has been documented that PIs are somewhat inhibitory of cellular proteases, such as cathepsins D and E. The K_i ratios for HIV PR and "unwanted targets" vary significantly between PIs. RTV shows the most significant inhibition of cathepsins D and E (Kempf et al., 1995). Both cathepsins are able to cleave resistin *in vitro* and have been implicated in the regulation of its degradation. HIV PIs have been evaluated for their ability to inhibit resistin cleavage. RTV or SQV, but not NFV, IDV, or ATV prevented cleavage (Geese and Ranade, 2008), which generally correlates with the severity of side effects caused by these PIs. Another possible mechanism of insulin resistance may be through inhibition of Akt signaling by PIs (Ben-Romano et al., 2003, 2004). The role of Akt in mediating insulin effects has been demonstrated in Akt2 knockout mice, which are insulin-resistant and have higher fasting and post-prandial glucose levels than do wild-type mice (Cho et al., 2001). Prolonged exposure of adipocytes and other cell types to nelfinavir resulted in impaired Akt phosphorylation and insulin-stimulated glucose uptake. An interesting outcome of these studies was the possible repositioning of PIs as anticancer therapies, since activation of the Akt pathway promotes cell proliferation and tumor formation (Altomare and Testa, 2005). It has been shown that some PIs can inhibit proliferation and cause cell death in more than 60 tested cell lines, and cytotoxic effects of chemotherapeutic agents such as docetaxel could be enhanced when combined with PIs (Gupta et al., 2007; Ikezoe et al., 2004; Yang et al., 2006). The potency of PIs in cellular anticancer assays is usually $\geq 10\,\mu M$, with NFV being the most potent. NFV is currently in clinical trials evaluating its ability to reduce solid tumors and also as a radiation/chemotherapy sensitizer (http://www.Clinicaltrial.gov). The Akt signaling pathway that originally linked PIs to cancer may not be the only mechanism of inhibition of malignant cell lines. The PI effects do not always correlate with Akt inhibition; and other off-target effects of PIs such as proteasome inhibition, ER stress, or induction of unfolded protein response may contribute to putative anticancer properties.

23.3.7 Approaches to the Design of HIV PIs with Improved Resistance Profiles

As was indicated in Section 23.3.1, the evolution of PIs revolved around improving potency, PK properties, and resistance profile. While even the first marketed PI, saquinavir, is quite potent as an inhibitor of the protease and as an antiviral agent, its poor PK properties made it a relatively ineffective drug. The quandary that medicinal chemists have been trying to deal with over the years is how to maintain high potency against the target in a molecule that has a good PK profile when the number of hydrogen bond interactions and lipophilic interactions required for this potency is high, leading to high molecular weights and generally breaking many of the "rule of 5" and other guidelines for good PK properties. In a similar vein, the more interactions that are required for tight binding, the more chance the target enzyme can mutate to make the drug less effective, leading to resistance.

Several approaches have been employed in recent years to design PIs with better resistance profiles. One classical approach can be termed structural diversity. That is, it stands to reason that the more dissimilar the new inhibitor is structurally from earlier inhibitors, the more likely it is to have a different resistance profile. In fact, the interactions that the inhibitor makes with the target need to be different, but that is likely to be the case. Noteworthy in this respect are the nonpeptide based inhibitors. While many such compounds were explored, the ones that have received the most attention are cyclic ureas, pyrone-based inhibitors, and lysine-derived inhibitors.

The cyclic ureas were first described in 1994 (Lam et al., 1994). These compounds were derived from a structure-based design program where the intent was to replace the conserved bound "flap" water in the enzyme that makes H-bonds to Ile50 and Ile50′ by a carbonyl in the inhibitor. Based on modeling of the earlier ritonavir type acyclic analogues DuPont–Merck scientists generated a variety of five-, six, and seven-membered ring templates with this carbonyl as well as hydroxy groups placed so as to interact with the catalytic Asps. The most potent of these analogues were seven-membered ring cyclic ureas decorated with either substituted or unsubstituted benzyls at each nitrogen and with one of the core hydroxyls removed. Several such analogues including DMP-323 (Fig. 23.5) were selective for HIV PR over other aspartyl proteases (Erickson-Viitanen et al., 1994). DMP-323 showed reasonable PK properties in animals and underwent clinical evaluation in humans, but was dropped due to poor bioavailability. This compound was significantly less potent against the I84V mutant protease (Nillroth et al., 1997). A second analogue, DMP450 (Fig. 23.5), which incorporated basic amine substituents on the two aryl groups, had greatly improved solubility and improved bioavailability in humans (Lam et al., 1996), but still had a suboptimal resistance profile (Hodge et al., 1996). Abbott researchers described a structurally related hydrazo analogue (Fig. 23.5) that was also potent (Sham et al., 1996), and subsequently various groups reported other types of analogues. However, it does not appear that any of these compounds made it to the clinic.

Pyrone-based nonpeptide inhibitors were discussed above in Section 23.3.1 in the context of the discovery of tipranavir.

A more recent discovery is the identification of lysine-based PIs. Lysine derivatives were identified as hits through screening of commercially available protected amino acids in an enzymatic assay with recombinant HIV PR. Subsequent rounds of chemical modification led to the discovery of a series of potent lysine sulfonamide PIs (Sevigny et al., 2006; Stranix et al., 2003, 2004, 2006). This series was licensed by Ambrillia (formerly Procyon) to Merck in 2006 and the lead compound, PPL-100, entered clinical trials as MK-8122, but was placed on developmental hold in 2008. This compound is a phosphate prodrug of PL-100 (Fig. 23.5), which has an antiviral IC_{50} of 16nM and subnanomolar potency in the enzyme assay (Stranix et al., 2006). PL-100 exhibited a favorable profile compared to reference PIs in the PhenoSense assay (Monogram Biosciences) against a panel of 63 MDR isolates (Dandache et al., 2007). *In vitro* selection studies resulted in generation of novel active site mutations, T80I and P81S, that do not cause cross-resistance with other PIs, along with K45R and M46I (Wu et al., 2006). The crystal structure of a close analogue of PL-100 in complex with WT HIV PR (Nalam et al., 2007) shows that the inhibitor makes direct hydrogen bonds with the flap, similar to substrates, cyclic ureas, and tipranavir and unlike most other PIs where this interaction is water-mediated. It forms hydrogen bonds mostly with main-chain or conserved atoms of the enzyme, which contributes to its favorable resistance profile.

Another approach to overcoming resistance is to design compounds that generate most of their binding energy with HIV PR by forming interactions with residues that cannot mutate. This principle is pretty much self-evident from the definition of resistance. Maximizing interactions with the protein main chain (hydrogen bonds) is one such approach. Freire describes this in terms of optimizing enthalpic binding (Ohtaka et al., 2004). Additional interactions are made with invariant residues or the invariant portion of a residue that mutates within a limited chemical space. For example, if a residue mutates from valine to alanine, a compound that made a hydrophobic interaction with the alanine might be able to make the same interaction with the alanine portion (CH) of the valine. Of course it would need to not be so bulky as to bump into the other carbons of the valine. While the details of the approaches researchers use to reach this endpoint may vary, they generally fit with these basic design principles. What is by no means so obvious is how to generate sufficient potency to have good antiviral properties while making only interactions with conserved substructures of the enzyme. From the discussion above, it becomes intuitively apparent that it is much easier to design a weak inhibitor with a good resistance profile than a potent compound with a good resistance profile. Additionally, a large inhibitor that makes enough of these invariant interactions to be potent may expand outside of the chemical space required for good PK properties.

Some of the approaches to tackle resistance that fit into these principles are described below. In some cases the favorable resistance profiles resulted from analyzing the SAR of many analogues. By correlating favorable resistance profiles with structural (X-ray or NMR) interactions and/or with thermodynamics of binding, hypotheses for the interactions responsible for the favorable resistance profiles could be generated.

One class of compounds that fits within the above parameters is the bis-THF-containing inhibitors originally described by Ghosh in 1996 and followed up on by many groups since then (see discussion on Darunavir in Section 23.3.1). Collaboration of Vertex and GSK on amprenavir analogues that incorporated the bis-THF resulted in the discovery of brecanavir (Fig.

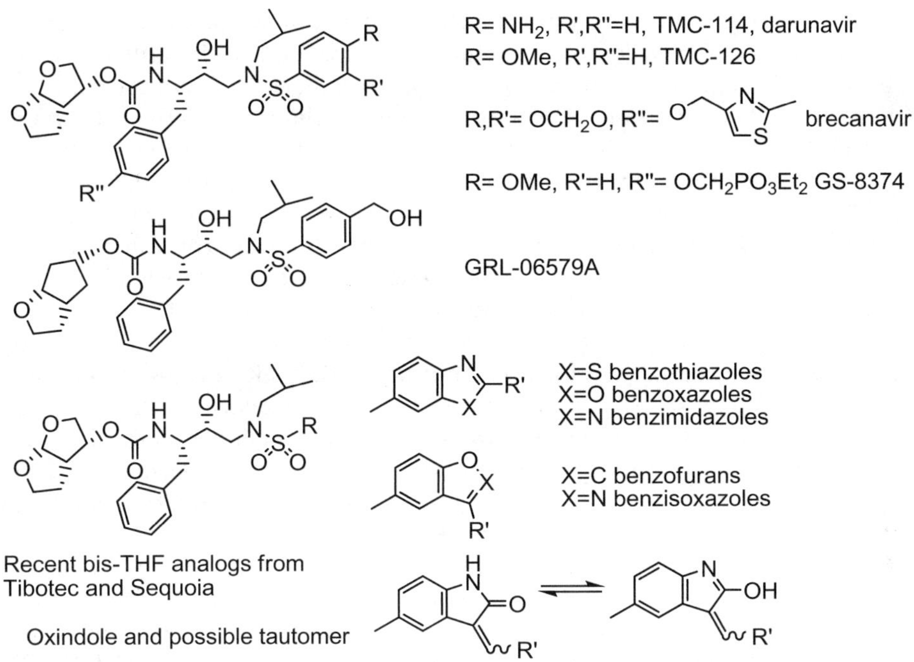

Figure 23.7. Chemical structures of selected bis-THF-containing HIV PIs and related analogues.

23.7), in which the methylenedioxy analog of darunavir was elaborated by a thiazolemethoxy substituent on the P1 benzyl (Ford et al., 2006; Hanlon et al., 2004; Hazen et al., 2007; Miller et al., 2006; Reddy et al., 2007b; Spaltenstein et al., 2005). This compound was very potent and had an excellent resistance profile. The thiazole was chosen because, besides its excellent potency and resistance profile, it was among the better analogues with respect to bioavailability; and additionally, blood levels in rats and dogs were dramatically increased in the presence of ritonavir. In clinical trials, blood levels were low in the absence of ritonavir, but were significantly boosted in its presence. The development of brecanavir was halted at the end of 2006 due to formulation problems.

Tibotec described replacement of the P2′ aryl group of darunavir by various bicyclic heteroaryls (Fig. 23.7). In these compounds the amino of TMC-114 or the ether oxygen of TMC-126 has been incorporated into a heteroaromatic ring to provide benzoxazoles, benzothiazoles, benzisoxazoles, benzimidazoles, and oxindoles (Surleraux et al., 2002a,b, 2003a,b, 2005a; Tahri and Wigerinck, 2004). X-ray crystal structures of 2-aminobenzoxazole and 2-aminobenzothiazole analogues showed that the azole N made an H bond with the NH of Asp30′ and the 2-amino group was interacting with the Asp30′ side-chain carboxyl. These interactions are similar to those made by the amino group of darunavir (Tie et al., 2004).

By analyzing multiple crystal structures of WT and mutant HIV PRs in complex with different inhibitors Sequoia Pharmaceuticals identified a series of conserved and critical polar interactions between the inhibitors and the enzyme. These interactions are directed to main-chain and active-site atoms of HIV PR, and they define a conserved atomic substructure that cannot be altered by mutations. This resulted in the identification of potent P2′ oxindole, indole, benzofuran, and benzisoxazole derivatives with excellent resistance profiles (Fig. 23.7). From among these, a substituted benzofuran analogue, SPI-256, was moved to clinical trials (Eissenstat et al., 2005; Eissenstat and Guerassina, 2005; Erickson et al., 2003, 2008; Wynne et al., 2008).

Recent reports from the Ghosh laboratory suggest that the bis-THF can be replaced by a hexahydrocyclopenta[*b*]furanyl group with no loss in potency. In this series a *p*-hydroxymethyl substituent on the benzenesulfonyl group (GRL-06579A) was more potent than the amino analogue (Ghosh et al., 2006b). The resistance profile was generally comparable to darunavir. X-ray crystal structure analysis showed H-bonding of the tetrahydrofuran O to Asp29NH and H-bonding of the hydroxymethyl group to Asp30′NH.

Schiffer and co-workers have proposed and published extensively on the substrate envelope hypothesis [reviewed recently in Nalam and Schiffer (2008)]. The structural analysis of peptide substrates corresponding to the cleavage site in Gag and Gag-Pol in complex with inactive D25N HIV PR revealed that peptides with different sequences adopt a similar shape upon binding (Prabu-Jeyabalan et al., 2002). The consensus volume occupied by the substrates has been termed the substrate envelope (King et al., 2004a). It has been proposed that inhibitors that bind HIV PR within the substrate envelope should exhibit a favorable resistance profile. The rationale is that the reduction in affinity of such inhibitors by resistant mutations will lead to a reduction of affinity for substrate and reduced viral viability. It has been noted that the primary HIV PR mutations do not extensively contact substrates, as opposed to inhibitors (King et al., 2004a), providing a structural rationale for the hypothesis. The analysis of currently approved PIs and newly designed compounds revealed a negative correlation between favorable resistance profile and protrusions outside the substrate envelope (Chellappan et al., 2007; King et al., 2004a,b; Prabu-Jeyabalan et al., 2006; Reddy et al., 2007a). Most recently a library of PIs based on the amprenavir hydroxyethylaminosulfonamide scaffold was designed using computational approaches, where the ability to conform to the substrate envelope was incorporated into the scoring function (Altman et al., 2008; Chellappan et al., 2007). Although the binding constants for the two best inhibitors were two to four orders of magnitude worse than clinical PIs, they were less affected by mutations, as judged by smaller mutant to WT K_i ratios, supporting the hypothesis. Potency could be improved (Altman et al., 2008), but generally only with a sacrifice in resistance profile.

Flexibility is a concept that is also often discussed in the context of resistance. Inhibitors that can adjust their interactions to changes in the target protein structure (mutants) will have better resistance properties. Freire described it in terms of the peptide substrate having a higher flexibility than the synthetic inhibitors and thus being more able to adapt to backbone rearrangements or subtle conformational changes induced by mutations in the protease (Luque et al., 1998). Conceptually, there are a variety of ways that inhibitor "interaction flexibility" might be achieved. One is by using inhibitors with large conformational flexibility that can change their conformation to optimize interactions with the mutated target. Unfortunately, the larger the number of rotatable bonds in a molecule, the less likely it is to have good PK properties (Veber et al., 2002). In addition, such compounds pay an entropic penalty on binding, resulting in lower potency, since the binding conformation is one of many possible available conformations. If conformational flexibility is relatively low, and the conformations required for tight binding to mutants are among the most stable, then the entropic penalty should be low. This effect may contribute to the excellent properties mentioned above for the bis-THF, which has only a small number of possible orientations of the bicyclic ring (Nivesanond et al., 2008). In recent studies aimed at the design of PIs based on the substrate envelope hypothesis, it was noted that not all envelope-respecting inhibitors exhibit favorable resistance properties, while some inhibitors that exceeded the envelope do, suggesting that other factors such as inhibitor flexibility may play an important role.

Another way that flexibility in interactions can be achieved is by having redundant or "adjustable" interacting groups. Removal (or weakening) of one interaction could be compensated for by introducing or increasing the strength of another, resulting in minimal net change in binding potency. A different approach would be to have a group in the inhibitor that can interact in multiple ways with the protein residue and thus adapt to mutation. For example, a lipophilic moiety incorporating a hydrogen-bonding group might be able to adjust its binding to an Ala to Ser mutant. An example of a group with both hydrophobic and hydrogen bonding properties is the β-methanesulfonyl-l-valine moiety used by LG Chemicals as a P2 substituent (Park et al., 1996). Another way that interaction flexibility might be accomplished is by an adjustable hydrogen bond network in the inhibitor. For example, if the inhibitor has stable tautomeric states such as keto–enol or imine–enamine that replace a hydrogen bond acceptor by a hydrogen bond donor (or vice versa), it could potentially accommodate a shift in the protein structure interaction from an amide NH of the backbone (or side chain) to the carbonyl. Such a tautomeric equilibrium could contribute to the excellent resistance profiles of some PIs with oxindole substituents at P2′ (Fig. 23.7).

In an attempt to make a PI prodrug with preferential intracellular retention, Gilead scientists synthesized several analogues of UIC-94003 (TMC-126) with phosphonic acid derivatives attached to the para-position of the P1 benzyl group (Cihlar et al., 2006). The original clinical candidate GS-9005 entered clinical trials in 2004, but was dropped due to low oral bioavailability and poor antiviral effect. A second phosphonate-containing analogue, GS-8374, has been the subject of recent presentations and publications. Although the phosphonate moiety was designed to be fully solvent-exposed, with no interaction with enzyme (confirmed by crystallographic study), it unexpectedly improved the inhibitor resistance profile. It was proposed that the solvation of the exposed phosphonate provides an anchor point for the inhibitor and that this "solvent anchoring" increases its degrees of freedom in the active site and helps the inhibitor adapt to the changes of the active site caused by mutations. Isothermal titration calorimetry measurements of inhibitor interactions revealed greater entropic compensation for the loss of enthalpy upon binding to mutant HIV PRs for the phosphonate-containing inhibitors than for UIC-94003, resulting in improved K_d ratios. GS-8374, when assayed using the PhenoSense Assay (Monogram Biosciences) against a panel of 24 resistant mutant viruses, exhibited a mean fold change of 6.2, comparable to tipranavir, but it is 90-fold more potent against WT. In comparison, darunavir and brecanavir had mean fold changes of 30 and 23 against the same panel (Callebaut et al., 2007).

The potency of PIs is usually described by their K_i, which is equivalent to the equilibrium dissociation constant. The latter is a ratio of off-rate (k_{off}) to on-rate (k_{on}). For PIs and for most enzyme inhibitors, the off-rate is the dominating factor in the equation; that is, the improvement of binding of an inhibitor (decrease in K_i) generally results from a decrease in off-rate. Similarly, the increase in off-rate for a given inhibitor is the main contributor to its decreased binding affinity toward drug-resistant mutants. The importance of the k_{off} and drug-target residence time for *in vivo* systems has recently been reviewed (Copeland et al., 2006; Tummino and Copeland, 2008). Some very potent PIs are such tight binders that their off-rates are very slow; numbers in the range of 100 h have been reported (Dierynck et al., 2007). On the other hand, the estimated half-life of viral maturation and virion release in cell culture is only several hours (Kaplan et al., 1994). Therefore a significant increase in off-rate due to mutations might be tolerated for such potent inhibitors before the effect on antiviral IC_{50} could be detected. In other words, even though binding of a very potent inhibitor to the mutant enzyme is weaker, it potentially may be outside the realm of impact on the viral kinetics. Thus increasing an inhibitor's potency against WT by means of decreasing its off-rate may be an important factor in raising its genetic barrier for resistance (Gulnik and Eissenstat, 2008). In terms of drug design, one potential way to achieve an infinitely long off-rate is through irreversible inhibitors. Medicinal chemists generally shy away from irreversible inhibitors because of the possibility of toxicity associated with the inhibitor reacting with proteins other than those being targeted. Several groups have worked on irreversible HIV PIs. The most advanced was LB71350, a clinically studied compound developed by the Korean company LG Biotech (Choy et al., 1996). This compound incorporated an epoxide into its core that was shown to alkylate Asp25 of the protease. The presence of two active-site aspartates acting in concert provides some inherent selectivity versus off-target alkylation: One of the catalytic Asps is situated perfectly to activate the epoxide oxygen by protonation while the other can nucleophilically attack the epoxide carbon to form an ester (Kona, 2008). This compound was structurally closely related to other PIs and did not show an unusually good resistance profile *in vitro*.

Another approach to avoiding resistance associated with current PIs is to interact with the protease in some other way. The different approaches to the design of HIV PIs that bind outside the active site have yet to bear fruit (as measured by antiviral activity) and will thus only be discussed very briefly. Significant progress in improving the potency of inhibitors targeting the dimer interface has been made [recently reviewed in Bannwarth and Reboud-Ravaux (2007)]. A series of recent studies identified mutations in HIV PR that destabilized dimer and allowed the NMR structure determination of the folded PR monomer (Bannwarth and Reboud-Ravaux, 2007; Ishima et al., 2003, 2007; Louis et al., 2007), and the study of the interactions of monomer with an interface peptide inhibitor (Frutos et al., 2007). These studies suggested that precursors of HIV PR form a much weaker dimer, which may allow the design of inhibitors targeting monomer prior to maturation and formation of the more stable dimer (Ishima et al., 2007; Louis et al., 2007). Binding to the monomeric precursor could be a potential explanation for the disruption by darunavir and tipranavir of HIV PR dimerization in cells observed using a FRET- based assay (Koh et al., 2007). The possibility of targeting PR flexibility through allosteric inhibitors has been suggested [for recent review see Hornak and Simmerling (2007)]. Folding inhibitors of HIV PR that act by trapping the protein in a partially denatured state by targeting highly conserved local elementary structures, "nuclei of folding," have also been proposed (Bonomi et al., 2007; Broglia et al., 2007, 2008). A short peptide with low micromolar activity that may inhibit the enzyme through this mechanism was reported (Broglia et al., 2006), although no additional proof of the proposed mode of inhibition was presented.

The real confirmation for the usefulness of any of the hypotheses described above would be the successful design of new inhibitors of HIV PR (or other HIV enzymes) that are not close structural analogs of current inhibitors. Unfortunately, to date most of these hypotheses have been used to explain the profiles of compounds that had already been synthesized.

23.4 CONCLUDING REMARKS

After the identification of HIV as the virus causing AIDS at the beginning of the 1980s significant efforts were made to understand the molecular biology and virology of HIV. This led to rapid identification and characterization of viral enzymes as important drug design targets essential for infectivity. The onset of HIV PR screening was accelerated by the existence of renin inhibitor programs at several pharmaceutical companies, since HIV PR is a member of the same family of aspartic proteinases. In addition, the rapid progress in structural biology of HIV PR provided guidance for rational optimization of the hits. In fact, HIV PIs are pointed to as the first major success of structure-based drug design. These efforts resulted in the nine HIV PIs presently approved by the FDA for the treatment of HIV/AIDS. HIV PIs are an important component of HAART and have contributed significantly to the improvement of antiretroviral regimens. Although PIs have a high genetic barrier to resistance, viral resistance to PI-containing regimens remains a major problem that is exacerbated by the emergence of viruses cross-resistant to the whole PI class. Suboptimal pharmacokinetic profiles are characteristic of all PIs, presumably because of the large number of (hydrophobic) interactions required to achieve high potency. However, the pharmacokinetic profiles of PIs can usually be significantly improved by pharmacokinetic boosters such as ritonavir. PK boosting allows for a reduction in the frequency of dosing by increasing plasma half-life of PIs, and it also reduces interpatient variability. However, ritonavir boosting comes at the cost of toxicity added on top of the side effects associated with the PIs. There is a persistent clinical need for HIV PIs with improved resistance and side effect profiles, and significant design efforts have been mounted in academia and the pharmaceutical industry to address this need. In addition, the design of novel CYP inhibitors with improved safety profiles as PK boosters for PIs and other HIV drugs has been initiated.

ACRONYMS

APV	Amprenavir
ATV	Atazanavir
CA	Capsid protein
CYP	Cytochrome P450
DRV	Darunavir
ER	Endoplasmic reticulum
FDA	Food and Drug Administration
HAART	Highly active antiretroviral therapy
HIV-1	Human immunodeficiency virus type 1
HIV-LD	HIV-associated lipodystrophy syndrome
IDV	Indinavir
IN	Integrase
LPV	Lopinavir
MA	Matrix protein
NC	Nucleocapsid protein
NFV	Nelfinavir
NMR	Nuclear magnetic resonance
NNRTI	Non-nucleoside reverse transcriptase inhibitor
PDB	Protein data bank
PI	Protease inhibitor
PK	Pharmacokinetics

PR	Protease
RH	Ribonucleoprotease domain of RT
RT	Reverse transcriptase
RTV	Ritonavir
SAR	Structure–activity relationship
SQV	Saquinavir
SREBP	Sterol regulatory element binding protein
TF	Trans-frame region
TPV	Tipranavir
WT	Wild type

ACKNOWLEDGMENTS

We wish to acknowledge John Erickson and Abelardo Silva for stimulating discussions.

REFERENCES

Acosta EP, Kakuda TN, Brundage RC, Anderson PL, Fletcher CV. Pharmacodynamics of human immunodeficiency virus type 1 protease inhibitors. *Clin Infect Dis* 2000;30(Suppl 2):S151–S159.

Altman MD, Ali A, Kumar Reddy GS, Nalam MN, Anjum SG, Cao H, Chellappan S, Kairys V, Fernandes MX, Gilson MK, Schiffer CA, Rana TM, Tidor B. HIV-1 Protease inhibitors from inverse design in the substrate envelope exhibit subnanomolar binding to drug-resistant variants. *J Am Chem Soc* 2008;130:6099–6113.

Altomare DA, Testa JR. Perturbations of the AKT signaling pathway in human cancer. *Oncogene* 2005;24:7455–7464.

Autar RS, Wit FW, Sankote J, Sutthichom D, Kimenai E, Hassink E, Hill A, Cooper DA, Phanuphak P, Lange JM, Burger DM, Ruxrungtham K. Ketoconazole is inferior to ritonavir as an alternative booster for saquinavir in a once daily regimen in Thai HIV-1 infected patients. *AIDS* 2007;21:1535–1539.

Baldwin ET, Bhat TN, Liu B, Pattabiraman N, Erickson JW. Structural basis of drug resistance for the V82A mutant of HIV-1 proteinase. *Nat Struct Biol* 1995;2: 244–249.

Bannwarth L, Reboud-Ravaux M. An alternative strategy for inhibiting multidrug-resistant mutants of the dimeric HIV-1 protease by targeting the subunit interface. *Biochem Soc Trans* 2007;35:551–554.

Barbaro G. Highly active antiretroviral therapy-associated metabolic syndrome: pathogenesis and cardiovascular risk. *Am J Ther* 2006;13:248–260.

Barrios A, Rendon AL, Gallego O, Martin-Carbonero L, Valer L, Rios P, Maida I, Garcia-Benayas T, Jimenez-Nacher I, Gonzalez-Lahoz J, Soriano V. Predictors of virological response to atazanavir in protease inhibitor-experienced patients. *HIV Clin Trials* 2004;5:201–205.

Bastard JP, Caron M, Vidal H, Jan V, Auclair M, Vigouroux C, Luboinski J, Laville M, Maachi M, Girard PM, Rozenbaum W, Levan P, Capeau J. Association between altered expression of adipogenic factor SREBP1 in lipoatrophic adipose tissue from HIV-1-infected patients and abnormal adipocyte differentiation and insulin resistance. *Lancet* 2002;359:1026–1031.

Beaulieu PL, Anderson PC, Cameron DR, Croteau G, Gorys V, Grand-Maitre C, Lamarre D, Liard F, Paris W, Plamondon L, Soucy F, Thibeault D, Wernic D, Yoakim C, Pav S, Tong L. 2′,6′-Dimethylphenoxyacetyl: a new achiral high affinity P(3)-P(2) ligand for peptidomimetic-based HIV protease inhibitors. *J Med Chem* 2000;43:1094–1108.

Becker SL. The role of pharmacological enhancement in protease inhibitor-based highly active antiretroviral therapy. *Expert Opin Investig Drugs* 2003;12:401–412.

Ben-Romano R, Rudich A, Torok D, Vanounou S, Riesenberg K, Schlaeffer F, Klip A, Bashan N. Agent and cell-type specificity in the induction of insulin resistance by HIV protease inhibitors. *AIDS* 2003;17:23–32.

Ben-Romano R, Rudich A, Tirosh A, Potashnik R, Sasaoka T, Riesenberg K, Schlaeffer F, Bashan N. Nelfinavir-induced insulin resistance is associated with impaired plasma membrane recruitment of the PI 3-kinase effectors Akt/PKB and PKC-zeta. *Diabetologia* 2004;47:1107–1117.

Bold G, Fassler A, Capraro HG, Cozens R, Klimkait T, Lazdins J, Mestan J, Poncioni B, Rosel J, Stover D, Tintelnot-Blomley M, Acemoglu F, Beck W, Boss E, Eschbach M, Hurlimann T, Masso E, Roussel S, Ucci-Stoll K, Wyss D, Lang M. New aza-dipeptide analogues as potent and orally absorbed HIV-1 protease inhibitors: candidates for clinical development. *J Med Chem* 1998;41:3387–3401.

Bonomi M, Gervasio FL, Tiana G, Provasi D, Broglia RA, Parrinello M. Insight into the folding inhibition of the HIV-1 protease by a small peptide. *Biophys J* 2007;93: 2813–2821.

Braun PWH, Hoffman D, Daumer M, Ehret R, Korn K, Thiele B, Burg T, Sturmer M, Wiesmann F, Kaiser R. Clinically relevant resensitization of PI saquinavir and atazanavir by L76V in multidrug-resistant HIV-1-infected patients. *Antiviral Therapy* 2007;12:abstract 129.

Brik A, Wong CH. HIV-1 protease: mechanism and drug discovery. *Org Biomol Chem* 2003;1:5–14.

Broglia RA, Provasi D, Vasile F, Ottolina G, Longhi R, Tiana G. A folding inhibitor of the HIV-1 protease. *Proteins* 2006;62:928–933.

Broglia RA, Tiana G, Sutto L, Provasi D, Perelli V. Low-throughput model design of protein folding inhibitors. *Proteins* 2007;67:469–478.

Broglia R, Levy Y, Tiana G. HIV-1 protease folding and the design of drugs which do not create resistance. *Curr Opin Struct Biol* 2008;18:60–66.

Callebaut C, Stray K, Tsai L, Xu L, He GX, Mulato A, Priskich T, Parkin N, Lee W, Cihlar T. Profile of GS-8374, a Novel Phosphonate-containing HIV PI: *in vitro* antiretroviral activity, toxicity, and resistance. Abstract 491. *CROI*, Los Angeles, 2007.

Caron M, Auclair M, Vigouroux C, Glorian M, Forest C, Capeau J. The HIV protease inhibitor indinavir impairs sterol regulatory element-binding protein-1 intranuclear localization, inhibits preadipocyte differentiation, and induces insulin resistance. *Diabetes* 2001;50:1378–1388.

Carr A. HIV protease inhibitor-related lipodystrophy syndrome. *Clin Infect Dis* 2000;30(Suppl 2):S135–S142.

Chellappan S, Kiran Kumar Reddy GS, Ali A, Nalam MN, Anjum SG, Cao H, Kairys V, Fernandes MX, Altman MD, Tidor B, Rana TM, Schiffer CA, Gilson MK. Design of mutation-resistant HIV protease inhibitors with the substrate envelope hypothesis. *Chem Biol Drug Des* 2007;69:298–313.

Chen Z, Li Y, Schock HB, Hall D, Chen E, Kuo LC. Three-dimensional structure of a mutant HIV-1 protease displaying cross-resistance to all protease inhibitors in clinical trials. *J Biol Chem* 1995;270:21433–21436.

Cheng YS, Yin FH, Foundling S, Blomstrom D, Kettner CA. Stability and activity of human immunodeficiency virus protease: comparison of the natural dimer with a homologous, single-chain tethered dimer. *Proc Natl Acad Sci USA* 1990;87: 9660–9664.

Cho H, Mu J, Kim JK, Thorvaldsen JL, Chu Q, Crenshaw EB 3rd, Kaestner KH, Bartolomei MS, Shulman GI, Birnbaum MJ. Insulin resistance and a diabetes mellitus-like syndrome in mice lacking the protein kinase Akt2 (PKB beta). *Science* 2001;292:1728–1731.

Choy N, Lee CS, Park C, Choi H, Son Y-C, Moon KY, Jung WH, Kim CR, Yoon H, Kim SC. Irreversible and orally absorbable HIV-1 protease inactivator. *Korean J Med Chem* 1996;6:309–316.

Cihlar T, He GX, Liu X, Chen JM, Hatada M, Swaminathan S, McDermott MJ, Yang ZY, Mulato AS, Chen X, Leavitt SA, Stray KM, Lee WA. Suppression of HIV-1 protease inhibitor resistance by phosphonate-mediated solvent anchoring. *J Mol Biol* 2006;363:635–647.

Clemente JC, Moose RE, Hemrajani R, Whitford LR, Govindasamy L, Reutzel R, McKenna R, Agbandje-McKenna M, Goodenow MM, Dunn BM. Comparing the accumulation of active- and nonactive-site mutations in the HIV-1 protease. *Biochemistry* 2004;43:12141–12151.

Clemente JC, Coman RM, Thiaville MM, Janka LK, Jeung JA, Nukoolkarn S, Govindasamy L, Agbandje-McKenna M, McKenna R, Leelamanit W, Goodenow MM, Dunn BM. Analysis of HIV-1 CRF_01 A/E protease inhibitor resistance: structural determinants for maintaining sensitivity and developing resistance to atazanavir. *Biochemistry* 2006;45:5468–5477.

Collins JR, Burt SK, Erickson JW. Flap opening in HIV-1 protease simulated by "activated" molecular dynamics. *Nat Struct Biol* 1995;2:334–338.

Condra JH, Schleif WA, Blahy OM, Gabryelski LJ, Graham DJ, Quintero JC, Rhodes A, Robbins HL, Roth E, Shivaprakash M, et al. *In vivo* emergence of HIV-1 variants resistant to multiple protease inhibitors. *Nature* 1995;374:569–571.

Copeland RA. Tight binding inhibition. In: *Evaluation of Enzyme Inhibitors in Drug Discovery*. Hoboken, NJ: John Wiley & Sons, 2005, pp. 178–213.

Copeland RA, Pompliano DL, Meek TD. Drug-target residence time and its implications for lead optimization. *Nat Rev Drug Discovery* 2006;5:730–739.

Craig JC, Duncan IB, Hockley D, Grief C, Roberts NA, Mills JS. Antiviral properties of Ro 31-8959, an inhibitor of human immunodeficiency virus (HIV) proteinase. *Antiviral Res* 1991;16:295–305.

Culm-Merdek KE, von Moltke LL, Gan L, Horan KA, Reynolds R, Harmatz JS, Court MH, Greenblatt DJ. Effect of extended exposure to grapefruit juice on cytochrome P450 3A activity in humans: comparison with ritonavir. *Clin Pharmacol Ther* 2006; 79:243–254.

Dandache S, Sevigny G, Yelle J, Stranix BR, Parkin N, Schapiro JM, Wainberg MA, Wu JJ. *In vitro* antiviral activity and cross-resistance profile of PL-100, a novel protease inhibitor of human immunodeficiency virus type 1. *Antimicrob Agents Chemother* 2007;51:4036–4043.

Darke PL, Jordan SP, Hall DL, Zugay JA, Shafer JA, Kuo LC. Dissociation and association of the HIV-1 protease dimer subunits: equilibria and rates. *Biochemistry* 1994;33:98–105.

Das A, Prashar V, Mahale S, Serre L, Ferrer JL, Hosur MV. Crystal structure of HIV-1 protease *in situ* product complex and observation of a low-barrier hydrogen bond between catalytic aspartates. *Proc Natl Acad Sci USA* 2006;103:18464–18469.

Davies DR. The structure and function of the aspartic proteinases. *Annu Rev Biophys Biophys Chem* 1990;19:189–215.

De Meyer S, Azijn H, Surleraux D, Jochmans D, Tahri A, Pauwels R, Wigerinck P, de Bethune MP. TMC114, a novel human immunodeficiency virus type 1 protease inhibitor active against protease inhibitor-resistant viruses, including a broad range of clinical isolates. *Antimicrob Agents Chemother* 2005;49:2314–2321.

De Meyer S, Vangeneugden T, van Baelen B, de Paepe E, van Marck H, Picchio G, Lefebvre E, de Bethune MP. Resistance profile of darunavir: combined 24-week results from the POWER trials. *AIDS Res Hum Retroviruses* 2008;24:379–388.

Degawa-Yamauchi M, Bovenkerk JE, Juliar BE, Watson W, Kerr K, Jones R, Zhu Q, Considine RV. Serum resistin (FIZZ3) protein is increased in obese humans. *J Clin Endocrinol Metab* 2003;88:5452–5455.

Desai MC, Hong AY, Liu H, Xu L, Vivian RW. Preparation of peptidomimetics as modulators of pharmacokinetic properties of therapeutics by inhibiting cytochrome P450 monooxygenase. WO 2008010921, 2008.

DHHS. Guidelines for the use of antiretroviral agents in HIV-1-infected adults and-adolescents, 2008, pp. 1–139.

Dierynck I, De Wit M, Gustin E, Keuleers I, Vandersmissen J, Hallenberger S, Hertogs K. Binding kinetics of darunavir to human immunodeficiency virus type 1 protease explain the potent antiviral activity and high genetic barrier. *J Virol* 2007;81: 13845–13851.

Dixit V, Hariparsad N, Li F, Desai P, Thummel KE, Unadkat JD. Cytochrome P450 enzymes and transporters induced by anti-human immunodeficiency virus protease inhibitors in human hepatocytes: implications for predicting clinical drug interactions. *Drug Metab Dispos* 2007;35:1853–1859.

Dorsey BD, Levin RB, McDaniel SL, Vacca JP, Guare JP, Darke PL, Zugay JA, Emini EA, Schleif WA, Quintero JC, et al. L-735,524: the design of a potent and orally bioavailable HIV protease inhibitor. *J Med Chem* 1994;37:3443–3451.

Dowell P, Flexner C, Kwiterovich PO, Lane MD. Suppression of preadipocyte differentiation and promotion of adipocyte death by HIV protease inhibitors. *J Biol Chem* 2000;275:41325–41332.

Doyon L, Croteau G, Thibeault D, Poulin F, Pilote L, Lamarre D. Second locus involved in human immunodeficiency virus type 1 resistance to protease inhibitors. *J Virol* 1996;70:3763–3769.

Drewe J, Gutmann H, Fricker G, Torok M, Beglinger C, Huwyler J. HIV protease inhibitor ritonavir: a more potent inhibitor of P-glycoprotein than the cyclosporine analog SDZ PSC 833. *Biochem Pharmacol* 1999;57:1147–1152.

Dunn BM. Structure and mechanism of the pepsin-like family of aspartic peptidases. *Chem Rev* 2002;102:4431–4458.

Dunn BM, Gustchina A, Wlodawer A, Kay J. Subsite preferences of retroviral proteinases. *Methods Enzymol* 1994;241:254–278.

Eissenstat M, Duan D. Benzofuransulfonamide derivatives, processes for preparing them, pharmaceutical compositions containing them, and their use as inhibitors of cytochrome P 450. WO 2008022345, 2008.

Eissenstat M, Guerassina T. Preparation of diamino-mono-ol dipeptide isostere core based resistance-repellent retroviral protease inhibitors. WO 2005087728, 2005.

Eissenstat M, Delahanty G, Topin A, Rajendran GR. Preparation of heterocyclyl sulfonylaminobenzylhydroxypropylcarbamates as HIV protease inhibitors. WO 2005110428, 2005.

Elston RS, Hall D, Shapiro J, Bethell R, Kohlbrenner V, Mayers D. Deselection of the I50V mutations occurs in the clinical isolates during aptivus/r (tipranavir/ritonavir)-based therapy. *Antiviral Therapy* 2006;11:abstract 92.

Erickson J, Neidhart DJ, VanDrie J, Kempf DJ, Wang XC, Norbeck DW, Plattner JJ, Rittenhouse JW, Turon M, Wideburg N, et al. Design, activity, and 2.8 A crystal structure of a C2 symmetric inhibitor complexed to HIV-1 protease. *Science* 1990;249:527–533.

Erickson JW. The not-so-great escape. *Nat Struct Biol* 1995;2:523–529.

Erickson JW, Burt SK. Structural mechanisms of HIV drug resistance. *Annu Rev Pharmacol Toxicol* 1996;36:545–571.

Erickson JW, Gulnik SV, Ghosh AK, Hussain KA. Multidrug-resistant retroviral protease inhibitors and associated methods. WO 9967254, 1999.

Erickson JW, Eissenstat M, Silva A, Gulnik S. Broad spectrum microbial and neoplasmic protein inhibitors. WO 2003057173, 2003.

Erickson JW, Eissenstat M, Silva AM, Afonina E, Gulnik S. Design and crystal structure of SPI-256: an experimental HIV protease inhibitor with a high genetic barrier to resistance. In: *ICAAC*, Washington, DC, Abstract 1266, 2008.

Erickson-Viitanen S, Klabe RM, Cawood PG, O'Neal PL, Meek JL. Potency and selectivity of inhibition of human immunodeficiency virus protease by a small nonpeptide cyclic urea, DMP 323. *Antimicrob Agents Chemother* 1994;38:1628–1634.

Ernest CS, 2nd, Hall SD, Jones DR. Mechanism-based inactivation of CYP3A by HIV protease inhibitors. *J Pharmacol Exp Ther* 2005;312:583–591.

Fassler A, Bold G, Capraro HG, Cozens R, Mestan J, Poncioni B, Rosel J, Tintelnot-Blomley M, Lang M. Aza-peptide analogs as potent human immunodeficiency virus type-1 protease inhibitors with oral bioavailability. *J Med Chem* 1996;39:3203–3216.

Fellay J, Marzolini C, Decosterd L, Golay KP, Baumann P, Buclin T, Telenti A, Eap CB. Variations of CYP3A activity induced by antiretroviral treatment in HIV-1 infected patients. *Eur J Clin Pharmacol* 2005;60:865–873.

Fichtenbaum CJ, Gerber JG. Interactions between antiretroviral drugs and drugs used for the therapy of the metabolic complications encountered during HIV infection. *Clin Pharmacokinet* 2002;41:1195–1211.

Flentge CA, Randolph JT, Huang PP, Klein LL, Marsh KC, Harlan JE, Kempf DJ. Synthesis and evaluation of bifunctional inhibitors of cytochrome P-450 3A. *Abstracts of Papers, 236th ACS National Meeting, Philadelphia, United States, August 17–21, 2008*, MEDI-299.

Foisy MM, Yakiwchuk EM, Hughes CA. Induction effects of ritonavir: implications for drug interactions. *Ann Pharmacother* 2008;42:1048–1059.

Ford J, Meaden ER, Hoggard PG, Dalton M, Newton P, Williams I, Khoo SH, Back DJ. Effect of protease inhibitor-containing regimens on lymphocyte multidrug resistance transporter expression. *J Antimicrob Chemother* 2003;52:354–358.

Ford SL, Reddy YS, Anderson MT, Murray SC, Fernandez P, Stein DS, Johnson MA. Single-dose safety and pharmacokinetics of brecanavir, a novel human immunodeficiency virus protease inhibitor. *Antimicrob Agents Chemother* 2006;50:2201–2206.

Freedberg DI, Ishima R, Jacob J, Wang YX, Kustanovich I, Louis JM, Torchia DA. Rapid structural fluctuations of the free HIV protease flaps in solution: relationship to crystal structures and comparison with predictions of dynamics calculations. *Protein Sci* 2002;11:221–232.

Freskos JN, Bertenshaw DE, Getman DP, Heintz RM, Mischke BV, Blystone LW, Bryant ML, Funckes-Shippy C, Houseman KA, et al. (Hydroxyethyl)sulfonamide HIV-1 protease inhibitors: identification of the 2-methylbenzoyl moiety at P-2. *Bioorg Med Chem Lett* 1996;6:445–450.

Frutos S, Rodriguez-Mias RA, Madurga S, Collinet B, Reboud-Ravaux M, Ludevid D, Giralt E. Disruption of the HIV-1 protease dimer with interface peptides: structural studies using NMR spectroscopy combined with [2-(13)C]-Trp selective labeling. *Biopolymers* 2007;88:164–173.

Furfine ES, Baker CT, Hale MR, Reynolds DJ, Salisbury JA, Searle AD, Studenberg SD, Todd D, Tung RD, Spaltenstein A. Preclinical pharmacology and pharmacokinetics of GW433908, a water-soluble prodrug of the human immunodeficiency virus protease inhibitor amprenavir. *Antimicrob Agents Chemother* 2004;48:791–798.

Gabriely I, Ma XH, Yang XM, Atzmon G, Rajala MW, Berg AH, Scherer P, Rossetti L, Barzilai N. Removal of visceral fat prevents insulin resistance and glucose intolerance of aging: an adipokine-mediated process? *Diabetes* 2002;51:2951–2958.

Gallant JE. Protease-inhibitor boosting in the treatment-experienced patient. *AIDS Rev* 2004;6:226–233.

Gatanaga H, Suzuki Y, Tsang H, Yoshimura K, Kavlick MF, Nagashima K, Gorelick RJ, Mardy S, Tang C, Summers MF, Mitsuya H. Amino acid substitutions in Gag protein at non-cleavage sites are indispensable for the development of a high multitude of HIV-1 resistance against protease inhibitors. *J Biol Chem* 2002;277:5952–5961.

Geese WJ, Ranade K. Polymorphisms of human resistin gene and uses for identifying HIV-1 protease inhibitors with reduced metabolic affects. US 2008004209, 2008.

Ghosh AK, Thompson WJ, McKee SP, Duong TT, Lyle TA, Chen JC, Darke PL, Zugay JA, Emini EA, Schleif WA, et al. 3-Tetrahydrofuran and pyran urethanes as high-affinity P2-ligands for HIV-1 protease inhibitors. *J Med Chem* 1993;36: 292–294.

Ghosh AK, Thompson WJ, Fitzgerald PM, Culberson JC, Axel MG, McKee SP, Huff JR, Anderson PS. Structure-based design of HIV-1 protease inhibitors: replacement of two amides and a 10 pi-aromatic system by a fused bis-tetrahydrofuran. *J Med Chem* 1994;37:2506–2508.

Ghosh AK, Kincaid JF, Walters DE, Chen Y, Chaudhuri NC, Thompson WJ, Culberson C, Fitzgerald PM, Lee HY, McKee SP, Munson PM, Duong TT, Darke PL, Zugay JA, Schleif WA, Axel MG, Lin J, Huff JR. Nonpeptidal P2 ligands for HIV protease inhibitors: structure-based design, synthesis, and biological evaluation. *J Med Chem* 1996;39:3278–3290.

Ghosh AK, Kincaid JF, Cho W, Walters DE, Krishnan K, Hussain KA, Koo Y, Cho H, Rudall C, Holland L, Buthod J. Potent HIV protease inhibitors incorporating high-affinity P2-ligands and (R)-(hydroxyethylamino)sulfonamide isostere. *Bioorg Med Chem Lett* 1998;8:687–690.

Ghosh AK, Ramu Sridhar P, Kumaragurubaran N, Koh Y, Weber IT, Mitsuya H. Bis-tetrahydrofuran: a privileged ligand for darunavir and a new generation of hiv protease inhibitors that combat drug resistance. *Chem Med Chem* 2006a;1: 939–950.

Ghosh AK, Sridhar PR, Leshchenko S, Hussain AK, Li J, Kovalevsky AY, Walters DE, Wedekind JE, Grum-Tokars V, Das D, Koh Y, Maeda K, Gatanaga H, Weber IT, Mitsuya H. Structure-based design of novel HIV-1 protease inhibitors to combat drug resistance. *J Med Chem* 2006b;49:5252–5261.

Ghosh AK, Dawson ZL, Mitsuya H. Darunavir, a conceptually new HIV-1 protease inhibitor for the treatment of drug-resistant HIV. *Bioorg Med Chem* 2007;15: 7576–7580.

Ghosh AK, Chapsal BD, Weber IT, Mitsuya H. Design of HIV Protease Inhibitors Targeting Protein Backbone: An Effective Strategy for Combating Drug Resistance. *Acc Chem Res* 2008;41:78–86.

Gianotti N, Soria A, Lazzarin A. Antiviral activity and clinical efficacy of atazanavir in HIV-1-infected patients: a review. *New Microbiol* 2007;30:79–88.

Gong YF, Robinson BS, Rose RE, Deminie C, Spicer TP, Stock D, Colonno RJ, Lin PF. *In vitro* resistance profile of the human immunodeficiency virus type 1 protease inhibitor BMS-232632. *Antimicrob Agents Chemother* 2000;44:2319–2326.

Gonzalez de Requena D, Gallego O, Valer L, Jimenez-Nacher I, Soriano V. Prediction of virological response to lopinavir/ritonavir using the genotypic inhibitory quotient. *AIDS Res Hum Retroviruses* 2004;20:275–278.

Granfors MT, Wang JS, Kajosaari LI, Laitila J, Neuvonen PJ, Backman JT. Differential inhibition of cytochrome P450 3A4, 3A5 and 3A7 by five human immunodeficiency virus (HIV) protease inhibitors *in vitro*. *Basic Clin Pharmacol Toxicol* 2006;98: 79–85.

Grant SK, Deckman IC, Culp JS, Minnich MD, Brooks IS, Hensley P, Debouck C, Meek TD. Use of protein unfolding studies to determine the conformational and dimeric stabilities of HIV-1 and SIV proteases. *Biochemistry* 1992;31:9491–9501.

Greenblatt DJ, von Moltke LL, Daily JP, Harmatz JS, Shader RI. Extensive impairment of triazolam and alprazolam clearance by short-term low-dose ritonavir: the clinical dilemma of concurrent inhibition and induction. *J Clin Psychopharmacol* 1999;19:293–296.

Griffiths JT, Phylip LH, Konvalinka J, Strop P, Gustchina A, Wlodawer A, Davenport RJ, Briggs R, Dunn BM, Kay J. Different requirements for productive interaction between the active site of HIV-1 proteinase and substrates containing -hydrophobic*hydrophobic- or -aromatic*pro- cleavage sites. *Biochemistry* 1992;31:5193–5200.

Gulnik SV, Eissenstat M. Approaches to the design of HIV protease inhibitors with improved resistance profiles. *Curr Opin HIV AIDS* 2008;3:633–641.

Gulnik SV, Suvorov LI, Liu B, Yu B, Anderson B, Mitsuya H, Erickson JW. Kinetic characterization and cross-resistance patterns of HIV-1 protease mutants selected under drug pressure. *Biochemistry* 1995;34:9282–9287.

Gulnik S, Erickson JW, Xie D. HIV protease: enzyme function and drug resistance. *Vitam Horm* 2000;58:213–256.

Gupta V, Samuleson CG, Su S, Chen TC. Nelfinavir potentiation of imatinib cytotoxicity in meningioma cells via survivin inhibition. *Neurosurg Focus* 2007;23:E9.

Gutmann H, Fricker G, Drewe J, Toeroek M, Miller DS. Interactions of HIV protease inhibitors with ATP-dependent drug export proteins. *Mol Pharmacol* 1999;56: 383–389.

Haas DW, Arathoon E, Thompson MA, de Jesus Pedro R, Gallant JE, Uip DE, Currier J, Noriega LM, Lewi DS, Uribe P, Benetucci L, Cahn P, Paar D, White AC Jr, Collier AC, Ramirez-Ronda CH, Harvey C, Chung MO, Mehrotra D, Chodakewitz J, Nguyen BY. Comparative studies of two-times-daily versus three-times-daily indinavir in combination with zidovudine and lamivudine. *AIDS* 2000;14:1973–1978.

Hanlon MH, Porter DJ, Furfine ES, Spaltenstein A, Carter HL, Danger D, Shu AY, Kaldor IW, Miller JF, Samano VA. Inhibition of wild-type and mutant human immunodeficiency virus type 1 proteases by GW0385 and other arylsulfonamides. *Biochemistry* 2004;43:14500–14507.

Haubrich RH. Resistance and replication capacity assays: clinical utility and interpretation. *Top HIV Med* 2004;12:52–56.

Hazen R, Harvey R, Ferris R, Craig C, Yates P, Griffin P, Miller J, Kaldor I, Ray J, Samano V, Furfine E, Spaltenstein A, Hale M, Tung R, St Clair M, Hanlon M, Boone L. *In vitro* antiviral activity of the novel, tyrosyl-based human immunodeficiency virus (HIV) type 1 protease inhibitor brecanavir (GW640385) in combination with other antiretrovirals and against a panel of protease inhibitor-resistant HIV. *Antimicrob Agents Chemother* 2007;51:3147–3154.

Hertel J, Struthers H, Horj CB, Hruz PW. A structural basis for the acute effects of HIV protease inhibitors on GLUT4 intrinsic activity. *J Biol Chem* 2004;279: 55147–55152.

Hertogs K, Bloor S, Kemp SD, Van den Eynde C, Alcorn TM, Pauwels R, Van Houtte M, Staszewski S, Miller V, Larder BA. Phenotypic and genotypic analysis of clinical

HIV-1 isolates reveals extensive protease inhibitor cross-resistance: a survey of over 6000 samples. *AIDS* 2000;14:1203–1210.

Hirosumi J, Tuncman G, Chang L, Gorgun CZ, Uysal KT, Maeda K, Karin M, Hotamisligil GS. A central role for JNK in obesity and insulin resistance. *Nature* 2002;420:333–336.

Hodge CN, Aldrich PE, Bacheler LT, Chang CH, Eyermann CJ, Garber S, Grubb M, Jackson DA, Jadhav PK, Korant B, Lam PY, Maurin MB, Meek JL, Otto MJ, Rayner MM, Reid C, Sharpe TR, Shum L, Winslow DL, Erickson-Viitanen S. Improved cyclic urea inhibitors of the HIV-1 protease: synthesis, potency, resistance profile, human pharmacokinetics and X-ray crystal structure of DMP 450. *Chem Biol* 1996;3:301–314.

Hoefnagel JG, Koopmans PP, Burger DM, Schuurman R, Galama JM. Role of the inhibitory quotient in HIV therapy. *Antivir Ther* 2005;10:879–892.

Hoetelmans RM, Reijers MH, Weverling GJ, ten Kate RW, Wit FW, Mulder JW, Weigel HM, Frissen PH, Roos M, Jurriaans S, Schuitemaker H, de Wolf F, Beijnen JH, Lange JM. The effect of plasma drug concentrations on HIV-1 clearance rate during quadruple drug therapy. *AIDS* 1998;12:F111–F115.

Hong L, Zhang XC, Hartsuck JA, Tang J. Crystal structure of an *in vivo* HIV-1 protease mutant in complex with saquinavir: insights into the mechanisms of drug resistance. *Protein Sci* 2000;9:1898–1904.

Hoog SS, Towler EM, Zhao B, Doyle ML, Debouck C, Abdel-Meguid SS. Human immunodeficiency virus protease ligand specificity conferred by residues outside of the active site cavity. *Biochemistry* 1996;35:10279–10286.

Hornak V, Okur A, Rizzo RC, Simmerling C. HIV-1 protease flaps spontaneously open and reclose in molecular dynamics simulations. *Proc Natl Acad Sci USA* 2006;103:915–920.

Hornak V, Simmerling C. Targeting structural flexibility in HIV-1 protease inhibitor binding. *Drug Discov Today* 2007;12:132–138.

Hsu A, Granneman GR, Bertz RJ. Ritonavir. Clinical pharmacokinetics and interactions with other anti-HIV agents. *Clin Pharmacokinet* 1998;35:275–291.

Hsu A, Isaacson J, Brun S, Bernstein B, Lam W, Bertz R, Foit C, Rynkiewicz K, Richards B, King M, Rode R, Kempf DJ, Granneman GR, Sun E. Pharmacokinetic-pharmacodynamic analysis of lopinavir–ritonavir in combination with efavirenz and two nucleoside reverse transcriptase inhibitors in extensively pretreated human immunodeficiency virus-infected patients. *Antimicrob Agents Chemother* 2003; 47:350–359.

Hyland LJ, Tomaszek TA Jr, Meek TD. Human immunodeficiency virus-1 protease. 2. Use of pH rate studies and solvent kinetic isotope effects to elucidate details of chemical mechanism. *Biochemistry* 1991a;30:8454–8463.

Hyland LJ, Tomaszek TA Jr, Roberts GD, Carr SA, Magaard VW, Bryan HL, Fakhoury SA, Moore ML, Minnich MD, Culp JS, et al. Human immunodeficiency virus-1 protease. 1. Initial velocity studies and kinetic characterization of reaction intermediates by 18O isotope exchange. *Biochemistry* 1991b;30:8441–8453.

Ikezoe T, Hisatake Y, Takeuchi T, Ohtsuki Y, Yang Y, Said JW, Taguchi H, Koeffler HP. HIV-1 protease inhibitor, ritonavir: a potent inhibitor of CYP3A4, enhanced the anticancer effects of docetaxel in androgen-independent prostate cancer cells *in vitro* and *in vivo*. *Cancer Res* 2004;64:7426–7431.

Ishima R, Freedberg DI, Wang YX, Louis JM, Torchia DA. Flap opening and dimer-interface flexibility in the free and inhibitor-bound HIV protease, and their implications for function. *Structure* 1999;7:1047–1055.

Ishima R, Torchia DA, Lynch SM, Gronenborn AM, Louis JM. Solution structure of the mature HIV-1 protease monomer: insight into the tertiary fold and stability of a precursor. *J Biol Chem* 2003;278:43311–43319.

Ishima R, Torchia DA, Louis JM. Mutational and structural studies aimed at characterizing the monomer of HIV-1 protease and its precursor. *J Biol Chem* 2007;282: 17190–17199.

Janneh O, Jones E, Chandler B, Owen A, Khoo SH. Inhibition of P-glycoprotein and multidrug resistance-associated proteins modulates the intracellular concentration of lopinavir in cultured CD4 T cells and primary human lymphocytes. *J Antimicrob Chemother* 2007;60:987–993.

Johnson VA, Brun-Vezinet F, Clotet B, Gunthard HF, Kuritzkes DR, Pillay D, Schapiro JM, Richman DD. Update of the drug resistance mutations in HIV-1: spring 2008. *Top HIV Med* 2008;16:62–68.

Jordan SP, Zugay J, Darke PL, Kuo LC. Activity and dimerization of human immunodeficiency virus protease as a function of solvent composition and enzyme concentration. *J Biol Chem* 1992;267:20028–20032.

Kaldor SW, Kalish VJ, Davies JF, 2nd, Shetty BV, Fritz JE, Appelt K, Burgess JA, Campanale KM, Chirgadze NY, Clawson DK, Dressman BA, Hatch SD, Khalil DA, Kosa MB, Lubbehusen PP, Muesing MA, Patick AK, Reich SH, Su KS, Tatlock JH. Viracept (nelfinavir mesylate, AG1343): a potent, orally bioavailable inhibitor of HIV-1 protease. *J Med Chem* 1997;40:3979–3985.

Kaletra, prescribing information. In: *Physicians' Desk Reference*, 62nd ed. Montvale, NJ: Thomson Healthcare Inc., 2008, pp. 456–466.

Kannisto K, Sutinen J, Korsheninnikova E, Fisher RM, Ehrenborg E, Gertow K, Virkamaki A, Nyman T, Vidal H, Hamsten A, Yki-Jarvinen H. Expression of adipogenic transcription factors, peroxisome proliferator-activated receptor gamma co-activator 1, IL-6 and CD45 in subcutaneous adipose tissue in lipodystrophy associated with highly active antiretroviral therapy. *AIDS* 2003;17:1753–1762.

Kaplan AH, Zack JA, Knigge M, Paul DA, Kempf DJ, Norbeck DW, Swanstrom R. Partial inhibition of the human immunodeficiency virus type 1 protease results in aberrant virus assembly and the formation of noninfectious particles. *J Virol* 1993;67:4050–4055.

Kaplan AH, Manchester M, Swanstrom R. The activity of the protease of human immunodeficiency virus type 1 is initiated at the membrane of infected cells before the release of viral proteins and is required for release to occur with maximum efficiency. *J Virol* 1994;68:6782–6786.

Katoh E, Louis JM, Yamazaki T, Gronenborn AM, Torchia DA, Ishima R. A solution NMR study of the binding kinetics and the internal dynamics of an HIV-1 protease-substrate complex. *Protein Sci* 2003;12:1376–1385.

Kellam P, Larder BA. Recombinant virus assay: a rapid, phenotypic assay for assessment of drug susceptibility of human immunodeficiency virus type 1 isolates. *Antimicrob Agents Chemother* 1994;38:23–30.

Kempf DJ, Codacovi L, Wang XC, Kohlbrenner WE, Wideburg NE, Saldivar A, Vasavanonda S, Marsh KC, Bryant P, Sham HL, et al. Symmetry-based inhibitors of HIV protease. Structure-activity studies of acylated 2,4-diamino-1,5-diphenyl-3-hydroxypentane and 2,5-diamino-1,6-diphenylhexane-3,4-diol. *J Med Chem* 1993;36: 320–330.

Kempf DJ, Marsh KC, Denissen JF, McDonald E, Vasavanonda S, Flentge CA, Green BE, Fino L, Park CH, Kong XP, et al. ABT-538 is a potent inhibitor of human immunodeficiency virus protease and has high oral bioavailability in humans. *Proc Natl Acad Sci USA* 1995;92:2484–2488.

Kempf DJ, Marsh KC, Kumar G, Rodrigues AD, Denissen JF, McDonald E, Kukulka MJ, Hsu A, Granneman GR, Baroldi PA, Sun E, Pizzuti D, Plattner JJ, Norbeck DW, Leonard JM. Pharmacokinetic enhancement of inhibitors of the human immunodeficiency virus protease by coadministration with ritonavir. *Antimicrob Agents Chemother* 1997;41:654–660.

Kim EE, Baker CT, Dwyer MD, Murcko MA, Rao BG, Tung RD, Navia MA. Crystal structure of HIV-1 protease in complex with VX-478, a potent and orally bioavailable inhibitor of the enzyme. *J Am Chem Soc* 1995;117:1181–1182.

King JR, Acosta EP. Tipranavir: a novel nonpeptidic protease inhibitor of HIV. *Clin Pharmacokinet* 2006;45:665–682.

King NM, Melnick L, Prabu-Jeyabalan M, Nalivaika EA, Yang SS, Gao Y, Nie X, Zepp C, Heefner DL, Schiffer CA. Lack of synergy for inhibitors targeting a multidrug-resistant HIV-1 protease. *Protein Sci* 2002;11:418–429.

King NM, Prabu-Jeyabalan M, Nalivaika EA, Schiffer CA. Combating susceptibility to drug resistance: lessons from HIV-1 protease. *Chem Biol* 2004a;11:1333–1338.

King NM, Prabu-Jeyabalan M, Nalivaika EA, Wigerinck P, de Bethune MP, Schiffer CA. Structural and thermodynamic basis for the binding of TMC114, a next-generation human immunodeficiency virus type 1 protease inhibitor. *J Virol* 2004b;78:12012–12021.

Klei HE, Kish K, Lin PF, Guo Q, Friborg J, Rose RE, Zhang Y, Goldfarb V, Langley DR, Wittekind M, Sheriff S. X-ray crystal structures of human immunodeficiency virus type 1 protease mutants complexed with atazanavir. *J Virol* 2007;81:9525–9535.

Klein LL, Chen H-J, Yeung MC, Flentge CA, Randolph JT, Huang PP, Hutchinson DK, Kempf DJ. Thiazolyl-carbamate derivatives as cytochrome P450 oxidase inhibitors and their preparation, pharmaceutical compositions and use in the treatment of viral infection. WO 2008027932, 2008.

Koh Y, Matsumi S, Das D, Amano M, Davis DA, Li J, Leschenko S, Baldridge A, Shioda T, Yarchoan R, Ghosh AK, Mitsuya H. Potent inhibition of HIV-1 replication by novel non-peptidyl small molecule inhibitors of protease dimerization. *J Biol Chem* 2007;282:28709–28720.

Kohl NE, Emini EA, Schleif WA, Davis LJ, Heimbach JC, Dixon RA, Scolnick EM, Sigal IS. Active human immunodeficiency virus protease is required for viral infectivity. *Proc Natl Acad Sci USA* 1988;85:4686–4690.

Kona J. Theoretical study on the mechanism of a ring-opening reaction of oxirane by the active-site aspartic dyad of HIV-1 protease. *Org Biomol Chem* 2008;6:359–365.

Kontijevskis A, Wikberg JE, Komorowski J. Computational proteomics analysis of HIV-1 protease interactome. *Proteins* 2007;68:305–312.

Korant BD, Strack P, Frey MW, Rizzo CJ. A cellular anti-apoptosis protein is cleaved by the HIV-1 protease. *Adv Exp Med Biol* 1998;436:27–29.

Koudriakova T, Iatsimirskaia E, Utkin I, Gangl E, Vouros P, Storozhuk E, Orza D, Marinina J, Gerber N. Metabolism of the human immunodeficiency virus protease inhibitors indinavir and ritonavir by human intestinal microsomes and expressed cytochrome P4503A4/3A5: mechanism-based inactivation of cytochrome P4503A by ritonavir. *Drug Metab Dispos* 1998;26:552–561.

Kovalevsky AY, Liu F, Leshchenko S, Ghosh AK, Louis JM, Harrison RW, Weber IT. Ultra-high resolution crystal structure of HIV-1 protease mutant reveals two binding sites for clinical inhibitor TMC114. *J Mol Biol* 2006a;363:161–173.

Kovalevsky AY, Tie Y, Liu F, Boross PI, Wang YF, Leshchenko S, Ghosh AK, Harrison RW, Weber IT. Effectiveness of nonpeptide clinical inhibitor TMC-114 on HIV-1 protease with highly drug resistant mutations D30N, I50V, and L90M. *J Med Chem* 2006b;49:1379–1387.

Kovalevsky AY, Chumanevich AA, Liu F, Louis JM, Weber IT. Caught in the Act: the 1.5 A resolution crystal structures of the HIV-1 protease and the I54V mutant reveal a tetrahedral reaction intermediate. *Biochemistry* 2007;46:14854–14864.

Kozisek M, Bray J, Rezacova P, Saskova K, Brynda J, Pokorna J, Mammano F, Rulisek L, Konvalinka J. Molecular analysis of the HIV-1 resistance development: enzymatic activities, crystal structures, and thermodynamics of nelfinavir-resistant HIV protease mutants. *J Mol Biol* 2007;374:1005–1016.

Krafft GA, Wang GT. Synthetic approaches to continuous assays of retroviral proteases. *Methods Enzymol* 1994;241:70–86.

Krausslich HG, Facke M, Heuser AM, Konvalinka J, Zentgraf H. The spacer peptide between human immunodeficiency virus capsid and nucleocapsid proteins is essential for ordered assembly and viral infectivity. *J Virol* 1995;69:3407–3419.

Kumar GN, Rodrigues AD, Buko AM, Denissen JF. Cytochrome P450-mediated metabolism of the HIV-1 protease inhibitor ritonavir (ABT-538) in human liver microsomes. *J Pharmacol Exp Ther* 1996;277:423–431.

Kumar GN, Dykstra J, Roberts EM, Jayanti VK, Hickman D, Uchic J, Yao Y, Surber B, Thomas S, Granneman GR. Potent inhibition of the cytochrome P-450 3A-mediated human liver microsomal metabolism of a novel HIV protease inhibitor by ritonavir: a positive drug–drug interaction. *Drug Metab Dispos* 1999;27: 902–908.

Kuzmic P. Kinetic assay for HIV proteinase subunit dissociation. *Biochem Biophys Res Commun* 1993;191:998–1003.

Kuzmic P. Program DYNAFIT for the analysis of enzyme kinetic data: application to HIV proteinase. *Anal Biochem* 1996;237:260–273.

Lafont V, Armstrong AA, Ohtaka H, Kiso Y, Mario Amzel L, Freire E. Compensating enthalpic and entropic changes hinder binding affinity optimization. *Chem Biol Drug Des* 2007;69:413–422.

Lam PY, Jadhav PK, Eyermann CJ, Hodge CN, Ru Y, Bacheler LT, Meek JL, Otto MJ, Rayner MM, Wong YN, et al. Rational design of potent, bioavailable, nonpeptide cyclic ureas as HIV protease inhibitors. *Science* 1994;263:380–384.

Lam PY, Ru Y, Jadhav PK, Aldrich PE, DeLucca GV, Eyermann CJ, Chang CH, Emmett G, Holler ER, Daneker WF, Li L, Confalone PN, McHugh RJ, Han Q, Li R, Markwalder JA, Seitz SP, Sharpe TR, Bacheler LT, Rayner MM, Klabe RM, Shum L, Winslow DL, Kornhauser DM, Hodge CN, et al. Cyclic HIV protease inhibitors: synthesis, conformational analysis, P2/P2′ structure–activity relationship, and molecular recognition of cyclic ureas. *J Med Chem* 1996;39:3514–3525.

Lamarre D, Croteau G, Wardrop E, Bourgon L, Thibeault D, Clouette C, Vaillancourt M, Cohen E, Pargellis C, Yoakim C, Anderson PC. Antiviral properties of palinavir, a potent inhibitor of the human immunodeficiency virus type 1 protease. *Antimicrob Agents Chemother* 1997;41:965–971.

Lapatto R, Blundell T, Hemmings A, Overington J, Wilderspin A, Wood S, Merson JR, Whittle PJ, Danley DE, Geoghegan KF, et al. X-ray analysis of HIV-1 proteinase at 2.7 A resolution confirms structural homology among retroviral enzymes. *Nature* 1989;342:299–302.

Lenhard JM, Furfine ES, Jain RG, Ittoop O, Orband-Miller LA, Blanchard SG, Paulik MA, Weiel JE. HIV protease inhibitors block adipogenesis and increase lipolysis *in vitro*. *Antiviral Res* 2000;47:121–129.

Lima VD, Gill VS, Yip B, Hogg RS, Montaner JS, Harrigan PR. Increased resilience to the development of drug resistance with modern boosted protease inhibitor-based highly active antiretroviral therapy. *J Infect Dis* 2008;198:51–58.

Liu F, Boross PI, Wang YF, Tozser J, Louis JM, Harrison RW, Weber IT. Kinetic, stability, and structural changes in high-resolution crystal structures of HIV-1 protease with drug-resistant mutations L24I, I50V, and G73S. *J Mol Biol* 2005;354: 789–800.

Liu F, Kovalevsky AY, Louis JM, Boross PI, Wang YF, Harrison RW, Weber IT. Mechanism of drug resistance revealed by the crystal structure of the unliganded HIV-1 protease with F53L mutation. *J Mol Biol* 2006;358:1191–1199.

Louis JM, Nashed NT, Parris KD, Kimmel AR, Jerina DM. Kinetics and mechanism of autoprocessing of human immunodeficiency virus type 1 protease from an analog of the Gag-Pol polyprotein. *Proc Natl Acad Sci USA* 1994;91:7970–7974.

Louis JM, Clore GM, Gronenborn AM. Autoprocessing of HIV-1 protease is tightly coupled to protein folding. *Nat Struct Biol* 1999a;6:868–875.

Louis JM, Wondrak EM, Kimmel AR, Wingfield PT, Nashed NT. Proteolytic processing of HIV-1 protease precursor, kinetics and mechanism. *J Biol Chem* 1999b; 274:23437–23442.

Louis JM, Ishima R, Torchia DA, Weber IT. HIV-1 protease: structure, dynamics, and inhibition. *Adv Pharmacol* 2007;55:261–298.

Luque I, Todd MJ, Gomez J, Semo N, Freire E. Molecular basis of resistance to HIV-1 protease inhibition: a plausible hypothesis. *Biochemistry* 1998;37:5791–5797.

Maguire MF, Guinea R, Griffin P, Macmanus S, Elston RC, Wolfram J, Richards N, Hanlon MH, Porter DJ, Wrin T, Parkin N, Tisdale M, Furfine E, Petropoulos C, Snowden BW, Kleim JP. Changes in human immunodeficiency virus type 1 Gag at positions L449 and P453 are linked to I50V protease mutants *in vivo* and cause reduction of sensitivity to amprenavir and improved viral fitness *in vitro*. *J Virol* 2002;76:7398–7406.

Mahalingam B, Wang YF, Boross PI, Tozser J, Louis JM, Harrison RW, Weber IT. Crystal structures of HIV protease V82A and L90M mutants reveal changes in the indinavir-binding site. *Eur J Biochem* 2004;271:1516–1524.

Mallon PW. Pathogenesis of lipodystrophy and lipid abnormalities in patients taking antiretroviral therapy. *AIDS Rev* 2007;9:3–15.

Mammano F, Trouplin V, Zennou V, Clavel F. Retracing the evolutionary pathways of human immunodeficiency virus type 1 resistance to protease inhibitors: virus fitness in the absence and in the presence of drug. *J Virol* 2000;74:8524–8531.

Martinez-Cajas JL, Wainberg MA. Protease inhibitor resistance in HIV-infected patients: molecular and clinical perspectives. *Antiviral Res* 2007;76:203–221.

Mascolini M. Ritonavir plus saquinavir: two trials with different results. *AIDS Treat News* 1996:5–6.

Matayoshi ED, Wang GT, Krafft GA, Erickson J. Novel fluorogenic substrates for assaying retroviral proteases by resonance energy transfer. *Science* 1990;247: 954–958.

McQuade TJ, Tomasselli AG, Liu L, Karacostas V, Moss B, Sawyer TK, Heinrikson RL, Tarpley WG. A synthetic HIV-1 protease inhibitor with antiviral activity arrests HIV-like particle maturation. *Science* 1990;247:454–456.

McTernan CL, McTernan PG, Harte AL, Levick PL, Barnett AH, Kumar S. Resistin, central obesity, and type 2 diabetes. *Lancet* 2002;359:46–47.

Meek TD, Rodriguez EJ, Angeles TS. Use of steady state kinetic methods to elucidate the kinetic and chemical mechanisms of retroviral proteases. *Methods Enzymol* 1994;241:127–156.

Mervis RJ, Ahmad N, Lillehoj EP, Raum MG, Salazar FH, Chan HW, Venkatesan S. The gag gene products of human immunodeficiency virus type 1: alignment within the gag open reading frame, identification of posttranslational modifications, and evidence for alternative gag precursors. *J Virol* 1988;62:3993–4002.

Mildner AM, Rothrock DJ, Leone JW, Bannow CA, Lull JM, Reardon IM, Sarcich JL, Howe WJ, Tomich CS, Smith CW, et al. The HIV-1 protease as enzyme and substrate: mutagenesis of autolysis sites and generation of a stable mutant with retained kinetic properties. *Biochemistry* 1994;33:9405–9413.

Miller JF, Andrews CW, Brieger M, Furfine ES, Hale MR, Hanlon MH, Hazen RJ, Kaldor I, McLean EW, Reynolds D, Sammond DM, Spaltenstein A, Tung R, Turner EM, Xu RX, Sherrill RG. Ultra-potent P1 modified arylsulfonamide HIV protease inhibitors: the discovery of GW0385. *Bioorg Med Chem Lett* 2006;16: 1788–1794.

Miserez AR, Muller PY, Spaniol V. Indinavir inhibits sterol-regulatory element-binding protein-1c-dependent lipoprotein lipase and fatty acid synthase gene activations. *AIDS* 2002;16:1587–1594.

Mitsuya H, Maeda K, Das D, Ghosh AK. Development of protease inhibitors and the fight with drug-resistant HIV-1 variants. *Adv Pharmacol* 2008;56:169–197.

Motwani B, Khayr W. Pharmacoenhancement of protease inhibitors. *Am J Ther* 2006;13:57–63.

Moyle G. Use of HIV protease inhibitors as pharmacoenhancers. *AIDS Read* 2001;11: 87–98.

Moyle GJ, Back D. Principles and practice of HIV-protease inhibitor pharmacoenhancement. *HIV Med* 2001;2:105–113.

Murata H, Hruz PW, Mueckler M. Indinavir inhibits the glucose transporter isoform Glut4 at physiologic concentrations. *AIDS* 2002;16:859–863.

Muzammil S, Ross P, Freire E. A major role for a set of non-active site mutations in the development of HIV-1 protease drug resistance. *Biochemistry* 2003;42:631–638.

Muzammil S, Armstrong AA, Kang LW, Jakalian A, Bonneau PR, Schmelmer V, Amzel LM, Freire E. Unique thermodynamic response of tipranavir to human immunodeficiency virus type 1 protease drug resistance mutations. *J Virol* 2007;81:5144–5154.

Myint L, Matsuda M, Matsuda Z, Yokomaku Y, Chiba T, Okano A, Yamada K, Sugiura W. Gag non-cleavage site mutations contribute to full recovery of viral fitness in protease inhibitor-resistant human immunodeficiency virus type 1. *Antimicrob Agents Chemother* 2004;48:444–452.

Nalam MN, Peeters A, Jonckers TH, Dierynck I, Schiffer CA. Crystal structure of lysine sulfonamide inhibitor reveals the displacement of the conserved flap water molecule in human immunodeficiency virus type 1 protease. *J Virol* 2007;81: 9512–9518.

Nalam MNL, Schiffer CA. New Approaches to HIV protease inhibitor drug design II: testing the substrate envelope hypothesis to avoid drug resistance and discover robust inhibitors. *Curr Opin HIV AIDS* 2008;3:642–646.

Navia MA, Fitzgerald PM, McKeever BM, Leu CT, Heimbach JC, Herber WK, Sigal IS, Darke PL, Springer JP. Three-dimensional structure of aspartyl protease from human immunodeficiency virus HIV-1. *Nature* 1989;337:615–620.

Nguyen AT, Gagnon A, Angel JB, Sorisky A. Ritonavir increases the level of active ADD-1/SREBP-1 protein during adipogenesis. *AIDS* 2000;14:2467–2473.

Nijhuis M, Schuurman R, de Jong D, Erickson J, Gustchina E, Albert J, Schipper P, Gulnik S, Boucher CA. Increased fitness of drug resistant HIV-1 protease as a result of acquisition of compensatory mutations during suboptimal therapy. *AIDS* 1999;13:2349–2359.

Nijhuis M, van Maarseveen NM, Lastere S, Schipper P, Coakley E, Glass B, Rovenska M, de Jong D, Chappey C, Goedegebuure IW, Heilek-Snyder G, Dulude D, Cammack N, Brakier-Gingras L, Konvalinka J, Parkin N, Krausslich HG, Brun-Vezinet F, Boucher CA. A novel substrate-based HIV-1 protease inhibitor drug resistance mechanism. *PLoS Med* 2007;4:e36.

Nillroth U, Vrang L, Markgren PO, Hulten J, Hallberg A, Danielson UH. Human immunodeficiency virus type 1 proteinase resistance to symmetric cyclic urea inhibitor analogs. *Antimicrob Agents Chemother* 1997;41:2383–2388.

Nivesanond K, Peeters A, Lamoen D, Van Alsenoy C. Conformational analysis of TMC114, a novel HIV-1 protease inhibitor. *J Chem Inf Model* 2008;48:99–108.

Noor MA. The role of protease inhibitors in the pathogenesis of HIV-associated insulin resistance: cellular mechanisms and clinical implications. *Curr HIV/AIDS Rep* 2007;4:126–134.

Northrop DB. Follow the protons: a low-barrier hydrogen bond unifies the mechanisms of the aspartic proteases. *Acc Chem Res* 2001;34:790–797.

Ohtaka H, Muzammil S, Schon A, Velazquez-Campoy A, Vega S, Freire E. Thermodynamic rules for the design of high affinity HIV-1 protease inhibitors with adaptability to mutations and high selectivity towards unwanted targets. *Int J Biochem Cell Biol* 2004;36:1787–1799.

Ohtaka H, Freire E. Adaptive inhibitors of the HIV-1 protease. *Prog Biophys Mol Biol* 2005;88:193–208.

Olsen DB, Stahlhut MW, Rutkowski CA, Schock HB, vanOlden AL, Kuo LC. Non-active site changes elicit broad-based cross-resistance of the HIV-1 protease to inhibitors. *J Biol Chem* 1999;274:23699–23701.

Olson DP, Scadden DT, D'Aquila RT, De Pasquale MP. The protease inhibitor ritonavir inhibits the functional activity of the multidrug resistance related-protein 1 (MRP-1). *AIDS* 2002;16:1743–1747.

Orans J, Teotico DG, Redinbo MR. The nuclear xenobiotic receptor pregnane X receptor: recent insights and new challenges. *Mol Endocrinol* 2005;19:2891–2900.

Osborne TF. Sterol regulatory element-binding proteins (SREBPs): key regulators of nutritional homeostasis and insulin action. *J Biol Chem* 2000;275:32379–32382.

Palmer S, Shafer RW, Merigan TC. Highly drug-resistant HIV-1 clinical isolates are cross-resistant to many antiretroviral compounds in current clinical development. *AIDS* 1999;13:661–667.

Park C, Choi H, Son Y-C, Lee CS, Choy N, Koh JS, Lee TG, Kwon YD, Kim SC, Yoon H. beta-Methanesulfonyl-L-valine as a novel, unnatural amino acid surrogate for P2 in the design of HIV protease inhibitors. *Bioorg Med Chem Lett* 1996;6:585.

Partaledis JA, Yamaguchi K, Tisdale M, Blair EE, Falcione C, Maschera B, Myers RE, Pazhanisamy S, Futer O, Cullinan AB, et al. *In vitro* selection and characterization of human immunodeficiency virus type 1 (HIV-1) isolates with reduced sensitivity to hydroxyethylamino sulfonamide inhibitors of HIV-1 aspartyl protease. *J Virol* 1995;69:5228–5235.

Paterson DL, Swindells S, Mohr J, Brester M, Vergis EN, Squier C, Wagener MM, Singh N. Adherence to protease inhibitor therapy and outcomes in patients with HIV infection. *Ann Intern Med* 2000;133:21–30.

Pauwels R, De Clercq E, Desmyter J, Balzarini J, Goubau P, Herdewijn P, Vanderhaeghe H, Vandeputte M. Sensitive and rapid assay on MT-4 cells for detection of antiviral compounds against the AIDS virus. *J Virol Methods* 1987;16:171–185.

Peranteau AG, Kuzmic P, Angell Y, Garcia-Echeverria C, Rich DH. Increase in fluorescence upon the hydrolysis of tyrosine peptides: application to proteinase assays. *Anal Biochem* 1995;227:242–245.

Perloff MD, von Moltke LL, Fahey JM, Daily JP, Greenblatt DJ. Induction of P-glycoprotein expression by HIV protease inhibitors in cell culture. *AIDS* 2000;14:1287–1289.

Perrin V, Mammano F. Parameters driving the selection of nelfinavir-resistant human immunodeficiency virus type 1 variants. *J Virol* 2003;77:10172–10175.

Petropoulos CJ, Parkin NT, Limoli KL, Lie YS, Wrin T, Huang W, Tian H, Smith D, Winslow GA, Capon DJ, Whitcomb JM. A novel phenotypic drug susceptibility assay for human immunodeficiency virus type 1. *Antimicrob Agents Chemother* 2000;44:920–928.

Pettit SC, Moody MD, Wehbie RS, Kaplan AH, Nantermet PV, Klein CA, Swanstrom R. The p2 domain of human immunodeficiency virus type 1 Gag regulates sequential proteolytic processing and is required to produce fully infectious virions. *J Virol* 1994;68:8017–8027.

Pettit SC, Everitt LE, Choudhury S, Dunn BM, Kaplan AH. Initial cleavage of the human immunodeficiency virus type 1 GagPol precursor by its activated protease occurs by an intramolecular mechanism. *J Virol* 2004;78:8477–8485.

Pettit SC, Lindquist JN, Kaplan AH, Swanstrom R. Processing sites in the human immunodeficiency virus type 1 (HIV-1) Gag-Pro-Pol precursor are cleaved by the viral protease at different rates. *Retrovirology* 2005;2:66.

Planken SP, Sutton SC, Chen R. Preparation of pyrazole derivatives as cytochrome P450 inhibitors. WO 2008004096, 2008a.

Planken SP, Sutton SC, Tran TH. Preparation of substituted benzylpyrazoles for inhibiting CYP450 enzyme system. WO 2008004100, 2008b.

Polgar L, Szeltner Z, Boros I. Substrate-dependent mechanisms in the catalysis of human immunodeficiency virus protease. *Biochemistry* 1994;33:9351–9357.

Porter DJ, Hanlon MH, Carter LH 3rd, Danger DP, Furfine ES. Effectors of HIV-1 protease peptidolytic activity. *Biochemistry* 2001;40:11131–11139.

Prabu-Jeyabalan M, Nalivaika E, Schiffer CA. Substrate shape determines specificity of recognition for HIV-1 protease: analysis of crystal structures of six substrate complexes. *Structure* 2002;10:369–381.

Prabu-Jeyabalan M, Nalivaika EA, King NM, Schiffer CA. Viability of a drug-resistant human immunodeficiency virus type 1 protease variant: structural insights for better antiviral therapy. *J Virol* 2003;77:1306–1315.

Prabu-Jeyabalan M, King NM, Nalivaika EA, Heilek-Snyder G, Cammack N, Schiffer CA. Substrate envelope and drug resistance: crystal structure of RO1 in complex with wild-type human immunodeficiency virus type 1 protease. *Antimicrob Agents Chemother* 2006;50:1518–1521.

Prasad JV, Para KS, Tummino PJ, Ferguson D, McQuade TJ, Lunney EA, Rapundalo ST, Batley BL, Hingorani G, Domagala JM, et al. Nonpeptidic potent HIV-1 protease inhibitors: (4-hydroxy-6-phenyl-2-oxo-2H- pyran-3-yl)thiomethanes that span P1-P2′ subsites in a unique mode of active site binding. *J Med Chem* 1995a;38: 898–905.

Prasad JVNV, Lunney EA, Ferguson D, Tummino PJ, Rubin JR, Reyner EL, Stewart BH, Guttendorf RJ, Domagala JM, et al. HIV protease inhibitors possessing a novel, high-affinity, and achiral P1′/P2′ ligand with a unique pattern of *in vitro* resistance. Importance of a conformationally-restricted template in the design of enzyme inhibitors. *J Am Chem Soc* 1995b;117:11070–11074.

Prot M, Heripret L, Cardot-Leccia N, Perrin C, Aouadi M, Lavrut T, Garraffo R, Dellamonica P, Durant J, Le Marchand-Brustel Y, Binetruy B. Long-term treatment with lopinavir–ritonavir induces a reduction in peripheral adipose depots in mice. *Antimicrob Agents Chemother* 2006;50:3998–4004.

Qari SH, Respess R, Weinstock H, Beltrami EM, Hertogs K, Larder BA, Petropoulos CJ, Hellmann N, Heneine W. Comparative analysis of two commercial phenotypic assays for drug susceptibility testing of human immunodeficiency virus type 1. *J Clin Microbiol* 2002;40:31–35.

Rajala MW, Qi Y, Patel HR, Takahashi N, Banerjee R, Pajvani UB, Sinha MK, Gingerich RL, Scherer PE, Ahima RS. Regulation of resistin expression and circulating levels in obesity, diabetes, and fasting. *Diabetes* 2004;53:1671–1679.

Reddy GS, Ali A, Nalam MN, Anjum SG, Cao H, Nathans RS, Schiffer CA, Rana TM. Design and synthesis of HIV-1 protease inhibitors incorporating oxazolidinones as P2/P2′ ligands in pseudosymmetric dipeptide isosteres. *J Med Chem* 2007a;50:4316–4328.

Reddy YS, Ford SL, Anderson MT, Murray SC, Ng-Cashin J, Johnson MA. Safety and pharmacokinetics of brecanavir, a novel human immunodeficiency virus type 1 protease inhibitor, following repeat administration with and without ritonavir in healthy adult subjects. *Antimicrob Agents Chemother* 2007b;51:1202–1208.

Reyataz, prescribing information. In: *Physicians' Desk Reference*, 62nd ed. Montvale, NJ: Thomson Healthcare Inc., 2008, pp. 910–920.

Riddle TM, Kuhel DG, Woollett LA, Fichtenbaum CJ, Hui DY. HIV protease inhibitor induces fatty acid and sterol biosynthesis in liver and adipose tissues due to the accumulation of activated sterol regulatory element-binding proteins in the nucleus. *J Biol Chem* 2001;276:37514–37519.

Roberts NA, Martin JA, Kinchington D, Broadhurst AV, Craig JC, Duncan IB, Galpin SA, Handa BK, Kay J, Krohn A, et al. Rational design of peptide-based HIV proteinase inhibitors. *Science* 1990;248:358–361.

Robinson LH, Myers RE, Snowden BW, Tisdale M, Blair ED. HIV type 1 protease cleavage site mutations and viral fitness: implications for drug susceptibility phenotyping assays. *AIDS Res Hum Retroviruses* 2000;16:1149–1156.

Rose RE, Gong YF, Greytok JA, Bechtold CM, Terry BJ, Robinson BS, Alam M, Colonno RJ, Lin PF. Human immunodeficiency virus type 1 viral background plays a major role in development of resistance to protease inhibitors. *Proc Natl Acad Sci USA* 1996;93:1648–1653.

Rose RB, Craik CS, Stroud RM. Domain flexibility in retroviral proteases: structural implications for drug resistant mutations. *Biochemistry* 1998;37:2607–2621.

Schutt M, Meier M, Meyer M, Klein J, Aries SP, Klein HH. The HIV-1 protease inhibitor indinavir impairs insulin signalling in HepG2 hepatoma cells. *Diabetologia* 2000;43:1145–1148.

Scott JD. Simplifying the treatment of HIV infection with ritonavir-boosted protease inhibitors in antiretroviral-experienced patients. *Am J Health Syst Pharm* 2005;62:809–815.

Seelmeier S, Schmidt H, Turk V, von der Helm K. Human immunodeficiency virus has an aspartic-type protease that can be inhibited by pepstatin A. *Proc Natl Acad Sci USA* 1988;85:6612–6616.

Sekar VJ, Lefebvre E, De Pauw M, Vangeneugden T, Hoetelmans RM. Pharmacokinetics of darunavir/ritonavir and ketoconazole following co-administration in HIV-healthy volunteers. *Br J Clin Pharmacol* 2008;66:215–221.

Sevigny G, Stranix B, Tian B, Dubois A, Sauve G, Petropoulos C, Lie Y, Hellmann N, Conway B, Yelle J. Antiviral activity and cross-resistance profile of P-1946, a novel human immunodeficiency virus type 1 protease inhibitor. *Antiviral Res* 2006;70:17–20.

Shafer RW, Schapiro JM. HIV-1 drug resistance mutations: an updated framework for the second decade of HAART. *AIDS Rev* 2008;10:67–84.

Shafer RW, Winters MA, Palmer S, Merigan TC. Multiple concurrent reverse transcriptase and protease mutations and multidrug resistance of HIV-1 isolates from heavily treated patients. *Ann Intern Med* 1998;128:906–911.

Sham HL, Zhao C, Stewart KD, Betebenner DA, Lin S, Park CH, Kong XP, Rosenbrook W Jr, Herrin T, Madigan D, Vasavanonda S, Lyons N, Molla A, Saldivar A, Marsh KC, McDonald E, Wideburg NE, Denissen JF, Robins T, Kempf DJ, Plattner JJ, Norbeck DW. A novel, picomolar inhibitor of human immunodeficiency virus type 1 protease. *J Med Chem* 1996;39:392–397.

Sham HL, Kempf DJ, Molla A, Marsh KC, Kumar GN, Chen CM, Kati W, Stewart K, Lal R, Hsu A, Betebenner D, Korneyeva M, Vasavanonda S, McDonald E, Saldivar A, Wideburg N, Chen X, Niu P, Park C, Jayanti V, Grabowski B, Granneman GR, Sun E, Japour AJ, Leonard JM, Plattner JJ, Norbeck DW. ABT-378, a highly potent inhibitor of the human immunodeficiency virus protease. *Antimicrob Agents Chemother* 1998;42:3218–3224.

Silha JV, Krsek M, Skrha JV, Sucharda P, Nyomba BL, Murphy LJ. Plasma resistin, adiponectin and leptin levels in lean and obese subjects: correlations with insulin resistance. *Eur J Endocrinol* 2003;149:331–335.

Silva AM, Cachau RE, Sham HL, Erickson JW. Inhibition and catalytic mechanism of HIV-1 aspartic protease. *J Mol Biol* 1996;255:321–346.

Skulnick HI, Johnson PD, Howe WJ, Tomich PK, Chong KT, Watenpaugh KD, Janakiraman MN, Dolak LA, McGrath JP, Lynn JC, et al. Structure-based design of sulfonamide-substituted non-peptidic HIV protease inhibitors. *J Med Chem* 1995;38:4968–4971.

Smith AB, 3rd, Hirschmann R, Pasternak A, Yao W, Sprengeler PA, Holloway MK, Kuo LC, Chen Z, Darke PL, Schleif WA. An orally bioavailable pyrrolinone inhibitor of HIV-1 protease: computational analysis and X-ray crystal structure of the enzyme complex. *J Med Chem* 1997;40:2440–2444.

Snasel J, Pichova I. The cleavage of host cell proteins by HIV-1 protease. *Folia Biol (Praha)* 1996;42:227–230.

Spaltenstein A, Kazmierski WM, Miller JF, Samano V. Discovery of next generation inhibitors of HIV protease. *Curr Top Med Chem* 2005;5:1589–1607.

Steppan CM, Bailey ST, Bhat S, Brown EJ, Banerjee RR, Wright CM, Patel HR, Ahima RS, Lazar MA. The hormone resistin links obesity to diabetes. *Nature* 2001;409:307–312.

Stoll V, Qin W, Stewart KD, Jakob C, Park C, Walter K, Simmer RL, Helfrich R, Bussiere D, Kao J, Kempf D, Sham HL, Norbeck DW. X-ray crystallographic structure of ABT-378 (lopinavir) bound to HIV-1 protease. *Bioorg Med Chem* 2002;10:2803–2806.

Storch CH, Theile D, Lindenmaier H, Haefeli WE, Weiss J. Comparison of the inhibitory activity of anti-HIV drugs on P-glycoprotein. *Biochem Pharmacol* 2007;73: 1573–1581.

Stranix BR, Sauve G, Bouzide A, Cote A, Sevigny G, Yelle J. Lysine sulfonamides as novel HIV-protease inhibitors: optimization of the N-epsilon-acyl-phenyl spacer. *Bioorg Med Chem Lett* 2003;13:4289–4292.

Stranix BR, Sauve G, Bouzide A, Cote A, Sevigny G, Yelle J, Perron V. Lysine sulfonamides as novel HIV-protease inhibitors: N-epsilon-disubstituted ureas. *Bioorg Med Chem Lett* 2004;14:3971–3974.

Stranix BR, Lavallee JF, Sevigny G, Yelle J, Perron V, LeBerre N, Herbart D, Wu JJ. Lysine sulfonamides as novel HIV-protease inhibitors: N-epsilon-acyl aromatic alpha-amino acids. *Bioorg Med Chem Lett* 2006;16:3459–3462.

Surleraux DLNG, Vendeville SMH, Verschueren WG, De Bethune M-PTMMG, De Kock HA, Tahri A, Erra Sola M. Preparation of 2-(substituted-amino)benzoxazole sulfonamides as broadspectrum HIV protease inhibitors. WO 2002081478, 2002a.

Surleraux DLNG, Wigerinck PTBP, Getman D, Verschueren WG, Vendeville S, De Bethune M-P, De Kerpel JOA, Moors SLC, De Kock HA, Voets MCJ. Broad-spectrum 2-(substituted-amino)-benzothiazolesulfonamide HIV protease inhibitors. WO 2002083657, 2002b.

Surleraux DLNG, Vergouwen BJB, De Kock HA. Broad-spectrum substituted benzisoxazole sulfonamide HIV protease inhibitors, preparation thereof, pharmaceutical compositions, diagnostic kits, and combinations with other antiretroviral agents. WO 2003097616, 2003a.

Surleraux DLNG, Wigerinck PTBP, Voets MCJ, Vendeville SMH, De Kock HA, Vergouwen BJB. Preparation of broad spectrum substituted benzimidazolesulfonamide HIV protease inhibitors. WO 2003076413, 2003b.

Surleraux DL, de Kock HA, Verschueren WG, Pille GM, Maes LJ, Peeters A, Vendeville S, De Meyer S, Azijn H, Pauwels R, de Bethune MP, King NM, Prabu-Jeyabalan M, Schiffer CA, Wigerinck PB. Design of HIV-1 protease inhibitors active on multidrug-resistant virus. *J Med Chem* 2005a;48:1965–1973.

Surleraux DL, Tahri A, Verschueren WG, Pille GM, de Kock HA, Jonckers TH, Peeters A, De Meyer S, Azijn H, Pauwels R, de Bethune MP, King NM, Prabu-Jeyabalan M, Schiffer CA, Wigerinck PB. Discovery and selection of TMC114, a next generation HIV-1 protease inhibitor. *J Med Chem* 2005b;48:1813–1822.

Tahri A, Wigerinck PTBP. Broad spectrum substituted oxindole sulfonamide HIV protease inhibitors. WO 2004016619, 2004.

Tang J, Hartsuck JA. A kinetic model for comparing proteolytic processing activity and inhibitor resistance potential of mutant HIV-1 proteases. *FEBS Lett* 1995;367:112–116.

Thaisrivongs S, Tomich PK, Watenpaugh KD, Chong KT, Howe WJ, Yang CP, Strohbach JW, Turner SR, McGrath JP, Bohanon MJ, et al. Structure-based design of HIV protease inhibitors: 4-hydroxycoumarins and 4-hydroxy-2-pyrones as non-peptidic inhibitors. *J Med Chem* 1994;37:3200–3204.

Thaisrivongs S, Watenpaugh KD, Howe WJ, Tomich PK, Dolak LA, Chong KT, Tomich CC, Tomasselli AG, Turner SR, Strohbach JW, et al. Structure-based design of novel HIV protease inhibitors: carboxamide-containing 4-hydroxycoumarins and 4-hydroxy-2-pyrones as potent nonpeptidic inhibitors. *J Med Chem* 1995;38:3624–3637.

Thaisrivongs S, Janakiraman MN, Chong KT, Tomich PK, Dolak LA, Turner SR, Strohbach JW, Lynn JC, Horng MM, Hinshaw RR, Watenpaugh KD. Structure-based design of novel HIV protease inhibitors: sulfonamide-containing 4-hydroxycoumarins and 4-hydroxy-2-pyrones as potent non-peptidic inhibitors. *J Med Chem* 1996;39:2400–2410.

Tie Y, Boross PI, Wang YF, Gaddis L, Hussain AK, Leshchenko S, Ghosh AK, Louis JM, Harrison RW, Weber IT. High resolution crystal structures of HIV-1 protease with a potent non-peptide inhibitor (UIC-94017) active against multi-drug-resistant clinical strains. *J Mol Biol* 2004;338:341–352.

Tie Y, Boross PI, Wang YF, Gaddis L, Liu F, Chen X, Tozser J, Harrison RW, Weber IT. Molecular basis for substrate recognition and drug resistance from 1.1 to 1.6 angstroms resolution crystal structures of HIV-1 protease mutants with substrate analogs. *FEBS J* 2005;272:5265–5277.

Tisdale M, Myers RE, Maschera B, Parry NR, Oliver NM, Blair ED. Cross-resistance analysis of human immunodeficiency virus type 1 variants individually selected for resistance to five different protease inhibitors. *Antimicrob Agents Chemother* 1995;39:1704–1710.

Todd MJ, Semo N, Freire E. The structural stability of the HIV-1 protease. *J Mol Biol* 1998;283:475–488.

Tomasselli AG, Heinrikson RL. Specificity of retroviral proteases: an analysis of viral and nonviral protein substrates. *Methods Enzymol* 1994;241:279–301.

Tummino PJ, Ferguson D, Hupe D. Competitive inhibition of HIV-1 protease by warfarin derivatives. *Biochem Biophys Res Commun* 1994;201:290–294.

Tummino PJ, Copeland RA. Residence time of receptor–ligand complexes and its effect on biological function. *Biochemistry* 2008;47:5481–5492.

Turner SR, Strohbach JW, Tommasi RA, Aristoff PA, Johnson PD, Skulnick HI, Dolak LA, Seest EP, Tomich PK, Bohanon MJ, Horng MM, Lynn JC, Chong KT, Hinshaw RR, Watenpaugh KD, Janakiraman MN, Thaisrivongs S. Tipranavir (PNU-140690): a potent, orally bioavailable nonpeptidic HIV protease inhibitor of the 5,6-dihydro-4-hydroxy-2-pyrone sulfonamide class. *J Med Chem* 1998;41: 3467–3476.

Tyndall JD, Pattenden LK, Reid RC, Hu SH, Alewood D, Alewood PF, Walsh T, Fairlie DP, Martin JL. Crystal structures of highly constrained substrate and hydrolysis products bound to HIV-1 protease. Implications for the catalytic mechanism. *Biochemistry* 2008;47:3736–3744.

Vacca JP, Dorsey BD, Schleif WA, Levin RB, McDaniel SL, Darke PL, Zugay J, Quintero JC, Blahy OM, Roth E, et al. L-735,524: an orally bioavailable human immunodeficiency virus type 1 protease inhibitor. *Proc Natl Acad Sci USA* 1994;91:4096–4100.

Vacca JP, Condra JH. Clinically effective HIV-1 protease inhibitors. *Drug Discovery Today* 1997;2:261–272.

Van't Klooster GAE, Wigerinck PTBP, De Meyer S, Baert LEC, De Kock HA. Use of a sulfonamide compound for improving the pharmacokinetics of a drug. WO 2006108879, 2006.

Vara Prasad JVN, Para KS, Lunney EA, Ortwine DF, Dunbar JB Jr, Ferguson D, Tummino PJ, Hupe D, Tait BD, et al. Novel series of achiral, low molecular weight, and potent HIV-1 protease inhibitors. *J Am Chem Soc* 1994;116:6989–6990.

Vazquez ML, Bryant ML, Clare M, DeCrescenzo GA, Doherty EM, Freskos JN, Getman DP, Houseman KA, Julien JA, Kocan GP, et al. Inhibitors of HIV-1 protease containing the novel and potent (*R*)-(hydroxyethyl)sulfonamide isostere. *J Med Chem* 1995;38:581–584.

Veber DF, Johnson SR, Cheng HY, Smith BR, Ward KW, Kopple KD. Molecular properties that influence the oral bioavailability of drug candidates. *J Med Chem* 2002;45:2615–2623.

Vega S, Kang LW, Velazquez-Campoy A, Kiso Y, Amzel LM, Freire E. A structural and thermodynamic escape mechanism from a drug resistant mutation of the HIV-1 protease. *Proteins* 2004;55:594–602.

Vermeiren H, Van Craenenbroeck E, Alen P, Bacheler L, Picchio G, Lecocq P. Prediction of HIV-1 drug susceptibility phenotype from the viral genotype using linear regression modeling. *J Virol Methods* 2007;145:47–55.

von Moltke LL, Durol AL, Duan SX, Greenblatt DJ. Potent mechanism-based inhibition of human CYP3A *in vitro* by amprenavir and ritonavir: comparison with ketoconazole. *Eur J Clin Pharmacol* 2000;56:259–261.

Vora S, Marcelin AG, Gunthard HF, Flandre P, Hirsch HH, Masquelier B, Zinkernagel A, Peytavin G, Calvez V, Perrin L, Yerly S. Clinical validation of atazanavir/ritonavir genotypic resistance score in protease inhibitor-experienced patients. *AIDS* 2006;20:35–40.

Vourvahis M, Kashuba AD. Mechanisms of pharmacokinetic and pharmacodynamic drug interactions associated with ritonavir-enhanced tipranavir. *Pharmacotherapy* 2007;27:888–909.

Walmsley S. Protease inhibitor-based regimens for HIV therapy: safety and efficacy. *JAIDS* 2007;45(Suppl 1):S5–S13.

Weber IT, Kovalevsky AY, Harrison RW. Structures of HIV protease guide inhibitor design to overcome drug resistance. *Frontiers in Drug Design and Discovery* 2007;3:45–62.

Wiegers K, Rutter G, Kottler H, Tessmer U, Hohenberg H, Krausslich HG. Sequential steps in human immunodeficiency virus particle maturation revealed by alterations of individual Gag polyprotein cleavage sites. *J Virol* 1998;72:2846–2854.

Williams JW, Morrison JF. The kinetics of reversible tight-binding inhibition. *Methods Enzymol* 1979;63:437–467.

Wlodawer A, Vondrasek J. Inhibitors of HIV-1 protease: a major success of structure-assisted drug design. *Annu Rev Biophys Biomol Struct* 1998;27:249–284.

Wlodawer A, Miller M, Jaskolski M, Sathyanarayana BK, Baldwin E, Weber IT, Selk LM, Clawson L, Schneider J, Kent SB. Conserved folding in retroviral proteases: crystal structure of a synthetic HIV-1 protease. *Science* 1989;245:616–621.

Wondrak EM, Nashed NT, Haber MT, Jerina DM, Louis JM. A transient precursor of the HIV-1 protease. Isolation, characterization, and kinetics of maturation. *J Biol Chem* 1996;271:4477–4481.

Wood E, Hogg RS, Yip B, Moore D, Harrigan PR, Montaner JS. Superior virological response to boosted protease inhibitor-based highly active antiretroviral therapy in an observational treatment programme. *HIV Med* 2007;8:80–85.

Wu JJ, Dandache S, Stranix BR, Panchal C, Wainberg MA. The HIV-1 Protease Inhibitor PL-100 has a high genetic barrier and selects a novel pattern of mutations. *Antivir Ther* 2006;11:S152.

Wynne B, Holland A, Ruff D, Guttendorf R. A first-in human study evaluation the safety, tolerability, and pharmacokinetics (PK) of SPI-256, a novel HIV protease inhibitor (PI) administered alone and in combination with ritonavir (RTV) in healthy adult subjects. In: *ICAAC*, Washington, DC, 2008, Abstract 1265.

Xie D, Gulnik S, Collins L, Gustchina E, Bhat TN, Erickson JW. Thermodynamics and proton uptake for pepstatin binding to retroviral and eukaryotic aspartic proteases. *Adv Exp Med Biol* 1998;436:381–386.

Xie D, Gulnik S, Gustchina E, Yu B, Shao W, Qoronfleh W, Nathan A, Erickson JW. Drug resistance mutations can effect dimer stability of HIV-1 protease at neutral pH. *Protein Sci* 1999;8:1702–1707.

Yan Q, Hruz PW. Direct comparison of the acute *in vivo* effects of HIV protease inhibitors on peripheral glucose disposal. *JAIDS* 2005;40:398–403.

Yang Y, Ikezoe T, Nishioka C, Bandobashi K, Takeuchi T, Adachi Y, Kobayashi M, Takeuchi S, Koeffler HP, Taguchi H. NFV, an HIV-1 protease inhibitor, induces growth arrest, reduced Akt signalling, apoptosis and docetaxel sensitisation in NSCLC cell lines. *Br J Cancer* 2006;95:1653–1662.

Yeh RF, Gaver VE, Patterson KB, Rezk NL, Baxter-Meheux F, Blake MJ, Eron JJ Jr, Klein CE, Rublein JC, Kashuba AD. Lopinavir/ritonavir induces the hepatic activity of cytochrome P450 enzymes CYP2C9, CYP2C19, and CYP1A2 but inhibits the hepatic and intestinal activity of CYP3A as measured by a phenotyping drug cocktail in healthy volunteers. *JAIDS* 2006;42:52–60.

Yoshimura K, Kato R, Kavlick MF, Nguyen A, Maroun V, Maeda K, Hussain KA, Ghosh AK, Gulnik SV, Erickson JW, Mitsuya H. A potent human immunodeficiency virus type 1 protease inhibitor, UIC-94003 (TMC-126), and selection of a novel (A28S) mutation in the protease active site. *J Virol* 2002;76:1349–1358.

Youle M. Overview of boosted protease inhibitors in treatment-experienced HIV-infected patients. *J Antimicrob Chemother* 2007;60:1195–1205.

Yusa K, Harada S. Acquisition of multi-PI (protease inhibitor) resistance in HIV-1 *in vivo* and *in vitro*. *Curr Pharm Des* 2004;10:4055–4064.

Zeldin RK, Petruschke RA. Pharmacological and therapeutic properties of ritonavir-boosted protease inhibitor therapy in HIV-infected patients. *J Antimicrob Chemother* 2004;53:4–9.

Zhang D, Chando TJ, Everett DW, Patten CJ, Dehal SS, Humphreys WG. *In vitro* inhibition of UDP glucuronosyltransferases by atazanavir and other HIV protease inhibitors and the relationship of this property to *in vivo* bilirubin glucuronidation. *Drug Metab Dispos* 2005;33:1729–1739.

Zhang YM, Imamichi H, Imamichi T, Lane HC, Falloon J, Vasudevachari MB, Salzman NP. Drug resistance during indinavir therapy is caused by mutations in the protease gene and in its Gag substrate cleavage sites. *J Virol* 1997;71:6662–6670.

Zhang ZY, Poorman RA, Maggiora LL, Heinrikson RL, Kezdy FJ. Dissociative inhibition of dimeric enzymes. Kinetic characterization of the inhibition of HIV-1 protease by its COOH-terminal tetrapeptide. *J Biol Chem* 1991;266:15591–15594.

Zhou H, Pandak WM Jr, Lyall V, Natarajan R, Hylemon PB. HIV protease inhibitors activate the unfolded protein response in macrophages: implication for atherosclerosis and cardiovascular disease. *Mol Pharmacol* 2005;68:690–700.

Zhou H, Gurley EC, Jarujaron S, Ding H, Fang Y, Xu Z, Pandak WM Jr, Hylemon PB. HIV protease inhibitors activate the unfolded protein response and disrupt lipid metabolism in primary hepatocytes. *Am J Physiol Gastrointest Liver Physiol* 2006; 291:G1071–G1080.

Zhou J, Yuan X, Dismuke D, Forshey BM, Lundquist C, Lee KH, Aiken C, Chen CH. Small-molecule inhibition of human immunodeficiency virus type 1 replication by specific targeting of the final step of virion maturation. *J Virol* 2004;78:922–929.

Ziermann R, Limoli K, Das K, Arnold E, Petropoulos CJ, Parkin NT. A mutation in human immunodeficiency virus type 1 protease, N88S, that causes in vitro hypersensitivity to amprenavir. *J Virol* 2000;74:4414–4419.

INDEX

Enzyme Inhibition in Drug Discovery and Development, edited by Lu and Li
Copyright © 2010 John Wiley & Sons, Inc.